STEREOCHEMISTRY
CONFORMATION AND MECHANISM

STEREOCHEMISTRY
CONFORMATION AND MECHANISM

ELEVENTH
Multi Colour
EDITION

P S Kalsi
Professor of Eminence
Shoolini University, Solan
Visiting Professor
Gujarat Forensic Sciences University
Gandhinagar
Former Visiting Professor of Chemistry
Indira Gandhi National Open University (IGNOU)
New Delhi
Former Dean of Colleges
Punjab Technical University, Jalandhar
Former Professor and Head
Department of Chemistry
College of Basic Sciences and Humanities
Punjab Agricultural University
Ludhiana

NEW AGE TECHNO PRESS

An Imprint of

NEW AGE INTERNATIONAL (P) LIMITED, PUBLISHERS
LONDON • NEW DELHI
Bangalore • Chennai • Cochin • Guwahati • Hyderabad • Kolkata • Mumbai
Visit us at www.newagepublishers.com

Published by New Age International (P) Ltd., Publishers
First Edition: 1990
Eleventh Multi Colour Edition: 2022

GLOBAL OFFICES

- **New Delhi** **NEW AGE INTERNATIONAL (P) LIMITED, PUBLISHERS**
7/30 A, Daryaganj, New Delhi-110002, (INDIA), **Tel.:** (011) 23253472, 23253771
E-mail: contactus@newagepublishers.com • Visit us at www.newagepublishers.com
- **London** **NEW AGE INTERNATIONAL (UK) LTD.**
27 Old Gloucester Street, London, WC1N 3AX, UK
E-mail: info@newacademicscience.co.uk • Visit us at www.newacademicscience.co.uk

BRANCHES

- **Bangalore** 37/10, 8th Cross (Near Hanuman Temple), Azad Nagar, Chamarajpet, Bangalore- 560 018
Tel.: (080) 26756823, **E-mail: bangalore@newagepublishers.com**
- **Chennai** 26, Damodaran Street, T. Nagar, Chennai-600 017, **Tel.:** (044) 24353401
E-mail: chennai@newagepublishers.com
- **Cochin** CC-39/1016, Carrier Station Road, Ernakulam South, Cochin-682 016
Tel.: (0484) 4051304, **E-mail: cochin@newagepublishers.com**
- **Guwahati** Hemsen Complex, Mohd. Shah Road, Paltan Bazar, Near Starline Hotel
Guwahati-781 008, **Tel.:** (0361) 2513881, **E-mail: guwahati@newagepublishers.com**
- **Hyderabad** 105, 1st Floor, Madhiray Kaveri Tower, 3-2-19, Azam Jahi Road, Near Kumar Theater
Nimboliadda Kachiguda, Hyderabad-500 027, **Tel.:** (040) 24652456
E-mail: hyderabad@newagepublishers.com
- **Kolkata** RDB Chambers (Formerly Lotus Cinema) 106A, 1st Floor, S N Banerjee Road
Kolkata-700 014, **Tel.:** (033) 22273773, **E-mail: kolkata@newagepublishers.com**
- **Mumbai** 142C, Victor House, Ground Floor, N.M. Joshi Marg, Lower Parel, Mumbai-400 013
Tel.: (022) 24927869, 24915415, **E-mail: mumbai@newagepublishers.com**
- **New Delhi** 22, Golden House, Daryaganj, New Delhi-110 002, **Tel.:** (011) 23262368, 23262370
E-mail: sales@newagepublishers.com

C-22-01-12962

Printed in India at Goyal Offset Works (P) Ltd., Sonepat, Haryana.
Typeset at In-house, Delhi.

NEW AGE INTERNATIONAL (P) LIMITED, PUBLISHERS
7/30 A, Daryaganj, New Delhi-110002
Visit us at **www.newagepublishers.com**
(CIN: U74899DL1966PTC004618)

Dedication

The book is dedicated with all my love and regards to Prof. Dr. S. Sivaram, an internationally known distinguished scientist. He is currently an Honorary Professor and INSA Senior Scientist at IISER Pune. Earlier he was a CSIR-Bhatnagar Fellow and Director of CSIR-NCL Pune (2002-10). He is decorated with prestigious awards from different learned National and International Scientific Societies and Organisations. Above all, he is singularly top placed in the country for these honours.

Prof. Sivaram is a dedicated academician with a vision to upgrade the educational system in the country through world class teaching and research and also holds the precious view that teaching and research must go hand-in-hand.

Foreword

INDIAN INSTITUTE OF SCIENCE EDUCATION AND RESEARCH
(An Autonomous Institution, Ministry of Human Resource Development, Govt. of India)
Dr. Homi Bhabha Road, Pune 411008

IISER PUNE

Dr. S. Sivaram, DSc, FNA, FASc, FNAE, FTWAS
INSA Senior Scientist and Honorary Professor

It is indeed a privilege and honour to pen this Foreword for the eleventh edition of the book titled "Stereochemistry: Conformation and Mechanism" authored by Professor P.S. Kalsi.

Stereochemistry is an old subject, yet very relevant, for understanding and appreciation of contemporary chemistry and biology. Rational stereoselective organic synthesis, chiral catalysis, elucidation of absolute configuration of natural products, conformational analysis for systematizing interpretation of chemical knowledge and understanding reaction mechanisms through the prism of stereochemical principles have enriched chemistry in manifold ways. Ability to synthesize single enantiomers in high enantiomeric excess has led to high atom economic synthesis eliminating time consuming resolution of enantiomers and wastage of half the starting materials used for synthesis. Much of the advancement in our understanding of stereochemistry is due to the emergence of powerful analytical tools such as high-resolution NMR spectroscopy, single crystal X-Ray crystallography, circular dichroism and optical rotatory dispersion methods as well as chiral liquid chromatography. Computational modelling and simulation tools, 3D visualization methods as well as augmented reality have provided chemists with amazing new methods to understand the role of stereochemistry in determining functional properties of molecules. The tragedy of S-(-)-Thalidomide is the fact that S- (-)-Amlodipine is the active form of the most successful anti-hypertension and coronary heart disease drug and that we make use of only *isotactic*-polypropylene, not *atactic* -polypropylene (consumption of over sixty million tons per annum) is a testimony to the fact that stereochemistry impacts us everyday of our life.

Stereochemistry also connects organic chemistry to modern biology. Most of the biologically important molecules display interesting stereochemical features. The fact that molecules that form the building block of life are all derived from L-amino acids has puzzled chemists and biologists alike and stimulated several interesting experiments in pre-biotic chemistry as well as raised interesting questions on the origins of life. Thus, stereochemistry, defines, both, form and functions of molecules.

Professor Kalsi's eleventh edition, lives up to the reputation of its earlier edition and provides a reader with a lucid introduction to the subject. It is well laid out, copiously illustrated and brought out in two colours. Each chapter includes many examples to illustrate the topic covered. The figures are nicely drawn and easy to understand. Apart from presenting stereochemical aspects of organic molecules, the role of stereochemistry in understanding mechanisms of organic reactions and synthesis has been highlighted throughout the text. This makes the study of organic chemistry pleasurable and provides a rational way to understand organic chemical transformations, rather than resorting to rote memory which students often resort to. All chapters have been updated with new material. A comprehensive listing of stereochemical terms, assignment of configurational descriptors from different projections of the molecules is a new added attraction. A large number of new problems have been added which provides a student to judge his understanding of the subject. This edition also brings out the important connection between the principles of stereochemistry to modern biology. This is a very welcome feature indeed.

This book will be very useful to students of organic chemistry as their first introduction to the subject and also to practicing professionals who need to brush up their fundamentals once in a while during their working life. The splendour and aesthetics of organic chemistry is obvious to only those who can visualize molecules in all their three dimensions. Yet most students learn organic chemistry using only two-dimensional drawings, either on paper or on a blackboard and find it difficult to understand concepts such as conformation, configuration, chirality and stereogenic elements as applied to molecules. Professor Kalsi has performed a yeomen service by bringing an appreciation of these principles in a simple manner to the readers of the book. It is my fond hope that this book should become an integral part of the book shelf of, both, students and professionals alike.

Dr. S. Sivaram

Preface to the Eleventh Edition

NEW TO THIS EDITION–GENERAL

This new eleventh edition extends the transformation started in the tenth edition to a significant extent. My major aim is to build and maintain an intellectual foundation of stereochemistry for the students taking organic chemistry as their profession. Apart from presenting stereochemical aspects of organic molecules, the role of stereochemistry in understanding mechanisms of organic reactions and synthesis has been further highlighted by adding new examples. The author has made an attempt to give students more confidence in predicting stereochemical aspects, to simplify the learning and make it easier for them to recognise similarities and differences among related stereochemical situations.

The major comments of a huge number of students and teachers using the tenth edition have been to define the major stereochemical manuplations by taking one compound as an example. This problem has been addressed by taking the example of bromochlorofluoromethane to explain chirality–writing of enantiomers, wedge and dash structure, Fischer projections, *R* and *S* notation, mental manipulation of tetrahedra, converison of Fischer projection into a Wedge-Dash representation and drawing a particular enantiomer. This problem has been addressed by rewriting the first seven pages.

This practice in fact was followed earlier as in the case of tartaric acid page 104 and has been strengthened in other examples as well.

- **Mastering skills**

  Proper understanding of stereochemistry requires the development of intellectual skills, which indeed are "survival skills" for a student. To achieve this goal extensive new material was added in boxes in all the chapters in the tenth edition. This practice continues in the new edition as well with replacement of old material with new one where the situation demanded as e.g., in Box 1.1 itself.

BOX 1.1
MENTAL MANIPULATION OF TETRAHEDRA
WHICH TWO BONDS ARE PLACED IN THE PLANE OF PAPER
(STUDY WITH MODELS WILL HELP)

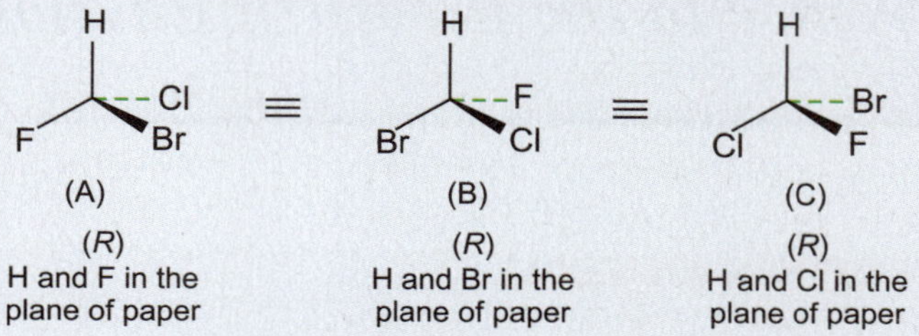

SCHEME B1.1

Consider one enantiomer e.g., (R) bromochlorofluromethane. In a tetrahedral structure only two bonds can be in the plane of the paper at any one time. In (Scheme B1.1), the manipulations simply show as to which two bonds are in the plane of paper.

- **Organic Chemistry Today-Application to Biological Sciences**

  The most dramatic developments at the beginning of 21st century are a result of collaboration between organic chemists and biologists. A good working knowledge of molecular biology is indeed a must for the survival skill for modern organic chemists and thus, organic chemistry has a long way to go. The applications of organic chemistry in general and organic stereochemistry in particular provides a bridge with biochemistry courses. The new addition provides more material by taking examples from biological systems.

NEW TO THIS EDITION–SPECIFIC

- *Chapter 1.* At several places the material has been extensively modified and rewritten. New problems and discussion have been incorporated.

  The first seven pages have been rewritten under the topic intoduction. Cahn-Ingold-Prelog *R*, *S* notation has been introduced early along with the basic concepts of enantiomers and diastereomers. The Fischer projections, methods to convert Fischer projections to Wedge and Dash formulas, have been detailed by taking the same compound–bromochlorofluoromethane.

  Chirality has also been introduced by simple example from cyclic compound-1, 2-dimethylcyclopentane which introduces enantiomers, diasteroomers and meso compound early in the study of chirality.

  New material on relating conformations in cyclohexanes and relationship between stereochemical and constitutional isomers has been added.

  New thought provoking examples on assignment of *R* and *S* configuration have also been added.

  New material with new examples on helical chirality are added too.

- *Chapter 2.* New material on *pro-R* and *pro-S* distinction by enzymes has been added.
- *Chapter 3.* The mechanistic details of Mitsunobu reaction which is the best method for inverting the configuration of an alcohol or subjecting a 2° alcohol to nucleophilic substitution (without elimination) has been rewritten for better understanding.
- *Chapter 4.* More material has been added on the conformations of acetaldehyde, ethane and related systems.

  Some more intersting examples on anomeric effect which strongly influence the conformational equilibria is pyranoses with C-1 electronegative atom have been added.

  Additional examples of molecules which are forced to adopt twist boat conformations and *trans*-decalin systems have been added.
- *Chapter 5.* In this chapter, the discussion on E2-syn-elimination has been rewritten for better understanding. More interesting examples of the involvement of *anti*-stereochemistry and predominance of Saytzeff product during E2-elimination are added.
- *Chapter 6.* More problems have been added on oxymercuration, and addition of HX-hydrohalogenation along with bromination to olefins.
- *Chapter 7.* New material has been added on Peterson reaction and Julia olefination. The mechanisms with the explanation of the stereochemical outcome of the reactions have been presented.

Ludhiana **P.S. Kalsi**

Preface to the First Edition

The French chemist Louis Pasteur (1850) was the first to reveal that optical activity at the molecular level was due to an asymmetric placement of atoms in a molecule. He further recognized that enantiomers had equal and opposite effects on polarized light. vant Hoff and Le Bel in 1874 (Nobel Prize) proposed the tetrahedral model for the carbon as the cause of molecular dissymmetry and the resultant optical rotation. This was the beginning of stereochemistry which forms a massive and an integral part of organic chemistry.

In recent years, stereochemistry has not only expanded in depth but, with the development of new kinds of nomenclature and concepts, it has also almost taken on an entirely new perspective. These developments offer knowledge of great interest to students and the present book incorporates this knowledge not only for the uninitiated student but for the expert as well.

The intention of this book, is to give equal weightage to various aspects of stereochemistry viz., chirality, stereochemical course of various reactions, molecular rearrangements and methods of assigning configurations. Though treated lightly, but several examples from the major special areas of organic chemistry: Terpenoids, steroids, alkaloids and carbohydrates have been presented to explain various stereochemical, mechanistic and conformational aspects.

The book runs into various chapters each of which is reasonably complete in itself. The first chapter describes chirality and some related aspects to provide a necessary background for understanding mechanism and conformations presented in subsequent chapters and is important in its own right. Some emphasis has also been laid on spectral methods in solving stereochemical problems.

I have tried to develop an internal order of almost all the basic concepts, to help the students to be able to apply the knowledge gained to new situations. Special attention is paid in illustrating the introduction to each chapter in some depth. The student may thus, not find himself in a situation where his understanding a concept requires knowledge that he has not gained. Special attention is paid to serve the needs of students in a way that will enhance their comprehension of the subject of stereochemistry.

It is now recognized that Hückel-Möbius approach can be gainfully and effectively used to explain pericyclic reactions, therefore, I have introduced this as the only approach to make several aspects of these reactions more understandable.

Problems and their solutions are provided at the end of each chapter and these are designed to test the students' understanding of each topic up to that point. Chapter summaries are presented in the appendix.

Advances in a scientific field are often reported in somewhat condensed communications, and find elaboration in papers and books after a lapse of some years. My acknowledgement, therefore, is to many authors who have preceded me in this task of authorship and from whose efforts I have drawn innumerable facts, ideas and theories. Particularly to some of these sources, mentioned in references, I am indebted to them.

Ludhiana

P.S. Kalsi

Acknowledgements

At the very outset, I would like to express my heartfelt gratitude to (late) Prof. S.C. Bhattacharyya under whom I got my research training (1960-1964), leading to Ph.D degree, at National Chemical Laboratory, Pune. I have also had the good fortune to work under (late) Prof. S.S. Deshapande (Ph.D., London, Contemporary of Prof. C.K. Ingold) at Master's level, at Holkar Science College, Indore. Prof. S.S. Deshapande was well known for his intellectual honesty and power in the country. I remain their student even today and there is no easy way to adequately acknowledge my debt to them.

I acknowledge the role of National Chemical Laboratory, Pune, in shaping my career and acumen. My training at NCL was the most demanding and exciting experience of my life which laid in me, a strong foundation and desire to write different texts in the area of organic chemistry.

Prof. K. Venkataraman, former Director of National Chemical Laboratory, Pune, played a remarkable role in my professional life. I have also been lucky enough to benefit from the intellect of many remarkable scientists like Prof. Dr. Parkash P. Wadgaonkar, NCL Pune and Prof. A.K. Singh, IIT Bombay. They have all been a great source of inspiration.

Ludhiana **P.S. Kalsi**

Contents

Chirality

1
CHAPTER

1.1 WHAT IS CHIRALITY ?

(A) Introduction

Kekulé described correctly the tetravalent nature of carbon. Chemistry was, however, still viewed in a two-dimensional way until 1874. In that year, J. van't Hoff and Le Bel added a third dimension to carbon. They proposed that the four bonds of carbon are not randomly oriented but have a specific spatial orientation. van't Hoff (Nobel Prize 1901) further advanced the idea of tetrahedral carbon atom. A chirality center is a tetrahedral carbon atom that is bonded to four different groups. A molecule that contains one chirality center is chiral and can exist as a pair of enantiomers (Fig. 1.1). Chirality centers are often designated with an asterisk (*).

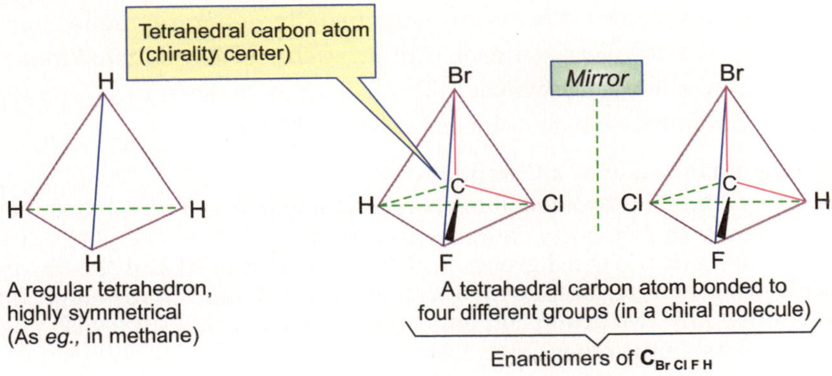

Fig. 1.1

A flying wedge representation of a tetrahedral carbon atom is shown in (Fig. 1.2). The conventions used to show the three-dimensionality are:

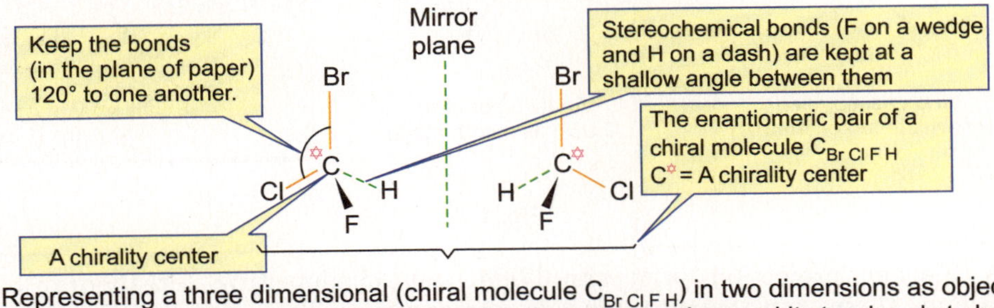

Representing a three dimensional (chiral molecule $C_{Br\,Cl\,F\,H}$) in two dimensions as object and its nonsuperimposable mirror image (enantiomers). Here carbon and its two bonds to bromine and chlorine are kept in the plane of paper. F is on a wedge while H is on a dash.

Fig. 1.2

- The central carbon atom is taken to lie in the plane of paper together with any two bonds. The bonds to carbon are represented by continuous lines of normal thickness.
- The thick wedge represents a bond coming out of the plane of paper towards one's eyes, while dashed line represents a bond receding into the plane away from the viewer (Fig. 1.2). In short one rays that (F) is on a wedge and (H) is on a dash.
- Note that any three atoms must lie in the plane at any given time for a molecule.

The tetrahedral geometry of carbon and a few other atoms (pyramidal) *e.g.*, N, P, Si and S and trigonal geometry of sp^2 hybrid carbon are key to the study of organic stereochemistry. The regular tetrahedron (Fig. 1.1) is highly symmetrical with four equivalent vertices. When these are occupied by four different achiral atoms or groups, symmetry disappears, and a three dimensional four points chiral complex is obtained ($C_{\boxed{Br}\boxed{Cl}\boxed{F}\boxed{H}}$) which is non superimposable on its mirror image (Fig. 1.2).

Significantly if one of the groups is chiral the symmetry is maintained as in *meso*-tartaric acid (see, scheme 1.2*a*).

The tetracovalent carbon (the fifth point) placed at the center of the tetrahedron is termed a chirality center whose presence usually generates molecular chirality. A unique feature of a chirality center is that an exchange of any two ligands reverses its chirality to give a new stereoisomer. This *e.g.*, may be seen in $C_{\boxed{Br}\boxed{Cl}\boxed{F}\boxed{H}}$ when an exchange gives the mirror image—the enantiomer (Fig. 1.3).

(B) Writing the Enantiomer of a Compound

One of the enantiomers of bromochlorofluoromethane (*e.g.*, I Fig. 1.3) can be drawn in any way by keeping the chirality center with any two bonds in the plane of paper *e.g.*, H and Br and then put the third group on the wedge, this is Cl and the fourth F on the dash. This gives one enantiomer, its mirror image is the second enantiomer. Interchanging the position of any two groups on one enantiomer in a molecule with one chirality center also gives its enantiomer (Fig. 1.3).

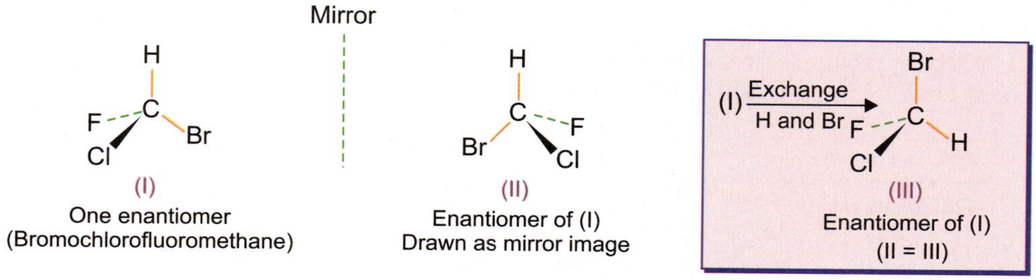

Fig. 1.3

A Fischer projection is a simplified way of depicting tetrahedral carbon atoms and their substituents in two dimensions. The molecule is drawn in the form of a cross, the central carbon being at the point of intersection. The horizontal lines signify bonds directed *toward* the viewer; the vertical lines are pointing away (Fig 1.3*a*), further details are in (Schemes 1.28*a* and 1.28*b*).

When one draws enantiomers in a Fischer projection one can draw the first enantiomer by placing the four atoms or groups bonded to the chirality center in any order (A, Fig. 1.3*a*). The second enantiomer can be drawn by interchanging two of the atoms or groups. It does not matter which two are interchanged. It is best to interchange the groups on the two horizontal bonds, because then the enantiomers look like mirror images as well. Thus in (A, Fig. 1.3*a*) interchanging Cl with Br gives its enantiomer (B, Fig. 1.3*a*).

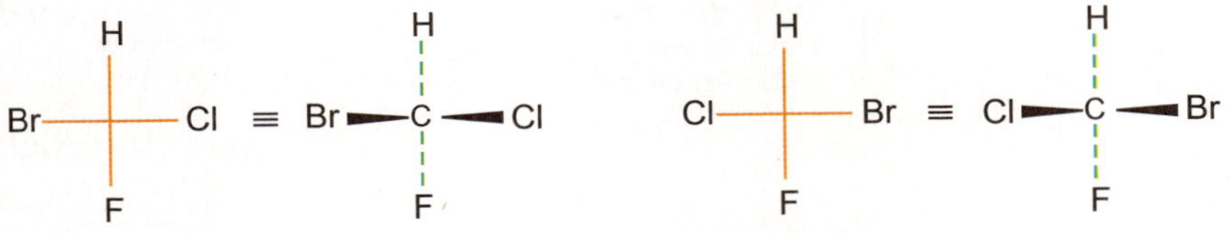

Fig. 1.3*a*

A useful working procedure to assign absolute configuration when the group of lowest priority is towards you. There is no need to turn or reorient the molecule to put it away, leave it as such but apply the *R*/*S* system backward. Thus consider the Fischer projection of one of the enantiomers of bromochlorofluoromethane (I, Fig. 1.3*b*) and an flying wedge representation of an enantiomer of 2-butanol (II, Fig. 1.3*b*).

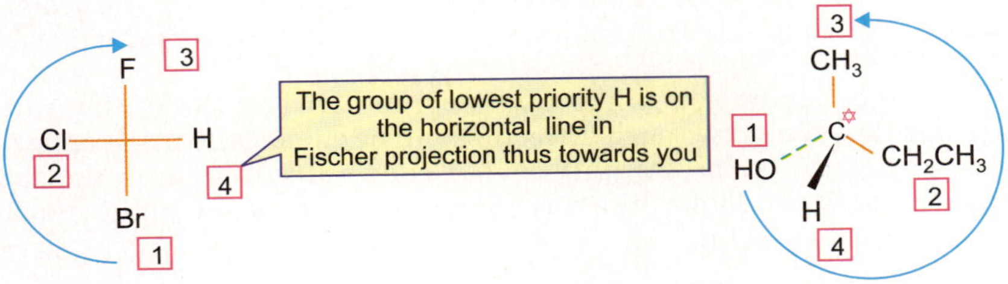

Fig. 1.3*b*

(C) Cahn–Ingold–Prelog R, S Notation (Assignment of absolute configuration to enantiomers)

The Cahn–Ingold–Prelog system ranks substituents at the chirality center in order of decreasing atomic number. In the case of e.g., bromochlorofluoromethane the order is Br > Cl > F > H. View the molecules so that the bond to the lowest ranked substituent–in this case H–points away from us and determine if the sequence of decreasing priority of the other three substituents is clockwise (*R*) or counterclockwise (*S*). One may orient the molecule to put (H) at the back and then assign configuration (Fig. 1.3*c*). Remember turning or tumbling the compound drawn in three dimensions does not affect the configuration of an enantiomer. (Compare this situation with Schemes 1.3*a*, 1.20*d* and 1.20*e*)

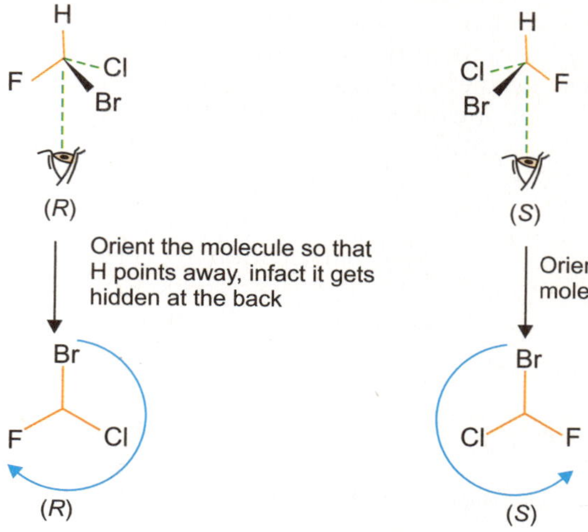

Fig. 1.3c

Three important terms used in stereochemistry.

- Configuration
- Absolute configuration and
- Relative configuration (For details see Figs 1.5*a* and 1.5*b*)

Configuration is the fixed spatial ordered arrangement of atoms in a molecule as in (I, Fig. 1.3).

Absolute configuration is the fixed spatial arrangement of the atoms in a chiral molecule that distinguishes it from its mirror image and its stereochemical description (*e.g.*, *R* or *S* I and II Fig. 1.3)

Relative configuration is the relationship between the configurations of two chiral molecules. When two such molecules can be chemically interconverted without breaking any bonds to the chirality center, they are said to have the same relative configuration independent of the direction of rotation of the plane of polarized and the (*R*) and (*S*) designation.

BOX 1.1
MENTAL MANIPULATION OF TETRAHEDRA WHICH TWO BONDS ARE PLACED IN THE PLANE OF PAPER (STUDY WITH MODELS WILL HELP)

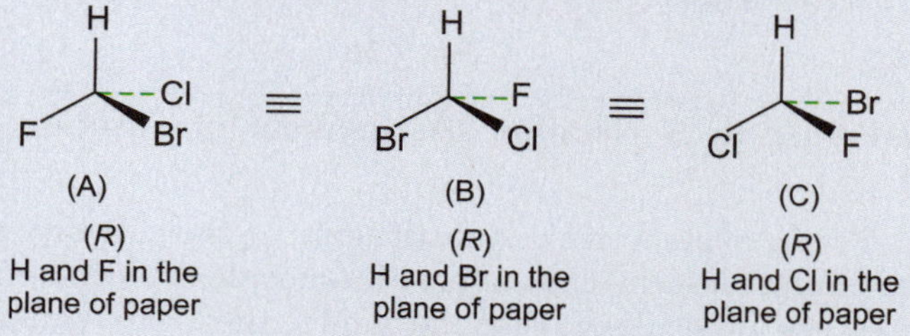

SCHEME B1.1

Consider one enantiomer e.g., (R) bromochlorofluromethane. In a tetrahedral structure only two bonds can be in the plane of the paper at any one time. In (Scheme B1.1), the manipulations simply show as to which two bonds are in the plane of paper.

Continued...

All the structures (A–C Scheme B1.1) represent the same molecule (R) – bromochlorofluoromethane, showing its different projections.

Naming the compound systematically including R and S system defines the identity of the compound. Switching of groups is not allowed since this inverts the stereochemistry of a chirality center.

For every chirality center in a Fischer projection, the horizontal lines are considered to be coming out of the page, and the vertical lines are considered to be going behind the page.

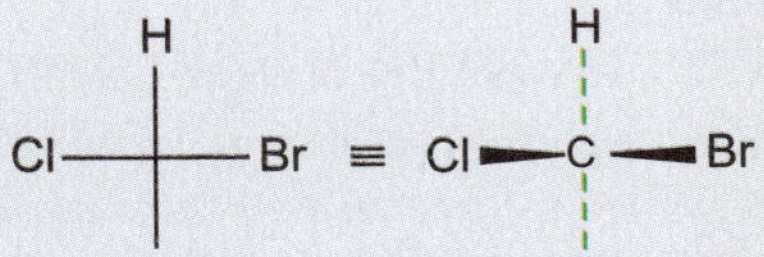

(A)

Fischer projection

(S)-Bromochlorofluoromethane

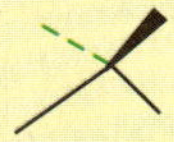

Flying wedge representation of a chirality center

SCHEME B 1.1a

A Fischer projection represents, all horizontal lines as wedges and all vertical lines as dashes (Scheme B1.1a).

So far one has displayed a chirality center as a flying wedge representation with two bonds in the plane of paper and one dash and wedge (Scheme B1.1a).

The chiral center of a Fischer projection can easily be converted into this more familiar type of drawing. One chooses one horizontal line e.g., the one with Br in the Fischer projection (A) and redraws it as a wedge (A′); then choose one vertical line e.g., the one with F and draw it as a dash (A′). The answer will be the same, regardless of which two lines are chosen (Scheme B1.1b).

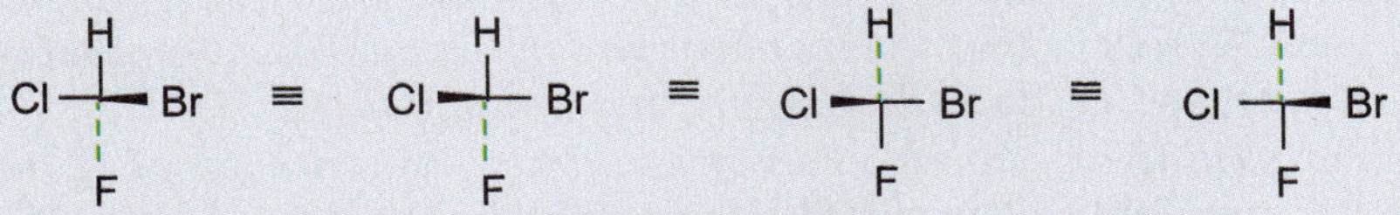

(A′)

(S)-Bromochlorofluoromethane

Convertion of Fischer projection into a drawing with two bonds in the plane of paper one on a wedge and one on a dash

SCHEME B 1.1b

The configuration at the chirality center in all the structures (Scheme B 1.1b) is S in all the drawings.

With a Fischer projection with several chirality centers, this procedure is repeated for every chirality center.

BOX 1.2
SIGNIFICANCE OF REDRAWING THE TETRAHEDRA WITH ANY TWO BONDS IN THE PLANE OF PAPER

In several situations e.g., while determining the R and S configurations of a chirality center, redrawing the tetrahedra keeping the bonds with priority 1 *and* 2 *in the plane of paper helps. Since the path* $1 \rightarrow 2 \rightarrow 3$ *determines the configuration (see, Scheme 1.20m and Box 1.13, page 76).*

BOX 1.3
DRAWING CHIRAL MOLECULES (STUDY WITH MODELS WILL HELP)

There can be several different ways to represent a chiral molecule on a two-dimensional page. Thus the four different projections represent the same enantiomer of (R)–2–butanol (Scheme B 1.2, also see Scheme 1.20 m and Box).

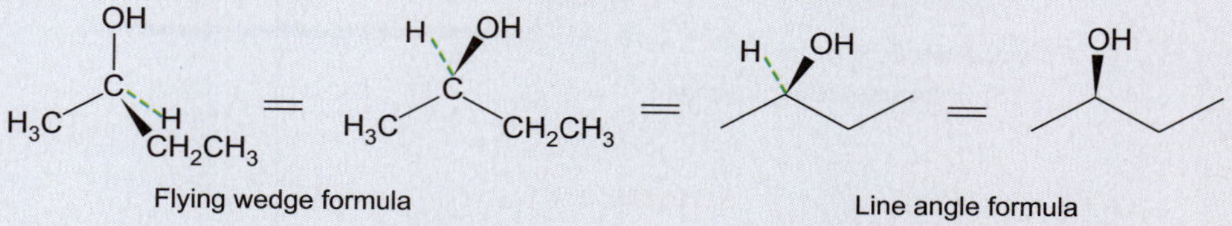

(1) (2) (3) (4)

(*R*)–2-butanol

Structures represent the same molecule (*R*)–2–butanol drawn differently.

SCHEME B 1.2

Consider the following points:

- *The compound (1, Scheme B 1.2) represents the flying wedge formula.*
- *Turning (1) in space gives (2). In (2, scheme B 1.2), the situation is similar to that already studied in (Scheme B 1.1), i.e., with tetrahedra only two bonds can be in the plane of paper at any one time. Thus in (1, Scheme B 1.2, OH and* CH_3*) are in the plane of paper and in the equivalent projection (2, Scheme B 1.2 now* CH_3 *and* CH_3CH_2 *are in plane of paper).*
- *For writing an abbreviated representation, one transforms (2 into 3 Scheme B 1.2) which is often called line angle or line wedge formula.*
- *Representation (4) is obtained by ignoring H on the chirality centre which is often not drawn.*
- *Writing the enantiomer of (4) is to draw its mirror image (Scheme B 1.3).*
- *The enantiomer of R-2-butanol(4) can also be drawn by interchanging the position of any two groups around the chirality center (e.g., H and OH here). This operation gives the projection of (S)-2-butanol (shaded, Scheme B 1.3), here now group of lowest priority (H) is on the thick wedge which as per usual practice is not shown. This operation is similar to that discussed in (Fig. 1.3).*

Continued...

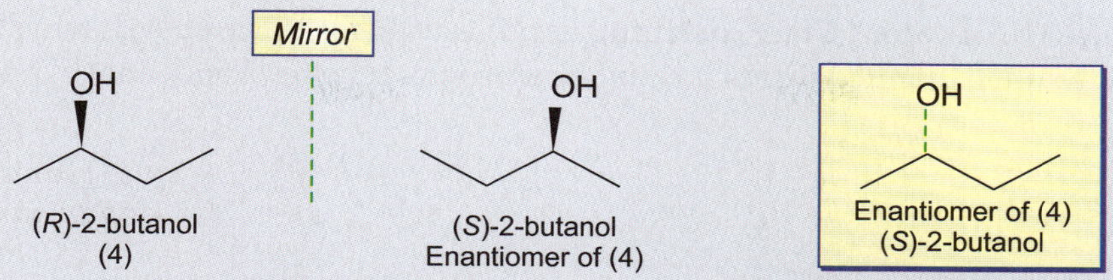

The (H) on C2 is on a dash and not shown

Written by interchanging any two of the groups e.g., (H and OH) to the chirality centre in (4) (In this projection H is towards one's eyes and would be on a wedge which as per usual practice is not shown)

SCHEME B 1.3

BOX 1.3*a*
DRAWING AN ENANTIOMER *e.g.*, (*S*) BROMOCHLOROFLUOROMETHANE WITH A DESIRED CONFIGURATION

- *A compound with one chirality center can have only two enantiomers. These may be drawn in several ways by keeping the chirality center in the plane of paper with any two bonds. Then the third group will be on the wedge and the fourth on a dash.*
- *The* ***Cahn–Ingold–Prelog system*** *ranks substituents at the chirality center in order of decreasing atomic number. Here the order is Br > Cl > F > H.*
- *The following procedure is applied to draw a particular enantiomer e.g., (S)–Bromochlorofluoromethane. (A, Scheme B 1.1, Also see, scheme 1.20e)*

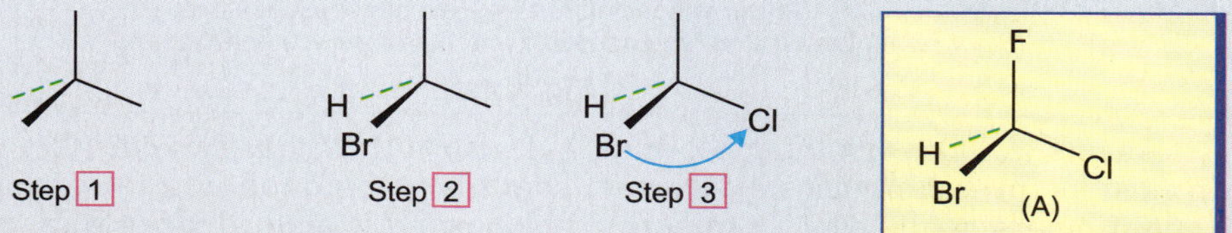

Step 1
Draw the bonds around the chirality center

Step 2
Put the group with lowest priority on the dash and with highest priority on any remaining bond

Step 3
Draw an arrow counterclockwise from the group with highest priority to the next available bond and put the group of next highest priority on that bond

(S)-Bromochlorofluoro-methane

Step 4
Add the remaining substituent on the last available bond

SCHEME B 1.3a

(D) The Term Chiral, Achiral and Meso

The word chiral (Greek word *Chier*, meaning hand) is used for those objects which have right-handed and left-handed forms, *i.e.*, molecules which have "handedness" and the general property of "handedness" is termed chirality. An object or a molecule which is not superimposable upon its mirror image is chiral. Enantiomers occur only with those compounds whose molecules are chiral.

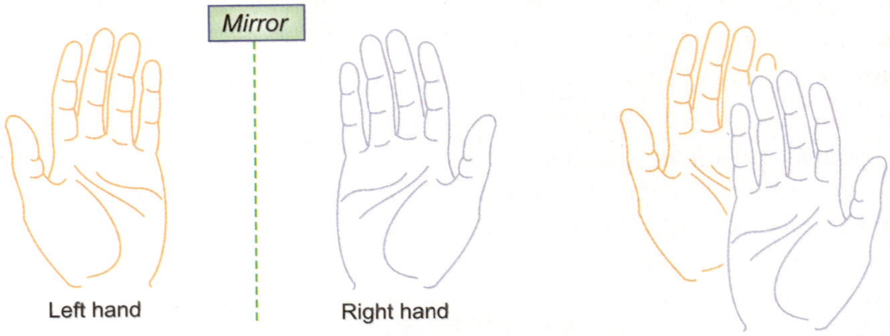

The mirror image of left hand is a right hand　　Left and right hands are not superimposable

Fig. 1.4

The *trans* isomer of 1,2-dimethylcyclopentane is *chiral* because it is *not superposable* on its mirror image, (Fig 1.4*a*). Enantiomers do not exist for achiral molecules.

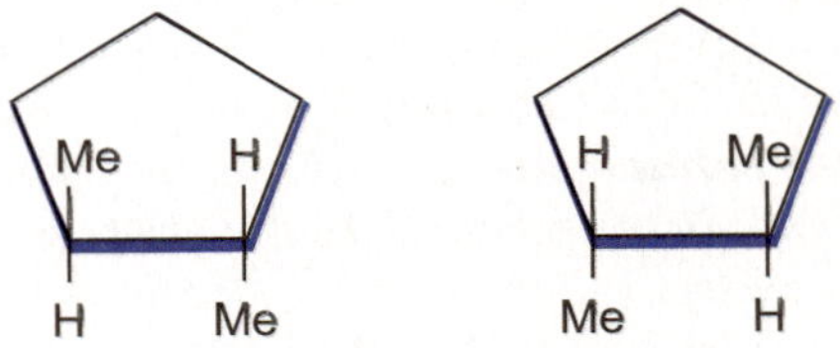

(Enantiomers)
Mirror images of *trans*-1,2-dimethylcyclopentane,
they are not superposable and therefore, are enantiomers

Fig. 1.4a

As one would learn later 1,2-dimethylcyclopentane has two chirality centers and can exist as three stereoisomeric forms. The *trans* compound (Fig 1.4*a*) exists as a pair of enantiomers, while *cis*-1, 2-dimethylcyclohexane is achiral since it is superposable on its mirror image (Fig 1.4*b*), moreover, since it has a plane of symmetry. An achiral molecule is superposable on its mirror image.

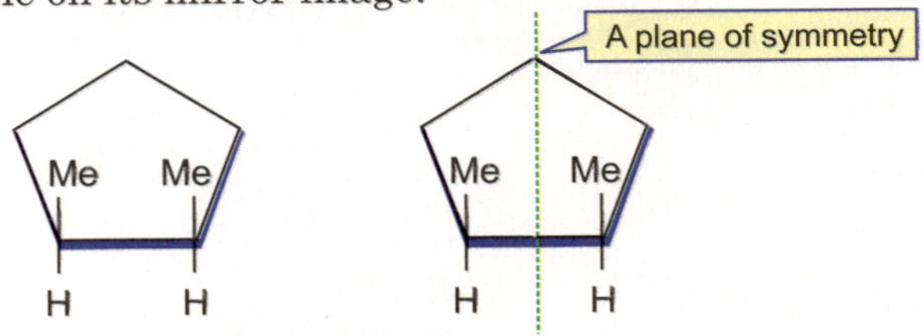

(Achiral *meso*)
Mirror images of *cis*-1,2-dimethylcyclopentane
these mirror images are superposable.
Thus this compound is achiral (*meso*) and has a plane of symmetry

Fig. 1.4b

The universal test for chirality of a molecule, or any object, is the nonsuperposability of the molecule or object on its mirror image.

Human hands are related to each other as object and its mirror image typical of enantiomers (Fig. 1.4). Thus *e.g.*, one's right hand will fit into a right glove and not into a left glove. The term chiral is used to describe molecules of enantiomers since these are related the same way that a right hand is related to a left hand. A structurally chiral system is the human body itself, the heart lies to the left of center and the liver to the right. From an evolution point of view most people are right-handed. Several plants display chirality by way they wind around supporting structures and thus represent as a left-handed or a right-handed helix.

All but one of the 20 amino acids, the components of proteins are chiral and left-handed. The molecules of almost all natural sugars on the other hand are right handed. The enantiomers display physiological differences, one enantiomeric form of a terpenoid limonene smells like oranges while the other enantiomer smells like lemons. One's nose is capable of distinguishing between enantiomers, *i.e.*, the receptor sites for the sense of smell are chiral (see, Scheme 1.2*b*).

Generally, otherwise, enantiomers have identical properties in a symmetrical environment, enantiomers react at the same rate with achiral compounds. Their properties may however, differ in an unsymmetrical environment, and enantiomers may react at different rates with other chiral compounds (see Scheme 1.2*b*). This difference is reflected in a compound being biologically active, while its enantiomer is not.

(E) Enantiomers and Diastereomers

Stereoisomers can be studied as two major classes :

- ***Enantiomers which are stereoisomers and are nonsuperposable mirror images of each other.*** (See, Fig. 1.4*a*). All other stereoisomers are diastereomers.
- ***Diastereomers are stereoisomers which are not mirror images of each other.***

Cis and *trans* isomers of cycloalkanes and *cis* and *trans* alkene isomers (not drawn here) are examples of stereoisomers which are diastereomers. (Fig 1.4*c*). *Cis* and

cis-1,2-Dimethylcyclopentane

trans-1,2-Dimethylcyclopentane

Cis and *trans* cycloalkane isomers are diastereomers.

Fig. 1.4c

trans isomers of 1,2-dimethylcyclopentane have the same molecular formula (C_7H_{14}), the same sequence of connectivity of their atoms, but different arrangements of their atoms in space. Either both methyl groups are attached to the same face of the ring, or the two methyl groups are attached to opposite faces of the ring. Moreover, the positions of the methyl groups cannot be interconverted by conformational changes. Thus, these compounds are stereoisomers, and since they are stereoisomers are not mirror images of each other, they can be further classified as diastereomers.

The alkene isomers, *cis*-and *trans*-2-butene are stereoisomers which are also diastereomers (Scheme 1.1), since they are not mirror images of each other. Moreover, they are not interconvertible due to large barrier of rotation of carbon-carbon double bond.

- ***An achiral molecule is superposable on its mirror image.***

The *cis* and *trans* isomers of 1,2-dichloroethene are both achiral because each isomer is superposable on its mirror image.

(F) The Terms Asymmetric Center, Chirality Center and Stereogenic Center (Stereocenter)

Several terms are used to designate *e.g.*, a sp^3 hybridised carbon atom bonded tetrahedrally to four different substituents in a chiral molecule: asymmetric center, or a chirality center. From the time of Le Bel and vant Hoff an atom with four different substituents was called an asymmetric atom. This is indeed so, since compounds with one such atom are truly asymmetric as they lack symmetry. There are molecules which have atoms with four different substituents and which also have various symmetry elements, including planes of symmetry as in *meso* tartaric acid. Use of the term asymmetric atom in these type of cases may be confusing if at all. An asymmetric center is sp^3 hybridised carbon attached with four different groups where the interchange of two groups reverses it stereochemistry (I, see, Fig. 1.3). An asymmetric carbon atom is the most common example of a chirality center. Chirality center is the IUPAC term for any atom holding a set of ligands in a spatial arrangement which is non-superimposable on its mirror image.

A stereogenic center or in short a stereocenter is an atom having groups of such nature that an interchange of any two groups will produce a different stereoisomer. The stereocenters involved in *cis-trans* isomerism may however, not be chirality centers (Fig. 1.5). All stereocenters may not be tetrahedral, thus the carbon atoms of *cis*- and *trans*-2-butene provide an example of trigonal planar (sp^2 hybridized) stereocenters, since an interchange of groups at either atom leads to a stereoisomer (a diastereomer). This means that although all chirality centers (asymmetric centers) are stereocenters, not all stereocenters are chirality centers.

The cyclobutane derivative (III, Fig. 1.5) is not chiral since it contains a mirror plane and there are no chirality centers. However, changing the configuration of the either carbon atom where the halogen atom joins the ring gives a different (achiral) diastereomer *trans*-isomer. The centers are thus stereogenic (as they give rise to other stereoisomers) even though they are not chirality centers.

Unlike (III), the cyclobutane derivative (IV) is chiral, it has two stereocenters which are chirality centers as well.

All three alternative terms, asymmetric center, chirality center or stereogenic center (stereocenter) are in current use as *e.g.* in the case of carbon atoms which have four different substituents. The term chiral is applied to the whole molecule to mean being capable of existing as nonsuperimposable object and mirror image forms. Chirality is the property of the whole molecule and cannot be localized in a particular atom or a center. Thus the use of the term chiral carbon is somewhat inappropriate, the term stereogenic carbon is conceptually much better.

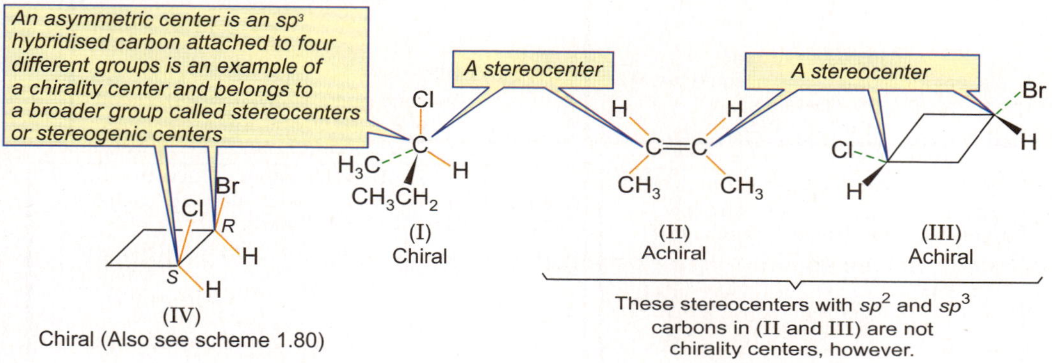

Fig. 1.5

1.2 ISOMERS

Isomers are defined as different compounds that have the same molecular formula. These are named depending on the difference which may be constitutional or configurational (stereoisomeric) (Scheme 1.1).

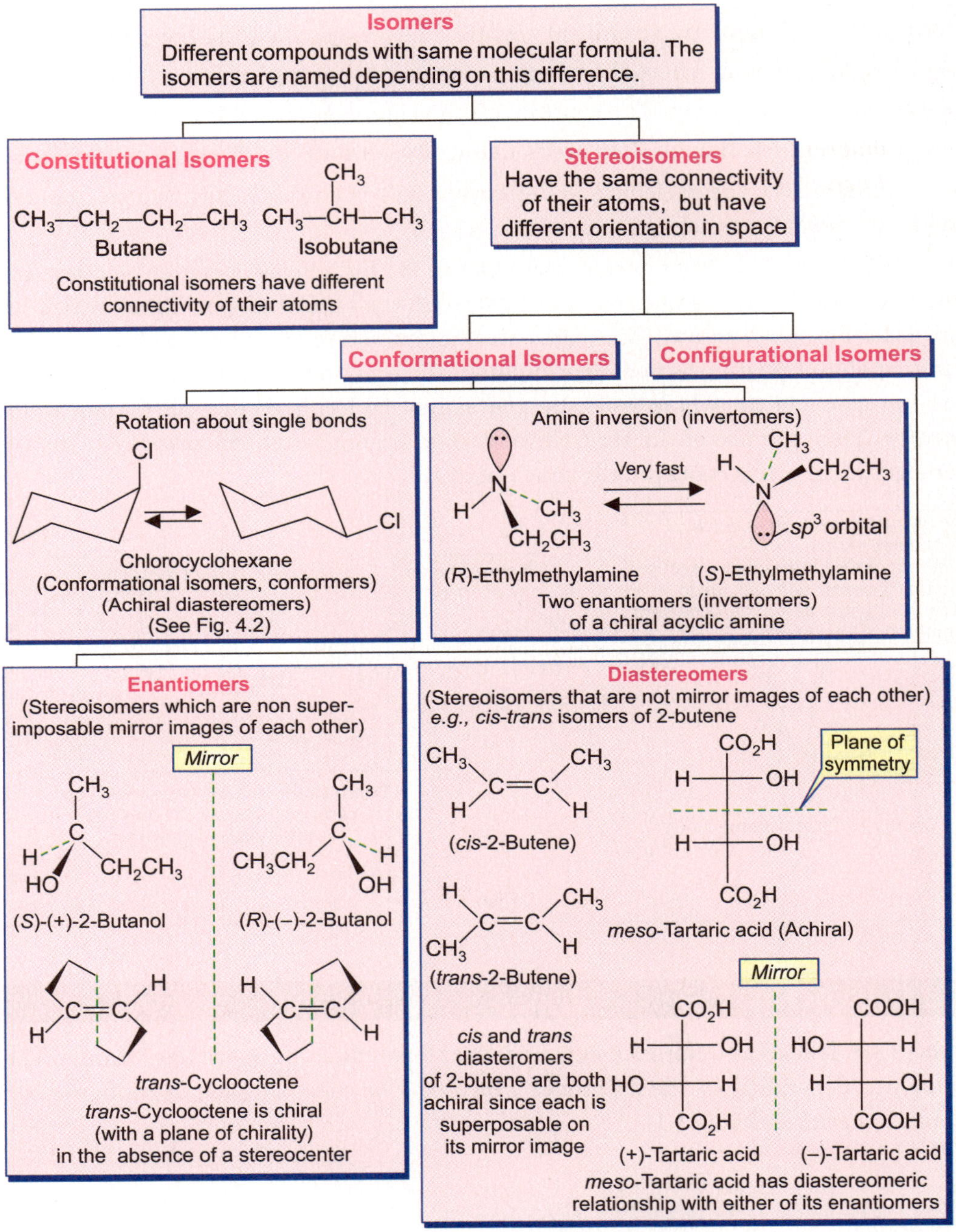

Isomers
Different compounds with same molecular formula. The isomers are named depending on this difference.
Constitutional Isomers
$CH_3—CH_2—CH_2—CH_3$
Butane
$CH_3—CH(CH_3)—CH_3$
Isobutane
Constitutional isomers have different connectivity of their atoms
Stereoisomers
Have the same connectivity of their atoms, but have different orientation in space
Conformational Isomers
Configurational Isomers
Rotation about single bonds
Cl
Chlorocyclohexane
(Conformational isomers, conformers)
(Achiral diastereomers)
(See Fig. 4.2)
Amine inversion (invertomers)
Very fast
sp^3 orbital
(R)-Ethylmethylamine
(S)-Ethylmethylamine
Two enantiomers (invertomers) of a chiral acyclic amine
Enantiomers
(Stereoisomers which are non superimposable mirror images of each other)
Mirror
(S)-(+)-2-Butanol
(R)-(–)-2-Butanol
trans-Cyclooctene
trans-Cyclooctene is chiral (with a plane of chirality) in the absence of a stereocenter
Diastereomers
(Stereoisomers that are not mirror images of each other) e.g., cis-trans isomers of 2-butene
(cis-2-Butene)
(trans-2-Butene)
Plane of symmetry
meso-Tartaric acid (Achiral)
Mirror
cis and trans diasteromers of 2-butene are both achiral since each is superposable on its mirror image
(+)-Tartaric acid
(–)-Tartaric acid
meso-Tartaric acid has diastereomeric relationship with either of its enantiomers

SCHEME 1.1

BOX 1.3b
RELATING CONFORMATIONS IN CYCLOHEXANES

- *How to know the relation of (A with I and II scheme B1.3b). Assign a numbering system (which does not follow IUPAC rules), and then determine the location and three-dimensional orientation of each substituent on the chair conformation. In case of (A, scheme B1.3c) there is a methyl group at C-1, which is up, and the ethyl group at C-2, is down:*

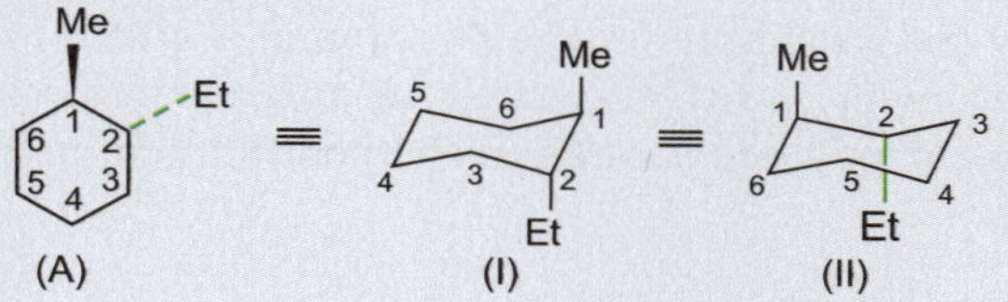

SCHEME B 1.3b

- *While assigning the numbers to the chair conformation, the first number can be placed anywhere on the ring. For example, it would be perfectly correct to draw the chair as in (II, scheme B 1.3c)*

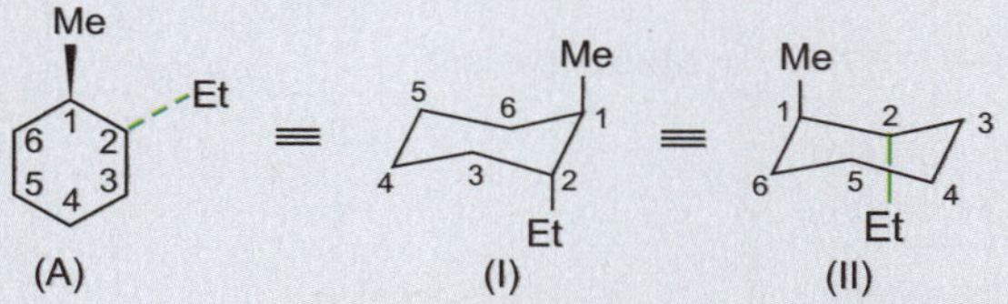

SCHEME B 1.3c

- *If one goes clockwise around the ring, one encounters the methyl group first and the ethyl group second. Moreover, in both chair conformations, the methyl group is up and the ethyl group is down. Therefore, these are identical conformations.*
- *A ring flip (scheme B1.3d) causes all equatorial groups to become axial and vice versa. The ring flip does not change whether the group is up or down.*

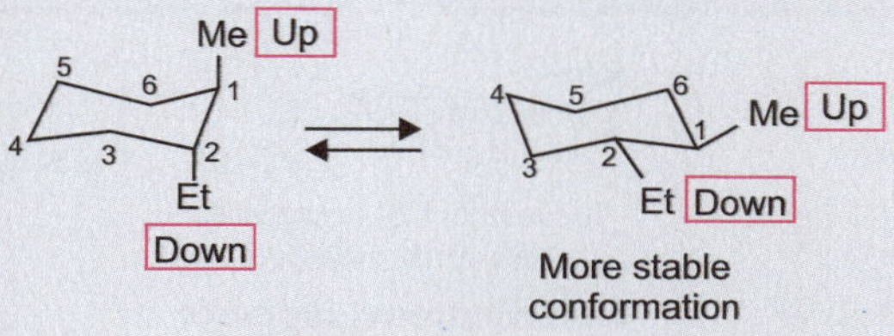

SCHEME B 1.3d

(A) Constitutional Isomers

Constitutional isomers differ in the connectivity of atoms including bond multiplicity; disregarding configuration and conformation. The equivalent older term structural isomers is obsolete and redundant. All isomers are structural isomers and structure includes constitution, configuration and conformation, as determined by X-ray diffraction.

Glucose and fructose with the same molecular formula $C_6H_{12}O_6$ (Scheme 1.1*a*) are constitutional isomers, glucose is an aldehyde while fructose is a ketone. E2-elimination is regioselective since more of one constitutional isomer is formed, compounds with different bond order, 1-pentene and 2-pentene are also called constitutional isomers.

Some other examples of constitutional isomers include:

- *Tautomers.* Isomers of different energies which are interconvertible *via* a low energy barrier, the isomerization involves atom or group migration.
- *Proton tautomers (Prototropy).* Enol-keto isomerization is an example of prototropy (*i.e.*, a change in the position of a proton) and the interconversion of the tautomers, 2-hydroxypyridine and pyridone is an example (Scheme 1.1*a*) of prototropy *i.e.*, change in the position of a proton and involves proton tautomers.

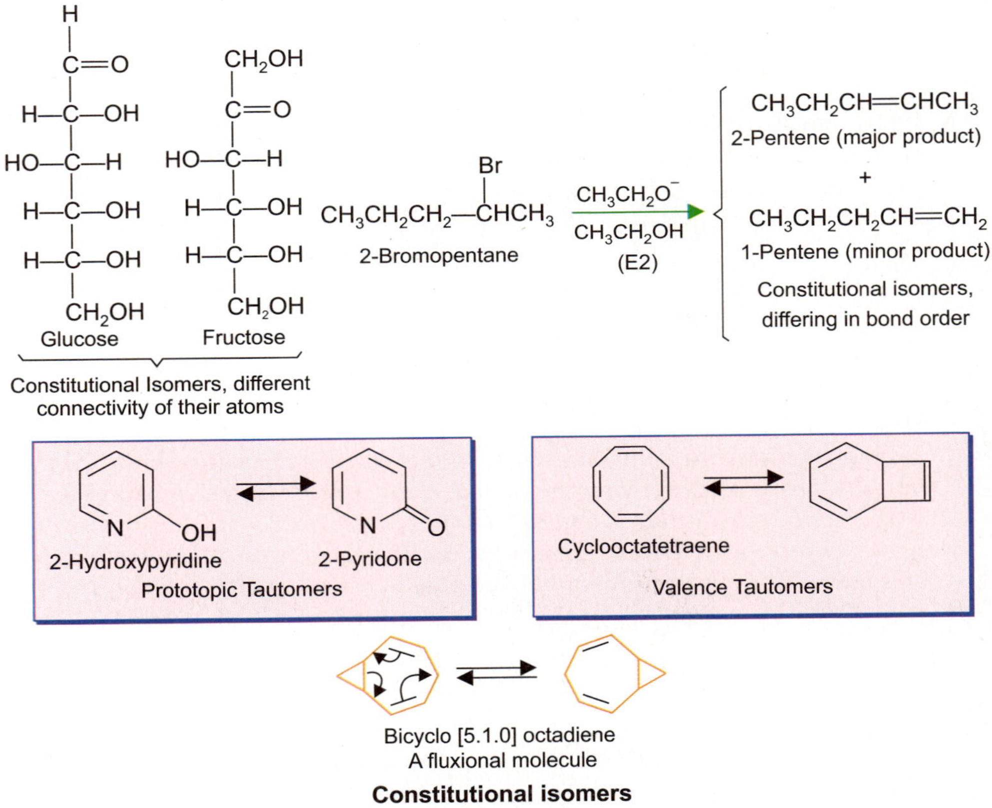

SCHEME 1.1a

- *Valence isomers.* These isomers or degenerate species that are interconvertible by reorganization of some of the bonding electrons. The interconversion is accompanied by atom movement and not atom migration. Thus valence isomers are not tautomers. Valence isomers can be separately identified and in case these have the same structure (degenerate species) the individual atoms can be separately identified. The interconversion of cyclooctatetraene and its bicyclic isomer is an example of valence isomers (Scheme 1.1*a*).

- *Fluxional molecules.* Molecules which undergo rapid degenerate rearrangement *i.e.,* a rearrangement into indistinguishable molecules *e.g., via* bond reorganization (Scheme 1.1*a*).
- *Allylic isomers.* These isomers result due to a rearrangement involving the allyl (propenyl group) and is recognized separately. These rearrangements may occur *via* the intermediate formation of a delocalized ion or a radical. The loss of X at one end and its return to the opposite end of the allyl system (Scheme 1.1*b*) leads to the overall rearrangement. The allylic isomers arise due to a 1,3-rearrangement.

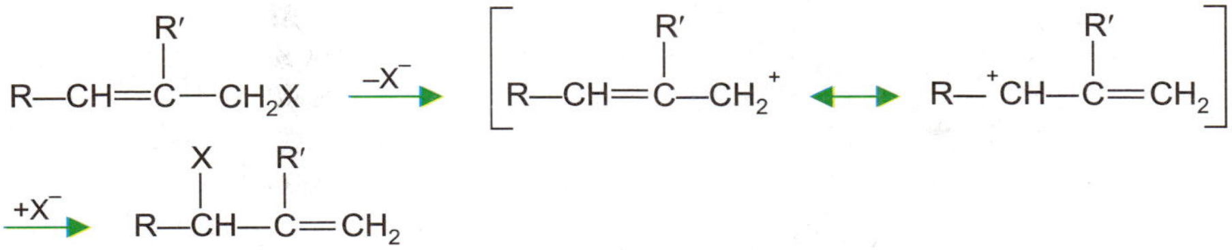

Rearrangement of allylic halides

SCHEME 1.1b

BOX 1.4
DECIDING STEREOCHEMICAL AND CONSTITUTIONAL RELATIONSHIPS IN CYCLOHEXANES

Consider the position of attachment for each substituent, determine the stereochemical relationship of the substituents (cis or trans), and then evaluate the conformational relationships (axial, equatorial, staggered, gauche, eclipsed, anti, syn-periplanar). When the point of attachment of substituents is different, then the compounds are constitutional isomers.

When the stereochemical relationships of substituents are different, then the compounds are stereoisomers (geometric isomers). If the conformational relationships are different, then the substances are conformers.

PROBLEM: ***Indicate the relationship among the compounds (A–C) in each set (I and II, Scheme B 1.4).***

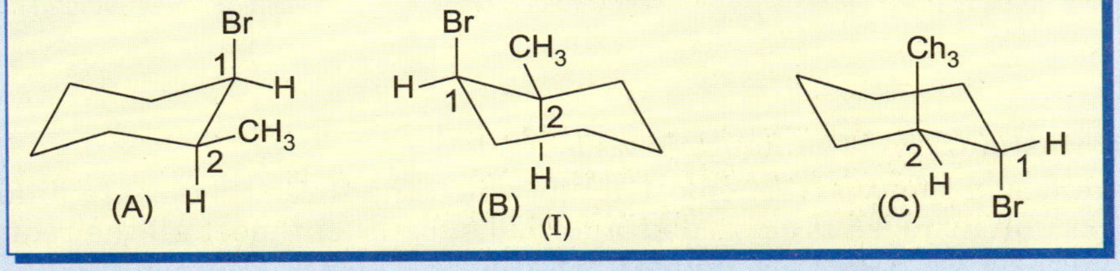

(I)

Continued...

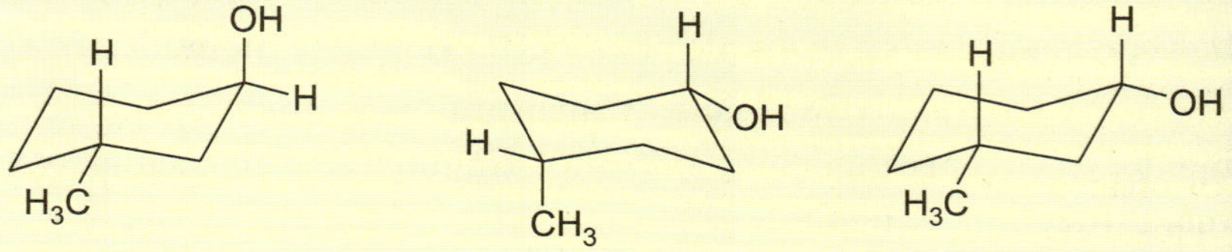

(II)

SCHEME B 1.4

Answer

The attachment points of the substituents (in set, I) are the same (1, 2 Scheme B 1.5) and their orientations are cis in compounds ***A*** *and* ***B.*** *Thus* ***A*** *and* ***B*** *are the same compound. The compounds (A and C) are stereoisomers (geometric isomers).*

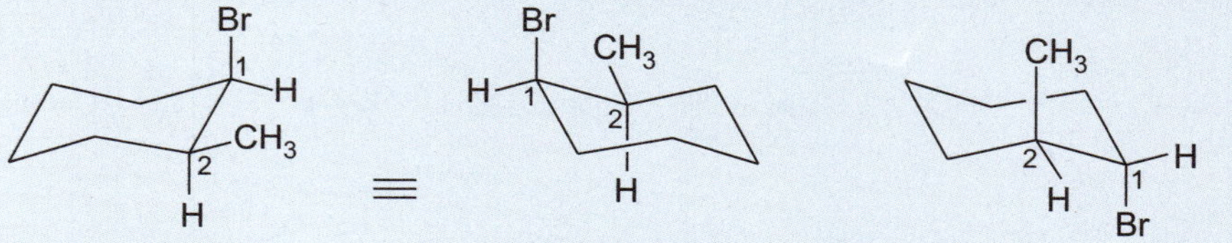

A	**B**	**C**
1-Bromo (axial) 2-Methyl (equatorial) *cis* (Both substituents are up)	1-Bromo (axial) 2-Methyl (equatorial) *cis* (Set-I)	1-Bromo (axial) 2-Methyl (axial) *trans* (CH_3 is up, Br is down)

SCHEME B 1.5

The attachment points for the substituents in (set, II) are the same in compounds ***A*** *and* ***C*** *(C1 and C3 Scheme B 1.6) but compound* ***B*** *has the substituents attached at C1 and C4. Compounds* ***A*** *and* ***C*** *are stereoisomers which differ in the stereochemical relationships of their substituents (****A*** *is trans, and* ***C*** *is cis). Therefore* ***A*** *and****C*** *are stereoisomers (geometric isomers), and A and B are constitutional isomers.*

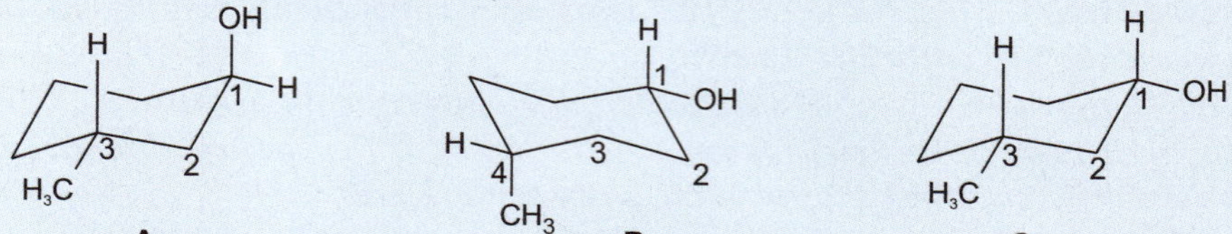

A	**B**	**C**
1-Hydroxy (axial) 2-Methyl (equatorial) *trans*	(Set-II) 1-Hydroxy (equatorial) 4-Methyl (axial) *cis*	1-Hydroxy (equatorial) 2-Methyl (equatorial) *cis*

SCHEME B 1.6

(B) Conformational and Configurational Isomers

Acyclic amines of the type (Scheme 1.1*c*) in which the three groups are different and the lone pair on nitrogen is classed as a formal substituent meet all the requirements of a chirality center. However, no optical activity is observed in amines of this type, even though these are chiral. This is due to very rapid pyramidal inversion (energy barrier to inversion is small ~25 kJ mol^{-1}) which interconverts enantiomers (called invertomers), further details are in (schemes 1.5).

Inversion is, however, slower for third-row elements. Thus, phosphines, (R_3P) with high energy barrier ~150 kJ mol^{-1} and sulfoxides ($R_2S{=}O$) as well can be obtained in optically active form.

In the case of asymmetrically substituted phosphines, the optically active isomers can often be isolated, and this is particularly so when one of the ligands is aromatic. This type of phosphines are integral components of chiral catalysts. An ethane–1, 2–diphosphine (DIPAMP) has two stereogenic phosphorus centers (scheme 1.1*c* also see scheme 6.58). Amine inversion is an example of conformational isomers (invertomers), while biphenyl enantiomers may be isolated due to restricted rotation. Such isomers are called atropisomers.

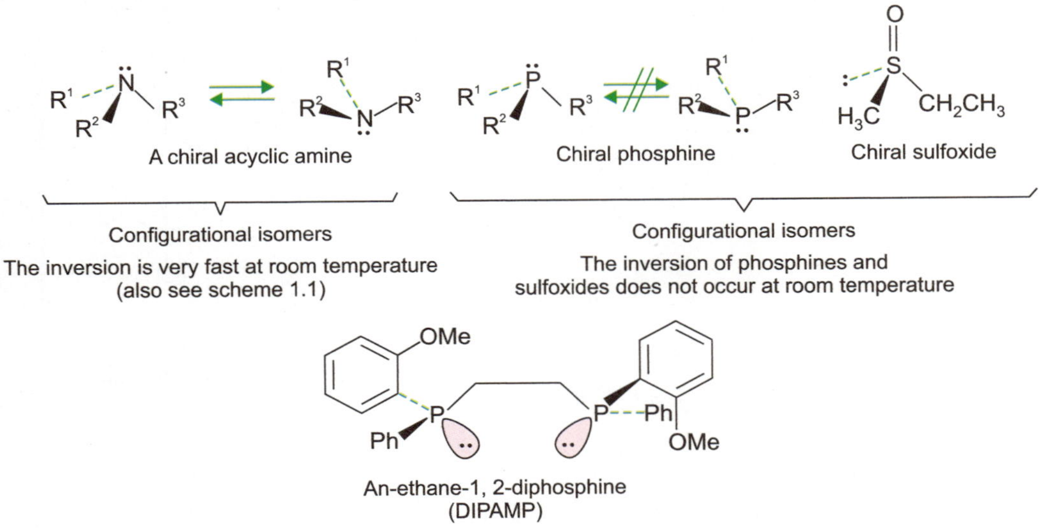

SCHEME 1.1c

The Distinction between Conformation and Configuration

- *Based on symmetry criteria, stereoisomers may be enantiomers (related as object and nonsuperimposable mirror image) and when these are not so related, these are termed diastereomers. The diastereomers occur in compounds with more than one stereocenter, cyclic compounds and compounds with double bonds (Scheme 1.1).*
- *Based on energy barrier criterion, configurational isomers are separated by a high energy barrier > 100 kJ mol^{-1}. The two enantiomers of 2-butanol (Scheme 1.1) are separated by high energy barrier since their interconversion involves a σ bond breaking. They represent configurational enantiomers. In the isomerisation of cis- and trans-2-butene a π bond is disrupted which requires an appreciable amount of energy; they are also configurational diastereomers. The stereoisomers which on the other hand are separated by comparatively low energy barrier < 60 kJ mol^{-1} so that interconversion is easy under ordinary conditions, are called conformational isomers. In some crowded molecules e.g., biphenyls however, rotation about a single bond may be sufficiently restricted to give stable and isolable conformers known as atropisomers which are configurational isomers (Scheme 1.1e).*

Continued...

The distinction between conformation and configuration is in fact subtle and not agreed upon universally. The acyclic amine inversion (Scheme 1.1) has a typically low energy barrier (33.5 kJ/mol) and may be considered either a configurational or a conformational change. These invertomers are however, better considered as conformational enantiomers. However, these arguments do not apply to chiral phosphines (Scheme 1.1c) where inversion is associated with high energy (~ 150 kJ/mol) to result in configurational isomers.

Rotation around a single bond may be easy to give conformational isomers e.g., in n-butane (Scheme 1.1d). Gauche butane is chiral. Two enantiomers interconvert by a conformational change too rapidly (conformational enantiomers). Anti butane is achiral and either of the gauche butanes is its diastereomer (conformational diastereomers). Gauche butane is an example of racemization.

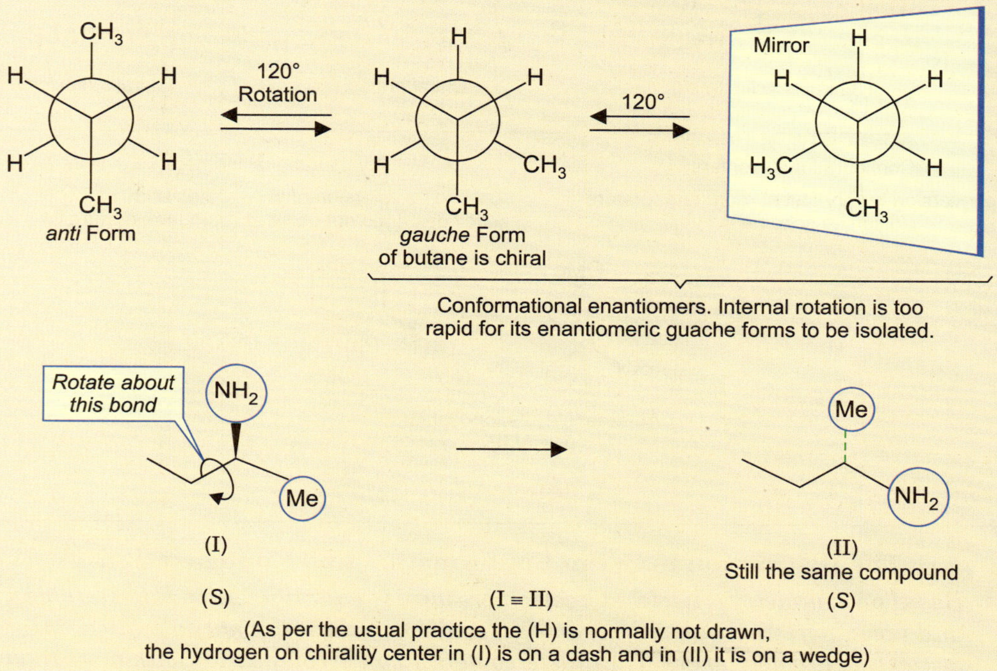

SCHEME 1.1d

Similarly chlorocyclohexane (Scheme 1.1) represents a pair of conformational diastereomers. Infact, chlorocyclohexane shows conformational isomerism at room temperature while configurational isomerism at –150°C (see Fig. 4.2). Another interesting example of conformational enantiomers is in (Scheme 4.33).

Configuration does not change with conformation

A compound can have infinite number of conformations but only one configuration—configuration does not change with conformation. When e.g., one of the bonds is rotated (I Scheme 1.1d) one moves from one thick bond to one dotted bond (or from one dotted bond to one thick bond). This conformational change has no effect on the chirality center. Examine e.g., the two different conformations of the same compound meso- tartaric acid (I and III, Scheme 1.41).

(C) Stereoisomers

When the isomers have the same sequence of covalent bonds, but differ in the relative disposition of their atoms in space, then the difference is stereoisomeric (scheme 1.1) (also see Fig. 1.2). Some examples are, enantiomers, diastereomers (epimers, anomers), conformational isomers (atropisomers and invertomers).

(D) More on Enantiomers—Optical Isomerism

Simple organic molecules

Consider a simple molecule *e.g.*, a compound with an sp^3 hybridized carbon with four different substituents as in 2-butanol (Scheme 1.1). The molecule cannot be superimposed on its mirror image and such molecules are said to be chiral (or handed). The pair of butanol molecules are termed enantiomers (from the Greek 'enantio' meaning opposite) which are defined as pair of molecules related as non-superimposable mirror images. The enantiomers of 2-butanol are drawn (Scheme 1.1) in the three dimensional projection formulas (a procedure to draw these projections is depicted in Scheme 1.15).

Another example of enantiomers is in D- and L-glucose drawn now in another projection (Scheme 1.2) called a Fischer projection, (a procedure to draw these projections is detailed in (Fig 1.3*b*) and (Scheme 1.17). The D-sugars have the OH group on the highest number chirality center (bottom stereocenter) on the right in the Fischer projection. The unnatural L-series of sugars are the enantiomers of the natural D-series. For the assignment of D and L configuration to sugars in their cyclic form (see Scheme 1.2*f*).

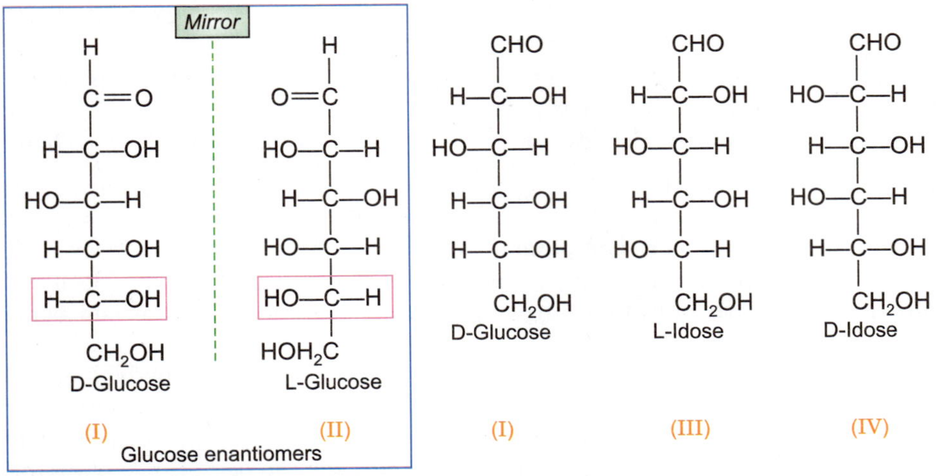

The mirror image (enantiomer) of D-glucose (I) is L-glucose (II) with inverted stereochemistry at every chirality center. By simply changing the stereochemistry at only one chirality center (bottom) of D-glucose does not, however, give L-glucose, it however, converts a D-sugar (D-glucose) into a L-sugar, which is L-idose (III) an enantiomer of natural D-sugar D-idose (IV).

To recognise glucose (The placement OH, OH, H, OH) from the highest chirality center to the lowest chirality center must be recognised. If this arrangement is on the right hand side of Fischer projection, the sugar is D-glucose, if it is on left hand side the sugar is L-glucose.

Note that inverting some but not all, chirality centers gives a diastereomer. Thus D-glucose and D-idose are diastereomers.

SCHEME 1.2

Compounds can be chiral and thus exist as a pair of enantiomers in the absence of chirality centers as in *trans*-cyclooctene (Scheme 1.1, further details are in Schemes 1.136–1.138). Another example of compounds which are chiral in the absence of chirality centers are biphenyl derivatives (Scheme 1.1*e*, further details are in Sec. 1.16, B).

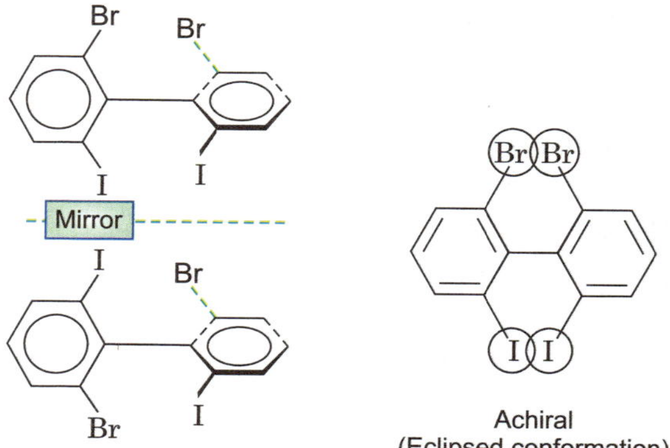

Chiral biphenyl (enantiomeric pair)
(Staggered conformation)
Chirality is due to the presence of an axis of chirality
(I)

Achiral
(Eclipsed conformation)
too crowded to exist

> *Stable conformational isomers (atropisomers) exist in compounds, e.g., biphenyls due to steric strain between the ortho substituents. These isomers become chiral when both rings are unsymmetrically substituted. The biphenyl then gets locked in one of the two chiral enantiomeric staggered conformations. This biphenyl would have been achiral if a symmetric (planar) high energy eclipsed conformation could be achieved (an impossible situation.)*

> *Unlike, gauche butane, conformational enantiomers in suitably substituted biphenyls can be isolated.*

SCHEME 1.1e

Necessary condition for chirality

Normally, the sufficient condition for chirality in an organic compund is the presence of a chirality center. This however, is not essential condition. The essential condition is however, the asymmetry of the molecule as a whole as in chiral biphenyl (scheme 1.1e)

Thus compounds of the type C_{abcd} exist in enantiomeric forms and are described as chiral and the carbon atom with four different achiral atoms or groups as substituents is called a chirality center. The phenomenon of enantiomers is also known as optical isomerism. An important property of compounds of type C_{abcd} *e.g.*, a molecule with one tetrahedral atom with four different groups attached to it is enantiomerism as in butanol (Scheme 1.1).

An important property of such enantiomers (*i.e.*, a chiral tetrahedral model) is that on interchanging any two groups at the chirality center converts one enantiomer into another. In addition to the compounds of the type C_{abcd} with one chirality center *e.g.*, 2-butanol (Scheme 1.1) which fulfill the conditions for the occurrence of enantiomeric pairs, several other structural situations may give rise to enantiomerism. These include compounds with more than one chirality center (as in glucose, Scheme 1.2).

Chirality centers other than carbon e.g., based on tetracoordinate silicon and tricoordinate sulfur or phosphorus also exist (see, schemes 1.5 and 1.6).

Pyramidal inversion at N is so fast that chiral amines cannot be resolved. Phosphines however, undergo pyramidal inversion slowly and can be resolved.

Chiral Organic Compounds

- *The presence of a chirality center is a sufficient condition for molecular chirality.*
- *Atoms other than carbon can be chirality centers. Examples include those based on tetracoordinate silicon and tricoordinate sulfur or phosphorus as the chirality center. In principle, tricoordinate nitrogen can be a chirality center in compounds of the type N(x, y, z), where x, y, and z are different, but inversion of the nitrogen pyramid is so fast that racemization occurs virtually instantly at room temperature.*
- *The essential condition for chirality is the absence of plane of symmetry and centre of symmetry (S_n) in a molecule. A chiral compound can have a rotational axis of symmetry C_n.*
- *Several molecules display chirality (optical isomerism) in the absence of chirality centers e.g., chiral biphenyls. Thus a compound is chiral if it is not superimposable on its mirror image.*
- *Organic stereochemistry is mainly based on tetrahedral geometry of carbon which is absolutely central to its study. Study of stereochemistry is also based on atoms like N, P, Si and S and to a lesser extent on the trigonal geometry of sp^2 hybrid carbon (see Scheme 1.44)*

(E) *Meso* Compounds

In tartaric acid (Scheme 1.2*a*) one has two chirality centers. A molecule with two chirality centers can give rise to a maximum of four stereoisomers (2^n as in 2-bromo-2-butanol, see Scheme 1.28*c*). However, if the two chirality centers carry an identical set of substituents, the number of stereoisomers is less than 2^n, since there will be a *meso* compound as in tartaric acid (Scheme 1.2*a*). The *meso* compound *e.g.*, tartaric acid has a plane of symmetry and is achiral.

Chirality and Chirality Centers

- *Chirality is a necessary and sufficient condition to generate enantiomerism and requires the absence of S_n (alternating axis of symmetry of any order).*
- *The presence of a chirality center usually imparts molecular chirality. A unique feature of such a chirality centre is that exchange of any two ligands inverts its chirality to yield a new stereoisomer. When all the ligands are achiral, the exchange gives an enantiomer, however, if one or more of the ligands are chiral, a diastereomer will be formed. This is seen in the case of tartaric acid. When one interchanges the groups on one chirality center in meso- tartaric acid (see, Scheme 1.2a) an enantiomer of tartaric acid is formed and the two have diastereomeric relationship. (inverting some, chirality centers but not all gives a diastereomer).*
- *Thus an organic molecule with one chirality center must be chiral, however, molecules with two or more chirality centers are not all chiral.*

Continued...

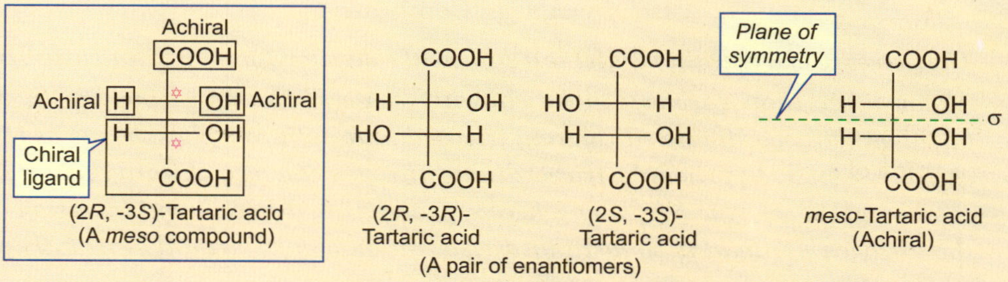

Stereoisomers of tartaric acid in Fischer projections—Interchange of H and OH at the bottom chirality center in *meso* tartaric acid gives an enantiomer of Tartaric acid (2*R*, 3*R*) with which it has diastereomeric relationship.

SCHEME 1.2a

A *meso* compound

A meso compound is an achiral compound (not optically active) which contains chirality centres and has an internal plane of symmetry). It has two chirality centres with opposite stereodescriptors. The latter condition renders the molecules optically inactive because there is internal compensation; the rotation caused by the R chirality centre is cancelled by that caused by the S (also see Fig. 1.4b).

Chirality centers *e.g.*, in tartaric acid stereoisomers are assigned *R* and *S* configurational descriptors, so as to specify stereochemical features of each stereoisomer. The enantiomer of (+) tartaric acid is its nonsuperimposable mirror image (–) tartaric acid and these constitute an enantiomeric pair. Notice that pairs of enantiomers (as expected) have opposite configuration at every chirality center.

A racemic (±) mixture has a sharp melting point which differs from the pure enantiomers. Because a pair of enantiomers exhibit the same chemical and physical properties, they cannot be separated by ordinary chemical or physical methods. The most general technique for resolving a pair of enantiomers is to subject them to a reaction with a chiral reagent to obtain a pair of separable diastereomeric products.

Table 1.1: Physical properties of stereoisomers of tartaric acid

Acid	Melting point, °C	$[\alpha]_D^{25°C}$	Solubility, g/100 g H_2O at 15°C
(2R, 3R)-(+)-Tartaric acid	171	+ 12.7°	139
(2S, 3S)-(–) Tartaric acid	171	– 12.7°	139
(2R, 3S)-Tartaric acid	140	0°	125
(±)-Tartaric acid	206	0°	20.6

(F) Complex Organic Molecules and Biomolecules

Except for few low molecular weight organic compounds, the organic substances found in living systems both animals and plants are chiral. No doubt these molecules (with several stereocenters) can theoretically exist as a number of stereoisomers, almost invariably only one stereoisomer is found in nature. Naturally occurring alkaloid brucine has several stereocenters which are located in fused ring systems, however, nature makes only one enantiomer (–)-brucine. Naturally occurring amino acids (with the exception of achiral glycine) are chiral. There are two possible enantiomers (optical isomers) for each amino acid, but only one of them (L-form) exists in the body. Enzymes are proteins which are derived from chiral amino acids, thus an enzyme is also chiral and can exist as enantiomers, however only one enantiomer exists naturally (since an amino acid exists only as one enantiomer these will construct only one mirror image form of the enzyme). Thus enzymes provide a chiral environment.

Enzymes catalyzed reactions are stereospecific and stereoselective

- *For definition of terms, stereospecific and stereoselective (see Sec. 1.13).*
- *Enzymes are chiral and enantiomerically pure.*
- *Enzymes display stereospecificity and stereoselectivity (see Sec. 2.3 D).*
- *All stereospecific reactions are necessarily stereoselective, however, the converse is not true.*

(G) Properties of Enantiomers

Each enantiomer of a pair has the same physical and chemical properties in achiral environments with the important exceptions of their interactions with (*i*) plane polarized light (optical activity) and (*ii*) chiral reagents. When plane polarized light is passed through the solution of each enantiomer (in the same solvent, using the same cell and same concentration), then the plane of polarized light is rotated in opposite directions by the same amount as in the enantiomers of tartaric acid (see Table 1.1)

Plane polarized light is in fact an equal mixture of left and right circularly polarized light which propagates through space as left handed and right handed helices respectively. Due to the chirality of the circular components of the plane polarized light the two enantiomers of a compound react with it differently.

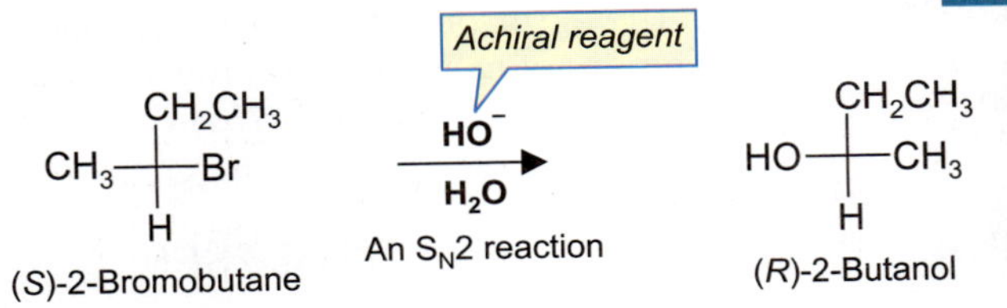

Achiral reagents e.g. hydroxide ion react with enantiomers [reaction only with (S)-enantiomer is shown] at the same rate. Chiral reagents discriminate between them.

Continued...

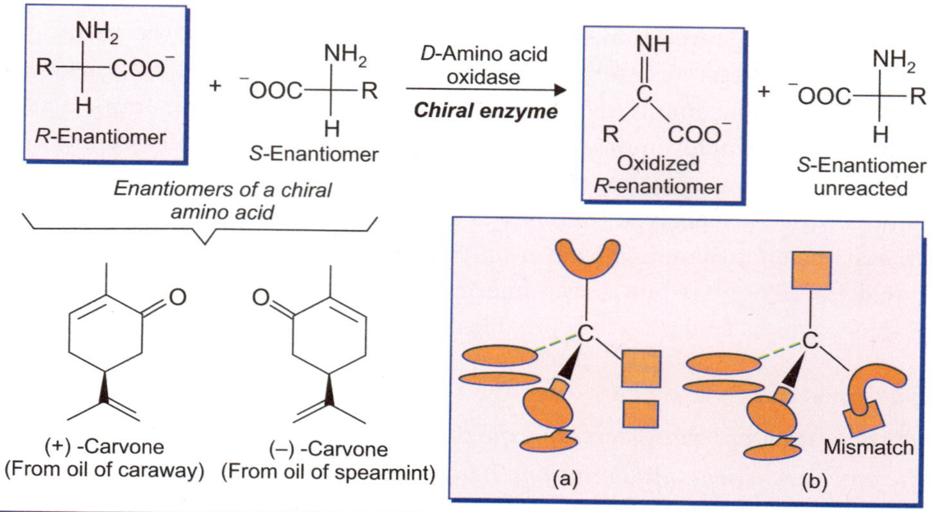

The individual enantiomers of carvone smell different, since the receptor sites in the nose are chiral. (a) An enzyme surface is capable of interacting with one of the enantiomers of e.g., 2-bromobutane at three binding sites. (b) The other enantiomer does not fit the same binding sites and therefore, ignored by the enzyme.

Thus the enzymes usually act on only one of the two enantiomers–a property known as stereospecificity–and in the case of prochiral molecules with only one of the enantiotopic ligands or faces–a property known as stereoselectivity (see schemes 2.14 and 2.18 for terminology).

SCHEME 1.2b

Each enantiomer shows the same chemical reactivity with achiral reagents *i.e.*, enantiomers react with achiral reagents at the same rate. Thus *e.g.* (*S*)-2-bromobutane reacts with achiral hydroxide ion to give (*R*)-2-butanol (Scheme 1.2*b*) by an S_N2 mechanism. The rate of this reaction is found to be the same with the enantiomeric (*R*)-2-bromobutane with hydroxide ion to give (*S*)-2-butanol.

When, however, the reagent is chiral *e.g.*, an enzyme, the two enantiomers will react at different rates. The enzyme D-amino acid oxidase reacts only with one of the enantiomers of a chiral amino acid, the (*R*)-enantiomer, the (*S*)-enantiomer remaining unchanged (Scheme 1.2*b*). Another example is found during the kinetic resolution of amino acids (See, Scheme 1.85).

Receptors are proteins which are chiral and thus these will bind one of the enantiomers better than the other *i.e.*, one enantiomer binds with a particular receptor whereas the other does not. Receptors located on the exterior of nerve cells in the nose are thus able to differentiate odours. The enantiomers of carvone (Scheme 1.2*b*) smell different since each fits into a different receptor. (Also, see Scheme 1.67*d*)

BOX 1.5

ENZYMES DISTINGUISH BETWEEN TWO ENANTIOMERS

The pocket in the enzyme chymotrypsin is lined with non-polar amino acids and for this reason chymotrypsin cleaves on the C–side of the amino acids with flat nonpolar side chains (Phe, Tyr, Trp). Thus chymotrypsin catalyzes the hydrolysis of the peptide bond on the C-side of amino acids which have aromatic six-membered rings (Phe, Tyr, Trp). The pocket in chymotrypsin thus matches with the binding of

Continued...

the side chains of the subtrate (Scheme B 1.7). The fit of the amino acid residue in the pocket in the active site of chymotrypsin is hydrophobic where the phenyl group fits perfectly (I). Moreover, the shape of the chiral pocket in the active size is uniquely matched with L-phenylalanine moiety of the substrate (The lock and key fit). The cavity however, does not match (II) with the "wrong" enantiomer, the D-enantiomer.

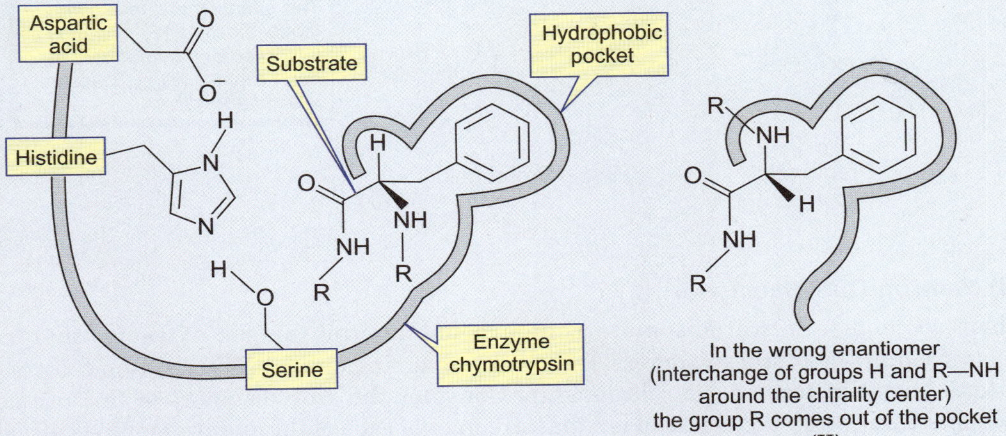

(I) (II)

SCHEME B 1.7

In summary chiral substances react only with substances that match their own chirality. This forms the basis for an enzyme to distinguish between two enantiomers of a compound, during enzyme catalyzed reactions. The enzyme first positions a molecule at the binding site on its surface (via, hydrogen bonds, electrostatic attractions, dispersion forces or even covalent bonds). An enzyme with specific bonding sites for three of the four groups on a stereocenter can make a distinction between a molecule and its enantiomer (or one of its disastereomers). From (Scheme 1.2*b*), one can see that one enantiomer can be absorbed at its three bonding sites while the other cannot. The enzyme lactate dehydrogenase *e.g.*, brings about the conversion of only one of the enantiomers of lactic acid to pyruvic acid ($CH_3COCOOH$, Scheme 1.2*b*).

BOX 1.6
CHIRAL DRUGS
IMPORTANCE OF RESOLUTION OF RACEMIC MIXTURES AND STEREOSELECTIVE SYNTHESIS

There are many examples of biologically active compounds including drugs where the enantiomers have distinctly different effects. The preparation of enantiomerically pure drugs, therefore, is one factor that makes stereoselective synthesis and the resolution of racemic drugs (separation into pure enantiomers) major areas of research today.

Moreover, in a number of cases, one of the enantiomers of a drug has been found to act as a blocker of the biological receptor site, thus diminishing the activity of the

Continued...

other enantiomer. Worse, one of the enantiomers may have a completely different, and sometimes toxic effect (Scheme B 1.8).

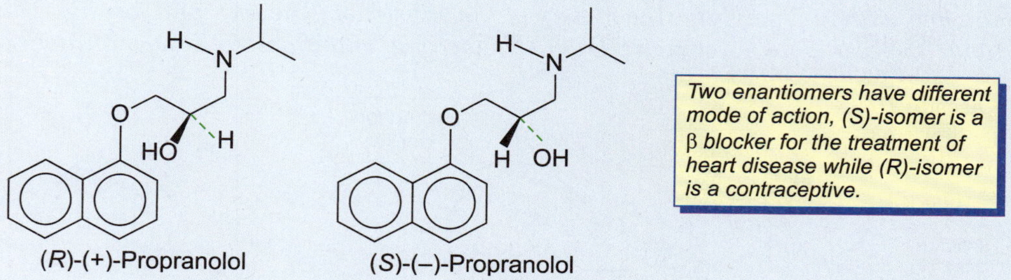

SCHEME B 1.8

(H) More on Diastereomers

Diastereomers are stereoisomers that are not mirror images of each other *i.e.*, diastereomers are stereoisomers that are not enantiomers. Thus *cis*-and *trans*-2-butene have diastereomeric relationship. Consider the stereoisomers of tartaric acid (Scheme 1.2*a*), *meso*-tartaric acid is a diastereomer of each of the enantiomers. As already said the pairs of enantiomers have the opposite configuration at every stereocenter *i.e.*, enantiomeric tartaric acids are (2*R*, 3*R*) and (2*S*, 3*S*). While a diastereomer *i.e.*, *meso*-tartaric acid has one stereocenter of common configuration while the other of opposite configuration (2*R*, 3*S*, Note that inverting some, but not all, of the chirality centers in a stereoisomer gives a diastereomer.

Consider another example from a cyclohexane derivative (I, Scheme 1.2*c*). One can draw one of its diastereomers (II) by changing the configuration at one of the chirality centres and its enantiomer (I*a*) by changing the configuration at every center.

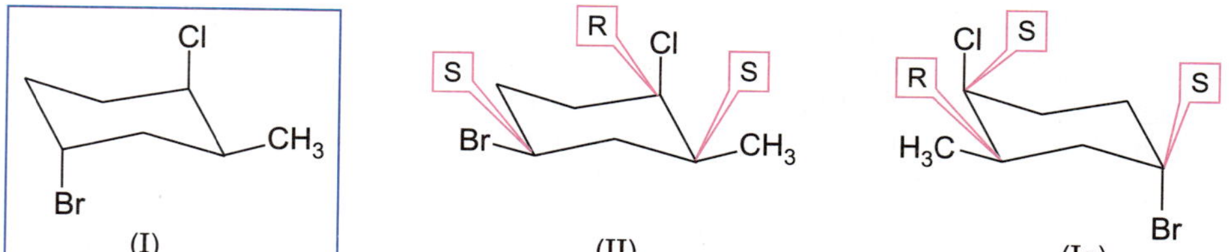

(II) Diastereomer of (I) (Only one center changed)

(I*a*) Enantiomer (mirror image of I) (All centers changed)

SCHEME 1.2c

Diastereomers are nonenantiomeric stereoisomers which have two or more chirality centers and differ in the projection of at least one of them. Thus galactose, glucose and mannose are all diastereomers (Scheme 1.2*d*).

Epimers are diastereomers

When two diastereomers differ in the configuration at only one chirality center (of several) then these are called epimers. The term is quite general, however, it is rarely used for molecules with only two chirality centers. The chirality center at which the difference in configuration occurs is said to be the epimeric center. Glucose and galactose are epimers at C4. Similarly glucose and mannose are epimers at C2 (Scheme 1.2*d*). Glucose and talose (as well as galactose and talose) are not epimers because they differ in the projections at more than one chirality center. Interesting examples of epimers and enantiomers are in (Scheme, 1.2*e*), involving bicyclo [2.2.1] heptane and [2.2.2] octane system. α-D-glucose and β-D-glucose are anomers which are epimers in the carbohydrate series that differ only at the hemiacetal carbon atom (see, Scheme 1.2*f*).

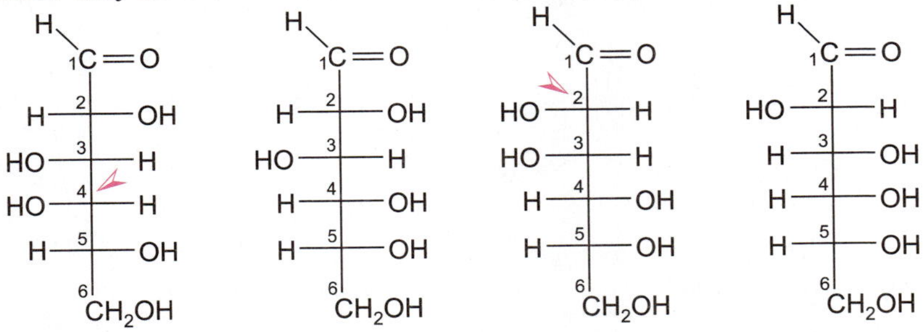

D-Galactose D-Glucose D-Mannose D-Talose

D-galactose, D-glucose and D-mannose are diastereomers, D-galactose and D-glucose are C4 epimers while D-glucose and D-mannose are C2 epimers. D-glucose and D-Talose, however, are not epimers (though these are diastereomers) since these differ in configuration at more than one chirality centre (C2 and C3).

SCHEME 1.2d

BOX 1.7

ISOMERS OF METHYLBICYCLO [2.2.1] HEPTANE

- *Carbons 1, 2 and 7 are different*
- *At C1 and C7 only one possible way of attaching CH_3 group (Plane of Symmetry)*
- *At C2 it could either point toward the short bridge exo or longer bridge endo*
- *Total stereoisomers 4 (three chirality centers, see Scheme 1.67h)*
- *Exo and endo stereoisomers are diastereomers (an epimeric pair)*

Constitutional isomers

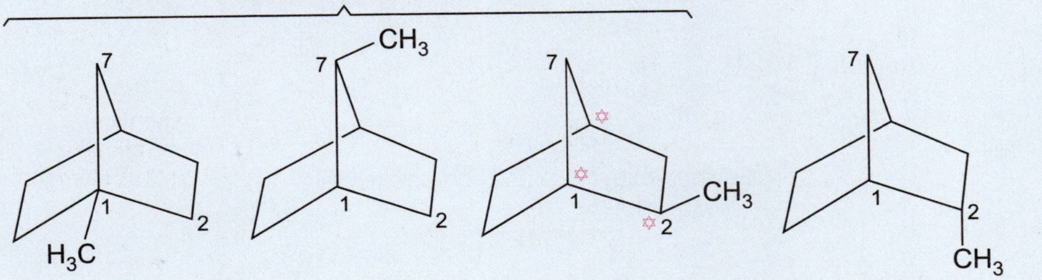

(I) *exo*-2-Methylbicyclo-[2.2.1]heptane

(II) *endo*-2-Methylbicyclo-[2.2.1]heptane

Configurational isomers (epimers)
Each (I and II) has a mirror image (not shown) which increases the total number of stereoisomer to four

SCHEME B 1.9

BOX 1.8
ISOMERS OF METHYLBICYCLO [2.2.2] OCTANE

Consider the structures of [2.2.2] octane system.

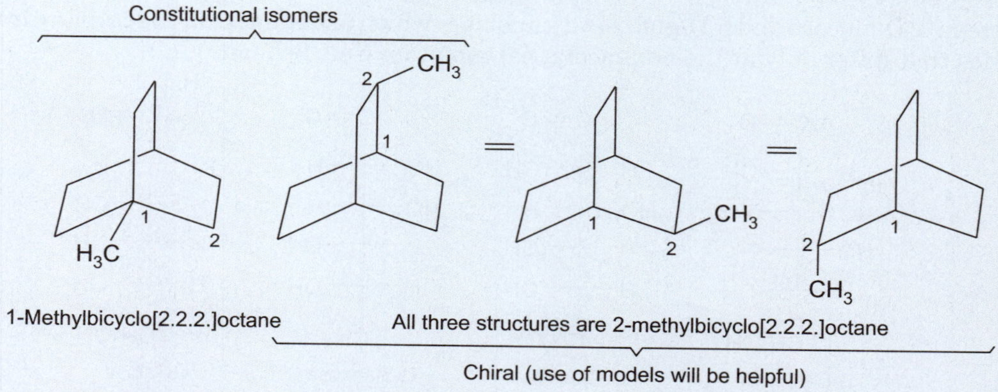

SCHEME B 1.10

- *There are only two different carbons C1 and C2 and two constitutional isomers (Scheme B 1.10).*
- *Unlike methylbicyclo [2.2.1] heptane, there are no exo and endo diastereomers as the two bridges are identical.*
- *2-Methylbicyclo [2.2.2] octane is chiral solely due to the chirality centre at C2 and exists as enantiomeric pair. (Scheme B 1.11).*

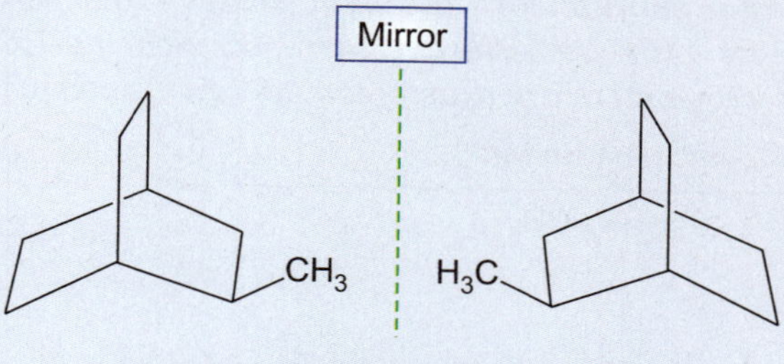

2-Methylbicyclo [2.2.2] octane is chiral

SCHEME B 1.11

More examples of enantiomers, diastereomers, epimers are found in methylbicyclo [2.2.1] heptane system and methylbicyclo [2.2.2] octane systems which are bridged ring systems. The prefix *epi* is a convenient way to name chiral structures that are epimer of molecules with non systematic names. For example ambrox (V, Scheme 1.2*e*) is a whale product which was used in perfumery. Its synthetic isomer that is epimeric at the 9-position which has a better potential as a perfume was named 9-*epi*-ambrox, a designation which completely defines its structure.

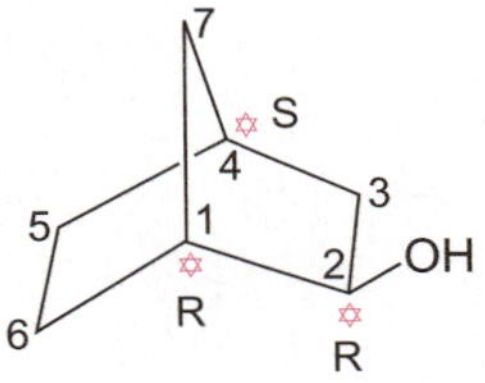

exo
(I)

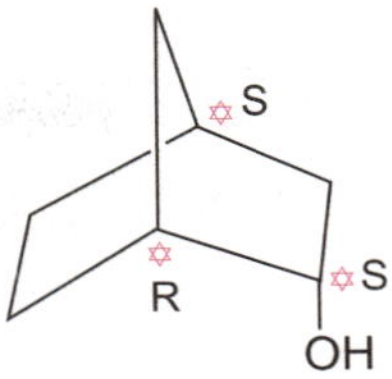

endo
(II)

(Epimeric pair)

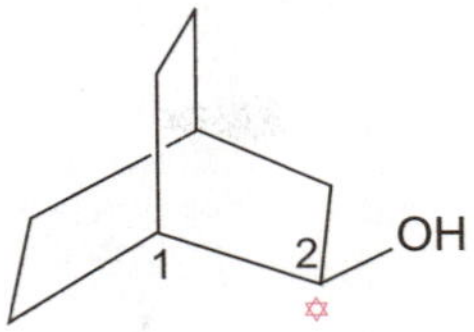

(III)

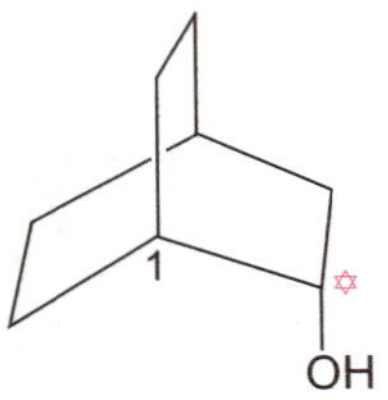

(IV)

(Enantiomeric pair)

The [2.2.1] heptane system in (I and II) has three chirality centers. These alcohols fulfil the criterion of epimers having different configurations at the hydroxyl bearing carbon C2 (Further details are in (Sec. 4.16)).

The [2.2.2] octane system has only one chirality center and therefore, can exist only as an entiomeric pair (More details are in Sec. 4.16) and (Box. 1.8)

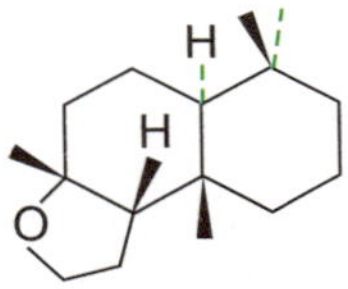

(V)

Ambrox

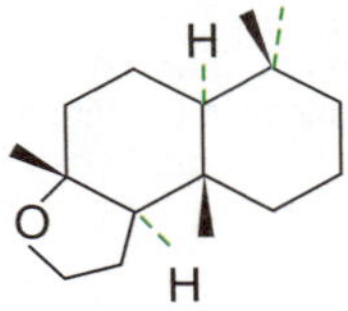

(VI)

9-*epi*-Ambrox

SCHEME 1.2e

Anomers are epimes (diastereomers) which differ in absolute configuration at the anomeric center during the formation of a cyclic hemiacetal of a carbohydrate

The term is primarily used in carbohydrate chemistry to indicate the two epimers at the hemiacetal or acetal carbon of a sugar in its cyclic form. The OH group at C5 of open chain form of glucose attacks the aldehyde (carbon C1) to form a hemiacetal. A new stereocenter at C1 is generated and a pair of diastereomers is formed. These diastereomers which differ in the configuration at C1 (called anomeric carbon) are called anomers (Scheme 1.2*f*). Thus α-D-glucopyranose and β-D-glucopyranose are diastereomers. When the H atom on the highest numbered chirality center is down (in a pyranose form), the sugar is D. With a D-sugar if the OH at C1 is down the sugar is α and if up the sugar is β. Also see (Scheme 1.20*g*), and for mechanism of epimerisation (see, page 31).

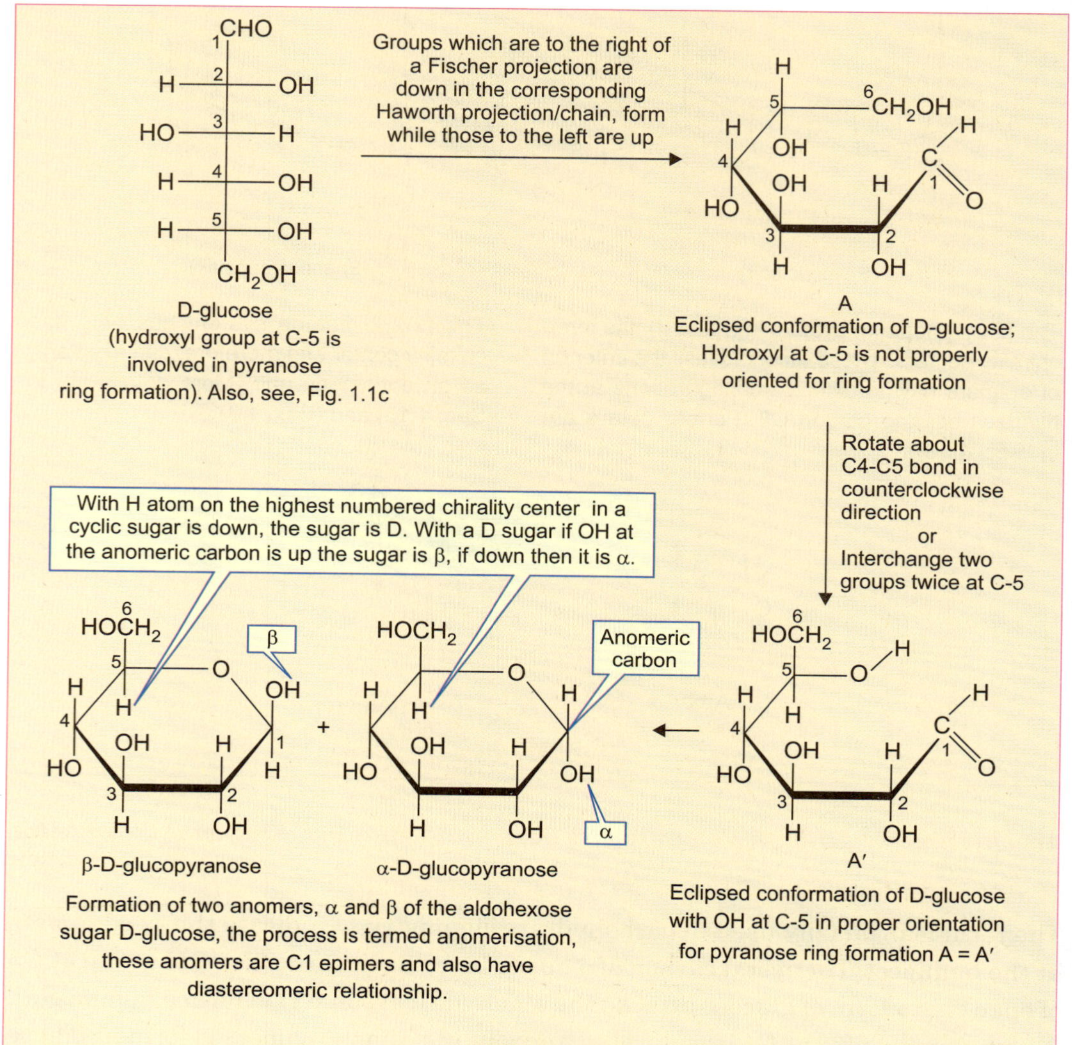

SCHEME 1.2f

In the case of glucose, α-D-glucose and β-D-glucose are interconnected fairly rapidly in solution through the open chain form *i.e.*, the Fischer projection. The β form predominates (β : α ~ 2 : 1) since in β-D-glucose all the substituents are in the equatorial positions (see, scheme B 1.12, Box 1.9). When crystalline α-D-glucose is dissolved in water, the initial rotation of the solution $[\alpha]_D^{20} = 111°$, falls gradually to an equilibrium value of 52.5° (The β-form has $[\alpha]_D^{20} = 19°$. The phenomenon is known as mutarotation.

CHAPTER 1

BOX 1.9
ANOMERISATION OF FISCHER PROJECTION OF D-GLUCOSE

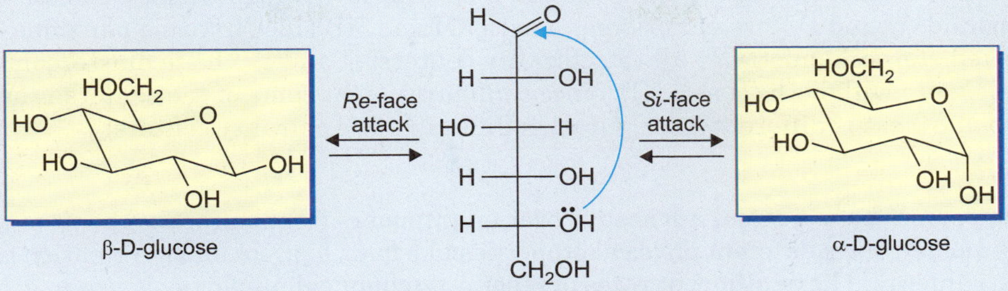

β-D-glucose

D-Glucose Fischer projection

α-D-glucose

α–D-glucose and β-D-glucose are diastereomers. They are also epimers and anomers.

SCHEME B 1.12

- *The descriptors Re and Si define the two heterotopic faces of a trigonal planar prochirality center (see Scheme 2.15) will three constitutionally different groups (with C1 in glucose, these are H, O and the rest of the five carbon chain of glucose.*
- *When the two groups attached with the carbonyl group are same as in HCHO and CH_3COCH_3 the faces of the carbonyl group are homotopic. When the groups are different as in CH_3CHO, the faces are enantiotopic and when one of the group contains one or more chirality centres, the faces are diastereotopic. Therefore, the faces of the carbonyl group in the Fischer projection of D-glucose are diastereotopic and addition to these faces gives diastereomers. Thus α-D and β-D glucose molecules are diastereomers (epimers at C1–anomers).*

Although *cis-* and *trans* isomers of alkenes are diastereomers (see, Scheme 1.1) that are achiral, the majority of diastereomeric compounds are chiral compounds however, which have more than one chirality center.

EXERCISE 1.1

Indicate the relationship-enantiomeric or diastereomeric among the products of the Diels-Alder reaction (scheme 1.3).

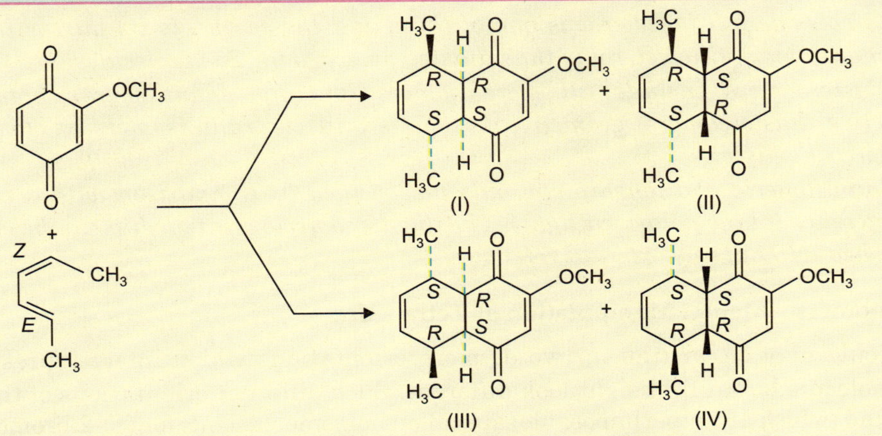

SCHEME 1.3

ANSWER. For the mechanism and stereochemical outcome of this reaction (see, schemes 8.48 and 8.49). The reaction takes place at the electron poor double bond of the dienophile and as expected the products will have rings with *cis*-fusion. The compounds (I and IV) are enantiomes as also (II and III) since in these pairs one has inverted stereochemistry at all the chirality centers. (I and IV) have diastereomeric relationship with both (II and III), since configuration at some of the centers is same while at others is different (configuration differs at two chirality centers).

Properties of Diastereomers

Diastereomers have a major advantage over enantiomers from a practical point of view. Diastereomers have different physical properties like m.p., b.p., solubility, retention times and R_f values and have different rates of reactions (chemical properties) even in achiral environments. Standard techniques like crystallization distillation or chromatography can therefore, be used to separate diastereomeric mixtures. For example, *meso* tartaric acid (specific rotation = 0) melts at 140°C as against either of the enantiomers (m.p. 171°C) and has different value of pK_a. Use of this difference in properties of diastereomers is made in the resolution of racemic mixture (See, Scheme 1.79). Another example is of erythroses and threoses (See, Scheme 1.37*j*).

Enantiomers and Diastereomers

One may note the following points :

- *Enantiomers (mirror images of each other) are related by symmetry elements of the second kind, i.e., σ plane, i, and S_n axis while diastereomers are not related by any such symmetry element.*
- *Since a molecule (or an object) can have only one mirror image, enantiomers can exist only in pairs. On the other hand, structural conditions permitting, a molecule can have any number of diastereomers, e.g., meso tartaric acid has two diastereomers (diastereomeric relationship with either of the enantiomers).*
- *Two stereoisomers cannot be enantiomers and diastereomers at the same time, i.e., enantiomeric and diastereomeric relationships are mutually exclusive.*
- *Diastereomers may be (or may not be) chiral in which case each of the diastereomer will show enantiomerism. Thus cholesterol with 8 stereocenters has 256 stereoisomers. Cholesterol is one of these and the second is its mirror image (enantiomer). Cholesterol is thus diastereomeric with 254 molecules.*
- *Diastereomers include all stereoisomers (but for enantiomers), optically active diastereomers, geometrical isomers and cis-trans isomers of classical stereochemistry.*
- *Enantiomeric relationship can only be specified by comparison with a chiral reference (plane polarized light). The diastereomeric relationship can be established without any external reference.*

(I) Homomers – Mental Manipulation of Tetrahedra

A molecule may be written in two or more orientations which infact represent the same molecule but at first sight look different. Such different orientations of the same compound (which are superimposable) are called homomeric. Thus for (*S*)-2-bromobutane one can write many equivalent orientations (scheme 1.3*a*)

For a related example (see, page 4). In these examples the chirality center is placed at the plane of the paper with any two bonds. Note that switching of groups is not allowed, since it inverts the configuration of the chirality center.

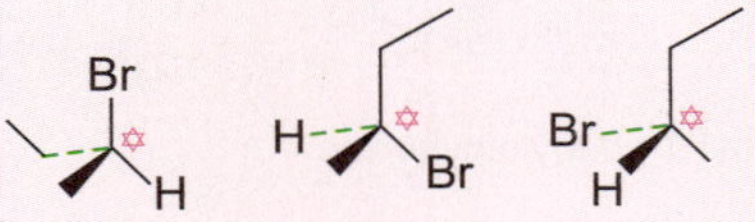

Ia = Ib = Ic

Identical structures
(S)-2-Bromobutane (*HOMOMERS*)

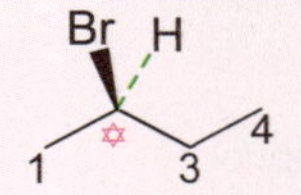

(II)

(*R*)-2-Bromobutane

Identity of a compound can be established from systematic nomenclature including (R) and (S) system. Similar name shows that compounds are same. The (identical) homomeric projections differ only as to which two bonds are in the plane of paper at any one time (see Scheme B1.1c)

SCHEME 1.3a

(J) Homochiral Molecules

The term 'homochiral' is sometimes used to mean 'optically pure' (enantiomeric purity). But this use to mean optically pure may be avoided since 'homochiral' has another meaning; if two molecules have the same absolute stereochemistry they can be said to be homochiral, *e.g.*, the right hands of a group of people are homochiral. Thus if one says 'those two amino acids are homochiral', then one has two different meanings depending on how one uses 'homochiral'. 'Enantiometrically pure' is unambiguous.

(K) Configuration is the Fixed Spatial Ordered arrangement of Atoms/groups in a molecule.

The actual three-dimensional arrangement of groups around a stereocenter is termed its configuration and the term configurational isomers is synonymous with stereoisomers. One way to designate the configuration of geometrical isomers is to use the terms *cis* or *trans* (or better *E*, *Z* nomenclature), other configurational descriptors are *D* and *L* and *R* and *S*.

The term conformations (conformational isomers, conformers or rotamers) refer to various shapes that a molecule can adopt by rotation about single bonds *e.g.*, an eclipsed or a staggered conformation of ethane and different conformations of chlorocyclohexane (see, Scheme 1.1). Unless it is held rigid by a small ring or double bonds, a molecule could have an infinite number of conformations, however, only one configuration. Thus, 2-butanol has two stable enantiomers (*R*)-2-butanol and (*S*)-2-butanol (see, Scheme 1.1) each exists as a dynamic mixture of conformations *e.g.*, in the ethyl group. Thus normally 2-butanol is considered to exist in only two enantiomeric forms (Scheme 1.1) as again shown in the planar projection formula (Scheme 1.3*b*). In doing so, one ignores the otherwise important conformational details. Using the concept of residual stereoisomerism, 2-butanol is considered to have only two residual stereoisomers (the two enantiomers) since fast rotation at room temperature reduces the number of isolable species. In case with amine invertomers as well *e.g.*, in N-methyl-1-phenylethyl amine (Scheme 1.3*b*) although there are two stereocenters at carbon and nitrogen, it has only two residual stereoisomers (residual enantiomers with different configuration only at the benzylic carbon).

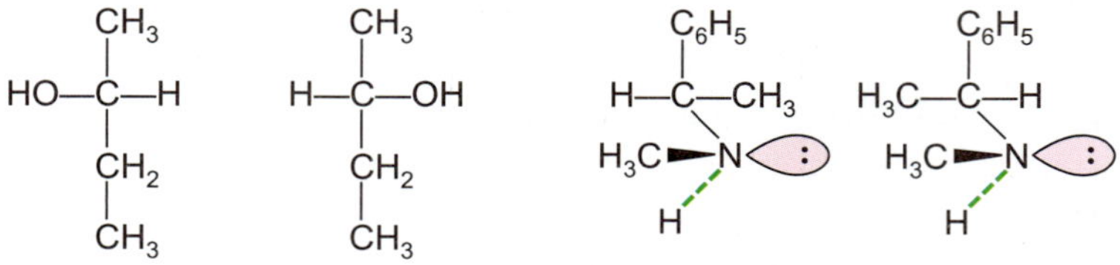

(*R*)-2-Butanol (*S*)-2-Butanol *N*-Methyl-1-phenylethylamine

"Residual stereoisomers"

SCHEME 1.3b

Whether two isomeric structures constitute different configurations or merely different conformations depend on how readily they interconvert. For example, the different conformations of ethane derivative interconvert millions of times per second at 25°C, while the *Z* and *E* configurations of ethylene derivatives do not interconvert spontaneously even at 200°C. Since in principle it should be possible to stop conformational interconversion by sufficient cooling, the operational distinction between configurations and conformations usually is based on whether the structures interconvert at room temperature (25°C, for a good example see Fig. 4.2).

(L) Absolute Configuration Differentiates Between Enantiomers

Apart from configuration, there are further two terms absolute configuration and relative configuration. Absolute configuration is the spatial arrangement of the atoms in a chiral molecule in a way that distinguishes it from its mirror image and its stereochemical description (*R* or *S*, *M* or *P*, *D* or *L*) as in the case of *e.g.*, 2-butanol enantiomers (Scheme 1.1) and glyceraldehyde enantiomers (Fig. 1.5*a*).

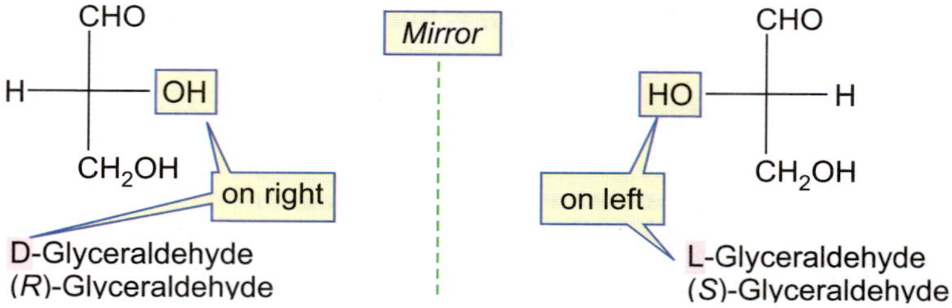

Absolute configuration of glyceraldehyde designated as D and L or *R* and *S* differentiates between right handed enantiomer and left handed enantiomer.

Fig. 1.5a

Thus absolute configuration differentiates between enantiomers *i.e.*, which is right handed enantiomer and which is left handed enantiomer.

There is no simple relation between the sign of rotation (+)- or (–) (Fig. 1.5*b*) and the absolute configuration. Thus D-glyceraldehyde (*R* configuration) is dextrorotatory while D-glyceric acid (*R* configuration) is levorotatory.

(M) Relative Configuration

If (+)-glyceraldehyde is converted to glyceric acid using a reaction that is known to put the COOH group where the CHO was, then the two compounds have the same relative configuration at their chirality centers. The oxidation reaction (Fig. 1.5*b*) does not break a bond to the chirality center in (+)-glyceraldehyde and

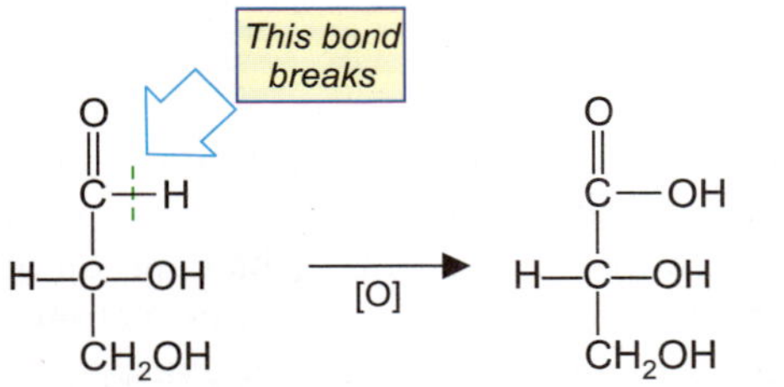

(+)-Glyceraldehyde is D-(+)-Glyceraldehyde is R(+)-Glyceraldehyde

D-(–)-Glyceric acid

(+)-Glyceraldehyde is converted chemically into (–)-glyceric acid without breaking any bond to the chirality center. Thus the two compounds have the same relative configuration. Since absolute configuration of (+)-glyceraldehyde (X-ray diffraction) is known to be (D-) as shown, (–)glyceric acid has also the same absolute configuration and still belongs to D-series.

Relating configurations, (+) glyceraldehyde is related to (–) glyceric acid. Both have the same relative configuration since in this oxidation reaction no bond to the chirality center is broken. This reaction thus proceeds with retention of configuration.

Fig. 1.5b

proceeds with retention of configuration. Thus the product glyceric acid of this conversion though levorotatory has the same relative configuration as that of (+)-glyceraldehyde. Since now the absolute configuration of D-(+)-glyceraldehyde (X-ray diffraction) is known, the absolute configuration of glyceric acid (still a member of D-series) is also known.

More on Relative Configuration

Relative configuration is the relationship between configurations of two chiral molecules. When two such molecules are chemically interconverted without breaking any bonds to the chirality center, these are said to have the same relative configuration (Fig. 1.5c) independent of the direction of rotation of the plane polarized light and independent of the (R and S) designation (also see, Scheme 1.29).

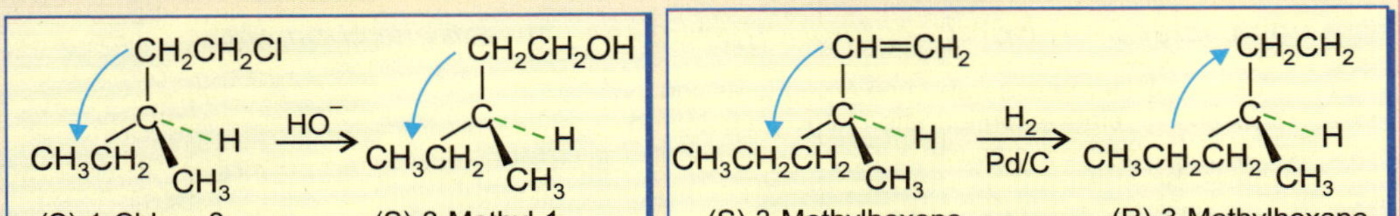

(S)-1-Chloro-3-methylpentane | (S)-3-Methyl-1-pentanol | (S)-3-Methylhexene | (R)-3-Methylhexane

(A)
Only OH substitutes for Cl, other groups maintain their relative positions.

(B)
Like in (A), the groups have maintained their relative positions, but their relative priorities (Cahn-Ingold-Prelog) have changed.

The reactants and products in sets (A) and (B) have same relative configurations, since in either transformation no bond to the chirality center is broken.

Fig. 1.5c

(N) Retention of Configuration

Retention of configuration is the conversion of one molecule into another with the same relative configuration. The term retention indicates that the spatial arrangement of atoms or groups at a chirality centre remains unaltered relative to a reference group. Thus, in a nucleophilic substitution reaction the group undergoing substitution serves as the reference group.

(O) Inversion of Configuration

Inversion of configuration is the conversion of one molecule into another which has the opposite relative configuration. Inversion of configuration does not require a change in designated absolute configuration, similarly a change in designated absolute configuration does not require an inversion of configuration. The reaction (III, See Scheme 1.28*g*) proceeds with inversion although (*R*) reactant gives (*R*)-product.

Relative configuration at a stereocenter is also the relation with that of any other stereocenter in the same molecule.

BOX 1.10

RELATIVE CONFIGURATION AT A CHIRALITY CENTER AS A RELATION WITH ANY OTHER CHIRALITY CENTER IN THE SAME MOLECULE

For cyclic systems with two substituents the stereodescriptors cis and trans are used while r, c and t are used if the ring has three or more substituents.

A reference group denoted by symbol 'r' is specified with respect to which the other substituents are described as cis or trans (A, scheme B 1.13).

The r-group is so chosen that it is attached to the lowest numbered ring member according to IUPAC rules. The common practice is to abbreviate the names, by replacing cis and trans by c and t respectively.

Following the rules the compound (I, scheme B1.13) is properly named.

- *The pair of enantiomers (I and II, Scheme B 1.13) represent a cyclic system with three substituents. These have the absolute configuration 1R, 2S, 3S and 1S, 2R, 3R while 2t, 3c-dibromocyclohexan-1r-ol represents only the relative configuration of the substituents in both (Scheme B 1.13 also see scheme 1.67b). Thus the nomenclature r, c and t cannot differentiate between enantiomers.*

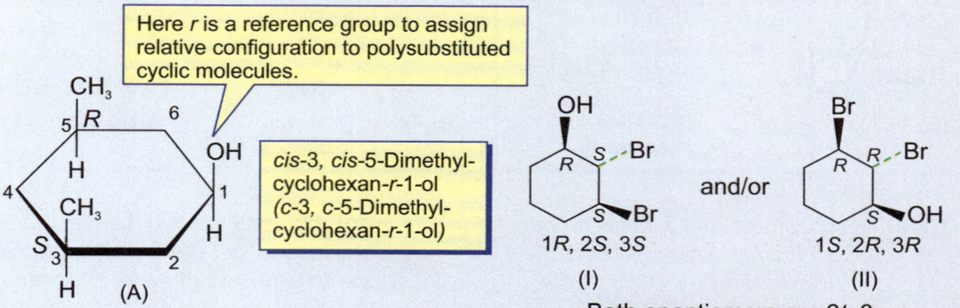

Both enantiomers are 2*t*, 3*c*-dibromocyclohexan-1r-ol

SCHEME B 1.13

1.3 MORE ON CHIRAL MOLECULES-ENANTIOMERISM (OPTICAL ISOMERISM)

Enantiomers occur only with those compounds whose molecules are chiral. A chiral molecule can be defined as one that is not superposable on its mirror image.

(A) The Term Achiral

Objects and molecules which are superimposable on their mirror images *e.g.* a cup (Scheme 1.3*c*) is achiral. An internal plane of symmetry is a hypothetical plane which bisects an object or a molecule into mirror-reflective halves. An object or a molecule with an internal plane of symmetry is achiral (can be superposed on its mirror image). Thus a cup is achiral since it can be divided into two equal halves by its plane of symmetry. Similar is the case with *meso*-tartaric acid (drawn as Fischer projection or perspective formula Scheme 1.3*c*). A hand is chiral since a plane cannot split it into two equal halves. Similarly either of the (+)- or (–)-enantiomers of tartaric acid is chiral.

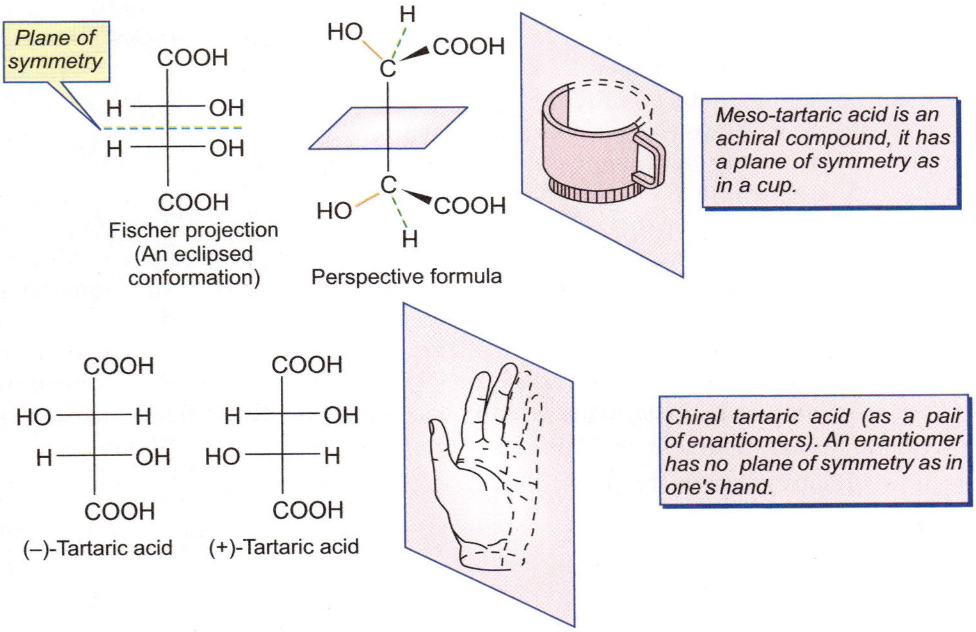

SCHEME 1.3c

(B) Chiral Molecules

(*i*) Nonsuperimposability of a Structure on Its Mirror Image

The ultimate criterion for chirality *i.e.,* existence of enantiomers in a molecule is the nonsuperimposability of a structure and its mirror image. [*Thus enantiomers must be isomers and mirror images as well, i.e., they must not be superimposable*].

(*ii*) Elements of Symmetry and Chirality—Rotation/Reflection Axis

Chirality—A Necessary and Sufficient Condition

Only those molecules are chiral which do not have an alternating axis of symmetry S_n (i.e., absence of a rotation/reflection axis). The chiral molecules can, however, have an axis of rotation (C_n).

An alternative approach to decide if or not a structure is chiral *i.e.*, capable of existing in enantiomeric forms (optically active forms) is to examine the symmetry of a molecule. A structure which lacks an alternating axis of symmetry is chiral and is not superimposable on its mirror image and therefore, can exist in optically active forms. Symmetry operations are discussed in detail (see Sec. 1.8), however, here the common practice is to look for a plane of symmetry (a one fold alternating axis of symmetry S_1) and a center of symmetry which is equivalent to S_2 will be discussed while higher subscripts for S_n are rare (A molecule is said to have an *n* fold alternating axis of symmetry S_n if rotation of 360°/*n* about an axis followed by reflection in a plane perpendicular to that axis brings the molecule into a position indistinguishable form the original).

meso-Tartaric acid is achiral even though it has two chirality centers since it has a plane of symmetry (see, Schemes 1.2*a* and 1.3*a*).

When one cannot detect a mirror plane of symmetry in a molecule, the molecule is not necessarily chiral. Thus the cyclobutane derivative (I, Scheme 1.3*d*) has no internal mirror plane of symmetry, yet the mirror image is superposable on the original molecule (when *e.g.*, the mirror image is rotated the two structures are superposable. Unlike a three dimensional formula which does not change on rotation, the two dimensional representations of three dimensional molecules like Fischer projections can only be manipulated in specified ways). In fact (I, Scheme 1.3*d*) has a center of symmetry *i.e.*, it contains an S_2 axis (an inversion center see, Sec. 1.9). A molecule is said to have a center of symmetry or an inversion center if all straight lines that can be drawn through the center of the molecule meet identical atoms (or points) at the same distance from the center. In other words inversion of all atoms (or points) in the molecule through the point leads to an arrangement indistinguishable from the original. There can be only one such inversion center in the molecule.

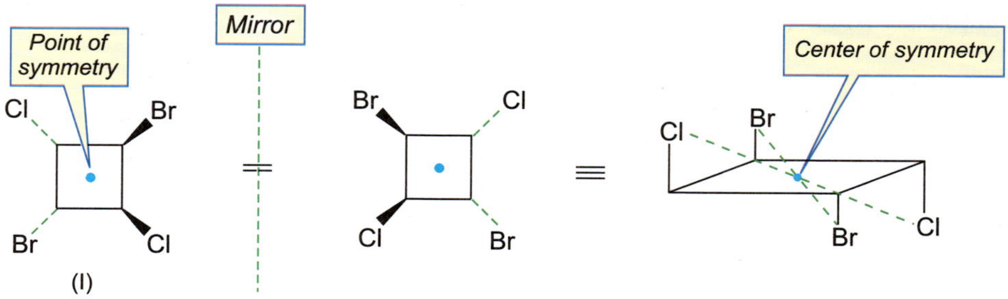

In this achiral cyclobutane derivative no plane of symmetry is detectable. Infact the compound has a center of symmetry (point of symmetry) and the two structures are superimposable on their mirror images.

SCHEME 1.3d

Molecules with a Plane or a Centre of Symmetry and Axis of Rotation

The mirror plane of symmetry (S_1) is the most frequently occuring rotation reflection axis in organic molecules. Examples of further compounds which have this axis are in (I and II Scheme 1.3e), cis-1, 2-dichlorocyclohexane and meso-2,3-dichlorosuccinic acid are achiral even though each has two chirality centers. Dichlorocyclohexane dicarboxylic acid (III) on the other hand is achiral even though it has four chirality centers due to the presence of a point of symmetry—presence of S_2 axis which is somewhat rare in organic chemistry. A molecule can, however, be chiral if it has only an axis of rotation. Both trans-1,2-dichlorocyclohexane and dichlorosuccinic acid (IV and V respectively, Scheme 1.3e) have a two fold axis of rotation (C_2) as the only symmetry element and both are chiral.

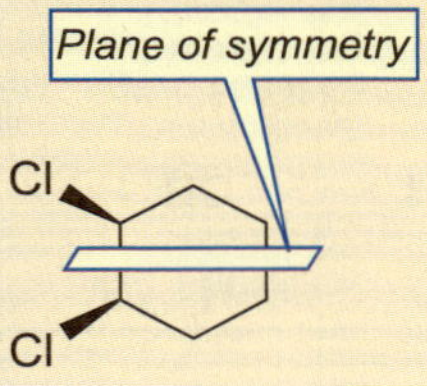

(I) Achiral

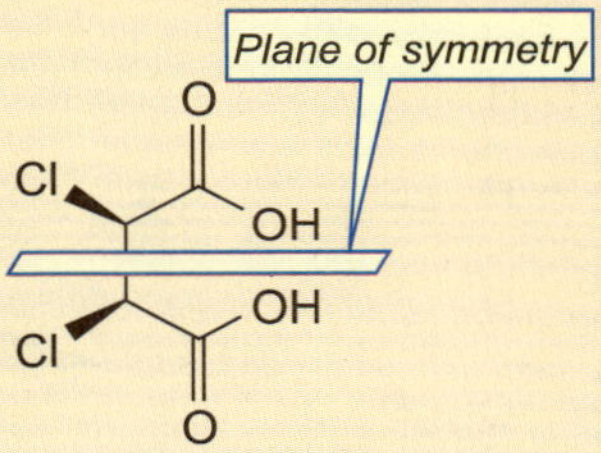

(II) Achiral

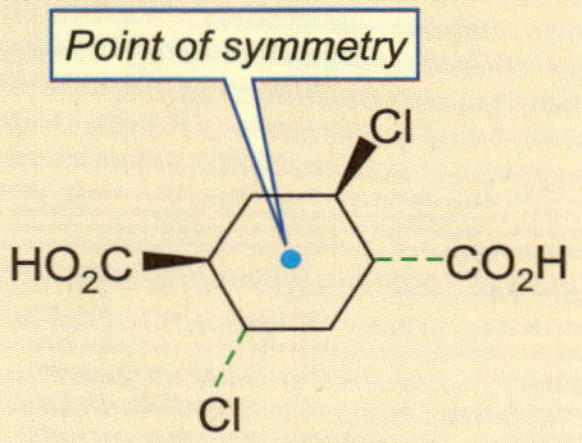

(III) Achiral

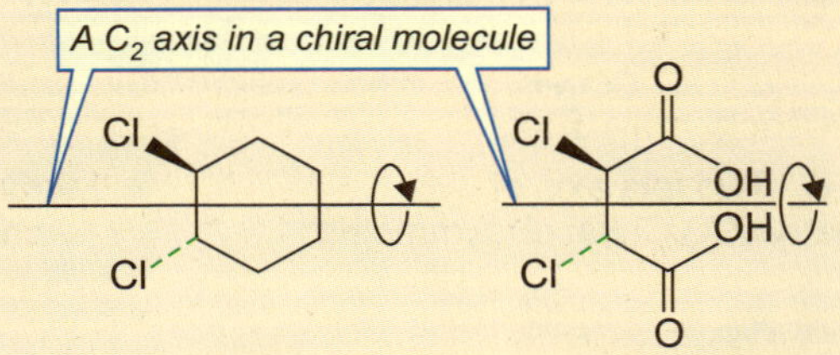

(IV) Chiral (V) Chiral

Some Stereoisomers of Inositols

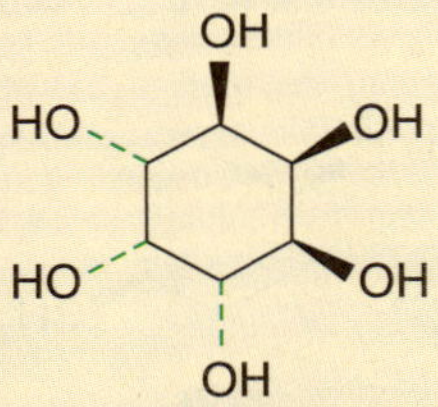

(VI) *meso*

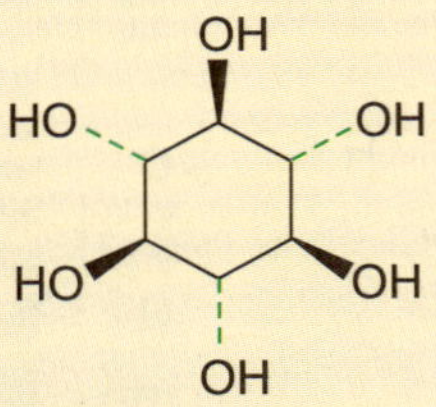

(VII) *meso*

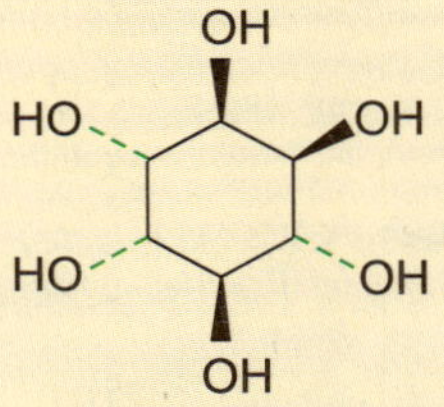

(VIII) Chiral

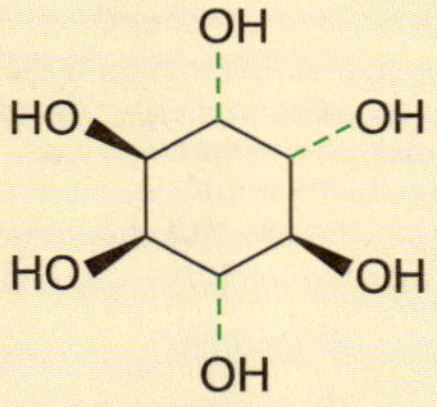

(IX) Enantiomer of (VIII)

Inositols exist as nine stereoisomers (see Box 1.11), seven of these are meso and one chiral which exists as an enantiomeric pair. If out of these isomers one considers four inositol isomers where three OH groups are up and three OH groups are down, then two are meso isomers VI and VII and one enantiomeric pair (VIII and IX).

SCHEME 1.3e

EXERCISE 1.2

The diacids derived from D-glucose, D-galactose and D-mannose (scheme 1.3f) have interesting symmetry properties. Explain.

ANSWER

The diacid (I, Scheme 1.3f) has no symmetry elements and is thus chiral asymmetric compound. The diacid (II) has a C_2 axis it is chiral dissymmetric compund, while (III) is achiral meso compund. The example of (III) shows that meso compounds can exist in molecules with more than two chirality centers.

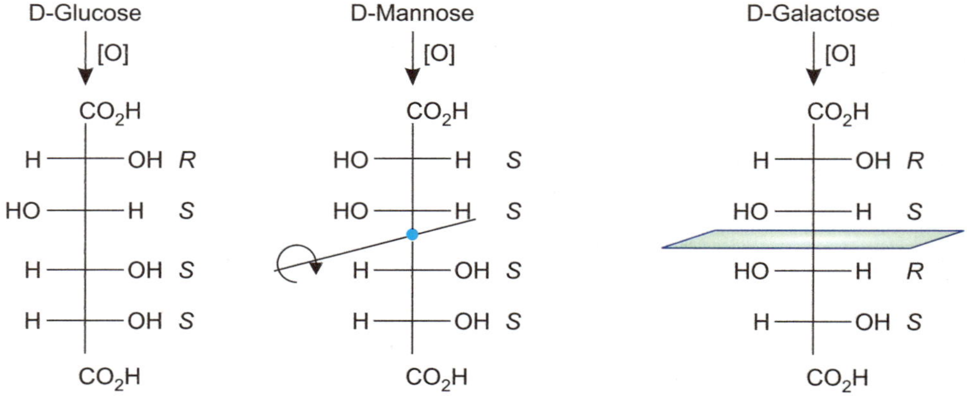

(I) No symmetry Chiral asymmetric

(II) Two fold axis of symmetry (C_2) Chiral dissymmetric

(III) Mirror symmetry Achiral *meso*

SCHEME 1.3f

EXERCISE 1.3

Pick out chiral and achiral structures from the following (Scheme 1.3g).

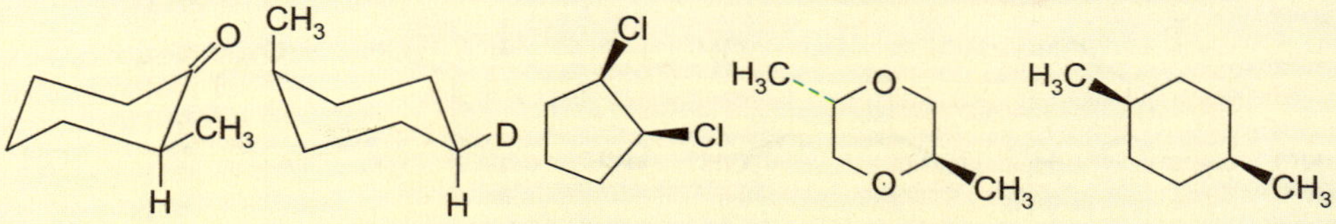

(I) (II) (III) (IV) (V)

SCHEME 1.3g

ANSWER. *(I) chiral ; (II) achiral ; (III) achiral (a plane of symmetry) ; (IV) achiral has no plane of symmetry but has a center of symmetry (Scheme 1.3h models will help) ; (V) achiral, has a plane of symmetry. It is always helpful to indicate H atoms in the projections.*

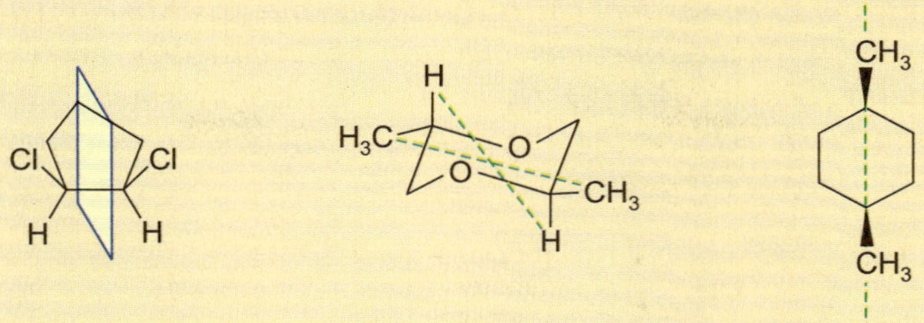

(III) (IV) (V)

SCHEME 1.3h

(*iii*) Presence of a Chirality Center

A molecule C_{abcd} is necessarily chiral when $a \neq b \neq c \neq d$ and none of these ligands are themselves chiral. In some cases a chiral ligand can generate mirror symmetry on a molecule C_{abcd} (*meso* compounds). The ligands can be alkyl, aryl, heteroatomic or their combination. A chirality center in a molecule is stared for emphasis.

All molecules with a single chirality center (point group Cl) must display optical activity. A minor difference among the four groups is sufficient to generate chirality as in (I, Scheme 1.4) which is optically active. The chirality of this center is retained even when two substituents show as small a difference as a hydrogen atom and its isotope deuterium as in 1-butanol-1-*d*, in II and epoxide III (Scheme 1.4).

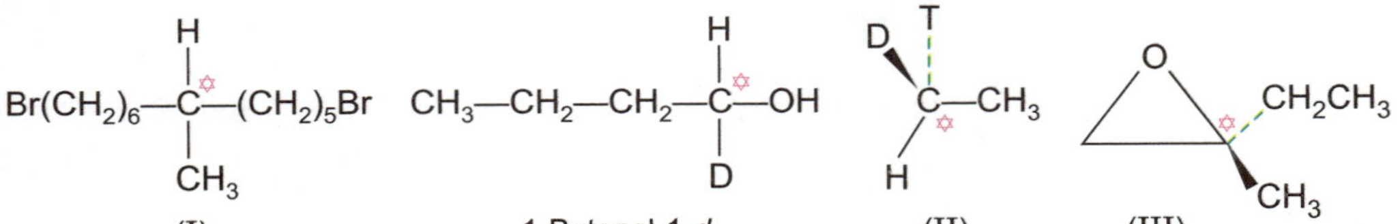

(I) 1-Butanol-1-*d* (II) (III)

SCHEME 1.4

BOX 1.11
MESO COMPOUNDS AND ENANTIOMERS IN COMPOUNDS WITH SEVERAL CHIRALITY CENTERS–INOSITOLS

Naturally occurring inositols are 1, 2, 3, 4, 5, 6-hexahydroxycyclohexanes that are found in all cells. One may compare myo inositol with compounds, (I and III, scheme B1.14)

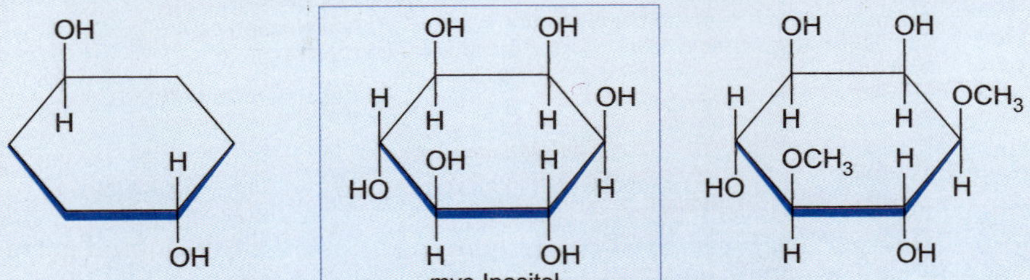

(I) *myo*-Inositol (II) (III)

A growth factor in animals

SCHEME B 1.14

Continued...

Inositol contains six equivalent chirality centers and exists in eight diastereomeric forms. myo Inositol is the meso stereoisomer having six chirality centres and a mirror plane of symmetry (Scheme B 1.15). No mirror plane of symmetry is present in the related compound (III, Scheme B 1.14), and since there is no chirality center in (I), thus it cannot be a meso compound.

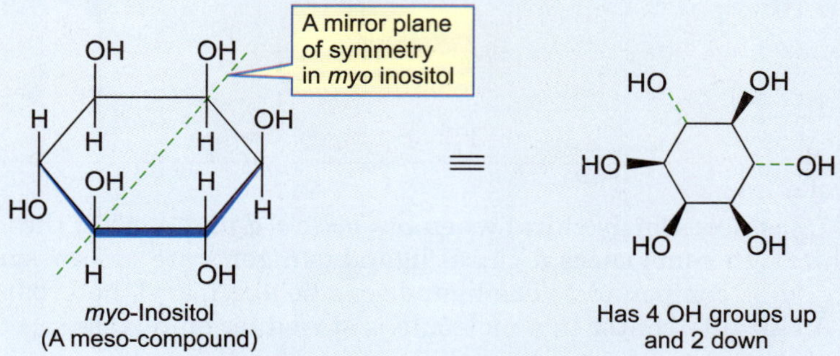

SCHEME B 1.15

Apart from myo inositol (Scheme B 1.14), the other six isomers are also meso having one or more σ planes. Interestingly one of the meso inositol isomers (A, Scheme B 1.16) has both an inversion centre as well as three σ planes. Only one σ plane is shown for simplicity.

Thus of the eight inositol isomers seven are meso (two of these are in Box 1.11) and eighth is chiral and exists as a enantiomeric pair (Scheme B 1.16, also see Scheme 1.3e). Inositol, therefore exists as total of nine stereoisomers.

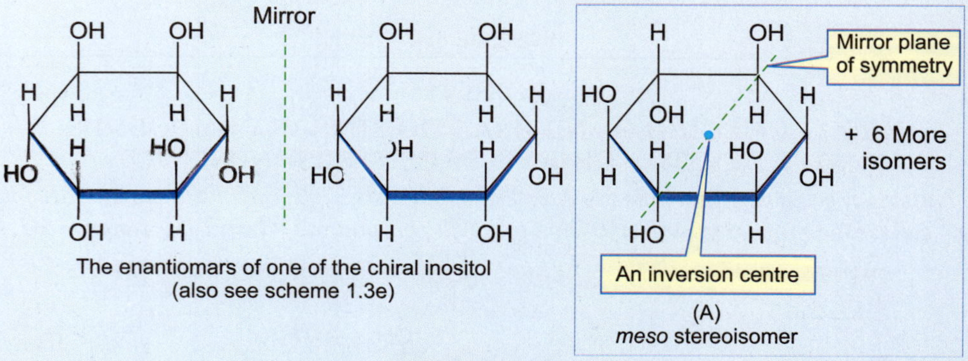

SCHEME B 1.16

(*iv*) Chirality Centers Other than Carbon Atom

(a) Compounds with tetrahedral arrangement of bonds: Silicon and germanium like carbon have a tetrahedral arrangement of bonds and display chirality provided the four groups are different. Some of the tetrahedral centres are Si (Silanes) and Ge (Germanes) derivatives a well as N in quaternary salts (shown as enantiomers) and phosphine oxides (Scheme 1.5).

In sulfones sulfur bonds tetrahedrally, however, as two of the groups have to be oxygen, normally the sulfones do not display chirality. An optically active sulfone (I, Scheme 1.5) has however, been prepared in which both the oxygens are different one being ^{16}O and the other ^{18}O. An adamantane derivative with four different bridgehead substituents becomes chiral and optically active as in (A, Scheme 1.5) which has been resolved. The molecule is completely asymmetric (Cl point group) and represents a kind of expanded tetrahedron. The center of chirality is represented by a dot in the

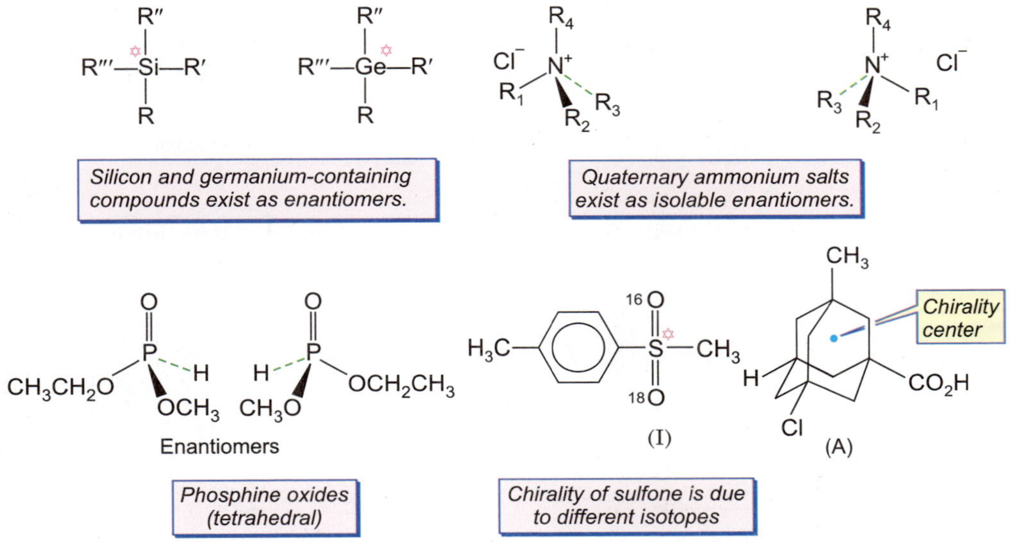

SCHEME 1.5

unoccupied space of the adamentyl carbon framework. This is an example to show that a chirality center in a molecule may not necessarily lie on an atom.

(b) Amines, phosphines and sulfoxides: Atoms with pyramidal; bonding may be chiral and therefore, can give rise to optical activity provided the atom has three different groups attached to it. The unshared pair of electrons then represents the fourth group (Scheme 1.6). Sulfur exhibits pyramidal bonding in sulfoxides, sulfonium salts etc., many of which have been resolved. An interesting example of a sulfoxide which is chiral due to two alkyl groups differing only in ^{12}C versus ^{13}C is (I, Scheme 1.6). In these compounds sulfur atom maintains a rather rigid geometry and does not interconvert between the two enantiomeric forms.

Sulfides such as (I, Scheme 1.7) can be oxidized to sulfoxides or sulfones, depending on the type and amount of oxidizing agent used. Since sulfoxides are pyramidal, diastereomeric (exo and endo) sulfoxides (II) and (III) are formed on mild oxidation of (I). Further oxidation gives the same sulfone, (IV), in each case.

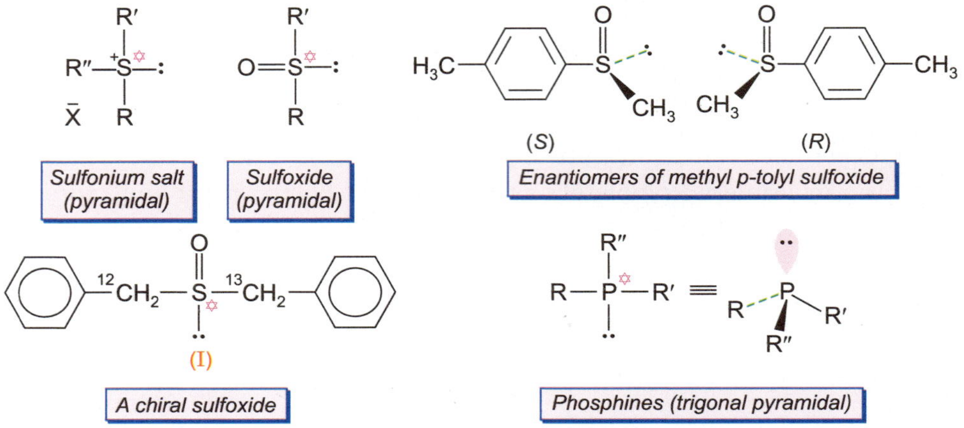

SCHEME 1.6

Note that sulphoxides with two different groups attached always have a chirality center (Scheme 1.7)

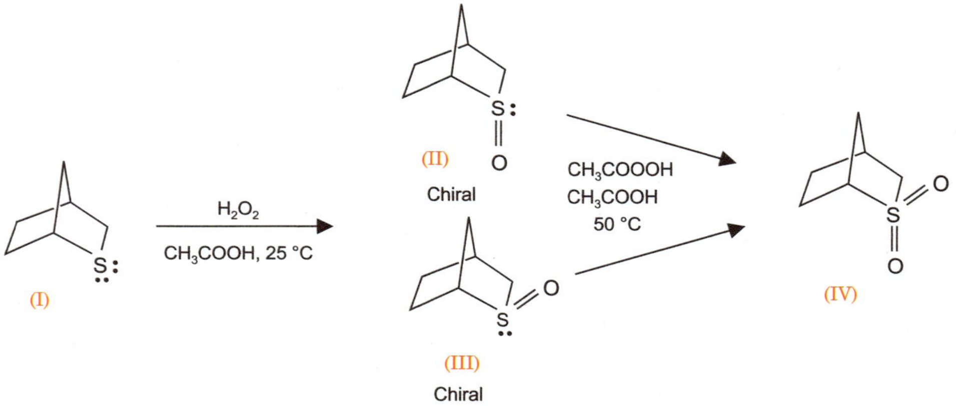

SCHEME 1.7

Chiral amines have pyramidal geometry, however, a nitrogen atom with a pair of nonbonded electrons rapidly turns inside-out at room temperature. This is called amine inversion (See, Scheme 1.1*c*). One may note that amines in which the chirality is due to the presence of a stereogenic carbon atom can however, be resolved into enantiomers (Scheme 1.8).

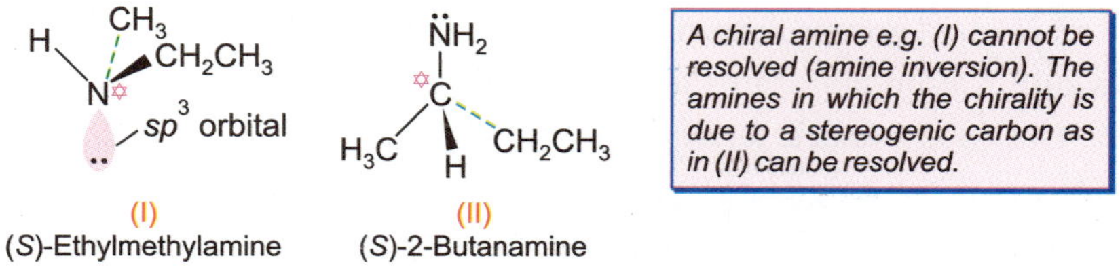

SCHEME 1.8

Due to rapid inversion, the individual amine enantiomers cannot be isolated. The two amine structures provide an interesting example of conformational enantiomers.

The pair of nonbonded electrons are required for amine inversion, therefore, quaternary ammonium ions, with four bonds to nitrogen with no nonbonded electrons, do not invert. Some quaternary salts do racemize. The tertiary amine so formed inverts configuration rapidly and recombination with the alkyl halide gives a racemic salt. The less nucleophilic the counter ion X, the more stable will be the quaternary salt configurationally.

Phosphorus is in the same group of the periodic table as nitrogen and tricoordinate phosphorus compounds e.g., phosphines like amines are trigonal pyramidal. Interestingly phosphines undergo pyramidal inversion comparatively very slowly than amines and similar is the case with chiral tricoordinate sulfur compounds like sulfoxides.

No bonds are broken during the change and the configurational nomenclature is retained as the change can be described as S → R or R → S. The amine interconversion is as an inversion (turning inside-out of an umbrella) and to avoid confusion, the enantiomers are termed invertomers (scheme 1.8a). When the lone pair is donated to a substituent, giving the $[NR^1 R^2 R^3 R^4]^+$ species, this rapid inversion is prevented. The tetra-alkyl ammonium salts with four different alkyl groups display optical activity.

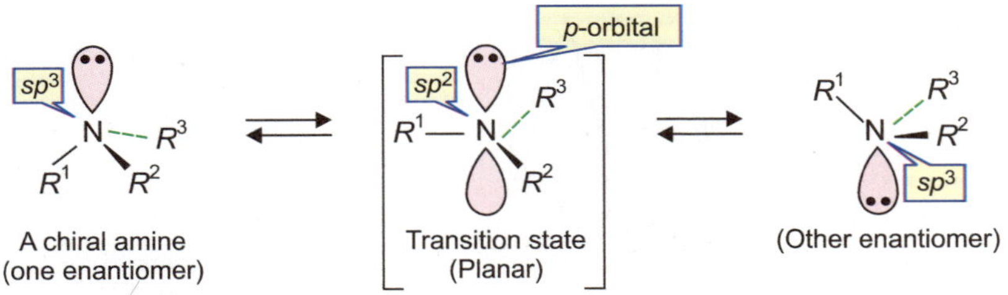

Amine inversion (very fast) takes place through a transition state in which the sp^3 nitrogen becomes an sp^2 nitrogen. The three groups bonded to the sp^2 nitrogen are coplanar in the transition state with bond angles of 120°, and the lone pair is in a *p* orbital. The "inverted" and "non-inverted" amine molecules are enantiomers. This is another example of racemisation.

SCHEME 1.8a

The unshared electron pair on the nitrogen of ammonia (and amines) occupies an approximately sp^3–hybridized orbital. This orbital has 75% 2*p* character. In the transition state for inversion (scheme 1.8*a*) the central atom is sp^2–hybridized and the unshared pair occupies a 2*p* orbital. A relatively small amount of energy is needed to add another 25% *p* character to the unshared pair; so, the inversion energy barrier is small, and inversion is fast.

When the central atom is from the third or greater period as in phosphines the unshared electron pair occupies an orbital with a high degree of *s* character, it takes significant energy to convert an electron pair in a 3*s* orbital to one in a 3*p* orbital. Therefore, the inversion barrier for these atoms is larger, and inversion is slow.

Distinction Between Conformation and Configuration

The enantiomers of a chiral acyclic amine may be best described as conformational isomers (barrier to inversion is small (~ 25 kJ mol^{-1}). This argument however, does not apply to a chiral phosphine which are configurationally stable at room temperature (high energy barrier 150 kJ mol^{-1}, see scheme 1.1*c*).

When the nitrogen atom is present at a ring junction in bridged ring systems, pyramidal inversion is no longer possible without bond cleavage. With proper substitution the tricoordinate nitrogen becomes a stable center of chirality, as in Troger's base (Scheme 1.8*b*). Similarly brucine (see, Scheme 1.80), an alkaloid contains a stereogenic nitrogen in addition to stereogenic carbons.

Optical stability is the measure of resistance of a pure enantiomer toward racemization under a given set of conditions. One type of nitrogen atom that inverts particularly slowly is a nitrogen atom in a three-membered ring. The planar transition state involving sp^2 nitrogen for flapping of some aziridines is sufficiently stained (energy barrier quite high 73 kJ mol^{-1}) for the isolation of individual enantiomers (Scheme 1.8*b*). In addition to nitrogen being a part of three membered ring, another factor which imparts configurational stability in amines in the presence of an adjacent atom with at least one lone pair of electrons e.g. Cl, O (Scheme 1.8*b*).

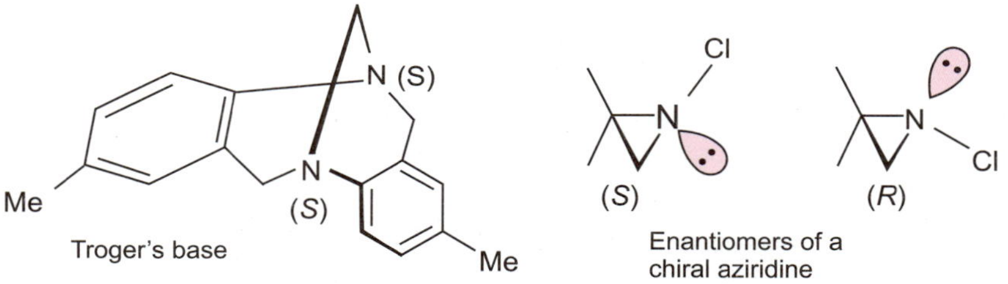

SCHEME 1.8b

Chiral carbanions racemize readily while carbocations and carbon free radicals being close to planarity are achiral independent of their substituents (Scheme 1.9).

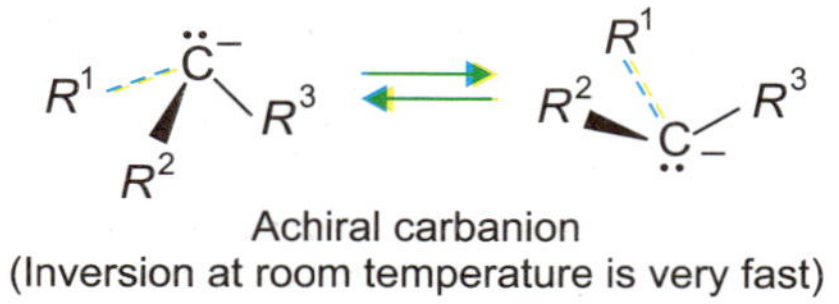

The carbocation and free radicals are planar with sp^2 hybridized central carbon. A carbanion has a sp^3 hybridized central carbon and is isoelectronic with amines and undergoes inversion very easily.

SCHEME 1.9

(*v*) Molecules with Chirality Axis (Stereoaxis)—Conformational Enantiomers and Diastereomers

Several molecules which do not have a chirality center are, however, capable of holding unlike pairs of groups is mutually perpendicular planes to make such molecules like allenes (VI, Scheme 1.10, C_2 symmetry) and spirans (VII) chiral and to exist in nonsuperimposable mirror image forms (Scheme 1.10). These molecules contain a chirality axis, which in the case of allenes (in allenes the central carbon is *sp* bonded), includes the allene system C=C=C. The geometry of the 1,2-diene system forces the ligands attached to the termini of the allene to occupy mutually perpendicular planes. When each terminus carries a pair of unlike groups (though the sets may be identical) the compound becomes chiral. The same conditions apply to spirans.

(VI)

A pair of enantiomers of 2,3-pentadiene (a chiral molecule)

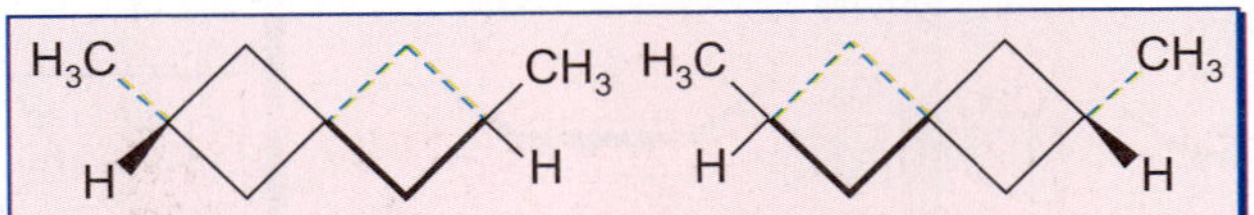

(VII)

A pair of enantiomers of 2,6-dimethylspiro [3,3] heptane

SCHEME 1.10

Suitably substituted biphenyls also have a chirality axis. Biphenyls are twisted from the coplanar (achiral) conformation as a result of steric strain between the *ortho* substituents on adjacent phenyl rings. In case both rings are unsymmetrically substituted as in (VIII, Scheme 1.11) a biphenyl has a chirality axis and exists in enantiomeric forms.

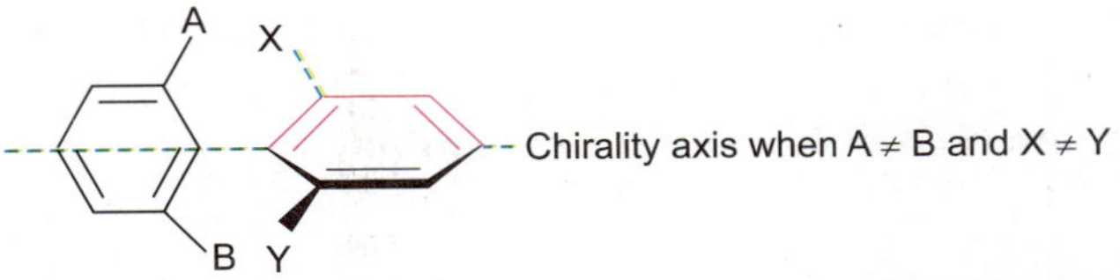

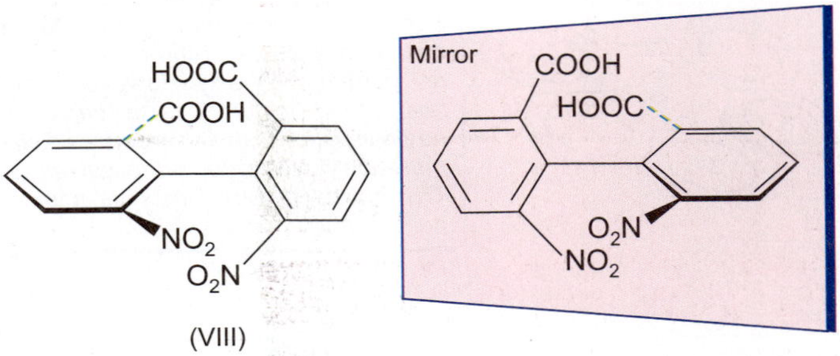

(VIII)

Correctly substituted ortho-biphenyls are chiral. Rotation about the junction bond will recemize a pure enantiomer.

SCHEME 1.11

(*vi*) Molecules with a Chirality Plane (Stereoplane)

Optically active *trans*-cyclooctene (Scheme 1.12) has been prepared and characterized at room temperature. The chirality of *trans*-cyclooctene is due to the presence of a plane of chirality, since there are two asymmetric arrangements of the tetramethylene bridge either below or above the plane of the carbon-carbon double bond. *Trans*-cyclooctene is locked into one of these enantiomeric conformations (also see, Schemes 1.52 and 1.53).

Bridged aromatic rings like paracyclophanes have a plane of chirality which is specified by the substituted benzene ring and the atoms directly bonded to it.

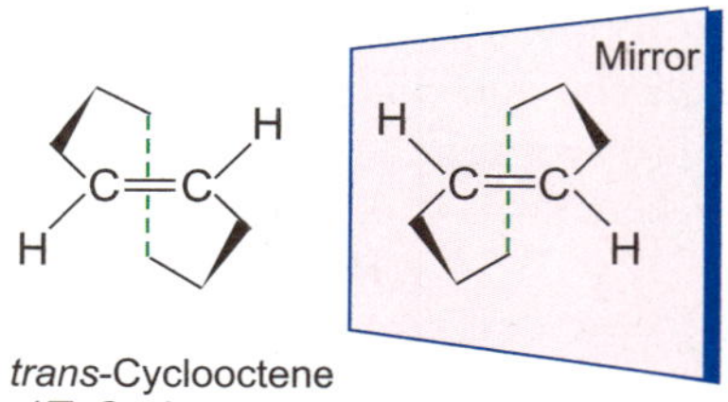

trans-Cyclooctene is chiral without a chirality center or a chirality axis but contains a stereoplane (chirality plane). *trans*-Cyclooctene has either tetramethylene bridge (above one enantiomer) or below (second enantiomer) the plane of trigonal atoms (chiral plane). This chirality plane is defined by double bond and its four directly attached atoms.

trans-Cyclooctene
(*E*)-Cyclooctene
(I) (II)

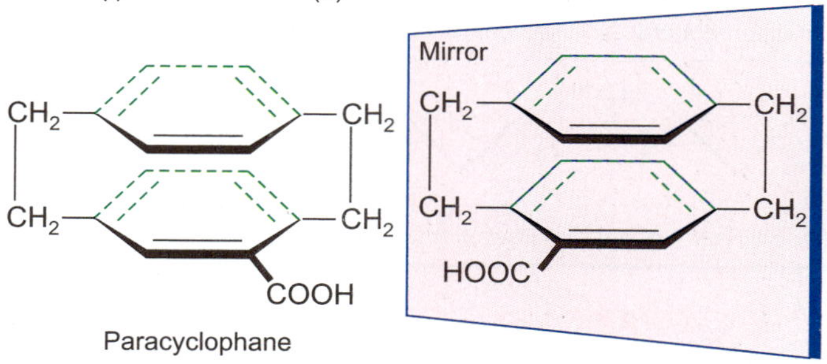

A bridged paracyclophane (suitably substituted to generate chirality) is chiral due to the presence of a chirality plane.

Paracyclophane
(I) (II)

SCHEME 1.12

(*vii*) Molecules with Helical Shape

Several molecules *e.g.*, hexahelicene (Scheme 1.13) display chirality due to helical shape, it has no stereocenter, and has no substituents to make the molecule chiral. In case the molecule was to be planar, it would suffer unbearable overcrowding between the terminal rings *a* and *a*′.

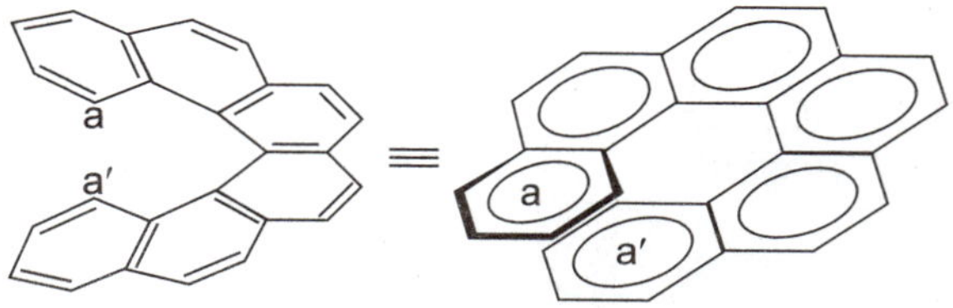

$[\alpha]_D = -3600\ (CHCl_3)$

Fusing benzene rings in a linear fashion as in naphthalene, anthracene etc. does not generate chirality. However, if rings are added to produce a curve, the addition of the sixth ring produces a chiral molecule-hexahelicene (helical shape). The molecule thus exists as a pair of helical enantiomers.

SCHEME 1.13

(*viii*) Chirality Due to Restricted Rotation of Other Types

Metallocenes substituted with at least two different groups on one ring become chiral and have been resolved (I, Scheme 1.14). Other metallic complexes also become chiral if geometry is suitable (II, Scheme 1.14). Thus fumaric acid-iron tetracarbonyl has been resolved into optically active forms.

Cyclooctatetraene is a conjugated eight-carbon ring system (Scheme 1.14) the molecule is not planar but is tub-shaped. A regular octagon has angles of 135° while sp^2 angles are most stable at 120°. In order to avoid this angle strain the molecule assumes a nonplaner achiral conformation. 1,2,3,4-Tetramethyl-cyclooctatetraene is chiral due to its tub shape and no plane of symmetry.

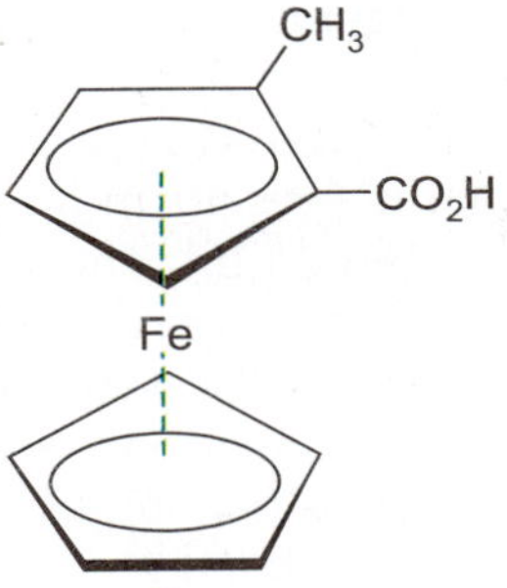

(I)

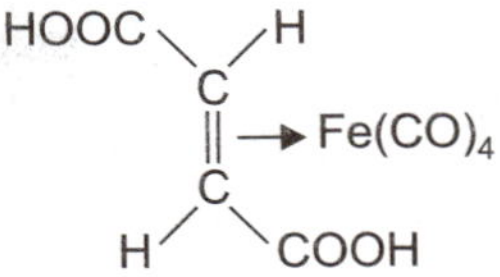

(II)

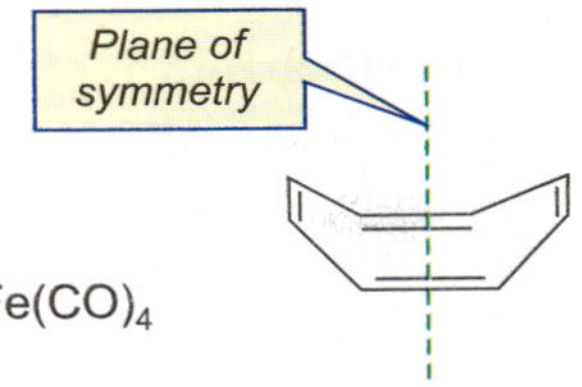

Cyclooctatetraene (Achiral due to the presence of plane symmetry)

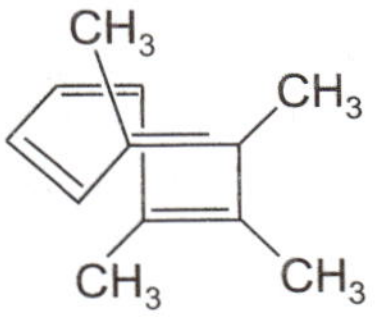

1, 2, 3, 4-Tetramethyl-cyclooctatetraene (Chiral, the plane of symmetry is lost)

SCHEME 1.14

(*ix*) Stereogenic Units in Organic Molecules (Stereogenecity)

A stereogenic unit is a structural unit whose presence in a molecule generates stereoisomerism. The most frequently encountered stereogenic units are chirality centres, chirality axes, and chirality planes and double bonds which can display *cis-trans* (*E*/*Z*) isomerism. Chirality centers like the C atom in (I, Scheme 1.14*a*), with four different ligands or *pyramidal* atoms with three different ligands represent the most common chiral units. The C=C=C unit in (II) constitutes a chirality axis, about which there is a chiral distribution of ligands. In (III, Scheme 1.14*a*) the carboxyl substituted benzene ring constitutes a chirality plane: exchange of the —CO_2H with any one of the H atoms drawn explicitly affords the enantiomer.

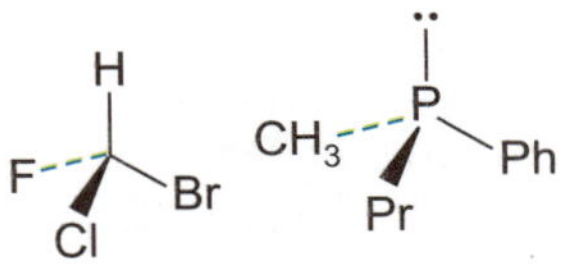

Molecules with a chirality center (Chiral)
(I)

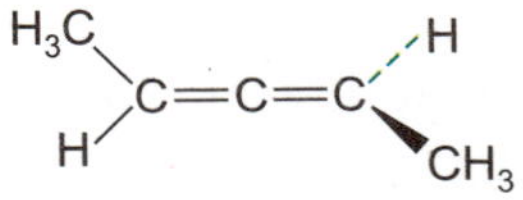

Molecule with a chirality axis (Chiral allene)
(II)

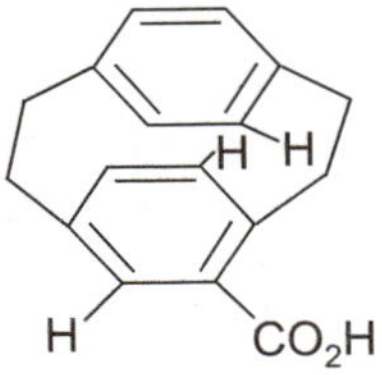

Molecule with a chirality plane (Chiral paracyclophane)
(III)

Chiral molecules with one stereogenic unit

SCHEME 1.14a

A double bond capable of showing *cis*, *trans* (*Z*/*E*) isomerism is also a stereogenic unit (I, Scheme 1.14*b*). Similarly some tri- and tetra-coordinate atoms lying on symmetry planes (II, Scheme 1.14*b*) represent stereogenic units (see, Fig. 1.5).

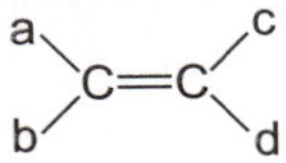

(I)

π-Diastereoisomerism
$a \neq b$ and $c \neq d$
(Achiral)

(II)

Two stereogenic centers (e.g. exchange of H with OH gives the *cis* diastereomer). Achiral due to these on a plane of symmetry.

Achiral molecules with one stereogenic unit

SCHEME 1.14b

Apart from the molecules discussed above, several stereogenic units may be present in the same molecule (Scheme 1.14*c*). In the naturally occurring lactone (I), one can detect four stereogenic units. The carbons at C-6 and C-8 are chirality centers, C-10, 1, 2 defines a chirality axis; the C-4, 5 double bond which is present in a large ring can display *Z*/*E* isomerism (see, Scheme 1.44) is a stereogenic unit; the C=C at C-7, however, cannot display *cis*/*trans* stereoisomerism since it is present in a five membered ring. Similarly in maxacalcitol (II, Scheme 1.14*c*) there are nine stereogenic units. Of these six are chirality centers; two are in the form of carbon-carbon double bonds whose configuration is shown. The last (9th) stereogenic unit is the single bond between the *E* and Z double bonds which can display partial double bond character, and its *s-trans* conformation is indicated.

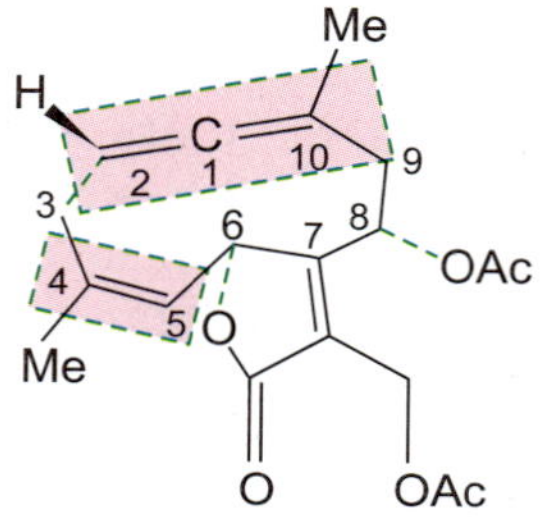

(I)

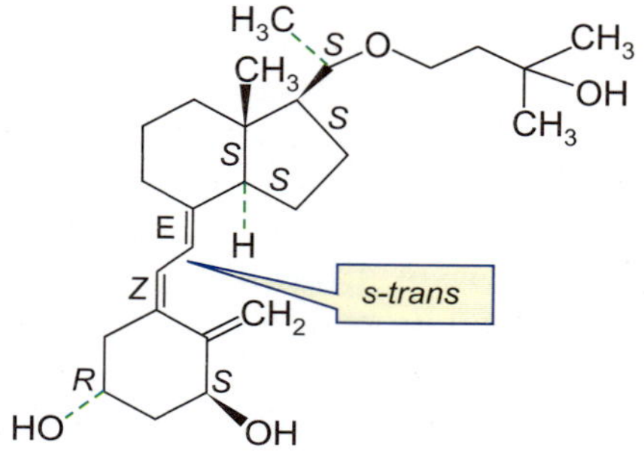

(II)

Molecules with several stereogenic units

SCHEME 1.14c

(*x*) A Useful Rule to know the Maximum Number of Stereoisomers–Stereogenic Units

- In compounds where stereoisomerism is due to chirality centers the total number of stereoisomers will not exceed 2^n, where n = number of chirality centers.
- Two identically substituted chirality centers however, give rise to only three stereoisomers. This is seen in the case of tartaric acid, since one of the stereoisomers is a *meso* compound superimposable with its mirror image.
- However one calculates the number of possible stereoisomers from 2^n where n= the number of stereogenic units. Consider *e.g.*, 3–penten–2–ol (Scheme 1.14*d*) which contains two stereogenic units *i.e.*, a double bond capable of showing *cis*, *trans* (*Z*/*E*) isomerism and a chirality center. Thus, there are formally $2^n = 2^2 = 4$ isomers.

Consider the molecule of vitamin A (1.14*d*), although it contains 5 double bonds only those in the side chain (4) can be considered. Thus the number of stereoisomers is $2^4 = 16$. Because of the ring strain the double bond in the six-membered ring can only have the *Z* configuration [Unlike the side chain double bonds, the double bond in the ring is not the stereogenic unit since it cannot display (*Z*/*E*) isomerism].

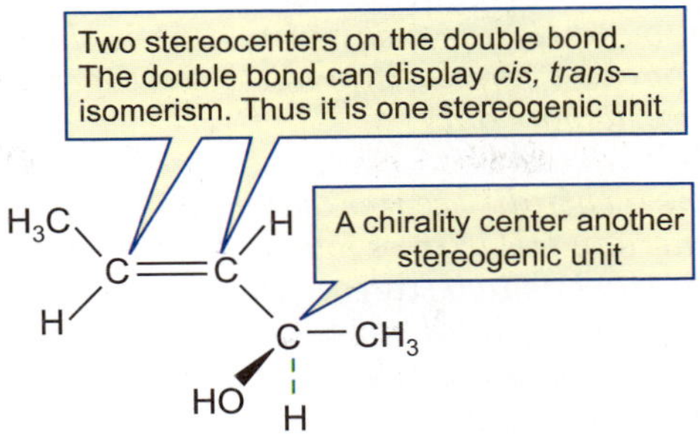

3-Penten-2-ol

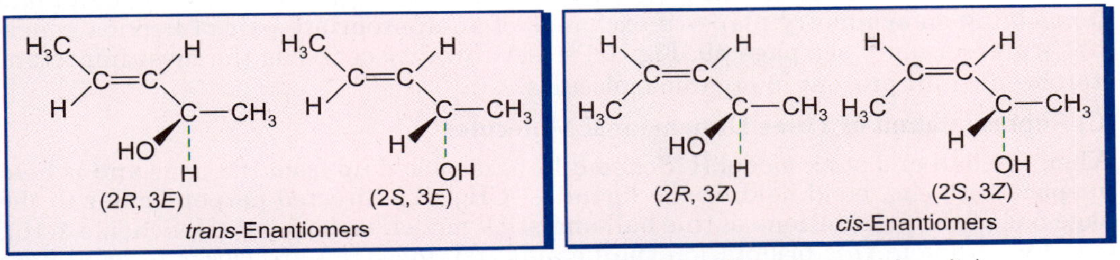

Four stereoisomers of 3-penten-2-ol (Two stereogenic units)

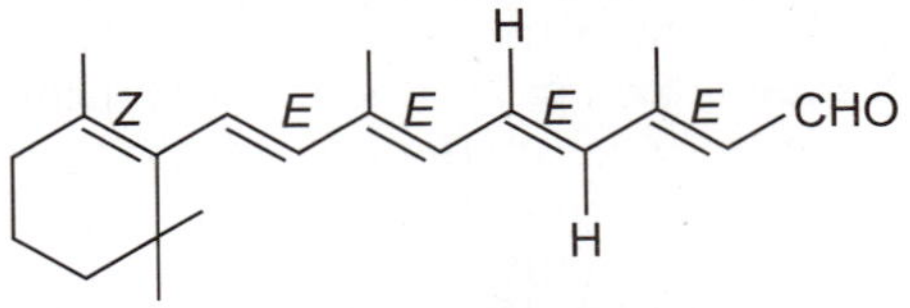

Vitamin A (Four stereogenic units)

SCHEME 1.14d

EXERCISE 1.2*a*

Predict the number of stereoisomers in the compounds (Scheme 1.15)

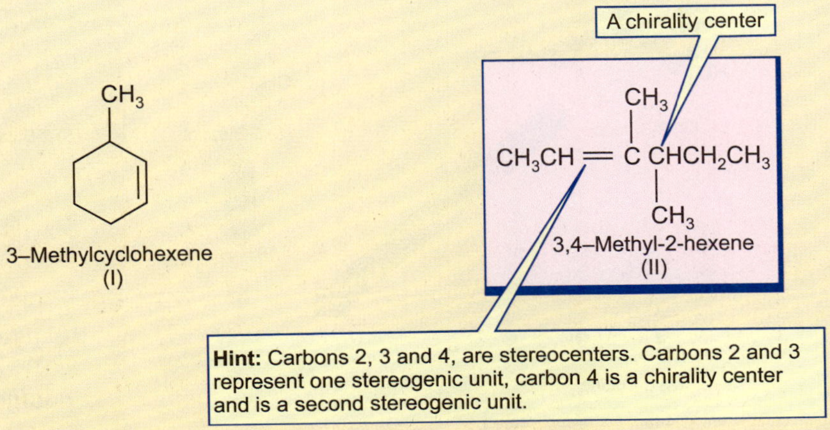

SCHEME 1.15

Continued...

ANSWER. *The compound (I) has a chirality center, the double bond being in the six membered ring can only be cis i.e., it is not a stereogenic unit. Thus this compound exists as an enantiomeric pair only.*

In (II) the double bond can show cis-trans isomerism and is therefore, a stereogenic unit, (II) also contains a chirality center which represents second stereogenic unit. The number of stereoisomers in (II) is thus four (2^n n = stereogenic units which is 2). Note this the C = C unit in (II) has two stereocenters but it represents one stereogenic unit. For calculating the number of total stereoisomers, these are stereogenic units and not stereocenters which are considered. The four stereoisomers are: (2Z, 4S), (2E, 4S), (2Z, 4R) and (2E, 4R).

A molecule is said to have a stereogenic unit (e.g., a chirality centre, axis, or plane, or a double bond showing *cis-tran*-isomerism or some tri- or tetra-coordinate atoms lying on symmetry planes if exchange of an appropriate pair of ligands generates a stereoisomer see page 49, Fig. 1.14*b*). A chirality center is the most important stereogenic unit present in organic molecules.

(C) Representation of Three Dimensional Molecules

When the ball and stick model (I, Scheme 1.16) is picked up from the page and is held in space with *e.g.*, bond holding the ligand (—CH_2CH_3) directed perpendicular to the page one of the orientations of this ball and stick model, is depicted (II, Scheme 1.16). One finds that in this orientation the ligands (H) and (OH) are closer to one's eyes (heavy wedges) while the ligand (CH_3) is directed away (broken wedge) and this three-dimensional picture (III) in which the bond holding the ligand (– CH_2CH_3) is placed in plane of the page is shown by a solid line is yet another useful way to depict chiral structures. These operations are equivalent to rotation of *e.g.*, your right hand (a chiral object). One of many orientations, may have thumb closer to you (equivalent to a wedge) or directed away from you equivalent to a dash— whatsoever may be the orientation, the right hand remains right hand. Thus a three-dimensional model of an enantiomer will retain its identity in any orientation you may depict it. In short, a right hand glove will remain right hand glove, whatever is the orientation.

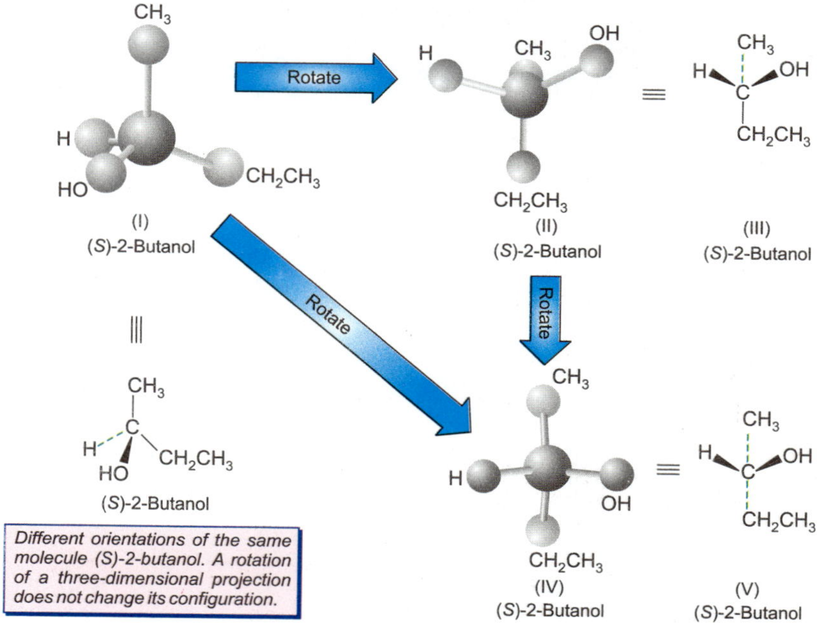

SCHEME 1.16 (Also see, Scheme 1.32)

A yet another orientation *e.g.*, (V, Scheme 1.16) of S-2-butanol is also drawn. This is equivalent to tilting the model (II) in space to an orientation (IV) when two groups are close to one's eyes while two are away [The knowledge of projection (III) in assigning *R* and *S* designations is in (Scheme 1.20*k*)].

(*i*) Fischer Projections

Fischer developed a two-dimensional plane projection formulas for three-dimensional molecules. These projections are particularly useful for carbohydrates and amino acids, and can also be used to represent many chiral molecules. Fischer projections are particularly useful for compounds with two or more chirality centers. The carbon chain in a Fischer projection is always drawn from top to bottom.

One must remember that Fischer projections represent eclipsed conformations and cannot tell anything about the preferred conformation of a molecule.

A Fischer projection uses a cross to represent the chirality center; the atom that is the chirality center is often omitted in which case it is symbolized by the interaction point of the two lines of a cross. All the bonds are drawn as solid lines with the understanding that horizontal bonds point toward the observer and vertical bonds point away. The carbon chain is drawn along the vertical line of the Fischer projection usually with the most highly oxidized end carbon atom at the top as in D-glucose (Scheme 1.18*a*). In (R)-1, 2-propanediol, the hydroxymethylene group being more oxidized is placed at the top (Scheme 1.17). In the case of 2-butanol, neither end carbon is more oxidized than the other and in this type of cases one simply keeps the carbon chain along the vertical.

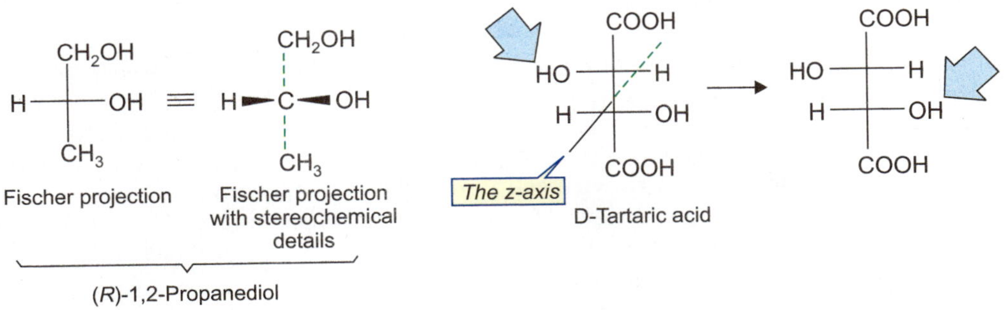

SCHEME 1.17

In D-tartaric acid *e.g.*, the two carbons at the end of the vertical chain have the same oxidation state. It, however, does not matter which carbon atom is situated at the bottom of the chain. When one turns D-tartaric acid through 180° about the Z-axis (Scheme 1.17), the same arrangement of atoms or groups is found.

> *For reference, the position of one OH group in D-tartaric acid (Scheme 1.17) is marked by an arrow before and after rotation. One may recall, a Fischer projection can be rotated through 180° in the plane of the paper without loosing identity. However, one cannot turn the projection through 90° or 270° which gives the mirror image. The projection cannot be taken out of the plane of the paper and rotated.*

Consider 2,3–dibromobutane (Scheme 1.18) drawn in Fischer projections and as a wedge dash representations. The molecule has a enantiomeric pair and a *meso* compound (The situation is similar to that in tartaric acid see, Scheme 1.2).

Consider first the enantiomeric pair (A and B) drawn in Fischer projection which is an eclipsed conformation.

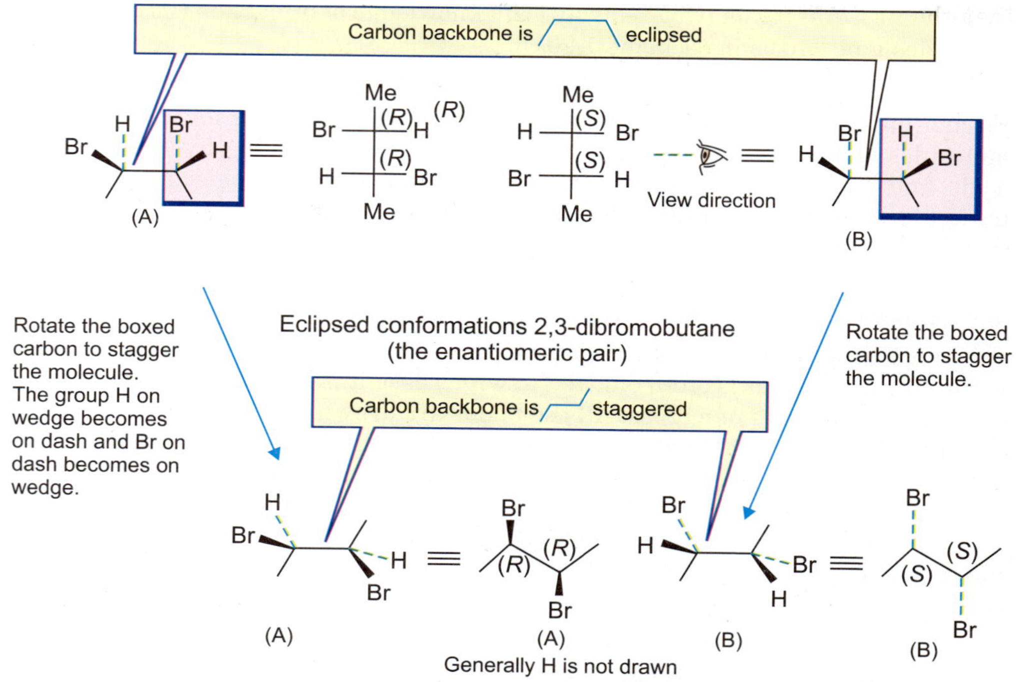

Staggered conformations 2, 3–dibromobutane (the enantiomeric pair)

SCHEME 1.18 (Compare with Schemes sec. 1.6)

- To transfer the groups from the Fischer projection to wedge formula, the wedge formula must be drawn with eclipsed carbon frame work.
- The Fischer projection is viewed from one side and depending on whether the group is closer to the view position or away it is put on a wedge or a dash respectively.
- To stagger the conformation one carbon is rotated.

In the meso compound the images (*C* and *C*′, Scheme 1.18*a* are) superimposable and in fact it is improper to draw these as two structures since they represent the same molecule (*i.e.*, *C* ≡ *C*′).

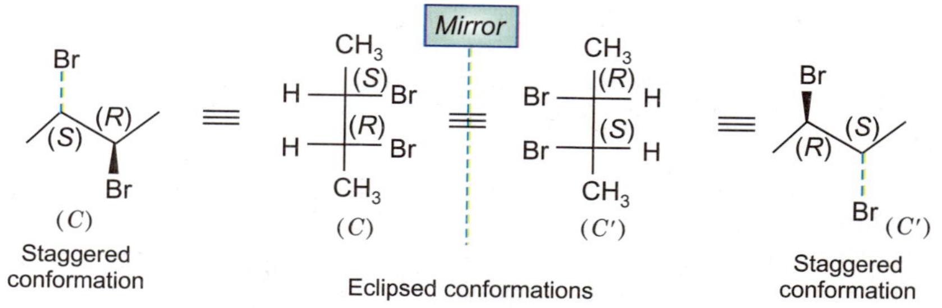

Meso 2,3–dibromo butane, (*C* and *C*′) represents the same molecule.

- Rotation of (*C*), 180° in the plane of paper gives same (*C*′)
- Mirror reflection of (*S*) configuration is (*R*) and of (*R*) is (*S*).

SCHEME 1.18a

When one rotates the Fischer projection by 90°, the vertical bonds become horizontal and the horizontal bonds become vertical. The viewer is confused and assumes that the horizontal bonds come forward and the vertical bonds go back. One sees a different view of the molecule (actually, the enantiomer of the original molecule). A 90° rotation of the Fischer projection in the plane of the paper (gives the enantiomeric structure) is not allowed (see Scheme 1.28*a*). Similarly its 180° rotation out of the plane of paper (gives the enantiomeric structure) is not allowed (see Scheme 1.28*a*).

A rotation of Fischer projection in the plane of paper by 180° (gives the same structure) and is a allowed process (see Scheme 1.28*a*).

Disadvantages of Fischer Projection

- These projections can be turned/rotated only in certain specified ways (see Scheme 1.28*a*).
- In compounds with more than one chirality center, a Fischer projection implies an eclipsing relationship of the groups attached to the two centers. One knows, however, that staggered conformations are more stable.
- Many organic reaction mechanisms (*e.g.* E2 reaction) necessitate that one should write the structures in the staggered form while only in few cases *e.g.*, pyrolytic eliminations an eclipsing relationship is required.
- In summary, Fischer projections are useful since, these allow to represent the stereoisomers in two dimensions, unfortunately, however, these are in fact the relatively unstable eclipsed conformations.
- Perspective (three-dimensional) formulas are not only more accurate representations but also at the same time these depict the molecule in its more stable staggered configuration.

More on Realistic Representation of Stereostructures of Molecules

Mention here is made of three conventions to draw stereostructures of compounds which avoid difficulties mentioned regarding Fischer projections.

(*ii*) The Flying Wedge Projection (Zig-zag Conformation, Masamune Projection)

The longest chain (backbone) is drawn as a "zig-zag" in the plane of the paper as a staggered arrangement (continuous lines) as shown for glucose (Scheme 1.18*a*). The substituents (other than H) in front and behind the plane are drawn with solid wedge lines and dotted lines respectively (Masamune projection). A "zig zag" can also be drawn in an eclipsed form, to clarify some stereochemical problems.

The Fischer projection was devised long before when nothing was known about the preferred conformations of molecules. It is now usual to follow a convention for acyclic molecules devised by Masamune.

Consider the following points:

- If one considers D-glucose, each chirality center is given (*R* or *S*) descriptor along with the positional number and D-glucose will be properly designated as (2*R*, 3*S*, 4*R*, 5*R*)-2, 3, 4, 5, 6-pentahydroxyhexanal. The H atoms are assumed and may not be drawn. For the method to convert Fisher projection to dashed-wedged line structure (*i.e.*, a staggered zig-zag see, Scheme 1.39*a*). A Fischer projection of D-(+)-glyceraldehyde is compared with the wedge formula and the wedge formula is drawn as zig-zag structure (Scheme 1.18*b*).

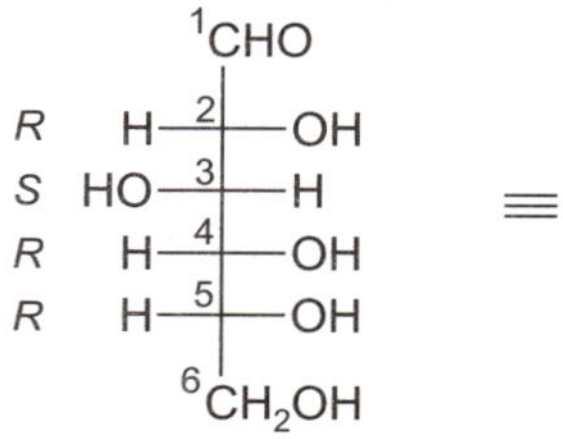

(I)
D-Glucose Fischer projection
(Eclipsed)

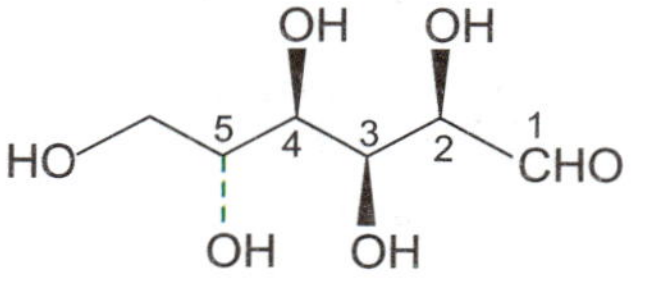

(II)
D-Glucose as "zig-zag" orientation
(Also, see scheme 1.40)
(Masamune projection)
(Staggered)

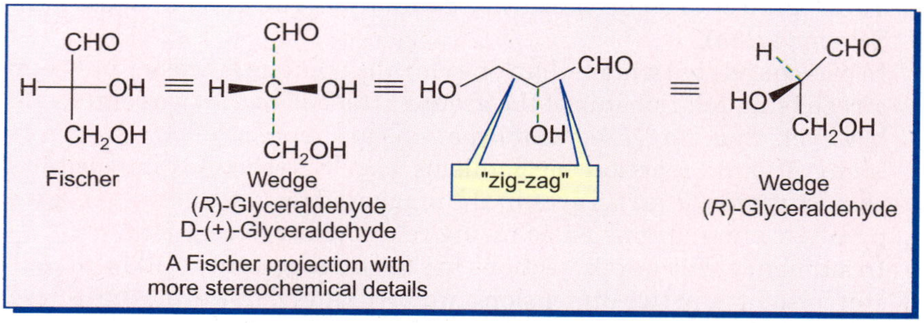

SCHEME 1.18b

- *Meso* stereoisomer(s) may not be easily detected while dealing with symmetric molecules. Thus in such a molecule with two chirality centers the Masamune projection will have a center of symmetry *i* if the stereocenters are adjacent (or separated by an even number of C atoms) but a plane of symmetry σ if they are separated by an odd number of C atoms in the main chain (Scheme 1.18*c*).

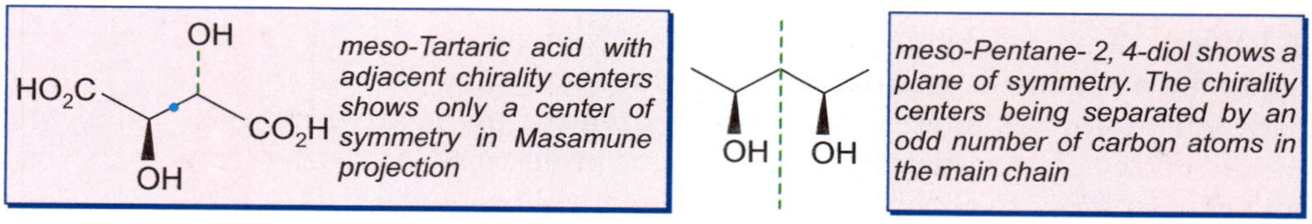

SCHEME 1.18c

- Masamune formulae for the designation of stereochemical relationship of diastereomers has been developed by aldol chemists. The carbon backbone is extended in the plane of the paper in the horizontal direction as a horizontal staggered zig zag shown with ligands (other than H) with bonds coming above or going below the plane of zig zag chain. If two substituents extend in the same direction, it is called *syn* and when these extend in opposite directions the stereochemical relationship is called *anti* (Scheme 1.18*d*, Also see Scheme 4.9).

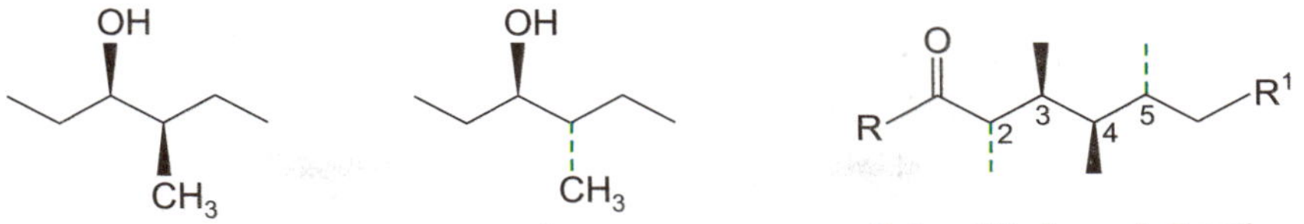

syn   *anti*   2, 3-*anti*-3, 4 *syn*, 4, 5-*anti* (Masamune)

SCHEME 1.18d

BOX 1.12
AN EXAMPLE OF STEREOCHEMICAL ANALYSIS USING *SYN* AND *ANTI* DESIGNATION OF STEREOCHEMICAL RELATIONSHIPS

While analysing a molecule regarding its stereochemical analysis, one looks for symmetry in the starting material to identify axes, planes or centres of symmetry. Diol (I, Scheme B 1.17) has an axis of symmetry whereas diol (II) has a plane. Diol (I) is C2 symmetric and chiral; while (II, Scheme B 1.17 is achiral).

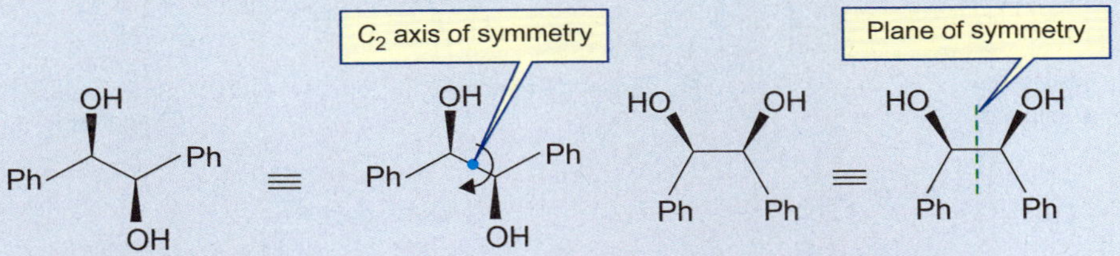

(I) *Syn*-Diol C_2 symmetric (Chiral)

Axis of symmetry coming towards one's eyes

(II) *Syn*-Diol

Plane of symmetry Achiral (*meso*)

Consider (I):
- It has two chirality centers, exists as enantiomers.
- Thus there could be four maximum stereoisomers
- Since groups attached to the chirality centers are identical, there is a possibility of having lesser number (possibility of existence of a *meso* compound).
- It is C_2 symmetric (chiral)
- An axis of rotation C_n, does not interfere with the chirality of a compound.
- The situation is just like that in enantiomers of tartaric acid.

Consider (II):
- The compound is related to *meso* tartaric acid.

SCHEME B 1.17

When one redraws the *syn*-diol (II, Scheme B 1.17) by rotating one carbon, one gets equivalent (II b, Scheme B 1.18) where one has *anti* relationship between the hydroxyl groups. In (II b) the molecule has now a center of symmetry, the molecule is achiral whatever way it is drawn.

Recall that plane of symmetry can be best seen in an eclipsed conformation, and center of symmetry in a staggered conformation (See, Scheme 1.41).

Continued...

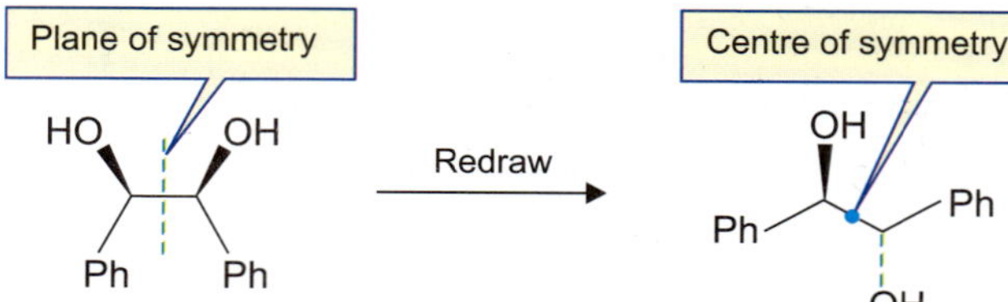

'*syn*' Diol
(II) Achiral
Scheme B 1.17

anti-diol
(II b) Achiral

SCHEME B 1.18

(*iii*) The "Sawhorse" Representation

In the "sawhorse" representation, the bond between the two carbon atoms (each carbon may be viewed as a letter 'Y') *e.g.* ethane (Scheme 1.18*e*) is drawn diagonally and is

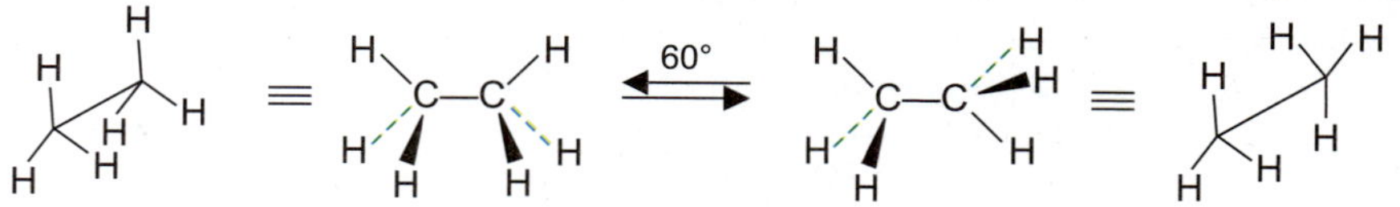

Sawhorse (Eclipsed)

Flying-wedge structure

Sawhorse (Staggered)

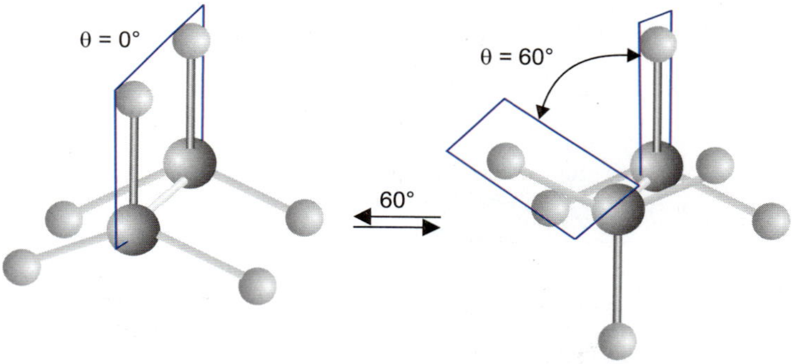

(Eclipsed)

(Staggered)

Ball and stick model showing dihedral angle θ

In ethane a 60° rotation of one carbon converts an eclipsed conformation (least stable, substituents on adjacent carbons as close as possible) to a staggered conformation when the substituents are as far apart as possible (In a staggered conformation, one sees that each C-H bond bisects the angle between the C-H bonds on the adjacent carbon in the middle) Torsional strain is the destabilization due to increased electron-electron repulsion in eclipsed bonds.

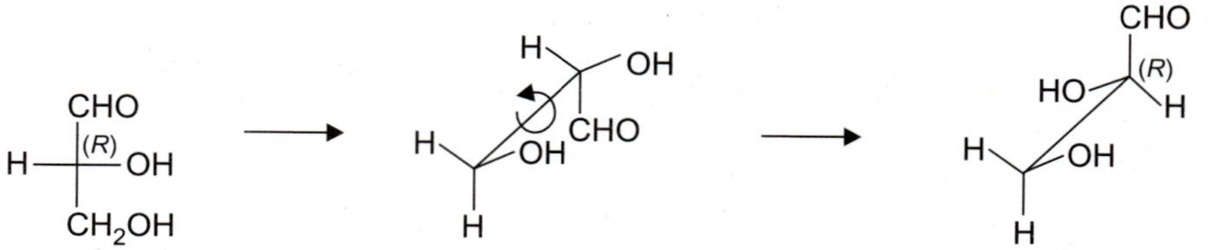

(D)-Glyceraldehyde
Fischer

(D)-Glyceraldehyde
Sawhorse (eclipsed)

(D)-Glyceraldehyde
Sawhorse (Staggered)

SCHEME 1.18e

slightly elongated. This projection shows the spatial relationship of atoms in a two-carbon system, which can be depicted, in the eclipsed or in staggered conformation as shown in ethane. It is just a simplified version of a ball and stick model where dihedral angle has been shown.

Various projections of (D)–glyceraldehyde have been presented in (scheme 1.18*b*). It can be converted into sawhorse projection, keeping in mind that it is an eclipsed conformation (also see scheme 1.32).

Sawhorse projection is useful to give a clear stereochemical picture of E2 elimination and addition reactions (Scheme 1.18*e*). Considering the sawhorse projection of products, it is clear that the ionic addition *e.g.* of bromine to a double bond is *anti* addition to give *trans*-products. Thus *trans*-2-butene on ionic addition of bromine gives only *meso* product (Scheme 1.18*f*). Further details of bromination of alkenes is in (Problems 1.23, 1.24 and Scheme 1.70).

A plane of symmetry can be best seen in an eclipsed conformation, while a center of symmetry can be best seen in a staggered conformation.

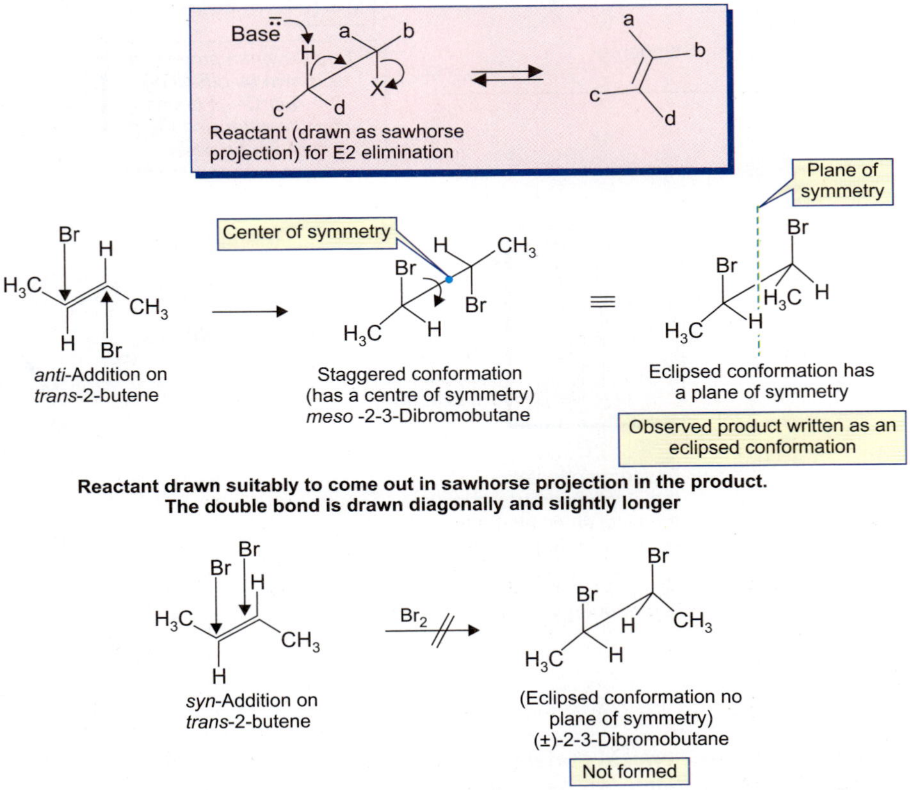

SCHEME 1.18f

(*iv*) Newman Projection

In this projection the molecule is viewed along the bond joining the two carbon atoms, the front carbon atom is shown by three lines (three bonds) which come together in a 'Y' shape. The back carbon is shown by a circle with three bonds pointing out from it (Scheme 1.18*g*). The Newman projection indicates a conformation beautifully, whether eclipsed or staggered. It is also used to depict carbonyl compounds.

The projection gives a clear picture on which side an incoming nucleophile will attack as in Cram's rule ((1) Scheme 1.18*h*, also see Scheme 2.24).

One views a Newman projection that places the carbonyl in a vertical position with the oxygen pointing up, the large group L can be pointing either straight out to the left or straight out to the right (both have the necessary 90° dihedral angle, Scheme 1.18*h*).

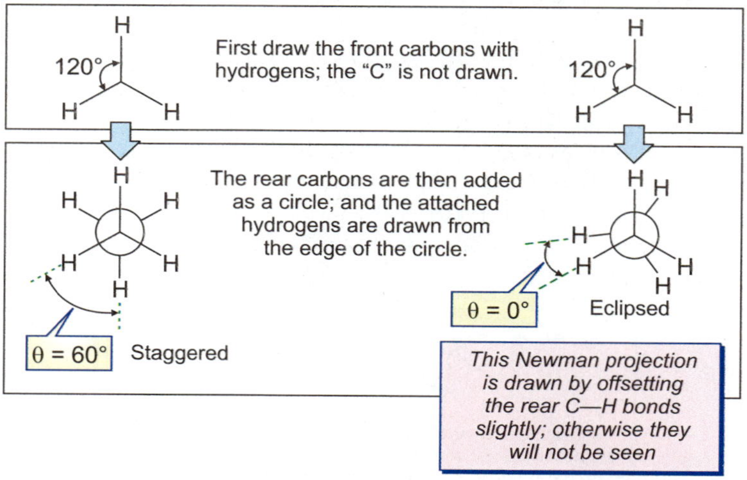

SCHEME 1.18g

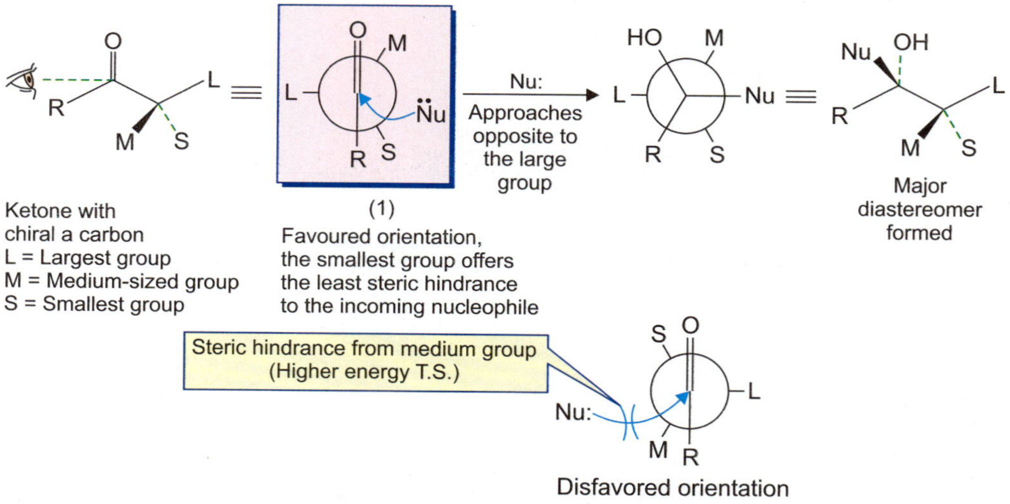

Stereochemical preferences predicted from Cram's rule using Felkin-Ahn model, from a Newmann projection

SCHEME 1.18h

When the aldehyde or ketone adopts a correct conformation, the nucleophile will predominantly attack the carbonyl from the side *opposite* the largest group to minimize steric strain in the transition state (at an angle approximately 109° from the C═O bond).

EXERCISE 1.2*b*

Which of the following compounds (I–III, scheme 1.18i) represents a meso compound?

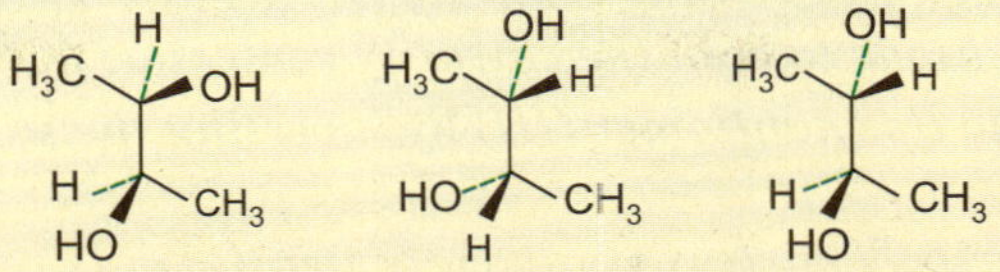

(I) (II) (III)

SCHEME 1.18i

ANSWER. The compound (III) is *meso. Strategy 1. When the carbon backbone of a compound is drawn in a staggered form look for a center of symmetry as in (III, scheme 1.18j).*

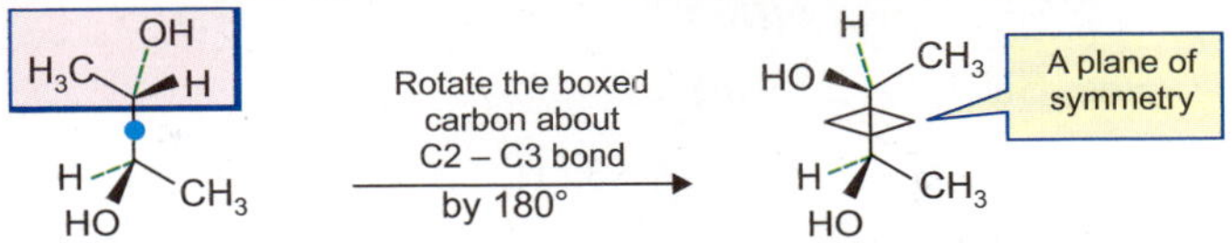

(III, with a center of symmetry is *meso*) (III, with a plane of symmetry is *meso*)

SCHEME 1.18j

Strategy 2. Rotate the boxed carbon around C2 – C3 bond to bring the carbon backbone in an eclipsed form. Now the plane of symmetry is clearly seen.

EXERCISE 1.2*c*

Write all the stereoisomers of 2,3-dihydroxybutanoic acid (scheme 1.18k).

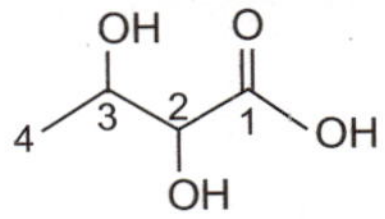

2,3-Dihydroxybutanoic acid

SCHEME 1.18k

ANSWER. *The compound is drawn with the carbon backbone in a*

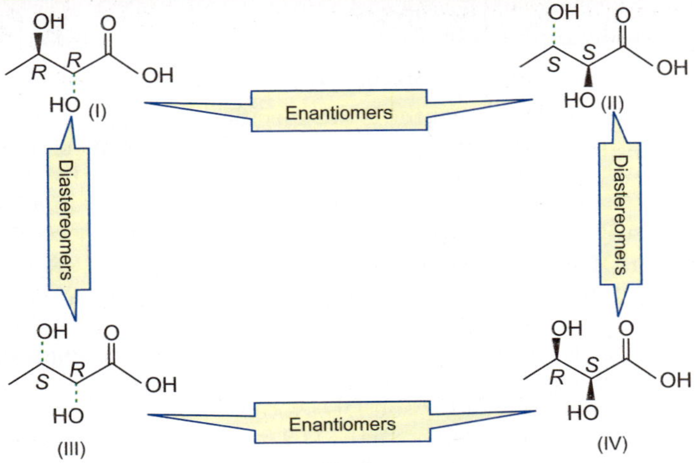

SCHEME 1.18l

Continued...

staggered Zig–Zag fashion.

The following points may be noted :

- *One may start e.g., by drawing both the OH groups at C2 and C3 on wedges and assign configurations to both the chirality centers as in (IV, scheme 1.18l)*
- *Convert (IV) into its enantiomer (III) by inverting the stereochemistry at both the centers C2 and C3.*
- *Inverting the stereochemistry only at one center (C3 in IV) gives its diastereomer (II).*
- *Write the enantiomer of (II) by inverting the stereochemistry at both the centers in (II) to give its enantiomer (I).*

EXERCISE 1.2*d*

Pick up the (2R, 3R) dihydroxy butanoic acid from (Scheme 1.18l) and convert it into Sawhorse and Fischer projections.

ANSWER: (See, Scheme 1.18 *m*)

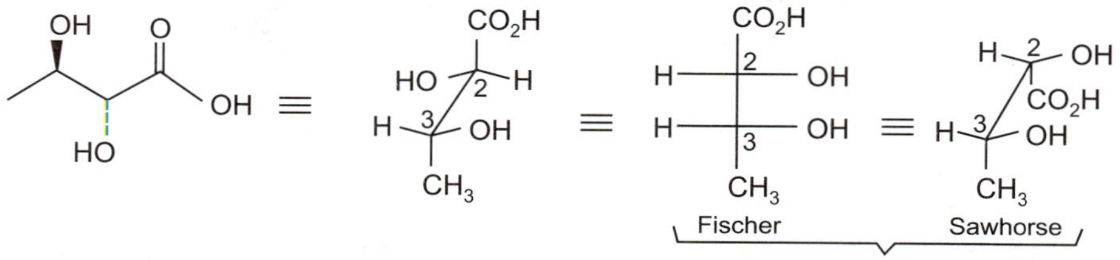

(2*R*, 3*R*) Stereoisomer of 2,3-dihydroxybutanoic acid staggered zig-zag conformation

Staggered Sawhorse conformation

Projection of the eclipsed sawhorse conformation correctly defines the location of groups on C2 and C3 in eclipsed Fischer projection. Eclipsed sawhorse projection is obtained from staggered sawhorse projection by rotation about the C2-C3 bond.

SCHEME 1.18m

1.4 CONFIGURATIONS AND THEIR SPECIFICATIONS *D-L* AND *R-S* SYSTEMS

(A) The D- and L-Designations—Configuration of Stereocenters

The configuration of a chirality center is the specification of the relative spatial arrangement of the four groups attached to it. Absolute configuration specifies their order so as to distinguish the two enantiomers and thereby define their chirality *i.e.*, the specification of (*R*) or (*S*) at each stereocenter (and dissymmetric grouping) in a molecule. The term relative configuration specifies the relation of the two chirality centers to each other.

The former DL system to denote absolute configuration is not without faults and, therefore, this system is seldom used today except for some classes of compounds like carbohydrates and amino acids. The (+) enantiomer of glyceraldehyde (See, scheme 1.1*c*) has its OH group on the right of the Fischer projection (D-configuration) while the (–) enantiomer has it on the left (L-configuration). The D- and L-glyceraldehydes are the standard reference compounds for the DL system of absolute configuration. Thus the D-series of sugars are those which have the OH group on the highest numbered chirality center (*i.e.*, the bottom chirality center) on the right in the Fischer projection and L-series have it on left (Scheme 1.18*b*). An aldose is degraded which results in the loss of the aldehyde carbon atom to afford a smaller sugar. Finally one obtains a triose which is then identified by comparing with glyceraldehyde enantiomers and assigned proper absolute configuration. D and L configurations are not related to the optical rotations of the sugars to which they are applied *i.e*, some sugars are D-(+)-while others are D-(–), thus D # *d* and L # *l*.

D-(+)-glucose means that dextrorotatory glucose has the same absolute configuration at the highest number chirality center as that of D-glyceraldehyde (the OH group on the bottom stereocenter on the right in Fischer projection Scheme 1.18*b*) and the configuration relative to the highest number chirality center at the other chirality centers must be memorized. Thus D—L system has the disadvantage of specifying the configuration of only one chirality center.

Some weaknesses of D-L nomenclature: Several difficulties were faced while working with D, L nomenclature. In the case of amino acids, L-(–) serine (Scheme 1.19) is taken as a reference. A further complication is introduced in assigning configuration to a molecule containing both a hydroxyl as well as an amino group. The 2-amino-2-deoxymannonic acid (A, Scheme 1.19) may well be assigned D-configuration with reference to D-glyceraldehyde and L-configuration with reference to serine (C2). No doubt the complication was removed by employing subscripts '*g*' for glyceraldehyde and '*s*' for serine so that the compound could be assigned either of configurations Dg or Ls. Not easy to do in the laboratory, but in biological systems the terminal CH_2OH group can be oxidized enzymatically without oxidation of the aldehyde group. The product glucuronic acid is important in animal systems because many toxic substances are excreted in the urine as glucuronides, derivatives of this acid.

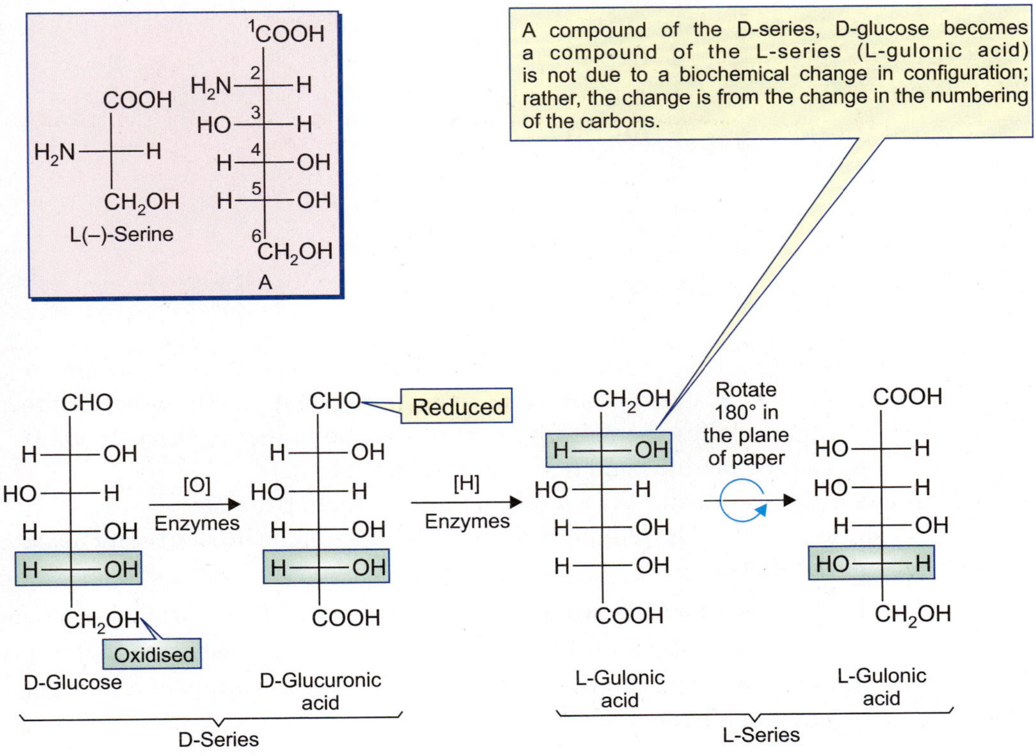

SCHEME 1.19

(B) The R and S Designations from Perspective Formulas – Three Dimensional Drawing

The system now widely used is the Cahn-Ingold-Prelog (R-S) system (which relates all compounds with only one set of rules) in which the four groups attached to the chirality center are numbered 1, 2, 3 and 4 and are ranked according to a set of sequence rules so that they can be assigned priorities for arrangement in a sequence 1 > 2 > 3 > 4. In a simple case the priorities follow the order of decreasing atomic number of the atom that is directly attached to the chirality center *i.e.*, Br(35), Cl(17), S(16), O(8), N(7), C(6) and H(1). The application of the rule can be illustrated with 2-butanol enantiomers (Scheme 1.20) which has a sequence OH > C_2H_5 > CH_3 > H.

The oxygen has the highest priority (1) and the hydrogen the lowest atomic number priority (4). To make the priority assignment for the methyl and ethyl groups where the atom that is directly attached to the chirality center is a carbon atom the next sets of atoms in the groups are examined. Thus, the ethyl group takes precedence (C, H, H) as compared with methyl (H, H, H).

In Table 1.2, atoms and groups are listed with increasing priority with the exception of a lone pair. H has the lowest priority. The following points may be noted:

Table 1.2: Atoms and groups with increasing priority

1—H	10—$CH{=}CH_2$	19—NH_2	28—SOR
2—D	11—$C(CH_3)_3$	20—$NHCH_3$	29—SO_2R
3—CH_3	12—C≡CH	21—NO	30—Cl
4—CH_2CH_3	13—C_6H_5	22—NO_2	31—Br
5—$CH_2(CH_2)NCH_3$	14—CH_2OH	23—OH	32—I
6—CH_2—$CH{=}CH_2$	15—CHO	24—OCH_3	
7—CH_2—C≡CH	16—COR	25—OC_6H_5	
8—$CH_2C_6H_5$	17—$CONH_2$	26—F	
9—$CH(CH_3)_2$	18—CO_2H	27—SH	

1. Double bond configuration : (*i*) *Z* has higher priority than *E* and (*ii*) *cis* has higher priority than *trans*, where *cis* and *trans* refer to the relationship between the substituent that includes the chirality center and the *higher* priority group at the other end of the double bond (*e.g.*, see Scheme 1.46*a*).
2. *l* (like) has a higher priority than *u* (unlike), and *r* higher than *s*.
3. Enantiomeric groups : *R* has higher priority than *S* and *M* is higher than *P* (*e.g.*, see, Scheme 1.139*c*).

- If the four atoms attached to the chirality center are different, priorities depend on atomic numbers, the atom of higher atomic number is assigned higher priority. If two atoms are isotopes of the same element, the atom of higher mass number has the higher priority (I, Scheme 1.19*a*).

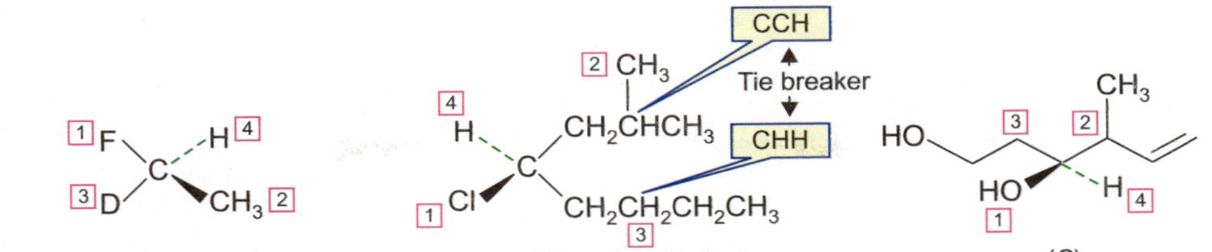

(*R*)-1-Deuterio-1-fluoroethane (I) | (*R*)-4-Chloro-2-methyloctane (II) | (*S*) (III)

SCHEME 1.19a

- If the relative priorities of two groups cannot be determined as above, these are decided by a comparison of priority atoms at the first point of difference (II, Scheme 1.19*a*). In the case of (III, Scheme 1.19*a*) of the two carbon chains attached to its chirality center, the chain with a secondary carbon is of higher priority than the one with primary carbon.

  In the case of (I, 1.19*b*) of the four chains attached to chirality center, butyl has priority over ethyl group. In the case of "oxygen chains" the first point of difference is reached at the carbon attached to oxygen. Similarly priorities are given in compound (II and III Scheme 1.19*b*). Often it may be necessary to proceed atom by atom till a point of difference is obtained.

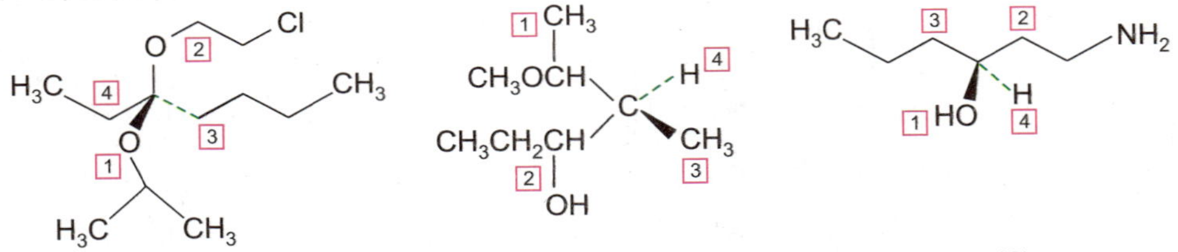

(*R*) (I) | (*S*) (II) | (*S*) (III)

SCHEME 1.19b

- In the case of double or triple bond, both atoms are considered to be duplicated or triplicated. This is shown in (Scheme 1.19*c*).

  For example, in glyceraldehyde (scheme 1.19*c*) the –OH group has the highest priority of all and the O, O, H of –CHO takes priority over the O, H, H of $-CH_2OH$. The complete sequence is thus $-OH > -CHO > -CH_2OH > -H$. In 1-amino-2-methyl-1-phenylpropane (I, Scheme 1.19*c*), therefore, the C, C, C, of phenyl take priority over the C, C, H of isopropyl, but not over N, which has a higher atomic number. The sequence is thus $NH_2 > C_6H_5 > C_3H_7 > H$.

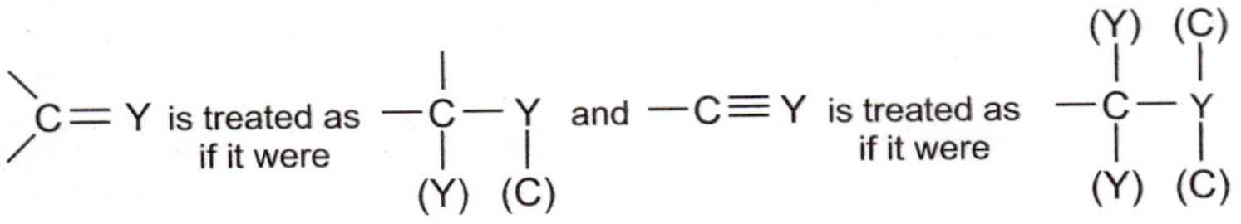

The symbols in parentheses are duplicate or triplicate representations of the atoms at the other end of the multiple bond.

Continued...

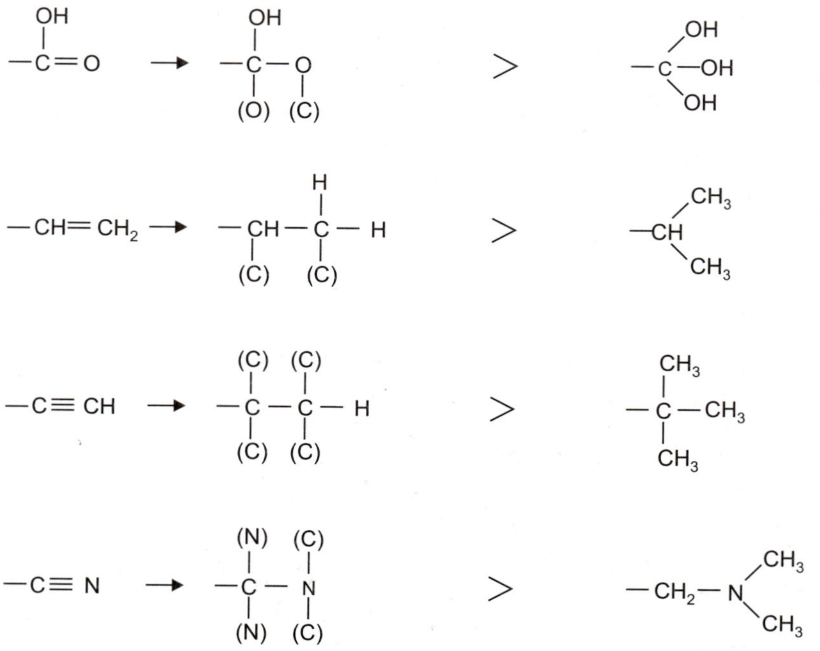

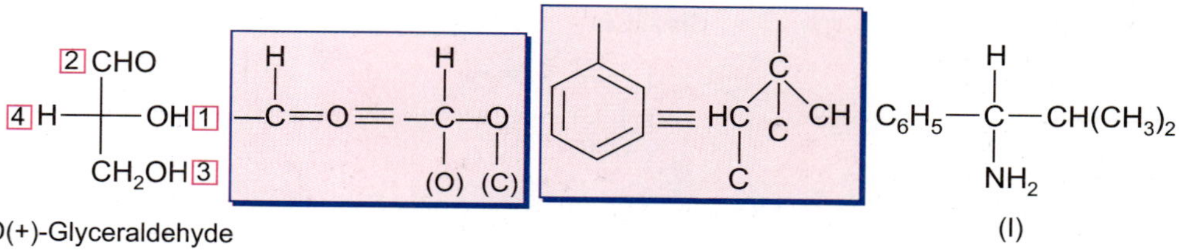

SCHEME 1.19c

- A R configured group has priority over S, for an example of the application of this rule (see Scheme 1.31a).
- One can assign priorities to the groups as given (see table 1.1) by considering the first point of difference and two more examples are in (Scheme 1.19d). In the first set, the

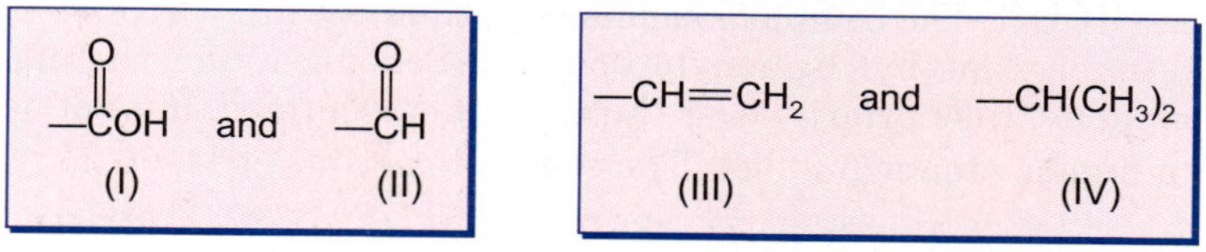

SCHEME 1.19d

point of difference is O of the OH in COOH group compared with H in the CHO group, therefore COOH group has higher priority (Scheme 1.9*e*). In the second set C-1 in each group has a similar pattern of atoms; *i.e.*, C(C, C, H), carbon bonded to two carbons and a hydrogen. For the vinyl group, bonding at carbon 2 is C(C, H, H). For the isopropyl group at carbon 2 it is C(H, H, H). The vinyl group is higher in priority than is the isopropyl group.

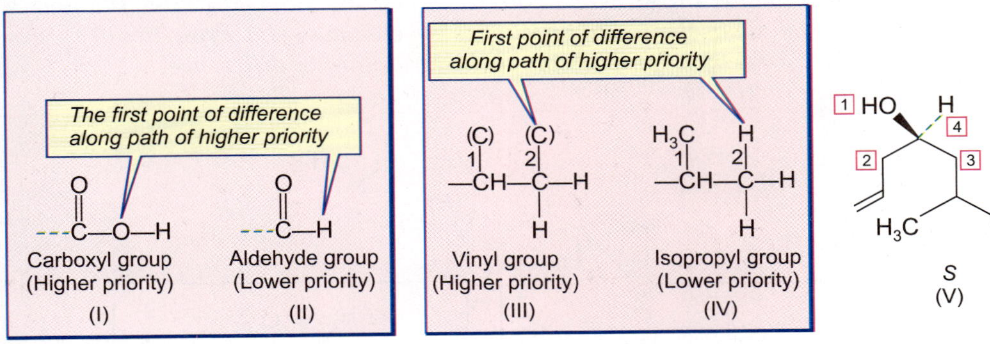

SCHEME 1.19e

EXERCISE 1.2e

Assign (R) or (S) configuration to each of the following compounds:

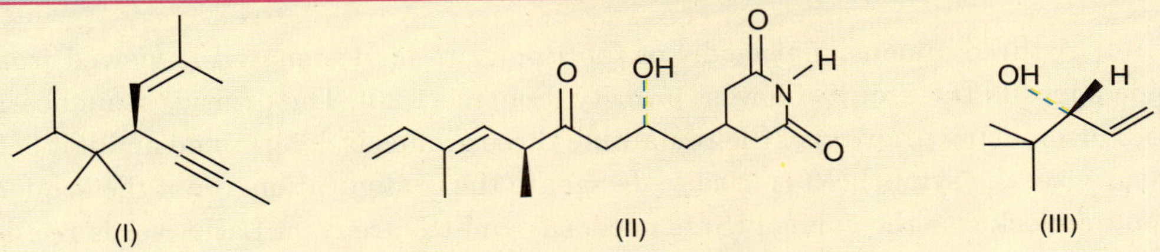

SCHEME 1.19f

ANSWER (*I*). *Looking to the carbon atoms attached directly to the chirality center*

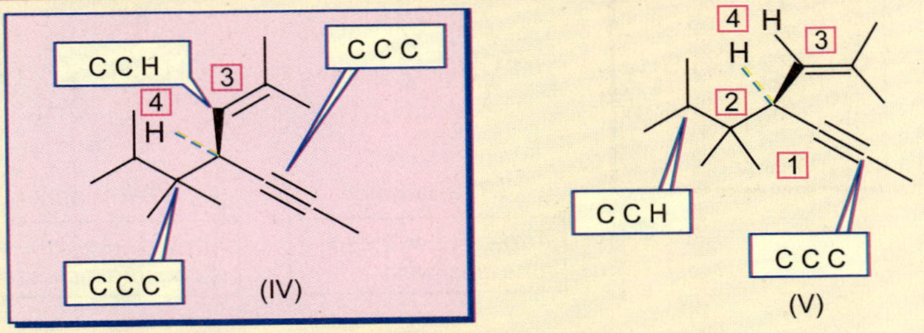

SCHEME 1.19g

Continued...

one gets the priorities shown in (IV, Scheme 1.19g) of these two atoms one is H, thus it gets the fourth priority. The carbon attached to the double bond gets the third priority. To assign the first and second priorities one must move one carbon away from the chirality center as in (V scheme 1.19g) and then make a list of three atoms connected to each of these positions. The carbon atom which is the part of the triple bond wins the tie-breaker. Thus (I) has (R) configuration.

(II) In this compound there are two chirality centers and at each center the fourth priority group is H which as per usual practice is not shown in (II). Priorities are given as shown (scheme 1.19h) by a comparison at the first point of difference.
(III) This compound is (R).

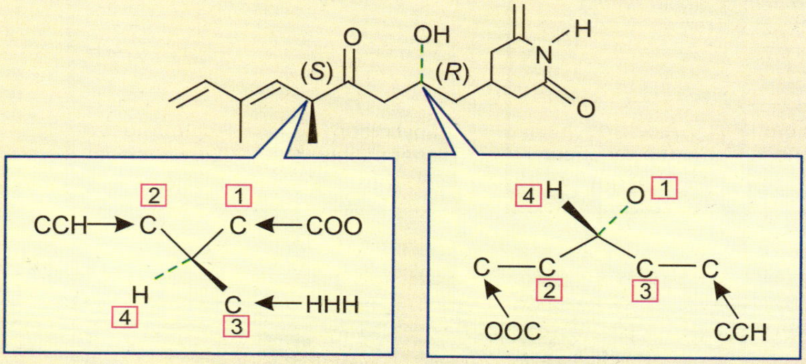

SCHEME 1.19h

A three dimensional model of the isomer to be designated is viewed from the side opposite the group of lowest priority (Scheme 1.20). The priority sequence 1 → 2 → 3 (decreasing priority) of the remaining three groups is determined and if it is found clockwise, the symbol (*R*) is used to designate the configuration and if the sequence is counterclockwise the symbol (*S*) is used. (*R*) and (*S*) are from Latin words *rectus* and *sinister* meaning right and left respectively.

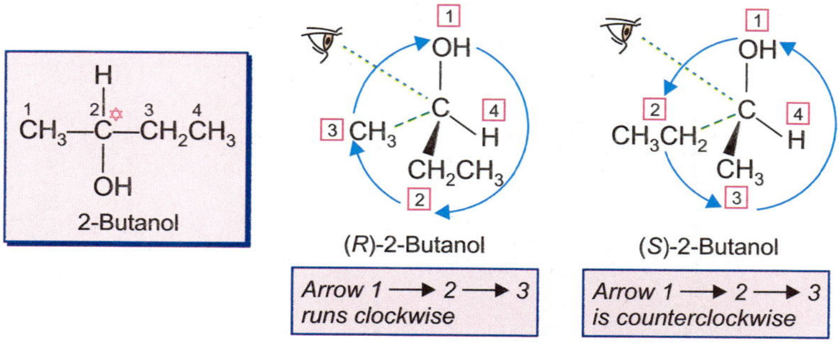

SCHEME 1.20

In the case of *R*-2-butanol and *S*-2-butanol (Scheme 1.20) the two enantiomers as drawn at first glance do not look like the mirror reflections. The interchange of any two groups around a stereocenter in a three dimensional drawing or a Fischer projection (with one stereocenter) gives the enantiomer. Therefore, one immediately views the drawings as enantiomers differing in the position of CH_3— and CH_3— CH_2— groups. The mirror image drawing of *S*-2-butanol is *R*-2-butanol and is drawn suitably in (Scheme 1.20*a*).

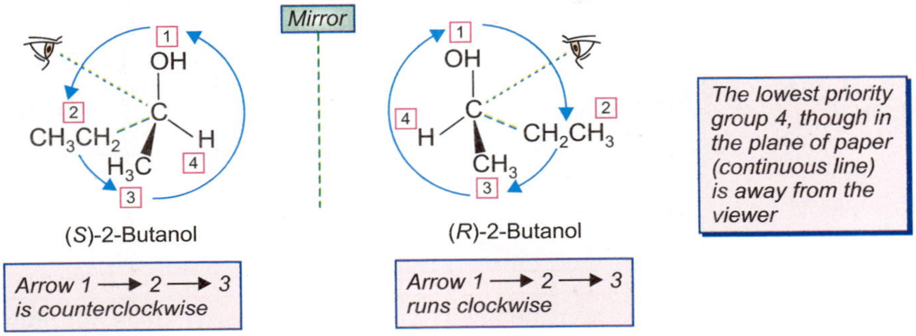

SCHEME 1.20a

One may position the same molecule, on paper or by using molecular model set to simply give different views of the same molecule. Some effort will then be needed to establish their identity. This may be done mentally or by manipulation on paper, and the later process is easy. Thus, yet another view of *R*-2-butanol is presented (Scheme 1.20*b*), that this and the drawing of (*R*)-2-butanol (Scheme 1.20*a*) represent the same molecule can be proved by interchanging the position of any two groups in one of these drawings twice (This process leads to retention of configuration at the stereocenter). Thus in drawing (Scheme 1.20*b*) first exchange 1 with 3 and then 2 with 4 and the arrangement thus obtained is identical to the one in (Scheme 1.20*a*). Identical names including *R* and *S* designation show that different orientations of a compound are infact same *i.e.*homomeric. This procedure shows that *e.g.*, different drawings of 2-butanol (Schemes 1.16, 1.20-1.20*c*) represent the same enantiomer (*S*)-2-butanol.

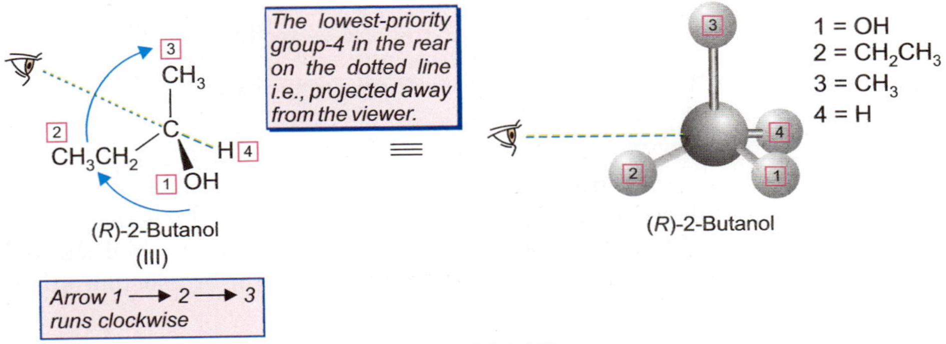

SCHEME 1.20b

To assign absolute configuration as *R* or *S* the three dimensional drawing, of the molecule has to be viewed with the lowest priority group at the back. The compounds (I-III, Scheme 1.20*c*) are already drawn (oriented) so that each orientation can be viewed either from left or right with the priority group 4 already oriented toward the rear.

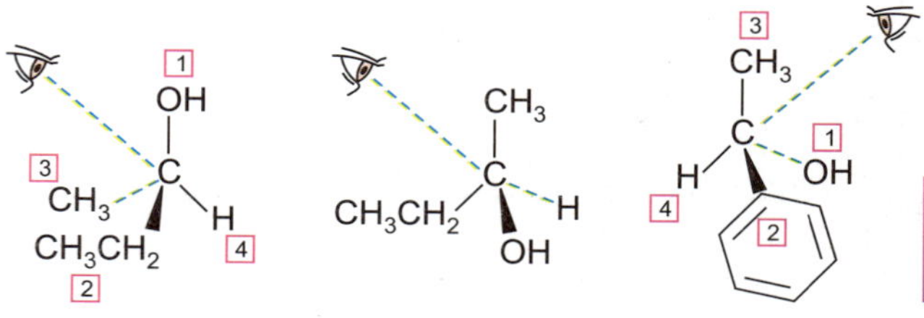

Compounds (I) and (II) can be viewed from left while (III) can be easily viewed from right

(I) (*R*)-2-Butanol (II) (*R*)-2-Butanol (III) (*R*)

SCHEME 1.20c

One may have many options when the priority 4 group as drawn does not point away from one's eye. One can reorient the substituents so that the priority group 4 points away from one's eyes and the orientation of the compound can be visualized easily for the assignment to *R* and *S* descriptor to the chirality center. Consider the situations (Schemes 1.20*d* – 1.20*f*).

- When the group of lowest priority is drawn at the top, one can look along the C—H bond from below the orientation. Alternatively one can tilt the orientation so that it can be viewed conveniently for the assignment of configuration (Scheme 1.20*d*)

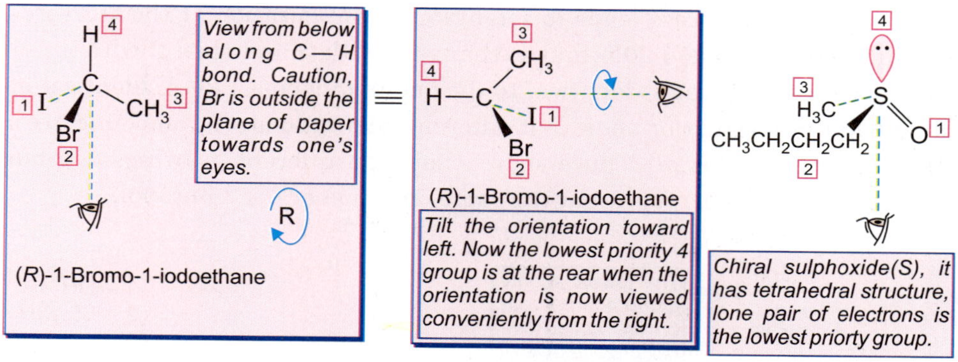

SCHEME 1.20d

- When the lowest priority group is oriented towards one's eyes, one may imagine holding the molecule by the group at top (1, Scheme 1.20e) and rotating it so that the group of lowest priority is positioned away and can be viewed through the chirality center as in (*S*) alanine. Another example is of bromochlorofluoroiodo methane (Scheme 1.20*e*).

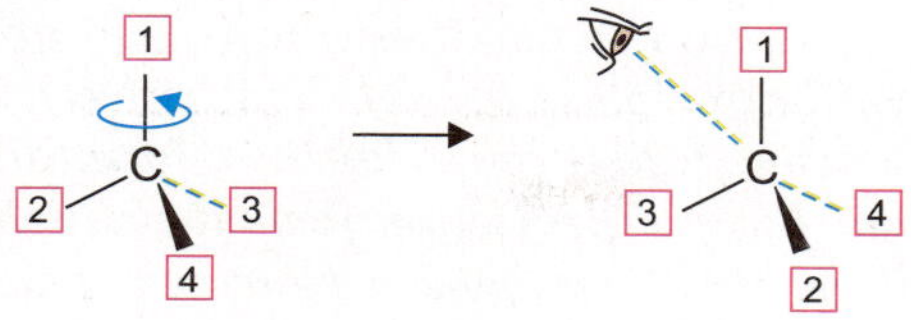

(I) Rotation

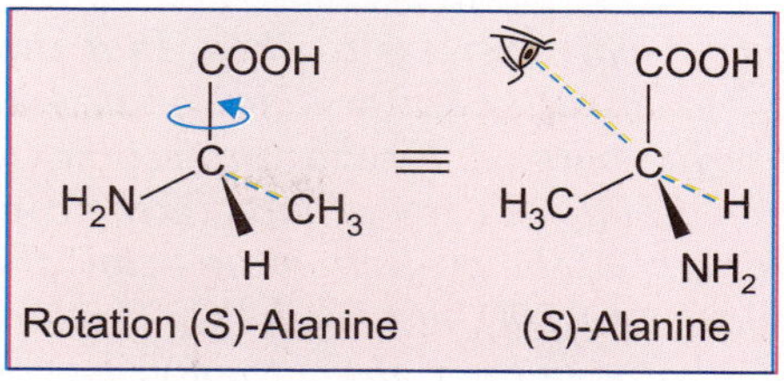

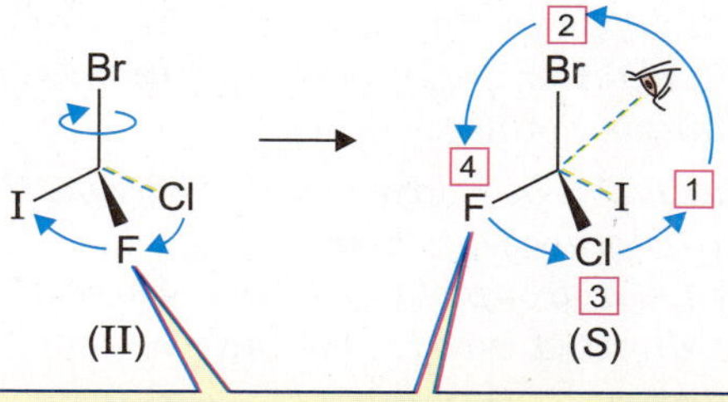

Holding the molecule (an enantiomer of bromochlorofluoroiodo methane) by the Br and rotating it as shown (II) *puts the lowest priority group (F) in the plane of paper. Looking down the C—F bond gives (S) configuration.*

When the lowest priority group is towards one's eyes, then turn the molecule to put the group of priority 4 at the rear for the assignment of configuration.

SCHEME 1.20e

- When the lowest priority group is towards one's eyes one may adopt alternative option. Interchange of the position of any two groups twice at the chirality center retains its configuration and also puts group of lowest priority away from the viewer (scheme 1.20*f*).

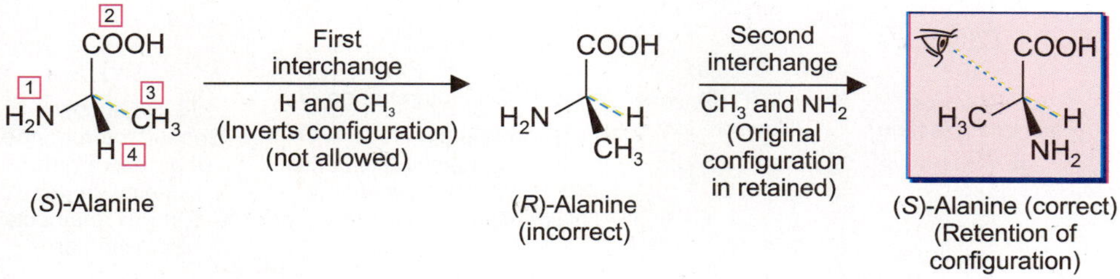

SCHEME 1.20f

One can assign R or S configuration directly from the Fischer projection by following the procedure given in (see, Scheme 1.26b).

Another useful exercise to show that interchanging the position of any two groups around a chirality center twice retains its configuration is the relation between a-D-glucose in common Fischer projection (I, Scheme 1.20g) and Haworth formula (II, Scheme 1.20d). All the groups in the common Fischer projection formula which are on the right of the projection are below the plane of the hexagon in Haworth structure (see, Scheme 1.20f), but for the C5 H which is below the plane of hexagon even though it is not lying on the right of the Fischer projection (I, scheme 1.20g). In fact one would find that interchanging two groups twice (first H with CH_2OH and then H with ether link) give the equivalent structure (modified Fischer projection IV, scheme 1.20g) in which H at C5 comes to right of the projection. Six membered ring exists in (energy minimum) chair form and is not flat.

The, flat Haworth formula can be relaxed to chair by putting one carbon (C4) up and other (C1) down as far a-D-glucose (Scheme 1.20g). In the chair form the substituents, maintain their identity i.e. a group closer to the observer remains closer and the one directed away remains directed away. One may remember that there are two chair forms possible interconverted by ring flip.

The following is the procedure for converting a Haworth projection of glucose into a chair conformation.

- *Place the ring oxygen at the back right-hand corner. The primary alcohol group (—CH_2OH) is the largest of the substituents and thus would be more stable in the equatorial position (C5 position).*

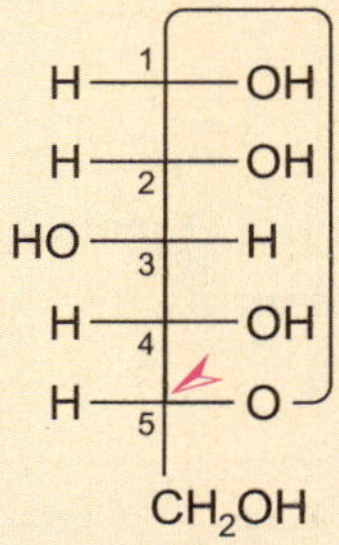

(I)
α-D-Glucose
(Fischer Projection)

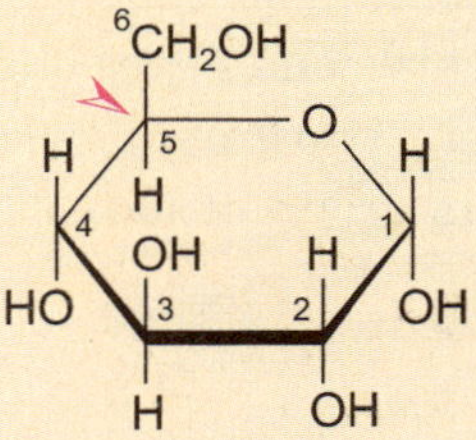

(II)
α-D-Glucose
(Haworth formula)

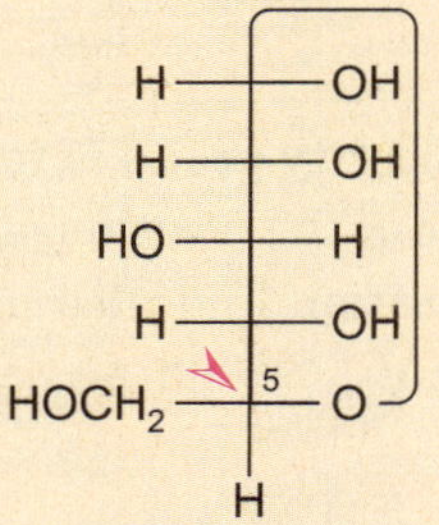

(III)
One interchange of groups (H and CH_2 OH) around chirality center C-5 in (I) inverts its configuration (not allowed)

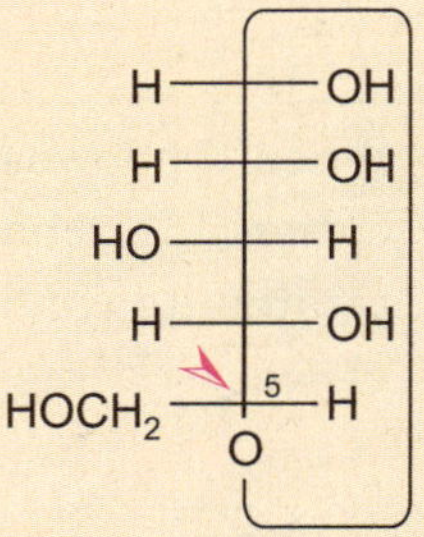

(IV)
α-D-Glucose
(Second interchange of groups around chirality center C-5 in (III) retains its configuration)

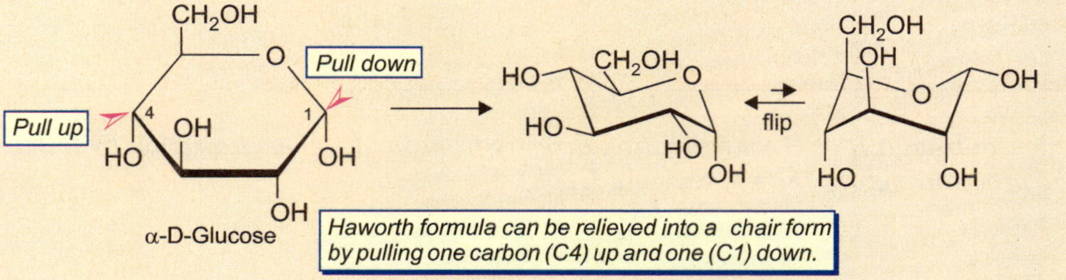

SCHEME 1.20g

Continued...

- *The OH group at C4 is trans to the —CH_2OH group (see Haworth projection), this is thus placed at the equatorial position (1, 2-diequatorial substituents are trans to one another (see Scheme 4.25).*
- *The C3 OH group is also trans to the C4 OH group, thus the C3 OH group is also placed in the equatorial position.*
- *Moving around the ring, thus gives the conformation of α-D-glucose.*
- *The axial positions at C5, C4, C3 and C2 are all occupied by hydrogens (not shown) which require little space and thus experience little steric strain. There is more on conformation of sugars (Sec. 4.15).*

(C) When the Chirality Center is the Part of the Ring

- When the chirality center is part of the ring, the paths around the ring are considered as independent substituents. **[For determining the retative priority (1 to 4) see the summary of sequence rules].** For the assignment of (R or S) descriptors the lowest priority group has to be directed away (Scheme 1.20*h*) and thus the absolute configuration of 4-methylcyclohexene (Scheme 1.20*h*) is (*R*).
- When in a system the group of lowest priority is instead directed toward the viewer *i.e.*, it is on the thick line, the structure is rotated 180° so that it gets directed away and the usual procedure is then applied. In structure (I, Scheme 1.20*h*) at both the stereocenters C1 and C2 the group of lowest priority *i.e.*, H (not shown as per usual convention) is directed towards ones eyes. Its identical view (II) is obtained by its rotation by 180° around the axis which passes through C1 and the middle of C3–C4 bond. The configuration as assigned from (II, Scheme 1.20*h*) with H now directed away from one's eyes comes out to be (1*S*, 2*R*).

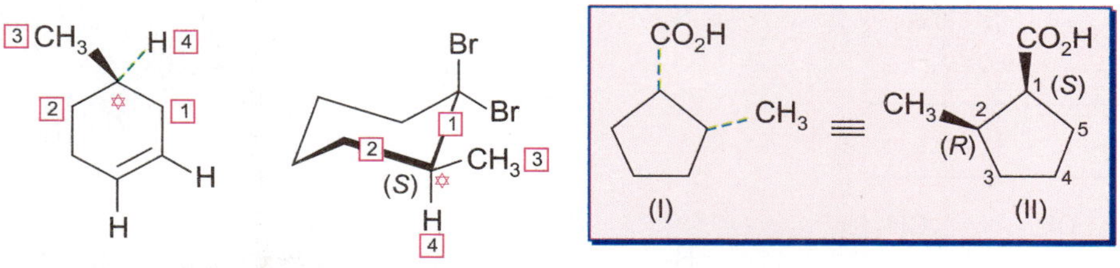

(*R*)-4-Methylcyclohexene 1, 1-Dibromo-2-methylcyclohexane

SCHEME 1.20h

(D) When the Group of Lowest Priority is instead towards One's Eyes

One may not turn or flip the molecule—to get the group of lowest priority away from one's eyes. Keep the orientation with the lowest priority group towards you, but now apply the R/S rule backward (I, Scheme 1.20*i*). In the molecule of camphor (two chirality centers) only the chirality center C1 containing CH_3 group is considered. The lowest priority group CH_3 is already away in (projection II, Scheme 1.20*i*) and when viewed from front the configuration at this stereocenter comes out to be (R). In the equivalent projection (II*a*, Scheme 1.20*i*) now the methyl group of lowest priority is instead towards one's eyes. To assign configuration to this stereocenter, apply the R/S rule backwards on (II*a*, Scheme 1.20*i* also see Scheme 1.20*j*).

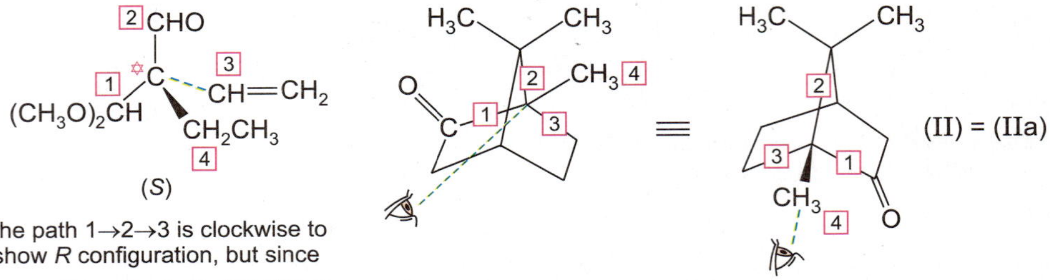

The path 1→2→3 is clockwise to show *R* configuration, but since group of lowest priority CH_3CH_3 is towards one's eyes apply *R*, *S* system backwards. Thus the correct configuration of the

(I)

From the view position CH_3 group is oriented away

(*R*)

(II)

1*R*-Camphor

From the view position the CH_3 group is close to the observer

(*R*)

(IIa)

1*R*-Camphor

SCHEME 1.20i

Recall that in [2.2.1] heptane system, the bridge can only be formed by *cis* fusion due to geometrical constraints, both the bridgehead groups must be *cis* to each other as shown in (I, Scheme 1.20*j*). For assigning *R* and *S* configuration, by looking from the front bridgehead carbon, the ligand on the rear bridgehead carbon (is away from the viewer) and may be indicated by a dotted line (III, Scheme 1.20 *j*). Other structures are given for practice.

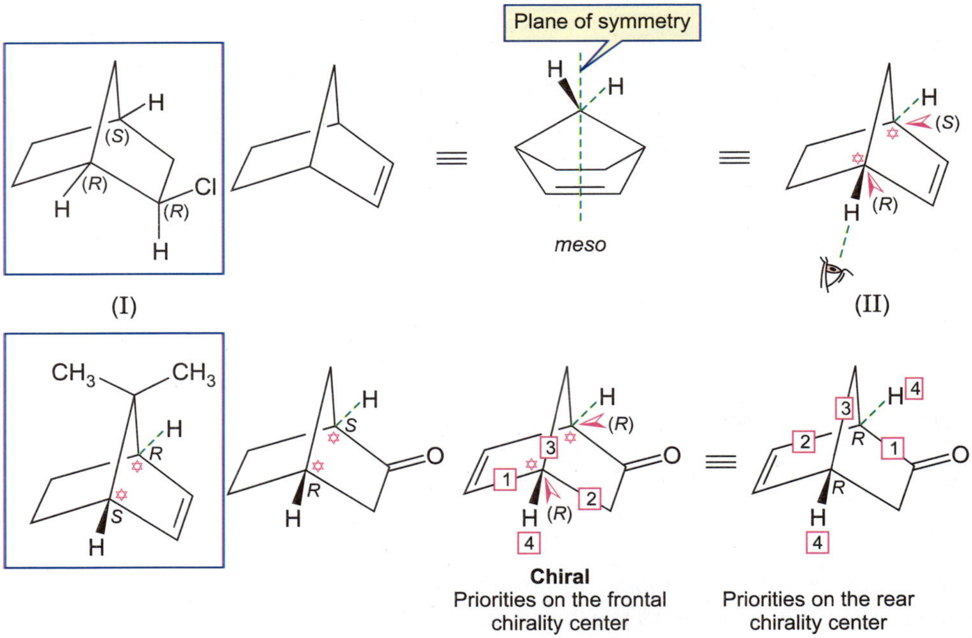

(I) (II)

(III) (IV) (V)

SCHEME 1.20j

(E) Some Typical Situations

To assign configuration at C-2 in sawhorse projection (I, Scheme 1.20*k*), view from the right side of the structure is correct as in (III, scheme 1.20*c*). However, the assignment of configuration to C-3 in (I, Scheme 1.20*k*) presents difficulty. Evidently the group of lowest priority at C-3 (H) is towards one's eyes. So for assignment of configuration at C-3 one works with *R*/*S* rule backwards. Alternately, the conversion of (I) to eclipsed sawhorse conformation (Ia) by rotation around C2-C3 bond puts the group of lowest priority away from one's eyes to be viewed conveniently from the opposite side to give the correct configuration. The situation in epoxide (II, Scheme 1.20*k*) is similar.

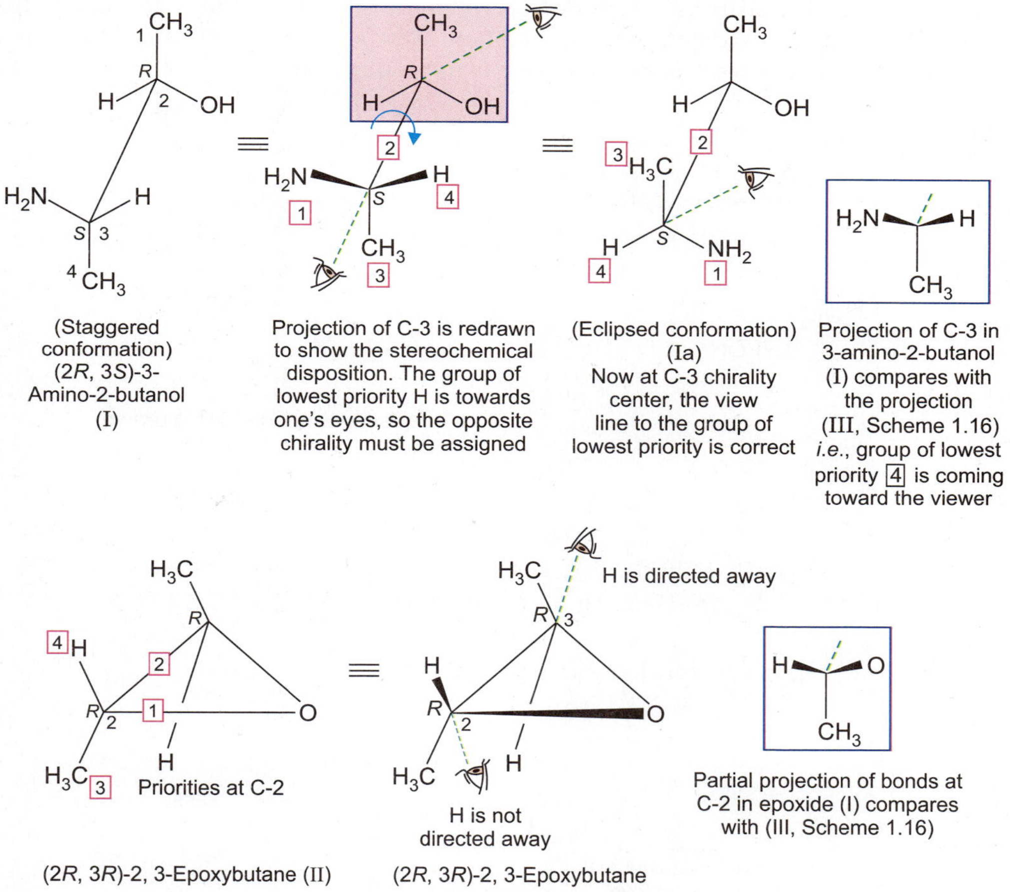

SCHEME 1.20k

Same considerations help to assign configurations from a Newman projection of a compound (Scheme 1.20l)

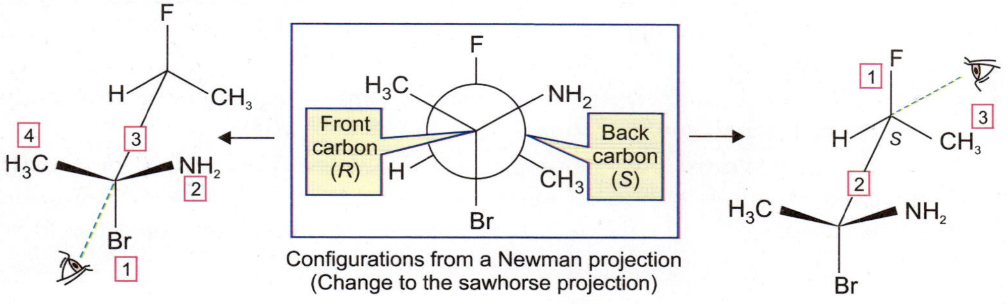

SCHEME 1.20l

In the epoxide (I, Scheme 1.20 m) there are two chirality centers, consider only C–2. Many students make the mistake of choosing the anticlockwise path 1 → 2 → 3 to assign wrong (*R*) configuration. The epoxy ring is however, drawn below the plane of paper thus, the path 1→ 2 → 3 is clockwise. (Also see configuration at C–3 in glucose (Exercise 1.5).

When the structure is reoriented by tilting it 90° out of plane, the bonds from the chirality centers to oxygen come in the plane of paper (continuous lines, Ia, Scheme 1.20m) and group of lowest priority moves away, configuration at C–2 is (*S*) (See Box 1.1, page 4).

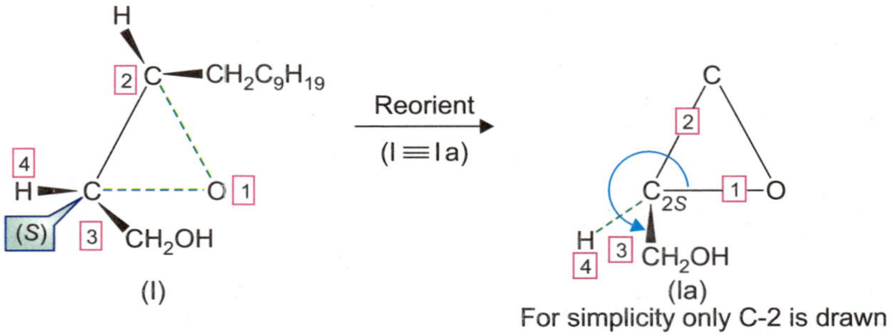

SCHEME 1.20m

BOX 1.13

DRAWING STEREOSTRUCTURES – WHICH TWO BONDS ARE IN THE PLANE OF PAPER–ASSIGNMENT OF R AND S CONFIGURATION MADE EASY

Intramolecular epoxidation (Scheme B 1.19):

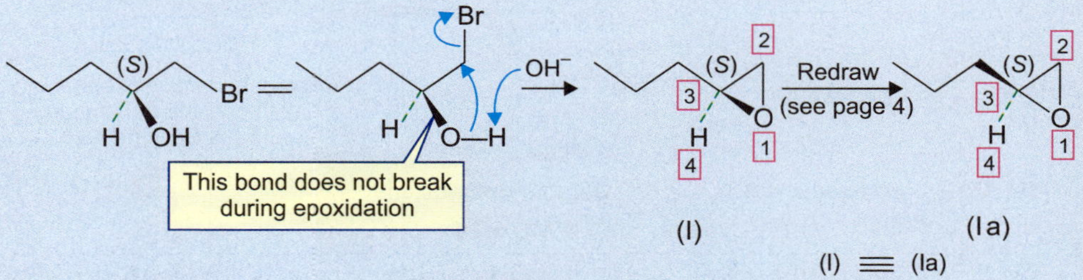

SCHEME B 1.19

- *No bond to the chirality center is broken; configuration of the carbon is retained, in the epoxide (I, Scheme B 1.19).*
- *Epoxide (I) has two bonds (methylene carbon on epoxy ring and n–propyl group in the plane of paper, configuration 1 → 2 → 3, S is in order).*
- *Equivalent way to draw the same epoxide is now to keep the bonds with priority* 1 *and* 2 *in the plane of paper (continuous lines) to get equivalent orientation (Ia). The configuration 1 → 2 → 3 from (Ia) anticlockwise also comes out to be the same (S).*

Continued...

Intramolecular epoxidation (Scheme B 1.20) now involves cleavage of the bond to the chirality center, thus the configuration of chirality center bonded to halogen will change. Now the assignment of configuration from (II, Scheme B 1.20) is not that straightforward. In (II Scheme B 1.20) one of the bonds to the chirality center (attached with oxygen, priority 1*) is below the plane of paper while the other priority* 2 *is in the plane.*

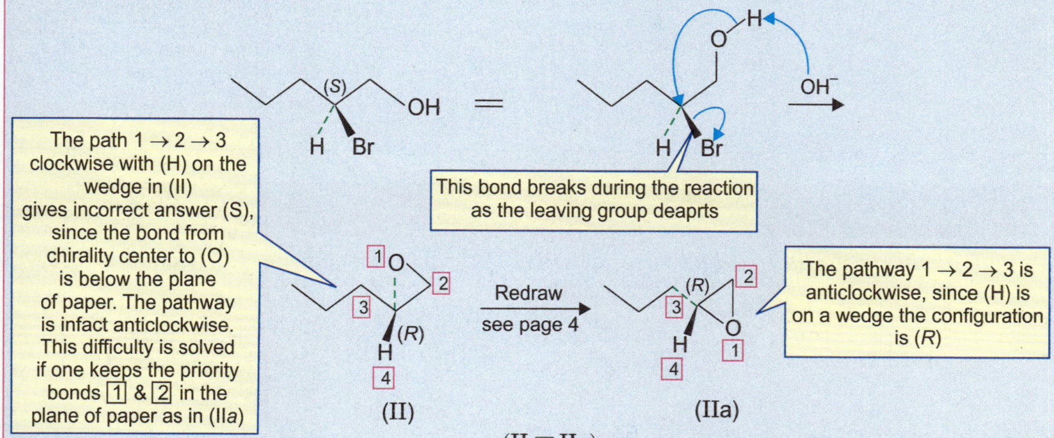

(II ≡ IIa)

SCHEME B 1.20

Thus assignment of configuration from (II, scheme B 1.20) is attended with same difficulties as observed in (Scheme 1.20m). If one redraws the structures (II) to an equivalent by now keeping priority bonds 1 *and* 2 *in the plane of paper as in (IIa), the pathway 1 → 2 → 3 is clearly anticlockwise to give the correct configuration (R).*

(F) The *R* and *S* Designations to Compounds Containing Nitrogen, Phosphorus and Sulphur

One may recall that amines and phosphines are approximately tetrahedral with the fourth ligand being the lone pair which is the lowest priority group. Consider the chiral phosphine (Scheme 1.21) in which the electron lone pair and the C_6H_5 group are in the plane of the paper (continuous lines) while pentafluorophenyl group C_6F_5 is projected in front of the plane while the methyl group is projected away. One may assign absolute configuration either by viewing the orientation from below (see Scheme 1.20*d*) or tilt the orientation to conveniently view it from the side with the lowest priority 4 group (electron lone pair) directed away as done with a compound with a chirality center (see scheme 1.20*d*). It is also easy to use the option in (Scheme 1.20*f*).

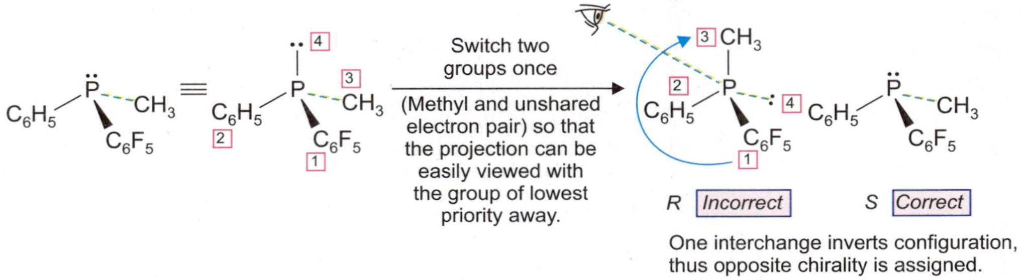

SCHEME 1.21

EXERCISE 1.3

Assign absolute configuration to the chirality centers in the compounds (Scheme 1.22).

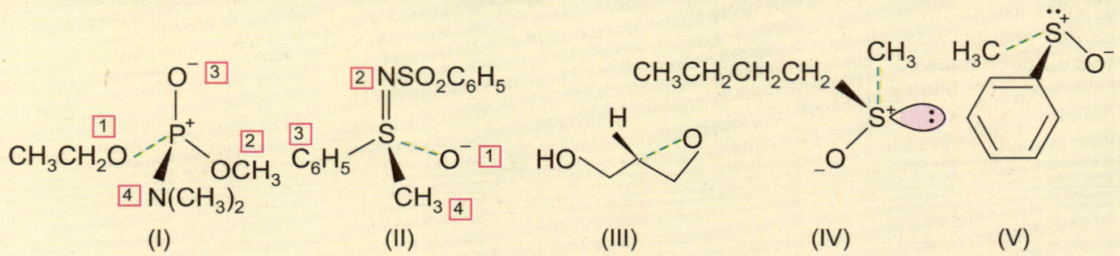

SCHEME 1.22

ANSWER. *(I) R ; (II) R ; (III) S ; (IV) S ; (V) S.*
*[**Hint.** (Both I and II) are S as drawn since priority group 4 is coming forward so the actual configuration is R in both.]*

Relationship between possible stereoisomeric structures

Determine the R and S configuration at each chirality center. For two chirality centers R, R and S, S are enantiomers of each other; R, S and S, R are enantiomers of each other. Any other combination is a pair of diastereomers.

EXERCISE 1.4

Discuss the stereochemical features in the compounds ((I)–(III) Scheme 1.23). Use of models will help.

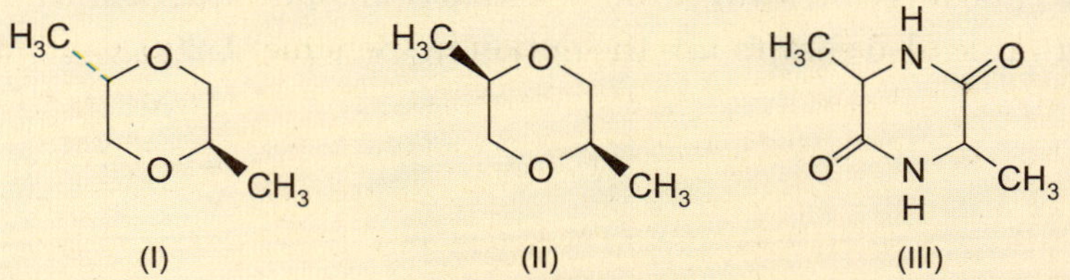

SCHEME 1.23

ANSWER. *(I) is achiral with a center of symmetry (see Scheme 1.3 e); (II) is chiral and since at every chirality center the group of 4 priority is below the projection, configuration is directly assigned (Scheme 1.24), its enantiomer has inverted configuration at every chirality center. The compound (III) has one ring and two chirality centers with identical substituents (showing the presence of meso structure), the amide group is flat. There are three stereoisomers an enantiomeric pair and a meso compound.*

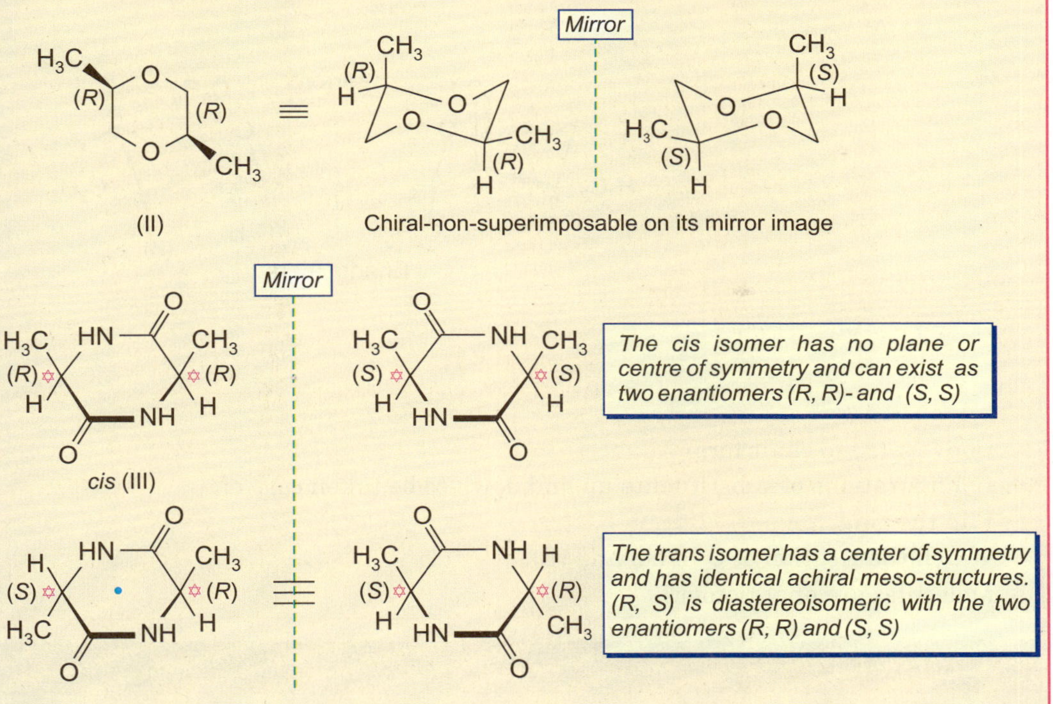

SCHEME 1.24

EXERCISE 1.4A

What is the stereochemical relationship between (A and B, Scheme 1.24a), constitutional isomers, homomers, enantiomers or diastereomers.

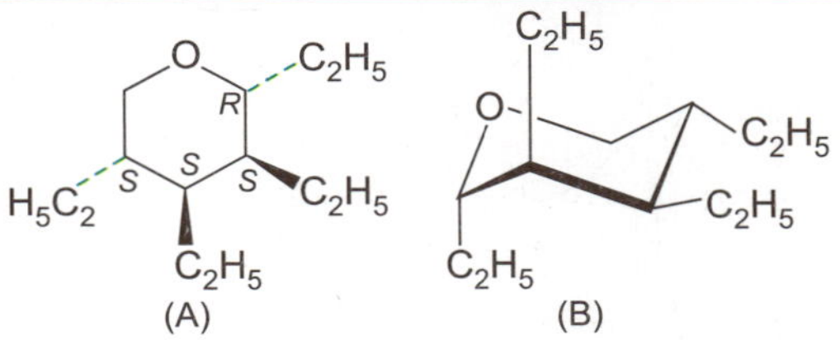

Continued...

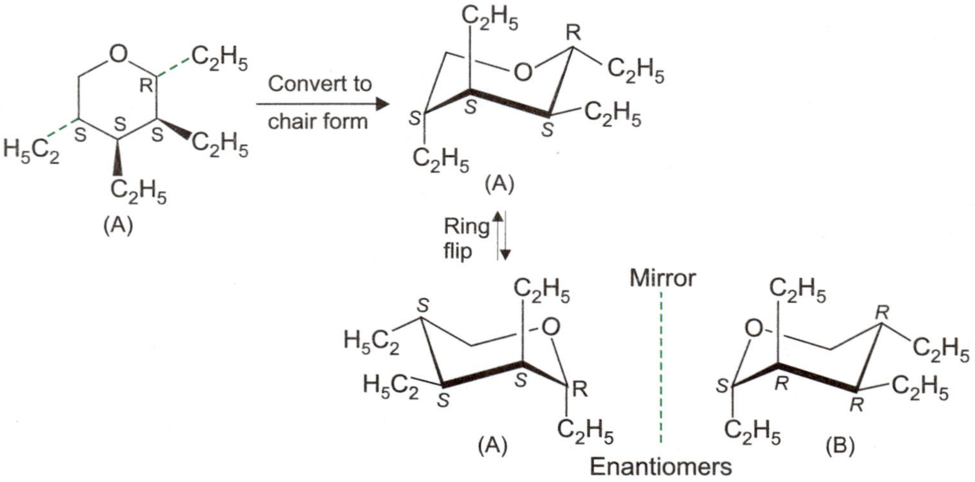

SCHEME 1.24a

ANSWER.

- Convert (A) to chair form.
- Configurations of substituents up and down to be taken care of
- Flip the ring.
- Watch the location of O in (A) and (B).
- Molecules are enantiomers.

EXERCISE 1.5

Assign configuration to the compounds (Scheme 1.25)

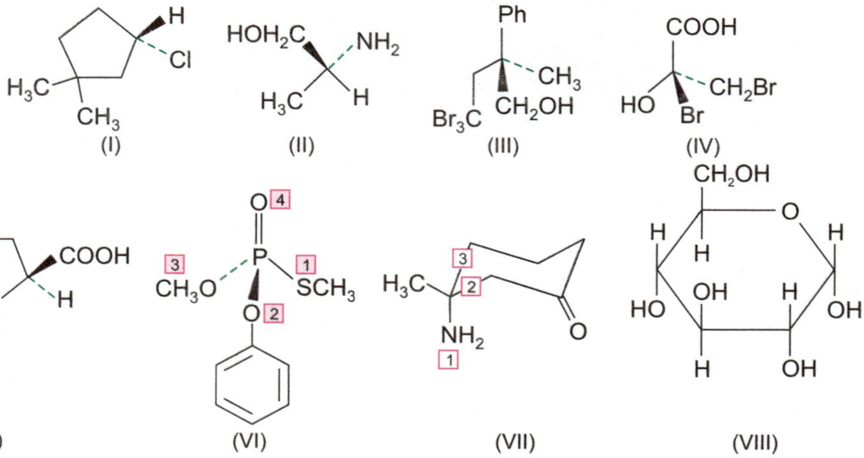

SCHEME 1.25

ANSWER. *(I) S; (II) R; (III) S, [hint bromines are too remote to be decisive]; (IV) S, (V) S; (VI) S; (VII) R, hint give priorities NH_2 is 1, CH_3 is 4 since 4 is towards one's eyes reverse the answer obtained from 1 → 2 → 3; (VIII) [hint priorities at C1 are OR > OH > C (OH) > H, the path 1 → 2 → 3 gives R, however since group of lowest priority H is forward the correct configuration at C1 is S (Scheme 1.26)]; priorities at C2 are OH > C(OH)OR > C(OH)C > H and following the same procedure as for C1, configuration at C2 is R.*

A Caution is needed in (VIII) at C3 (priorities are OH > CH(OH) COR > CH(OH) COH > H) group of lowest priority H is down, note that OH at C3 is not in the plane of paper but towards one's eyes so the path 1 → 2 → 3 is counterclockwise and should be correctly followed. The configuration at C3 is S which is more clearly derived from the conformational formula. The configuration at other centers C4 and C5 is as indicated.

Caution for configuration at C3 compared with C1 and C2. At C3 (H) is away the priority [1] group (OH) is not in the plane of paper but out of plane of paper towards you. The path 1 ⟶ 2 ⟶ is infact anticlockwise.

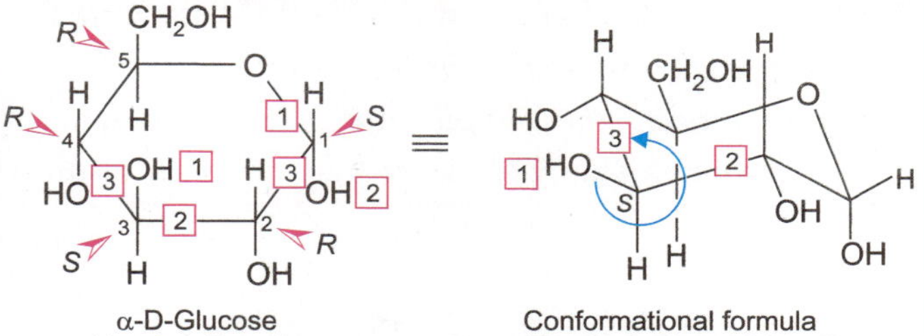

α-D-Glucose
Haworth formula
(VIII)

Conformational formula
(Configuration at C3 is *S*)

SCHEME 1.26

EXERCISE 1.6

Give the stereochemical relationship between the pair of compounds (Scheme 1.27). (Hint. Consult material in Box 1.4)

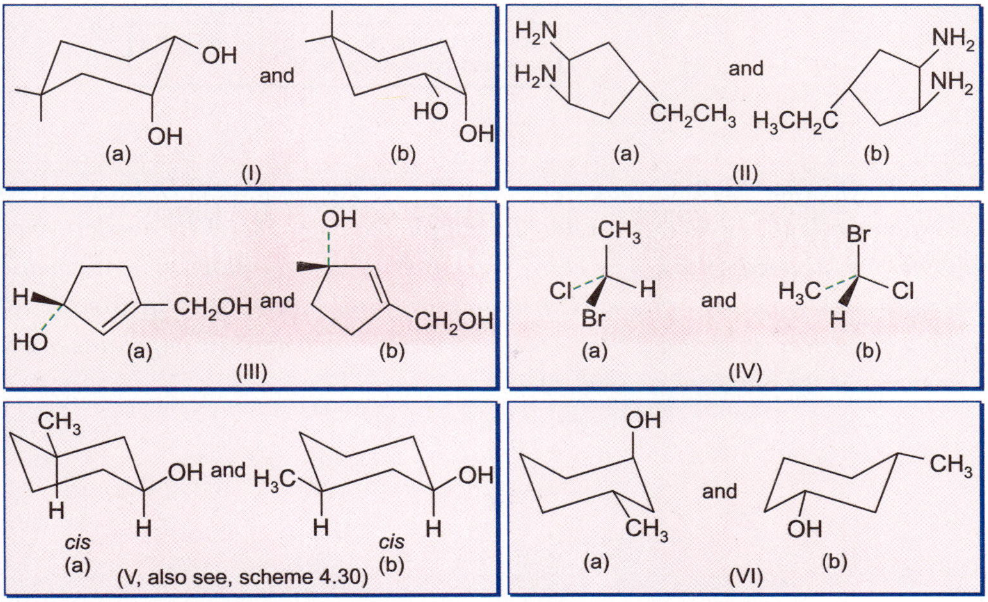

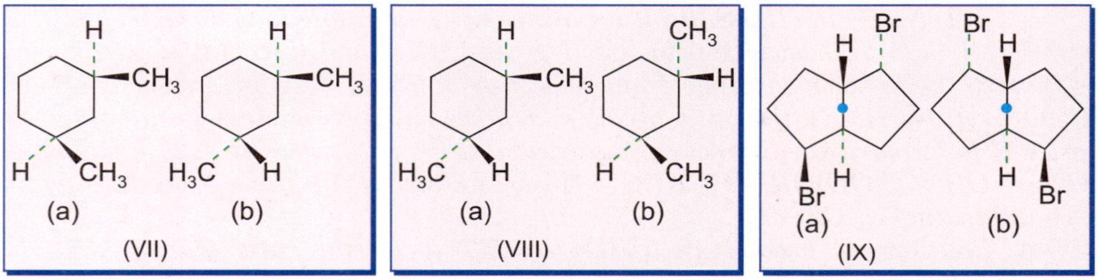

SCHEME 1.27

ANSWER. *(I) Conformational isomers (both 1 S, 2 R, configuration does not change with conformation) ; (II) same; (III) enantiomers [turn the molecule (b) clockwise so that the two CH_2OH groups in (a) and (b) are next to each other, these come out to be non-superimposable mirror images]; (IV) enantiomers [(a) is R while (b) is S]; (V) enantiomers, firstly the configuration of (a) is 1S, 3R in (b) it is 1R, 3S, i.e. configuration at each center is inverted. Moreover, one may draw the planar formulas of both*

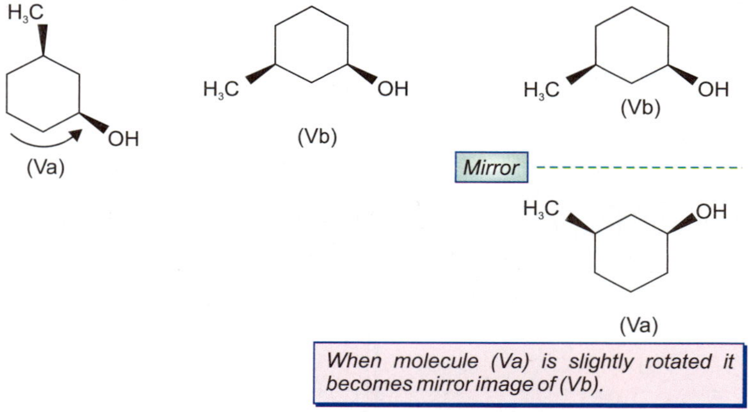

SCHEME 1.28

ignoring H (Scheme 1.28) to show that these are mirror images; (VI) diastereomers (a) is trans while (b) is cis (for a related example see page 15); (VII) diastereomers, configuration inverted at only one chirality center; (VIII) enantiomers configuration inverted at both chirality centers; (IX) same compound rotate the structure (b) 180° around a horizontal axis and it becomes the structure (a), both the compounds are the same meso (see the center of symmetry).

(G) The R and S Designations from Fischer Projections—A Comparison with a Three Dimensional Projection

A simple and most widely accepted procedure used now to assign R and S designations to the chirality centers when a compound is written in Fischer projection involves only two operations.

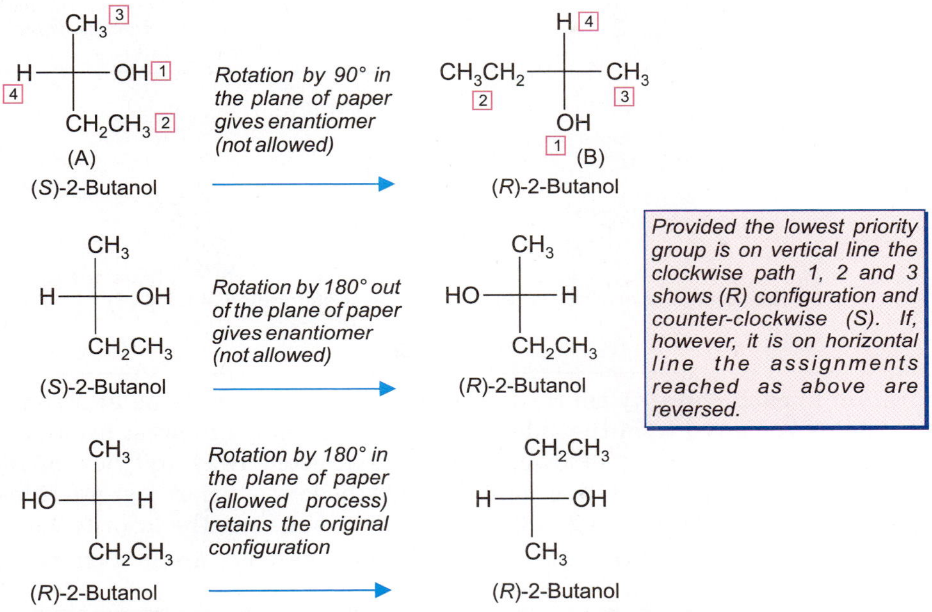

SCHEME 1.28a

Assign the priority order to the ligands and trace a semicircle joining 1 → 2 → 3 ignoring '4' the lowest priority group. When '4' is on the vertical line in the Fischer projection (no matter if it is at the top or at the bottom) the sequence gives the correct designation, if however, 4 is on the horizontal line the sequence gives the wrong answer and should be reversed. Thus, *e.g.*, in (*S*)-2-butanol (A, Scheme 1.28*a*) H is on the horizontal line, therefore, the designation arrived at from the sequence 1 → 2 → 3 is to be reversed. Rotation of a Fischer projection by 90° in the plane of paper is not allowed since it gives the enantiomer. Thus rotation of Fischer projection of (*S*)-2-butanol by 90° in the plane of paper gives (*R*)-2-butanol (B, Scheme 1.28*a*) in which now H is on the vertical line and the designation arrived at from the sequence 1 → 2 → 3 in (A, Scheme 1.28*a*) gives the correct designation (*R*). One can confirm from proper nomenclature including (R and S) designation that rotation of a Fischer projection by 180° out of plane of paper gives the enantiomer and is not allowed, however, its rotation by 180° in the plane of paper maintains the configuration and is allowed.

Consider one of the stereoisomers of 3-bromo-2-butanol *e.g.*, one enantiomer of *erythro*-3-bromo-2-butanol in the Fischer projection. Assign priorities to the groups to determine the configuration at C2. The "wrong" designation (*R*) arrived at from the sequence 1 → 2 → 3 is however, to be reversed to (*S*) since the group of lowest priority is on the horizontal line. Similarly the configuration at C3 comes out to be (*R*) and the otherwise shorthand name (*erythro*) of this stereoisomer in IUPAC terms comes out to be (2*S*, 3*R*)-3-bromo-2-butanol. Similarly in a three dimensional wedge representation (perspective formula) with two chirality centers the

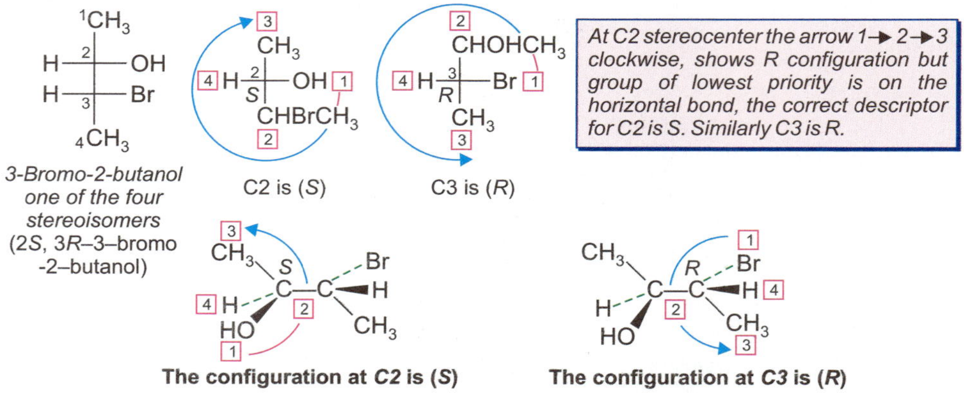

SCHEME 1.28b

configuration to each chirality center is assigned separately and one at a time. The chirality center is viewed from the side opposite to the group of lowest priority. Thus at C2 (Scheme 1.28*b*) the group of lowest priority is directed away and viewing it from right hand side gives the correct descriptor *S*. At C3 the lowest priority group is instead towards ones eyes, the arrow 1 → 2 → 3 is *S* as drawn, but since the priority 4 group is coming forward so the correct configuration is *R*. More examples are in (Scheme 1.20*j*).

1.5 STEREOISOMERISM RESULTING FROM MORE THAN ONE CHIRALITY CENTER–MORE INFORMATION

(A) *Threo* and *Erythro* Nomenclature (A Comparison with *Meso* and (±) Forms)

A molecule with two adjacent chirality centers, when there are two groups which are common to each carbon while the third is different *i.e.*, $C_{abx} - C_{aby}$ gives rise to *erythro* and *threo* diastereomers. Since erythrose and threose are diastereomers, other dia-stereomeric pairs of molecules which have two adjacent chirality centers are designated as *erythro* or *threo* depending on whether similar groups are on same side (*erythro*) or on opposite sides (*threo*) of the Fischer projection respectively. Thus *e.g.*, 3-bromo-2-butanol with two chirality centers has (2^2) four stereoisomers, which are named (Scheme 1.28*c* also see Schemes 1.37*d*, 1.38, 1.38*a*, 6.51 and 6.52).

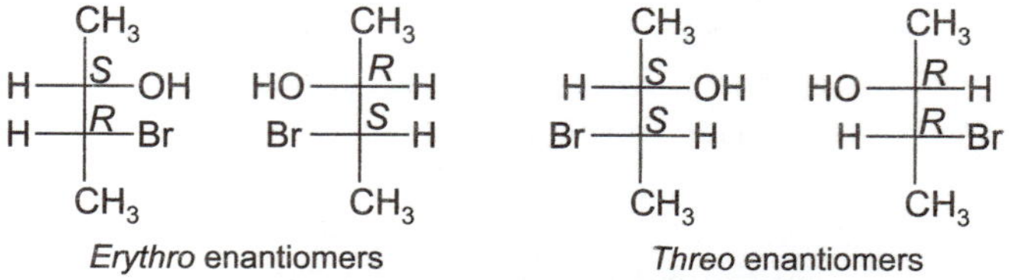

Four stereoisomers of 3-bromo-2-butanol

SCHEME 1.28c

Erythro and *threo* is the short hand method employed by organic chemists to name appropriate compounds. However, if one has to refer to a particular enantiomer of the *erythro* or *threo* pair, (*R* and *S*) notation is used with the name of each. *Erythro* and *threo* diastereomers can be identified in analogy with erythrose and threose (see, Scheme 1.37*d*).

The terms *erythro* and *threo* are generally applied only to those molecules which do not have symmetric ends (Scheme 1.28*d*). When however, the ends are symmetric as *e.g.* in 2, 3-dibromobutane and tartaric acid the terms *meso* and (+) are preferred since the use of these terms shows if or not a diastereomer has an enantiomer. A use of the prefixes *erythro* and *threo* is made to name dissymmetric molecules where the ends are different. For the formation of *erythro* or *threo* products during addition to *e.g.* an olefin (see, Problem 6.23 and Schemes 6.51 and 6.52).

Further details are in Scheme 1.37*d*.

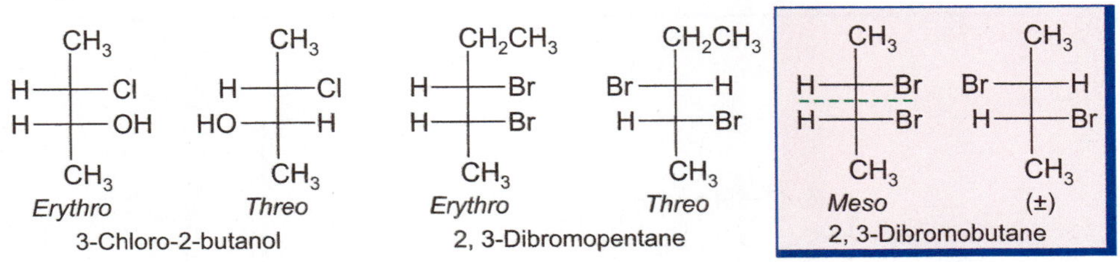

SCHEME 1.28d

Unlike a Fischer projection a three dimensional structure like a wedge projection can however, be manipulated (without loosing its identity) in any way one wishes *i.e.*, turning rotating or tumbling (see, Scheme 1.20*d*). A useful check for identity of two structures (drawn as Fischer projection or three dimensional drawing) would be to name each compound including its (*R—S*) designation. In case the names come out to be the same then the structures are the same. It is indeed important that one should be able to assign absolute configuration to a chirality center so as to understand the structures drawn in different ways.

The terms cis, syn, erthro in predicting stereochemical out come of reactions which generate two chirality centers are in (Schemes 6.51, 6.52 and problem 6.23).

For a related system of nomenclature (u and l), see (Schemes 1.43c–1.43f)

(B) Relationship between Possible Stereoisomeric Forms

- Recall that quickest method to find relationships is by determining *R*, *S* configuration at each chirality center to reach the conclusion (See Exercise 1.6).
- In Fischer projections rotation through 180° in the plane of paper of a projection for comparison with others is helpful. Also placing as many similar groups as possible in identical positions vertical or horizontal in one projection compared to the other by interchanging the position of two substituents at the same chirality center for comparison is also helpful (Practice Exercise 1.7). On the basis of *R*, *S* configuration for comparison, the same conclusion is reached, thus *e.g.*, (I*a* and I*b*, Scheme 1.28*e*) have both *S* configuration and are thus identical.

- Never operate on more than one stereocenter at a time while interconverting Fischer projections.

EXERCISE 1.7

Show if the structures in each pair (scheme 1.28c) are identical, enantiomers or diastereomers.

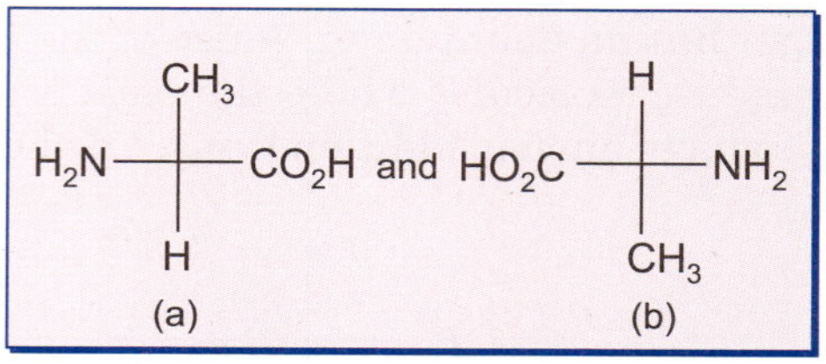

(I)

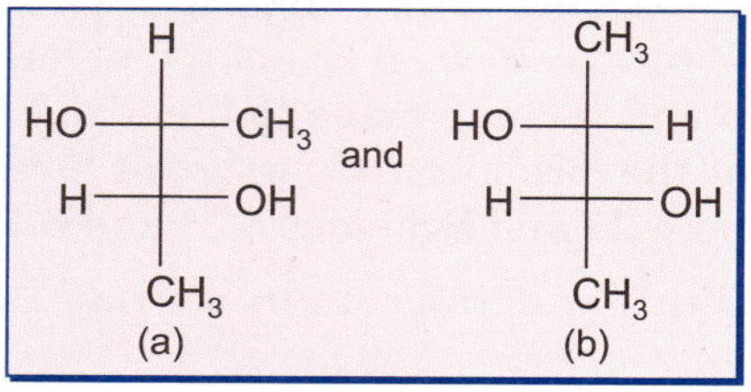

(II)

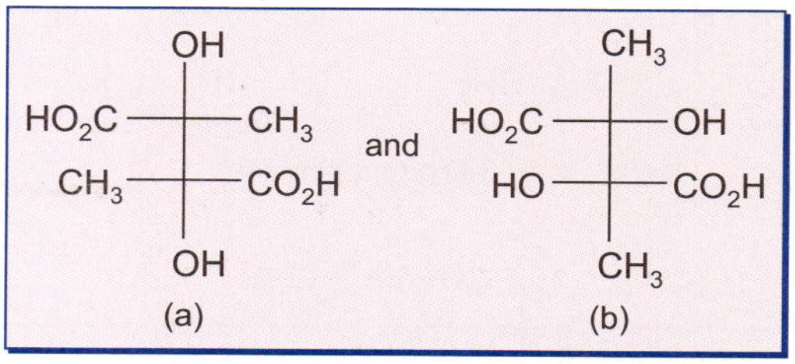

(III)

SCHEME 1.28e

ANSWER. *(I) Identical (180° rotation in the plane of paper of either structure gives the other) ; (II) diastereomers (working hint for such problems is to place as many corresponding (similar) groups as possible in identical positions and then decide relationship. Interchanging the position of two groups twice does not change the configuration of a chirality center. In (II) at the top carbon the methyl groups are not in the same position in (a) and (b). When one does two interchanges at this chirality center (first CH_3 with H and then H with OH without disturbing the bottom carbon the equivalent structure of (IIa) becomes (IIc) which is a meso compound. Thus (IIa = IIc) is diastereomeric with (IIb). (III) By doing a similar manipulation, the structures come out to be enantiomers (interchanging a pair of groups once on each stereocenter e.g., in (III a) CH_3 with OH on top chirality center and again OH with CH_3 at the bottom chirality center inverts configuration at every chirality center i.e., produces a mirror image and regenerates (IIIb). Thus (IIIa and IIIb) represent an enantiomeric pair.*

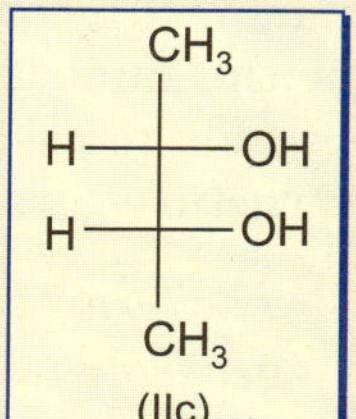

(IIc)

(C) The Comparison Between *D*, *L* and *R*, *S* Conventions

An important caution about the sequence rule (*R*/*S*) is that chemical and biogenetic families are not necessarily correlated *e.g.*, the natural α-amino acids can be represented as structure (I, Scheme 1.28*f*) in Fischer projections (the L-configuration).

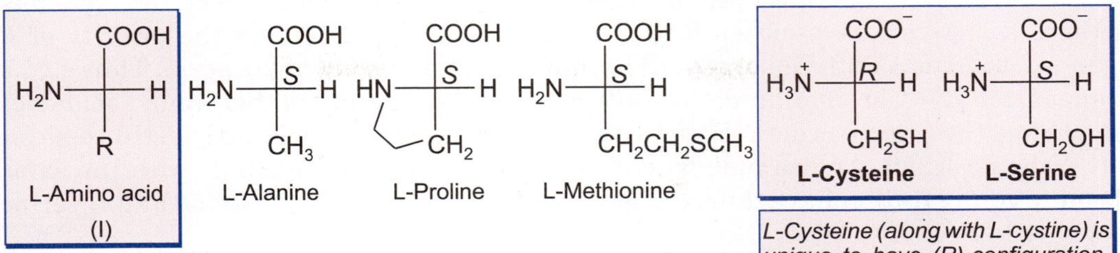

All protein amino acids belong to ***L-series.*** *Most of these have* ***S configuration.***

L-Cysteine (along with L-cystine) is unique to have (R)-configuration. The CH_2SH group of cysteine has priority over COO^- since S has a greater atomic number than O.

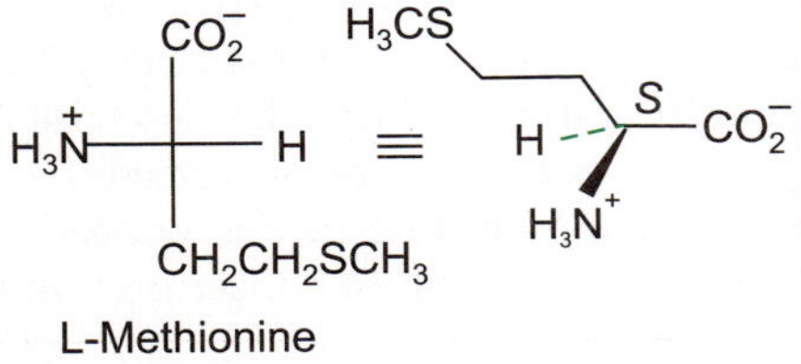

The priority order of the atoms and groups directly attached to the chirality center :
$\overset{+}{H_3N}$— > —CO_2^- > —$CH_2CH_2SCH_3$ > H—
Sulfur is one atom farther removed from the chirality center, and so C—O outranks C—C—S.

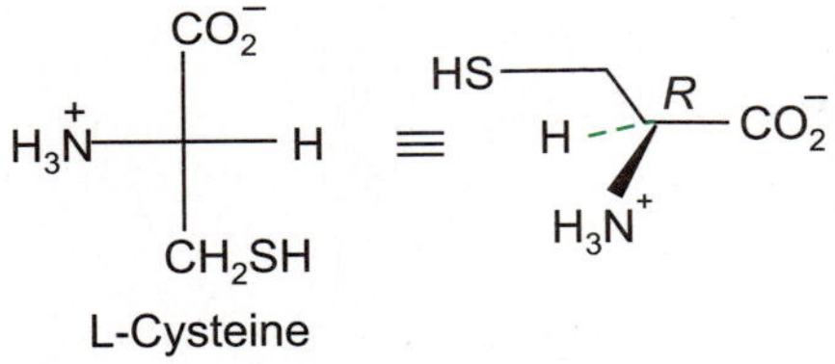

The priority order of the atoms and groups directly attached to the chirality center:
$\overset{+}{H_3N}$— > $HSCH_2$— > —CO_2^- > H—

SCHEME 1.28f

The reason that L-cysteine has the *R* configuration and all the other L-amino acids have the *S* configuration lies in the fact that the $-CH_2SH$ substituent is the only side chain that outranks $—CO_2^-$ according to the CIP rules. Recall that rank order is determined by atomic number at the first point of difference, and –C–S outranks –C–O. In all the other amino acids, $—CO_2^-$ outranks the substituent at the chirality center. This is the reason for the survival and utility of local systems *e.g.*, D and L for carbohydrates and amino acids and the α, β nomenclature for steroids.

(D) Relationship between Configuration and Configurational Nomenclature (*R* and *S*, D and L)

There is no necessary connection between the two systems of nomenclature–the D, L and *R, S* systems. The D, L system is based on chemical transformation of a compound to give either D- or L-glyceraldehyde. Thus the compounds which can be correlated with D-glyceraldehyde belong to D-series and those correlated with L-glyceraldehyde belong to L-series.

Amino acids from proteins, thus belong to L-series (I, scheme 1.28*f*) *e.g.*, L-analine and the heterocyclic amino acid L-proline. The use of D and L-system, continues since it shows the essential similarity between all the protein amino acids all belonging to L-series. However, the *R*, *S* system suffers from the disadvantage since the members of the L-series can have either *R* or *S* configuration depending on the priority of *R* ligand (I, scheme 1.28*f*). Generally all L-amino acids have *S*-configuration. Thus *e.g.*, a change from R = CH_3 in alanine to a different alkyl group in another amino acid does not change the priorty order NH_2 > COOH > R > H and the configuration still remains *S* (I, Scheme 1.28*f*). An example where two closely related compounds with the same absolute configuration have different descriptors (*R* or *S*) is of L-cysteine and L-serine (Scheme 1.28*f*).

Maintaining the configuration of the molecule means that the four groups bonded to the chirality center maintain their relative positions. It doesn't, however, necessarily mean that an *S* reactant will always give an *S* product or a *R* reactant will always give a *R* product. When in a reaction a bond to a chirality center is broken, the resulting configuration of the chirality center depends on the mechanism of the reaction. Thus *e.g.*, when the hydroxide ion reacts with (*S*)-2-bromobutane in a one step (S_N2) reaction (eq. I, Scheme 1.28*g*), the product of this reaction (*R*)-2-butanol has the opposite relative configuration and the reaction is said to have proceeded by an inversion of configuration. However, inversion of configuration does not require a change in designated (*R* or *S*) absolute configuration, nor does a change in designated absolute configuration require an inversion of configuration.

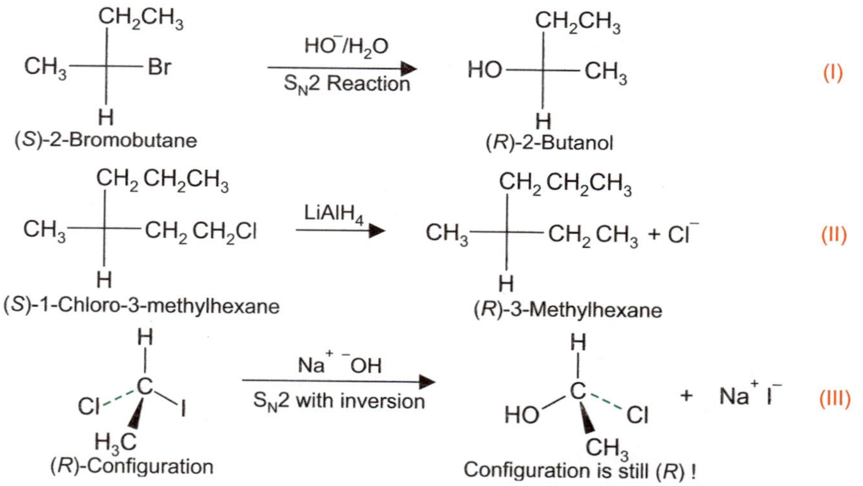

SCHEME 1.28g

In reaction (eq. II, Scheme 1.28*g*) the relative position of the groups as present in the reactant is maintained in the product (*i.e.*, relative configuration of both is same). However, the *S* reactant gives *R* product, the change in the priorities of the groups as defined by CIP rules and not a change in their position (inversion of configuration) is what leads an *S* reactant to become the *R* product. Again the (*R*) reactant (eq. III, Scheme 1.28*g*) on S_N2 reaction gives a product with inversion of configuration which however, is still (*R*). Thus there is no guarantee that inversion will change an (*R*) chirality center to (*S*) chirality center. This example further shows that among the halogens an iodide ion is the best leaving group.

(E) Correlation between Configuration and the Direction of Rotation of Plane Polarized Light

The direction of rotation of plane-polarized light is often specified and incorporated with the names of optically active compounds. Each optically active compound has a characteristic specific rotation. The enantiomers of a compound (mirror image relationship) rotate the plane of polarized light to the same amount but in opposite directions (Scheme 1.29).

If the configuration of a chiral compound *i.e.*, an enantiomer is known to be either (*S* or *R*) it, however, does not tell the direction that the enantiomer will rotate the polarized light. Some (*R*) molecules rotate polarized light to the right and are therefore, dextrorotatory (+) and some rotate it to the left and are thus levorotatory (–). One can assign (*S* or *R*) configuration to an enantiomer of a compound by looking at the structure and applying CIP rules. The only way one can tell if a compound is (+) or (–) is by examining the compound with a polarimeter. Thus *e.g.*, (*S*)-lactic acid and (*S*)-sodium lactate have the same configuration yet one is dextrorotatory and the other is levorotatory (Scheme 1.29).

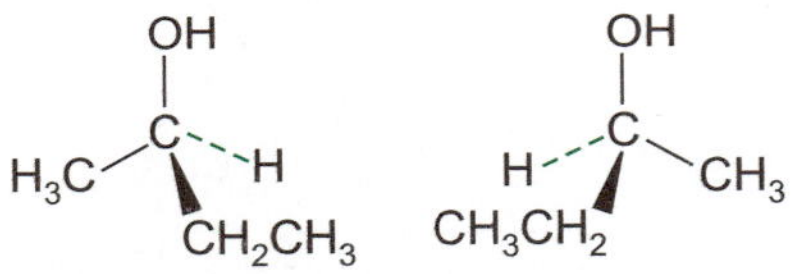

(*R*)-(–)-2-Butanol $[\alpha]_D^{25} = -13.52$ (*S*)-(+)-2-Butanol $[\alpha]_D^{25} = +13.52$

A pair of enantiomers of 2-butanol

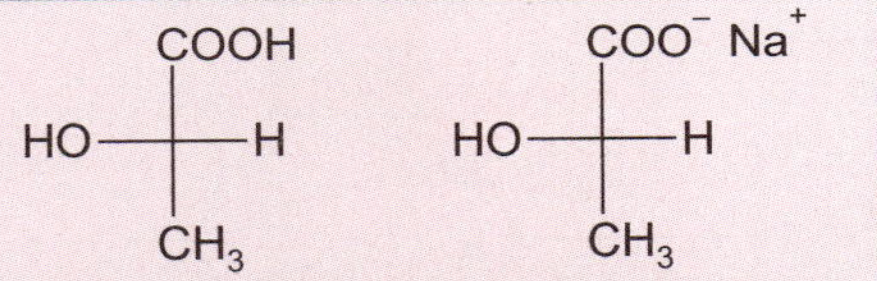

(*S*)-(+)-Lactic acid $[\alpha]_D^{25\,°C} = +3.8$ **(*S*)-(–)-Sodium lactate** $[\alpha]_D^{25\,°C} = -13.5$

Compounds with same configuration

SCHEME 1.29

The enantiomer of (*S*)-(+) lactic acid is (*R*)-(–) lactic acid. A way to determine which enantiomer has which configuration is using X-ray crystallography and the analysis of crystals of pure enantiomers.

In summary there is no correlation between the configurations of enantiomers and the direction in which these rotate the plane polarized light. There is also no correlation between (*R* and *S*) designation and the direction of rotation of plane polarized light.

(F) *Meso* Compounds

The maximum number of stereoisomers for a compound with *n* chirality centers is 2^n. However, some compounds with *e.g.*, two such centers have only three stereoisomers. In such cases, the four substituents bonded to one chirality center are identical to the four substituents bonded to the other chirality center. An example is of tartaric acid which has a pair of enantiomers and one stereoisomer is the *meso* compound. *Meso* compounds contain chirality centers as well as a plane of symmetry as shown in *meso*-tartaric acid (see Scheme 1.2*a*). A compound with a plane of symmetry is achiral (can be superposed on its mirror image). Thus *meso*-compounds will not be optically active even though they have chirality centers.

The *meso* compounds *e.g. meso*-tartaric acid have two features in common: (*i*) a plane of symmetry and (*ii*) atleast two chirality centers with opposite stereodescriptors. The latter condition renders the molecule optically inactive since there is internal compensation, the rotation due to (*R*) chirality center is cancelled by that caused by (*S*) chirality center.

Meso compounds also exist in cyclic structures. Thus compound (I, Scheme 1.30) has a plane of symmetry which passes through C1, C7 and C4 and the chirality centers C2 and C6 have opposite configuration thus it represents a *meso* compound. The compound (II, Scheme 1.30), however, does not have a plane of symmetry and configuration is also *S* on both C2 and C6, this compound is thus not *meso*. Another example is in cyclohexane derivatives (III and IV, Scheme 1.30), (III) has a plane of symmetry passing through C1 an C4). Similar arguments hold for the stereoisomers of 1,2 dibromocyclobutane.

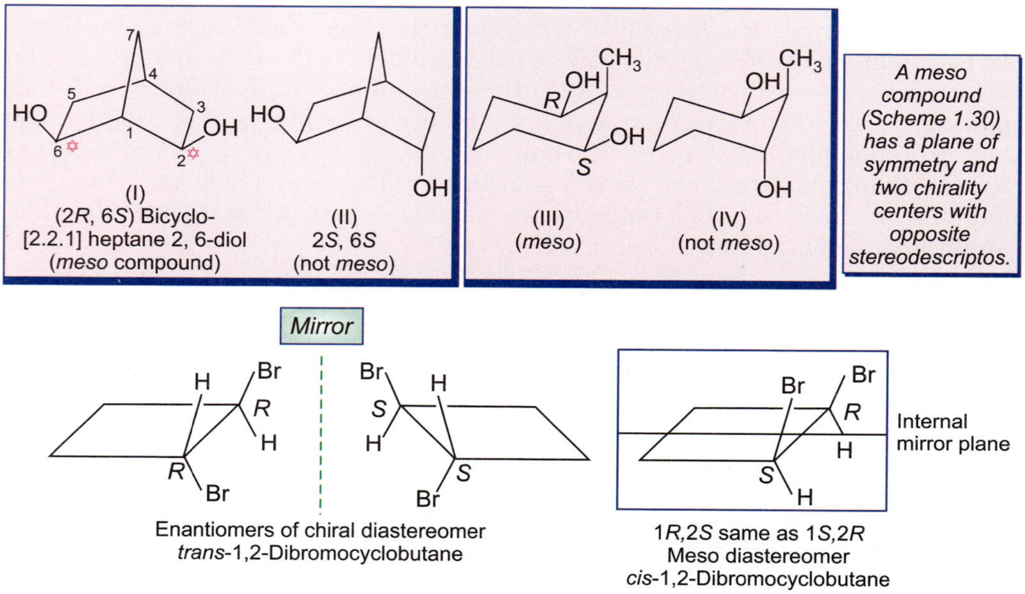

SCHEME 1.30

(G) Pseudoasymmetric Centers (Pseudochirality Center)

Among all the possible stereoisomers of 2,3,4-tribromopentane (Scheme 1.31), one has a pair of *meso*-compounds and an enantiomeric pair. The *meso* (I) and (II) have three stereocenters each, out of which only two C2 and C4 are chirality centers.

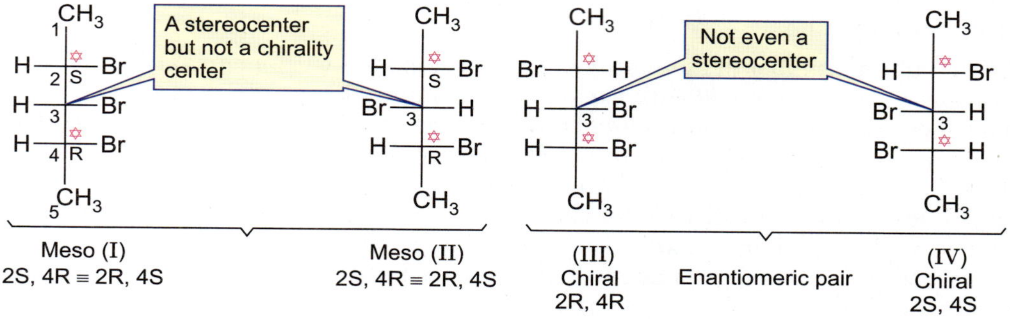

SCHEME 1.31

A chirality center (asymmetric center) is "any atom holding a set of ligands in a spatial arrangement which is not superimposable on its mirror image" and must have four different groups on it. However in (I and II, Scheme 1.31) C-3 has two groups that are identical (except for their stereochemistry). Thus C-3 is not a chirality center. But it is a stereocenter alright. In structure-(I) interchanging the H and Br at C-3 gives structure (II, a diastereomer). This type of stereocenter which has two identical enantiomorphic groups and two other non-identical groups and lies on a plane of symmetry is termed a pseudoasymmetric center (also see, Scheme 1.67*c*).

As two of the ligands on the pseudochirality center are the same, in terms of atomic number (but for their configuration) a further criterion is invoked, that is an *R*-configured group has priority over *S*-configured group (*R* before *S* in the alphabet). Thus it is possible to determine the configuration of a pseudochirality centre by application of the sequence rules. To distinguish a pseudochirality centre from a chirality centre the stereodescriptors *r* and *s* are used (Scheme 1.31*a*).

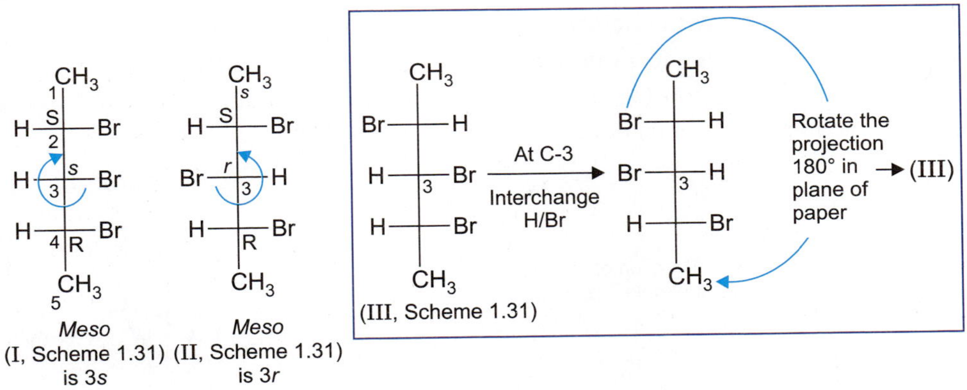

SCHEME 1.31a

In the enantiomers (III and IV) C3 is not even a stereocenter since *e.g.*, in (III) the interchange of ligands at C-3 followed by rotation of the projection by 180° in the plane of paper keeps the two structures unchanged.

The Term Chirotopic

The term chirotopic has been coined by Mislow and Siegel 1984 (as shown by local symmetry) a chirotopic atom is one which resides in a chiral environment e.g., all the five atoms of CHFClBr bromochlorofluoromethane are chirotopic, though only C atom is a chirality center. Thus any point in a chiral molecule is chirotopic, however, even in an achiral molecule there may be several chirotopic points or atoms. In meso-tartaric acid e.g., all atoms are chirotopic, but for the center of symmetry (midway between C2 and C3 which is achirotopic). All points on a plane of symmetry in a molecule are achirotopic.

Stereogenicity and Chirotopicity

It has been suggested that stereogenicity and chirotopicity are two different and distinct characters of a stereocenter. These characters not only overlap, but are closely associated in organic chemistry. In the active isomers (III and IV, Scheme 1.31) C-3 is chirotopic but non-stereogenic, while in the meso forms C-3 is achirotopic but stereogenic. Moreover, application of CIP rules when applied to pseudoasymmetric centers become less straightforward. One has already seen cases e.g., amino acids where according to D-L system of nomenclature all naturally occuring amino acids belong to L-series while on the basis of CIP nomenclature majority of them belong to S configuration while some to R (see, Scheme 1.28f).

1.6 INTERCONVERSION BETWEEN FISCHER AND THREE DIMENSIONAL (WEDGE) FORMULAS

(A) Compounds with One Chirality Center

Consider the following points regarding D-glyceraldehyde.

- Orient the wedge formula (or model) in a way so that two groups *e.g.*, H and OH are facing you (on wedges) and two groups CHO and CH_2OH are away from you (on dashes) as in (Ia and Ib scheme 1.32).

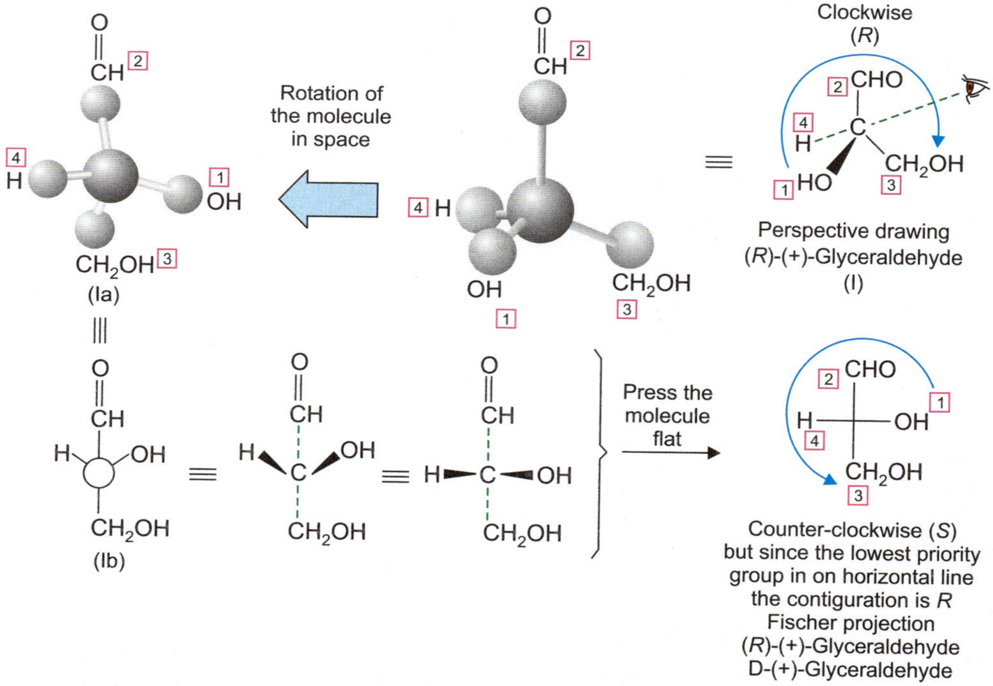

(A Simple Mental Exercise) Conversion of Dashed-Wedge Line Structure of D-Glyceraldehyde into Fischer Projection–compare with (scheme 1.34)

SCHEME 1.32

- Now simply press the molecule flat in the plane of the paper to get the Fischer projection (In the Fischer projection—thus obtained the sequence $1 \rightarrow 2 \rightarrow 3$ is counterclockwise but; since the group of lowest priority is on horizontal line, the designation thus arrived is reversed). Similar procedure when used for (*R*)-(–)-lactic acid is given (scheme 1.33).

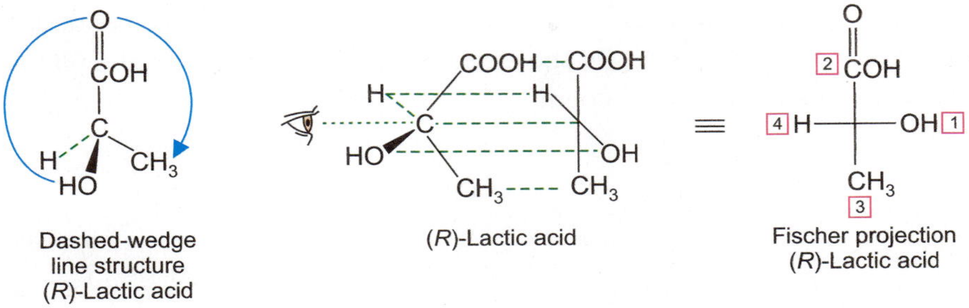

SCHEME 1.33

- **A simple mental exercise to convert a dashed-wedge line structure into a Fischer projection.**

Step 1. Draw the groups attached to continuous lines of the dashed-wedge line structure on the ends of a vertical line, as one sees these *i.e.*, in (I, scheme 1.34) CH_3 at the top and CHO at the bottom.

Step 2. Imagine mentally the group on the thick wedge to be in between the continuous lines of the wedge line structure so as to be either on the right-hand side or left-hand side of the horizontal line of the Fischer projection depending on as to how dashed-wedge structure is drawn. In (I, scheme 1.34), the continuous lines in the plane of paper are drawn with opening

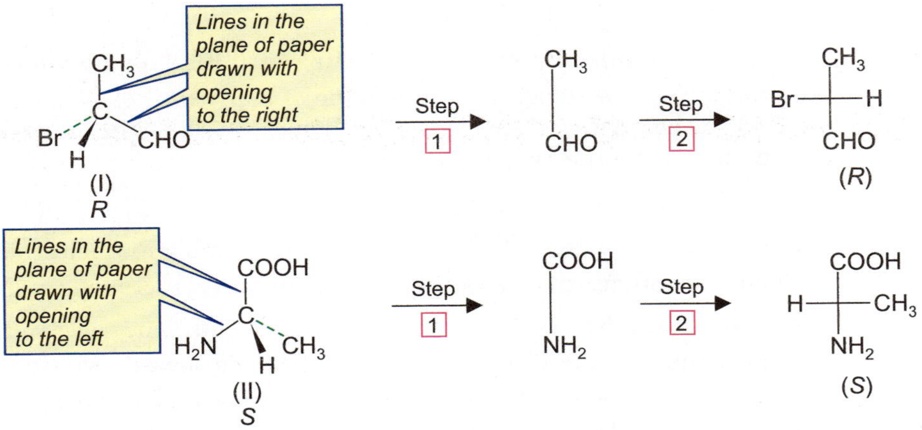

A simple mental exercise to safely convert any dashed-wedged line structure into Fischer projection

SCHEME 1.34

on the right, thus H on thick wedge in (I, scheme 1.34) when imagined to be in between continuous lines must be drawn on the right of Fischer projection. In (II, scheme 1.34) the continuous lines in the plane of paper are drawn with opening on the left, thus H on the thick wedge when imagined to be in between continuous lines, now must be drawn on the left of the Fischer projection.

When the dashed-wedge line structure is drawn with the continuous lines drawn at the bottom (III, scheme 1.35), the structure may be rotated in the plane of paper to put the continuous lines with opening either to the left or right and apply the method described above.

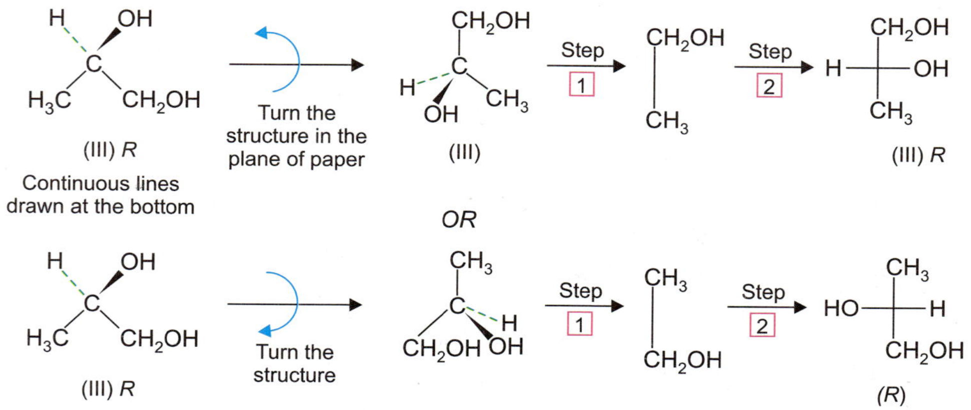

A simple mental exercise to safely convert any dashed-wedged line structure into Fischer projection

SCHEME 1.35

Summary–Conversion of flying wedge structure into Fischer projection

In fact rotate any flying wedge structure in the plane of paper (*i.e.*, II, scheme 1.34 or III, scheme 1.35) to the arrangement as in (I, Scheme 1.34, lines in the plane of paper drawn with opening to the right) and apply the same procedure as used for (I, Scheme 1.34) in every case.

Often, the conversion of a dashed-wedge line structure into a Fischer projection assists in assignment of configuration as in the case of sulphoxides (Scheme 1.36).

Unsymmetrical sulfoxides are chiral, the oxygen and lone electron pair play the role of substituents. The electron lone pair has the lowest priority while the oxygen the highest. For *R* and *S* designation, the molecule has to be viewed from the side opposite to the lone pair, (see scheme 1.20d), or convert the dashed-wedge line structure into its Fischer projection (as shown in Schemes 1.34 and 1.35) and then assign the configuration. The bonds with lone pair and S = O are taken to be in the plane of paper and drawn as continuous lines.

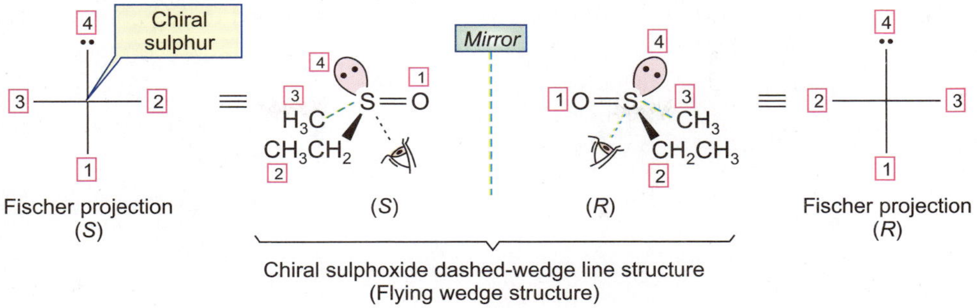

SCHEME 1.36

- A simple mental exercise to convert a Fischer projection into dashed-wedge line structure.

Step 1. Draw the groups of the vertical line on Fischer projection as continuous lines of the dashed-wedge line structure with opening on the right (Scheme 1.37).

Step 2. Then put the group on the right-hand side of horizontal line on the thick wedge.

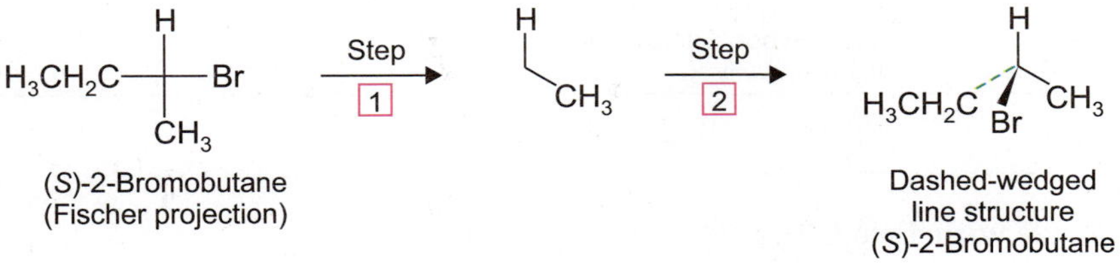

SCHEME 1.37

When the groups of the vertical line of Fischer projection are drawn as continuous lines of the dashed-wedge line structure with opening on the left, then instead the group on the left-hand side is put on the thick line (Scheme 1.37*a*). These projections are same as discussed (see Scheme 1.34).

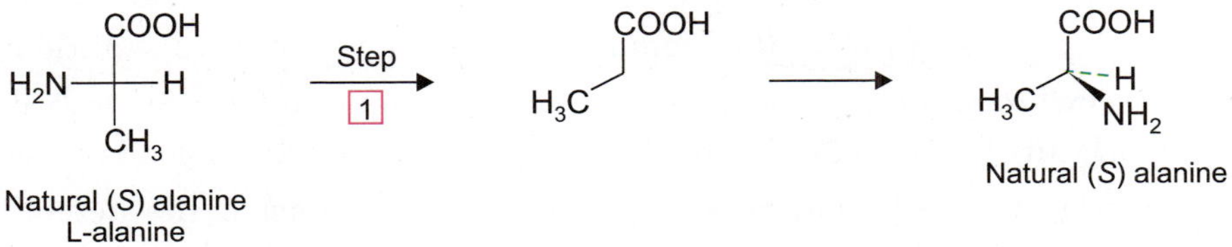

SCHEME 1.37a

Similarly unnatural D-alanine can be drawn as discussed (see, Scheme 1.37) as depicted in (Scheme 1.37*b*).

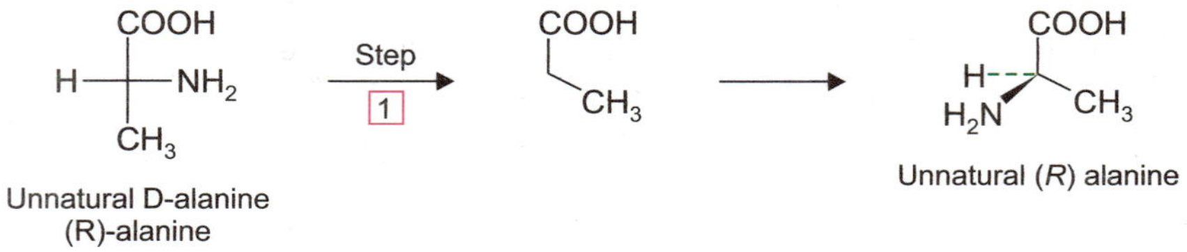

SCHEME 1.37b

With chiral amine oxides and related systems, the oxygen atom is taken as the fourth substituent (Scheme 1.37*c*).

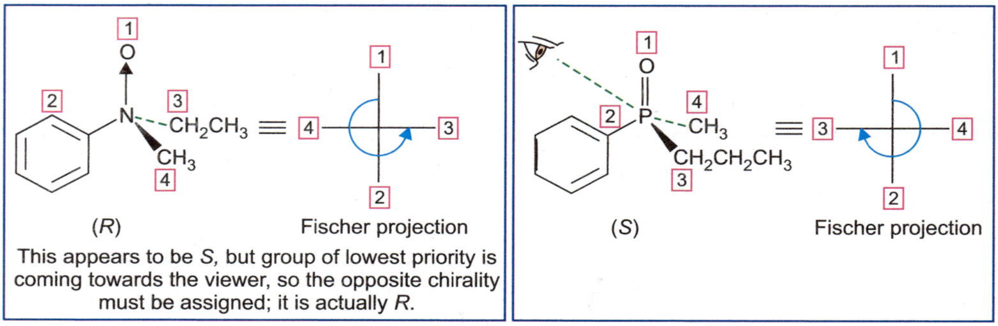

SCHEME 1.37c

(B) Compounds with Two Chirality Centers

(1) *Interconversion of Fischer into Sawhorse Projections*

Consider one of the four stereoisomers of 3-bromo-2-butanol:

- 3-Bromo-2-butanol has two chirality centers, therefore, it can exist in four stereoisomeric forms (see scheme 1.28*c*).
- Since the four substituents bonded to one chirality center are not identical to those of the other there is no *meso* stereoisomer.
- The four stereoisomers can be drawn in Fischer projections (see scheme 1.28*c*) and exist as a pair of *erythro* enantiomers and a pair of *threo* enantiomers.
- The *erythro* and *threo* nomenclature is a shorthand (informal) method of nomenclature based on Fischer projections to distinguish between diastereomers.
- The proper IUPAC nomenclature would be to name each of the stereoisomers with (*R* and *S*) designation (see scheme 1.28*b*).
- If one picks up one of the *erythro* enantiomers of 3-bromo-2-butanol (*e.g.*, 2*S*, 3*R* stereoisomer I, scheme 1.37*d*) its Fischer projection can be converted into "sawhorse", representation through the eclipsed form (II) to the staggered form.

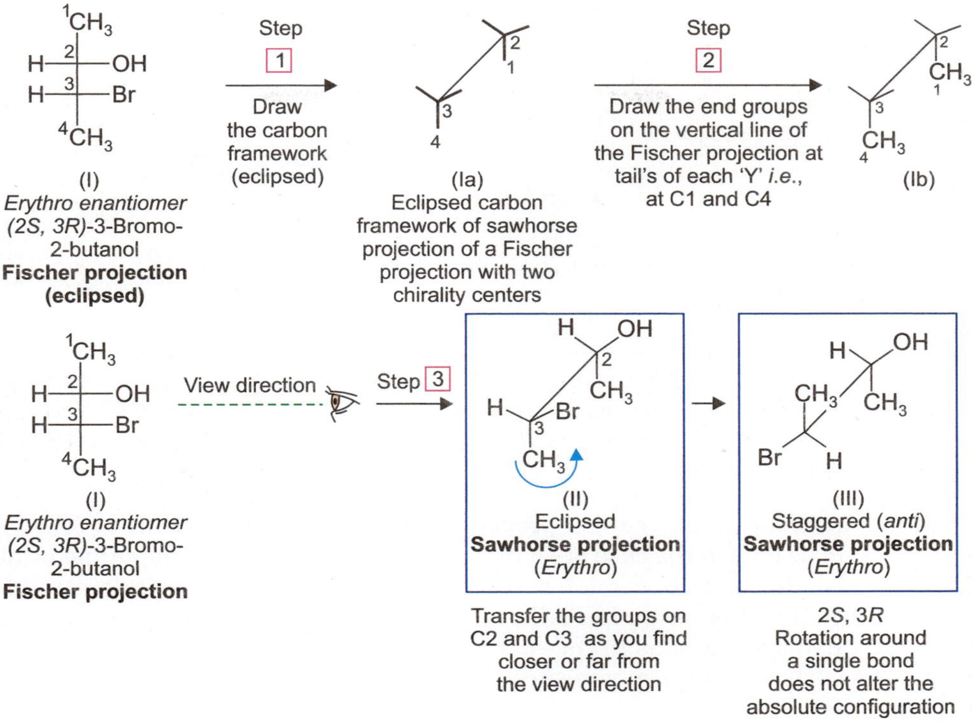

SCHEME 1.37d

- In the Fischer projection (I, scheme 1.37*d*) the substituents on the two chirality centers are eclipsed. To transform it into its sawhorse projection one must write first these, projections in their eclipsed form. For a beginner it may be helpful to do it stepwise. In the sawhorse representation in its eclipsed form, in fact each chirality centers may be looked up as the letter 'Y'. As a first step write the carbon backbone in eclipsed form shown by bold lines as in (Ia, scheme 1.37*d*) and put the end groups (CH_3, CH_3) on the vertical line of the Fischer projection on the "tails of each Y" (Ib). Then transfer the substituents on the horizontal lines of the Fischer projection on the "head" of each 'Y' (C_2 and C_3) on the right-hand or the left-hand as they appear on the Fischer projection from the view side to give (II). Then stagger the substituents by rotation of one carbon keeping the other carbon static.

EXERCISE 1.8

When fumarate reacts with D_2O in the presence of the enzyme fumarase only one stereoisomer of deuterated malate is formed (scheme 1.37e). Explain if this a syn or anti-addition.

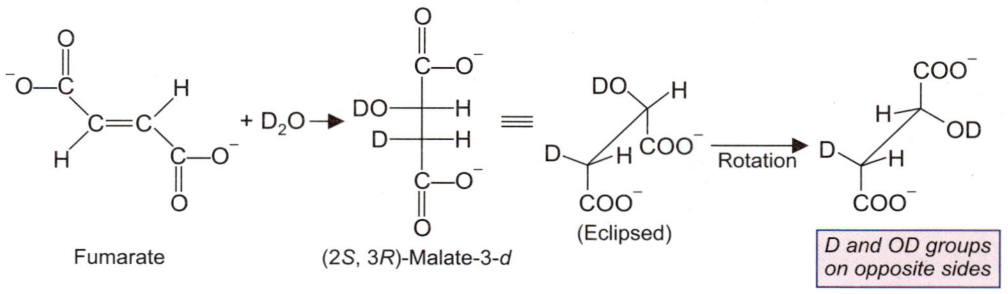

SCHEME 1.37e

ANSWER. *Convert the Fischer projection of the product into a three dimensional perspective drawing (first eclipsed form then stagger it so as to bring two COO^- groups anticoplanar i.e., these two to have 180° angle as in the starting material). Examine the relationship of the –D and –OD groups. Thus the addition is anti (Scheme 1.37e).*

It is important to take care of proper numbering of carbon atoms to follow this interconversion as *e.g.*, in the case of enantiomeric pair of 2, 3-dibromobutane (scheme 1.37*f*), otherwise this interconversion may lead to the enantiomer. One may count from the top. The last chirality center in Fischer projection (3) may be considered as the front carbon in sawhorse formula as in (Ia, scheme 1.37 *f*). If instead from (I), C2 is written as the frontal carbon of the sawhorse projection, then one actually is writing the enantiomeric structure (II).

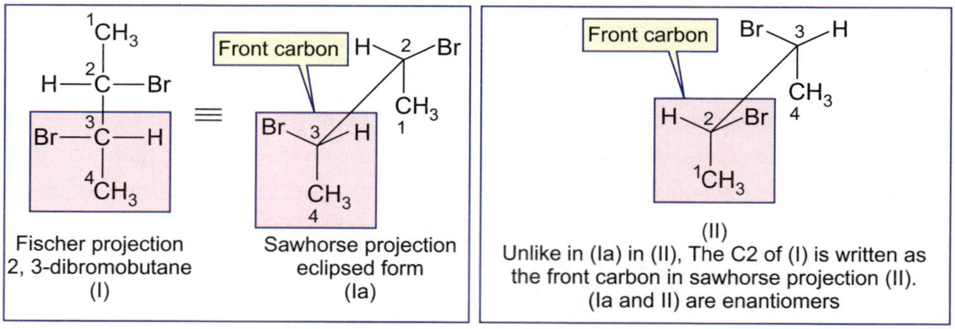

SCHEME 1.37f

(2) *Conversion of Fischer Projection into Dashed-wedge Line Projection*

Consider the following points:

- When one looks to a Fischer projection perpendicular to the plane of the page then both the groups on the horizontal lines (solid wedges) are towards the viewer. But if viewed from one side *e.g.*, right-hand side then the groups on the right-hand side are closer to the viewer than the groups on the left-hand side (scheme 1.37*g*).
- A consideration of the Fischer projection of D-erythrose (viewed from right-hand side) adds both the OH groups on thick wedges and hydrogen atoms on dashed lines, on the eclipsed carbon framework of Fischer projection to give (I, scheme 1.37*g*).

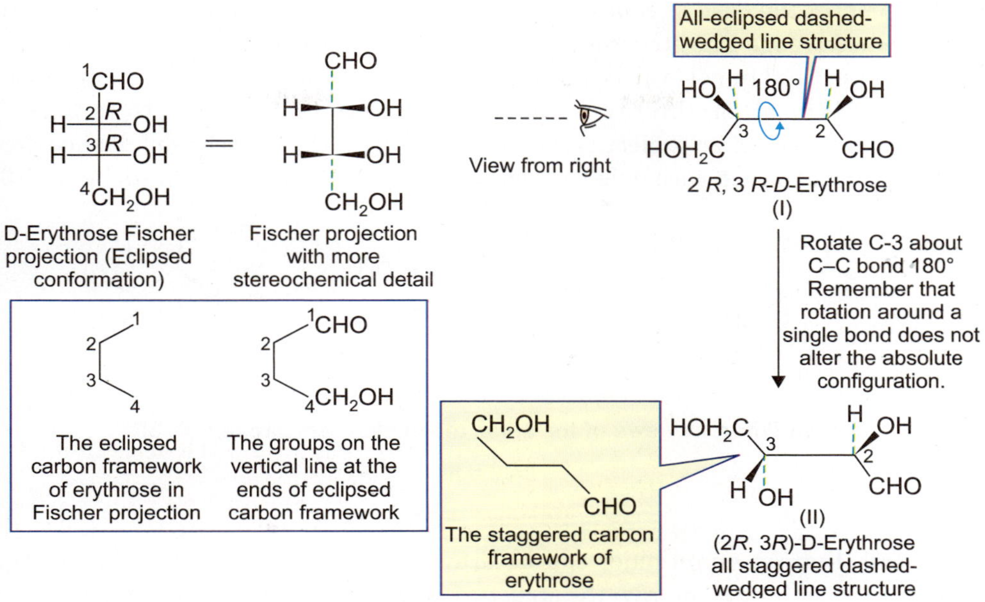

SCHEME 1.37g

- One can view such an all-eclipsed form from a molecular model and it is seen to have a roughly circular shape.
- To get the all-staggered conformation rotate either carbon *e.g.*, carbon C-3 around C-2, C-3 bond by 180°. By doing so the $-CH_2OH$ group which is in the plane of paper downwards goes upwards. In doing, so the group towards one's eyes (solid wedge) goes away (dashed line) and vice versa to afford (II, scheme 1.37*g*).
- The same result is obtained by looking to the Fischer projection of D-erythrose from the left-hand side (scheme 1.37*h*).

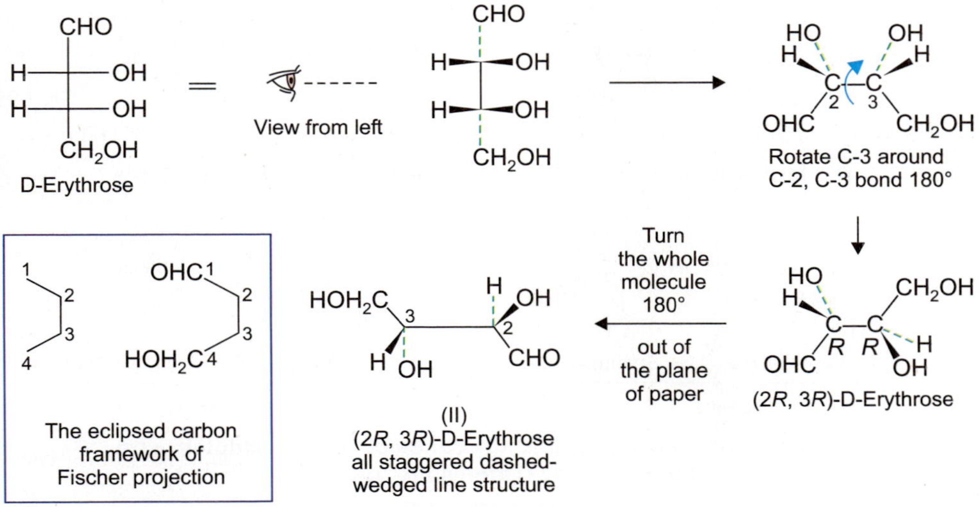

SCHEME 1.37h

As seen in the case of compounds with one chirality center, where a stereoisomer (in dashed-wedge line structure) can be drawn in several ways (see scheme 1.36), similarly there could be several views of the same compound with two or more chirality centers. Considering D-erythrose one can depict the structure in several ways depending on which side the Fischer projection is viewed to write the all-eclipsed dashed-wedge line structure and which carbon is rotated to stagger the arrangement (Scheme 1.37*i*).

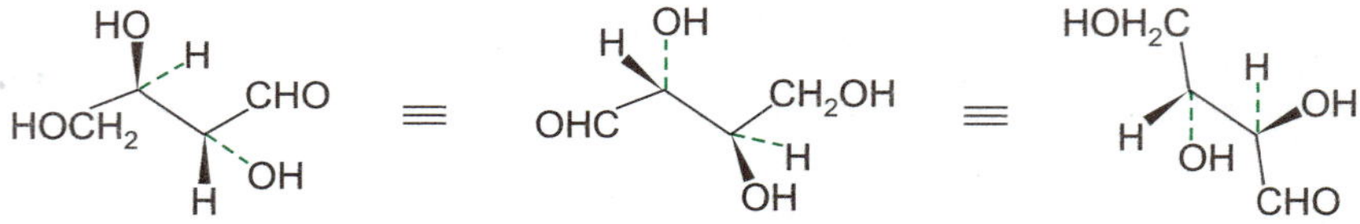

(I) (II) (III)

Three different views of the same molecule D-erythrose (2R, 3R)

I = II = III

SCHEME 1.37i

Recall, that an enantiomer can be drawn by inverting the stereochemistry at every chirality center *i.e.*, redraw the structure and replace all dashes with wedges and all wedges with dashes as in the case of erythrose enantiomers (Scheme 1.37*j*); while inverting the stereochemistry at only one of the stereocenters gives its diastereomer. Thus D-erythose and D-throse are diastereomers.

One can recognise *erythro* or *threo* structure (see, Scheme 1.28*c*) from the staggered dashed-wedge line structure. If one succeeds to eclipse every like group H with H; OH with OH and the unlike groups as well (CHO with CH_2OH), the configuration is *erythro*. If, however, one can eclipse only either any of the identical pair or unlike groups at any time then the configuration is *threo* (see Scheme 1.38).

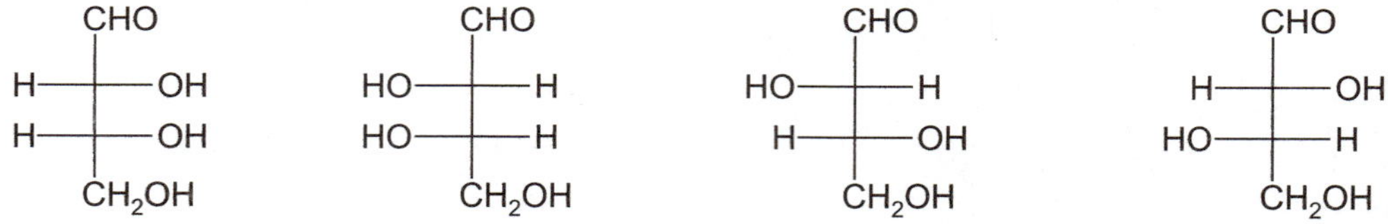

D-Erythrose L-Erythrose D-Threose L-Threose

The four aldotetroses drawn as their eclipsed Fischer projections

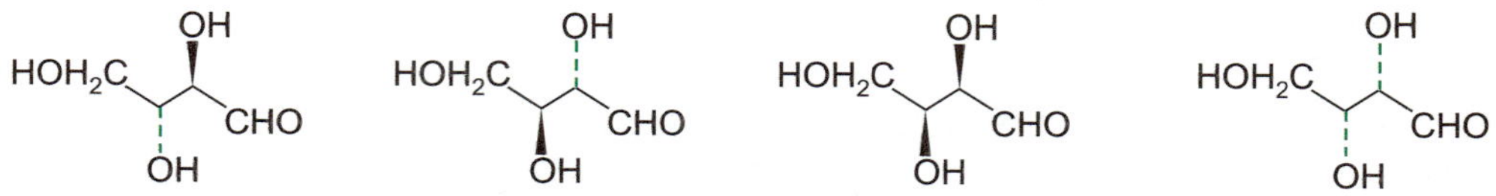

D-(−)Erythrose L-(+)Erythrose D-(+)-Threose L-(+)-Threose

Enantiomers Enantiomers

Staggered dashed-wedge line structures of the four aldotetroses. For simplicity H atoms are not drawn

SCHEME 1.37j

(3) *Conversion of the Dashed-wedge Line Structure into Fischer Projection*

A Fischer projection is an energy maximum eclipsed conformation. Thus the first step in such an exercise is to draw the molecule in its energy maximum eclipsed arrangement. View from one side and transfer the groups on to Fischer projection depending on whether these look close or far from the view side (Scheme 1.38). Consider the following points regarding compound (I, Scheme 1.38). The compound is written in a staggered arrangement (see, the zig-zag of continuous lines connecting CHO with CH_2OH).

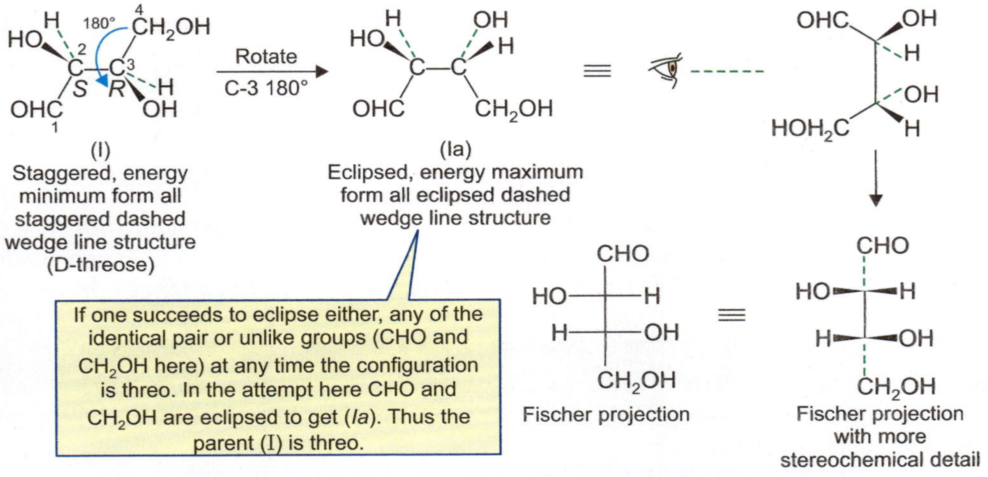

Conversion of a dashed-wedge line structure into Fischer projection

SCHEME 1.38

Rotate either C-2 or C-3 to convert this arrangement into eclipsed conformation (Ia, scheme 1.38). Here C-3 is rotated 180°, CH_2OH from the top comes down and the groups on this carbon which were towards one's eyes in (I) go away and the groups away from one's eyes come towards one's eyes (*i.e.*, groups on a wedge and a dash change places).

EXERCISE 1.8A

The erythro and threo diastereomers of quaternary ammonium salts are presented in Fischer projections (scheme 1.38a). Identify the erythro-diastereomer from the Newman projections (I and II, scheme 1.38a).

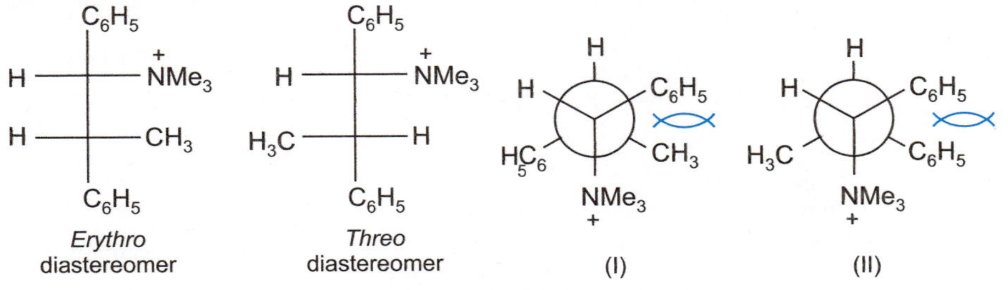

SCHEME 1.38a

Continued...

ANSWER. *In (II), it is possible to eclipse each of the like substituents, H with H, C_6H_5 with C_6H_5, and also the 'unlike' substituents CH_3 and $\overset{+}{N}Me_3$, the configuration is erythro.*

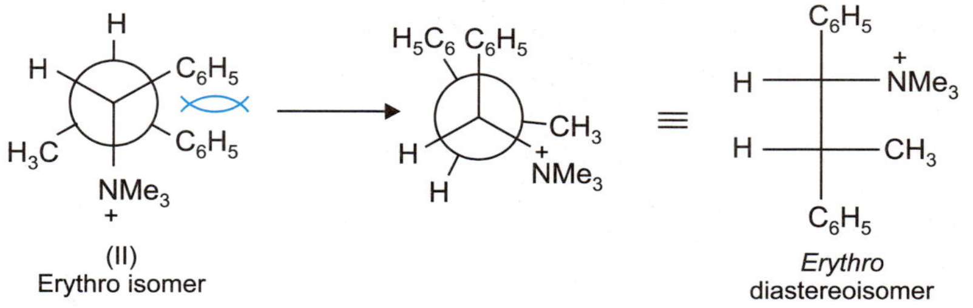

SCHEME 1.38b

The configurations at C(2) and C(3) in (I, scheme 1.38a) are now such that in e.g., Newman projections one can eclipse only (a) the H atoms, or (b) the C_6H_5 groups or (c) the unlike substituents CH_3 and N^+Me_3 at any one time. Thus (I) represents the threo isomer.

EXERCISE 1.8B

Convert the following perspective formulas (dashed-wedged line structures) (scheme 1.38c) to Fischer projections.

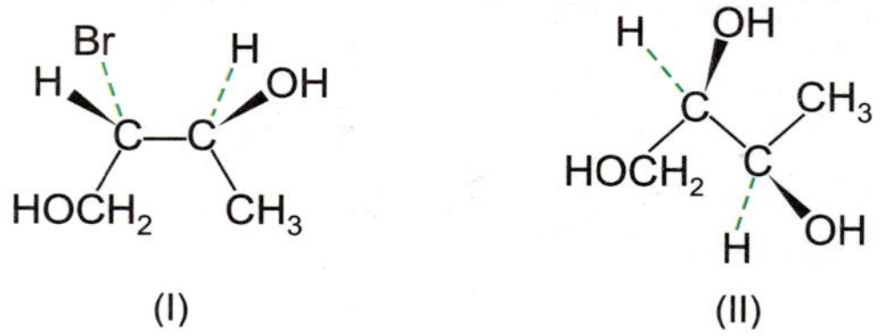

SCHEME 1.38c

ANSWER. *Arrangement in (I) is already in an eclipsed conformation rotating it by 90° in the plane of paper gives the equivalent (Ia scheme 1.38d). Viewing this arrangement from one side e.g., left-hand side gives the Fischer projection (Ib).*

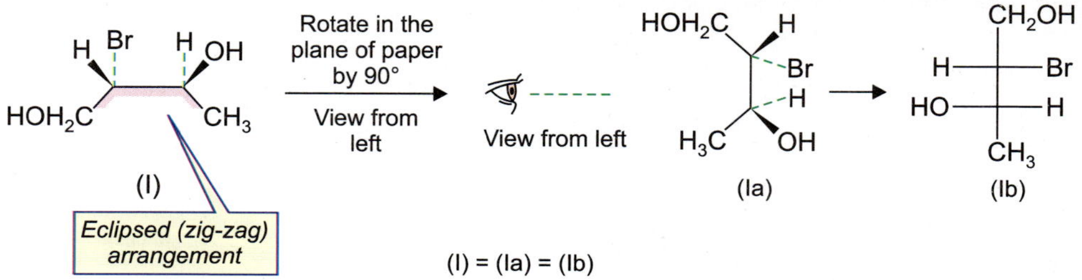

SCHEME 1.38d

Continued...

In (II, scheme 1.38c) the arrangement is a staggered conformation, thus it has to be first converted into an eclipsed conformation (IIa scheme 1.38e) and then the groups are placed on to a Fischer projection which gives (IIb).

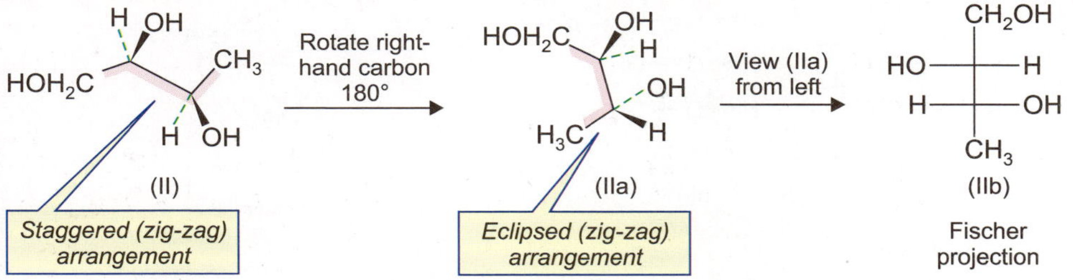

(II) = (IIa) = (IIb)

SCHEME 1.38e

(C) Compounds with Three Chirality Centers–Conversion of a Fischer Projection into Dashed-wedge Line Structure

Consider a compound (I, scheme 1.39) with three chirality centers the procedure is just the same as discussed above. In the first step again one views the Fischer projection from one side *e.g.*, from right and the groups on the Fischer projection are transferred on to the eclipsed carbon framework (I*b*, scheme 1.39). For convenience the eclipsed projection (I*b*) is rotated in the plane of paper to get (I*c*, scheme 1.39), keeping the central chirality center intact rotate the adjacent carbons to get the desired arrangement (I*d*).

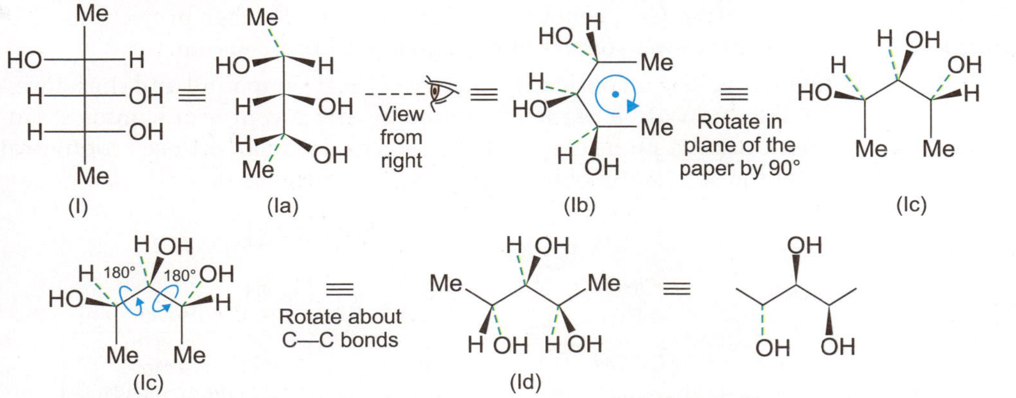

Conversion of a Fischer projection with three chiralitycenters into a dashed-wedge line structure

SCHEME 1.39

(D) Compounds with Four Chirality Centers

Consider the Fischer projection of D-glucose (I, scheme 1.40), following the procedure of (scheme 1.39), one arrives at all eclipsed dashed-wedged line structure (Ic, scheme 1.40). Now in this case (a compound with four chiralitycenters) one rotates two alternate carbons by 180° (C-3 and C-5 in this example) to get the staggered conformation (I*d*). One may note that if the whole arrangement (I*d*, scheme 1.40) is rotated 180° out of the plane of paper one get the equivalent arrangement (Ie).

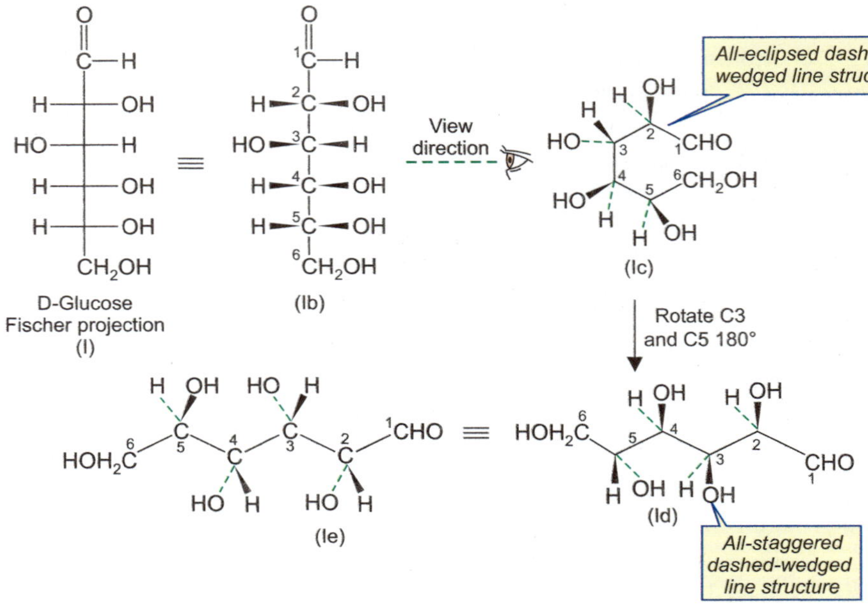

SCHEME 1.40

(E) Determining Relationship Among Stereoisomers–Stereoisomeric Tartaric Acids Drawn in Different Orientations

One will see that there can be several correct ways of depicting a molecule *e.g.*, in dashed-wedge line notation (see, schemes 1.3*a*) as well as Fischer projections. Firstly an example of one of the stereoisomers of a tartaric acid is discussed.

- *Meso*-tartaric acid is an optically inactive achiral compound and therefore, one must be able to locate a plane of symmetry and a center of symmetry in the given orientation. A plane of symmetry can be seen in an eclipsed configuration while a center of symmetry in a staggered conformation.

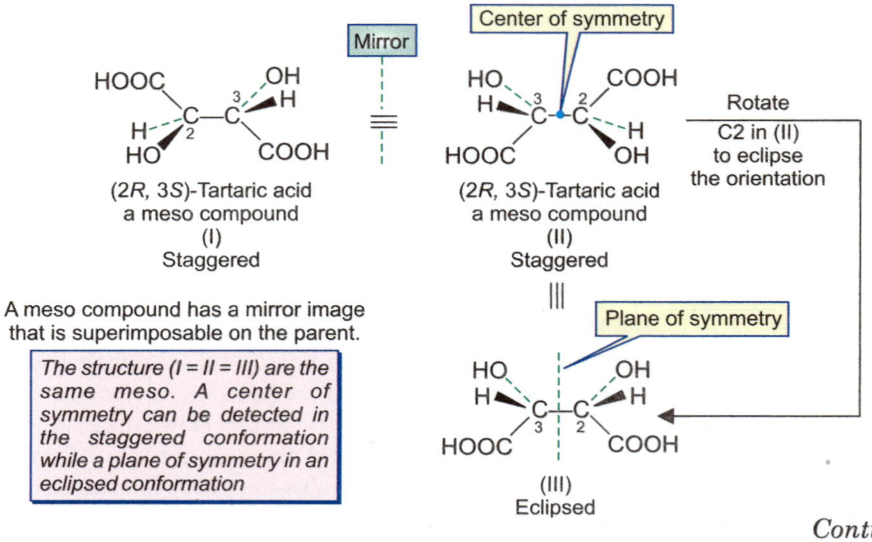

Continued...

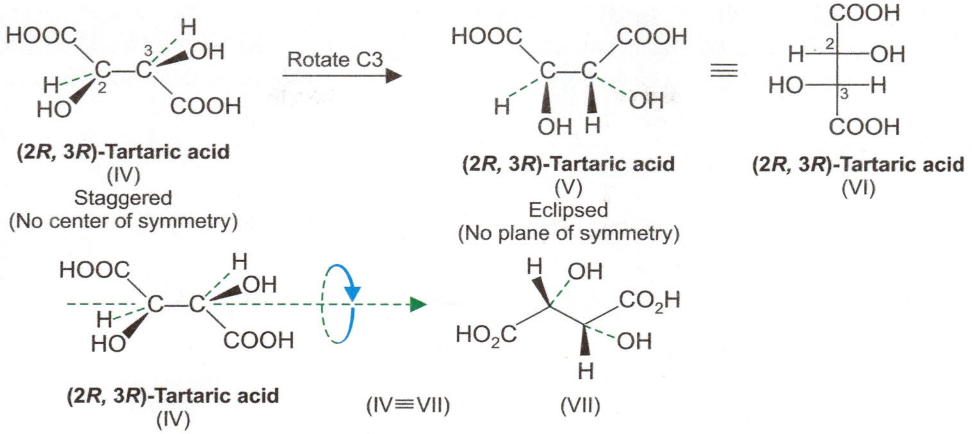

SCHEME 1.41

- If one focuses on the conformations of one of the enantiomers of tartaric acid (2R, 3R IV or V, Scheme 1.41) one in principle may use any conformation staggered (IV) or eclipsed (V, Scheme 1.41) as the situation may demand. One cannot detect a center or a plane of symmetry in (IV or V) respectively. [The projection (IV, Scheme 1.41) can be obtained by switching the position of groups at one of the chiralitycenters *e.g.*, C3 in (I, Scheme 1.41). The projections (I and IV) are diastereomes, recall that diastereomers, (with two chiralitycenters) have same configuration on one chiralitycenter and inverted configuration on the second. The enantiomers have inverted stereochemistry at every chiralitycenter.
- The orientations (IV and VII scheme 1.41) are identical since rotation of the structure through a horizontal axis passing through C2–C3 in (IV) by 180° out of the plane of paper gives (VII). Recall unlike a Fischer projection which cannot be lifted out of the plane of paper a perspective drawing does not lose its identity on this operation.
- One can recognize a *meso* compound from among the orientations drawn *e.g.*, in the Newman projection by rotation about the C—C bond. That stereoisomer is *meso* if one succeeds in getting an eclipsed conformation in which all the substituents are matched up (Scheme 1.42) one can also recognize a center of symmetry in the staggered conformation of *meso*-tartaric acid.

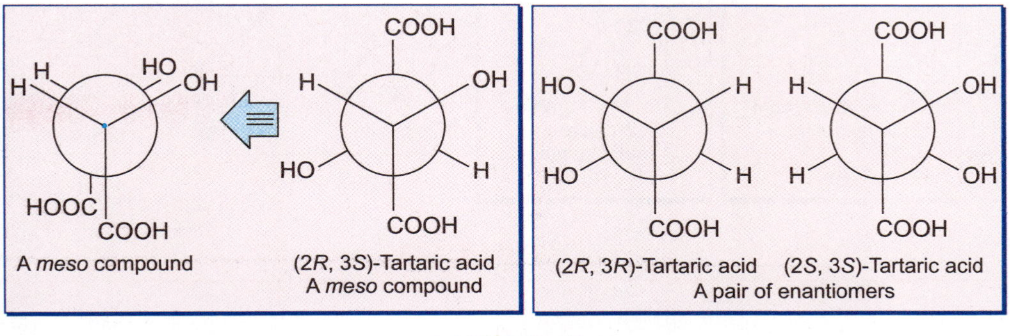

SCHEME 1.42

EXERCISE 1.9A

From the Newman projections (scheme 1.42a) point out which are identical; enantiomers; meso; or diastereomers.

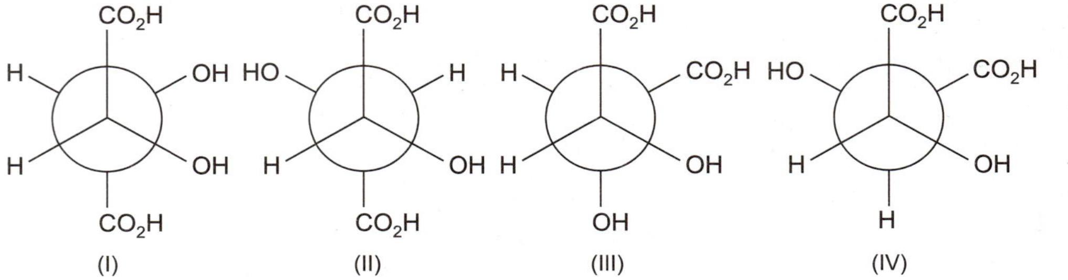

Different Newman projections of tartaric acid

SCHEME 1.42a

ANSWER

- *Conformations (I and IV) and (II and III) are identical and each is 2R, 3S.*
- *There are no enantiomeric pairs.*
- *Projections (II and III) are meso conformers of tartaric acid.*
- *Projections (I and IV) of tartaric acid are diastereomeric with projections (II and III).*

EXERCISE 1.9B

Write the wedge and dash structures for all the possible isomers of 3-bromo-2-chloro-4-methylpentane (for a related problem see scheme B 1.3a).

ANSWER. *There are two chirality centers in (I, scheme 1.43). There will be thus four stereoisomers. Since the chirality centers do not have identical set of substituents and the molecule does not have an internal mirror plane of symmetry, there is no possibility of the existance of a meso* stereoisomer. *The easiest way to solve such a problem is to*

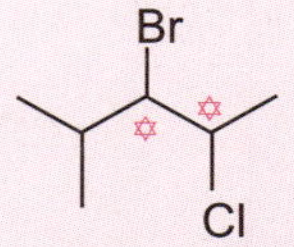

3-Bromo-2-chloro-4-methylpentane

(I)

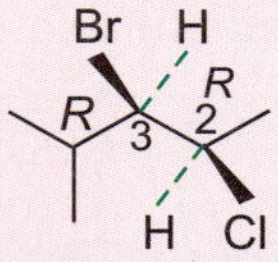

(2*R*, 3*R*)-3-bromo-2-chloro-4-methylpentane

(II)

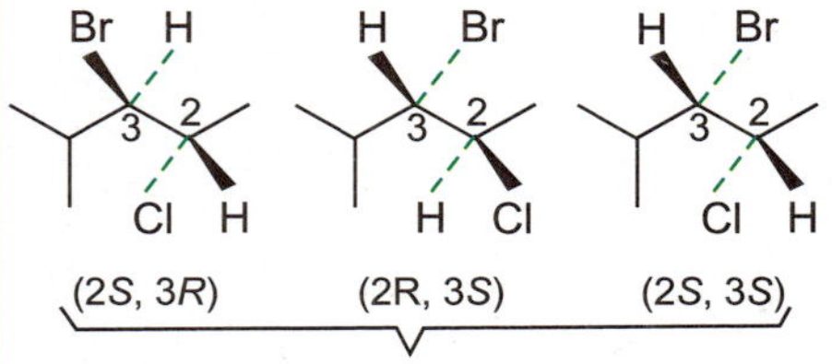

SCHEME 1.43

draw the staggered carbon framework of the compound and then assign R and S configuration by adding H atoms on their chirality centers on dashes i.e., below their carbons (II). The other stereoisomers can be drawn from (II) by switching the configurations at each of the chirality centers one at a time and switching the R/S designations to match (III, scheme 1.43).

EXERCISE 1.10

What is the stereochemical relationship between the configurational isomers (I–III scheme 1.43a)?

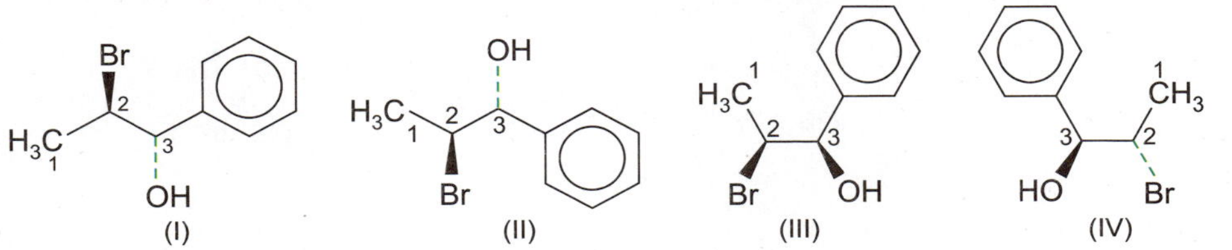

SCHEME 1.43a

ANSWER. *One may follow several approaches as already discussed, one may consider here two approaches* [A] *and* [B] .

[A] *Assign R and S descriptors to each stereocenter.*

(I) is 2R, 3S

(II) is 2S, 3R

(III) is 2S, 3R

Priorities	
C2	Br > CH (OH) Ph > CH_3 > H
C3	OH > CH(Br) CH_3 > Ph > H

Thus (I and II) are enantiomers, since at every chirality center the configuration is inverted, while (II and III) are same.

If one redraws the molecule (IV) to ((IV b), scheme 1.43 b) so that the carbon backbone as in ((IV)b) now mirrors the backbone as in (Ia and Ia') one finds that these isomers are not mirror images of each other due to chirality centers, this means that these are diastereomers. Moreover, the configurational descriptors in (IV) are 2S, 3S, while in (Ia) these are 2R, 3S and since inverting some, but not all chirality centers gives diastereomers, (IV) is a diastereomer of both (Ia and Ia').

[B] *Redraw the projections by rotation of one chirality center or the whole projection. One may show the H atoms also for clarity (scheme 1.43b).*

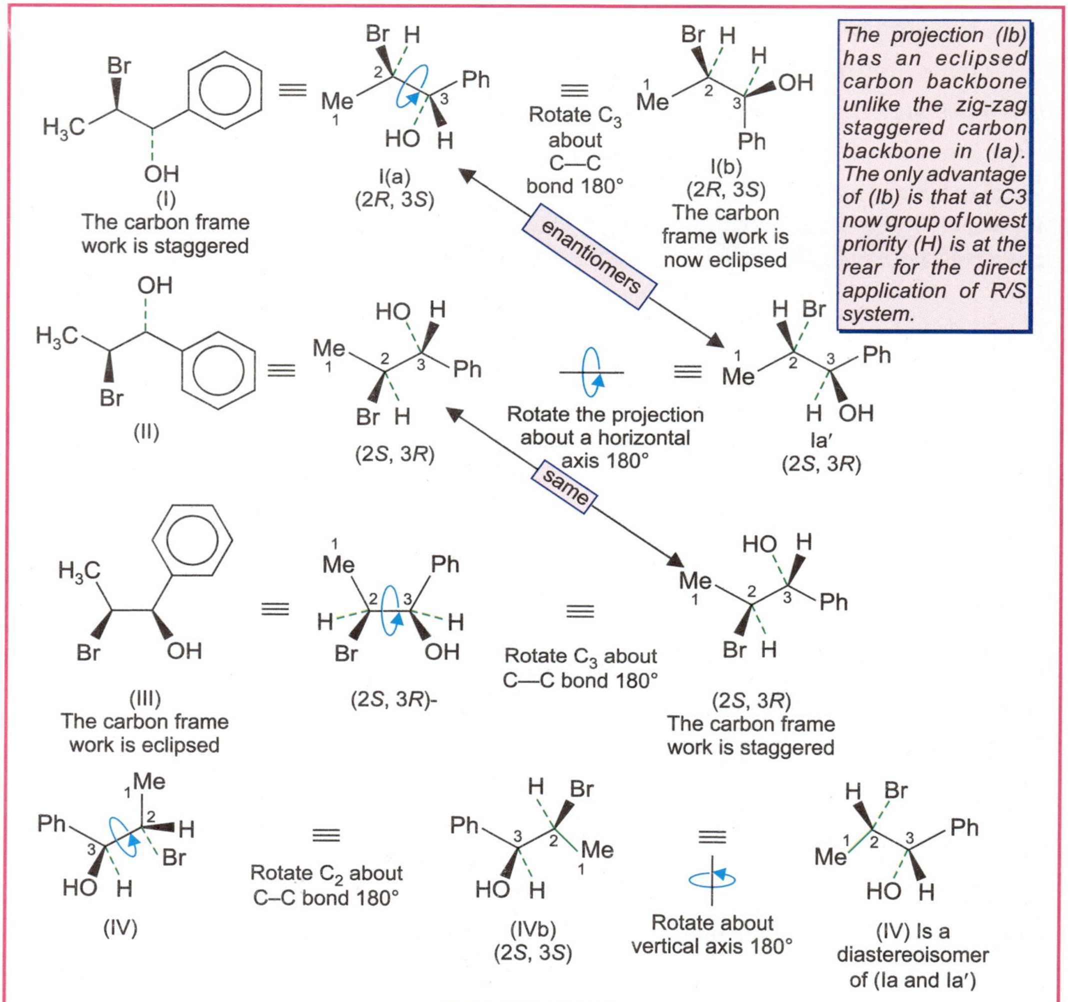

SCHEME 1.43b

More on Relative Configurations and Configurational Descriptors

- ***D, L are configurational descriptors for carbohydrates and amino acids.*** *Their use for other types of chiral compounds is obsolete.*
- ***d, l Characterization of enantiomers.*** *This differentiation of enantiomers by the sign of their optical rotation at a specified wavelength (normally 589 nm, sodium D line emission) is obsolete. Now (+) or (–) for dextrorotatory and levorotatory respectively are used. The prefix dl is still used for racemates, however the better used designations are (±) or rac.*

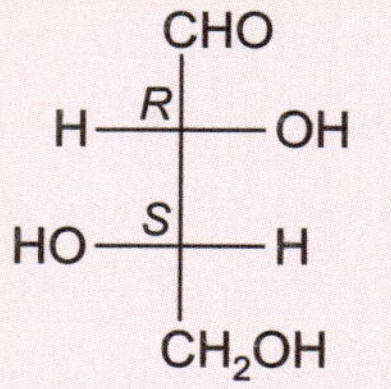

2R, 3S-Threose
u-2,3,4-Trihydroxybutanal
SCHEME 1.43c

Continued...

- ***Stereodescriptors representing different chirality centers.*** *These are separated by the use of commas, thus L-threose is represented as (2R, 3S)–threose along with the usual positional number (Scheme 1.43c). The pair of descriptors RS is used for a racemate with one chirality center. Two pairs of stereodescriptors imply a racemate with two like (RS, RS) as in erythrose or two unlike (RS, SR) chiralitycenters as in threose and in other systems (Scheme 1.43c).*
- ***Stereodescriptors l and u.*** *These are the stereodescriptors used for compounds with two chiralitycenters. One uses the symbol l (for like) and u (for unlike) when the CIP descriptors for the two chiralitycenters are like (R, R or S, S or RS, RS) or unlike (R, S or S, R or RS, SR) respectively (scheme 1.43d). This notation is unambiguous as the CIP system itself but is not convenient, because it requires assignment of the CIP descriptors for both chirality centers before the l or u descriptor is assigned. However, most chemists want an "at a glance" designation for a system of two chirality centers.*

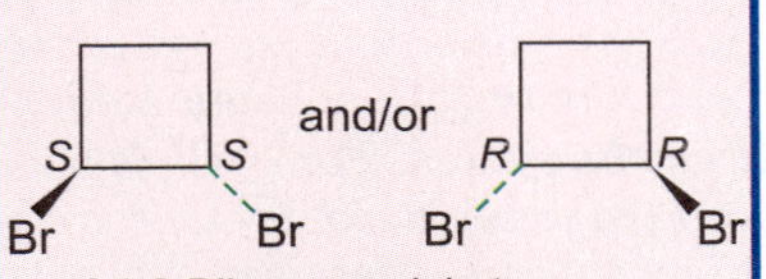

l-1,2-Dibromocyclobutane

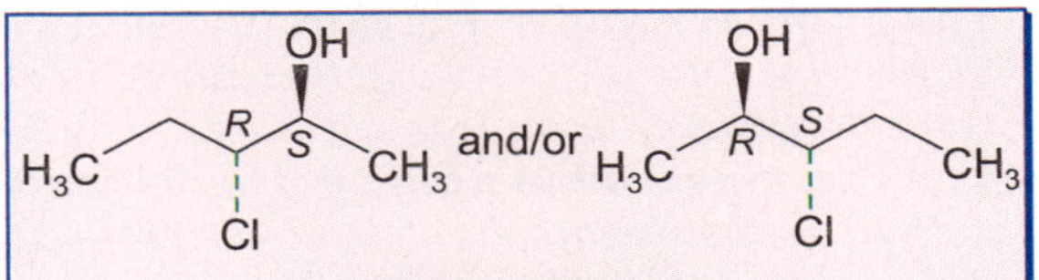

u-3-Chloropentan-2-ol

SCHEME 1.43d

- ***R* and S* nomenclature.*** *Alternatively, in a convention employed by Chemical Abstracts, the two enantiomers (2R, 3R)- and (2S, 3S)-tartaric acid may be specified by (2R*, 3R*). The symbols (2R*, 3R*) point that if C(2) has R configuration, so does C(3), and likewise that if C(2) has S configuration, so does C(3). meso-Tartaric acid can formally be depicted as having (2R*, 3S*) configuration which shows that if C(2) has R configuration, C(3) has the S and vice versa.*

  For a pure enantiomer of unknown absolute configuration the stereodescriptors R and S* or rel are used.*

  Thus one calls (±) threose as u–2, 3, 4–trihydroxybutanal (scheme 1.43c) and the use of these soft descriptors l and u is extended to other like and unlike pairs as well e.g., MM or PP have the symbol l while MP has the symbol u.
- ***Stereodescriptors l and u for compounds with three or more chirality centers.*** *When two constitutionally "identical" groups differ in their configuration, the priorty rules state that an (R) group has a higher priorty than one that in (S) [R before S in alphabet]. While dealing with compounds e.g. D-glucose with the configuration 2R, 3S, 4R, 5R (scheme 1.18a) the center with the lowest number is compared with each of the others and racemic glucose is, ull.*
- ***Priority between groups with l and u configuration.*** *A CIP sub-rule states that of the two constitutionally similar groups the one with an l configuration e.g., R, R or S, S has priority over the groups with a u configuration. Thus the compound (I, scheme 1.43e) is assigned a proper stereodescriptor at each of the five chirality centers.*

- ***The prefixes threo and erythro.*** *This is an easy "at a glance" nomenclature for compounds with two chirality centers and was devised several years ago. This useful nomenclature can run into problems (II scheme 1.43e).*

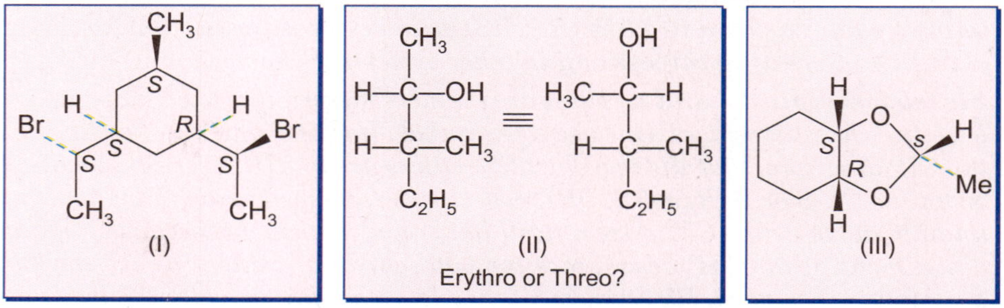

SCHEME 1.43e

- ***The prefixes r and s.*** *The chirality centers are designated by R and S and also by r and s depending on whether they are chirality or pseudochirality centres, respectively. An example of a pseudochirality center is in (III, scheme 1.43e).*
- ***The r, c, t method to name polysubstituted carbocycles.*** *For cyclic systems with two substituents the stereodescriptors cis and trans are used, while r, c and t are used if three or more substituents are present on the ring (detailed discussion is in scheme 1.67b).*
- ***Summary.*** *The soft descriptors (threo, erythro; syn, anti etc) are the shorthand methods used by organic chemists and have a danger of running into misuse. The relative configurations thus are only reliable when the orientations display 3-D relationships.*

A Good Exercise for the Use of u and l Nomenclature
An Example from Cram's Rule

When a chirality center in a substrate is adjacent to a carbonyl group, the stereochemical course of nucleophilic addition follows the Cram's rule (for full details see schemes 1.18 g and 2.23b – 2.28b). Consider the preferential cyanohydrin formation from (±)-2-methylbutanal with HCN (scheme 1.43f). One explains the outcome by saying that u-configured pair of enantiomers are the major products. The minor product being the l-configured racemate of 2-hydroxy-3-methylpentanenitrile.

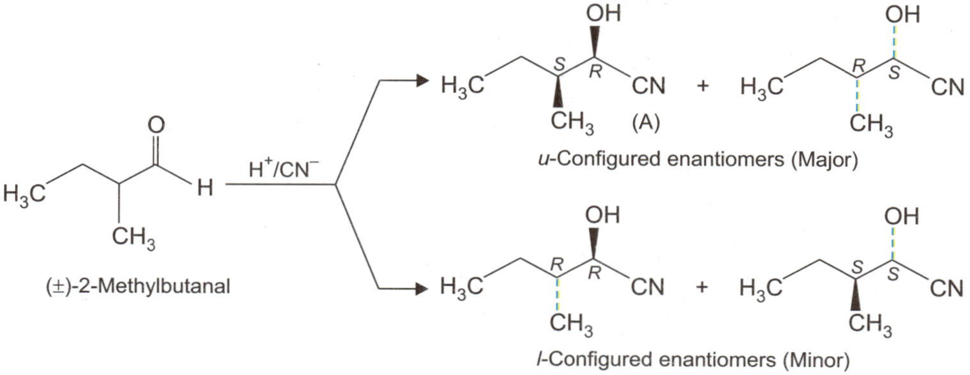

SCHEME 1.43f

1.7 SOME ASPECTS OF GEOMETRIC ISOMERISM (pi DIASTEREOISOMERISM AND TORSIONAL CHIRALITY IN CARBON-CARBON DOUBLE BONDS)

In case the rotation about a bond joining two mutivalent atoms is restricted, then suitable substitution leads to isomerism. The most common method of preventing rotation about a bond is either the formation of a multiple bond, as in π-diastereoisomerism or by the incorporation of the bond in a ring system as in σ-diastereoisomerism. Isomerism arises in both cases to give isomers which represent different chemical entities known as geometrical isomers. Isomers with similar groups on the same side of the molecule are known as *cis* and those with similar groups on opposite sides are known as *trans*.

This class of diastereomers includes *cis, trans*, *E*, *Z* as well as *syn* and *anti*-isomers.

(A) Geometric Isomerism on Double Bonds

For an alkene to show geometrical isomerism the two groups on one end of the double bond must be different and the two groups on the other end of the double bond must also be different. That is, in terms of the structure ((I), scheme 1.44), A must be different from B, and D must be different from E. In such a situation both of the carbons of the double bond are said to be trigonal planar stereocenters (for details see Fig. 1.5).

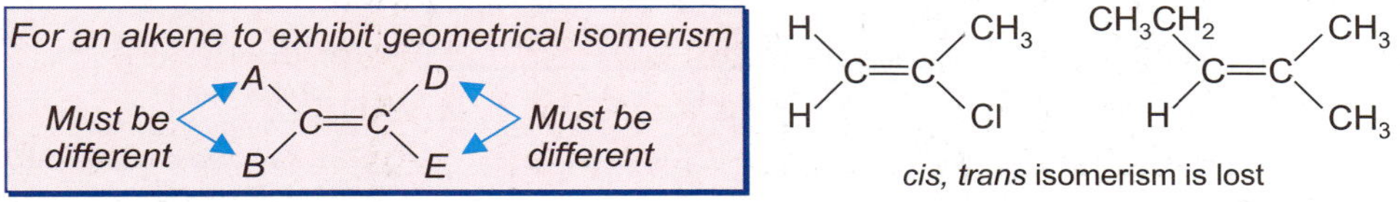

SCHEME 1.44

(B) The E-Z System of Nomenclature of Alkenes and Nitrogen Containing Compounds

Now, the *E-Z* system which is based on the priorities of groups on the Cahn-Ingold-Prelog convention, is used to designate diastereomeric alkenes. The terms *cis* and *trans* are unambiguous only when used to designate the stereochemistry of disubstituted alkenes. If, however, the alkene is trisubstituted or tetrasubstituted these terms either become ambiguous or cannot be applied at all. For example, in the case of the alkene 1-bromo-1-chloropropene (Scheme 1.45) it is not possible to designate it as *cis* or *trans* since no *pair* of groups are the same.

In the *E-Z* system the two groups attached to each carbon of the double bond are arranged in order of priority as in 1-bromo-1-chloropropene (scheme 1.45), Br > Cl and CH_3 > H. If the two groups of higher priority are located on the same side of the double bond the alkene is designated *Z* isomer (from the German word *Zusammen* meaning together). If on the other hand, the two groups of higher priority are on opposite sides of the double bond then the alkene is designated *E* isomer (from the German word, Entegegen meaning opposite). In compound (I, scheme 1.45) one carbon of the double bond has OCH_3 and COOH, OCH_3 being of higher priority than COOH. The second carbon has two ring residues, one of these residues has higher priority over the other and the compound (I, scheme 1.45) therefore, represents the *Z* isomer and similarly compound (II, Scheme 1.45) is also the *Z* isomer (Cl > $CH_2CH_2CH_3$ and CH_3 > H). If an alkene has more than one double bond, the stereochemistry about each double bond is to be specified (III scheme 1.45). Cumulenes with odd number of double bonds (*V*, scheme 1.45) display π-diastereomerism and *E* and *Z* isomers are possible, thus 1-chloropenta-1,2,3-triene is *Z*-isomer. The smallest bridgehead alkene stable under normal conditions is bicyclo (3.3.1) non-1-ene. This molecule contains a *trans*-cyclooctene system (see, scheme 1.52).

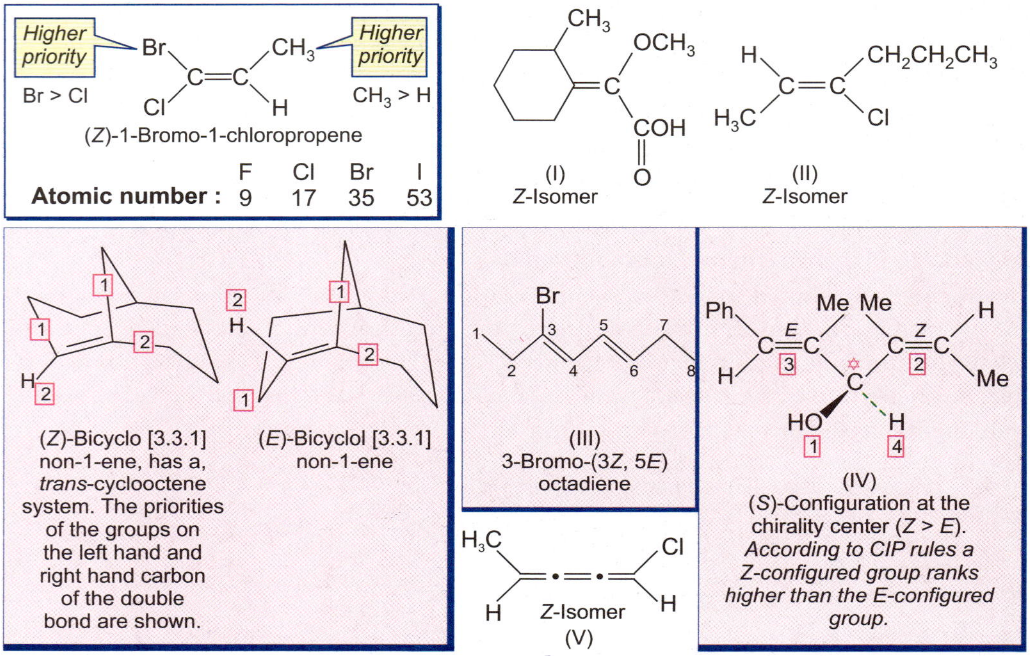

SCHEME 1.45

Between the two diastereomeric groups, one having *Z* configuration gets priority over the group having *E* configuration (CIP) rules, thus the chirality center in (IV, scheme 1.45) is assigned (*S*) configuration.

Diastereomeric oximes and other nitrogen containing compounds are also now given (*E*, *Z*) nomenclature (scheme 1.46). The term *syn* is similar in meaning to *Z* (or *cis*) and the term *anti* is similar in meaning to *E* (or *trans*) are used in naming certain classes of compounds (scheme 1.46).

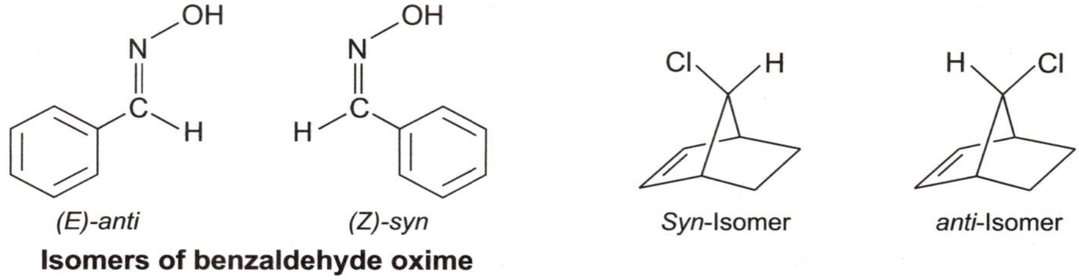

SCHEME 1.46

When two constitutionally identical groups differ in their configuration the priority rules state that R group has priority over S group, and a Z configured group ranks higher than the E configured group.

A Typical Case of *E/Z* Isomerism

Consider 4-methylhepta-2, 5-diene which can exist as four configurational isomers since both double bonds could be *E*, *E* or *Z*, *Z*. These can also have different configuration (scheme 1.46*a*) and in that case C4 would become a chirality center. This can be assigned a stereodescriptor by CIP rules since a *Z* configured group has higher priority than *E* group.

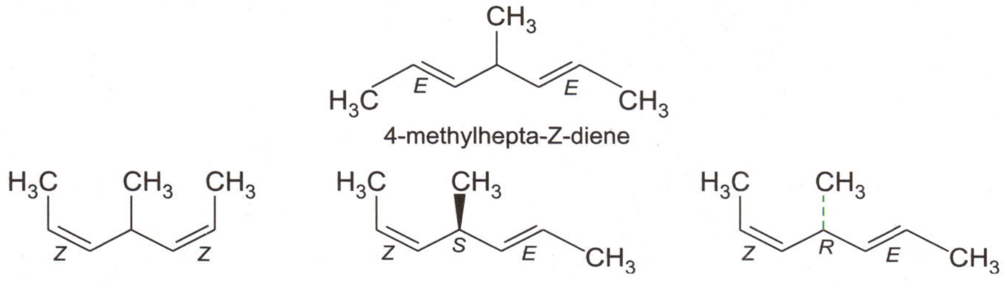

SCHEME 1.46a

In molecules of the type A=CH—CH=B, the fully eclipsed conformation with a dihedral angle of 0° is termed *s-cis*, while the conformation with a dihedral angle of 180° is called *s-trans*. In butadiene e.g., (scheme 1.47), the *s* preceding the *cis* and *trans* refers to *cis* and *trans* with respect to a single bond possessing some double bond character which provides the torsional barrier nearly 25 kJ mol^{-1} (which is much higher than in ethane) to interconversion. This is entirely a conformational effect. In Diels-Alder reaction the diene must be able to react in the *s-cis* conformation and any structural feature that interferes with the attainment of *s-cis* conformation inhibits the reaction (also see scheme 1.14*c*).

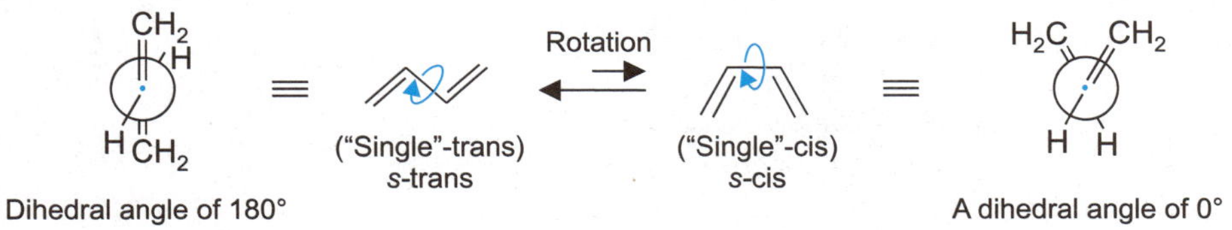

SCHEME 1.47

Geometrical isomerism is also shown by cumulenes provided the number of double bonds in odd (see, scheme 1.114).

A still higher energy barrier (about 89 kJ mol^{-1}) is found in the rotation around C—N bond, the double bond character of the amide bond is due to the localization of the nitrogen lone pair of electrons. Thus due to restricted rotation there can be two distinct isomers which are termed *E* and *Z*. The substituents of highest priority in *N*-methylbenzamide defining the isomerism are the oxygen and the N-methyl groups, when these are on the same side the descriptor *Z* is used and when these are on opposite sides the structure is *E* (scheme 1.48).

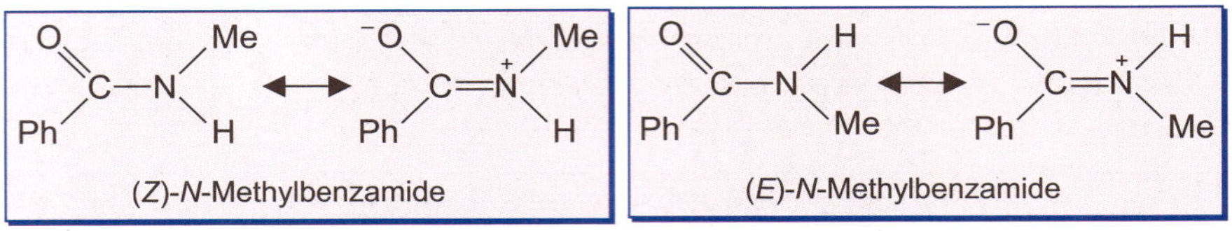

SCHEME 1.48

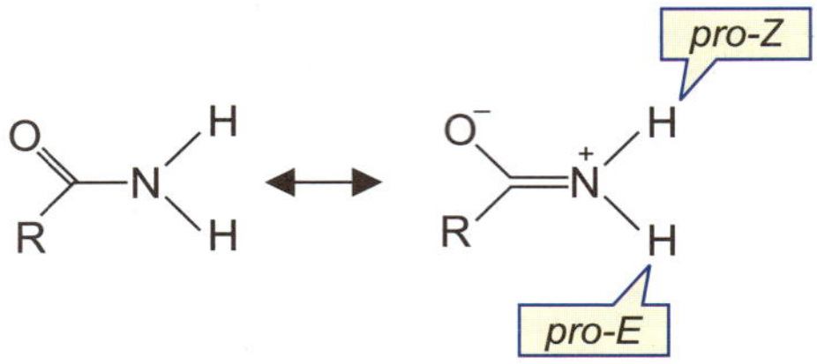

SCHEME 1.49

The rotation is however, too fast to allow their isolation, although these can be distinguished by low temperature NMR. Thus the absorption of amide protons at room temperature occurs at different fields in ^{1}HNMR spectrum (a unified signal is however, observed at raised temperature when rotation around the C–N bond starts to occur scheme 1.49). The amide hydrogen atoms (scheme 1.49) are diastereotopic and thus are distinguished by the descriptors *pro-Z* and *pro-E* (see, Sec. 2.3).

Significantly the barriers in peptidylprolines as in (scheme 1.50) are so important for protein folding that to bring about such a change in proline isomers in nature there is a widespread enzyme (peptidylprolyl-*cis*, *trans*- isomerase PPI) to bring about this rotation (scheme 1.50).

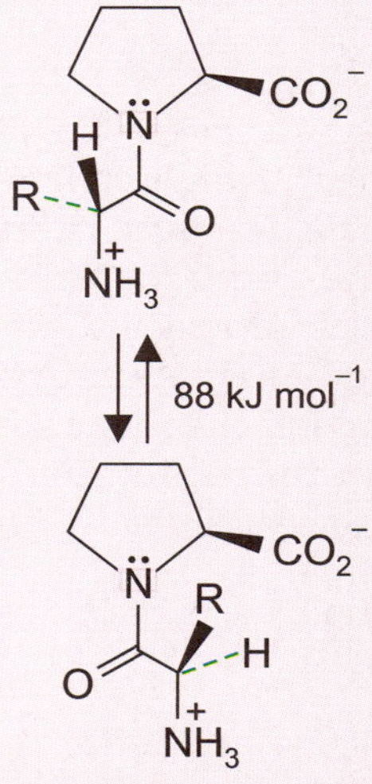

SCHEME 1.50

> *Any two species in equilibrium cannot ordinarily be separated at room temperature unless these are separated by a minimum energy barrier of about 100 kJ mol^{-1}.*

On the other hand there are compounds in which nearly free rotation is possible around a formal C = C double bonds. In these cases the double bonds are seriously weakened *via* resonance which decreases the double bond character to allow easier rotation. Thus *e.g.*, the compound (scheme 1.51) has a barrier to rotation of 92 kJ mol^{-1} compared to a simple alkene ~ 260 kJ mol^{-1}. Such alkenes are called “push pull” alkenes. In these push pull alkenes the barrier to rotation is close to the value found for some C—C single bonds.

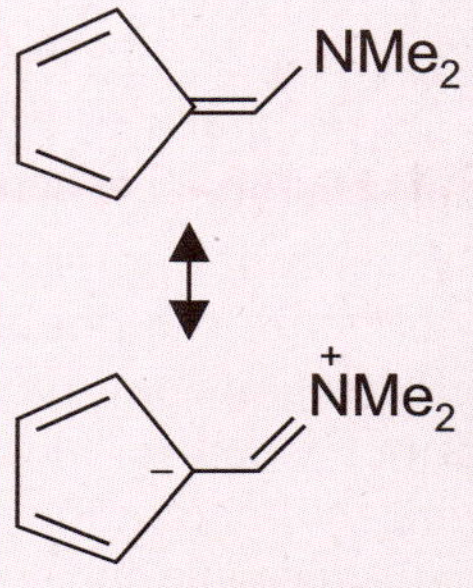

SCHEME 1.51

(C) Stereoisomerism—Geometric (cis-trans) Isomerism in Monocyclic Compounds (A Summary)

(i) Cis-Trans Isomerism and Chirality-Cycloalkenes

In the cycloalkenes from cyclopropene to cyclooctene the configuration about the double bond in each case is *cis* (scheme 1.52). A *trans*-double bond *e.g.*, cannot be geometrically accommodated *e.g.*, in a six membered ring.

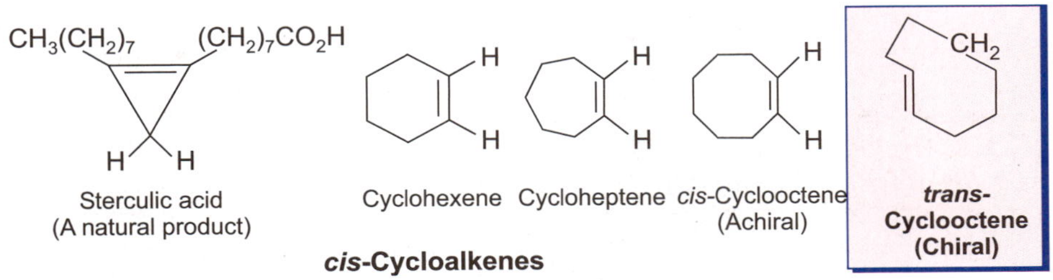

SCHEME 1.52

Cyclohexene is a *cis* alkene, and necessarily so, as the chain with four sp^3 hybrid carbons is not long enough to link C(1) and C(2) with formation of a *trans* alkene. A hypothetical *trans*-cyclohexene (I, scheme 1.53) would be too highly strained to exist at room temperature.

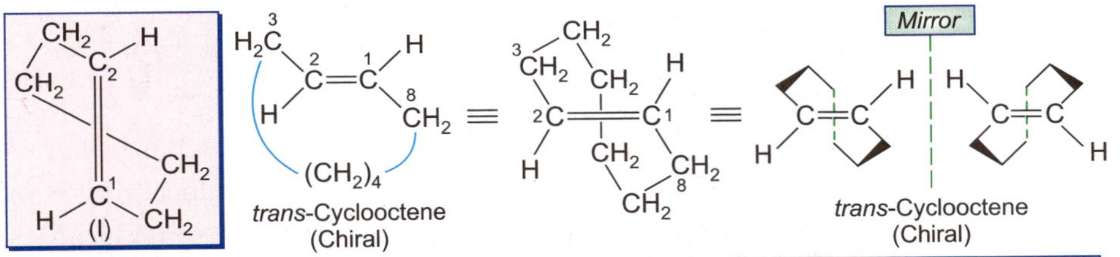

trans-Cyclohexene is too strained to exist

trans-Cyclooctene is chiral and a strained molecule which cannot achieve a planar symmetric conformation. It is optically active, the two enantiomers may be regarded to have helical shape. cis-Cyclooctene has achiral diastereomer trans-cyclooctene.

SCHEME 1.53

EXERCISE 1.10a

Assign E/Z nomenclature to the compounds (scheme 1.54)

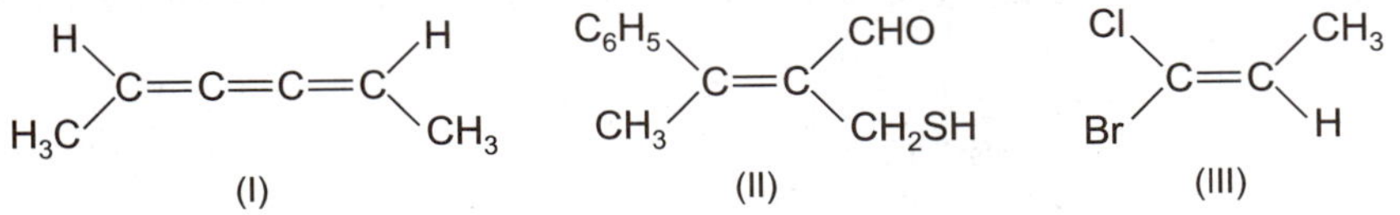

SCHEME 1.54

ANSWER. ***(I)**, (Z)-hexa-2, 3, 4-triene ; **(II)**, E ; **(III)** E.*

EXERCISE 1.11

Assign E/Z nomenclature to the geometric isomers of compounds (scheme 1.55).

Cyclohexene

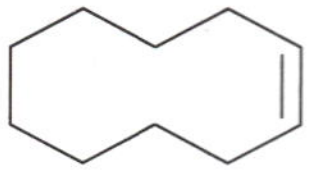

Cyclodecene

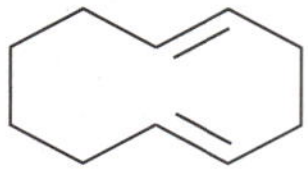

1, 5-Cyclodecadiene

SCHEME 1.55

ANSWER. *Cyclohexene has no E isomer which would be too strained to exist. The E/Z isomers for other compounds are in (scheme 1.56).*

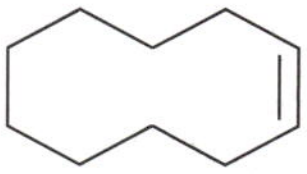

(*Z*)-Cyclodecene

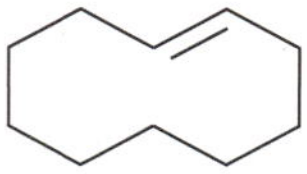

(*E*)-Cyclodecene

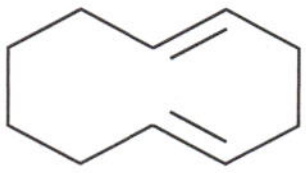

(1*E*, 5*E*)-1,5-Cyclodecadiene

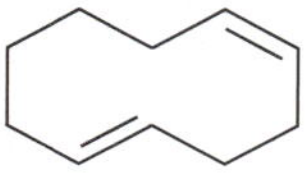

(1*Z*, 5*E*)-1,5-Cyclodecadiene

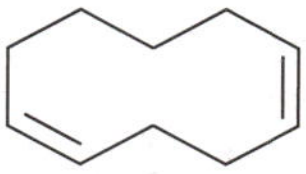

(1*Z*,5*Z*)-1,5-Cyclodecadiene

SCHEME 1.56

To date *trans*-cyclooctene is the smallest *trans* cycloalkene that has been prepared in pure form which is stable at room temperature (the *cis-trans*-cyclooctenes represent a diastereomeric pair). In *trans-cyclooctene* (scheme 1.53) there is considerable strain and it is less stable than its *cis* isomer by 9.1 kcal/mol. *Trans*-cycloheptene and *trans*-cyclohexene have been detected, but these are so unstable that these have never been isolated in pure form under ordinary conditions. *Trans*-cyclooctene is a chiral molecule with chirality plane. The two trigonal carbons and the atoms which are directly attached to these are in a plane and the polymethylene bridge is skewed in the third dimension. Its two enantiomers have been isolated. In one of the enantiomers, the four carbon methylene chain lies above the double bond on one side and below it on the other. ***(Further details of chirality of trans-cyclooctene are in schemes 1.136-1.137).***

(ii) Geometric and Stereoisomerism in Cycloalkanes (Alicyclic Compounds)

The presence of a ring like a double bond prevents rotation and when properly substituted makes *cis-trans* isomerism possible in cycloalkanes. The substituted carbons need not be adjacent (as also is the case in acyclic compounds see scheme 1.43). Rotation around the sigma bonds forming the ring would require the attached atoms or groups to pass through the center of the ring. This process is prevented from happening unless the ring contains eight or more carbon atoms, depending on the size of the substituents. Thus *cis-trans* isomers can exist in rings and properly designated (scheme 1.57).

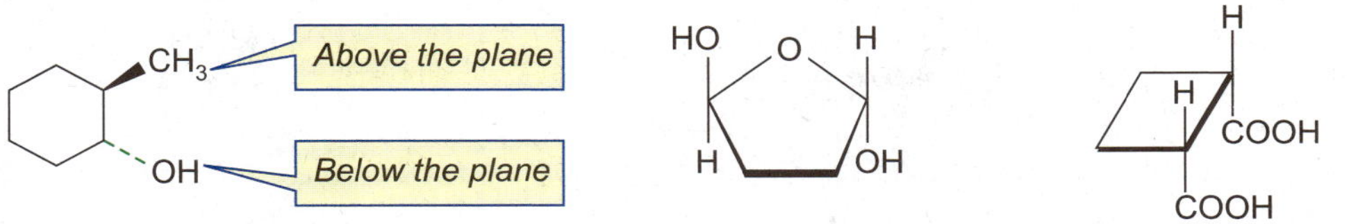

A *trans* compound

trans (OH groups on opposite sides of the ring)

cis (COOH groups on the same side of ring)

SCHEME 1.57

The stereochemistry of cyclic compounds with two or more chirality centers is basically similar to that in acyclic compounds. The shape of these rings and the stereochemical features of cyclohexane derivatives will be discussed in more detail in chapter 4.

A way of showing how groups are attached to the ring, is using a dotted line to show a group below the plane of the ring and solid wedge to represent a group above the plane (scheme 1.57). A part of the molecule when shown by thick line(s) is to orient it in a way so that this thick lined portion is projected towards ones eyes, this is often done to depict stereochemical features more clearly. The *cis* and *trans* isomers in cycloalkanes are in fact diastereomers.

(a) Cyclopropanes

A monosubstituted cyclic compound *e.g.*, chlorocyclopropane has no chirality center, possesses a plane of symmetry and is achiral (scheme 1.58). Similarly 1, 1-dichlorocyclopropane is achiral. 1,1,2-trichlorocyclopropane has however, one chirality centre and should exist as nonsuperposable mirror images (*i.e.*, enantiomers, scheme 1.58). A 1, 2-disubstituted compound (two chirality centers, exists in four forms (*i.e.* when both the substitutents are different): these are a pair of *cis*-enantiomers and a pair of *trans*-enantiomers when the substituents are the same *e.g.*, as in 1, 2-dichlorocyclopropane, the *cis* form has a plane of symmetry and thus represents a *meso*-type achiral compound which has superposable mirror image (scheme 1.59).

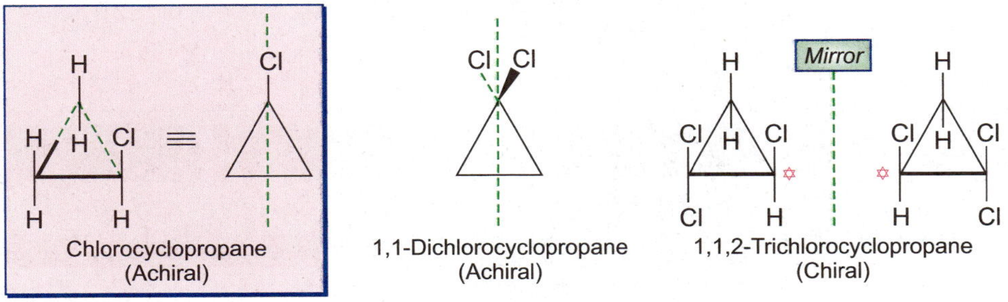

Chlorocyclopropane (Achiral)

1,1-Dichlorocyclopropane (Achiral)

1,1,2-Trichlorocyclopropane (Chiral)

SCHEME 1.58

In the case of *trans*-1, 2-dichlorocyclopropane, the mirror image is not superposable on the original. It is, therefore, a chiral molecule and two resolvable enantiomers exist.

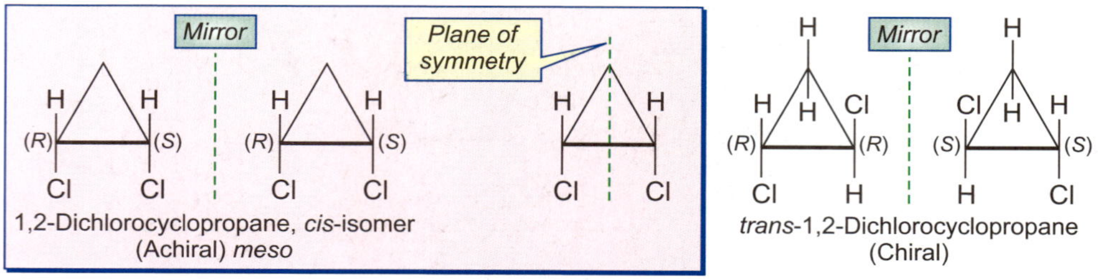

SCHEME 1.59

(b) Cyclobutanes

As with cyclopropanes, monosubstituted and 1, 1-disubstituted cyclobutanes exist in only one achiral form, whereas 1, 2-disubstituted cyclobutanes exist as a pair of enantiomers and a *meso* form (when the two substitutents are same). Both the *cis* and the *trans* 1, 3-disubstituted derivatives possess a plane of symmetry so that there are only two geometric isomers and none is optically active. (This is true of any disubstituted cycloalkane containing an even number of carbon atoms, where the substituents are on the opposite sides of the ring). One can also detect a center of symmetry in the case of *trans*-1, 3-dimethylcyclobutane which is at the center of the ring (this, however, is not shown, also see scheme 1.67*e*).

The mirror image of *e.g.*, *trans*-1, 3-dimethylcyclobutane is thus superimposable on the original scheme 1.60). This one can see by simply rotating the structures. Further proof that it is an achiral compound is due to the fact that it has a plane of symmetry. The plane of symmetry is lost in (I, scheme 1.60) which is chiral.

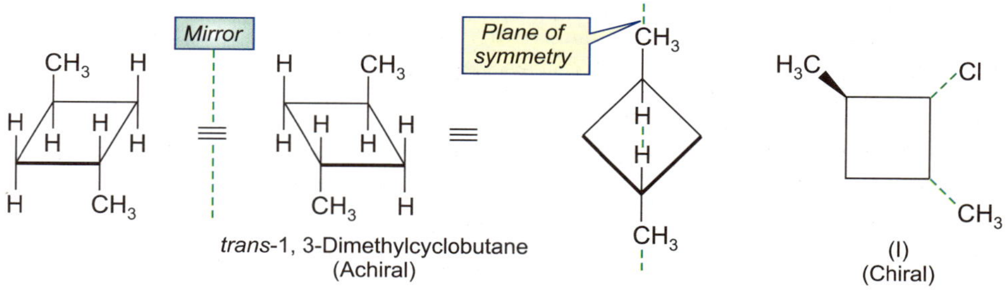

SCHEME 1.60

EXERCISE 1.12

Draw projections to show the presence of geometric cis-trans isomers and enantiomers for (I and II scheme 1.60a).

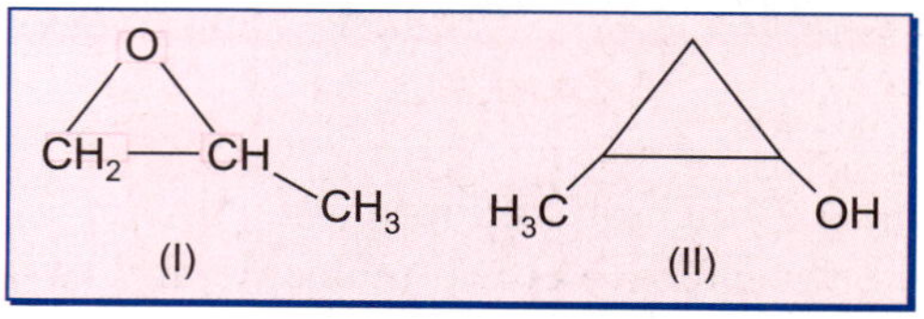

SCHEME 1.60a

ANSWER. *In (I) there is one chirality center (the four different substituents being H, O, CH_2 and CH_3). It has no geometric isomers, the enantiomers are in (scheme 1.61). Compound II has two geometric isomers cis and trans and each of these (chiral) has a non-superimposable mirror image thus it has four stereoisomers.*

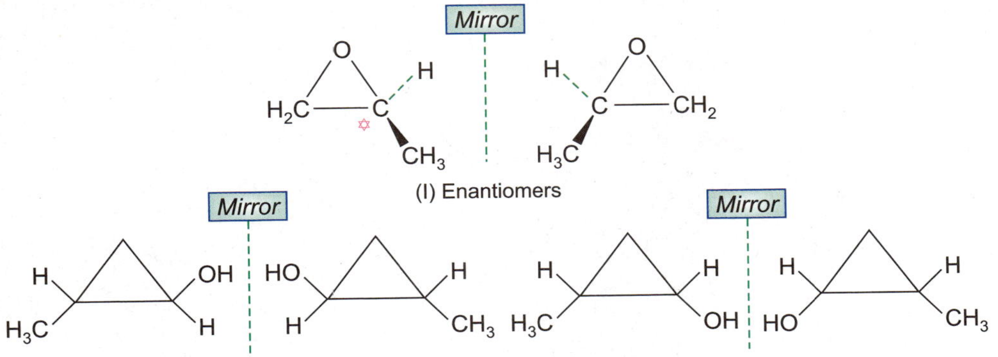

(I) Enantiomers

(II) *trans* (II) *cis*

SCHEME 1.61

(c) *Cyclopentanes*

The discussion may be taken up with cyclopentane ring, which like cyclopropane and cyclobutane is essentially planar. The planar projection formulas of cyclohexane derivatives can obviously not be considered as entirely correct representations. The planar representations are however, useful to symbolize and to count the various possible stereoisomers. Cyclohexane derivatives thus present some problems and a full discussion will be taken up in Chapter 4.

1,2-Dimethylchlopentane has two chirality centers and exists in three stereoisomeric forms (scheme 1.62). A structure with two stereocenters does not have always four possible stereoisomers (2^2). Sometimes there are only three. In these situations, both the chirality centers are identically substituted (*i.e.*, groups on the two chirality centers are the same), this situation is indeed similar with tartaric acid, (see scheme 1.34). The *trans*-diastereomer (1, 2-dimethylcyclopentane) has an enantiomer, but the *cis*-diastereomer has an internal mirror plane of symmetry and it is the *meso*-compound and is achiral. A compound which contains two (or more chirality centers) but is superimposable with its mirror image is a *meso* compound (tartaric acid with two identically substituted chirality centers has also a *meso* compound (see scheme 1.41).

Introduction of the third methyl group to give 1, 2, 3-trimethylcyclopentane introduces interesting stereochemical features. Structures (3 and 4) represent an enantiomeric pair, while all other pairs are diastereomers. The *meso* structures can be distinguished from the C_2, CH_3 group which may be viewed as *cis* or *trans* to the other two methyl groups taken together as a pair. The situation here is the same as in dimethylcyclohexanols (see, scheme 1.67*c*).

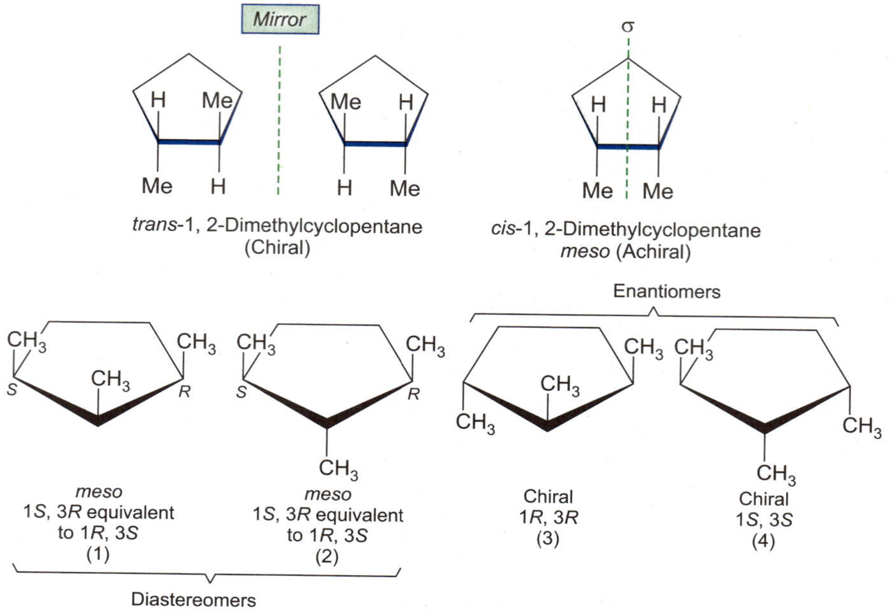

Stereoisomers of 1,2,3-trimethylcyclopentane

SCHEME 1.62

(d) Cyclohexanes (an introduction)

Unlike the essentially planar cyclopentane derivatives *e.g.*, 1, 2-dimethylcyclopentane discussed in (scheme 1.62), examine the formula of 1, 2-dimethylcyclohexane which also has two chirality centers. One may expect 2^2 *i.e.* four stereoisomers. There are indeed four (one can however, isolate only three stereoisomers for further details see chapter 4). The *trans*-1, 2-dimethylcyclohexane like *trans*-1, 2-dimethylcyclopentane exists as a pair of enantiomers. Significantly the *cis*-isomer (scheme 1.63) has also two enantiomers, since when this isomer is drawn in the chair form one can not detect a plane

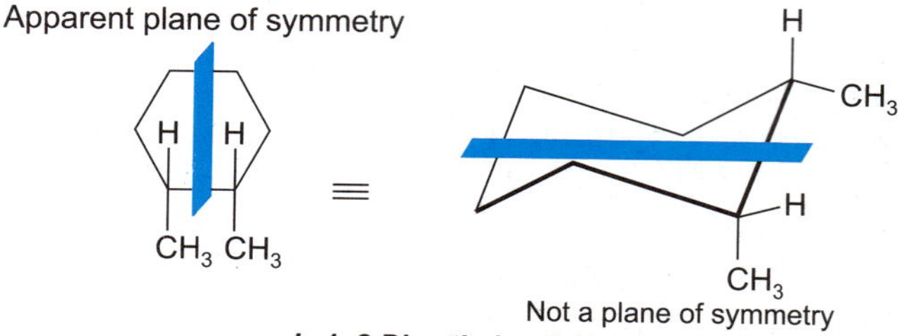

***cis*-1, 2-Dimethylcyclohexane**

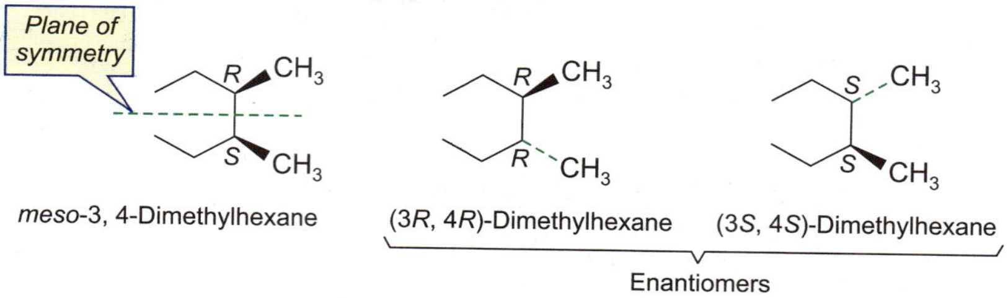

SCHEME 1.63

of symmetry. However, when one draws the *cis* isomer (scheme 1.63) as a hexagon it apparently looks to have a plane of symmetry and thus like *cis*-1, 2-dimethylcyclopentane one may be deceived to conclude initially that it should exist as a *meso* compound. It is thus important to remember that the plane of symmetry must be apparent in the case of a cyclohexane, whether the molecule is drawn as a hexagon or a chair conformation. However, it is not necessary to study the conformations to know if a cyclic compound is optically active. A cyclic compound cannot be optically active if its planar structure has a plane of symmetry (it may be meso or racemic), for further details (see schemes 4.32–4.34).

These stereochemical complications are removed when in *cis*-1, 2-dimethylcyclohexane the ring is cleaved at the C-C bond opposite the one connecting the chirality centers (scheme 1.63). The resulting acyclic compound 3, 4-dimethylhexane, has three stereoisomers a *meso* compound and two enantiomers.

cis and *trans* Isomers are diastereomers

In summary, therefore, the cis, trans or geometric isomers are diastereomers which differ in chemical and physical properties. The structural features leading to geometric isomerism are the presence of a suitably substituted double bond (carbon-carbon, carbon-nitrogen, nitrogen-nitrogen) or the presence of a suitably substituted cyclic structure. The cis/trans notation is used to designate the configuration in carbocyclic structures as well. In addition, chirality of stereocenters has to be evaluated and designated by (R, S) notation. Thus e.g., (Ia, scheme 1.64) is trans-2-methylcyclohexanol, and it is correctly named as (1R, 2R)-2-methylcyclohexanol since its mirror image (enantiomer) (1S, 2S)-2-methyl cyclohexanol is also the trans isomer. The cis-isomer (IIa) is also chiral and has a diastereomeric relationship with the trans isomer, the two enantiomers are (IS, 2R)-2-methyl cyclohexanol and (IR, 2S)-2-methylcyclohexanol (scheme 1.64).

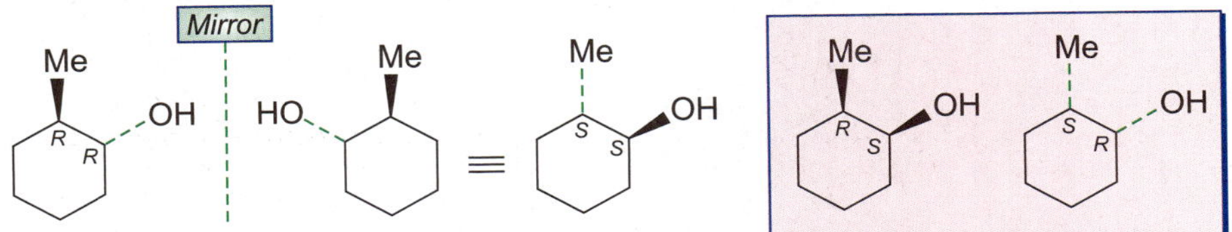

(1*R*, 2*R*)
(Ia)

(1*S*, 2*S*)
(Ib)

trans-2-Methylcyclohexanol

(1*S*, 2*R*)
(IIa)

(1*R*, 2*S*)
(IIb)

cis-2-Methylcyclohexanol

cis and trans-2-methylcyclohexanols are two diastereomers and each in turn gives two enantiomers

SCHEME 1.64

(e) Cyclohexanes as Planar Structures

Cyclohexane rings are chair shaped rather than being flat. Some stereochemical analysis regarding the number of chirality centers, detection of plane of symmetry and whether a molecule would display optical activity or not can be done with drawings using planar rings. Careful analysis of one chair conformation of *cis*-1,2-dimethylcyclohexane (see, scheme 1.63) shows that it is chiral—it is not superimposable on its mirror image. The ring flipped conformation, however, represents the enantiomer of the original conformation (see, scheme 4.33).

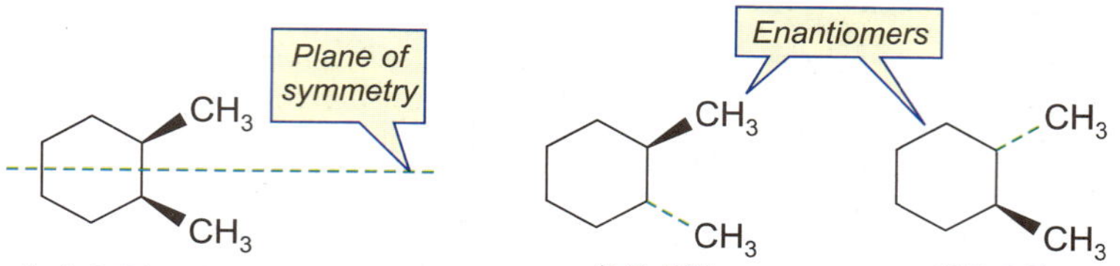

cis-1, 2-Dimethylcyclohexane (1*R*, 2*S*) or (1*S*, 2*R*) (Optically inactive)

(1*R*, 2*R*) (1*S*, 2*S*) *trans*-1, 2-Dimethylcyclohexane (Optically active)

SCHEME 1.65

The conformers of *cis*-1,2-dimethylcyclohexane interconvert rapidly at room temperature, thus the molecule represents a mixture of both conformational enantiomers (a recemic mixture) and is not optically active. By other arguments the same molecule may also be considered as a *meso* compound. The chirality centers are identical (same substituents) and the configurations at two chirality centers are opposite. A consideration of the planar structure of *cis*-1, 2-dimethylcyclohexane (scheme 1.65) shows that it must not be optically active since it has a plane of symmetry. One however, cannot detect a plane of symmetry in the planar ring of *trans*-1,2-dimetylcyclohexane (scheme 1.65), it is thus expected to be chiral and optically active. The following points may be noted:

- A flat (planar) structure of *e.g.*, *cis*-1,2-dimethylcyclohexane can be drawn in two different ways. This depends if one views the conformation from top or bottom. Thus the two structures (I and II, scheme 1.66) are equivalent. One can confirm it—if (I) is turned 180° out of the plane of paper one gets (II). One may note that during this operation a group closer to one's eyes (on thick wedge) goes away (on dotted line) and vice versa. The structures thus remain identical.

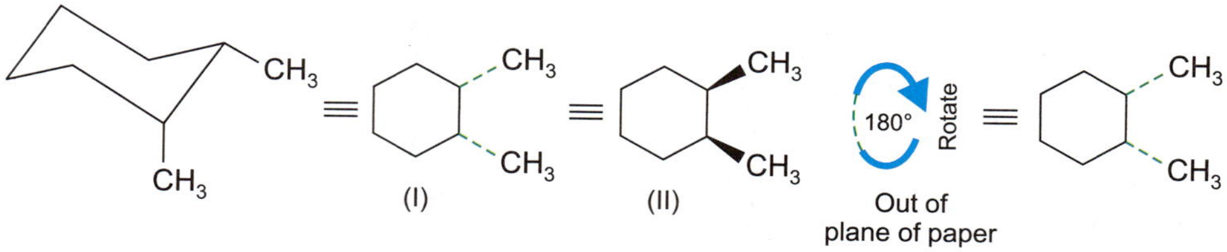

SCHEME 1.66

- If the planar structure of a cyclic compound has a plane of symmetry, the compound cannot be optically active. This is seen in *cis*-1,2-dimethylcyclohexane. In such a situation the planar structure is superimposable on its mirror image (scheme 1.66*a*).

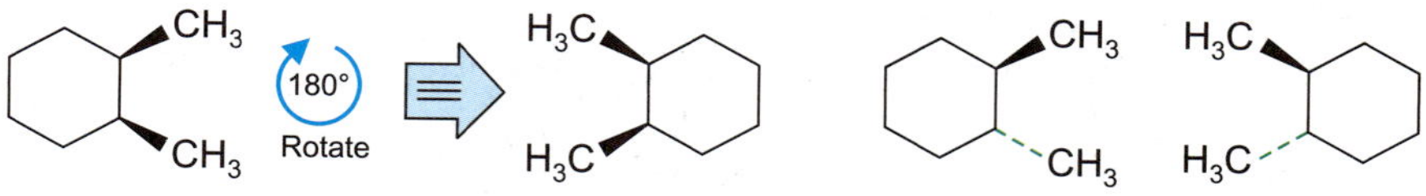

Mirror image (identical) Structures-*cis*-1, 2-dimethylcyclohexane

(I) *trans*-1, 2-Dimethylcyclohexane- not superimposable on its mirror image

SCHEME 1.66a

- A planar structure of *trans*-1,2-dimethylcyclohexane is expected to be optically active (no plane of symmetry) since the planar structure and its mirror image are not superimposable (I, scheme 1.66*a*) (for further details see scheme 4.40).

EXERCISE 1.13

Draw the planar structures of all possible stereoisomers of 1, 3-dimethylcyclohexane and 2, 4-dimethylcyclobutane carboxylic acid. Show their stereochemical relationship by assigning R, S nomenclature to chirality centers.

ANSWER. *It is in scheme 1.67. A chirality center in a meso compound that sits on the mirror plane is chiral when four different groups are attached to it. When two structurally "identical" groups differ in their configuration, the priority rules state that an (R) group has a priority over (S). Thus the carbon atom bearing the COOH group in (I) has the (S) configuration, and the (R) configuration in (II, scheme 1.67). In fact, the carbon atom bearing the COOH group in (I and II) is a pseudochirality center and more appropriately it is designated by s and r respectively (see, schemes 1.31 and 1.67c).*

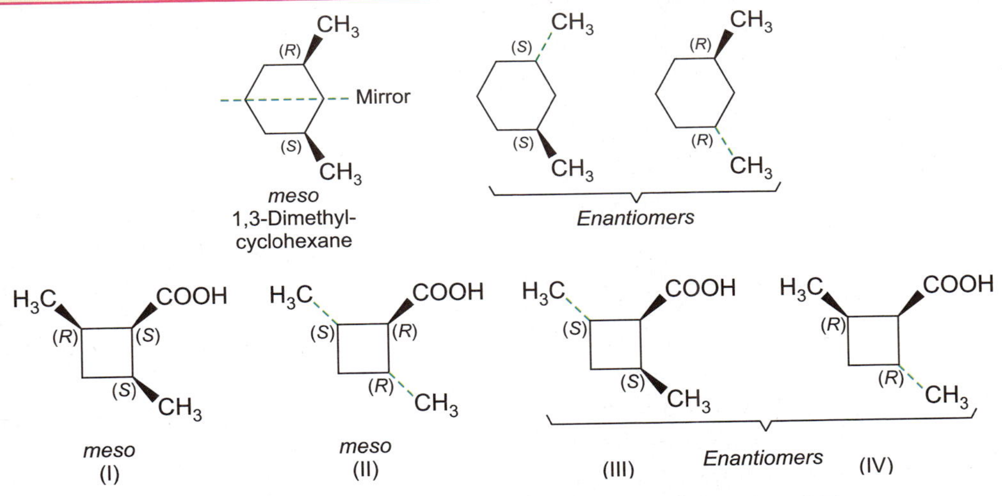

Stereoisomers of 2,4-dimethylcyclobutane carboxylic acid

SCHEME 1.67

Cyclohexanol (like *e.g.*, a monosubstituted cyclopropane) is an achiral compound, it has a plane of symmetry which is perpendicular to the page and that passes through the hydrogen and hydroxy group attached to it, and C4. Note how one side of the ring mirrors the other side. The plane of symmetry is again detectable in both the chair conformations in one of which OH is equatorial while in other it is axial (scheme 1.67*a*). The two chair conformations of a monosubstituted cyclohexane *e.g.*, 1-chlorocyclohexane actually have diastereomeric relationship, since they are stereoisomers that do not have a mirror image relationship. However, the energy barrier for ring flip of one cyclohexane chair conformation to another is of the order of 10 kcal/mole. Thus, at room temperatures, they are in rapid equilibrium. In fact, the situation is analogous to that seen with butane where the two enantiomeric gauche conformations are in rapid equilibrium (see scheme 1.1*c*). In 2-cyclohexene-1-ol (scheme 1.67*a*) one side of the ring no longer mirrors the other, so it no longer has a plane of symmetry. It is a chiral molecule. In this compound C1 is a chirality center the groups being H, OH and two different ring residues.

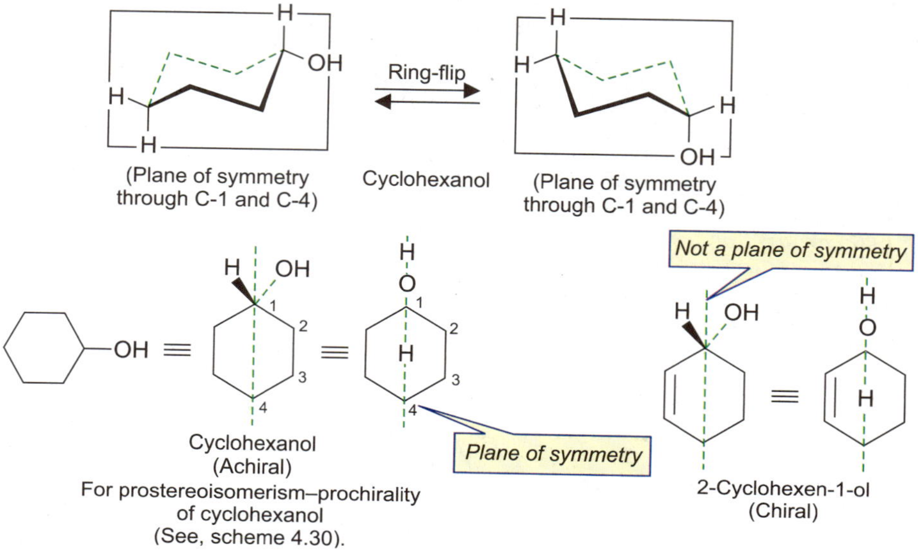

SCHEME 1.67a

(iii) Nomenclature of Poly-Substituted Cycloalkanes–r, c and t Method–Relative Configuration

Cyclic stereoisomers with only two differently substituted carbons are named either *cis* or *trans*. The *cis-trans* nomenclature does not, however, suffice for compounds with more than two substituents. For these compounds, a system is used in which the configuration of each group is given with respect to a reference group designated by *r*, followed by a hyphen, before its locant. The relation of the other substituents relative to the reference substituent is then expressed by adding *c* or *t* (for *cis* or *trans*) before their locant. Thus, compound (III and its enantiomer IIIa scheme 1.67*b*) are *c*–3, *t*-5-dimethylcyclohexan-*r*-1-ol; the alternate direction of numbering would yield *t*-3, *c*-5- dimethylcyclohexan-*r*-1-ol and in these type of cases the IUPAC recommends choosing that numbering which gives a *cis*-attachment at the first point of difference. The same situation is involved in naming compound (IV, scheme 1.67*b*).

Thus, the *cis* or *trans* relationship of the methyl groups to the hydroxyl group (chosen as reference) is specified. However, until recently, there had been ambiguity in naming such compounds. For example (I, scheme 1.67*b*) could unambiguously be given the prefix *cis, cis* but (II, scheme 1.67*b*) might be *trans, trans* or perhaps *cis, trans*. For a chiral compound use of *RS* system to each chirality center would help in assigning names but this system cannot be applied here, in (I scheme 1.67*b*) C1 is pseudochirality center, while in (III and IIIa) it is not even a stereocenter (see, scheme 1.67*c*).

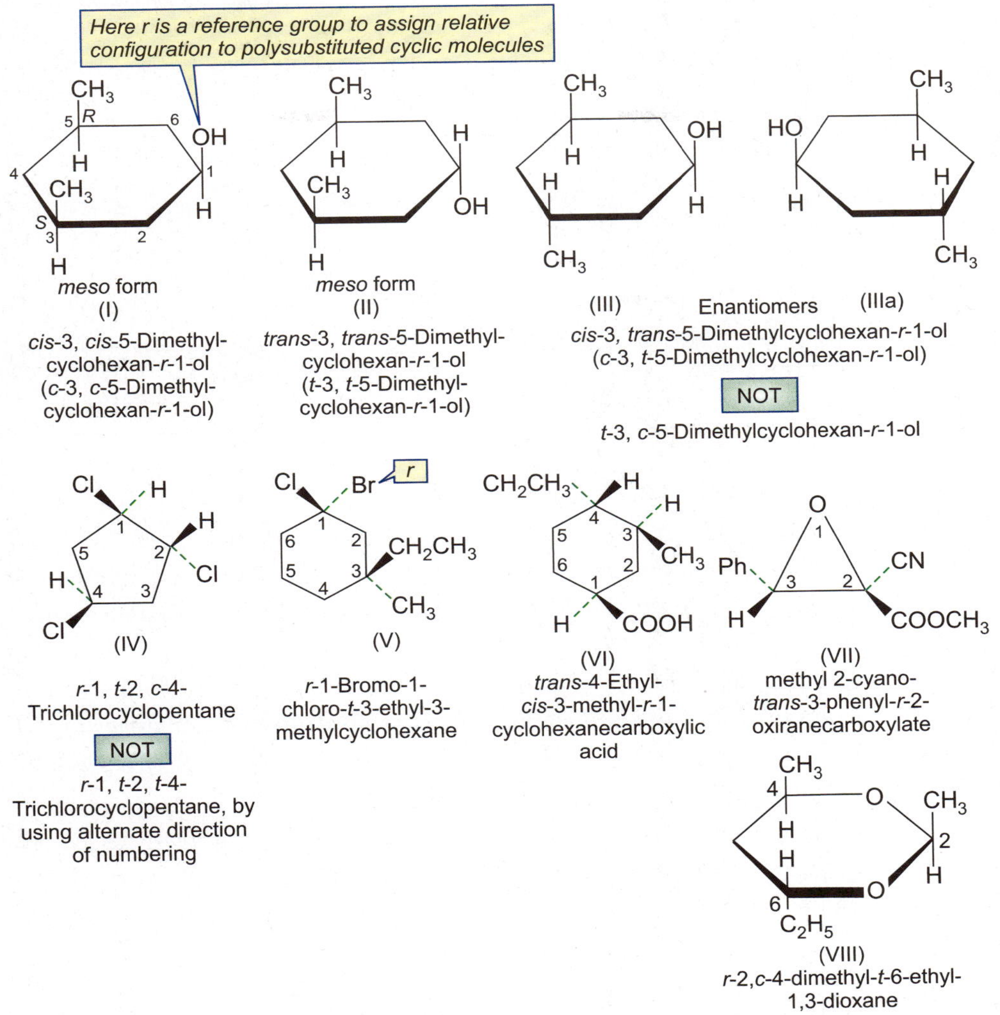

Assignment of the relative configuration to a molecule–the *r*, *c*, *t* method.

SCHEME 1.67b

The *c* also gets preference over *t* when alternative possibility is there (IV, scheme 1.67*b*).When two different substituents are present on each centre (C-1 and C-3) as in V, scheme 1.67*b* sequence, rule is used to choose the reference (*r*) substituent and also the substituent on the other centre, that is, Br > Cl, Et > Me. Another example is of (VIII, scheme 1.67*b*).

More on Stereochemistry of Dimethylcyclohexanols

Dimethylcyclohexanols (scheme 1.67c) exist only as four stereoisomers.

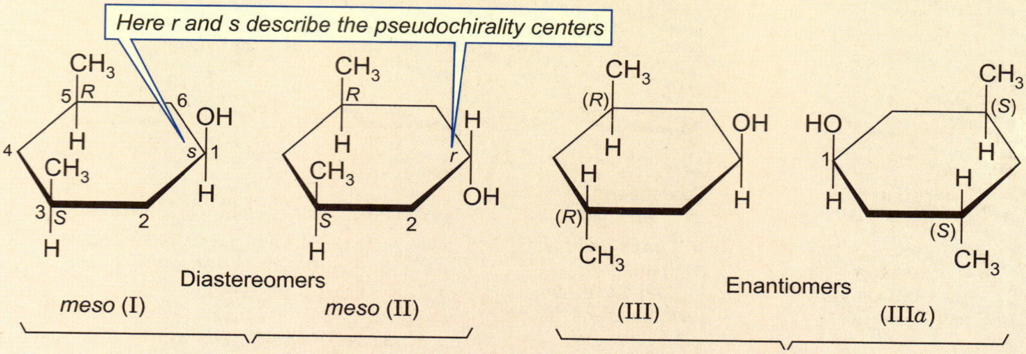

SCHEME 1.67c

The structures (I and II, scheme 1.67c) are meso diastereomers (In I and II C-1 is a stereocenter but not a chirality center) it is a pseudochirality center and can be distinguished from the configurational descriptors used for the pseudochirality centre (also see scheme 1.31).

If one considers e.g., one of the enantiomers (IIIa, scheme 1.67c) both C-3 and C-5 have S configuration, thus no configurational descriptor can be given to C-1. Moreover C-1 is not a stereocenter [C-1 is non-stereogenic, exchange of H and OH (followed by ring flip) regenerates the (IIIa) itself].

To distinguish a pseudochirality center from a chirality center, the stereodescriptors r and s are used. Remember r in combination with c and t is also used to define the relative configuration if three or more substituents are attached to the ring (see, scheme 1.67b).

(iv) Representation of Stereoisomers in Cyclic Compounds

The enantiomer of carvone (I, scheme 1.67*d* also see, scheme 1.2*c*) is its mirror image (II) or (II*a*) and the later is drawn by inverting the stereochemistry in (I) at the stereocenter [one may note that if one rotates (II) by 180° out of the plane of paper one gets (II*a*), the group which was shown as thick wedge goes away from one's eyes and is thus shown as hatched *i.e.*, dotted wedge] and vice versa. A three dimensional figure does not loose its identity on rotation, however, a Fischer projection can only be rotated in specified ways. Thus if an enantiomer of carvone is written as in (II*a*, scheme 1.67*d*), to assign the stereodescriptor, to the chirality center, the compound must be viewed from the side opposite the group of least priority. One then rotates the structure to get the proper orientation (II, scheme 1.67*d*) and the path 1 → 2 → 3 is clockwise to show that the enantiomer is *R*.

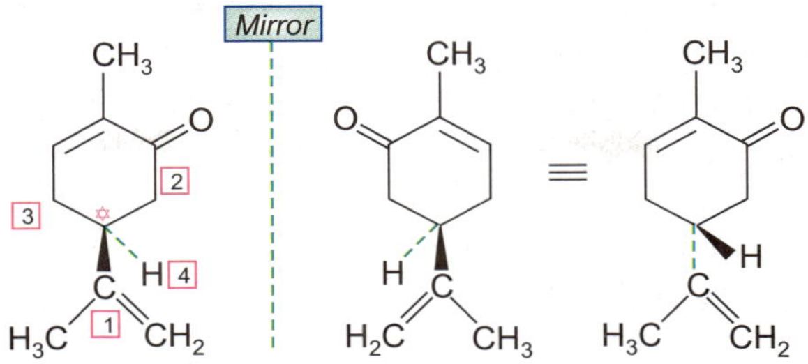

(I)
(*S*)-(+)-Carvone
(Caraway)

(II)
(*R*)-(–)-Carvone
(Spearmint)

(IIa)

SCHEME 1.67d

One has already learnt, that if the group of lowest priority is not directed away from ones eyes, as in (IIa, scheme 1.67d). It may not be necessary to turn the structure. Keep it as such, however apply the R/S system of nomenclature backward i.e., the descriptor thus obtained must be reversed.

(v) The Number of Stereoisomers in Cyclic Compounds

Consider cholesterol a constitutionally unsymmetrical compound, as expected, with eight chirality centers it can exist as $2^8 = 256$ stereoisomers. One only considers tetrahedral stereocenters although a double bond is also present in the molecule. The double bond in cholesterol being in a six-membered ring can only be *cis*, thus it does not contribute to the number of stereoisomers (compare with vitamin A Scheme 1.14*d*). Thus cholesterol (scheme 1.67*e*) has an enantiomeric relationship with one of these stereoisomers and diastereomeric relationship with the rest 254 stereoisomers.

However, the number of stereoisomers in a cyclic compound can be less than expected (due to increased symmetry than their acyclic counterparts) or when some diastereomers cannot form due to steric reasons.

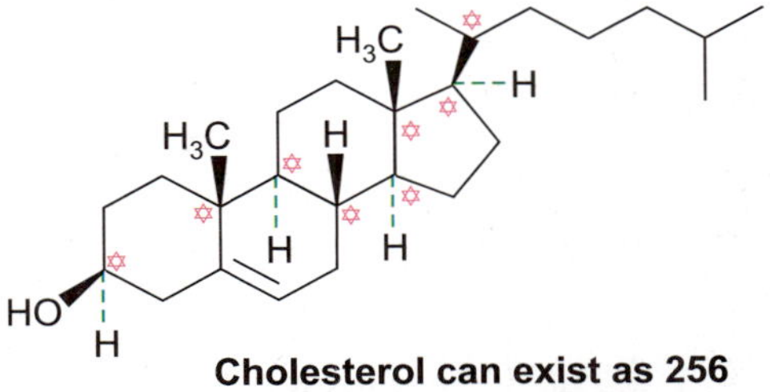

Cholesterol can exist as 256 stereoisomers as expected

SCHEME 1.67e

(*a*) With increased symmetry: It is easy to recognize when a compound *e.g.*, with two chirality centers has a stereoisomer that is meso. *A compound with the same four atoms or groups bonded to two different chirality centers will have three stereoisomers: one will be a meso compound, and the other two will be enantiomers (see tartaric acid, scheme 1.1).*

Consider 1-bromo-2-hydroxycyclobutane. The compound with two chirality centers exists as two diastereomers (*cis* I and *trans* II scheme 1.68*f*) and each has an enantiomer (not drawn here) and thus one has in all expected 4 stereoisomers.

With increased symmetry, 1-bromo-3 hydroxycyclobutane with two non-equivalent stereocenters exists only in two diasteromeric forms (III and IV scheme 1.67*f*). In the case of this molecule both C1 and C3 are therefore, stereocenters but not chirality centers (compare with Fig. 1.5).

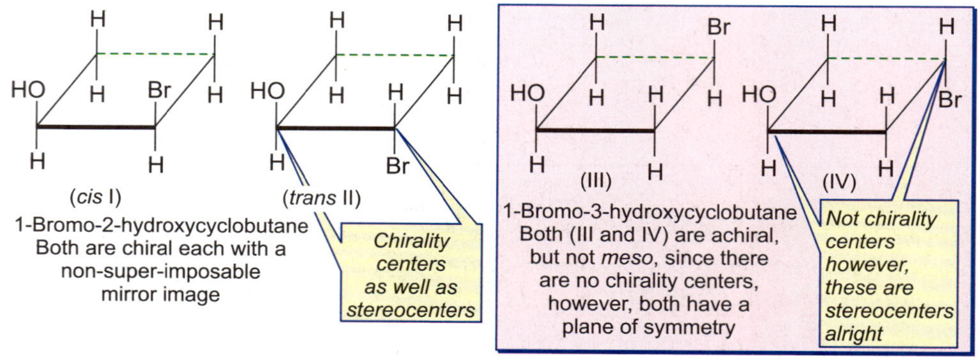

SCHEME 1.67f

Twistane is a highly symmetrical molecule (scheme 1.67*g*) with four equivalent chirality centers, however, it exists only in two enantiomeric forms. (One can easily detect a main C_2 axis in this chiral compound, as one shall see below, the presence of one or more C_n axes does not interfere with the chirality in a molecule). In fact twistane has three mutually perpendicular C_2 axes, the third C_2 axis is perpendicular to the plane of paper. With symmetry the number of stereoisomers fall, thus twistane exists only in two stereoisomeric forms even though there are four chirality centers.

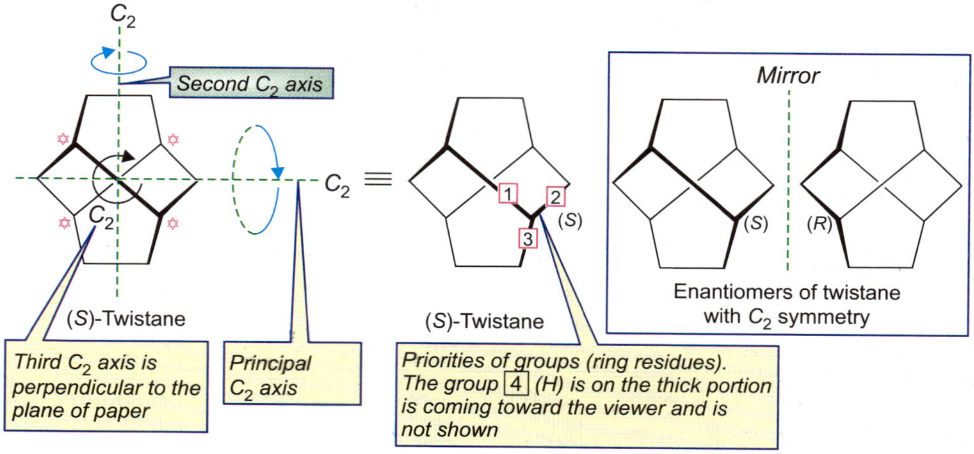

SCHEME. 1.67g

Occurance of boat and twist-boat form in stable molecules

The chair form of cyclohexane is generally the most stable conformer, however, under certain conditions the boat or the twist-boat can be stabilised in preference to the chair. Other factors such as formation of intramolecular H-bond (see scheme 4.47) tend to favour the formation of the flexible conformers. Fixation of chair form in adamantane (see, scheme 4.56), fixation of twist form in twistane (see, scheme 4.23) and boat form in norbornane and camphor (see scheme 4.54) are other examples.

(*b*) When some diastereomers cannot form: Due to steric reasons (configurational restriction), when some diastereomers cannot form the number of stereoisomers is less. In camphor the bridge can only be formed by *cis* fusion, both the bridgehead methyl group and hydrogen must be *cis* to each other and camphor with two chirality centers (two bridge head carbons) exists only as a (±) pair, and these two chirality centers behave as a single element of chirality (scheme 1.67*h*). Similarly norbornane monosubstituted in position 2 (I scheme 1.67*h*) has three chirality centers and again for steric reasons carbons 1 and 4 behave as a single element of chirality and only four stereoisomers exist for such derivatives. These stereoisomers are (+) and (–) *endo* and (+) and (–) *exo*. Similar is the case with α-pinene. (For further details, see scheme 4.54)

The smallest bridged system reported so far in which the bridge is *trans* is the [4.3.1] system in the *trans* ketone (II, scheme 1.67*h*). In this case there are four isomers, since both the *trans* and the *cis* are pairs of enantiomers.

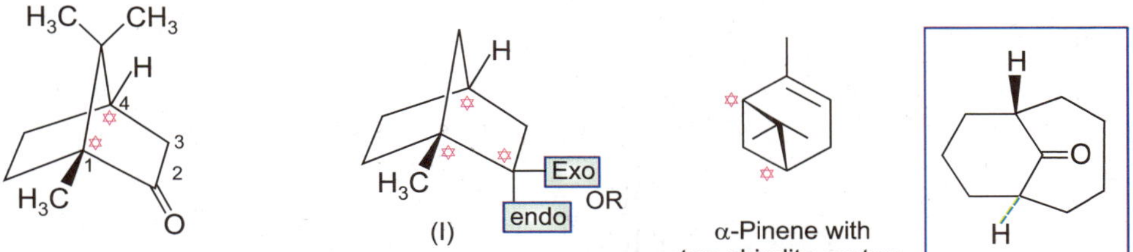

Camphor with two chirality centers exists only as a(±) pair

Norbornnane mono-substituted in position 2 has three chirality centers, but exists only as four stereoisomers

α-Pinene with two chirality centers exists only as a(±) pair

SCHEME 1.67h

1.8 SYMMETRY ELEMENTS, OPERATIONS, POINT GROUPS AND STEREOCHEMICAL PROPERTIES

A molecule which shows optical activity, or can be resolved into optical antipodes, is known as a chiral molecule while one which does not show optical activity is an achiral molecule. A chiral molecule usually, though not always, contains at least one chirality center *e.g.*, stereogenic carbon. Optical activity can also result from lack of molecular symmetry and molecules which do not have chirality center, such as some biphenyls, can also be optically active. Two definitions of chirality may be considered; a molecule is described as chiral if it cannot be superimposed on its mirror reflection, or, alternatively, if it does not possess an alternating axis of symmetry. These definitions complement each other, the first projects a pictorial aspect while the second a mathematical approach.

The first definition can be explored by the use of a set of molecular models and a mirror while the second by the detection of symmetry elements in a molecule and by carrying out symmetry operations on it. A molecule is said to have symmetry elements when certain parts of the molecule can be interchanged with other parts so that after the interchange the appearance of the molecule is indistinguishable from the original. The symmetry operations are the ways of interchanging parts of a molecule.

(A) A Simple or Proper Axis of Symmetry (C_n)

When an imaginary line (axis) can be drawn through a molecule so that rotation by 360°/n gives the molecule indistinguishable from the original, then that molecule is said to have a rotation axis (C_n of order n).

For example, (*E*)-1, 2-dichloroethene has a simple axis of rotation that passes through the midpoint of the molecule and is perpendicular to the plane described by the atoms of the molecule. Rotation through 180° about the axis leads to an arrangement identical to the original. The (*E*)-1, 2-dichloroethene is said to have two-fold axis of rotation. The net result of this C_2 operation is interchanging the position of two hydrogen atoms and two chlorine atoms in the (*E*)-1, 2-dichloroethene molecule. As the hydrogen atoms and chlorine atoms are indistinguishable so the new arrangement is indistinguishable from the old (the H and Cl which were close to one's eyes initially shown by thick wedges go away from you, now shown by hatched wedges. The molecule has also a center of symmetry). If however, one of the hydrogen atoms was replaced by deuterium the molecule would no longer have C_2 axis.

Boron trifluoride has, in a similar way an axis passing through its midpoint, about which a 120° rotation produces an orientation indistinguishable from the original (Scheme 1.68).

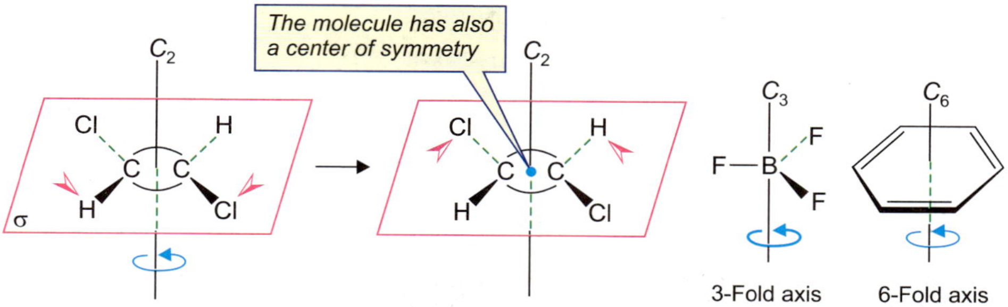

(E)–1, 2–Dichloroethene has a 2-fold axis of symmetry (C_2) and a center of symmetry as well.

SCHEME 1.68

The planar forms of cyclobutane and cyclopentane can be shown to have four-fold and five-fold simple axis of rotation, respectively, and benzene has a six-fold (C_6) simple axis of rotation. All linear molecules have a C_∞ axis as in O=C=O and CH=CH an equivalent arrangement is always obtained whatever be the angle of rotation.

(B) Plane of Symmetry (σ)

A molecule possesses a plane of symmetry if:

(*i*) All the atoms of the molecule are in the same plane. For example, (*E*)-1, 2-dichloroethene possesses a plane of symmetry as well which is the plane of the molecule and includes all the atoms (scheme 1.68). The designation sigma (σ) for planes of symmetry comes from the German word spiegel (meaning mirror).

Thus, all planar molecules posses atleast one plane of symmetry, identical with the molecular plane. Linear molecules have an infinite number of σ planes which intersect along C_{∞}. The planes of symmetry which are perpendicular to the principal axis are termed σ_h (*h* for horizontal) whereas those containing the principal axis are marked σ_v (v for vertical).

(*ii*) A molecule has a plane of symmetry if an imaginary double-sided mirror imagined to be inserted through the molecule reflects both the halves so that the new arrangement is indistinguishable from the original molecule. In other words, if a plane (mirror-plane) can be passed through the molecule, so that it divides the molecule into two symmetrical halves, one half-reflecting the other (scheme 1.69).

The symmetry elements present in *e.g.*, a reactive intermediate help in predicting the stereochemical outcome (scheme 1.69).

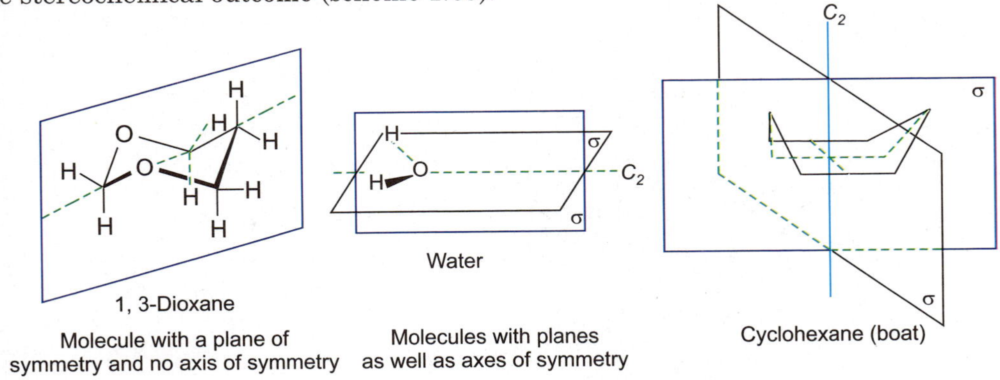

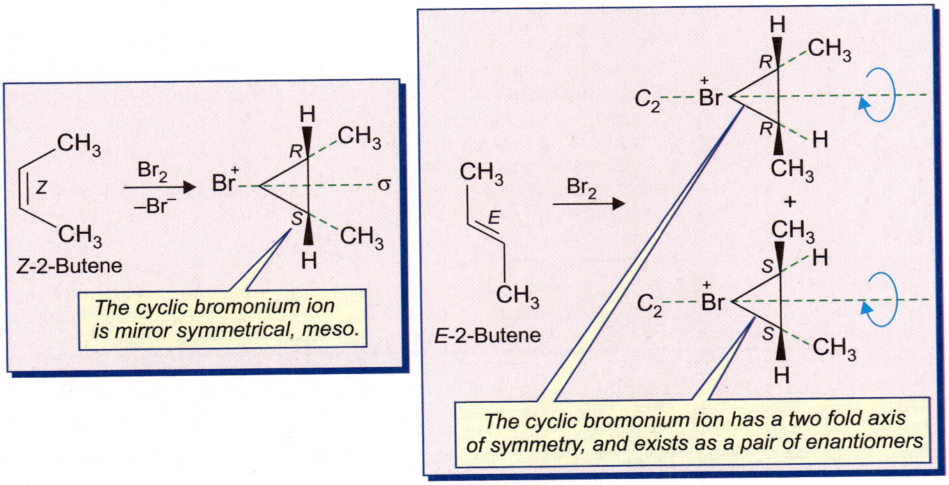

SCHEME 1.69

BOX 1.14

PREDICTION OF STEREOCHEMICAL OUTCOME FROM SYMMETRY ELEMENTS

The addition of bromine to an olefinic π bond is generally anti, stereospecific. Cis-2-butene (Z-2-butene) adds bromine e.g., from the top face (scheme B1.21) to give the intermediate bromonium ion (bridged ion) which is achiral i.e., it has a plane of symmetry along with bromide ion [reaction at the other face of the alkene bottom face not shown, however, would give the same bromonium ion, the two faces of (Z)-2-butene are overall homotopic]. The bromide ion then attacks the bromonium ion at either carbon from bottom (S_N2) type of displacement at equal rates to yield the two enantiomers in equal amounts i.e., as the racemic form.

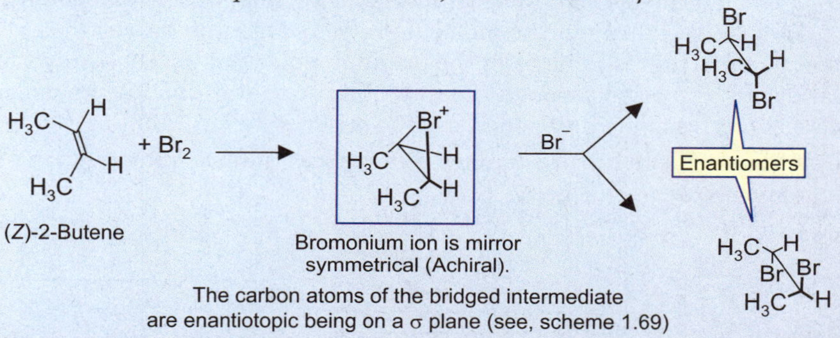

Bromonium ion is mirror symmetrical (Achiral).

The carbon atoms of the bridged intermediate are enantiotopic being on a σ plane (see, scheme 1.69)

*Anti-addition of Br_2 to (Z)-2-butene gives a racemic mixture of (2R, 3R) and (2S-3S), 2, 3-dibromobutane through the **achiral bromonium ion** in which olefin geometry is maintained. Either of the two equal anti approaches of Br^- on the bromonium ion carbons gives enantiomers. The original olefin geometry changes, thus cis-olefin gives (±)-2, 3-dibromobutane.*

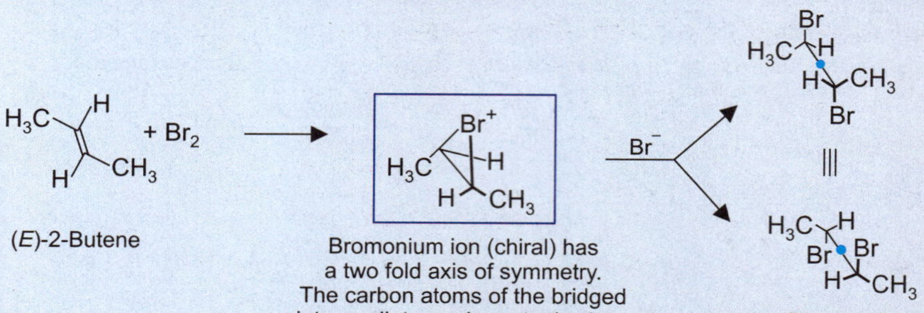

Bromonium ion (chiral) has a two fold axis of symmetry. The carbon atoms of the bridged intermediate are homotopic since these are exchangeable through C_2 axis

meso-Staggered projection with a center of symmetry

*Anti-addition of Br_2 to (E)-2-butene gives meso-2, 3-dibromobutane through the **chiral bromonium ion** (in which the olefin geometry is preserved). Either of the anti-attack by Br^- changes the original olefin geometry. Thus trans-olefin gives a meso product as shown by the presence of center of symmetry.*

According to one common definition a reaction is stereospecific if the reactant can exist as two stereoisomeric forms e.g., (Z)-2 butene and (E)-2-butene and each gives distinctly different stereoisomeric products. Thus reaction of 2-butene with bromine is a stereospecific addition.

SCHEME B 1.21

trans-2-Butene reacts with bromine e.g., from the top face to give a chiral bromonium ion with a C_2 symmetry and bromide ion [reaction at the other face (bottom) would yield the enantiomer of the bromonium ion, this attack has not been shown however, the two faces of trans-2-butene are everall enantiotopic.] When the bromonium ion is opened by S_N2 type displacement at either carbon, the same achiral meso compound is formed (also, see Problems 1.23 and 1.24).

A plane of symmetry is equivalent to a one-fold alternating axis of symmetry (S_1). Some molecules have symmetry planes in addition to simple axes of rotation, while some others have a symmetry plane as their sole symmetry-element (scheme 1.69). Water has two mutually perpendicular σ-planes both containing the C_2 axis and intersecting along it. All planar molecules such as water, must contain at least one symmetry plane which is the molecular plane.

If one considers the boat conformation of cyclohexane (scheme 1.69) one can identify two perpendicular mirror planes of symmetry which intersect on the C_2 symmetry axis.

(C) Center of Symmetry (Point of Symmetry)

A molecule is said to have a center of symmetry if all straight lines that can be drawn through the center of the molecule meet identical atoms at equal distance from the center.

One has seen that in an acyclic compound a plane of symmetry can be best seen in its eclipsed form, when it is written in Fischer, sawhorse, a Newmann or a wedge projection (see scheme 1.41). One can however, detect a center of symmetry (point of symmetry when such a center is present in the given compound) by writing the staggered conformation. The same is true for cyclic compounds as well *e.g.*, in staggered chair conformation of 1,2-dimethyl-cyclohexane one cannot detect a plane of symmetry (see, scheme 1.63), while it is present in the eclipsed boat conformation (see scheme 4.34). Thus *meso*-tartaric acid is shown to have a center of symmetry in its staggered (*anti*) conformation (scheme 1.41). One has already seen a C_2 axis and a plane of symmetry (σ) in (*E*)-1, 2-dichloroethene which also has a center of symmetry (see scheme 1.68). A center of symmetry is equivalent to a two-fold alternating axis of symmetry S_2. The objects with a point of symmetry (center of symmetry) are termed centrosymmetric.

(D) Alternating Axis of Symmetry (Rotation Reflection Axis) S_n

A molecule has an *n*-fold alternating axis of symmetry (S_n) if the molecule when turned 360°/*n* about an axis followed by reflection in a plane perpendicular to that axis brings the molecule in a position indistinguishable from the original.

It turns out that S_1 is the same as plane of symmetry (σ), S_2 is the same as center of symmetry (*i*) and higher descriptors of S_n are rare. Thus 1, 3-dichloro-2, 4-difluorocyclobutane (I, scheme 1.70) has a center of symmetry as its only symmetry element. When (I, scheme 1.70) is rotated 180° about the axis passing though the center of the molecule and then reflected in a mirror perpendicular to this axis, an arrangement superimposable on the original is obtained. The multiplicity *i.e.*, the foldedness of the alternating axis is given by the extent of the rotation in Ist step *i.e.*, 360°/180° = 2; thus (I, scheme 1.70) has a two fold axis (S_2).

BOX 1.15

A useful Mental Exercise with Meso 2, 3-dibromobutane

One can detect a plane of symmetry in the eclipsed Fischer projection (I, drawn with more stereochemical details) but not in its staggered conformation (II, written in the related style which instead has center of symmetry).

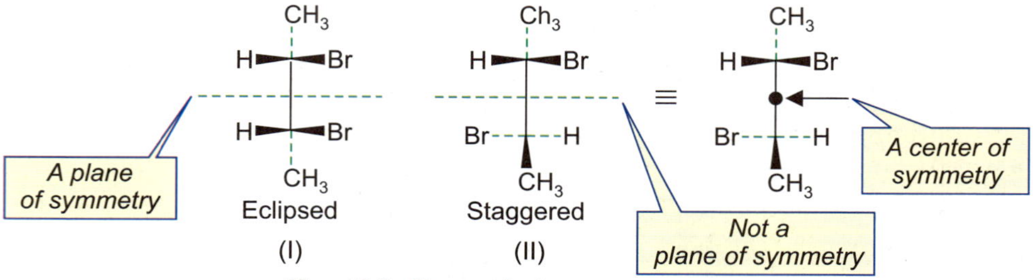

***Meso*-2,3-dibromobutane**

SCHEME B 1.22

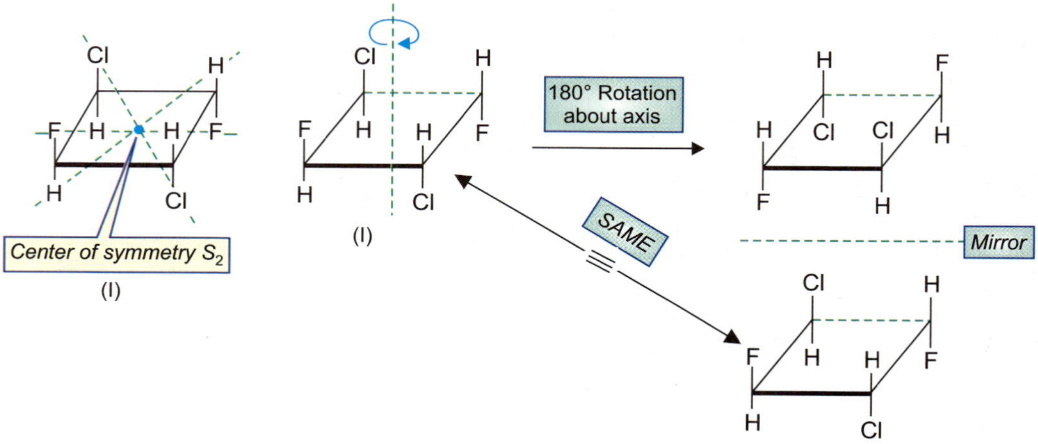

Demonstration of center of symmetry S_2 in (I)

SCHEME 1.70

For detecting an alternating axis of symmetry, the two operations may be reversed and the net result remains the same, as shown for *meso* tartaric acid drawn in sawhorse projection (Ia, scheme 1.71). The following points may be considered:

- In the sawhorse projection drawn in the eclipsed form (I*a*, scheme 1.71, plane of symmetry is clearly visible) *C*3 is rotated to get the staggered form (I*b*, scheme 1.71) where now the center of symmetry is also clearly seen.

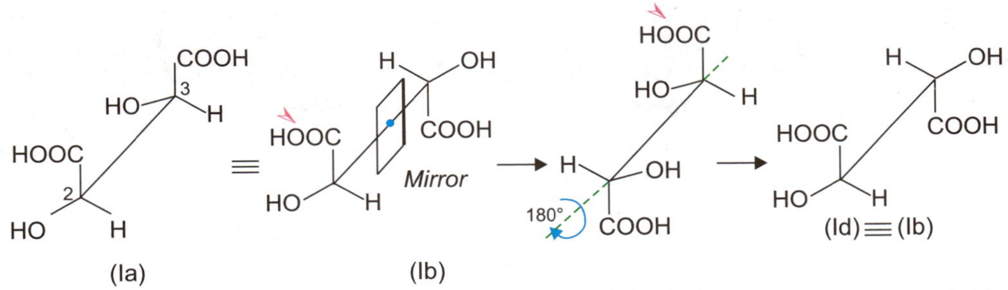

Presence of S_2 axis in *meso*-tartaric acid (drawn as sawhorse staggered conformation)

SCHEME 1.71

- Reflect the conformation (I*b*, scheme 1.71) in a mirror placed at the center of the C2-C3 axis and at right angles to it.
- The orientation (I*c*) is thus obtained. The frontal *C*2 carbon shown by a solid thick arrow in (I*b*) will take a position in the rear shown by partly filled arrow in the reflection.
- When (I*c*, scheme 1.71) is rotated 180° around the axis perpendicular to the mirror an orientation (I*d*) is obtained which is identical (superposable) on the original.
- The presence of an S_2 axis is also seen by drawing a wedge projection of *meso*-tartaric acid (see, scheme 1.77).

Under the S_n operation the equivalent atoms are carried from one side of the reflection plane to the other in an alternating sequence, hence the name alternating axis.

The cyclobutane derivative (I, scheme 1.72) has no plane of symmetry but it is superimposable with its mirror image (II). The molecule however, has a center of symmetry, therefore, it is achiral.

For an S_n operation, a molecule is rotated around an axis by 360°/*n* followed by reflection in a plane perpendicular to the axis. The order of these two operations may be reversed and the same result would be obtained. These reversed operations when applied to (I, scheme 1.72) confirm the presence of S_2 axis in it.

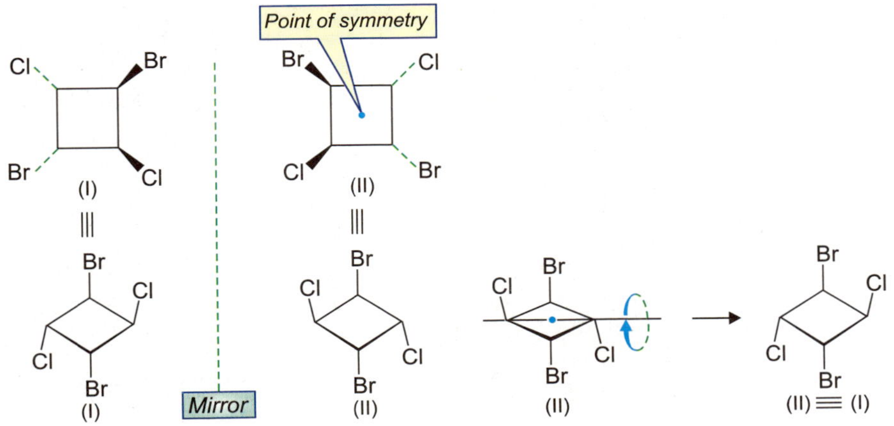

Demonstration of alternating axis of symmetry (S_n)

SCHEME 1.72

1.9 SYMMETRY ELEMENTS, ASYMMETRY, CHIRALITY A COMBINED LOOK

(A) Determination of Chirality

An achiral molecule is identical and can be superimposed upon its mirror image, whereas a chiral molecule is one which cannot be superimposed on its mirror image. The property of chirality can thus be described by constructing the models of two molecules under study and look for the existance of non-superimposable mirror image.

One may describe the chirality in terms of symmetry elements of the molecule under study. When the molecule has a center of symmetry (C_i) or a plane of symmetry (σ) or a n-fold alternating axis of symmetry (S_n) the mirror images of the molecule are superimposable and the molecule is achiral (optically inactive).

Olean (scheme 1.73) is an attracant for the olive fly. The molecule has no plane or center of symmetry and is chiral. One enantiomer of this compound attracts females while the other attracts males. This can be seen from its three dimensional model. The molecule is however, not superimposable on its mirror image for the assignment of configuration (See, page 192).

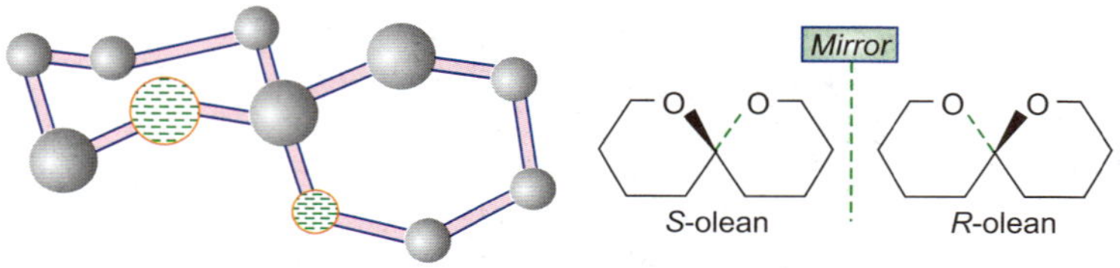

Olean
There is no plane of symmetry
(Chiral)

Two enantiomers of olean

SCHEME 1.73

Care must be used if this approach is adopted; for example, the spiro compound (a, scheme 1.74) possesses neither a center nor a plane of symmetry, but it cannot exist in enantiomeric forms because it possesses a four-fold alternating axis of symmetry (S_4). Indeed (*a*) is superimposable on its mirror image.

To show that (*a*) does contain a four-fold alternating axis of symmetry, an orientation of (*a*, Scheme 1.74) is rotated through 90° about the co-axis of both rings to get (*b*). Reflection of (*b*) through the central plane (*i.e.*, through the *N* atom) perpendicular to this axis yields a molecule identical with (a, scheme 1.74). During the process of reflection the upper half of (*b*) coincides with the lower half of (*a*) and the lower half of (*b*) with the upper half of (*a*).

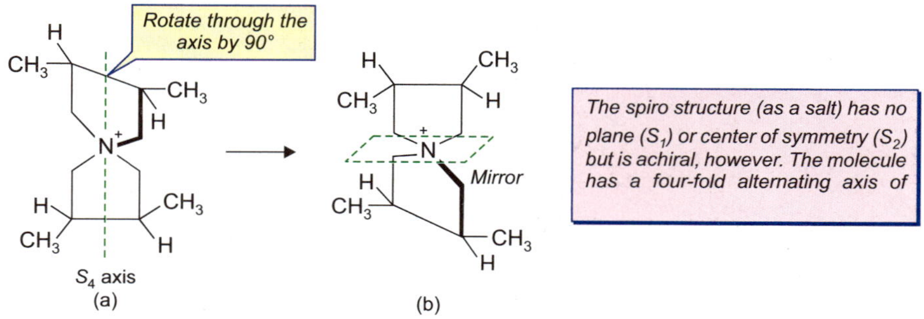

SCHEME 1.74

It is not a necessary condition for chirality that a molecule should have no symmetry elements. A necessary and sufficient condition for a molecule to have a non-superimposable mirror image is that it must not contain a plane, a centre or a *n*-fold alternating axis of symmetry. However, these requirements do not preclude the presence of a simple axis of symmetry in a chiral molecule. The presence of one or more C_n axes in a compound does not interfere with the chirality of a compound.

There are several examples of compounds *e.g.* (I and II, Scheme 1.75) with non-superimposable mirror images; (chiral compounds), in which a symmetry element (C_2) is detectable. The allenes (see, scheme 1.77*d*) provide yet another example of chiral molecules with C_2 axes.

(B) The Terms Chiral, Asymmetric and Dissymmetric

One uses three terms (chiral, asymmetric and dissymmetric) for molecules which show enantiomerism (optical isomerism, *i.e.*, chirality). An asymmetric compound is that (chiral) compound which lacks all elements of symmetry.

All molecules contain an infinite number of C_1 axes i.e., if one rotates anything 360° (360°/1 = 360°) passing through it in any direction, the result is the same arrangement in the same three dimensional orientation which it had originally.

A chiral compound which has one chirality center *i.e.*, a single carbon atom which is attached to four different groups is always asymmetric. Thus bromochlorofluoromethane is an asymmetric compound (see Fig. 1.3) and an asymmetric compound lacks an C_n axis as well.

The term chiral is synonymous with dissymmetric, however, the term chiral is now widely used. When a chiral compound has a simple axis of symmetry (usually a C_2 axis) and no other symmetry element is present the compound is then termed dissymmetric. Thus, the chiral molecules (I and II scheme 1.75) are dissymmetric. The molecule (I scheme 1.75) has no chirality center, it has no plane or center of symmetry.

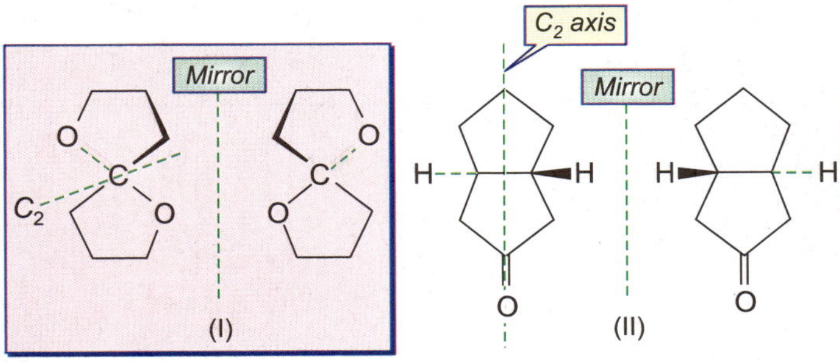

A chiral compound cannot have a plane or center of symmetry, but it can have a simple axis of rotation (this is also so e.g. in (+)-tartaric acid (see scheme 1.76)

The presence of one or more C_n axes in a compound does not interfere with the chirality of a compound.

SCHEME 1.75

Tartaric acid (scheme 1.76, drawn in eclipsed form) may be taken as yet another example to study its chirality in terms of symmetry elements. This acid exists as a pair of enantiomers (I and II scheme 1.76) each of which is diastereomeric with the *meso* (optically inactive) form (III). It is easy to write the three stereoisomers of tartaric acid in a solid and dashed wedge formula (a sort of zig-zag, see scheme 1.38*e*) in eclipsed conformation. Start by writing *meso* isomer (plane of symmetry). The second stereoisomer II is obtained by interchanging the position of groups at one chirality center. The third stereoisomer is the mirror image of (II, scheme 1.76), so invert the stereochemistry at each chirality center in II to get III. One can easily detect a C_2 axis in *e.g.*, the enantiomer (I, scheme 1.76). This molecule is thus a dissymmetric (chiral) molecule. A plane of symmetry is easily detected in the meso-form (III). One can also detect a center of symmetry (C_i) in the staggered *meso* form of tartaric acid (scheme 1.77). That a center of symmetry is equivalent, to a two fold alternating axis of symmetry (S_2) is again proved in (scheme 1.77) by following the usual operations one has already seen the presence of an S_2 axis in *meso*-tartaric acid when drawn in sawhorse projection (see, scheme 1.71).

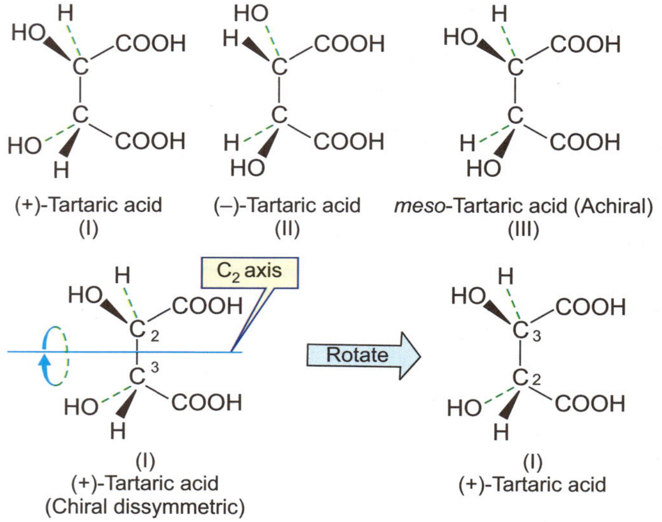

Rotation of (+)-tartaric acid (I) through the axis (as shown) by 180° gives the original, to show that (+)-tartaric acid has a simple two fold axis of symmetry C_2. [During the process of rotation C_3 from bottom in (I) occupies the top position in new arrangement, i.e. all the groups towards ones

SCHEME 1.76

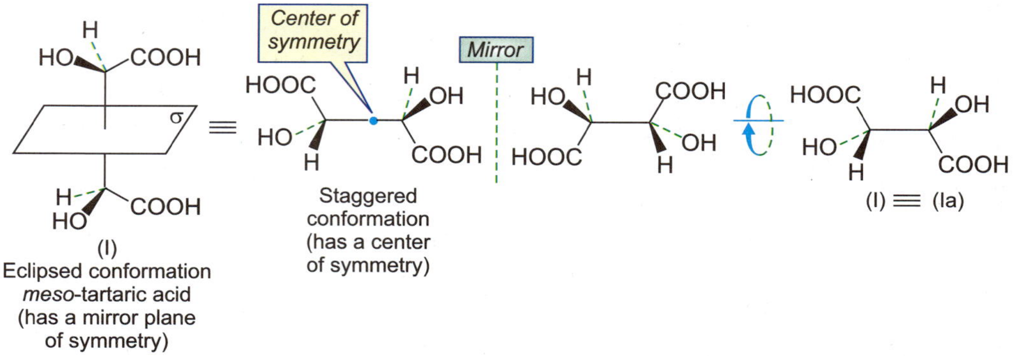

Demonstration of the presence of S_2 axis in *meso* tartaric acid drawn in wedge formula

SCHEME 1.77

(C) Point Group Classification

Point group classification of an organic molecule, a model or an object is based on symmetry classification according to the symmetry elements, axes, planes or alternating axis of symmetry it contains.

A molecule that lacks any symmetry element (other than the infinite number of C_1 axes passing through the center of mass which all molecules possess) is asymmetric. All asymmetric molecules are necessarily chiral. Thus e.g., 1,1,2-trichloropropane is an asymmetric molecule (scheme 1.77b)

A molecule that cannot be superimposed on its mirror image is dissymmetric i.e., chiral. Those molecules have one or more rotational axes of any fold as the only symmetry elements. Thus trans-1,2-dichlorocyclopropane (scheme 1.77a) is chiral and dissymmetric, it has a C_2 axis.

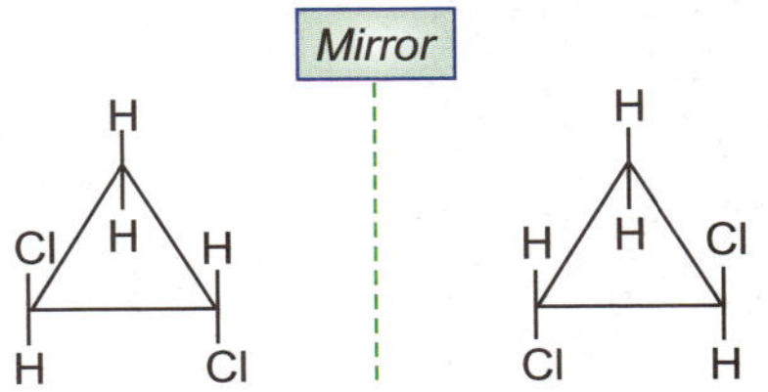

SCHEME 1.77a

(*i*) Chiral Point Groups

- **C_1 Point group.** When no symmetry element is present in a molecule (expect for C_1), the molecule is then asymmetric and belongs to C_1 point group. All organic molecules with one chirality center carbon belong to C_1 group (I and II, scheme 1.77*b*).

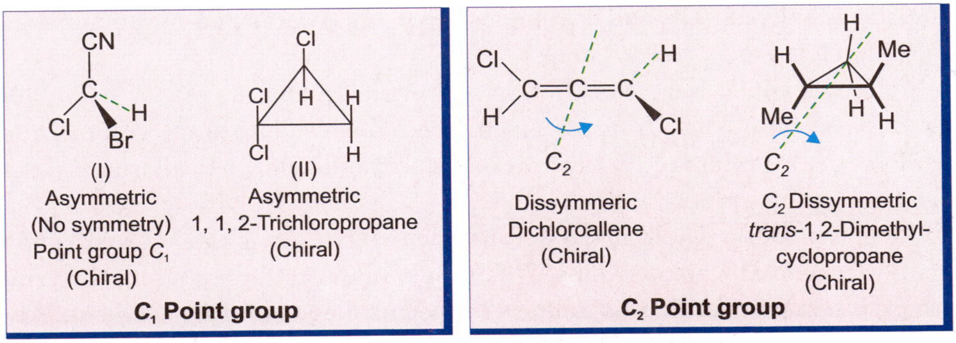

SCHEME 1.77b

- **C_n (n > 1) point group.** The molecules having one rotation axis only build the point group C_n. Several molecules belonging to point group C_2 having one C_2 as the only element of symmetry are common. Such molecules (chiral) are disymmetric but not asymmetric (scheme 1.77*b*). Thus dichloroallene is chiral and belongs to C_2 point group.

- **D_n ($n > 1$) point group (D stands for dihedral symmetry).** In this (chiral) point group in addition to C_n there are nC_2 axes perpendicular to main C_n axis. One may detect-a C_2 axis and two C_2 axes perpendicular to this in the chiral molecule of twist boat and in twistane, thus both belong to D_2 point group (Scheme 1.77*c* also see, scheme 1.67*g*).

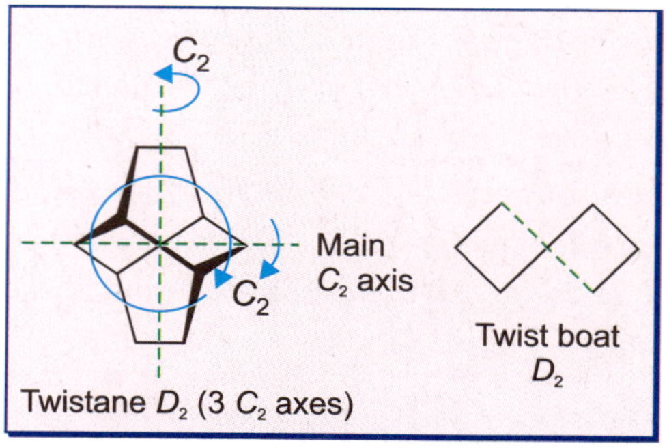

SCHEME 1.77c

(*ii*) Achiral Point Groups

The molecule and its mirror image will be identical when the molecule belongs to one of the achiral point groups. Any point group with one or more of the elements σ, *i* or S_n is achiral. Several example have already been discussed *e.g.*, σ plane is present in *cis*-1,2-dichlorocyclopropane (see, scheme 1.59) point group C_s; center of symmetry *i* is present in (I, scheme 1.72) point *group* C_i; S_n rotation, reflection axis S_4 is present in spiro structure (scheme 1.74) point group S_4.

In several cases (scheme 1.77*d*) achiral point groups may be built out of combination of both σ planes and C_n axes. With one C_n and $n\sigma$ planes intercepting at C_n the point groups are C_{nv} and the planes are σ_v *e.g.*, water and pyridine belongs to C_{2v} and chloroform to C_{3v} point group.

Molecules which have one C_n axis and one σ_h plane (σ_v planes being absent) belong to the group C_{nh} *e.g.*, *trans*, 1-2-dichloroethylene belong to C_{2h} group. Thus (*E*)-1,2-dichloroethene (scheme 1.68) with a C_2 axis perpendicular to the horizontal σ plane belongs to C_{2h} point group.

When one adds σ planes to the axes present in D_n (C_n axis and nC_2 axes) one leads to D_{nd} and D_{nd} and D_{nh} point groups. When the σ planes intersect along the principal axis, these bisect pairs of C_2 axis and are termed as diagonal in D_{nd} , and an example is of allene (scheme 1.77*d*). In case there is a horizontal plane which is accompanied by *n* vertical σ planes which include rather than bisect the *n* horizontal C_2 axes, then point group is called D_{nh}.

Thus benzene belongs to D_{6h} point group since it has one C_6, six C_2, six σ_v and one σ_h, similarly, acetylene has $D_{\infty h}$ symmetry *i.e.*, cylindrical symmetry.

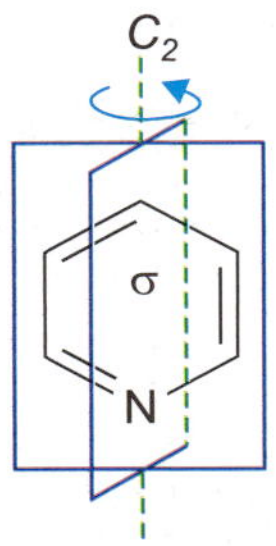

C_{2V} point group. The C_2 axis lies along the intersection of the 2 vertical σ planes.

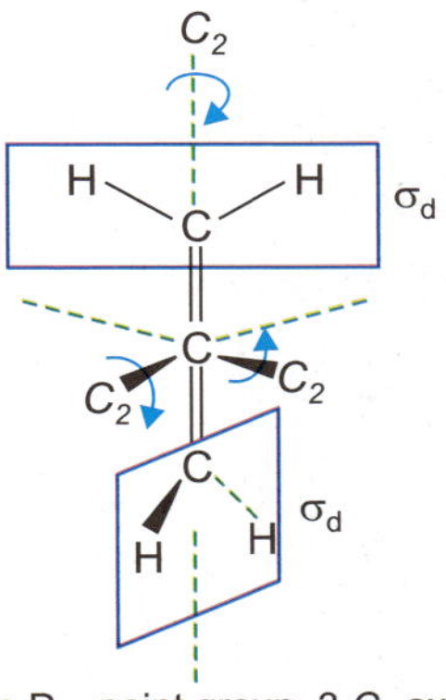

Allene D_{2d} point group. 3 C_2 axes with the 2 σ planes intersecting the principle (vertical) C_2 axis.

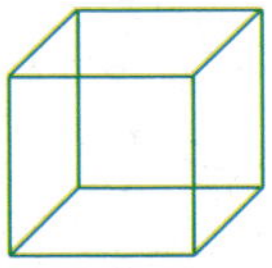

Cubane

O_h point-group. 3 C_4, 4 C_3 and 6 C_2 axes and 9 σ planes

SCHEME 1.77d

Lastly mention may be made of the point groups T_d, which is applicable to regular tetrahedral molecules like, CH_4, CCl_4 etc. and O_h, octahedral point group to which the well known compound cubane belongs (scheme 1.77*d*). The final high symmetry point group is K_h which is applicable to objects having all symmetry elements. Molecules, however, cannot have K_h symmetry and this point group is applicable only to single isolated atoms.

1.10 RESOLUTION OF RACEMIC MIXTURES

It is important not to confuse *meso* compound with racemates both of which do not show optical activity. A *meso* compound is a single achiral substance while racemate is a 50:50 combination of a pair of enantiomers with the mixture having zero optical rotation.

Enantiomers have similar physical and chemical properties in achiral environment while diastereomers display different physical and chemical behaviour even in achiral environments. In a pair of enantiomers the individual components are mirror-images, all the interatomic distances in both molecules remain the same. Physical and chemical properties depend on the atoms in a molecule and their interatomic distances. A pair of diastereomers no doubt contain identical atoms, and moreover each atom is joined to the same atoms in each molecule however, there is a difference in arrangement of the atoms in space. Consequently, diastereomers differ in energy content, and they, therefore, show different chemical and physical properties. The energy difference between a pair of diastereomers therefore, provides the basis of a method for separating pairs of optical isomers (enantiomers) which are the constituents of a racemic mixture.

The term resolution is used to describe the separation of a racemic mixture into its enantiomeric constituents. One can obtain an optically active compound from a racemic mixture using two methods:

(*a*) via resolution of individual enantiomers from a racemic mixture and

(*b*) by carrying out an asymmetric synthesis (stereoselective synthesis) using a chiral reagent or a catalyst.

There are examples of molecules like drugs (See page 25) where the enantiomers have distinctly different effects. Thus, resolution of racemic mixtures is one of the major areas of research today.

The individual enantiomers have identical physical and chemical properties and can only be distinguished by a chiral environment. Plane polarized light is such a chiral environment and one enantiomer is dextrorotatory and one is levorotatory. The diastereomers are however, different substances and have different properties. The resolution involves the conversion of a pair of enantiomers of a racemic mixture into a pair of diastereomers with an optically active reagent. The diastereomers are then separated by making use of their different properties *i.e.*, crystallization, solubility, chromatography etc.

The optically active reagents used for resolution are mostly obtained from biological sources where these occur as only one enantiomer. Thus, though brucine has six chirality centers and can theoretically exist as $2^6 = 64$ stereoisomers nature only makes one enantiomer (–)-brucine. Instances are there when more than one stereoisomer is found but these rarely exist together in the same biological system. Fortunately nature does not waste its precious energy and resources unnecessarily, only one of the stereoisomers is produced and utilized by any given organism. The chiral substances react only with substances that match their own chirality (two enantiomers of carvone have different odors, see scheme 1.67d).

Each diastereomer is then treated separately with a second reagent to regenerate the resolving reagent and the pure enantiomer.

The following methods can be employed for the resolution of racemic mixtures.

(A) Mechanical Separation—Crystallization Method—Resolution via Conglomerates

Direct crystallization of an optically active compound from a racemate may be achieved provided the crystals of the latter form as conglomerates.

A conglomerate is an exactly 50 : 50 mixture of crystals of pure (+)-enantiomer with crystals of the pure (–)-enantiomer. When crystallization of a racemate leads to the formation of a conglomerate, then by definition one has achieved spontaneous resolution.

In very rare cases the two enantiomers of a racemic mixture (racemate) will crystallize separately. The racemic sodium, ammonium salt of tartaric acid crystallized by slow evaporation of an aqueous solution is a conglomerate and Pasteur could pick out the crystals of (+)- and (–)-enantiomers, since the crystals have enantiomeric relationship (scheme 1.78).

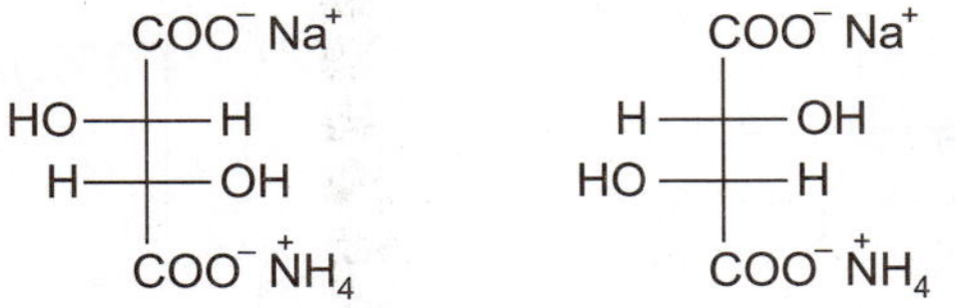

Sodium ammonium tartrate *Left-handed crystals* — Sodium ammonium tartrate *Right-handed crystals*

Rarely, (as in the case of sodium ammonium tartrate), the two enantiomers of a racemic mixture may crystallize separately and then separated mechanically.

SCHEME 1.78

Pasteur's work was a unique combination of sensitive judgement, a firm hold on the problem, good luck and hard work. Pasteur's salt was a conglomerate only at temperatures below 25°C (work carried out in Paris).

(B) Resolution Through Preferential Crystallization

Biologically active α-methyl-L-dopa *e.g.*, is obtained in optically active form by circulating a supersaturated solution of the racemate through two chambers, each of which contains seed crystals of the respective enantiomers. One may also bring about crystallization of one of the enantiomers of a racemate while the other is kept in a supersaturated state, if the crystal of one form is not available (scheme 1.78*a*). This later method has been used for the resolution of chloramphenicol an antifungal compound. It is sometimes possible to induce selective crystallization by seeding with a crystal of an optically active form of another molecule.

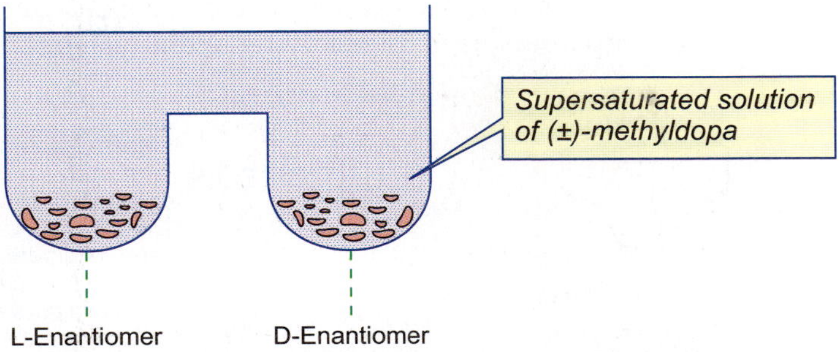

Diagrammatic scheme for the resolution of racemic methyldopa

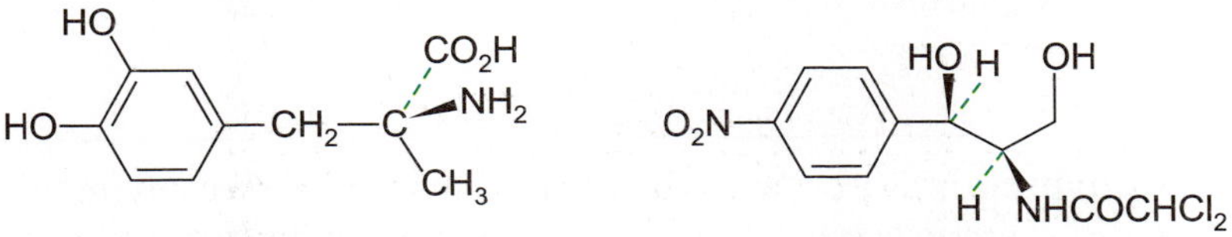

α-Methyl-L-dopa — Chloramphenicol

SCHEME 1.78a

(C) Resolution Through the Formation of Diastereomers

A racemic mixture of an alcohol *e.g.*, 2-butanol reacts with pure (+)-tartaric acid (the resolving agent as a naturally occurring pure enantiomer) to give diastereomeric esters (scheme 1.79). These esters are separated by chromatography. The resolving agent is then cleaved from the

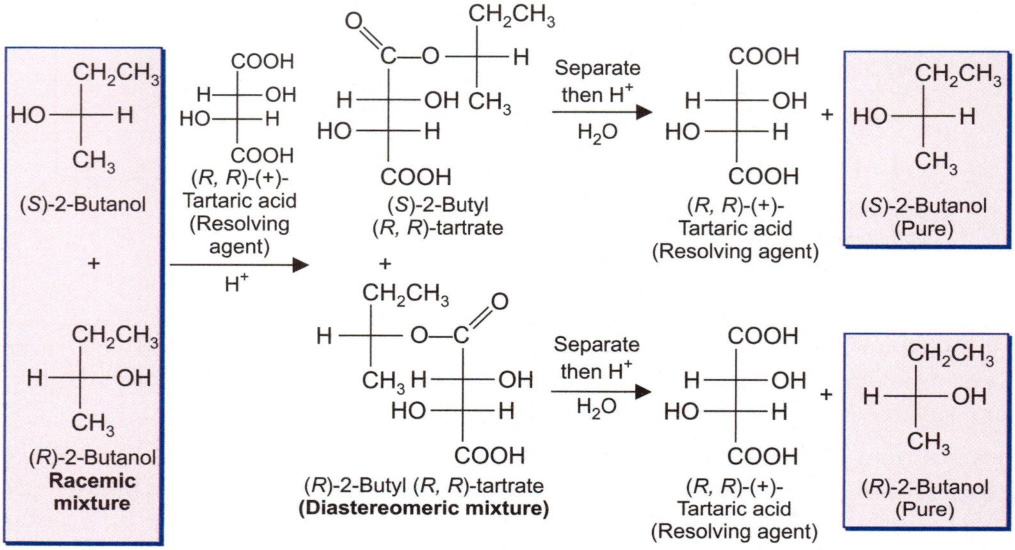

SCHEME 1.79

separated enantiomers of 2-butanol by hydrolysis to give (*R*)-2-butanol and (*S*)-2-butanol in their pure form. The recovered (+)-tartaric acid would probably be thrown away since it is cheap (available from any winery) and nontoxic.

Organic acids (*e.g.*, tartaric acid) and bases (*e.g.*, brucine) are important groups of naturally occurring compounds used as resolving agent (scheme 1.80).

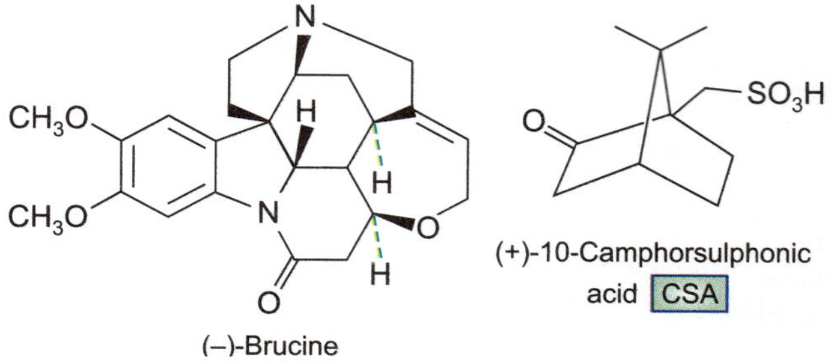

Most resolving reagents are naturally occurring. Nature makes only one enantiomer, (even though several stereocenters may be present). Racemic mixture on reaction with this pure enantiomer gives a mixture of separable diastereomers.

SCHEME 1.80

The most important criterion is the proper choice of the resolving agent. Thus a racemic mixture of organic acids is often separated into its pure enantiomers by using the naturally occurring (optically active—one enantiomer) alkaloid brucine (a base). The diastereomeric mixture of salts formed initially is separated by crystallization from which pure enantiomeric acids are separated by treatment with HCl (scheme 1.81).

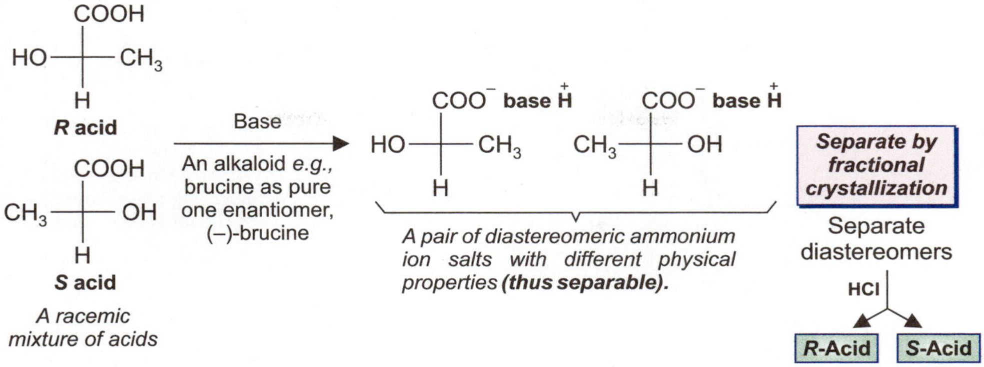

SCHEME 1.81

A racemic alcohol (neutral compound) may also be resolved by first attaching an "acidic" handle. The alcohol is first converted into; a half ester of phthalic acid by heating with phthalic anhydride. The half ester is then typically resolved as an acid by using e.g., the optically active alkaloid brucine (scheme 1.82). The optically active enantiomers of half-acid-half ester are then hydrolyzed under basic conditions to give two optically active enantiomers of the alcohol.

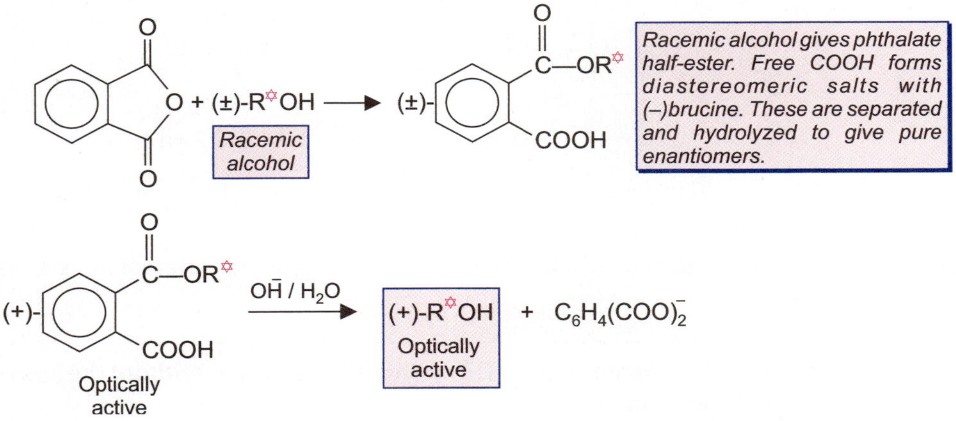

SCHEME 1.82

(D) Resolution Through the Formation of Molecular Complexes

Instead of forming stable salts or covalent compounds between the substrates (a racemic mixture) and the resolving reagents, it has now become more and more common to use weak interactions (chromatography, complexes) as a means of resolving enantiomers. In some cases it is possible to make molecular complexes. These complexes should be easy to form and easy to decompose. For example the optically active fluorenone derivative (I, scheme 1.83) TAPA has been used to resolve hexahelicene. (–)-TAPA and (–)-hexahelicene crystallize whereas the diastereomeric π-complex of (+)-hexahelicene remains in solution.

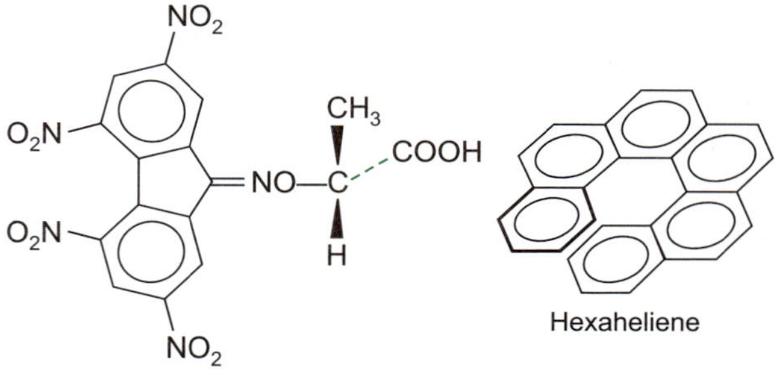

(I)
TAPA

Diastereomeric, charge transfer π complexes between the optically active π acid TAPA [α (2,4,5,7-tetranitro-9 fluorenylideneaminooxy)] propionic acid and chiral Lewis bases (e.g., hexahelicene) are often strong to allow resolution.

SCHEME 1.83

Compounds in which conventional functionality is absent, like chiral alkenes, arenes, sulfoxides and phosphines can be resolved by the use of a metal centered complex to give diastereomeric derivatives with *e.g.*, a recemic alkene like *trans*-cyclooctene. The coordination of the twisted carbon-carbon double bond of *trans*-cyclooctene is more stronger than ethene with platinum in the square planar complex (I, scheme 1. 83a) containing a chiral unit S-1-phenylethyl amino moiety. After the diastereomeric complexes are separated, their treatement with KCN is attendend with the displacement of a pure enantiomer (from *e.g.*, II scheme 1.83*a*) by CN^-.

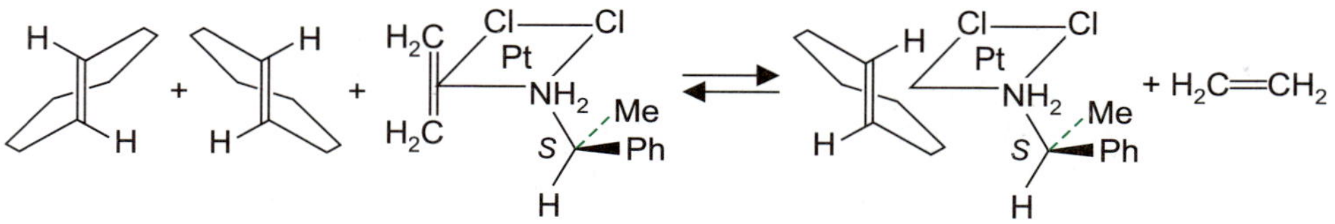

(I) (II)

Resolution of *trans*-cyclooctene via a platinum complex (only one diastereomer is shown)

SCHEME 1.83a

(E) Resolution by Chiral Chromatography (Differential Absorption) without the Necessity of Isolating the Diastereomers

Absorption and Adsorption

Absorption is the intimate mixing of the atoms or molecules of one phase with the atoms or molecules of a second phase, while adsorption is a process in which molecules (either gas or liquid) adhere to the surface of a solid.

Component enantiomers of the racemic mixture can be separated by chromatography on an optically active support *e.g.* silica in which surface hydroxyl groups have been derivatized by treatment with (*R*)-phenylglycine. In such a technique covalent chemical bonds are not formed. Advantage is taken of the formation of partial bonds—complexes— as the pair of enantiomers passes over an optically active support. The optically active support which is also called chiral stationary phase (X*) forms complexes with the enantiomers A and A′ These complexes are diastereomers with different physical properties, including bond strengths of A ... X and A′...X. Both AX and A′X are in equilibrium with the free enantiomers, and these equilibria is different for the two diastereomeric complexes. Thus A and A′ will move through the column at different rates and elute at different times.

Thus the enantiomers are separated by their interaction with the enantiomerically pure chiral stationary phase by the temporary formation of diastereomers.

Troger's base [for structure see (Scheme 1.8*b*)] cannot be resolved by using an enantiomer of an acid as a resolving agent since it is decomposed in contact with acids, it was resolved by chromatography on lactose powder. Several widely used chiral adsorbents (chiral stationary phases) are derivetized natural polymers *e.g.*, of cellulose (scheme 1.83*b*) and other fully synthetic chiral polymers.

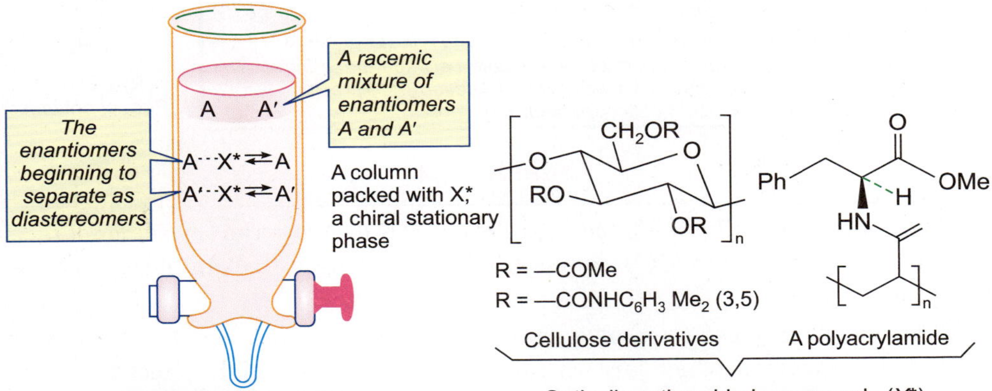

Separation of enantiomers through chromatography on a chiral stationary phase *X**.

Optically active chiral compounds (*X**) used as chiral stationary phases in their enantiomerically pure forms.

SCHEME 1.83b

One may like to learn here an example of asymmetric induction as well (this asymmetric induction (Scheme 1.83c) is infact one of the procedures to carry out asymmetric synthesis detailed in chapter 2. Thus achiral (I, Scheme 1.83c) on reaction with optically active (II, N-benzyl-3-(hydroxymethyl)-4-methyl-1, 4-dihydropyridine] gives (S)(+)-mandelic acid (97.5%) while (R) (–)-mandelic acid is produced to the extent of only 2.5%. In this process where a chiral reagent (II) has given up its chirality to another (i.e., I) while (III) becomes achiral is termed self immolative.

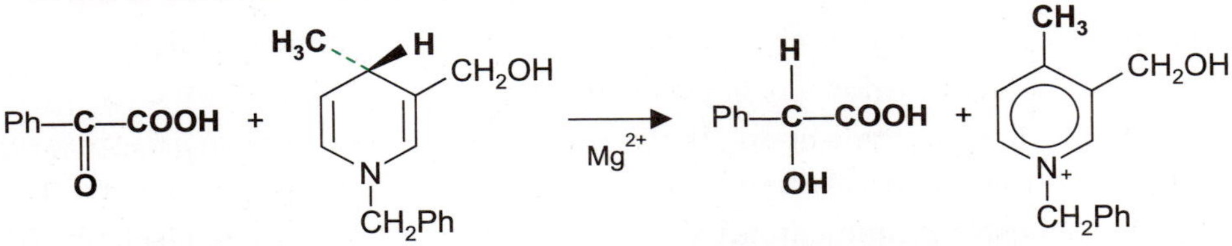

(I) (II) (*S*)-(+)- *S*-(+)-Mandelic acid (III)

SCHEME 1.83c

(F) Biochemical Methods of Resolution – Use of Enzymes

The laboratory reduction of 2-butanone (an achiral substrate) gives only a racemic product of 2-butanol. However, the reduction with an enzyme (a chiral reagent) gives an optically pure (*R*)-2-butanol (scheme 1.84, this is a typical example of an asymmetric synthesis, also See scheme 2.60*a*).

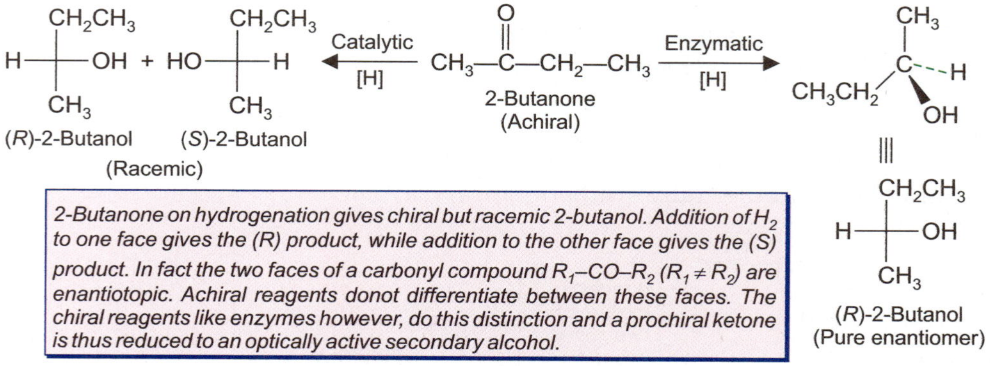

SCHEME 1.84

The mould *Penicillium glaucum*, preferentially destroys the (+) isomer of racemic ammonium tartrate while the (–) isomer is left intact. This is an example of asymmetric destruction.

The use of enzymes to resolve enantiomers has become very popular only very recently. The enzymes which are chiral materials display high selectivity for one enantiomer of the racemic mixture. Enzymes are soluble in aqueous solution, thus it is practically very difficult to have reasonable concentration of organic substrates in the aqueous medium to achieve resolution at good rate. However, a variety of esterases lipases function well in organic solvents and thus the major stumbling block is removed.

It is easy to acetylate a racemic alcohol and treat the racemic esters with a lipase. One enantiomer is hydrolyzed to the alcohol while the other remains as an ester. These are then separated by chromatography. This recent method of resolution of a racemic alcohol is gaining importance and is far superior to the technique discussed under scheme 1.82.

Role of the enzyme esterase to resolve the non-steroidal pain reliever naproxen into its enantiomers (*S*-naproxen is biologically active) is interesting. The following points may be noted:

- The ethyl esters of naproxen crystallize in two enatiomeric forms.
- Both (*R*)- and (*S*)-form of crystals are insolube in water.
- Both form of crystals in an alkaline solution are treated with an esterase. The (*S*)-ester is selectively hydrolyzed and goes into solution as the sodium salt.
- The (*R*)-ester remains unaffected and filtered (*i.e.*, separated) (scheme 1.84*a*).
- The (*R*)-ester is racemized (via the enol, the compound has an acidic α-hydrogen) and treated again with esterase.

The biochemical methods named **kinetic resolution** have found important application to resolve (±)-amino acids. A (±)-amino acid is acylated and acylated racemic mixture is treated with an enzyme "acylase 1" (hog kidney acylase). This enzyme is capable of hydrolyzing amide links (– NHAc) of L-amino acids only (stereospecificity) (scheme 1.85). Thus at the end a free L-enantiomer with the D-enantiomer which is still acetylated is obtained. This mixture is now easily separated by usual methods. Because the resolution of the enantiomers depends on the difference in the rate of reaction of the enzyme with the two *N*-acetylated compounds, the method is called kinetic resolution. Another example of kinetic resolution of chiral allylic alcohols is with sharpless reagent (see, scheme 2.52).

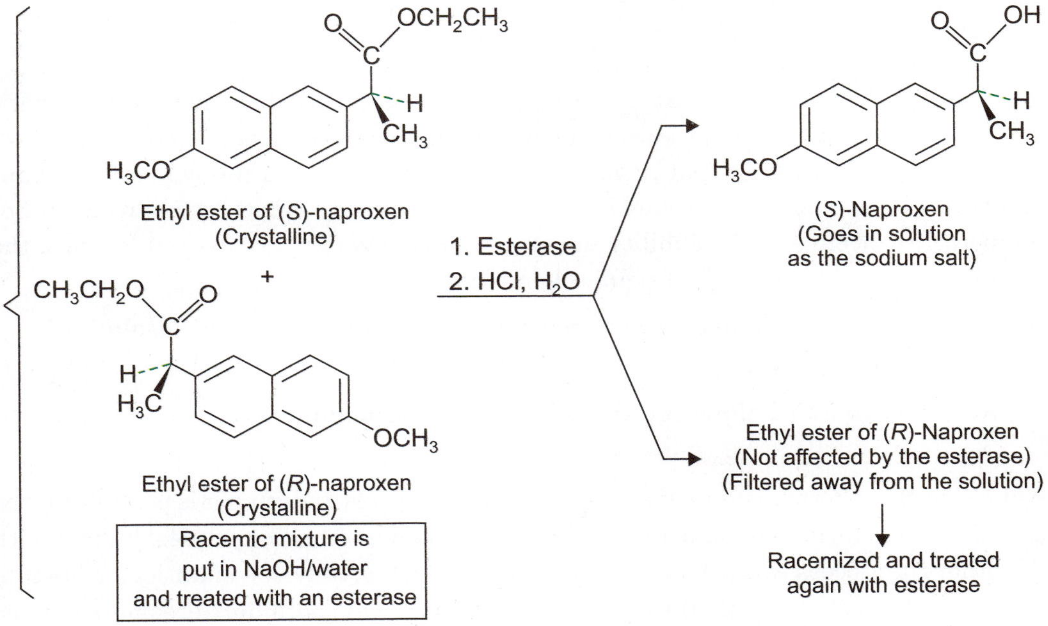

SCHEME 1.84a

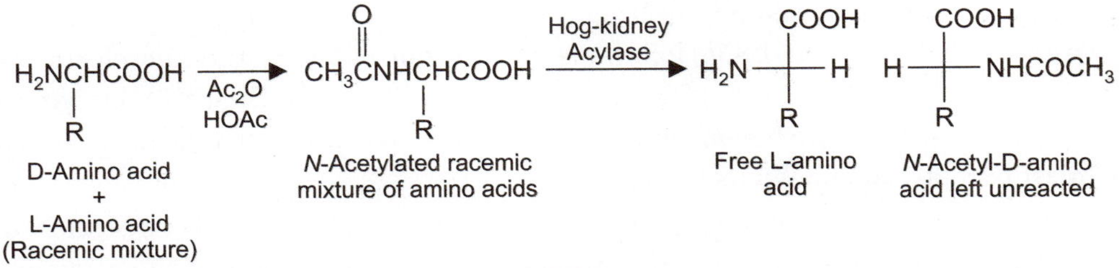

SCHEME 1.85

Another example is the reaction of racemic acetylphenylalanine with p-toluidine catalyzed by the enzyme papain gives the *p*-toluidide of acetyl-L-phenylalanine and leaves unchanged D-phenylalanine (scheme 1.85*a*).

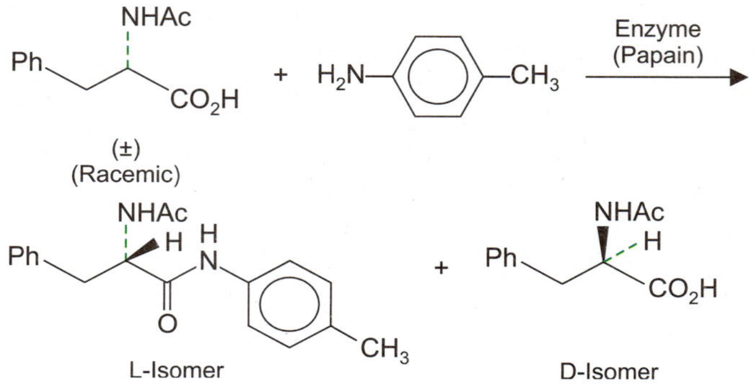

SCHEME 1.85a

Apart from using only natural enzymes for resolution of chiral molecules, genetic engineering techniques are now used and it is possible to develop new esterases or other enzymes in the laboratory that react specifically with single enantiomers of important substrates. New properties are engineered into enzymes such as better stability at higher temperature or enhanced solubility in organic solvent/water mixtures to increase the practical value of enzymes for organic chemistry.

(G) Asymmetric Transformation of Racemates and Total Spontaneous Resolution (see Schemes 1.108b–1.108d)

(H) Resolution via Chiral Recognition and Inclusion Compounds

(*i*) Use of chiral crown ethers

A chiral host can form diastereomeric inclusion compounds and often it is possible for the host to form an inclusion compound with one enantiomer of a racemic guest but not with the other—a process termed *chiral recognition i.e.*, when only one enantiomer fits into the chiral host cavity, while the other does not. Generally, both diastereomers may be formed, however, one is formed more readily than the other. If one removes the guest, it has already been partially resolved. Thus chiral crown ether (chirality due to restricted rotation, scheme 1.85*b*) partially resolves *rac*-methyl phenylglycinate HPF_6. When the aqueous solution of *rac*-amine salt is mixed with a chloroform solution of optically active crown ether (scheme 1.85*b*) followed by separation of layers, the chloroform layers contained about twice as much of the complex formed between the chiral crown ether and (*R*) amine salt. (In inclusion compounds there is no bonding between the host and the guest except van der Waals forces).

The chiral crown ethers of the type (scheme 1.85*b*) have been attached to polystyrene resin to perform chiral chromatographic separation of enantiomers of a-amino acids. As expected the two enantiomers move down the column which is packed with this optically active (enantiomerically pure) support form diastereomeric complexes of different stability and solubility. The enantiomer which forms a more stable complex keeps bound to the column while the other comes out from the column first.

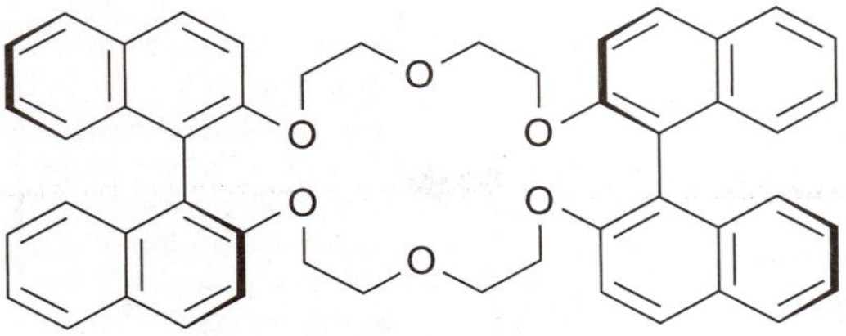

A ***chiral*** crown ether

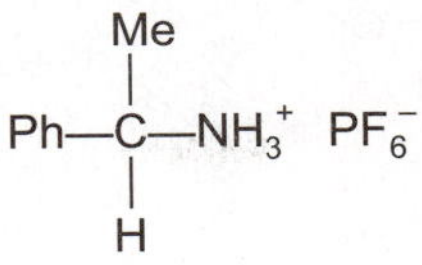

rac-Methyl phenyl glycinate HPF_6

Inclusion of one enantiomer may be preferential in the chiral cavity of crown ether. Diastereomeric inclusion compounds may have difference in stability or solubility.

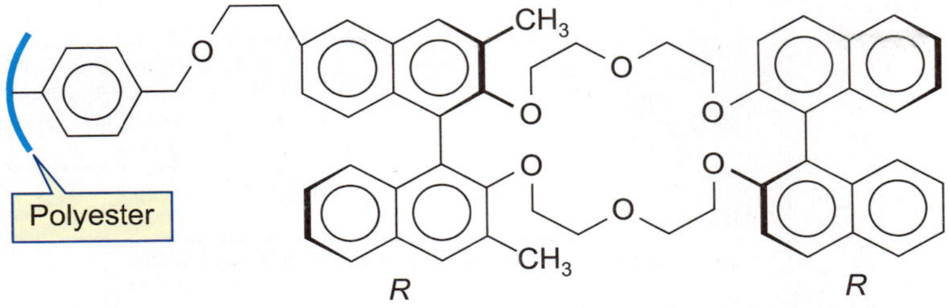

A chiral (*R*, *R*)-binaphthyl crown ether attached to a polystyrene resin

SCHEME 1.85b

In summary the following points may be noted:

- Preferential inclusion of one of the enantiomers may be due to favourable hydrogen bonding and van der Waals attraction.
- Proper match between the size and shape of the guest molecule with that of the host cavity (*e.g.*, the size and shape of the guest molecule can vary with a conformationally flexible molecule, there may also be play in the space in the cavity).
- The diastereomeric inclusion compounds are separable due to their differential stability or solubility.

(ii) Use of cyclodextrins

Cyclodextrins (natural origin, chiral) are made up of six, seven or eight glucose units connected in a large ring are termed respectively α, β or γ cyclodextrins. These have a shape of a tub with primary OH groups projecting from the narrow side of the tub while the secondary OH groups from the wide side (scheme 1.85*c*). The cyclodextrins are used for the resolution of a variety of compounds *via* inclusion complexes.

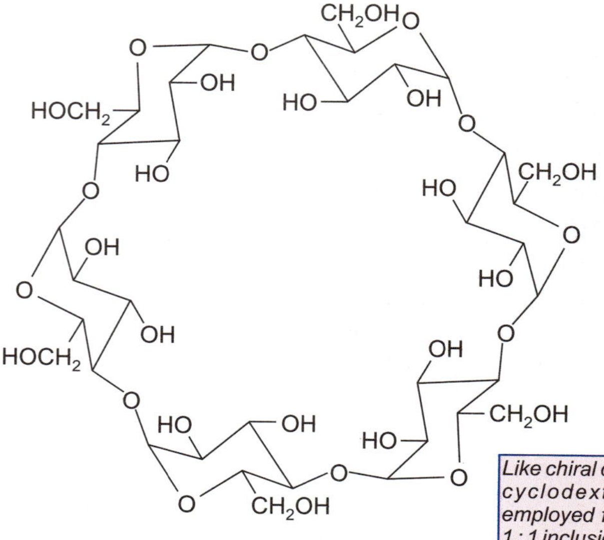

α-Cyclodextrin

The cyclodextrins are macrocyclic glucose oligomers with six (α) seven (β) or eight (γ) glucopyranose units linked α-1,4. These are degradation products of starch with a tub like shape. These have non polar, hydrophobic cavities. The OH groups of sugar rings make the exterior hydrophilic and water soluble.

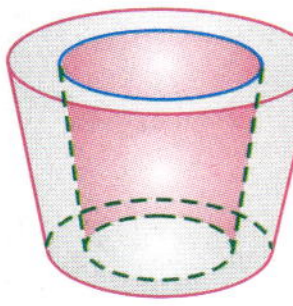

A cyclodextrin is represented like a pale type figure

Like chiral crown ethers and cryptands, cyclodextrins have also been employed for resolution via crystalline 1 : 1 inclusion complexes in solution.

SCHEME 1.85c

One distinguishes two broad classes of inclusion compounds. The type discussed in above two examples involving chiral crown ethers and cyclodextrins is termed 'Cavitates' in which the guest molecule i.e., the chiral substrate to be resolved is partially or entirely enclosed within a second chiral substance having a chiral cavity. The second type is termed clathrates where the guest molecules are surrounded by several molecules of the resolving agent to form a cage or channel.

(*iii*) Use of TOT in resolution

TOT (tri-*O*-thymotide) (Scheme 1.86) provides a remarkable recent example of resolution by inclusion compound formation. TOT crystallizes in the absence of any guest (a solvent) as a racemic compound [(*P*)-(+)-right handed propeller and (*M*)-(–)-left handed propeller]. In the presence of a guest (*e.g.*, benzene) it however, crystallizes as a conglomerate (formation of clathrate inclusion complexes). In such a conglomerate the crystals may have TOT in either the (*P*)-(+)- or (*M*)-(–)-form and consequently the

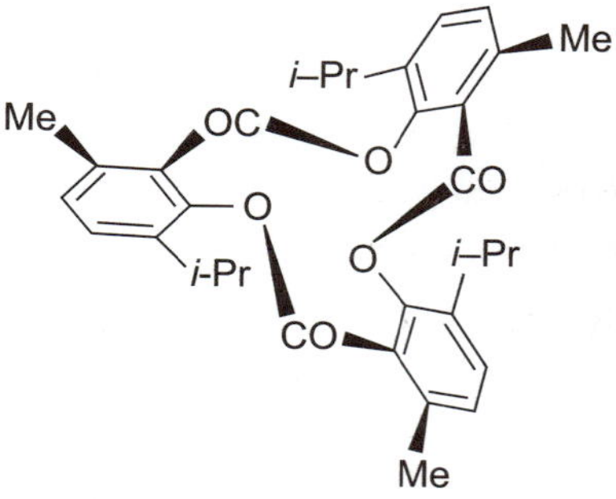

TOT (tri-*O*-Thymotide)

TOT (tri-O-Thymotide) is racemic when guest free. In the presence of a guest it forms clathrate inclusion complexes which are conglomerates, with TOT as (P)-(+)-right handed propeller or (M)-(–)-left handed propeller.

SCHEME 1.86

cavities become chiral. The guest containing crystals on dissolution separately at room temperature lose their optical activity due to interconversion (enantiomerization) of TOT enantiomers via bond rotations. Thus the channel clathrates between TOT and chiral guests are used in resolution.

(*iv*) Other methods of resolution

Optically active diisopinocamphenyl borane (see scheme 2.39) can be used to resolve racemic olefins. The reagent adds to one enantiomer and the other is unchanged. Chiral allylic alcohols have been resolved with chiral epoxidizing agents made from tartarate complexes of titanium (see scheme 2.52). One enantiomer is epoxidized while the other remains unchanged.

1.10A MEASUREMENT OF OPTICAL ACTIVITY

Light is a wave phenomenon in which vibrations take place at right angles to the direction in which the light travels. Infinite number of planes pass through the line of propagation and in ordinary light vibration take place in all these planes. Plane polarized light is light in which vibrations take place in only one plane, and this is realised by passing ordinary light through a polarizer which forms an important component of a polarimeter (scheme 1.87). Plane polarized light is a vector sum of left and right circularly polarized light, which propagates through space as left and right handed helices. These two forms of light represent two enantiomers, and therefore, the two enantiomers of a compound interact with it in opposite ways.

An optically active compound is one which rotates the plane of polarized light. An optically active substance which rotates the plane-polarized light to the right (clock-wise) is said to be dextrorotatory often abbreviated by the letter *d*, and its optical rotation is given a (+) sign. A substance is said to be levorotatory (often abbreviated by the letter *l*) if it rotates plane-polarized light to the left (counterclockwise), and its optical rotation is given a (–) sign.

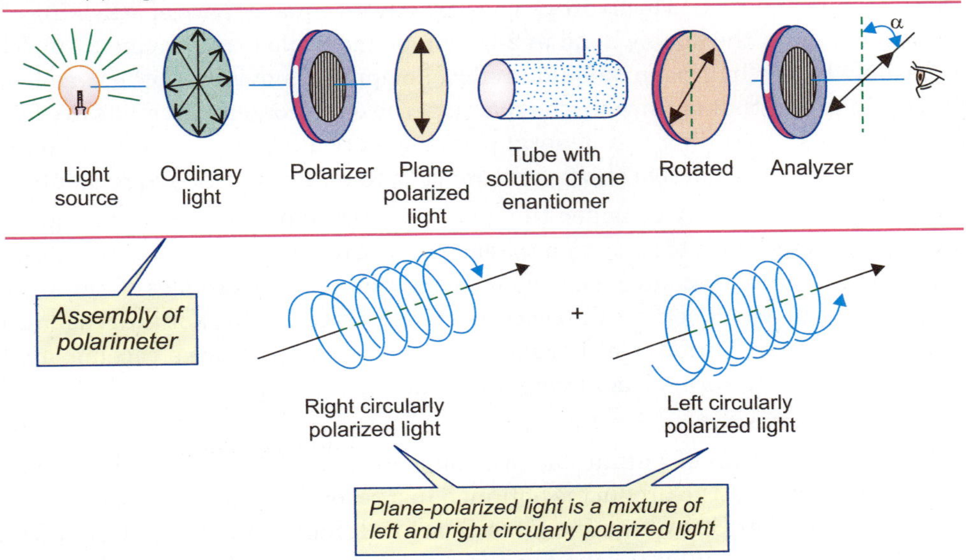

SCHEME 1.87

Optical rotation is a function of concentration, sample thickness, temperature, wavelength of polarized light, etc. It is usually recorded in the literature in terms of specific rotation $[\alpha]^t_\lambda$.

where t = temperature measurement in °C

λ = wavelength of polarized light (usually sodium D line, 5893 Å)

α = observed angle of rotation in degrees

l = sample thickness in decimeters

c = concentration of solution in g/100 ml

1.11 RACEMIC MIXTURE AND RACEMIZATION

Racemic mixture or a racemate is an equimolar mixture of two enantiomers. Since a racemic mixture contains equal numbers of *dextrorotating* and *levorotating* molecules, the net optical rotation is zero. A racemic mixture is often symbolized by (±) or (*dl*).

The process whereby a pure enantiomer is converted into a racemic mixture is called racemization. Racemization may be accomplished in a trivial sense by simply mixing equal amounts of two pure enantiomers. Racemization may also result from the following chemical interconversions.

(A) Formation of Enantiomers (Racemization During Reactions That Create Chirality Centers)

Several organic reactions can yield a chiral product from an achiral starting material. Thus the addition of hydrogen to the carbon-oxygen double bond of 2-butanone (achiral compound) in the presence of a catalyst creates a chirality center. The two enantiomers of 2-butanol (achiral compound) however, are produced in equal amounts—the product is racemic (see, Scheme 1.84). The hydrogen has exactly an equal chance of attacking above or below the plane of the double bond in 2-butanone molecule (see, scheme 2.16). There is no reason why the hydrogen should prefer one approach over the other ; as long as there is nothing else that is chiral in the reaction, the enantiomeric products are formed in equal amounts. According to a general principle optically active products cannot be formed when optically inactive substances react with optically inactive reagents.

One should carefully consider the stereochemistry of the reactants and follow through the mechanism of the reaction to reach the correct products. One may focus the step that creates the stereocenter. The following are some of the examples of racemization during reactions in which neither the reactants (C=O, C=C, C^+ etc.) nor the reagents (H_2, Br_2, etc.) are chiral. Addition to either face is equivalent to produce a racemic mixture (these processes are described as having no enantioselectivity).

(*i*) Mechanism Involving a Radical Intermediate—Bromination of Butane

The radical bromination of butane (achiral molecule) at C2 yields a chiral molecule (scheme 1.88), however, the product is obtained in a racemic form. Abstraction of either methylene hydrogen at C2 by bromine gives an achiral radical, Reaction of Br_2 with this radical is equally likely at either the top or the bottom face, a situation which gives a racemic mixture of products (the C2 hydrogens are enantiotopic).

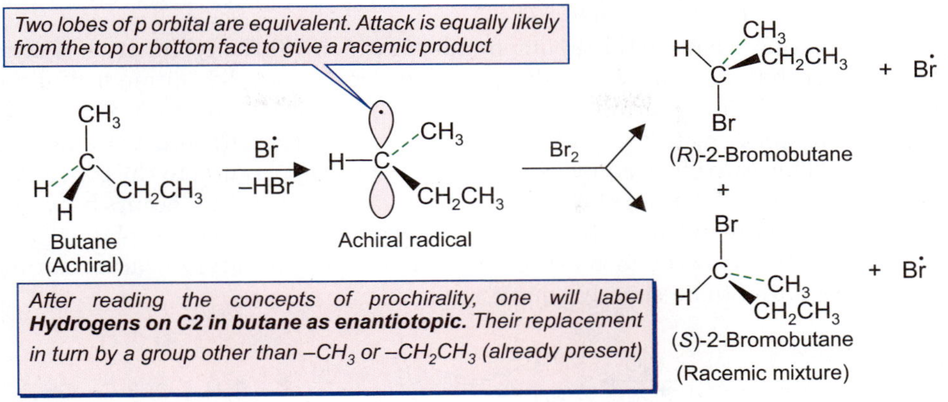

SCHEME 1.88

(*ii*) Mechanism Involving a Carbocation—Addition of HBr to 1-Butene

This addition proceeds via the intermediate formation of a planar carbocation (scheme 1.89) to give again a racemic mixture of products.

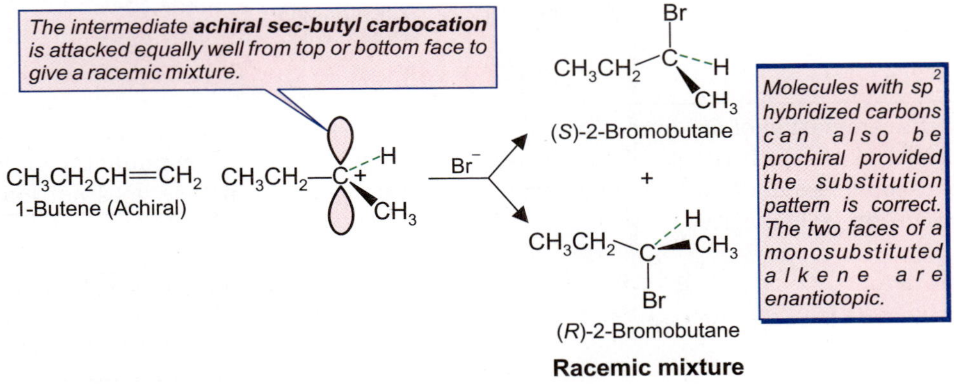

SCHEME 1.89

(*iii*) Mechanisms Involving Cyclic Intermediates

An example is found during *e.g.*, epoxidation of *E*-2-butene (see, scheme, 1.101).

(B) Racemization from One Enantiomer

(a) By Rotation About a Single Bond

The biphenyls and related compounds, in which optical activity is due to the restriction of rotation about a single bond, racemize when enough thermal energy is employed for the energy barrier between the enantiomers to be surmounted at a practicable rate (see Scheme 1.117). Amine inversion is yet another example of recemization.

Cyclic compounds which exist in enantiomeric conformations *e.g.*, *cis*-1, 2-dimethyl-cyclohexane (see Scheme 4.33) undergo racemization *via* ring inversion and apparently do not involve any achiral intermediate or transition state.

(b) Via an Enol or Enolate Anion

Racemization occurs in those compounds, in which a carbonyl function is attached to a stereocenter that also carries a hydrogen. When (*R*)-3-phenyl-2-butanone is dissolved in aqueous ethanol that contains NaOH or HCl, the optical rotation of the solution gradually drops to zero, to yield a racemic mixture of the (*R*) and (*S*) enantiomers (scheme 1.90). This rate of racemization is found to be proportional to the concentration of ketone and the concentration of NaOH or HCl. Racemization thus, occurs by way of the intermediate enol form in which the former stereogenic carbon becomes planar (achiral). As racemization involves the formation of the enol form, the rates of racemization and enolization are found to be exactly equal.

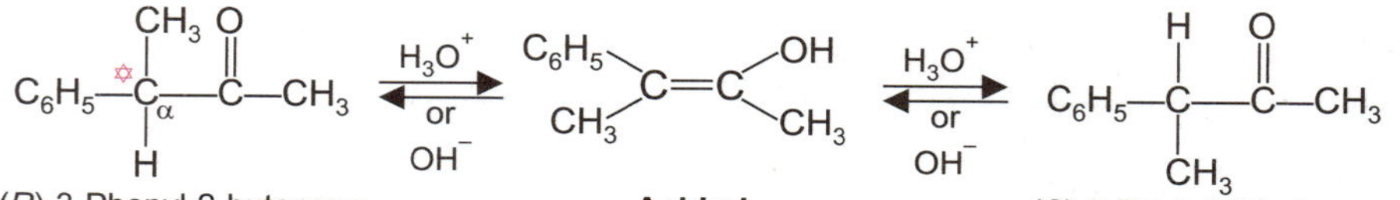

(*R*)-3-Phenyl-2-butanone **Achiral** (*S*)-3-Phenyl-2-butanone

A compound with H-containing stereocenter next to carbonyl group (α-position) gives a flat and achiral enol in an acid or base solution (labile nature of H in the α-position of carbonyl). Approach of the electrophile e.g., H_3O^+ from either face gives a mixture of enantiomers-a racemate.

$$\text{Rate} = k[\text{Ketone}][H^+] \quad \text{or} \quad k'[\text{Ketone}][OH^-]$$

SCHEME 1.90

Racemization of an optically active ketone occurs only if the stereocenter is α to the carbonyl group. If the aldehyde or ketone is chiral because of asymmetry at some other carbon, the enol form is also chiral, enolization in such a case does not lead to racemization (scheme 1.91).

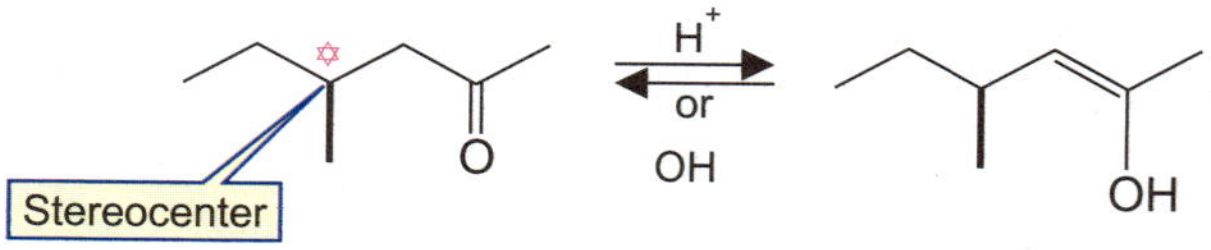

WheninachiralketonetheH-containing stereocenter is not in the α-position, racemization does not occur in acid or basesolution.

(*S*)-4-Methyl-2-hexanone (**Chiral**) **(Stillchiral)**

SCHEME 1.91

EXERCISE 1.14

Groups others than the carbonyl group also activate the adjacent protons which are easily removed by base, examples are nitro compounds, sulfones and nitriles. Thus the optically active nitrile (I) is expected to racemize when treated with base. However, (I) undergoes deuterium exchange 4000 times faster than it racemizes on treatment with sodium methoxide in deuteriomethanol. Explain.

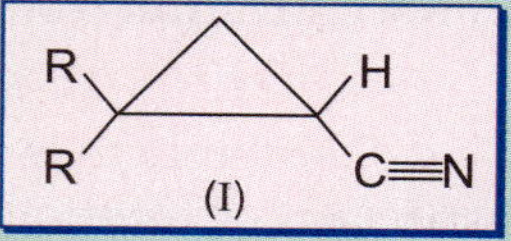

ANSWER. *The inversion of carbanion (II) formed by the abstraction of a proton from the position adjacent to the nitrile group with base is difficult since the nitrile group has to become coplanar with the cyclopropane ring and this flipping process is unfavourable since excessive strain energy will be involved. (scheme 1.91a).*

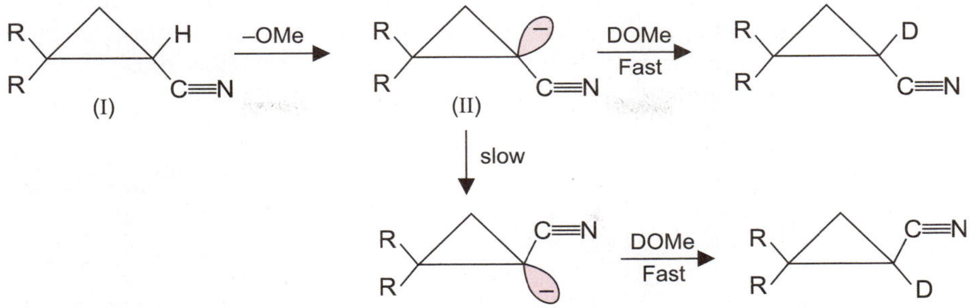

SCHEME 1.91a

Racemization is often a facile process which is troublesome as well as it has nuisance value. The presence of both the enantiomers of thalidomide in a drug formulation led to birth defects in the children of some women who took the substsance during pregnancy. Even if the pure (*R*) form was given to patients, the mutagenic (*S*)-isomer was formed by racemization under physiological conditions because the stereocenter is α to a carbonyl group (scheme 1.92). Similarly amino acids can undergo racemization due to the presence of a stereocenter α to the carbonyl group in these compounds. In proteins, the peptide bonds render this structural feature prone to racemization, however, it does not occur under physiological conditions.

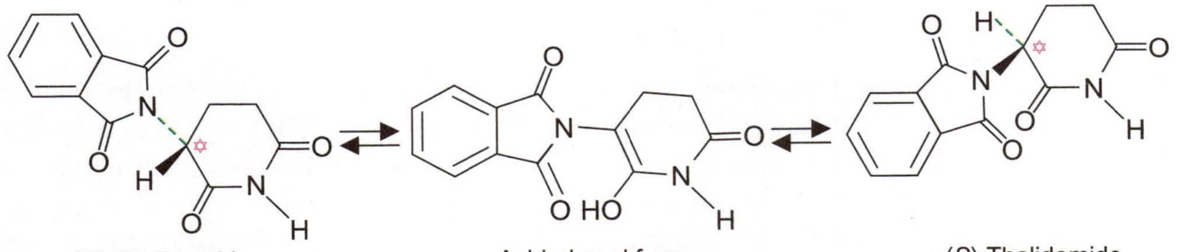

SCHEME 1.92

(C) *By Substitution Reactions*

(*i*) S_N2 reactions

Normally in an S_N2 reaction the incoming nucleophile (*e.g.*, OH^- in scheme 1.28*g*, eq. I) initiates the reaction from the back while Br^- leaves the molecule from the front in a concerted process. The chirality center, thus undergoes an inversion of configuration with respect to the stereochemistry of substrate. In case the incoming nucleophile and the out going leaving group are same as in the case of reaction of optically active 2-iodoctane (Scheme 1.93) with sodium iodide, the reaction becomes reversible and an equilibrium is set up between the two enantiomers (I and II, scheme 1.93) leading to racemization. Conclusive proof that inversion of configuration occurs during an S_N2 reaction was provided by studying the reaction between optically active 2-iodooctane and radioactive iodide ions. This is one of the simplest possible types of bimolecular substitution reacton, and involves replacement of iodide ions by radioactive iodide ions so that product and starting material are chemically identical. The process also involves inversion of configuration, and is thus accompanied by a loss of optical activity. It was seen that the rate of loss of optical activity is twice the rate of incorporation of radioactive iodide ions [*i.e.*, the inversion rate is half the rate of racemization since for every pair of S isomers only one is inverted to give the racemate].

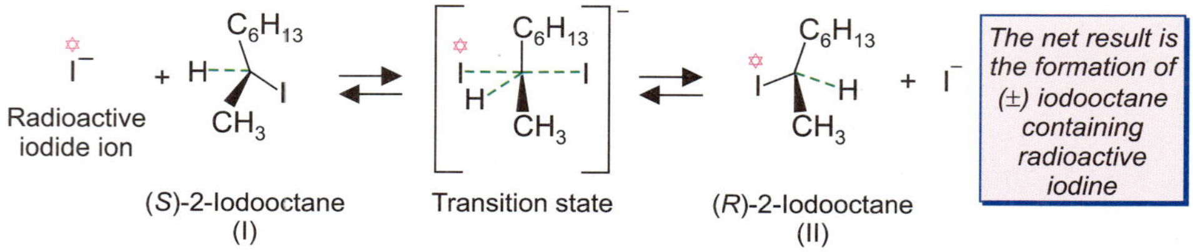

(Racemization via an S_N2 reaction)

SCHEME 1.93

In case the process involved an achiral intermediate, like a carbocation, the rate of racemization will be equal to the rate of incorporation of radioactivity. On the other hand, in an S_N2 reaction every substitution (eq. I, scheme 1.28*g*) involves inversion. Racemization is, therefore, complete when half of the material gets inverted (and has incorporated radioactivity) so that the rate of racemization is twice the rate of incorporation of radioactivity. This experiment provides the most convincing proof to date that an S_N2 reaction is accompanied by inversion of configuration.

(*ii*) S_N1 reaction

Although many S_N1 reactions proceed with racemization, many others result in more inversion of configuration in the product than retention. This is due to initial ion pair formation where the leaving group blocks the front side of the carbocation to favour inversion. It is only when the leaving group diffuses away leaving the carbocation in free form that the nucleophile can attack equally well from either side to result in equal amounts of inversion and retention (see scheme 3.26).

(*iii*) During a molecular rearrangement

Optically active monoterpenoid camphene undergoes racemization on treatment with acid. The reaction of (I, scheme 1.94) with a proton leads to the formation of a carbocation which undergoes a methyl shift with its bonding pair of electrons to give another carbocation, the loss of a proton gives (II) which is the mirror image of (I, scheme 1.94).

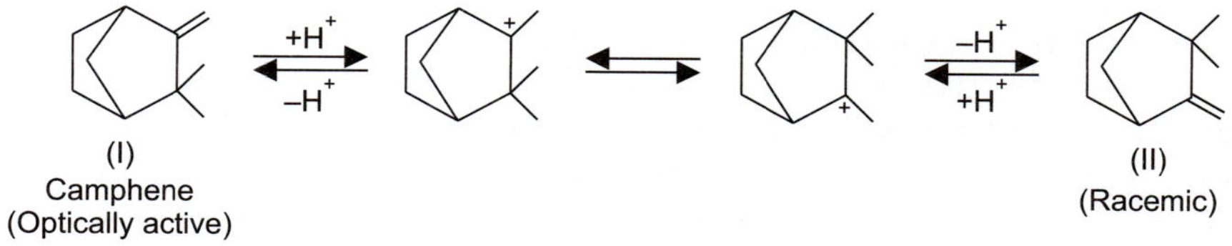

Racemization of monoterpenoid camphene on treatment with acid

SCHEME 1.94

1.12 SOME STEREOCHEMICAL REACTIONS NEAR A CHIRALITY CENTER (FORMATION OF DIASTEREOMERS)

Several reactions *e.g.*, electrophilic addition of HBr to monosubstituted alkene (see, scheme 1.89, achiral reactants) or the radical bromination of butane at C2 (see, scheme 1.88) introduce chirality in the molecule, however, the product obtained is racemic. In these reactions a planar sp^2 hybridized and therefore, achiral intermediates (a carbocation and a radical respectively) are formed. The two faces are equally susceptible to attack on two equivalent reaction sites to result in racemic products.

The presence of a chirality center in the starting substrate affects the outcome of the reaction to give an unequal mixture of product diastereomers. Thus the addition of HBr to (*R*) enantiomer of 4-methyl-l-hexene proceeds through a carbocation (scheme 1.95).

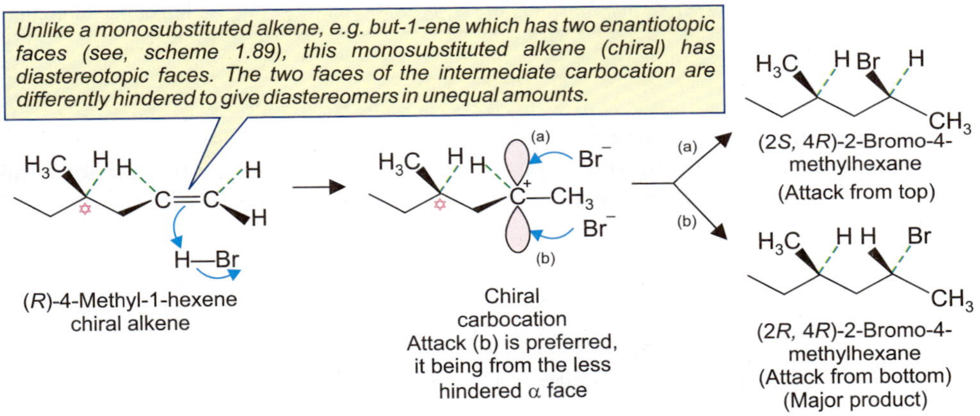

Formation of unequal mixture of diastereomers

SCHEME 1.95

Because there is a chirality center in the carbocation intermediate, it is a chiral species and therefore, it does not have a plane of symmetry. One face of the carbocation will be more sterically hindered than the other, the incoming bromide ion will therefore, have greater access to the less sterically hindered face. As a result of this the two diastereomers will be formed in unequal amounts. The reaction is stereoselective, precisely this is a diastereoselective reaction since the stereoisomer produced in excess is a diastereomer (for further details see under asymmetric synthesis—also see schemes (1.104–1.106). The reaction (scheme 1.95) creates a new chirality center, thus two stereoisomers are formed. These stereoisomers are diastereoisomers since, one of the chirality centers (initially present) has the same configuration in both while other (newly created) has the opposite configuration. This also is the basis of Crams rule to addition to diastereotopic ketones (see, scheme 2.23*b*).

Similarly, during acid catalyzed addition of water to 1, 4-dimethyl-cyclohexene (the presence of a chirality center) the two faces of the intermediate carbocation are not identical (scheme 1.96). Attack of water from the axial direction is hindered by the two axial hydrogen atoms and is therefore, slower to give the minor diastereomer (I, scheme 1.96) and the addition of water to the equatorial face dominates and the major product is the diasteromer (II, scheme 1.96) where the hydroxyl group is *trans* to the methyl group at C4. Both the diastereoisomers are achiral (plane of symmetry, also see schemes 7.6 and 7.7).

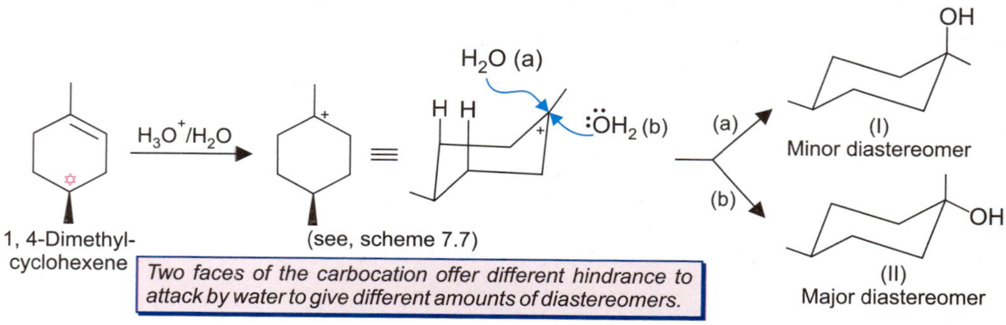

SCHEME 1.96

The chlorination of (*S*)-2-bromobutane (scheme 1.97) at C3 again gives a mixture of diastereomers in unequal amounts based on arguments similar to that for scheme 1.95 *[The methylene protons close to a stereocenter are generally diastereotopic, (see scheme 2.10)]*.

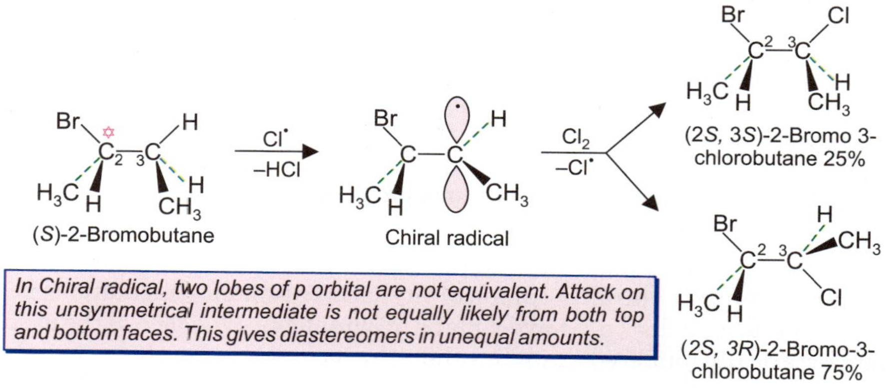

SCHEME 1.97

The radical chlorination of optically active 2-bromobutane *e.g.*, the (*S*) enantiomer at two terminal methyl groups at C1 or C4 will produce optically active products since the original chirality center remains intact. The chlorination at C2 would lead to loss of optical activity since now the bond to chirality center is cleaved to give a planar achiral sp^2 hybridized radical. Its chlorination will lead to racemization (compare with scheme 1.88).

Conformational control and diastereoselectivity during electrophilic additions to a double bond in acyclic systems may be provided by an $A^{1,3}$ strain, (see scheme 4.67). In an acyclic system (scheme 1.98) there must be a facial preference for attack, An $A^{1,3}$ strain generates a conformational control as found during hydroboration of the alkene (scheme 1.98). The hydration of a double bond via hydroboration involves anti-

Markovnikov addition of the B–H bond; *cis* addition of the B–H bond; addition of the B–H bond from the *less hindered side* of the double bond, and oxidation with retention of configuration (scheme, 1.98, also see scheme 6.24).

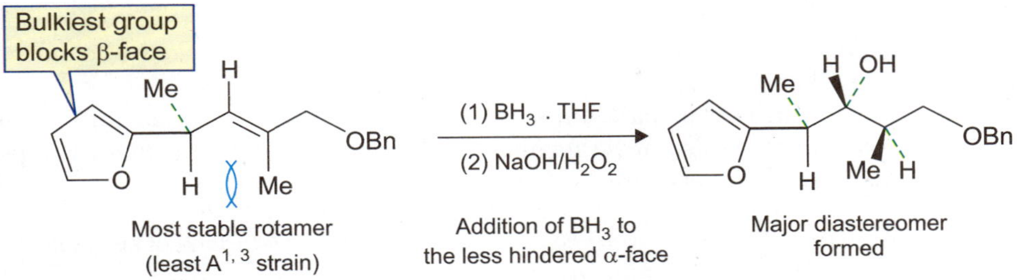

SCHEME 1.98

Effect of a given chirality in a substrate on addition reactions (Diastereoselectivity)

- *The chirality center already present can be R, S or both (i.e., in last possibility, the starting material is racemic.)*
- *In case the existing chirality center has R configuration (i.e., the starting material is enantiometrically pure R stereoisomer as in scheme 1.95), the products of addition (diastereomers) will have either the R, R, or R, S configuration. Each diastereomer will be optically active since only one enantiomer is formed.*
- *In case the existing chirality center was with S configuration the products (diastereomers) formed on addition would be S, S or S, R. These also would be optically active and infact represent the enantiomers of the diastereomeric pair formed from the starting material in which the existing chirality center had R configuration.*
- *In case the starting material is a racemic mixture, all the four stereoisomers would be formed i.e., product will be optically inactive. The outcome of the reaction will still be two diastereomers (each is formed as a pair of enantiomers) and the same diastereoselectivity will be observed.*

In summary, therefore, during *e.g.*, addition reactions to an olefin which has a chirality center elsewhere in the molecule, if optically active starting materials are employed the resulting diastereomers will be optically active. The racemic starting materials yield optically inactive diastereomeric mixture, in both situations the diastereomers will be formed in unequal amounts.

1.13 STEREOSELECTIVE AND STEREOSPECIFIC REACTIONS

(A) Stereospecificity – A Common Definition

When a reactant can exist in stereoisomeric forms and if one stereoisomeric reactant gives one stereoisomeric product while the other gives a different stereoisomeric product exclusively or abundantly, the reaction is called stereospecific.

Alkenes react with peroxy acids via, a **syn** addition, and it is stereospecific epoxidation. 2-Butene can exist in stereoisomeric forms-*cis*- and *trans*-2-butene which are diastereomers. Epoxidation of *trans*-2-butene gives a racemic mixture. The faces of the double bond are enantiotopic. The addition of oxygen to either face of the alkene generates an enantiomer (Scheme 1.99). *cis*-2-Butene with identical faces (homotopic through rotational axis of symmetry) gives only (*meso*) *cis*-2,3-dimethyloxirane, no matter which face of the alkene accepts the oxygen atom.

This epoxidation is thus called a stereospecific reaction (epoxidation) since the reactant can exist in stereoisomeric forms, *cis*-diastereomer gives a *meso* compound while *trans*-diastereomer gives a racemic mixture (Scheme 1.99).

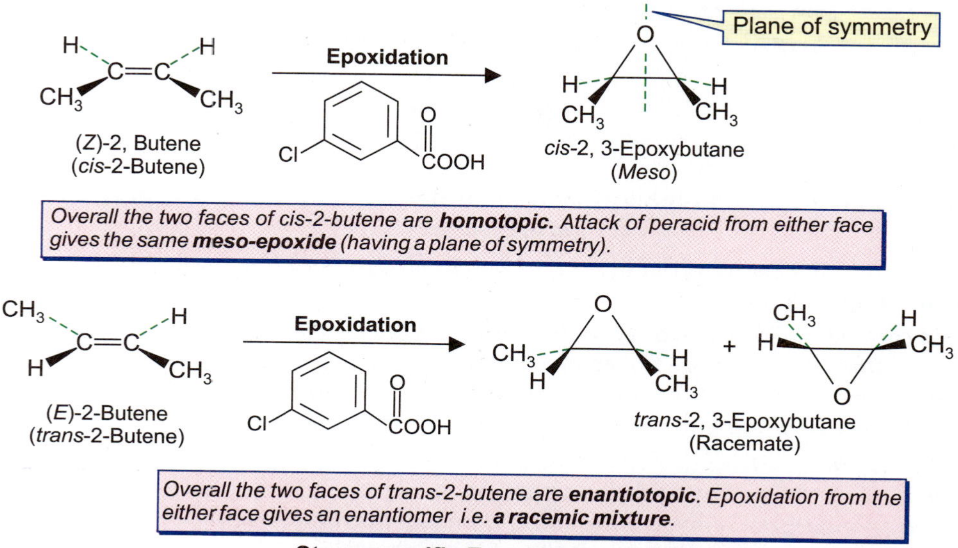

Stereospecific Epoxidation of Alkenes

SCHEME 1.99

(B) Stereoselectivity – A Common Definition

A **stereoselective reaction** is one in which the reactant is not necessarily chiral (as in the case of an alkyne) but in which the reaction generates predominantly or exclusively one stereoisomeric form of the product.

Synthesis of *trans*-2-butene can be achieved by treating 2-butyne with lithium in liquid ammonia. The *anti* addition of hydrogen by this reaction gives the *trans* product and not the other diastereomer *cis*-2-butene (Scheme 1.100).

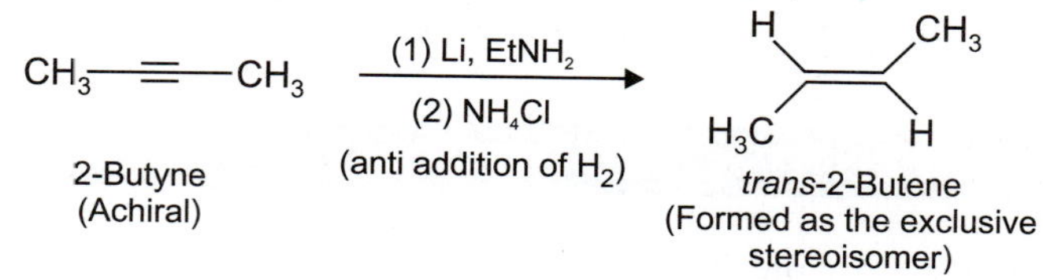

A stereoselective reduction of an alkyne

SCHEME 1.100

The enzymes when in contact with prochiral molecules, react only with one of the enantiotopic ligands or faces, a property called stereoselectivity. For an example, glycerol undergoes phosphorylation exclusively at the *pro-R* hydroxymethylene group with adenosine triphosphate (ATP) in the presence of an enzyme-glycerol kinase (scheme 1.101).

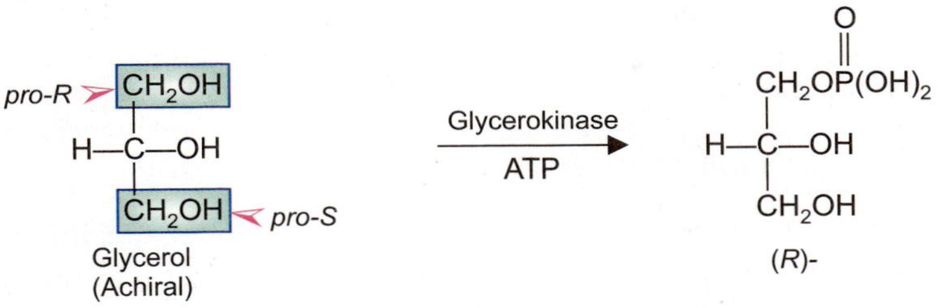

SCHEME 1.101

(*i*) Bromination of Cyclohexene – Could be Called a Stereoselective or Stereospecific Reaction

Cyclohexene can only exist in one stereoisomeric form with *cis*-geometry of the double bond. When cyclohexene is brominated, the product is one stereoisomeric product *trans*-1-2-dibromo-cyclohexene (scheme 1.102) no cis-isomer is formed. One may compare this stereoselective bromination of cyclohexene with stereospecific bromination of 2-butene stereoisomers (see page 132). Moreover, the bromination proceeds through the cyclic bromonium ion, if the intermediate were a simple carbocation both *cis* as well as *trans*-1, 2-dibromocyclohexane would have formed. Thus the reaction mechanism for this stereoselective reaction is the same as in the case of stereospecific bromination of *cis*-*trans* diastereomers of 2-butene. For this reason some chemists call it a stereoselective addition while others call it stereospecific and both are correct.

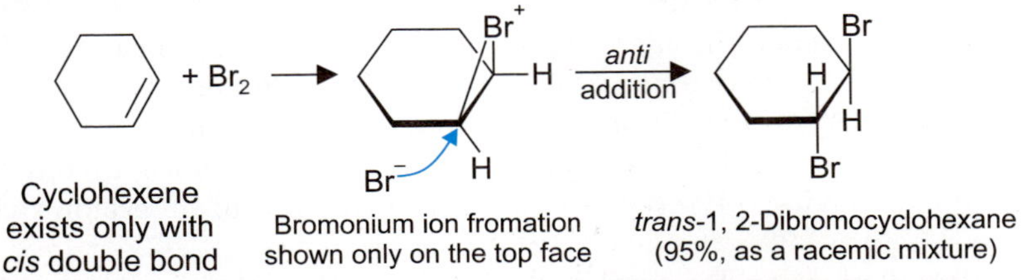

anti-Addition of bromine via bromonium ion which is attacked by Br^- from the opposite face (bottom face) (i.e., the two bromine atoms in the product are on opposite faces of the ring) gives only trans-1, 2-dibromocyclohexane as a racemic mixture. No cis-isomer is formed.

A stereoselective reaction

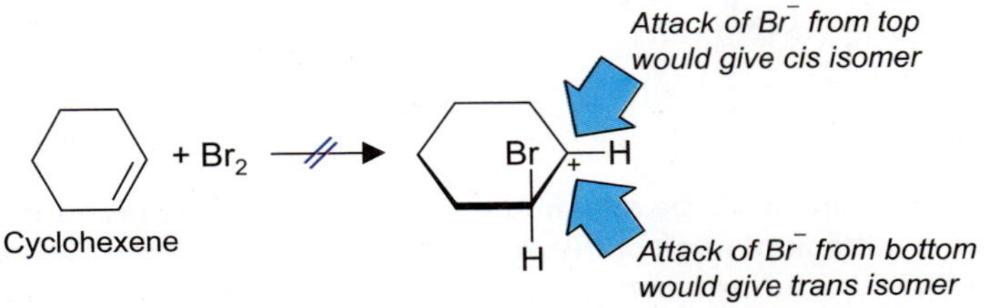

(A carbocation intermediate not formed)

SCHEME 1.102

A Closer Look at Stereospecific and Stereoselective Reactions—A Tricky Situation

- *There appears to be confusion over the terms stereoselective and stereospecific.*
- *The common definitions given above suffer from the disadvantage e.g., a reaction which is stereospecific e.g., for two stereoisomeric alkenes must be called stereoselective for a small ring alkene (e.g., cyclohexene) which has no stereoisomers, however, the mechanism is same.*
- *Thus a reaction carried on a compound that has no stereoisomers cannot be stereospecific, however, at most stereoselective (see, scheme 1.100).*

(C) More Satisfactory Definitions—Stereoselectivity and Stereospecificity

- One may look at the mechanism, when there is a choice in the stereochemical pathway *e.g.*, addition to the less hindered face of the double bond the reaction is stereoselective. When there is no choice the reaction is stereospecific *e.g.*, S_N2 displacement with inversion of configuration has no choice in the reaction pathway which must lead with inversion of configuration. Thus all S_N2 reactions are stereospecific.

If there is a choice ***in the stereochemical pathway,*** *whether that be 5 to 1, or 1000 to 1, the reaction is* ***stereoselective,*** *if there is* **no choice** *then* ***the reaction is simply stereospecific.***

- Since carbocations are planar, it might be expected that optically active starting materials should give racemic products on S_N1 reaction when a completely free carbocation is formed. The incoming nucleophile would be expected to attack equally readily from both sides of the cation. Because both enantiomers are favoured the S_N1 reaction thus is *not stereoselective* since neither enantiomer is formed predominantly.

  The S_N1 reactions may attain stereoselectivity in practice. A mixture of racemization and inversion is observed, suggesting that the leaving group is still in the vicinity of carbocation either free or as an ion pair. This inhibits attack from the same side as the departing group and therefore, favours attack from the opposite side, giving rise to a greater amount of inversion than retention (see scheme 3.39).
- During a reaction by an S_N1 mechanism, the intermediate cation (scheme 1.103) may be attacked from either side by a nucleophile (there is no longer the necessary inversion one sees with S_N2). Attack occurs selectively from the less hindered face of the cation (scheme 1.103).

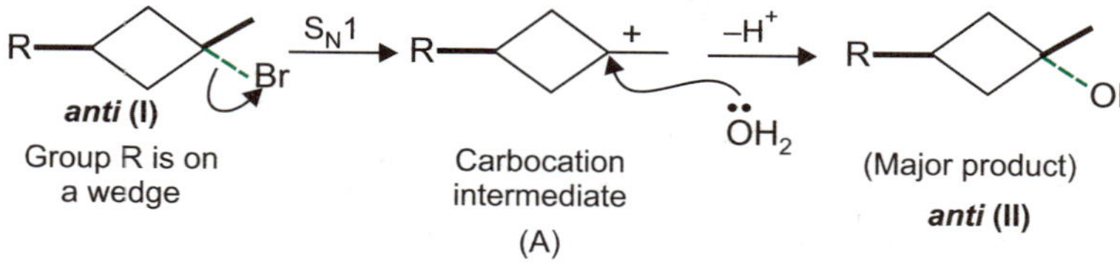

Carbocation intermediate could be attacked from either side. The predominant product is however, formed by attack of nucleophile from less hindered side *i.e.*, away from R group (on a wedge) so that OH ends up away from R on the dash. The reaction is stereoselective.

SCHEME 1.103

With stereoselectivity, there is indeed a possibility that different isomers will give the *same* product (Compare this with stereospecific reactions). The intermediate cation (A) is the same whether the starting material is *syn* (I) or *anti* (I) and must give the same product *anti* (II).

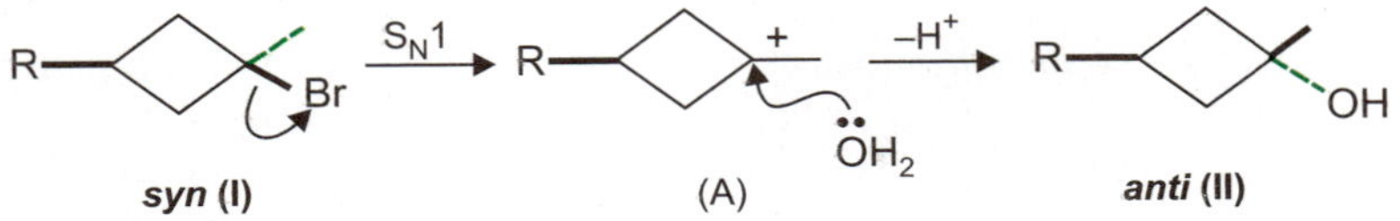

SCHEME 1.103a

- The S_Ni reaction (see scheme 3.56) provides a good example of a stereoselective reaction.
- In a stereoselective reaction the molecule *selects* a particular reaction pathway for some reason like steric hindrance, kinetics or thermodynamics but the molecule is not forced by mechanism to react one way only.
- A stereospecific reaction can also be stereoselective (however, the converse is not true). Thus, *syn* addition describes stereospecificity in the catalytic hydrogenation of alkenes, while the preference for addition to the less hindered face of the double bond describes stereoselectivity (see schemes 1.105 and 1.106). Similarly one describes that epoxide formation on a steroid (scheme 1.104) is stereoselective on the less hindered face of the C = C via a stereospecific *syn* addition. The ring opening of the epoxide by the nucleophile (Br^-) occurs stereospecifically (S_N2 pathway) to give overall *anti* addition.
- For an another example of a compound which has stereoselective and stereospecific components (see scheme 6.82).

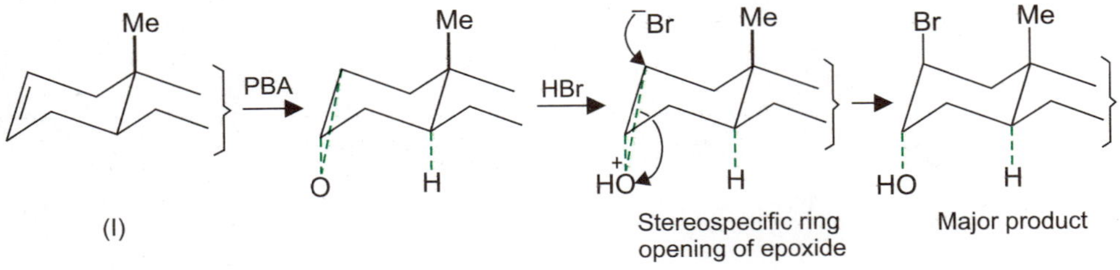

Epoxidation of steroid is stereospecific, but with the methyl group shielding the β-face, the reaction is also stereoselective. Thus the reaction has both the stereoselective and stereospecific components.

SCHEME 1.104

- Thus stereoselective reactions proceed via the most favorable pathway (kinetic control) or via the pathway which gives the most stable stereoisomer as the major product (thermodynamic control, (see scheme 1.106*b*)).
- This difference in the definitions is, however, seldom maintained in practice and both terms are used synonymously.

EXERCISE 1.15

Write the structure of the product(s) from the epoxidation of 1-butene. Assign R and S descriptors to stereocenters. Is the reaction stereoselective?

ANSWER. *The faces of a monosubstituted alkene are enantiotopic. Epoxidation of 1-butene by attack of the peracid from either face of the alkene affords a racemic mixture of epoxides (scheme 1.104a). The reaction is not stereoselective but is stereospecific syn-addition, since the mechanism is fixed (concerted formation of the two new carbon-oxygen bonds).*

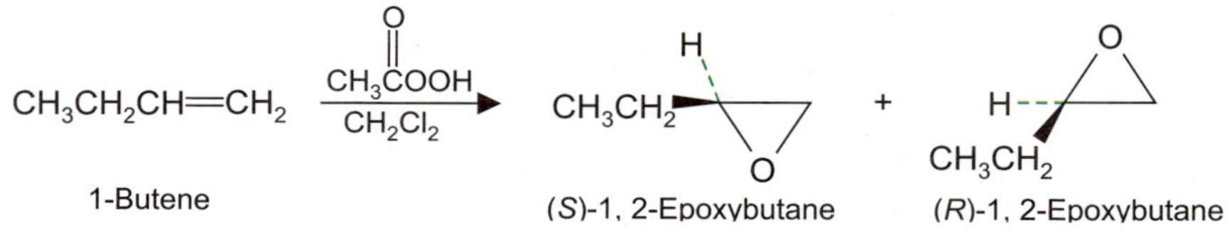

SCHEME 1.104a

(D) Hydrogenation a Common Stereoselective *syn* Addition of Hydrogen

A reaction which introduces a second chirality center into a starting reactant which already has one, may not give equal quantities of two possible diastereomers (see scheme 1.96). The catalytic hydrogenation of 2-methyl (methylene) cyclohexane gives both *cis* and *trans*-1, 2-dimethylcyclohexane, but in unequal amounts (scheme 1.105). More cis-1, 2-dimethyl-cyclohexane is formed than *trans*. The less hindered face of the double bond approaches the catalyst surface, and this is the face to which hydrogen is added (a *syn* Addition). Hydrogenation of 2-methyl (methylene) cyclohexane occurs preferentially at the side of the double bond opposite that of the methyl group to give *cis* stereoisomer as the major product. Thus it is a stereoselective reduction since there is a choice in the stereochemical pathways.

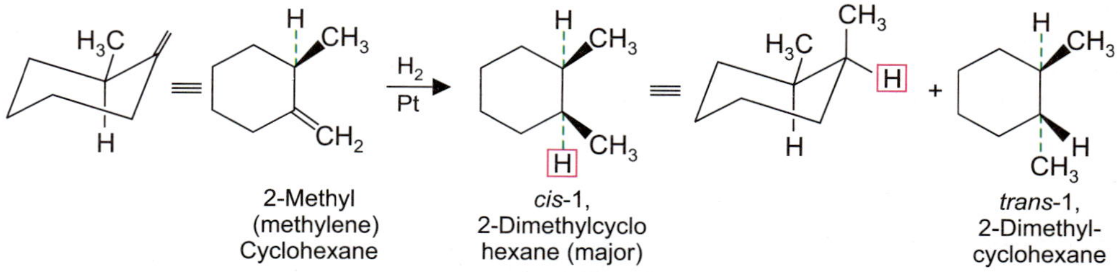

A stereoselective reaction

SCHEME 1.105

The catalytic hydrogenation of α-pinene is also a stereoselective *syn* addition of hydrogen. The observed stereoselectivity depends on the mode of alkene approach to the catalyst surface. The two hydrogens (shown by arrows) are added on the less hindered side of α-pinene. *Cis*-pinane and *trans*-pinane are common names which show that relationship between the pair of methyl groups on the bridge and the third methyl group (scheme 1.106). (Enzyme-catalyzed reactions are also completely stereoselective as shown in scheme 2.19).

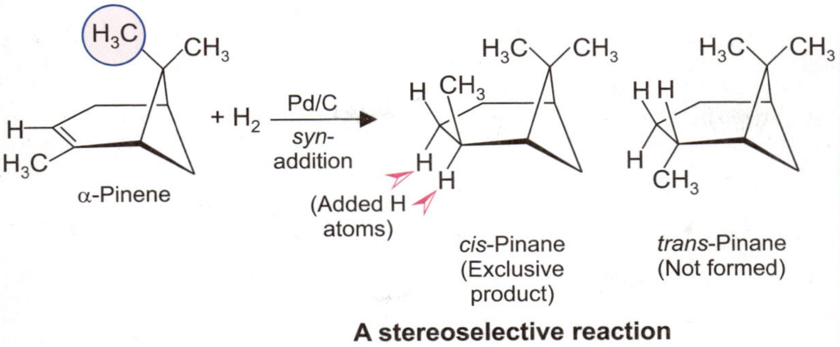

The methyl group which lies over the double bond of α-pinene shields its top face, preventing a close approach to the surface of the catalyst. Hydrogenation of α-pinene, therefore, takes place preferentially from the bottom face of the double bond

SCHEME 1.106

EXERCISE 1.16

Bromination of cyclohexene has been designated a stereoselective reaction (see, scheme 1.104). However, some chemists also term it stereospecific. Explain.

ANSWER. *A reaction in which a single starting material can give two or more stereoisomeric products but yields one of them in greater amounts than the other (or even to the exclusion of the other) is said to be stereoselective. However, this reaction may also be called stereospecific, since there is no choice in the stereochemical pathway. This pathway involves the formation of a bromonium ion on one face, followed by its S_N2 opening. Thus the reaction process is mechanistically constrained to proceed in a stereochemically defined manner (just like S_N2 reaction) and is equally well defined as stereospecific.*

EXERCISE 1.17

The hydroboration-oxidation of an alkene brings about the net syn-addition of the elements of H—OH to the double bond. By convention this reaction is stereospecific syn addition (scheme 1.106a) defined reaction pathway. Predict the stereostructure of the product of hydroboration of α-pinene (for structure of α-pinene see scheme 1.106).

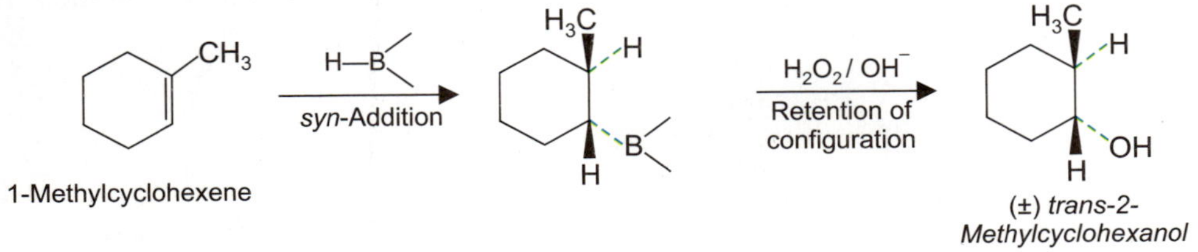

SCHEME 1.106a

ANSWER. *Recall that a stereospecific reaction can also be stereoselective. Thus, though syn addition during the catalytic hydrogenation of an alkene describes stereospecificity, while the preference for addition to the less hindered side of the double bond reflects stereoselectivity. Hydroboration-oxidation of α-pinene like hydrogenation also occurs from the less hindered bottom face and therefore, it is a stereoselective reaction (Scheme 1.106b, note that here α-pinene is drawn slightly differently than in scheme 106).*

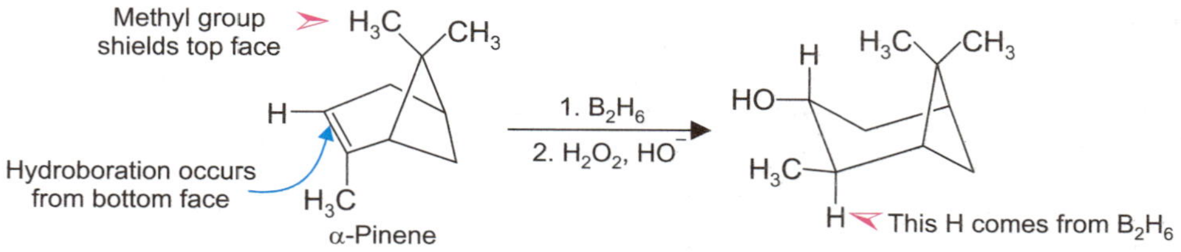

SCHEME 1.106b

More on Stereoselective and Stereospecific Reactions

Often one comes across S_N1 reactions which though involve carbocations, but which favour inversion due to shielding effect of the departing anion, such reactions (see, scheme 3.39) are termed to proceed with low stereoselectivity (i.e., some enantioselectivity). The S_Ni reaction is said to proceed with stereoselective retention of configuration (see scheme 3.56) at the stereocenter.

Stereospecific participation by the σ electrons and stereospecific assistance to ionization by π electrons is observed during the non-classical carbocation formation in norbornane systems (see, scheme 3.70 and 3.79).

(E) How much Stereoselectivity is Possible?

This depends on several factors :

- The nature of substrate
- The type of reaction
- The reagents involved.

In the laboratory, chemists employ enantiomerically pure reagents or catalysts to convert an achiral compound into one enantiomer of product (asymmetric synthesis). In nature, enzymes perform this enantioselectivity, for example the enzyme-catalyzed oxidation of dopamine to (–)-norepinephrine. The chiral reaction environment created by the enzyme gives rise to 100% stereoselectivity in favor of the enantiomer (scheme 1.107).

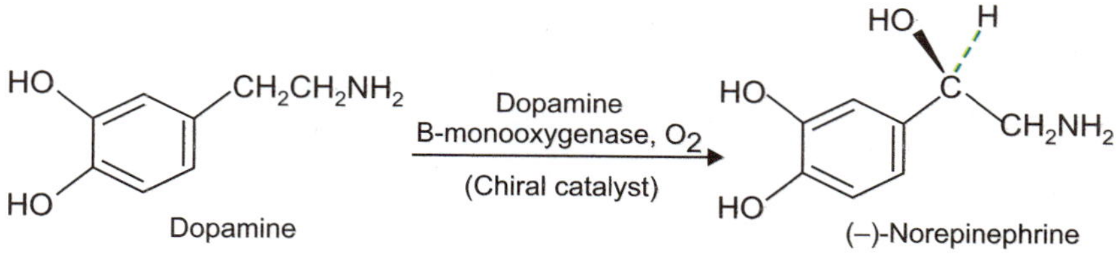

SCHEME 1.107

(F) An Unusual Stereoselective S_N1 Reaction

The high-energy carbocation intermediate formed during an S_N1 displacement (scheme 1.108) reacts with the first nucleophilic species it comes across, regardless of which lobe of the carbocation *p* orbital the nucleophile approaches.

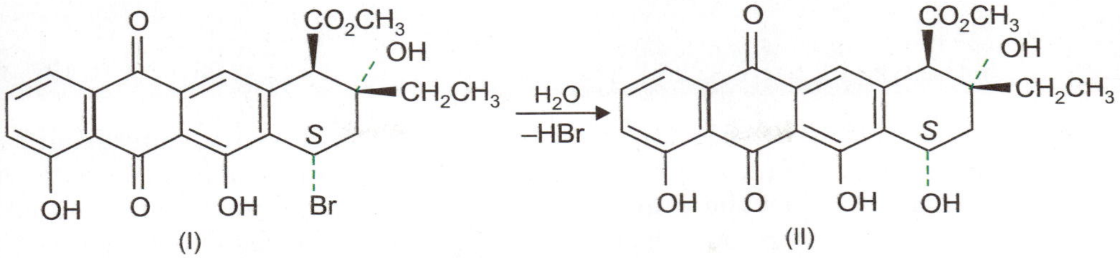

An unusual stereoselectivity during S_N1 displacement

SCHEME 1.108

A secondary halide (I, scheme 1.108) with a good leaving group (bromide), a highly polar protic solvent but a poor nucleophile (water) are the conditions ideal for an S_N1 reaction—and displacement of bromide by water occurs with over 90% retention of configuration to give (II). Approach of a nucleophile to the top lobe of *p* orbital of the carbocation involved is partly blocked by the ethyl group. Moreover, the hydroxy group on the bottom face of the ring "directs" nucleophilic addition of a water molecule from below by hydrogen bonding (scheme 1.108*a*).

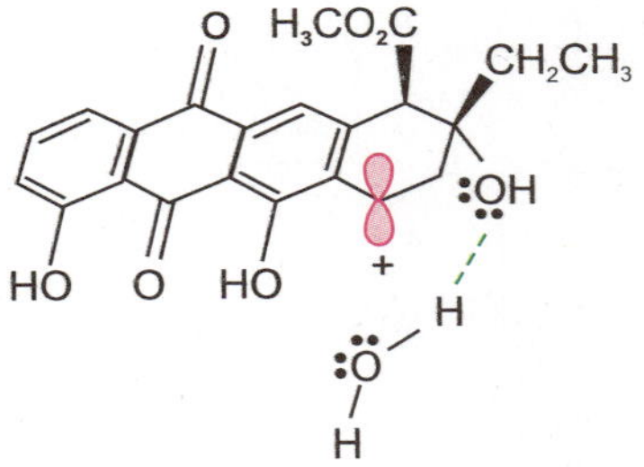

Aclarubicin

SCHEME 1.108a

The reaction gains importance since the compound aclarubicin belonging to anthracycline antibiotics is a component of widely used antilumour drugs.

1.14 ENANTIOMERIC EXCESS – OPTICAL PURITY

Optical activity can be measured in a mixture of enantiomers if these are present in unequal amounts. Using the value of the measured rotation, one can calculate the composition of such a mixture. For example, if a solution of (+) alanine from a fossal displays an [α] of only 4.25 (*i.e.* one-half of the value for the pure enantiomer), one can conclude that 50% of the sample is pure (+)-isomer while the other 50% is racemic. It is said to have 50% enantiomer excess. Because the racemic portion consists of equal amounts of (+) and (–), the actual composition of the sample is 75% (+) and 25% (–). The 25% (–) enantiomer cancels the rotation of a corresponding amount of the (+) enantiomer. This mixture is called 50% (*i.e.*, 75%–25%) optically pure. The observed optical rotation is one-half that of the pure dextrorotatory enantiomer. Optical purity can be found from the following relationship:

$$\%\ \text{Optical purity} = \left(\frac{[\alpha]_{\text{observed}}}{[\alpha]}\cdot 100\right) = \text{Enantiomer excess}$$

The enantiomeric excess can also be determined from NMR spectroscopy (see, scheme 1.150).

1.15 OPTICAL ACTIVITY-STEREOISOMERISM IN COMPOUNDS WITHOUT A STEREOGENIC CARBON—(AXIAL CHIRALITY)

The chiral compounds discussed so far contains one or more chirality centers and have their chirality specified at one or more such centres. As an additional and a specific case of chirality one may consider the suitably substituted adamantanes with four different groups at the bridgehead positions which are chiral and therefore, display optical activity. The adamantane derivative (scheme 1.109) may be regarded as the formal analogue of 2-bromopropionic acid and is completely asymmetric (belonging to C_1, point group). This has four nonequivalent chirality centers and due to highly symmetrical structure of its parent hydrocarbon adamantane however, it exists only as a (±) pair. In fact, the four substituents form a regular tetrahedral arrangement and the arrangement has a center of chirality (shown by a dot) in the unoccupied space of the molecular framework of the adamantane. Two points come to light, firstly the center of chirality in a molecule may not always lie on an atom (as *e.g.*, on carbon in the case of 2-bromopropionic acid). Secondly, the number of stereoisomers may be lesser than calculated from 2^n (see Schemes 1.67*f* and 1.67*h*).

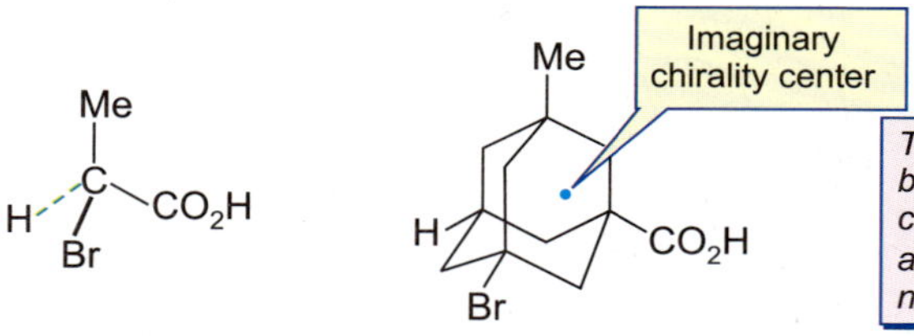

2-Bromopropionic acid

3-Methyl-5-bromoadamantane-carboxylic acid (Chiral) (For chiral adamantanes with a chirality axis (see Scheme 1.126)

The adamantane derivative has four bridgehead substituents and four chirality centers. It has however, only a (±) pair. A chirality center thus may not always reside on an atom.

SCHEME 1.109

However, in some compounds with nonsuperimposable mirror images it is not possible to identify a chirality center and it then becomes necessary to focus our attentions on a larger portion of the molecule. In the same sense as we have spoken of chirality in the case of compounds with a chirality center, we may in the case of some other chiral compounds speak of axial chirality, planar chirality and helicity.

Whilst for models with a chirality center all the four groups have to be different, for ones with an axis of chirality (stereoaxis) a smaller number of differences are sufficient. As one will see later on; for models with a plane of chirality (stereoplane), even one difference is enough.

Thus, the presence of a chirality center is not a necessary however, it is a sufficient condition for molecular dissymmetry. The overall chirality of a molecule can be factorized into three elements, chirality centers, stereoaxes, and stereoplanes, while still another element of chirality is helicity.

Several important classes of chiral compounds have one or more chirality centers. Thus, for example, a chirality center X can be detected in a molecule when the four different ligands *a*, *b*, *c* and *d* of a central atom X are located on corners of a tetrahedron (I, scheme 1.110). In case this center is replaced by a linear grouping *e.g.*, C=C=C, the tetrahedron becomes elongated *i.e.*, extended along the axes of the grouping. On speculative elongation of this tetrahedron as in (II), the chirality center is extended to produce a stereoaxis. In this type of an extended tetrahedron on which axial chirality is based, the conditions for chirality are much less stringent when compared to a regular tetrahedron. Indeed a reference to (II) shows that an extended tetrahedron (with lesser symmetry) will be chiral if the pair of ligands at one end of the axis and the pair at other end constitute two different ligands, *i.e.*, the minimum condition for chirality is that the ligand $a \neq b$ (scheme 1.110).

Elongated tetrahedron approach can be applied to a variety of compounds *e.g.*, allenes, spiranes, biphenyls etc. which are chiral not due to the presence of a chirality center but a stereoaxis. The application to an allene (scheme 1.110) shows, that an allene *e.g.*, 2, 3-pentadiene will be chiral if (minimum) the two substituents at each end are different *i.e.*, ($a \neq b$, scheme 1.110).

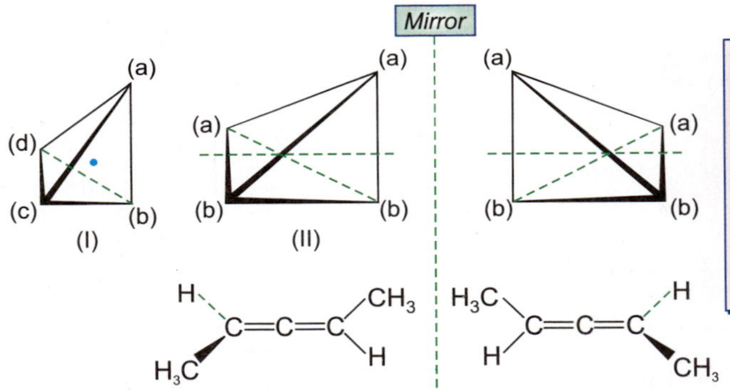

Enantiomers of 2, 3-pentadiene

A chirality center, C_{abcd} in a usual tetrahedral compound ((I) C_1 asymmetric) becomes a stereoaxis (II) on notional elongation of a tetrahedron (C_2 symmetry). Minimal conditions for chirality in compounds with stereoxis e.g., an allene is ($a \neq b$).

SCHEME 1.110

The allenes with three or four different substituents are also chiral, the prediction that suitably substituted allenes could be chiral was made by van't Hoff in 1875 and was verified in I935 by Maitland and Mills.

Several of these axially chiral compounds have a two fold axis of symmetry (C_2) and therefore, cannot be termed asymmetric (see sec. 1.9*c*).

One has already seen that one of the enantiomers of tartaric acid e.g. (+)-enantiomer though chiral has also a C_2 axis (see scheme 1.76). The presence of one or more C_n axes in a compound does not interfere with its chirality. For these reasons the term asymmetric carbon has been replaced by stereogenic carbon (a stereocenter). Interestingly, however, the term asymmetric synthesis still survives. An asymmetric center (carbon) is better termed chirality center. An asymmetric center is sp^3 hybridised carbon attached with four different groups where the interchange of two groups produces an enantiomer. In short chirality can exist in a molecule with an axis of symmetry, but it cannot exist in a compound with a plane of symmetry, a center of symmetry or an alternating axis of symmetry.

A shrewd eye for symmetry will detect a C_2 axis *e.g.,* in the substituted allene 1, 3-dichloropropa-1, 2-diene in its Newman projection ; this C_2 axis passes through the center of the molecule and rotation of the molecule about this axis by 180° gives an identical molecule (II, scheme 1.111). The arrangements (Ia and Ib, scheme 1.111) are enantiomers. The enantiomer (Ib, scheme 1.111) has an equivalent representation in (Ic) and the later is obtained by simply exchange of groups in (Ia) in left end carbon. Thus exchange of ligands at either of the terminal atoms across the stereoaxis reverses the chirality (Chirality axis which is present in allenes and biphenyls is a stereogenic unit). In a stereogenic unit an exchange of ligands generates a stereoisomer. The structure (Ib) when rotated 90° around C=C=C axis gives the equivalent structure (Ic, scheme 1.111).

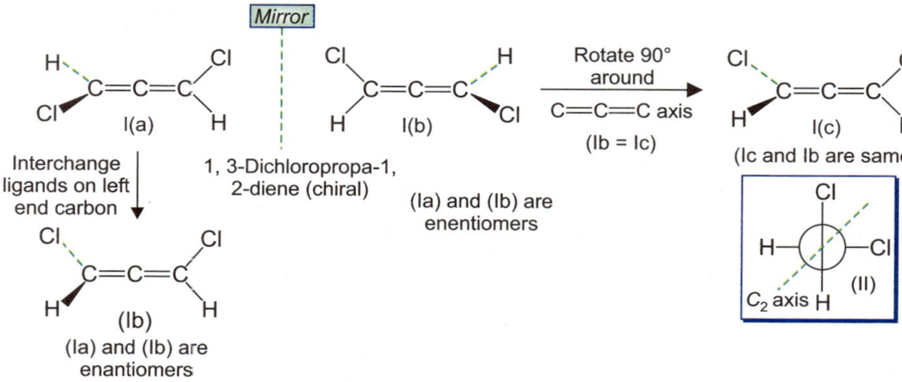

SCHEME 1.111

The allenes of the general formula Cab = C = Cab have a C_2 axis (point group C_2). In case the three or four of the end groups of an allene are different the C_2 axis disappears and then the molecule becomes totally asymmetric belonging to C_1 point group (See scheme 1.112).

(A) Optical Isomerism of Allenes Spiranes and Related Compounds (Axial Chirality)

In the spatial arrangement of the cumulative double bonds of allene, the four substituents of the allene grouping are situated at the apexes of an imaginary tetrahedron, elongated. The following points may be noted:

- In an allene, the central carbon atom is *sp* hybridized and linear, and the two outer carbon atoms are sp^2 hybridized and trigonal. The central, *sp* hybrid carbon atom must therefore, use different *p* orbitals to form the *pi* bonds with the two outer carbon atoms. The two unhybridized *p* orbitals on a *sp* hybrid carbon atom are perpendicular, so the two *pi* bonds must also be perpendicular (scheme 1.112).
- In allene (I, 1, 2-propadiene, symmetry group D_{2d}) the C=C=C is a potential chiral axis. The planes defined by H(Cl)H and H(C3)H are mutually perpendicular.

- Allene itself is achiral since it has two planes of symmetry. In order to generate chirality, the two planes of symmetry must be eliminated. When unlike substituents are added at each end of the C=C=C unit, the substituted allene becomes chiral. This is so in 2, 3-pentadiene (scheme 1.112) substituent at one end (H and CH_3 ; H ≠ CH_3) and the other end (H and CH_3 ; H ≠ CH_3).
- An allene with unlike substituents only at one end as in (scheme 1.112) leaves one symmetry plane and the allene is achiral.
- The allene (IV, scheme 1.112) with all the four end groups different (this is also so when three end groups are different) loses the C_2 axis and becomes asymmetric belonging to C_1 point group.

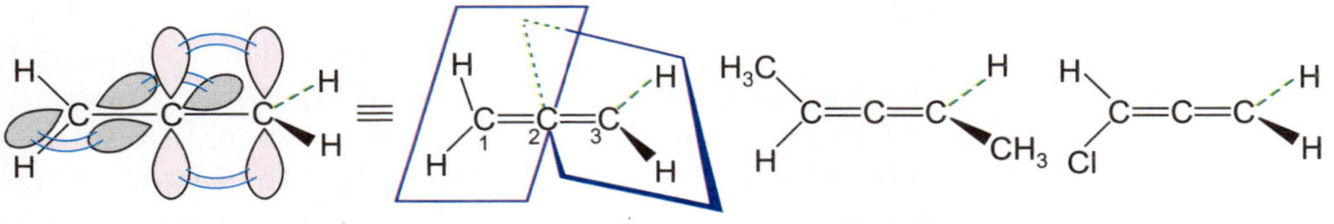

(I) Allene (propa-1, 2-diene) (Achiral) | (II) 2-3-Pentadiene (Chiral) | (III) (Achiral)

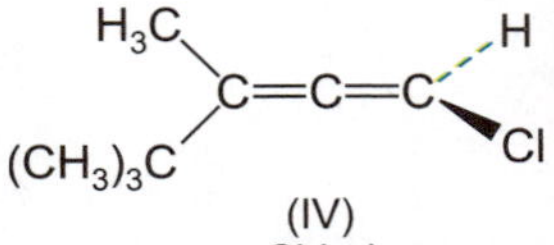

(IV) Chiral (Asymmetric)

Substituted allenes [minimum each pair of ligands at end carbons different abC=C=Cab (a ≠ b)] are chiral. Chiral allenes e.g., 2, 3-pentadiene have a C_2 axis. The interchange of groups at either end reverses the chirality - gives an enantiomer (see, scheme (1.111). The stereogenic unit in chiral

SCHEME 1.112

The cumulated bonding systems (compounds with two or more successive double bonds) of the type (I, Scheme 1.113) with an even number of double bonds do not have a plane of symmetry or a center of symmetry and therefore, must be chiral, resolvable into enantiomers.

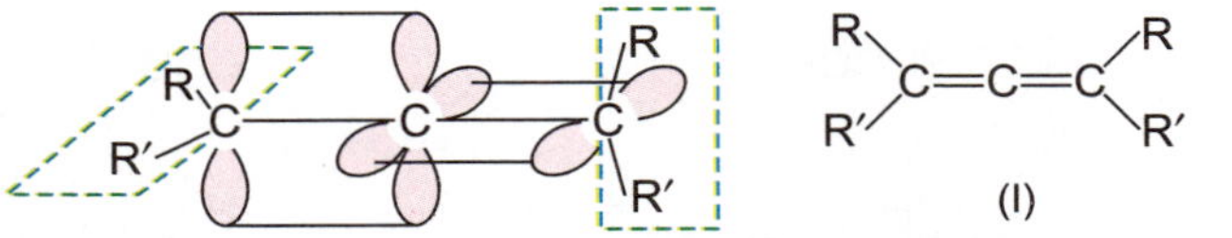

(I)

A cumulene (even number of double bonds) shows enantiomerism if R ≠ R′
The four substituents lie in two perpendicular planes

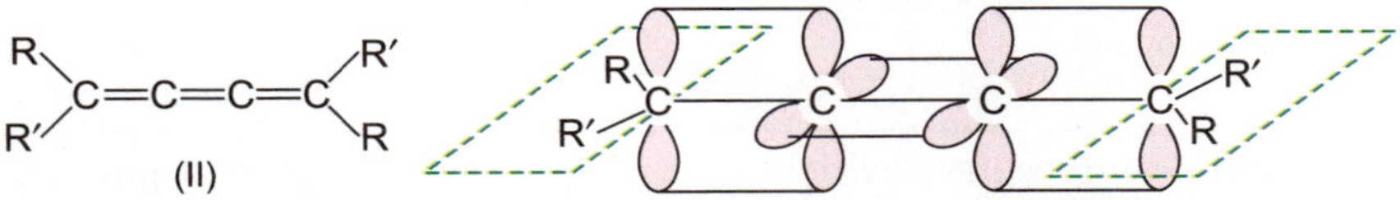

(II)

A cumulene (odd number of double bonds) shows geometrical isomerism.
The four substituents lie in the same plane, These cumulenes display π-diastereomerism and *E, Z* nomenclature is applicable (see, Scheme 1.45)

SCHEME 1.113

Interestingly the compounds with odd number of cumulated double bonds instead *display Z-E* (geometrical) isomerism. When the allene chain of compound (I, scheme 1.113) is extended by one more double bond (introduction of another *sp*-hybridized carbon atom) one then gets a system (II, scheme 1.113) in which the substituted groups at the two ends of the cumulated chain now lie in the same plane and geometrical isomerism is shown.

The replacement of one double bond in an allene by a ring gives alkylidenecycloalkanes (sometimes referred to an hemispiranes) does not alter the basic geometry of the system of allenes and suitably substituted compounds, therefore, exist in optically active form *e.g.*, (I and II, scheme 1.114). Related compounds in which sp^2 -carbon is replaced by nitrogen, *e.g.*, compound (III, scheme 1.114) has also been obtained as enantiomers.

The overall framework in (IV, scheme 1.114) is planar (a cumulene with odd number of double bond, the five membered ring is equivalent to a double bond). The compound displays *cis, trans* isomerism, due to two chirality centers, the system can display a pair of enantiomers as well. The total number of stereoisomers is four and not eight the compound (IV) infact has three stereogenic units and expected to display eight stereoisomers (see, schemes 1.15 and 1.14*d*) from the relation 2^n where n = stereogenic units. The number of stereoisomers is less than expected, since the ligands attached to the chirality centers are similar, thus some stereoisomers become *meso* (compare tartaric acid).

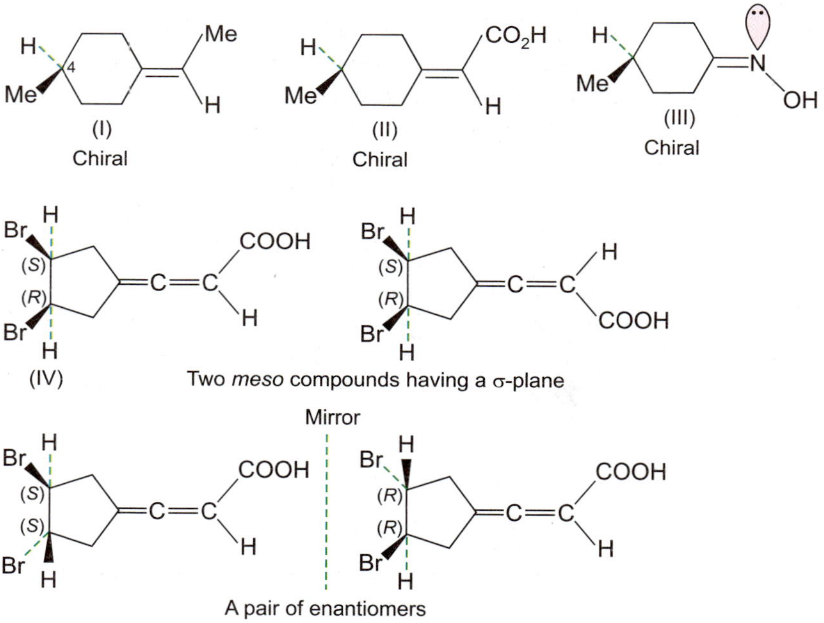

SCHEME 1.114

Recall that a stereocenter which has two identical enantiomorphic ligands and two other non-identical ligands, and which lies on a plane of symmetry, is termed a pseudochirality center (pseudoasymmetric centre). The stereodescriptors in this case are given as lower case italic letters to differentiate such centres from true chirality centres (See, Schemes 1.31 and 1.67*c*). Because two of the ligands on the pseudoasymmetric centre are the same, in terms of atomic number, a further criterion is invoked and that is that an R stereogenic centre has priority over an S (R before S in the alphabet also refer to problem 1.45).

From these considerations the allene (I, scheme 1.114*a*) is then said to have a pseudochirality axis (pseudoasymmetric axis). Since the aliphatic side chains in (I, scheme 1.114*a*) are mirror symmetric, both the atoms (positions 3 and 5) have opposite configuration. Thus the allene has a pseudochirality axis which is attached to two enantiomorphic residues at one end and two different groups at the other. Of these enantiomorphic residues, the R-configured branch has priority over the S-configured branch. Therefore, the r configuration is assigned to the allene.

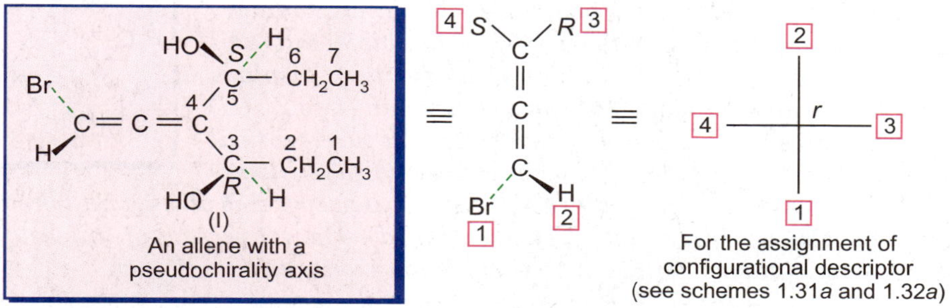

3*R*, 4r$_a$, 5*S*)-4-(2-bromoethenylidene) heptane-3, 5-diol)

SCHEME 1.114a

The replacement of both double bonds in an allene by ring systems gives a spiran. If neither ring contains a plane of symmetry, *spiro* compounds become chiral. Examples are (IV) and *S*-(+)-*spiro* [3, 3] hepta-1, 5-diene (V) (Scheme 1.114*b*).

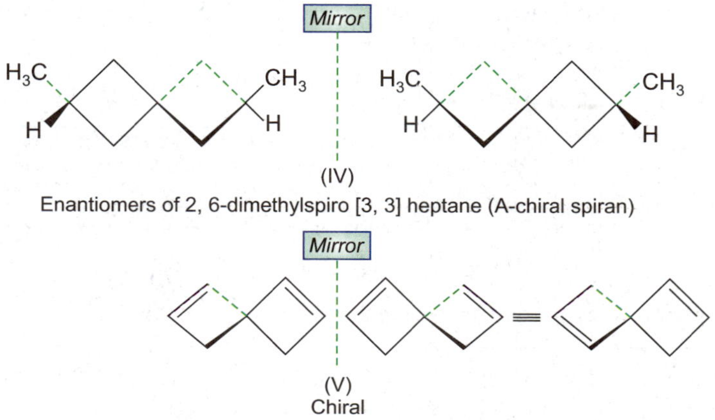

SCHEME 1.114b

EXERCISE 1.18

Which of the following compounds (Scheme 1.114c) is expected to be chiral ?

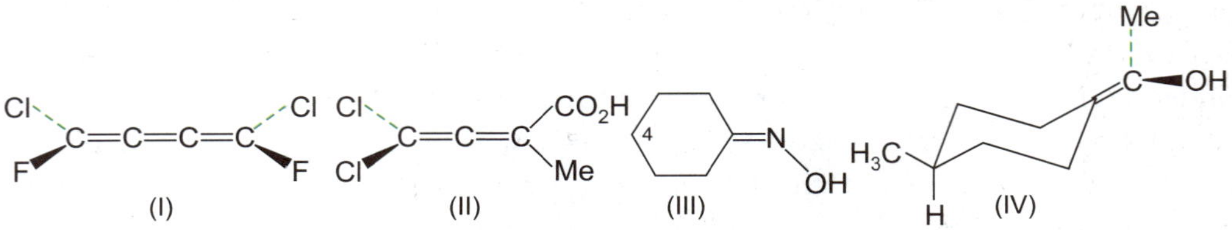

SCHEME 1.114c

ANSWER. *Compound (I) is achiral (the molecule contains three adjacent π bonds, odd number of double bonds and thus the molecule behaves like an alkene and displays geometrical isomerizm and the configuration in (I) is Z ; (II) is achiral since in this allene one carbon carries two identical substituents ; (III) is achiral since there are no substituents at C4 of the six-membered ring ; (IV) is chiral, however it is an enol and is convertible to more stable achiral ketone (Scheme 1.114d). The conversion of this ketone under acid or base catalysis will give back (IV) along with its enantiomer.*

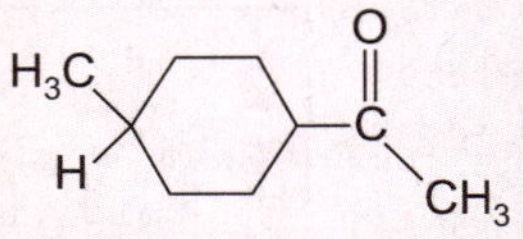

SCHEME 1.114d

(B) Optical Isomerism in Biphenyls (Atropisomerism)

(a) Optically Active Biphenyl Derivatives

The two conformations of *meso*-2, 3-dichlorobutane are nonsuperimposable mirror reflections of each other (scheme 1.115). It is, however, not possible to separate the enantiomeric forms since rotation about the central C—C bond occurs very rapidly resulting in the interconversion of the two conformations. When the barrier to rotation about a C—C exceeds about 80kJ per mole in some suitably substituted compounds, the rotation at room temperature is slow enough to allow isolation of the two optically active isomers. This is so in the case of some biphenyls.

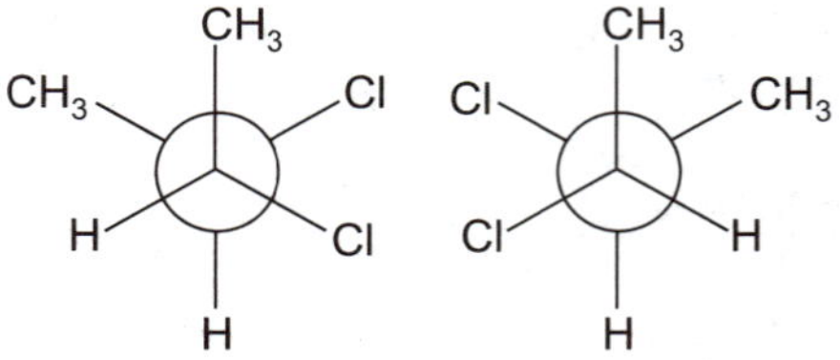

***meso*-2-3-Dichlorobutane** *has two conformations which are non-superimposable mirror images. The two enantiomers can not be isolated due to fast rotation about central C-C bond.*

SCHEME 1.115

In the crystal, both benzene rings of biphenyl lie in the same plane (I, Scheme 1.116). However, in solution and vapour phase the two rings are twisted with respect to each other by an angle of 45° due to steric interactions between the 2, 2′ and 6, 6′ pairs of hydrogens (II, Scheme 1.116). These interaction effects are further enhanced by *ortho* substituents larger than hydrogen so that:

- The rotation about the bond linking the two phenyl rings does not occur due to steric hindrance between the bulky *ortho* substituents.
- The two rings lie in different planes which are perpendicular, to make it impossible for the molecule to achieve a conformation in which the two aromatic rings are coplanar and this conformation has a plane of symmetry and this rules out chirality. This is so in a symmetrical structure shown for example by the planar formula of O, O′ difluorodiphenic acid (see scheme 1.120).
- Resolvable (chiral) biphenyls must contain two different *ortho* substituents on each ring and these make each ring unsymmetrical (H is taken as a substituent) and two rings are held in perpendicular planes.

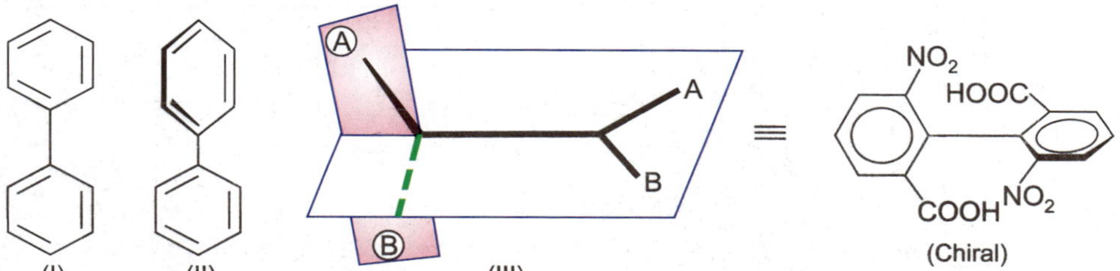

(I) Crystal (II) Solution
Bipnenyl

(III)
Substituted biphenyl with different bulky *ortho* substituents A and B on each ring

A = NO_2
B = COOH

SCHEME 1.116

- Each ring in a biphenyl individually represents a plane of symmetry irrespective of *ortho* substituents (and other groups). Consider the biphenyl which can be schematically represented (III, scheme 1.116) depicting the rings in perpendicular planes with two bulky ortho substituents (A) and (B) on each ring (A ≠ B). Such a biphenyl is chiral since neither of these planes can bisect the other into two equal halves.
- The biphenyl (III, scheme 1.116) has a mirror image (scheme 1.117) and can be resolved into its enantiomers and each enantiomer is stable indefinitely. The nitro and carboxylic groups are so bulky that they cannot pass by each other.

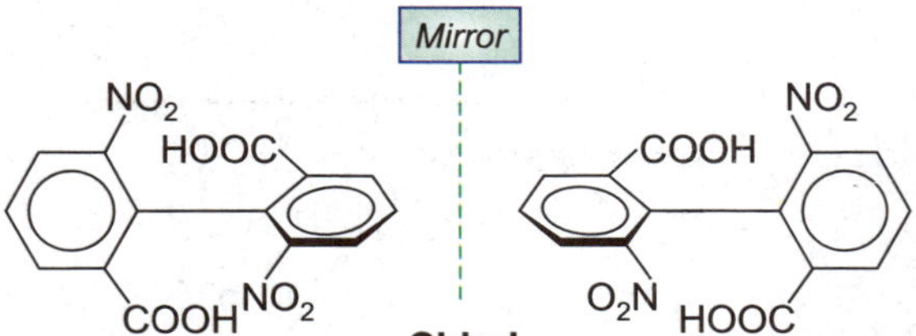

Chiral
(Optically active)

Bulky O-substituents restrict free rotation, the rings remain orthogonal (perpendicular to each other). When the O-substituents on either side of inter-ring bond are different the biphenyl becomes chiral.

SCHEME 1.117

- None of the rings should be symmetrically substituted, so that the molecule cannot have a plane of symmetry. Thus, the biphenyl (scheme 1.117) is chiral (A ≠ B in either pair, COOH ≠ NO_2 of the *ortho* substituents. The biphenyl (scheme 1.118) is, however, achiral. In this case *e.g.*, ring (II) is symmetrically substituted (A = B = COOH). A plane of ring (I) drawn perpendicular to ring (II, scheme 1.118) is a plane of symmetry and since it bisects the plane of ring (II) into two equal halves, thus the biphenyl (scheme 1.118) is achiral.

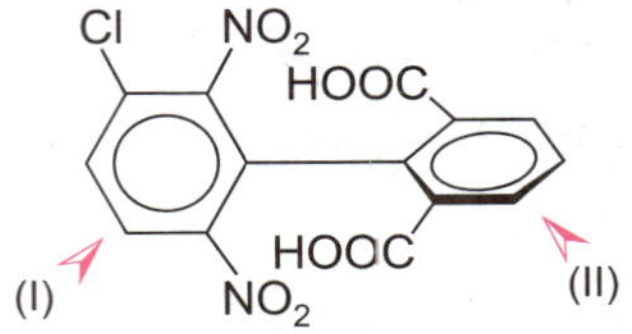

Achiral

If one or both rings in a biphenyl have identical ligands, it is achiral. Each ring individually represents a plane of symmetry irrespective of O-groups. When held in mutually perpendicular planes, one plane of symmetry should not bisect the other (or vice versa) into two equal halves.

SCHEME 1.118

Atropisomerism

The chirality in biphenyls is generated by restricted rotation around a single bond provided each ring has an appropriately substituent pattern. Isolable stereoisomers resulting from restricted rotation about single bonds are called atropisomers while rotamers are stereoisomers obtained by rotation about a single bond.

The term atropisomer is derived from the Greek "without turning." Atropisomers may be configurationally stable at room temperature when the hindrance to rotation may be rather severe. Racemization can take place when the sizes of the ortho-substituents are such that these can slip past each other and the two rings become co-planar.

In order to display optical activity the substituents in the *ortho* position must be large enough to prevent the two rings from becoming coplanar (the rotational energy barrier must be high enough so that interconversion of enantiomeric conformers does not occur). Thus *e.g.*, all attempts to resolve diphenic acid (scheme 1.119) have failed.

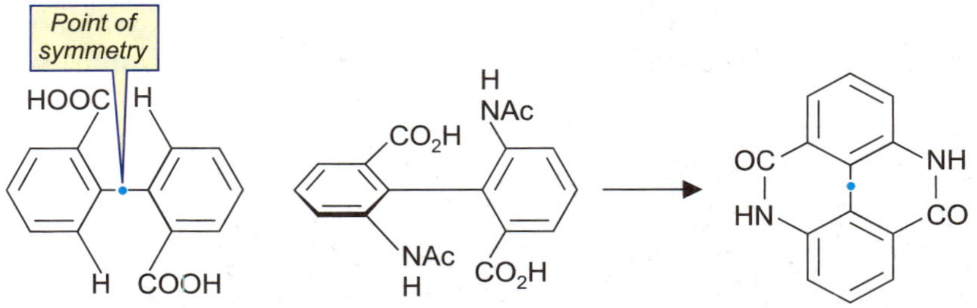

Biphenyl-2-2′-dicarboxylic acid (Diphenic acid) Achiral

(I) Chiral (Optically active)

(II) (Achiral)

SCHEME 1.119

The process of slipping a small hydrogen past the carboxylic acid group is very facile so that racemization of enantiomers occurs very rapidly through the planar form. In the planar from the center of symmetry is clearly seen.

Interestingly the diamide (I, scheme 1.119) is optically active and is resolvable. The activity is lost on hydrolysis since the resulting dilactam (II, scheme 1.119) is forced to be planar and a point of symmetry can be easily detected in it.

In the case of compounds with a chirality center, or in the case of allenes, one enantiomer may interconvert into the other only if a chemical bond is ruptured and formed again, *i.e.*, by means of a chemical reaction. In the case of atropisomers, however, the other enantiomer may be formed if the substituent in one *ortho*-position is successful in "pushing through" past the smaller *ortho*-substituent on the second ring. The stability of the optical activity of atropisomers may thus serve as a measure of the size (effective volume) of the substituents. When the bulky nitro groups (scheme 1.117) are replaced by the smaller fluorine atoms the resulting compound, 6, 6′-difluorobiphenyl-2, 2′-dicarboxylic acid can still display optical activity (scheme 1.120). However, the compound racemizes readily, *i.e.*, the enantiomers are readily interconverted. The process involves squeezing fluorines past the adjacent carboxyl groups *via* the planar conformation. Once they reach the planar conformation the chirality is lost and racemization results. This transition state is congested and requires the bending of bonds. The process takes energy and is measurably slow.

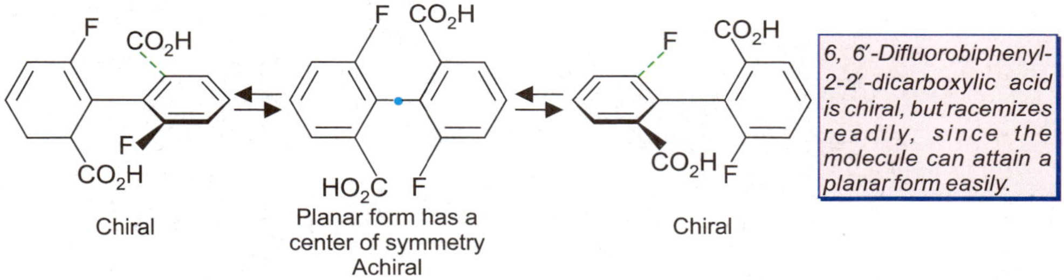

SCHEME 1.120

- Although the hydrogen atom is quite small, optically active compounds exist with two or even three *ortho*-positions of the biphenyl occupied by hydrogen. In the second case the benzene ring that is unsubstituted in the *ortho*-positions must have a substituent in a *meta*-position. The *meta*-substituent has no influence on rotation but it creates the necessary chirality of the molecule ; 3-bromobiphenyl-2′-trimethyl-arsonium iodide (I, scheme 1.121) may serve as an example.
- In the case of (I, scheme 1.121) the rotation around the pivotal bond is only moderately restricted to allow interconversion between two enantiomeric atropisomers in solution and thus mono-*ortho*-substituted biphenyls in general are not resolvable, but show mutarotation when chiral additives are mixed. Thus the (+) camphorsulfonate of the arsonium salt (II, scheme 1.121) shows mutarotation (first order asymmetric transformation, see scheme 1.108) in solution. Here the chiral counterion (counter ion, an ion associated with another ion of opposite charge) (+) camphorsulfonate discriminates between the readily interconvertible atropiosomers of the biphenyl to give two unequally populated

diastereomeric salts. Thus an interconversion of diastereomers occurs in solution since the two diastereomers (*i.e.*, diastereomeric salts) are not equally stable.

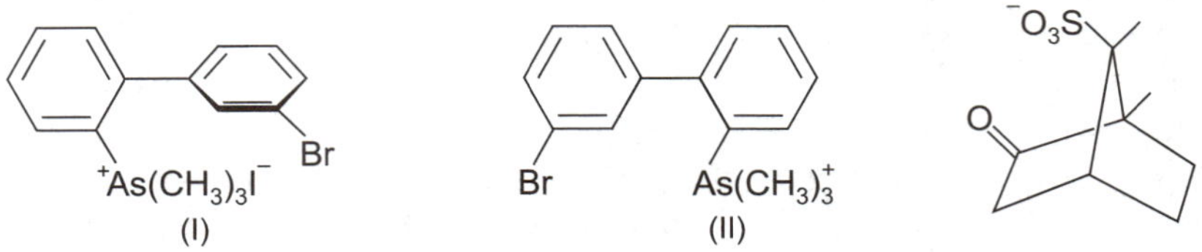

(I) (II)

SCHEME 1.121

- In addition to the bulk of the *ortho* substituents, the substituents in the *meta*-position tend to enhance racemization barriers by what is called as "buttersing effect", *i.e.*, by preventing the outward bending of an *ortho* substituent, which would otherwise occur in the transition state (coplanar conformation) for racemization. This bending would allow the ortho substituents to slip past each other more readily. Thus the rate of racemization of the 3-nitroderivative (I, scheme 1.122) is much lower compared with the 5′-nitroderivative (II, scheme 1.122).

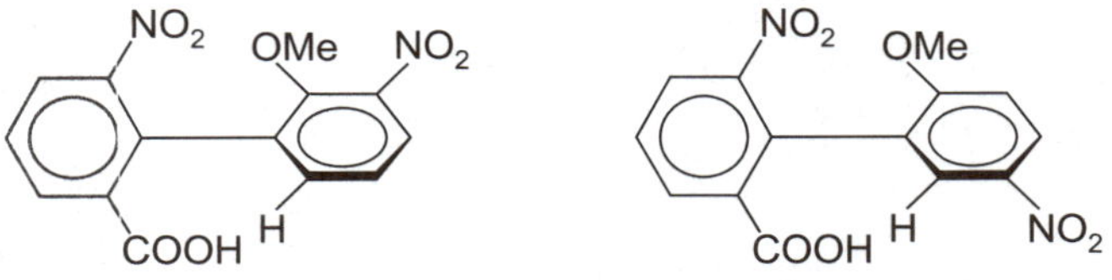

(I) (II)

A "buttressing effect,"

SCHEME 1.122

The apparent order of steric hindrance produced by different groups (as gauged by racemization rates of differently *ortho*-substituted biphenyls) in Br > CH_3 > Cl > NO_2 > COOH > OCH_3 > F > H. This order roughly parallels van der Waals radii of atoms and groups.

EXERCISE 1.19

(1) What is the stereogenic unit in the chiral biphenyl (I, scheme 1.122a) ? (2) What would happen if rotation about the junction bond occurs? (3) What happens if e.g., the ligands in ring A are exchanged?

ANSWER. *(1) A stereoaxis. (2) It would lead to racemization (scheme 1.122a). (3) As the definition of the stereogenic unit requires, exchange of an appropriate pair of ligands generates a stereoisomer. Exchange of ligands in ring A of (I, scheme 1.122a) gives its enantiomer (II). Formation of (II) from (I) is also a result of rotation of rings around the junction bond (racemization).*

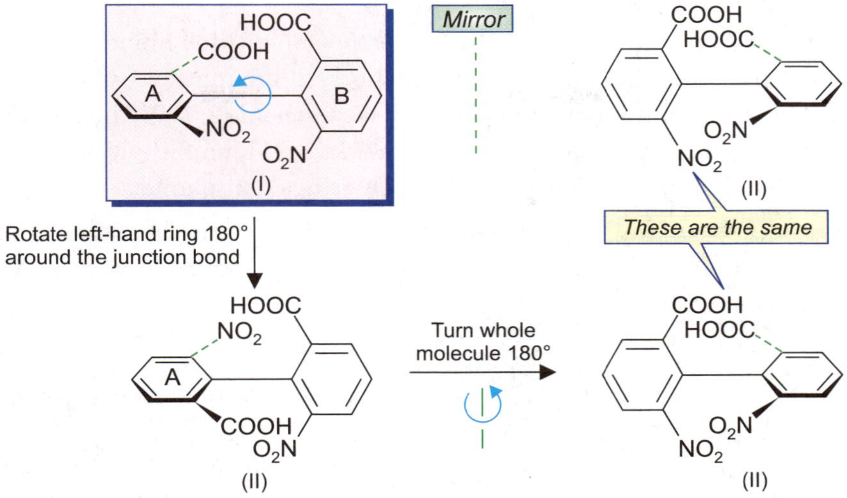

SCHEME 1.122a

(b) Bridged Biphenyls

When a 2, 2′-bridge of the type shown in (I, scheme 1.123) with $n = 1$, one has a disubstituted fluorene, this being a planar molecule, is not resolvable. When $n = 2$, the compound is a disubstituted 9,10-dihydrophenanthrene. The nonplanar six membered ring can give rise to atropisomerism and the compound is resolvable, provided the other two *ortho-* positions are substituted with bulkyl groups *i.e.*, methyl (II, scheme 1.123), otherwise they slip through the plane readily (please consider models). In case n is larger than 2 the bridged biphenyls lead to atropisomers irrespective of the bulk of two *ortho* substituents. In these cases the non-planarity is maintained by the puckering of the rings which in planar configuration suffer from angle strain and non-bonded interactions. An example of such a compound is (III, scheme 1.123), however, these bridged biphenyls undergo easy racemization since the angle strain and steric interaction in medium rings is not very large.

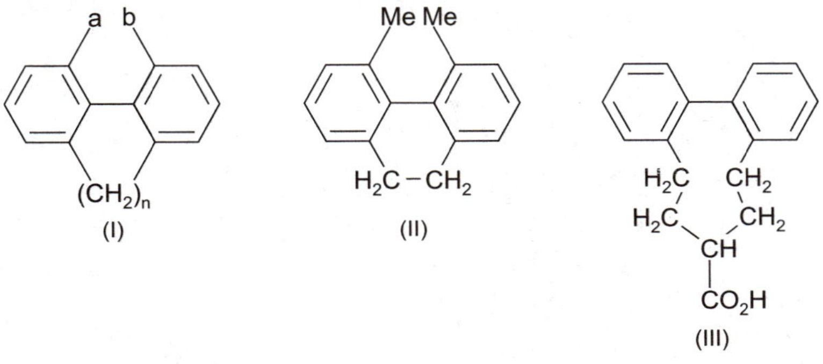

SCHEME 1.123

(c) Atropisomerism in Compounds Other Than Biphenyls

1, 1′-Binaphthyl may be looked as a di-*ortho*-substituted biphenyl (for structure see Scheme 1.108c) and has been resolved. The 2, 2′-binaphthol (scheme 1.124) is a constituent of a useful chiral reagent BINAL -H (see, scheme 2.53*f*). The rotation around the pivotal bond is hindered due to steric factors. In 1, 1′-binaphthyl itself *e.g.*, the peri H's provide enough steric hinderance to keep the rings non-planar.

One or both of the phenyl groups may be replaced by other aromatic or heteroaromatic rings, appropriately substituted *N*-phenylpyrrole and *N*, *N*′-bipyrryl (scheme 1.124) are resolvable.

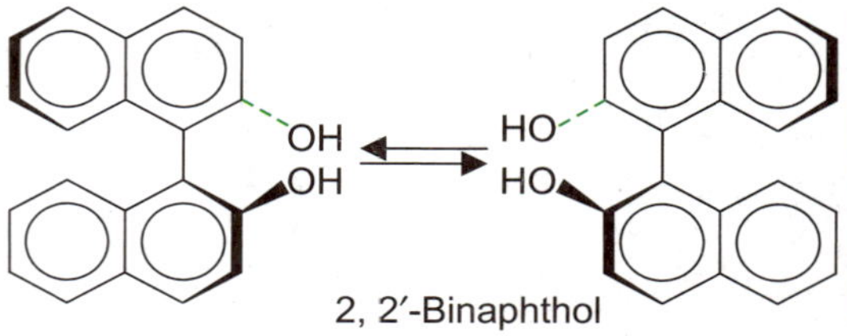

2, 2′-Binaphthol

2, 2′-Binaphthol and other related compounds display chirality (axial) due to restricted rotation around the bond joining the aromatic units. The rings remain orthogonal (perpendicular) to one another and the structure is able to exist in two enantiomeric forms (atropisomerism). 1, 1′-Binaphthyl is constituent of several chiral reagents (see schemes 2.54 and 6.59b)

SCHEME 1.124

Atropisomers may also be displayed by styrenes (I, scheme 1.125) or another situations with restricted rotation about single bonds (scheme, 1.125). High torsional barriers to rotation about sp^2–sp^3 and sp^3–sp^3 in (II and III respectively, scheme 1.125) make these compounds chiral and resolvable.

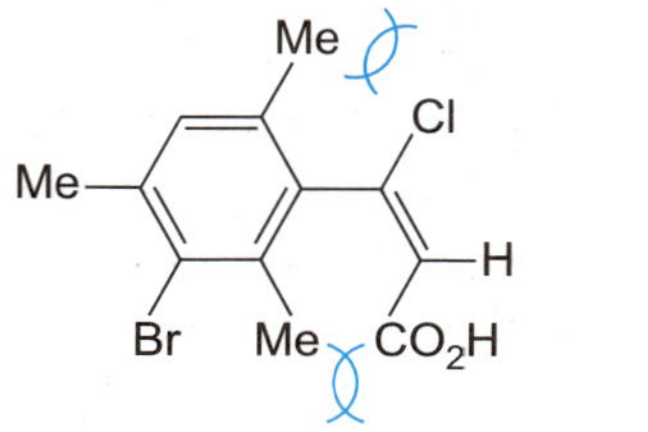

(I)

C=C Not planar with aryl ring, the compound is resolvable

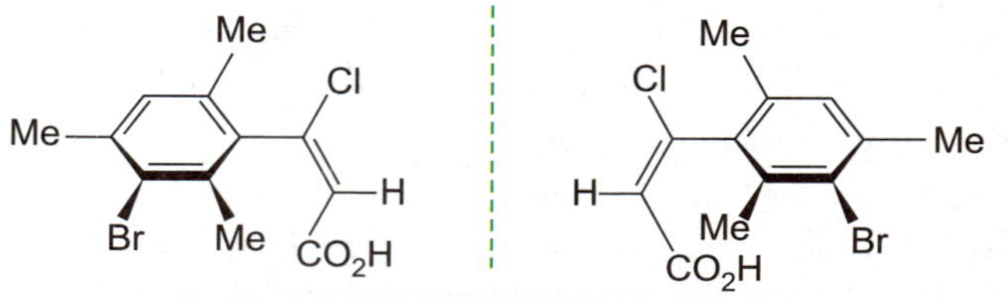

Steric crowding would be generated if the molecule had planar conformation

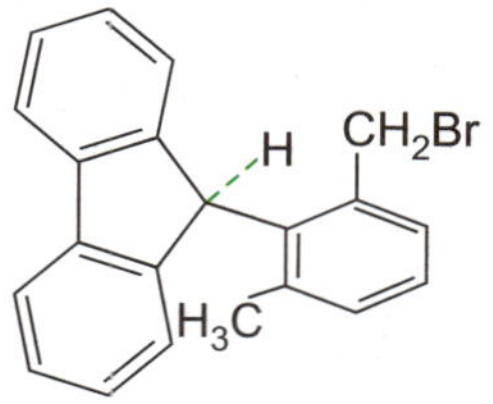

(II)

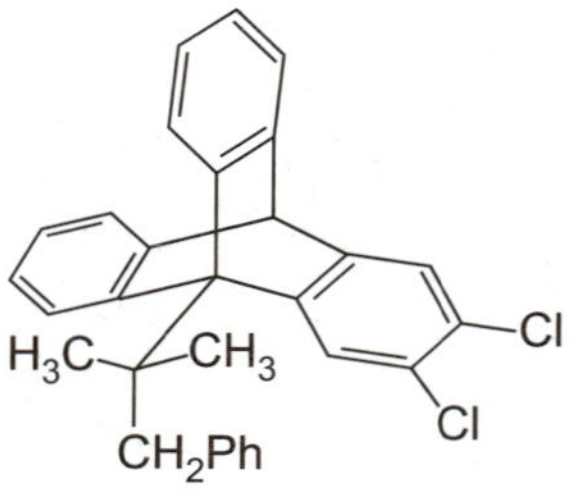

(III)

SCHEME 1.125

Optical Activity of Adamantanes

An example of a cyclic compound e.g., an adamantane derivative where the center of chirality is not located on any atom but in the unoccupied space inside the molecule has been presented (see, scheme 1.109). Adamantane has only two different carbons, the CH and CH_2 groups. Symmetry makes the four methine groups equivalent and the six methylene groups equivalent ((I) and (II) scheme 1.126). This makes (III and IV) equivalent (see the relative location of substitutents in both III and IV). The compound with two chirality centres is achiral with a plane of symmetry.

One may note that dimetrically opposite methylene planes in an adamantane are perpendicular to one another and with proper substitution axial chirality is generated as is so in allenes. One such example is of an optically active adamantoid is (V, Scheme 1.126) which has axial chirality. In this case C2 and C6 methylenes are not planar but lie in different planes which are perpendicular just like in biphenyl, and with proper substitution (as in a chiral biphenyl) become chiral. There is more material on adamantanes (see, scheme 4.56)

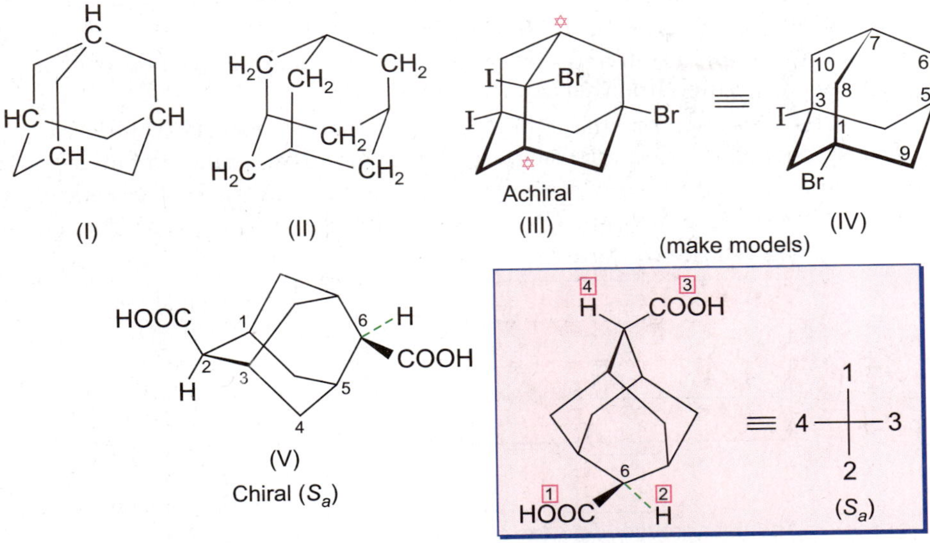

Apply the procedure detailed in (Schemes 1.132 and 1.132b)

SCHEME 1.126

(C) Nomenclature of Compounds with Chirality Axes (Axial Chirality)

Two ways may be employed for assigning R/S configuration to compounds with a chirality axis. One may adopt any of these. The second method is developed (P.S. Kalsi) as a problem solving hint and is the easiest.

Stereodescriptors to Denote the Absolute Configuration

Generally, for tetrahedrally or trigonal pyramidally coordinated chirality centres, the two possibilities are denoted by the stereodescriptors R and S. For compounds with chirality axes or chirality planes, the stereodescriptors R_a and S_a, R_p and S_p are used, respectively. Helical chirality is denoted by the descriptors M and P.

For the compounds with chirality axes or chirality planes the stereodescriptors R and S may be sufficient. For the sake of more clarity the prefixes R_a, S_a for compounds with chirality axes and R_p and S_p for compounds with chirality planes are used indicating that the stereochemical descriptor refers to a chirality axis or a chirality plane.

First Method

(*i*) Allenes

For the assignment of configurational nomenclature to axially chiral molecules *e.g.*, allenes, the orbital picture (see scheme 1.112) is projected to a Newman formula *i.e.*, (I → Ia scheme 1.127). The following points may be noted when the procedure is applied to (*S*)-2, 3-pentadiene.

- The molecule can be viewed from either end of the stereoaxis to give the same descriptor. The near groups are given priority over far groups.
- For better visualization it is always better to put the groups nearest the view direction on a thick line (horizontal or vertical).
- Thus in the projection formula of (*S*)-2, 3-pentadiene (scheme 1.127), the horizontally placed, (nearest the view direction) CH_3 and H are to be numbered 1 and 2 and the remaining vertically placed (rear) ligands are simply assigned priorities according to sequence rules *i.e.* CH_3 and H are numbered 3 and 4 respectively. The sequence 1 → 2 → 3 gives the configurational descriptor *S* (anticlockwise).

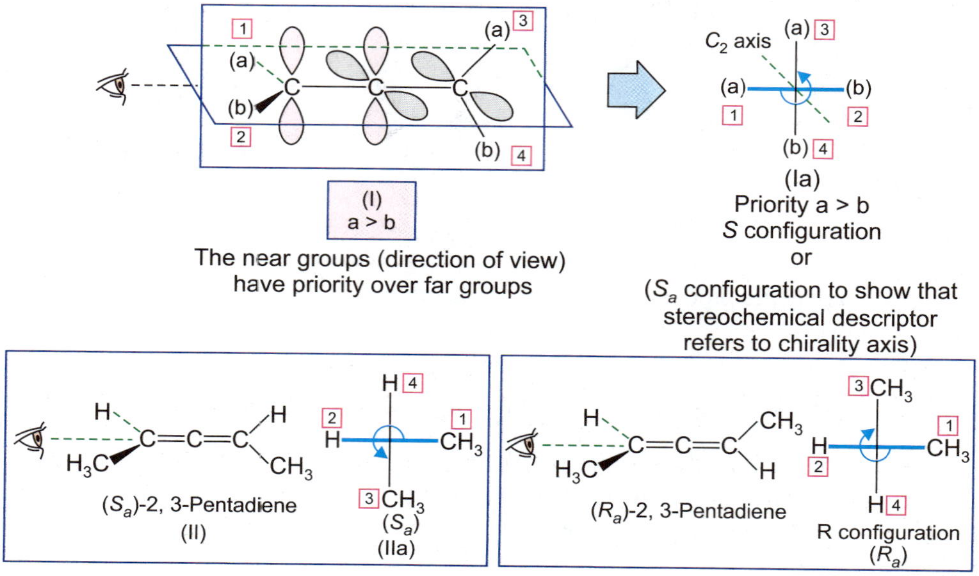

SCHEME 1.127

- Interchange of the two geminal groups at one end in these molecules gives the enantiomer, *e.g.*, (*R*)-2, 3-pentadiene is obtained from the (*S*)-enantiomer (scheme 1.127) and is properly named as shown.
- As already said, it makes no difference which end the structure is viewed, the stereodescriptor will remain unchanged. Consider again (*S*)-2-, 3-pentadiene now viewed from other direction (scheme 1.128) as compared to that in (scheme 1.127). Now the far groups (view, scheme 1.127) become the near groups and consequently get priority. These are put on the thick vertical line and the configuration again comes out to be (*S*).

A beginner may note (ignoring the priority numbers) that the position of vertical groups in Newman projection (IIa, scheme 1.127 and III, scheme 1.128 remains the same from either view direction (front or rear) however, the groups on horizontal line have interchanged positions—the H on the horizontal line in (view II, scheme 1.127) is on the left but will come on the right when viewed from the other direction.

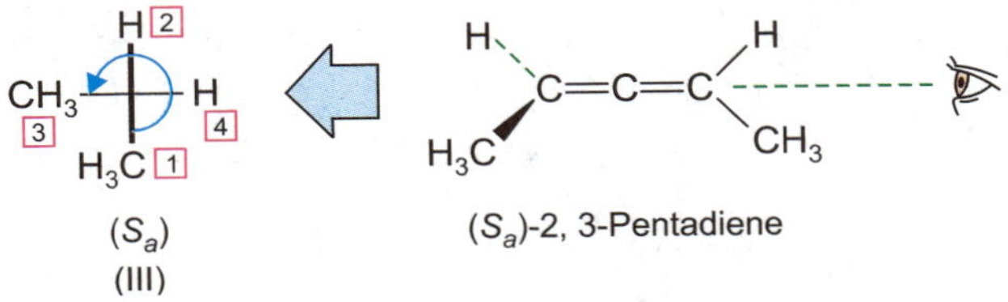

SCHEME 1.128

- For practice one may assign stereodescriptors to the two enantiomers of 2, 3-pentadiene drawn in different orientation (scheme 1.129).

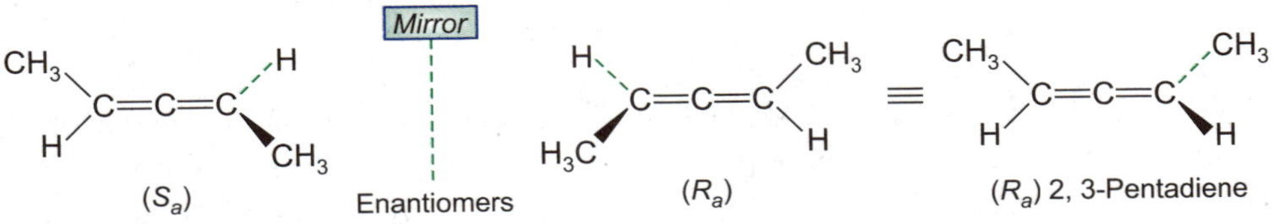

SCHEME 1.129

(*ii*) Spiranes and Alkylidene-Cycloalkanes

These compounds possessing chirality axes are assigned configurational descriptors in the same way as discussed for allenes and also by simpler method (Kalsi, 2004, see scheme 1.132 *a*). As in allenes, the interchange of the two germinal groups in these molecules also leads to enantiomers (II compared to I, scheme 1.130). In alkylidene-cycloalkanes, one of the double bond of allene is replaced by a ring, while in spiranes both the double bonds of an allene are replaced by rings.

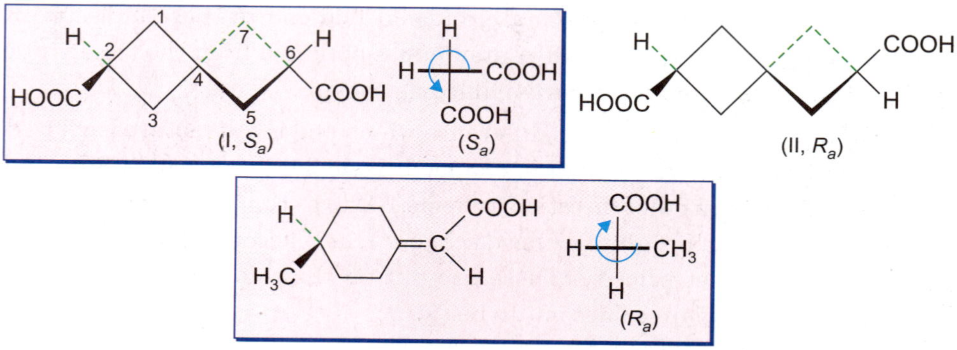

SCHEME 1.130

(*iii*) Biphenyls

In the case of biphenyls, however, it is important to note that the four *ortho*-carbons are sequenced properly according to CIP rules. Thus in the biphenyl (scheme 1.131) in the left ring the sequence is C-2 > C-6 > (OCH_3 > C-H) while in the other ring C2′ > C6′ (N > C) and therefore, configuration is S when the biphenyl (scheme 1.131) is viewed from either direction.

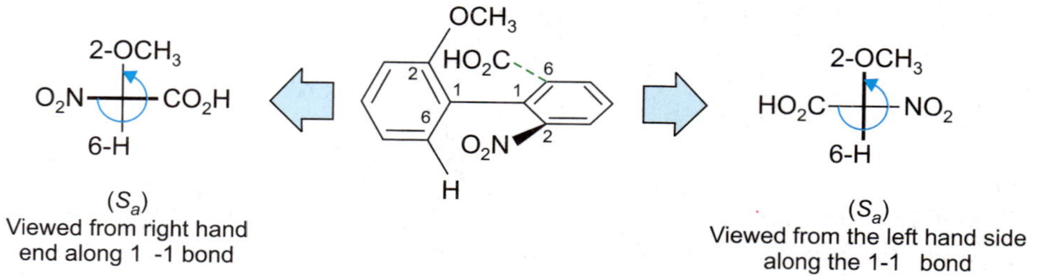

SCHEME 1.131

When in a biphenyl the C2 and C6 in a ring are attached to identical atoms, the priority order of the *ortho* carbon atoms is then determined through exploration around the ring or side chain. Thus in the biphenyl (scheme 1.132) the 2-H adjacent to 3-Br precedes 6-H.

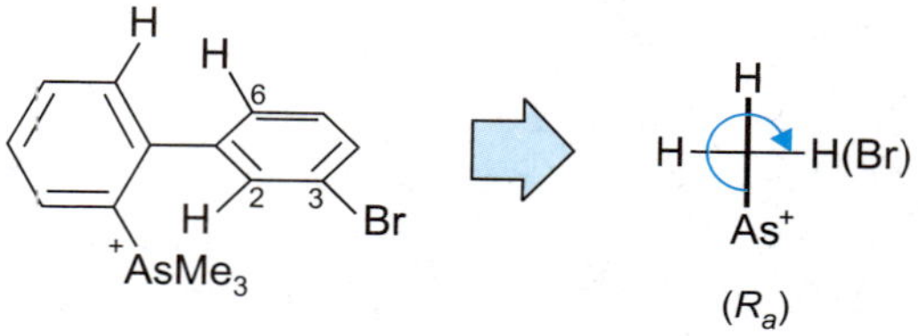

SCHEME 1.132

Second Method

A simple mental exercise to assign R/S configuration to compounds with axial chirality (Allenes, spiranes and biphenyls)

P.S. Kalsi: *A Simple Mental Exercise for Assigning Configurational Descriptors (R/S) to Molecules with a Stereoaxis (Chiral axis)*, Int. J. Chem. Sci. 2(3), 2004, 285–290.

This is the simplest exercise developed by author and can be used quickly to assign configuration to any compound with a stereoaxis and is used by adopting the following procedure:

- Turn the molecule to bring the groups in the plane of paper (continuous lines) at the top while the pair of groups which are perpendicular to this plane *i.e.*, stereochemical bonds (with groups on a wedge and a dash) come at the bottom (I-II, scheme 1.132*a*).
- Always view the molecule from bottom, thus the bottom groups take precedence over the other pair at the top (II, scheme 1.32*a*) and assign priorities to all the four groups.
- Draw a cross (III, scheme 1.132*a*) and put the groups (as their priorities) on the top of (II) on the horizontal line as you see these in (II). Thus Br(3) is on the right hand side while CH_3(4) is on the left.

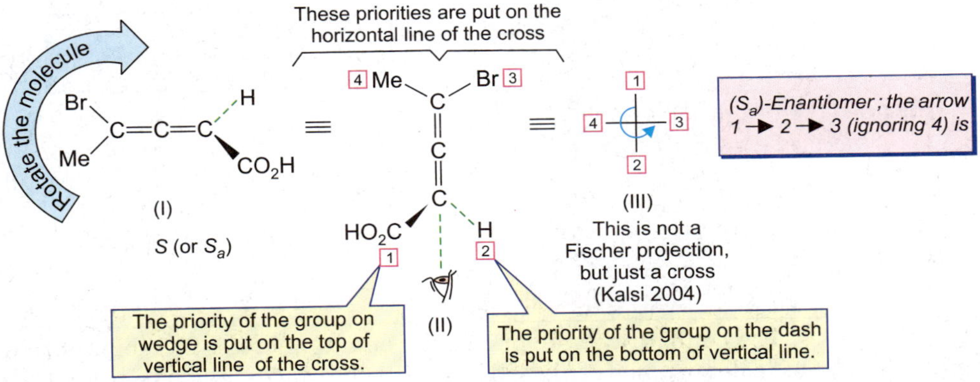

SCHEME 1.132a

- Put the groups of the bottom pair (perpendicular to the plane of paper) on the vertical line of the cross with the group on the thick line (*i.e.*, the group on the wedge) at the top while the group on the dash *i.e.*, the group on dotted line) at the bottom of the vertical line.
- Ignoring the priority group (4), the sequence 1 → 2 → 3 clockwise shows *R* configuration while 1 → 2 → 3 anticlockwise shows *S* configuration.

Example 1. Consider (scheme 1.132*b*) the enantiomer considered here is the same as (I, scheme 1.132*a*). Thus if the orientation (scheme 1.132*b*) is rotated by adopting the procedure above. It comes out that this stereoisomer is *S*-configured and thus (I, scheme 1.32*a* and this stereoisomer (scheme 1.32*b*) represent the same compound and are thus homomers.

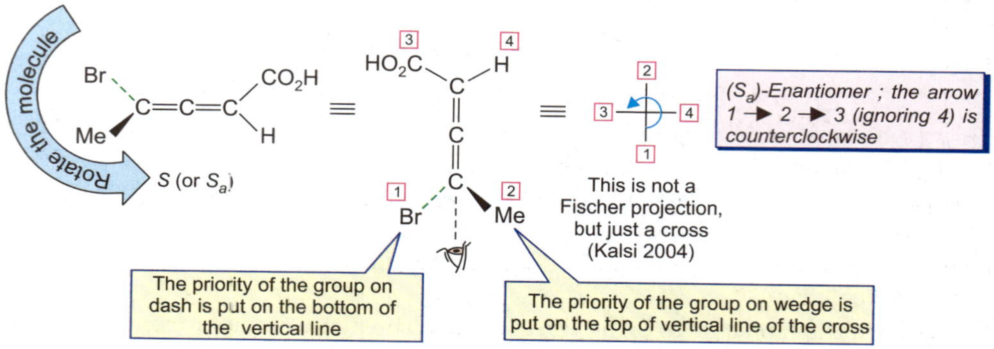

SCHEME 1.132b

Example 2. Consider one of the enantiomers of a chiral biphenyl by considering the relevant *ortho* groups (scheme 1.132c). One turns the molecule so as to put the ring A (in the plane of paper) at top while ring B which is perpendicular to ring A at the bottom. Looking from bottom and assignment of priorities to the *ortho* groups shows that it is *S* enantiomer. The NO_2 group in ring B is on a thick portion of the molecule so it is at the top of vertical line of the cross.

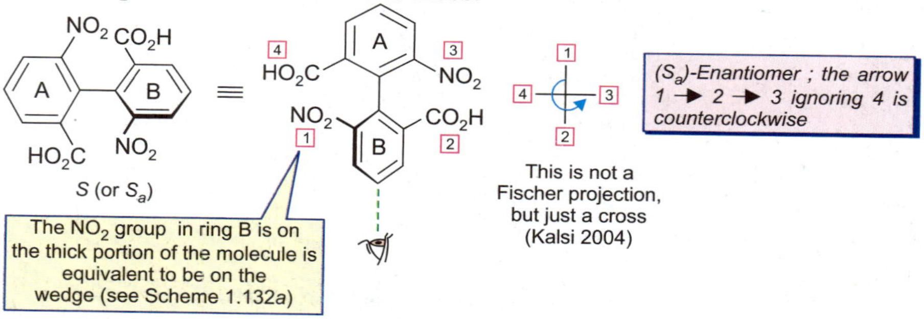

SCHEME 1.132c

Example 3. The configuration of one of the enantiomers of 2, 2′-binaphthol (I, scheme 1.132*d*) is to be determined. This will be with reference to the carbons C2, C9 and C2′ and C9′. One looks along C1 and C1′ bond, by first turning the orientation so that as drawn, the ring system A (in the plane of the paper) is at top while the ring system B perpendicular to it is at bottom (I → II, scheme 1.132*d*). As in other cases the molecule is viewed from bottom and priorities are assigned. The configuration comes out to be *R*.

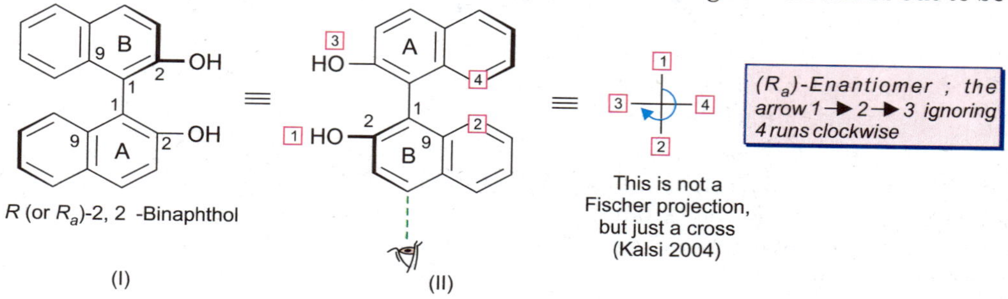

SCHEME 1.132d

In alkylidenecycloalkanes *e.g.*, (I, scheme 1.132*e*) one of the double bonds of an allene is replaced by a ring and the basic geometry does not change and these compounds display enantiomerism.

A number of cyclic oximes, semicarbazones and phenylhydrazones (II, R = OH, $NHCONH_2$, and NHPh scheme 1.132*e*) have also been resolved, in these compounds sp^2–carbon is replaced by nitrogen.

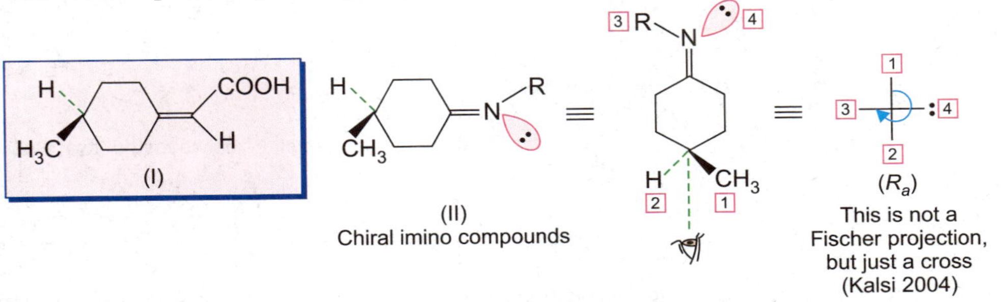

SCHEME 1.132e

(D) Helical Descriptors M and P for Molecules with Chirality Axes

The perpendicular segment of the substituents about the chirality axis means that such compounds like, allenes, biphenyls, alkylidenecycloalkanes, etc. have very short helical segments and their configuration may be denoted as *P* or *M* (as for compounds which are chiral due to helical structure). For the assignment of helix descriptors only the groups of highest priority nearer to the view direction and the far are considered. Thus, the considered, ligands are 1 and 3. In case the turn from priority front group 1 to priority rear group 3 is clockwise, the stereodescriptor is *P*, if anticlockwise it is *M*.

One has seen that the smallest carbocyclic ring that can incorporate a double bond with trans-geometry is eight membered (see schemes 1.53 and 1.136). Similarly the smallest sized ring which can contain an allene has nine carbons. One enantiomer of cyclonona-1, 2-diene ((I), scheme 1.133) has the S_a configuration. Interestingly here a molecule with planar chirality *trans* cyclooctene has been converted to one with axial chirality.

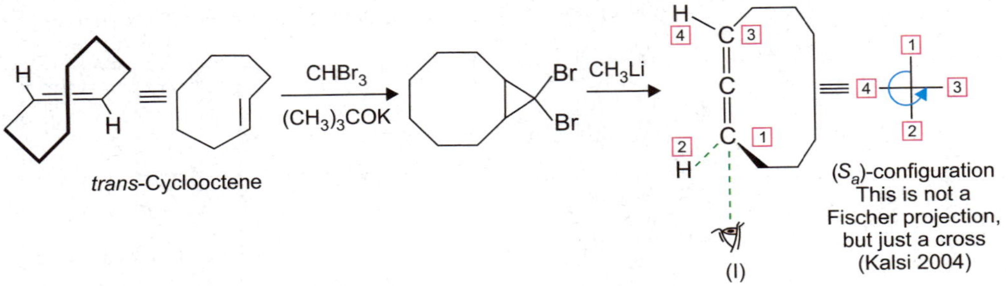

Assignment of stereodescriptor to (I) with a chirality axis

SCHEME 1.133

For the assignment based on helical chirality rewrite the cross derived from (I, scheme 1.133). For the assignment of helical chirality only the groups of highest priority nearer to the view direction and far are considered (scheme 1.133*a*). Moving from priority front group 1 to the priority near group 3 is clockwise (scheme 1.133*a*). Thus the stereodescriptor is *P*.

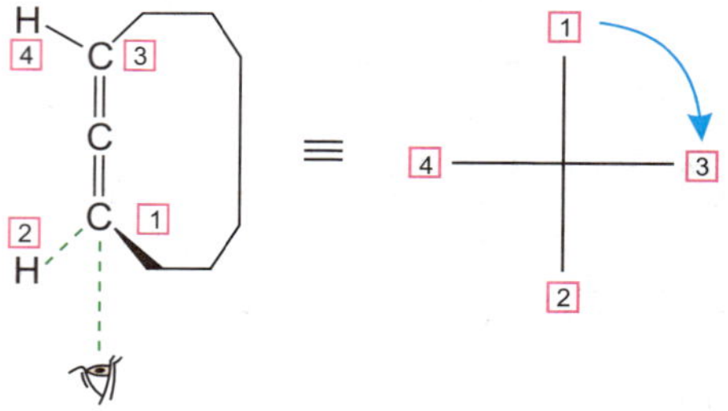

(I) *P*

There is a correlation between the two types of stereodescriptors

$R_a \rightarrow M$; $S_a \rightarrow P$

The compound (I) has the stereodescriptor S_a based on chirality axis (see, scheme 1.133). The configuration may also be denoted as *P* based on helical chirality.

SCHEME 1.133a

EXERCISE: 1.19*a*

(a) Assign configuration to (I, scheme 1.133b) based on axial chirality and helicity.

(b) Write the projection of compound (II scheme 1.133b) which represents S_a enantiomer. Also write the structure of R_a enantiomer.

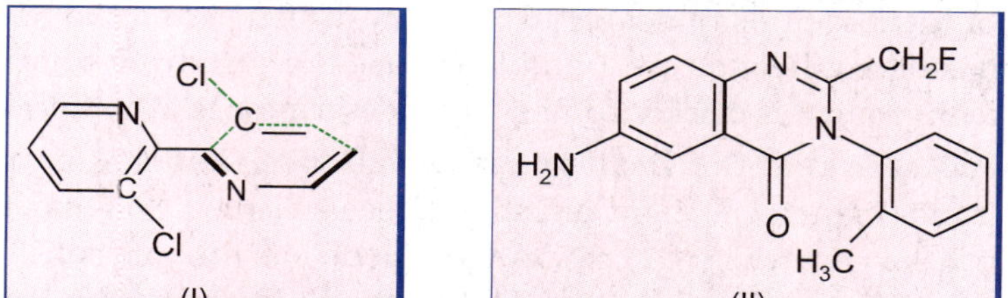

(I) (II)

SCHEME 1.133b

ANSWER: *(a) Rewrite the structure (I) for the application of the procedure (See, scheme 1.133a).*

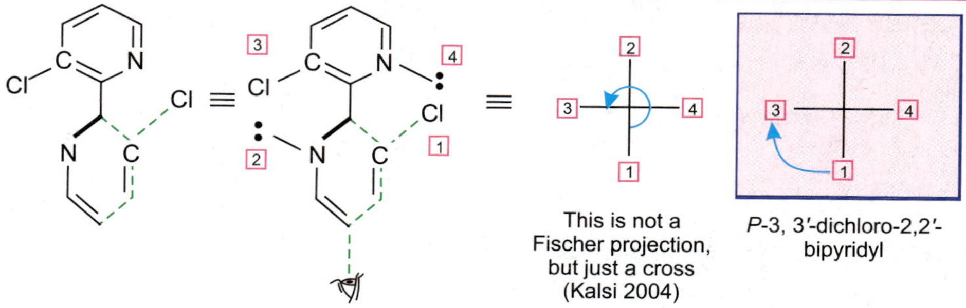

(S_a)-3, 3′ dichloro-2,2′-bipyridyl

This is not a Fischer projection, but just a cross (Kalsi 2004)

P-3, 3′-dichloro-2,2′-bipyridyl

SCHEME 1.133c

(b) The compound has to be rewritten for a view along the chirality axis (III, scheme 1.133d). Assign the priorities, put them on a cross (as in schemes 1.132a and 1.132c).

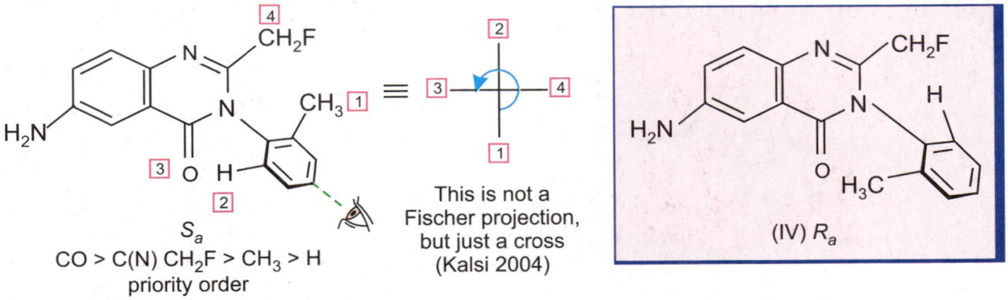

The arrow 1 ⟶ 2 ⟶ 3 (ignoring 4) runs counterclockwise. The exchange of CH_3 group with the H atom give R_a enantiomer (IV, scheme 1.133*d*).

SCHEME 1.133d

BOX 1.15
OPTICALLY ACTIVE SPIRANES WITH CHIRALITY CENTRES-SPIRANES WHICH ARE CENTRALLY CHIRAL

It may (wrongly) appear that *e.g.*, in a spirane (I, scheme 1.130) there are two chirality centres C2 and C6. Their chirality, however, cannot be specified since during sequencing, branches of equal priority meet before their priority is determined. In some spiranes which contain two equivalent ring systems the central atom common to both the rings is truly a chirality center as in 5, 5′-spiro-bishydantoin (I, scheme B 1.22). The four legands attached to this chirality centre are sequenced as below:

- Start from any of the two equivalent ring systems *e.g.*, from ring A

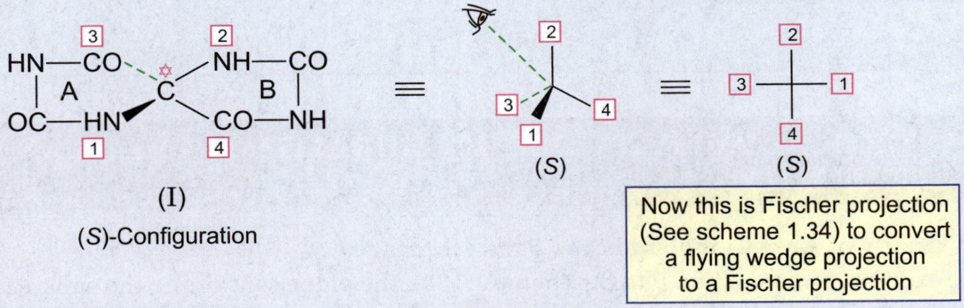

SCHEME B 1.22

- Assign priority [1] to NH (Sequence rules) of ring A
- Priority [2] is then assigned to NH of ring B
- Priority [3] is given to CO of ring A, since exploration from it lead to more preferred [1] NH

Continued...

When a molecule can display both central and axial chirality the former has precedence for configurational nomenclature.

Similarly for (II, scheme B 1.23) one starts from right hand side ring

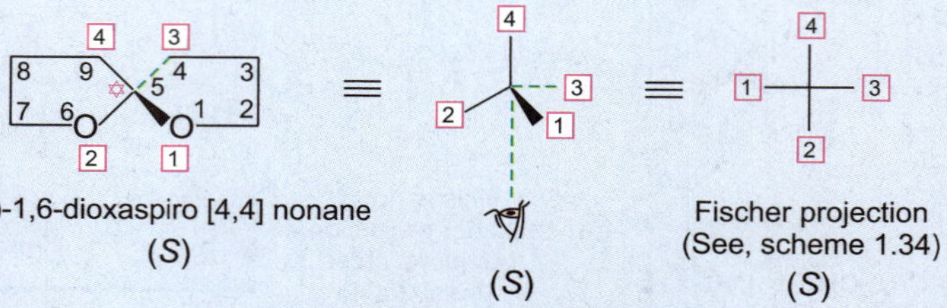

The four ligands attached to the chirality center C5 are sequenced as:

O at position 1 > oxygen at position 6 > C4, CH_2 > C9, CH_2

SCHEME B 1.23

Apart from central or axial chirality, a spirane may contain one or more chirality centres at the rings in which case diastereomers would occur. Thus when the dione (III, scheme B 1.24) is reduced, three diastereomeric diols are formed which are chiral and resolvable.

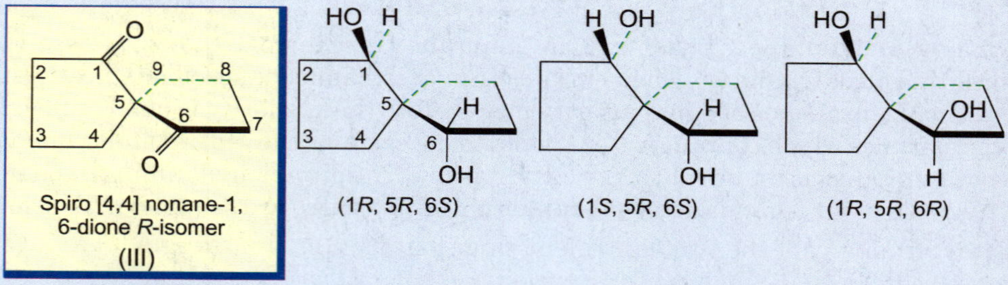

SCHEME B 1.24

1.16 OPTICAL ACTIVITY DUE TO CHIRALITY PLANE (PLANAR CHIRALITY)

(A) Chirality of Ansa Compounds and Paracyclophanes

If one examines a compound like (I, scheme 1.134), the plane containing the substituted benzene ring and the two oxygen atoms specifies a chirality plane. The molecule is chiral, though it does not have a chirality center or a chirality axis. The other enantiomeric form of (I) will have the bridge on the opposite side of the chiral plane.

Consider the molecule 2-bromo-hydroquinone without the polymethylene bridge. The plane containing the substituted benzene ring and the two oxygen atoms now represents the plane of symmetry and this attains the status of a chiral plane only when the polymethylene bridge is incorporated into 2-bromohydroquinone. Thus, the derivatives of hydroquinone, are optically active having planar chirality.

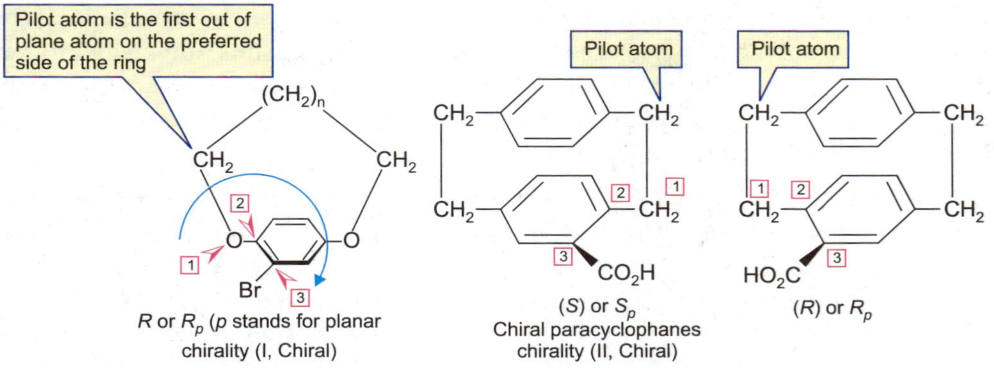

SCHEME 1.134

These derivatives of hydroquinone are called *ansa* compounds (ansa-handle in *Latin*, in ansa compounds, the two *para* positions of an aromatic ring are attached to hetero-atoms which in turn are connected through a polymethylene chain). In all chiral compounds of this type the aromatic ring is dissymmetrically substituted (*i.e.*, two-dimensionally chiral) and the polymethylene bridge can be either above or below the plane of the aromatic ring to give enantiomers.

The methylene ring is perpendicular to the plane of the benzene ring ; substituent – Br, prevents the rotation of the benzene nucleus inside the large ring. The compound (I, Scheme 1.134 $n = 8$) is resolvable, however when $n = 9$ the racemization is very easy since now the out of plane movement of the benzene ring is no longer restricted due to steric factor and the two enantiomeric forms are not configurationally stable.

A simple paracyclophane with one aromatic ring which resembles an ansa compound (I, scheme 1.134) has been resolved. In other paracyclophanes the two benzene rings are approximately parallel (arranged one above the other). To have enantiomerism at least one of the rings must be dissymmetrically substituted, (II, scheme 1.134) and when the methylene chains are small, the substituted ring cannot rotate within the cyclophane system so that interconversion of enantiomers (racemization) does not occur, and this is a form of atropisomerism. When each methylene bridge contains four methylene groups the compound cannot be resolved. Evidently, rotation of the substituted phenyl ring is now rapid for racemization.

(B) Nomenclature of Compounds with Planar Chirality

To assign configuration to compounds with chiral planes the sequence rule is extended by choosing a pilot atom. This is a priority atom directly linked to the plane but is not itself in the plane, thus it is first out of the plane atom. The pilot atom is chosen from that side of the plane which is most preferred (CIP rules). In the case of (I, scheme 1.134), the preferred side of the plane is the one which has the *ortho* bromine and therefore, the methylene group on the left hand side is the pilot atom. When both sides are equivalent, the pilot atom can be chosen from any side of the plane.

Then one classifies the adjacent three atoms (again chosen by precedence if there is a choice) of the plane as these are encountered along the bonds and given numerals 1, 2 and 3 or *a, b, c*. Now viewed from the pilot atom if the order $1 \rightarrow 2 \rightarrow 3$ describe a clockwise array in the chiral plane, the configuration is *R* or R_p if the array is counter-clockwise, the configuration is *S* or S_p.

In the molecule (I, scheme 1.134) *e.g.*, this path is O—C—C (Br), which is clockwise and thus, the configuration is (*R*), or R_p to indicate that stereochemical descriptor refers to a chirality plane.

To assign helix nomenclature to compounds with planar chirality one uses the pilot atom plus atoms *a*, *b* and *c* specified above. If the motion in moving to the pilot atom from the plane of chirality is clockwise then the configuration is (*P*) and if it is anticlockwise then the configuration is (*M*) as shown in (I, scheme 1.135).

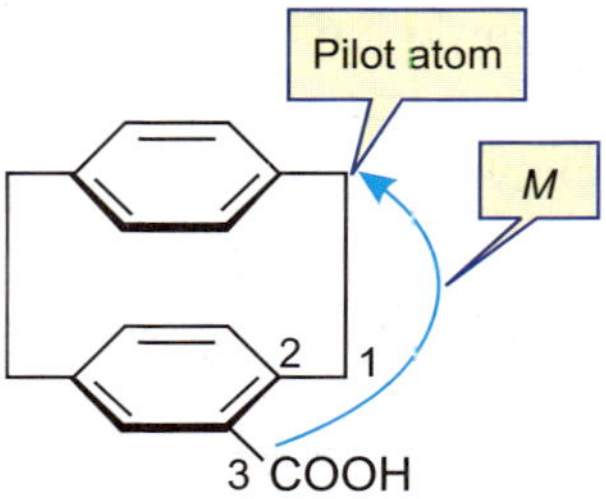

(I)
(*S*) or S_p planar chirality
(See, scheme 1.134)
(*M*) Helical chirality (Motion from chiral plane to pilot atom anticlockwise)

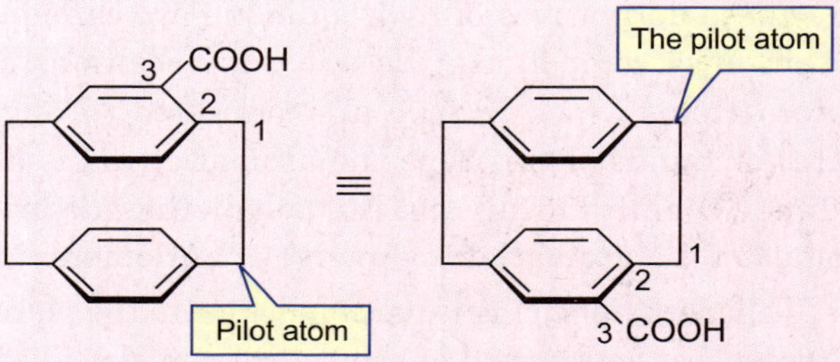

(II) (II*a*)

Reorient the projection (II) to bring the pilot atom towards you.

SCHEME 1.135

Note that the chiral plane $1 \rightarrow 2 \rightarrow 3$ is to be viewed with the pilot atom pointing toward the observer as in (I or II*a* scheme 1.135). Thus the view is to be from the pilot atom. It is then seen that R_p compounds correspond to *P* and S_p correspond to *M* which is opposite to the correlation in axial chirality (see scheme 1.133*a*).

(C) *Trans* (*E*) Cyclooctene

Another special example of planar chirality is *trans*-cyclooctene (see, Scheme 1.53). The introduction of a *trans*-double bond into rings containing six carbon atoms or less is not possible since it would introduce large strain in the molecule. *Trans*-cycloheptene has only been detected with spectrometers, it has a short lifetime and has not been isolated.

Trans-cyclooctene, on the other hand, has been isolated. The ring is large enough to accommodate the geometry required by a *trans* double bond, and thus, *cis*-and *trans*-forms of cyclooctene have independent existence (see, Scheme 1.54) *trans*-Cyclooctene is chiral due to planar chirality the two double bonded carbon atoms, as well as the two hydrogens and the two carbons directly adjacent constitute the chiral plane and polymethylene bridge (4 methylene chain) is skewed in the third dimension. One may recall that *cis* and *trans* 2-butenes on the other hand have trigonal stereocenters; these molecules are not chiral since each has a plane of symmetry (the plane of the molecule).

The interconversion (racemisation) of the two enantiomers in *E*-cycloctene (Scheme 1.136) requires the swinging of the tetramethylene chain over and below the plane of the trigonal atoms (chiral plane) and this swinging is opposed by ring strain.

In the case of higher homologues the mobility of the polymethylene chain increases and consequently the rotational barrier decreases. As a result *e.g.*, in the case of *trans*-cyclononene the optically active form can be detected only at – 80°C while *trans*-cyclodecene is an extremely mobile system.

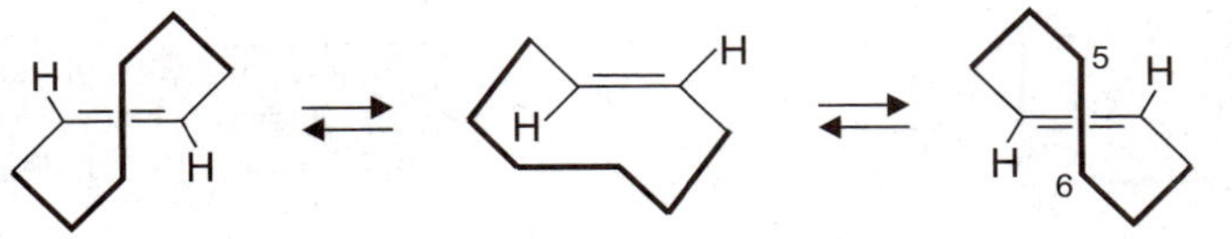

(I) (*R*)-Enantiomer or R_p (*p* stands for planar chirality) (*R*)-Crown

(*S*)-Chair or S_p

(II) (*S*)-Enantiomer or S_p (*S*)-Crown

Racemisation of (*E*)-cyclooctene involves the conformational change (*R*)-crown ⇄ (*S*)-chair ⇄ (*S*)-crown.

Racemisation of (*E*)-Cyclooctene

SCHEME 1.136

The molecule of *trans*-cyclooctene (*i.e.*, *E*-cycloctene) has a C_2 axis which passes through the center of the double bond and bisecting C5, C6 bond and the molecule thus belongs to point group C_2 (II, Scheme 1.136).

To assign chirality descriptors to the enantiomers of *E*-cycloctene (as in other molecules with planar chirality a pilot atom is selected first). In *E*-cycloctene, the chiral plane contains the two double-bonded carbon atoms, as well as the two hydrogens and the two carbons directly adjacent (these are the carbons of the methylene groups). Two equivalent pilot atom therefore, exist as shown by arrows in the (*S*)-enantiomer (Scheme 1.137), in such cases, the pilot atom may be chosen from any side. Starting from the either pilot atom one classifies the atoms of the plane and the sequence starts from the first in-plane atom and continues through atoms in the plane, always following the path leading to the more preferred atom. This is indicated in the structure by letters *a*, *b* and *c* ; viewed from the pilot atom if the order $a \rightarrow b \rightarrow c$ is clockwise, the configuration is (*R*) or R_p and if it is anticlockwise, the configuration is (*S*) or S_p (Scheme 1.137 where *p* stands for chirality plane). The configuration comes out to be (*S*) irrespective of the chosen pilot atom since in each case the paths are C—C=C. The path $a \rightarrow b \rightarrow c$ is to be explored from the pilot atom and for better visualization, the pilot atom should be brought to the front with the double bond on the back.

If an enantiomer of *trans*-cyclooctene is drawn as in (I, Scheme 1.137) with the double bond in front and then the pilot atom is at the back. However, to assign configuration the molecule is to be viewed from the pilot atom. An easy procedure is to turn the molecule so that the rotated (equivalent) orientation (Ia, Scheme 1.137) can be easily viewed from the pilot atom which is now in the front. The configuration thus comes out to be (*S*).

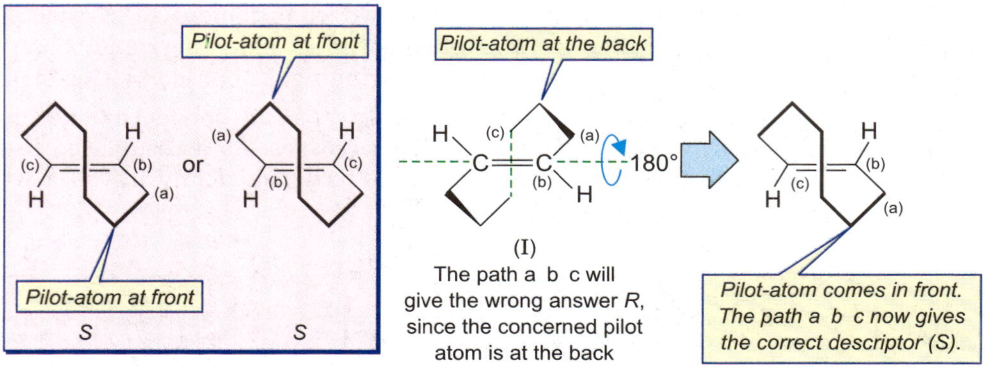

SCHEME 1.137

Just like axial chirality, planar chirality can be looked as a kind of helicity and specified as *P* or *M*. To assign *P* or *M* nomenclature the pilot atom is included with atoms *a*, *b* and *c*. It comes out that the R_p enantiomers (chiral plane nomenclature) correspond to *P* while S_p correspond to *M i.e.*, opposite to the correlation in axial chirality.

Thus to assign a helical stereodescriptor to compounds with planar chirality *e.g.* (*E*)-cyclooctene the relationship of the pilot atom to the chiral plane containing atoms C_a, C_b and C_c is considered (I, Scheme 1.137*a*). The motion in moving to the pilot atom from the plane of chirality in (I, scheme 1.37*a*) is clockwise thus it is (*P*). Similar considerations on (II) show that it is the enantiomeric structure (*M*).

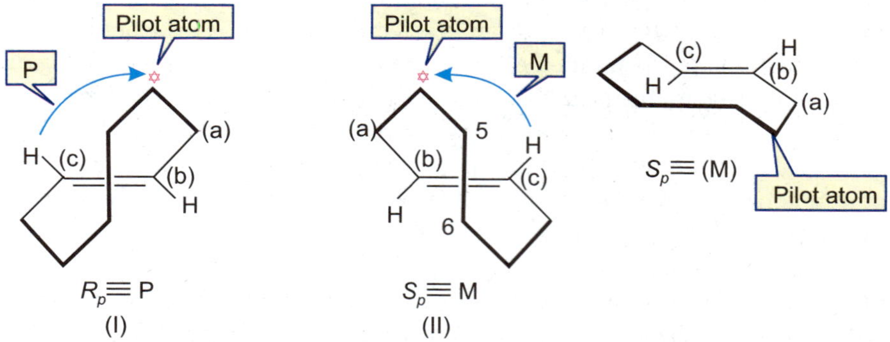

(*E*) – Cyclooctene–Helical descriptor is assigned on the basis of relationship of the pilot atom (✡) to the chiral plane containing atoms C_a, C_b and C_c.

SCHEME 1.137a

1.17 OPTICAL ACTIVITY OF COMPOUNDS DUE TO HELICITY

Helices (Scheme 1.138) are chiral objects, a right-handed helix (a clockwise rotation when viewed along the axis and moving from the front to the rear) is designated P (plus) while a left-handed helix is M (minus). The concept of helicity provides a simple method for designating the chirality of compounds with this feature. The terminal benzene rings in hexahelicene (Scheme 1.138 also see, Scheme 1.13), for example, cannot occupy the same plane without coming in serious conflict with one another. Therefore, the molecule is forced to adopt a nonplanar shape in which one side of the molecule must lie above the other because of crowding.

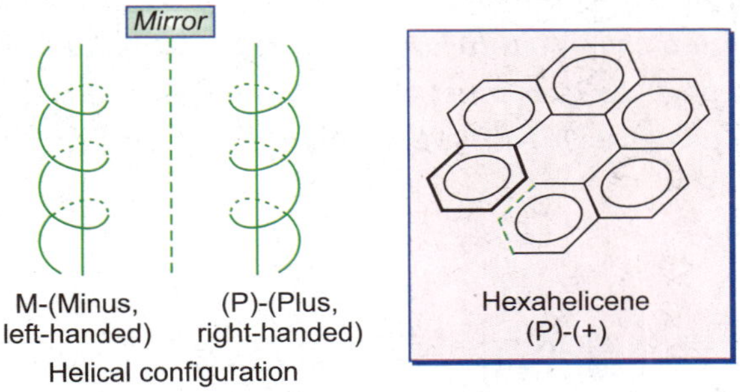

SCHEME 1.138

Hexahelicene (Scheme 1.138) is chiral by virtue of its helical shape which could be either left-or-right-handed in orientation. The entire molecule is infact less than one full turn of the helix, but this is enough to generate chirality in hexahelicene, it has been resolved into remarkably stable enantiomers which display spectacularly high optical activity and correspond to 'right' and 'left-handed' spirals. In hexahelicene and the middle rings (3 and 4) lie in a plane, while the terminal rings (1 and 6) fall above and below the plane respectively.

EXERCISE 1.20

Assign configuration to the compounds (I and II, Scheme 1.139).

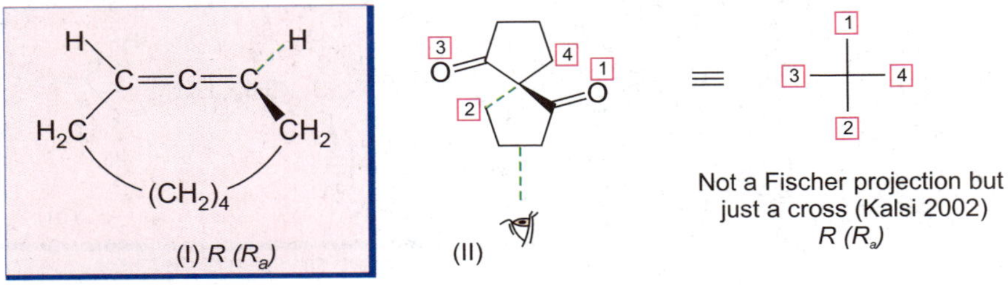

SCHEME 1.139

ANSWER. *(I) Turn the molecule in the plane of paper to get the projection (see Scheme 1.132a) to be viewed from bottom. This is an axially chiral allene and this enantiomer has R (R_a) configuration, its enantiomer is in (Scheme 1.133).*

The compound (II, Scheme 1.139) is a spirane and could be considered axially chiral and applying the procedure (See, Scheme 1.132 a) its configuration comes out be R which should be specified properly as R_a (for axial chirality, Scheme 1.139).

Recall that spiranes like this are also recognised to be centrally chiral. The configuration of this type of spiranes. (See page 192) starts with either carbonyl and proceeds as already discussed.

The four ligands attached to the chirality center are sequenced as shown (Scheme 1.139a). This compound comes out to have (S) configuration and is the enantiomer of (III, Scheme B 1.24 page 192). It must be noted that when a compound can display both central as well as axial chirality, the former has precedence for configurational nomenclature.

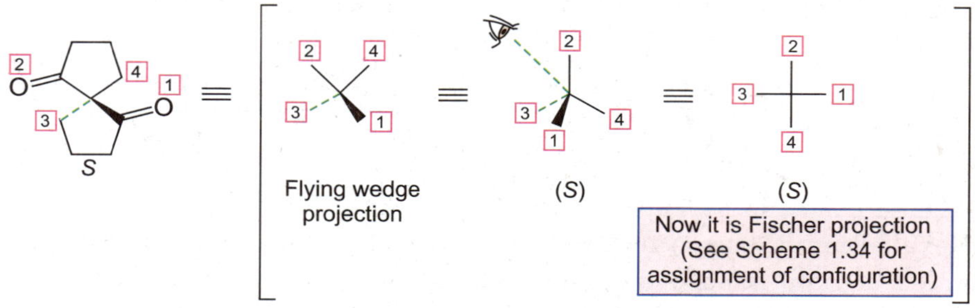

SCHEME 1.139a

EXERCISE 1.21 (*a*)

(a) Assign configuration to the chirality center in (I, Scheme 1.139b), (b) how (II and III, Scheme 1.139b) are related, enantiomers; diastereomers; homomers or constitutional isomers?

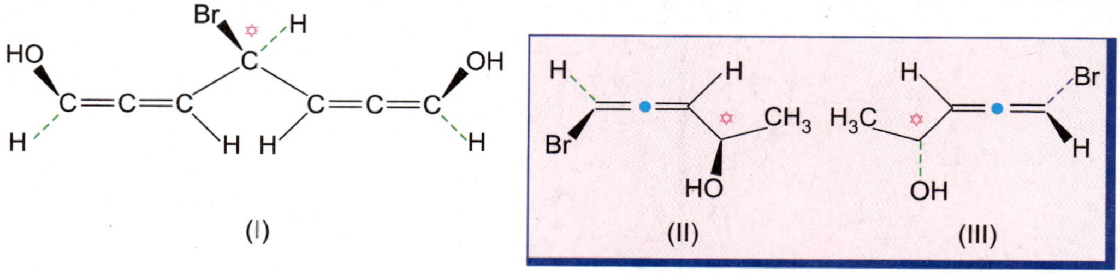

SCHEME 1.139b

ANSWER. (I) First assign axial chirality descriptors to the two allenic moieties, since these are two of the ligands on the chirality center. In (II) on the chirality center, the second priority ligand is substituted allenyl moiety. Both the chirality centers in II and III are of same configuration (*R*), both the allenes (II and III) on the basis of axial chirality are *Sa*, thus (II and III) are homomers.

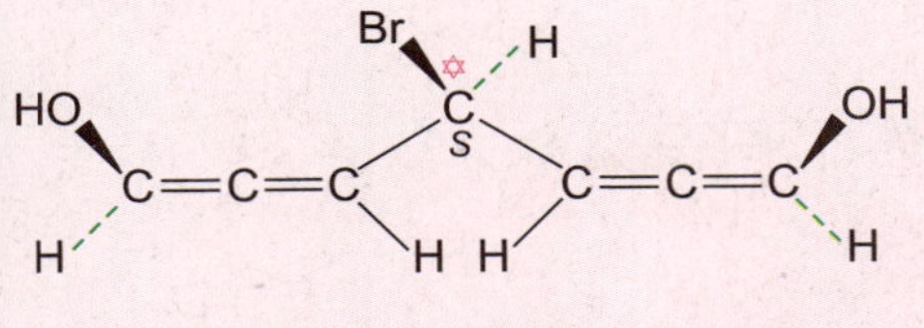

(I)

Shaded allene is (*R*) *i.e.* R_a
and non shaded allene is (*S*) *i.e.*
S_a from (Scheme 1.133a, Kalsi 2002).
Thus the chirality center is (*S*) CIP rules *R* > *S*.

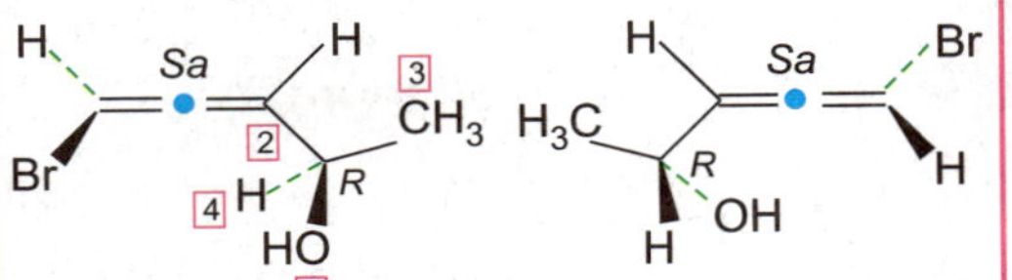

(II) (III)

SCHEME 1.139c

Aromatic compounds represent another category, in which suitable substitution forces the molecule to buckle from the most favourable, planar arrangement. For example, phenan-threne, a planar molecule, when substituted in positions 4 and 5, for instance, by methyl groups becomes somewhat skewed and exists as two enantiomers. Consequently, it has been possible to isolate the optically active (I, Scheme 1.140). Conformational helicity is also encountered in benzophenanthrene in which again the terminal rings and substituents are in different planes to make the molecule exist as two helical enantiomers (II, Scheme 1.140). Conformational helicity is encountered in some secondary structures of polypeptides, and many proteins have significant portions of their chains stabilized in the α-helix, which can be either left-or right-handed.

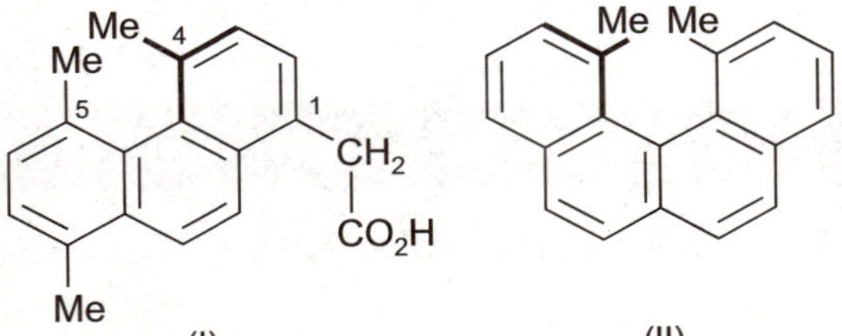

(I) P-(+) (II) *M*-(–)

Phenanthrenes and benzophenanthrenes are normally expected to be planar. However, due to molecular overcrowding the ring structures assume a helical shape.

SCHEME 1.140

1.18 MOLECULAR STEREOISOMERISM

Consider [14] annulene (annulenes are large ring compounds with conjugated double bonds which may exist in *cis* and *trans* configuration) in which hydrogens in the ring with four *trans* double bonds cause steric strain. The molecule is thus somewhat destabilized and somewhat nonplanar and quite reactive. The molecule of [14] annulene, however does show the presence of diamagnetic ring current. In each of the structures, four transannular hydrogens overlap with each other to give two conformational diastereomers which differ in the mode of overlapping. The two forms are separable (on silica gel column impregnated with silver nitrate) but are easily interconvertible (Scheme 1.141).

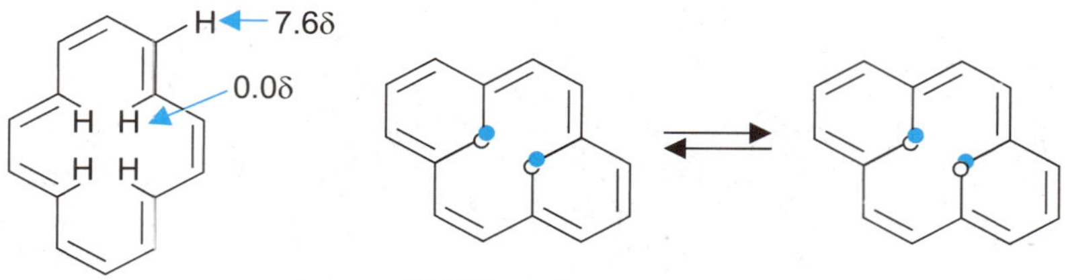

[14] Annulene

SCHEME 1.141

Annulenes

These are monocyclic conjugated polyenes and were synthesized to test Hückel's rule. Their name uses a bracketed numerical prefix to indicate the number of carbons followed by the word annulene. [4n] Annulenes have alternating short (double) and long (single) bonds and are antiaromatic. The expected aromatic character of [4n + 2] annulenes is rendered low by angle and van der Waals strain, and is regained when the ring contains 18 or more carbons.

1.19 CHIRALITY IN METALLIC COMPLEXES

Several metallocenes substituted with atleast two different groups on one ring become chiral and have been resolved (see, Scheme 1.14). Fumaric acid-iron tetracarbonyl has also been resolved while its corresponding maleic acid compound has a plane of symmetry and is therefore achiral.

1.20 NMR SPECTROSCOPY AND STEREOCHEMICAL INFORMATION

A working knowledge of NMR spectroscopy is assumed. One can get both conformational and configurational information:

- The chemical environment around a proton(s) can be determined from chemical shifts.
- The intensity of a peak points to the number of chemically equivalent protons.
- Steric disposition of neighbouring protons and their number can be determined from J values.
- Coupling constant J and chemical shift data give information regarding the relative stereochemistry of diastereomers.

(A) Coupling Constant and Relative Configuration

Karplus curve (Scheme 1.142) gives relationship between the vicinal coupling constant and the dihedral angle. This information is useful to assign stereochemistry in rigid cyclohexane

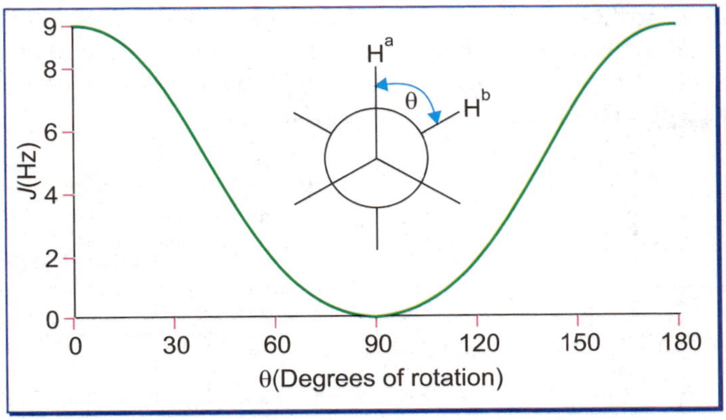

Effect of vicinal coupling constant $3J_{H^a—C—C—H^b}$ (Hz) on the dihedral angle. For θ of 0°, 60°, 90° and 180° the values of J are around 10, 3, 0 and 10 Hz respectively.

SCHEME 1.142

---

derivatives (Scheme 1.143). In the case of disubstituted alkenes the coupling between the *trans*-alkene protons is generally greater for the *E*-isomer than for *Z* isomer (Scheme 1.144) which also shows the dependence of the coupling constant on the angle between hydrogens attached to sp^2 hybridized carbons. When one considers the stereostructures (I and II, Scheme 1.144) the steric disposition of H_a and H_b are reflected by the values of coupling constants. Small coupling constant in (I) is indicative of dihedral angle between H_a and H_b close to around 90° while in (II) it should be around 180°.

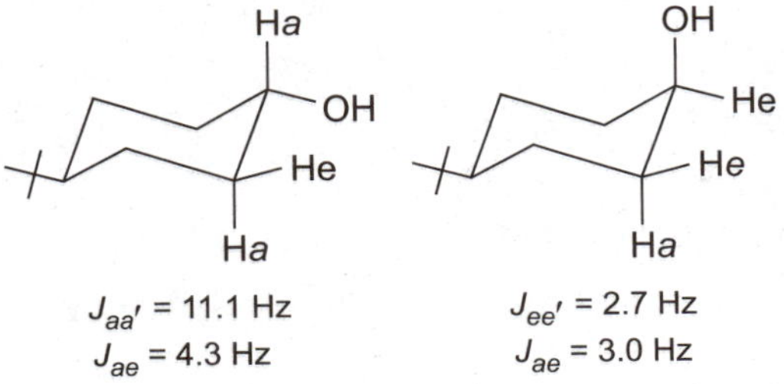

Dihedral angle (θ) between two axial hydrogens is 180° while the axial-equatorial and equatorial-equatorial dihedral angles are both 60°.

SCHEME 1.143

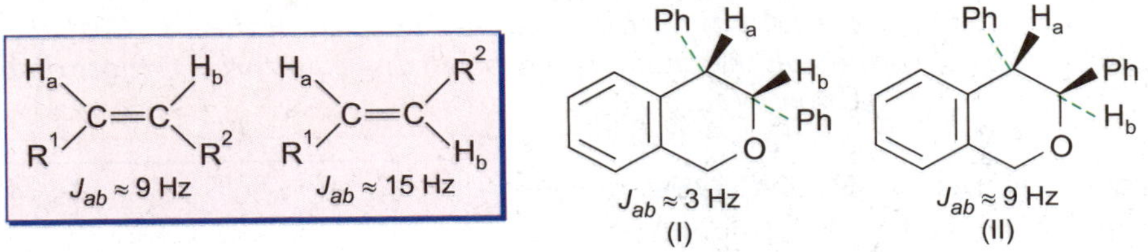

SCHEME 1.144

(B) Chemical Shift and Configuration

The anisotropy of neighbouring bonds or functional groups is a through-space phenomenon and is seen in (I, Scheme 1.145), $\delta H^a = 1.77$ is the effect of neighbouring methylene group while $\delta H^b = 1.95$ is the effect of carbonyl group.

A proton or a group of protons which are coplanar with an aromatic ring, C=C or C=O groups are deshielded while the protons which are held above their π electron clouds are shielded. In cyclohexanes the equatorial proton is always deshielded due to C—C bond anisotropy (Scheme 1.145).

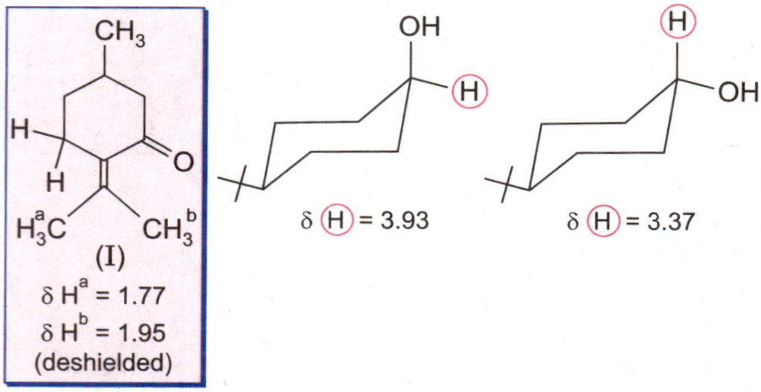

Stereochemistry in cyclohexane derivatives can be assigned since an equatorial proton is always found down field (~ 0.5δ) than the axial proton on the same carbon.

SCHEME 1.145

EXERCISE 1.22

How from ¹H NMR chemical shift data one can assign stereostructure in the pairs (I and II, bicyclo [2.2.1] hept-2-en-7ols and α and β-D glucose III and IV Scheme 1.146).

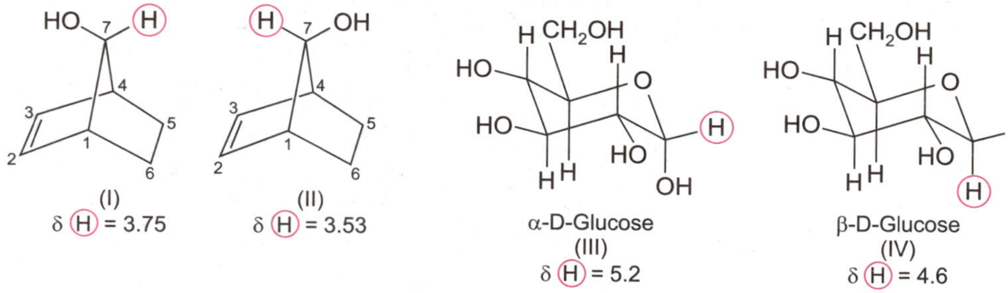

SCHEME 1.146

ANSWER. *In the anti isomer (II) the proton at C7 is held over the double bond it is therefore, shielded compared to that in (I). In the glucose molecule the proton on the anomeric carbon is the most deshielded and can be detected in the ¹H NMR spectrum. However, in (III) the proton on the anomeric carbon being equatorial is more down field compared to the same proton in (IV).*

(C) Nuclear Overhauser Effect (NOE)

When two nuclei which have different chemical shifts, but are close in space *e.g.*, H_a and H_b in (I, Scheme 1.147, but not in II) and the molecule is simultaneously irradiated with the ratio frequency υ_a which is the resonance frequency of H_a during recording of NMR spectrum one observes two things. Firstly the signal corresponding to H_a disappears because

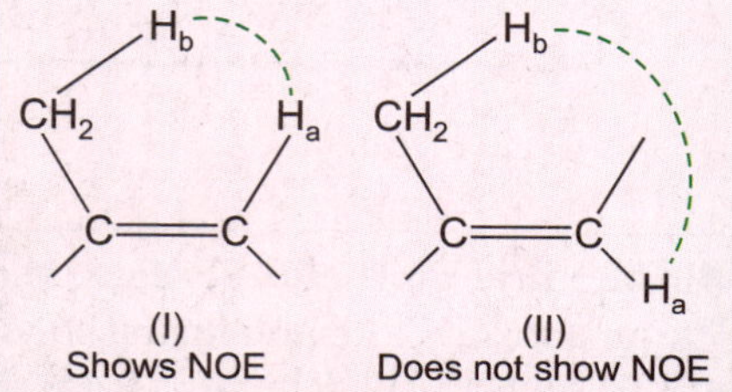

SCHEME 1.147

of saturation while the signal of H_b is enhanced in intensity because of increased spin relaxation. One use of NOE is *e.g.*, to determine the configuration of more heavily substituted alkenes (Scheme 1.147). One has already seen that disubstituted alkenes can be distinguished as *E* or *Z* based on coupling between alkene protons which is generally greater for the *E* isomer.

> *A spin excited nucleus H_a undergoes spin relaxation via the transfer of its spin energy to an adjacent nucleus H_b over short distances (2 – 4Å) and does not involve coupling.*

(D) Use of Shift Reagents–Enantiomer Recognition

Shift reagents were introduced in 1969, for spreading out NMR absorption patterns without increasing the strength of the applied magnetic field. Addition of shift reagents forms labile molecular associates with electron-donating polar groups such as OH, C=O, and NH_2 to bring about large changes in the chemical shifts of protons (or carbons in ^{13}C-NMR). The shifts known as lanthanide-induced shifts (LIS) are inversely proportional to the third power of the distance of the nuclei from the lanthanide and provide a sensitive method to know the relative distance of various groups and atoms from the complexation site. A common shift reagent, which is a paramagnetic metal [*e.g.*, an europium (III) salt] derivatized to generate solubility in an organic solvent, associates with polar functional groups in a molecule, to bring about a downfield shift of the resonance frequency of protons in the locality. If the ligand around the transition metal is chiral, a fluorinated camphor derivative for example Eu(hfc)$_3$, then the two enantiomers of the chiral compound will form diastereoisomeric complexes with the organometallic agent and some protons may then have different chemical shifts. Thus the interaction of the chiral europium shift reagent with the two enantiomers will give different signals in their NMR spectra *e.g.*, in a chiral carbonyl compound (Scheme 1.148). The role of the CLSR is to interact with the two enantiomers in a different manner, through the formation of weak diastereomeric complexes. This differential interaction together with the shift properties of the CLSR allows the ^{1}H absorptions of the respective enantiomers to be measured by integration of these signals with certainty. The diastereomeric excess (*de*) of one diastereomer X over another Y is given by the expression (Scheme 1.149). Enantiomeric excess can also be measured by NMR spectroscopy and thus 90 : 10 ratio of two enantiomers gives an enantiomeric excess of 80%.

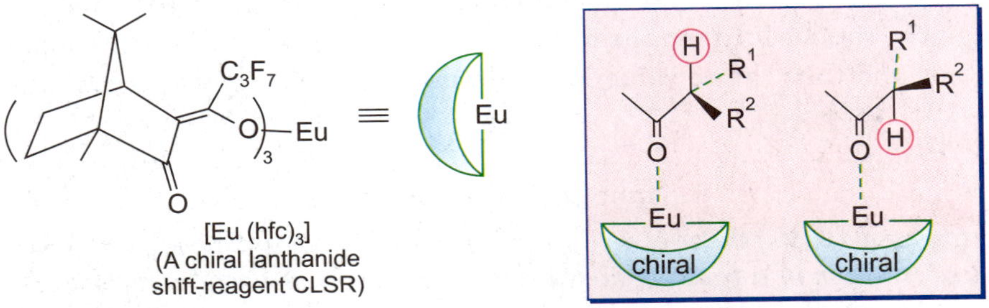

The signals of circled hydrogen atoms will be shifted down field and will have different chemical shifts

SCHEME 1.148

$$\text{d.e.} = \frac{\%\ \text{Diastereoisomer X} - \%\ \text{Diastereoisomer Y}}{\%\ \text{Diastereoisomer X} + \%\ \text{Diastereoisomer Y}}\ \%$$

$$(\text{e.e.}) = \frac{\%\ \text{Enantiomer X} - \%\ \text{Enantiomer Y}}{\%\ \text{Enantiomer X} + \%\ \text{Enantiomer Y}}\ \%$$

SCHEME 1.149

(E) Role of Mosher's Acid (MTPA)

In another method a mixture of enantiomers is reacted with an optically pure substance to give a new covalent bond and two different diastereomers and these will have different chemical shifts. The optically pure compound used for this purpose should have easily identifiable features in the NMR spectrum. One such widely used compound to determine the ratio of two enantiomeric alcohols in a mixture is Mosher's acid (3, 3, 3-trifluoro-2methoxy-2-phenylpropanoic acid MTPA). Mosher's acid gives a pair of diastereomers *e.g.*, with a racemic alcohol 1-phenylethanol (Scheme 1.150). The diastereomeric ratio can be assessed from the ^{1}H NMR spectrum by integration of the methoxy proton absorptions in the diastereomers or by integration of the CF_3 groups of the diasteromeric pair.

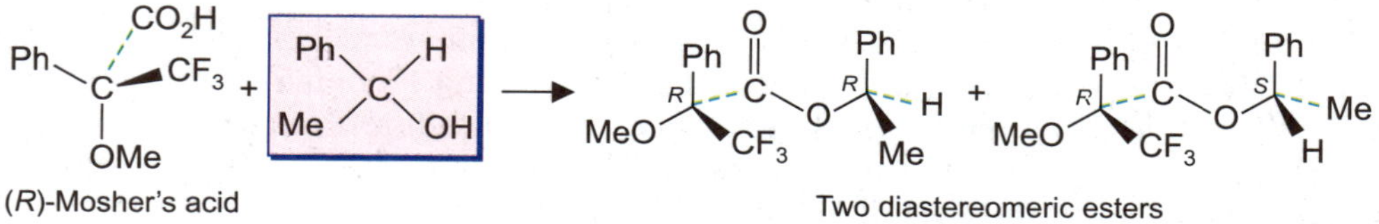

SCHEME 1.150

Thus by employing the esters of this type it is possible to find the ratio of two enantiomeric alcohols in the given mixture. The integrated intensity values of the two signals gives the ratio between the amounts of each enantiomer. If the ratio is 1:1 one is having a racemic mixture, if the ratio is 9:1 then the mixture consists of 90% of one enantiomer and 10% of the other and thus the *ee* value is the difference between these percentage values *i.e.*, 90 – 10% =80% *ee*.

Asymmetric Reaction and *ee*

Earlier (around 1970's) the ee of an asymmetric reaction was determined by measuring the optical rotation of a racemic mixture (see, sec. 1.14). A useful development to find out the enantiomeric purity of a mixture is gas or liquid chromatography with chiral supports. This object can also be achieved by NMR spectroscopy when enantiomers give separate signals with a chiral lanthanide shift reagent (CLSR), or via reaction of a completely or partially racemic alcohols with Mosher's acid. The integrated intensity value of distinct signals in the diastereomers involved in either case gives the ratio between the amounts of each enantiomer.

PROBLEMS

1.1. Water has a two fold axis of symmetry (C_2), while benzene has a C_6 axis, comment on the molecules with C_∞ and C_1 axes.

1.2. Explain briefly a relation between elements of symmetry and optical activity.

1.3. Depict the symmetry plane on the structure of *cis*-1,2-dichlorocyclopropane molecule. Does molecule of H_2O also have a symmetry plane?

1.4. How can you demonstrate a highly symmetrical molecule like methane to have multiple S_4 axes? Relate it with a cube.

1.5. Label the following pairs of compounds as homomers, constitutional isomers, enantiomers or diastereoisomers. Assign *R*/*S* configuration to the stereogenic centers in (I and II).

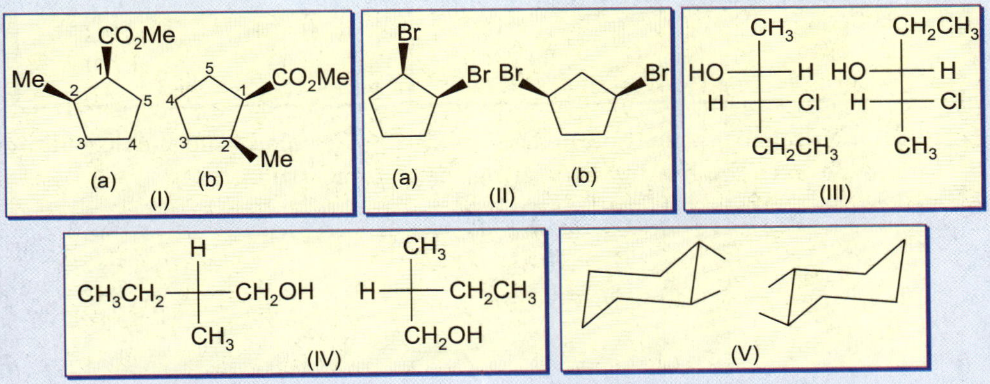

1.6. Given the following Fischer projections, indicate : (*a*) relation of one projection with other, (*b*) their names and assignment to (*R*) or (*S*).

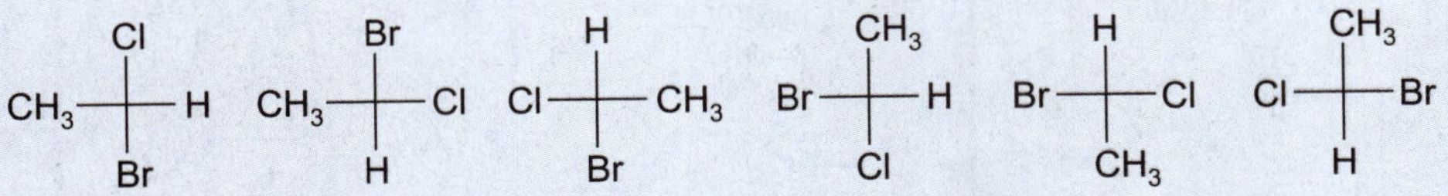

1.7. How many stereoisomers can exist for each of the compounds (I—IV)?

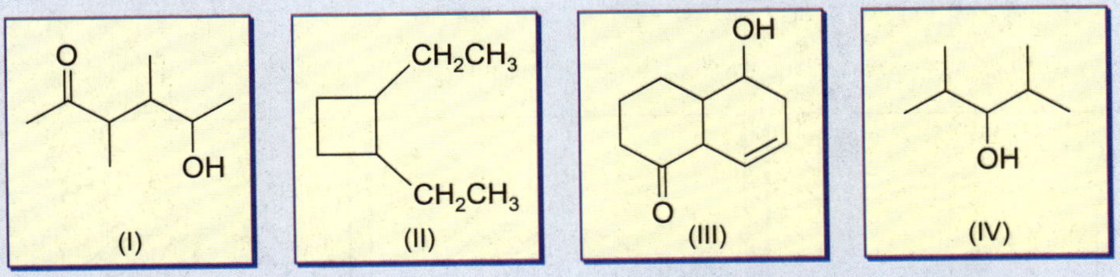

1.8. Write the Fischer projection for (*R*)-2-iodobutane and convert it into its "Wedge and dotted line" representation.

1.9. How are the following conformations of 2, 3-dibromobutane related with one another?

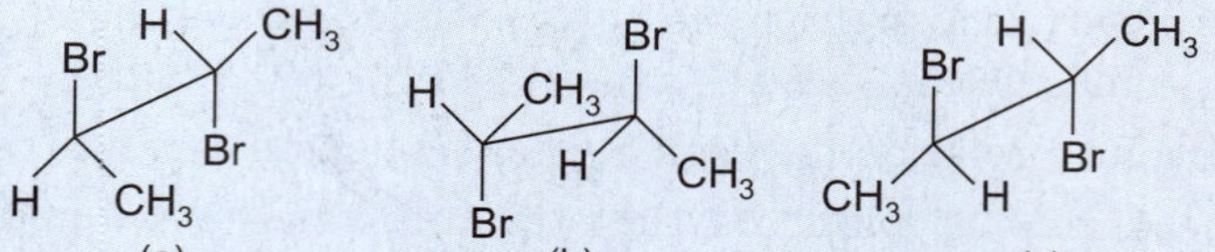

(a) (b) (c)

1.10. What is the stereochemical relationship, (enantiomers or diastereomers) between each of the pair (I and II)?

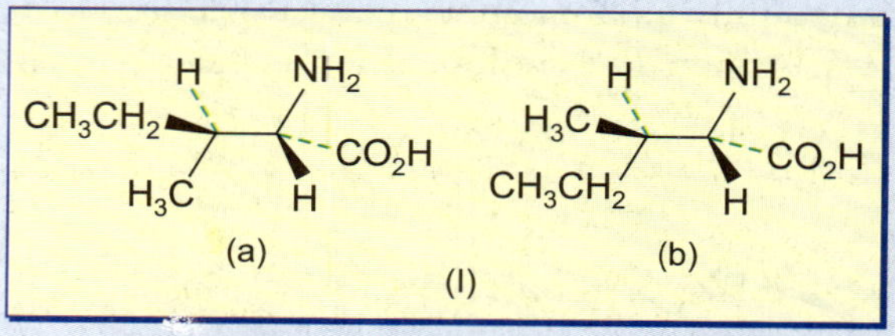

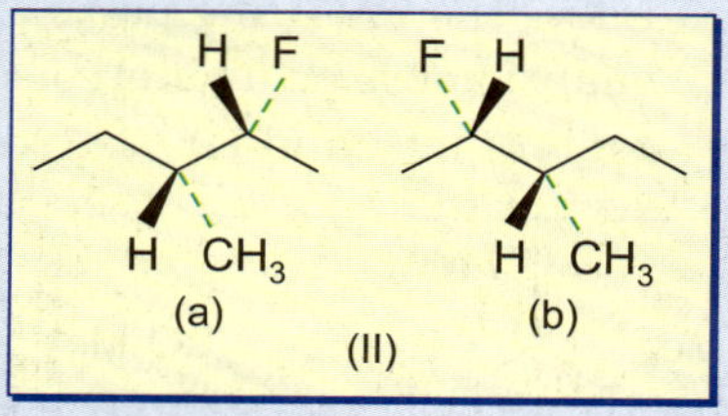

1.11. Butane on monochlorination gives a mixture of 1-chloro and 2-chlorobutane. Comment on their chirality by writing their stereostructures.

1.12. The following biphenyl is chiral. Designate its configuration.

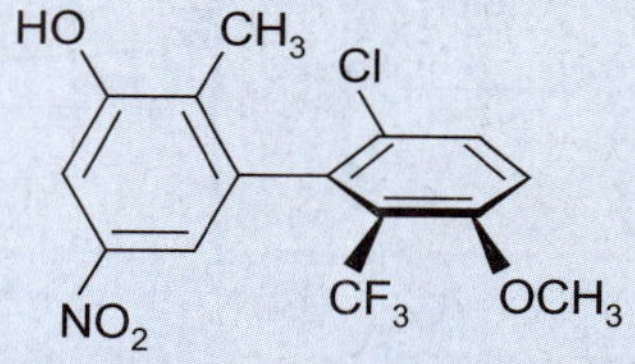

1.13. Comment on the chirality (optical isomerism) of the following biphenyls.

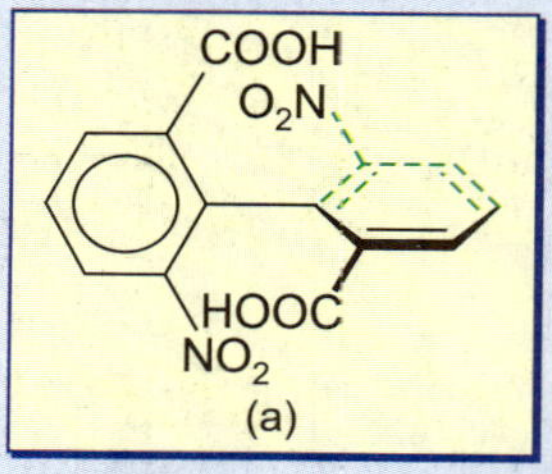

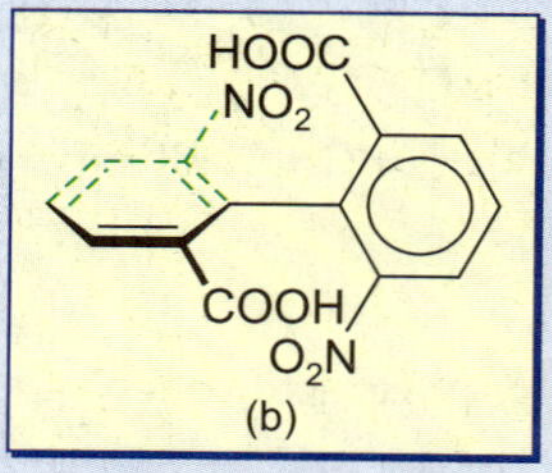

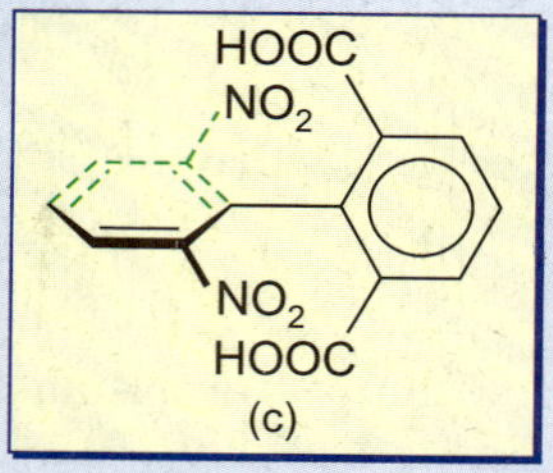

1.14. Which of the following molecules are chiral?

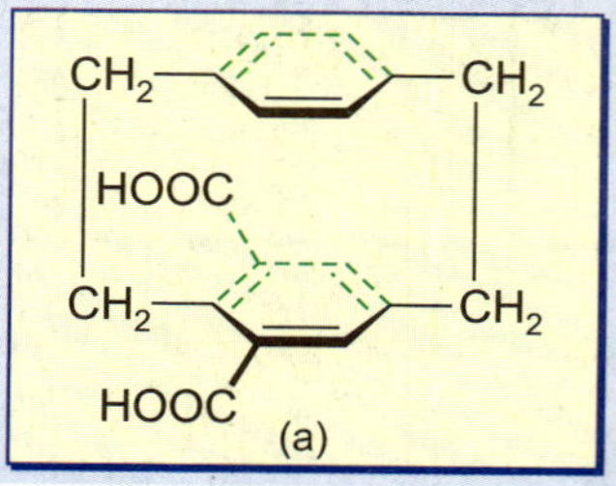

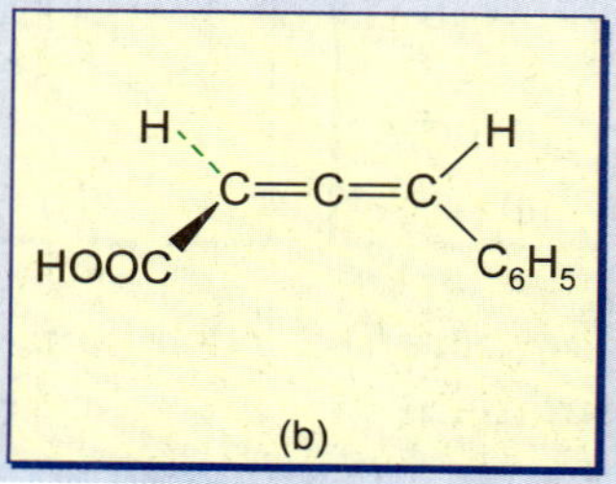

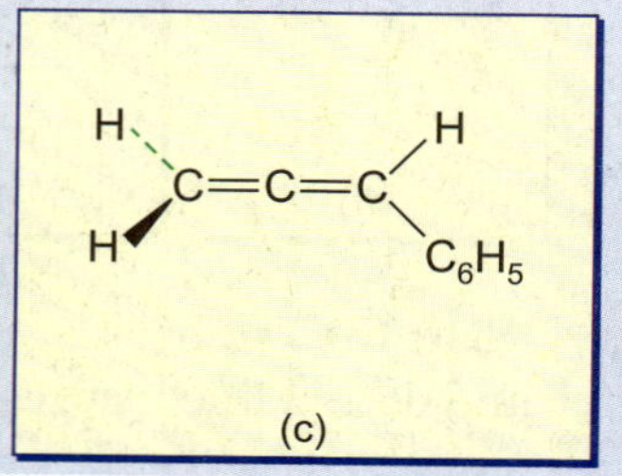

1.15. Comment on the identity and chirality of the following structure of penta-2, 3-diene.

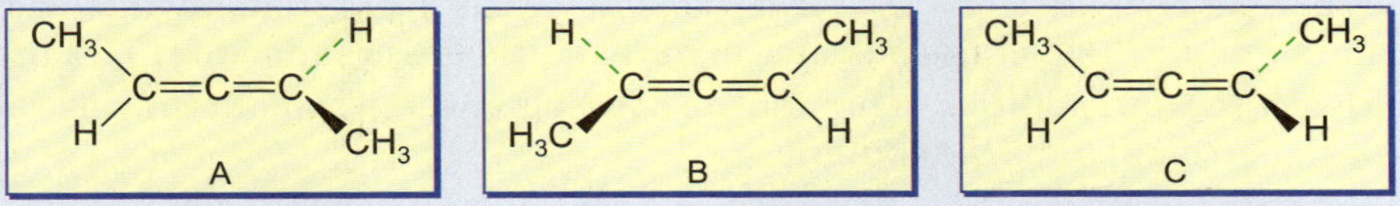

A B C

1.16. Label each compound as either having center, axes or planes of symmetry. How the chirality of compounds 2 depends on *m*, *n*, and *x*? Assign the order of priorities to individual ligands in compounds 4 and 5.

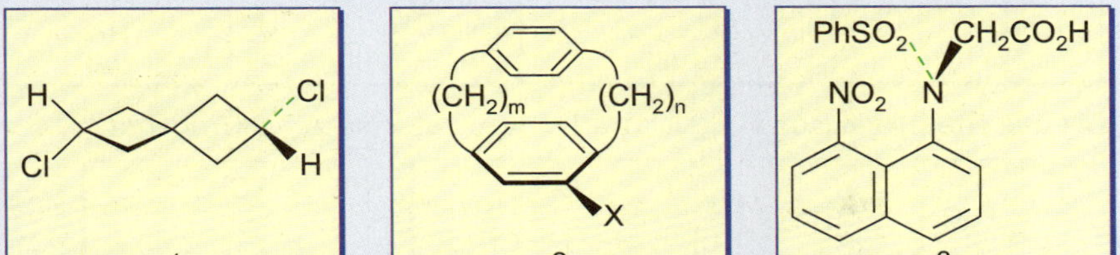

1 2 3

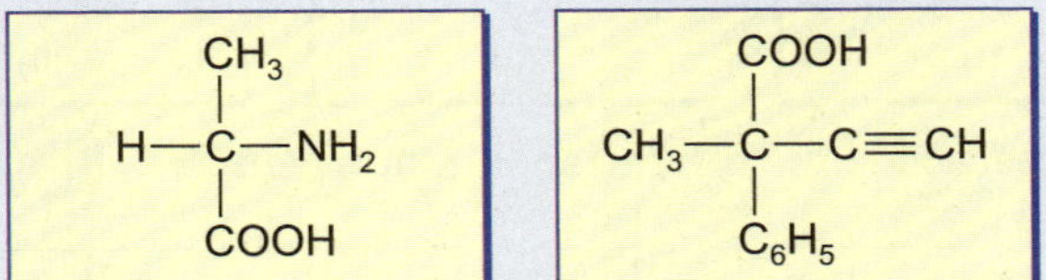

4 5

1.17. Give a general definition of chirality and support it with examples.

1.18. The two enantiomers of a chiral amine readily interconvert by a process known as nitrogen inversion (Scheme 1.142). In the case of quaternary ammonium compounds, such inversion is not possible, and chiral ions may be separated into enantiomers which are relatively stable. The optically active allylethylmethylphenyl-ammonium halides undergo racemization slowly in solution. Rate of racemization is temperature dependent and is faster with the iodide than with the bromide. Give a mechanism for this racemization.

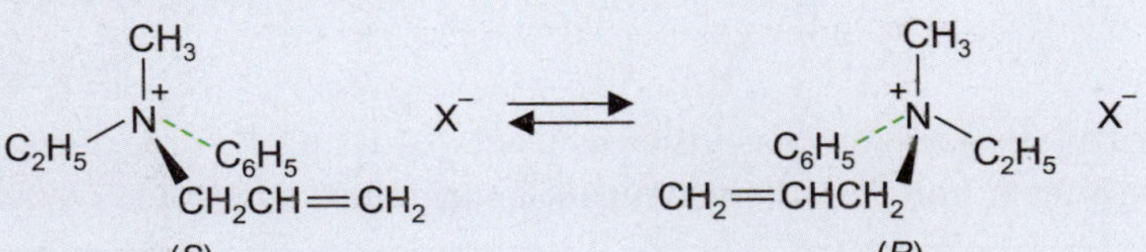

(*S*) (*R*)

1.19. Write the structure of the lowest molecular weight alkane, which is chiral. Depict its isomeric structure as well.

1.20. Write the structures of the geometric isomers of 1,2-cyclopentanediol and comment on their stereochemistry.

1.21. What are the essential conditions for a compound to be chiral? The presence of a stereocenter is not always essential for a compound to exhibit chirality. Explain.

1.22. Why is it not possible under ordinary conditions to resolve amines even though three different groups are attached to nitrogen atoms?

1.23. What factors determine the formation of stereoisomers from an alkene during bromination? Depict the outcome of *anti* addition of bromine to (*E*)-2-butene.

1.24. Depict the *anti* addition of bromine to (*Z*)-2-butene and explain the relationship with outcome from (*E*)-2-butene.

1.25. How the projections (I—IV) are related with one another?

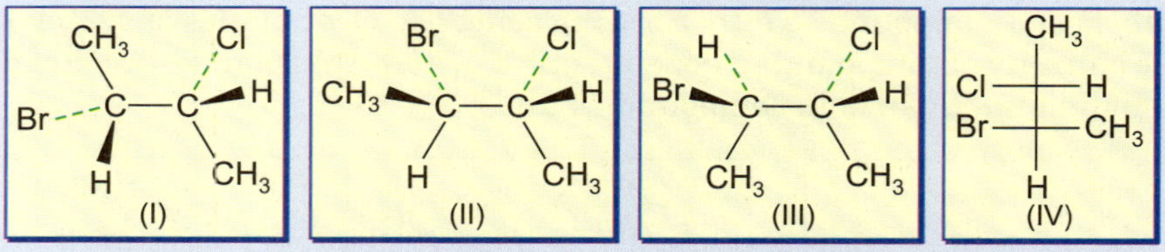

1.26. How are the projections (I, II, and III) are related to one another?

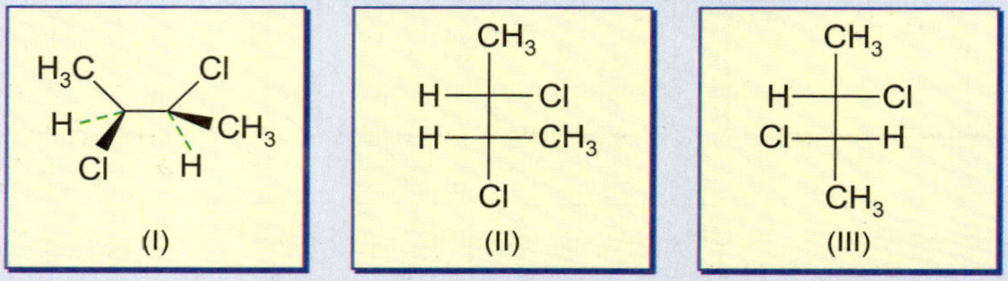

1.27. Identify the stereochemical relationship between the pairs of compounds 1 and 2.

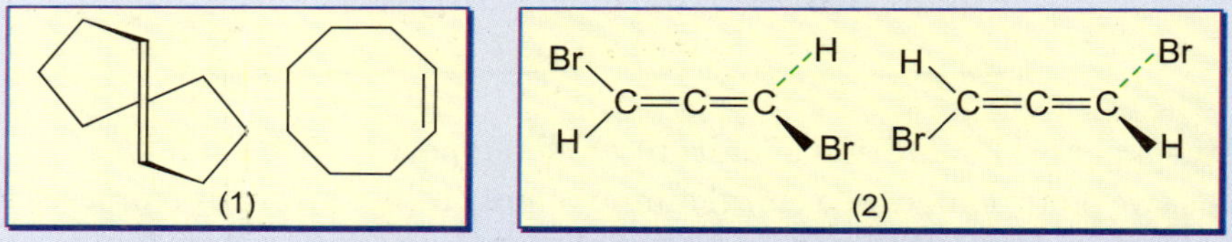

1.28. What is the relationship between (I and II)? Decide by interconverting one projection into other?

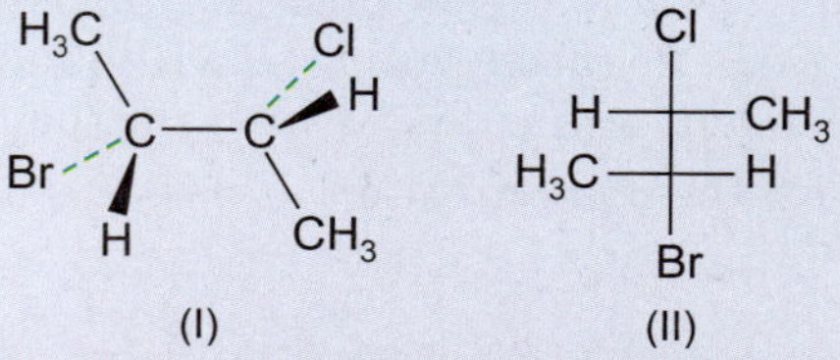

1.29. Discuss the stereochemical relationship between the following pairs of compounds, by assigning *R* and *S* nomenclature (See problem 1.31).

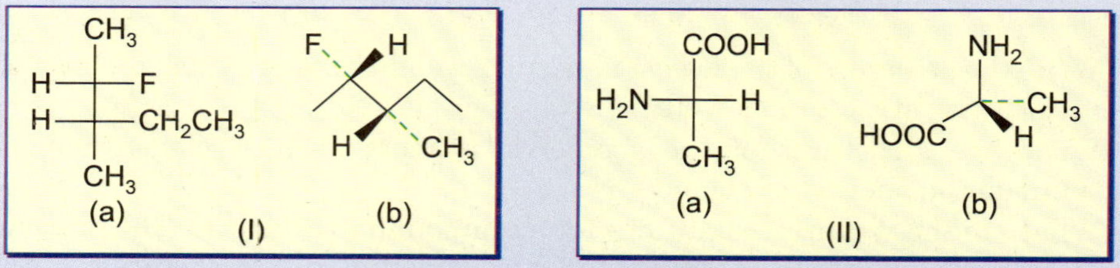

1.30. Given the priorities designate the structures below as *R* or *S*.

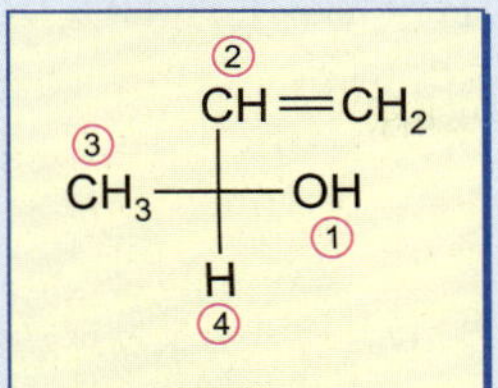

(I)

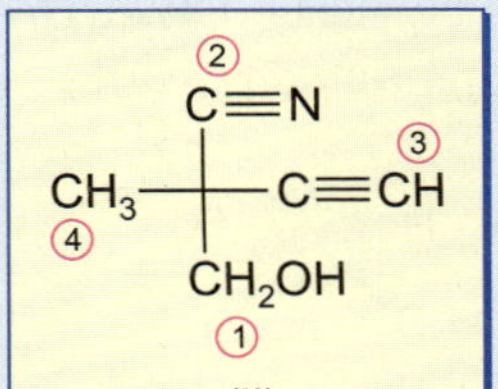

(II)

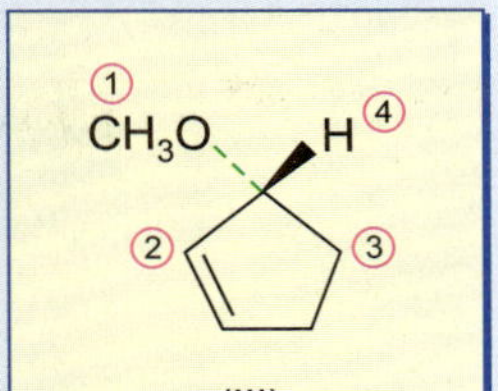

(III)

1.31. Draw the structure of (*R*)-3-ethylcyclohexene.

1.32. Assign configurational descriptors to the following compounds

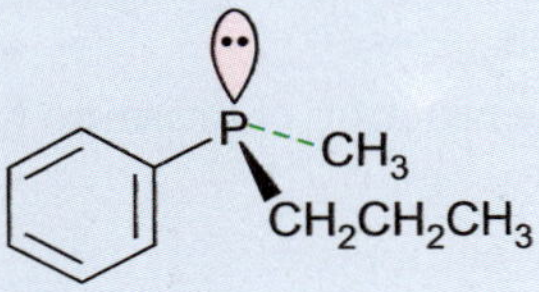

(I)

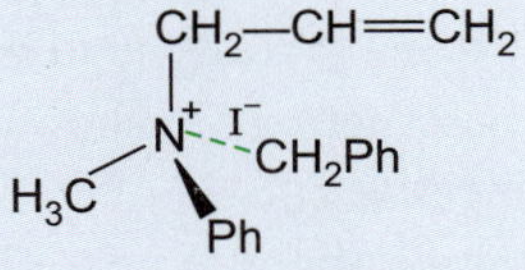

(II)

1.33. Interconvert the Fischer projections of (*R*) and (*S*)-2-butanol into their flying wedge projections and *vice-versa*.

1.34. How the following structures are related to each other: constitutional isomers, enantiomers, diastereomers or identical compounds?

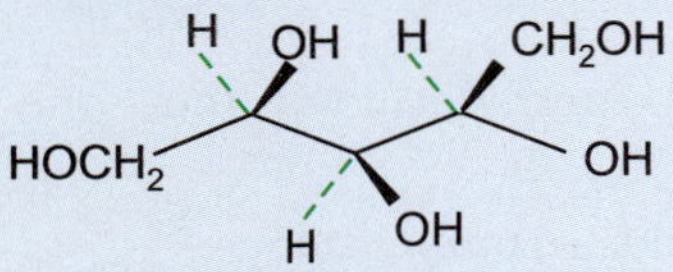

(I)

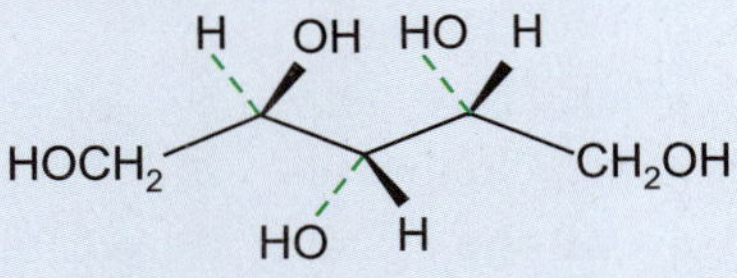

(II)

1.35. What is the relationship between the compound (A and B)?

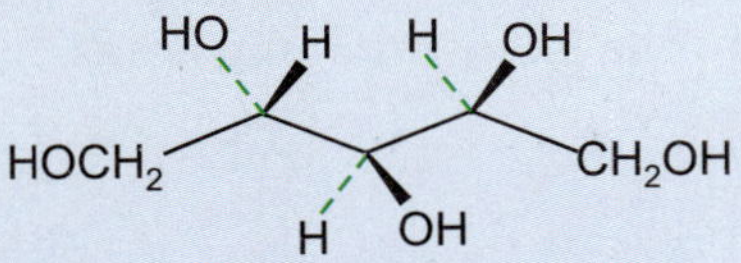

(A)

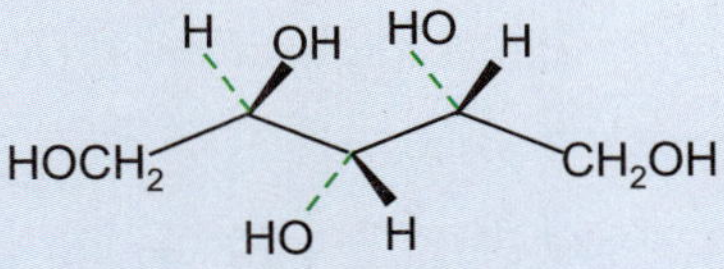

(B)

1.36. How the compounds (I and II) are related ?

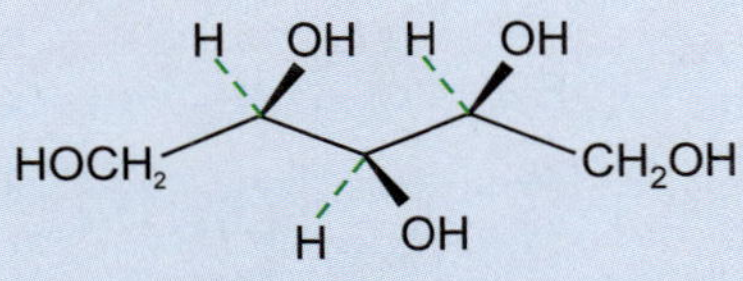

(I)

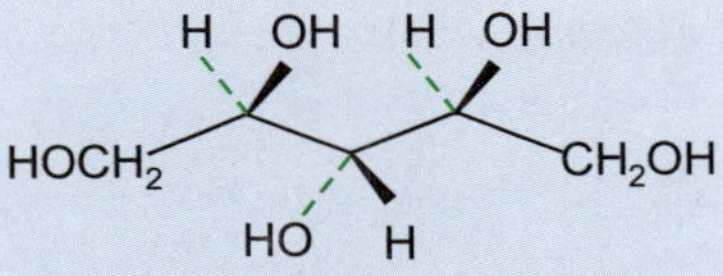

(II)

1.37. Comment on the chirality of the naturally occurring antibiotic mycomycin.

$$HC{\equiv}C{-}C{\equiv}C{-}CH{=}C{=}CH{-}CH{=}CH{-}CH{=}CH{-}CH_2COOH$$

1.38. Give a stereodescriptor to the following chiral compound.

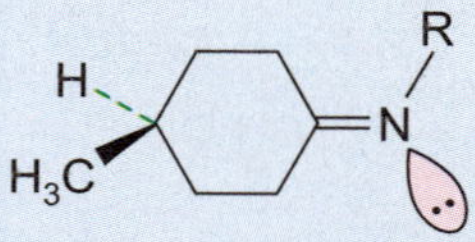

1.39. Write the structure of *meso* tartaric acid in Newman projection and translate it into Fischer projection.

1.40. Maleic acid is reacted with osmium tetroxide followed by hydrolysis. Write the stereostructure of the product, in two different ways.

1.41. Show that *meso* tartaric acid contains an S_2 axis.

1.42. Write stereostructures to show as to why the reaction of bromine with cyclohexene is stereoselective.

1.43. Provide systematic names to the following compound:

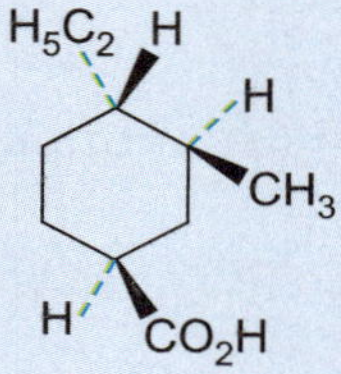

(I)

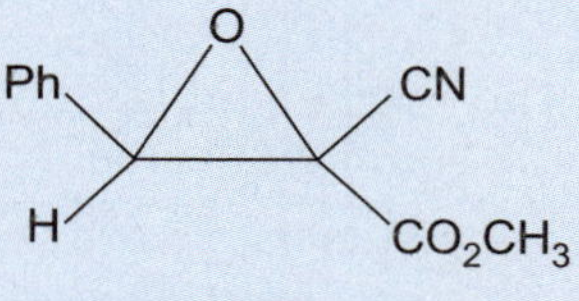

(II)

1.44. Discuss the chirality of 1-bromo-3-methylcyclobutane and 1-bromo-3 methylcyclohexane.

1.45. Draw all the possible stereoisomers from the compound in the box.

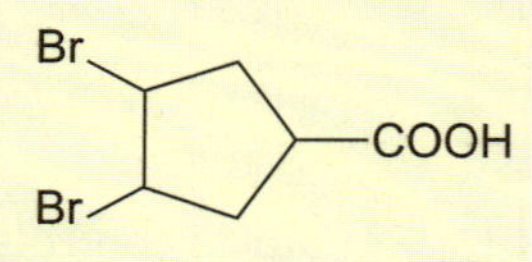

1.46. The molecules of water and phenanthrene belong to the same point group. Explain.

1.47. (*a*) Assign configuration, if (*R*) or (*S*) to the following compounds.

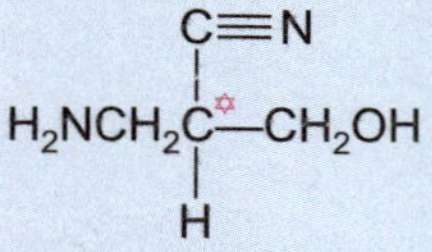

(I)

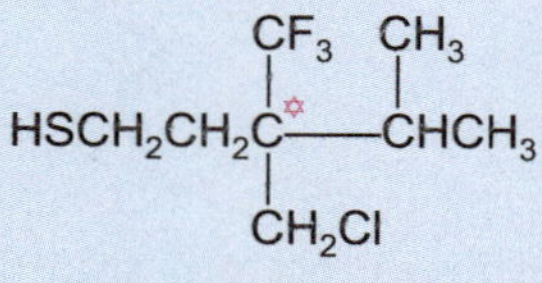

(II)

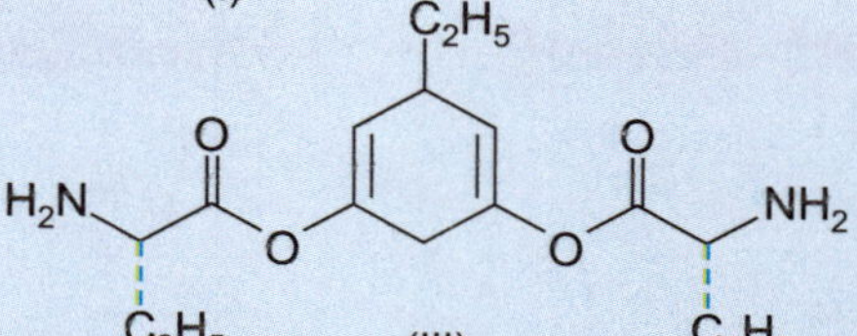

(III)

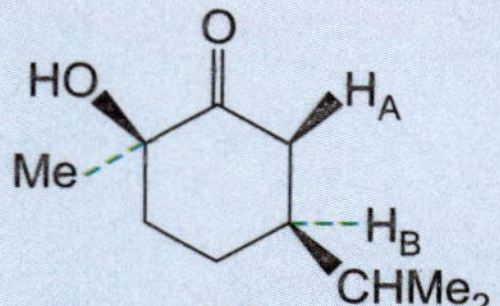

(IV)

(*b*) When the cyclohexanone (IV) is dissolved in benzene J_{AB} is 3Hz, but in methanol it is 11 Hz. Explain.

ANSWERS TO SELECTED PROBLEMS

1.1. Linear molecules like acetylene have a C_∞ axis. In this case even an infinitesimal rotation (360°/∞) about this axis leads to an orientation indistinguishable from the original. The trivial onefold axis C_1 is never considered, as all molecules have an infinite number of C_1 axes.

1.2. If a molecule is not superimposable on its mirror image, then it can display optical activity. Such molecules are (*i*) asymmetric *i.e.*, these do not have any symmetry elements (except a trivial axis, C_1) or (*ii*) dissymmetric (chiral) these possess proper axes but no alternating axis of symmetry *i.e.* improper axis. Thus, if a molecule has an S_n axis, then it is not optically active. If it has no S_n axis, then it is optically active. If a C_n axis is the only symmetry element present in a molecule then it is optically active.

1.3.

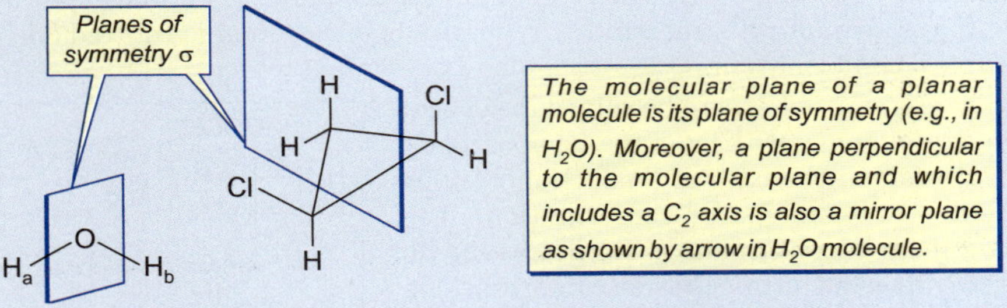

1.4. The axis depicted in 1 passes through the middle of the molecule and bisects opposite HCH angles. If one considers the clockwise rotation of the molecule by 90° around this axis and subsequent reflection in a plane perpendicular to it, one gets an orientation of the molecule superimposable on the original. One can easily depict not only the same axis by inscribing the tetrahedral model of methane in a cube 2, but the other S_4 axes as well. The S_4 axis shown bisects the opposite sides of the cube. A cube has six equivalent faces, therefore, there are three S_4 axes. Notably each S_4 in methane molecule is also a C_2.

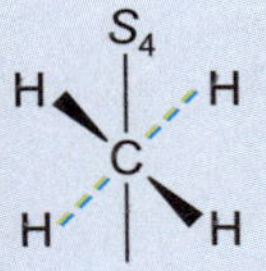

1
Methane

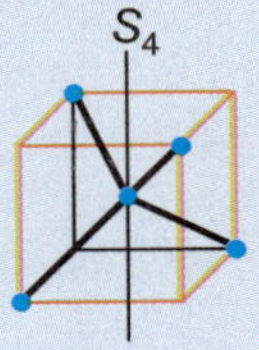

2
Cube

1.5. (*i*) Enantiomers, when *e.g.*, (*b*) is rotated in the plane of paper towards left it becomes mirror image of (*a*). The configurations of (*a*, 1*S*, 2*R*) and (*b*, 1*R*, 2*S*) also prove this.

(*ii*) Constitutional isomers, both are *meso*, (*a*, 1*R*, 2*S*) while (*b*, 1*R*, 3*S*).

(*iii*) Constitutional isomers.

(*iv*) Enantiomers, one is *R* other is *S* (interchanging two groups twice on the stereocenter generates a mirror image projection of the other).

(*v*) Enantiomers.

1.6. These structures are all equivalent, a structure being written by interchanging any two pairs of groups. The compound is (*R*)-1-bromo 1-chloroethane.

1.7. (I), eight; (II), three (one is *meso* since the substituents on the two stereocenters are the same); (III), eight; (IV), none (not a chiral molecule).

1.8.

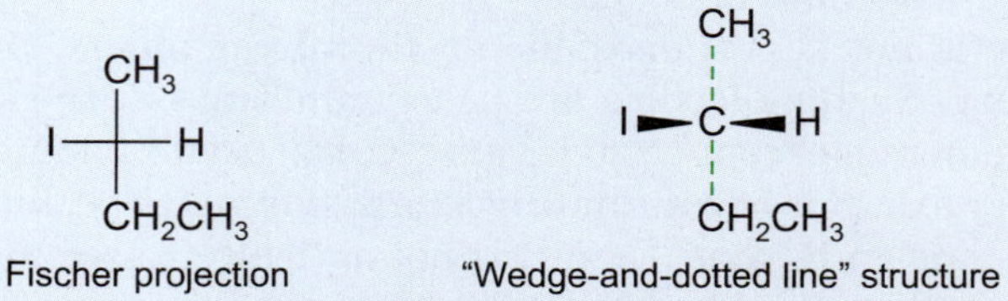

1.9. Conformations (*a*) and (*b*) are an enantiomeric pair while (*c*) is the *meso*-form. [**Hint.** The sawhorse projections are drawn in staggered conformations. Since the groups on each stereocenter are same there must be *meso* compound and a center of symmetry is seen in (*c*). When in (*c*) one carbon is rotated one gets an eclipsed conformation now with a plane of symmetry].

1.10. This type of problem can be solved in several ways and some of these have been already discussed. In the case of (I) both molecules are drawn as staggered conformations (see the zig-zag of continuous lines). These can be first converted into eclipsed conformations and then translated on to Fischer projections to see the relationship. These come out to be diastereomers. Even as such if one looks at the dashed-wedged line formulas in both (I*a* and I*b*). The configuration at the right hand carbon is same but differs only at the left hand carbon to show that these are diastereomers.

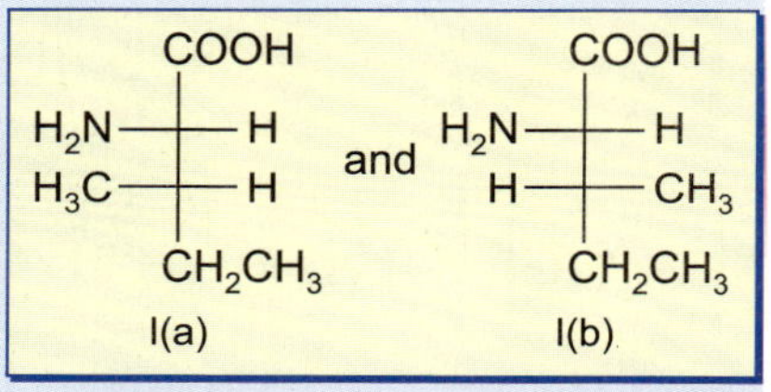

In the case of (II*b*) its rotation out of the plane of paper (180°) gives an enantiomeric projection (inverted stereochemistry at each stereocenter). Thus II*a* and II*b* are enantiomers.

1.11. 1-Chlorobutane is achiral. The 2-chlorobutane formed in this reaction is a racemic mixture *i.e.*, an equimolar mixture of (*R*)-2-chlorobutane and (*S*)-2-chlorobutane. The reactive intermediate in the reaction leading to 2-chlorobutane is the *sec*-butyl free radical, which is almost planar and therefore, is achiral and reacts with Cl_2 on either side of the molecule. Also see (Scheme 1.87).

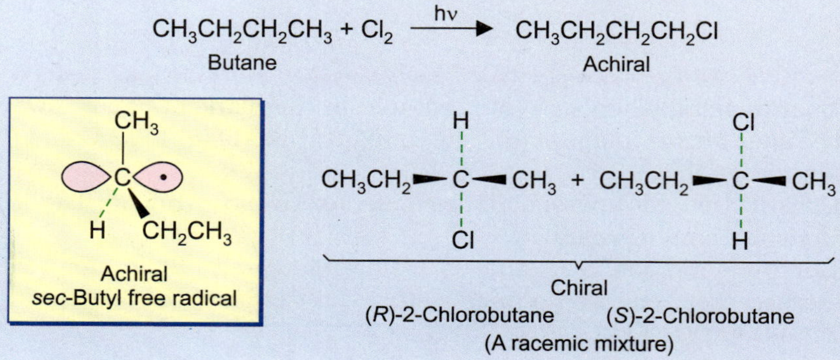

1.12. (*R*)-Configuration.

1.13. The biphenyls (*a*) and (*b*) have the two differently substituted *ortho* positions on each ring, the groups being indeed large to pass each other on rotation about the central bond. Thus, these represent chiral molecules; a pair of enantiomers. Compound (*c*) is a structural isomer with a plane of symmetry, therefore, it is not chiral even though in this case the rotation is equally restricted.

1.14. Only (*b*).

1.15. It is chiral allene. A and B are enantiomers and B and C are equivalent.

[**Hint:** B rotated around the axis of allenic carbons clockwise gives C].

1.16. Compounds of the type 2 (paracyclophanes) have been resolved when *m* and *n* are fairly small ($m = 3$, $n = 4$ and $X = CO_2H$). It is then that the benzenoid rings cannot rotate at a significant rate. When the connecting methylene chains are long, rotation is so fast that the enantiomers rapidly equilibrate.

4.	NH_2,	COOH,	CH_3,	H ;	5.	COOH,	C_6H_5,	C ≡ CH,	CH_3
	1	2	3	4		1	2	3	4

1.18. Due to reversible dissociation into an amine and an alkyl halide (see Scheme 1.5). The recombination of the amine with the alkyl halide gives a racemic salt. The less nucleophilic the counterion X^-, more configurationally stable is the quaternary salt, nucleophilic reactivities follow the order $I^- > Br^- > Cl^- > F^-$.

1.23. Two factors. Configuration of the alkene and the *anti* stereochemistry of addition (see, Scheme 1.18 *e*). Anti addition of Br_2 to either face of (*E*)-2-butene gives the same meso-2,3-dibromobutane.

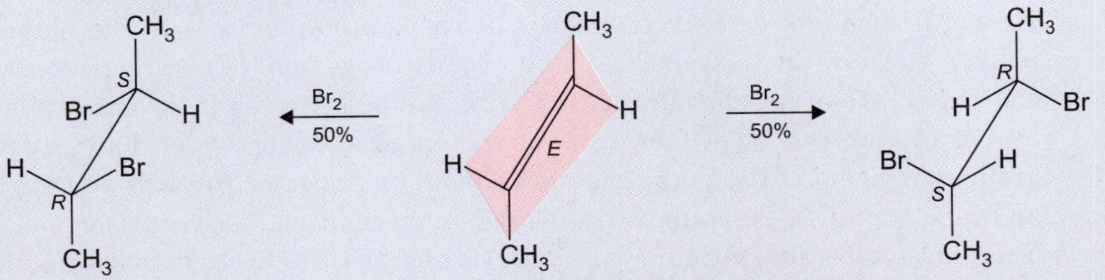

meso (*E*)-2-Butene *meso*

1.24. The two faces of (*E*)-2-butene are enantiotopic which yields two enantiomeric cyclic bromonium ions. The subsequent nucleophilic attack by Br^- is independent as to which bromonium ion is attacked (The chirality centers being homotopic), in either case the same *meso* product is formed (nucleophilic attack leads to inversion at one of the chirality centers.

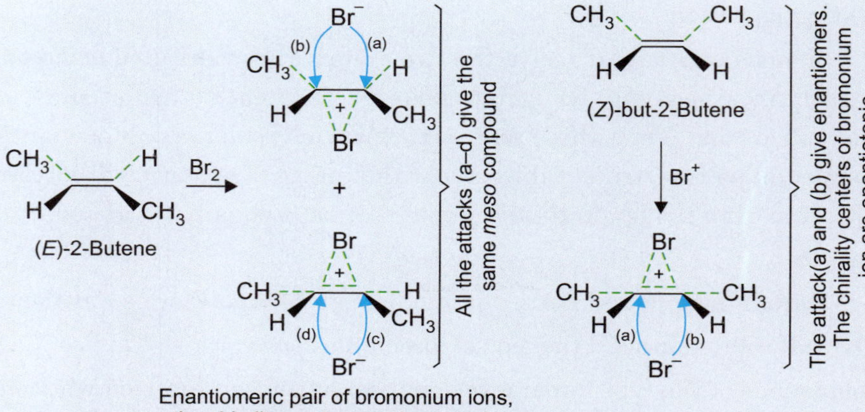

Enantiomeric pair of bromonium ions, the chirality centers are homotopic

Anti addition of Br_2 to (*Z*)-2-butene gives a racemic mixture of (2*R*, 3*R*)- and (2*S*, 3*S*)-2, 3-dibromobutane. The two faces of (*Z*)-2-butene are homotopic. The attack on either face gives the same bromonium ion.

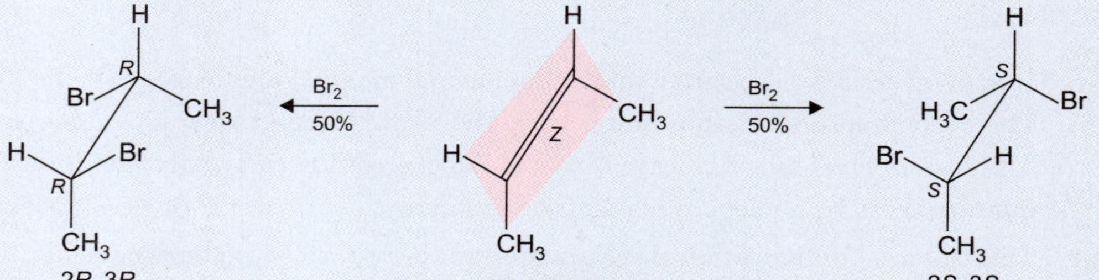

1.25. All the structures represent the same compound. Orientation (I) is the standard zig-zag (staggered) carbon chain (look at the continuous lines in the plane of paper). To know the relation of (I) with Fischer projection (IV) firstly (I) must be converted into eclipsed conformation. This can be achieved in several ways (*i*) by *e.g.* any rotation by 60° towards you by this procedure, the left hand methyl group will go out of the plane of page and will be projected towards you to give an orientation (II). This can be translated on an eclipsed Fischer projection. (*ii*) You may rotate the left hand carbon in (I) towards you by 180° to get another equivalent orientation (III) which can be translated on to Fischer projection (V).

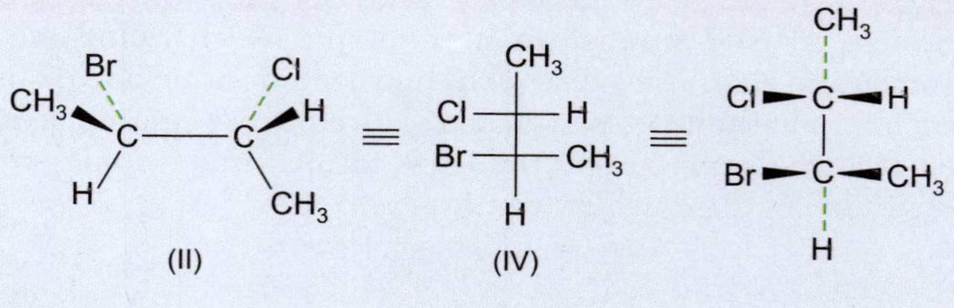

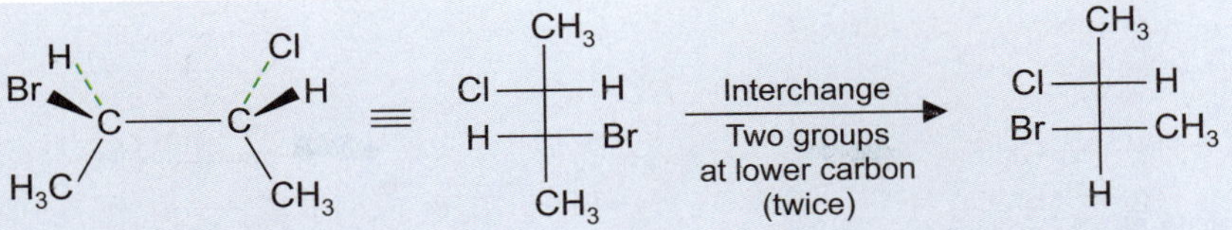

(III) (V) (V = IV) (IV)

In (V) if one switches first (CH_3, Br) and then (Br, H) on the lower stereocenter, its configuration remains unchanged and it comes out to be same as (IV).

1.26. All the orientations represent the same compound.

[**Hint:** The compound is already an eclipsed conformation which is translated on to the Fischer projection (II). The Fischer projection (III) is obtained by switching two groups twice and also provides a more conventional drawing with carbon chain vertical].

1.27. (*a*, 1) Diastereomers. (2) Same enantiomer of a chiral allene.

[**Hint:** One interchange of groups at a geminal carbon of a chiral allene gives an enantiomer, two such changes will leave the structure unchanged (see, Scheme 1.128).

1.28. Two structures are identical. One may convert the dashed-wedge line perspective formula which is presented here in a staggered zig-zag like structure (see the continuous lines in the plane of paper in a dashed-wedged line formula whether it is staggered or eclipsed) into the Fischer projection. Since Fischer projection is eclipsed, one must first derive an eclipsed conformation from the dashed-wedge line structure by rotating either stereogenic carbon and then transferring the groups on the Fischer projection (as done in Scheme 1.37*k*).

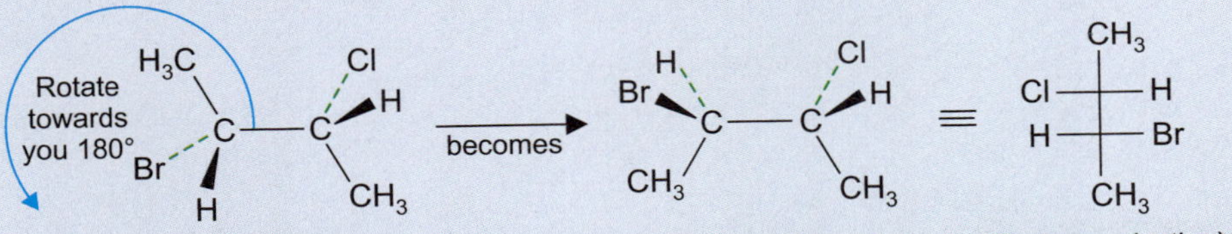

(Staggered projection) (Eclipsed projection) (Fischer projection) (III)

Now for determining the relationship of (III) with (II, problem 1.31) place as many similar groups in projection (III) in identical positions to that in (II). Recall that interchanging the position of two groups twice around a chirality center retains its configuration while interchanging only once inverts it.

Top chirality center

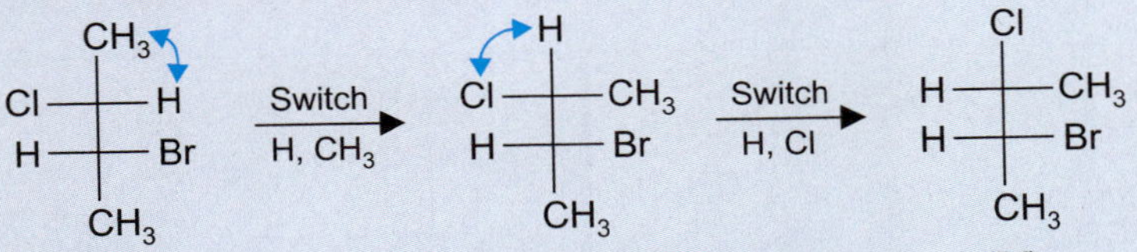

(III) (IV) (III IV)

Bottom chirality center

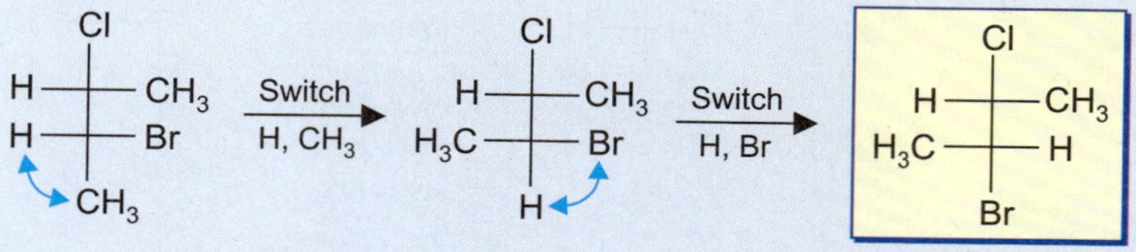

(V)
(III = IV = V)

1.29. Pair (I) represents identical compounds (2*S*, 3*S*)-2-fluoro-3-methylpentane. Pair (II*a* is *S*, while II(*b*) is *R*). The two compounds are therefore enantiomers.

1.30. In Fischer projection (I) the group of lowest priority is on the vertical line thus the path $1 \rightarrow 2 \rightarrow 3$ anticlockwise gives the correct description. In II, the group of lowest priority is on the horizontal line, the path $1 \rightarrow 2 \rightarrow 3$ clockwise (*R*) is to be reversed and the configuration is *S*. In the cyclic structure III, the group of lowest priority is not away. In order to do that, rotate the molecule 180° out of the plane of paper (now H will be away from you—OCH_3 towards you and the double bond will move to the right hand side), after this path $1 \rightarrow 2 \rightarrow 3$ is clockwise, the compound is *R* [or the path in III $1 \rightarrow 2 \rightarrow 3$ *i.e.*, *S* must be reversed to get the correct configuration *R* since the group of lowest priority is towards one's eyes].

1.31.

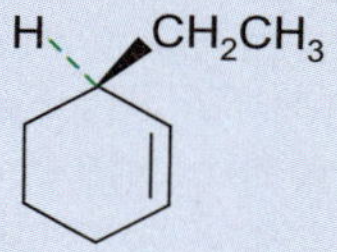

1.32.

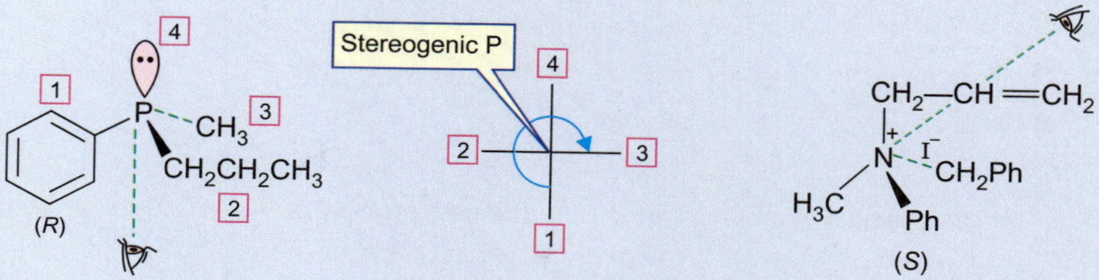

(I) Fischer projection (Ia) (II)

View from bottom (see, Scheme 1.20*d*), or convert into Fischer projection by following the procedure (Scheme 1.34).

1.33. Read the material (Schemes 1.34 to 1.37*c*) for solving this problem.

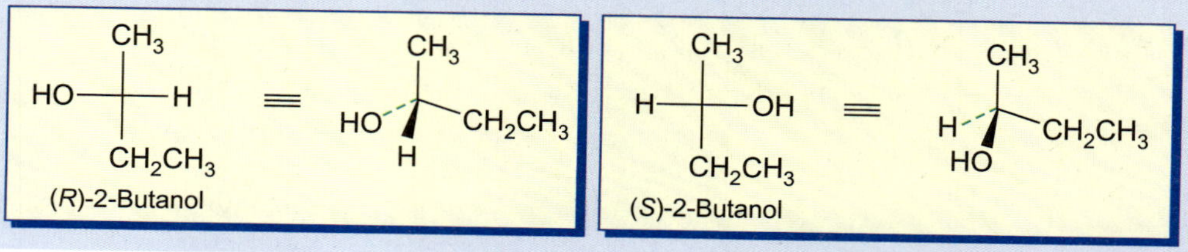

1.34. The compounds (I and II) are identical.

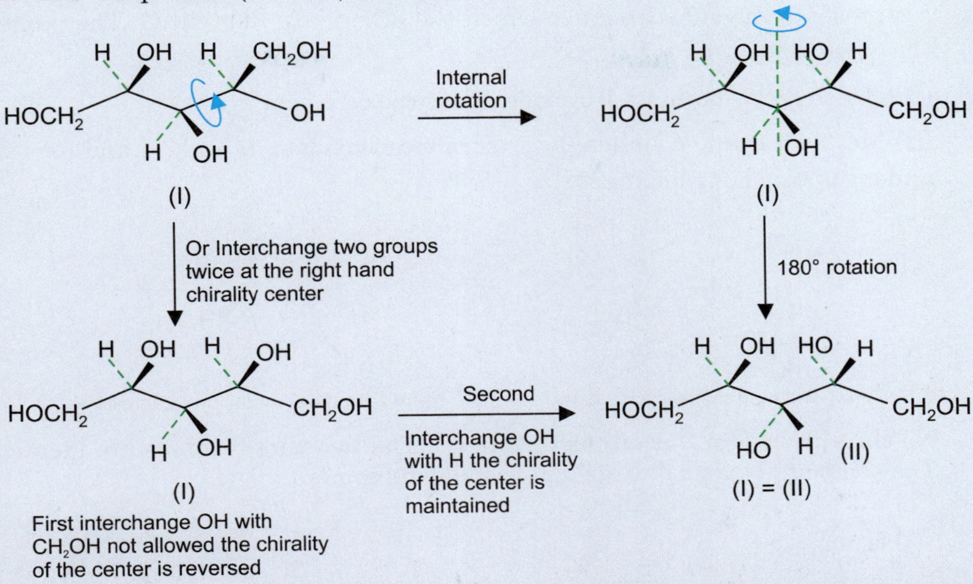

1.35. The compounds A and B are enantiomers, since at each chirality center one replaces all dashes with wedges and all wedges with dashes (interchange the places of two groups at each chirality center).

1.36. Both are *meso* and are diastereomers (differing in the stereochemistry at the central chirality center).

1.37. It is a chiral allene with the grouping of the type abC=C=Ccd. The chirality arises due to nonplanar arrangement of four groups about an axis called a chiral axis (stereoaxis).

1.38. Follow the modified simple procedure given in (sec. 1.16).

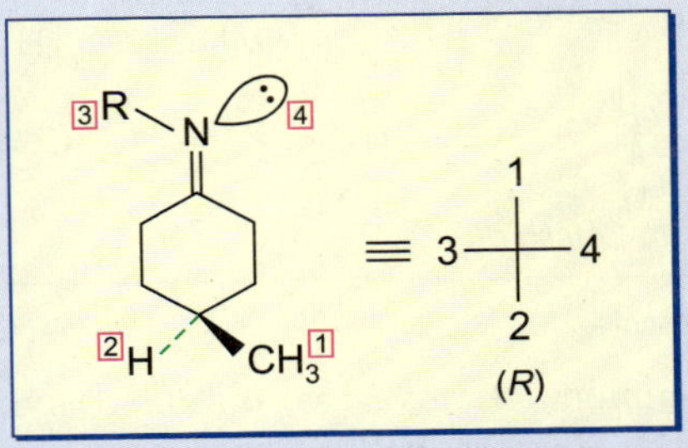

1.39. See Schemes 1.41 and 1.42.

1.40. See page 544.

1.41. See Scheme 1.76.

1.42. See Scheme 1.102.

1.43. When there are more than two substituent-bearing atoms in a ring, the symbol *r* represents the substituent to which the others are referenced. The name of the compounds are:

(I) *t*-4-Ethyl-*c*-3-methyl-*r*-1-cyclohexanecarboxylic acid,

(II) Methyl 2-cyano-*t*-3-phenyl-*r*-2-oxiranecarboxylate, (*t* and *c* stand for *trans* and *cis* in this nomenclature).

1.44.

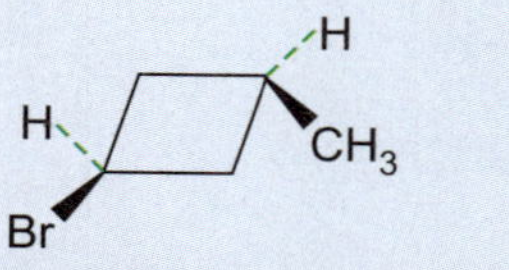

cis-1-Bromo-3-methylcyclobutane

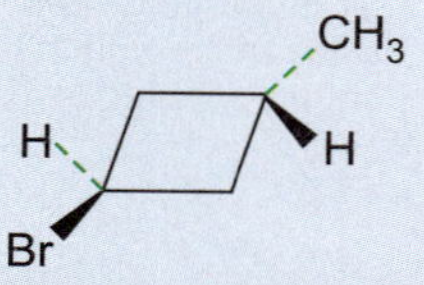

trans-1-Bromo-3-methylcyclobutane

No chirality center, for either C_1 and C_3. The two ring residues are identical. The compound exists only as a pair of diastereomers.

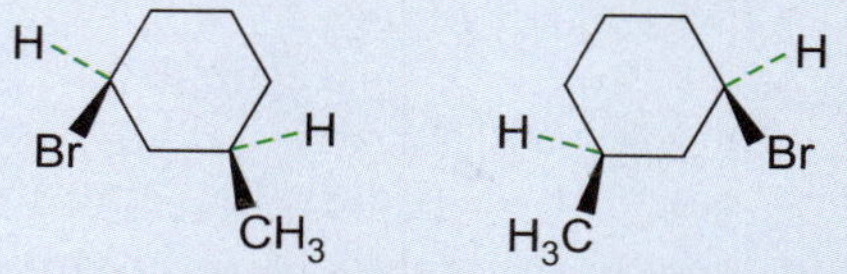

cis-1-Bromo-3-methylcyclohexane

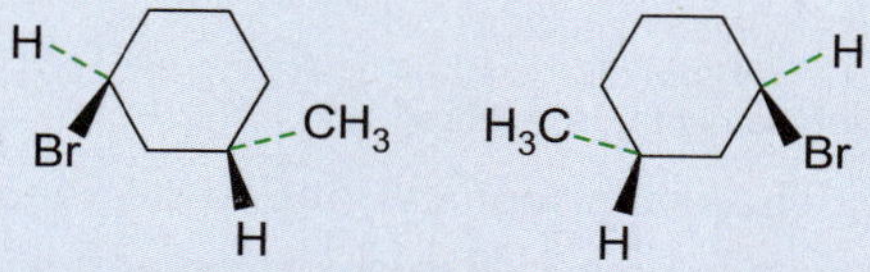

trans-1-Bromo-3-methylcyclohexane

The compound has two chirality centers, for either C_1 and C_3. The two ring residues are different. The compound exists as four chiral stereoisomers.

1.45. Writing the stereostructures of the carbon framework shows that it has two chirality centers and a pseudochirality center (Three stereocenters in all also see, Scheme 1.68*c*). The four stereoisomers are shown.

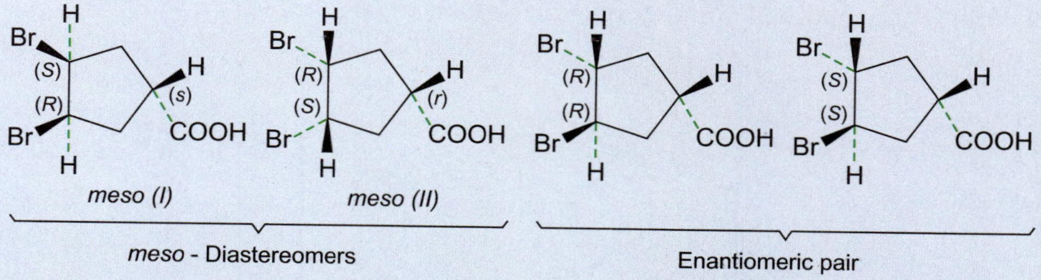

1.46. Since both of these molecules have two planes of symmetry and a C_2 axis, the point group is C_{2v}, (the C_2 axis and two planes of symmetry include the axis).

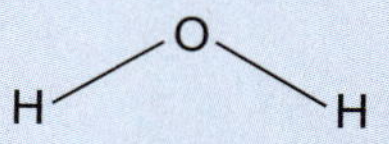

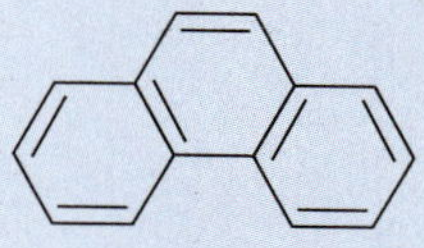

1.47. (*a*) The priority of groups in the compounds (I and II) are as shown. Therefore (I) is (*S*) and (II) is also (*S*). In the case of compound (III), there are two chirality centers in the side chain and a pseudochirality center (pseudoasymmetric center)in the ring.

$\boxed{1}\ CH_2OH > \boxed{2}\ C{\equiv}N > \boxed{3}\ CH_2NH_2 > \boxed{4}\ H$

Priority sequence
(I)

$\boxed{1}\ CH_2Cl > \boxed{2}\ CF_3 > \boxed{3}\ CH_3\overset{CH_3}{\overset{|}{C}}H > \boxed{4}\ CH_2CH_2SH$

Priority sequence
(II)

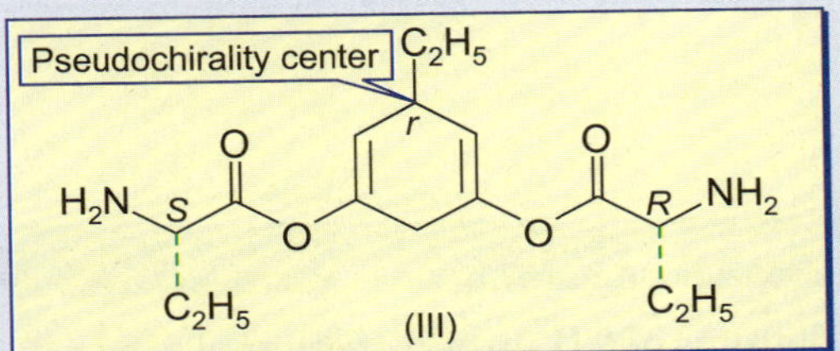

(III)

When a carbon carries two groups which are different and other two are enantiomorphic and lies on a mirror plane of symmetry, then this carbon is termed pseudochirality center.

(*b*) In methanol both protons must be axial (See, Scheme 1.142), the large isopropyl group is equatorial (V), the polar *OH* group is stabilised by hydrogen bonding. In benzene, the stabilisation effect is absent and therefore, the ring flips to attain stabilisation with carbonyl group by intramolecular hydrogen bonding. The loss of putting isopropyl group axial is compensated by hydrogen bonding (VI).

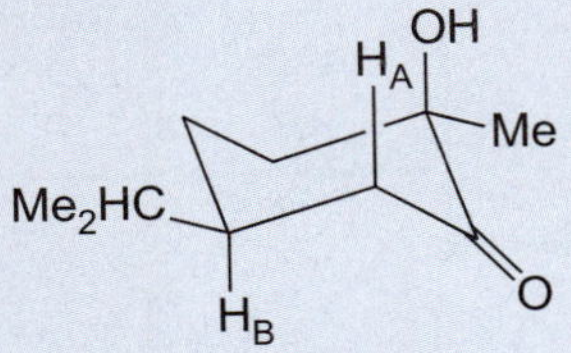

(V)
In methanol

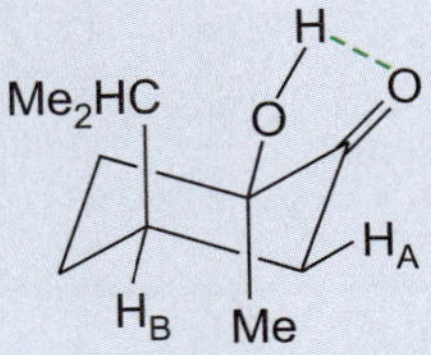

(VI)
In benzene

Prochirality—Prostereoisomerism and Asymmetric Synthesis

2

CHAPTER

2.1 INTRODUCTION – NOMENCLATURE

Isotopic labelling shows that the two H atoms of CH_2 group of ethanol are very different *i.e.*, heterotopic (Scheme 2.1). These heterotopic ligands are enantiotopic since *e.g.*, on isotopic replacement chiral entities formed are enantiomers. Moreover the ligands *i.e.*, the two H atoms of the CH_2 group in ethanol which are related by a mirror plane are enantiotopic. The two H atoms in the CH_2 group of ethanol behave differently in their enzyme catalysed rates of oxidation.

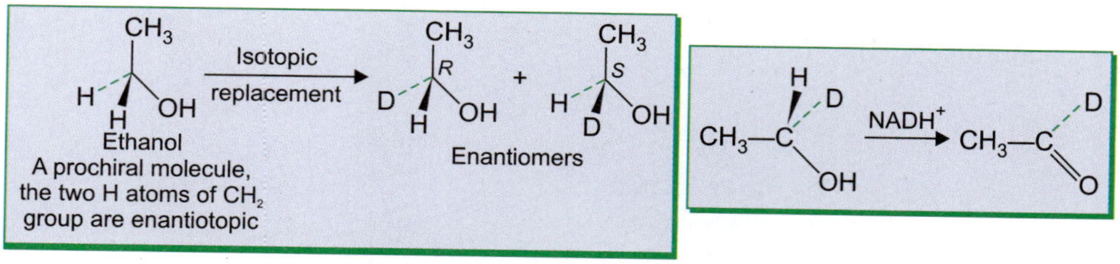

SCHEME 2.1

(A) Prochiral Molecules

Prostereoisomerism/prochirality is the property of certain molecules due to which these can be converted into stereoisomers (enantiomers or diastereomers).

A tetrahedrally bonded atom (of an achiral compound) of the general formula C_{abc_2} (as in propionic acid or ethanol) which becomes a chirality center C_{abcd} and the compound becomes chiral on replacement of one of the identical groups with a different group *d* is called a prochiral center and the molecule as prochiral. This situation is available in the molecules of propionic acid and ethanol.

In a molecule like propionic acid (Scheme 2.2) the carbon atom marked C_α is not a chirality center, the carbon C_α can be made a chirality center by replacing one or other of the two apparently identical ligands (in this case the ligands are two H atoms of the methylene group) in turn by a different ligand *e.g.*, OH to give enantiomeric products (*S*)-lactic acid and (*R*)-lactic acid. Thus propionic acid is a prochiral molecule and with the same reasoning ethanol (Scheme 2.1) is also a prochiral molecule.

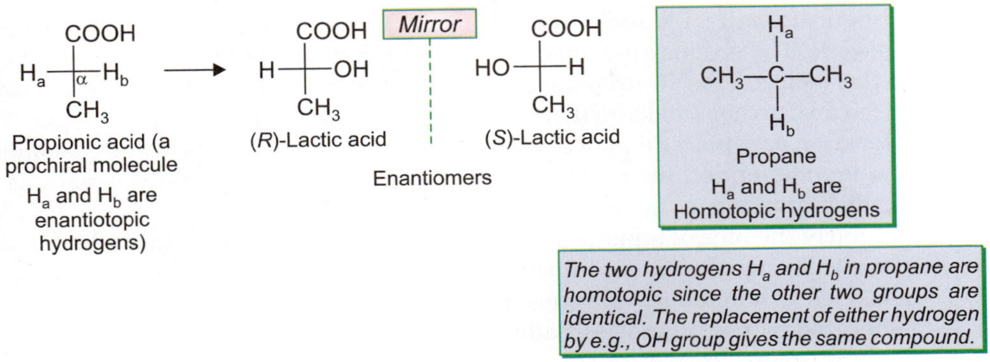

An achiral molecule is called prochiral if it can be converted in a single chemical reaction into a chiral molecule

SCHEME 2.2

(B) Homomorphic Ligands (Topicity)

The two apparently identical ligands in propionic acid (Scheme 2.2) *i.e.* H atoms of the methylene group are called "homomorphic" from Greek *homos* meaning same and *morphe* meaning form, these are indistinguishable when considered in isolation but distinguishable when part of a molecule.

The following points be noted:

- In the case of homomorphic atoms, these must of the same element *e.g.*, two H or two Br atoms.
- In the case of homomorphic groups these must have the same constitution and configuration. Thus in compound (I, Scheme 2.2*a*), C4 is a prochiral center since apart from H and OH it is attached with two other groups (ring residues) which are constitutionally and configurationally identical.

In ketone (II, Scheme 2.2*a*), the groups attached to the carbonyl group are not homomorphic since though constitutionally similar, these are not configurationally identical (also see, Scheme 2.20*b*).

(I)
C4 is prochiral
The two ring residues attached to C4 are homomorphic

(II)
The two ligands attached to the carbonyl group are not homomorphic

SCHEME 2.2a

- Two or more homomorphic ligands may be distinguishable when part of a molecule that are either equivalent (homotopic) or not equivalent (heterotopic from Greek *heteros* meaning different and topos meaning place *i.e.*, enantiotopic or diastereotopic).
- Topicity is thus the spatial relationship between constitutionally and configurationally identical (homomorphic) atoms or groups of atoms in a molecule. The groups are homotopic to mean that they occupy equivalent places in the molecule as in propane (Scheme 2.2). The groups may be heterotopic which are further subdivided into enantiotopic or diasterotopic groups, such that alteration or replacement of one or other of the ligands leads to stereoisomeric compounds [enantiomers as in ethanol (Scheme 2.1) or diastereomers as in 1,2-dichloropane (Scheme 2.10)].
- Similarly the idea of topism is also applied to *spaces* on either face of a trigonal atom which may become occupied by an incomming atom or group and are again classified as homotopic, enantio—and diastereotopic based on the environment generated by the ligands already attached to such an atom as *e.g.*, in pyruvic acid (Scheme 2.2*b*).
- Thus the groups or faces in a molecule can be homotopic or heterotopic. The heterotopic groups or faces are further sub-divided into enantiotopic and diastereotopic groups or faces, and the topicity is designated by the stereo-descriptors *pro-R, pro-S, Re, Si, pro-E* and *pro-Z* (see, Scheme 1.49).
- Thus topicity describes the relationships of two or more homomorphic ligands (or faces) which constitute a set. A ligand by itself cannot be called homotopic or heterotopic.
- Stereoheterotopic ligands (or faces) may be differentiated by chemical, biochemical or NMR spectral methods. Stereoselective synthesis has it base in different behaviour of heterotopic groups or faces towards chemical reactions.

(C) Stereoheterotopic Ligands

The two (apparently identical) H atoms of methylene group in propionic acid and ethanol (Schemes 2.1 and 2.2) are called stereoheterotopic groups. Stereoheterotopic ligands may be enantiotopic (as in propionic acid), diastereotopic (as in 2-butanol see, Scheme 2.13*a*) or a molecule may have only homotopic ligands (as in propane, Scheme 2.2).

(D) Heterotopic Faces

When the carbonyl group in pyruvic acid (Scheme 2.2*b*) is reduced by the addition of hydride from *e.g.* $NaBH_4$ the enantiomers of lactic acid are again obtained by the equally feasible addition to either the front or the rear face. The sp^2 hybrid carbon of carbonyl group in pyruvic acid is called prochiral and is said to have two heterotopic faces which are enantiotopic.

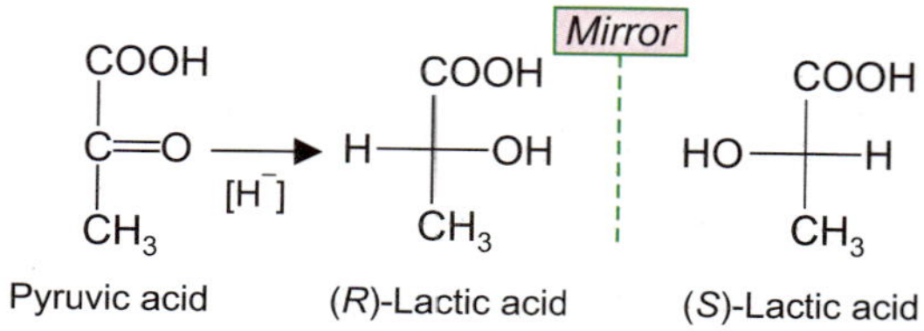

Pyruvic acid is a prochiral molecule. Chirality is introduced by nucleophilic attack, [H⁻]. The product formed by attack on top face is the mirror image of that formed by attack on the bottom face. The two faces of the carbonyl group in pyruvic acid are termed enantiotopic.

SCHEME 2.2b

2.2 HOMOTOPIC AND HETEROTOPIC LIGANDS AND FACES

Two criteria—substitution/addition or symmetry may be employed to determine the topic relationships of homomorphic ligands and faces and one of these tests is however, sufficient. Unfortunately the still commonly used term "equivalent" lacks precision—the two methylene hydrogen atoms of propionic acid (Scheme 2.2) are equivalent when detached (*i.e.*, these are homomorphic) from the system of propionic acid. These protons are not equivalent due to their placement in propionic acid, where, these are heterotopic. Heterotopic ligands and faces may be either enantiotopic or diastereotopic.

(A) Homotopic Ligands and Faces

(*i*) Substitution – Addition Criterion

Two homomorphic ligands are homotopic if substitution (replacement) of each one of them in turn by another atom or group leads to the same structure (the replacement ligand must be different not only from the original one, but also from all other ligands attached to the same carbon). Thus the three methyl hydrogen atoms in acetic acid or toluene are homotopic because replacing any one of these by *e.g.*, chlorine gives the same chloroacetic acid or benzyl chloride respectively. The two methine hydrogen atoms in (*R*, *R*)–(+)–tartaric acid (I, Scheme 2.2*c*) or in its enantiomer are homotopic since replacement of either of them *e.g.*, by deuterium gives the same (2*R*, 3*R*)-tartaric-2-*d* acid. This example shows that chirality does not interfere in having homotopic ligands.

Two faces of a double bond are homotopic provided addition to either face gives the same product. Thus the two faces of *e.g.*, *cis*-2-butene are homotopic (epoxidation is used as the test reaction) since epoxidation on either face gives the same *meso* product (see, Scheme 1.102). Similarly the two faces in formaldehyde (and also in acetone) are homotopic (see Scheme 2.5*a*).

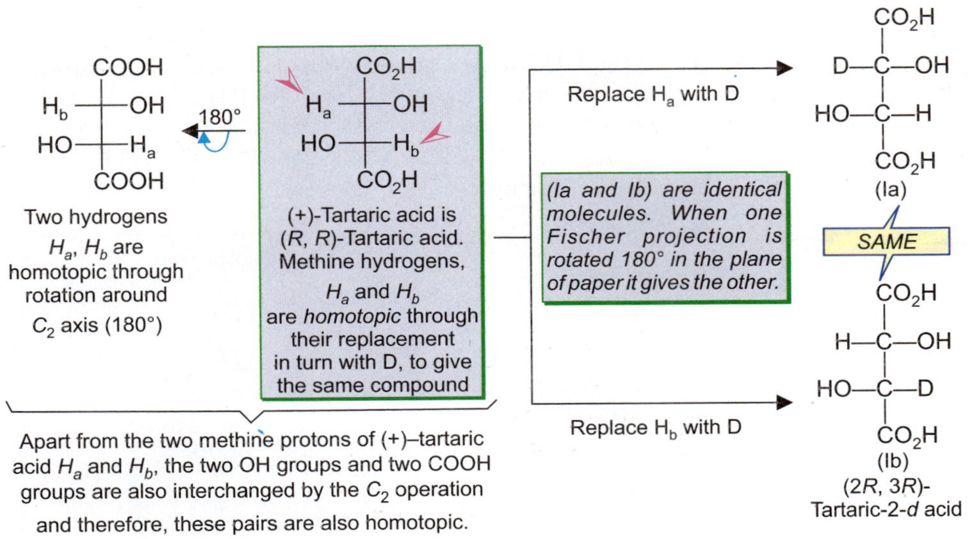

SCHEME 2.2c

(*ii*) Symmetry Criterion

Ligands or faces of a double bond are homotopic if they can interchange places through a rational axis (C_n). In (+) tartaric acid, the two methine protons (H_a and H_b) and the two OH and COOH groups are interchangeable through rotation around a C_2 axis. These pairs of ligands (Scheme 2.2*c*) are thus homotopic.

In a molecule *e.g.*, acetic acid two successive 120° rotations of a methyl group around the axis (C_3 rotational axis, Scheme 2.3) allow each hydrogen to take the position of either of the other two (with no structural change). Thus in a rapidly rotating methyl group of acetic acid all the hydrogens are rendered equivalent and are thus homotopic. The methyl group is not stationary, if it was so then *e.g.*, in one of its eclipsed conformations, the hydrogens would have been rendered heterotopic which is not the case.

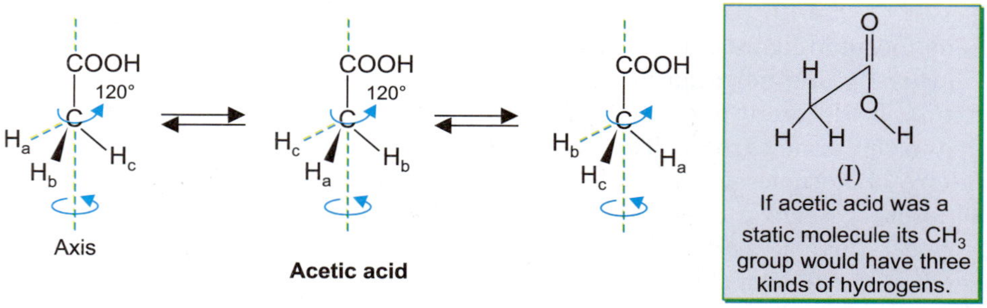

SCHEME 2.3

In the eclipsed conformation of acetic acid (I, Scheme 2.3) provided it existed as shown, then one of the protons which is eclipsed with the carbonyl group is rendered different from others, it being in the shielding zone of C = O group would be a shielded proton in ^1H NMR.

In molecules of the type CH_2Y_2, the two hydrogens are homotopic as in propane (Scheme 2.2) and also in malonic acid CH_2 $(COOH)_2$. The two methine protons (as well as the two OH and the two COOH groups) of (+)–tartaric acid are interchangeable through a C_2 axis (see Scheme 2.2*c*). These pairs of ligands are therefore, homotopic. Similarly the faces of double bonds, carbonyl

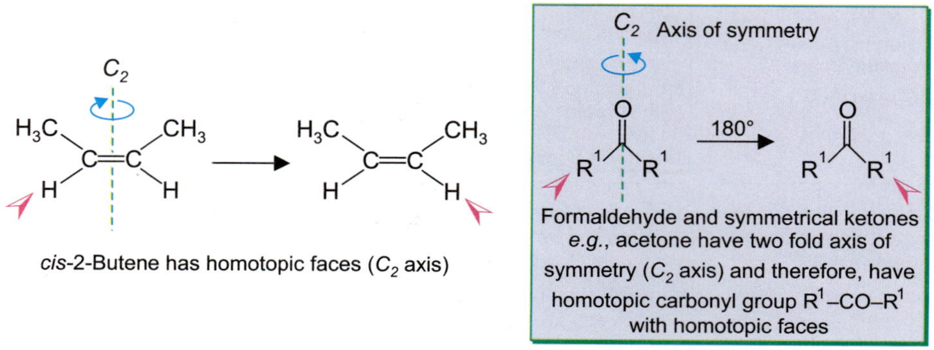

SCHEME 2.4

(compounds) or carbocations are homotopic if these interchange by a C_2 axis. This is so *e.g.*, in *cis*-2-butene, formaldehyde and symmetrical ketones (Scheme 2.4). The diol (I, Scheme 2.5) has a two fold axis of symmetry (C_2 to show that the two hydroxyl groups are homotopic. This fact is proved experimentally since on monoacetylation the diol just gives only one product.

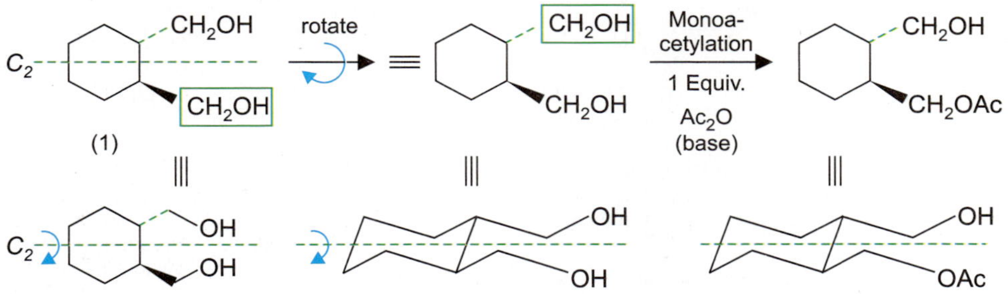

SCHEME 2.5

EXERCISE 2.1

Demonstrate that the two faces of formaldehyde (methanal) are homotopic. Are the faces in cyclopropanone and trans-2, 3-dichlorocyclopropanone homotopic ?

ANSWER. *Approach of a reagent e.g., a nucleophile to add a new ligand not already present in the molecule (which should not be either H or O in the case of e.g., methanal) yields the same product (Scheme 2.5a). Thus approach of a reagent e.g., EtLi to either face F and F′ of methanal affords the same product. Moreover, the faces are also exchangeable through C_2 axis to make the faces homotopic. Based on similar arguments the faces in both cyclopropanone and trans-2, 3-dichlorocyclopropanone are also homotopic.*

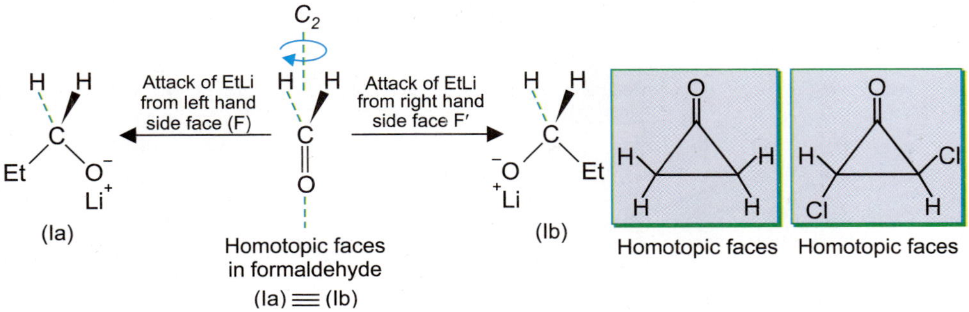

SCHEME 2.5a

(B) Enantiotopic Ligands and Faces

(i) Substitution – Addition Criterion

Two heterotopic ligands are enantiotopic if their replacement in turn by a different achiral ligand gives rise to enantiotopic products. In the case of propionic acid (CH_3CH_2 COOH the protons of the methylene group are enantiotopic (see, Scheme 2.2). The protons H_1 and H_2 in *meso-* tartaric acid are enantiotopic since their replacement in turn by a test group D gives rise to enantiomeric structures (Scheme 2.6). Interestingly *meso*-tartaric acid is a rare example of a molecule with heterotopic ligands but no discernible prochiral atom or other element of prochirality.

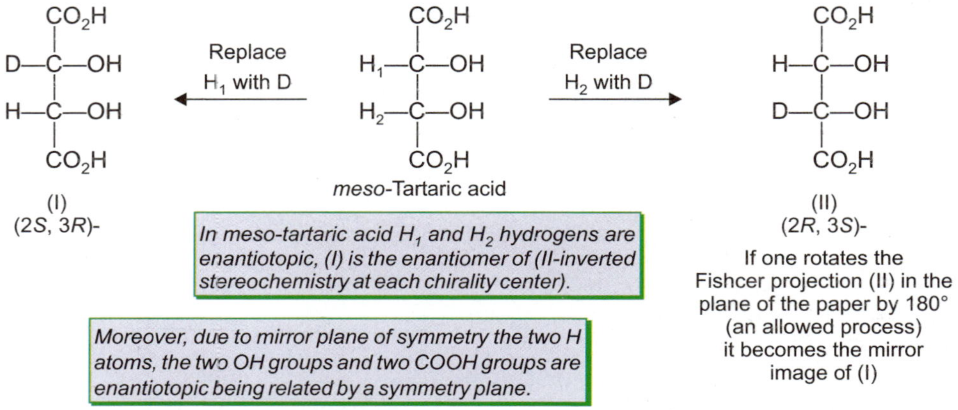

SCHEME 2.6

Based on similar criterion, the faces F and F′ of *e.g.*, aldehydes (other than formaldehyde) *e.g.*, ethanal are enantiotopic (Scheme 2.7). Nucleophilic attack from one face (F) gives a product that is the mirror image of that formed by the attack of the same nucleophile from the other face (F′).

These molecules with sp^2 hybridized carbons which become chiral (to give a 50 : 50 mixture of enantiomers *via* addition from either top face or bottom face) are also prochiral molecules. The two faces of the carbonyl group in these compounds are termed enantiotopic (Scheme 2.7). Consider the case of *trans*-2-butene which has enantiotopic faces since epoxidation to either face gives an enantiomer (see, Scheme 1.102.)

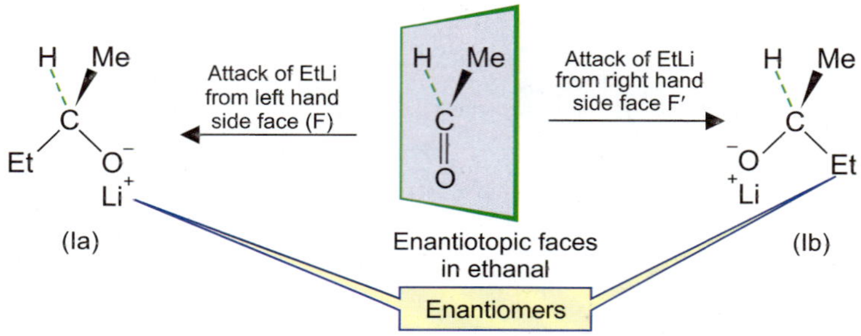

SCHEME 2.7

(*ii*) Symmetry Criterion

The enantiotopic ligands and faces cannot be interchanged by a simple axis of symmetry C_n (Scheme 2.7*a*) (operation of symmetry element of first kind) but are interchangeable by σ (plane of symmetry), center of symmetry *i* or alternating axis of symmetry S_n (operation of a symmetry element of second kind). Thus the two H atoms of CH_2 group in ethanol are enantiotopic since these are related by a symmetry plane (see Scheme 2.1) and not by a (simple) symmetry axis.

When R^1 does not equal R^2, as in aldehydes (Scheme 2.7*a*) and unsymmetrical ketones, the carbonyl compound looses C_2 axis (Scheme 2.7*a*), the faces of the carbonyl compound are now known as enantiotopic.

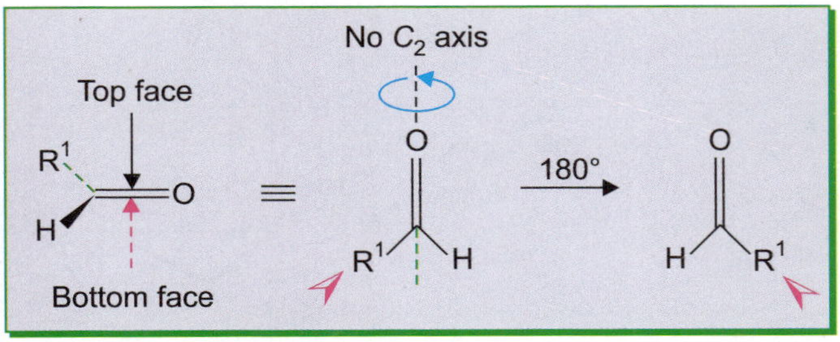

Unlike in (scheme 2.4) when in aldehydes and unsymmetrical ketones the groups attached to the carbonyl group are different, the faces of the carbonyl group become enantiotopic.

SCHEME 2.7a

In a generalized carbonyl compound R^1—CO—R^2 with the molecular plane at right angles to the plane of the paper, it is flat, it has two enantiotopic faces, one as viewed from bottom and the other from the top. The enantiotopic faces of this type of carbonyl compounds are also interchangeable through a σ plane which coincides with the plane of the molecule.

In *meso*-tartaric acid a σ plane of symmetry (see Scheme 2.6) is obvious which shows that two H atoms, two OH groups and two COOH groups are enantiotopic. Enantiotopic ligands are also interchangeable through the operation of a center of symmetry *i*. The center of symmetry is also obvious in the stable staggered conformation of *meso*-tartaric acid (see Scheme 1.77).

The two faces of *trans*-2 butene are enantiotopic (presence of plane of symmetry, and center of symmetry) in contrast to that in *cis*-2-butene (plane of symmetry, but C_2 also) where the two faces are homotopic (see, Scheme 1.99).

EXERCISE 2.2

Identify the indicated hydrogens/ligands in the compounds (Scheme 2.8) as homotopic or enantiotopic.

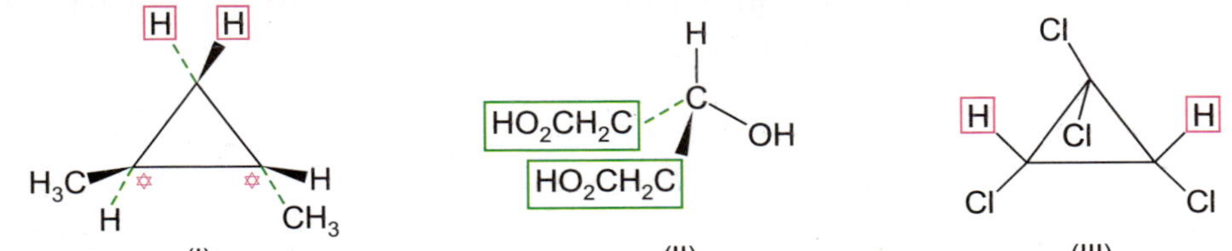

(I) (II) (III)

trans-1, 2-Dimethylcyclopropane

SCHEME 2.8

ANSWER. *In (I) the hydrogen atoms of the methylene group are homotopic since C_2 symmetry axis interchanges them (IV, Scheme 2.8a). Moreover, replacement with X of either*

Continued....

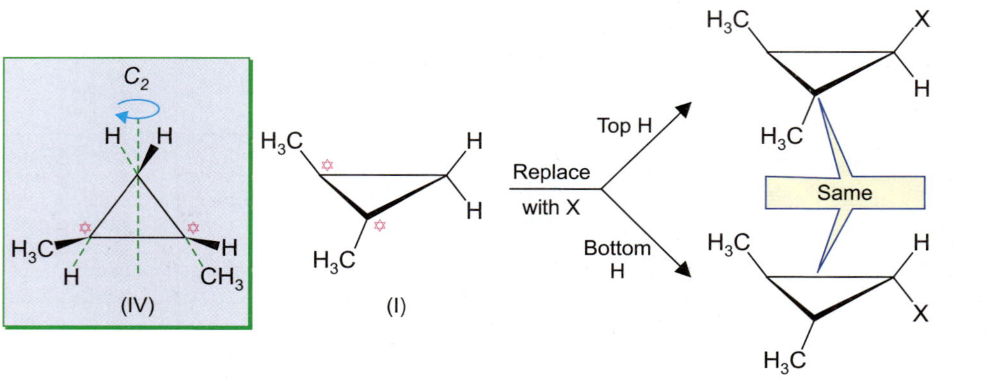

SCHEME 2.8a

of these hydrogen atoms in turn gives the same compound (turn either one over, it becomes superimposable on the other). In (II) mono esterification gives an enantiomeric pair ((V) and (VI), Scheme 2.8b) thus the two ligands are enantiotopic. In (III) since the protons cannot be interchanged by a simple symmetry axis but are related by a mirror plane of symmetry (VII, Scheme 2.8b) these are enantiotopic.

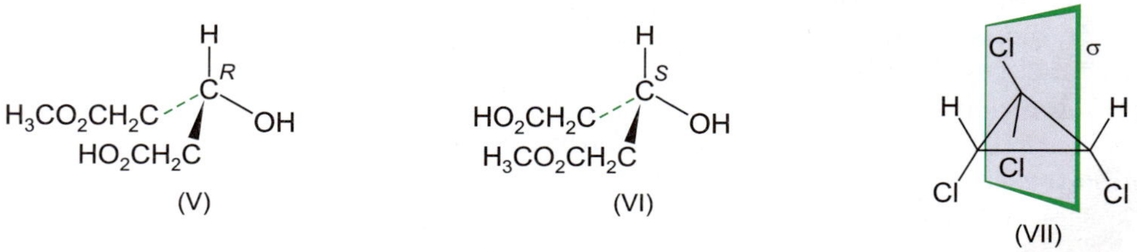

SCHEME 2.8b

(C) Diastereotopic Ligands and Faces

(*i*) Substitution – Addition Criterion

When the replacement of either of the two (homomorphic) ligands by a different achiral test ligand gives diastereomers then such (homomorphic) ligands are termed diastereotopic. The addition criterion is similarly applied to recognize diastereotopic ligands and faces. In propene (Scheme 2.9) replacement of H_a and H_b in turn gives diastereomeric alkenes (*cis*- or *trans*) thus H_a and H_b are diastereotopic hydrogen atoms.

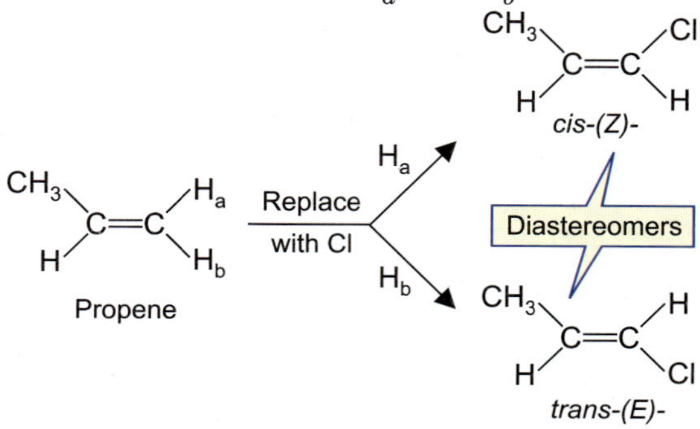

In propene replacement of H_a and H_b in turn by Cl gives a cis-trans pair of diastereomeric alkenes. Thus H_a and H_b in propene are diastereotopic.

SCHEME 2.9

BOX B 2.1
PROCHIRAL CENTER / PROSTEREOCENTER

In cyclobutanol (an achiral molecule, Scheme B 2.1), the ligands H^e and H^f are diastereotopic which on substitution in turn by OH group give rise to diastereomeric (cis and trans) products. Protons H^e and H^f in cyclobutanol are diastereotopic, their replacement in turn by "OH" gives cis-and trans-diols which are diastereomeric. Since in eithr cis-or trans diol chirality is not introduced at C3 (plane of symmetry in the diols) C3 in Cyclobutanol is not a prochiral center but is a prosterocenter (i.e., proachiral center). Thus C3 in cyclobutanol displays prostereoisomerism and not prochirality.

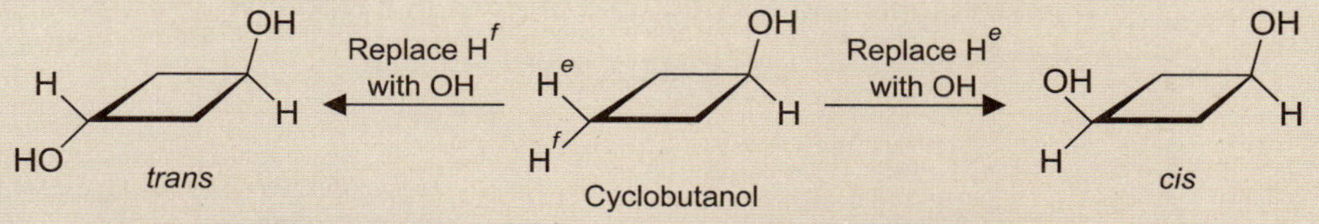

SCHEME B 2.1 (For another example see, Scheme 4.30)

Two geminal methylene protons or other (homomorphic) ligands (*e.g.*, hydrogens of methyl groups) in chiral molecules when *e.g.*, adjacent to a chirality center) are usually diastereotopic (as in 1, 2-dichloropropane (Scheme 2.10).

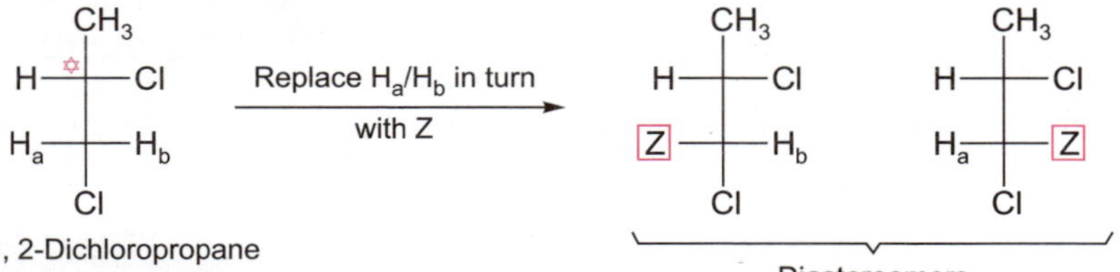

Geminal methylene protons, adjacent to a chirality center are usually diastereotopic

SCHEME 2.10

The geminal protons close to a chirality axis in the chiral biphenyl (I, Scheme 2.10*a*) *e.g.*, the methylene protons in each of the CH_2OH groups in ring A and B are diastereotopic since there is no mirror plane of symmetry bisecting H_A–C–H_B angle. However, when the biphenyl is achiral (II) based on same arguments. The geminal hydrogens of CH_2OH in ring A of (II) are diastereotopic while the geminal hydrogens of CH_2OH group of ring B of (II) are enantiotopic. Similarly, H_a and H_b in (III) are diastereotopic.

Cyclobutanol (Scheme 2.11) presents another case of a compound having both enantiotopic and diastereotopic protons.

The two faces of a carbonyl group close to a chirality center are diastereotopic. Thus addition of HCN to methyl α-phenethylketone (Scheme 2.12) gives diastereomers. Similarly achiral 4-*t*-butylcyclohexanone has two diastereotopic faces since addition of hydride ($NaBH_4$) can occur from the axial or the equatorial side to give diastereomers (Scheme 2.13).

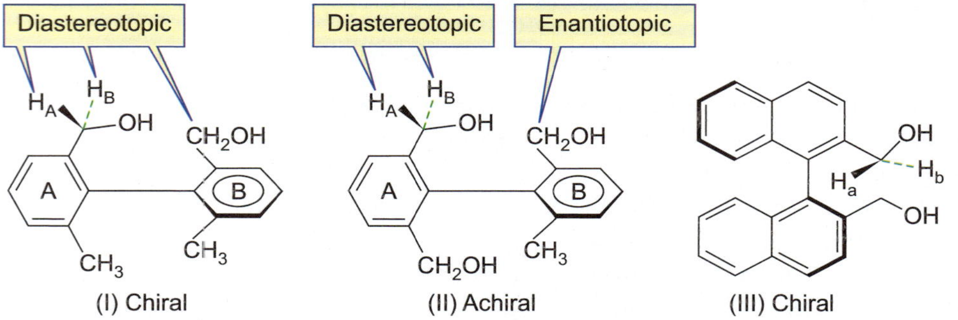

Diastereotopic and enantiotopic protons in a biphenyl

SCHEME 2.10a

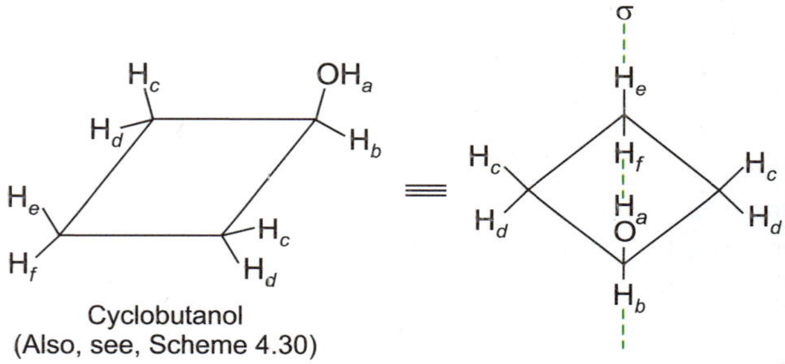

Cyclobutanol
(Also, see, Scheme 4.30)

In cyclobutanol protons H_e and H_f are diastereotopic. Cyclobutanol has also a plane of symmetry. Protons H_c are mirror images of each other and are enantiotopic, similarly protons H_d are also enantiotopic. The relationship between each H_c and H_d pair is again diastereotopic.

SCHEME 2.11

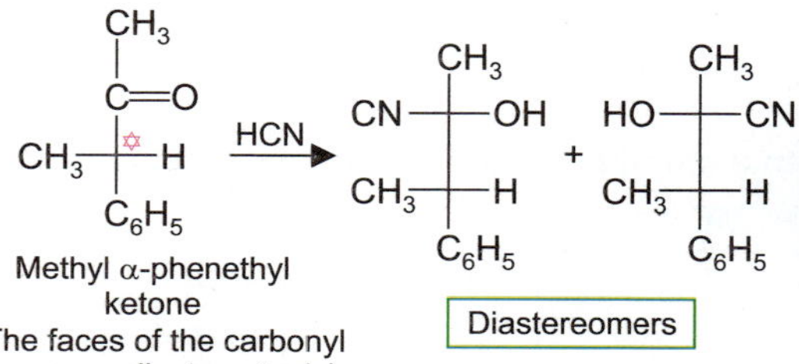

Methyl α-phenethyl ketone
(The faces of the carbonyl group are diastereotopic)

When in aldehydes (other than formaldehyde) and unsymmetrical ketones R-CO-R′ (R≠R′) the groups R, R′ contain one or more chirality centers, the faces of the carbonyl group become diastereotopic. Addition to one face gives a diastereomer and addition to the other face gives another diastereomer.

SCHEME 2.12

(*ii*) Symmetry Criterion

The diastereotopic ligands or faces cannot be related by any symmetry relation *i.e.*, C_n (axis of symmetry) or S_n (alternating axis of symmetry). One example is taken to explain this fact. 2-Butanone has enantiotopic faces since these faces are interchangeable through a σ plane coinciding with the plane of the molecule, however, when the C═O group is adjacent to a chirality center as in the case of a ketone (Scheme 2.12), then its two faces are no longer interchangeable through a σ plane. Thus the faces of the ketone (Scheme 2.12) are diastereotopic.

Cyclohexanone itself has two equivalent faces however, its monosubstituted derivatives have two diastereotopic faces as in achiral 4-methyl cyclohexanone (I, Scheme 2.13) and chiral 3-methyl and 2-methyl cyclohexanone (II and III Scheme 2.13).

Consider 4-*t*-butylcyclohexanone (Scheme 2.13) is an achiral molecule, the faces of the carbonyl group are diastereotopic - addition of hydride to the two faces (in turn) gives two (achiral) diastereomeric products. Here also, since chirality is not introduced at carbonyl carbon, this carbon is prostereocenter rather than prochiral – (a case of prostereoisomerism rather than prochirality).

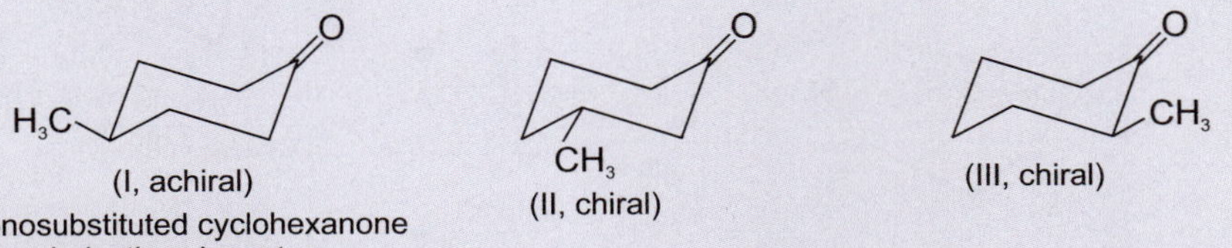

Monosubstituted cyclohexanone derivatives have two diastereotopic faces

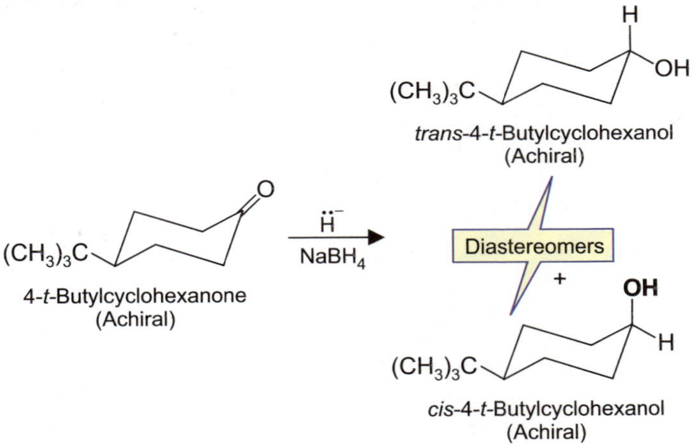

SCHEME 2.13

EXERCISE 2.3

The indicated hydrogens (Scheme 2.13a) are diastereotopic? Explain.

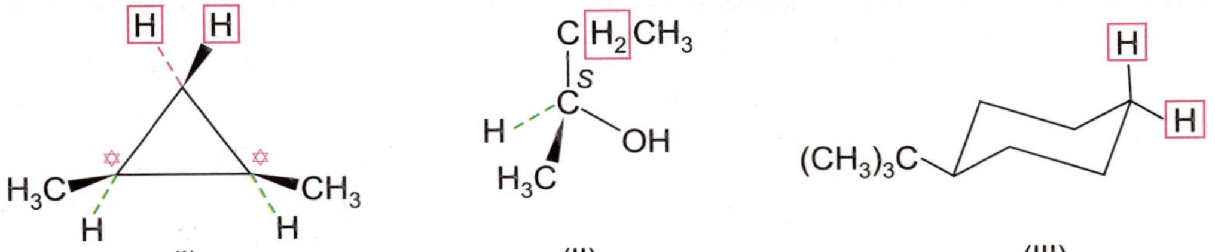

(I) *cis*-1, 2-Dimethylcyclopropane

(II) *S*-2-Butanol

(III)

SCHEME 2.13a

Continued...

ANSWER. *In (I) the methylene protons are close to a chirality center (infact two chirality centers) and are thus diastereotopic. These are also not related by simple symmetry axis. Moreover, their replacement by a group X gives diastereomers (Scheme 2.13b). In (II) one has both a chirality center and a prochiral center. The methylene protons are diastereotopic as shown by isotopic replacement (Scheme 2.13c). The indicated hydrogens in (III) are diastereopic since a replacement test gives diastereomers (Scheme 2.13c).*

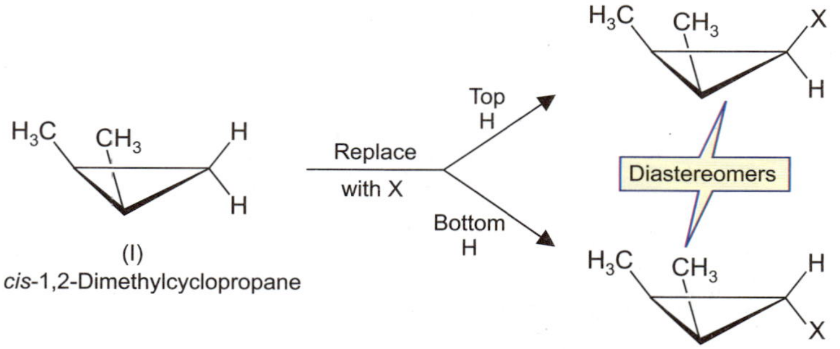

SCHEME 2.13b

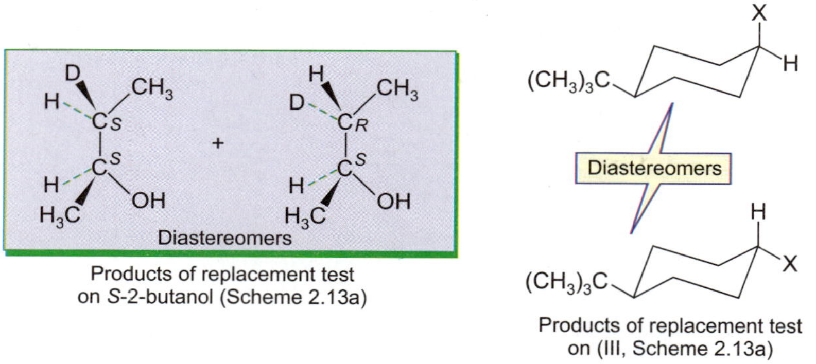

SCHEME 2.13c

A summary of topic relationship

Topicity	Substitution-addition criterion	Symmetry criterion	Difference
Homotopic	Identical product	Ligands related through C_n and faces by C_2 axis	No difference by any method
Enantiotopic	Enantiomeric products	Ligands (faces) related through σ, *i*, or S_n	Distinguishable, in principle, in chiral media (NMR), by chiral reagents, and enzymes
Diastereotopic	Diastereomeric products	Ligands and faces not related by any symmetry element	Distinguishable, in principle, by all methods

Homomorphic ligands will be homotopic if the operation of the symmetry axis make the nuclei in question interchange places. Thus in 1,3-dioxolane (I, Scheme 2.13*d*), the hydrogen atoms at *C*(2) are homotopic since they are interchanged by operation of the C_2 axis. On the other hand,

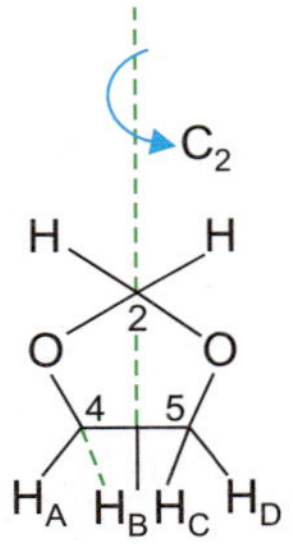

1, 3-Dioxolane

(I)

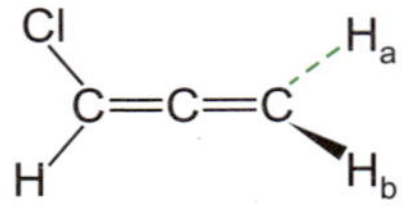

A monosubstituted allene where H_a and H_b are enantiotopic. The allene has a prochiral axis. (Also see, Scheme 1.114a)

(II)

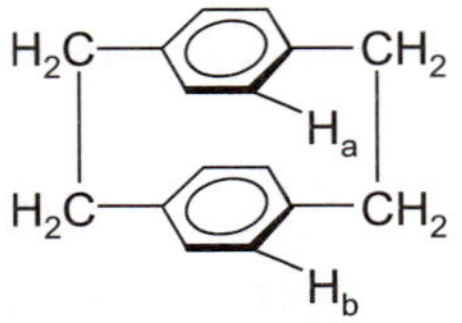

A paracyclophane with enantiotopic protons (H_a and H_b) and a prochiral plane

(III)

SCHEME 2.13d

the geminal hydrogen atoms at *C*(4), or *C*(5), are not interconverted by the C_2 symmetry operation and are thus heterotopic (H_A with respect to H_B and H_C with respect to H_D). However, H_A and H_D are homotopic (as are H_B and H_C), since these are also interchanged by the C_2 axis.

In monochloroallene (II) and paracyclophane (III) only one pair of enantiotopic H's is shown) H_a and H_b are enantiotopic, since replacement of either in turn with *e.g.*, D gives enantiomers (also see, Schemes 1.114*a* and 1.112). Moreover, in both (II and III) the hydrogens H_a and H_b exchange positions through σ plane.

2.3 NOMENCLATURE OF HETEROTOPIC LIGANDS AND FACES AND SOME STEREOSELECTIVE REACTIONS

Recall that the groups or faces in a molecule are classified as homotopic or heterotopic. The heterotopic groups or faces are further sub-divided into enantiotopic and diastereotopic groups or faces, and the topicity is designated by the stereodescriptors *pro-R, pro-S, Re, Si, pro-E* and *pro-Z*.

(A) *pro-R, pro-S* Nomenclature

(*i*) Molecules with Prochiral Center(s)

To name the ligands at a prochiral center *e.g.*, H_a and H_b in ethanol (Scheme 2.14), the ligand to be labelled is arbitrarily assigned a higher CIP priority over the other.

If H_a is arbitrarily preferred over H_b in the sequence rule, the sequence is $OH > CH_3 > H_a > H_b$ and the (hypothetical) configurational symbol to the prochirality center in ethanol will be *R*, thus H_a is designated *pro-R* and by default H_b becomes *pro-S*. The same answer would have come out if H_b was given precedence over H_a, in that case the sequence would have been $OH > CH_3 > H_b > H_a$. The hypothetical configurational symbol for ethanol is then *S*, hence H_b is *pro-S*.

It might be noted that the same result would have been obtained by replacing first, one hydrogen and then the other by deuterium since deuterium has priority over hydrogen. Replacement of H_a by D gives (*R*)-ethanol-1-*d* and hence H_a is *pro-R*; similarly, replacement of H_b by deuterium gives (*S*)-ethanol-1-*d* and hence H_b is *pro-S*. Another example is found in propionic acid (I, Scheme 2.14), it is always better to assign a descriptor to H_a first *i.e.* a group directed towards you then automatically H_b of lowest priority will be away from you *i.e.*, an arrangement for the direct assignment of CIP nomenclature. Similarly in (NADH, II, Scheme 2.14) the assignment can be made easily. For the assignment of *pro-R* and *pro-S* descriptors to (A, Scheme 2.14) use the information explained in (Scheme 1.20*k*).

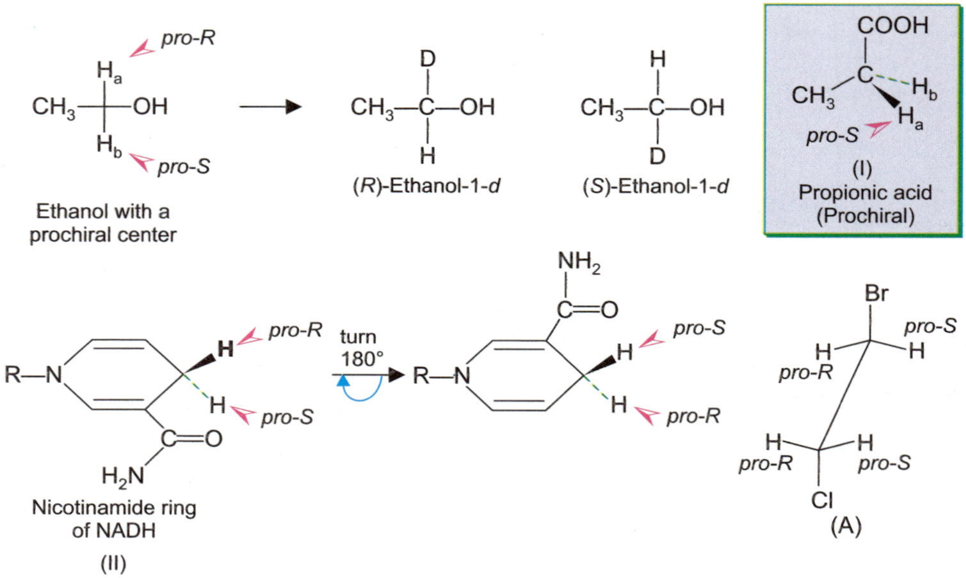

Chiral centers can be labelled *R* or *S* not only in enantiomers but also in many diastereomers, so the designations *pro-R* and *pro-S* are not confined to enantiotopic ligands but may also be used for a number of diastereotopic ligands as well.

SCHEME 2.14

Cyclohexanol is another interesting example which combines the concepts of prochirality and proachirality *i.e.*, prostereoisomerism (see, Scheme 4.30).

BOX 2.2
PROCHIRALITY CENTER AND PROPSEUDOCHIRALITY CENTER

Citric acid molecule (Scheme B 2.2) has a plane of symmetry perpendicular to the page and passing through the central carbon atom. Via this plane the two protons H_a interchange and two protons H_b interchange and are thus enantiotopic.

However, there is no plane of symmetry passing between the protons of each methylene group and this renders protons H_a and H_b on each CH_2 group diastereotopic. Moreover, replacement of e.g., H_a on one carbon by D generates two chirality centers in the molecule. This molecule is diastereomeric with the one in which H_b on the same carbon is replaced by D (Scheme B 2.2).

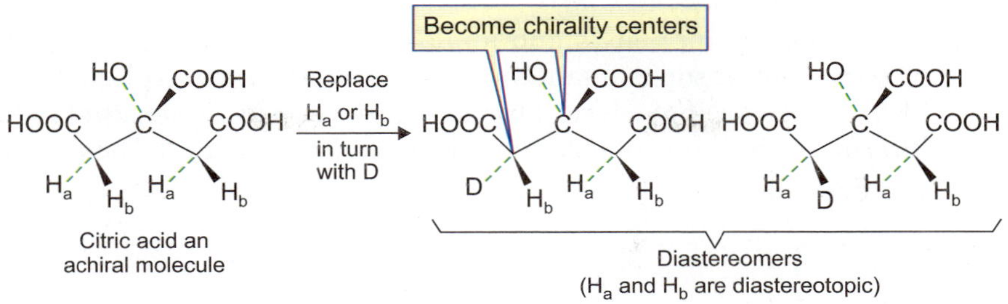

SCHEME B 2.2

The central carbon atom in citric acid is prochiral (Scheme B 2.3). The identical groups attached to this carbon are –CH_2COOH. To label these as pro-R and pro-S one can start from either of these identical groups e.g., the one in a shaded box, one thus gives it arbitrarily higher priority over the other. The priorities for CIP nomenclature are as shown. Thus the ligand CH_2COOH in the shaded box is labelled pro-R. Following the same procedure the ligand CH_2COOH in the unshaded box come out to be pro-S.

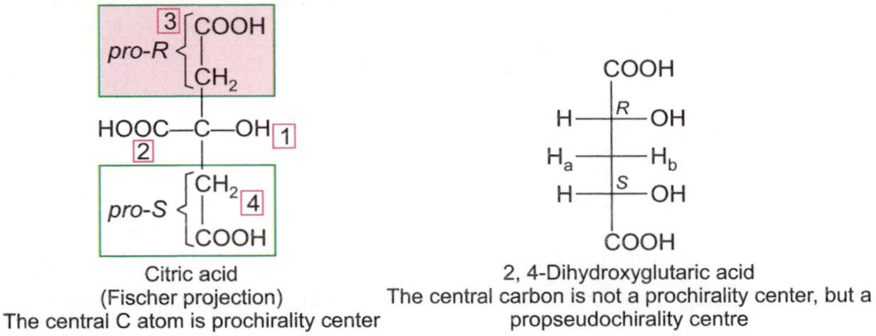

SCHEME B 2.3

Compared with citric acid the central carbon atom in 2, 4-dihydroxyglutaric acid is not a prochirality center. It can however, be considered as a propseudochirality center (Scheme B 2.4) since substitution of a hydrogen atom leads to a pseudochirality center [when a carbon carries two groups which are different (H and D), and two groups which are enantiomorphic and lies on a mirror plane of symmetry is called a pseudochirality center]. Thus the hydrogen atoms on the central carbon atom of 2, 4–dihydroxyglutaric acid can be distinguished as pro-s and pro-r.

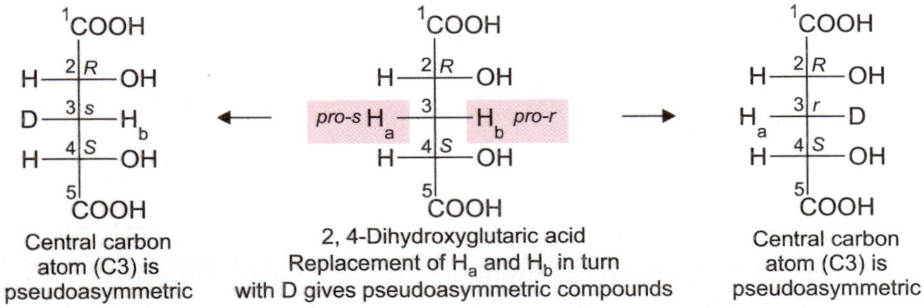

SCHEME B 2.4

(*ii*) Molecules with Prochiral Axis

The molecules like allenes, biphenyls and analogues when suitably substituted may have a pair of stereoheterotopic ligands *e.g,* H_A and H_B in a biphenyl derivative (I, Scheme 2.14*a*). To assign topic descriptor to H_A its priority is elevated over that of H_B and then *R*/*S* configuration is determined (Kalsi 2004, Scheme 1.133*a*) which leads to *pro-R* or *pro-S* designation.

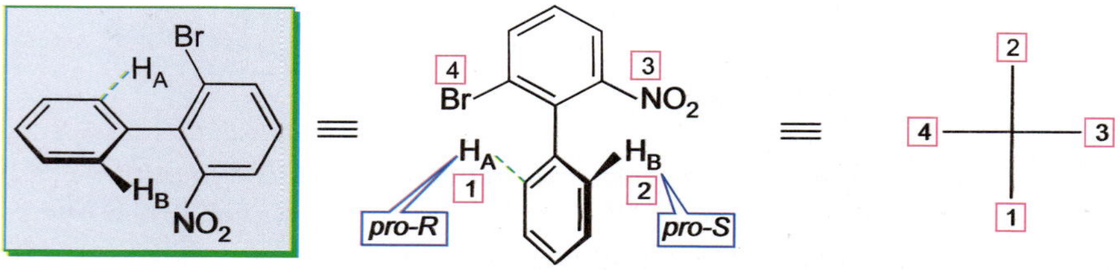

(I)
Molecule with a prochiral axis

Rotate (I) so that the stereochemical bonds are at the bottom (See, Scheme 1.132a)

Not a Fischer projection (Kalsi 2004 see Scheme 1.132a) just a cross. Path 1→2→3 to assign configuration to the hypothetical chiral axis in (I) with elevated priority to H_A is *R*, thus H_A is *pro-R*, and H_B will be *pro-S*

SCHEME 2.14a

(*iii*) Molecules with Prochiral Plane

Consider the **paracyclophane (I, Scheme 2.14b, also see Scheme 2.13*d*), to give a descriptor** to *e.g.*, H_b, it is given priority over H_a **so that the upper methylene C becomes the** pilot atom, and the path 1 → 2 → 3 is determined (**from the pilot atom towards ones** eyes) to assign hypothetical configuration to the chiral plane which comes out to be S thus H_b is *pro-S*. Similarly when H_a is considered, the bottom methylene becomes the pilot atom and H_b becomes *pro-R*. Caution, when the bottom methylene group is the pilot atom (with H_a of imaginary higher priority over H_b) the projection shall have to be turned 180°, so that it could be viewed from the pilot atom towards ones eyes.

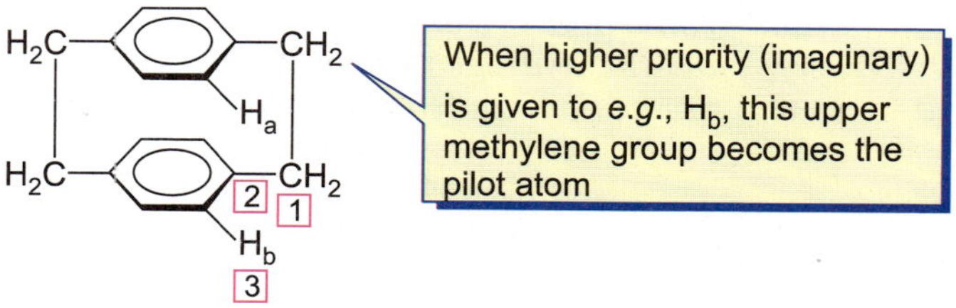

(I)

SCHEME 2.14b

(B) Heterotopic Faces of Carbonyl Compounds (*Re*, *Si* Nomenclature)

The groups and faces in a molecule(s) which are enantiotopic or diastereotopic are collectively termed as heterotopic (stereoheterotopic). In the case of carbonyl compounds

(R^1–CO–R^2) when the groups $R^1 = R^2$, as in acetone and formaldehyde the faces are homotopic. When R^1 and R^2 are different, the two faces of the trigonal center are different enantiotopic and then the carbonyl carbon is called prostereogenic carbonyl carbon. [This carbon will become a chirality center when attacked by a group unlike any of the groups already bonded to it and the addition product will be a pair of enantiomers, see Scheme (1.84) and Scheme 2.16.]

A different situation arises when the groups R^1 and R^2 contain one or more chirality centers. Then the faces of carbonyl group become diastereotopic (see, Schemes 2.23*b*, 2.22 and 7.9).

The faces of the carbonyl group are differentiated by the *Re*–*Si* nomenclature. The groups around the carbonyl group are given priorities (CIP rules for *R, S* nomenclature). If going from the group of highest priority of the group of lowest priority around the face of the carbonyl group, the path is clockwise, the face is *Re* and if it is anticlockwise, the face is *Si* (Scheme 2.15). The closest face of the carbonyl group *e.g.*, in acetophenone (I, Scheme 2.15) is *Si* face. To explore the other face one either goes behind the page or rotates the molecule (180° out of the plane of paper *e.g.*, to get the view of the *Re* face (I*a*, Scheme 2.15). Both the faces may be viewed simultaneously as shown for

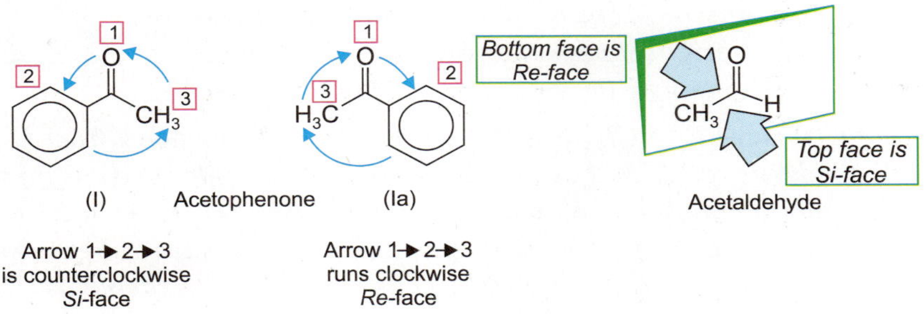

SCHEME 2.15

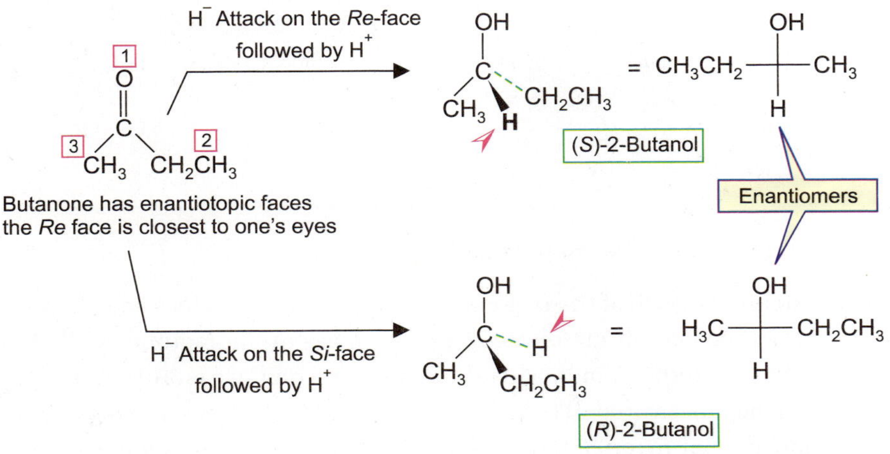

SCHEME 2.16

acetaldehyde (Scheme 2.15). Thus 2-butanone during its reduction with hydride (Scheme 2.16) presents both the *Re* face (*i.e.* added H is closest to ones eyes and is shown by a wedge) to give (*S*)-2-butanol as well the *Si* face (*i.e.*, added H is away from ones eyes and is shown by a dash) to give (*R*)-2-butanol. Significantly during enzymatic reduction, 2-butanone, presents specifically only one of the faces *i.e.*, the *Si* face to give only one enantiomer *i.e.*, (*R*)-2-butanol. One may however, note that as the CIP rules operate, there is no relationship between the face attacked by the nucleophile and the stereodesignator of the chirality center in the product. In the preceding example (Scheme 2.16) hydride addition to the *Re* face gives (*S*)-enantiomer whereas hydride addition to the *Si* face gives the (*R*)-enantiomer.

EXERCISE 2.4

Depict the top face in methyl acetate as presented (Scheme 2.16a) as Re or Si.

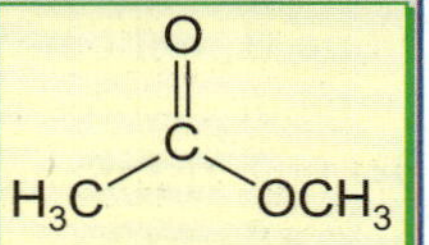

Methyl acetate
SCHEME 2.16a

ANSWER. *This is a general structure R′-CO-R² where one of the groups contains an oxygen. This situation requires the assignment of priority between the carbonyl oxygen and the R-group oxygen to assign Re, Si faces of such molecules.*

Considering CIP rules R′ = CH_3 has the least priority. One knows from CIP rules that double bonds are considered as two single bonds and triple bonds as three. Thus one may represent the carbonyl oxygen of methyl acetate as (I, Scheme 2.16b) where the new carbon atom has no further substituent and is thus taken to be attacked to a 'phantom atom' of atomic number zero (II). Thus (I) is to be considered as (III). This shows that methoxy oxygen has priority over the carbonyl oxygen and therefore, the face presented is Si (IV, Scheme 2.16b).

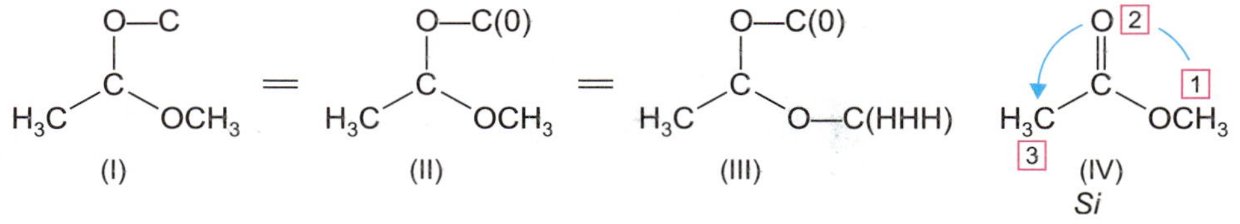

SCHEME 2.16b

(C) Faces of Alkenes (*Re, Si* Nomenclature)

In the case of an alkene each of the trigonal carbon atoms has its face specified separately (Scheme 2.17). In the case of maleic acid the two faces are homotopic while in fumaric acid these are enantiotopic. A monosubstituted alkene has two enantiotopic faces. Thus in the case of monosubstituted alkenes *e.g.*, but 1-ene (or other alkenes of the type abC = Cdd) only one symbol *Re* or *Si* will define the either face (I, Scheme 2.17).

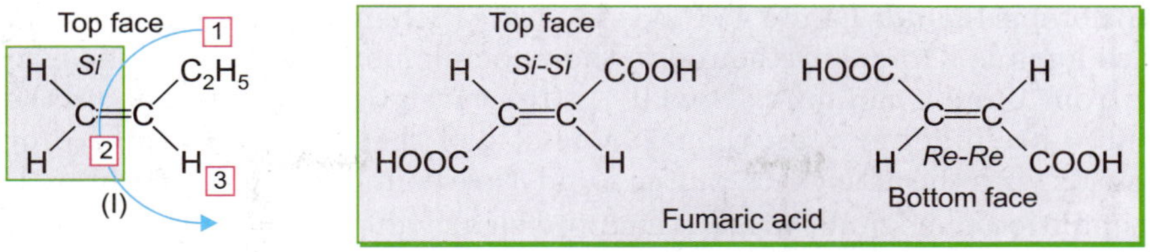

SCHEME 2.17

(D) Asymmetric Enzymatic Transformations (Heterotopic Ligands and Faces and Biochemical Reactions)-Stereoselectivity and Stereospecificity of Enzymes

Unlike the achiral chemical reagents, the chiral enzymes are capable of discriminating between two ligands and faces which are enantiotopic, as well as between two enantiomers (Scheme 2.18). The enzymes act on only one of the two enantiomers—a property known as stereospecificity (see, Scheme 1.2b) and in the case of prochiral molecules with only one of the enentiotopic ligands or faces—a property known as stereoselectivity.

(*i*) Enzyme Mediated Oxidation

The *pro-S* and *pro-R* hydrogens cannot be distinguished by an achiral chemical reagent, however, these hydrogens can be distinguished *e.g.*, during oxidation with enzymes as in the case of ethanol only *pro-R*-hydrogen being removed (Scheme 2.18).

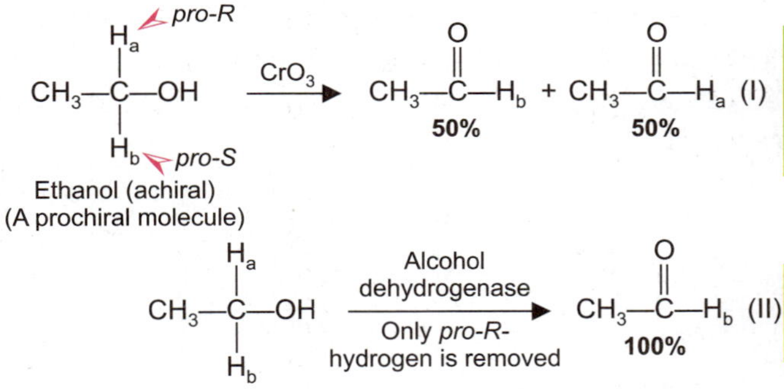

The pro-R and pro-S hydrogen can not be differentiated with achiral chemical reagents. Chiral reagents e.g., enzymes, however, distinguish between these.

Stereoselective oxidation of ethanol with the enzyme alcohol dehydrogenase.

SCHEME 2.18

(*ii*) Enzymes can distinguish between Prochiral Groups

This is explained by considering the phosphorylation of glycerol (Scheme 2.18*a*).

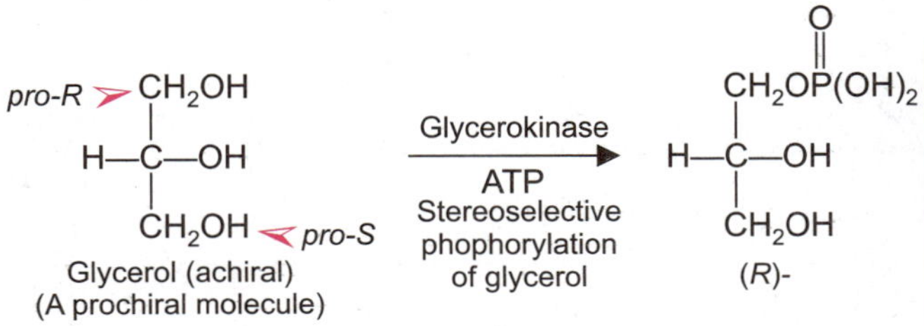

During the phosphorylation of glycerol the pro-R hydroxymethylene group is esterified exclusively by adenosine triphosphate (ATP) in the presence of an enzyme (glycerol kinase). Here also the enzyme distinguishes between the two enantiotopic primary alcoholic groups of glycerol.

SCHEME 2.18a

(*iii*) *pro*-R and *pro*-S Distinction via Enzymes

The enzymes react with one of the enantiomers, since an enantiomer can be absorbed on the enzyme surface via three points attachment (see Scheme 1.2*b*). Similarly the two enantiotopic ligands (1 and 4 of the substrate (Scheme 2.18*b*) can be differentiated, since once ligands 2 and 3 are bound to their complementary sites on the enzyme only ligand 1 from 1 and 4 can bind. Thus if 1 is the catalytically active site on the enzyme and 2 and 3 are binding sites, then 1 but not 4 of the substrate can be brought into juxtaposition with the active site 1 after 2 and 3 are bound with binding sites. Therefore, the two enantiotopic ligands are distinct and clearly distinguishable (Scheme 2.18*b*).

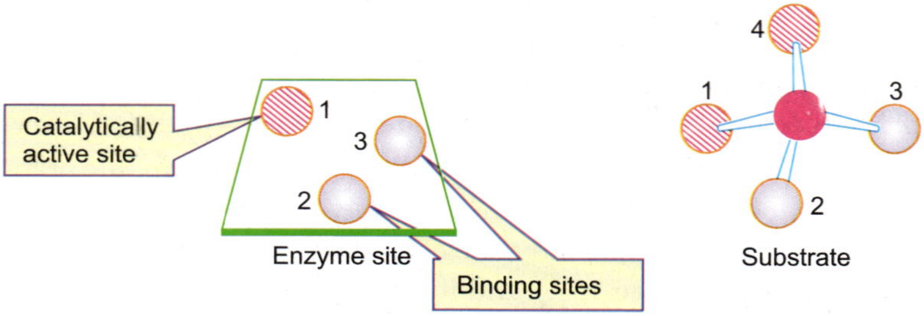

SCHEME 2.18b

(*iv*) Fumarase Catalyzed Hydration and Dehydration

The enzyme fumarase catalyzes the hydration of fumaric acid to malic acid (Scheme 2.19) to yield only the *S*-enantiomer. In this reaction the hydroxyl group is added to the *Si* face of one of the carbon atoms of the double bond. This reaction is reversible with pronounced stereochemical requirements, thus neither the *cis* isomer of fumaric acid (maleic acid) nor the *R* enantiomer of malic acid can serve as substrates in this fumarase catalyzed hydration-dehydration equilibrium.

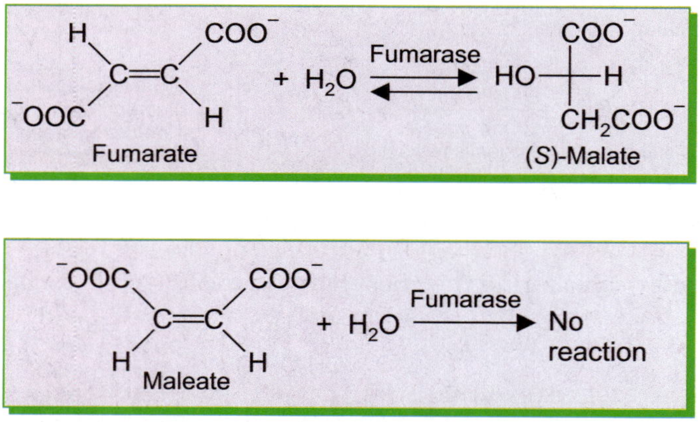

SCHEME 2.19

(*v*) Enzyme Mediated Reduction—Bakers' Yeast

When the enzyme alcohol dehydrogenase converts acetaldehyde to ethanol using the coenzyme NADH they discriminate between the two faces of the trigonal planar carbonyl substrate so one of the two possible stereoisomeric forms of the tetrahedral product predominates. The preference of several NADH dependent enzymes for either *Re* or *Si* face of their respective substrates is known. One of the most widely used is yeast alcohol dehydrogenase (YAD). For example acetaldehyde is reduced to ethanol with yeast alcohol dehydrogenase (YAD) Bakers yeast in the presence of the hydride donating coenzyme NADH. Acetaldehyde has two enantiotopic faces (*Re* and *Si*) and NADH has two diastereotopic hydrogens (*pro-R* and *pro-S*. Since R attached to N is chiral). It is established that during the reduction of acetaldehyde (Scheme 2.20) the *pro R* H from NADH is transferred only to the *Re* face of acetaldehyde and this hydrogen becomes *pro-R* in ethanol. In the reverse reaction (*S*)-ethanol-1-*d* give back the original deuterated acetaldehyde to show that the *pro-R* carbinol H is abstracted back by NAD^+ in oxidative step. Note that NAD^+ has two faces (*Re* and *Si*) and either of these can accept a hydride, significantly only the *Re* faces of both acetaldehyde and NAD^+ are involved and only *Pro-R* of both ethanol and NADH participate in these reactions.

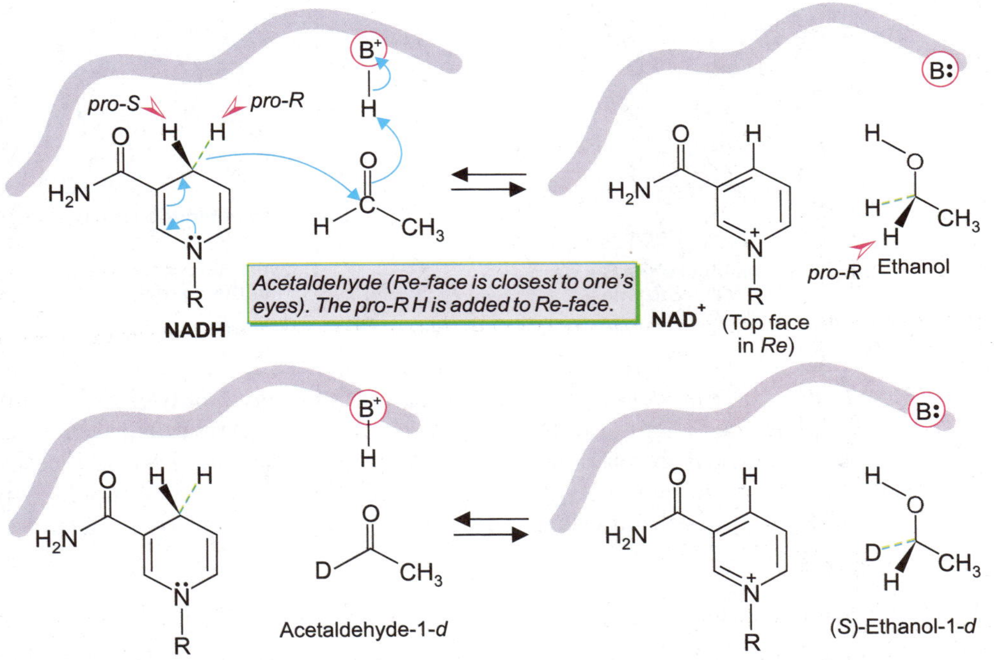

SCHEME 2.20

The evidence that only the *Re* face of acetaldehyde is involved is that when acetaldehyde-1-*d* is reduced *S*-ethanol-1-*d* is formed exclusively (the *pro-R* H *is transferred to the Re face (this H is on thick wedge in the product).*

- The complete stereoselectivity of this reduction shows that the three dimensional array involving the substrate, the enzymes binding sites and the coenzyme associated with it are oriented in a way so that one of the two diastereomeric transition states is greatly favored over the other and consequently one face of the substrate is involved in hydride transfer.

  An enzyme has a specific binding site for the coenzyme and after binding, the enzyme blocks one side of the coenzyme. When the enzyme blocks the B-side of NADH, the substrate will then bind to A-side leading to the transfer of H_a

hydride ion (*pro-R* hydrogen). When the A-side of the coenzyme is blocked by the enzyme, then the substrate binds to the B-side and not the H_b hydride ion (*pro-S*) will be transferred. A large number of dehydrogenases are known which can either transfer *pro-R* or *pro-S* hydrogen of NADH (Scheme 2.20*a*).

- Transfer of the hydride ion concomitant with protonation of the oxygen atom form an acidic side chain of an appropriate amino acid results in reduction of the ketone or aldehyde to an alcohol. This process is similar to the laboratory method that uses a hydridic reagent followed by protic workup.

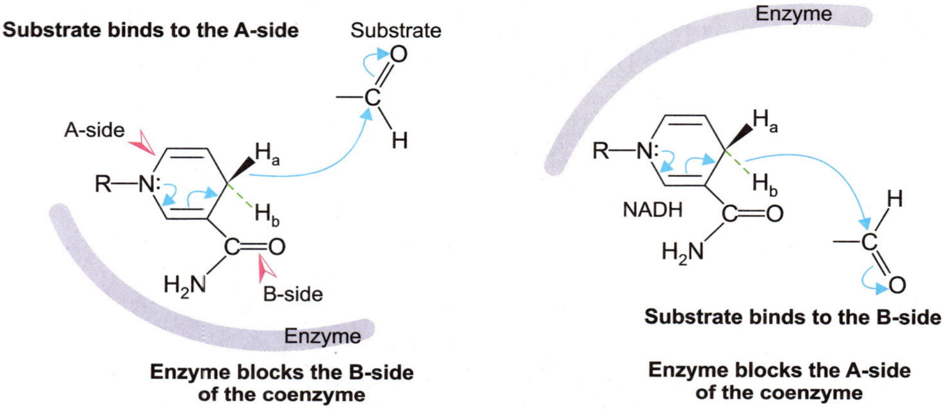

SCHEME 2.20a

- Recall that enzymatic reduction of achiral 2-butanone gives only one enantiomer of 2-butanol (see Scheme 1.85). Biological reduction of pyruvic acid catalyzed by the enzyme lactate dehydrogenase give only (*S*)-lactic acid. During biological reduction, pyruvic acid presents only one of its faces to the chiral enzyme while achiral hydride addition occurs on both faces to give a racemic mixture (see Scheme 2.1*a*).

(E) The *Re Si* Nomenclature

One uses descriptors *Re* and *Si* to define the two heterotopic faces of a trigonal planar prochirality center to which three constitutionally different groups are attached (I and II, Scheme 2.20*b*). When, however, the two of the three groups are enantiomorphic, decriptors *re* and *si* are used. Thus the top face of (2*R*, 4 *S*)-2, 4-dibromopentan-3-one is the *Re* side (*R* > *S*).

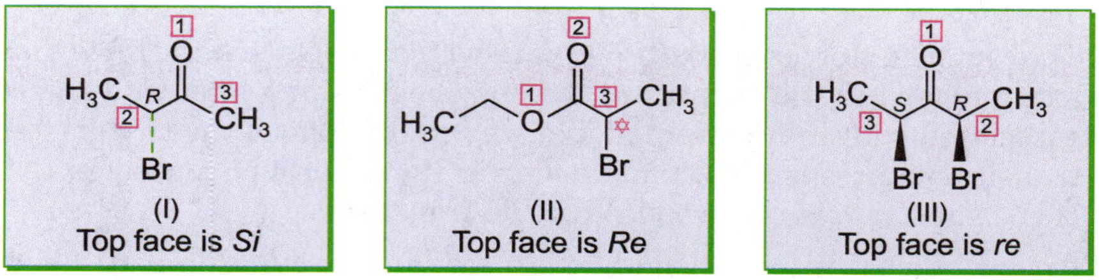

SCHEME 2.20b

(F) Reactivity of Diastereotopic Ligands with Achiral Chemical Reagents and Enzymes

The diastereotopic hydrogens however, as expected have different reactivities both with chemical reagents and enzymes. Thus for E2 elimination from 2-bromobutane (with H^a and H^b as diastereotopic protons) the substituents to be eliminated (*i.e.* H and Br) must be *anti* to each other. There are two staggered conformers (Scheme 2.21) in which this arrangement is possible. One of these conformers, (Ia, Scheme 2.21) is however, less stable due to more crowding and therefore, more of the *trans*-2-butene is formed (preferential elimination of H^b and Br).

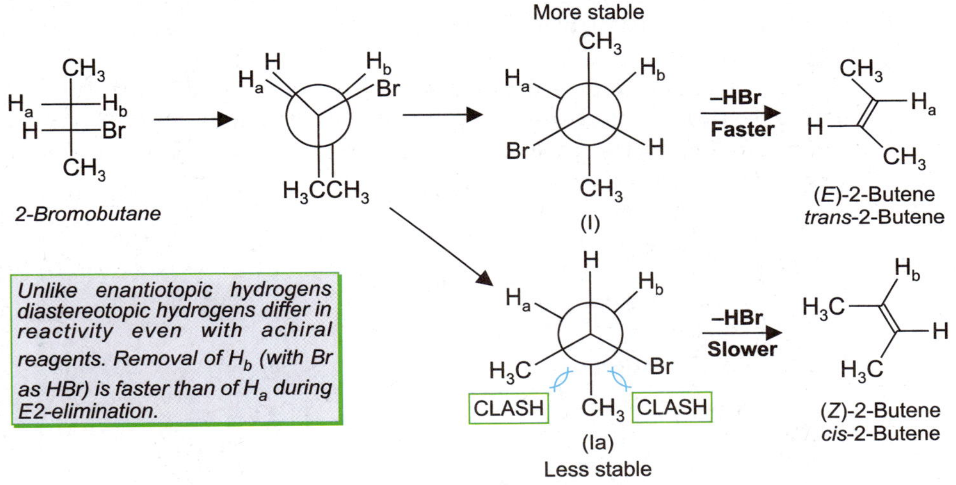

SCHEME 2.21

2.4 ASYMMETRIC SYNTHESIS – STEREOSELECTIVE SYNTHESIS (STEREOSELECTIVITY)

1. Asymmetric Synthesis – An Introduction

Optically active compounds, before 1940 could be obtained in stereoisomerically pure form only by isolation from natural sources or via the resolution of a racemic mixture, or by a few laboratory controlled enzymic reactions.

In general the chemical reactions give products which contain either chiral centres, axes, or planes, however, the isolated material is only the racemic form. This is due to the fact that reactants, reagents or solvents are achiral and are themselves racemic. In the absence of a chiral influence, a reaction producing enantiomers gives them in equal amounts (racemic mixture) *via* the transition states of identical energies. These reactions, therefore, take place at identical rates to give equal amounts of the enantiomers.

2. Principle of Asymmetric Synthesis

For the preferential formation of one stereoisomer (either enantiomer or diastereomer) over the other, either the reactant, or the reagent, or the solvent must be the pure enantiomeric form. The chiral agent must play an active part in the reaction and has to be integral to the transition state, so that two diastereoisomeric transition states are formed. Consequently one stereoisomer is produced more rapidly than the other. In summary an asymmetric synthesis involves competing reactions with diastereoisomeric transition states which takes place at different rates.

The following points may be noted:

- Two conditions need to be met for a successful asymmetric synthesis. The first condition of an asymmetric synthesis is the presence of a prochiral unit in the substrate molecule *i.e.,* presence of enantiotopic or diastereotopic groups or faces. If one considers the reduction of a prochiral ketone with lithium aluminum hydride (Scheme 2.21*a*) attack of the achiral reagent at either enantiotopic face leads to two enantiomeric transition states of identical energy and probability, and the product of the reaction is a racemic mixture of the two enantiomeric alcohols. Such a reaction cannot be called an asymmetric synthesis because the second condition is not met, namely, asymmetric induction. **Asymmetric induction** refers to the extent of excess of one enantiomer over the other achieved in an asymmetric synthesis.

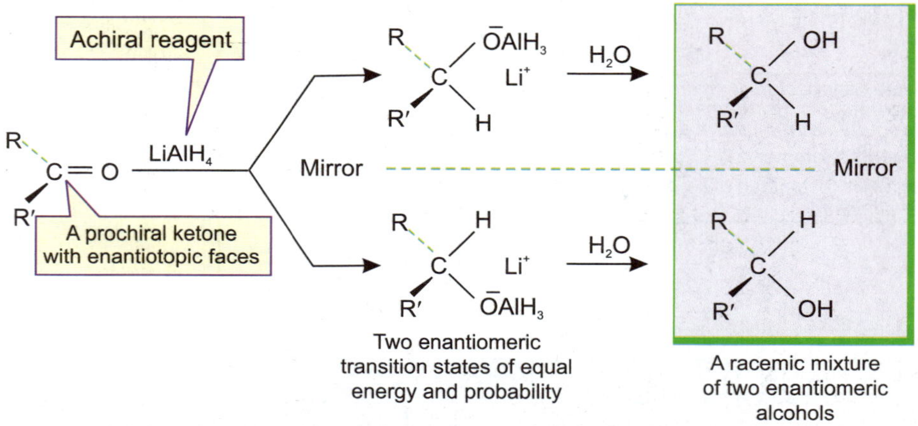

Not an example of symmetric synthesis since asymmetric induction is not observed.

SCHEME 2.21a

- The second condition of asymmetric induction is brought about in the reaction pathway by the presence of an *element of chirality* which plays an active role in the reaction: chiral reagent, chiral solvent, chiral catalyst, circularly polarized light, or element of chirality in the substrate molecule itself. By "active role" it is meant that the element of chirality is part and parcel of the transition states, which, as a result, become diastereoisomeric. The diastereomeric transition states differ in their thermodynamic properties.
- Diastereotopic ligands and faces undergo reactions, with achiral or chiral reagents, through diastereomeric transition states which differ in all their thermodynamical properties and two or more diastereomeric products are formed at different rates and therefore in different amounts. Such reactions whether carried out under kinetical or under thermodynamical control are stereoselective.

3. Principle Categories of Asymmetric Synthesis

(a) The Use of Chiral Substrates (First Generation Methods)

These methods (first generation methods) require the use of enantiomerically pure natural products *e.g.*, stereoids, terpenoids, alkaloids, amino acids etc. (Chiral pool, Scheme 2.21*b*).

Some Examples of Molecules in the Chiral Pool:

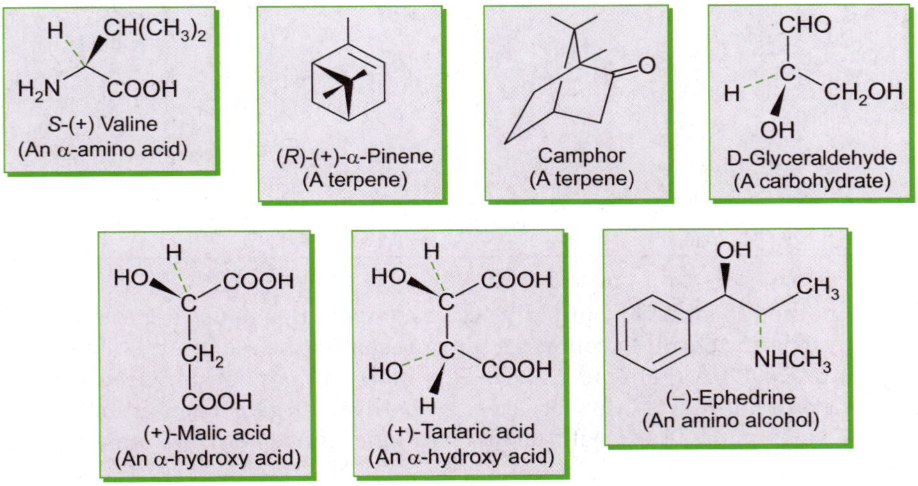

SCHEME 2.21b

(*i*) Stereoselective reduction of cholestan-3-one (Diastereoselectivity)

Cholestan-3-one on reduction with lithium aluminum hydride gives exclusively the equatorial alcohol (I, Scheme 2.22) by attack of the reagent from the less hindered face (*α* face) of the molecule. The alcohol (I, Scheme 2.22) is the more stable of these two diastereomers (I and II, Scheme 2.22). The conformationally fixed steroid molecule has sterically hindered β face due to angular methyl groups. It is an example of diastereoselective reduction.

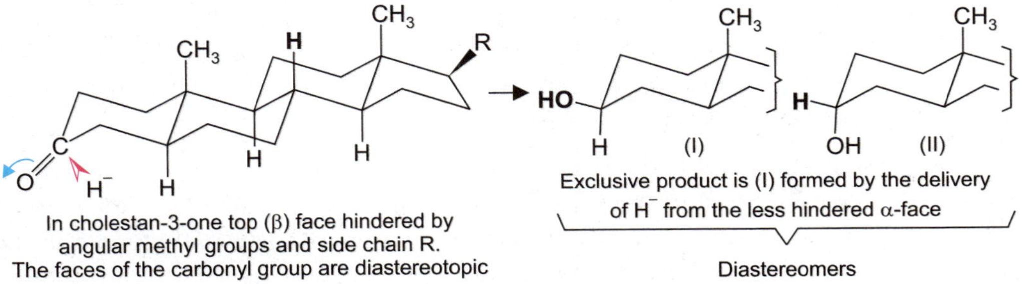

SCHEME 2.22

(*ii*) Hydroboration of (–)-isopulegol

In this case as well, the new chirality center is generated under the influence of chirality centers already present and in particular the one bearing a hydroxyl group. The interaction of the hydroxyl group (nucleophilic group) with the electron deficient borane directs the hydroboration from the rear of the molecule (Scheme 2.23).

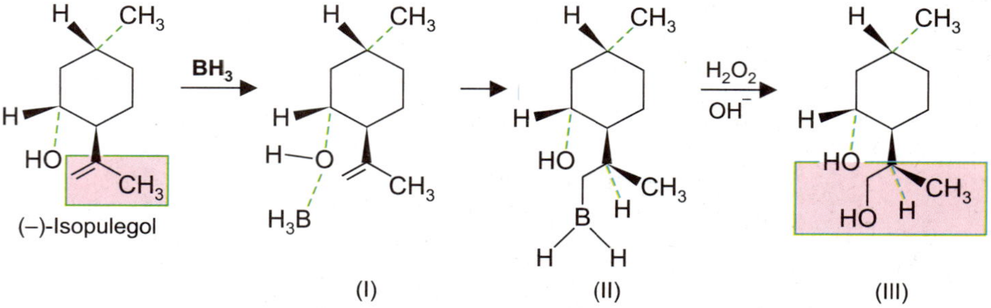

SCHEME 2.23

(*iii*) Conversion of L-tyrosine into L-DOPA

A different method to make an optically active compound is to use a method which does not effect any existing chirality center already present in the reactant. L-Tyrosine is thus converted into L-DOPA (Scheme 2.23*a*). This method may be compared with a chiral starting material that reacts by a stereospecific pathway to give an enantiomerically pure product. These examples represent useful strategies since both the starting materials are natural products (from chiral pool).

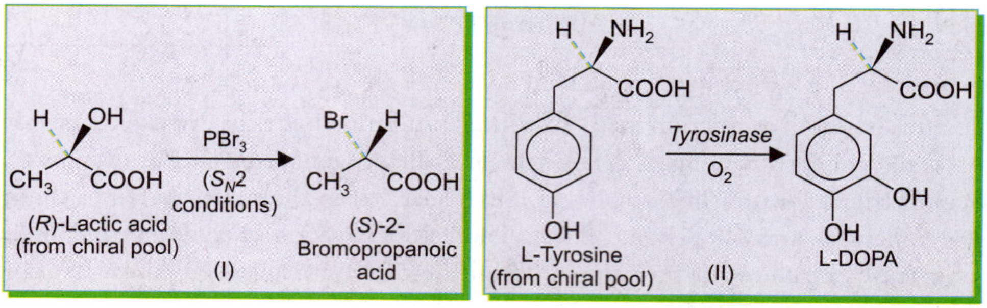

SCHEME 2.23a

A chiral material from the chiral pool may lead to a straightforward method to make an enantiomerically pure product (asymmetric synthesis). It may react in a stereospecific pathway (I) or without effecting any existing chirality center (II, Scheme 2.23*a*).

(*iv*) Nucleophilic attack on acyclic chiral carbonyl compounds (Cram's rule—the Felkin-Ahn modification, a diastereoselective synthesis via diastereoface differentiation)

The ketone (I, Scheme 2.23*b*) has a chirality center near the carbonyl group and thus the faces of the carbonyl group are diastereotopic. This chirality center will influence the approach of the attacking reagent to the prochiral carbonyl. Thus the approach of the reagent to one of the faces of this type of a molecule will be easy compared to the other leading to diastereoselectivity.

A nucleophile *e.g.*, will approach (I, Scheme 2.23*b*) from the face which presents less steric hindrance and this is evidently over the "hydrogen face" (*i.e.*, *Si* face) to give (II, Scheme 2.23*b*). However, if the nucleophile approaches the more hindered "methyl face" (*i.e.*, *Re* face) the product would be (III). The following points may be noted:

- Provided the conformation of (I, Scheme 2.23*b*) is frozen as shown the preference would be for the formation of (II).

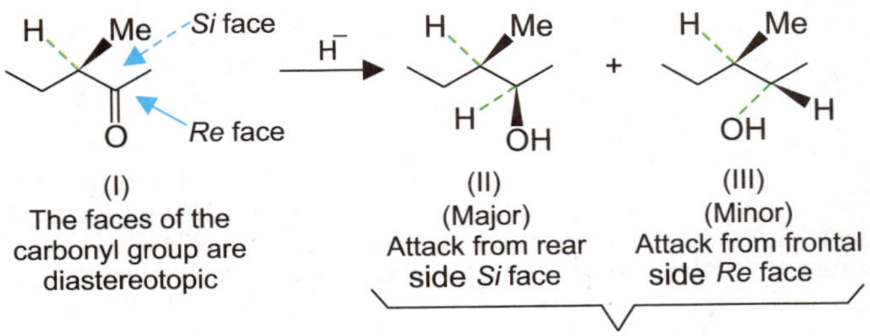

With the locked confirmation as in (I) the hydride would prefer to attack from less hindered rear face of C=O (side of hydrogen Si face) to be at the rear (not shown in the figure) rather than more hindered side of methyl (Re face)

SCHEME 2.23b

- The acyclic compounds of the type being discussed are, however, not locked into a particular conformation. The conformation adopted by the substrate may be locked (in the presence of the reagent *i.e.*, attacking nucleophile) in the diastereomeric transition state leading to products.
- Thus for the diastereoselectivity the facial and orientational preference has to be addressed and at the same time the conformation of the substrate in the diastereomeric transition state must be known.
- For certain additions to the carbon oxygen double bond of chiral aldehydes and ketones Cram's rule is useful to predict as to which diastereomer of the two will predominate.

The four groups attached to the chirality center (at α carbon) are COR, S, M and L where S, M and L designate small, medium and large group respectively. The oxygen of the carbonyl orients itself so as to be between the small and the medium sized groups. A typical Cram's model is shown as its Newmann projection (Scheme 2.24). Thus L being the largest group is oriented as far away as possible from the carbonyl group. In this orientation L will however, eclipse the group R. The rule is that the incoming group preferentially attacks on the side of the plane containing the small group (*i.e.*, the less hindered face). (Scheme 2.24). Based on Cram's model reaction of (I, Scheme 2.25) with Grignards reagent gives (II, Scheme 2.25) as the major product.

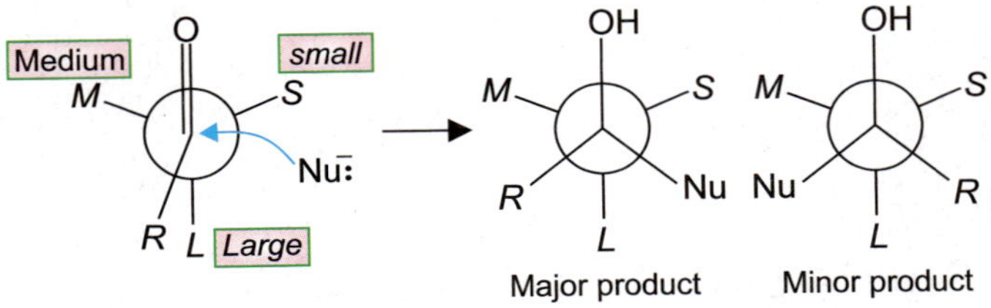

Crams rule predicts the nucleophilic attack on acyclic diastereotopic aldehydes and ketones. Group L orients trans to the carbonyl oxygen and nucleophile prefers to enter from less hindered side.

SCHEME 2.24

The Felkin-Ahn developed a variation based on theoretical calculations in terms of steric intersections and predicts more fully, the results of this chiral induction and is preferred. The Felkin Ahn (FA) model differs from Cram's rule in the conformation adopted by the carbonyl compound. The following points may be noted:

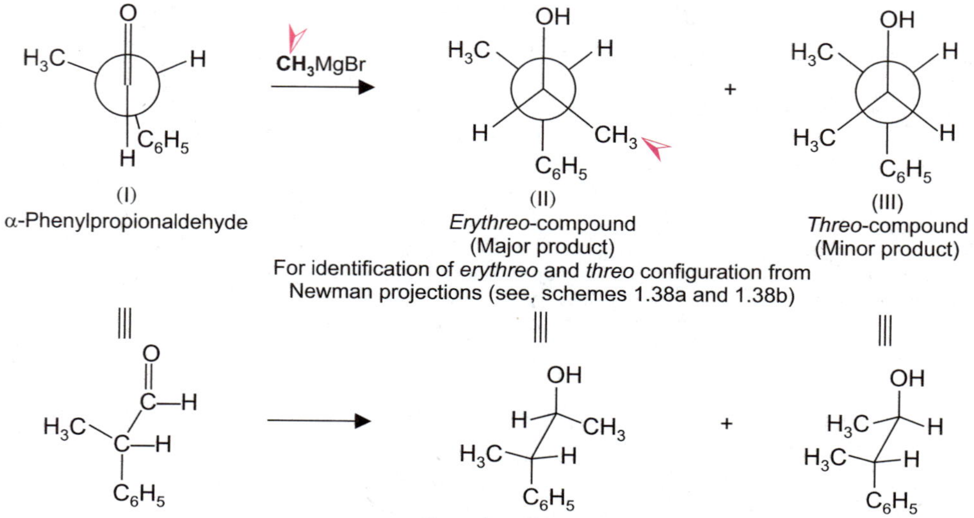

Cram's rule

SCHEME 2.25

- In the FA model the C—L bond is positioned perpendicular to the carbonyl group (Scheme 2.26). This is unlike Cram's rule where L is assumed to be antiperiplanar to the C=O group. This arrangement removes unfavorable eclipsing interactions between L and R.
- There are two such conformations (I and II, Scheme 2.26).
- The nucleophile approaches the carbonyl carbon in a plane perpendicular to that of the –CO– fragment from the side opposite the C—L bond and at an obtuse angle (angle of 107°) to the carbonyl group C=O which corresponds nearly to the tetrahedral angle of Nu—C—O in the product (IV, Scheme 2.26 for further explanation (see, Scheme 7.4).
- The consideration of the reactive conformations (I and II, Scheme 2.26) shows that (I) is of lower energy due to less steric interaction between the nucleophile and the smallest group S to give (III, Scheme 2.26) as the major product.

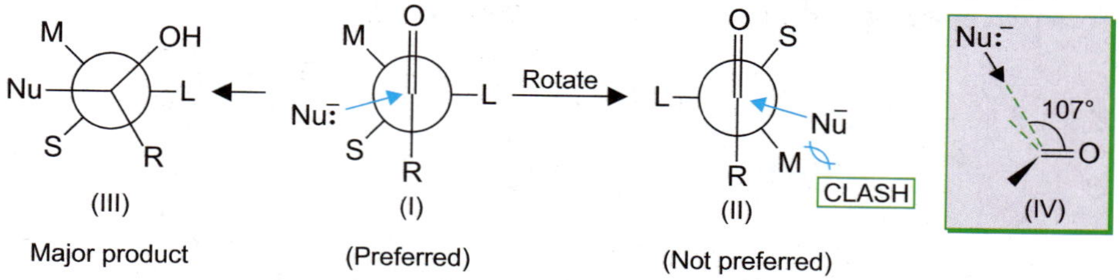

Cram's rule using updated Felkin-Ahn model

SCHEME 2.26

Thus the reduction of (*S*)-3-phenylbutanone (*S*)-PhCHMe-COMe (in which the sizes of the groups decrease in the order Ph > Me > H) with lithium aluminium hydride gives a product which predominates in the diastereomer predicted by Cram's rule (Scheme 2.27).

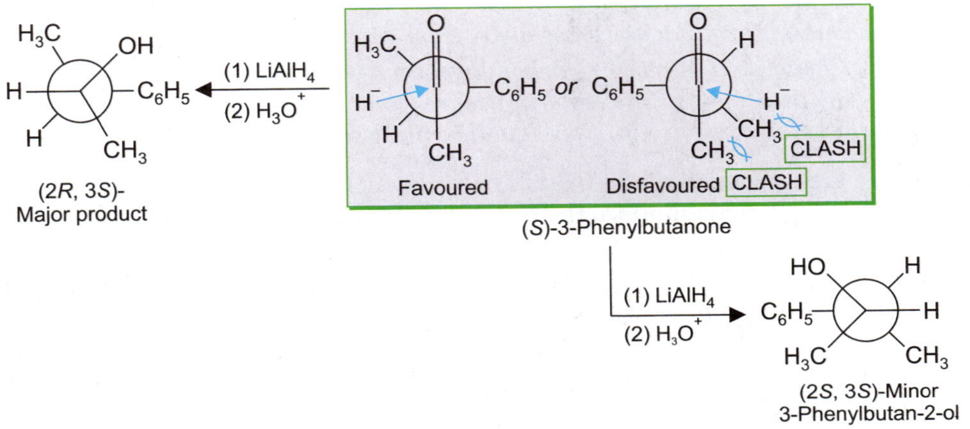

SCHEME 2.27

The preference for one product stereoisomer over the other depends on the difference in sizes between S and M. In case S and M are very similar, there is likely to be little difference in their interactions with the incoming nucleophile (as well as R) and the diastereoselectivity of such molecules is likely to be poor.

BOX 2.3
ERYTHRO/THREO FROM STAGGERED SAWHORSE OR NEWMAN PROJECTIONS

If as in (II, Scheme 2.25) it is possible to eclipse each of the like substituents H with H, CH_3 with CH_3 and unlike substituents OH and C_6H_5 the configuration is erythro (Scheme B 2.5)

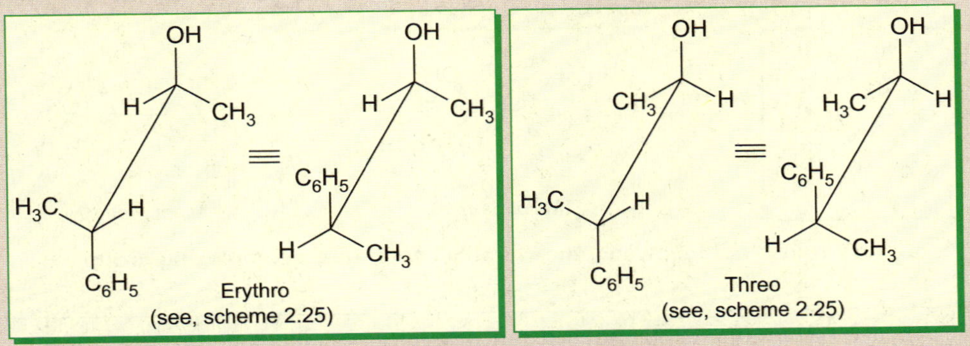

SCHEME B 2.5

The configurations at the two chirality centers in threo configuration are such that now in sawhorse projection one can eclipse only the H atoms, or the CH_3 groups or the unlike substituents OH and C_6H_5 (as is so in the present case) at any one time.

Summary of Felkin-Ahn model

- *Label the largest and the smallest groups.*
- *Place the largest group perpendicular to carbonyl in Newman projection. The two remaining groups on the chirality center can then be arranged in two ways (i) the smaller group towards or away from carbonyl oxygen.*
- *The nucleophile adds to the carbonyl group at an optimum angle of around 107°. The attack of the nucleophile occurs from the face of the carbonyl group from the small sized group, since it will be hindered by the medium sized group.*

There are situations when Cram's rule may not be followed since the conformations of the carbonyl compounds in the transitions state are no longer dependent mainly on steric factors.

In such cases one of the substituents on the chirality center is an alkoxy, hydroxyl or other complexing group. Lithium cations *e.g.*, coordinate effectively with oxygen atoms and thus reduction of ketones of the type (I, Scheme 2.28) with lithium aluminium hydride proceeds through a relatively rigid chelate compound of type (II, Scheme 2.28). Thus the reagent brings about the reduction by first acting as a conformational lock (also see Scheme 7.17) by coordination with both the methoxy oxygen and the ketone oxygen. A high degree of stereoselectivity is observed since the hydride attacks the chelate from the less hindered side (*i.e.* away from the phenyl group, in this case it is the *Re* face which offers the least resistance to the approach of the reagent) but not necessarily in the Cram's sense.

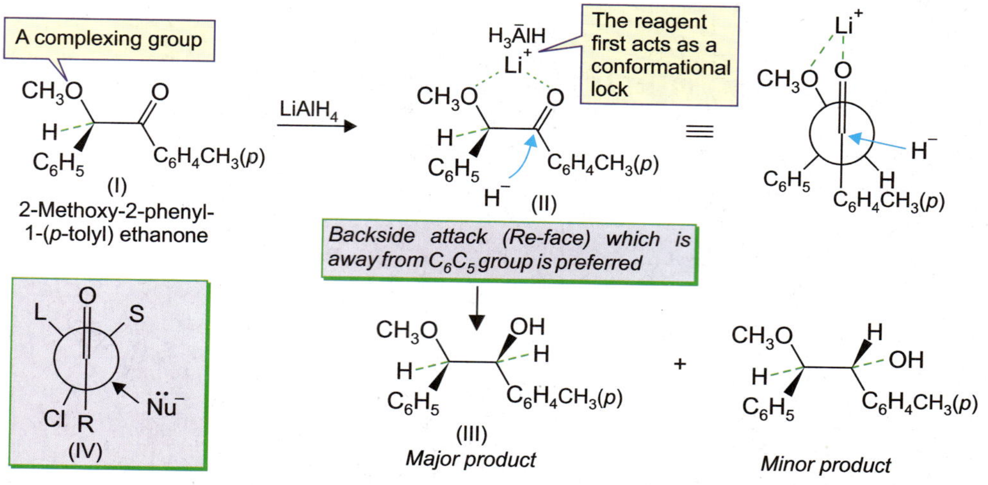

Cram's rule is not followed if the chirality center has a complexing group

SCHEME 2.28

In cases when one of the substituents is strongly electronegative (*e.g.*, chlorine) the preferred transition state corresponds to conformation (IV, Scheme 2.28 due to the tendency of the negatively polarized oxygen and chlorine atoms to be as far apart as possible. Again in such cases Cram's rule is not necessarily followed.

EXERCISE 2.5

Write the structure of major product of addition of a nucleophile (Nū:) to the ketones (I and II, Scheme 2.28a).

ANSWER. *In the case of (I) first identify the largest, medium and smallest groups. Write the proper Felkin-Ahn model (Scheme 2.28a) conformation. The nucleophile will add to the carbonyl group at an obtuse angle with C=O from the small sized group.*

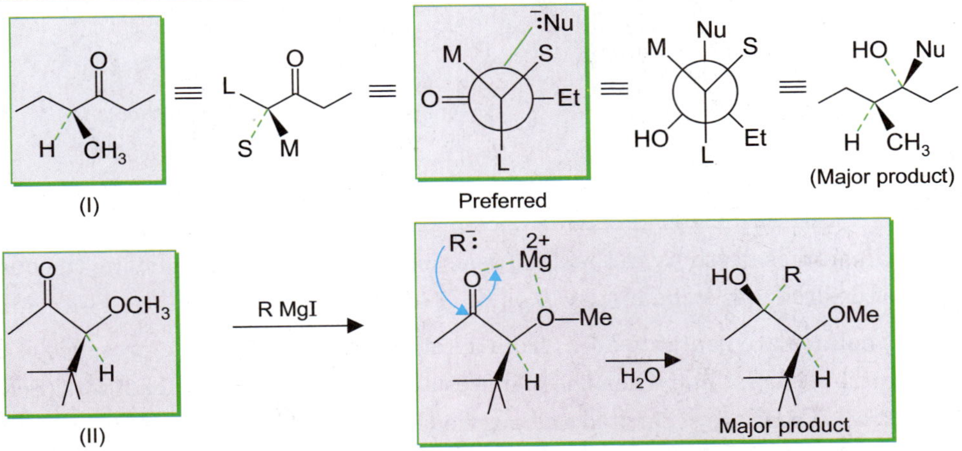

SCHEME 2.28a

In the case of (II, Scheme 2.28a) since there is a coordinating methoxyl group at the chirality center which is α to the ketone, chelation shown in (III, Scheme 2.28b) locks the molecule and now the nucleophile is delivered from the less hindered side i.e., away from the bulky t-butyl group.

(b) Diastereoselectivity in Aldol Reactions (Directed Aldol Reaction)

(*i*) Introduction

The following points may be noted: [for introduction to aldol condensation see (Sec. 7.10)].

- The aldol reaction creates two chirality centers from achiral starting materials and in a most general case, there are four stereoisomers of the aldol product (Scheme 2.29). Thus *syn* or *anti* diastereomers are produced, each as a pair of enantiomers.
- Regarding the stereochemistry of the reaction, therefore, one has to control diastereoselection *i.e.*, whether (racemic) *syn* or (racemic) *anti* product is formed as the major product. Secondly one has also to aim at enantioselection *i.e.*, formation of one of the four possible stereoisomers.

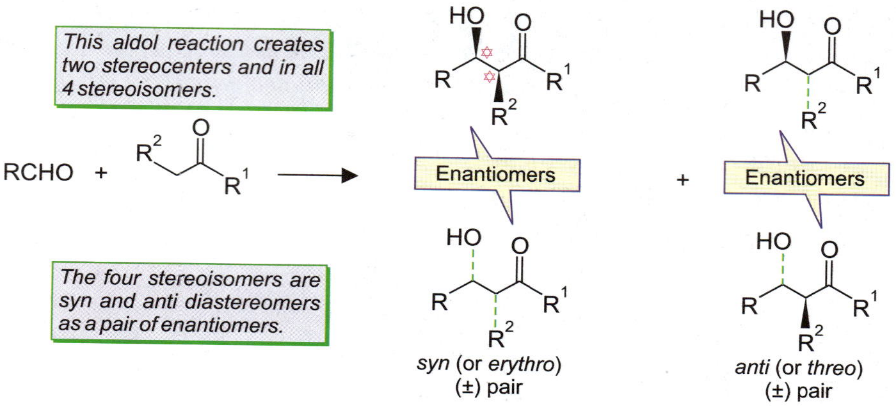

SCHEME 2.29

- Diastereoselectivity in the aldol reaction is achieved by employing the enolate of desired stereochemistry (*E* or *Z*).
- Enolates are generated *e.g.*, from a ketone and a base in the presence of chlorotri-methylsilane when the enolates are trapped as silyl enol ethers (Scheme 2.30). These are separated and purified by chromatography and then converted into pure (*Z*)- or (*E*)-enolate with fluoride ion (nucleophilic substitution at the silicon atom by fluoride ion, fluoride ion has a high affinity for silicon, since Si-F bonds are very strong).

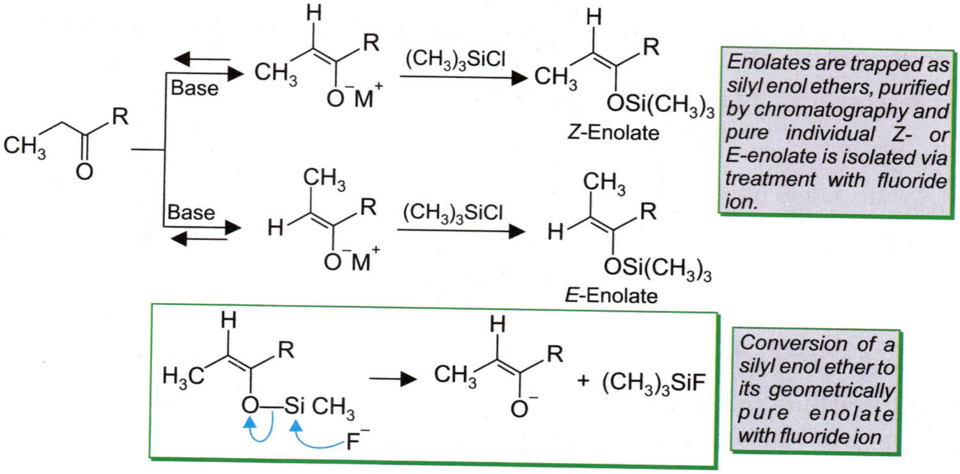

SCHEME 2.30

- Mukaiyama found that titanium tetrachloride ($TiCl_4$) and other transition metal compounds can induce silyl enol ethers to attack carbonyl compounds, producing Aldol-like products (Scheme 2.30*a*, also see Scheme 7.52*a*).

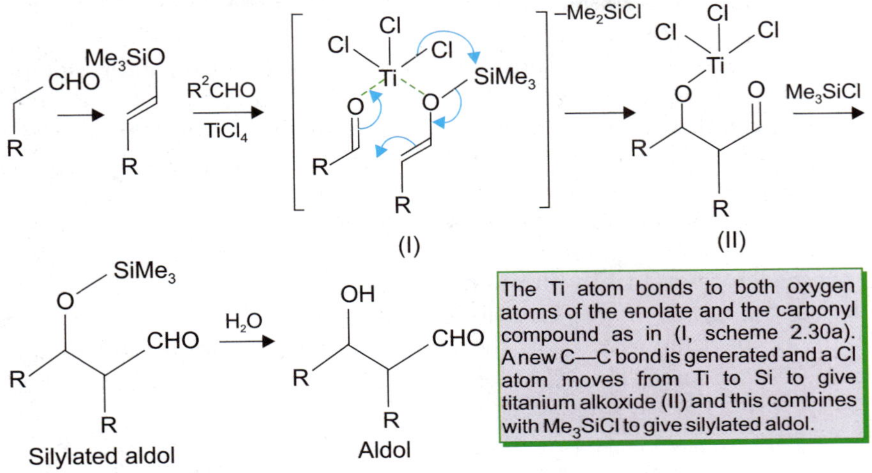

SCHEME 2.30a

- Methods are available to produce either *E* or *Z* enolates in pure forms.
- *Z* enolates give mainly 2, 3-*syn* aldols while the *E*-enolates give the 2, 3-*anti* aldols (Scheme 2.31).

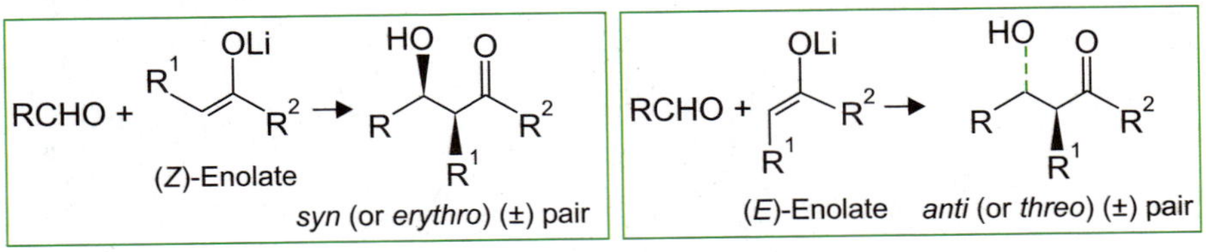

SCHEME 2.31

- Greater diastereoselectivity in aldol reactions is achieved by employing boron enolates as the carbon nucleophiles. The boron oxygen bonds are shorter than lithium oxygen bonds, and consequently the steric interactions in the transition state are magnified to result in greater stereoselectivity. Z-vinyloxyboranes *e.g.*, are readily prepared by reacting ketones with a dialkylboron-trifluoromethane-sulphonate (triflate) and a mild base diisopropylethylamine (in these enolates the boron atom is bonded to the oxygen atom of the ketone) and these react with aldehydes to give *syn* diols in higher yields (Scheme 2.32).

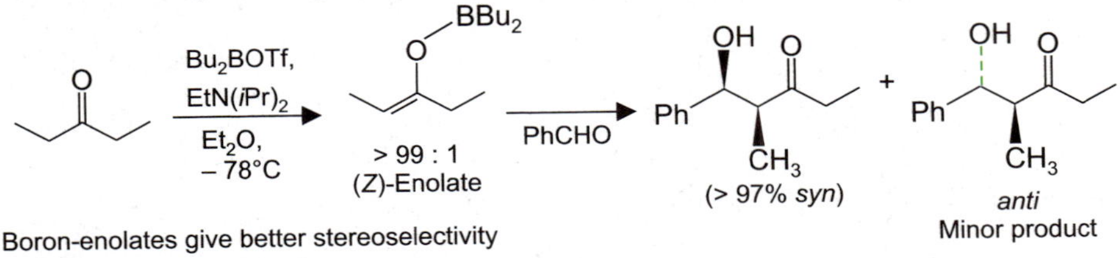

SCHEME 2.32

(*ii*) The diastereoselectivity of the aldol reaction—A chair like transition state (Also see, Scheme 7.13)

The diastereoselectivity is achieved by the reaction (*i.e.* formation of the new C—C bond) proceeding via a chair like six membered transition state in which the ligated metal atom is bonded to the oxygen atom of the aldehyde as well to that of the enolate. If the geometry of the enolate is fixed, the only variable is the orientation of the aldehyde, and therefore, one deals with transition states of different stabilities. This is so since the electron rich enol double bond and the electrophilic carbonyl carbon atom can be brought into close proximity via two transition states *e.g.*, in the case of *Z* enolate (I, Scheme 2.33). One of the transition states III is disfavored due to 1, 3-non-bonded interactions between the substituents and thus the reaction takes place largely *via* transition state (II, Scheme 2.33) to give *syn* aldol. Similar arguments show that the reaction with the *E*-enolate proceeds preferentially through the transition state (IV, Scheme 2.33). The aldol condensation can be made enantioselective (see, Scheme 2.35), by introducing chirality into one partner which can then be removed, thus aldol condensation now selects for one enantiomer. This is an example of double asymmetric synthesis (see, Scheme 2.35).

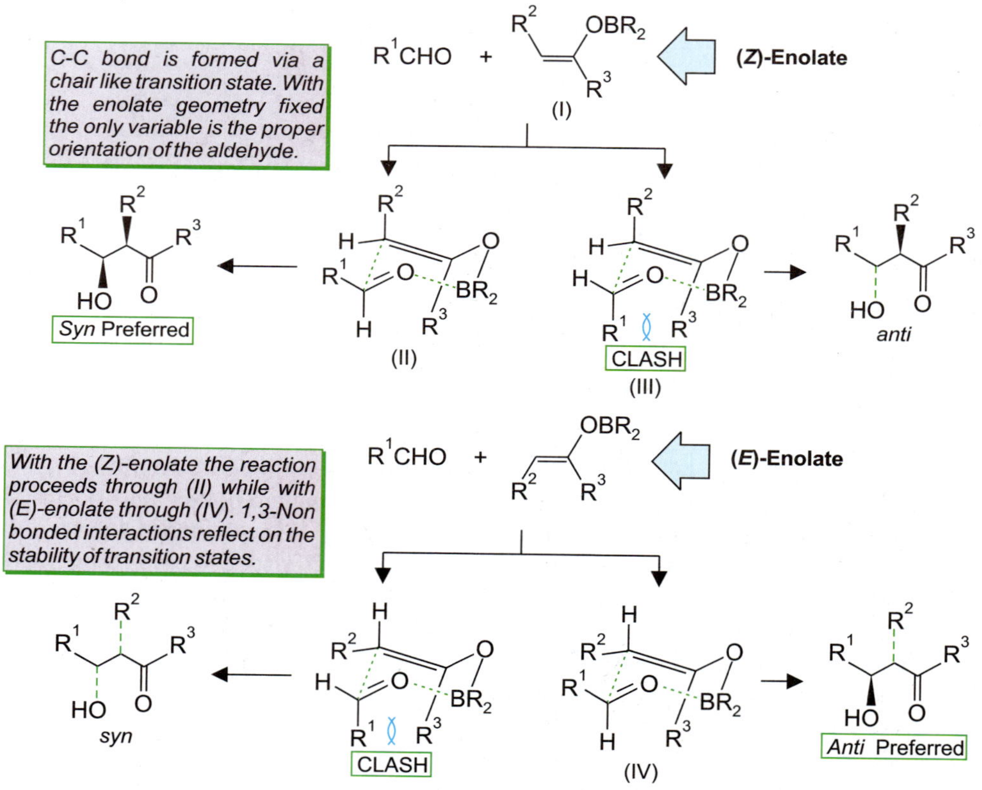

SCHEME 2.33

(c) The Use of Chiral Auxiliaries (Second Generation Methods)

During asymmetric synthesis a chiral auxiliary (a reagent) is attached chemically to the achiral substrate to give a chiral intermediate. This is followed by the reactions of asymmetric synthesis, during which the auxiliary dictates the preferred stereochemistry by directing the course of further reactions by making available one preferred pathway for attack by the added reagents. At the end of the synthesis the chiral auxiliary is removed, and thus does not appear in the final product.

(*i*) Alkylation of chiral enolates (asymmetric α-substitution of a carboxylic acid and synthesis of optically active α-hydroxy acids)

In (Scheme 2.34) alkylation of propanoic acid is taken as an example which involves the following steps:

- The chiral pyrrolidine unit is derived from *S*-(–) proline.
- The carboxylic acid substrate (activated propanoic acid) reacts with (*S*)-prolinol a chiral auxiliary to give an amide.
- Deprotonation with LDA gives a chiral enolate entirely in the *Z* from and at the same time the hydroxylic hydrogen is also replaced by a lithium atom. The complex (I, Scheme 2.34) thus maintains its structure by the hydroxymethyl group of the chiral auxiliary by chelating in its lithiated form with the enolate OLi group.
- The chiral enolate is alkylated with butyl iodide which mainly occurs form the top face of the molecule (the less hindered direction of approach, *i.e.* the added alkyl group C_4H_9-group is shown by a thick wedge). Thus the electrophilic attack occurs on the *Si* face of the double bond (at C2 the site of electrophilic attack), the *Re*–face being hindered by the chiral auxiliary. The net result is the formation of (*R*)-2-butylpropanoic acid in 88% *ee* on hydrolysis.

In case one uses (*R*)-prolinol as the chiral auxiliary, it would then hinder the approach of the electrophile to the *Si* face at C2 and force it to approach instead the *Re* face to give (2*S*)-product.

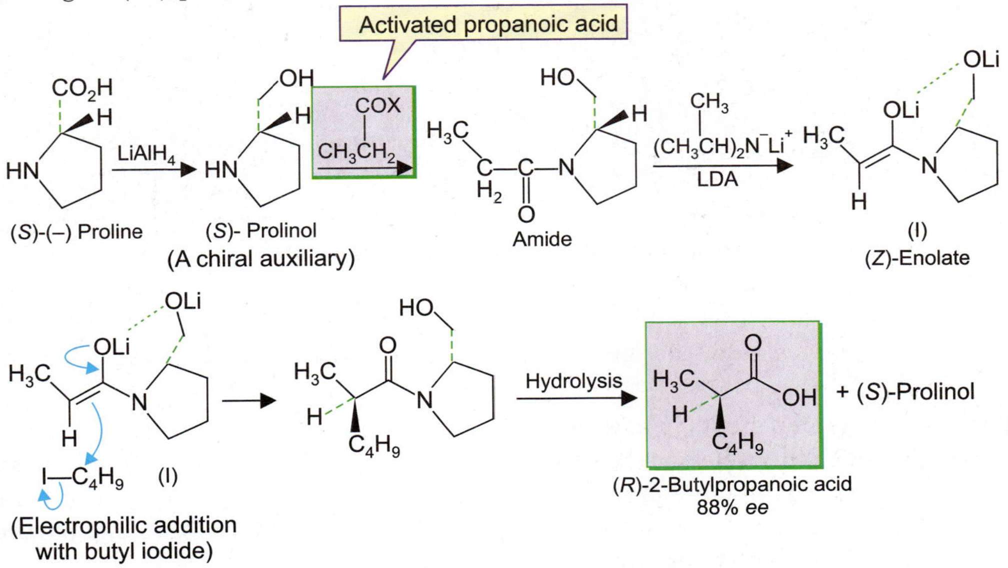

Asymmetric α-substitution of a carboxylic acid (propanoic acid)

SCHEME 2.34

Optically active α-hydroxy acids (I, Scheme 2.34*a*) are important intermediates in organic synthesis and occur in nature. An interesting method leading to their synthesis involves an asymmetric halolactonization (Scheme 2.34*a*). The symmetric synthesis starts with a suitable α, β–unsaturated acid which is converted into the acyl derivative (II, Scheme 2.34*a*) of *S*-proline (the chiral auxiliary). Reaction with *N*–bromosuccinimide gives a bromonium ion which lactonizes to a bromolactone (III, Scheme 2.34*a*). Its reduction with tributyltin hydride gives the lactone which on hydrolysis gives the optically active α-hydroxy acid. Thus tiglic acid gives (*R*-2-hydroxy-2-methylbutanoic acid by following this reaction sequence (Scheme 2.34*a*).

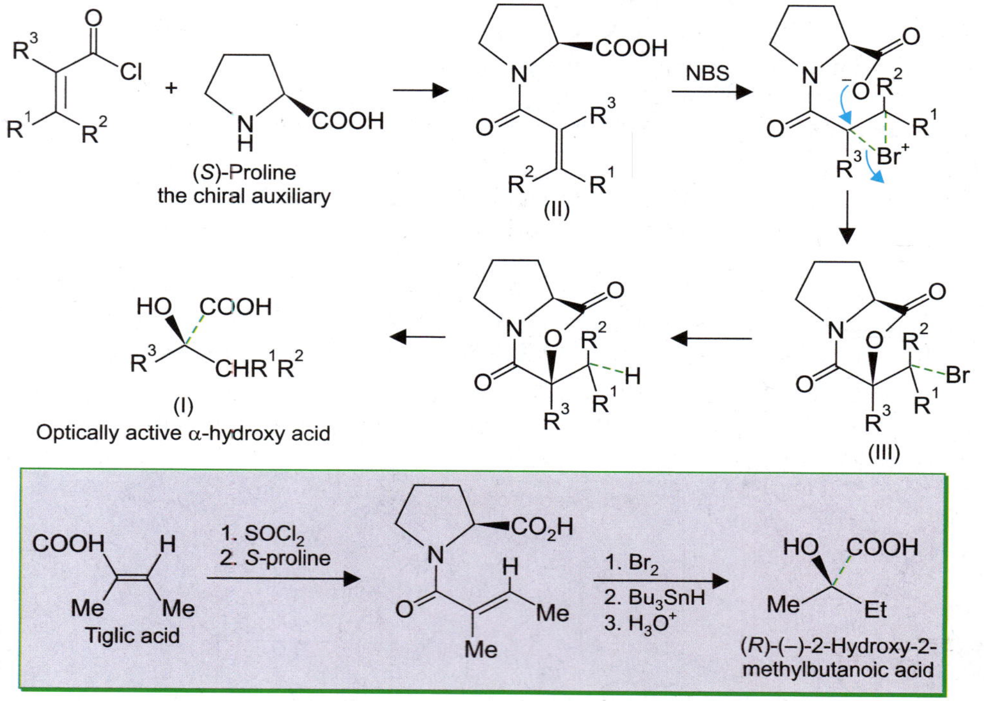

Asymmetric synthesis of α-hydroxy acids

SCHEME 2.34a

(*ii*) Enantioselective aldol condensation via chiral enolates (Double asymmetric synthesis)

When one of the components in aldol condensation is optically active, the reaction may becomes highly enantioselective in which case only one of the four stereoisomers (see Scheme 2.29) predominates. Boron enolates with chiral auxiliary group (chiral boron enolates) have been used with success. Thus chirality is introduced into one partner and after the reaction if is removed.

The propionamide (I, Scheme 2.35) as it *Z*-boron enolate reacts with achiral aldehydes to give *syn* aldols (II, Scheme 2.35) almost exclusively. On hydrolysis enantiomerically pure 3-hydroxy-2-methyl carboxylic acids (III, Scheme 2.35) are formed.

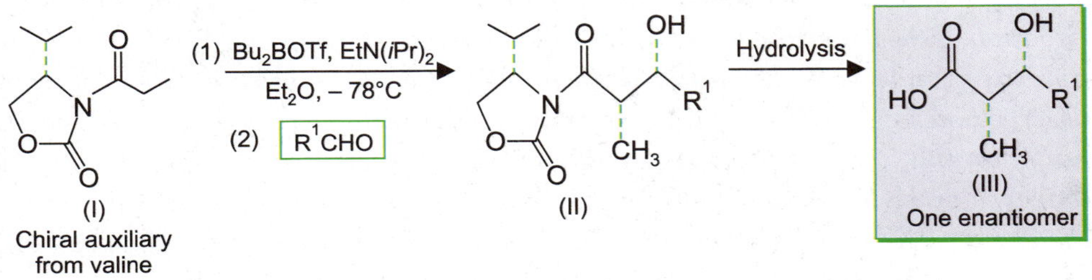

SCHEME 2.35

This type of a reaction where an optically active substrate (chiral enol derivative) reacts to generate both new chirality centers enantioselectively is called double asymmetric synthesis.

(*iii*) C-Alkylation via chiral acyl iron complexes (enantioselective synthesis of an acid)

In (Scheme 2.35*a*) cyclopentadienyl (triphenylphosphine iron) complex with attached acyl unit from R—CH_2COOH (I, Scheme 2.35*a*) is chiral. The following points may be noted.

- In this synthesis iron carbonyl is the chiral auxiliary.
- Butyl lithium transforms the acyl carbonyl group into the lithium enolate (II, Scheme 2.35*a*) with *E* geometry (methyl group on the same side as the enolate oxygen, atomic members of oxygen and iron are 8 and 28 respectively).

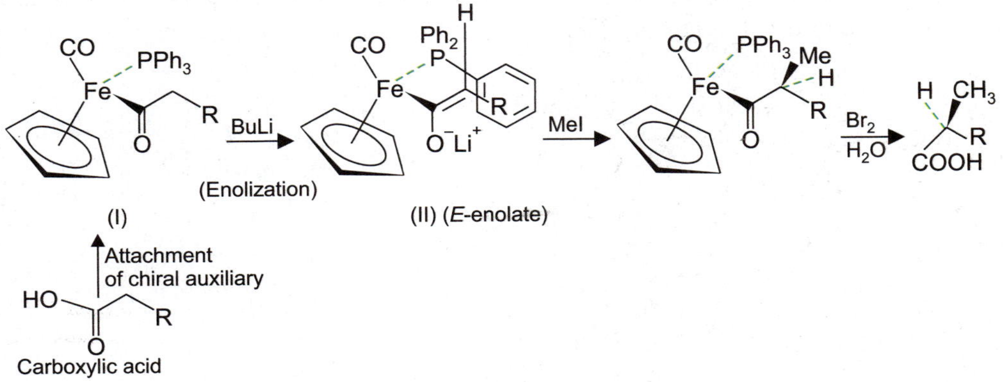

Use of an iron carbonyl as a chiral auxiliary

SCHEME 2.35a

- One of the phenyl groups of PPh_3 unit effectively blocks the approach of the electrophile from CH_3I from the *Re* face at C2 of (II, Scheme 2.35*a*). The alkylation thus occurs from the *Si* face (closest of the observer).

Enantioselective Synthesis of an Acid

The enantioselective synthesis of an acid (Scheme 2.35a) is an important application of coordination to metal centers. The enantiotopic faces can be rendered diastereotopic via ligation of the substrate to metal centers which are stereogenic. A chiral acyl iron complex can undergo C alkylation with high diastereoface selection. The outcome of observed stereochemistry of the final product (enantioselectivity) is based on the assumption that the intermediate enolate ion has E geometry, one face of which is effectively blocked by one of the phenyl groups on the iron complex. The alkylation e.g., of the enolate occurs from other side. The iron complex thus provides the necessary chirality and being sterically demanding blocks the approach of a reagent to one face.

- On one electron oxidation with bromine in water gives an alkylated acid in an enantiomerically pure form. Oxidation with bromine cleaves the C—Fe bond and generates the acid derivative.

(*iv*) Chiral aza-enolates-asymmetric synthesis of a-alkylcarboxylic acids

Chiral oxazolines [I, Scheme 2.36 (in an oxazoline C2 is in fact a masked carboxyl, group therefore, asymmetric alkylation of a substituent at C2 opens a route for the asymmetric synthesis of alkylated carboxylic acids)] on treatment with LDA (I, Scheme 2.36) gives the *Z*-enolate (II, Scheme 2.36). The major product (almost exclusive) on lithiation gives chelated compound (II, Scheme 2.36) in which lithium is held below, the plane of the ring by the methoxyl group. Alkylation of (II, Scheme 2.36) takes place from underside of the molecule as shown (III, Scheme 2.36) *i.e.*, on the side opposite the bulky phenyl substituent, (this direction is provided by the lithium and further shows that both the methoxy group and phenyl group are essential for the observed optical yield).

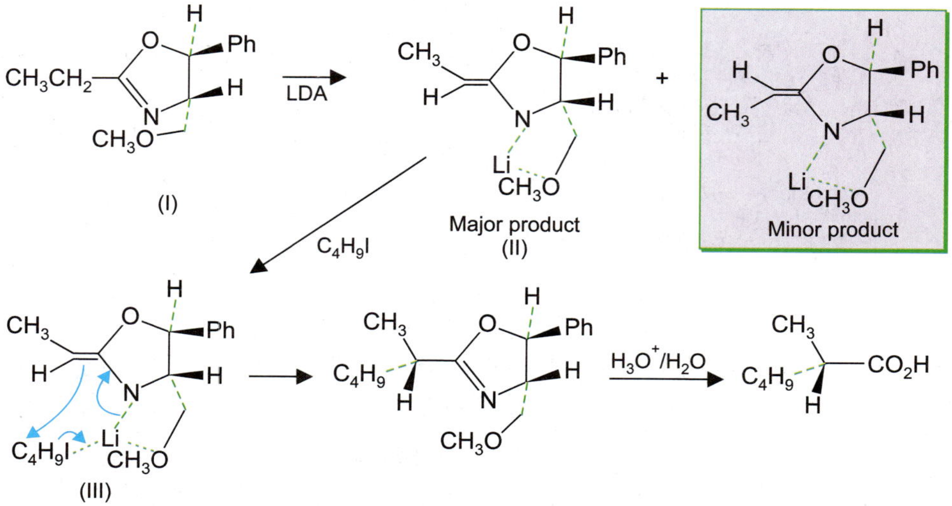

SCHEME 2.36

(v) Enantioselective alkylation of aldehydes and ketones via chiral hydrazones

S-1-Amino-2-methoxy-methylpyrrolidine is called SAMP and its enantiomer as RAMP and are prepared from *S*- and *R*- prolines respectively. They bear a chelating methoxy group (Scheme 2.37) and are employed as chiral auxiliaries in enantioselective alkylation of aldehydes and ketones.

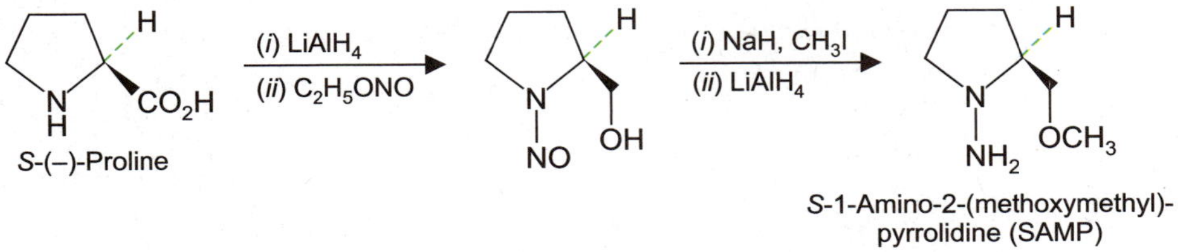

SCHEME 2.37

An easy way to bring about an asymmetric alkylation α to a carbonyl group is to just convert a ketone to a chiral imine or hydrazone followed by deprotonation with a strong base. Thus the hydrazone of 3-pentanone on treatment with LDA affords a complex (I, Scheme 2.38, the α-hydrogens of hydrazones are acidic). The resulting lithium aza-enolate adopts an orientation and is stabilized by forming a complex involving the lithium complex, the α-carbanion and the chiral auxiliary (at the methoxyl group). This chiral chelated carbanion is a nucleophile which will react *e.g.*, with alkyl halides (1-iodopropane in Scheme 2.38) *via* nucleophilic substitution to form exclusively only one of the two diastereomers (*S, S* Scheme 2.38) and not (*R, S*). On ozonolysis the *S, S* diastereomer (II, Scheme 2.38) afforded the *S*-enantiomer (99% of the product) of 4-methyl-3-heptanone, and the chiral auxiliary is recovered as the *N*-nitroso compound which may be reduced back to SAMP, RAMP and SAMP reacted with the same achiral carbonyl compound to permit enantioselective α-alkylation to occur in either direction.

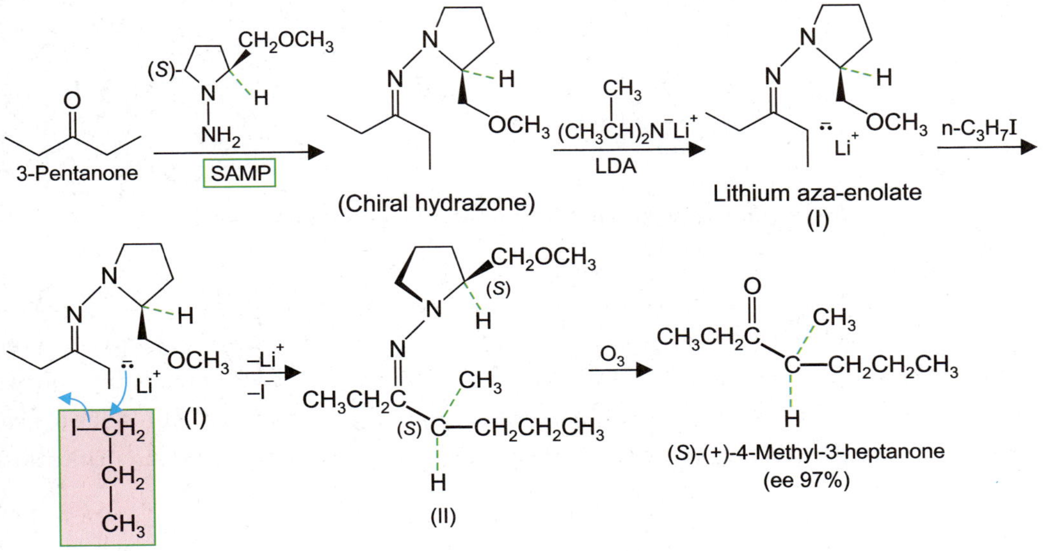

SCHEME 2.38

EXERCISE 2.6

How one can carry out asymmetric alkylation of cyclohexanone α to the carbonyl?

ANSWER. *This can be achieved by converting cyclohexanone into a chiral hydrozone by reaction with SAMP. Its conversion into lithium aza-enolate, followed by methylation and ozonolysis gives the desired product (Scheme 2.38a).*

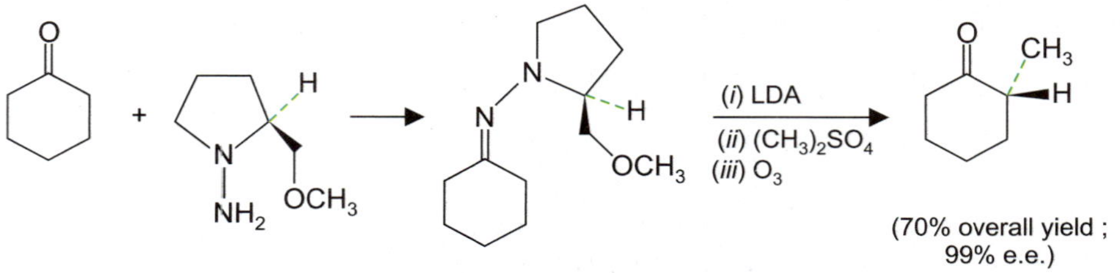

SCHEME 2.38a

***(vi)* Asymmetric Diels-Alder Reaction.** This is detailed in (Schemes 8.51–8.53)

(d) The Use of Chiral Reagents (Third-Generation Methods) and Chiral Catalysts

***(i)* Asymmetric hydroboration with diisopinocampheylboranes (an enantioselective reduction of alkenes to alcohols)**

Hydroboration of a naturally occurring terpenoid hydrocarbon α-pinene (which is chiral) gives chiral alkylboranes–monoisopinocampheylborane $(Ipc)BH_2$ or diisopinocampheylborane $(Ipc)_2$ BH as only one stereoisomer (Scheme 2.39), by attack of BH_3 from the less hindered α-side of α-pinene resulting in stereoselective addition.

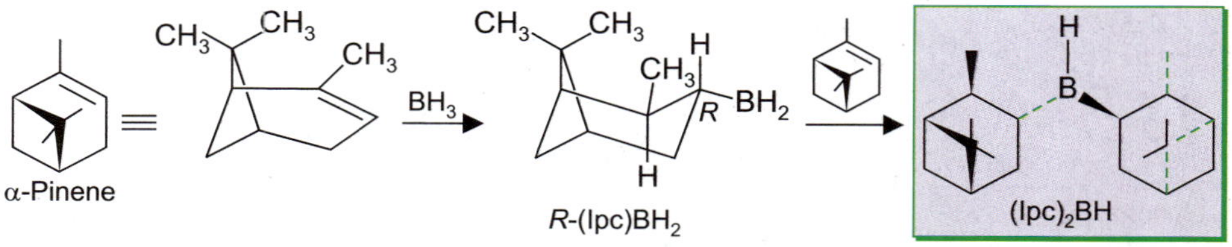

***R* specifies configuration at carbon atom bonded to boron**

SCHEME 2.39

The use of chiral borane $(Ipc)_2BH$ in its optically pure form on hydroboration of the prochiral (*Z*)-but-2-ene, followed by oxidation gives (*R*)-butan-2-ol with high optical purity is in (Scheme 2.40). Monoisopinocampheylborane (Ipc) BH_2 reacts with (*Z*)-but-2-ene to give (*S*)-butan-2-ol, and this reflects on the importance of careful reagent preparation/purification.

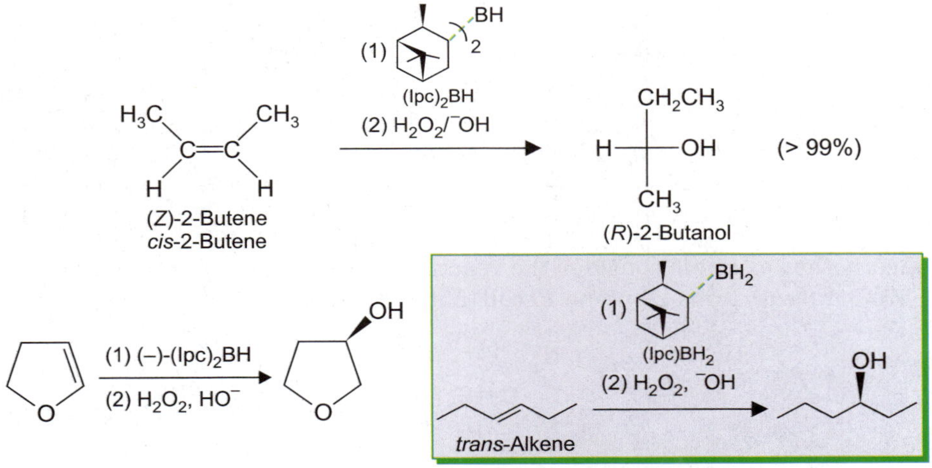

IPC_2 BH reacts with symmetrical *cis*-olefins with high stereoselection, while IPC BH_2 works well for enantioselective hydroboration of *trans*-alkenes

SCHEME 2.40

It has been shown that $(Ipc)BH_2$ is a very good enantioselective reagent for hydroboration of *trans*-alkenes, trisubstituted alkenes and 1-substituted cycloalkenes. $(Ipc)_2BH$ on the other hand, works very well with *cis* olefins with high enantioselection. The mechanism of the reaction between (*Z*)-but-2-ene (*cis*-2-butene) with $(Ipc)_2BH$ (Scheme 2.41) involves a different interaction of the two faces of the alkene with the chiral borane to give diastereomeric (triorgano) boranes (addition occurs preferentially to one face). A chiral alcohol is then produced after oxidative hydrolysis with hydrogen peroxide and base. The transition state leading to (*R*) alcohol (Scheme 2.41) is much lower in energy (one may consider the orientation of the alkyl groups on olefin with respect to $(Ipc)_2BH$ portion in the transition state leading to major diastereomer, (Scheme 2.41).

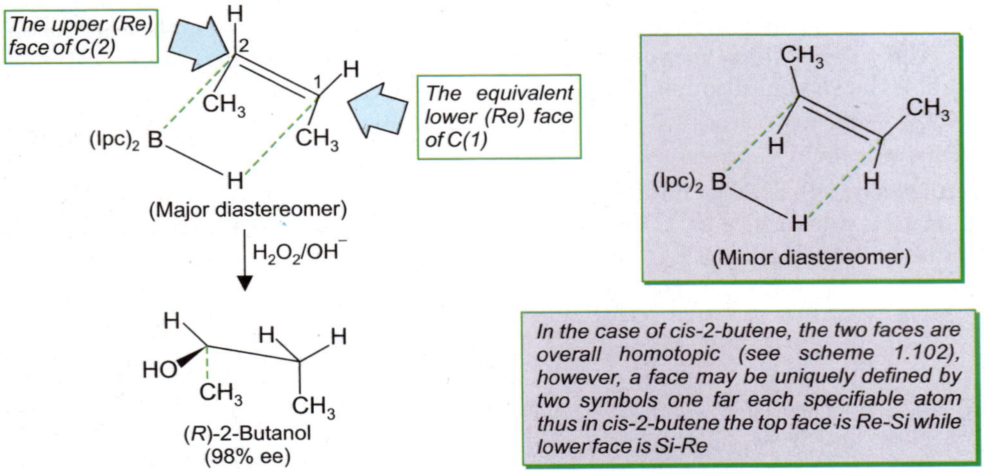

SCHEME 2.41

The results of Scheme 2.41 are interesting since cis-2-butene (a prochiral alkene) is devoid of any inbuilt structural features that could block the entry of a reagent to one face compared to other (camphor Scheme 7.8a has, however, such inbuilt structural features). $(IPC)_2 BH$ is a very bulky dialkyl chiral borane, when compared to cis-2-butene in which the comparison to resistance by attack of $(IPC)_2 BH$ is H versus methyl. The alcohol is formed by the attack of bulky borane to the upper (Re) face of C2 or the equivalent lower face (Re) of C1.

As another example consider the reaction of one enantiomer of a chiral borane (I, Scheme 2.41*a*) with *trans*-2-butene. Recall that hydroboration reaction is very sensitive to

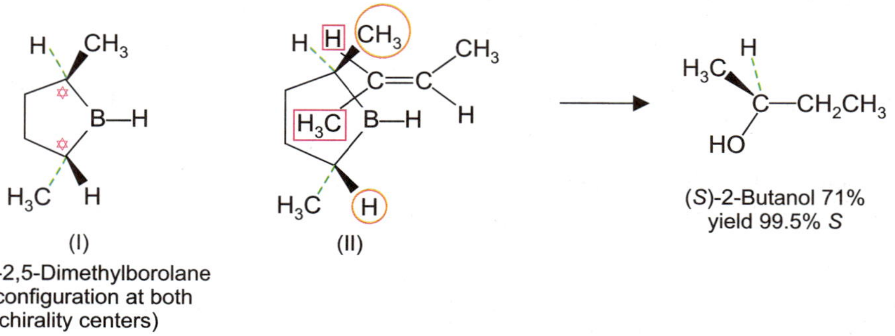

SCHEME 2.41a

steric effects. The chiral borane approaches the alkene in a manner so as to minimize steric interactions. As shown in the transition state (II, Scheme 2.41*a*), there are fewer steric interactions than if the other face of alkene had approached the chiral borane. A favourable geometry of the reaction involves *e.g.*, the approach of the alkene from above the borane. The smaller group on the alkene (The H atom in the square) interacts with larger group on the borane (The CH_3 in the circle) and the larger group on the alkene (the CH_3 in a square interacts with a smaller group on the borane (the H in the circle).

Note that, to be successful in producing a single enantiomer of the product, the borane must be enantiomerically pure. Any of the (*S*, *S*)-enantiomer that is present in the borane will result in the formation of an equal amount of the other enantiomer of 2-butanol, the (*R*)-enantiomer in this case (also see, Scheme 6.49).

(*ii*) Asymmetric reduction using chiral trialkylboranes (enantioselective reduction of aldehydes and ketones)

A useful reagent 9-BBN is made by the reaction of 1, 5-cyclooctadiene with BH_3. In this reaction BH_3 adds twice to 1, 5-cyclooctadiene, first intermolecularly and then intramolecularly (Scheme 2.42).

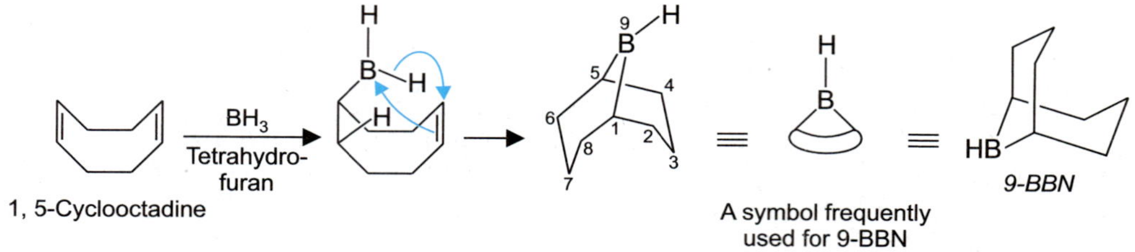

SCHEME 2.42

Hydroboration of (+) α-pinene with 9-BBN gives an extremely useful chiral trisubstituted borane which is commercially available and known as (*R*)-alpine borane (IPC. BBN, Scheme 2.43). It is an efficient enantioselective reducing agent and reduces a variety of carbonyl groups. In these reactions the hydride transferred during the reduction originates in the 9-BBN and (+)-α-pinene is its chiral carrier [(+)-α-pinene is regenerated during the process].

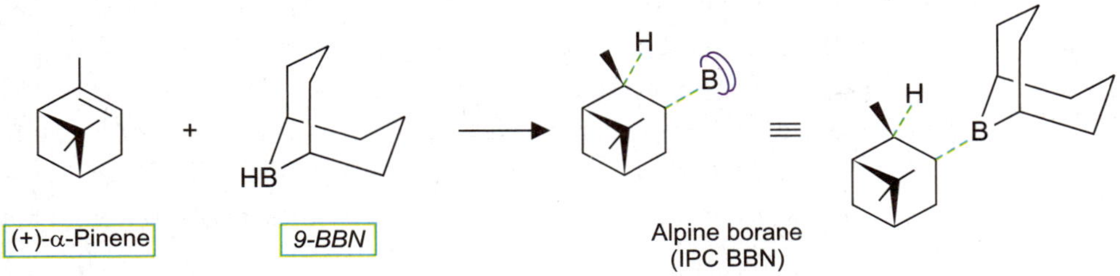

SCHEME 2.43

It is thought that the adduct transfers a hydride to the ketone via a six membered boat shaped cyclic transition state, where the larger group on the ketone (Ph of $PhCOCH_3$) preferentially lies away form the α-pinene moiety to make the steric congestion minimum (Scheme 2.44). The adduct of 9-BBN can be made with both (+) and (–)-α-pinene (the later is known as *S*-Alpine Borane) to synthesize both the enantiomers individually. Thus the reduction of unsymmetrical (prochiral ketones with enantiotopic faces) *e.g.*, acetophenone $C_6H_5COCH_3$ to optically active secondary alcohols is achieved.

Monodeuteriated and therefore, chiral primary alcohols can be made by this method by employing deuteriated analogue of Alpine Borane (Scheme 2.44, also see Scheme 7.16).

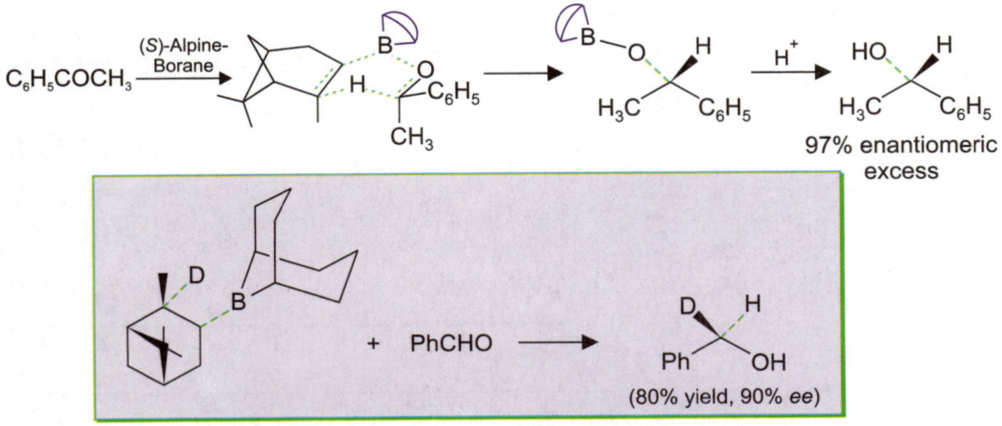

Stereoselective reductions of carbonyl groups

SCHEME 2.44

(*iii*) Use of chiral oxazaborolidine complex of borane (enantioselective reduction of ketones)

The borohydride (II/BH_3 Scheme 2.45) derived from the reaction of oxazaborolidine (I, Scheme 2.45) with diborane is an effective reducing agent and brings about an enantioselective reduction of ketones. A ketone may be reduced stoichiometrically with (I)/BH_3 (Scheme 2.45) or the reaction can also be carried out under catalytic conditions using a small amount of (I, Scheme 2.45) and excess BH_3/THF. Under stoichiometric conditions, only two of the three hydride ions attached to boron are reactive. The absolute configuration of the product can be predicted *e.g.* in the present case, the reagent which has the (*S*)-configuration at the secondary center produces the (*R*)-alcohol from the ketone, while the (*R*) reagent (prepared from D-proline) yields the (*S*)-alcohol. The reduction of acetophenone provides a specific example of this reaction (Scheme 2.46).

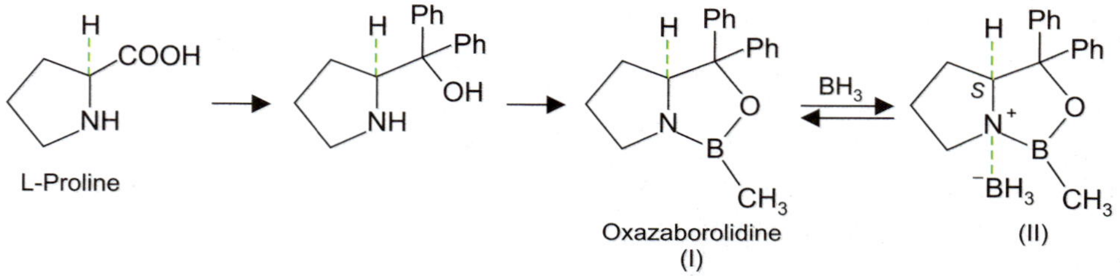

The excellent catalytic properties of (I) lie in the fact that it brings two reactants (BH_3) and a ketone in close proximity in a stereocontrolled manner

SCHEME 2.45

The reaction attains optimal stereoselectivity (a stereoselective reaction is one in which the reactant is not necessarily chiral ($C_6H_5COCH_3$) but it gives predominantly or exclusively one stereoisomeric product).

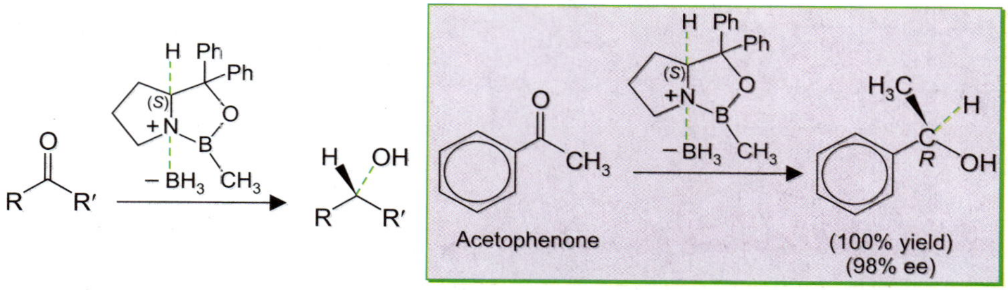

SCHEME 2.46

An enatiotopic ketone (*e.g.*, acetophenone) coordinates with electrophilic boron and the hydrogen atom transfer occurs *via* a six membered transition state in a way that the larger substituent (phenyl) attached to the carbonyl carbon atom is far removed from the heterocyclic complex as in (II, Scheme 2.47 rather than III).

The following points may be noted for the mechanism of this reduction.

- Borane coordinates to the nitrogen atom of the catalyst (see, Scheme 2.45).
- Next the carbonyl oxygen of the ketone (*e.g.*, acetophenone) coordinates to the electrophilic boron and the hydrogen atom transfer takes place.
- The high enantioselective, catalytic properties of oxazaborolidine additive is thus to properly orient the two reactants (BH_3 and the carbonyl compound) into close proximity in a stereocontrolled fashion so that only one face of the carbonyl group reacts (one boron atom acts as a Lewis acid to bind the basic oxygen atom of the carbonyl group while the other produces the hydride ion for the reduction of C=O bond).

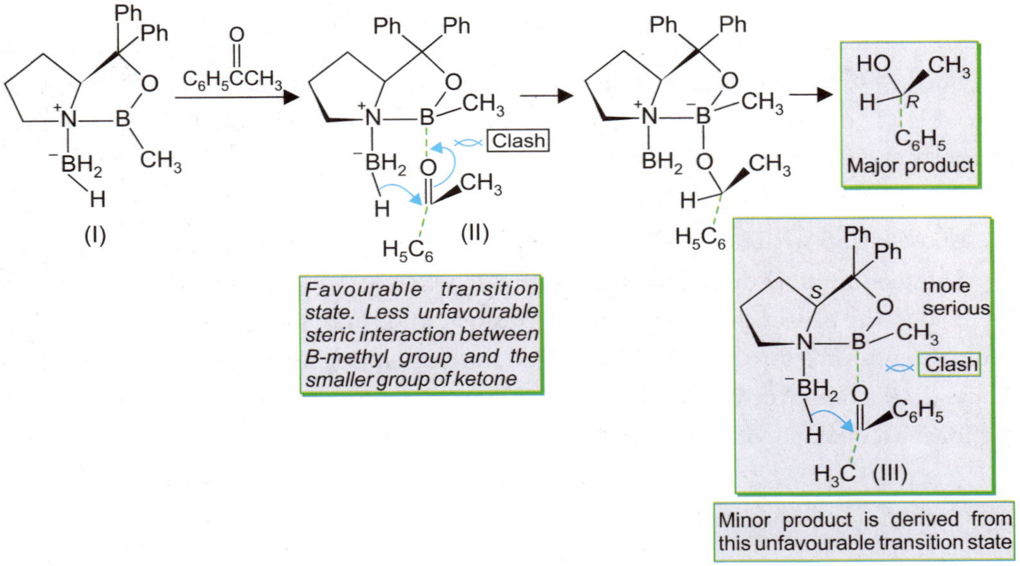

SCHEME 2.47

(*iv*) Sharpless asymmetric epoxidation (SAE) is an enantioselective reaction that oxidises alkenes to epoxides

The epoxidation of an alkene is achieved by reaction with perbenzoic acid or *m*-chloroperbenzoic acid (see Schemes 6.84–6.90). Chiral epoxides prepared this way lead only to recemates as no chiral inducing agents are present. The asymmetric synthesis of optically active epoxides is the need of the day, since these are required for the total synthesis of biologically active compounds.

In the sharpless asymmetric epoxidation (discovered 1981, Nobel prize 2001) the most useful substrate for oxidation is an allylic alcohol. The epoxidation is carried out with *t*-butylhydroperoxide (*t*-BuOOH) and titaniumtetraisopropoxide Ti (i-OPr)$_4$ and the procedure becomes highly enantioselective when enantiomerically pure tartrate esters (DET) are employed (Scheme 2.48).

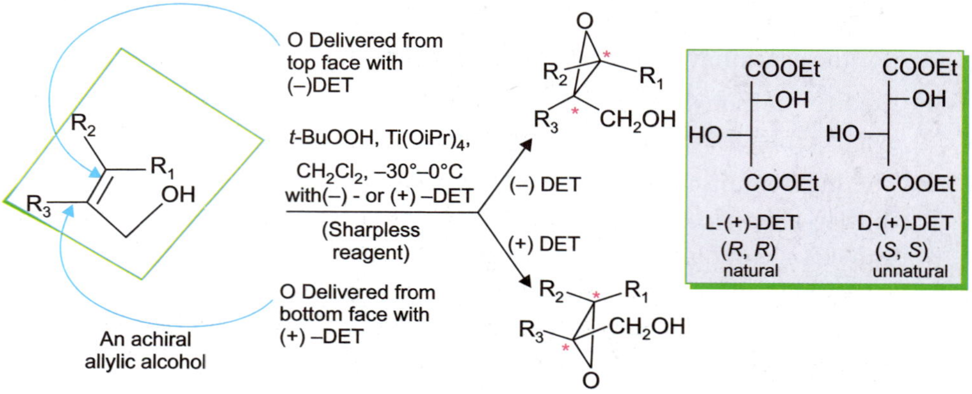

SCHEME 2.48

The key to the problem to determine as to which enantiomer is formed for a given isomer of DET, one draws the allylic alcohol in a plane with the CH_2OH group placed in the bottom right corner then:

- Epoxidation with (–)-DET adds an oxygen atom from above the plane.
- Epoxidation with (+)-DET adds an oxygen atom from below the plane.

The technique is one of the best methods whereby prochiral allylic alcohols are epoxidized in the presence of *t*-BuOOH $Ti(i\text{-}OPr)_4$, and (+)-or (–)-diethyl tartrate (DET) with high regio- and stereoselectivity to afford the corresponding optically active epoxides (Scheme 2.49, compare with problem 2.14*b*). Only double bonds with an allylic –OH group form an epoxide, others usually do not react.

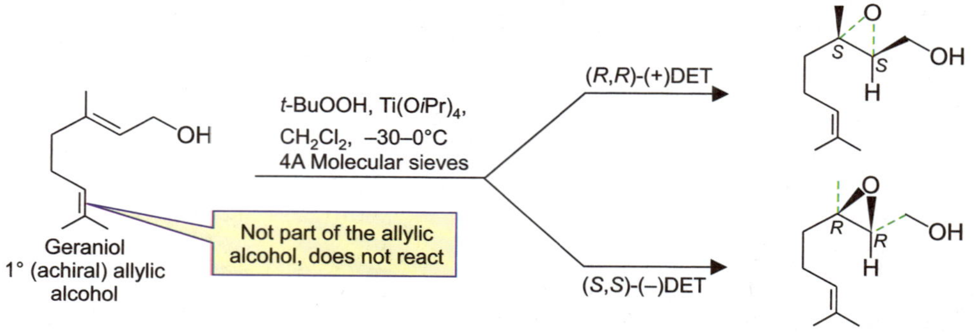

SCHEME 2.49

The sharpless reagent consists of three components: *tert-butyl* hydroperoxide, $(CH_3)_3COOH$; a titanium catalyst—usually titanium (IV) isopropoxide, $Ti[OCH(CH_3)_2]_4$; and diethyl tartrate (DET) (see Scheme 2.48). There are two different chiral diethyl tartrate isomers, labelled as (+)-DET or (–)-DET to indicate the direction in which they rotate polarized light (enantiomeric pair).

EXERCISE 2.7

Predict the major product of epoxidation of (I and II, Scheme 2.50c)

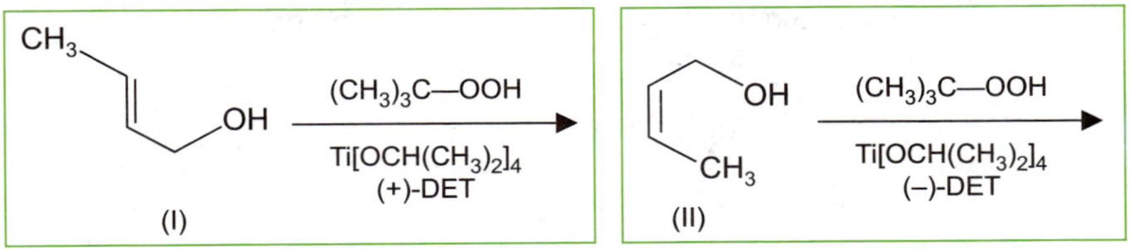

SCHEME 2.50

ANSWER. *To draw an epoxidation product:*

- *Draw the allylic alcohol with the OH group on the bottom right corner of the plane (see Scheme 2.48), one may re-draw the alkene if required.*
- *(+)-DET adds the O atom from below, and (–)-DET adds the O atom from above.*

(I) The OH group is drawn on the bottom right corner of the plane and (+)-DET is used, so the O atom is added from below (Scheme 2.50a)

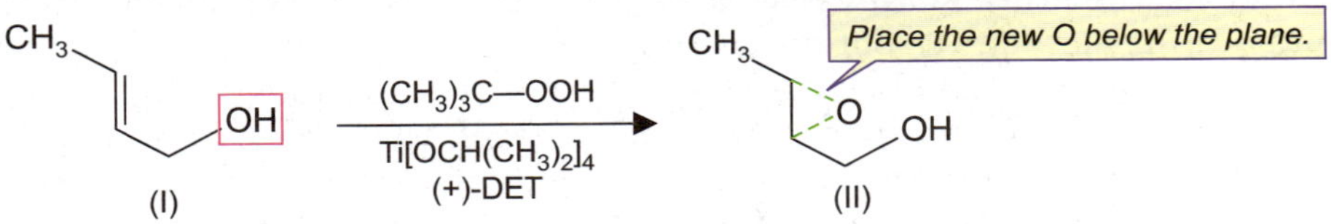

SCHEME 2.50a

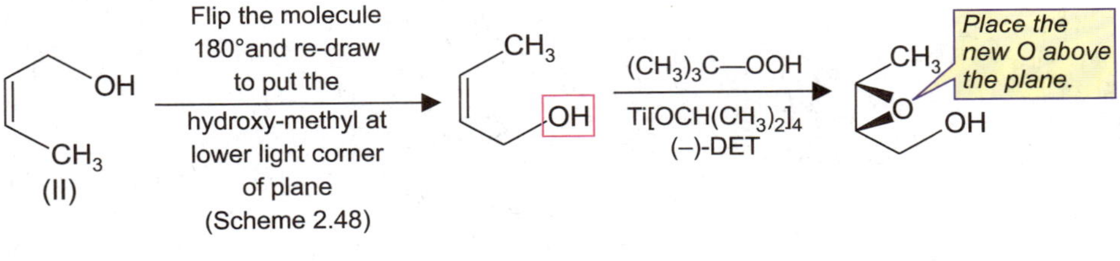

SCHEME 2.50b

Diethyl tartrate, the allylic alcohol, and the oxidant *t*-BuOOH displace the isopropoxide groups on titanium to form the active Ti-catalyst in a complex ligand exchange pathway. Thus titanium holds everything together *i.e.*, the tartrate ligand, the substrate, the allylic alcohol and the oxidant *t*-BuOOH. There are two titanium atoms and two unsymmetrically bound tartrate units. When the catalyst combines with the allylic alcohol and the hydroperoxide, the same titanium atom takes both of these and only one diastereotopic face of the alkene is now presented where the active oxygen atom is transferred (Scheme 2.51).

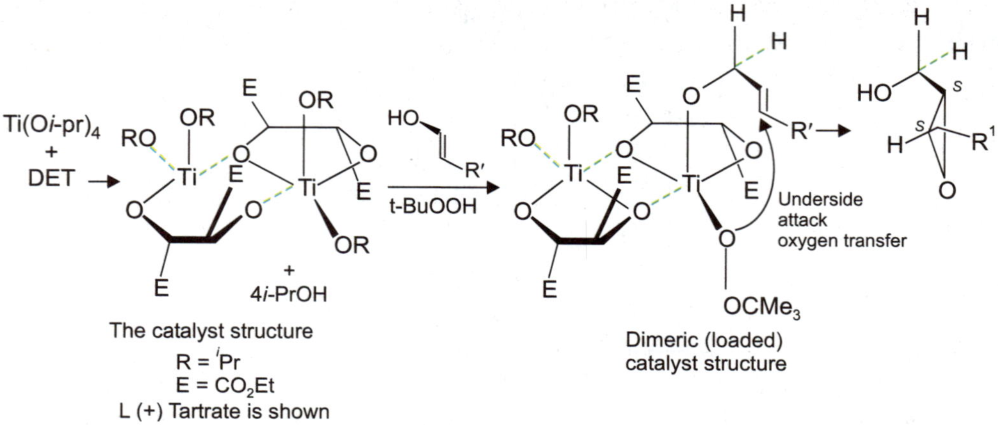

SCHEME 2.51

(v) Kinetic resolution of chiral allylic alcohols with the Sharpless reagent

The titanium catalyst is sensitive to pre-existing chirality in the substrate. As a result, the epoxidation of racemic secondary allylic alcohols with a given tartrate-titanium-isopropoxide combination occurs rapidly only with one of the enantiomers (Scheme 2.52). The other slower reacting enantiomer is left behind. This procedure thus, provides a method for obtaining two different optically active compounds a chiral epoxide and a resolved chiral allylic alcohol (a route to optically pure allylic alcohols). The reagents used for the kinetic resolution are essentially similar to those used for asymmetric epoxidation except that diisopropyltartrate esters (DIPT, Scheme 2.52) are often substituted for the diethylester analogs (DET, Scheme 2.52).

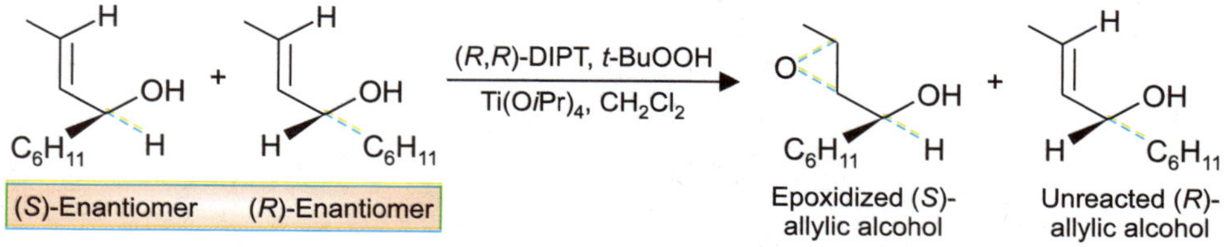

SCHEME 2.52

(vi) Asymmetric oxidation of sulfides to sulfoxides

The reagents used for asymmetric epoxidation are used for the asymmetric oxidation of sulfides to sulfoxides, however, in this water is actually needed in order to achieve good selectivity (Scheme 2.53, for chiral sulfoxides see Schemes 1.6 and 1.7).

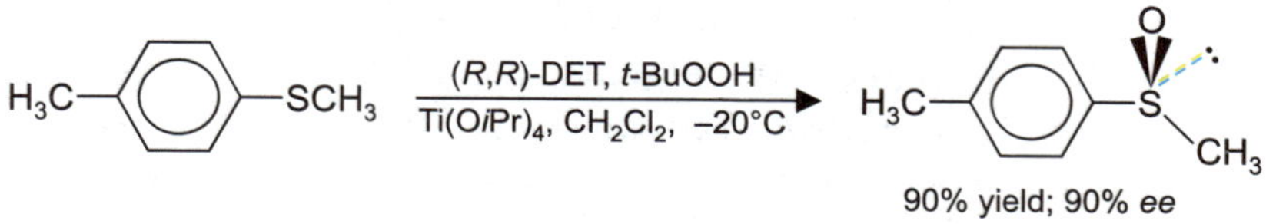

SCHEME 2.53

(vii) Enantioselective epoxidation *via* oxygen transfer catalysts – Jacobsen epoxidation

In 1990, Eric Jacobsen reported on a manganese-containing chiral catalyst which epoxidizes olefins with reasonably high *ee* values using low priced cooxidant NaOCl under mild conditions.

The catalyst is made by condensing 1,2-diaminocyclohexane and a phenolic-aldehyde to generate two carbon-nitrogen double bonds to give a ligand (I, Scheme 2.53*a*) which is referred to as "Salen". Salen binds manganese in the +3 oxidation state. This species (II, Scheme, 2.53*a*), reacts with NaOCl to form the oxidizing species, *cis*-alkenes show high selectivity with Mn (Salen*) (O) Cl reagent to give corresponding epoxides.

Chiral dioxiranes developed by Professor Y. Shi on the other hand work very well to epoxidize *trans* and trisubstituted alkenes. A dioxirane moiety is generated by the reaction of a keto group with hydrogen persulfate ion (HSO_4^-). The chiral dioxiranes derived from readily available sugar derivatives of D-fructose and L-Sarbose (Scheme 2.53*b*) where the pair of hydroxyl groups are blocked by ketal/acetal formation. The enantioselective epoxidation is carried out by treating an alkene with ketone FR* (derived from D fructose or L-sorbose and persulfate ion (Scheme 2.53*c*).

CHAPTER 2

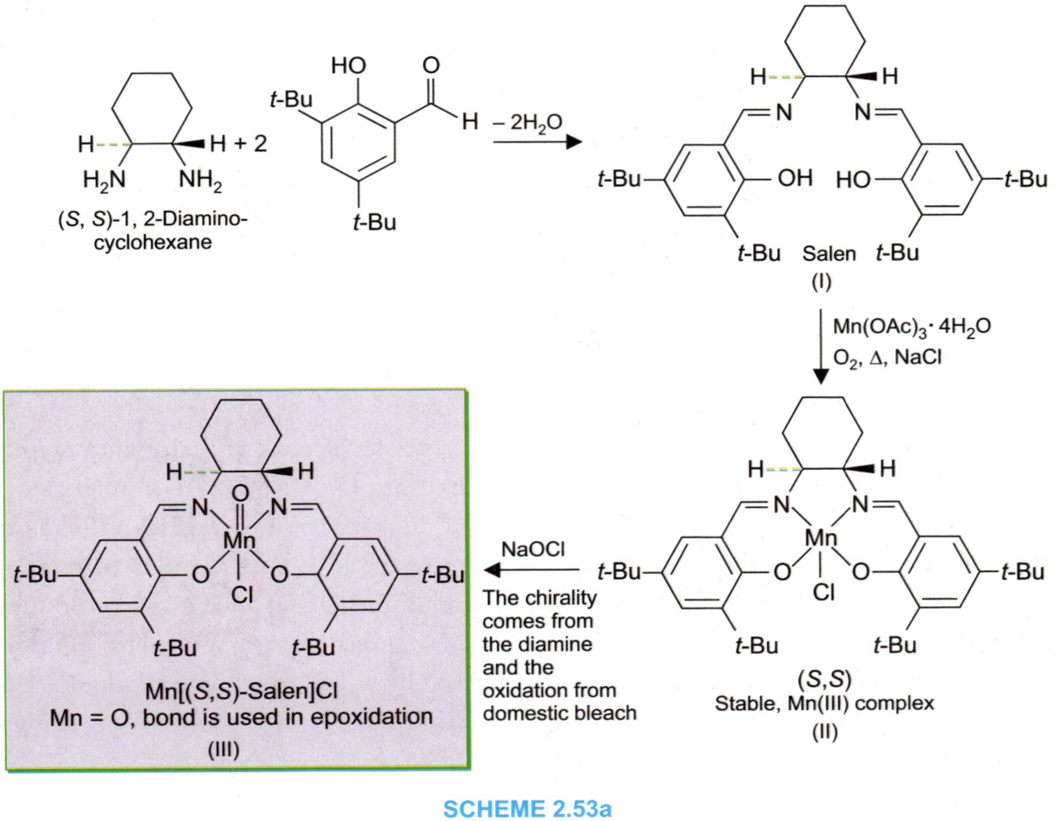

SCHEME 2.53a

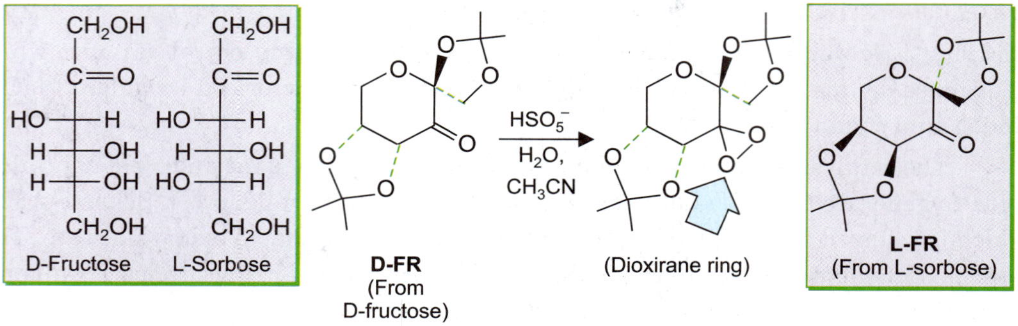

SCHEME 2.53b

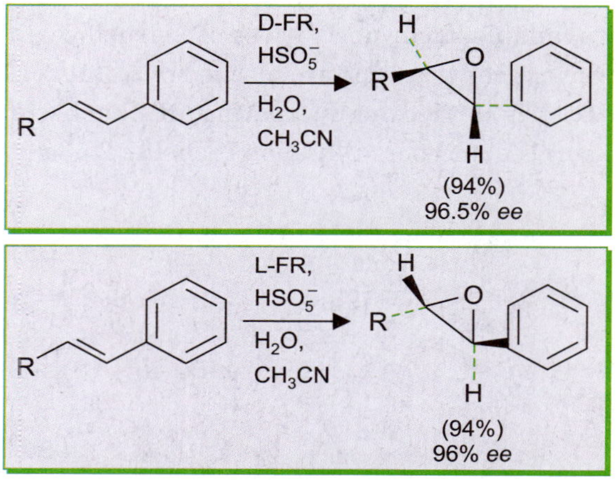

SCHEME 2.53c

(*viii*) Sharpless asymmetric dihydroxylation (AD) – The catalyst

It was discovered by Sharpless that prochiral alkenes on treatment will catalytic amounts of the solid, nonvolatile dipotassium osmate dihydrate ($K_2[Os^{VI}O_2(OH)_4]$), potassium ferricyanide $K_3Fe(CN)_6$, a stoichiometric oxidant, along with a chiral ligand leads to enantioselective *syn*-dihydroxylation. The enantiomerically pure ligands are easily available from cinchona alkaloids and are dihydroquinine (DHQ) and dehydroquinidine (DHQD). The use of phthalazine spacer (PHAL) superceded the monomeric alkaloid ligands, the $(DHQ)_2$ PHAL and $(DHQD)_2$PHAL where two cinchona alkaloid units are connected together via a linker molecule (PHAL) are known by the trade names *AD*-mix-α and *AD-mix*-β, respectively (Scheme 2.53*d*).

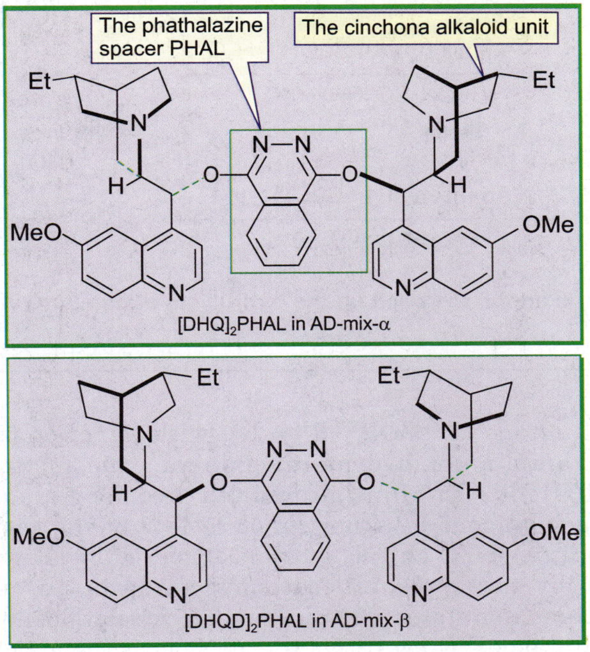

SCHEME 2.53d

(*ix*) Stereochemical outcome of asymmetric dihydroxylation (AD)

AD means asymmetric dihydroxylation, and α and β signify which face of the double bond undergoes addition.

The dihydroxylation reaction is general for many alkene types, and simply using the other reagent mix switches the stereochemical outcome of the transformation. The ee values are however, not identical, since $(DHQ)_2$PHAL and $(DHQD)_2$PHAL are *diastereomers,* not enantiomers.

A mnemonic device for predicting the stereochemical outcome of the diols from AD reactions is in Scheme 2.53*f*. The largest substituent on the alkene is placed at the lower left corner of the model (Scheme 2.53*e*). AD-mix-α dihydroxylates the bottom (α)-face of the alkene, while AD-mix-β induces dihydroxylation from the top (β)-face.

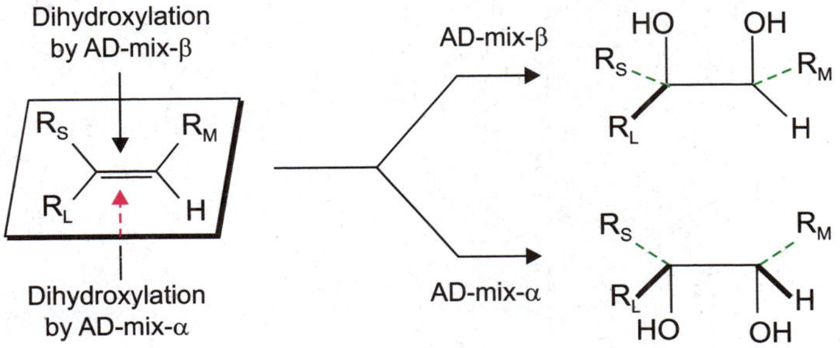

(Contd...)

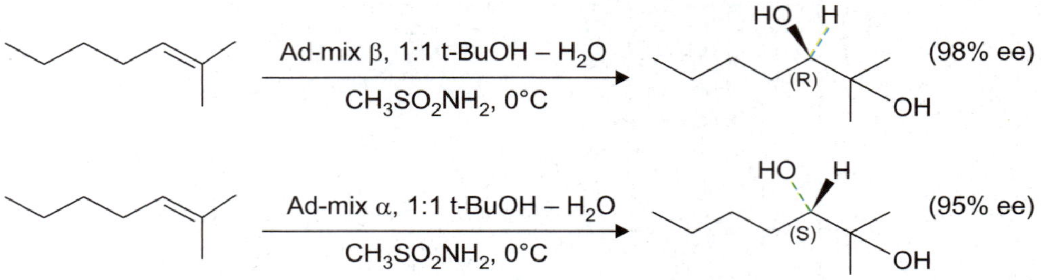

$MeSO_2NH_2$ is added to speed up the hydrolysis of osmium (VI) glycolate

SCHEME 2.53e

(x) Asymmetric reduction using BINAL-H

An unsymmetrical ketone (R^1 COR^2, $R^1 \neq R^2$, with enantiotopic faces) on reduction with lithium aluminium hydride gives a racemic mixture of the secondary alcohol $R^1CH(OH)R^2$. Lithium aluminium hydride is achiral and the approach of this reducing agent to either *Re* or *Si* face of the carbonyl carbon atom is equally facile. However, a chiral hydride donor will transfer an achiral unit H^-: enantioselectively to a prochiral carbonyl group of an enantiotopic ketone (R^1 COR^2) [via two transition states with diastereomeric relationships (of different energies) to give one enantiomer (*R*-) or (*S*-) of the alcohols as the major product].

A commonly used asymmetric reducing agent ('BINAL-H') is derived from optically pure binaphthol, ethanol and lithium aluminium hydride ((Scheme 2.53*f*). It reduces acetylenic ketones, aromatic ketones and α,β-unsaturated ketones with high optical purities. The (*R*)-BINAL-H selectively adds the hydride to the *Si* face of the substrate, while the (*S*)-reagent adds it to the *Re* face (Scheme 2.53*f*).

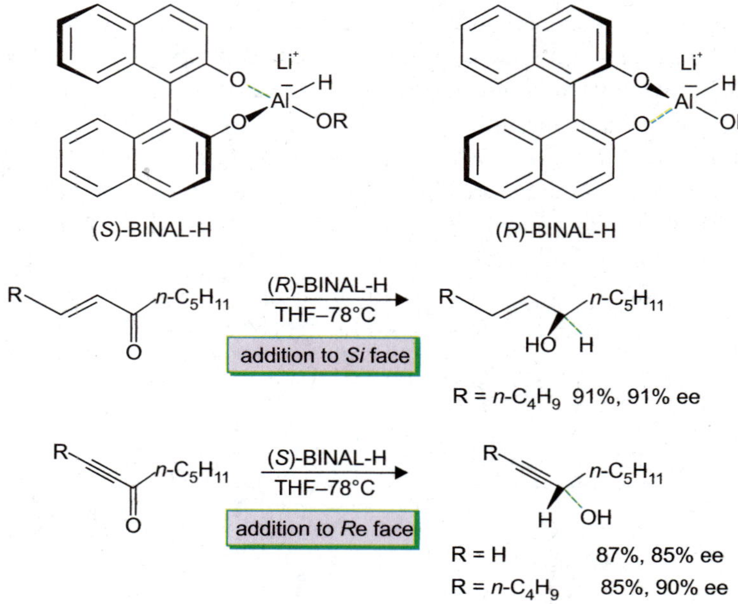

SCHEME 2.53f

Asymmetric reduction of ketones like butanone and 4, 4-dimethylpentan-2-one (DIMPO) is carried out (with high enantioselectivity) with (*R*, *R*)-or (*S*-*S*)-dimethylborolane. Generally the *RR*-reducing agent gives the *R*-alcohol while the *SS*-reagent yields the *S*-product.

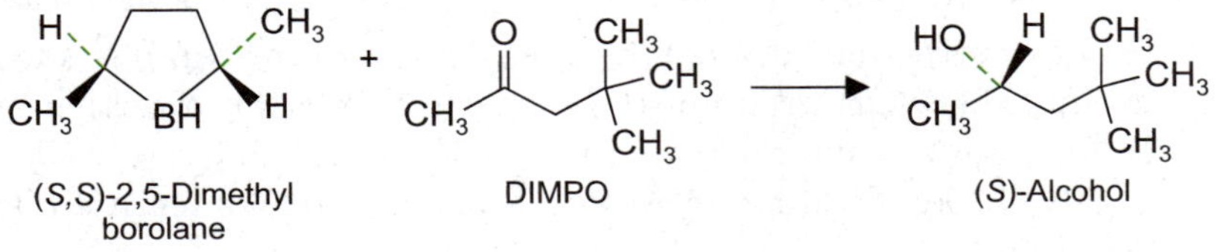

SCHEME 2.54

(e) Miscellaneous Enantioselective Synthesis

1. **Asymmetric Diels Alder reactions (for details see chapter 8)**
2. **Enantioselective catalytic hydrogenation asymmetric synthesis of L-alanine and L-dopa (for details see chapter 6)**
3. **Asymmetric dihydroxylation (for details see chapter 6)**

(f) Phase Transfer Catalysis (Fourth Generation Methods)

This is one of the examples from the fourth generation asymmetric synthesis involving asym-metric catalysis. An efficient enantioselective alkylation of 2-phenylindanone (I, Scheme 2.55) has been carried out in a two phase system (toluene-water)

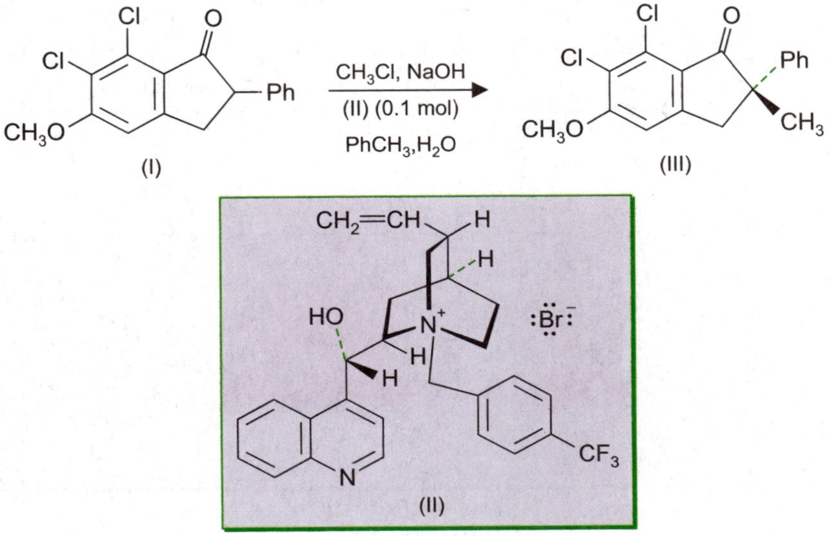

SCHEME 2.55

catalyzed by quaternary ammonium salt (the benzyl cinchoninium cation II, Scheme 2.55). This salt is derived from the alkaloid cinchonine (phase transfer conditions). The high enantioselectivity is believed to be the result of the following factors:

- The quinuclidine ring lies behind the plane of the indanone enolate.
- A highly specific complex is formed between the enolate and the catalyst.
- There is *pi* interaction between the benzyl group of the catalyst and the 2-phenyl group. The 9-OH group of the catalyst provides a directive handle (via H-bonding) and the backside approach of the alkylating agent CH_3Cl is comparatively blocked leading to the formation of (*S*)-(+)–2-methylindanone (III, Scheme 2.55).

(g) Enantioselective Intramolecular Aldol Condensation (Carbon-Carbon Bond Forming Reaction–Robinson Annelation Reaction)

The cyclopentan-1, 3-dione (a, Scheme 2.56) undergoes an intermolecular aldol condensation under the catalysis of a chiral base (*S*-proline) with high enantioselectivity. When during a Robinson annulation reaction (Scheme 2.56, overall process of addition and cyclization), the cyclization step is carried out in the presence of (*S*)-proline, the bicyclic hydroxydiketone (II, Scheme 2.56) is obtained in 94% optical purity. The following steps may be noted for this enantioselectivity:

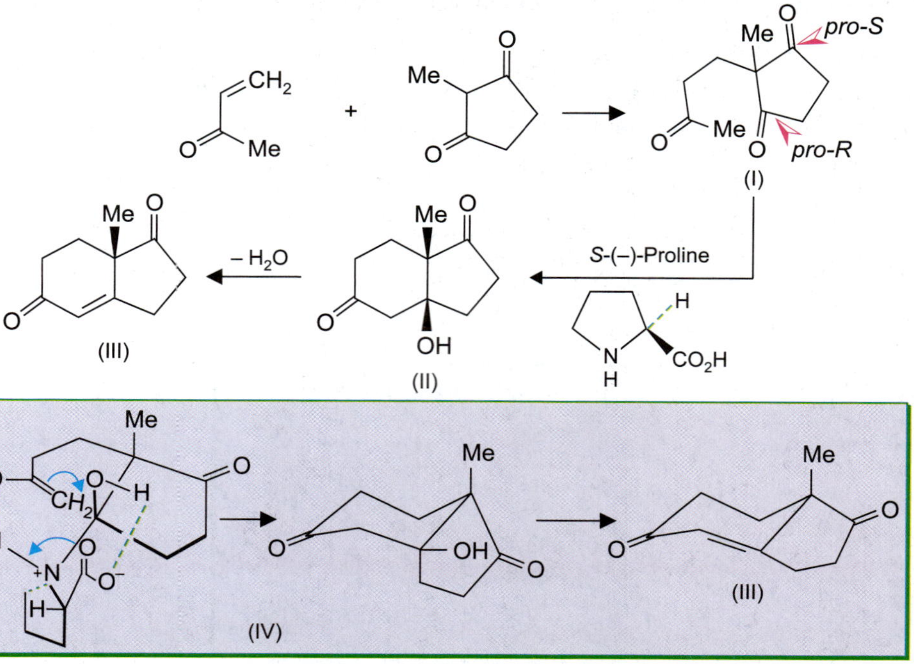

SCHEME 2.56

- The chiral catalyst distinguishes between the two carbonyl groups in the cyclopentane ring (I, Scheme 2.56) which are enantiotopic and distinguished by the chiral catalyst which, therefore, add preferentially to one of these. The two carbonyl groups on the cyclopentane ring are enantiotopic and designated *pro R* and *pro S*.

- Two hydrogen bonds in (IV, Scheme 2.56) impose a topographical rigidity on the conformation of the tricyclic transition state.
- The proline residue as a consequence is *trans* related to the angular methyl group and carbon-carbon bond formation will consequently take place from the side opposite to the methyl group to give the *cis*-fused hydroxy diketone (II, Scheme 2.56) which undergoes dehydration as usual.

(h) Self Regeneration of Stereocenters-α-Alkylation of Amino Acids

A chiral starting material *e.g.* *S*-proline (Scheme 2.57) is derivatized in such a way so that a new chirality center is created under the influence of the original chirality center as in (I, Scheme 2.57). The tetrahedral geometry of the original stereocenter is then destroyed (II, enolate formation with LDA). The stereochemical information is thus stored in the new chirality center and the original chirality center can then be regenerated stereospecifically as shown during α-alkylation of amino acids. Although the alkyl group gets attached from the lower face of the molecule on the same side of the bulky *t*-butyl group, this face is in fact less hindered than the upper face. A consideration of the model of the enolate (II, Scheme 2.57) shows that the bulky *t*-butyl group occupies a pseudo-equatorial position to make lower face less hindered.

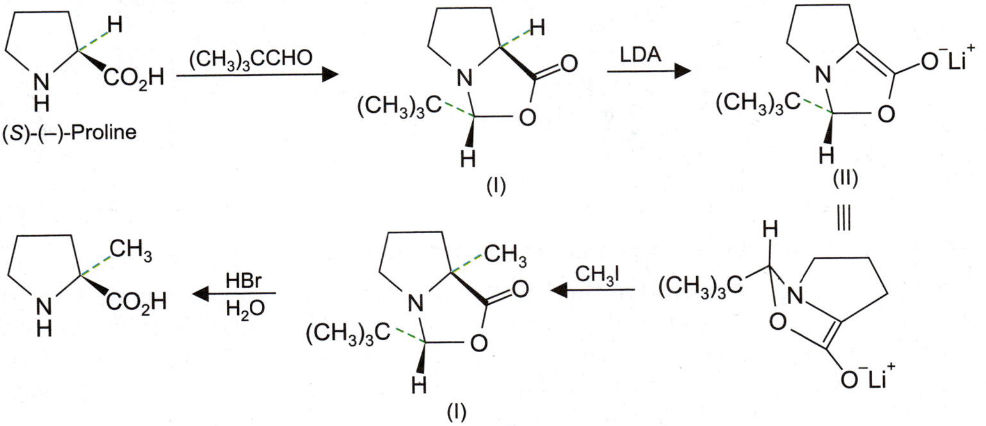

SCHEME 2.57

(i) Stereoselective Formation of a Double Bond (Conversion of Z-Cyclooctene into E-Cyclooctene)

The affinity of phosphorus for sulfur is utilized in synthesis of alkenes from 1, 2-diols *via* thiocarbonate intermediates (Scheme 2.58). The Z-cyclooctene (Scheme 2.59) *e.g.*, is converted into *trans*-diol (epoxidation via acid catalyzed hydrolysis) which on reaction with thiophosgene gives the cyclic thionocarbonate which on reaction with triethylphosphite gives *E*-cyclooctene (a highly stained compound). When the diol is resolved before the reaction, one can make enantiomerically pure *E*-cyclooctene. The key to this procedure is that in order to form the thionocarbonate the diol has to adopt a conformation (Scheme 2.59) which ensures the formation of the *trans*-double bond. One may recall that just as a double bond can be oxidized stereoselectively to *cis*- or *trans*-1, 2-diols, the diols may also be converted into the corresponding olefinic compounds stereoselectively.

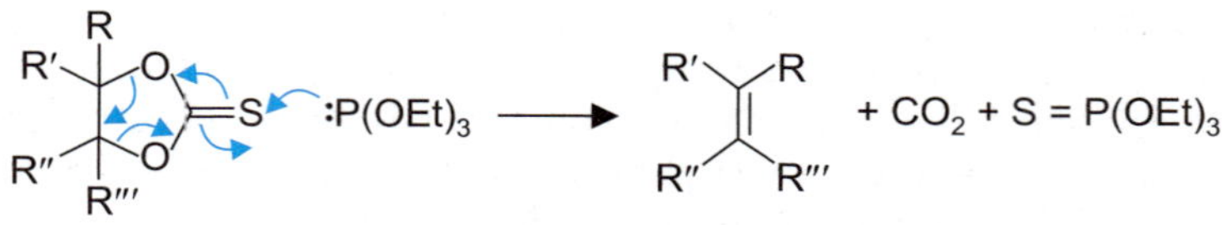

SCHEME 2.58

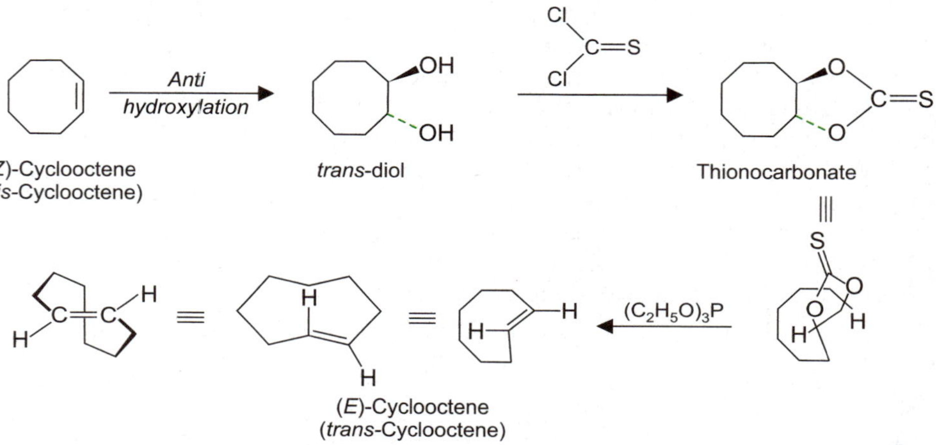

SCHEME 2.59

(j) Asymmetric Synthesis Using Circularly Polarized Light

Absolute chiral synthesis involve the formation of optically active compounds without the incorporation of other optically active compounds. One of the most interesting of some of the known examples of absolute chiral synthesis involves the light induced cyclization–(photoreactions) of 1,2-diarylethylenes to dihydrophenanthrene derivatives (oxidation with iodine and oxygen gives phenanthrene Scheme 2.60).

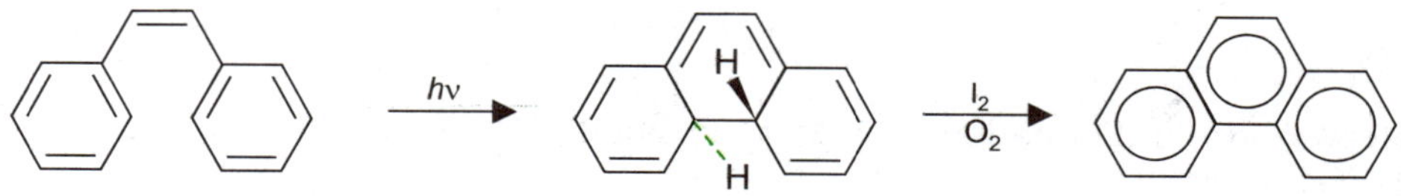

SCHEME 2.60

When 1,2-diarylethylene is subjected to these reactions in the presence of circularly polarized light to induce cyclization, optically active octahelicene is formed (study the chirality of hexahelicene). With right hand circularly polarized light the (–) enantiomer is formed in 2% optical purity, while with the left hand circularly polarized light the (+) enantiomer is formed to the same extent. The chiral reagent here is the circularly polarized light and the two forms of the product have a nonsuperimposable mirror image relationship to each other. The role of circularly polarized light is reminiscent of an optically active compound in a conventional resolution. It combines with individual enantiomers of the diarylethylene, forming a pair of excited states which are diastereomerically related and, are thus, formed and then decomposed at different rates (Scheme 2.60).

(k) Use of Bakers' Yeast, Other Enzymes and Proteins

Recall, that reductions can be catalyzed by enzymes *i.e.*, a prochiral ketone is reduced to optically active secondary alcohol (see, Schemes 1.85 and 2.20). The enzyme is only the chiral catalyst which, however, does not provide the hydrogen atom for reduction. The hydrogen atoms, instead are provided by the *relevant coenzyme e.g., NADH*. One may carry out the reduction using whole cells, like bakers yeast where both the enzyme and the coenzyme are provided by the organism. Thus ethyl acetoacetate (Scheme 2.60*a*) is reduced selectively to ethyl (*S*)-(+)-3-hydroxybutanoate using one of the reducing enzymes found in Baker's yeast alcohol dehydrogenase. It is one of the enzymes which are involved in the metabolism of D-glucose to ethanol. In this reaction enantiomeric excesses ranging from 70–97% have been achieved particularly when oxygen is excluded during the fermentation (anaerobic). However, a dependence on the size of the two groups has been discovered. Reduction of ethyl acetoacetate (Scheme 2.60*a*) with Baker's yeast yielded the (*S*) alcohol but reduction of ethyl β-ketovalerate instead gave the (*R*) alcohol. For the mechanism of this reaction (see Scheme 2.20).

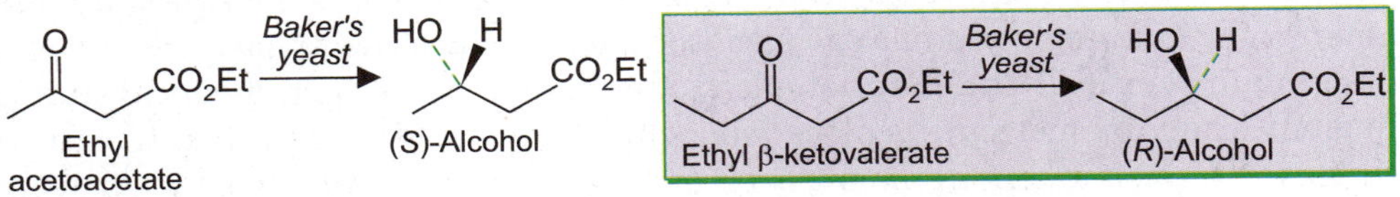

SCHEME 2.60a

Meso compounds undergo enantioselective reactions in the presence of enzymes, horse liver alcohol dehydrogenase (HLADH) selectively oxidises the diol (Scheme 2.60*b*) to the lactone.

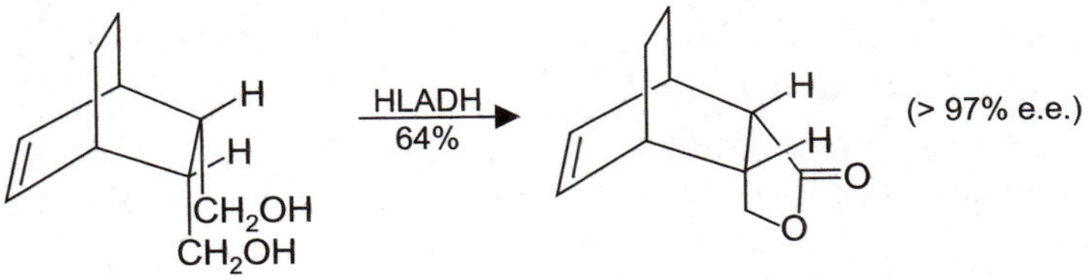

SCHEME 2.60b

2.5 STEREOHETEROTOPIC LIGANDS AND NMR SPECTROSCOPY

(*i*) Introduction

After learning the concepts of prostereoisomerism/prochirality, one can easily classify like atoms or groups of atoms in a molecule according to their symmetry relationships and the following points may be noted:

- The term *equivalent ligands e.g.*, equivalent hydrogens is still widely used for homotopic hydrogens, however, it lacks precision.
- All homotopic hydrogens (also called equivalent) are symmetry equivalent, however, not all symmetry equivalent hydrogens are homotopic (these could well be enantiotopic).
- In achiral conditions symmetry equivalent atoms are chemically equivalent *i.e.*, these undergo reactions at the same rate. These are called chemically equivalent (a term which again lacks precision).
- Under chiral conditions homotopic hydrogens *e.g.*, still cannot be distinguished, however, enantiotopic ligands *e.g.*, hydrogens can be distinguished. Thus in a chiral atmosphere enantiotopic hydrogens are not chemically equivalent (in glycerol *pro-R*-hydroxymethylene group is exclusively phosphorylated with ATP in the presence of an enzyme glycerol kinase (see Scheme 2.18*a*).

(*ii*) NMR Identification

(*a*) Homotopic Ligands

Ligands *e.g.*, hydrogens are homotopic (often termed chemically equivalent) if these can be interchanged by a rotational axis of symmetry. These reside in identical environments and cannot be distinguished by NMR it being an achiral tool. Homotopic protons, therefore, show chemical shift equivalence and are termed *isochronous*.

(*b*) Enantiotopic Ligands

Enantiotopic nuclei are related by a plane of symmetry (there is no rotational axis of symmetry). Enantiotopic protons are also *isochronous* since their geometrical environments remain identical. The enantiotopic protons, however, become chemical shift non-equivalent when the molecule is placed in a chiral environment and then these can be distinguished. Such an environment is created by using an optically active solvent or by placing, the material in the active site of an enzyme. Under these conditions a diastereotopic relationship is established. For example, the protons of bromochloromethane are enantiotopic (chemically equivalent as there are often called) since these are on a plane of symmetry containing C, Br and Cl (I, Scheme 2.61). In a chiral environment (represented by placing a hand a chiral object) to one side, (II Scheme 2.61), the protons remain no longer enantiotopic, a diastereotopic relationship is established and the two protons become chemically non-equivalent (The plane of symmetry is lost in the presence of a chiral environment *e.g.*, a hand).

The two methyl groups of dimethylsulfoxide (DMSO, III, Scheme 2.61) are enantiotopic and are chemical shift equivalent (*isochronous*) under normal achiral conditions. In the presence of a chiral additive, $PhCHOHCF_3$ with which sulphoxides form some kind of associates, the two methyl groups are seen as a doublet.

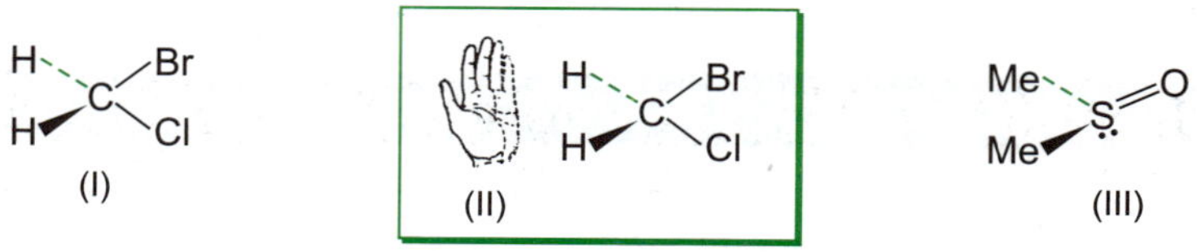

SCHEME 2.61

(c) Diastereotopic Ligands

Diastereotopic ligands are not symmetry equivalent *i.e.*, non-interchangeable by any symmetry operation. Diastereotopic protons *e.g.*, reside in different environments and always in principle display chemical shift differences. As an example, two similar groups, like two hydrogens or two methyl groups adjacent to a stereocenter become diastereotopic (chemically non-equivalent).

The ^{13}CNMR spectrum of S-(+)-3-methyl-2-butanol with two diastereotopic methyl groups (Scheme 2.62), as expected displays these as a closely spaced pair of peaks at 17.9 and 18.1 ppm (in all 5 peaks).

One may see that these two diastereotopic methyl groups are always non-equivalent even with free rotation. In any of the fixed staggered conformers (Scheme 2.62) one cannot detect a plane of symmetry. Similarly as expected the methylene protons adjacent to a stereocenter are diastereotopic (chemically non-equivalent). Thus, as expected the methyl ester of 2,3-dibromo-2-methylpropionic acid shows an AB quartet (Jab = 10 Hz) for the two diasteretopic protons H_a and H_b.

It may be mentioned that difference in chemical shifts of diastereotopic ligands is often small and may often remain undertectable (*accidental isochrony*).

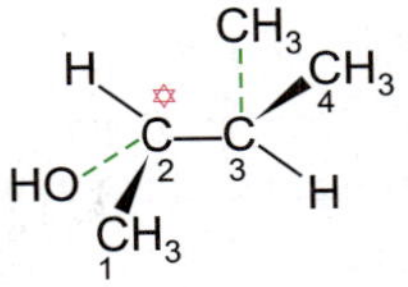

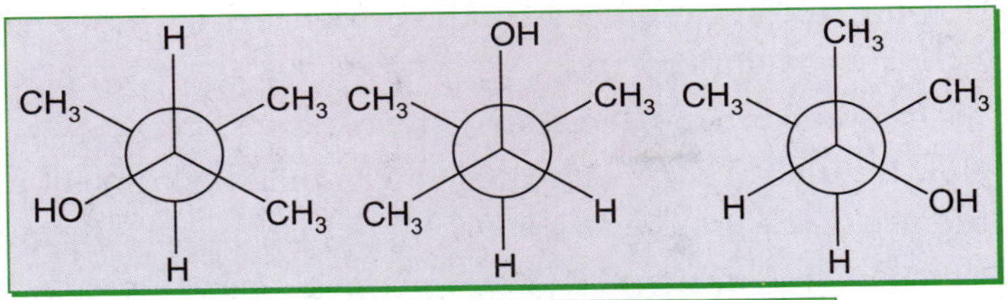

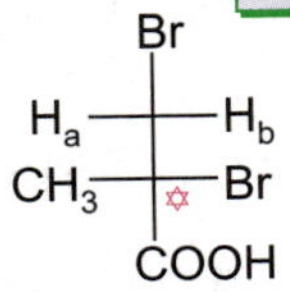

The methylene protons H_a and H_b in 2, 3-dibromo-2-methylpropionic acid being diastereotopic couple with one another and display themselves

SCHEME 2.62

PROBLEMS

2.1. Using symmetry considerations show if or not the two H's of CH_2Cl_2, CH_2ClF are homotopic or enantiotopic.

2.2. Predict the hydrogen atoms in *cis*-1, 2-dichlorocyclopropane and its *trans*–isomer as homotopic, enantiotopic or diastereotopic.

2.3. Label the groups/faces (where applicable) homotopic, enantiotopic or diastereotopic in the following compounds.

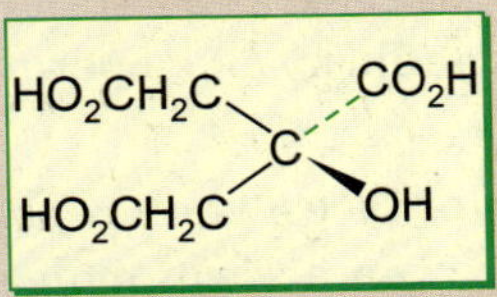

1

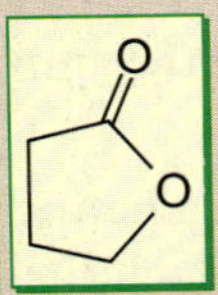

2

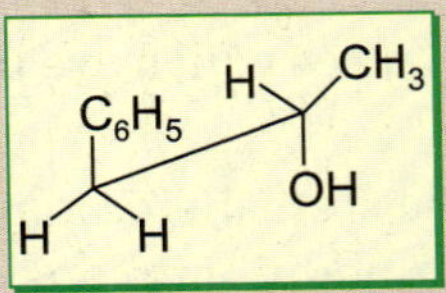

3

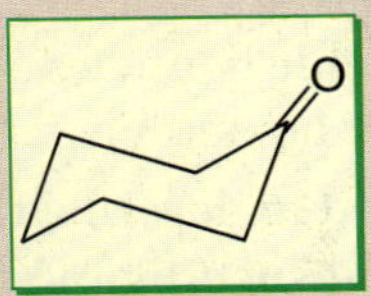

4

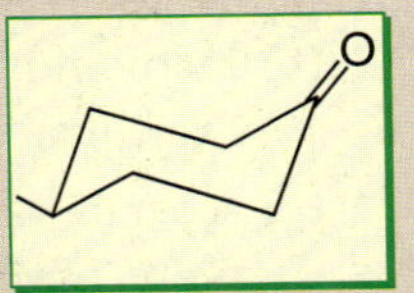

5

2.4. Label the methylene hydrogens of ethanol as *pro-R* and *pro-S* by assigning priorities to the ligands and depicting the path 1 → 2 → 3 as clockwise or anticlockwise.

2.5. Depict the two faces of acetophenone as *Re* or *Si*.

2.6. Show the hydride attack from lithium aluminium hydride from the *Re* face of 2-butanone.

2.7. The enzyme fumarase catalyzed hydration of fumaric acid gives (*S*)-malic acid. Depict the stereochemistry of the reaction.

2.8. Fill in the blanks.

(*i*) The difference between the Cram's rule and Felkin-Ahn model is the ...... of the carbonyl group.

(*ii*) When $R^1 \neq R^2$ the two faces of the carbonyl compound R^1—CO—R^2 are known as ......

(*iii*) Groups or faces which are enantiotopic or diastereotopic are collectively known as ......

(*iv*) Reaction on an allylic alcohol with *tert*-butylhydroperoxide catalyzed by titanium tetraisopropoxide in the presence of (+)- or (–)–diethyltartrate is called ...... epoxidation.

(*v*) *R*-alpine broane is made by the reaction of (+) α-piene and ....

2.9. How the role of chiral enolates determine alkylation of propanoic acid? Discuss the use of (*S*)-proline as a source for chiral enolate formation.

2.10. Using hydroboration–oxidation on the alkene (I), predict the faces of the *pi* bond as enantio- or diastereotopic.

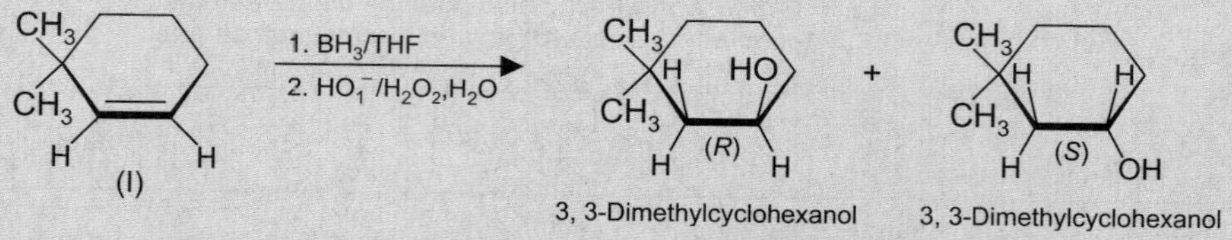

2.11. Label the statements true or false.

(*i*) If the rotation of a molecule about an axis by 360°/*n* followed by reflection in a plane at right angle to that axis gives a superimposable entity, the molecule is said to have a symmetry axis of order *n*.

(*ii*) Faces of a double bond which are not symmetry related are called diastereotopic faces.

(*iii*) Homomorphic ligands are both structurally and configurationally identical when detached.

(*iv*) Homomorphic ligands and faces which are interchangeable by a C_n operation are termed homotopic.

(*v*) A term used to describe the stereoheterotopic ligands or faces when replacement of one such ligand or addition to one such face in an achiral molecule to give chiral products is prochirality and not prostereoisomerism.

2.12. Indicate if the hydrogens marked H_a and H_b are homotopic, enantiotopic or diastereotopic, in the following compounds. Are the methylene protons in (V) enantiotopic and related by σ plane?

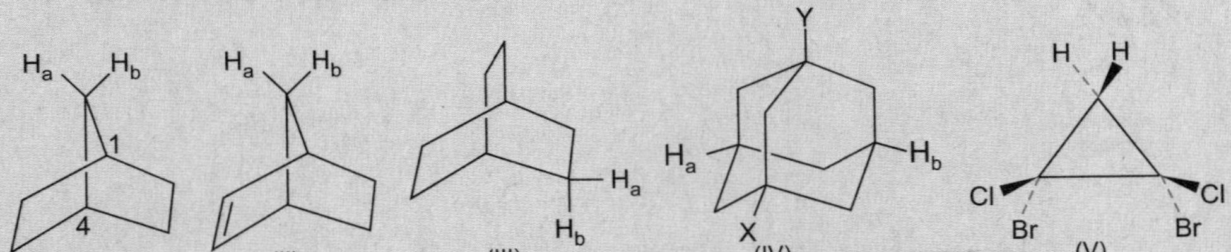

(I) (II) (III) (IV) (V)

2.13. (*a*) Why C-3 in compound (1) is a prochiral center, name the appropriate ligands on C-3 as *pro-R* and *pro-S*.

(*b*) Designate H_a and H_b in the allene (II) and in cyclohexane derivative (III) as *pro-R* and *pro-S*.

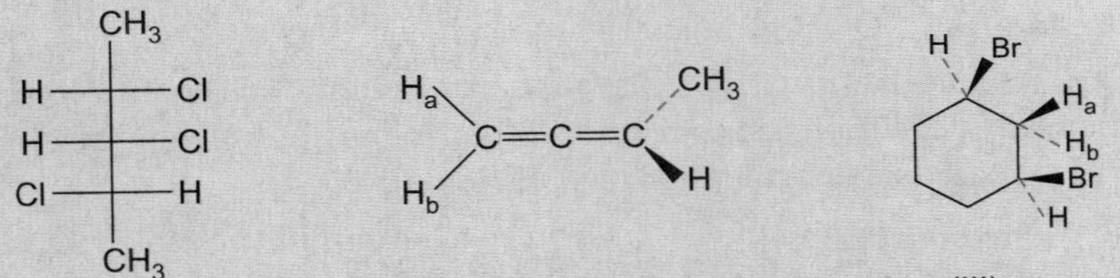

(I) (II) (III)

2.14. (*a*) How are the compounds (I and II) related to each other?

(*b*) Draw the expected products of Sharpless epoxidation of the following allylic alcohol (III) using (+)-diethyl tartrate as the chiral catalyst.

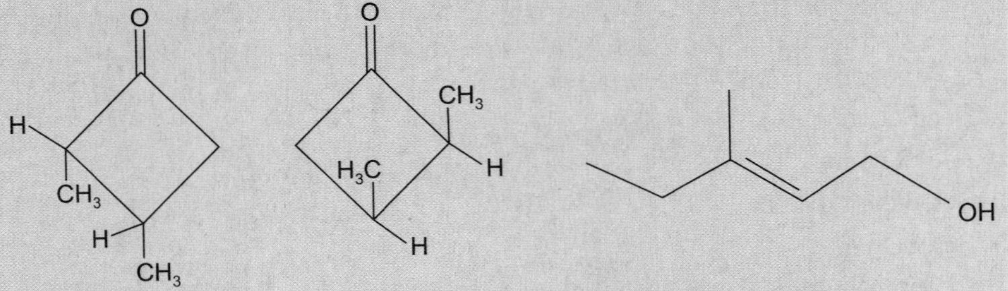

(I) (II) (III)

ANSWERS TO SELECTED PROBLEMS

2.1. 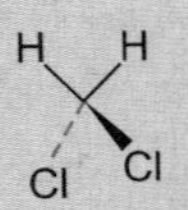 Homotopic protons by C_2 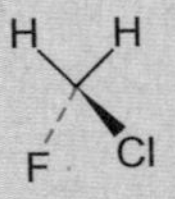 Enantiotopic protons by σ

2.2. The methylene protons in the *cis*-isomer are diastereotopic. These being close to a chirality center. The other pair of protons is enantiotopic being interchangeable by a plane of symmetry. In the *trans*-isomer methylene protons though close to a stereocenter are, however, homotopic (exception), these being interchangeable by a C_2 axis. The other pair is also homotopic for the same reason (also see, Scheme 2.10*a*).

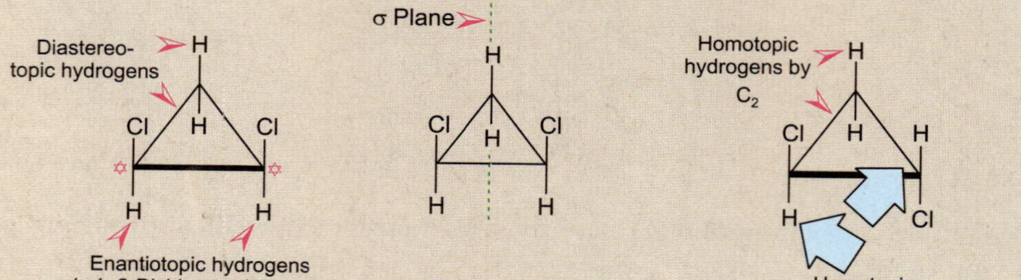

cis-1, 2-Dichlorocyclopropane

trans-1, 2-Dichlorocyclopropane

2.3. 1. The two-CH_2COOH groups are enantiotopic.

2. The two faces of the carbonyl group in the lactone are enantiotopic, as can be shown by reduction (compare with 2-butanone *i.e.*, an unsymmetrical ketone whose faces are enantiotopic (see, Scheme 1.85).

3. In 1-phenylisopropanol, C-2, is a stereocenter and the molecule is chiral C-1 is prochiral since replacement of the stereoheterotopic hydrogens by a group different from other ligands generates diastereoisomeric molecules, because of two centers of chirality. The two C-1 hydrogen atoms, therefore, are diastereotopic.

4. Cyclohexanone has homotopic faces.

5. Monosubstituted derivatives of cyclohexanone have two diastereotopic faces since addition to one face gives a *cis* product while to the other face gives the diastereometric *trans* product.

2.4. See Scheme 2.14.

2.5. See Scheme 2.15.

2.6. See Scheme 2.16.

2.7. See Scheme 2.19.

2.8. (*i*) Conformation; (*ii*) enantiotopic; (*iii*) stereoheterotopic, (*iv*) sharpless (*v*) 9-BBN.

2.9. See Scheme 2.34.

2.10. The hydroboration–oxidation gives enantiomeric product, the faces are therefore enenatiotopic.

2.11. (*i*) False; (*ii*) true; (*iii*) true; (*iv*) true; (*v*) false, (hint—prostereoisomerism is more general term since in some cases replacement of one or other stereoheterotopic ligands or addition to one or other heterotopic faces may give achiral diastereomers which do contain stereogenic but not chiral elements).

2.12. (I) Homotopic (interchangeable by rotational axis of symmetry); (II) diastereotopic (replacement by a test group would given *syn* and *anti* compounds which are diastereomers, the protons are not interchangeable by a simple axis of symmetry); (III) enantiotopic (gainfully if one makes a model and rotates it to put H^a and H^b nearest to the eyes these are on a mirror plane of symmetry); (IV) enantiotopic (the adamantane is an expanded tetrahedron and an adamantane with four different bridgehead substituents is chiral, see Schemes 1.5 and 1.110). No, the methylene protons in (V) are instead on the σ plane and diastereotopic (related problem Scheme 2.13*a*).

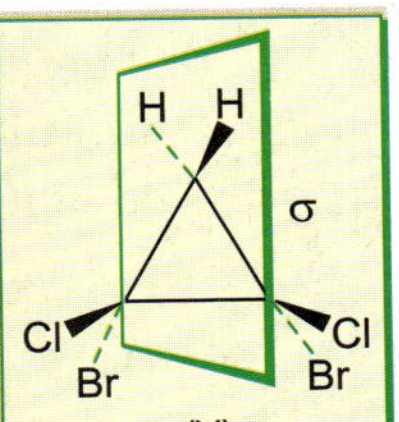

(V)

2.13. (*i*) The two ligands $CHClCH_3$ on C3 apart from H and Cl have the same constitution and configuration and are thus homomorphic (see, Scheme 2.2*a*) and enantiotopic.

(*ii*) The allene has a prochiral axis. To assign *pro-R* and *pro-S* descriptors to H_a and H_b, one of these *e.g.*, H_a is imagined to have elevated priority over H_b and then *R*/*S* configuration is determined (Kalsi 2004, Scheme 1.33*a*) to reach *pro-R* and *pro-S*.

(*iii*) *R* configuration has priority over *S* configuration.

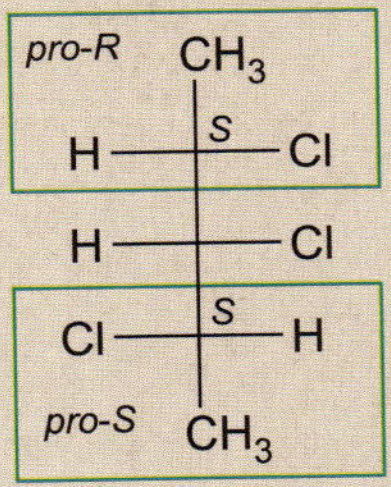

(I)

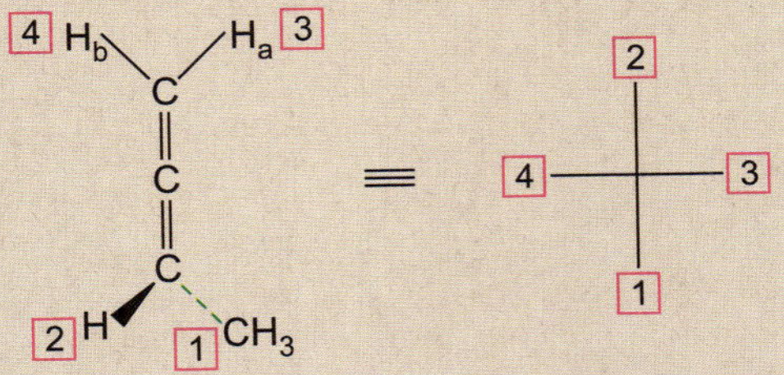

Rotated the molecule so that the stereochemical bonds are at the bottom (see scheme 1.133a Kalsi 2004)

Not a Fischer projection, just a cross, path 1→2→3 to assign configuration to the hypothetical chiral axis in (II) with elevated priority to H_a is *R* thus H_a is *pro-R*, and therefore, H_b is *pro-S*

(II)

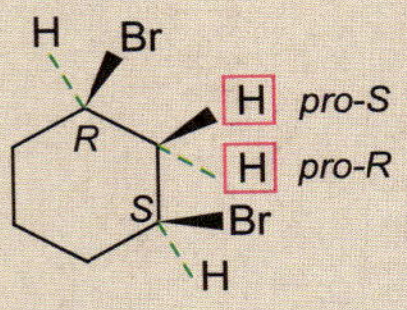

(III)
(*R*) has priority over (*S*)

2.14. (*a*) Both (I and II) represent the same compound (2*S*, 3*S*)-2,3-dimethylcyclobutanone, but display properly the two different faces of the trigonal center 1, 2 and 3 show the priority ratings in the Cahn-Ingold-Prelog.

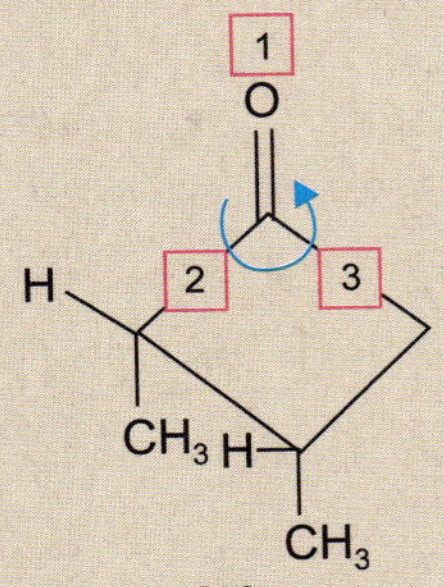

Si face
(I)

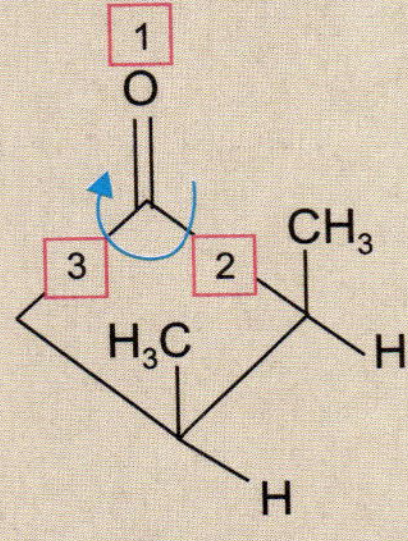

Re face
(II)

(b) The carbon skeleton of the allylic alcohol is first reoriented to match the orientation in (Scheme 2.48), the epoxidation is completed, and the carbon skeleton is reoriented to match the original drawing.

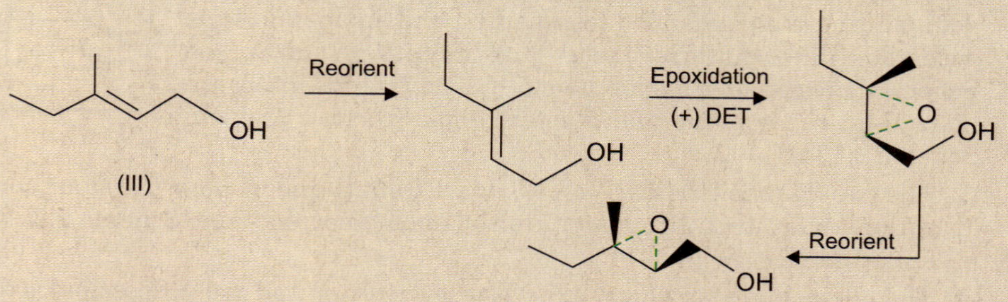

Aliphatic Nucleophilic Substitution

3

CHAPTER

3.1 INTRODUCTION

Nucleophilic substitution is one of the most versatile reactions in organic chemistry. For reaction to occur, the leaving group departs carbon with the pair of bonding electrons. The nucleophile, Nu:⁻, must contain at least one unshared pair of electrons and must be either neutral or negatively charged. One of the most commonly used families of leaving groups is the halides. Iodide is both a good leaving group and a good nucleophile. The structure of the alkyl group, the nature of the leaving group and the nucleophile are three major variables that control the mechanism of substitution and the formation of side products. This heterolytic reaction, involves two electron shifts and proceeds mostly by one of the two mechanisms. S_N1 (substitution nucleophilic unimolecular) and S_N2 (substitution nucleophilic bimolecular) which are distinguishable by both kinetic and stereochemical criteria.

EXERCISE 3.1

Why the rate of reaction (I) increases in the presence of very small amount of added NaI as in (II)?

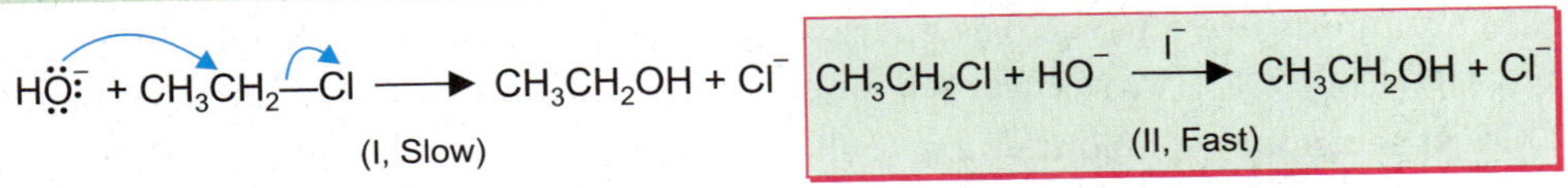

SCHEME 3.1

ANSWER. *The procedure takes advantage of iodide being an excellent nucleophile and an excellent leaving group. Iodide ion acts as a covalent catalyst (Scheme 3.1a). The first S_N1 reaction in the catalyzed reaction (III) is faster than the uncatalyzed reaction (I) since iodide ion is a better nucleophile than hydroxide ion (nucleophile in the uncatalyzed reaction). The second S_N2 reaction in the catalyzed reaction is again faster than the uncatalyzed reaction since iodide ion is a weaker base making it, a better leaving group than chloride ion (the leaving group in the uncatalyzed reaction). The iodide ion therefore, increases the rate of formation of ethanol by changing a relatively slow one-step reaction into a reaction with two relatively faster steps.*

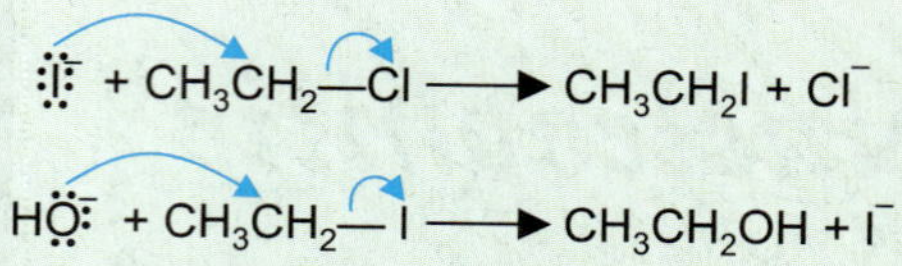

(III)

Mechanism for the iodide-ion-catalyzed reaction

SCHEME 3.1a

EXERCISE 3.2

On treatment of (S)-2-iodooctane (Scheme 3.2) with NaI in solution its optical activity is lost. Explain.

ANSWER. *I^- is both a good nucleophile and a good leaving group. Each displacement is thus attended with inversion at the chirality center. Since the process is fast it occurs multiple times at every substrate molecule which ultimately leads to racemization.*

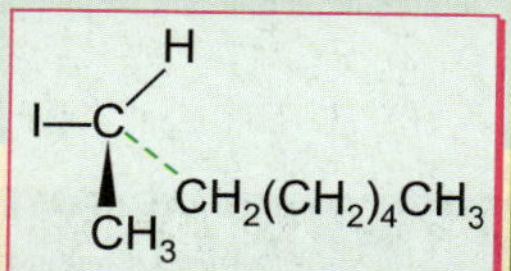

(*S*)-2-Iodooctane

SCHEME 3.2

For the information on stereospecificity and stereoselectivity of nucleophilic substitution reactions refer to (Schemes 1.99 –1.103a)

3.2 S_N2 REACTION MECHANISM AND EVIDENCE

(a) A Mechanism for the S_N2 Reaction – A Stereospecific Reaction

S_N2 reaction at a saturated carbon is a (concerted) one step process without an intermediate (Scheme 3.3), that always proceeds with inversion of configuration like the inversion of an umbrella inside out when caught in a wind.

The chemical kinetics of the reaction has been determined. The reaction displays second order kinetics. The value of the rate constant k_r depends on several factors including temperature and the free energy of the transition state (see Scheme 3.7). The S_N2 reaction is almost always observed for primary alkyl halides and related compounds.

Most secondary compounds *e.g.* (Scheme 3.3) undergo S_N2 reaction with inversion of configuration in which the substituents are alkyl groups. Thus S_N2 reaction is stereospecific and due to inversion of stereochemistry in this reaction, one can design ways to synthesise a desired stereoisomer. When iodide ion is used to displace the bromine in enantiomerically pure reactant (*S*)–(+)-2-bromobutane, it is converted to an enantiomerically pure product (Scheme 3.3). The negative iodide (nucleophile) pushes a pair of electrons into the partially positive carbon atom from the back side. The bromine begins to move away with the pair of electrons that had bonded it to the carbon. The transition state shows the formation and breaking of the bonds. The geometry of the transition state appears to be that in which the incoming and leaving groups are both weakly bonded to carbon in a linear fashion. The three remaining bonds of pentacoordinate transition state lie in a plane perpendicular to the two weak bonds.

Eventually the bond between the carbon and the leaving group is completely broken and the attacked carbon again becomes tetrahedral with inverted configuration.

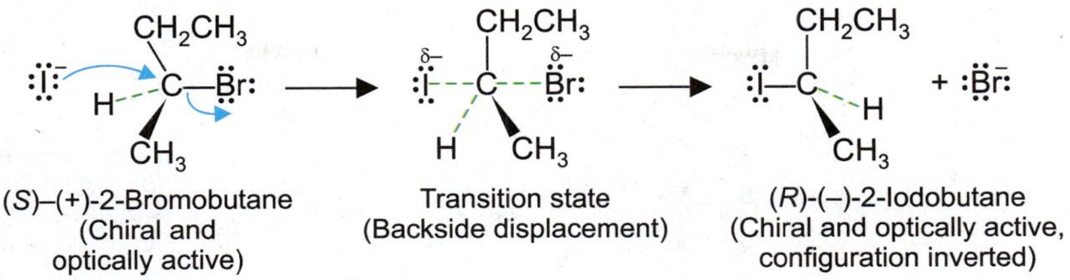

SCHEME 3.3

The potential energy profile for the S_N2 reaction (Scheme 3.4) shows that there is only one transition state and no intermediates between the reactants and products. The reactants are slightly higher in energy than the products, since the reaction is known to be exothermic. The transition state is much higher in energy since it involves a five–coordinate carbon atom with two partial bonds. The top of the energy hill corresponds to the transition state of a typical S_N2 reaction. The difference in free energy between the reactants and the transition state is the free energy of activation. The difference in free energy between the reactants and products is the free energy change for a reaction. A reaction which has a low free energy of activation will occur much faster than a reaction with higher one.

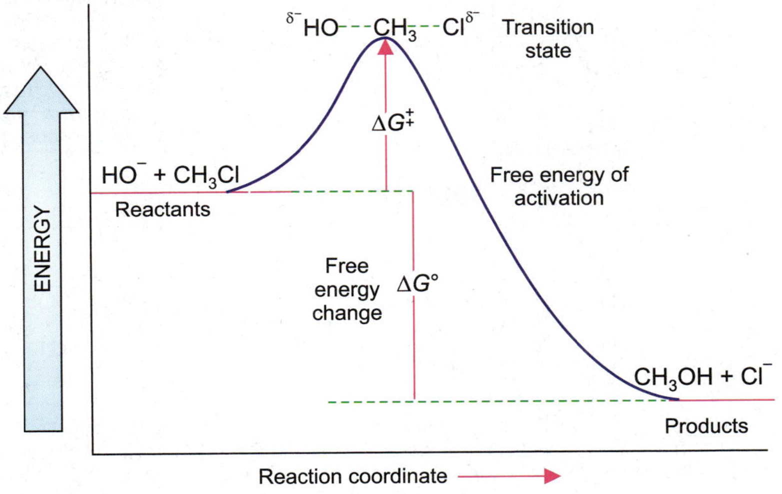

SCHEME 3.4

In the orbital picture of the transition state the carbon undergoing substitution is sp^2-hybridized and has a p orbital perpendicular to the plane of the bonds which partly overlaps with an orbital of the leaving group as well as with that of the incoming nucleophile (Scheme 3.5). As the nucleophile approaches the back side and attacks the small back lobe of the sp^3 hybrid orbital used by carbon to bind the halogen atom, the rest of the molecule becomes planar at the transition state by changing the hybridization at carbon to sp^2. The negative charge is no longer located entirely on the nucleophile, it is also partly located on the leaving group. When the reaction proceeds to products, the inversion motion is completed, the carbon returns to the tetrahedral sp^3 configuration, and the leaving group becomes a fully charged anion.

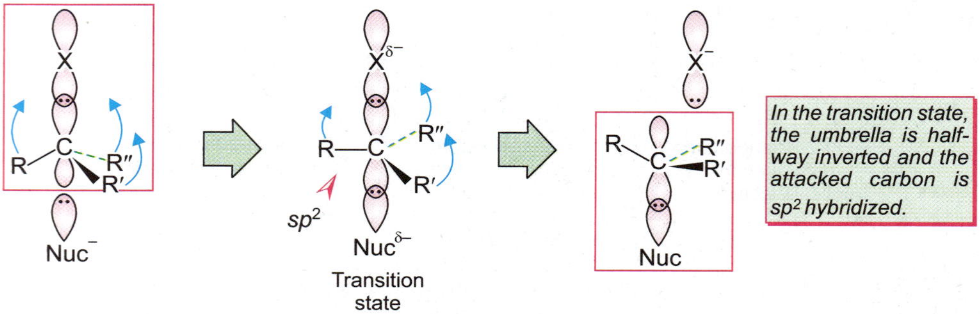

SCHEME 3.5

(b) The Stereochemistry of the S_N2 Reaction

In this one step reaction one could imagine that nucleophile could approach the substrate from the same side (front side) of the C—X bond leading to retention of configuration of the attacked carbon (Scheme 3.5*a*) or it could attack the carbon of substrate as away as possible from the C—X bond leading to inversion of configuration at the attacked carbon.

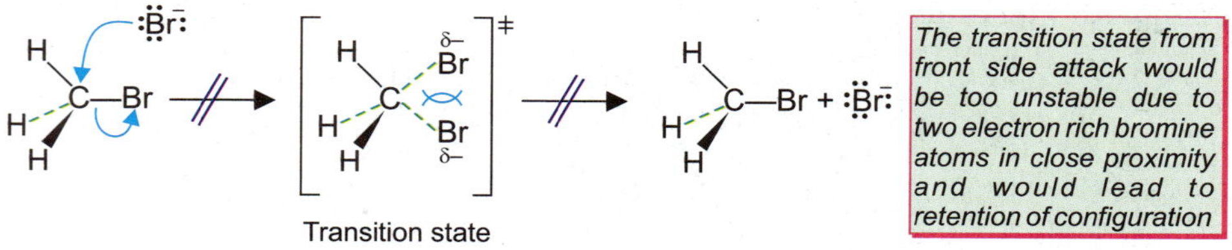

SCHEME 3.5a

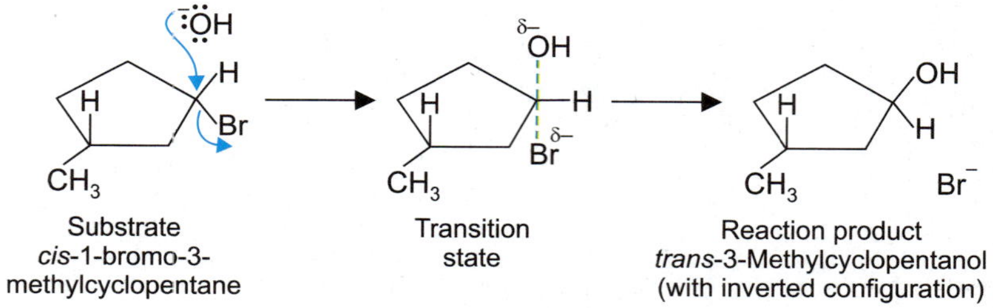

SCHEME 3.6

When the S_N2 reactions occur at a chirality center, there is always an inversion of configuration at the chirality center. This has been confirmed with acyclic compounds as well (see, Scheme 3.3). With cyclic compounds the nucleophile ends up being bonded to the opposite side of the ring from which the leaving group departs (a configurational inversion, Scheme 3.6). This changes the stereochemical relation between the substituents.

(c) The Structure of the Substrate–Requirement of Collinear Transition State

The rate of direct displacement *i.e.,* an S_N2 reaction is very sensitive to the steric bulk of the substituents present on the carbon undergoing such a reaction. This is expected since the degree of coordination (pentacoordinated transition state) increases at the reacting carbon atom. Thus from the steric point of view, the optimum substrate could be CH_3—X. Each replacement of hydrogen by a more bulky alkyl group (at α-carbon) should decrease the rate of reaction. Consequently, the order of reactivity of alkyl groups is expected to be methyl > primary > secondary > tertiary and this is observed. Table 3.1 gives the relative rates of typical S_N2 reactions. Methyl halides react most rapidly and tertiary halides react so slowly as to be unreactive by the S_N2 mechanism.

Neopentyl halides, are primary halides and even then these are unreactive in S_N2 reactions. This situation shows that steric hindrance effects are operative even if the β-carbon is substituted by alkyl groups. An activated complex at the transition state comprises the elements of a molecule of the alkyl halide and a hydroxide ion (eqn. I, Scheme 3.7).

If the methyl group of ethyl bromide is replaced by a *tert*-butyl group to give neopentyl bromide [2, 2-dimethylbromopropane], *i.e.*, the *primary* nature of the alkyl halide is kept unchanged, the rate constant of the new reaction under the same experimental conditions (eqn. II, Scheme 3.7) is very much smaller. In neopentyl halides, the *tert*-butyl group blocks the best pathway for rear side displacement of the leaving group (Schemes 3.7 and 3.8).

Table 3.1: Steric hindrance to the S_N2 reaction

Class of halide	Compound	Relative rate
Methyl	CH_3-Br (More reactive)	>1000
Primary (1°)	CH_3CH_2-Br	50
Secondary (2°)	$(CH_3)_2CH$-Br	1
Tertiary (3°)	$(CH_3)_3C$-Br(Less reactive)	<0.001
n-Butyl (1°)	$CH_3CH_2CH_2CH_2$-Br	20
Isobutyl (1°)	$(CH_3)_2CHCH_2$-Br	2
Neopentyl (1°)	$(CH_3)_3CCH_2$-Br	0.005

When the hydrogens of CH_3Br are successively replaced by -CH_3 groups, the rate of reaction with a given nucleophile becomes progressively slower. Steric hindrance is also pronounced even when the β-carbon is increasingly substituted by alkyl groups.

$$\underset{\text{Ethyl bromide}}{CH_3CH_2Br} + Na^+ OH^- \xrightarrow[T]{H_2O} CH_3CH_2OH + Na^+ Br^- \quad \text{...(I)}$$

Rate of reaction at temperature T = $k[CH_3CH_2Br]\,[OH^-]$

$$\underset{\text{Neopentyl bromide}}{(CH_3)_3CCH_2Br} + OH^- \xrightarrow[T]{H_2O} (CH_3)_3CCH_2OH + Br^- \quad \text{...(II)}$$

Rate of reaction at temperature T = $k[(CH_3)_3CCH_2Br]\,[OH^-]$

$$k_{(eqn\ I)} << k_{(eqn\ II)}$$

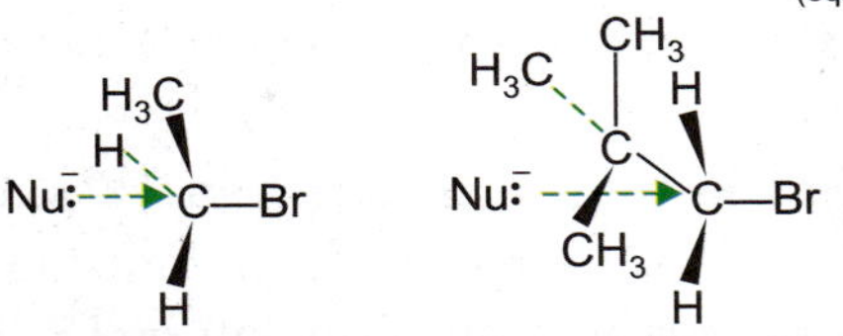

Ethyl bromide Neopentyl bromide

The S_N2 reaction is thwarated in neopentyl bromide due to difficulty of displacement since tert-butyl group shield the rear of C-Br bond.

SCHEME 3.7

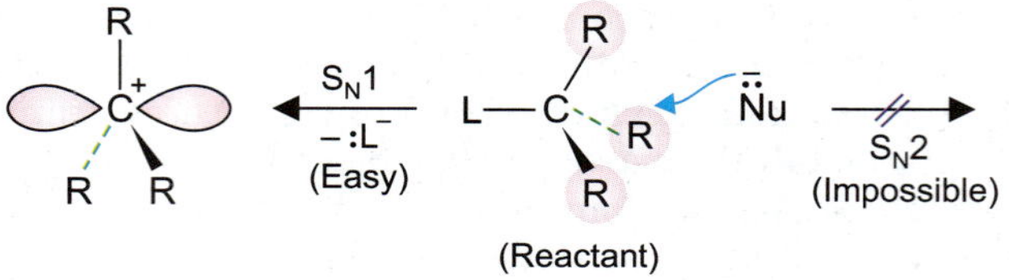

SCHEME 3.8

These examples show the importance of the geometry of the transition state for S_N2 reaction during the access of the nucleophile to the rear of the C—L bond. The nucleophile, the attacked carbon and the leaving group are collinear. In keeping with these requirements of the transition state, in several cyclic systems as well, the S_N2 reaction is either rendered effectively slow or completely stops. This happens (like in acyclic systems) if severe non-bonded interactions are met by the nucleophile as it attempts to gain access to the carbon which is bonded to the leaving group while maintaining the essential collinearity as in the case of neopentyl halides (Scheme 3.7). In the case of bicyclo [4.4.0] decane (I, Scheme 3.9) the axial angular methyl group hinders the approach of the nucleophilic reagent to the backside of the reaction centre. In the 1-substituted caged compounds (II, III and IV, Scheme 3.9), it is almost impossible for the nucleophile to approach the backside of C-1 from the "inside" of the molecule. Even if nucleophile could manage to attack from the rear it is not possible to invert the configuration of C-1.

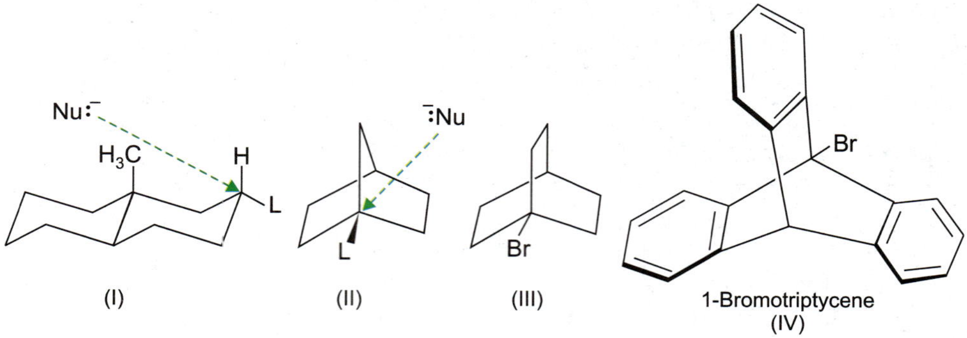

SCHEME 3.9

In an acyclic (noncyclic) haloalkane the hybridization around the halogen-bearing carbon is sp^3 with bond angles around 109°. In the transition state for an S_N2 reaction this carbon get rehybridized to sp^2 and the bond angles involving three unreacting groups around this carbon can easily expand to 120°.

Strong evidence has been provided that the transition state in an S_N2 reaction must be ideally linear. On base treatment of (I, Scheme 3.10 removal of a proton), the negatively charged carbon of the ion (II) does not attack the methyl group in an internal S_N2 process to produce (III, Scheme 3.10). The kinetic studies (reaction is bimolecular) have shown that

(III) is formed by instead an intermolecular process. The intramolecular (an internal S_N2 reaction) process though with a favourable entropy, does not occur since a linear transition state cannot be attained between the nucleophile, substrate carbon and leaving group.

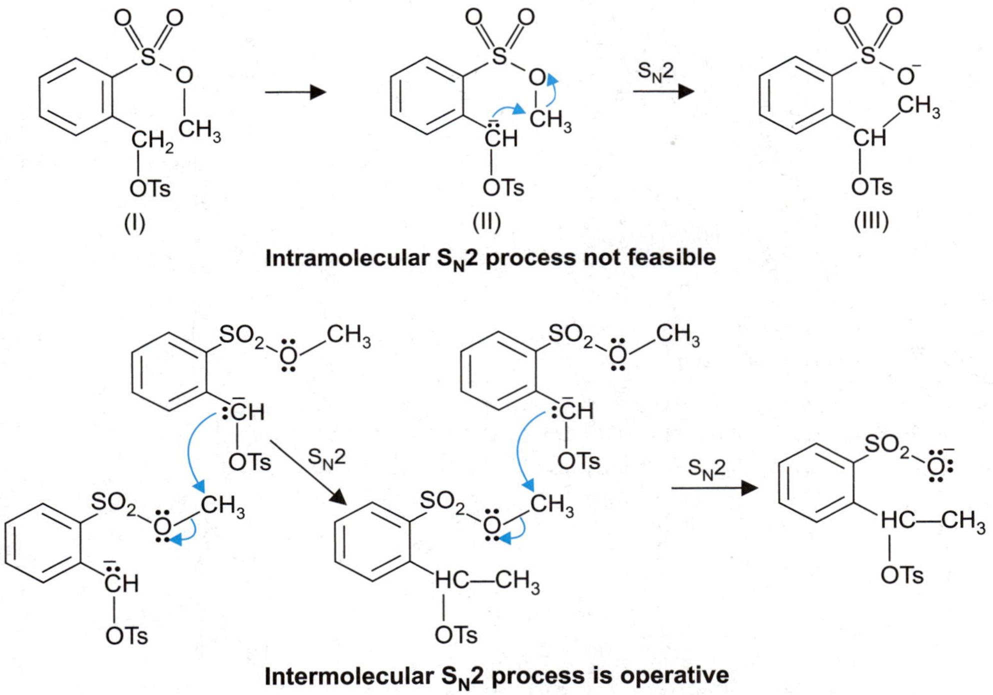

SCHEME 3.10

EXERCISE 3.3

Predict the products in the S_N2 reactions (Scheme 3.10a).

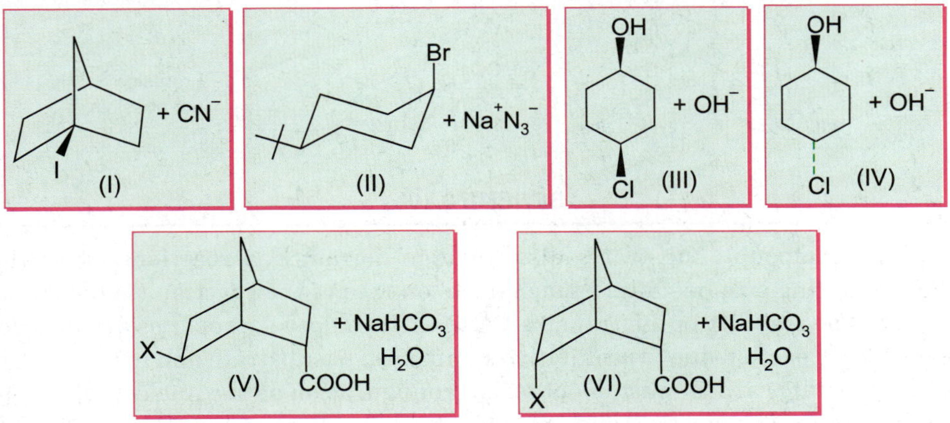

SCHEME 3.10a

ANSWER. *Substrate (I) is tertiary, it cannot undergo S_N2 reaction, the backside of the C–I bond is completely blocked (Scheme 3.10b).*

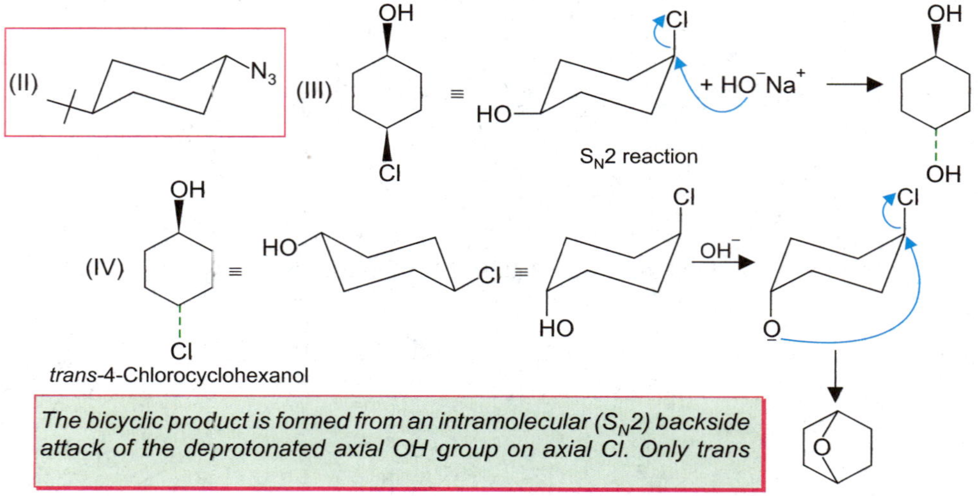

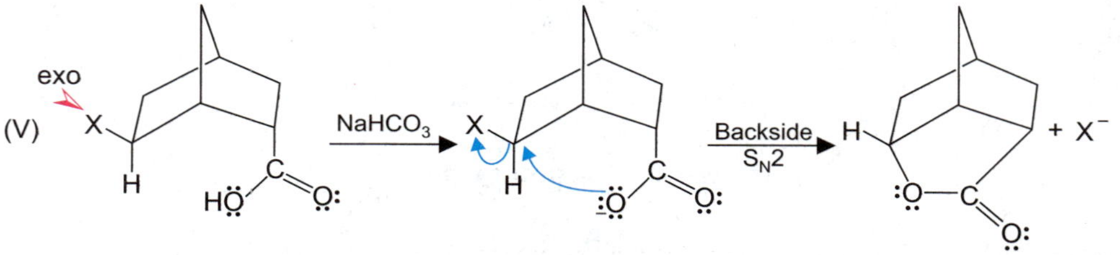

The carboxylate salt is in the correct position for backside S_N2 displacement of X to give a lactone.

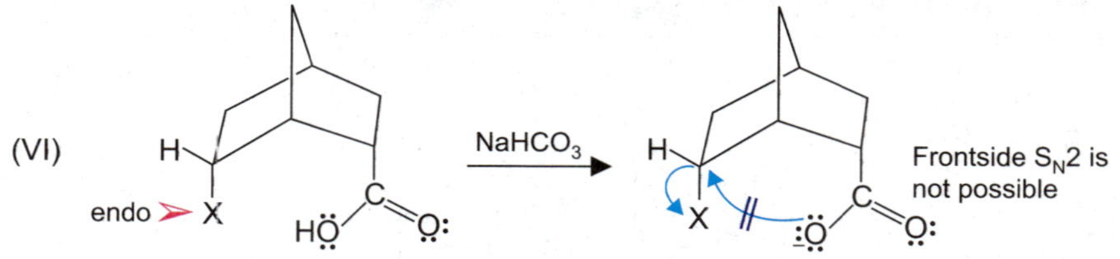

SCHEME 3.10b

Ring compound substrates also undergo normal S_N2 reactions (see Scheme 3.6), however, ring size has interesting effects on rate of the reaction (Table 3.2). The answer is in the stability of the transition state. The halocyclopropanes are unreactive to undergo S_N2 reaction due to prohibitive strain both in substrate and in the transition state. The reacting carbon must adopt an sp^2 configuration as the nucleophile replaces the leaving group. The normal bond angle for this hybridization would be 120°, and the cyclopropane ring cannot be distorted much from 60° (Scheme 3.11).

Table 3.2: Relative Reactivities of Cycloalkylbromides in S_N2 reaction with a Nucleophile (Nu:⁻), Iodide Ion

Compound	Relative rate
Cyclopropyl bromide	extremely slow
Cyclobutyl bromide	8×10^{-3}
Cyclopentyl bromide	1.6
Cyclohexyl bromide	1×10^{-2}
Isopropyl bromide (acyclic compound)	1.0

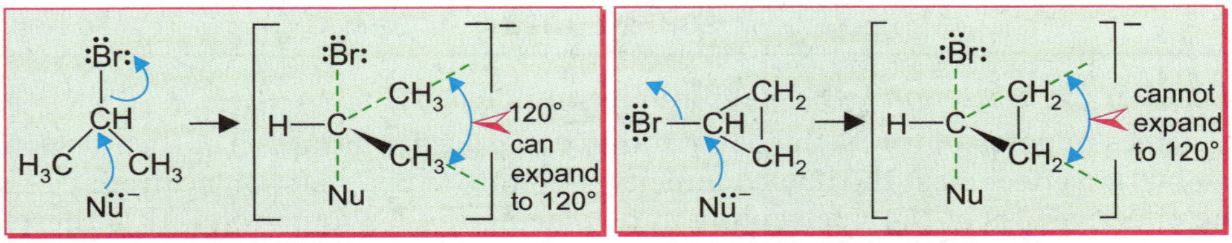

SCHEME 3.11

Halocyclohexanes although seemingly more capable of attaining sp^2 hybridization at the reacting carbon are slower in S_N2 reactions. When a nucleophile approaches an equatorial halide (I, Scheme 3.12) it faces steric hindrance by the two axial hydrogens at C–3 and C–5. The leaving axial group on the other hand faces steric hinderance to its exit (II, Scheme 3.12).

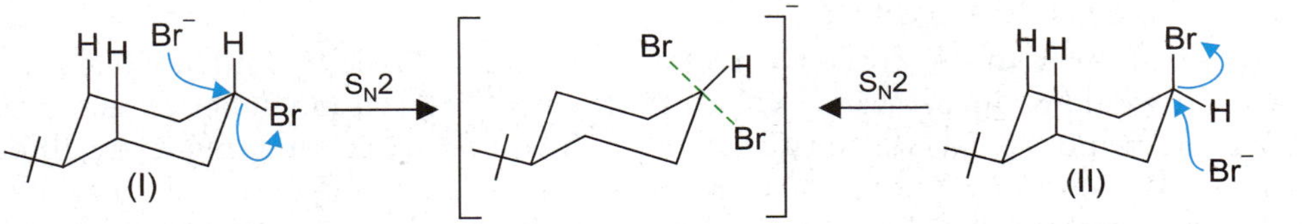

Approach by I⁻ hindered by axial hydrogens

(Same transition state)

Departure of I⁻ is now hindered by axial hydrogens

SCHEME 3.12

EXERCISE 3.4

Compared to equatorial cyclohexyl bromide, axial cyclohexyl bromide reacts faster in the S_N2 displacement although the transition state involved is the same (see, Scheme 3.12) in both cases. Explain.

ANSWER. *There is no energy difference between the transition states, these being of same energy. The axial iodo conformer (Br), however, increases its energy leading to lowering of the activation energy (Scheme 3.12a) for displacement in it compared to the equatorial cyclohexyl bromide.*

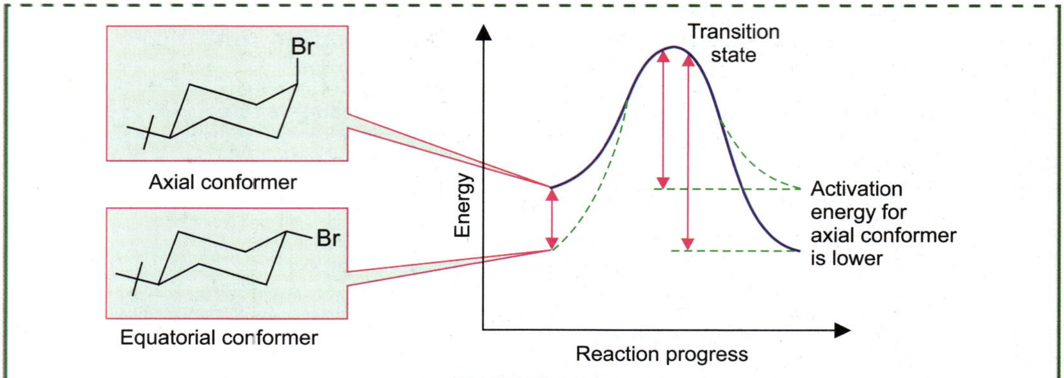

SCHEME 3.12a

(d) Effect of the Nucleophile/Nucleophilicity and Basicity

Two factors in particular *i.e.,* the polarizability (hard soft character) of the nucleophile and solvation effect most the nucleophilicity. The rate of S_N2 reaction is directly related to the effectiveness of the nucleophile in displacing the leaving group. It is therefore, seen that of a pair of nucleophiles containing the same reactive atom *e.g.,* methanol and methoxide ion, the species with a negative charge available for bonding is the more powerful nucleophile. Put in other words, of a base and its conjugate acid, the base is always more nucleophilic. Therefore, OH^- is a stronger nucleophile than H_2O, SH^- a stronger nucleophile than $(CH_3)_2$ SH and NH_2^- is stronger nucleophile than NH_3. This finding is reasonable since a stronger bond between the nucleophilic atom and carbon would mean a more stable transition state and thus a reduced activation energy. Because the S_N2 reaction is concerted the strength of the partially formed new bond will be reflected in the energy of the transition state, the more negative the attacking species, the faster the reaction should be.

There seems to be a good correlation between basicity and nucleophilicity. However, it would be incorrect to explain that methoxide is a much better nucleophile since it is much more basic. Basicity and nucleophilicity are two fundamentally different properties. Basicity (a measure of a thermodynamic phenomenon) is defined by the equilibrium constant for abstracting a proton (Scheme 3.13). On the other hand, nucleophilicity (a measure of a kinetic phenomenon) is defined by the rate of attack on an electrophilic carbon. The hydroxide ion (OH^-) is a stronger base than a cyanide ion (CN^-); at equilibrium it has a greater affinity for a proton (the *pKa* of H_2O is ~ 16, whereas the *pKa* of HCN is ~ 10).

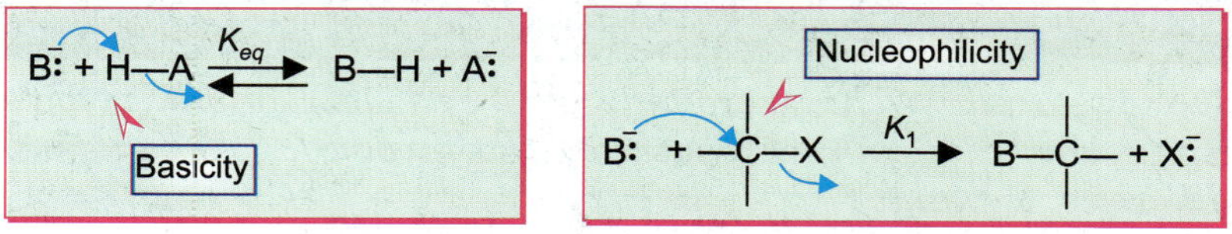

SCHEME 3.13

However, cyanide ion is stronger nucleophile, *i.e.,* it reacts more rapidly with a carbon having a leaving group than a hydroxide ion. In both cases (Scheme 3.13), however, a new bond is formed. When the new bond is to a proton, the species has reacted as a base; in case the bond is to a carbon it has reacted as a nucleophile. Moreover, since nucleophilicity is used to describe trends in the kinetic aspects of reactions, the relative nucleophilicity of a given species may differ from substrate to substrate.

A sterically bulky nucleophile is less reactive than a smaller one. As S_N2 reaction is not only sensitive to the bulky groups on the electrophile, but to the bulky groups on the nucleophile as well. This is due to non-bonded repulsions which would develop in the transition state. Thus although *t*-butoxide ion is a stronger base than ethoxide ion, but the bulky *t*-butoxide is a weaker nucleophile (Scheme 3.14). Basicity is little effected by steric hindrance, since the attack is on an unhindered proton.

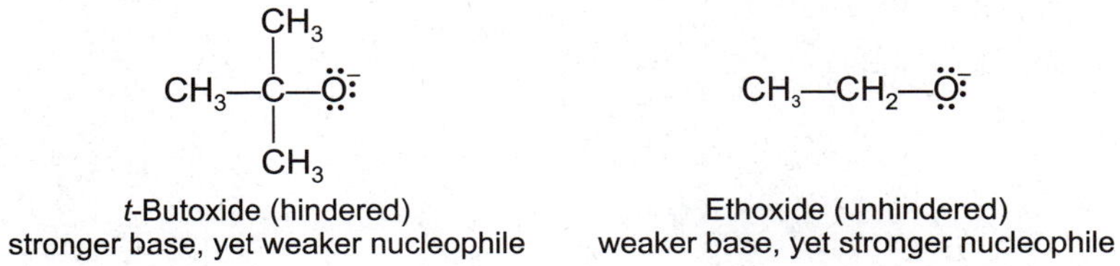

SCHEME 3.14

It is found that quinuclidine which is a tied-back cage compound reacts faster with ethyl iodide compared to triethylamine (Scheme 3.15). In these compounds the nucleophilic nitrogen atom is flanked by three two-carbon chains. In the cage compound, they are tied back and are not free to rotate. In triethylamine, they are free to rotate and thus effectively increase the bulk of the nucleophile. Thus the tied back cage compound quinuclidine reacts much faster with ethyl iodide than does triethylamine.

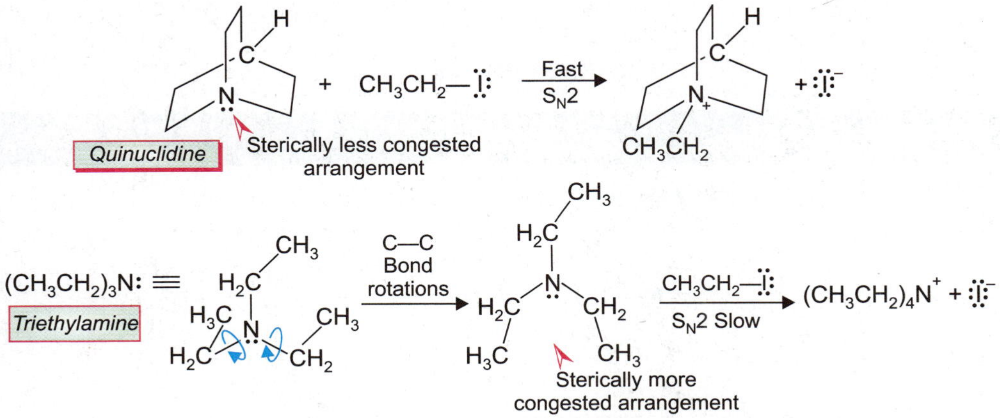

SCHEME 3.15

Nucleophilicity decreases on going from left to right in the periodic table. This follows the increase in electronegativity from left to right. Thus high electronegativity is unfavourable, due to tightly held electrons which are therefore, relatively less available for donation to the substrate for S_N2 process. Thus OH^- is more nucleophilic than F^-; NH_3 is more nucleophilic than H_2O.

CHAPTER 3

BOX 3.1
STERIC EFFECTS AND S_N2 AND E2 REACTIONS

Both S_N2 and E2 reactions are sensitive to steric effects, however, steric requirements for (E2, deprotonation) are far less severe compared to S_N2 reaction (Scheme B 3.1).

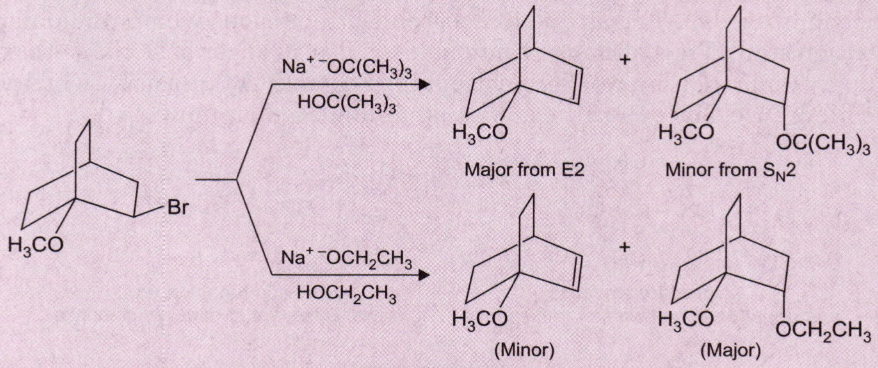

SCHEME B 3.1

When the aim is to generate an anion via an acid-base reaction, or to induce an E2 elimination reaction one may think to use of a sterically-hindered, non-nucleophilic base, such as diisopropylethyl amine (DIEA), 1, 8-diazabicyclo [5.4.0] undecen-7-ene (DBU), or 1, 5-diazabicyclo [4.3.0] non-5-ene (DBN Scheme B 3.2).

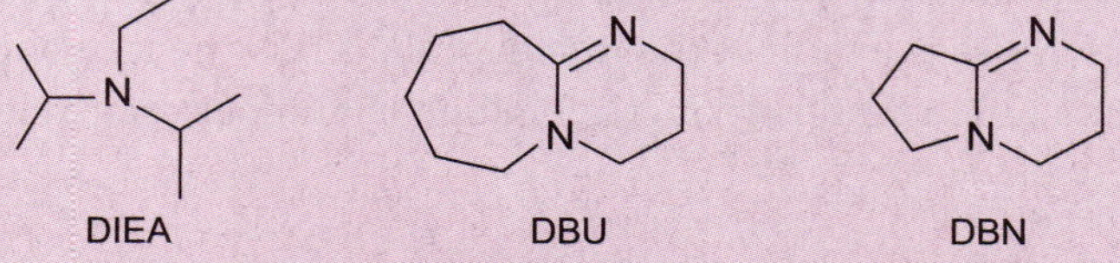

SCHEME B 3.2

EXERCISE 3.5

Explain the outcome of the following S_N2 reactions:

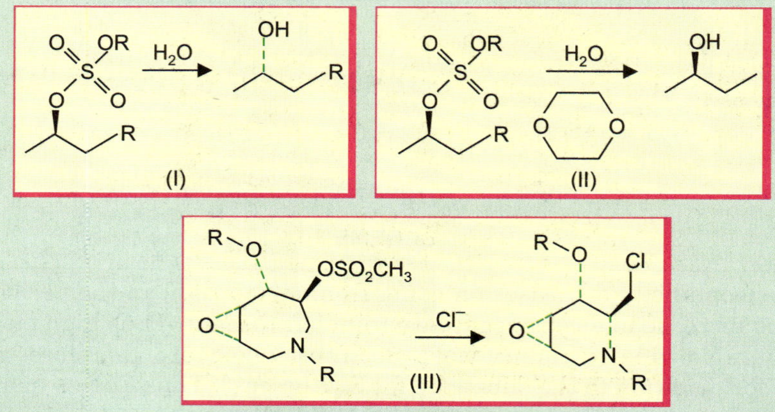

SCHEME 3.15a

Continued...

ANSWER.

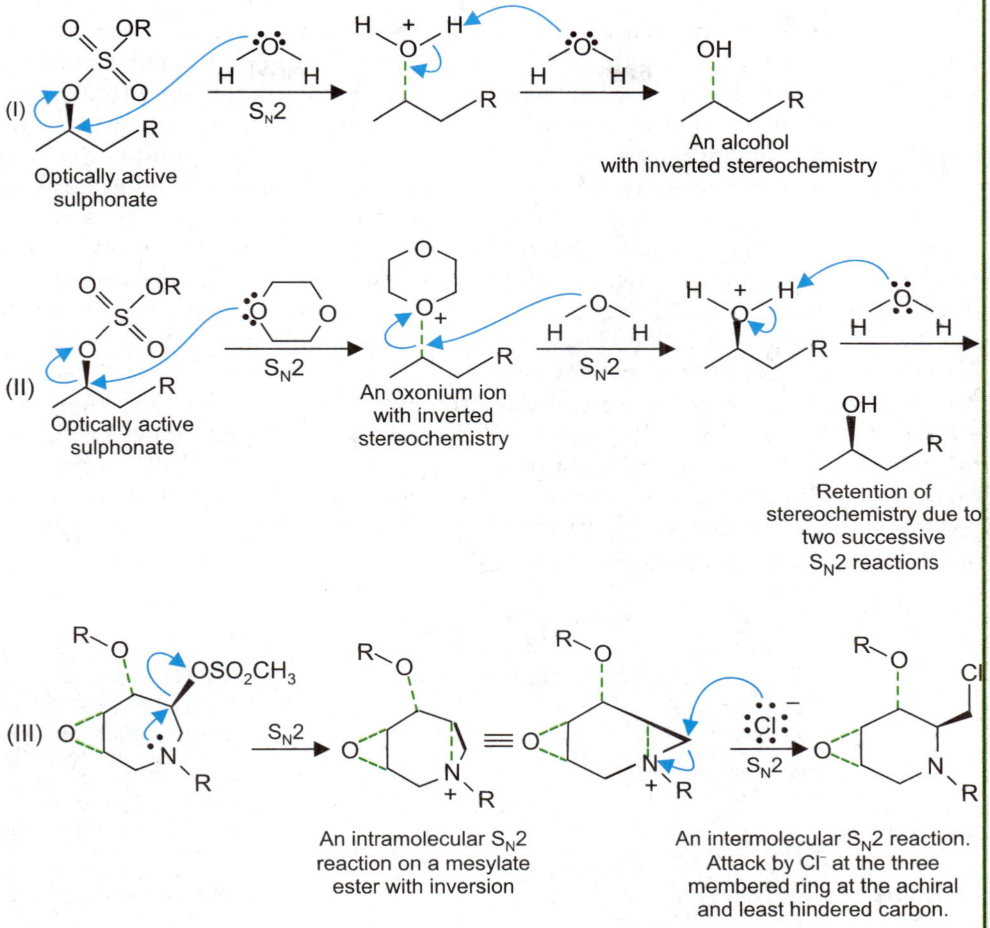

SCHEME 3.15b

Nucleophilicity increases going down the periodic table *i.e.*, Br^- is more nucleophilic than Cl^- and SH^- is more nucleophilic than OH^-. For the halide ions the nucleophilic reactivites follow the order (Scheme 3.16).

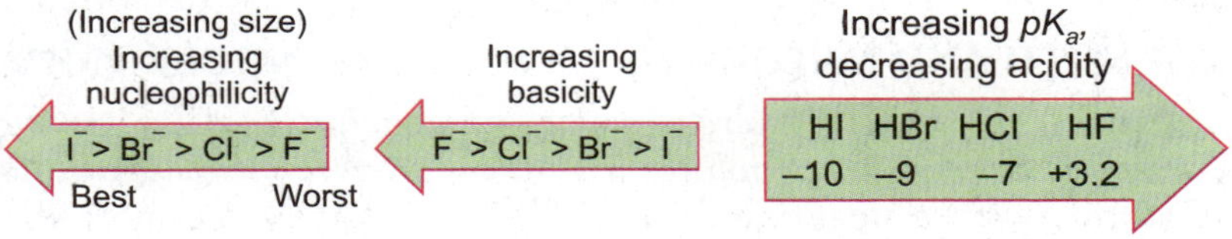

SCHEME 3.16

I^- is the best nucleophile while F^- is poorest. This order is the reverse of the basicity order of the halide ions. Thus one finds that the correlation between basicity and nucleophilicity does not hold for the halides. Going down a column in the periodic table the atoms become larger. One has more electrons at a greater distance from the nucleus. These electrons are more loosely held and the atom with this situation is more polarizable and more nucleophilic. The increase in polarizability (the ease with which the electron cloud on an atom can be distorted is called the polarizability of the atom) leads to an ease of the distortion of the electron could of the attacking atom of the nucleophile. This allows for more effective overlap in the transition state with the back lobe of the slowly rehybridizing sp^3 hybrid used to attain bonding to the leaving group. For these very reasons the basicity of the larger elements is relatively poor than smaller elements, since overlap with the hydrogen 1*s* orbital is poor. Thus, one can use successfully the extent of polarizability to explain as to why I^- is a better nucleophile than F^- (Scheme 3.17). It is only with I^- that the partial bond formation begins at relatively large distances as the electron cloud of the nucleophile gets "pulled" to the carbon atom where the S_N2 reaction occurs. Thus the activation energy of the substitution is decreased. This however, is not the case with F^- with has tightly bound electrons that cannot begin to form at C—F bond until the atoms are very close together. One may appreciate that the degree of nucleophilicity increases on going down in the periodic table even for uncharged nucleophiles *i.e.*, $H_2Se > H_2S > H_2O$ and $PH_3 > NH_3$. In these cases particularly, the fact that the increasing polarizability improves nucleophilic power explains the nucleophilicity trend. In these uncharged nucleophiles, the solvent effects must, however, be less pronounced.

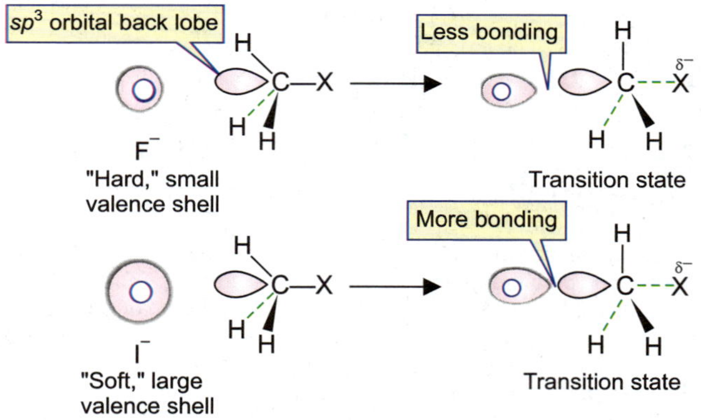

Fluoride ion and iodide ion as nucleophiles in the S_N2 reaction. Fluoride with tightly bound electrons which cannot begin to form a C—F bond until the atoms are close together. Iodide has more loosely bound outer electrons that start bonding earlier in the reaction.

SCHEME 3.17

BOX 3.2

STRENGTH AND NATURE OF LEWIS BASE–NUCLEOPHILE

- Generally, *nucleophilicity increases with increasing basicity, provided the comparison is between nucleophiles with a common nucleophilic atom*. Thus, a nucleophile with a negative charge (Nü⁻) will always be more basic, and therefore more nucleophilic than a neutral nucleophile (Nu). With the exception of amine and phosphorus nucleophiles, nucleophiles used in S_N2 reactions are anions, namely, they have a negative charge (Nü⁻). Recall that, the lower the pK_a is, the stronger the acid and the weaker the conjugate base. (Scheme B 3.3).

Continued...

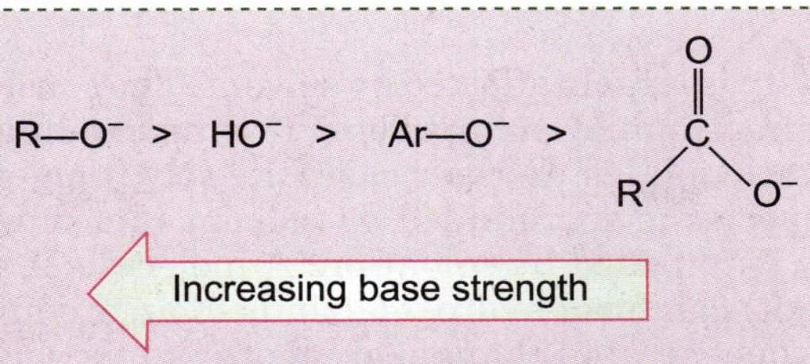

SCHEME B 3.3

- The size of the atom which donates the pair of electrons has a profound effect on its nucleophilicity. *The larger the atom, the greater is its nucleophilicity* (Scheme B 3.4).

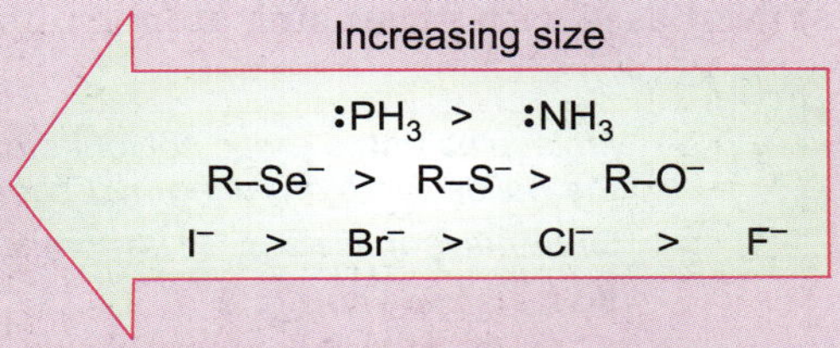

SCHEME B 3.4

- Ambident nucleophiles are capable of reacting at two sites (see Scheme 3.58–3.62).
- Alkoxides can serve the functions of both bases and nucleophiles (Scheme B 3.5)

$$\underset{pK_a\ 13}{EtOO{-}CH_2{-}COOEt} + CH_3CH_2O^- \rightleftharpoons EtOO{-}\ddot{C}H{-}COOEt + \underset{pK_a\ 16}{CH_3CH_2OH}$$

SCHEME B 3.5

- When a reactant has more than one acidic hydrogen is treated with a base, *the most acidic hydrogen reacts first*, since it leads to the formation of the more stable anion (Scheme B 3.6).

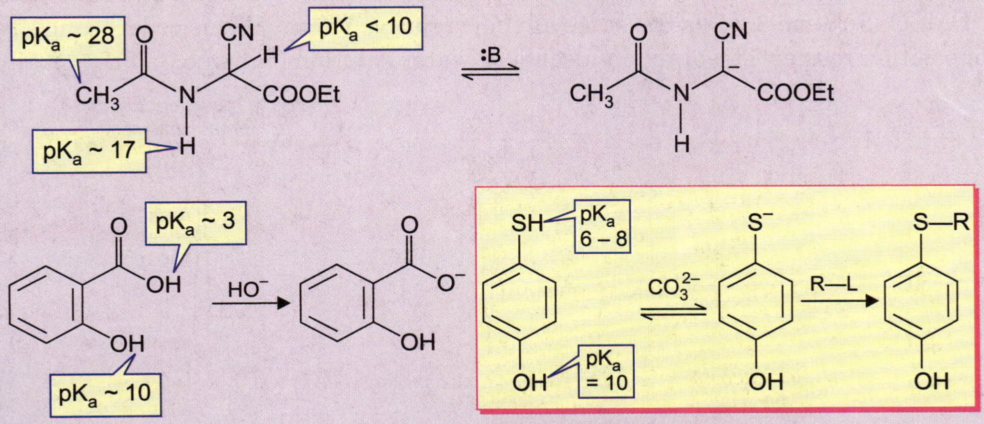

SCHEME B 3.6

(e) Effect of Solvent

The nucleophilic reactivity is largely affected by solvent. The consideration of the increase in polarizability of the nucleophile in improving the nucleophilic power is particularly useful for uncharged nucleophiles. In their case, the solvation effects *i.e.,* the extent to which a nucleophilic species is surrounded by solvent molecule, should be much less strong. Nucleophilicity is impeded by solvation. A molecule of a solvent *e.g.,* water or an alcohol (protic solvent) can form hydrogen bonds to an anionic nucleophile (Scheme 3.18). Smaller anions (concentrated change) are more tightly solvated than larger ones. This phenomenon, therefore, creates a solvent induced barrier to attack at the substrate. For a solvated anion to act as a nucleophile energy is required to "strip off" some of the solvent molecules. This energy is much larger in the case of a small strongly solvated ion *e.g.* F^- as compared to I^-. Generally, the polarizability of atoms and ions shows an increase with increasing atomic number going down in the periodic table, while solvation of ion (in protic solvents) decreases with increasing atomic number.

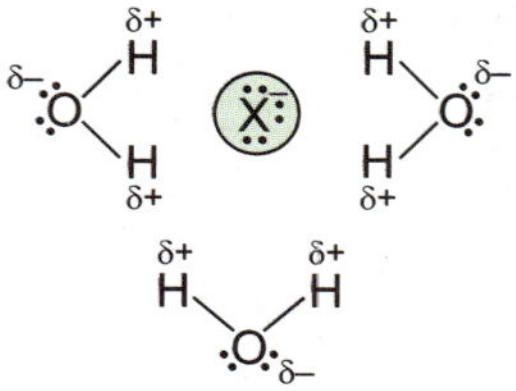

Molecules of the protic solvent e.g., water solvate the halide ion by forming hydrogen bonds to it. A small nucleophile e.g., F^- is more strongly solvated than a larger one e.g., I^-.

SCHEME 3.18

Nucleophilic displacement reactions involve ions either as nucleophiles or as products, as a consequence relatively polar solvents are required. A polar solvent has a relatively high dipole moment and dielectric constant. Both hydroxylic (water, methanol, ethanol) as well as polar aprotic solvents dimethyl sulfoxide DMSO, CH_3SOCH_3, dimethyl formamide DMF, $HCON(CH_3)_2$; hexamethylphosphoric triamide HMPT $[(CH_3)_2\,N]_3\,PO$ are used. Aprotic solvents cannot form hydrogen bonds (no hydroxy groups) thus they do not solvate anions to any appreciable extent. Consequently with anionic nucleophiles, reaction rates are far greater in the polar aprotic solvents. There is lack of stabilization of these nucleophiles by hydrogen bonding, these are therefore, almost "naked" to be highly reactive as nucleophiles. The relative order of nucleophilicity of halide ions in *e.g.* DMSO (Scheme 3.20) is the same as their relative basicity. This order is, however, in opposition to their strengths in alcohol or water solution (Scheme 3.19).

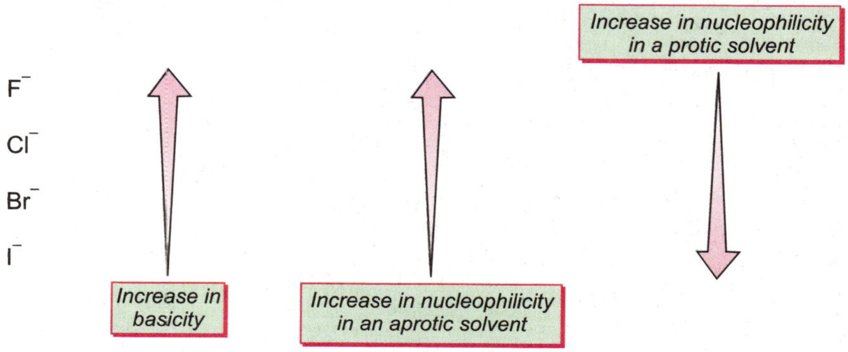

SCHEME 3.19

A polar aprotic solvent dissolves ionic compounds and it solvates cations, the way similar to protic solvents by orienting the negative end of its dipole around the cation (Scheme 3.20). It is however, unable to solvate the anion by H-bonding. Moreover, methyl groups in the case of DMSO *e.g.,* shield the S which is the positive end of the dipole, this prevents the solvation of the anion.

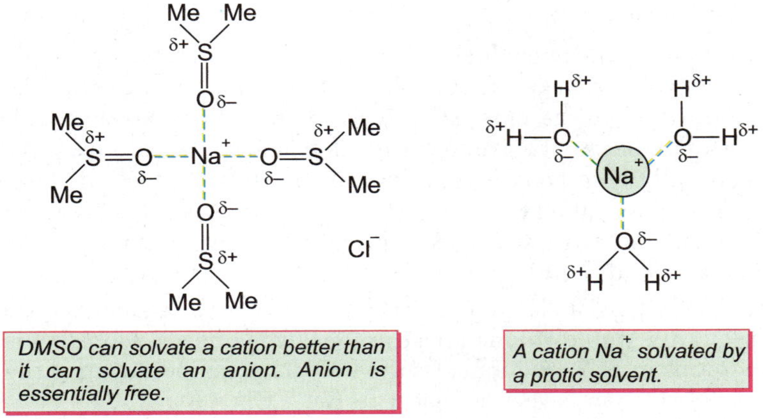

SCHEME 3.20

In summary, the stability of ions depends heavily on solvation. In the gas phase, therefore, fluoride the better base is also the better nucleophile. As the solvent polarity is increased some S_N2 reactions go faster, some slower. Thus *e.g.,* in reaction (I, Scheme 3.21) a change to a more polar solvent (from a non-polar hydrocarbon to a much more polar alcohol solvent) results in a faster reaction. Since one is considering rates, the structure of transition state must be considered. A more polar solvent will have little effect on the uncharged starting materials (eqn I, Scheme 3.21) but will stabilize the charge separated transition state. The result is a decreased activation energy. The reaction thus will go faster in a polar solvent like alcohol (a solvent like alcohol can solvate both positive and negative species, but aprotic solvents can only solvate positive species see Scheme 3.20).

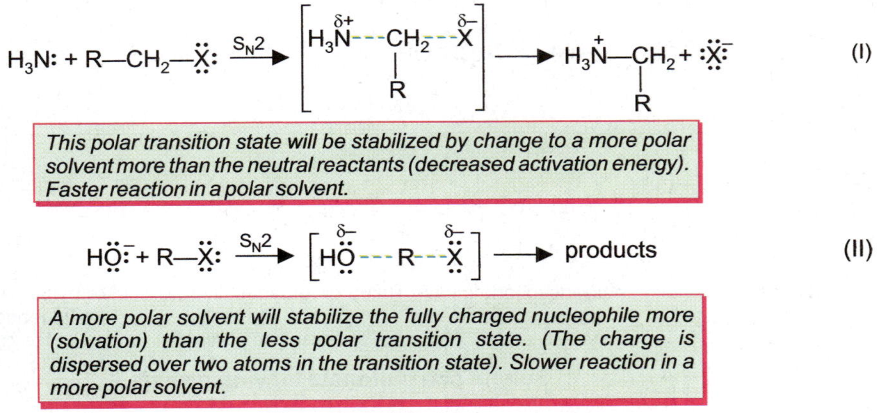

SCHEME 3.21

CHAPTER 3

In the reaction (eqn II, Scheme 3.21), the nucleophile bears a full negative charge. This charge is dispersed in the transition state over a larger area and therefore, in the transition state the negative charge is less concentrated. Thus in this case a change to a more polar solvent will stabilize the fully charged nucleophile (solvation) more than the transition state, the activation energy increases and the reaction slows.

(f) Effect of Leaving Group

A good leaving group is the one which becomes a stable ion after its departure. As most leaving groups leave as a negative ion, the good leaving groups are those ions which stabilize this negative charge most effectively. The weak bases do this best, thus the best groups are weak bases. If a group is a weak base *i.e.,* the conjugate base of a strong acid, it will generally be a good leaving group. In an S_N2 reaction the leaving group begins to grain negative charge as the transition state is reached. The more the negative charge is stabilized, the lower is the energy of the transition state; this lowers the energy of activation and thereby increases the rate of reaction.

The acids, HCI, HBr, HI and H_2SO_4 are all strong acids since the anions Cl^-, Br^-, I^- and HSO_4^- are stable anions, these anions (weak bases, see Scheme 3.9) are also good leaving groups in S_N2 reactions. Of the halogens, an iodide ion is the best leaving group and the fluoride ion is the poorest: $I^- > Br^- > Cl^- > F^-$.

The order of basicity is opposite: $F^- > Cl^- > Br^- > I^-$, the reason that alkyl fluorides are ineffective substrates in S_N2 reactions (Scheme 3.22) is related to the relatively low acidity of HF (pKa = 3) (Stronger the acid the weaker is its conjugate base, among the halogen acids, HI is the stronger acid see Scheme 3.16). Sulfonic acids R SO_2OH are similar to sulfuric acid in acidity and the sulfonate ion RSO_3^- (a weak base) is a very good leaving group. Alkanesulfonate ions, alkyl *p*-toluenesulfonates are therefore, very good substrates in S_N2 reactions (Scheme 3.23).

S_N2 Reaction			Reactivity
$HO^- + RCH_2I \longrightarrow RCH_2OH + I^-$	(Most reactive)		↑ Increase in reactivity
$HO^- + RCH_2Br \longrightarrow RCH_2OH + Br^-$			
$HO^- + RCH_2Cl \longrightarrow RCH_2OH + Cl^-$			
$HO^- + RCH_2F \longrightarrow RCH_2OH + F^-$	(Least reactive)		

SCHEME 3.22

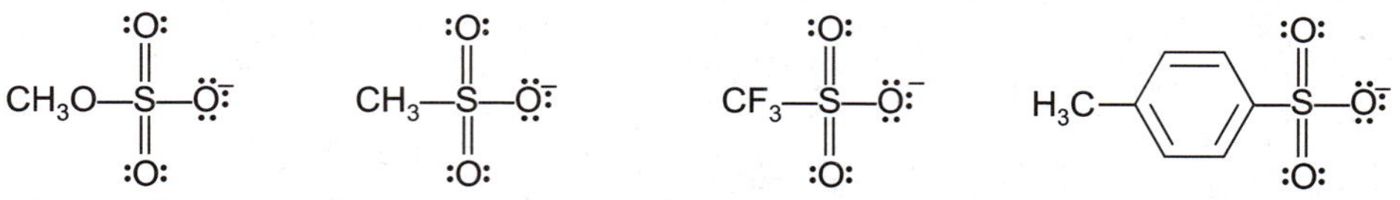

Methyl sulfate ion — Methanesulfonate ion (Mesylate ion) — Trifluoromethanesulfonate ion (Triflate ion) (a super leaving group) — 4-Methylbenzenesulfonate ion (*p*-Toluenesulfonate ion, tosylate ion)

Sulfate and sulfonate leaving groups

SCHEME 3.23

The triflate ion ($CF_3SO_3^-$) is one of the best leaving groups known, it is the anion of CF_3SO_3H which is a strong acid much stronger than sulfuric acid (Scheme 3.23). Consider the reaction (eqn. I, Scheme 3.24), the reverse of which though a nucleophilic reactions does not occur *i.e.,* ethyl alcohol and chloride ion do not react. This can be understood if one considers the leaving tendency of Cl^- in the forward reaction and the leaving tendency of OH^- in the reverse reaction in terms of their basicities. For some people it is easy to compare the acid strengths of the conjugate acids. HCl is a much stronger acid than H_2O *i.e.,* Cl^- is a much weaker base than OH^- (since stronger the acid, the weaker its conjugate base). Because it is a weaker base Cl^- is a better leaving group). As a consequence of this OH^- can displace Cl^- in the forward reaction but Cl^- cannot displace OH^- in the reverse reaction.

When the difference between the basicities of the nucleophile and the leaving group is not large, the reaction would be reversible, as in (eqn. II, Scheme 3.24). The *pKa* values of conjugate acids of the two leaving groups are not very different HBr–9 and HI–10.

$$CH_3CH_2Cl + OH^- \longrightarrow CH_3CH_2OH + Cl^- \qquad (I)$$

$$CH_3CH_2Br + I^- \rightleftharpoons CH_3CH_2I + Br^- \qquad (II)$$

SCHEME 3.24

(g) Phase Transfer Catalysis–Role of Crown Ethers

Nucleophilic reactivity is increased in the presence of a suitable crown ether. The crown ether 18-crown-6, *e.g.,* coordinates very effectively with potassium ions. Salts like KF, KCN and CH_3 COOK which are otherwise insoluble in non-polar solvents like benzene, in the presence of 18-crown-6 dissolve in it. Thus, the situation in the organic phase is the presence of relatively unsolvated anions to carry out nucleophilic substitution on an organic substrate. For example, KF which is both insoluble in benzene and unreactive to organic halides, brings about an efficient nucleophilic displacement on a halide in the presence of 18-crown-6 (Scheme 3.25). The role of a crown ether here is of a catalyst which brings the anion into the organic phase, and the process is termed phase transfer catalysis.

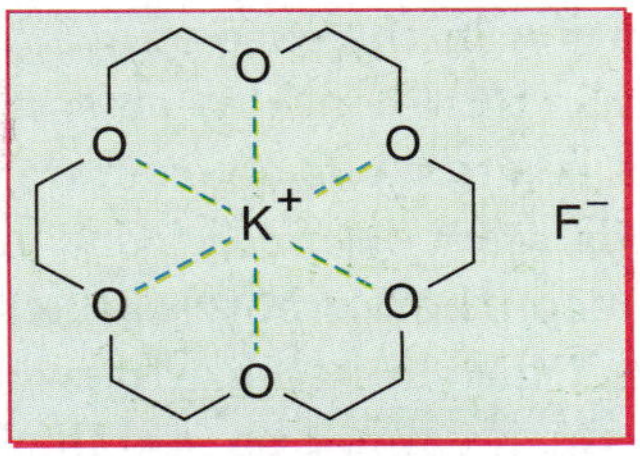

$$C_6H_5CH_2Cl + \overset{+}{K}\overset{-}{F} \xrightarrow[\text{Benzene}]{\text{18-Crown-6}} C_6H_5CH_2F + \overset{+}{K}\overset{-}{Cl}$$

100%

SCHEME 3.25

3.3 S_N1 REACTION

It is seen that when in a substrate at least one of the groups is aromatic there is some tendency for the leaving group in a substrate to leave on its own accord (basis for the S_N1 reaction) rather than it can be expelled by the incoming nucleophile (as is the case in an S_N2 reaction). In summary the substrates which contain good leaving groups and adjacent substituents like a phenyl group which can stabilize a positive charge mesomerically are prone to undergo S_N1 reactions. In these reactions there are two distinct steps, the first of which is the loss of the leaving groups, the second being the introduction of the nucleophile *i.e.*, the departure of the leaving group and the approach of the nucleophile are not linked. The first consequence of S_N1 reaction is that if the substrate is chiral, the product is a racemic mixture (Scheme 3.26). Thus when either enantiomer of chloride (Scheme 3.26) undergoes an S_N1 reaction the product is almost completely racemized, since the achiral planar intermediate carbocation is attacked both from left and right with equal probability.

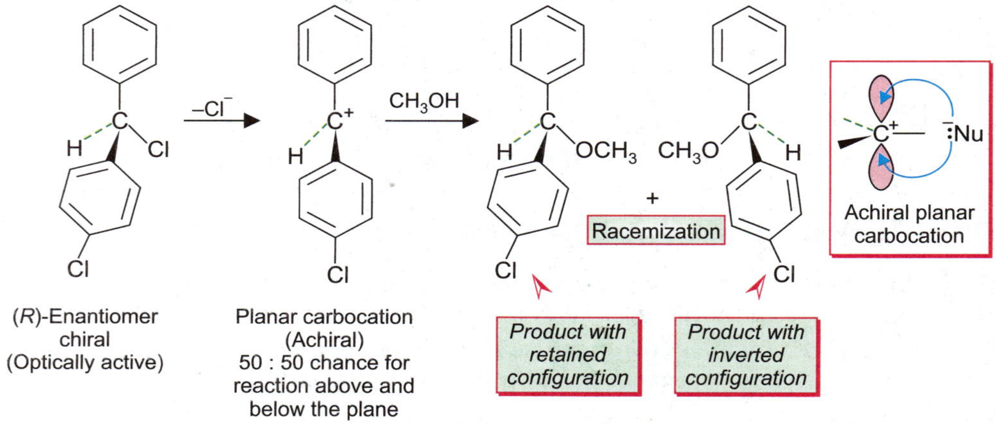

SCHEME 3.26

In the case of *tert.* substrates *e.g.*, *t*-butyl chloride the crowding prevents an incoming nucleophile from approaching the carbon bearing the leaving group from the rear (S_N2 conditions). Saturated aliphatic carbocations do not have either π bonds or unshared electron pairs and thus resonance effects cannot have much influence on their stabilization. However, the formation of a carbocation in the case of a 3° alkyl halide leads to relief of strain which is associated with the tetrahedral geometry and in the planar carbocation the three methyl groups adopt positions as apart from each other as possible (Scheme 3.27) when the cationic carbon atom becomes more highly substituted, additional possibilities for declocalization develop. The *tert*-butyl carbocation has three hyperconjugative interactions (Scheme 3.28) with which to stabilize the positive charge, thus it is more stable than the methyl cation which cannot be stabilized by hyperconjugation. Generally, *the more highly substituted the carbocation, the more stable it is.* The *tert*-butyl carbocation has actually been isolated at low temperature and its crystal structure determined.

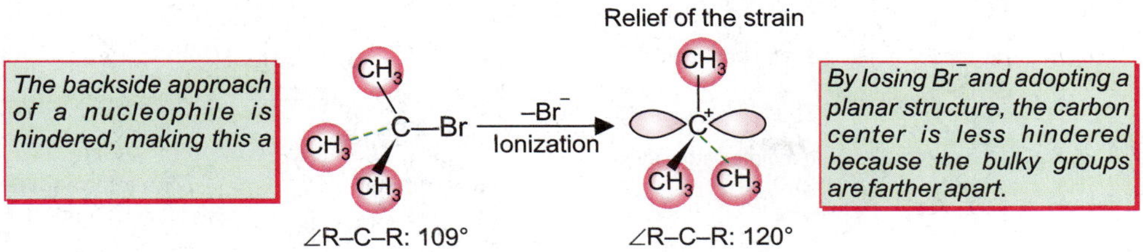

SCHEME 3.27

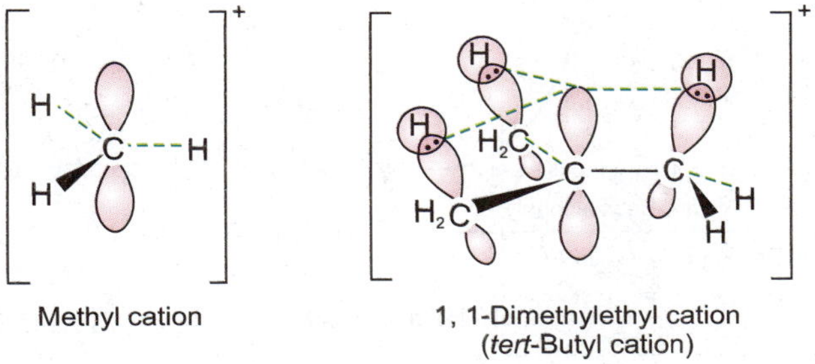

SCHEME 3.28

(a) The Mechanism of S_N1 Reaction

S_N1 rate = K_r[R—X]

S_N2 rate = K_r[R—X] [NuC:⁻]

SCHEME 3.29

It is a first order reaction–the rate of reaction depends on the concentration of the alkyl halide but not the concentration of nucleophile (Scheme 3.29). A carbocation is formed in the slow step of an S_N1 reaction. Because a tertiary carbocation is more stable and therefore, easier to form than a secondary carbocation which in turn is more stable than a primary carbocation, therefore, on successive replacement of methyl groups of *tert*-butyl bromide by hydrogens, the rate of the S_N1 reaction decreases progressively. This is opposite to the order of reactivity exhibited by alkyl halides in S_N2 reaction.

In step 2, the carbocation *e.g., tert*-butyl cation reacts with the nucleophile, when *e.g.*, H_2O is the nucleophile *i.e.*, during solvolysis in the third step, deprotonation gives the product. The mechanism reflects three different events (*i*) rate determining dissociation of the haloalkane to give a carbocation (Scheme 3.30) and a halide ion, nucleophilic attack by water on the carbocation to give an alkyloxonium ion. *Note that the loss of a proton is in fact a separate acid-base reaction that takes place after the substitution process is finished.* Each of the individual steps of the overall S_N1 reaction are in equilibrium, each one is an acid-base reaction, and that the carbocation functions are a Lewis acid in the first and second steps.

An energy diagram tracing the progress of the S_N1 reaction is shown (Scheme 3.29). The rate of the reaction reflects the activation energy required to form the carbocation intermediate. The activation energy required for the second step *i.e.*, the addition of the nucleophile to the carbocation is much smaller, thus step 2 is fast. The rate of the second step has no effect on the overall rate of reaction.

CHAPTER 3

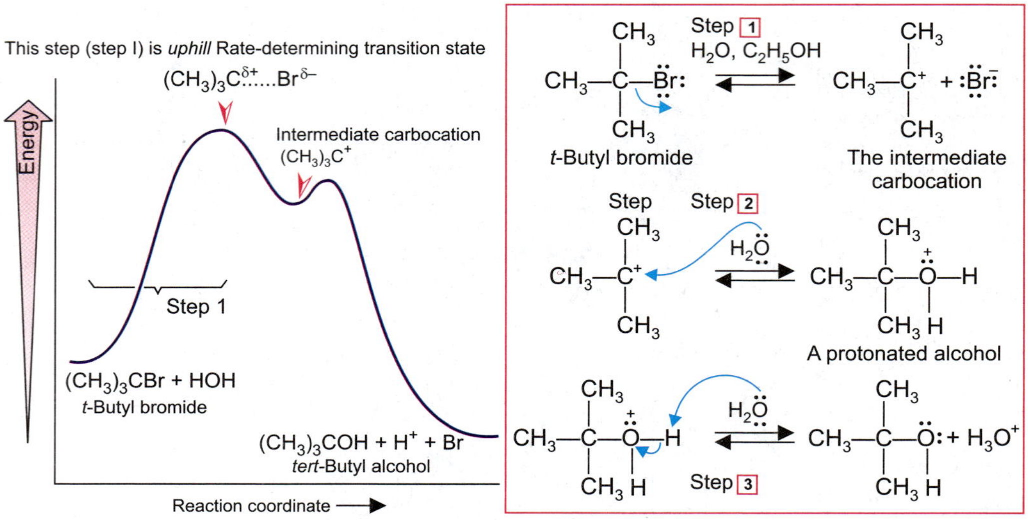

SCHEME 3.30

(b) The Effect of Structure of the Carbon Framework on S_N1 Reactions–Hammond Postulate

Carbon frameworks capable of stabilizing carbocations facilitate the S_N1 reactions, the stabilization generally increasing in the order $CH_3^+ < RCH_2^+ < R_2CH^+ \simeq C{=}C{-}C^+ < R_3C^+ < ArC^+$. Stable carbocations are planar and thus, compounds like *t*-butyl bromide (Scheme 3.30) undergo a facile replacement of bromine with the formation of the cation $(CH_3)_3C^+$, since it can assume a planar configuration as required by the sp^2 hybridized central carbon.

For an S_N1 reaction to proceed at a reasonable rate, the formation of relatively stable carbocation is essential since it is then that the free energy of activation for the slow step of the reaction will be low enough for it to proceed. [Recall (Scheme 3.30), this step *i.e.*, step 1 is *uphill* as far as free energy is concerned ($\Delta G°$ for this step is positive) as well as in terms of enthalpy ($\Delta H°$ is also positive). This step is, consequently endothermic. Hammond's postulate states–the transition state for a step which is uphill in energy should show a strong resemblance to the product of that step. Since the product of this step (an intermediate) is a carbocation, any feature that stabilizes it such as dispersal of the positive charge by electron-releasing groups (hyperconjugation) should also stabilize the transition state where the positive charge is generated (Scheme 3.31)].

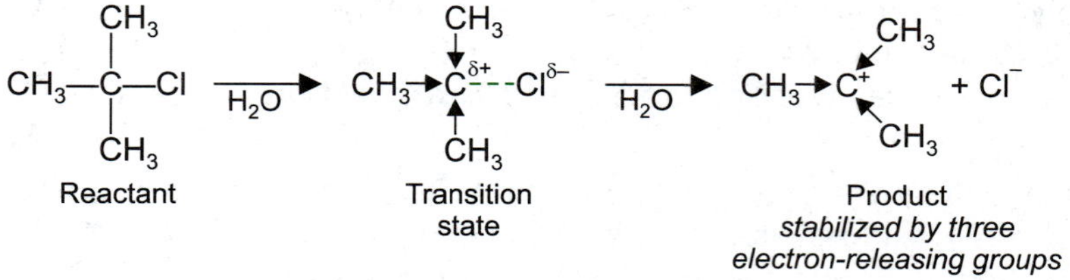

SCHEME 3.31

The carbocation intermediates also get stabilization from delocalization of their positive charge in to adjacent unsaturated groups. Thus benzylic and allylic halides undergo S_N1 reactions more rapidly than the related systems in which such carbocation stabilization is not available.

These arguments explain as to why the reaction of 3-bromocyclohexene (Scheme 3.32) with methanol (CH_3OH) is faster than the reaction of bromocyclohexane with methanol. Both the substrates are secondary halides, the reaction with methanol, a neutral nucleophile, will tend to occur by an S_N1 process. Although both carbocations are secondary, the one derived from 3-bromocyclohexene is also a resonance-stabilized allylic carbocation. Its resonance stabilization increases the rate of its formation. No such stabilization occurs in the reaction of bromocyclohexane (Scheme 3.32).

For a similar reason furfuryl chloride (a 1° alkyl halide) does not undergo S_N2 substitution. S_N1 reaction is promoted due to the resonance stabilisation of the cation formed by ionisation of the 1° chloride (Scheme 3.32)

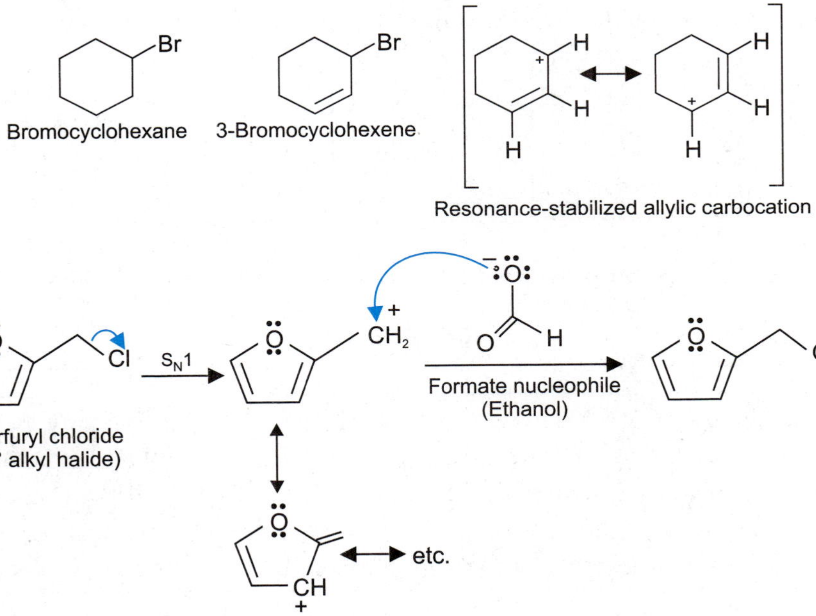

SCHEME 3.32

Consider the acetolyses of *cis*- and *trans*-4-*t*-butyl cyclohexyl tosylates (I and II, Scheme 3.33). In the case of *cis* compound I, the tosyl group is in the axial position and thus, suffers from steric 1:3 non-bonded interactions with the axial hydrogens. This leads to a faster reaction than with *trans*-equatorial isomer II by a factor of about 3.4.

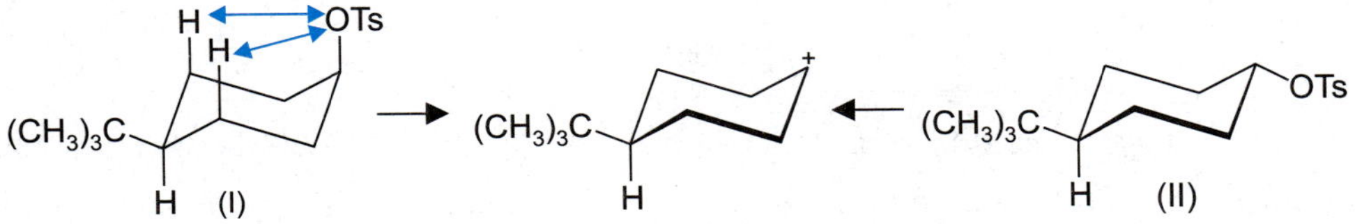

SCHEME 3.33

The S_N1 reactions fail or occur very slowly, if at all, at the bridgehead positions. S_N1 reactions proceed via planar carbocations and bridgehead carbons cannot become the seat of carbocations (S_N2 reactions fail, since the attack of the nucleophile from the rear is hindered). In case however, if the rings are large enough *e.g.,* in bridgehead compounds the strain due to the formation of the carbocation is somewhat relieved. Thus, [2.2.2] bicyclic systems *e.g.* (III, Scheme 3.34) undergo S_N1 reactions comparatively easier than the smaller bicyclic systems *e.g.,* the bicycloheptyl system (II, Scheme 3.34), however the reaction is still slower than with open chain compounds (I).

Similarly the bridgehead adamantyl tosylate (I, Scheme 3.35) does undergo an S_N1 reaction but more slowly. The formation of a carbocation at a bridgehead in (III, Scheme 3.35) is partly favored due to decrease in non-bonded interactions. The carbocation (III, Scheme 3.35) can be flattened without involving strain.

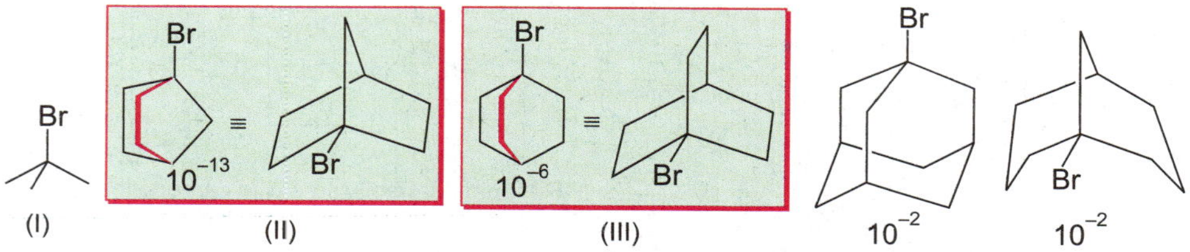

Relative rates of solvolysis of *t*-butyl and bridgehead bromides in H_2O-EtOH

SCHEME 3.34

An S_N2 reaction cannot occur and is completely ruled out at a bridgehead position like (I, Scheme 3.35, the nucleophile cannot attack from the rear, and the carbon bearing the leaving group cannot be inverted). Very interestingly the cubyl tosylate (II, Scheme 3.35) undergoes an S_N1 reaction much faster than expected when compared to the bridgehead adamantyl tosylate and in this compound the CCC bond angles are 90°.

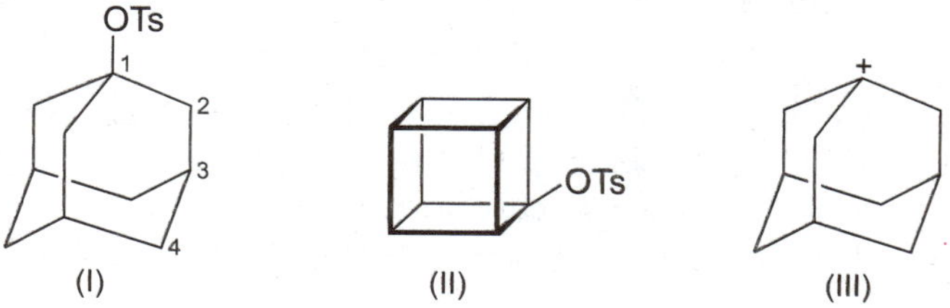

SCHEME 3.35

Interestingly enough, some nucleophilic substitution reactions which normally occur through carbocations do take place at [2.2.1] bridgeheads, provided the leaving group is of the type that it cannot itself function as a nucleophile. In the case of substrate (I, Scheme 3.36) chlorobenzene is the nucleophile.

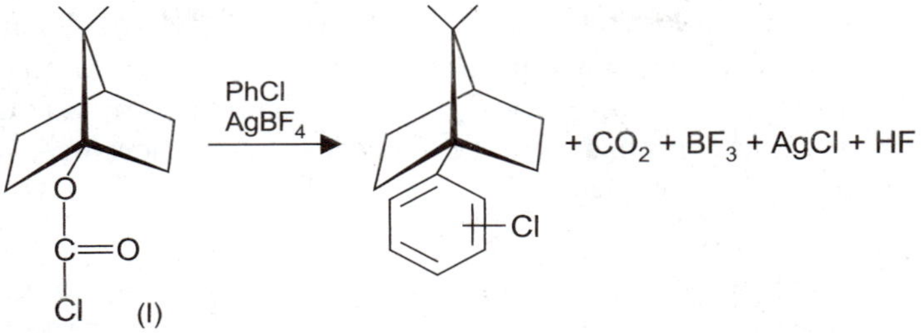

SCHEME 3.36

(c) Ion Pairs in S_N1 Mechanism

Since the carbocation carbon atom has sp^2 hybridization, it is expected to be planar, and therefore achiral. Equal amounts of enantiomeric products are expected during the reaction of a chiral substrate by attack at either side of the plane with equal probability resulting in complete racemization. Although many S_N1 reactions result in complete racemizations, some others do not. It is found that during many S_N1 reactions there is 5 to 20% inversion and in few cases a small amount of retention of configuration has been observed. These results have been explained by invoking the formation of ion pairs. Thus the S_N1 reactions proceed as shown (Scheme 3.37). The following points may be noted.

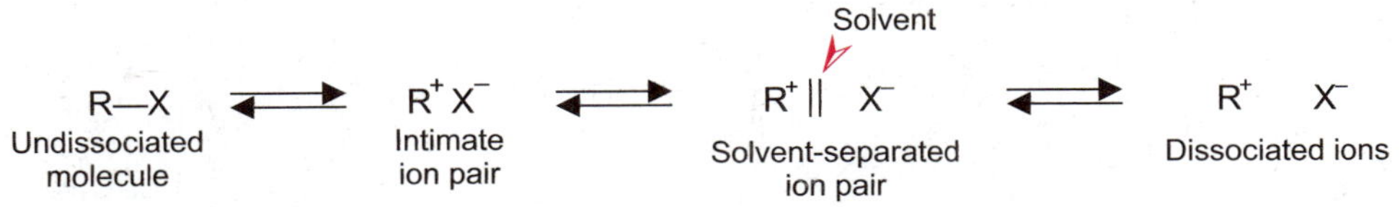

SCHEME 3.37

- In an intimate ion pair (also called contact or tight ion pair) R^+ does not behave like the free cation of the dissociated species (Scheme 3.37). There seems to be significant bonding between R^+ and X^- and the asymmetry is well maintained *i.e.*, individual ions retain their original stereochemical configuration. Thus during the course of S_N1 reaction, X^-

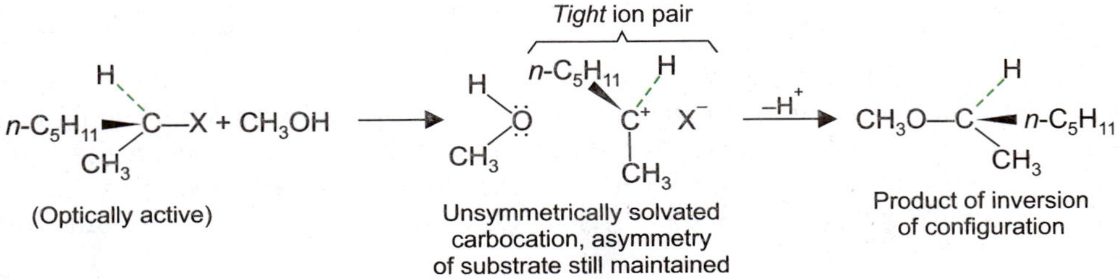

SCHEME 3.38

solvates the cation from the side it departs from, while solvent molecules near the intimate ion pair can solvate it only from the opposite side. The situation is like an unsymmetrically solvated carbocation, where the nucleophilic attack by a solvent molecule (on tight ion pair) gives inverted product (Scheme 3.38).

- In summary when strong ion pairing exists, one side of the carbocation is shielded from the incoming nucleophile (solvent during solvolysis) so, some enantioselectivity is observed. This process may therefore, be equally well represented (Scheme 3.39).

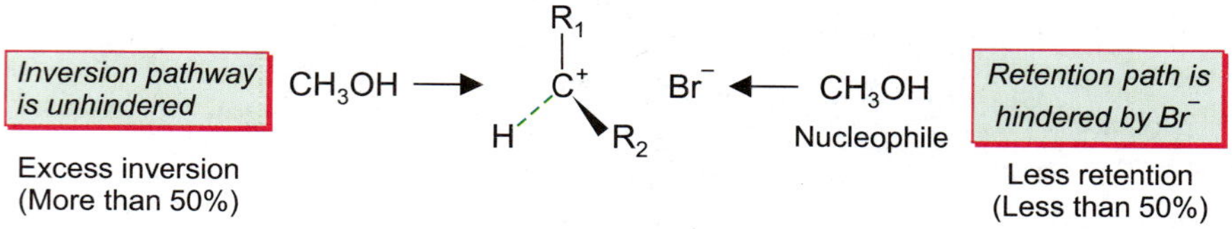

SCHEME 3.39

- If the nucleophile attacks carbocation of the solvent separated ion pair, the stereochemistry of the carbocation is not now maintained as tightly. The leaving groups is still in a position to partially block the approach of the nucleophile to that side of the carbocation. Thus more of the product with inverted configuration will be formed, leading to more racemization (perhaps total).
- If the free R^+ is formed, it is planar and the attack by nucleophile *e.g.,* solvent during solvolysis occurs with equal probability form either face to give complete racemization via symmetrically solvated carbocation (Scheme 3.40). One may describe this carbocation as prochiral.

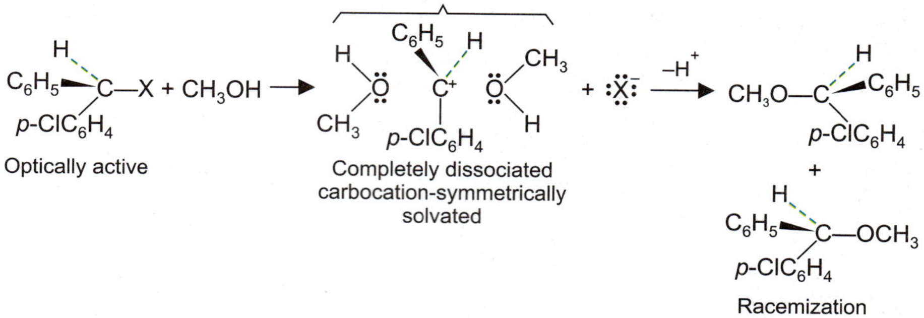

SCHEME 3.40

- Less stable *i.e.,* very short lived the carbocation, the more likely it is to be attacked before X^- separates from it and therefore, inversion of configuration is much more extensive (see, Scheme 3.38). When a substrate can give a stable carbocation (with long life) *e.g.,* by charge declocalization into the attached aromatic ring (see, Scheme 3.40), the anion hardly makes any contribution towards its stability.

During the hydrolysis of 1-chloro-1-phenylethane (Scheme 3.41) the benzene ring stabilizes the intermediate carbocation I and as a consequence the optically active starting material undergoes a predominant (98%) racemization. On replacing the benzene ring by a cyclohexane ring the ion (II, Scheme 3.41) formed in the reaction derives most of its stability from the intimate ion pair. Thus, it does not readily ionize to the free ion. Therefore, the product originates mostly by the nucleophilic attack on the ion pair, leading to almost exclusive inversion of configuration.

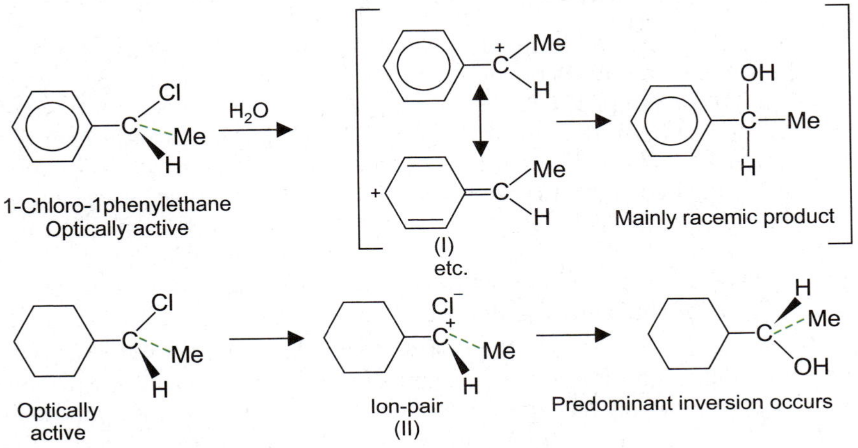

SCHEME 3.41

- The overall balance between racemization and inversion is very sensitive to several factors such as solvent and negative ion and the above two reactions provide extreme situations, most reactions coming between the two.
- Some cases, S_N1 reactions have been found to proceed with partial retention (20 to 50%) of configuration. Ion pairs have been involved to explain for example, the phenolysis of optically active α-phenylethyl chloride, an ether with retention of configuration is obtained and involves a four center mechanism (Scheme 3.42).
- It has been suggested that nucleophilic attack on an intimate ion pair can afford inverted product by mechanism that is borderline between S_N1 and S_N2 reaction.

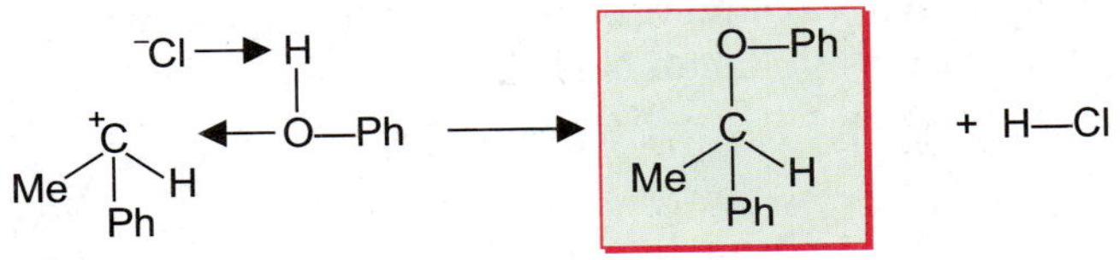

SCHEME 3.42

- The ion associated with another ion of opposite charge is termed counter ion or "gegenion" from the German meaning counter. Thus *e.g.*, in the case of a tight ion pair R^+X^-, R^+ is the counter ion of X^- and vice versa. A reaction in which the intimate ion pair recombines to afford the original substrate is called an internal return.
- *The special salt effect:* The effect produced by addition of an ion, that is generated in a reaction is termed common ion effect which leads to depression in the rate. Thus *e.g.*, the addition of chloride ion decreases the rate of solvolysis of diphenylmethyl chloride, the first step of the reaction $RCl \rightleftharpoons R^+ + Cl^-$ is the

rate determining. The presence of Cl^- from another source is subject to mass-law effect on the equilibrium and consequently on the overall rate of reaction.

- A change in the rate of a reaction due to the addition of a salt (electrolyte) that generally does not have an ion in common with reactants or product is termed *special salt effect*.
- The addition of a salt changes, the ionic strength of the solution which in turn affects the free energies of the ions which are being formed or destroyed. For example, addition of $LiClO_4$ (or LiBr) generally increases the rate of an S_N1 reaction (ions being formed $RX \rightleftharpoons R^+ + X^-$ in the rate determining step). An increase in the ionic strength of the solution favors their formation. One also looks at these results where the ClO_4^- (or Br^-) traps the solvent separated ion pair to afford $R^+ \parallel ClO_4^-$ and these species being unstable under these conditions give back the product. Thus, the amount of solvent separated ion pair that would have returned to the starting material is reduced leading to an increase in the overall reaction.

Recall the possibilities of inversion and racemization of the product during solvolysis reaction. One may note that the formation of an ion pair and subsequent internal return may effect the stereochemistry of the substrate RX. In some cases racemization of the optically active substrate RX occurs via internal return. However, partial or complete retention is also observed. These results can be explained (Scheme 3.43). Evidence for the involvement of ion pairs during racemization, via internal return is that such racemization is faster than solvolysis (entropy considerations).

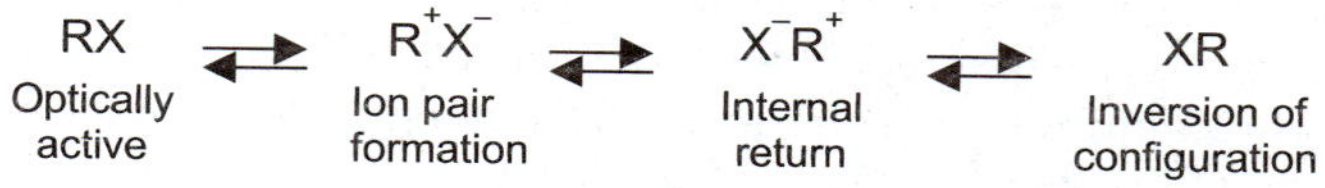

SCHEME 3.43

On acetolysis optically active *trans*-3-chloro-5-methylcyclohex-1-ene (I, Scheme 3.44) afforded racemic mixture of *cis*- and *trans*-acetates. This is due to the fact the intermediate allylic cation possesses a plane of symmetry. The rate at which optical activity was lost in this reaction was found to be faster by a factor of about four. Additionally, it was shown that the

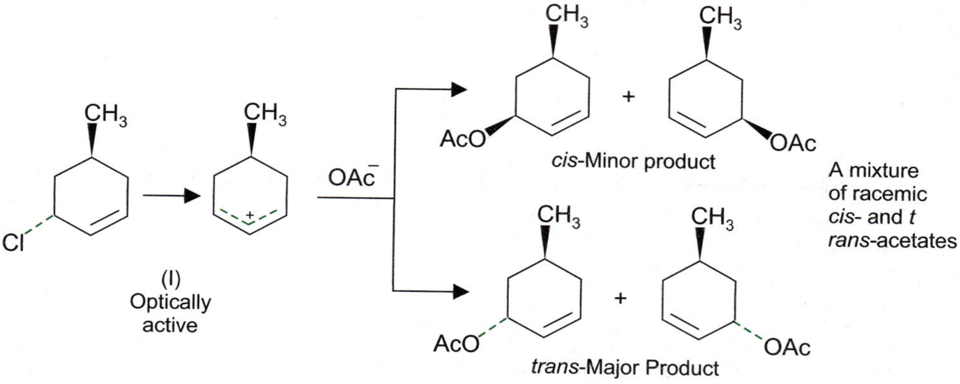

SCHEME 3.44

chloride racemized without even undergoing solvolysis (internal return) and the loss of its optical activity occurs within an ion pair (Scheme 3.45), the product being the racemic *trans*-3-chloro-5-methyl-cyclohex-1-ene. This is so since in the tight ion pair the ions are in close contact and therefore, prevent that movement of the chloride ion from one face of the cyclohexane ring to the other.

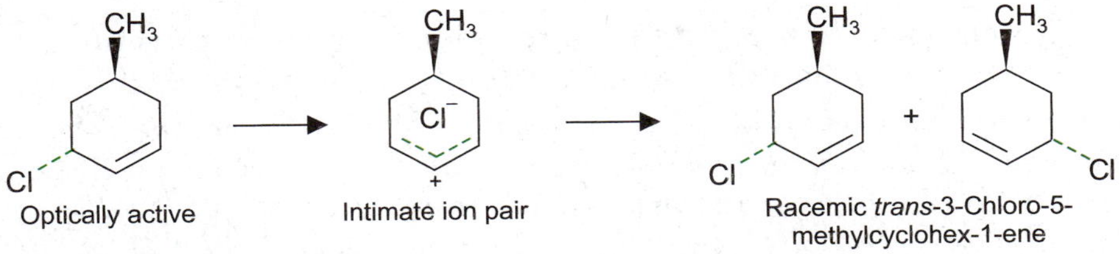

SCHEME 3.45

(d) Role of Solvent

In contrast to the use of aprotic solvents for S_N2 reactions, S_N1 reactions are enhanced by polar and protic solvents, due to the ability of a protic solvent to stabilize both cations and anions. The transition state leading to the intermediate carbocation and halide ion is stabilized more than the reactants (Scheme 3.46), thus the free energy of activation is lower. The transition state for this endothermic step is, where separated charges are getting developed and therefore, it assembles the ions that are ultimately formed.

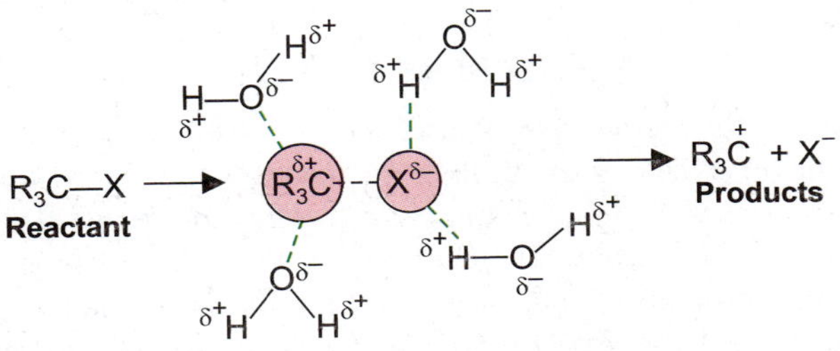

Transition state

Separated charges are developing.

Solvation stabilizes the transition state

SCHEME 3.46

(e) Effect of Leaving Group and Nucleophile on S_N1 Reactivity (a Comparison with S_N2 Reaction)

In an S_N1 reaction the carbocation formed in the slow step reacts rapidly in the second step with any uncleophile present in the reaction mixture. Thus increasing the rate of the fast step has no effect on the rate of the prior slow carbocation forming step. The rates of S_N1 reactions are unaffected by either the concentration or the identity of the nucleophile. A good nucleophile (HO^-, CH_3O^-) however, favors an S_N2 reaction over an S_N1 reaction.

An S_N1 reaction is favored by a poor nucleophile (H_2O, CH_3OH) not by increasing its rate , but by depressing the competing S_N2 reaction. Thus in summary, an S_N2 reaction is favored by a high concentration of a good nucleophile, while an S_N1 reaction is favored either by low concentration of a good nucleophile or by a poor nucleophile.

A closer look at S_N1 and S_N2 reactions–nature of solvent, nucleophile, substrate and leaving group

A few more examples are presented to understand broadly as to how the structure of the substrate, the nucleophile and the reaction conditions determine the S_N1 or S_N2 mechanism for a substitution reaction and how these conditions affect the reactivity within each mechanism. Thus compared to 100% inversion in the S_N2 mechanism (I, Scheme 3.46a) the S_N1 reaction with water proceeds with 2% enantiomeric excess.

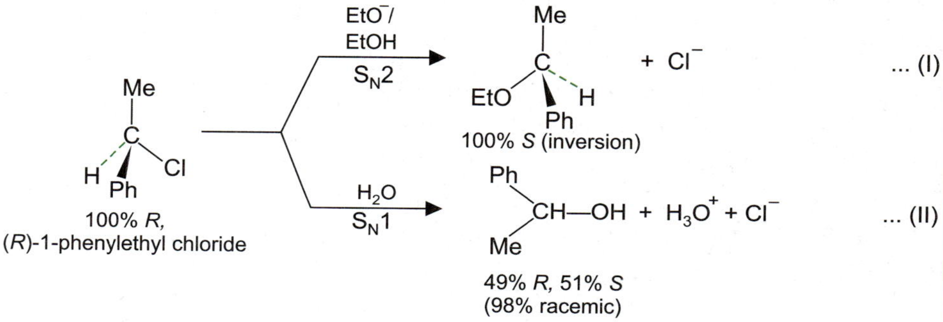

SCHEME 3.46a

On acetolysis the same S_N1 reactions produces an acetate mixture (42% R and 58% S i.e., 84% racemic with a 16% enantiomeric excess. Thus in S_N1 reaction there is normally an excess of inversion over retention. The nature of the solvent and the substrate play a role in the extent of formation of inverted product over product of retention. Unlike the acetolysis of (R)-1-phenylethyl chloride, (achiral) A on acetolysis, however, shows a far more stronger preference for inversion of configuration (Scheme 3.46b), however, some retention of configuration distinguishes these results from the outcome of an S_N2 reaction.

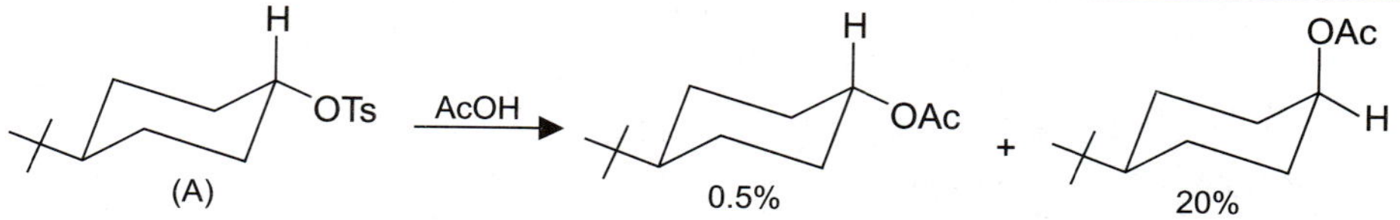

One may also compare the results with (problem 3.14).

SCHEME 3.46b

(f) The Nature of Leaving Group

Both in S_N2 as well as in an S_N1 reaction, the leaving group begins to acquire a negative charge as the transition state is reached. If this developing negative charge at the leaving group is stabilized, this also stabilizes the transition state *i.e.*, lowers its free energy.

Consequently the free energy of activation is lowered and the rate of reaction is increased. The relative abilities of various groups to leave carbon in S_N1 reactions are roughly the same as already discussed for S_N2 reactions. Of the halogens iodide ion is the best leaving group while fluoride ion is the poorest. Similarly *p*-toluenesulfonate ion OTs^- and triflate ion (trifluoro-methanesulfonate ion $CF_3SO_3^-$) are very good leaving groups.

(g) Two Important Preparative Reactions Involving S_N2 Reactions

- **Methyl esters from carboxylic acids and diazomethane.**

Whereas the alkylations take place in basic or neutral solutions, carboxylic acids can be directly methylated with diazomethane (Scheme 3.47). The actual nucleophile (the carboxylate ion) and the actual methylating agent are produced from the reaction partners by proton transfer.

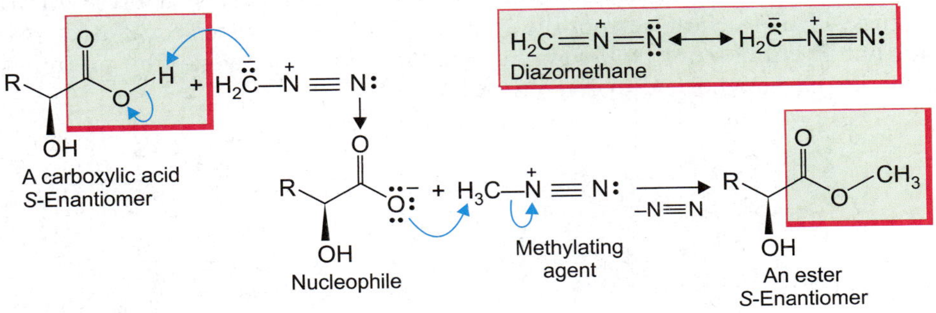

SCHEME 3.47

Diazomethane is a reactive compound with a strongly nucleophilic carbon atom. Resonance structures show that *the carbon atom in one form bears an electron pair and a negative charge*, so it can act as either a base or a nucleophile.

The methyldiazonium ion is a potent alkylating agent, since its methyl group is bonded to an excellent leaving group (N_2).

- **Mitsunobu reaction-inversion of configuration of a chirality center bearing an OH group.**

Mitsunobu reaction is one of the best methods for inverting the configuration of an alcohol or subjecting a 2° alcohol to nucleophilic substitution. Normally, E2 elimination competes with S_N2 substitution at 2° C(sp^3), but little or no elimination occurs in the Mitsunobu reaction.

In the Mitsunobu reaction a chiral 2° alcohol and a carboxylic acid are converted to an ester with inversion of configuration at the electrophilic C. The reaction requires a reagent combination, in which triphenylphosphine, Ph_3P, diethyl azodicarboxylate (DEAD), and an acid, PhCOOH react in the presence of the alcohol substrate (Scheme 3.47*a*).

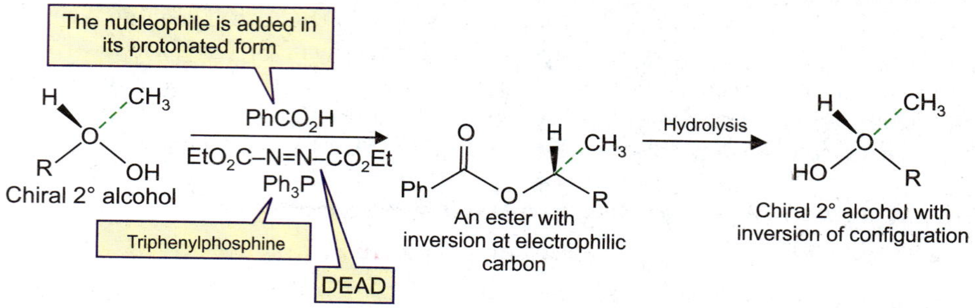

SCHEME 3.47a (Mitsunobu reaction)

The mechanism of the Mitsunobu reaction (Scheme 3.47*b*) involves addition of nucleophilic Ph_3P to the electrophilic N in DEAD. (The alcohol could add to DEAD instead, but P is far more nucleophilic.) The addition is preceded by protonation of DEAD by the carboxylic acid to give an adduct (A).

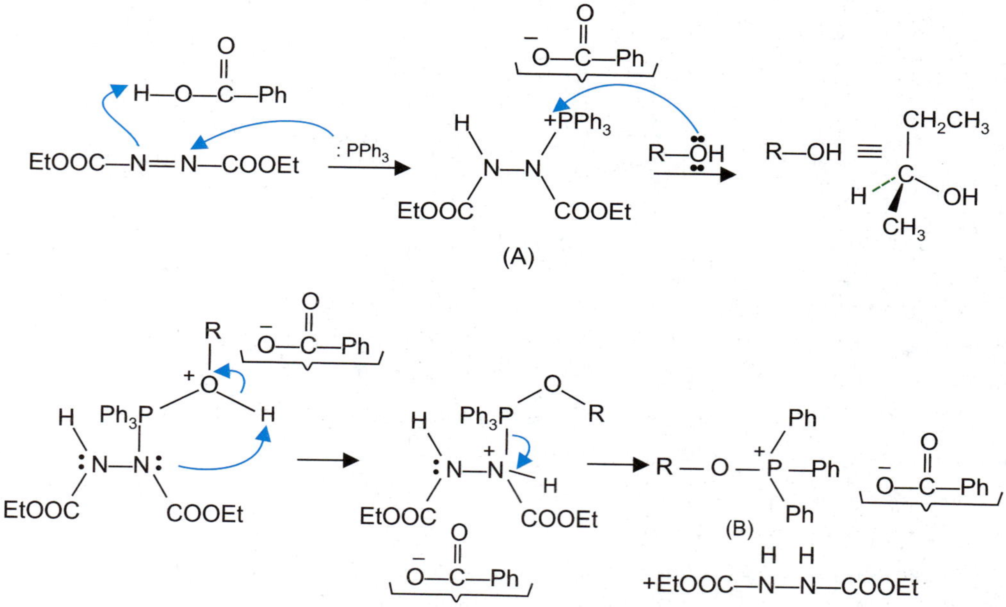

SCHEME 3.47b (Mechanism of Mitsunobu reaction)

The adduct (A) reacts in Step 2 with the alcohol—as a nucleophile—to yield the corresponding (alkoxy) triphenylphosphonium ion (B), $[RO–PPh_3]^+$. Phosphorus, being a period 3 element, can accommodate 10 electrons, and compounds with five bonds to the phosphorus atom are common.

In the actual substitution step (Scheme 3.47c) the nucleophile $PhCOO^-$ reacts with the electrophilic carbon atom that bears the leaving group, which is triphenylphosphine oxide, $Ph_3P{=}O$. Formation of the P=O bond pushes the Mitsunobu reaction to completion (Scheme 3.47c).

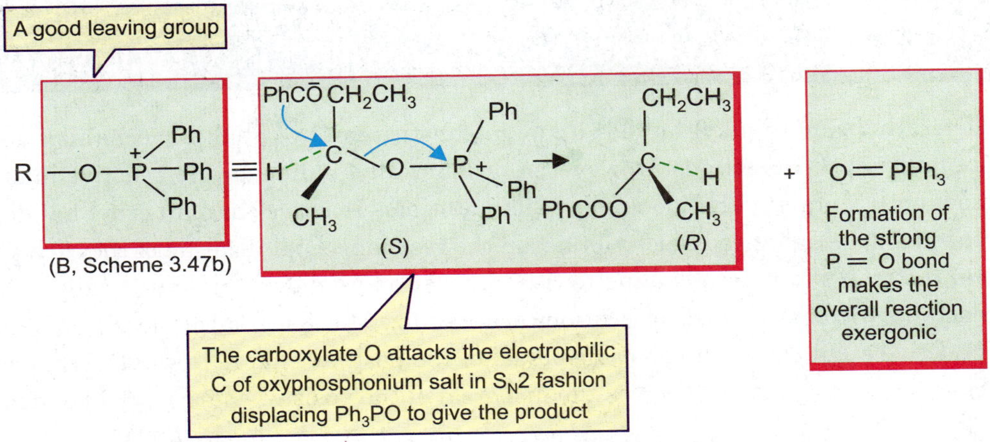

SCHEME 3.47c (Last step of Mitsunobu reaction)

The Mitsunobu reaction is thus not a simple S_N2 displacement of ^-OH from R by $PhCO_2^-$, because HO^- is a very poor leaving group in S_N2 reactions and will not leave under these mild conditions, and besides, Ph_3P and DEAD are required to generate a phosphorus – containing leaving group.

The alcohol is normally 1° or 2°, and the acids that are most effective have a $pK_a < 12$. Thus, carboxylic acids, phenols, thiols, thioacids, and inorganic compounds such as HCN and HN_3 are suitable reactants. Of particular value is carboxylate ion, normally a poor nucleophile is substitution reactions, which works well in the Mitsunobu reaction. More examples of this reaction are in Scheme 3.47*d*.

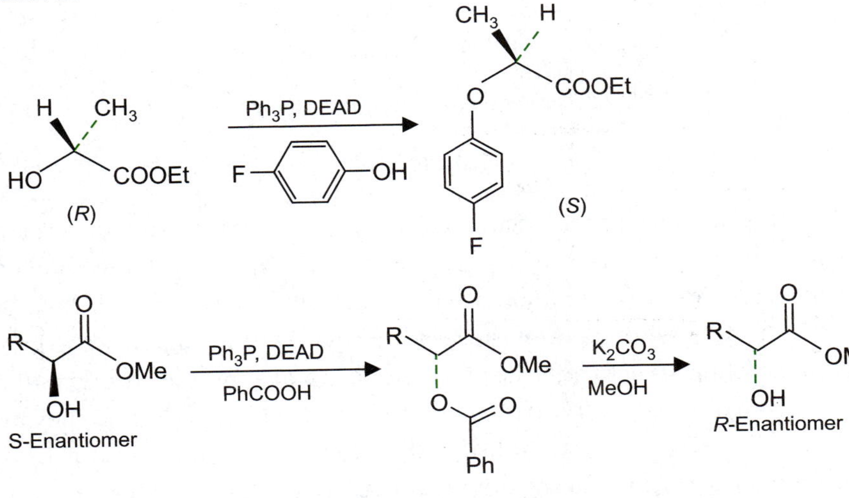

SCHEME 3.47d

3.4 NUCLEOPHILIC SUBSTITUTION OF ALLYLIC SYSTEMS—S_N1' AND S_N2' REACTIONS. REARRANGEMENT IN ALLYLIC SYSTEMS

Generally primary alkyl halides undergo substitution by the S_N2 mechanism and do not undergo S_N1 reactions. However, a primary allyl halide is very reactive in an S_N1 reaction *e.g.*, an allyl halide is more than 30 times reactive than an ethyl halide. The reason for the enhanced reactivity lies in the resonance stabilization of the carbocation and the transition state leading to the carbocation as shown for structurally isomeric reactants (Scheme 3.48). Carbocations are stabilized by dispersal of the positive charge. Inductive stabilization involves dispersal of the positive charge through sigma bonds, resonance-stabilization involves the dispersal of the positive charge by *pi*-bonds.

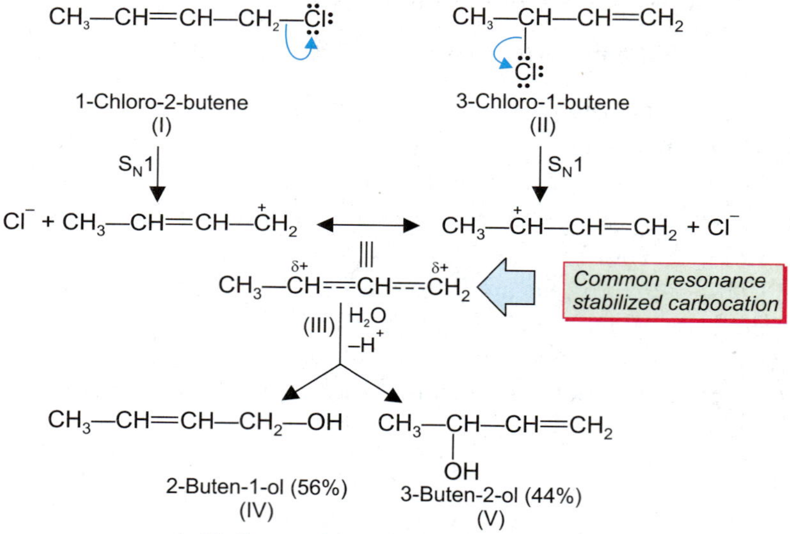

Allylic rearrangement-S_N1' mechanism

SCHEME 3.48

When the S_N1 mechanism occurs with an allylic rearrangement, it is often called the S_N1' mechanism. The structurally isomeric reactants (I and II, Scheme 3.48) ionize (at different rates) to give a common allylic carbonium ion (III) which then undergoes nucleophilic capture at both ends of the allylic system, but in this case the two ends are not equivalent. Thus structurally isomeric products are formed in unequal amounts. However, the same product mixture is obtained from both (I and II) as they are formed through a common resonance stabilized carbocation (III).

The symmetry of the parent allyl cation (Scheme 3.49) makes the two terminal CH_2 groups equivalent. Thus the nucleophile, *e.g.* water in the case of hydrolysis, will bond to either end. However, in the absence of isotopic labels, a single product is formed.

If the symmetry of the reactant is reduced by the site-specific introduction of an isotopic label, *e.g.* a ^{13}C atom (Scheme 3.49), the sites of nucleophilic attack are distinguished, and the two differently labelled products are obtained in equal amounts.

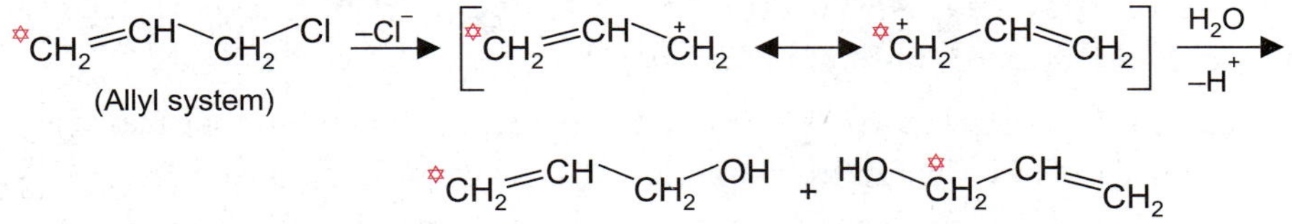

Allylic rearrangement in an S_N1' reaction detection of by isotopic labelling

SCHEME 3.49

S_N2' is substitution nucleophilic bimolecular with rearrangement, is a concerted nucleophilic displacement where the site of attack is at an atom other than the original point of attachment of the leaving group generally one multiple bond separated from the original point of attachment. This mechanism is kinetically similar with that of the S_N2 reaction and the S_N2' reaction normally occurs exclusively when the S_N2 process is sterically hindered (Scheme 3.50).

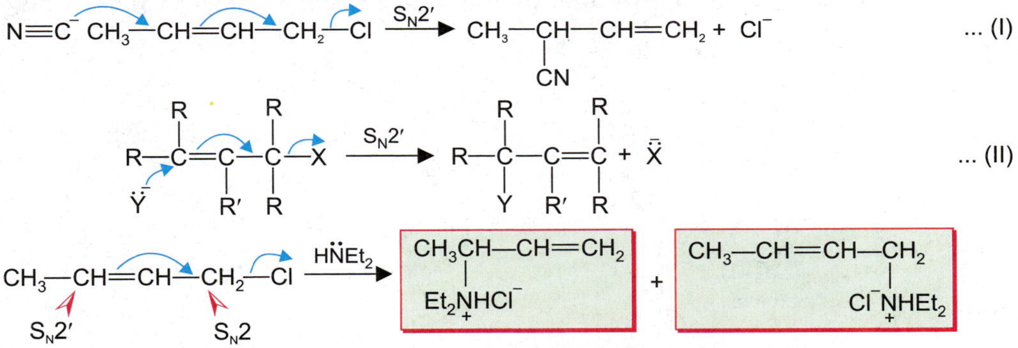

Examples of S_N2 and S_N2' mechanisms

SCHEME 3.50

3.5 NUCLEOPHILIC DISPLACEMENTS AT ALLYLIC HALIDES/ TOSYLATES

Allyl halides also undergo S_N2 reaction at faster rates than primary alkyl halides or even methyl halides. The reason for the greater S_N2 reactivity of allyl halides is that the allyl *pi* bond *pi* cloud reduces the energy of the transition state of an S_N2 reaction. In the transition state, the carbon undergoing reaction changes from the sp^3-hybrid state to the sp^2-hybrid state and has a *p*-orbital. This *p*-orbital forms partials bonds with both the incoming nucleophile and the leaving group. The entire grouping of atoms (transition state) have a negative charge. Adjacent *p*-orbitals, as in an allylic group, undergo, partial overlap with the transitional *p*-orbital. Thus, adjacent *p*-orbitals help delocalize the negative charge and thus lower the energy of the transition state (Scheme 3.51). Unlike S_N1 reactions on allylic systems which are prone to allylic shift of the double bond (low synthesis utility), allylic halides however, react cleanly (without rearrangement) by way of the S_N2 mechanism.

Reaction of (I, Scheme 3.51) proceeds readily at room temperature with carboxylate anion which is a weak nucleophile. The reaction is favoured because it is an intramolecular reaction (favoured) entropically, and secondly the reaction involves a highly reactive allyl bromide.

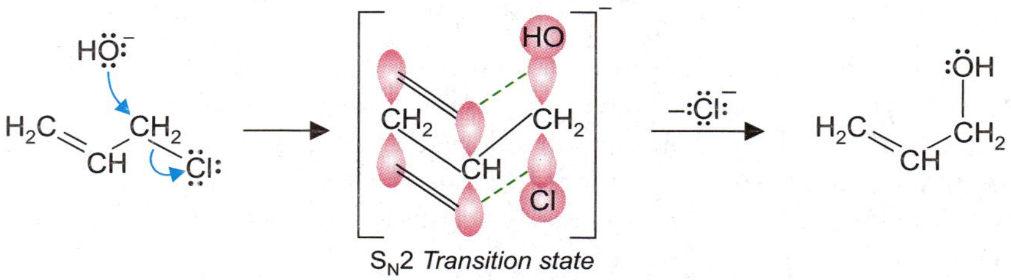

Stabilization of the transition state in an S_N2 reaction of allyl chloride

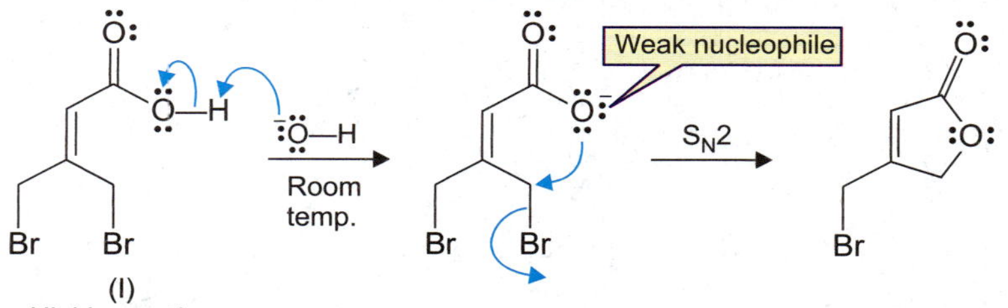

(I)
Highly reactive
allyl bromide

SCHEME 3.51

3.6 NUCLEOPHILIC SUBSTITUTION AT THE BENZYLIC POSITION

Benzylic halides readily undergo S_N2 reactions unless they are tertiary. Tertiary benzylic and allylic halides, like other tertiary halides, are unreactive in S_N2 reactions due to steric hindrance. This high reactivity during S_N2 displacement on the benzylic halide is due to the fact that the *p* orbital that partially bonds with the nucleophile and the leaving group also overlaps with the *pi* electrons of the ring (Scheme 3.52). This stabilizing conjugation lowers the energy of the transition state and this increases the reaction rate. Benzylic compounds also react rapidly by the S_N1 pathway because of the relative stability of benzyl cation (Scheme 3.52).

EXERCISE 3.6

Write the outcome of the reactions (Scheme 3.51a) and predict the reaction to be S_N1 or S_N2 in each case.

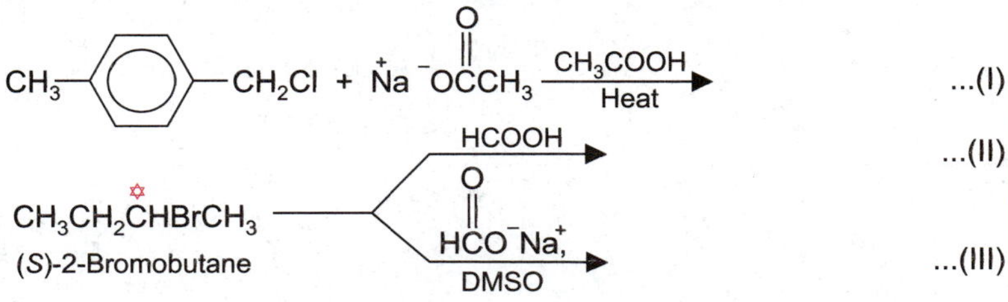

SCHEME 3.51a

ANSWER. *(I) The acetate ion is a weak nucleophile, however benzylic halide is a reactive substrate. The substitution can occur either by an S_N1 or S_N2 path to give*

$$CH_3—C_6H_4—CH_2—O\overset{O}{\overset{\|}{C}}CH_3.$$

(II) Racemization will occur via an S_N1 reaction, the solvent HCOOH (a carboxylic acid) is highly polar and protic the product being $CH_3CH_2(O\overset{O}{\overset{\|}{C}}H)CH_3$.

(III) The product will be the (R) enantiomer $CH_3CH_2CH(O\overset{O}{\overset{\|}{C}}H)\ CH_3$ formed via S_N2 reaction since now the nucleophile is good and the solvent is aprotic.

3.7 NUCLEOPHILIC SUBSTITUTION OF VINYLIC AND ARYL HALIDES

Vinylic halides *i.e.*, haloalkenes in which the halogen is directly attached to the double bond and aryl halides do not show reactivity in S_N1 and S_N2 reactions. S_N1 reactions of these halides are so slow that other kinds of reactions occur instead, such as addition to the multiple bond. The failure to undergo S_N2 reactions by these systems is due to the fact that the approach of the nucleophile to the back of sp^2 carbon is repelled the π electron cloud of double bond or the aromatic ring (Scheme 3.53).

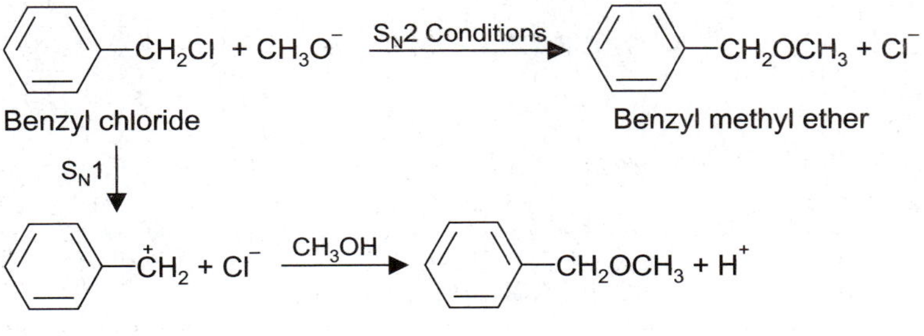

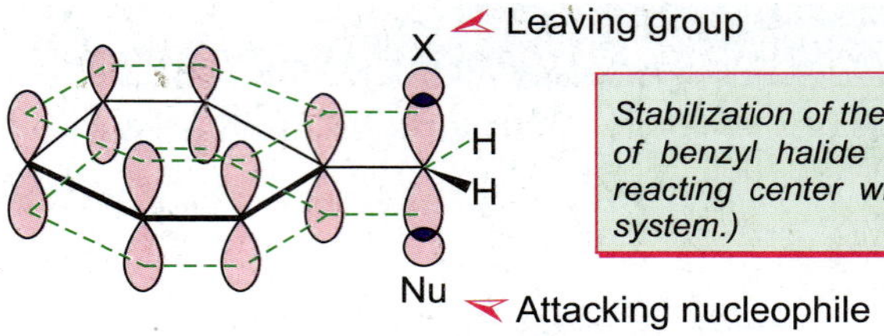

Stabilization of the S_N2 transition state of benzyl halide (conjugation of the reacting center with the benzene pi-system.)

SCHEME 3.52

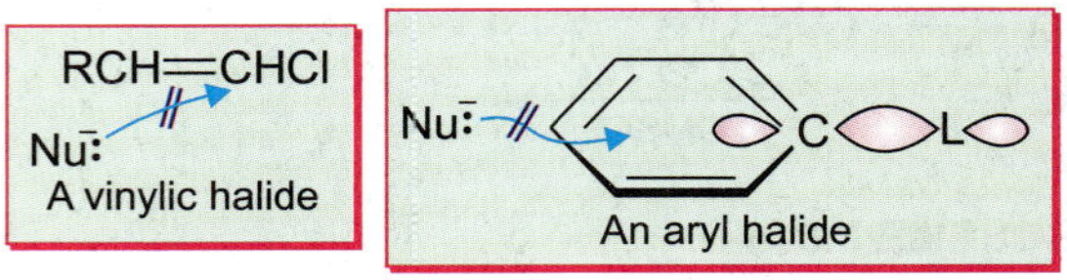

In the case of an aryl halide, the aromatic ring blocks the approach of the nucleophile to carbon at the side opposite the bond to the leaving group: Inversion of configuration is impossible.

SCHEME 3.53

The difference between the energy needed to form a vinyl cation and that needed for a simple primary carbocation is comparable to the difference between primary and secondary carbocations. Secondary carbocations are common intermediates in many reactions but simple primary carbocations are practically unknown in solution. Primary vinyl cations are similarly not known is S_N1 reactions. The decreased stability of a vinylic cation (Scheme 3.54) is due to the positive charge being on a more electronegative carbon. The positive charge in a vinylic cation is on an *sp* carbon, while the positive charge on an alkyl cation is on an sp^2 carbon; *sp* carbons are more electronegative than are sp^2 carbons, and hyperconjugation is less effective in stabilizing the charge on a vinylic cation.

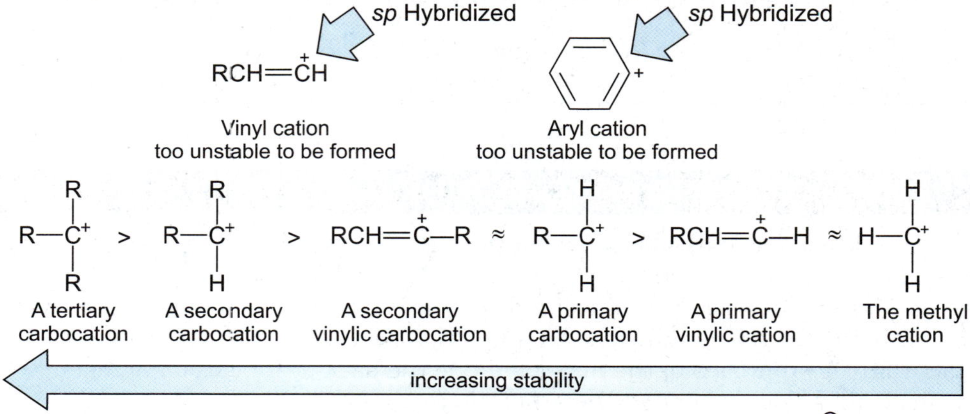

$$CH_3CH_2CH_2Cl + I^- \xrightarrow{S_N2} CH_3CH_2CH_2I + Cl^-$$

n-Propyl chloride

$$CH_3CH=CHCl + I^- \xrightarrow{S_N2} \text{No reaction}$$

1-Chloropropene

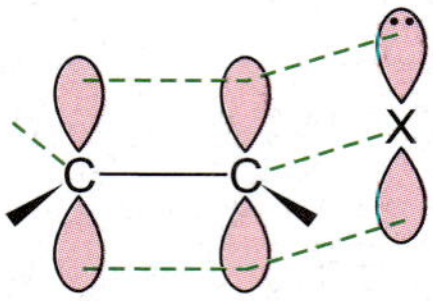

π-Orbital overlap in a vinyl halide

$$[CH_2=CH-\ddot{\underset{..}{X}}: \longleftrightarrow \bar{C}H_2-CH=\overset{+}{\underset{..}{X}}:]$$

SCHEME 3.54

This lack of reactivity is also explained by an increased difficulty in removing an atom with its pair of electrons from a bond to a vinyl orbital with its higher *s*-character than from a simple primary sp^3-orbital. The sp^2 carbon orbital involved in the vinyl halide bond (Scheme 3.54) is expected to produce a shorter and stronger bond than the ethyl sp^3-orbital. However, an additional factor leading to a still shorter and stronger bond is π overlap between the *p* orbital of the double bond and a lone-pair orbital of the halogen as shown. This type of π-overlap can also be represented by resonance involving Lewis structures. As a result of this overlap the C—X bond in a vinyl halide attains partial double bond character.

Vinylic Halides Do not Display S_N2 Reactivity

For a vinylic halide to display S_N2 reactivity and therefore, to reach the transition state (I, Scheme 3.55), the carbon in the carbon-halogen bond needs to be rehybridized from sp^2 to sp.

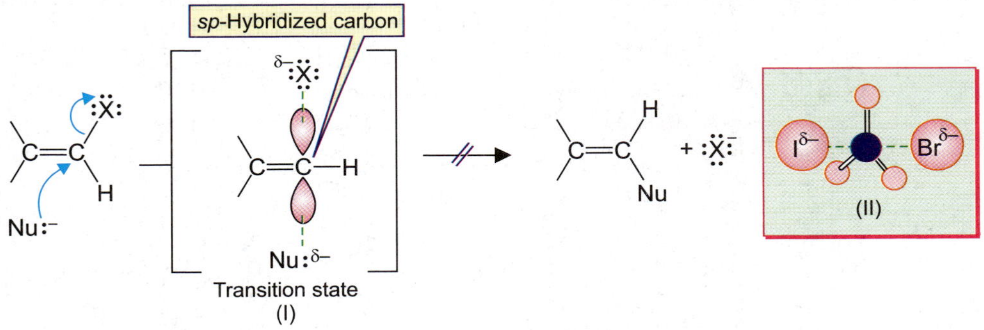

SCHEME 3.55

In an S_N2 reaction of an alkyl halide on the other hand, the hybridization change is sp^3 to sp^2 (II, Scheme 3.55). The sp hybridization state has very high energy and conversion of an sp^2-hybridized carbon into an sp-hybridized carbon needs around 21 kJ/mol (5 kcal/mol) more energy. This reduces the rate of S_N2 reaction of vinylic halides.

Secondly the attacking nucleophile (Scheme 3.55) would have to approach the vinylic halide at the backside of the halogen-bearing carbon and in the plane of the alkene. This would lead to clash between the nucleophile, leaving group and the alkene substituents R (Scheme 3.55a). This will tend to raise the energy of the transition state, and thus both hybridization and steric effects retard the S_N2 reaction to a point of no reactivity.

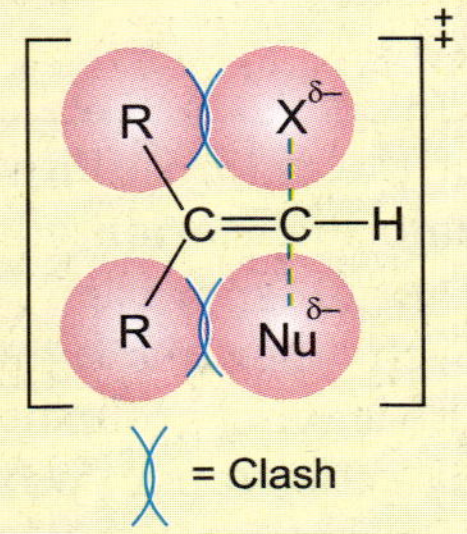

SCHEME 3.55a

3.8 THE S_Ni MECHANISM-TRANSFORMATION OF ALCOHOLS TO ALKYL CHLORIDES - A STEREOSELECTIVE RETENTION OF CONFIGURATION

Reaction of an alcohol with thionyl chloride converts the OH group into a good leaving group. Moreover, the reagent also provides a nucleophile (halide ion) to replace the leaving group. The oxygen atom of the alcohol replaces one of the chlorine atoms attached to sulfur producing an equivalent of HCl. In a solvent like diethyl ether most of the HCl formed during conversion of the alcohol into chlorosulfite is lost (HCl is a gas) and chloride for the substitution step comes from the decomposition

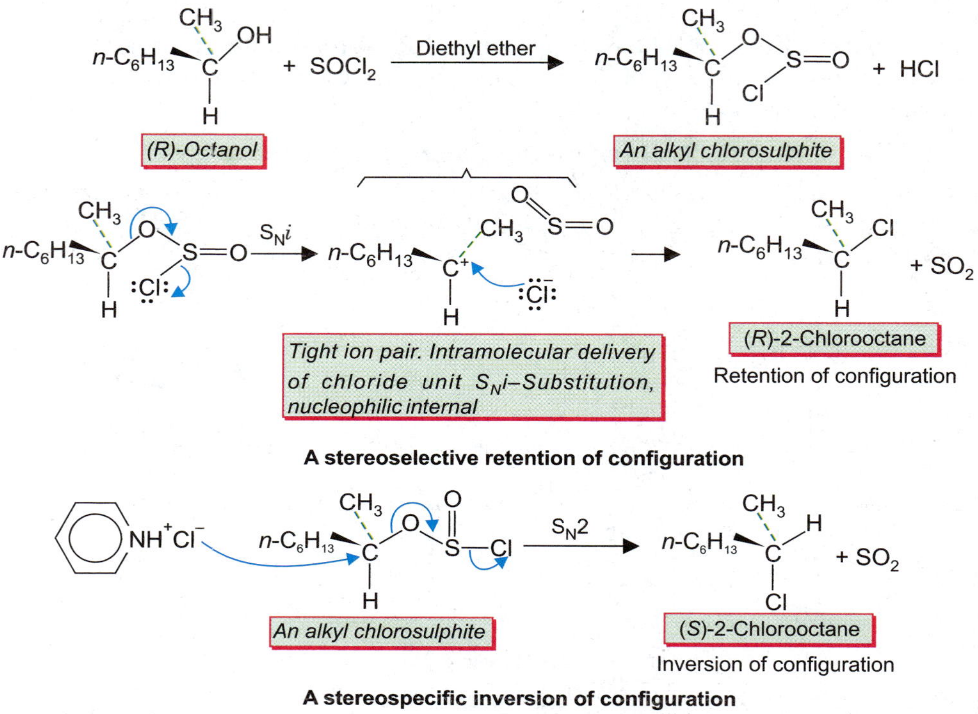

SCHEME 3.56

of the chlorosulfite $^{-}$O SOCl (The leaving group). As the sulfur dioxide departs, the chloride is oriented at the same side of the reacting carbon atom. In the nonpolar ether, charge separation is also unfavorable and the chloride and carbocation are attracted to each other in a tight ion pair. Collapse of the ion pair leads to the product, when part of the leaving group attacks necessarily from the front leading to retention of configuration in a uncommon process called S_Ni mechanism (Scheme 3.56). Thus in the presence of ether optically active 2-octanol reacts with thionyl chloride to give 2-chloroactane with retained configuration, the reaction is then said to proceed with stereoselective retention of configuration, which changes to stereospecific reaction in the presence of pyridine.

When, however, pyridine is used as a solvent pyridinium hydrochloride is formed which ensures that chloride ions exist in the reaction medium. The attack by the chloride ion on the electrophilic carbon atom of the alkyl chlorosulfite occurs from the back side by an S_N2 mechanism if the alcohol is 1° or 2° to give the product of inversion of configuration (Scheme 3.56). Thus now this reaction occurs through streospecific S_N2 mechanism. These reactions thus disclose as to how reaction conditions can affect the stereochemical outcome of a reaction.

Compared to ether, tetrahydrofuran or dioxane can solvate a charge better because the ethyl substitutents of diethyl ether provide steric hindrance, making it difficult for the nonbonding electron of the oxygen to approach the compound to be solvated. When the reaction is done in dioxane solution, an oxonium ion with inversion is formed from the solvent and the chlorosulfite ester. The oxonium ion then undergoes substitution by chloride. Two inversions are involved so that the result is overall retention (Scheme 3.56*a*).

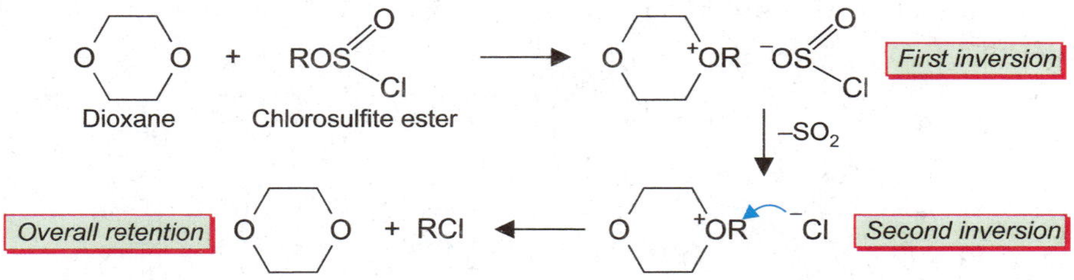

SCHEME 3.56a

One may see that the leaving group in these reactions is created by a concerted movement of electrons which produce the very stable sulfur oxygen double bond of SO_2 and displacement of chloride ion from the ester group. As expected the chlorosulfite esters of 3° alcohols react by an S_N1 mechanism (decomposition of the chlorosulfite ester to R^+, SO_2 and Cl^-) when Cl^- at tacks the carbocation.

3.9 MIXED S_N1 AND S_N2 REACTIONS

A majority of epoxide ring opening reactions occur with inversion of stereochemistry at one end to the ring supporting an S_N2 mechanism under basic or neutral conditions (this indeed is the major pathway). The less substituted carbon is attacked by the nucleophile (Scheme 3.56*b*).

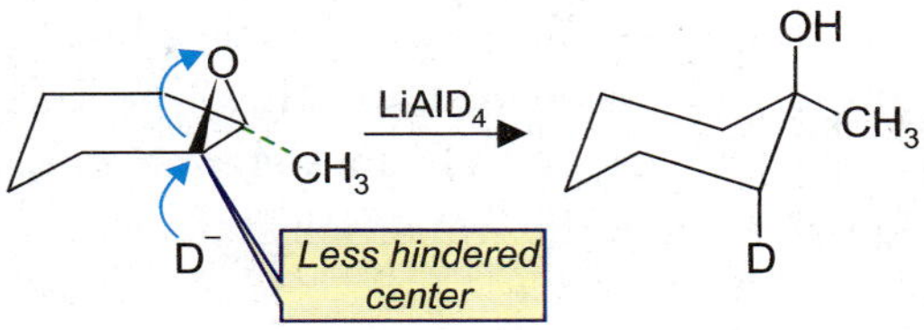

S_N2 ring opening of 1,2-epoxy-1-methylcyclohexane ($LiAlD_4$). Exclusive trans diaxial product formation (rules of an S_N2) reaction hold-attack at less hindered carbon with inversion).

SCHEME 3.56b

Under acidic conditions however, it is the protonated epoxide that undergoes the backside displacement *i.e.,* S_N2 type opening. The acid catalysis changes the regioselectivity, the attack is on the more hindered center. Consider the following points:

- The opening of epoxide ring in 1, 2-epoxy-1-methylcyclohexane with HBr, yields one isomer in which the Br and OH groups are *trans* (Scheme 3.56*c*) [*In fact when an epoxide ring is fused to a cyclohexane ring, the S_N2 opening always gives* 1, 2 *diaxial (trans product) rather than* 1, 2*-diequatorial product (trans product)*]. Thus inversion is observed at the stereocenter, the reaction is S_N2 type.

- Interestingly bromide ion (Br^-) attacks the more hindered epoxy carbon which may suggest an S_N1 mechanism.
- Unlike the carbocation intermediate of a true S_N1 reaction the transition state of this acid catalyzed ring opening is a bridged carbocation in which the oxygen atom is still blocking one face, the reaction therefore, takes place from the other side. As a consequence the stereochemistry is still retained and not lost completely which would have been the case if a free carbocation was involved.

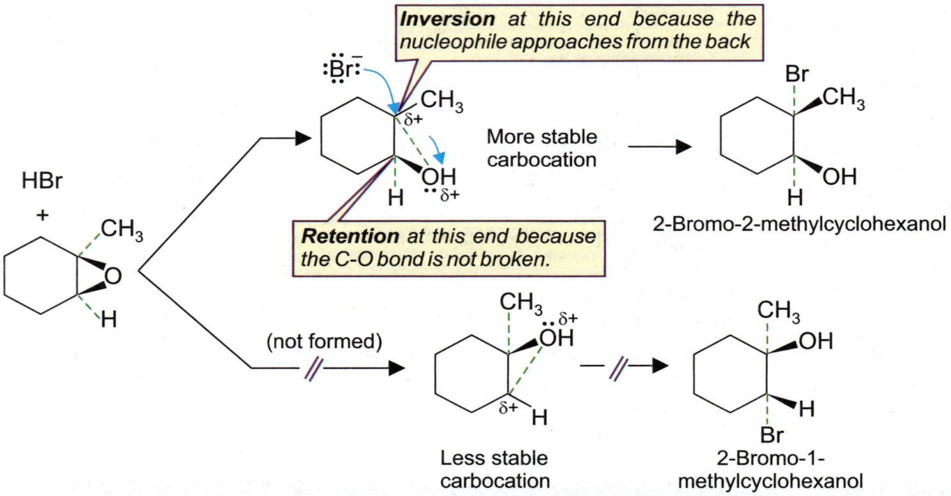

Acid catalyzed ring opening of 1, 2-epoxy-1-methylcyclohexane. A high degree of S_N1 carbocationic character in the transition state but S_N2 displacement.

SCHEME 3.56c

- In summary, the stereochemistry of the acid catalyzed ring opening of an unsymme-trical epoxide is S_N2-like since the attack of the nucleophile is from the side opposite the oxonium ion, bridged intermediate. However, the regiochemistry is S_N1 like, due to the partial carbocation character of the transition state—the uneven charge distribution in the transition state counteracts steric hindrance.

3.10 AMBIDENT NUCLEOPHILES-REGIOSELECTIVITY

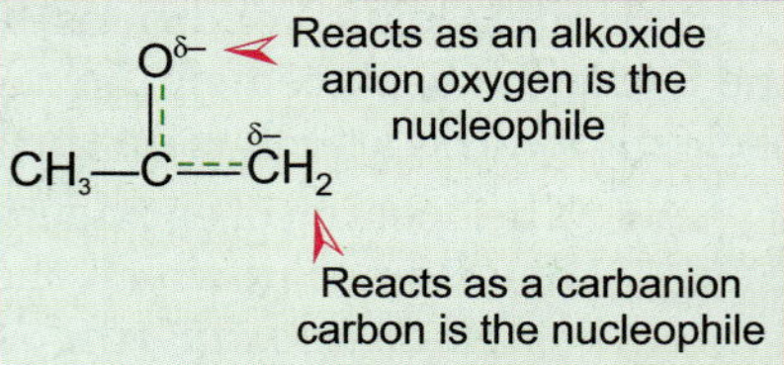

ambident nucleophile

SCHEME 3.57

Ambident nucleophiles (ambi = both, dent "teeth" Latin) are nucleophiles which have two nucleophilic sites ("two teeth") and thus are capable of reacting at two sites. This is shown *e.g.*, for the enolate anion derived from acetone (Scheme 3.57), which of these nucleophilic sites will react with an electrophile largely depends on the electrophile and on the reaction conditions. Protonation occurs preferentially on oxygen, since the resonance hybrid resembles more with the enolate since it is the more stable resonance contributor (the negative charge is on a more electronegative atom). Chlorotrialkylsilanes also *e.g.*, react exclusively at the oxygen atom of he enolate in reaction called silylation. Silylation is a nucleophilic substitution at the silicon atom by the oxygen atom of the enolate (Scheme 3.58). This preference is due to very strong oxygen-silicon bond compared to carbon-silicon bond. The trimethyl silyl enol ethers can be purified and converted back to the enolate—a process needed for directed aldol condensation. Silyl ethers are also employed for protecting hydroxyl groups.

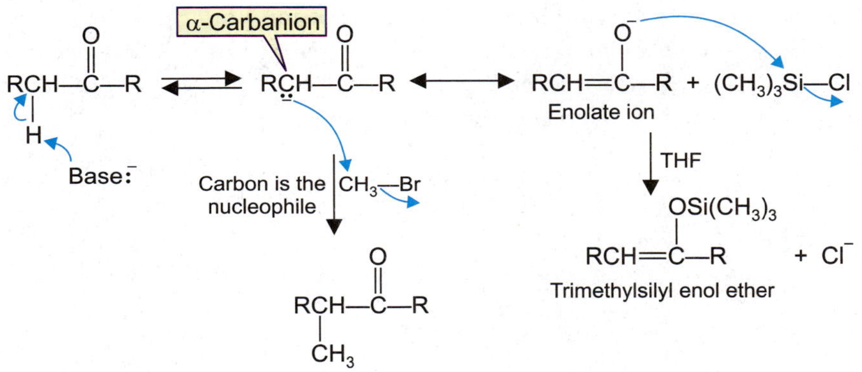

SCHEME 3.58

The anion formed after the removal of a proton with a base can also attack a saturated carbon to yield α-substituted products (*C*-alkylation, Scheme 3.58). Nitrogen and oxygen (first-row elememts) are of comparable nucleophilicities. The ratio of the products actually depends on the reaction conditions (Scheme 3.59). Methyl iodide reacts with cyanide ion to give methyl cyanide, the major product and small amounts of methyl isocyanide (Scheme 3.59).

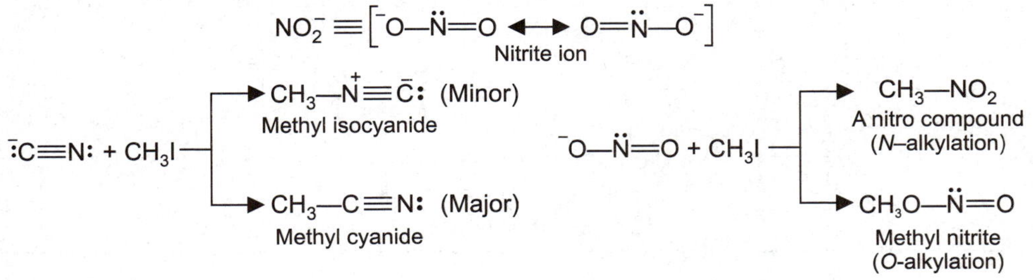

SCHEME 3.59

EXERCISE 3.7

Explain the outcome of reaction (Scheme 3.60).

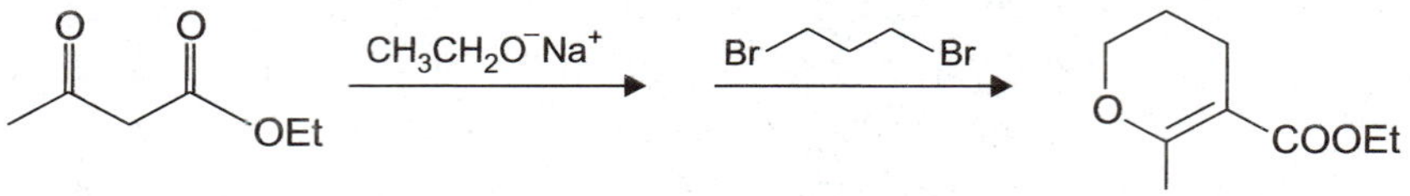

SCHEME 3.60

ANSWER. *Formation of the initial ambident anion (I, Scheme 3.61) is followed by C-alkylation. Formation of a second ambident anion undergoes O-alkylation, to give the cyclic product. The preferred C-alkylation would result in the formation of a 4-membered ring (ring strain), carbon anion is more nucleophilic compared to oxygen anion, since carbon is less electronegative than oxygen.*

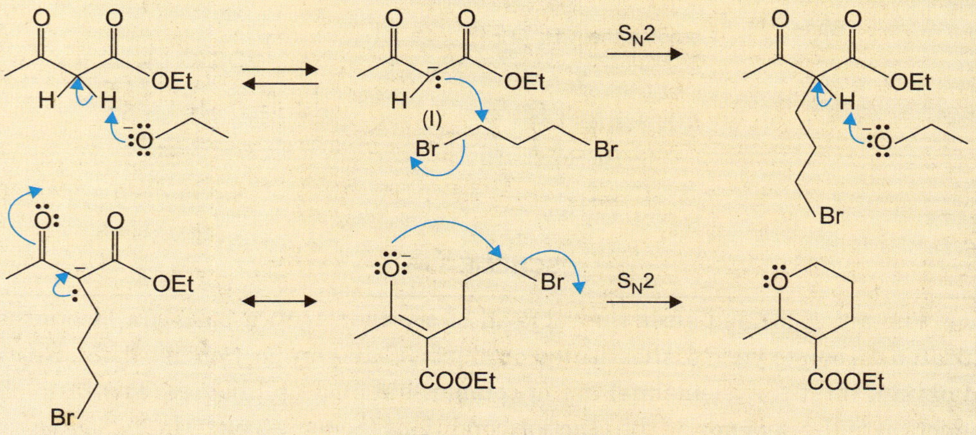

SCHEME 3.61

The outcome of reaction (Scheme 3.62) where the ambident anion (A) can react at either the N or S atoms is explained due to the greater nucleophilicity of sulphur.

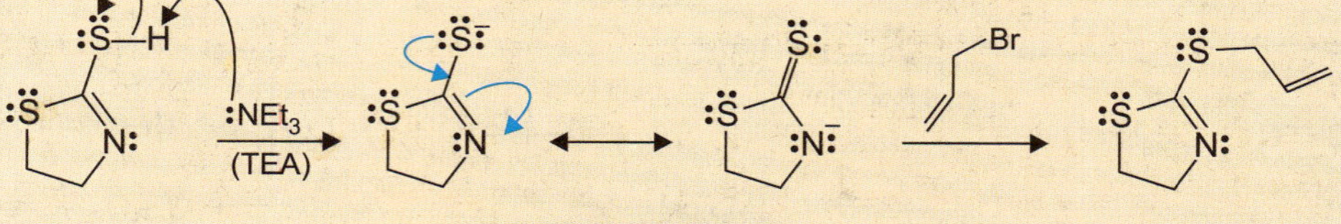

SCHEME 3.62

3.11 NEIGHBOURING GROUP PARTICIPATION—ANCHIMERIC ASSISTANCE

(a) Introduction

The following points may be noted:

- S_N2 reactions at chiral carbons give products with *inverted* configuration. In S_N1 reactions as well, which normally yield at least partially racemized products, it is usually easiest for the nucleophile to attack from the direction opposite to that from which the leaving group is departing. Thus the S_N1 products formed with inversion of configuration should always be more compared with retention.
- While most nucleophilic substitution reactions, whether they proceed by S_N1 or S_N2 mechanisms, result predominantly in inversion of configuration, that is not necessarily the case if the β-carbon-the carbon bonded to the carbon bearing the leaving group–has a substitutent with unshared pair of electrons or an aromatic ring *e.g.,* which can also act as a nucleophile. In these cases, nucleophilic substitutions frequently proceed mostly with retention of configuration and with enhanced rates *i.e.,* both the kinetics and the stereochemical outcome are strongly effected.
- Whether the substrate is an open chain compound or a cyclic molecule, the neighbouring group effects are the most evident when the leaving group and the neighbouring group can easily assume an (S_N2 type geometry) *trans* coplanar arrangement. The stereochemistry of epoxide formation from halohydrins with alkali is related to neighbouring group participation. In the case of cyclohexane derivative it proceeds best and at a fast rate if the groups involved are both axial. This is so, as the axial alkoxide is favorably located for a near attack (S_N2 type) on the carbon bearing the axial halogen (Scheme 3.62*a*). Halohydrins that cannot achieve a *trans*-coplanar *i.e.,* an *anti* orientation between the nucleophile and the leaving group infact do not give epoxides.

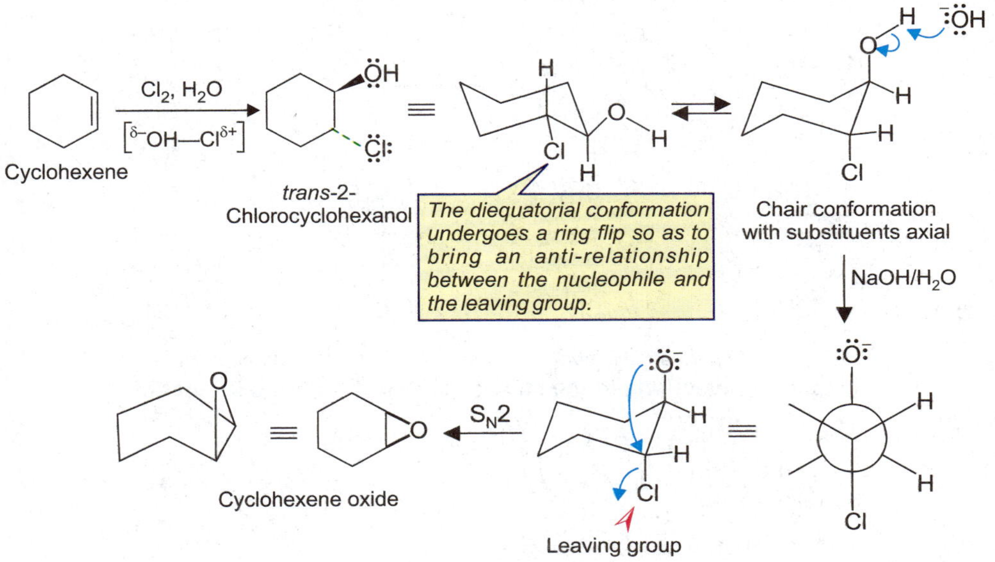

SCHEME 3.62a

- The neighbouring group participation infact is a two step process. In the first step the neighbouring group (acting as an internal nucleophile) attacks carbon at the reaction center (S_N2 attack) and the leaving group is lost to give a bridged intermediate. This is then attacked in the second step by an external nucleophile (another S_N2 attack) and the internal nucleophile goes back to where it came from, two consecutive S_N2 reactions lead to retention of configuration at the reacting carbon.

- The term bridged ion is synonymous with a nonclassical carbocation. A bromonium or a phenonium ion are infact classical ions, but are often (ambiguously) termed bridged ions. None of these involves a three-center two electron bonding.

(b) Examples of Neighbouring Group Participation

(i) An Acyclic (Open Chain) System of 3-Bromo-2-Butanol-Neighbouring Group Participation by a Halogen Atom

The following points may be noted:

- The molecule of 3-bromo-2-butanol has two chiralitycenters, thus it can have four stereoisomers *i.e.,* a pair each of *erythro-* and *threo-*enantiomers (Scheme 3.63, also see Scheme 1.33, one may see that there can be several ways of writing the structure of a chiral compound. Here each Fischer projection has an equivalent projection (180° rotation in the plane of paper) drawn under each.
- Fischer projections are eclipsed conformations and to study neighbouring group participation the concerned groups are to be *trans* and coplanar.
- Firstly one considers the pure (optically active) *threo-*enantiomer and its reaction with HBr to study the neighbouring group participation by the halogen atom (bromine).

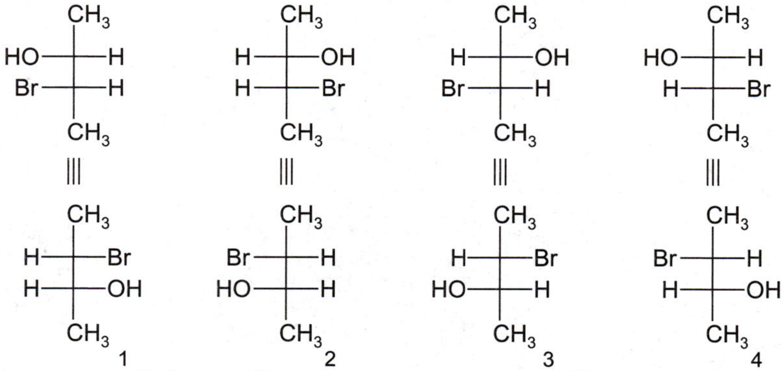

Erythro-enantiomers *Threo*-enantiomers

Fischer projections of the stereoisomers of 3-bromo-2-butanol

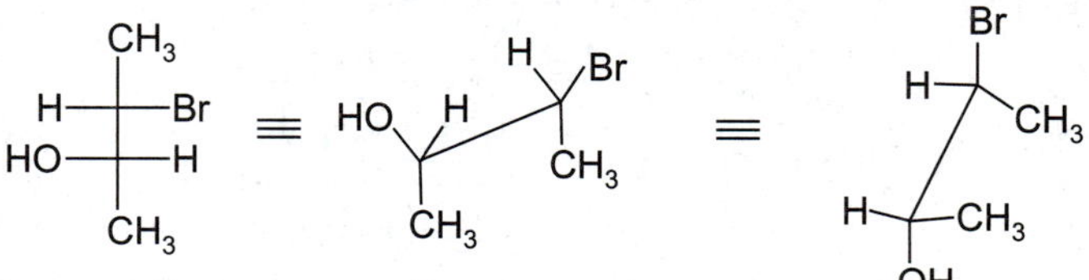

Threo-enantiomer optically active 3

3*a*

3*b*

SCHEME 3.63

- The Fischer projection of optically active *threo*-enantiomer (3, Scheme 3.63) is first transformed into sawhorse formula (eclipsed form 3*a*, Scheme 3.63) following the procedure already explained (see Scheme 1.37*d*). The carbon atoms C-2 and C-3 in 3*a* are then rotated appropriately so as to bring Br and OH in *anti* (*trans* coplanar) conformation (3*b*, Scheme 1.63).
- Reaction of optically active (3*b*, Scheme 3.64) with HBr lead to loss of optical activity to give racemic mixture (2*R*, 3*R* and 2*S*, 3*S* dibromobutane, compare this to the *anti*-addition of bromine to alkenes (see, Scheme 1.100).
- Thus in (3*c*, Scheme 3.64) the bromine atom acts as a neighbouring group displacing the leaving group (resembling intramolecular S_N2 reactions) to give a cyclic bromonium ion with inversion. The brominium ion is then opened by another S_N2 reaction by bromide ion (attack a, Scheme 3.64) with inversion. The net result is the formation of (5, scheme 3.64) with overall retention of configuration (*i.e.*, due to two successive displacements with inversion).
- Due to reactions with achiral reagents each end of the (symmetrical, achiral) cyclic bromonium ion (4, Scheme 3.64) is indistinguishable from the other end, therefore, exactly 50 per cent of each ring opening reaction occurs at each end. This results in the formation of enantiomers (5 and 6, Scheme 3.64) in exactly equal amounts. One may note that the bromonium ion (4, Scheme 3.64) has a plane of symmetry thus ends (C2 and C3) are enantiotopic.
- Similar arguments will explain the formation of *meso*-2-3-dibromobutane (*R*, *S*-2,3-dibromobutane) from optically active *erythro*-isomer (1 or 2, Scheme 3.63, compare with addition of bromine to *E*-2 butene Scheme 1.100).

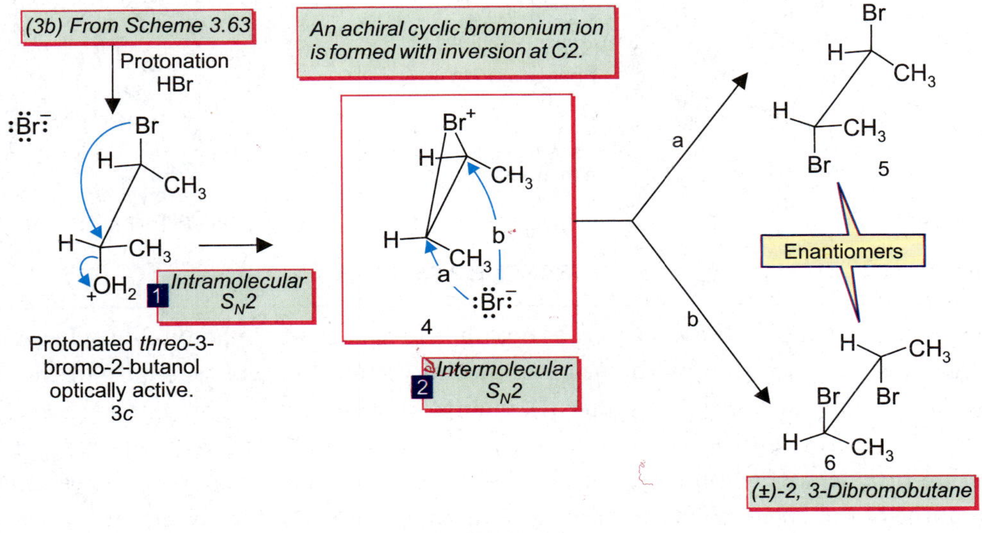

SCHEME 3.64

(ii) Effect of Neighbouring Phenyl Group—π Systems of Aromatic Rings

As the aromatic ring also has nucleophilic properties to a certain extent, one may expect that their presence in the β-position under certain conditions will influence the substitution by the formation of a cyclic ion (phenonium ion). The first to prove the validity of this assumption was Cram in the course of the reaction of stereoisomeric toluene-*p*-sulphonates of 3-phenyl-2-butanols (I and II, Scheme 3.65) with acetic acid. While reacting the optically

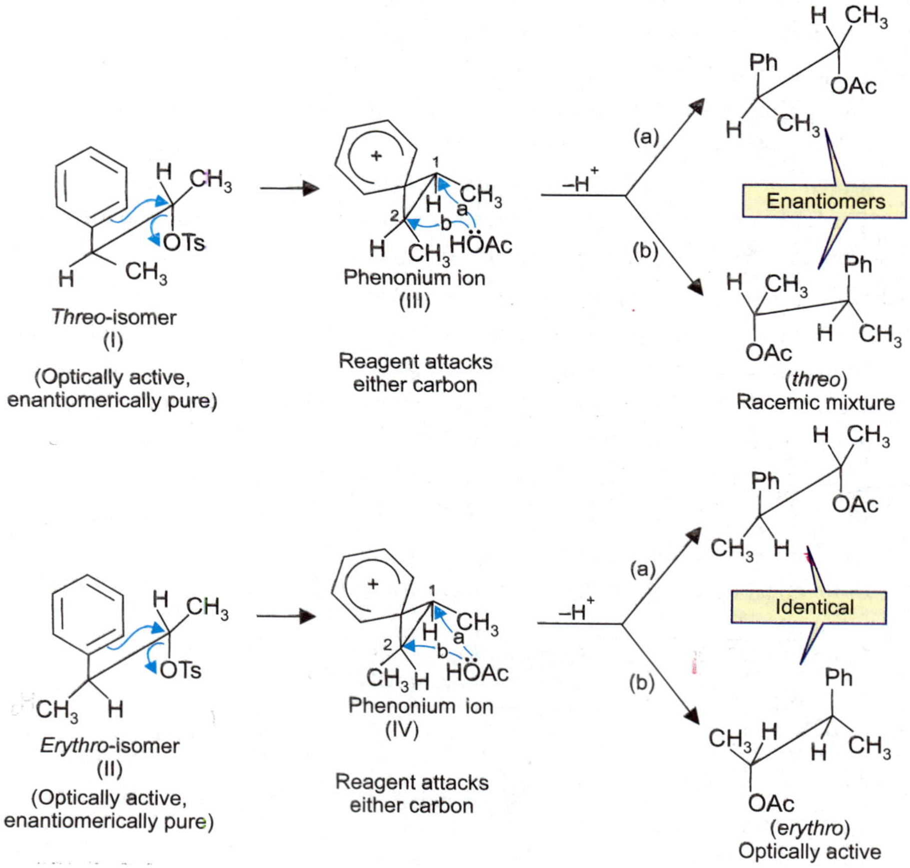

SCHEME 3.65

active *erythro*-isomer or *threo*-isomer, he obtained in both cases 3-phenyl-2-butyl acetates of the same configuration as the original alcohol. The difference was only in the fact that the optically active *threo*-isomer (I), yielded racemic product only and the *erythro* isomer (II) yielded optically active product. The absence of the *erythro*-isomer in the reaction products when one starts from the *threo*-isomer and *vice versa* shows that in the course of the formation of the phenonium ion a change of the configuration at C–1 takes place, and when it opens, the configuration changes either at C–1 or at C–2. The loss of optical activity in the case of the *threo*-isomer is caused by the formation of the absolutely symmetrical ion (III, Scheme 3.65). To enhance understanding, one may show these results by drawing the structures in other stereo-fashion (Scheme 3.66). In either case cyclic phenonium ion is formed. This ion from (I, Scheme 3.66) has a plane of symmetry and gives equal amount of enantiomers, while the phenoniun ion from II has no plane of symmetry but has a C_2 axis and thus attack at either end of three membered ring gives the same acetate (optically active) with the same configuration as the starting tosylate.

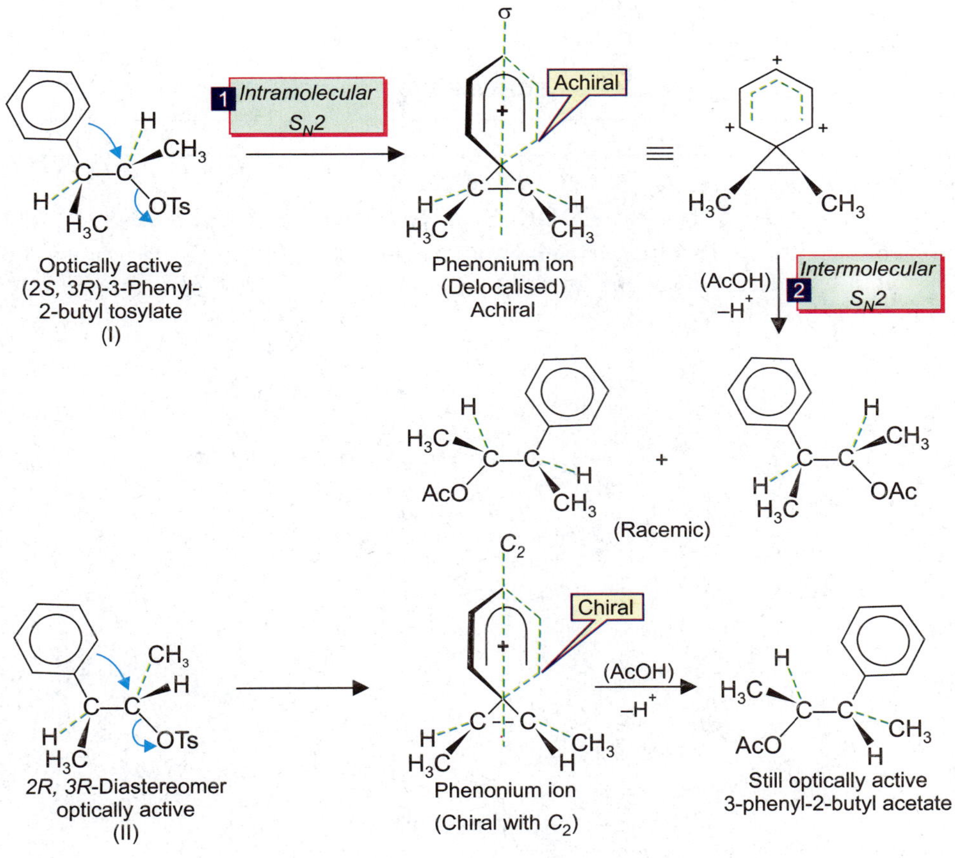

SCHEME 3.66

A further evidence for the involvement of phenonium ions was that in case acetolysis of both (I and II, Scheme 3.66) was stopped before completion the unreacted (I, Scheme 3.66) had completely racemized while (II, Scheme 3.66) had retained its optical activity. Thus in each case (shown here for one of the *threo* enantiomers Scheme 3.67), the phenonium ion and the tosylate ion existed as tight ion pairs. Internal return by tosylate ions (faster than attack by acetic acid) will give the starting tosylates with explainable stereochemistry.

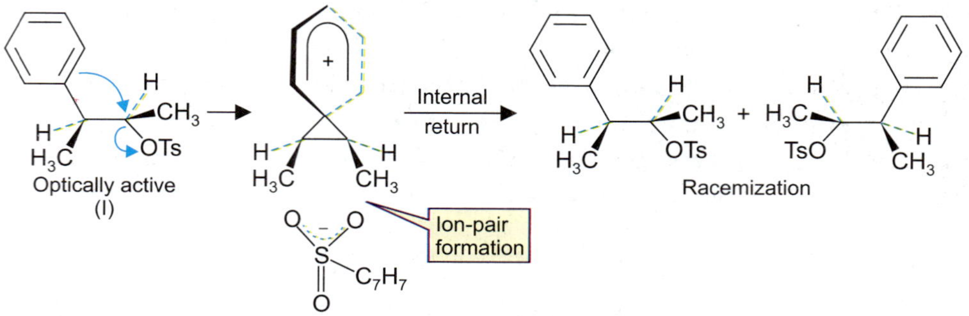

SCHEME 3.67

BOX 3.3
A DIFFERENT LOOK AT ARYL PARTICIPATION

Apart from studying the effect of neighbouring phenyl group as in (Schemes 3.65–3.67), one may look at it still differently (Scheme B 3.7). The overall reaction is as in (Scheme B 3.7).

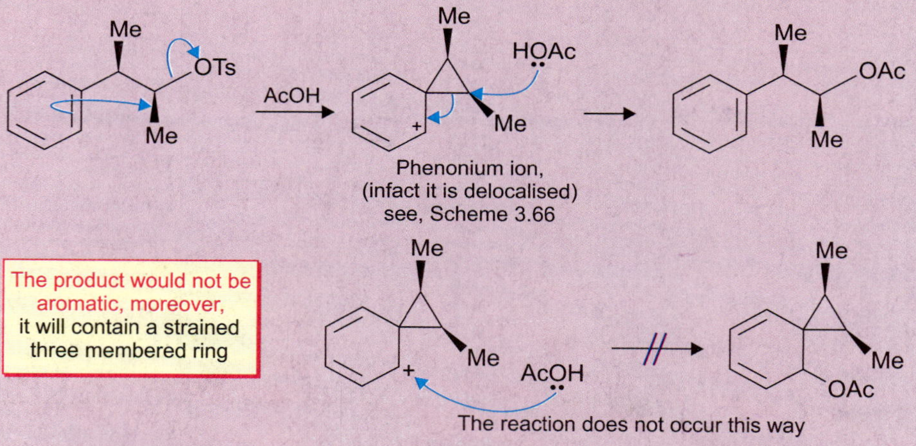

SCHEME B 3.7

First consider the chiral phenonium ion (Scheme B 3.8 which has no plane or center of symmetry but has a C2 axis. Thus starting from enantiomerically pure reactant one gets one pure enantiomer of the product (Scheme B 3.9).

Now consider the symmetrical achiral phenonium ion obtained from diastereomeric reactants to that employed in (Scheme B 3.10). Note that from either pure enantiomeric reactant the same achiral phenonium ion is obtained. Attack on each end of either of these achiral phenonium ions gives a different enantiomer (C or D) so from either enantiomer of starting material one gets the same racemic mixture of products. This loss of stereochemical information compares with an S_N1 reaction of enantiomerically pure compounds. Both reactions pass through an achiral intermediate.

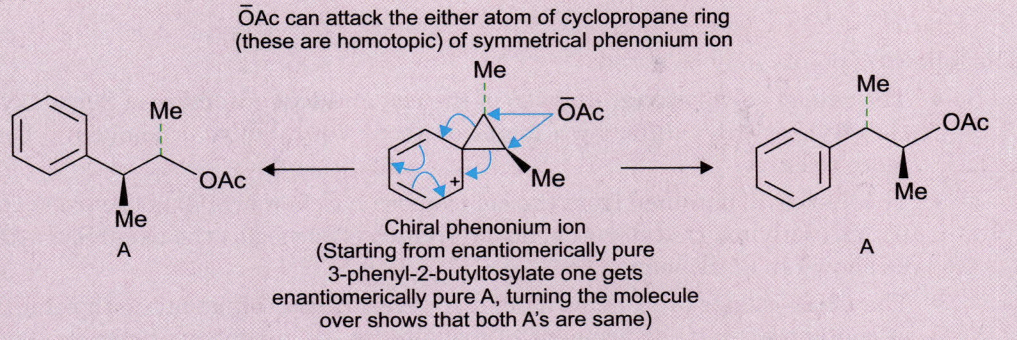

SCHEME B 3.8

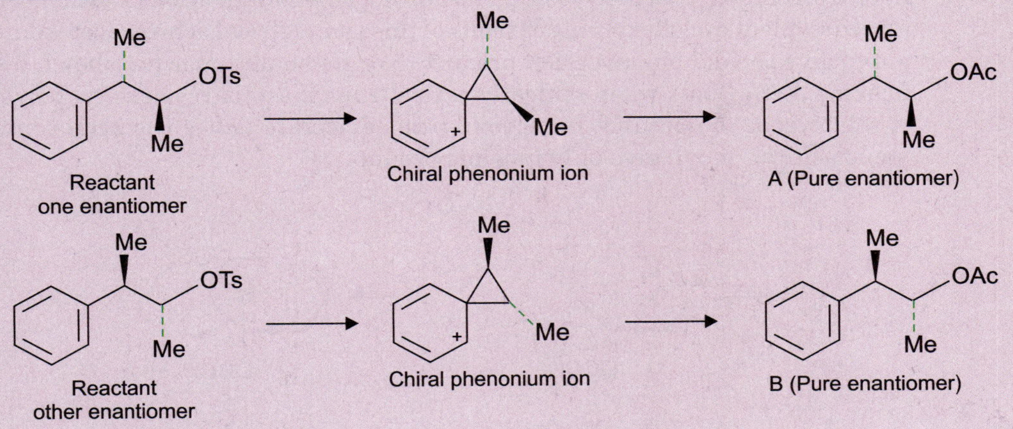

SCHEME B 3.9

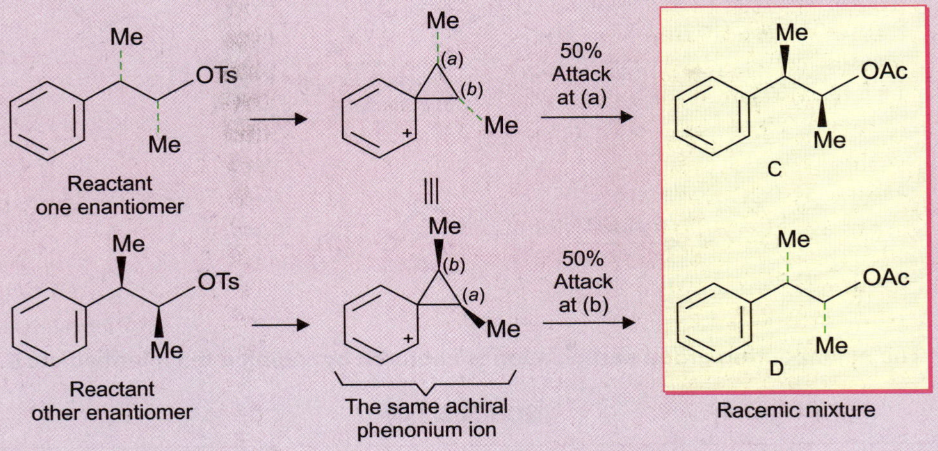

SCHEME B 3.10

(iii) Neighbouring Group Participation in a Cyclic System—Neighbouring Group Participation by Acetoxyl Group

The following points may be noted:

- The rate of solvolysis (acetolysis) of the *cis*- and *trans*-isomers of 2-acetoxycyclohexyl tosylates differ by a factor of about 700, the *trans*-compound being more reactive.
- The diacetate obtained from the *cis*-isomer (1, Scheme 3.68) is the *trans* compound (with inverted stereochemistry) which is formed by the usual S_N2 attack as shown in (2, Scheme 3.68).
- The *trans*-isomer (3, Scheme 3.68) also gives *trans* product but with retention of configuration. Displacement of tosylate by the neighbouring *trans*-placed acetoxy group gives a cyclic five membered acetoxonium ion with a resonance hybrid structure (4, Scheme 3.68). The acetate ions can then attack either of the two equivalent cyclohexyl ring carbons of this symmetrical achiral acetoxonium ion to give the double inversion product (5 or its enantiomer not shown here, Scheme 3.68). Thus when optically active *trans* 2-acetoxycyclohexyl tosylate is solvolyzed, the product is racemic *trans*-diacetate (since the acetoxonium ion is achiral, it can give only racemic product).

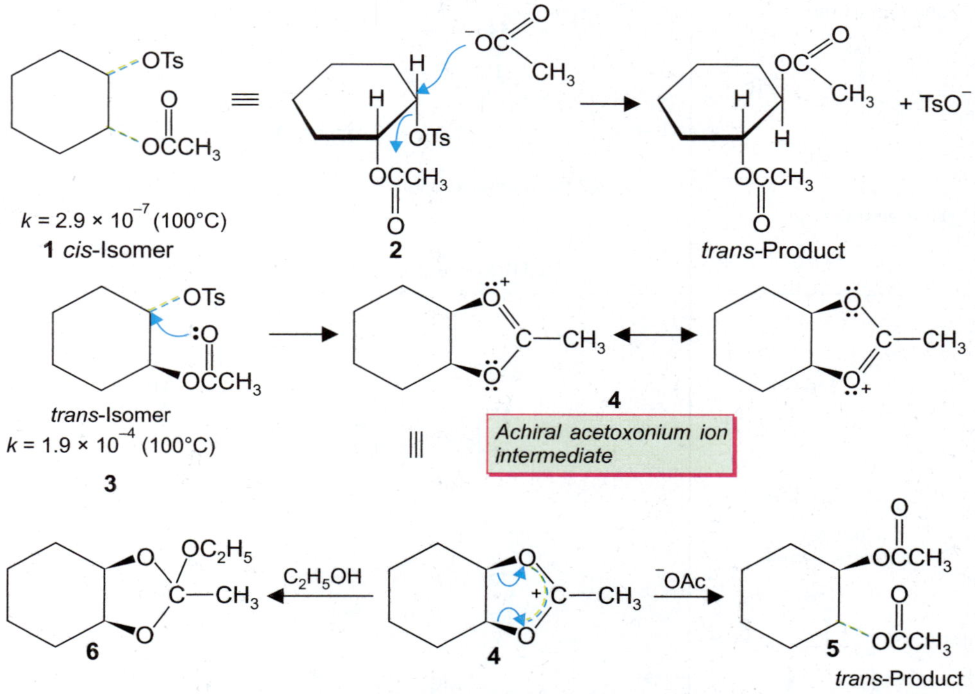

The neighbouring group participation is captured by trapping the intermediate 4.

SCHEME 3.68

- That acetoxonium ion is indeed the intermediate is confirmed by capturing it as a cyclic *ortho*-ester (6, Scheme 3.68) during solvolysis in the presence of ethanol. Ethanol adds to the "carboxyl carbon" of (4, Scheme 3.68) to give the *ortho*-ester.

BOX 3.4

LOSS OF ABSOLUTE STEREOCHEMICAL INFORMATION BUT PRESERVATION OF RELATIVE STEREOCHEMISTRY DURING NEIGHBOURING GROUP PARTICIPATION

One has observed two features during neighbouring group participation

- Acceleration in rate and
- Retention of stereochemistry.

Look at the achiral intermediate oxonium ion (4, Scheme 3.68), if a single enantiomer of the starting material is used, racemic product is formed through this achiral intermediate. Attack at one carbon atom gives one enantiomer; attack at the other gives the other enantiomer (Scheme B 3.11).

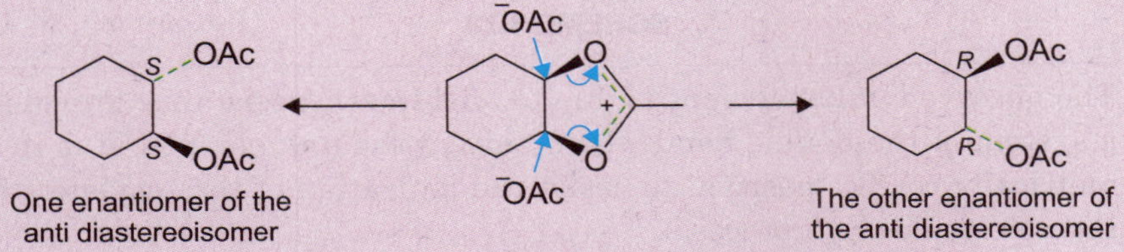

SCHEME B 3.11

(iv) *Neighbouring Group Participation by a Double Bond (Non-Classical Carbocations-Bridged ions)*

(a) Acetolysis of anti and syn 7-norbornenyl tosylate and the saturated compound

A graphic example of largest rate enhancement due to a neighbouring double bond is observed in the solvolysis of *anti*-7-norbornenyl tosylate (I, Scheme 3.69) which reacts 10^7 times as rapidly as its *syn*-isomer (II) and 10^{11} times as rapidly as 7-norbornyl tosylate (III). Moreover, *anti*-norbornenyl tosylate, reacts stereospecifically, with retention of configuration. (A stereo-specific assistance to ionization by π-electrons). The following points may be noted:

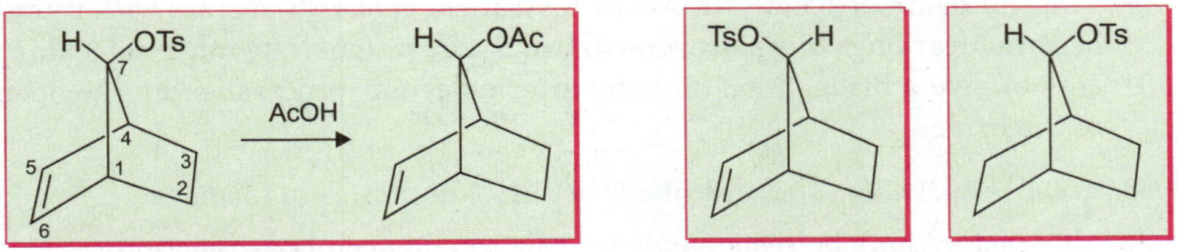

(I) 10^{11} Relative rates (II) 10^4 (III) 1

SCHEME 3.69

- These results are explained by the participation of π electrons of the double bond to give a carbocation of special stability—"the non classical" ion as an intermediate in the reaction of *anti*-7-norbornenyl derivative (Scheme 3.70).

The nonclassical carbocation is a delocalized cation in which the charge is distributed via a closed multicenter bonding. The most common situation involves three carbon atoms. Two of these are bonded to each other by a σ bond while the third is bonded to the other two by a two-electron three center-bond.

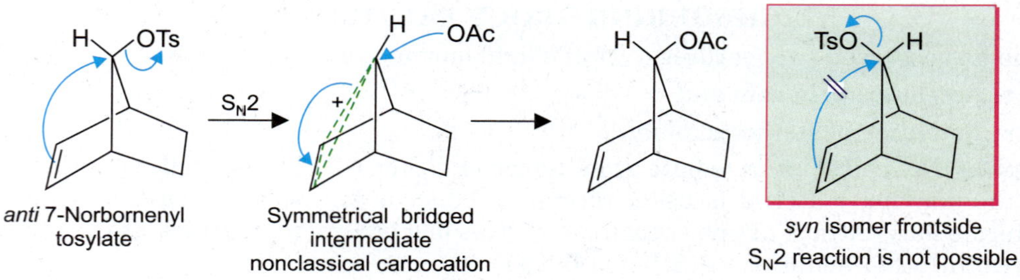

SCHEME 3.70

- The observed enhancement is due to anchimeric assistance provided by the π system of the double bond which only in the *anti* compound is in the geometrically correct position to assist the ionization of the tosylate (this is an intramolecular S_N2 process).
- In the case of *syn*-tosylate the *syn* double bond cannot assist in ionization, it being a front side S_N2 displacement—an unknown process (Scheme 3.70).
- The anchimeric assistance (neighbouring group participation) by the double bond leading to rate enhancement in the case of *anti*-tosylate is faster than the possible direct intramolecular reaction (*i.e.*, S_N2) and is thus the dominant process.
- The attack by acetic acid ($^-$OAc) to the symmetrical bridged intermediate is again another S_N2 process (now intermolecular S_N2 process) and must take place from the rear of the departing leaving group, here the two partial bonds to C–7 in the bridged intermediate (Scheme 3.70) act as a leaving group with respect to the incoming acetic acid as nucleophile.
- The end result of acetolysis of *anti*-tosylate to give *anti*-acetate with retention of configuration is therefore, due to two S_N2 reactions, one intramolecular process to give a bridged ion and one intermolecular process leading to opening of the bridge.

(b) Reasons for remarkably rapid acetolysis of anti-7-norbornenyl tosylate

The following factors may be responsible for this repaid acetolysis reaction:

- The developing cation at C–7 (leading to the formation of a symmetrical bridged intermediate, Scheme 3.70) is in a perfect position to interact with the center of the double bond–a region where electron density is highest.
- In comparison, in a homoallylic cation, the cationic carbon can interact with only one end of the double bond.

- In the norbornenyl system the five membered rings are distorted from planarity to large extent, thus C–7 is brought is close proximity to the double bond. Comparing this situation with 4–bromocyclo-pentene (Scheme 3.71) where the ring is almost planar does not display any rate enhancement on heating in water.

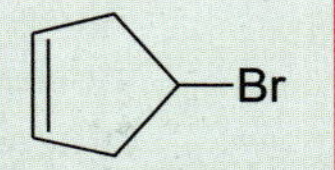

SCHEME 3.71

(c) Evidence–The acetolysis of anti-7-norbornenyl tosylate involves a symmetrical bridged ion intermediate (a highly delocalized non-classical structure) or a rapidly equilibrating mixture of cations.

The most famous controversy of modern times in organic chemistry (Winstein, Cram and H.C. Brown) concerned the question of bridged ions (non-classical carbocations) and unbridged or classical carbocationic intermediates invoked during the participation of neighbouring π systems or σ participation. The point of difference had been if the contributing structures are the contributing resonance forms to a single symmetrical bridged ion or these are unsymmetrical equilibrating cations each having its own separate existence (Scheme 3.72). Evidence for the nonclassical intermediate (A, Scheme 3.72) comes from the ^{1}HNMR study. It is shown that in the case of a relatively stable norbornadienyl cation (B, Scheme 3.72), the 5 and 6 protons are not equivalent to the 2, 3 protons, to show an interaction between the charged carbon and one double bond which indeed is an evidence for the presence of (A, Scheme 3.72).

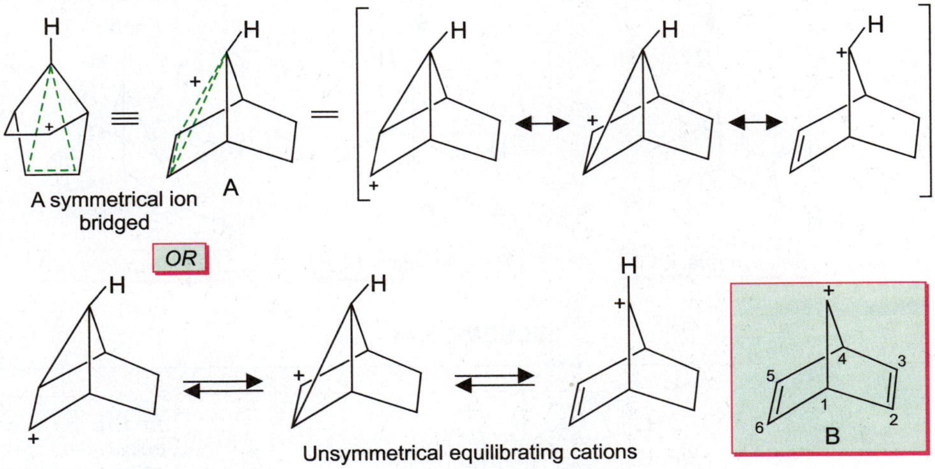

SCHEME 3.72

The following evidence seems to favour the formation of a non classical bridged ion (non-classical carbocation).

- Methyl groups were introduced on the double bond of *anti*-7-norbornenyl derivative– *p* nitrobenzoate of *anti*-bicyclo [2.2.1] hepten-7-ol. The relative rates of acetolysis increased as expected (Scheme 3.73) and it was found that the introduction of second mehtyl group on the double bond had a very large effect on solvolysis compared to the first methyl group. These observations showed that the transition state of this solvolysis reaction is the one in which both ends of the double bond share the positive charge at all times *i.e.*, a non classical bridged cation is involved (Scheme 3.73). Thus both methyl groups exert their stabilizing effect at the same time.

- In case, however, rapidly equilibrating carbocations were involved (Scheme 3.74) and therefore, a positive charge was generated at only one end of the double bond at any given time, a methyl group would assist the formation of the 7-norbornenyl cation. However, the effect of the second methyl group on the other carbon of the double bond should have had little effect (at the most the predicted rate would be only $2 \times 13.3 = 26.6$ and not that large, 148 as was in fact observed experimentally. Thus in the open (unbridged) cations the two methyl groups can never stabilize the cation at the same time—a fact observed in practice.

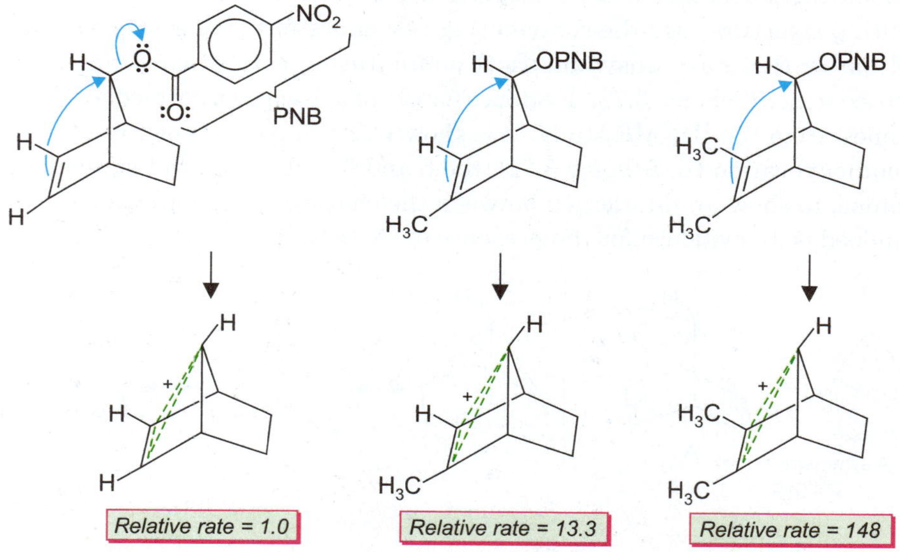

SCHEME 3.73

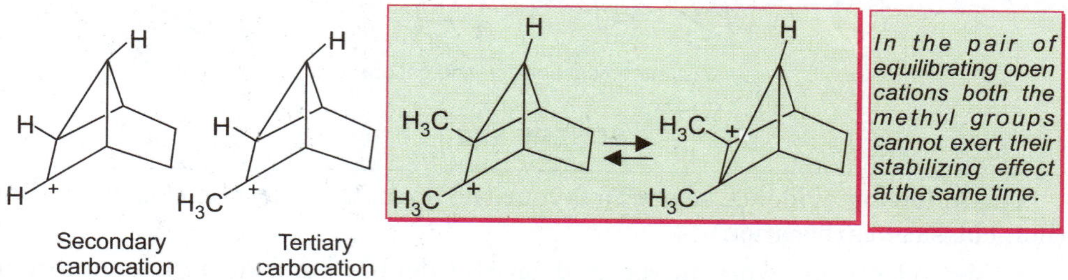

SCHEME 3.74

- The extent of participation of the carbon-carbon double bond in the ionization of *anti*-7-norbornenyl systems is significantly influenced by the substitution at C–7.

The placement of an aryl substitutent at C–7 diminishes the rate enhancing effect due to participation of the double bond. When an aryl group is present at C–7, the resulting benzyl-type stabilization of the potential carbocation at C–7 decreases the importance of participation by the double bond. The extent of stabilization further depends on the substitutents on the phenyl ring. For *p*-methoxyphenyl, phenyl and *p*-trifluoro-methylphenyl, the rate factor for the unsaturated relative to the saturated system is 3,40 and 3.5×10^4 respectively. The double bond has clearly a much important role to play as a neighbouring group with poorly stabilizing *p*-trifluoromethyl substituted system (Scheme 3.75).

- Significantly the solvolysis of *anti*-7-norbornenyl derivatives (Scheme 3.75) proceeded with retention of configuration to give product with original *anti*-configuration provided the phenyl group at C–7 was either unsubstituted or was substituted with electron withdrawing groups (Scheme 3.75). The mixture of *anti* and *syn* isomers was formed when the substitutents were *p*-dimethylamino or *p*-methoxy groups (Scheme 3.76).

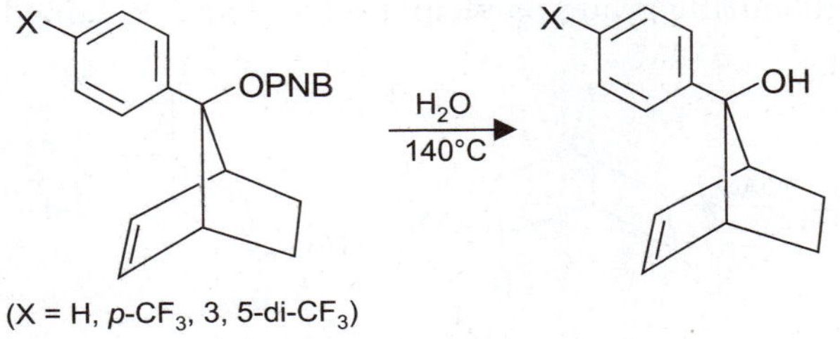

SCHEME 3.75

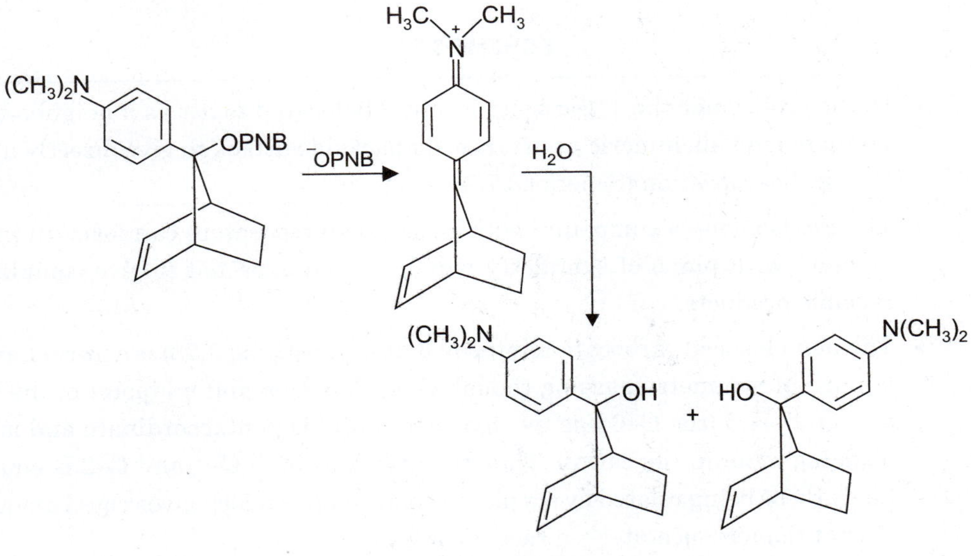

SCHEME 3.76

(v) Neighbouring Group Participation by σ Bonds–Anchimeric Assistance by Alkyl Groups the "Non Classical ion" Hypothesis

Evidence has been presented that C=C acts as a neighbouring group and that a non classical carbocation (a bridged cation) may be formed. Evidence is also available to show that a suitably located C—C in a substrate can also participate in the departure of the leaving group and non classical carbocations may be involved. (Stereospecific participation of of 1, 6 σ-electrons).

The following points may be noted:

- During the acetolysis of *exo*- and *endo*-norbornyl tosylates (Scheme 3.77) it is found that the solvolysis of *exo* isomer is 350 times faster than the *endo* isomer. Both the isomers give only the *exo* acetate (retention of configuration). However, optically pure *exo*-tosylate gives 100% racemic product while an optically pure *endo*-tosylate gives 93% racemic product. These observations are in keeping with neighbouring group participation and are explained:

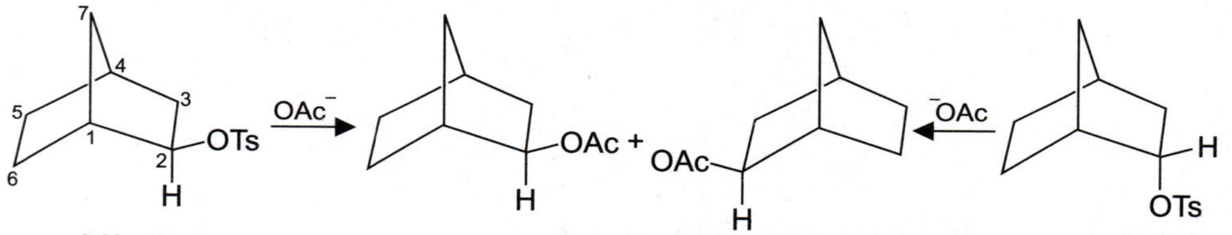

***exo*-2-Norbornyl tosylate**
Relative rate ~ 350
(Optically active)

exo Acetate–retention! and racemization!

***endo*-2-Norbornyl tosylate**
Relative rate = 1
(Optically active)

SCHEME 3.77

- In the *exo*-isomer the 1, 6 σ-bond is suitably located to act as a neighbouring group to lend anchimeric assistance via backside attack to give directly a non classical carbocation (Scheme 3.78).
- The *exo*-isomer—a compound with serveral stereocenters can form an intermediate with plane of symmetry which is then expected to give completely racemic products.
- The non-classical carbocation intermediate, (I, Scheme 3.78) is achiral having a plane of symmetry passing though C–4, C–5, C–6 and midpoint of the C–1 and C–2 bond (the C–6 has two hydrogens and is pentacoordinate and is the bridging atom in the cation, Thus the attack at both C–1 and C–2 is equally likely these being related by a σ plane (enantiotopic) which gives equal amounts of enantiomeric acetates—a racemic mixture.

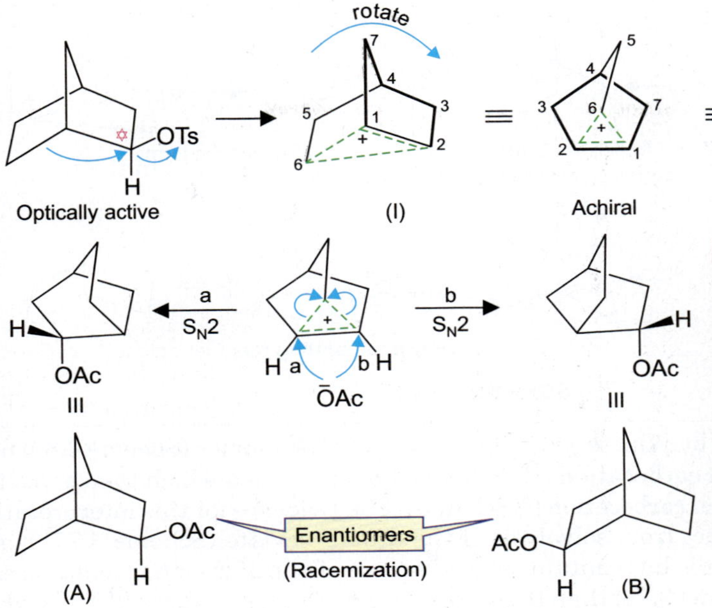

> *The species (1) is a nonclassical carbocation and involves three carbon atoms. Two of these are bonded to each other by a σ bond while the third is bonded to the other two by a two-electron three-center bond (also see, Scheme 3.70). Thus (1) is a three center two-electron bonding system.*

SCHEME 3.78

A Nonclassical Carbocation (a Bridged cation) is a Carbonium ion

The normal coordination number for a positively charged carbon atom without a multicenter bond is 3. When the coordination number of a positively charged carbon is greater than 3 it is considered hypervalent and then it truly represents a carbonium ion. Thus a carbonium ion is a hypervalent carbocation having a valence (coordination number) greater than 3. In nonclassical norbornyl cation (I, Scheme 3.78) C–6 has a coordination number of 5 and C–1 and C–2 have a coordination number of 4. Thus nonclassical norbornyl carbocation is an example of a carbonium ion.

- The attack occurs from the *exo*-side due to the cage structure of the non classical carbocation intermediate (I, Scheme 3.78). Moreover, the attack must occur from the direction opposite that of bridging interaction (S_N2) and this is *exo*-direction.
- Prof. H.C. Brown, however, put forward an interpretation as an alternative to Winstein's non-classical norbornyl cation (Scheme 3.79). The non-classical carbocation is stabilized, relative to secondary cation, by C—Cσ bond delocalization (+ve charge is spread). A rapidly equilibrating classical carbocation (Scheme 3.79) via 1, 2-shift where the inter conversion of classical carbocations was presumed to be repaid relative to capture by the nucleophile (from *exo*-side) would also explain the result. Most of the chemists generally now accent the intermediacy of a non-classical carbocation.

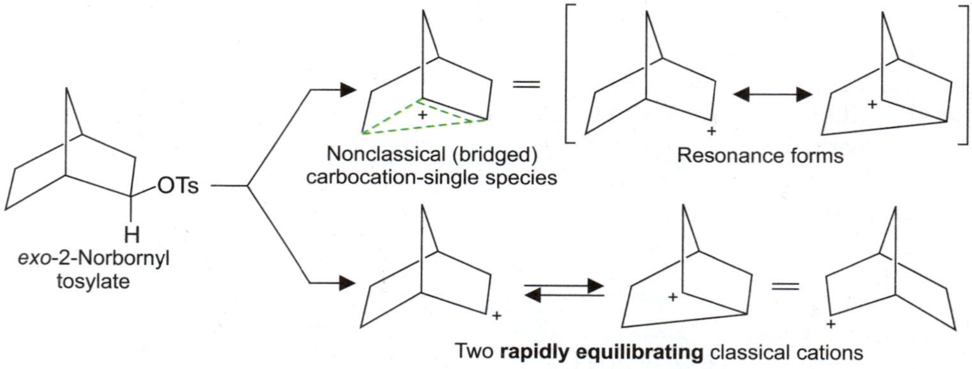

SCHEME 3.79

- It was suggested by Winstein that in the case of *endo*-isomer (Scheme 3.80) first an open classical carbocation (II, Scheme 3.80) is formed which then gives the same non-classical carbocation (I, Scheme 3.80). Evidence for this interpretation is that the product from solvolysis of the *endo*-brosylate (Scheme 3.77) is not completely racemic but contains slightly more of one of the enantiomeric *exo*-acetates (A Scheme 3.78) than (B) corresponding to some inversion. Thus when carbocation (II, Scheme 3.80) is formed some of it goes to give (A, Scheme 3.78) before it can collapse to the bridged ion (I, Scheme 3.80).

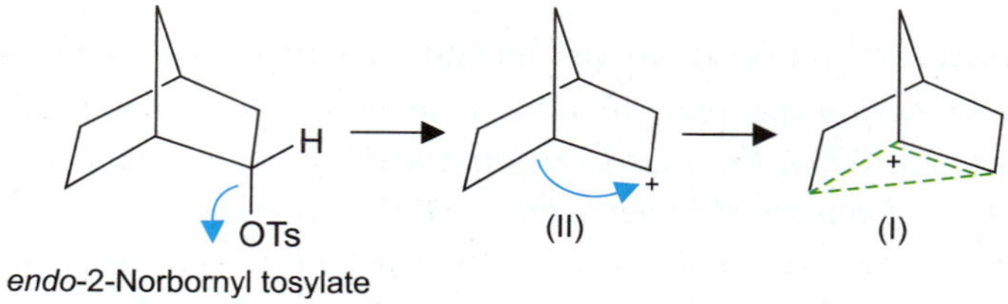

SCHEME 3.80

EXERCISE 3.8

What evidence is against a simple S_N2 displacement on optically active exo-2-norbornyl tosylate (Scheme 3.80a) without the σ bond participation?

ANSWER. *In fact the stereochemical outcome from optically active exo-2-norbornyl tasylate or its epimer (endo-2-norbornyl tosylate) on acetolysis forms the basis for the neighbouring group effect and involvement of a bridged non-classical carbocation. In both substrates optical activity is lost and moreover there is retention i.e., formation of only a racemic mixture of exo-tosylates*

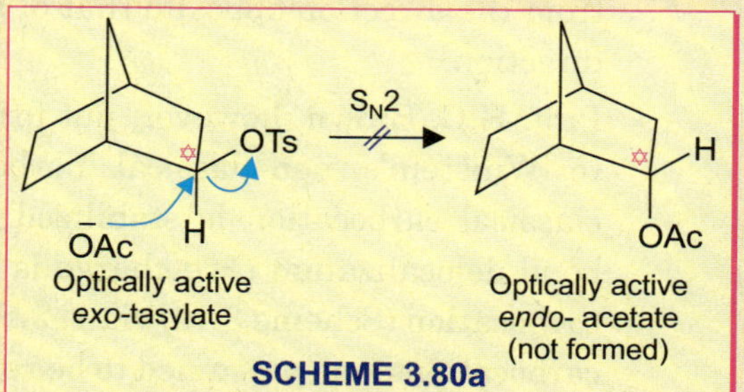

SCHEME 3.80a

Bridging imparts stabilization and nothing more, when other forms of stabilization are available, ions will be then open classical species. ^{13}CNMR spectroscopy distinguishes between equilibrating structures and bridged species. Thus 2-phenylnorbornyl cation (Scheme 3.81) has the classical structure. This benzylic cation gets stabilized by π-electrons of the benzene ring thus bridging has no added advantages. Firstly consider the resonance–stabilized carbocation (Scheme 3.82) the two carbons (shown by dots) which share the positive charges as expected are almost equivalent. In equilibrating ionic structures (two independent ions) such carbons differ by about 100 ppm.

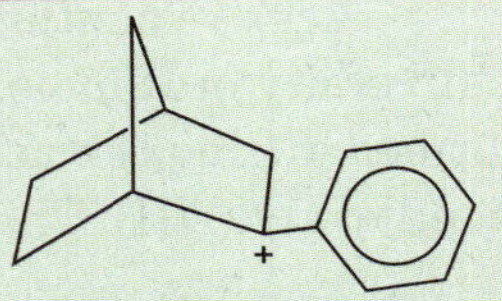

Classical carbocation

SCHEME 3.81

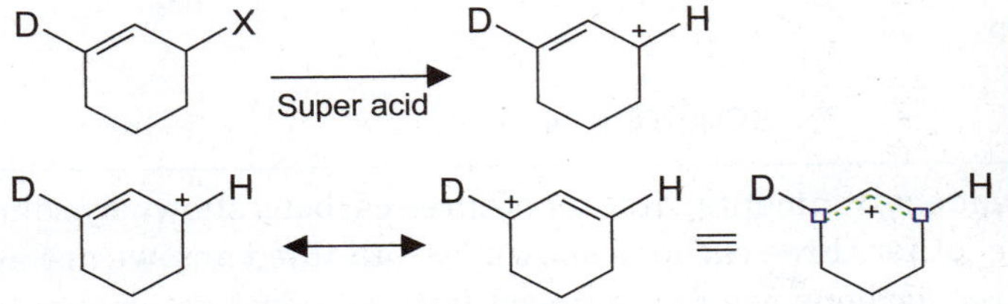

In ^{13}C NMR there is essentially no difference in the two carbons which share the positive charge.

SCHEME 3.82

That 2-norbornyl cation is indeed bridged has been shown by detecting it in a highly polar but non-nucleophilic solvent (super acid media, SbF_5-SO_2). The ^{13}C NMR showed very similar signals for the deuterated and undeuterated positively charged carbons (Scheme 3.83).

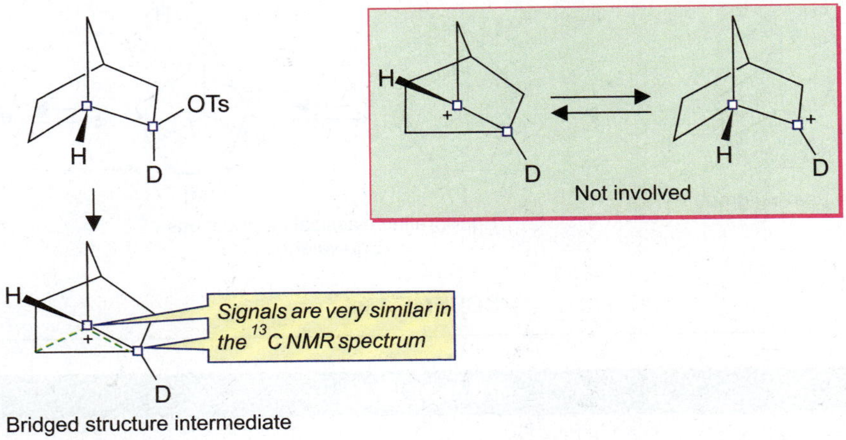

SCHEME 3.83

This evidence excludes the formation of an equilibrating pair of cations, in which case the two dotted carbons should have displayed widely separated signals in ^{13}CNMR (Use of this technique is complicated due to hydride shifts in this case however, the weight of evidence seems to strongly favour the bridged structure for norbornyl cation). [M. Saunders, Yale University].

Lastly mention may be made of ion (II, named after its discoverer R.M. Coates, University of Illinois, Scheme 3.84) studied at low temperature in super acid which is formed during the acetolysis of (I, Scheme 3.84) via the participation of cyclopropane σ bond to give (III). That this ion has structure (II, Scheme 3.84) with three different sets

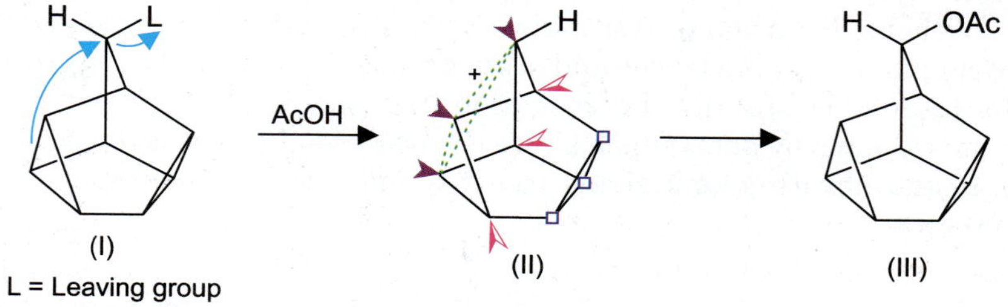

SCHEME 3.84

of carbons in three different environments (first set of three carbons shown by solid arrow share the positive charge, other three carbons shown by half filled arrows are adjacent to these while rest of three carbons are far removed from the first-set shown by dots) was shown since as expected it displayed three signals in its ^{13}CNMR spectrum thus the ion (II, Scheme 3.84) is bridged. In case rapidly equilibrating cations were involved the rearrangement is degenerate—every rearrangement leads in fact to the same cation shown for only two cations and then one could expect only one ^{13}CNMR signal (Scheme 3.85 I and I*a* are equivalent).

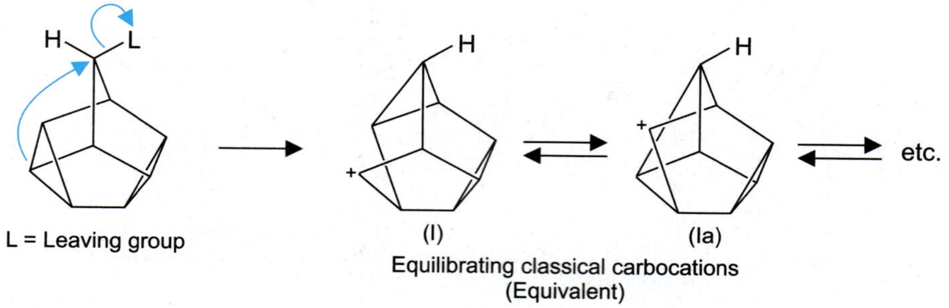

SCHEME 3.85

PROBLEMS

3.1. Comment on the following observations:

(*a*) Neopentyltosylate reacts at a slow rate with lithium iodide but resists reaction with lithium chloride while *n*-butyl tosylate reacts with either of the nucleophiles at reasonable rate.

(*b*) I-Phenylethyl chloride solvolyzes much faster than 2, 2-dimethyl-1-phenylpropyl chloride.

(*c*) Optically active *sec*-butyl chlorosulphite on warming in dioxan affords *sec*-butyl chloride with 97% retention, while the same reaction carried in isooctane leads to 43% inversion.

(*d*) Tri-*tert*-butylcarbinol solvolyzes much more rapidly in acid solutions as compared with *tert*-butyl alcohol.

(*e*) The yields of substitution product in these reactions of *sec*-butyl chloride with $(CH_3)_2N^-$, $(CH_3)_3CO^-$, CH_3O^-, and I^- increase in the order given.

3.2. During displacement of alkyl halides (R—X), the ratio of rates with two amines (Rate ratio = quinuclidine : triethylamine) is given below; with triethylamine it being always slower, how do you account for this rate difference and for its increase in the given series:

CH_3X	57
C_2H_5X	252
$(CH_3)_2CHX$	706

3.3. Name some ionizing and non-ionizing solvents.

3.4. Show the products with mechanism of following reactions:

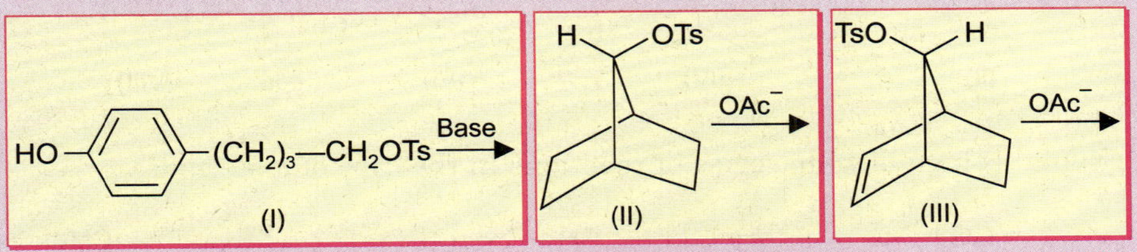

3.5. One of the compounds (I or II, reacts with Cl^- to give (III) which one is that. Give the mechanism of its formation.

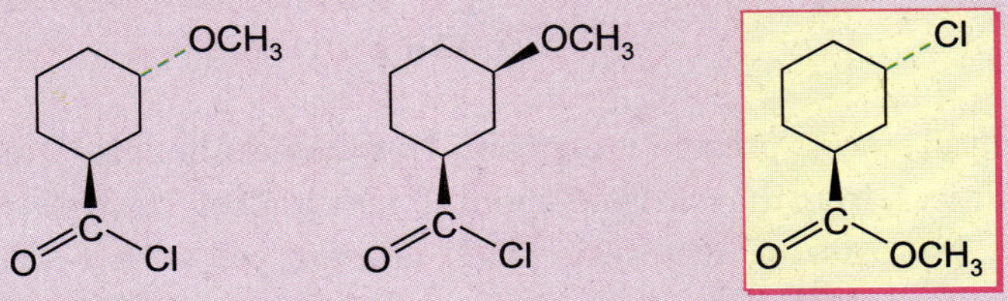

(I) (II) (III)

3.6. What conclusions do you draw from the fact the solvolysis of 1 in the presence of hydride ions ($NaBH_4$) affords small amounts of 2 and 3? (3 is a cyclopropyl derivative.)

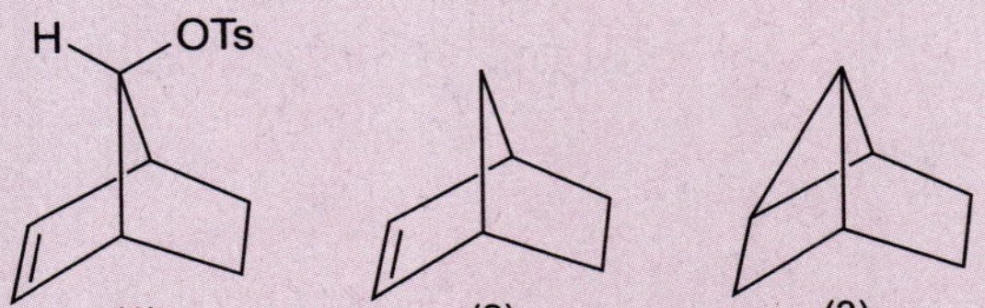

(1) (2) (3)

3.7. Primary halides of the type, $ROCH_2X$ apparently display S_N1 reactivity while most primary halides do not. Propose a resonance based explanation.

3.8. How can you explain the relative rates of solvolysis in ethanol of the following chlorides?

$C_6H_5CH_2Cl$	$C_6H_5CH(Cl)CH_3$	$(C_6H_5)_2CHCl$	$(C_6H_5)_3CCl$
(0.08)	(1)	(300)	(3×10^4)

3.9. Which compound in the each pair is expected to undergo a more rapid solvolysis on heating under reflux in ethanol?

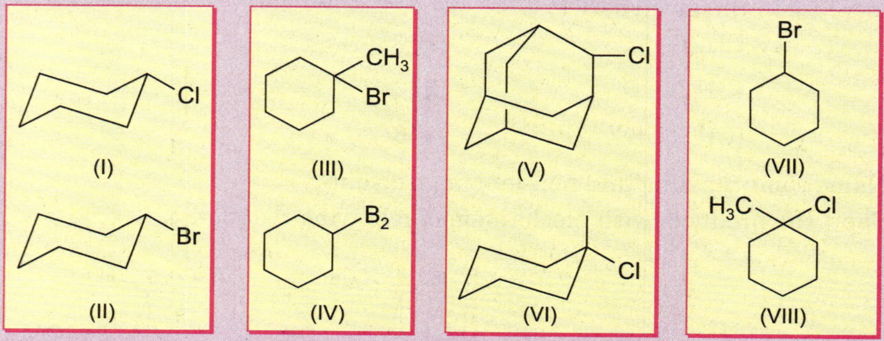

3.10. Explain the following series of reactions in the case of a steroid:

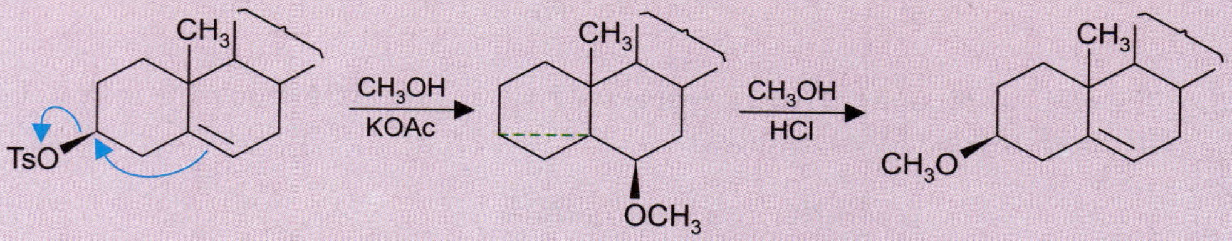

3.11. The neopentyl system present is neopentyl *p*-toluenesulfonate (I) on acetolysis gives products entirely from the rearrangement of the initially formed primary carbocations (Sec. 3.3). The rate of acetolysis of (I) is identical with I-adamantylcarbinyl *p* toluenesulfonate (II). However, during the acetolysis of (II) considerable product (III) is formed from the initially formed primary carbocation in addition to the product (IV) of rearrangement. Explain.

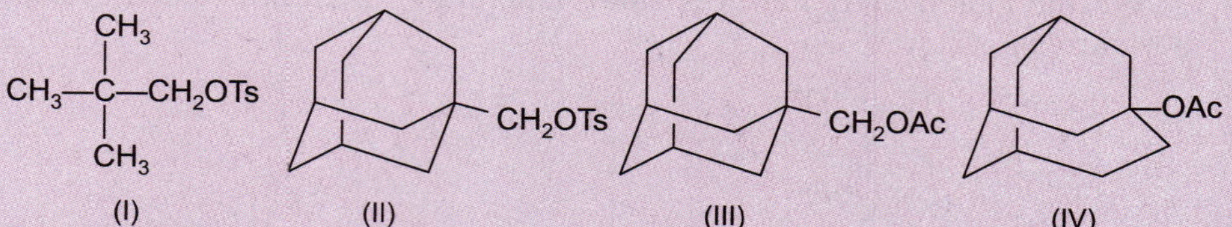

(I) Neopentyl system (II) (III) (IV)

3.12. Discuss the stereochemical outcome of the following reaction.

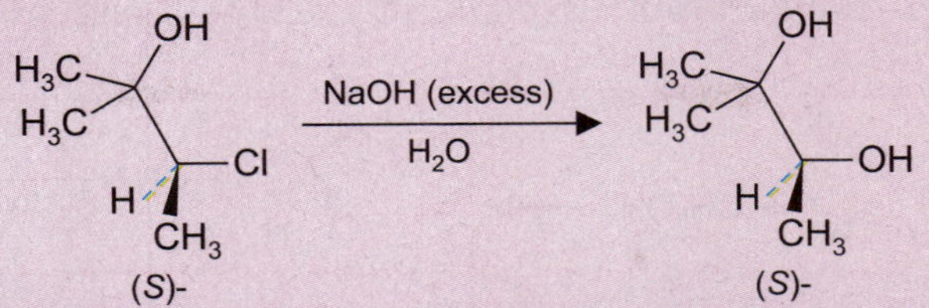

3.13. What is neighbouring group participation? Explain giving an example from a cyclohexane derivative.

3.14. Explain the following solvolysis of optically pure compounds.

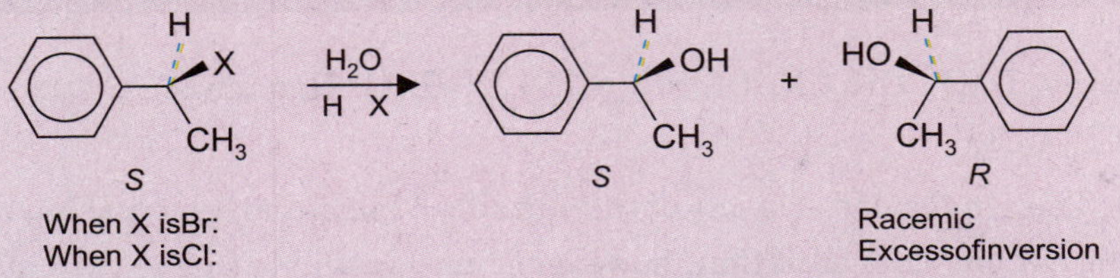

3.15. Explain as to why in the following *endo* tosylate the C–1, C–7 bond fails to provide anchimeric assistance to give the bridged ion?

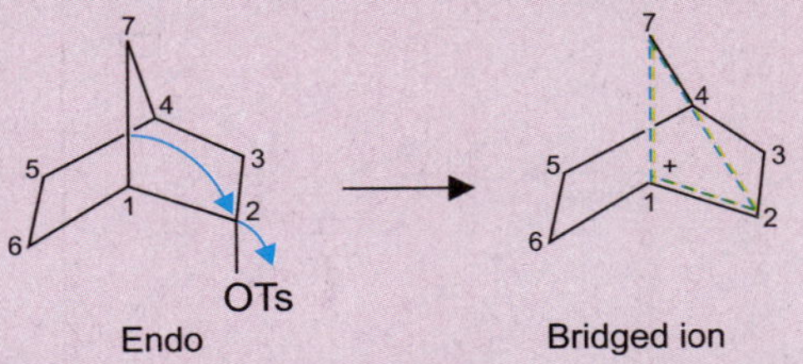

3.16. Why α bromopropionate on treatment with methanol does not undergo the normal S_N2 reaction?

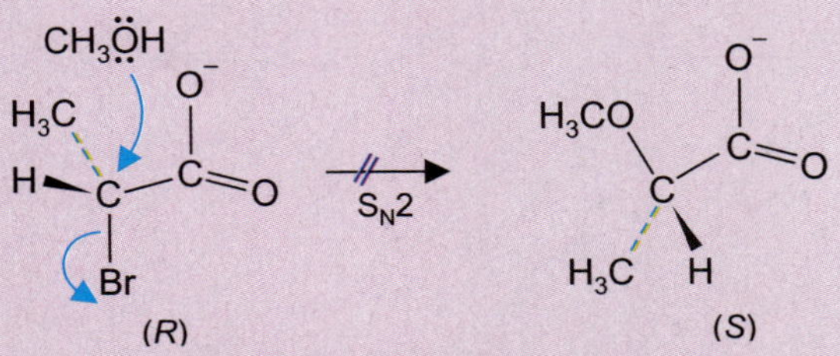

3.17. Predict whether the following reactions will proceed by S_N1 or S_N2 reaction pathway.

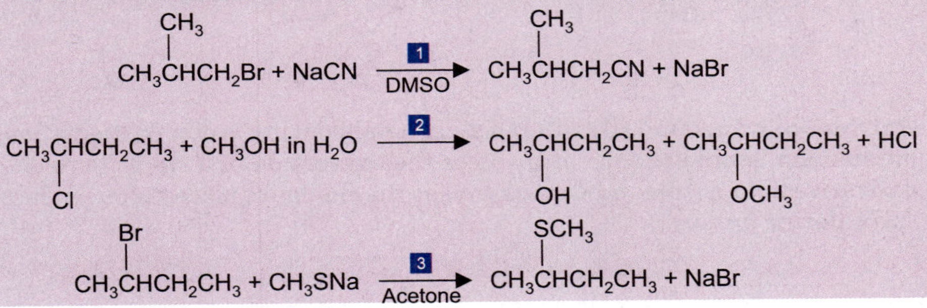

3.18. Solvolysis of (*a*) in methanol gives two ethers while reaction of (*b*) gives a product with inverted configuration. Explain the results along with stereochemical outcome.

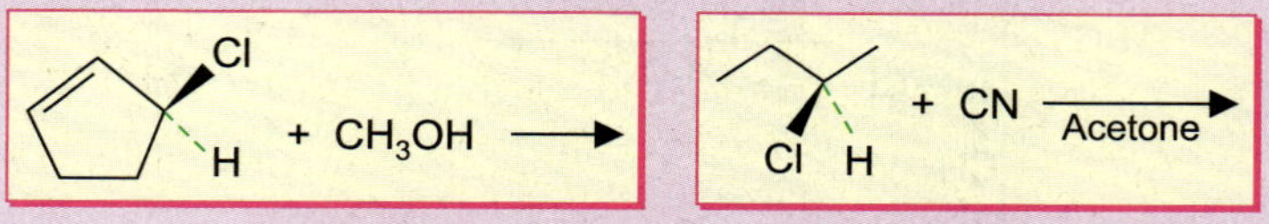

(a) (b)

ANSWERS TO SELECTED PROBLEMS

3.3. Ionising, DMF, CH_3CN, DMSO, $(CH_3)_2CO/H_2O$. Non-ionizing, ether, alkanes, $CHCl_3$, CCl_4.

3.4. (III). The *syn* tosylate via one of the suitable σ bonds can give a relatively stable allylic cation. This situation, however is not available in the case of (II). This situation explains the relative rates of acetolysis (Scheme 3.69).

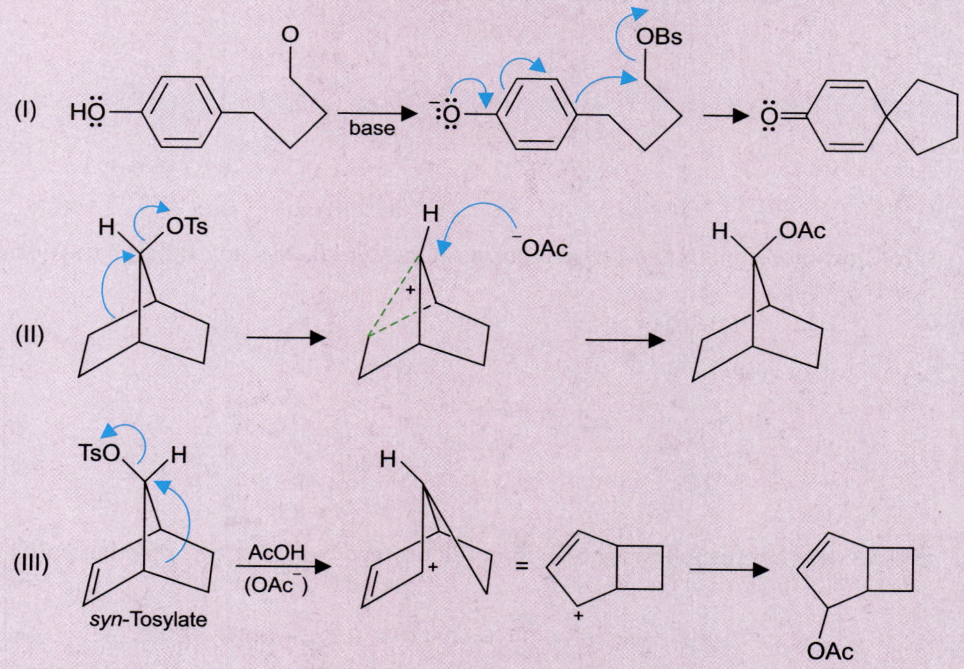

3.5. Only the *cis* compound (II) will react via neighbouring group participation. The mechanism is unique since it involves the conversion of a *cis* 1, 3-cyclohexane derivative in to a *trans*-product as well as the change of places between methoxy and chlorine groups.

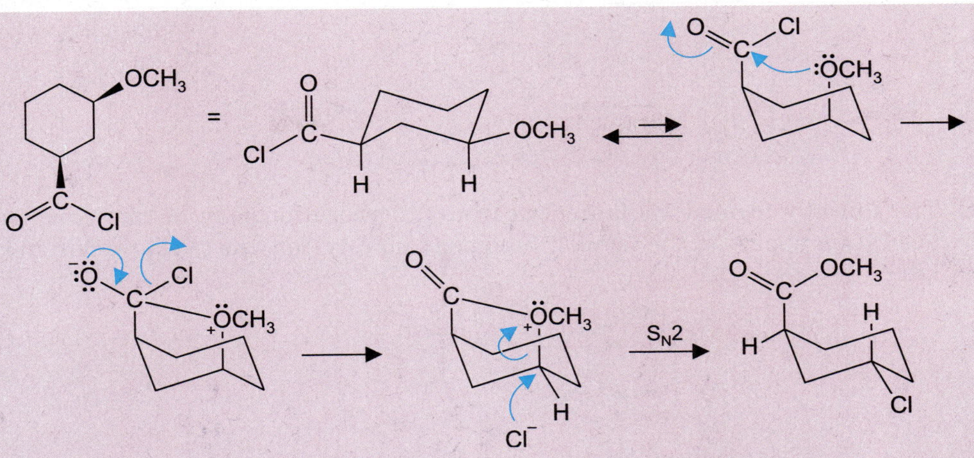

3.6. Neighbouring group participation by the suitably located *pi* electrons in the *anti* tosylate involving the non-classical intermediate, 1, some resonating structures have "cyclopro-pane-like" characters.

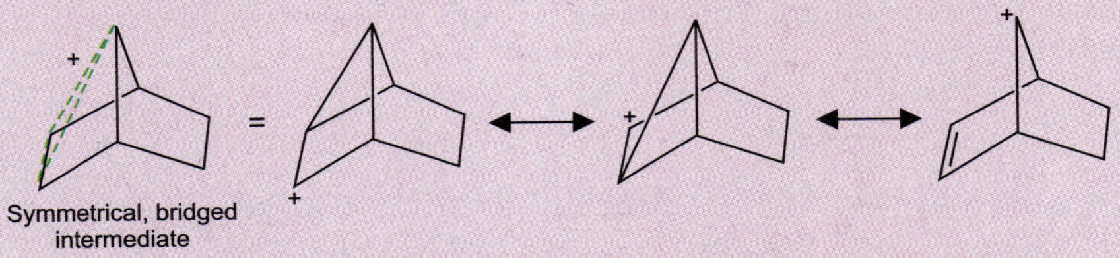

3.7. This is due to carbocation stabilization by a neighbouring hetero atom.

$$CH_3CH_2-O-CH_2Cl \longrightarrow Cl^- + \left[\begin{array}{c} CH_3CH_2-\ddot{O}-\overset{+}{C}H_2 \\ \updownarrow \\ CH_3CH_2-\overset{+}{\ddot{O}}=CH_2 \end{array}\right] \xrightarrow[C_2H_5OH]{-H^+} CH_3CH_2-O-CH_2OC_2H_5$$

3.9. Compound (II) since bromide is a better leaving group than chloride; compound (III) since it will generate a more stable 3° carbocation; compound (VI); compound (VIII).

3.10. The cylopropyl ring is formed by the participation of the double bond which displaces the leaving group with inversion at the stereocenter. The methoxy group then enters from the face of the starting reactant from which the leaving group departed. The reaction is known as the ***iso*-cholesterol** (or ***i*-cholesterol) rearrangement.** On adding acid, the *iso*-cholesteryl ether is converted back into a cholesteryl derivative. It has been shown that displacement reactions of cholesteryl tosylate are about 100 times as fast as reactions of cyclohexyl tosylate. The double bond participates in displacement of the tosylate group to form a homoallylic cation, in which the charge is shared between the two ends of the cation.

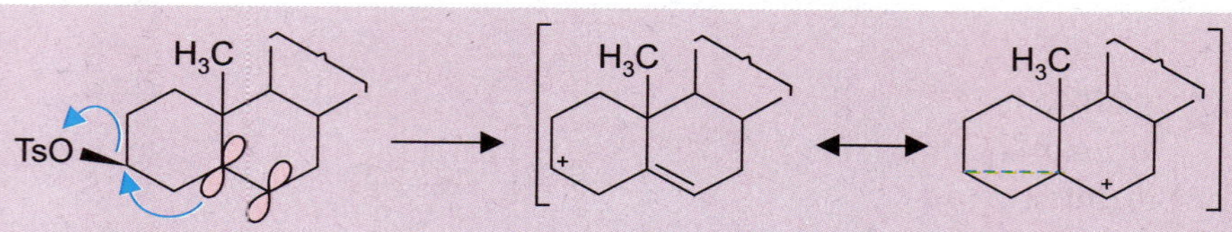

3.11. The initially formed 1-adamantylcarbinyl carbocation may be showing some hesitation to pass to the tertiary 3-homoadamantyl ion due to ring strain in the later.

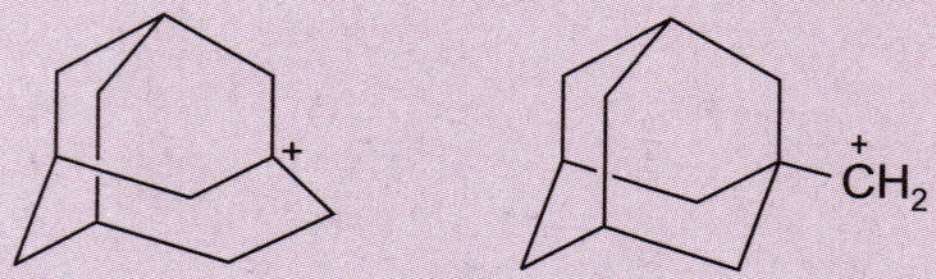

3-Homoadamantyl carbocation   1-Adamantylcarbinyl carbocation

3.12. During this substitution configuration is retained, thus it is not a direct S_N2 reaction (which will give inversion of configuration) nor a S_N1 reaction (which will give racemization). This reaction could be thus two S_N2 reactions at the reaction center.

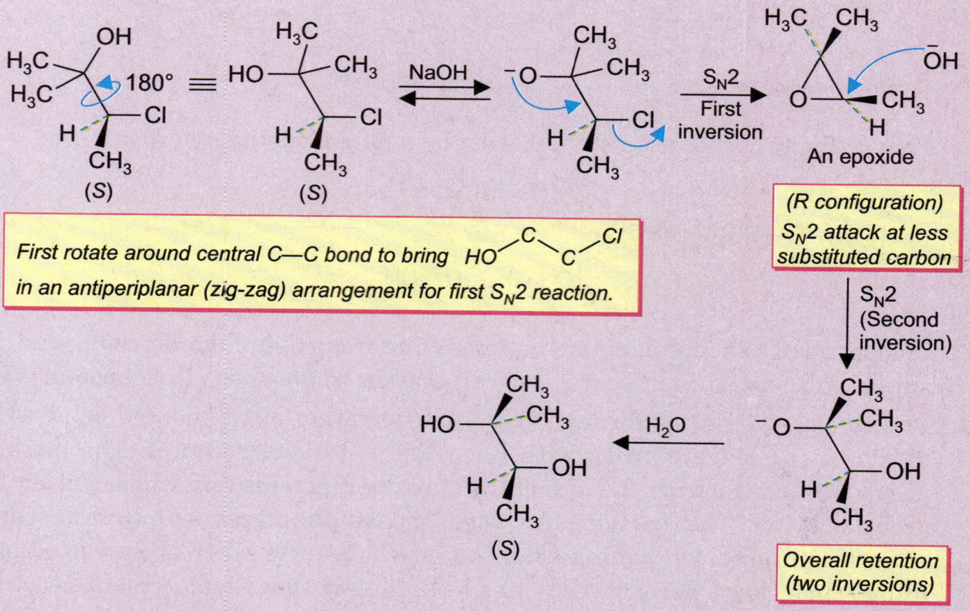

3.14. Chloride is a good but not the best leaving group (see Scheme 3.16), its exit to give the carbocation is somewhat sluggish. It thus remains in the vicinity of the carbocationic carbon to shield one face, the attack of the nucleophile thus for sometime occurs entirely from opposite face to give more of inversion. On the other hand bromine is a far better leaving group so it clearly departs without shielding the face from which it departs, in this case the racemization is complete.

3.15. The formation of the bridged ion would involve the contraction of one of the five membered rings of the norbornyl system to a four membered ring to cause strain.

3.16. Due to neighbouring group participation by the carboxylate ion to give an α lactone (intramolecular S_N2 reaction). α-Lactones are smallest cyclic esters, which cannot be isolated (reactive intermediates). In the second step methyl alcohol opens the α-lactone ring by a second S_N2 displacement. The overall result of this mechanism is net retention of configuration.

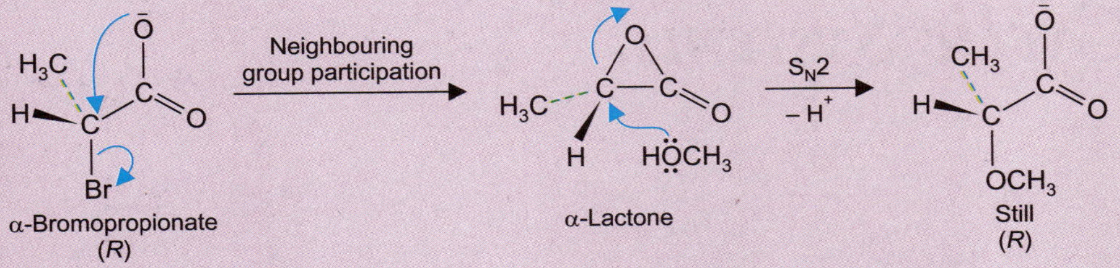

3.17. *Reaction* 1: S_N2 pathway, since the substrate is primary and cyanide ion is a good nucleophile. Dimethyl sulfoxide (DMSO), a polar aprotic solvent, is a particularly good solvent to bring about nucleophile assisted substitution reactions because its ability to solvate cations (in this case, Na^+) is good, while its ability to solvate anions (in this case, CN^-) is poor.

Reaction 2: S_N1 pathway since 2-chlorobutane is a secondary alkyl halide that can ionize to form a fairly stable secondary carbocation. The mixture of methanol and water is a good ionizing solvent. Both water and methanol are poor nucleophiles.

Reaction 3: S_N2 pathway bromide ion is a good leaving group on a secondary carbon. The sulfide ion is a strong nucleophile. Acetone, a polar aprotic solvent, is a good medium for S_N2 reaction.

3.18. Compound (*a*) is chiral and (*S*)-enantiomer, ionization of chlorine gives a resonance stabilized secondary allylic carbocation which is achiral and is expected to give almost a racemic mixture of ethers. Compound (*b*) is (*R*)-enantiomer and with a strong nucleophile and aprotic solvent gives S_N2 displacement with inverted configuration of the substrate.

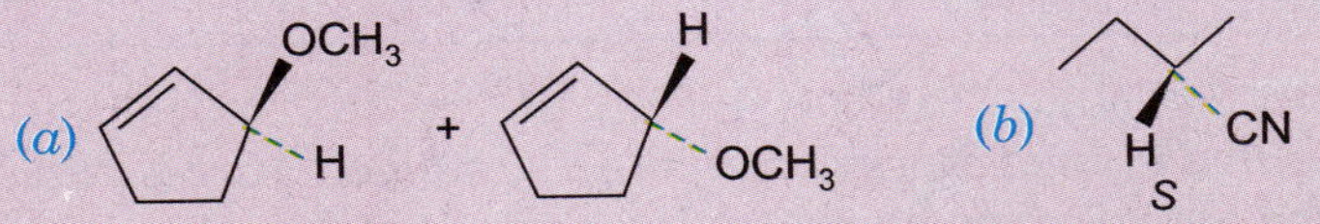

Conformations and Stereoisomerism of Acyclic and Cyclic Systems

4

CHAPTER

4.1 RESTRICTED ROTATION ABOUT SINGLE BONDS CONFORMATIONS OF ETHANE AND BUTANE

(a) Conformation, Configuration and Conformational Analysis

Whether two stereoisomeric structures represent different configurations or simply different conformations depends on how readily they interconvert. Various conformations of ethane derivatives interconvert millions of times per second at 25°C, however, the E and Z configurations of ethylene derivatives do not interconvert spontaneously even at 200°C. As, in principle, one could stop conformational interconversions by sufficient cooling, the operational distinction between configuration and conformation is based on if the structures interconvert at room temperature, 25°C. The various shapes that a molecule can adopt by rotation about single bonds are called conformations. Repulsive interaction between bonds (bonding electrons) on adjacent atoms lead to restricted rotation (torsional strain), while the steric strain in a molecule increases as the size of groups increases. A study of various conformations of a compound and their relative stabilities is called conformational analysis.

Conformation and Configuration

- *Stereoisomers separated by high energy barrier (> 100 kJ mol^{-1}) are isolable at room temperature and are called configurational isomers. Stereoisomers separated by relatively low energy barrier (< 60 kJ mol^{-1}) are easily interconvertible at ambient temperature and are called conformational isomers or conformers. Unless a conformation is held rigid by a small ring or double bonds, a molecule can have an infinite number of conformations, but only one configuration.*

(b) Dihedral Angle

The dihedral angle is an important stereochemical parameter, much more than bond length and bond angle which are structural parameters. When an ethane molecule rotates about its carbon—carbon bond, two extreme conformations can result: the staggered conformation and the eclipsed conformation (Scheme 4.1). An infinite number of conformations between these two extreme (skewed conformations) are also possible. The potential energy relationships between conformations, is conveniently discussed in terms of dihedral angle. A dihedral angle, (Greek letter θ), is the angle generated by two intersecting planes. In the Newman projection

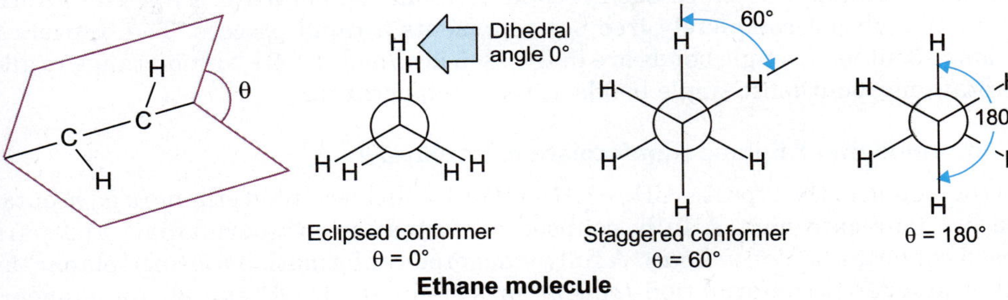

SCHEME 4.1

of the eclipsed conformation of ethane (Scheme 4.1), two H—C—C planes are shown, the angle at which these planes intersect (the dihedral angle) is 0° while in a staggered conformation the dihedral angle of the two H—C—C planes is 60°, or 180°.

(c) Conformations of Ethane (Conformational Analysis)

Two of the remarkable conformations of ethane are eclipsed conformation and staggered conformation.

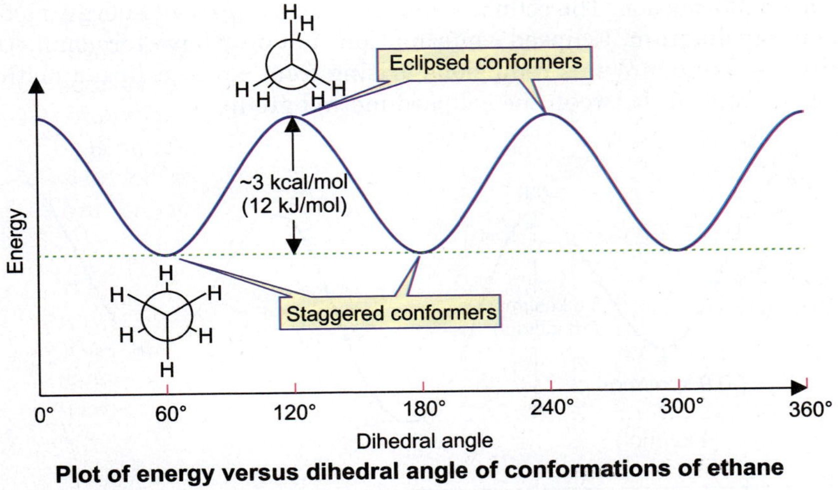

SCHEME 4.1a

The staggered conformations of ethane are the low energy forms while the eclipsed conformations represent transition states. The energy difference between the two conformers, which infact represents the barrier of interconversion between two staggered conformers is about 3 kcal/mol (12 kJ/mol) (Scheme 4.1*a*). Thus in an alkane in general, each pair of eclipsed bonds that it contains leads to an energy cost of about 1 kcal/mol (4.2 kJ/mol). This torsional strain is primarily due to electron pair repulsions when the bonds are eclipsed. The activation energy for rotation about the C—C bond in ethane is small, the thermal energy from the surroundings at the room temperature (20 kcal/mol, 83.7 kJ/mol) is sufficient to cause staggered conformations of ethane to interconvert millions of times each second at room temperature. Thus the internal rotation though not completely free but represents a rapid process. The barriers for rotations about most single bonds are in the (3–5 kcal/mol, 12–21 kJ/mol) range to allow fast rotations about most single bonds at room temperature.

(d) Conformations of Butane (Conformational Analysis)

For a compound of the type A—CH_2—CH_2—B (a 1,2-disubstituted ethane) *e.g.*, *n*-butane there are four extremes. A fully eclipsed conformation *syn*-periplanar) and partly eclipsed conformation *anticlinal*). A fully staggered conformation (*anti*periplanar) and another staggered conformation (*gauche* or *syn*-clinal). In ethane all the staggered conformations are equivalent, in *n*-butane on the other hand there are two different staggered conformations. In one of these (*synclinal*) there is a gauche relationship between methyl groups while in other conformation the relationship is *anti* (*anti*periplanar). Two groups are said to be *gauche* when the dihedral angle between them is 60° and in an *anti* conformation the groups are maximum distance apart, the dihedral angle between them being 180°. These staggered conformations (Scheme 4.2) are free of torsional strain however, the *anti*periplanar conformer is the most stable, because the methyl groups are far apart. The methyl groups in the *gauche* conformations are close-enough to each other, so that the van der Waals forces between them are *repulsive*. This repulsion causes the *gauche* conformations to have approximately 0.9 kcal/mol (3.8 kJ/mol) more energy than the *anti* conformation. The eclipsed conformations represent energy maxima in the potential energy diagram. Eclipsed conformations not only have torsional strain, they have additional van der Waals repulsions arising from either eclipsed methyl groups and hydrogen atoms or between the eclipsed methyl groups.

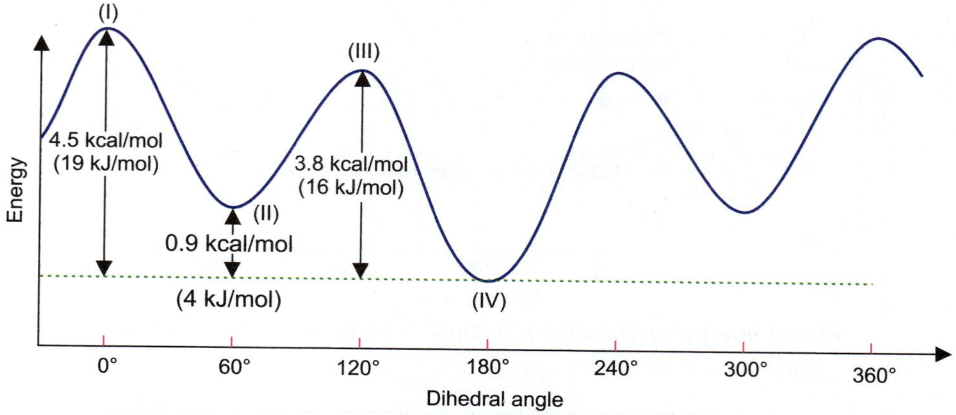

Plot of energy versus dihedral angle for conformations of butane

(Contd...)

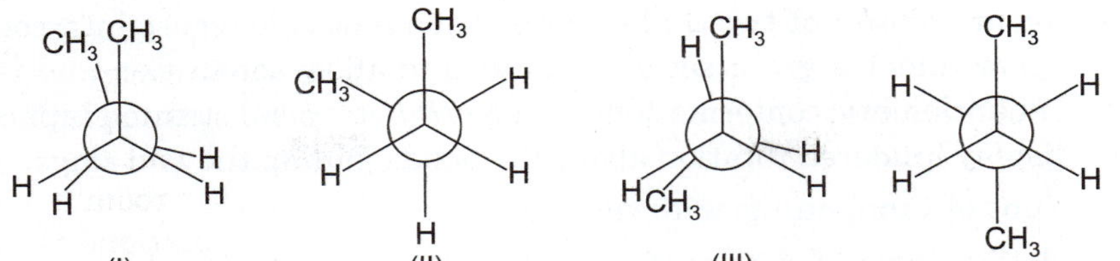

(I)	(II)	(III)	(IV)
(Synperiplanar, sp) fully eclipsed dihedral angle 0°	***(Synclinal, sc)*** *gauche* dihedral angle 60°	***(Anticlinal, ac)*** partially eclipsed dihedral angle 120°	***(Antiperiplanar, ap)*** *anti* dihedral angle 180°

SCHEME 4.2

While it takes only about 3 kcal/mol (12 kJ/mol) for ethane to rotate from staggered to eclipsed conformation, it takes about 5 kcal/mol (21 kJ/mol) for *n*-butane to rotate from *anti* to *syn*periplanar conformation (Scheme 4.2).

For butane and most of the molecules of the type A—CH_2—CH_2—A and A—CH_2—CH_2—B, the *anti* conformation is the most stable. In one group of compounds, containing small electronegative atoms like fluorine and oxygen *gauche* form is predominant. There is yet no explanation for this and the presence of intramolecular H bonding *e.g.*, in the case of 2-fluoroethanol as the reason has been ruled out.

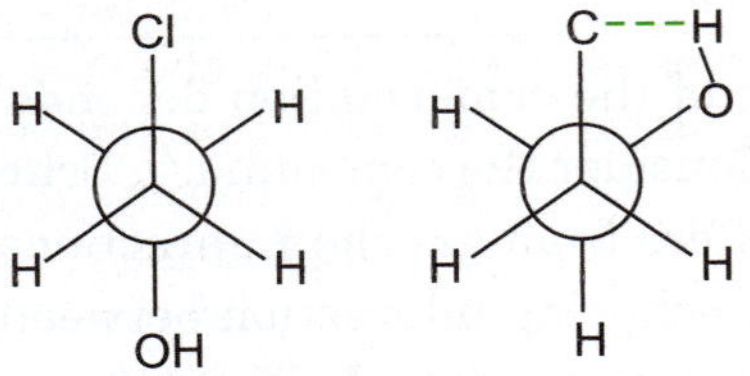

(A) (B)

It was earlier believed that formation of an intramolecular hydrogen bond in 2-fluoroethanol as seen in (B) explains its conformational preference over (A). This has been however ruled out Griffith R.C; Roberts, J.D. Tetrahedron Left 1974, 3499.

SCHEME 4.2a

In one case the two conformational isomers (I and II, Scheme 4.3) of a single aliphatic hydrocarbon have been isolated. The isomers have been separately crystallized and are stable at room temperature. The dihedral angles are, however, distorted from 60°, due to steric hindrance between bulky groups [for defining conformations (I and II, Scheme 4.3, from the dihedral angle consult Scheme 4.4].

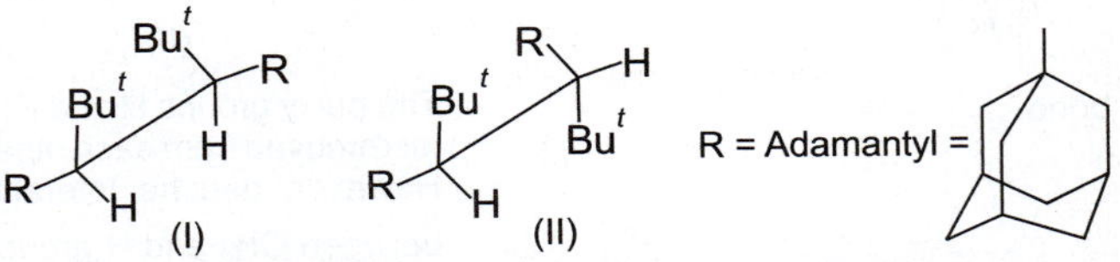

(I) (II)

3, 4-di (1-Adamantyl)-2, 2, 5, 5-tetramethylhexane

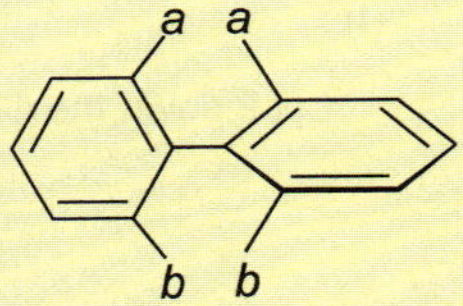

(III)

SCHEME 4.3

Atropisomers represent two conformations of a molecule whose interconversion is sufficiently slow under a given set of conditions to allow separation and isolation. Thus, the two enantiomeric conformations of the *ortho*–tetrasubstituted biphenyl (III, Scheme 4.3) display hindered rotation about the bonds joining the two rings.

(e) Conformations of Compounds with sp^3 – sp^2 Bonds

In acetaldehyde *e.g.*, the preferred conformation is the more stable eclipsing conformation (I, Scheme 4.3*a*) than the bisecting conformation (II). Similarly the propene moiety shows close similarities with the carbonyl compounds and the more stable eclipsed conformation is shown in (*III*, Scheme 4.3*a*).

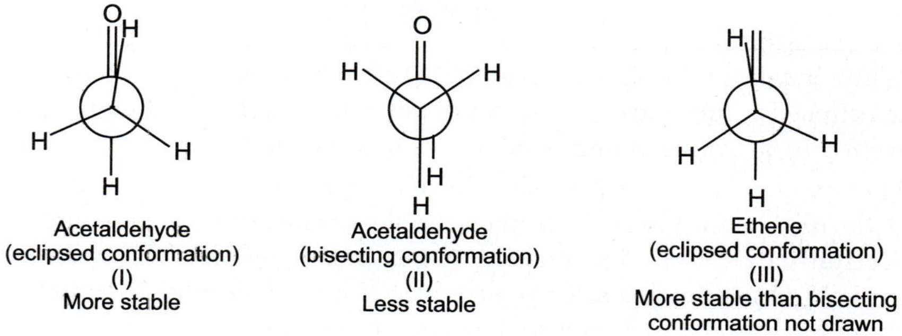

SCHEME 4.3a

Generally in such cases the predominance of the conformation depends on bulky substituents and the influence of the solvent. Consider the compound (I, Scheme 4.3*b*). One of its conformations (*II*, Scheme 4.3b) suffers from gauche interactions between the methyl group and the R group and also an eclipsing interaction between R′ and a hydrogen atom. The more significant of these interactions (CH_3/R) can be removed by rotating the front carbon atom counterclockwise (Scheme 4.3c) to get the most stable conformation (*III*).

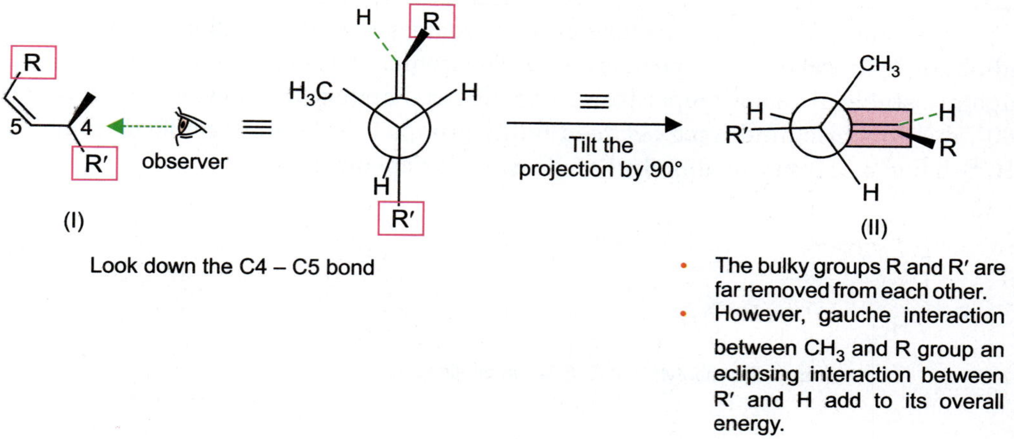

SCHEME 4.3b

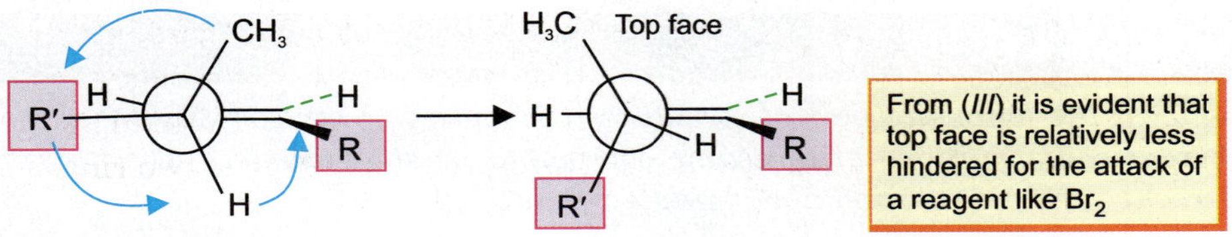

Rotate the frontal carbon in conformation (*II*, Scheme 4.3b)

Bottom face
(III)
Most stable conformation

SCHEME 4.3c

Unlike ethane and *n*-butane (where various conformations are easily recognized from the dihedral angle of atoms or groups these being symmetrical molecules) when the substituents on the two carbon are different, the following considerations are applied to define the conformations from the dihedral angle.

- In 2-bromopropanamide, all the ligands on the two carbons are different. The dihedral angle is then measured from one substituent on each carbon which is preferred by sequence rules *i.e.*, from oxygen and bromine as in eclipsed and *anti* conformations (Scheme 4.4).
- In 3-chloropentan-2 ol a hydrogen atom is common to both carbons, in such cases when an atom or a group is common to both carbons dihedral angle is measured between these two ignoring the higher priority of remaining substituents (Scheme 4.4).

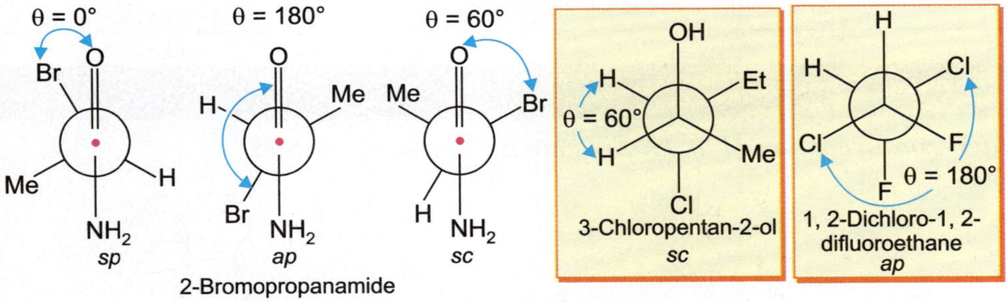

SCHEME 4.4

- When two atoms or groups are common to both the carbons, the pair of highest priority defines the dihedral angle as in 1, 2-dichloro-1, 2-difluoroethane (Scheme 4.4).

EXERCISE 4.1

Predict if the following pair of structures (Scheme 4.5) are constitutional isomers, different conformations of the same compound or stereoisomers which cannot be interconverted via rotation about single bonds.

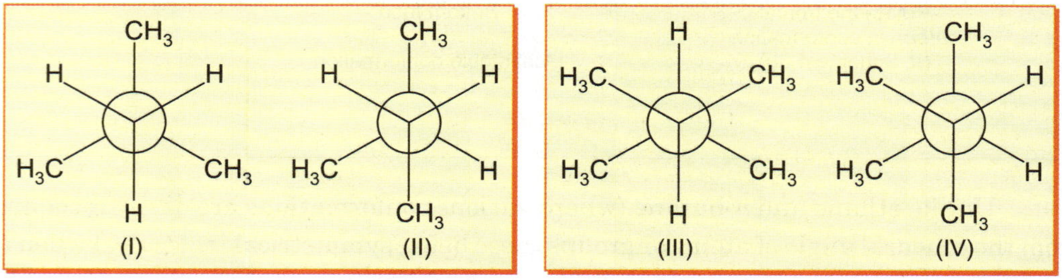

SCHEME 4.5

ANSWER. *The trick is to write the structures in the Lewis form to show their atomic connections (Scheme 4.6) to decide the problem. Thus (I) and (II) are constitutional isomers while (III) and (IV) are the different staggered conformations of the same molecule.*

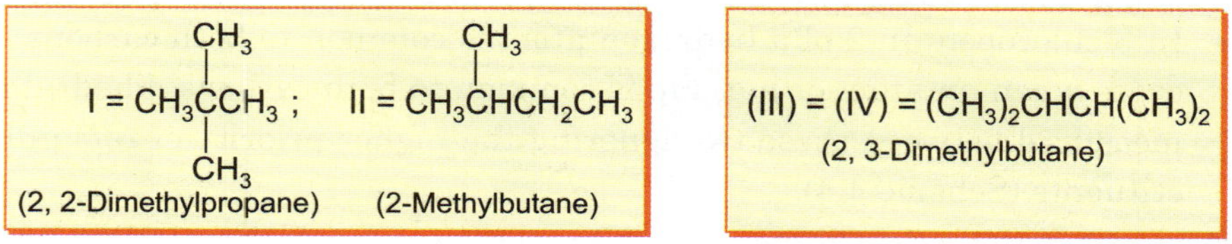

SCHEME 4.6

EXERCISE 4.2

Which conformation (I or II, Scheme 4.7) is more stable?

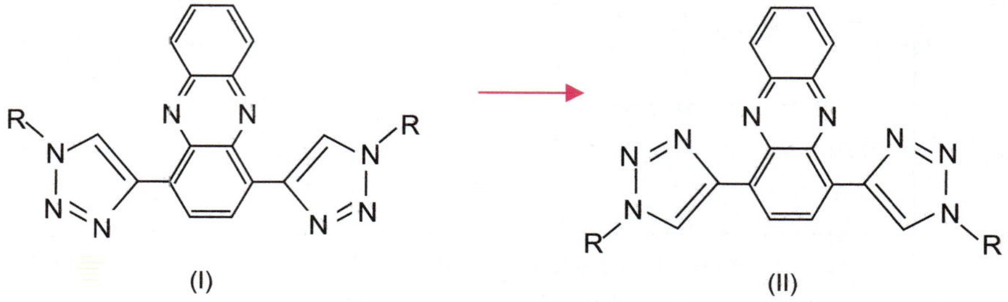

SCHEME 4.7

ANSWER. *Conformation (I, Scheme 4.7) is more stable. Conformation (II) can be obtained by rotation of σ bonds in (I) by 180°. In (II) the lone pairs are directed toward each other resulting in electrostatic repulsion (III, Scheme 4.7a)*

(Contd...)

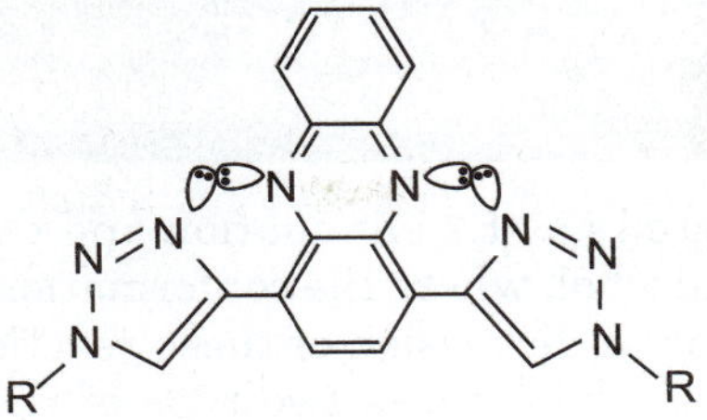

(III)

In (II, Scheme 4.7) the lone pair of electrons are directed toward each other to generate electrostatic repulsion as shown in (III, Scheme 4.7*a*). This repulsion is absent in (I, Scheme 4.7).

SCHEME 4.7a

4.2 ORIGIN OF CONFORMATIONAL ENERGY

An inspection of the model of ethane shows that the hydrogen atoms on adjacent carbon atoms are far apart making steric interaction negligible. The basic cause of the energy barrier could be studied only after accurate measurements of the energy difference between the maxima and minima for a number of molecules were made with the help of microwave spectroscopy.

Table 4.1: Conformational Energy Barriers of Various Compounds

Compound	*Energy barrier ($kJ\ mol^{-1}$)*	*Compound*	*Energy barrier ($kJ\ mol^{-1}$)*
	(*a*)		(*b*)
CH_3—CH_3	11.5	CH_3—CHO	4.8
CH_3—CH_2F	13.8	CH_3—$COCH_3$	3.6
CH_3—CH_2Cl	14.9	CH_3—COCl	5.7
CH_3—CH_2Br	14.9	CH_3—CO_2H	2.0

The results in Table 4.1 reveal the lack of importance of steric effects. The barrier is not significantly changed when hydrogen is replaced by any of the halogens. A difference between the barrier heights however, exists as given in the two columns of the table, and it is significant that the barriers (*a*) are for the rotation of an sp^3-hybridized carbon relative to an sp^3-hybridized carbon, while (*b*) are for the rotation of an sp^3-hybridized carbon relative to an sp^2-hybridized carbon. The bond hybridization has therefore, importance. The barrier must therefore, arise mainly from interaction of the bonding electrons on adjacent carbon atoms. In simple molecules like ethane, this indeed is the main source of the energy barrier. However, when substituent groups become bulky, steric effects make their contribution.

4.3 CONFORMATION AND CHEMICAL REACTIVITY IN ACYCLIC SYSTEMS (E2 ELIMINATION AND PINACOL REARRANGEMENT)

Under this section two reactions are discussed *i.e.,* E2 elimination and the molecular rearrangement (pinacol type) of a β-amino-alcohol, where the conformation adopted by the reacting molecule has significance. Before a discussion of these reactions is taken up, it is necessary to have some basic ideas about these two type of reactions. The requirement that the migrating and the departing groups in the starting compound are *anti* is termed as the stereoelectronic requirement. In the bimolecular (E2) elimination (Scheme 4.8), the elimination of HBr from an alkylbromide by treatment with base gives an alkene. This reaction has a stereoelectronic requirement; H and the Br atoms in the starting compound have to be *trans*- and coplanar (*anti* to each other, *i.e., anti* coplanar). *Anti*-elimination is preferred in the E2 reaction, the *syn*-elimination (when the leaving group and the hydrogen) are held *syn*-coplanar in an eclipsed conformation of a rigid compound are, however less common and are further discussed in chapter 5.

On treatment with a strong nucleophile, tertiary alkyl halides undergo elimination rather than substitution because steric hindrance prevents the backside approach required by the S_N2 reaction (Scheme 4.8). The nucleophile, on the other hand acts as a base and approaches the protons on the carbon atom adjacent to the one bearing the leaving group. One of these hydrogen atoms is removed by the base, the electrons in the C—H bond move to generate the π bond and expel the leaving group, a concerted process E2 elimination (since in the dehydrohalogenation presented here, the rate of reaction depends on the concentration of both substrate and base).

For an E2 reaction, the leaving groups are almost always *anti* in the transition state, which requires this conformation to be accessible if a reaction is to occur readily.

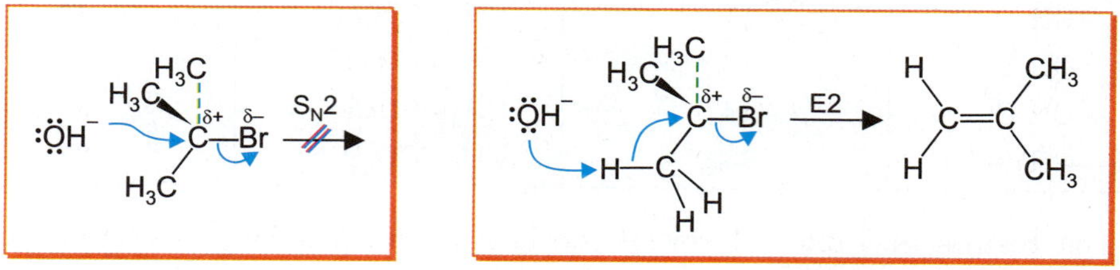

SCHEME 4.8

(a) E2 Elimination–Consideration of Appropriate Conformation (Detailed in Chapter 5)

The stereoelectronic requirement of E2 elimination reaction must involve only those (staggered) conformations of *a*-molecule in which the ligands to be eliminated attain an antiperiplanar arrangement. One starts from a given conformation of a compound in which a carbon atom(s) is rotated through different angles till a conformation is reached where the departing substituents become *anti* to each other.

Consider the E2-elimination from 1-bromo-1, 2-diphenylpropane, a compound with two chiralitycenters. It exists in four stereoisomeric forms, two stereoisomers as enantiomeric pair while the other two as diastereomeric pair. Consider one of the diastereomers drawn as Masamune projection (1, Scheme 4.9 also see Scheme 1.18*c*), where two phenyl group are *anti*-periplanar and in the plane of page (on continuous lines). For E2 elimination instead the proton and bromine must be *anti*-periplanar. Thus rotate both the carbons in (1) appropriately to get (1a) in which now one phenyl groups lies toward you on (thick line) and the other away from you on (dotted line, use of models is suggested). As the hydroxide attacks the C–H bond and eliminates Br^-, this arrangement is preserved and the two phenyl groups end up *trans* (the alkene is *E*).

The other diastereomer (2, Scheme 4.9) based on same reasoning gives *Z*-alkene. Each diastereomer reacts at a different rate. The diastereomer (2) reacts at a slower rate than (1). The reason for this slower rate is the attainment of *anti*-periplanar formation (from diastereomer 2) leads to gauche relationship, between the phenyl groups leading to steric interaction (For the same elimination reaction see, problem 4.10 and Scheme 5.20).

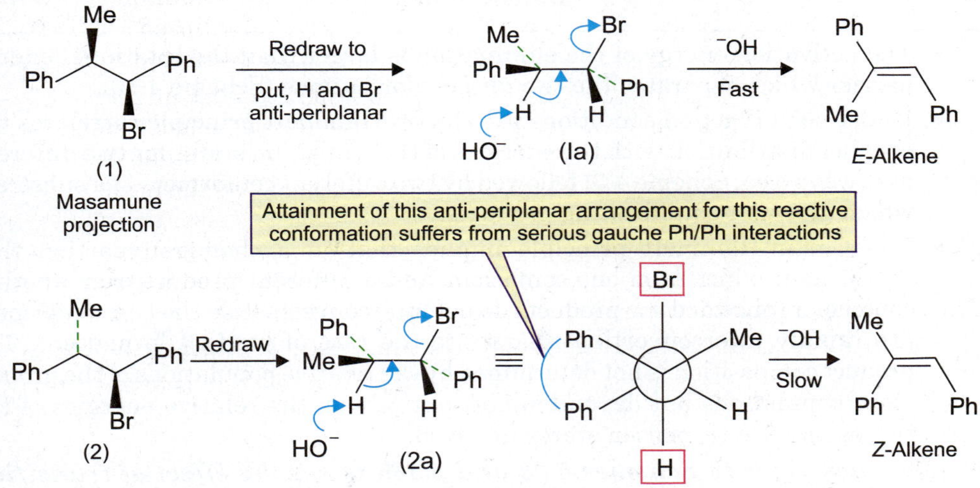

SCHEME 4.9

(b) Curtin-Hammett Principle

> *When the anti-conformation, either cannot be attained for geometrical reasons or if attained it would be highly sterically congested, a slower syn pathway for elimination is followed. (syn-elimination is observed in quaternary ammonium ions).*

When 2-bromobutane undergoes base induced elimination of hydrogen bromide, the reaction can occur through either of the two *anti*-conformers (I and II, Scheme 4.10) to give *cis* and *trans*-2-butene respectively, the latter predominates. The following points may be noted:

- The formation of *trans* product in larger proportion may be due to two reasons: The conformation of the reactant leading to it may be heavily populated than that which gives the minor (*cis*) product; the transition state which leads to the major (*trans*) product may be of lower energy than the one leading to minor product.

- Both the above factors or either of these may be responsible for the formation of *trans* product.

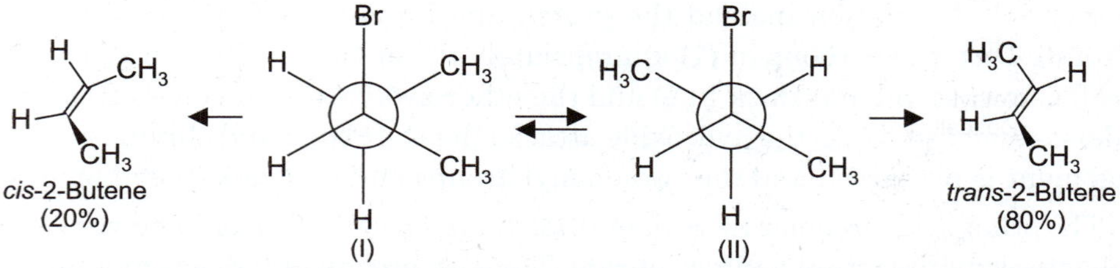

Cutrin-Hammett principle relates the product distribution to relative energies of the transition states from e.g., the conformers of 1-bromobutane (conformationally heterogeneous starting material) rather than the relative populations of the conformers. [The products must be non-equilibrating and the conformers be rapidly inter-convertable relative to the rate of product formation.]

SCHEME 4.10

- The activation energy of the elimination is larger than the rotational energy barrier which separates the two *anti* conformations (Scheme 4.10).
- Under such reaction conditions the Curtin-Hammett principle correlates the product distribution with the energies of the transition states for two different pathways (see, Scheme 4.3) followed by two different conformers of a substrate which give non-equilibrating products.
- The Curtin-Hammett principle implies that in a chemical reaction that gives one product from one conformer and a different product from another conformer (provided the products do not interconvert while the two conformers are rapidly interconverting relative to the rate of product formation). The product composition is not determined by the relative populations of the ground state populations but depend almost entirely on the relative energies of the representative transition states involved.

[***One may refer to Schemes 5.33 and 5.36b to see the effect of transition state energies on product distribution***].

(c) Stereochemistry of Wagner—Meerwein Rearrangement

(i) Introduction

The following points may be noted :

- A reaction proceeding through an intermediate carbocation can give products of substitution, addition of elimination with rearranged carbon skeleton.
- Stereochemical factor involved in molecular rearrangements depends on steric or conformational factors and the timing of the bond-breaking and bond-making processes. When the process is concerted, configuration will be retained in the migrating group and inverted at the migration terminus. When the migration origin becomes a carbocation, as in a Wagner type rearrangement the stereochemical identity is generally lost.
- In most Wagner type shifts the leaving group is generally believed to depart before the migration takes place.
- The configuration at the migration terminus in the product is dependent on the lifetime of the intermediate carbocation.

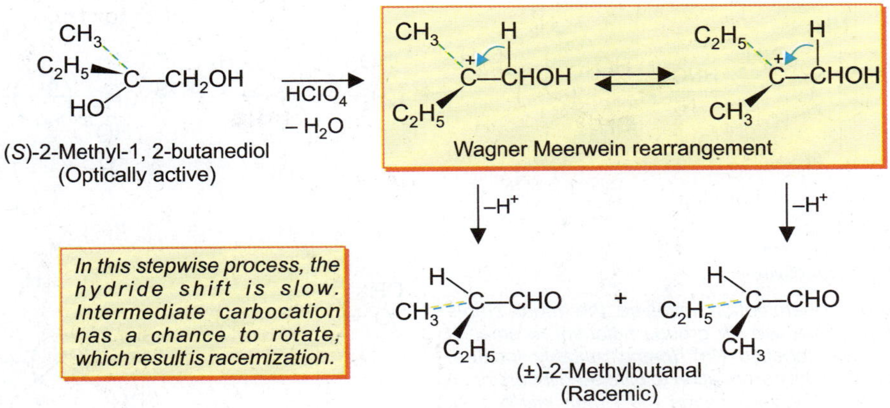

SCHEME 4.11

- Optically active 2-methyl-1,2-butanediol undergoes Wagner Meerwein rearrangement to give a racemic product (Scheme 4.11). The hydride shift is slow and the intermediate carbocation has a fair chance to rotate.
- In a related deamination process (Scheme 4.12), the reaction is very rapid (not much opportunity for the carbocation to rotate) and ends up in a product with significant inversion of configuration at the migration terminus.

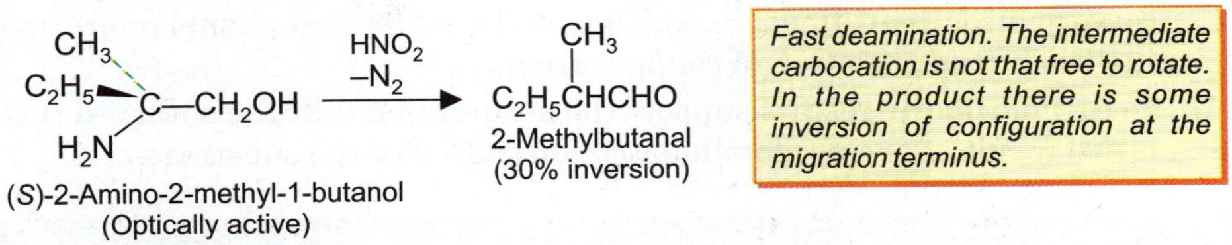

SCHEME 4.12

(ii) Pinacolic Deamination—the Role of the Most Stable Conformation

Deaminative rearrangement of one optically active isomer of 2-amino-1, 1-diphenyl-1-propanol proceeded with 90 per cent inversion. When one specific phenyl group was labelled with ^{14}C, the labelled group migrated to give the product of inversion at the migration terminus and the unlabelled group migrated so as to give retention (Scheme 4.13). The following points may be noted:

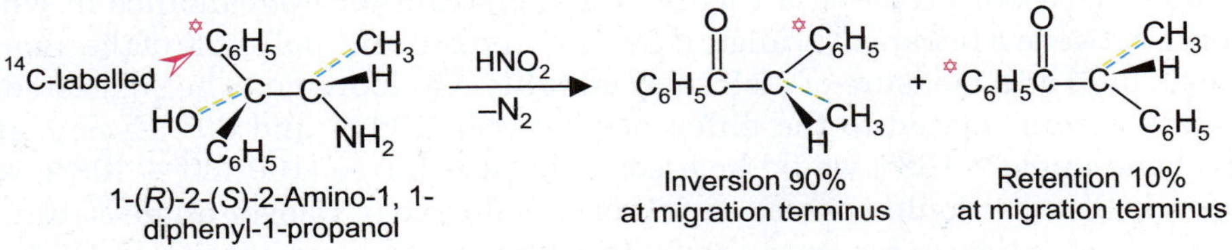

SCHEME 4.13

- The reaction proceeds involving the most stable staggered conformer where the two phenyl groups are adjacent to the smallest atom hydrogen (Scheme 4.14).

CHAPTER 4

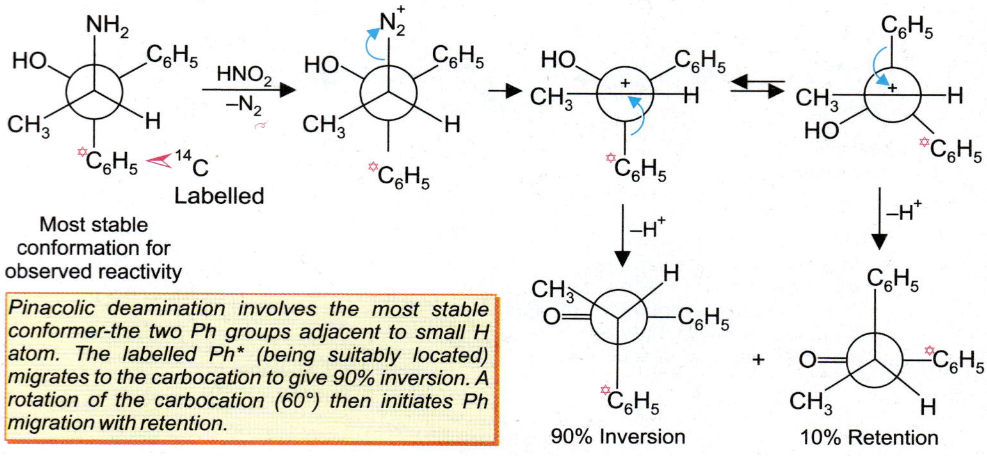

SCHEME 4.14

- The carbocation is produced by the loss of nitrogen gas and in the compound with ^{14}C labelled phenyl, the migration of the labelled phenyl group gives the major product of inversion (The labelled phenyl group is suitably located in the plane of vacant, *p* orbital of the carbocation).
- If the intermediate carbocation rotates by an angle of 60° (Scheme 4.14) a conformation is attained, in which now the unlabelled phenyl migration gives the product of retention of configuration.
- In conclusion the results support the assumption that the observed reactivity is the result of one predominant conformation of the substrate.

4.4 ANGLE AND PITZER STRAIN

Angle strain is also referred to as Baeyer strain. The normal angle between the carbon bonds in an alkane is about 111°; when the carbon atom forms part of a ring, the angle will be controlled by the geometric requirements of the ring. Three-membered ring system in cyclo-propane must be flat (since, mathematically, three points define a plane) so that the bond angles within the ring are forced to be 60° thereby causing strain. The amount by which the bond angle deviates from normal is a measure of the bond-angle strain.

Baeyer (1885) proposed a theory of angle strain for cycloalkanes in which the difference between a tetrahedral angle (109°.30′) and the internal angle of the appropriate polygon is used as a measure of molecular stability. Cyclopropane, he suggested, would have angle strain related to the difference between 109.5° and 60°. Cyclopentane (if planar, bond angle = 108°) would be essentially strain-free (109°.30′ – 108°), whereas larger cycloalkanes would possess an increasing degree of Baeyer angle strain.

The Baeyer theory, however, is not consistent with experimental data. Cycloalkanes larger than cyclopentane show only a small increase in strain energy, and very large cyclic molecules actually become almost strain-free. The problem lies in Baeyer's simple assumption that the rings are flat polygons (geometrical figures). C_3 and C_4 rings have large angle-strain, while the medium-size rings ($C_8 - C_{12}$) possess moderate strain, and large rings show only small strain energies. Though one often draws cyclic compounds as if they were planar geometric figures, their structures are not planar, however.

Pitzer strain is also referred to as bond-eclipsing strain of *gauche* and eclipsed conformations. Bond eclipsing has already been encountered in acyclic compounds, where it is generally relieved by rotation to give a staggered conformation,. In a cyclic compound a rotation may be impossible and therefore the strain resulting from eclipsing of bonds to neighboring carbon atoms will make a contribution, often substantial, to the strain in the ground state. The amount of strain due to the cyclization of a chain can be determined experimentally by measuring the heat of combustion per CH_2 group and comparing it to the value for acyclic analog. The two main contributions to the cyclic strain are the Baeyer strain and the Pitzer strain.

In small rings, the Baeyer strain is particularly significant, while it is less important or negligible in common rings (normal), in medium rings (8-to11-membered), and in large rings (12- and higher membered). In all cycloalkanes the Pitzer strain is operative and tends to be relieved by deviation from planarity of the carbon skeleton. In large rings, transannular interactions account for the relatively marked strain.

4.5 CYCLOALKANE RINGS OTHER THAN CYCLOHEXANE

(a) Conformation of Cyclopentane

If one considers a planar pentagonal structure of cyclopentane (Scheme 4.15), one would have C—C—C bond angles of 108°, a value so close to the normal tetrahedral angle of 109°.30′ that no significant strain effect would be expected. However, in such a structure all of the hydrogens are completely eclipsed and it would have considerable torsional strain. The molecule finds it energetically favourable to distort largely from a planar conformation to relieve this torsional strain even though it increases the angle strain. The actual structure is of "envelope" shape. The additional bond-angle strain involved in this arrangement (due to four carbon atoms in a plane is more than compensated by a decrease in eclipsed hydrogens. The out-of-plane methylene group is approximately staggered with respect to its neighbors.

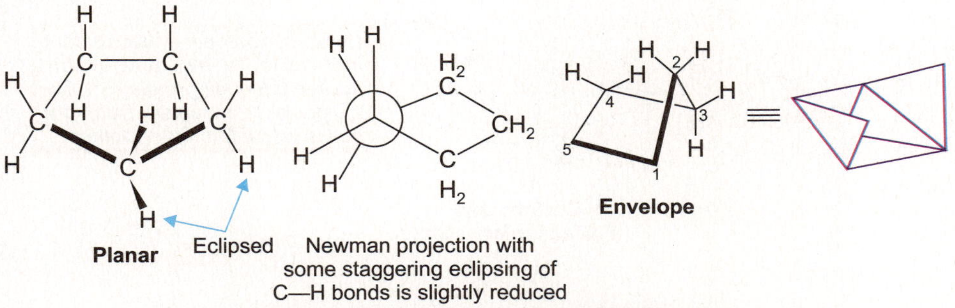

Different projections of cyclopentane

SCHEME 4.15

The envelope conformation of cyclopentane is dynamic. By twisting about the various C—C bonds, successive conformations are realised in which four carbons are in a plane and the fifth is out of plane.

(b) Conformation of Cyclopropane and Cyclobutane

In cyclopropane, one wonders as to how the three carbon atoms are connected in a ring without seriously violation the concept of bond angles. However, it is important to know that the highest electron density of the C—C bonds does not lie along the lines connecting the carbon atoms. Bonding electrons lie principally outside the triangular internuclear lines resulting in bent bonds (Scheme 4.16).

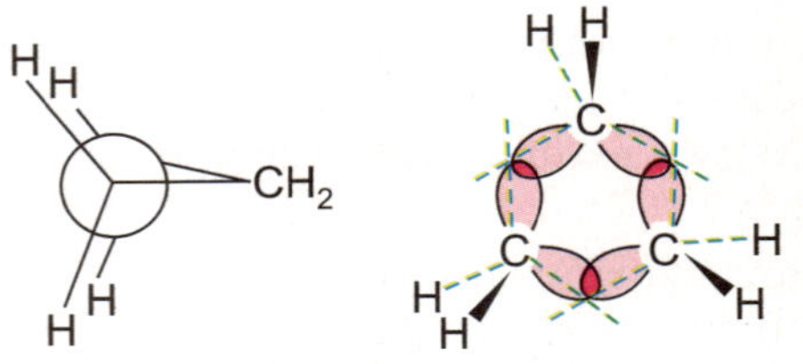

Cyclopropane

> ***Cyclopropane*** *suffers from both angle and torsional strain. Each C–C bond is eclipsed due to rigidity of the molecule. Three carbons form an equilateral triangle (bond angles 60°) Hybridization at each carbon is sp^3. Angle between sp^3 hybrid AO's is 109.5°.*

SCHEME 4.16

In cyclobutane, the internuclear angles of 90° are not as small as in cyclopropane. The C—C bonds are not so bent, and there is less strain per bond. However, there are four strained bonds rather than three and there are eight pairs of eclipsed hydrogen rather than six.

As three points define a plane, the carbon framework of cyclopropane must have a planar structure. However, cyclobutane can exist in a nonplanar conformation. Spectroscopic studies show that cyclobutane and many of its derivatives possess nonplanar structures in which one methylene group is bent (folded) at an angle of about 25° from the plane of the other three ring carbons (Scheme 4.17). In this conformation, some increase in bond angle strain is compensated by the reduction in the eclipsed hydrogen interactions. The strain that is present in the C7 to C10 compounds is not the result of angle strain (the molecules being puckered), but due to interfering hydrogen atoms (Scheme 4.17).

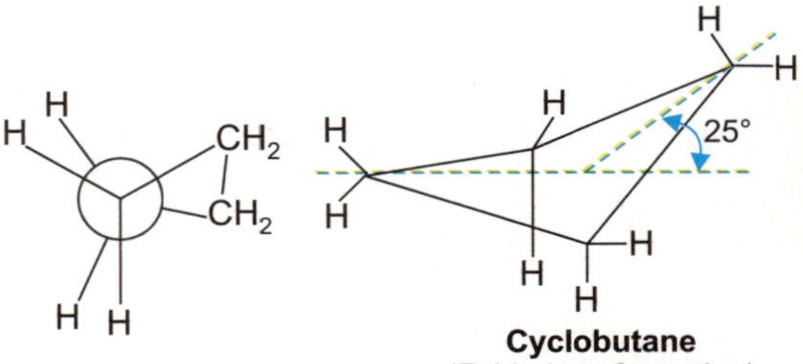

Cyclobutane
(Folded conformation)

> *Unlike cyclopropane, cyclobutane molecule is not planar but puckered (bending angle 25°). The molecule flips rapidly from one puckered conformation to other*

SCHEME 4.17

The heat of combustion is the amount of heat which is released when a compound is burned with an excess of oxygen in a sealed container (a *bomb calorimeter*). If the compound has extra energy due to ring strain, that extra energy is released in the combustion. The heat of combustion is usually measured by the temperature rise in the water bath surrounding the "bomb."

A cycloalkane can be represented by the molecular formula $(CH_2)_n$, (Scheme 4.17*a*).

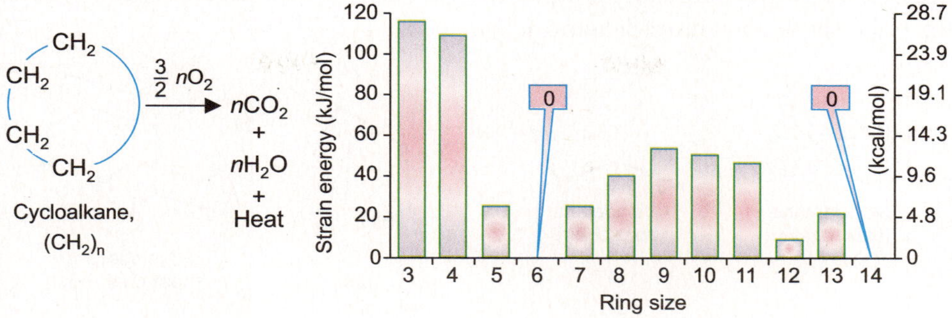

SCHEME 4.17a

The molar heat of combustion of cyclohexane is almost twice that of cyclopropane, simply because cyclohexane contains twice as many methylene (CH_2) groups per mole. To compare the relative stabilities of cycloalkanes, we divide the heat of combustion by the number of methylene (CH_2) groups. The result is the energy per CH_2 group. These normalized energies allow us to compare the relative amounts of ring strain (per methylene group) in the cycloalkanes. The cycloalkane strain energies are presented (Scheme 4.17*a*), it is seen that small and medium rings are strained, however, cyclohexane is strain free.

Baeyers strain theory is wrong since as per his assumption all cycloalkanes are not flat (*e.g.*, cyclohexane exists predominantly in a non-planar, puckered conformation), thus angle strain exists only in three and four membered rings. Most rings unlike three and four membered ring adopt puckered three dimensional conformations to allow the bond angles to be nearly tetrahedral. In the case of medium-ring (C_7–C_{11}) cycloalkanes, torsional strain due to H ↔ H eclipsing interactions on adjacent carbons and steric strain caused by the repulsion between nonbonded atoms which come too close are the most important factors. Thus, three kinds of strain contribute to the overall energy of a cycloalkane (see, Scheme 4.18).

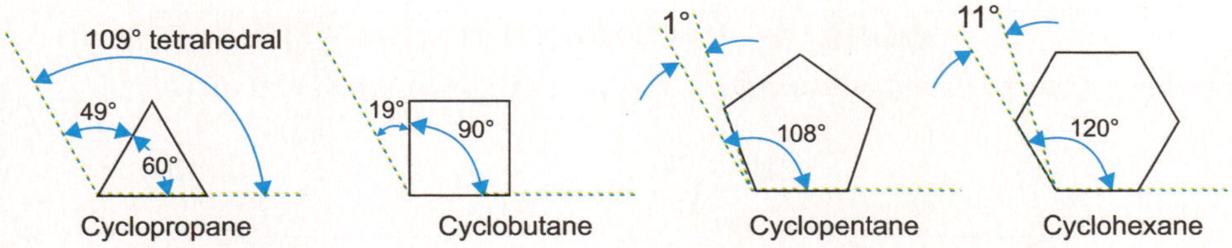

(c) Conformations of Some Medium Sized Rings

Cycloheptane, cyclooctane, and cyclononane also exist in nonplanar conformations. The small instabilities of these higher cycloalkanes appear to be due, primarily to torsional strain and van der Waals repulsions between hydrogens across rings. The nonplanar conformations of these rings, however, are essentially free of angle strain. Although not known with certainty, the most stable conformations of cycloheptane and cyclooctane appear to be those shown in (Scheme 4.18).

Cycloheptane (twist-chair)

Cyclooctane (boat-chair)

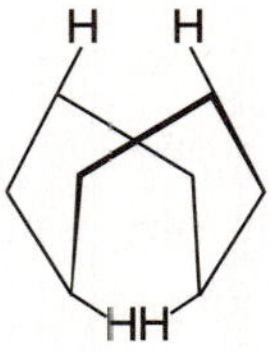

transannular interaction Cyclooctane (boat-boat)

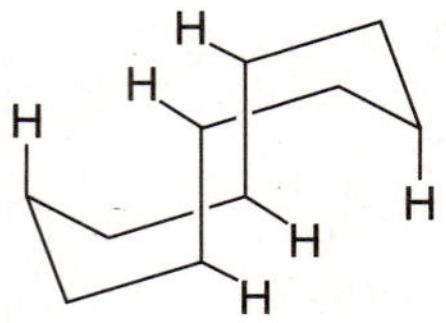

Cyclodecane (boat-chair-boat)

SCHEME 4.18

Puckered cyclooctane can assume no less than seven high symmetry forms which interconvert through inversions, pseudorotations or *via* their combinations. Cyclooctane is the smallest ring in which transannular interactions may occur. The boat-boat conformation has the carbon skeleton which corresponds to a fragment of the diamond lattice (compare with adamantane). A model of this conformation reveals that two hydrogen atoms compete for the same region of space. This is therefore, high-energy conformation. This type of transannular interactions can be quite severe in cyclooctane and other medium ring compounds.

A distinctive feature of medium sized rings is that they display transannular reactions. These reaction do no involve the neighboring atoms; and occur among atoms on opposite sides of the ring. An example of this type of anomalous reaction is *e.g.,* the formolysis of epoxy-cyclodecane to give *trans*-cyclodecane-1,6-diol by a transannular hydride shift (Scheme 4.19). Normally such a reaction from a cyclic or an acyclic epoxide affords a 1,2-diol. *Trans*-cyclooctene displays this reaction during the addition of chlorine, when 1,4-dichlorocyclooctane is obtained (10%) involving a 1,5-hydride shift (Scheme 4.19) in the ion in addition to the normal product.

LOOKING BACKWARD TO CHAPTER 1

Assignment of priorities and configuration in a cyclopropane derivative.

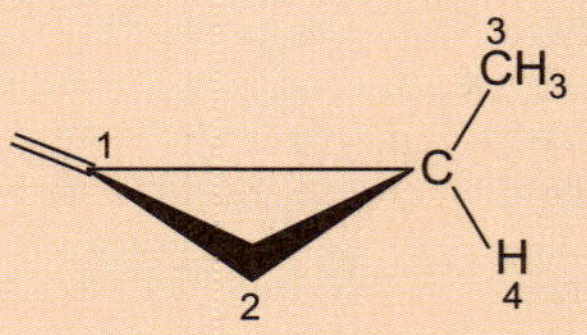

(*S*)

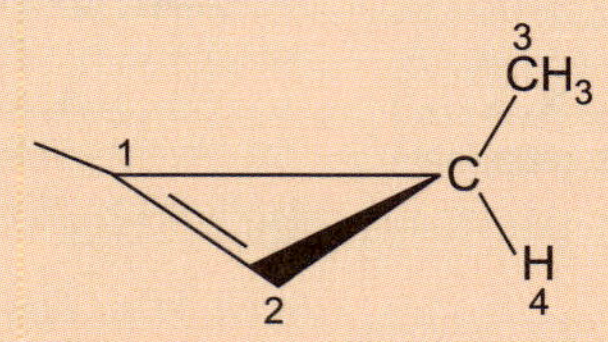

(*S*)

SCHEME 4.19

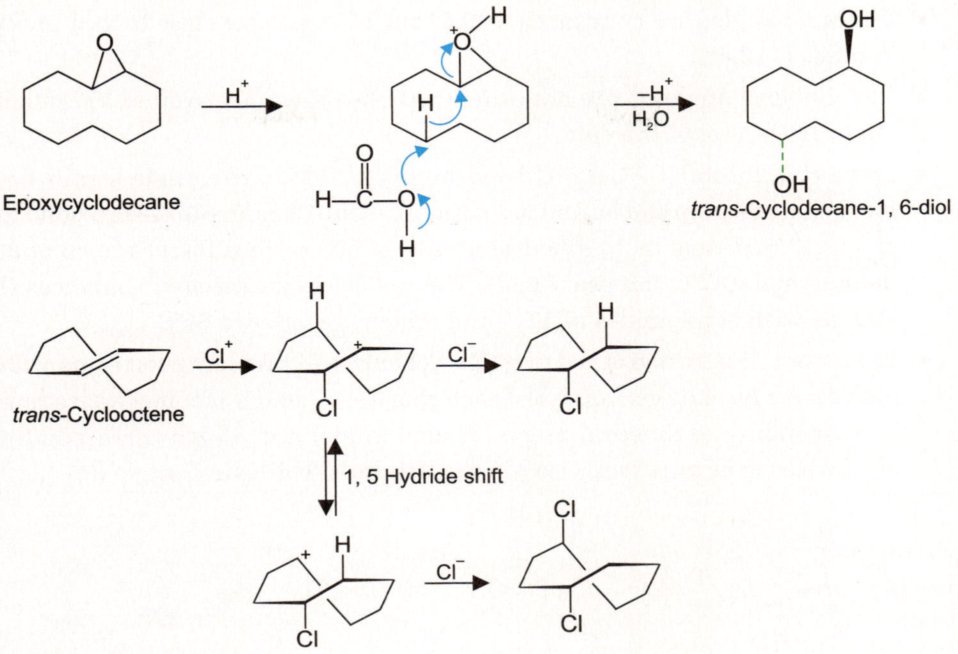

SCHEME 4.20

4.6 CONFORMATIONS OF CYCLOHEXANE (CONFORMATIONAL ANALYSIS)

The importance of Baeyer strain was first appreciated in 1890 by Sasche, who pointed out that two non-planar models of cyclohexane could be constructed in which all the bond angles can be 109.5°, so that the systems were free of Baeyer strain. One of these was a fairly rigid form, shaped roughly like a chair and the other was a flexible form whose most symmetrical form was shaped like a boat.

(a) The Chair Conformation of Cyclohexane

The following points may be noted:

- Cyclohexane molecule exists predominantly in a non-planar, puckered conformation the chair conformation (I, Scheme 4.21, conventional drawing).
- All bonds are staggered (Newman projection II, Scheme 4.21), therefore, Pitzer strain is minimized. However, a gauche butane like interactions between neighboring methylene groups leads to steric strain (II, Scheme 4.21).
- An ideal chair form would have dihedral angles of 60°, and C–C–C bond angles of 109.5°.

- The bond angles are not exactly 109.5° but 111°, a value close to that in acyclic alkanes (112.4°).
- The dihedral angles θ are not exactly 60°, but these are around 56° causing a slight flattening of the ring.
- Since the normal $C-CH_2-C$ bond angle is 112.4°, the angle strain for the ideal chair form would be about 4 kJ/mol (about 1 kcal/mol). Also, a bond angle of 112.4° corresponds to dihedral angles of 52°, with a Pitzer strain of 2.9 to 3.3 kJ / mol (0.7 to 0.8 kcal / mol). The real cyclohexane chair balances these strains with bond angles of 111° and dihedral angles of 56°.
- In another illustration of the ring (III, Scheme 4.21 providing some quantitative data for conformational analysis) each ring C—C bond is labelled with a number corresponding to dihedral angle (torsion angle) and a sign corresponding to clockwise (+) or anticlockwise (–) rotation for the dihedral angle (θ).

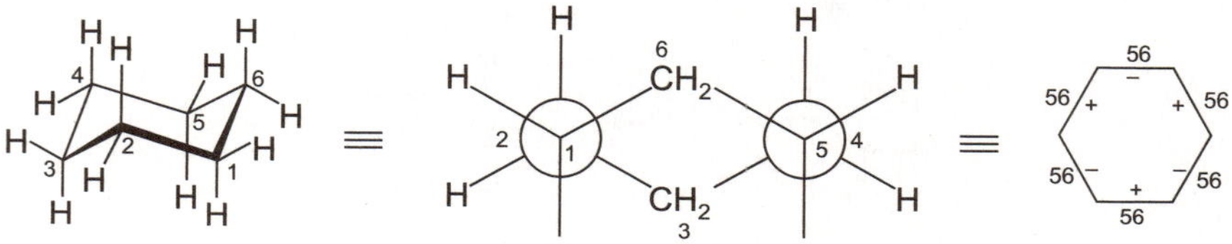

(I) Chair conformation (Has both a C_3 and S_6 axis and a center of symmetry)

(II) Newman projection (All bonds are staggered) The dihedral angle is not 60° but close to 56°

(III)

Cyclohexane chair

SCHEME 4.21

- As far as the symmetry of the chair conformation, mention may be made of a vertical axis passing through the center of the chair (not shown in Scheme 4.21, but see Scheme 4.24) which represents both a C_3 as well as an S_6 axis. The molecule has also a center of symmetry.

(b) The Boat Conformation of Cyclohexane

The following points may be noted:

- In the boat form of cyclohexane there is complete eclipsing of the hydrogens (torsional strain) attached to the carbon atoms forming the "side" CH_2-CH_2 bonds (shown for one side in III, Scheme 4.22) and the Newman projection (II, Scheme 4.22) shows two eclipsed butane units. Thus the two bonds labeled 0, 0 in (IV, Scheme 4.22) are completely eclipsed.

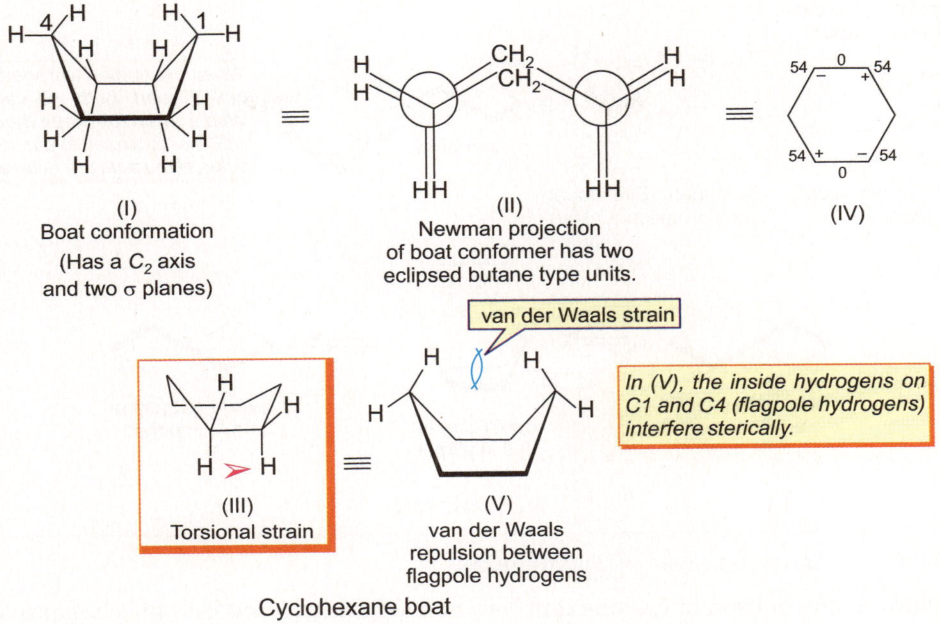

SCHEME 4.22

- The "inside" hydrogens on C–1 and C–4 ("flagpole hydrogens") interfere with each other sterically in transannular interaction (V, Scheme 4.22) and this is a second factor which leads to instability for boat form in comparison to chair. The "flagpole" hydrogens lie only 1.83 Å apart than their sum of van der Waals radii 2.5 Å.
- Thus a boat conformation is considerably less stable than a chair conformation because of the torsional strain associated with four pairs of eclipsed C—H interactions and the steric strain between the two "flagpole" hydrogens. The difference in energy between chair and boat conformations is approximately 29 kJ (6.9 Kcal) / mol (see Fig. 4.1).

(c) The Twist Boat Conformation of Cyclohexane

One has already seen that three forms of strain reflect largely on the energy of a conformation in a cycloalkane these are bond angle, eclipsing (torsional) and transannular (across the ring) strain. The chair conformation is at the minimum energy possible for cyclohexane system being free from bond angle and eclipsing strain. The only steric strain is due to gauche-butane like interactions between neighboring methylene groups.

The boat conformation is a higher-energy conformation due to eclipsing on its "side", carbons and due to transannular interaction between flagpole hydrogens.

The boat conformation is somewhat stabilized by moving apart the "flagpole" hydrogens. The new conformation called "twist boat" which is realized has minimized transannular interactions between "flagpole" hydrogens and at the same time torsional strain between the pair of side carbons of the initial boat are also relieved (Scheme 4.23). The symmetry of the twist boat is reduced and the molecule becomes chiral.

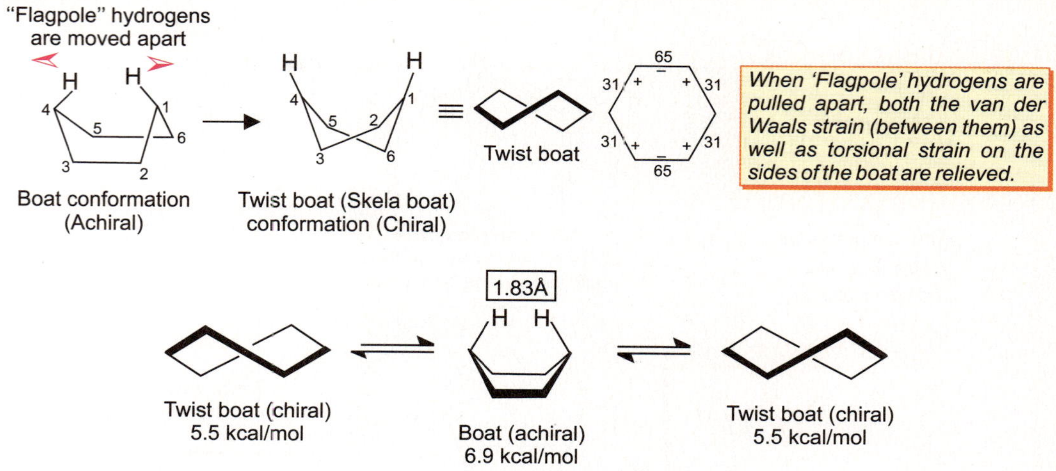

SCHEME 4.23

(d) Conformational Analysis of Cyclohexane

One chair conformation of cyclohexane can be easily converted into an alternate chair conformation. The details of this transformation and an estimate of the energies of different intermediates and transition states along the path of one chair to the alternate chair provides another example of conformational analysis.

The following points may be noted:

- The activation energy for cyclohexane ring inversion (Fig. 4.1) is 10 kcal/mol (42 kJ/mol). It is a rapid process with a half life of around 10^{-5} s at 25°C.
- The transition state conformation is called the half-chair. The half chair can be realized on moving the carbons C1, C2, C3 and C4 of a chair conformation into one plane and C6 above the plane and C5 below the plane. It contains many eclipsed carbon-hydrogen bond.
- The half chair conformer lies about 10 kcal/mol (42 kJ/mol) above the chair form. The 20 kcal/mol (83.7 kJ/mol) of energy that is available at room temperature provides plenty of energy to overcome this barrier. Thus the ring flipping is fast and occurs ~ 100,000 times per second at room temperature.
- During the inversion in cyclohexane, in the first step the chair conformation is converted into a twist boat conformation which lies 5.5 kcal/mol (23 kJ/mol) above the chair.
- The full boat lies atop of the barrier separating two twist forms and like the half chair it is also a transition state. The twist boat arrangement can pass through a second half chair to give the other cyclohexane.

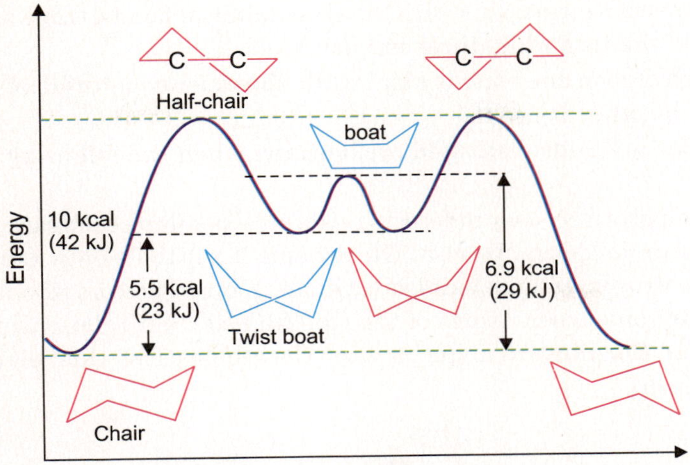

Fig. 4.1

4.7 EQUATORIAL AND AXIAL BONDS IN CHAIR FORM OF CYCLOHEXANE

The twelve C—H bonds in the chair form of cyclohexane are of two types. Six of these are parallel to the three-fold axis of symmetry of the chair. These are represented by vertical lines in the plane of the paper and are designated as axial (I, Scheme 4.24). The remaining six bonds are inclined at an angle of 109°.28′ to the three-fold axis and are described as equatorial.

The following points may be noted:

- Of the six axial bonds three are above (up) relative to the mean plane of the ring while three are down (below). Thus, the molecular model of cyclohexane can stand firmly on the table by the three upper or the three lower hydrogens (I, Scheme 4.24).
- On each face of chair conformer of cyclohexane there are three axial hydrogens and their orientation (up or down) alternates from one carbon atom to the next (I, Scheme 4.24).

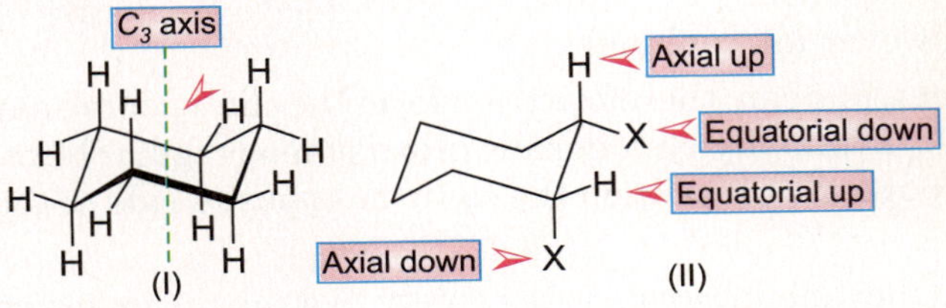

An equatorial bond is recognised (it makes a letter of "Z" with a ring residue). An equatorial bond can be labelled up or down by notionally extending it. It will clearly either go up or go down

SCHEME 4.24

- Similar situation obtains with the six equatorial bonds, three are up the mean plane of the ring while three are down.
- On each carbon one bond is axial while the other equatorial, of these one is up while the other is down.
- Consider a 1,2-disubstituted cyclohexane when the following points become clear.
- It is comparatively easy to recognize the axial bonds on each carbon and whether each is up or down. Thus in (II, Scheme 4.24) the relationship between an axial hydrogen and the axial substituent X on adjacent carbons is *trans*-since these are on opposite sides of the ring (in fact their relationship is *trans* and coplanar, the dihedral angle between these being 180° this relationship is also called *anti*).

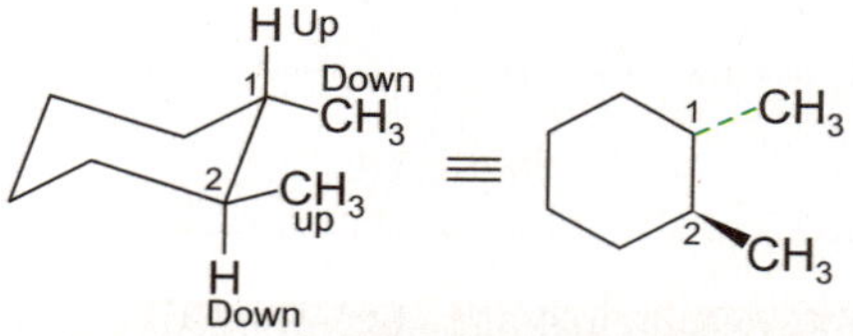

trans-1, 2-Dimethylcyclohexane

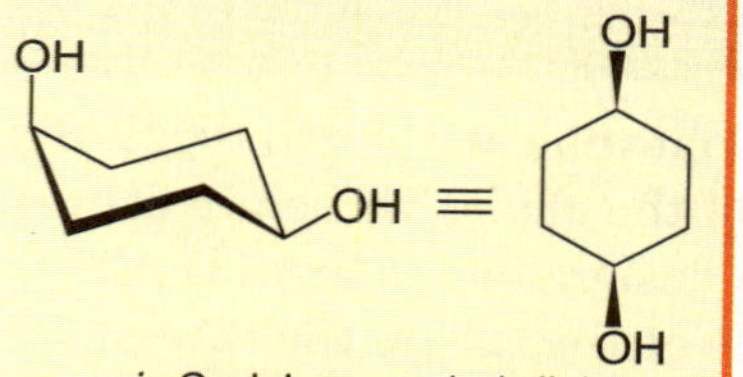

cis-Cyclohexane-1, 4-diol

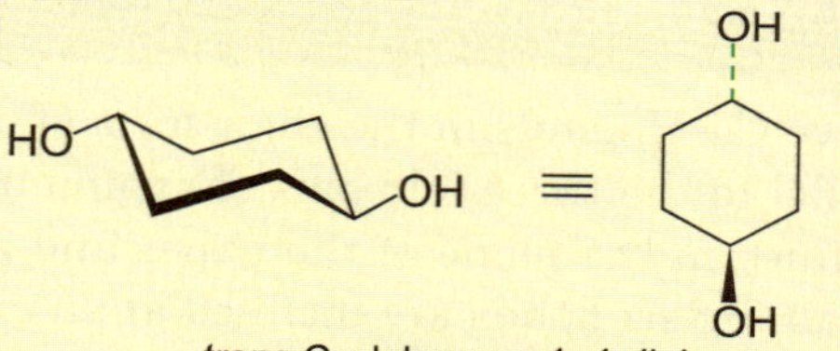

trans-Cyclohexane-1, 4-diol

SCHEME 4.25

- The directionality, is however, not so obvious for equatorial bonds. The two hydrogens (Scheme 4.25) are *cis* (axial hydrogen is up, notionally extending the C—H equatorial bond one finds that it as well goes up).
- The two X groups are *cis* as well, but now both are down (notionally extending the C—X equatorial bond one finds that it goes down).
- Both C—H and C—X equatorial bonds represent *trans*-relationship—one goes up while the other down (on notionally extending the two). These 1,2-diequatorial bonds are *trans*-but however not coplanar.
- Based on these argument, one can, therefore transform the chair conformation (Scheme 4.25) of *trans*-1, 2-dimethylcyclohexane into a planar regular hexagon and can know if the groups are present on the same or opposite side of the C_6 ring.

At room temperature cyclohexane interconverts rapidly to a mirror image chair conformation. As one chair form converts to the other, often referred to as ring inversion or ring flip, all the equatorial hydrogen atoms become axial and all the axial hydrogens become equatorial without losing their identity as up or down (Scheme 4.26).

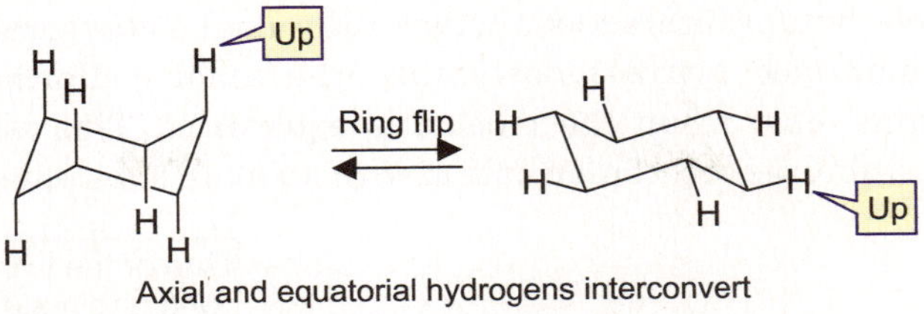

SCHEME 4.26

The equatorial-axial interconversion is so rapid at room temperature that all hydrogen atoms on cyclohexane can be considered equivalent. The energy barrier between the alternate chair conformations is only about 10 kcal/mol.

Ring flipping is slowed down below – 80°C and therefore, axial and equatorial hydrogen atoms can be differentiated as two different, sets of protons, each set giving a separate ^{1}HNMR signal.

4.8 CONFORMATIONS OF SUBSTITUTED CYCLOHEXANES

(i) Conformations of Methylcyclohexane

In cyclohexane itself the two chair forms have equal energy. When however, groups other than hydrogen are attached to the cyclohexane ring, the two chair forms are no longer equivalent in energy. In methyl cyclohexane *e.g.,* the methyl group occupies either equatorial or an axial position (Scheme 4.27). The following points may be noted:

- The two conformations of methyl cyclohexane are not equivalent. In the equatorial conformer the methyl group extends into space away from the rest of the molecule. In the axial conformer, the methyl substituent is close to other two axial hydrogens on the same side of the molecule *syn* axial hydrogens.
- This destabilizing influence arises due to 1, 3-diaxial interactions due to steric repulsions between an axial substituent, the methyl group in the present case and another axial substituent (including hydrogen atoms) located two carbon atoms away (Scheme 4.27).

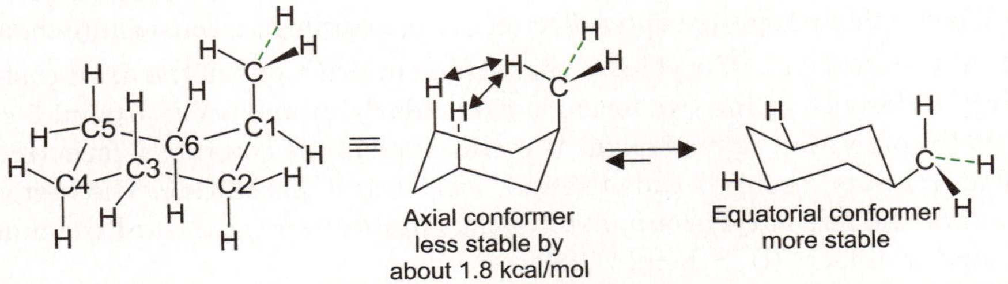

1, 3-Diaxial interactions are caused by steric repulsion between substituents two carbon atoms away.

SCHEME 4.27

- Two *gauche* butane interactions between the axial methyl group and the two ring C—C bonds (see arrows) destabilize the axial conformer while no such interactions exist when the methyl is equatorial. This is the second factor which destabilizes any conformation with an axial substituent (Scheme 4.27*a*).

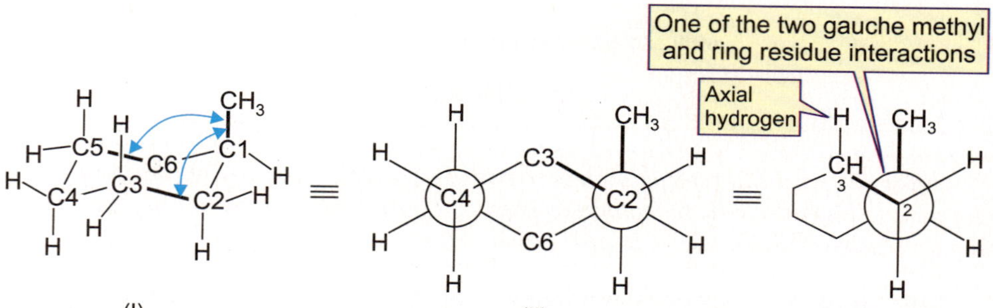

(I) (II) (III)

Two gauche interactions involving CH_3—C1—C2—C3 and CH_3—C1—C6—C5 are clearly seen in chair conformation (I). One can however, see only one of the gauche interactions in the Newman projections (II and III), the other can be seen in viewing these projections along 1–6 bond).

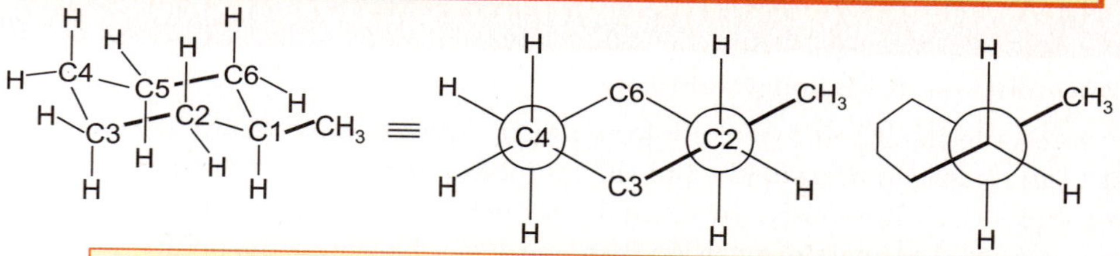

In the equatorial conformer the units CH_3—C1—C2—C3 and CH_3—C1—C6—C5 (shown by thick lines) have now anti orientation

SCHEME 4.27a

(ii) Larger Groups Compete for Equatorial Positions

Presence of a group larger than methyl on a cyclohexane could lead to even greater dominance of the equatorial form over the less stable axial conformation in the equilibrium mixture. The 1, 3-diaxial interactions responsible for the destabilizing gauche interactions will be further magnified and reflected in the equilibrium constant (Scheme 4.28).

Bulky groups *e.g.*, *t*-butyl have very strong interactions in the axial position and the energy difference in the two forms is particularly pronounced (around 5 kcal/mol) so that 99.9% of *tert*-butylcyclohexane is maintained in the equatorial form (II, Scheme 4.28). The *tert*-butyl group is thus used to "lock" the conformation. The better way to put it is that the *tert*-butyl group distorts the equilibrium far toward the much more stable equatorial form (II, Scheme 4.28).

When two bulky *t*-butyl groups are present on a chair cyclohexane, these are much less hindered when both occupy equatorial positions.

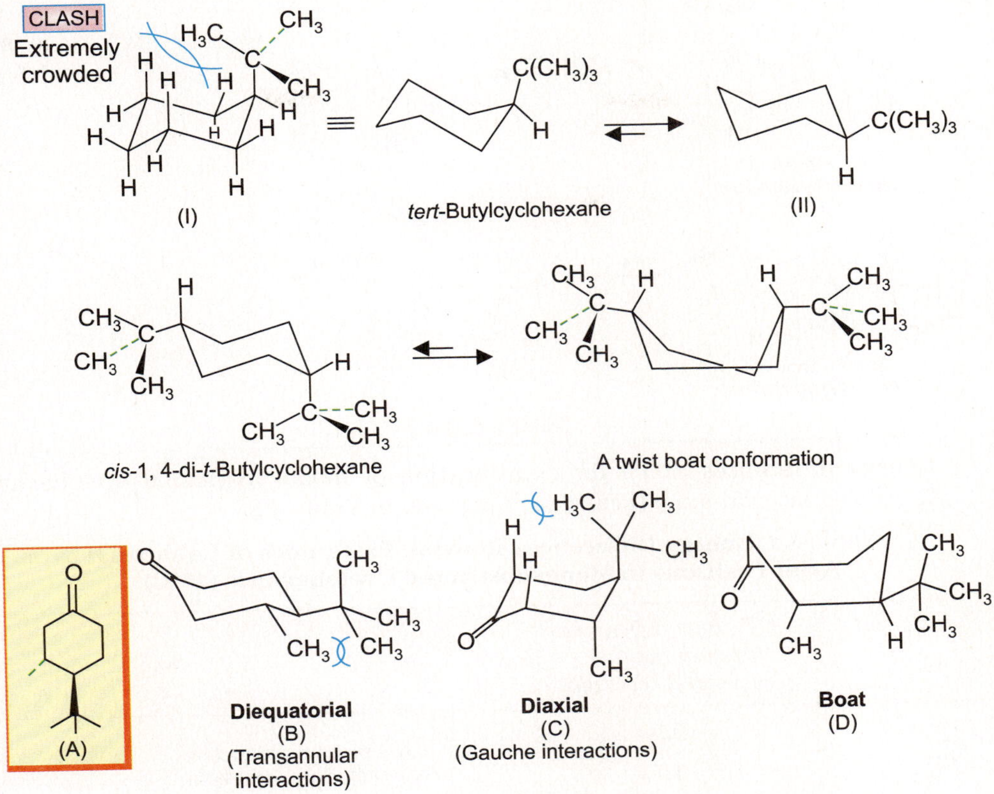

SCHEME 4.28

In a case of *cis*-1, 4-di-*t*-butylcyclohexane one group must be axial and other equatorial and none of the chair conformer allows both bulky groups to be equatorial. The molecule, therefore, adopts a twist boat conformation (Scheme 4.28).

Based on same arguments, the best conformation which can be drawn for compound (A, Scheme 4.28) is (D). Recall that carbonyl carbon is sp^2 hybridised, thus this is neither axial no equatorial.

BOX 4.1

TERTIARY BUTYL GROUP IS ALWAYS EQUATORIAL

A reason for cis-1-bromo-4-tert-butylcyclohexane to react at a much faster rate during E2 elimination than its trans-isomer is the requirement of trans-diaxial relationship in a cyclohexane derivative. This requirement is already present in cis-isomer (Scheme B 4.1), where bromine atom is axial with t-butyl group being always equatorial.

In the trans-isomer ring flip must place the bromine atom in the axial position first, but this is not possible since t-butyl group prevents it. Thus the trans-isomer, first attains a less stable boat conformation for the E2 reaction to proceed and therefore at a slow rate (see also Exercise 5.2).

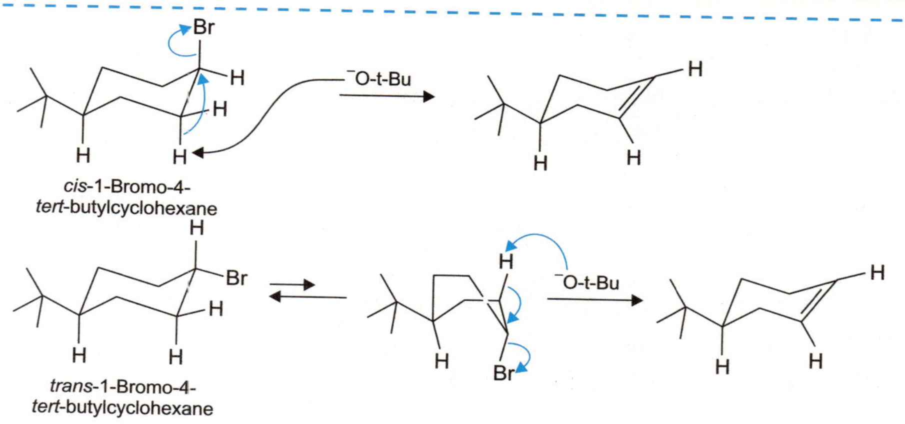

SCHEME B 4.1

Energy differences ($\Delta G°$) for equilibration of monosubstituted cyclohexanes, $C_6H_{11}X$. $\Delta G°$ (equatorial-axis) (kcal mol^{-1}) is given in Table 4.2.

Table 4.2: Energy Differences showing Preference of Equatorial over Axial Positions in Monosubstituted Cyclohexanes (25°C)

Substituent	*$-\Delta G°$, Free-difference between the two conformers, kcal/mol*	
– H	– 0.0	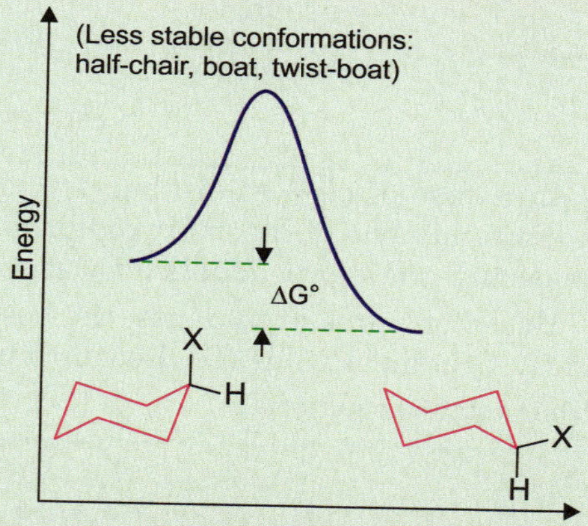
$-CH_3$	– 1.7	
– OH	– 1.0	
– COOH	– 1.4	
$-CH_2CH_3$	– 1.8	
$-CH(CH_3)_2$	– 2.1	
$-C_6H_5$	– 2.9	
$-C(CH_3)_3$	– 5.5 Very large	
– F	– 0.24	
– Cl, – Br, – I	– 0.5	

Generally, substituents larger than hydrogen prefer to be equatorial on a cyclohexane ring to avoid 1, 3-diaxial interactions. The conformational free energies – $\Delta G°$ for the ethyl group (1.8 kcal/mol) (7.5 kJ/mol) and isopropyl group (2.1 kcal/mol) (8.6 kJ/mol) are only slightly larger than that for the methyl group (1.7 kcal/mol), (7.1 kJ/mol) while that for the *tert*-butyl group (5.5 kcal/mol) (22.5 kJ/mol) is much larger. The ethyl and propyl groups can be rotated so that a hydrogen is pointed back over the ring to interact with the axial hydrogens, so their effective steric bulk is not much different from that of a methyl group. In contrast, the *tert*-butyl group is forced to have one of its methyl groups pointed over the ring, causing much more severe 1, 3-diaxial interactions (Scheme 4.28*a*).

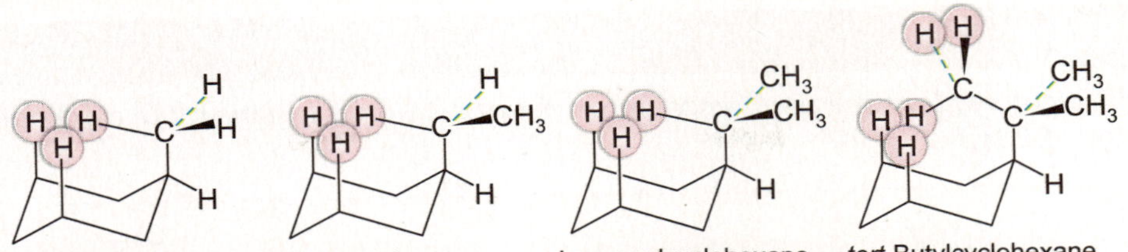

Methylcyclohexane Ethylcyclohexane Isopropylcyclohexane *tert*-Butylcyclohexane

SCHEME 4.28a

It is often said that the *tert*-butyl group locks the molecule into the form with the *tert*-butyl group equatorial. However, it is not a good way to put it. The molecule is not locked but the forms with the large *tert*-butyl group axial and equatorial are in rapid equilibrium with the equatorial form greatly predominating. Very few molecules are in the axial form at any one time and the equilibrium is mobile.

In iodocydohexane, *e.g.*, iodine has little preference for being axial or equatorial. Although iodine is a large atom, the C—I bond in much longer than C—H bond and therefore, 1, 3-diaxial interactions involving iodine are not significant.

(iii) The Favoured Conformation of a Disubstituted Cyclohexane with Unlike Substituents

To predict the relative stabilities of substituted cyclohexane derivatives, one may use the data (Table 4.2) and subtract the $\Delta G°$ value for the axial substituent from the $\Delta G°$ value for the equatorial group:

- If $\Delta G° < 0$, then the conformation drawn first is favoured.
- If $\Delta G° > 0$, then the conformation after a ring-flip is favoured.
- If $\Delta G° = 0$, then neither conformation is favoured.

For example consider *trans*-3-chlorocyclohexane-carboxylic acid (I, Scheme 4.28*b*) and draw its conformation which comes out to be (II).

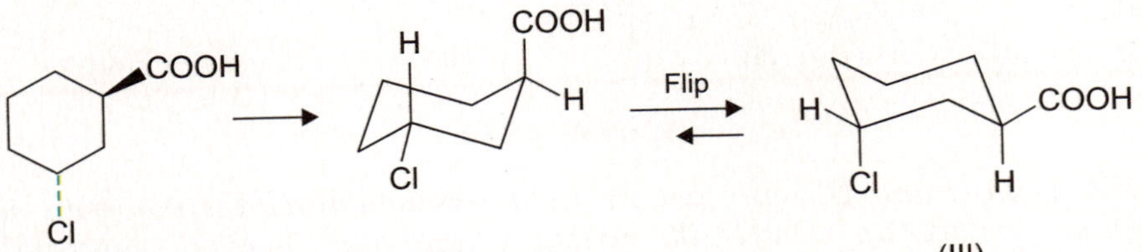

(I) (II) (III)

trans-3-Chlorocyclohexanecarboxylic acid   This chair form is favoured

SCHEME 4.28b

Calculation of the energy difference using the equation, $\Delta G° = \Delta G°_{(\text{eq substituent})} - \Delta G°_{(\text{ax substituent})}$ (Table 4.2) for this molecule, as drawn, the chlorine atom is equatorial (eq) and the acid group is axial (ax). Thus, $\Delta G° = \Delta G°_{(\text{Cl})} - \Delta G°_{(\text{COOH})} = (-0.5) - (-1.4) = +0.9$. Since $\Delta G° > 0$, the conformation formed after a ring flip (III) is the more stable.

EXERCISE 4.3

How one can explain the preference for an equatorial orientation in the compounds (Scheme 4.28b)?

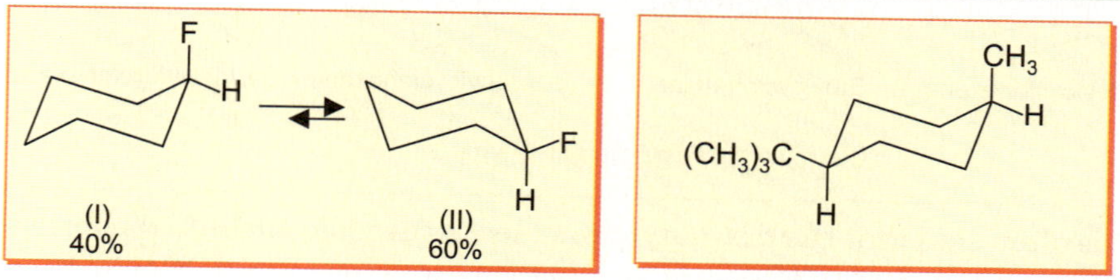

SCHEME 4.28c

ANSWER. *The tert-butyl group must adopt an equatorial orientation, consequently the methyl group is fixed in the axial position.*

A halogen atom does not require much space, moreover, F and H are nearly similar in size thus the preference of F to take up an equatorial orientation is less. The halogens F, Cl, Br, and I do not differ much in their preference for the equatorial position. As the size of the halogen increases in the order F < Cl < Br < I, so does the carbon-halogen bond distance, and the two effects tend to cancel.

EXERCISE 4.4

What is the relationship between the pairs (Scheme 4.28d)-different conformations of the same compound or stereoisomers which cannot be interconverted via rotation around single bonds?

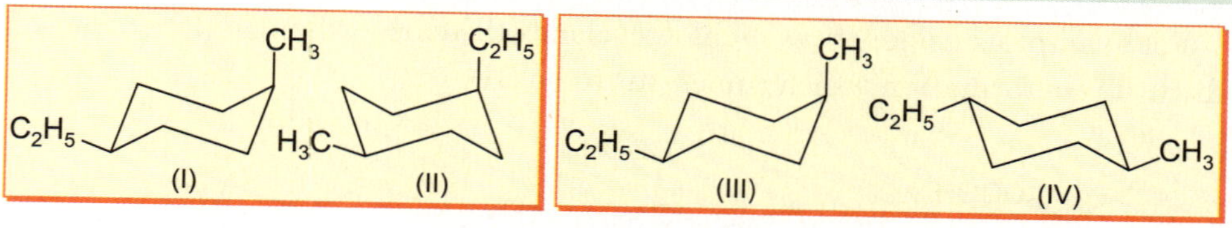

SCHEME 4.28d

ANSWER. *Both (I and II) represent cis-1-ethyl-4-methylcyclohexane (both methyl and ethyl groups are "up"). In (I), the methyl is axial and the ethyl equatorial. The orientations are opposite in (II). The two structures are ring-flipped forms of each other i.e., different conformations of the same compound.*

One may be deceived on the first look since the rings (III and IV) look like ring-flipped forms. However, in both structures ethyl group is equatorial (chair–chair interconversion converts all the equatorial bonds to axial and vice versa). In (III) the two substituents are cis while in (IV) these are trans, thus these are stereoisomers (diastereomers).

(iv) Conformations of Chlorocyclohexane (Isolation of Conformers)

Jensen in 1969 isolated the diastereomeric chlorocyclohexanes working at – 150°. The following points may be noted:

- The methine proton of chlorocyclohexane (which shows only a averaged ^{1}H NMR signal at room temperature) resolves into two sets of signals at – 115° C (Fig. 4.2).

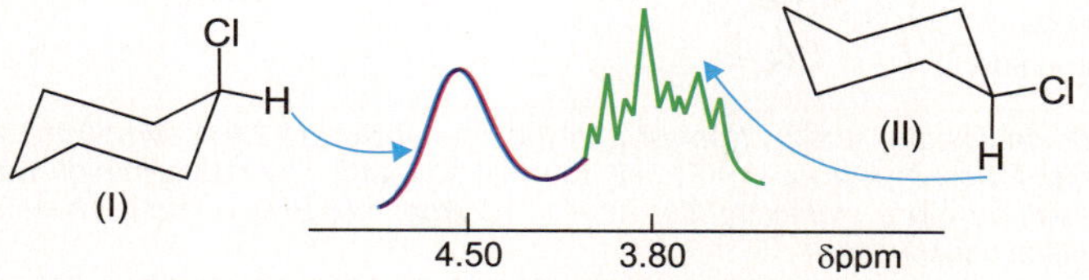

Fig. 4.2

- Diastereomeric chlorocyclohexanes are quite stable at – 150°C and can be isolated in their pure forms via fractional crystallization.
- In their pure forms the equatorial conformer displays only one signal (δ 4.50) while the axial conformer displays the other signal (δ 3.80).
- Each conformer gives its respective single spectrum (*i.e.,* only one signal either at δ 3.8 or 4.5 was observed) showing no impurity of the other diastereomer.
- On warming the solution to room temperature, the equilibrium mixture of axial and equatorial conformations of chlorocyclohexanes was reformed from each single isomer.
- Interestingly (unlike at room temperature) the conformers of chlorocyclohexane can be studied independently at – 115° by ^{1}H NMR which displays both the equatorial methine proton and the axial methine proton at different chemical shift values (Fig. 4.2). The broad singlet at δ 4.50 (broad singlet is due to equatorial proton while multiplets at δ 3.80 are due to axial proton. The equatorial proton is coupled with neighboring protons which are all gauche with lower value of coupling constant. The axial proton, however, is coupled to both axial (high coupling constant) as well as equatorial protons and gives well resolved multiplets.
- This is an example of conformational isomerism at 25° and configurational isomerism at –150°.

CHAPTER 4

4.9 CONFORMATIONAL ANALYSIS OF MONO AND DISUBSTITUTED CYCLOHEXANES – RELATIVE STABILITY OF CONFORMERS

The two forms of chair methylcyclohexane are in equilibrium with the equatorial conformer favored by a ratio of 95 : 5. By using the expression $\Delta G° = -1.36$ K (at 25°C) the difference in energy between the two conformers is 1.7 kcal/mol (7.1 kJ/mol). The energy difference, $\Delta G°$ between the axial and the equatorial isomers of several monosubstituted cyclohexanes has been measured (Table 4.2).

The data (Table 4.2) can be used to determine the relative stability of many substituted cyclohexane conformers. Thus one can analyze different conformations of 1-methyl-4-phenyl cyclohexane (Scheme 4.29) for relative stability. The conformation with both groups equatorial is much more stable than the other. Overall, the *trans*-isomer is more stable than the *cis*-isomer by the amount of strain due to the axial methyl group in the *cis*-isomer *i.e.,* by 1.7 kcal/mol (7.1 kJ/mol).

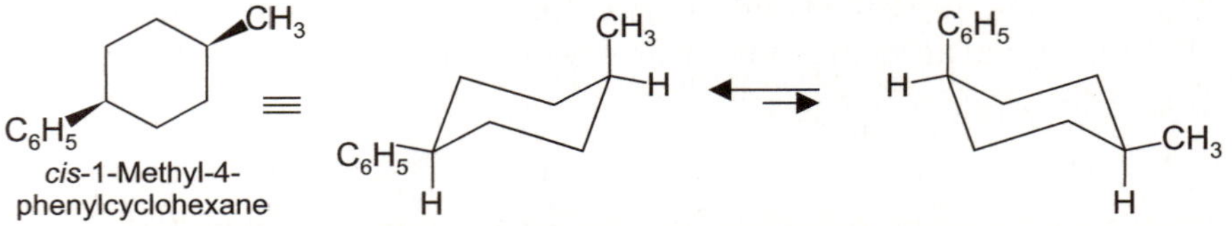

The axial destabilization energy (table 4.2) of the phenyl group [(2.9 kcal/mol) (12.1 kJ/mol)] is larger than for methyl group [(1.7 kcal/mol) (7.1 kJ/mol)]. The conformation with phenyl equatorial will be more stable by 2.9 – 1.7 = 1.2 kcal/mol (5.0 kJ/mol). Therefore it will predominate at equilibrium.

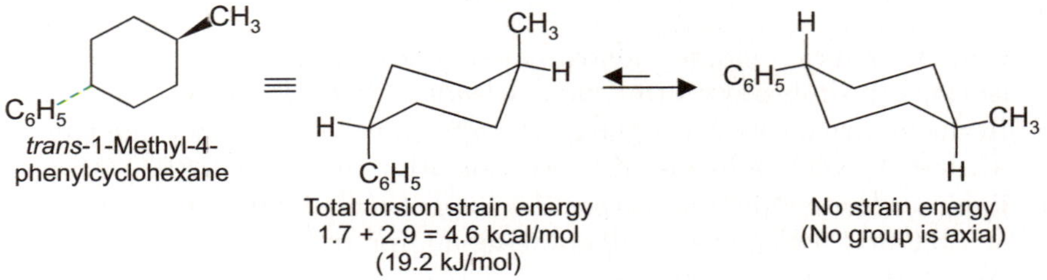

SCHEME 4.29

From the conformational analysis one can, therefore, predict the relative stability of its various conformers. In general a substituent (except for hydrogen) in the equatorial position produces the more stable (the preferred) conformation. For this reason β-D-glucose with five equatorial groups is more stable (Scheme 4.44) than α-D-glucose with four equatorial group and one axial group.

4.10 STEREOISOMERISM OF DISUBSTITUTED CYCLOHEXANES (*cis*-AND *trans*-ISOMERS AND CHIRALITY)

(A) Introduction (Summary)

Stereoisomerism of alicylic compounds (3-5-membered rings) has already been discusses (Schemes 1.65–1.67).

The following points may be noted for cyclohexane systems:

- When two or more substituents are introduced on a ring (infact any size ring) stereo-isomerism becomes possible, as already seen for 3-5 membered rings.
- It is often easy to know the number of stereoisomers of a cyclic compound by considering the ring to be flat (even though it may exist in a chair or other conformations).
- The existence of stereoisomers is independent of conformations. Different stereoisomers will have different conformations.
- The first step is therefore, to know the number of stereoisomers (from the flat structures), identification of all the stereoisomers and finally the conformations of each of the stereoisomers.
- To examine the chirality via the plane of symmetry, it is always better to examine both the planar (flat) structure as well as the conformational formula.

(B) Cyclohexane Derivatives—Stereoisomerism through Prostereoisomerism

Single substituent in a cyclohexane ring invariably prefers the equatorial position, but when a second substituent introduced one must consider if it is *cis*- or *trans*- to the first substituent, and whether it is on C_2, C_3 or C_4. In disubstituted derivatives of cyclohexane, the conformational preference will be for a chair containing both substituents equatorial, or, when this is not possible, for the bulkier of the substituents to be equatorial.

One can study the stereoisomerism in cyclohexane derivatives by considering *e.g.*, the prostereoisomerism of a mono-substituted cyclohexane derivative cyclohexanol (Scheme 4.30). Cyclohexanol is achiral (see Scheme 1.67*a*) however, it contains five centers of prochirality (carbons 1, 2, 3, 5 and 6). The two enantiotopic ligands of C-1 are the two edges of the ring and are designated *pro-R* and *pro-S* (Scheme 4.30). The other four prochiral centers (carbons 2. 3, 5 and 6) carry diastereotopic hydrogens, while carbon-4 has two diastereotopic hydrogen atoms and it itself is proachiral (replacement of either of the hydrogens leads to achiral diastereomeric 1, 4-disubstituted cyclohexane derivatives). Thus, for disubstituted cyclohexanes, stereoisomerism is possible.

Consider the enantiomers of *cis*-3-methylcyclohexanol (Scheme 4.30, also see Scheme 1.27), the pathway of going from C1 to C3 in one enantiomer is clockwise and in another it is anticlockwise.

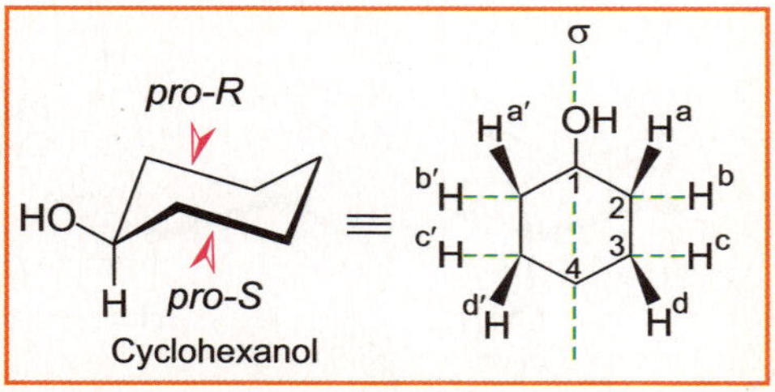

The two ring residues at C-1 are enantiotopic being on the mirror plane. H^a and $H^{a'}$ and also H^b and $H^{b'}$ are mirror images of each other and are enantiotopic, their replacement gives enantiomeric products. Similar is the case with H^c, $H^{c'}$ and H^d, $H^{d'}$. The CH_2 groups exhibit non-equivalent geminal hydrogens which are diastereotopic.
Thus H^a and H^b are diastereotopic, and same is the case with H^c and H^d. Carbon C–4 also contains two diastereotopic hydrogens, but C–4 is not a prochiral center, it is a proachiral center (see, Scheme B 2.1)

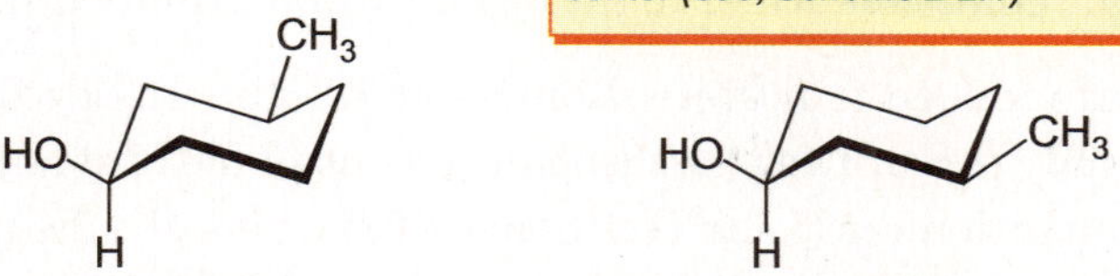

*I*R, 3*S* *I*S, 3*R*

Enantiomers of *cis*-3-methylcyclohexanols
(The methyl groups are occupying enantiotopic positions on the cyclohexanol framework)

SCHEME 4.30

(C) 1, 2-Disubstituted Cyclohexanes (With Similar Substituents)

1, 2-disubstituted cyclohexanes are probably the most challenging of the disubstituted cyclohexanes. Consider the following points:

- A consideration of the planar hexagon structures of *cis* or *trans* isomers shows that 1, 2-dimethylcyclohexane has two chiralitycenters, thus it can have four (2^n) stereoisomers (for the planar structure of *cis*-1, 2-dimethylcyclohexane and its study see (Schemes 1.63, 1.65 and 1.66).

CHAPTER 4

BOX 4.2
DRAWING DISUBSTITUTED CYCLOHEXANES

Two groups substituted on a cyclohexane ring may be either *cis* or *trans*. The *cis*- and *trans*-disubstituted rings are geometric isomers and cannot be interconverted at room temperature; however, either isomer may assume a variety of conformations. For example, consider some chair forms of *cis* and *trans*-1, 2-dimethylcyclohexane (Scheme B 4.2 also see Schemes 4.31 and 4.32).

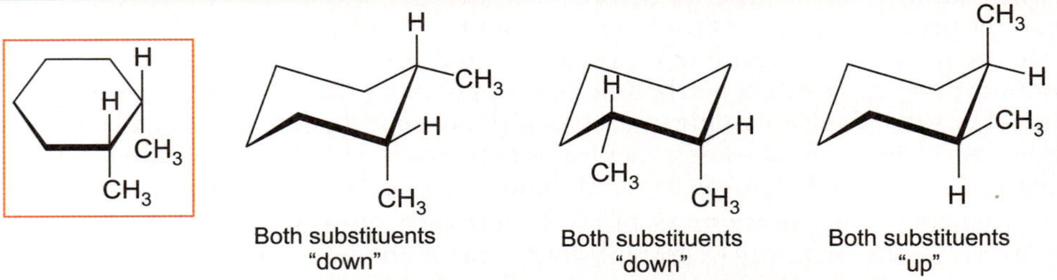

Some different representations of *cis*-1,2-dimethylcyclohexane

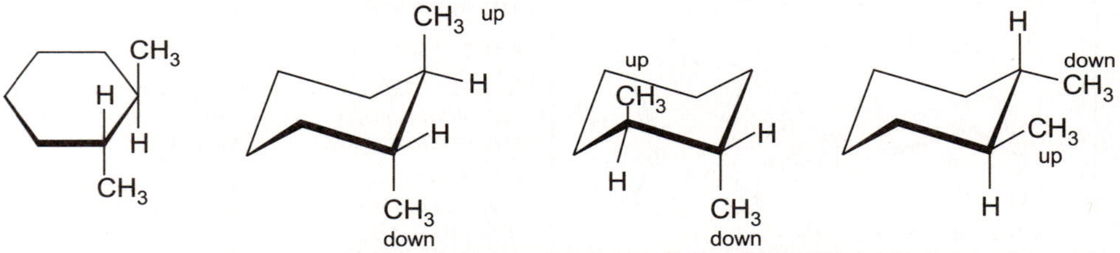

Some different representations of *trans*-1,2-dimethylcyclohexane

SCHEME B 4.2

Changes from one chair conformer to the other reverses their equatorial axial status.

- There are indeed four stereoisomers of 1,2-dimethylcyclohexane but one can isolate only three. The *trans* isomer (has no plane of symmetry) and exists as a pair of enantiomers (I and II Scheme 4.31) where the diequatorial conformer is the most stable. *cis*-1, 2-Dimethylcyclohexane (see Scheme 4.32) as such is also chiral but its enantiomers interconvert very rapidly to give a racemic mixture.
- In the case of *trans*-1, 2-dimethylcyclohexane (Scheme 4.31 I and II) are enantiomers. On ring flip *trans*–1, 2-dimethylcyclohexane converts the diequatorial stereoisomer (I) into the diaxial form (I*a*) and these two molecules are diastereomers (conformational diastereomers). Similarly (II and II*a*) are also conformational diastereomers.

A ring flip in trans-1, 2-dimethylcyclohexane (I, Scheme 4.31) will convert both the methyl groups into the less stable diaxial conformation. Remember a 1, 2-disubstituted cyclohexane will be cis if one group is axial and the other equatorial while it will be trans if either both the groups are equatorial or both axial.

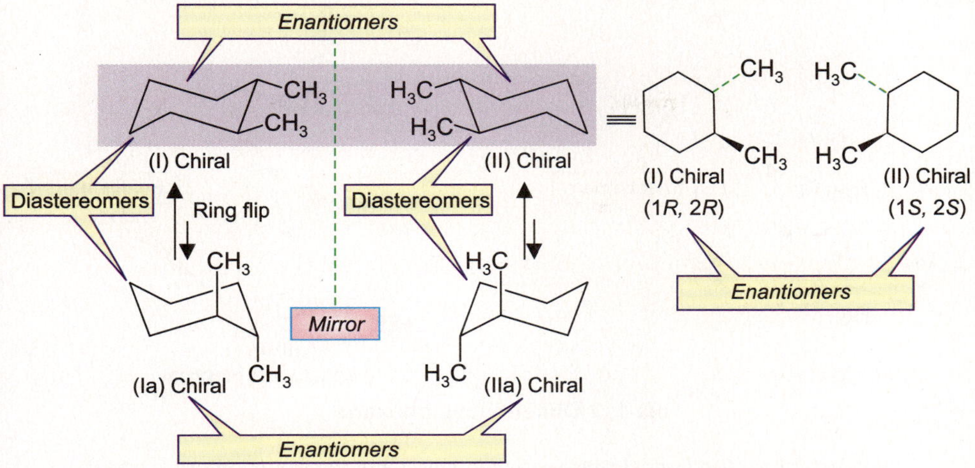

trans-1, 2-*Dimethylcyclohexane*

1, 2-Disubstituted derivatives may exist as four stereoisomers each having two distinct chair conformations (with similar or different substituents). The same is true for 1, 3-disubstituted derivatives. Note that only one of the two enantiomers is often represented (e.g., see shaded structures, Schemes 4.31 and 4.32).

SCHEME 4.31

- The *cis*-isomer of 1, 2-dimethylcyclohexane has no plane of symmetry when the chair form is considered (see, Scheme 1.63). However, the following arguments are necessary to understand.
- A planar structure of cyclohexane may be helpful when conformational details are not studied, moreover a given cyclic compound cannot be optically active if its planar structure has a plane of symmetry. Whether it is *meso* or racemic may be studied via its conformers (see, Schemes 4.32 and 4.34).
- The two mirror image conformational structures (I and II, Scheme 4.32) are not identical and are non superimposable. Therefore, both are chiral molecules and represent a pair of enantiomers.
- Both (I and II, Scheme 4.32) are however, interconvertible by ring flip. The inter-conversion of enantiomers by ring flip can be best shown by models. Alternatively as shown (I and II, Scheme 4.33), have a nonsuperimposable mirror image relationship, however if (I) undergoes a ring flip and the arrangement is rotated through an axis by 90° it becomes identical with its mirror image. Thus ring flip in *cis*-1 , 2-dimethylcyclohexane converts one enantiomer into the other (a process termed enantiomerization) *i.e.*, equilibration between the two conformations produces a racemic mixture.

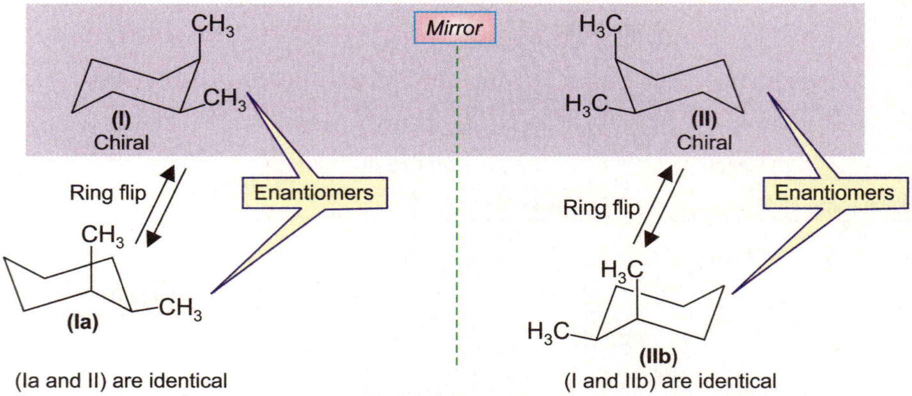

***cis*-1, 2-Dimethylcyclohexane**

As such *cis*-1, 2-dimethylcyclohexane is chiral, but the enantiomers (I and II) vary rapidly interconvert to give a racemic mixture.

SCHEME 4.32

Thus (I and II, Schemes 4.32 and 4.33) represent conformational enantiomers a situation similar to that found in gauche butane (see, Scheme 1.1*c*). Notice that the configurations of the chirality centers remain unchanged during the ring-flip as expected, since no bonds are broken during this conformational change. Thus *cis*-1, 2-dimethyl-cyclohexanes may exist as a racemic mixture or as a *meso* compound (see, Scheme 4.34).

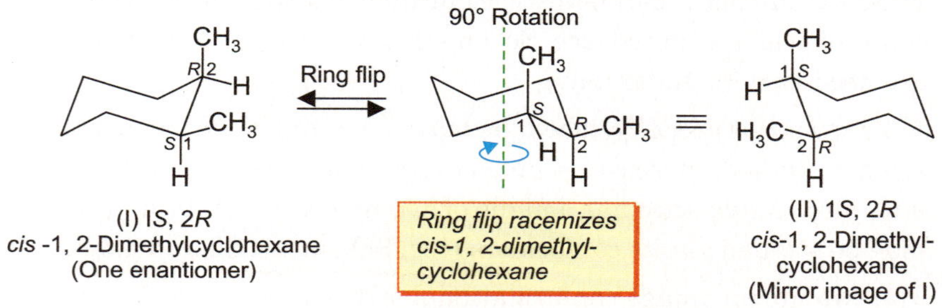

SCHEME 4.33

At normal temperatures therefore, only three isolable stereoisomers of 1, 2-dimethyl-cyclohexane exist, trans-isomer as a pair of enantiomers and cis-isomer as a racemic mixture.

1, 2-Dimethylcyclohexane–A closer look

The following points may be noted:

- *In the case of cis-1, 2-dimethylcyclohexane ring flip racemizes it i.e., on ring flip one enantiomer is converted into another, the compound may also be a meso compound (see, Scheme 4.34).*
- *In the case of trans-1, 2-dimethylcyclohexane a ring flip converts the diequatorial stereoisomer into the diaxial form. These two molecules are conformational diastereomers and not enantiomers. There is no racemization during its ring flip.*
- *In principle in the case of trans-1, 2-dimethylcyclohaxane both diaxial as well as diequatorial forms should be resolvable, both being chiral. However, the trans-diaxial-1, 2-dimethylcyclohexane cannot be isolated, since it simply flips into the more stable diequatorial form.*

EXERCISE 4.5

Depict the dihedral angles between the methyl groups in cis and trans-1, 2-dimethylcyclopropane and cis and trans-1, 2-dimethylcyclohexane (Scheme 4.33a). Which of these is resolvable ? What is the significance of this ?

ANSWER.

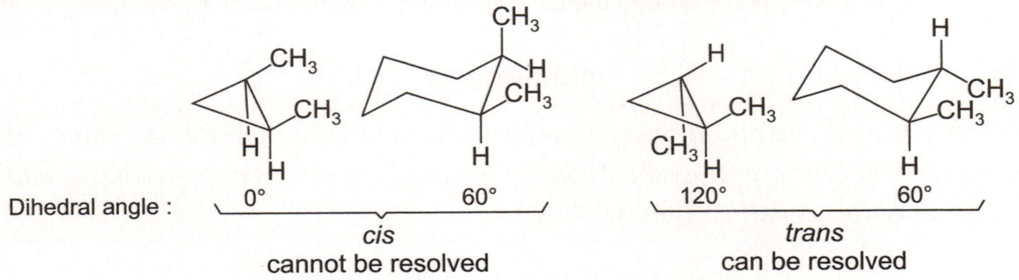

SCHEME 4.33a

Neither cis-1, 2-dimethylcyclopropane nor cis-1, 2-dimethylcyclohexane can be resolved, cyclopropane derivative being meso and the cyclohexane derivative flips into its mirror image. Their counterparts can however, be resolved. This has significance since planar cyclopropanes and nonplanar cyclohexanes behave similarly. Thus in deciding the problems of stereochemistry cyclohexane can be treated in their planar forms.

There is a further interesting and relevant point regarding cis-1, 2 dimethyl-cyclohexane and like molecules. The configuration of C1 and C2 (IS, 2R) in both (I and II, Scheme 4.33) remains the same and these configurations stay with these carbons whatever the conformation recall that a compound can have different conformations but same configuration sec. 1.2, 9). The substituents at the two chiralitycenters are the same, thus cis-1, 2-dimethylcyclohexane may be thought to represent a meso-compound. In fact, if a particular conformer of a molecule is chiral, the molecule is not necessarily chiral. A molecule cannot be chiral if it is in equilibrium with a structure (or a conformation) that is achiral. In case there is no achiral conformation then the molecule is chiral. Moreover a molecule cannot be chiral if any two of its conformers are enantiomers. The boat conformation of cis-1, 2-dimethylcyclohexane (Scheme 4.34) easily obtainable from either (I and II, Scheme 4.34) has a plane of symmetry passing between the stereocenters.

(Contd...)

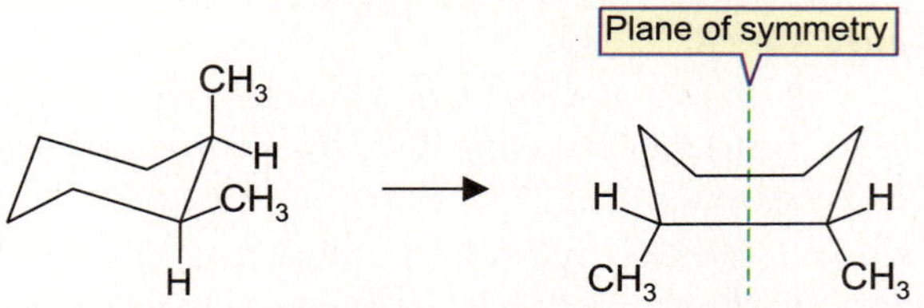

An enantiomer of *cis*-1, 2-dimethyl-cyclohexane (Chair form chiral) — *cis*-1, 2-Dimethylcyclohexane (Boat form achiral)

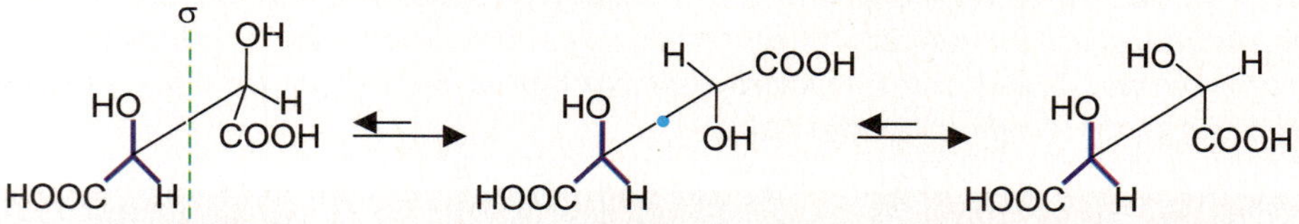

Synperiplanar (I) *meso*-Tartaric acid (Plane of symmetry)

Anticlinal (II) *meso*-Tartaric acid (Center of symmetry)

Anticlinal (III) *meso*-Tartaric acid chiral conformation but in equilibrium with (I and II)

The desired conformation (II or III, Scheme 4.34) can be otained from e.g., (I) by the rotation of one half of the molecule around the central carbon – carbon bond, by 120°.

SCHEME 4.34

Meso-Tartaric acid can adopt several conformations (I-III Scheme 4.34), some chiral and some achiral but since the chiral conformations is in equilibrium with the achiral ones, the molecule overall cannot be chiral.

(D) 1, 2-Disubstituted Cyclohexanes with Two Different Substituents

In case the two substituents are different in a 1, 2-disubstituted cyclohexane, then both *cis* as well as *trans* isomers are resolvable. The *trans* isomer exists in a diequatorial conformation, while the *cis* isomer in equatorial axial conformation with the bulkier of the groups predominantly in the equatiorial position.

One may consider the example of 1-isopropyl-2-methylcyclohexanes where *cis* as well as *trans* isomers exist. Unlike the *cis* dimethyl case (see, Scheme 4.32) these two compounds (I*a* and I*c*, Scheme 4.34*a*) are not enantiomeric. They are conformational diastereomers—different conformational isomers with different physical and chemical properties. In principle, each could be isolated and resolved. So there is a total of four possible *cis* isomers, the two shown (I*a* and I*c* Scheme 4.34*a*) and their mirror images.

^{13}C NMR SIGNALS IN 1, 1, DIMETHYLCYCLOHEXANE

Cyclohexane chairs interconvert at room temperature and above but this chair-chair flip can be stopped at low temperature. Consider one chair at low temperature and both chairs at high temperature. At low temperature there will be two different methyl groups one axial, the other equatorial. At high temperature, one has averaged structure and there will be only one methyl group.

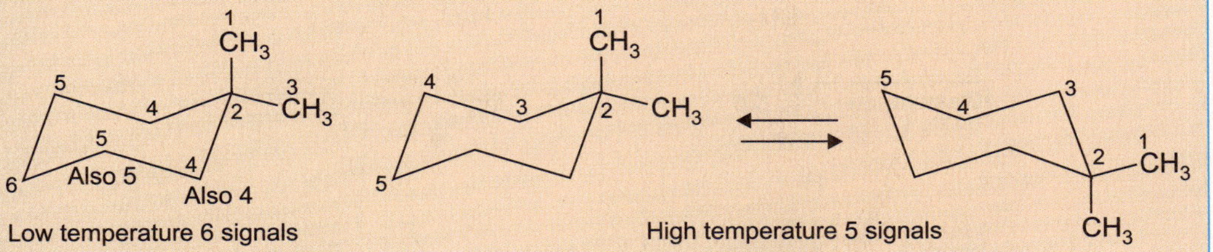

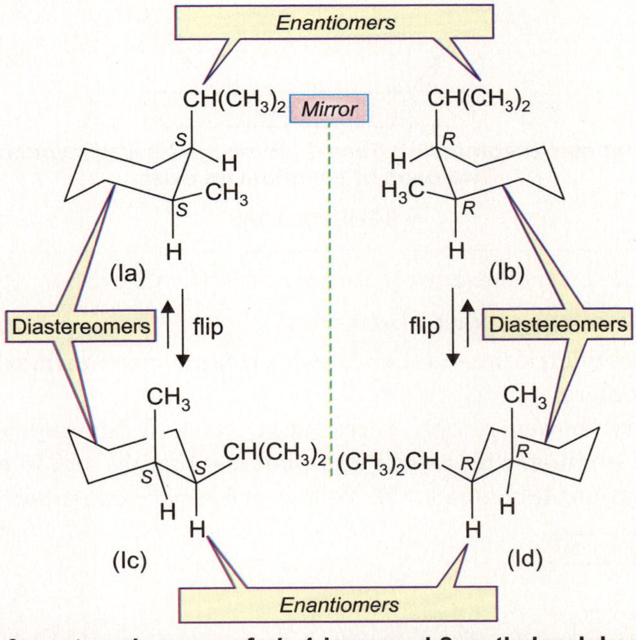

The four stereoisomers of *cis*-1-isopropyl-2-methylcyclohexane. Two pairs of enantiomers exist.

SCHEME 4.34a

Similarly the four stereoisomers of *trans*-1-isopropyl-2-methylcyclohexane are presented (Scheme 4.34*b*).

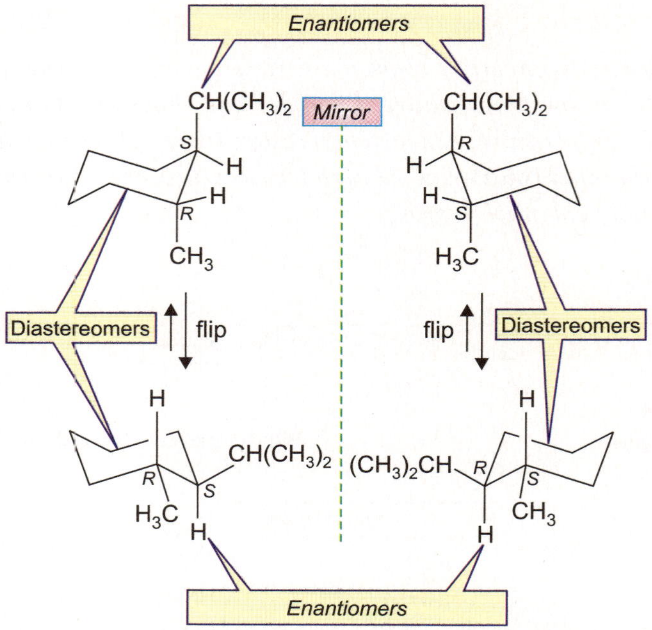

The four stereoisomers of *trans*-1-isopropyl-2-methylcyclohexane. Two pairs of enantiomers exist.

SCHEME 4.34b

(E) 1, 3-Disubstituted Cyclohexanes (with Identical Substituents)

Consider the following point-wise discussion:

- 1, 3-Dimethylcycloxane has two chirality centers, one can, therefore, expect ($2^2 = 4$) stereoisomers.
- In reality, however, only three exist, *cis*-1, 3-dimethylcyclohexane has a plane of symmetry both seen in a planar structure and in either of the chair conformations (Scheme 4.35). Thus it is a *meso* compound.

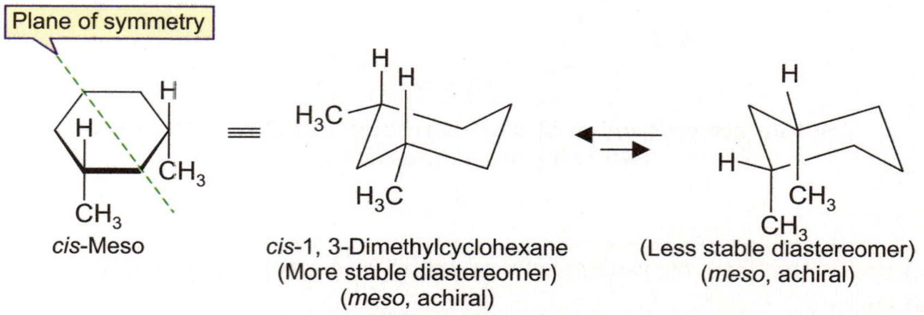

These two forms are diastereomers–each is a meso compound

SCHEME 4.35

- *trans*-1, 3-Dimethylcyclohexane is chiral (does not have a plane of symmetry) and exists as a pair of enantiomers (Scheme 4.36).

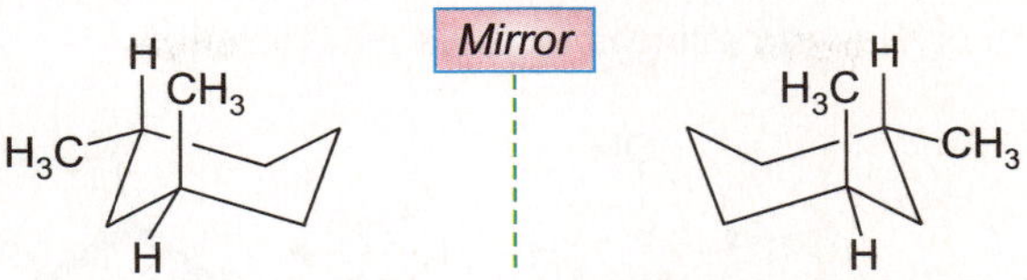

Enantiomers of *trans*-1, 3-dimethylcyclohexane

SCHEME 4.36

- In *trans*-1, 3-dimethylcyclohexane (like *cis*-1, 2-dimethylcyclohexane one group is axial and the other equatorial. One has already seen that in *cis*-1, 2-dimethylcyclohexane one enantiomer on ring flip changes into other *i.e.*, ring flip racemizes the compound.
- The ring flip in *trans*-1, 3-dimethylcyclohexane does not racemize this compound but regenerates the same enantiomer *i.e.*, ring inversion here (unlike in *cis*-1, 2-dimethyl-cyclohexane) only converts (+) into (+) and (–) into (–) enantiomers (Scheme 4.37).

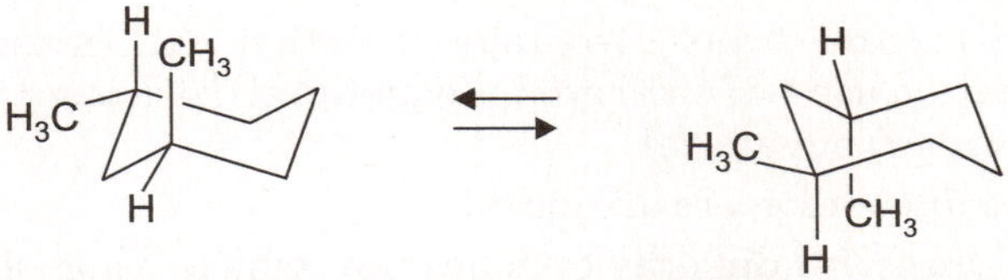

trans-1, 3-Dimethylcyclohexane

These two "forms" are identical–just turn this one over, "back-to-front".

***trans*-1, 3-Dimethylcyclohexane (chiral) flips into itself.**

SCHEME 4.37

- A preferred diequatorial 1, 3-*cis* isomer can adopt a diaxial conformation by ring flip if a situation so demands. This happens during *e.g.*, anhydride formation from *cis*-1, 3-cyclohexane dicarboxylic acid (Scheme 4.38).

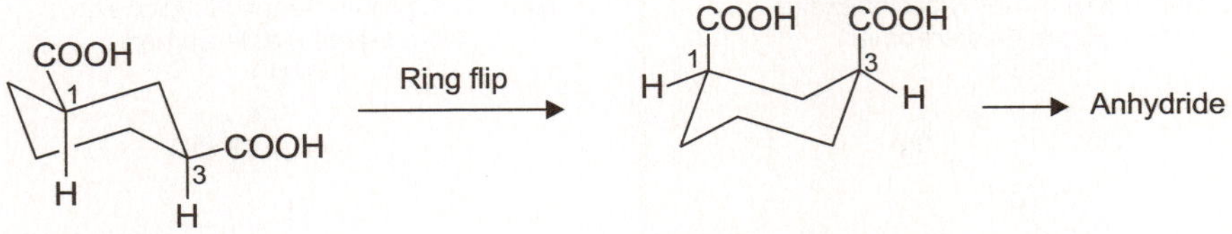

(I)
cis-1, 3-Cyclohexane-dicarboxylic acid
diequatorial conformer
(More stable)

(II)
cis-1, 3-Cyclohexane-dicarboxylic acid
diaxial conformer
(Less stable)

The -COOH groups in (I) are far apart for anhydride formation. Ring flip to give (II) brings them within reacting distance.

SCHEME 4.38

CHAPTER 4

The chair form of cyclohexane is the preferred conformation however other conformations are known and in some situations require a conformation other than chair (Scheme 4.38*a*). The lactone moeity functions as a stereochemical bridge, maintaining the hydroxyl and carboxylic acid groups in a *cis* relationship.

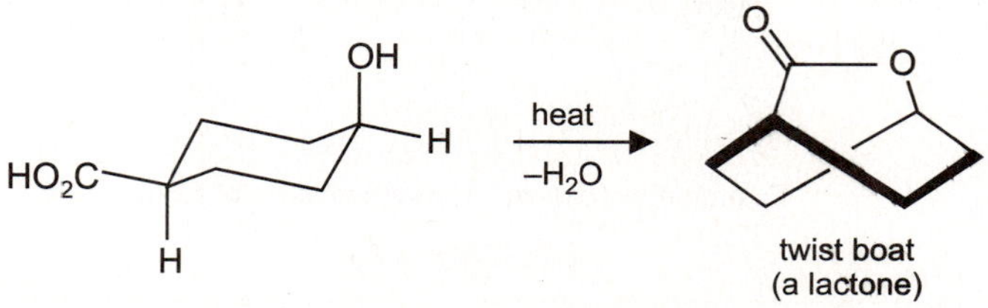

SCHEME 4.38a

(F) 1, 3–Disubstituted Cyclohexanes with Different Substituents

One can draw the possible chair forms of *cis*-and *trans*-1-isopropyl-3-methylcyclohexane (Scheme 4.38 *a*) *i.e.*, when the substituents are different:

- Note that unlike the achiral (*meso*) *cis*-1, 3-dimethylcyclohexane (see, Scheme 4.35), the *cis*-1-isopropyl-3-methylcyclohexane (Scheme 4.38*a*, with different substituents) is chiral.
- Both in *cis*-and *trans*-1-isopropyl-3-methyl cyclohexanes the most stable conformation has two alkyl groups equatorial (*cis*-isomer) while the large group (*trans*-isomer) equatorial.
- Both the diastereomers are chiral.
- Unlike *trans*-1-3-dimethylcyclohexane which ring flips on itself, a 1, 3-disubstituted cyclohexane with different substituents does not do so and four stereoisomers are available (Scheme 4.38*b*).

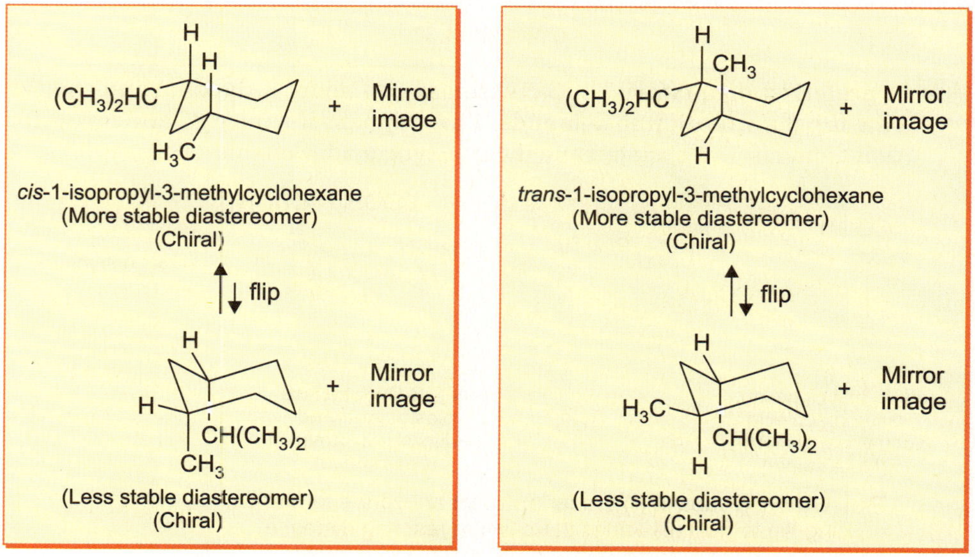

SCHEME 4.38b

(G) 1,4-Disubstituted Cyclohexanes (1,4-Dimethylcyclohexane)

The *cis* isomer exists in two identical conformations (axial equatorial and equatorial axial) while the *trans* in two non equivalent (equatorial, equatorial and axial, axial) conformations (Scheme 4.39, only one conformer in each case is shown). The vertical plane passing through C1 and C4 is a mirror plane of symmetry so all the conformers are achiral (even if the two substituents are different). The *cis* and *trans* isomers represent diastereomers.

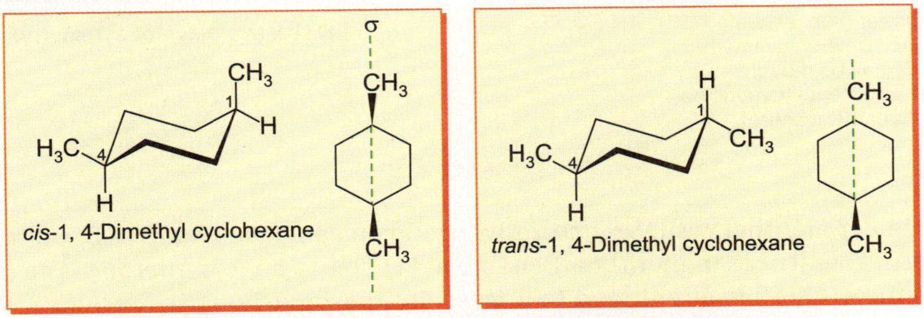

In a trans-1, 4-disubstituted cyclohexane either both substituents are diaxial or diequatorial.
In the *cis* isomer one group is axial while other equatorial.
Both *cis* and *trans* 1, 4-disubstituted cyclohexanes are achiral diastereomers, (plane of symmetry through C1 and C4)

SCHEME 4.39

(H) Summary

The chirality and nomenclature of different disubstituted cyclohexanes (with the more stable conformer and similar substituents) can be quickly reviewed from (Scheme 4.40). When one considers the planar structures of *cis*-1, 3-dimethylcyclohexane and *cis*-1, 2-dimethylcyclohexane (I and II respectively, Scheme 4,40) one finds that both have a plane of symmetry and both must be optically inactive. In compound (I) both the conformations (see Scheme 4.35) also have a plane of symmetry, thus these are achiral and *meso*. On the other hand (II) consists of two chiral conformations which are enantiomeric. Thus both (I and II, Scheme 4.40) have a plane of symmetry in their flat structures however, (I) represents a *meso* compound while (II) a racemic mixture of two separable enantiomers (low temperature). This difference is not clear from the planar structures which therefore, is one of the disadvantages of representing cyclohexanes in their flat form.

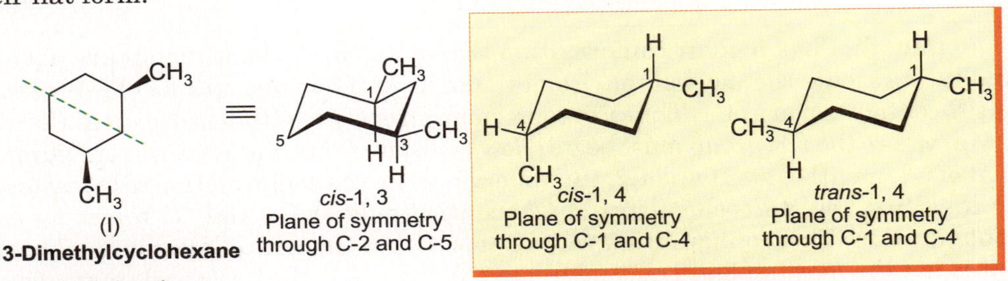

Optically inactive, its conformers (diequatorial and diaxial) have a plane of symmetry. It represents a *meso* compound

Achiral
(A plane of symmetry)

Continued...

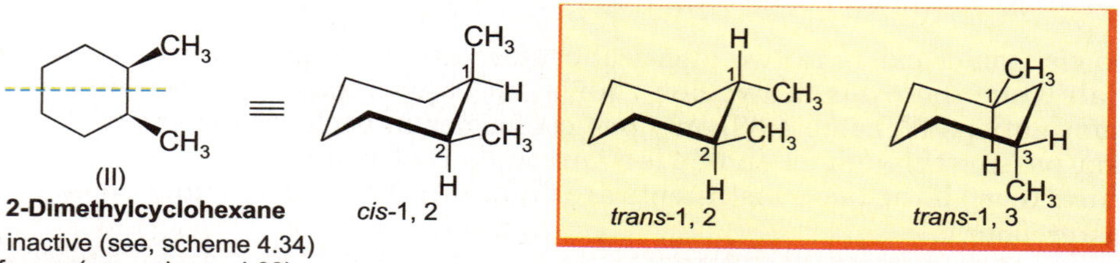

Optically inactive (see, scheme 4.34) its conformer (see, scheme 4.33) has no plane of symmetry, it may be thought to exist as an enantiomeric pair (*racemic* mixture).

Chiral **(No plane of symmetry)**

SCHEME 4.40

Consider menthol with three chirality centers it can have up to $2^3 = 8$ stereoisomers. The structural features in menthol (Scheme 4.41) do not have a mirror plane (like that in I, (Scheme 4.41) that might lead to a *meso*-stereoisomers.

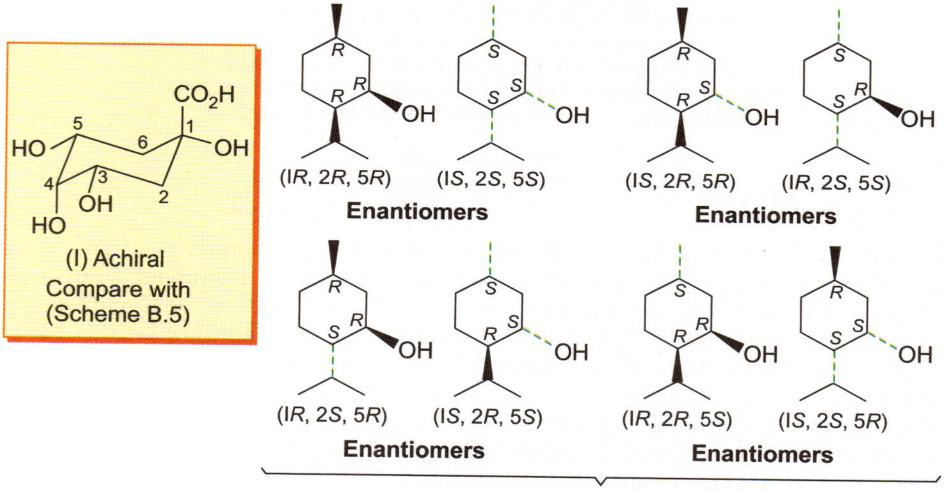

Menthol (with 3 chirality centers) is 2-isopropyl-3-methylcyclohexanol with 8 stereoisomers

SCHEME 4.41

Menthol has a chair conformation where all three substituents are equatorial and it represents the most stable isomer, moreover when one speaks of (–)-menthol it has *R*-configuration at C1. Thus the conformation of (–)-menthol is drawn as (I, Scheme 4.41a) where the OH group must be up. Now (+)-isomenthol has opposite configuration to that of (–)–menthol at C1 and C2, thus one can write the conformation of (+)-isomenthol by inverting the stereochemistry of (–)-menthol only at C1 and C2 to get its correct configuration. This configuration (A, Scheme 4.41*a*) though configurationally correct, but is not the most stable. The ring flip in (A) gives (A1) the preferred conformation of (+)-isomenthol. Isomenthol possesses *cis* alkyl groups, one of which therefore, must be axial in any chair conformation.

Only menthol the most stable isomar has a chair conformation in which all three substituents are equatorial. Neomenthol's two alkyl groups, like those in menthol, are

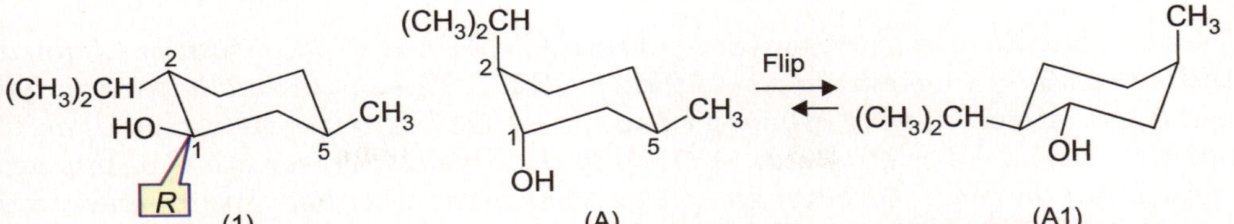

(1) (–)-Menthol

(A) (+)-Isomenthol

(A1) Most stable conformation of (+)-isomenthol

Inverted stereochemistry only at C1 and C2.
The configuration is correct, but it is high energy conformation

SCHEME 4.41a

trans and in a 1, 4 relationship in the ring. Neomenthol therefore has a conformation in which both alkyl groups are equatorial and only the relatively "small" hydroxy group is axial. The stability order is therefore: menthol > neomenthol > isomenthol

BOX 4.3
A FINE EXERCISE IN CONFORMATION AND STEREOCHEMISTRY IN A CYCLOHEXANE DERIVATIVE- ABIETIC ACID

The alcohol (II, Scheme B 4.3) obtained from abietic acid on dehydration with phosphorus pentachloride gave a triply unsaturated hydrocarbon (IV). Obviously, a primary alcohol attached to a quaternary carbon atom cannot be dehydrated in a simple fashion, the transformation was believed to involve the conversion of a hydroxymethylene group, attached to the nucleus, to a ethyl group via a Wagner change. The driving force for the Wagner shift under acid treatment of (II, Scheme B 4.3) is the non-bond 1 : 3-interactions between the angular methyl group at C10 and a methyl group at C4 (as shown in A).

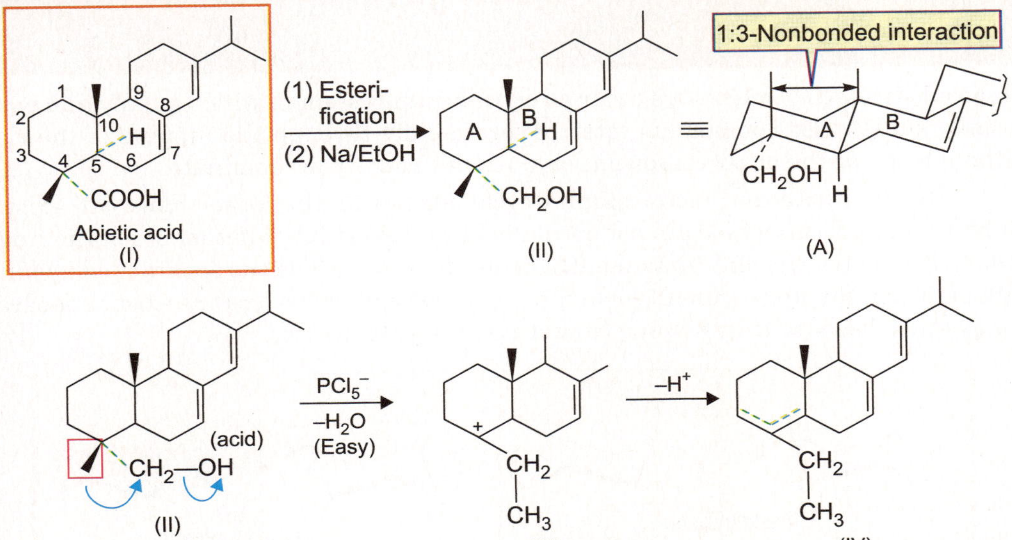

Abietic acid (I)

(II)

(A)

(II)

(III)

(IV) Triply unsaturated hydrocarbon

SCHEME B 4.3

(Contd...)

The tricarboxylic acid (corresponding to ring A) obtained by the oxidation of abietic acid with $KMnO_4$ is a meso compound (Scheme B 4.4). This shows that the tricarboxylic acid has a mirror plane of symmetry and the – COOH groups (meta placed) on C4 and C10 have a cis relationship. A study to the dissociation constants of this acid showed that the center carboxyl group was trans to the other two. The argument was based on the observation that the difference between pK_1 and pK_2 of cycloalkane-1,2-dicarboxylic acids is greater for the cis-than for the trans-acid. Hence rings A and B are trans fused.

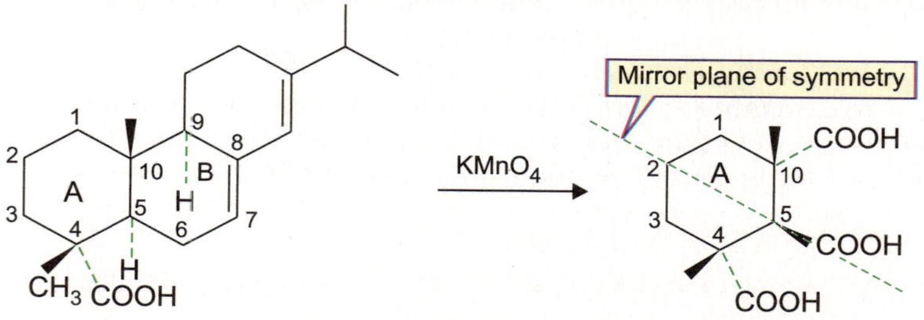

$[\alpha]_D$–110°
Abietic acid

Meso acid
$[\alpha]_D$–0°
A mirror plane of symmetry passes through C5 and C2, (see, Schemes 4.40 and 4.39).

SCHEME B 4.4

4.11 EQUILIBRIA OF DISUBSTITUTED CYCLOHEXANES AND RELATED SYSTEMS

Monosubstituted cyclohexanes occur as equilibrium system in which axially and equatorially substituted compounds interconvert rapidly through the flipping of the ring, although one or the other conformational isomer usually predominates.

Disubstituted and more complex cyclohexanes on the other hand can exist in stereoisomeric forms which are not normally interconvertible, whether or not the ring is able to flip. Thus, *cis*- and *trans*-menthane exist (Scheme 4.42) as separate compounds although ring-flipping occurs in each. Similarly, *cis*- and *trans*-decalins exist as separate compounds, because ring-flipping cannot interconvert the two.

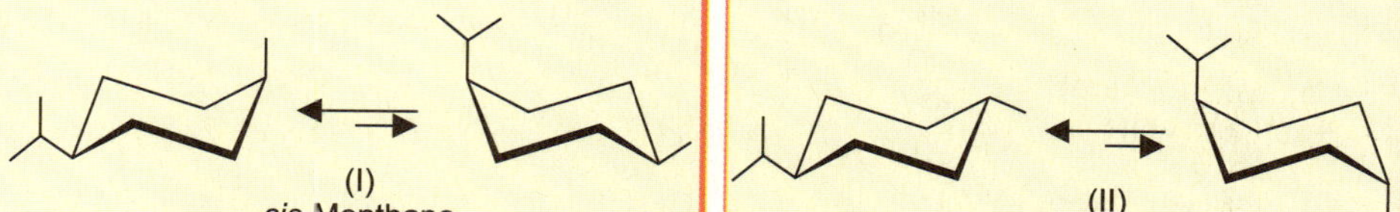

(I) *cis*-Menthane

(II) *trans*-Menthane

A cis-disubstituted cyclohexane remains *cis* even after ring flip and same is the case with the trans isomer.

SCHEME 4.42

However, the presence of some structural situations allows the interconversion of such stereoisomers under some conditions. For example, the incorporation of a carbonyl group in menthane gives *cis* and *trans*-isomers which are interconvertible

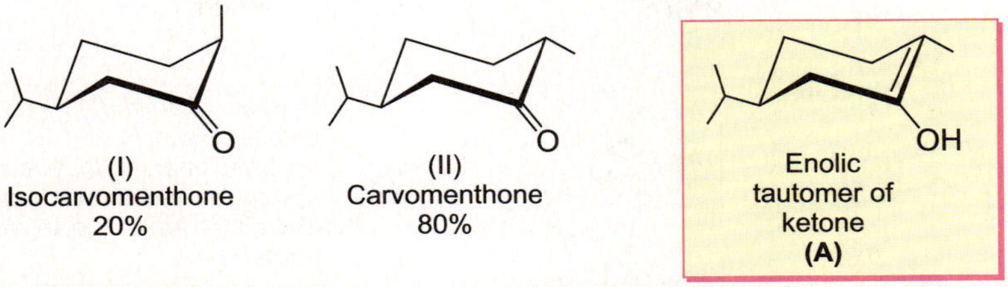

***cis*-Disubstituted cyclohexane can be converted to a *trans*-isomer via an enol.**

SCHEME 4.43

in both acidic and basic conditions *via* enol (A). The *trans* isomer, in which both alkyl substituents are equatorial, predominates (Scheme 4.43). This is the basis of 2-alkyl and 3-alkylketone effect (for details see sec. 4.18, also see 1-decalone, Scheme 4.61*a*).

Interconversion can also be brought about provided the ring system readily undergoes ring opening and ring-closure. The best known example is α-D-Glucose and β-D-glucose (see, Scheme 1.109).

CHAPTER 4

4.12 EFFECT OF CONFORMATION ON REACTIVITY – CYCLIC SYSTEMS – INTRODUCTION

One has already learnt about the effect of conformation on reactivity in acyclic systeme *e.g.*, during eliminatoin and Wagner type shifts (Schemes 4.10–4.14). Further examples are found in Chapter 5.

Several examples have also been discussed in cyclic systems, where to satisfy a particular mechanism a suitable conformation (may be of higher energy) is adopted as in (see Schemes 4.38 and B 4.1). Here one will learn of more examples dealing with stereo-electronic effects and that, ring formation, rearrangement and neighboring group participation are interrelated reactions with the stereoelectronic requirement that the groups involved must be *anti*periplanar.

This situation normally involves a diaxial orientation, however, a ring residue may also be *anti*periplanar to an equatorial leaving group. Consider the four diastereomeric 2-bromo-4-phenyl-cyclohexanols (I – IV) Schemes 4.44–4.46) the phenyl group is used to anchor the chair conformation and prevent it from flipping. In each case the stereoelectronic requirement for the groups involved in the reaction to be antiperiplanar is obvious. The following points may be noted:

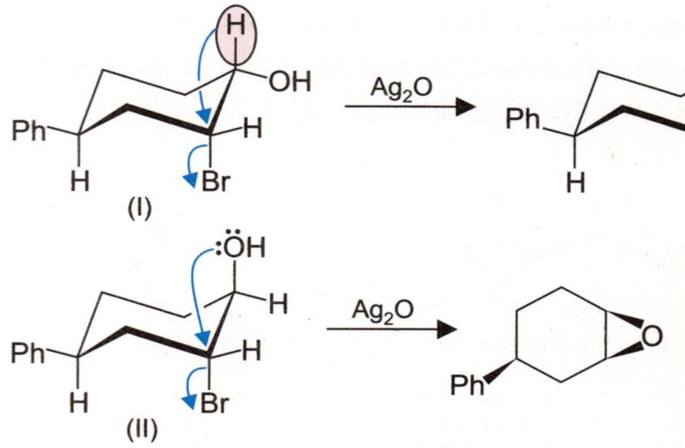

Hydride shift in (I) gives a ketone, since H and Br are antiperiplanar. In (II) reacting OH or O⁻ and Br are again antiperiplanar (epoxide formation).

SCHEME 4.44

- In all the reactions OH group provides the electronic push.
- In (I, Scheme 4.44) ketone is formed by hydride shift, the hydrogen involved is antiperiplanar to the loss of bromine induced by Ag_2O.
- In (II, Scheme 4.44) the reacting (OH or O^-) and leaving group (Br) are again antiperiplanar and an epoxide is formed on the top face of the ring from the OH group which is up.
- In (III, Scheme 4.45) equatorial Br has a ring residue with which it has *anti* periplanar arrangement. Thus the migration of a ring residue now gives a ring contracted product.

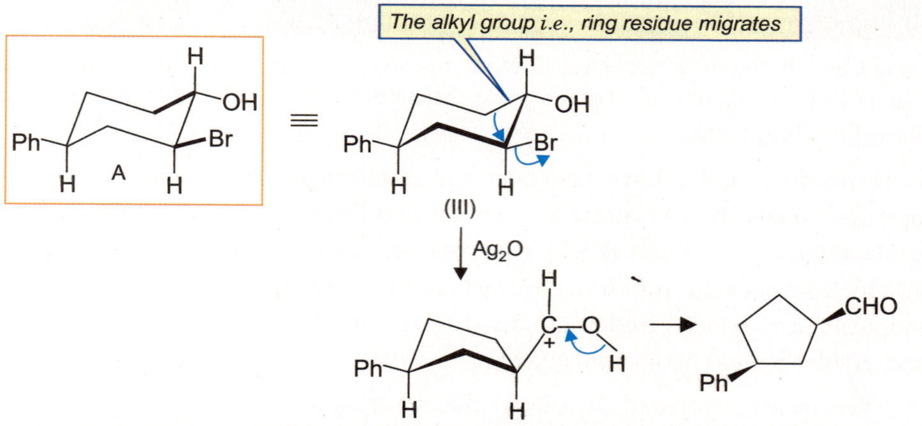

In (III) Ag_2O induces loss of Br (equatorial) and only a ring residue is antiperiplanar to this leaving Br which migrates to give ring contraction, *trans*-and coplanarity is shown between Br and migrating ring residue by bold bonds in (A).

Scheme 4.45

- In (IV, Scheme 4.46) proper stereoelectronic arrangement is not available though it is the stable starting material. The molecule thus undergoes a ring flip to less stable conformation (IV*a*, 4.46, with *syn*-axial phenyl and bromine). The product is a ketone, the one also obtained from (I, Scheme 4.44).

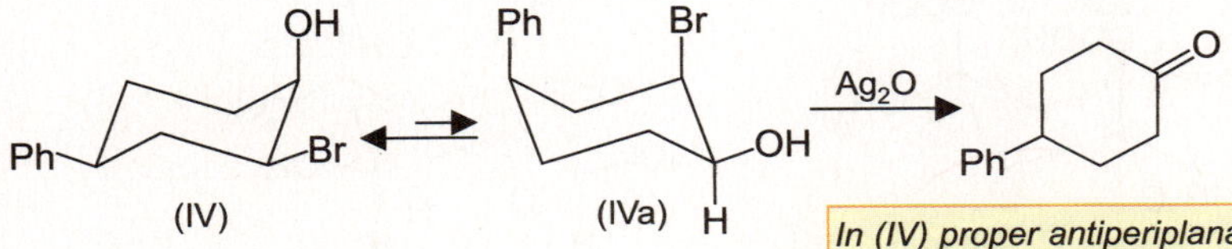

In (IV) proper antiperiplanar arrangement is not available, it reacts via alternate high energy conformation (IVa, obtained by ring flip) which provides this requirement. The cyclohexanone formed is the same as from (I, scheme 4.44)

SCHEME 4.46

4.13 SOME TYPICAL CYCLOHEXANES

Because of its bulk, the *t*-butyl group has effective preference for an equatorial position. When excessive strain is involved, a distortion of the cyclohexane ring occurs. For example, phenyl and *t*-butyl are both bulky groups, a crystal structure analysis of a compound that has a *cis*-4 phenyl-1-*t*-butyl cyclohexane structure shows that the ring has been stretched out somewhat but still has essentially a chair conformation with axial-phenyl and equatorial-*t*-butyl groups.

In *trans*-1, 3-di-*t*-butylcyclohexane (I, Scheme 4.47) a chair-cyclohexane ring would require one *t*-butyl group to be axial, in this compound the cyclohexane ring is twisted in order to avoid placing the *t*-butyl group is an axial position to attain a so-called "twist-form" or "skew-boat" structure (Scheme 4.47). This skew-boat form occurs in several compounds containing bulky groups but is not however, a significant conformation for cyclohexane itself.

Similarly in the case of *r*-1-*cis*-*di*-*t*-butyl-*cis*-*2*-*cis*-5-dihydroxycyclohexane (II, Scheme 4.47) an equilibrium exists between a chair and twist boat conformations (The twist boat conformation gets some stabilization from hydrogen bonding).

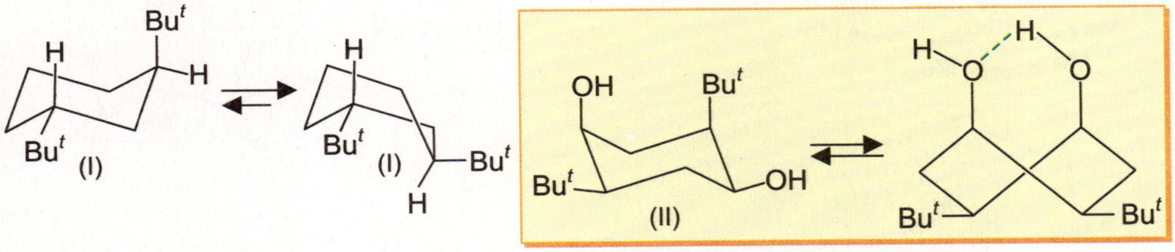

SCHEME 4.47

The most stable conformation of a cyclohexane derivative is the chair form with the larger groups in equatorial positions. Because a methyl group is larger than a carboxylic acid group, as indicated by the greater value of ΔG^{o} (Table 4.2) the methyl groups occupy the equatorial positions in Kemp's triacid. (Scheme 4.48).

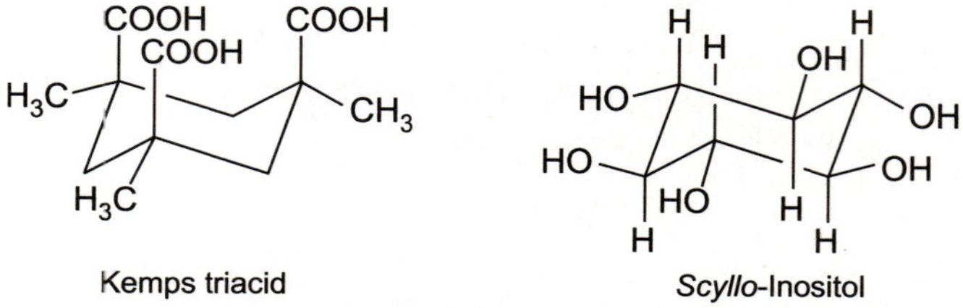

SCHEME 4.48

The most stable conformation of a cyclohexane derivative is the chair form in which groups larger than hydrogen atoms are in the equatorial positions. The most stable isomer of inositol (of 8 isomers see Schemes 1.3*e* and 1.4*a*–1.4*c*) is expected to have all six OH groups in the equatorial positions (Scheme 4.48).

An Israeli research group (1990) reported that the all-*trans* isomer of 1, 2, 3, 4, 5, 6-hexaisopropylcyclohexane was more stable in the conformation with all six isopropyl groups axial, not equatorial (I, Scheme 4.49). In the all-equatorial chair conformation (*II*, Scheme 4.49) there is severe steric repulsion due to the nearby isopropyl groups shown between a pair of neighbouring isopropyl groups (These interactions occur for each pair of neighbouring isopropyl groups.)

In the all axial chair conformation (I, Scheme 4.49) the C—H bond of each isopropyl group can all point towards each and the steric repulsion can be minimized only among H atoms shown only for the top face of the cyclohexane ring).

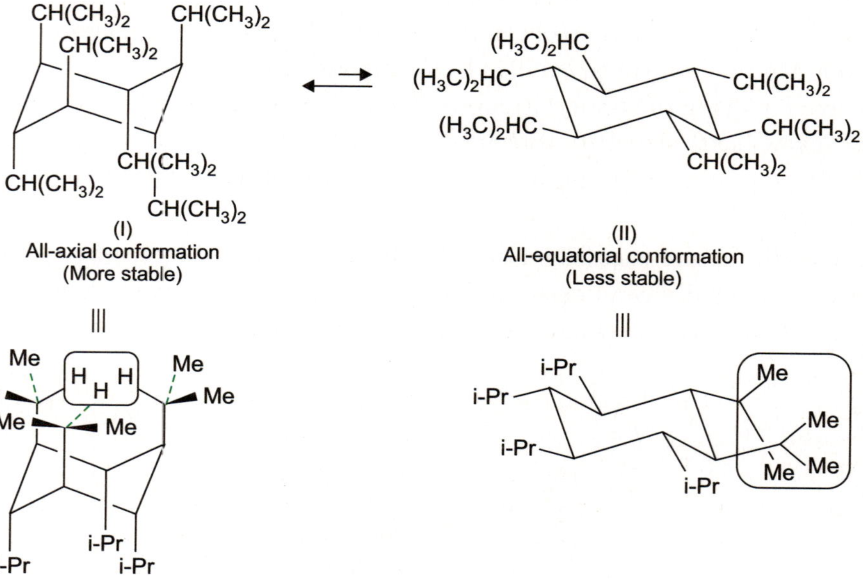

SCHEME 4.49

4.14 CONFORMATION OF HETEROCYCLES

A six membered ring containing an oxygen atom (tetrahydropyran) assumes a chair conformation. At all positions generally alkyl substituents prefer an equatorial arrangement (Scheme 4.50), however, an α-alkoxy substituent shows increased stability in the axial position. This effect is called the anomeric effect (see Scheme 4.53).

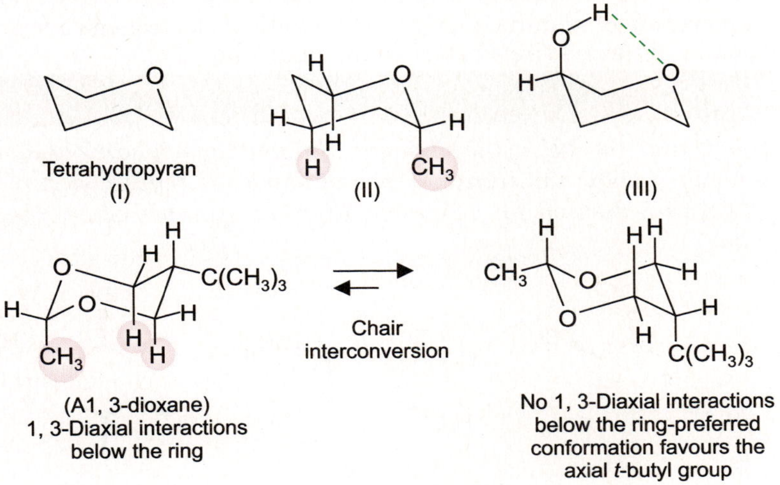

SCHEME 4.50

In (II, Scheme 4.50) substitution next to the oxygen atom leads to more severe 1, 3-diaxial interactions compared with cyclohexane because C—O bonds are shorter than C—C bonds and bring the axial groups closer. In the case of situations like (III, Scheme 4.51) the presence of an oxygen atom in the ring allows hydrogen bonding that can stabilize hydroxyl groups in the axial position. This is particularly important for sugars because they have several OH groups.

In *cis*-2-methyl-5-*tert*-butyl-1, 3-dioxane (Scheme 4.50) the preferred conformation has the bulky tert-butyl group axial and the methyl group equatorial. Since the divalent oxygen has no substituents, therefore, the 1, 3-diaxial interactions which are the main unfavorable interactions for axial substituents in cyclohexanes are absent.

It is consistenly seen that 5-alkyl substituents in 1, 3-dioxanes show a smaller equatorial preference than they do in cyclohexane. This decreased preference is due to decreased Van der Waals repulsions in the axial orientation, since there are now no hydrogens that are *syn*-axial to the 5-alkyl substituent. A 2-alkyl substituent, however, has a greater preference for the equatorial orientation in a 1, 3-dioxane than in cyclohexane, presumably because the decreased C—O bond length 1.43 Å (relative to C—C bond length 1.54 Å) brings an axial 2-alkyl group closer to the *syn*-axial hydrogens at C(4) and C(6) to result in an increased Van der Waals repulsion. Similarly, an axial 4-alkyl substituent in a 1, 3-dioxane suffers a greater Van der Waals repulsion with the axial hydrogen at C(2) that it does in cyclohexane.

A part from the usual factors that govern the equatorial versus axial orientation of substituents on a six-membered ring, two other factors are important when oxygen is present:

1. an equatorial OH is less crowded and better solvated by water than an axial one
2. the anomeric effect

The first of these is simple and cautions us to the fact that the relative energies of two species may be different in solution than in the solid state or the gas phase. Hydrogen bonding to water stablizes equatorial OH groups better than axial ones.

The anomeric effect stabilizes *axial* OH and other electronegative groups at the anomeric carbon in pyranose rings better than equatorial.

EXERCISE 4.6

Draw the conformations of each of the following dibromodimethyl cyclohexanes (Scheme 4.51). Bring about a chair interconversion and show the relationship of conformations in each case between original conformation and conformation after the chair interconversion.

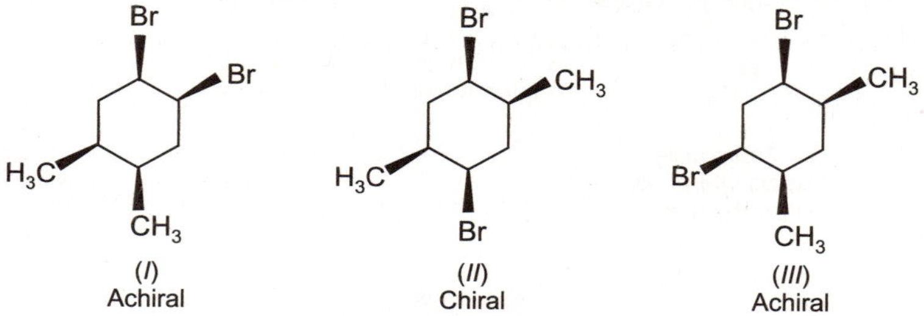

SCHEME 4.51

ANSWER.

(I)

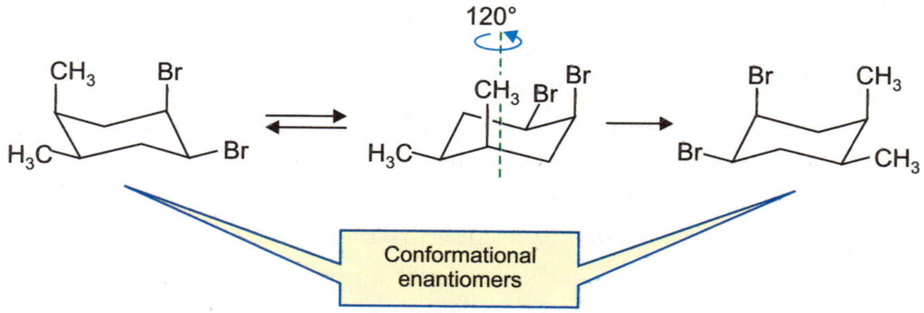

(II)

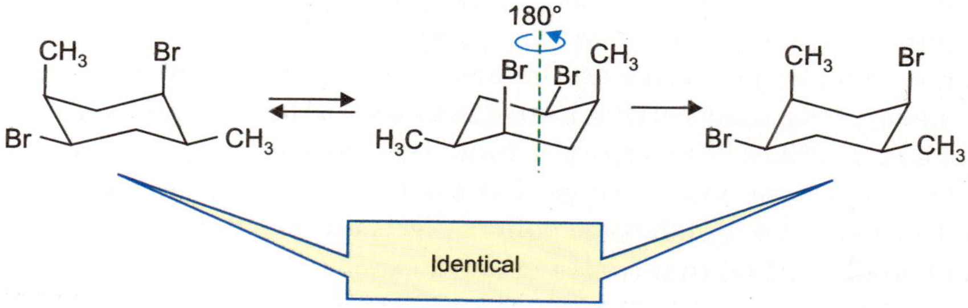

Continued...

(III)

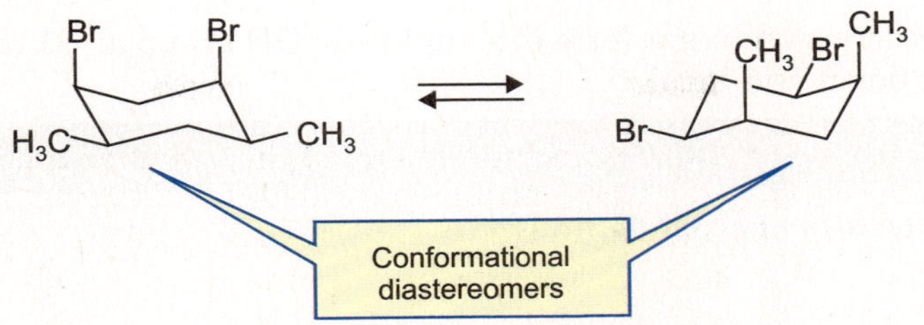

SCHEME 4.51a

4.15 CONFORMATION OF SUGARS AND ANOMERIC EFFECT

(A) Conformation of Common Sugars

One has already learnt drawing the Fischer, Haworth and "zig-zag" structures of D-glucose (see, Schemes 1.2*e*, 1.20*g* and 1.33*c*). The conversion of Haworth formula of α-D-glucose to chair conformation has been presented (Scheme 1.20*g*). Mechanism of mutarotation is given (Scheme 1.109).

The following points may be noted regarding drawing conformations of some sugars.

- The six membered ring which includes an oxygen atom is related to the heterocyclic ring tetrahydropyran (Scheme 4.52). A saccharide in its six membered form is called a pyranose.

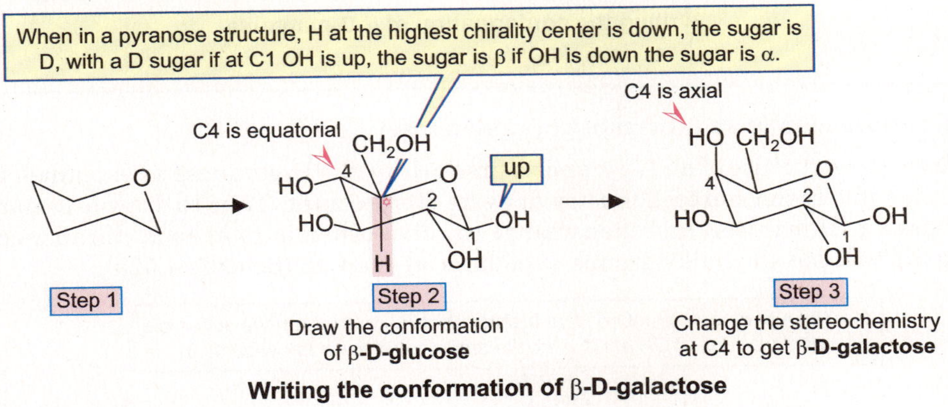

Writing the conformation of β-D-galactose

SCHEME 4.52

- At the starting point draw the chair conformation of tetrahydropyran.
- Secondly draw the conformation of β-D-glucose which is very easy to draw since all the OH groups are in equatorial position.
- From β-D-glucose it is easy to draw the chair conformation of any sugar provided one knows the epimeric relationship it has with β-D-glucose (β-D-mannose is its C2 epimer while β-D galactose is its C4 epimer). Thus *e.g.*, for drawing the

conformation of β-D-galactose change the stereochemistry at C4 from equatorial to axial position.

- In the corresponding α-form of a sugar the OH group at C1 (anomeric carbon) will be down and axial.

EXERCISE 4.7

Draw the conformation of α-D-mannose.

ANSWER

Step 1. *Draw the chair form of tetrahydropyran.*

Step 2. *Draw β-D-glucose by adding all OH groups in the equatorial position.*

Step 3. *Since D-glucose and D-mannose are C2 epimers change the stereochemistry at C2 (from equatorial to axial to get the conformation of β-D-mannose).*

Step 4. *In an α-anomer the OH group is axial at this carbon thus change the stereochemistry of OH from equatorial to axial at the anomeric carbon to get α-D-mannose (Scheme 4.52a).*

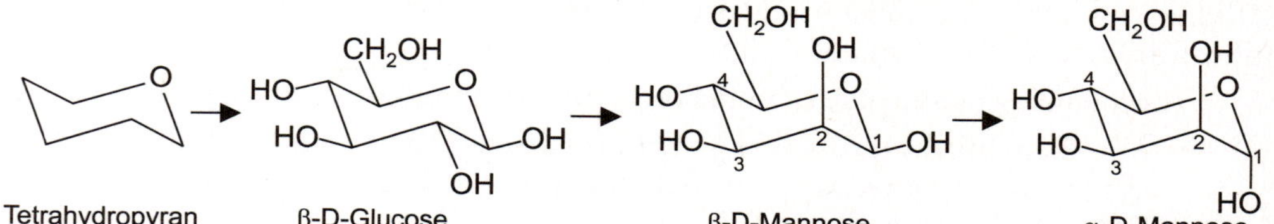

Tetrahydropyran Step 1 | β-D-Glucose Step 2 | β-D-Mannose Step 3 | α-D-Mannose Step 4

Writing the conformation of α-D-mannose

SCHEME 4.52a

(B) Conformation of an L-Pyranose (α-L-Mannose)

To draw the structure of an L-pyranose, firstly draw the D-pyranose as explained above and then invert the stereochemistry at every stereocenter. Thus to draw α-L-mannose first draw α-D-mannose and then change the stereochemistry at each chirality center. Ring flip will put the bulky groups in equatorial position (Scheme 4.52*b*).

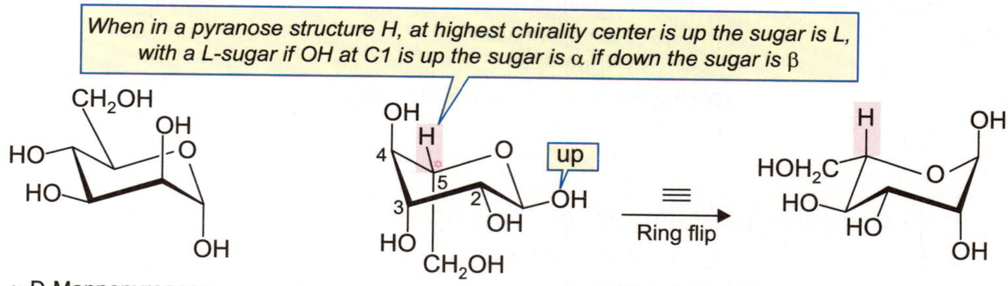

α-D-Mannopyranose α-L-Mannopyranose

SCHEME 4.52b

(C) Anomeric Effect

One has seen that when either pure anomer of D-glucose is dissolved in water it gives an equilibrium mixture of anomers where β-anomer (equatorial OH at C1) predominates (see, Scheme 4.44). Similar behavior is again displayed by D-mannose (via the corresponding open chain form) in which now, however, the α-anomer with axial hydroxyl at C1 predominates (I, Scheme 4.53) over the β-anomer (II).

This unusual favoring of the α (or the axial) orientation first observed in the case of sugars is called anomeric effect. In fact generally it is observed that during reactions which introduce oxygen or halogen group X (*e.g.*, OH, OMe, OCOMe and Cl) at the anomeric carbon atom of a pyranose ring, the major stereoisomer is the one in which the new group occupies the axial position.

This substituent (X) seems to exert two opposing effects. A steric effect which directs the substituent to the equatorial position, an equatorial group is less crowded and better solvated than an axial one. Secondly an electronegativity effect is the main cause which favors the axial position (III, Scheme 4.53). In the case of D-glucose (Scheme 4.44) the anomeric effect seems to play a secondary role to the steric difference of having the anomeric OH equatorial versus axial.

It has been suggested that the effect is operative mainly by a stabilizing interaction between the axial lone pair of electrons on the ring oxygen atom and the antiperiplanar, antibonding σ* orbital of the C–X bond. Thus some of the electron density moves from oxygen to the σ* antibonding orbital. This is possible only if the substituent X is axial (III, Scheme 4.53).

Alternatively, electrostatic repulsive forces between the dipoles due to the ring oxygen lone pairs and the exocylic oxygen or halogen may account in part for the observed axial preference. (IV and V, Scheme 4.53).

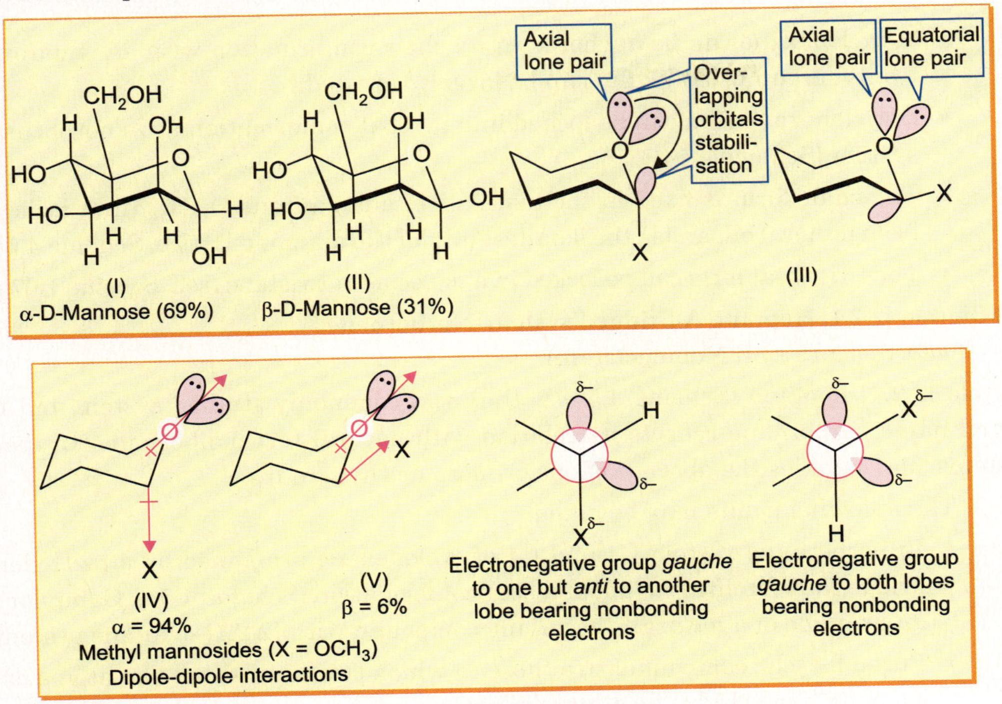

SCHEME 4.53

The anomeric effect may strongly inlfuence the *conformational* equilibria in pyranoses with a C–1 electronegative atom (oxygen or halogen). Thus the equilibrium mixture of the β-pyranosyl chloride derived from xylose triacetate is exclusivly a conformer in which chlorine is axial. These two conformations are not interconverted by mutarotation but by chair–chair interconversion that all four substituents occupy axial positions in the more stable conformer (also see, Scheme 4.49).

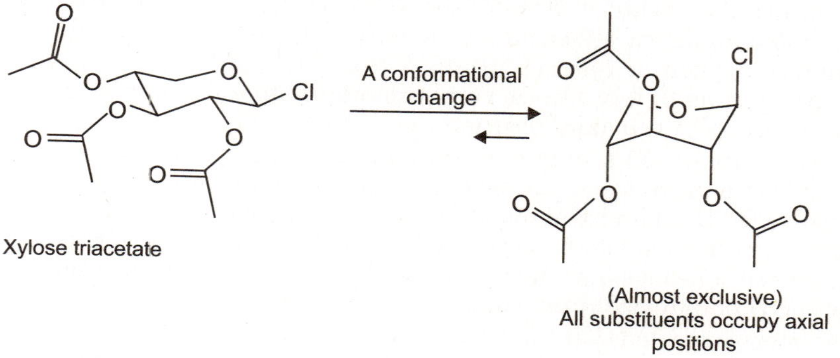

SCHEME 4.53a

4.16 POLYCYCLIC COMPOUNDS

Recall the following points:

- A cyclohexane can be anchored in its chair conformation with an equatorial *t*-butyl group when its flip can be largely arrested.
- A cyclohexane ring can be locked into its boat conformation as in norbornane system (Scheme 4.54).
- The chair form is fixed in the molecule of adamantane $C_{10}H_{16}$ (first isolated from mineral oil, and is the smallest unit of the diamond lattice, Scheme 4.56).
- The twist boat form can be locked in a molecule of twistane (see, Scheme 1.67 g).

(A) Bicyclo [2.2.1] Heptane A Bridged system–Norbornanes (*Exo, Endo, Syn* and Anti Nomenclature)

Bicyclo [2.2.1] heptane (Scheme 4.54) is the most commonly studied system and its trivial name norbornane reflects its relation with natural terpene bornane. In these systems 'nor' signifies the absence of one or more methyl groups.

The following points may be noted:

- In norbornane system C1 and C4 of cyclohexane are joined by a methylene bridge thus forcing it to adopt a boat conformation (Scheme 4.54). Camphor is an example of a naturally occurring compound with [2.2.1]–heptane system.
- Fused bicyclic compounds are those in which two rings share two atoms or a common bond. In bridged bicyclic compounds two rings share three or more atoms or a common face.

BOX 4.4
BRIDGEHEAD CARBONS

Bridgehead carbons are the carbons connected to each other by three different chain of carbons. In norbornane these carbons (stareed, Scheme 4.54) are linked by two carbon chains while the third chain C–7 has only one carbon. This may be viewed when two cyclopentane rings are imagined to close in and join to a norbornane system in which the two rings share three atoms (Scheme B 4.5).

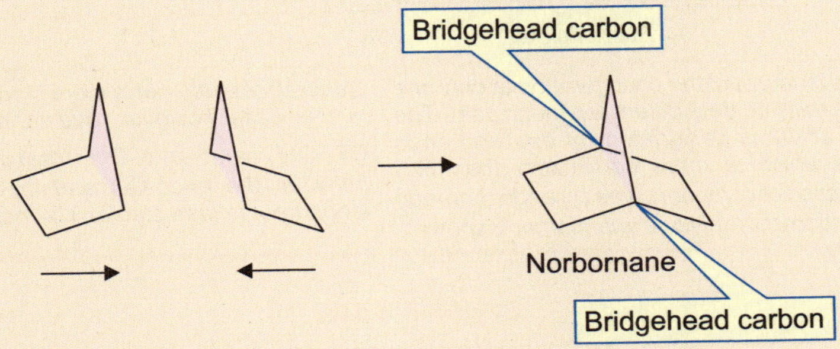

SCHEME B 4.5

- In norbornane (Scheme 4.54) the name bicycloheptane shows that it is a bicyclic compound containing a total of seven carbon atoms. The numbers in breackets represent the number of carbon atoms in the respective bridges, in order of decreasing size. The sum of the numbers inside the bracket is two less than the total number of carbons since the two bridgehead carbons C–1 and C–4 are not included.

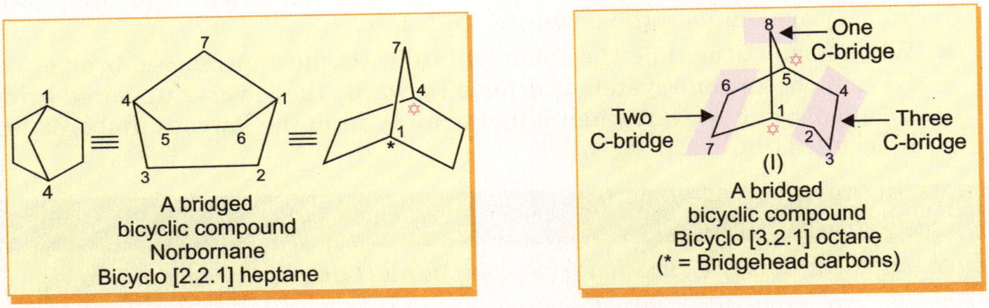

SCHEME 4.54

- The numbering of carbons beings from one bridgehead carbon and follows the largest bridge first (I, Scheme 4.54*a*).

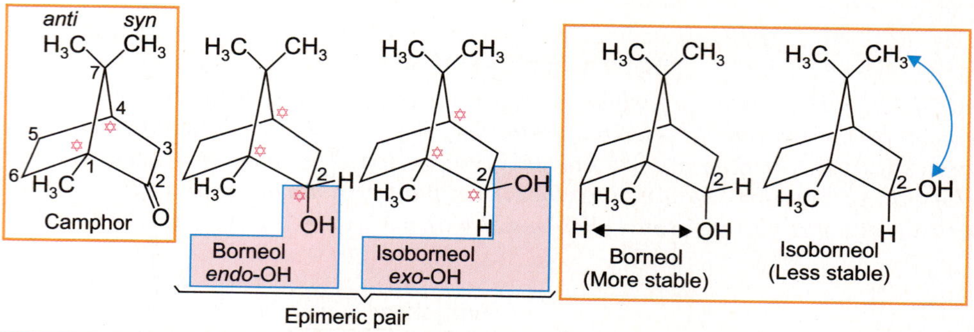

In borneol and isoborneol the configuration at only one of the three chirality centers differs (epimeric pair). The configuration at C2 is designated by exo and endo nomenclature which is more convenient than R/S system. Reduction of camphor gives (±) endo (borneol) and (±) exo isoborneol (only one enantiomer is shown).

Generally exo isomer is more stable than endo, in this case however, due to more serious C7–CH_3 and exo-OH interaction, than between the endo-OH and the endo C6–H isoborneol is less stable. The normal stability

SCHEME 4.54a

In camphor with a norbornane carbon skeleton (Scheme 4.54*a*) there is a choice as two bridges are the same. In this case numbering is fixed by the C = O group in one of the bridges. The numbering thus requires the lowest possible number to the carbon containing the functional group and this is C–2.

- The terms *exo* and *endo* are used for the substituents on a bridge which connects the bridgehead carbons. An *exo* substituent is the one which is on the same side of the smaller bridge and *endo* if it is opposite to the smaller bridge (Scheme 4.54*a*). This usage is clear from borneol and isoborneol, the reduction products of camphor. For details of reduction (see Scheme 7.8*e*). The configuration at C2 in borneols can be specified by *R* and *S* system of nomenclature, however, an easy system uses the terms *exo* and *endo*.
- When a substituent on the smallest bridge (highest numbered bridge) *i.e.*, at C7 in a norbornane system is directed towards the lowest numbered bridge it is called *syn*, however, when it points away from the lowest numbered bridge it is called *anti*.

EXERCISE 4.8

(i) How many epimers are expected from compound (I and II Scheme 4.54b)?
(ii) How one can prove the distinct position of methylene protons in norbornene?

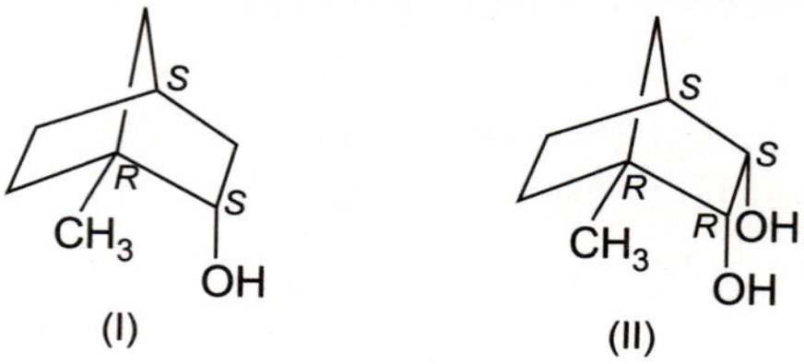

SCHEME 4.54b

Continued...

ANSWER. *(i) In the bridgehead compounds (Scheme 4.54b) due to geometrical constraints (see Scheme 1.67h), the number of stereoisomers is always one half of what would be expected if this restriction was not there. The epimers differ in the absolute configuration at only one of the chirality centers. Thus (I, Scheme 4.54b) with three chirality centers has only one epimer (A), and all the four possible stereoisomers are drawn for the norbornane derivative (I, Scheme 4.54b).*

For the same reasons the norbornane derivative (II, Scheme 4.54b) with four chirality centers has only two epimers (B and C).

Moreover, it is not possible to assign configuration to these systems from such stereodescriptors like exo, endo, syn and anti. The enantiomer of (exo A, Scheme 4.54c) is still exo.

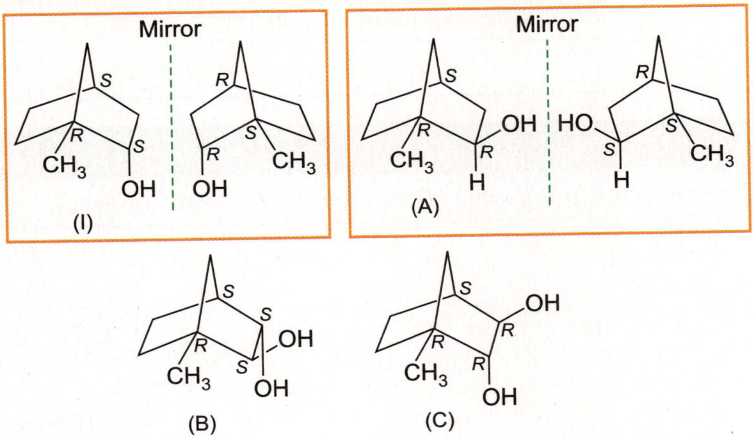

SCHEME 4.54c

One may note the homotopic and heterotopic hydrogens in the molecules (Scheme 4.54d) by recalling that generally:

- *When a carbon is bonded to two identical ligands and two different ligands, then two identical ligands are enantiotopic.*
- *When the other two ligands are also identical, the ligands are homotopic.*
- *When the two ligands are close to a chirality center, these are diastereotopic.*

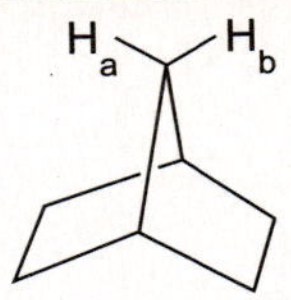

Homotopic (Related by C_n)

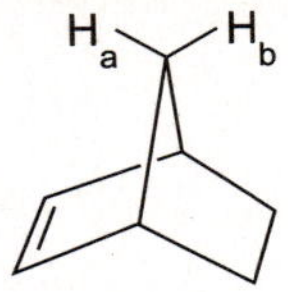

Diastereotopic (Not related by C_n methylene protons H_a/H_b close to two chirality centers) How to distinguish H_a/H_b by ^1H NMR?

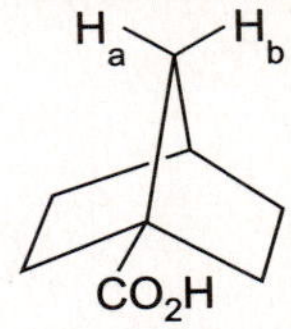

Enantiotopic (Not related by C_n. Methylene protons H_a/H_b not close to chirality centers. Carbon with H_a/H_b is bonded to two different groups)

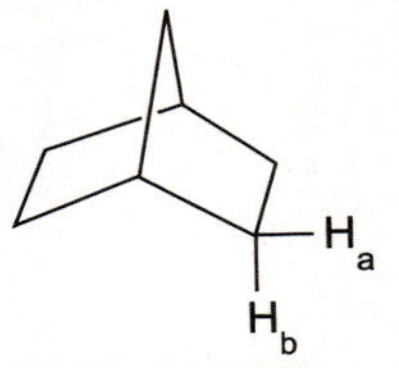

Diastereotopic (H_a is *exo* H_b is *endo*)

Continued...

(ii) The position of methylene protons in norbornene may be explained, since the syn and endo protons respectively are shielded with respect to the anti and exo protons. This is due to anisotropy of double bond. Protons situated over double bonds are more shielded than those in the plane of alkenes or carbonyl groups.

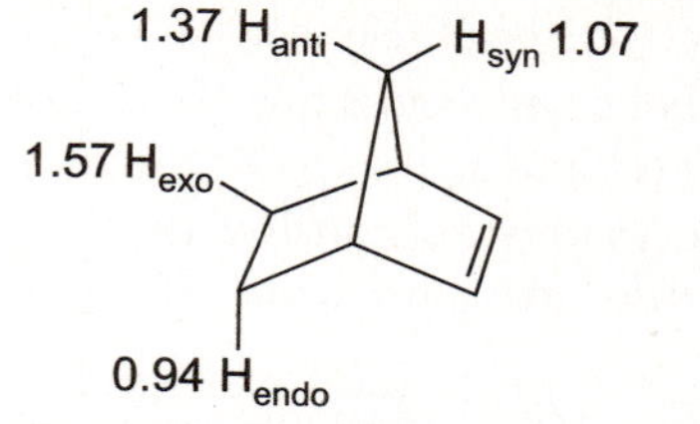

Norbornene (Chemical shifts δ ppm)

SCHEME 4.54d

EXERCISE 4.9

Explain the stereochemical outcome of the reaction (Scheme 4.54e).

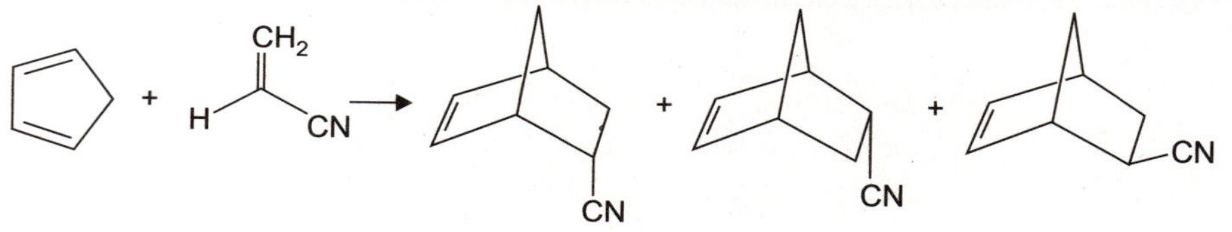

(1) (2) (3, Minor)

Major (Racemic mixture)

SCHEME 4.54e

ANSWER. *In a Diels-Alder reaction the less stable endo-product predominates. The endo-product is formed as a racemic mixture of (1 and 2, Scheme 4.54e). The compound (3) is formed as a minor product.*
The enantiomer (2) is drawn by placing the ligand at the enantiotopic position rather than a mirror image (Scheme 4.54f).

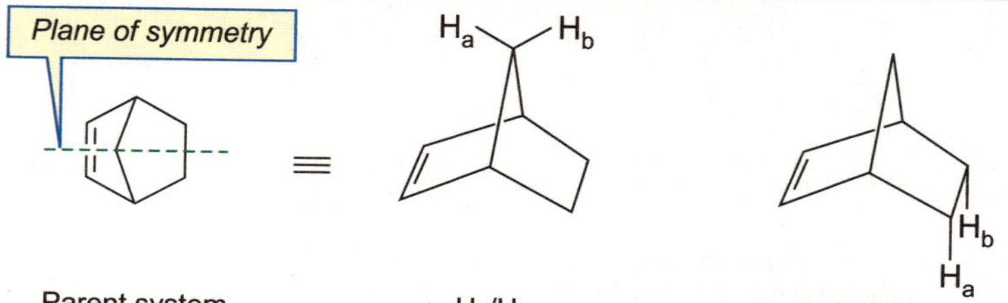

Parent system (*meso*) — H_a/H_b (Diastereotopic) — H_a/H_b (Enantiotopic) Being on a mirror plane of symmetry

SCHEME 4.54f

(B) Bicyclo [2.2.2] Octane System

This is another interesting bridged bicyclic system where 1, 4 positions of the cyclohexane boat are joined by an ethylene bridge. As is the case with bicyclo- [2.2.1]–heptane system, no *cis-trans* isomers can exist. It is only possible to fuse a two-carbon bridge to carbons 1 and 4 of cyclohexane ring if the two-carbon bridge is fused in a *cis* fashion (Scheme 4.55).

The following points may be noted:

- There are only two different carbons C1 and C2 in methylbicyclo [2.2.2] octane and two constitutional isomers.
- Unlike methylbicyclo [2.2.1] heptane, there are no *exo* and *endo* diastereomers as the two bridges are identical.
- 2-Methylbicyclo [2.2.2] octane is chiral and exists as an enantiomeric pair (Scheme 4.54*g*).
- Any two methylenic protons of bicyclo [2.2.2] octane are enantiotopic (not interchangeable through C_n but exchange through mirror plane of symmetry).

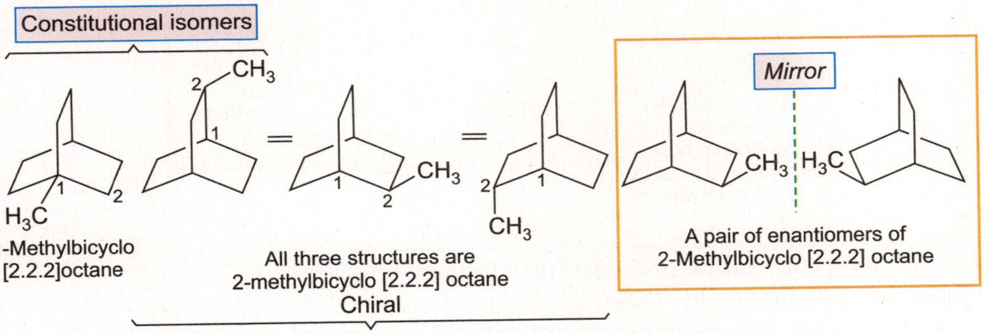

SCHEME 4.54g

This situation may be compared with methylbicyclo [2.2.1] heptane (see Box 1.7) where carbons 1, 2 and 7 are different (Scheme 4.54*h*).

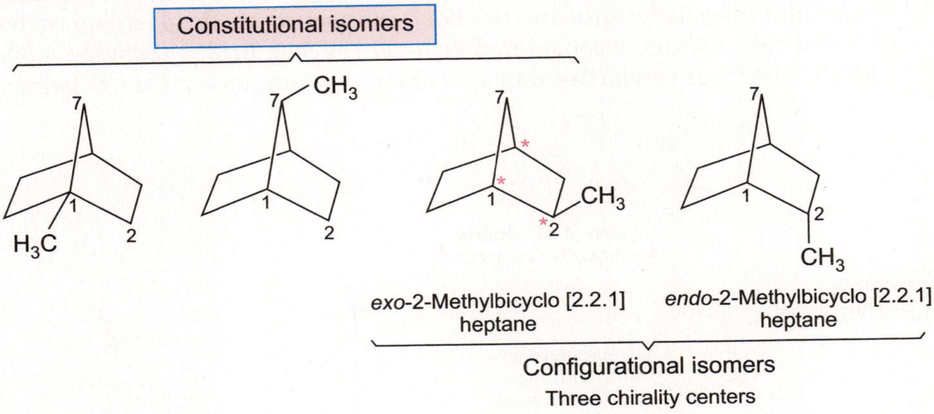

SCHEME 4.54h

Cyclic systems with two or more rings joined in the "fused" or "bridged' fashion may lead to some constraints on the number of stereoisomers and structural possibilities for a given system. The bridged-ring compound norbornane is a rigid molecule where the cyclohexane ring is forced in the boat conformation. Norbornane has fewer isomers than predicted, for it may be thought that a bicyclo [2.2.1] heptane structure can exist in which the bridging carbon is linked to the topside of the cyclohexane ring at C-1 and the bottomside at C-4. This system however, would be too strained to be isolated and only a single bicyclo [2.2.1] heptane is known. Similarly less strained bicyclo [2.2.2.] octane (Scheme 4.55) is only known in the *cis* fused form. On the other hand, bicyclo [4.4.1] undecane (Scheme 4.55) constitutes a system with a methylene group bridging a considerably larger and more flexible ring than cyclohexane. It exists in two isomeric forms. A consideration of molecular models shows this difference rather clearly, and reveals that the *trans* isomer of bicyclo [2.2.1] heptane cannot be constructed while the *trans* isomer of bicyclo [4.4.1] undecane is easily built (Scheme 4.55).

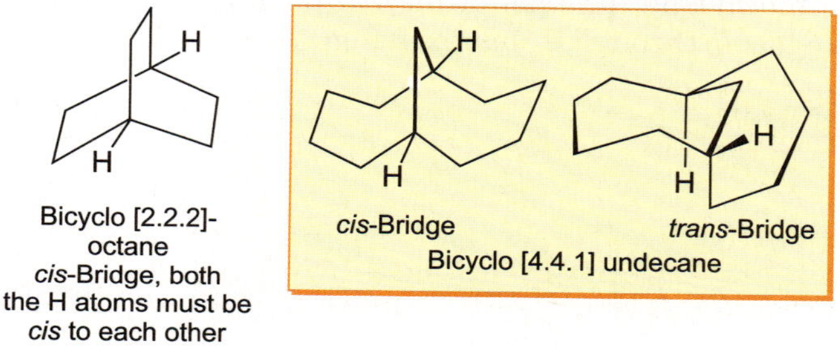

SCHEME 4.55

(C) Other Polycyclic Compounds–An Introduction

The following points may be noted:

- Two rings may be joined together in three general ways by sharing carbons. When one carbon is shared the mode of attachement is called spiro. In a spiro compound the two rings must lie in different perpendicular planes since the centeral carbon is approximately tetrahedral (Scheme 4.55*a*).
- When two rings can share a pair of adjacent carbons the arrangement is called a fused bicyclic compound. Three bridges connect the bridgehead carbons as in decalin (bicyclodecane) and the bond connecting the bridgehead carbons in a fused ring system is considered as a zero carbon bridge (Scheme 4.55*a*). In decalin the rings can be fused in a *cis* or *trans* arrangement (see, Scheme 4.59).

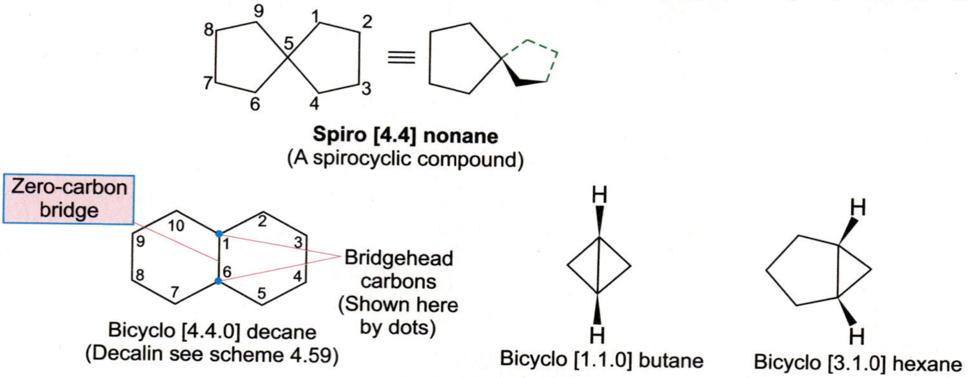

SCHEME 4.55a

- When the rings are small only *cis*- fusion is observed since *trans* fusion will introduce too much ring strain. Thus only *cis*- isomers of the compounds (Scheme 4.55*a*) are known.
- Recall that in a small bridged compound, the bridge can only be *cis*, but when the ring containing the bridge is large the bridge could as well be *trans* (see Scheme 4.55).

(D) Strained Carbocycles — Bredt's Rule

Cyclohexene is a strain-free compound with an estimated strain energy of 1 kcal/mol (4 kJ/mol). Only its *Z* geometrical isomer is known. Cyclooctene is the smallest cycloalkene of which *E* isomer has been isolated (see, Scheme 1.54).

The Bredt's rule states that a bridged bicyclic compound cannot have a double bond at a bridgehead position unless one of the rings contains atleast eight carbon atoms. Norbornene is a stable compound, since the double bond is away from the bridgehead carbon, however, its isomer (I, Scheme 4.55*b*) with a bridgehead double bond is unknown. When however, the number of ring atoms is larger in which the ring strain (the cause of non-formation of bridgehead double bonds) is lower, the presence of a bridgehead double bond is possible.

In the case of (II, Scheme 4.55*b*), however, the *trans* double bond is instead contained in a nine-membered ring and thus, the compound has been prepared and isolated. A further interesting example of a compound with a bridgehead double bond is of (III, Scheme 4.55*b*), the compound has a stereocenter (C5), is chiral and resembles *trans*-cyclooctene.

During β-eliminations in bicyclic systems, the bridgehead atoms are not involved, provided the ring bearing the incipient "*trans*" double bond has at least eight atoms.

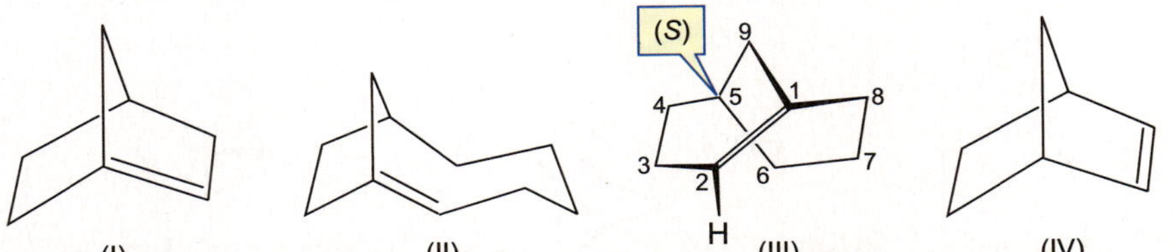

(I) Bicyclo [2.2.1] hept-1-ene (Unknown-violation of Bredt's rule)

(II) Bicyclo [5.2.1] dec-1-ene

(III) (*S*)-Bicyclo [3.3.1] non-1, (2)-ene

(IV) Norbornene a stable compound

SCHEME 4.55b

Adamantane (I, Scheme 4.56) is tricyclic and highly symmetrical structure with only two different type of carbons, CH and CH_2. Symmetry makes the four methine groups equivalent and six methylene groups equivalent (see, Scheme 1.126).

The following points may be noted regarding adamantane system.

- Its emperical formula is $C_{10}H_{16}$, (index of hydrogen deficiency shows it to be tricyclic).
- Carbons 1, 3, 5 and 7 are bridgehead carbons (Scheme 4.56).
- Removal of C-10 leaves a bicyclic [3.3.1] system
- Addition of C-10 gives the third ring.
- Adamantane is not strain free.

- It has more than four times the strain in cyclohexane.
- Framework structure forces tetrahedral angles.
- However C-CH_2-C do not want to be tetrahedral. (Angle in strain free propane molecule is 112.5°, strained cyclohexane has 111.5° angle, its further reduction to 109.5° causes strain in adamantane).

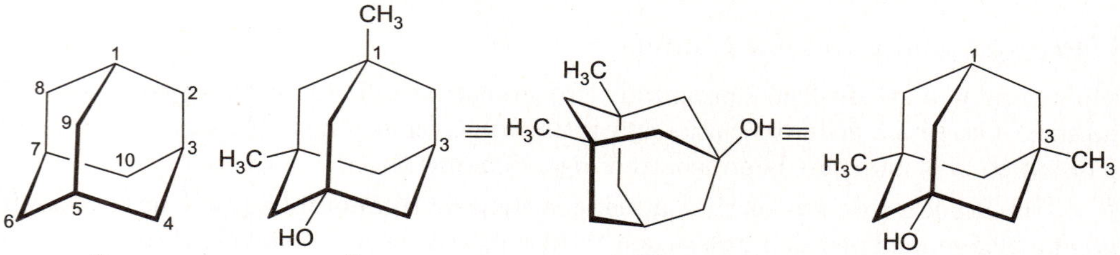

(I) (II (III) (IV)

All (achiral) same compound. The plane of symmetry is seen when (II) is redrawn as (III). The projection (IV) is obtained when C1–CH_3 is shifted to equivalent C3 position (note that four methine are equivalent and six methylene groups are equivalent).

SCHEME 4.56

An adamantane derivative with four different ligands at bridgehead positions becomes chiral and exists as a (±) pair (I, Scheme 4.57). An adamantane with diagonally opposite methylenes with proper substitution becomes chiral (chiral like a chiral biphenyl II, Scheme 4.57, also see, Scheme 1.126).

The presence of a C_2 axis and the chirality axis in an adamantane with a chirality axis is in (III, Scheme 4.57).

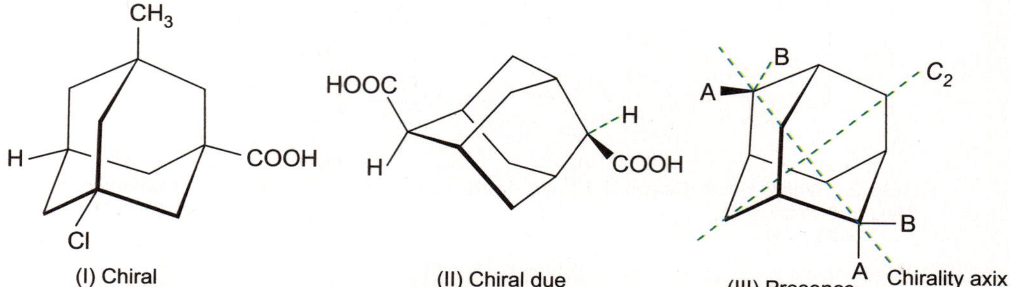

(I) Chiral
An adamantane derivative with four different groups at bridgehead carbons is chiral

(II) Chiral due to the presence of a chirality axis

(III) Presence of a chirality axis and a C_2 axis

SCHEME 4.57

A simple mental exercise in stereochemistry is in (Scheme 4.58)

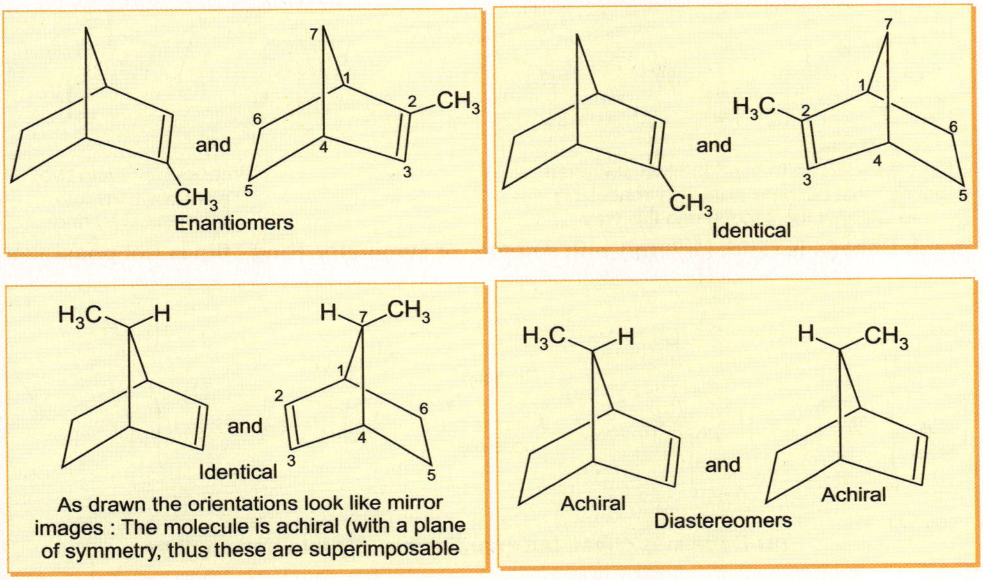

SCHEME 4.58

CHAPTER 4

BOX 4.5
PROCHIRALITY FACES OF CARBONYL COMPOUNDS REVISITED

R^1—C(=O)—R^2 The faces are homotopic if $R^1 = R^2$, enantiotopic if $R^1 \neq R^2$, diastereotopic if either R^1 and R^2 have a chirality center (Scheme B 4.6).

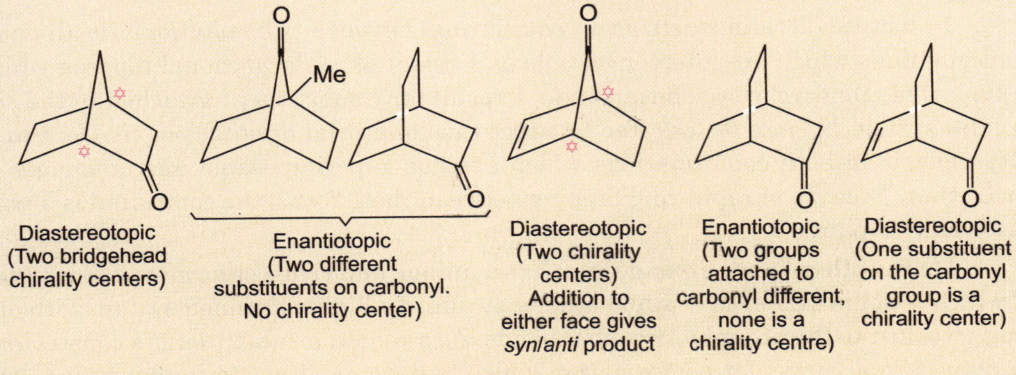

SCHEME B 4.6

(E) Fused-Rings: The Decalins

The compound bicyclo [4,4,0] decane, better known as decalin, exists in two diastereoisomeric forms depending on the way in which the two cyclohexane rings (A and B) are fused together (Scheme 4.59). It is also readily determined from the models that *cis* and *trans* decalins are diastereomers as they cannot be interconverted by bond rotations.

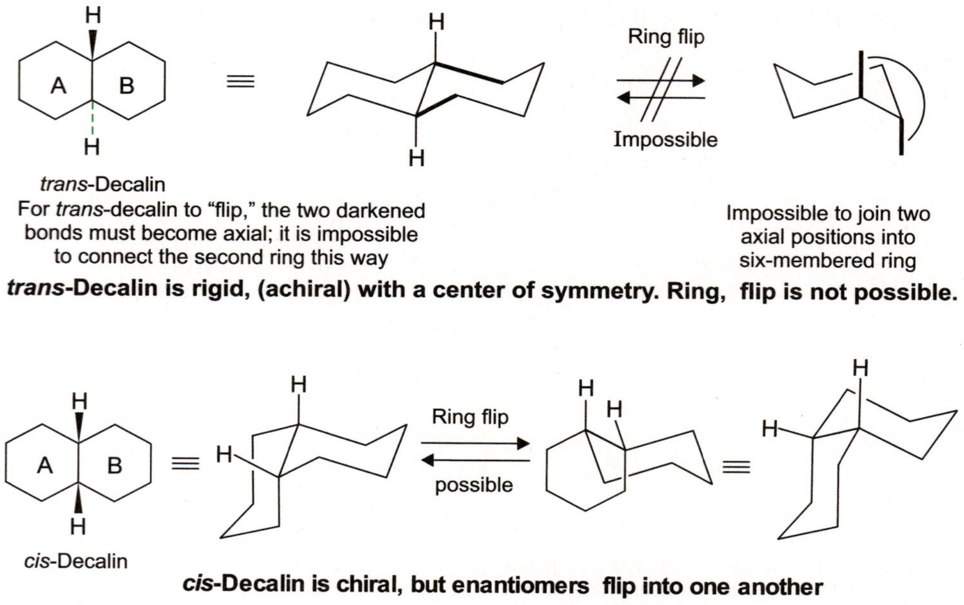

SCHEME 4.59

Decalin may be considered as a fusion of a four carbon chain *i.e.*, —CH_2 CH_2 CH_2CH_2— to the chair form of cyclohexane. *Trans*-decalin is fused with this four-carbon chain (equivalent to largest groups) in equatorial positions while the *cis*-decalin this fusion involves equatorial-axial positions (Scheme 4.59). As a result in *cis*-decalin the two hydrogens attached at the point of fusion between two rings lie on the same side of the ring while in *trans*-decalin, they are on opposite sides.

In fact, *cis*-decalin exists as an equilibrium between two enantiomeric all-chair conformations which are interconvertible as a result of conformational flipping which is typical of monocyclic cyclohexanes. As a result, any substituent attached to the *cis*-decalin system is more or less free to adopt the equatorial orientation. *cis*-Decalin is disymmetric in both conformations which are non-superimposable mirror images of each other. Because of rapid ring flipping between these forms the compound is a non-resolvable *dl*-pair.

On the other hand, *trans*-decalin has a unique and rigid conformation, inversion of which is not possible, as it would otherwise afford a highly strained system with one ring attached to the other by two axial bonds. As a result, a substituent is constrained to remain in a particular conformation which is dependent on its configuration. The rigid conformation of *trans*-decalin has been used in the same manner as *t*-butyl groups to study the relative rates of equatorial and axial substitution. The hindered approach to axial positions has been demonstrated *trans*-Decalin has a centre of symmetry and is therefore, optically inactive, It may be further noted that substituents located at the point of fusion between two rings (angular positions) in *cis*-decalins are axial with respect to one ring while equatorial with respect to the other. However in the case of *trans*-decalins these are axial with respect to both the rings. It may be remembered that simple rotation of groups about carbon—carbon bonds cannot bring about interconversion of *cis*-and *trans*-decalins. Therefore, in this respect decalins resemble the *cis*-and *trans* 1, 2-disubstituted cyclohexanes and as has already been said, in decalins the 1, 2-substituents are the two ends of four carbon chain.*trans*-Decalin as a result is stabler by about 2.7 kcal/mole (11.3 kJ/mole) as compared with *cis*-form.

The barrier to interconversion of *cis*- and *trans*- decalin can be decreased by heteroatom substitution in some cases *e.g.,* in azadecalin (Scheme 4.60). If a nitrogen atom occupies a bridgehead position *cis* and *trans* forms cannot be isolated. The rapid interconversion of the *cis* and *trans* forms in this case is the result of an easy inversion of configuration at nitrogen which eliminates the need for a *trans*-diaxial transition state.

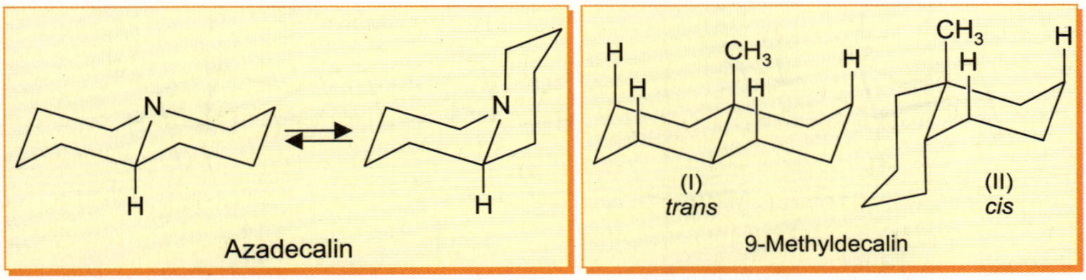

SCHEME 4.60

When an angular methyl group is introduced, the *cis*-form becomes slightly more stable than the *trans*-form. Steroids and several other natural products contain the 9-methyl decalin system. The introduction of an angular methyl group reduces the differences in energy between the decalins, because in the *trans*-compound (the methyl group is sterically compressed by four axial hydrogen atoms) and by only two (C_2 and C_4) in the case of *cis*-isomer (Scheme 4.60).

As discussed above in a *trans*-decalin system, neither ring can undergo a ring-flip to give a different chair conformation. Both rings are however free to adopt a higher energy boat conformation as shown for ring (B, Scheme 4.60*a*).

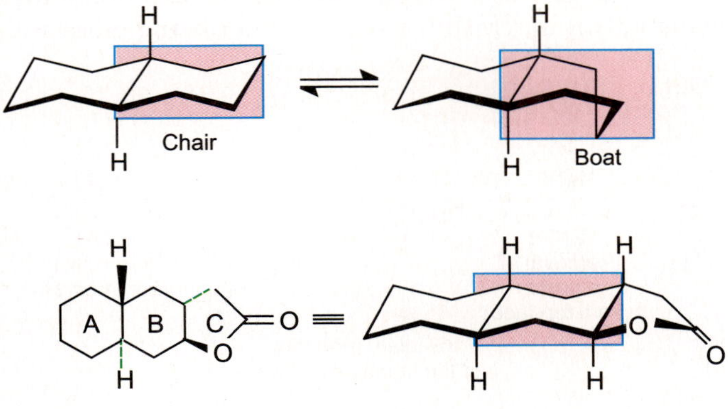

(I)

SCHEME 4.60a

The ring (B) of *trans*-decalin can be locked into a chair conformation by incorporating a ring (C, as in I, Scheme 4.60*a*)

When the rings in decalin system are not substituted other than at bridgeheads, the *trans* isomers are always achiral having a σ plane vertical to the common bond. The chiral *cis* isomers exist as unresolvable racemic mixture. On the introduction of a carbonyl group, one gets a 1- or 2- decalone, the two bridgehead carbons become chiral and both the *cis*- and *trans*-decalones exist as resolvable (±)-pair (Scheme 4.61) non steroid conformations.

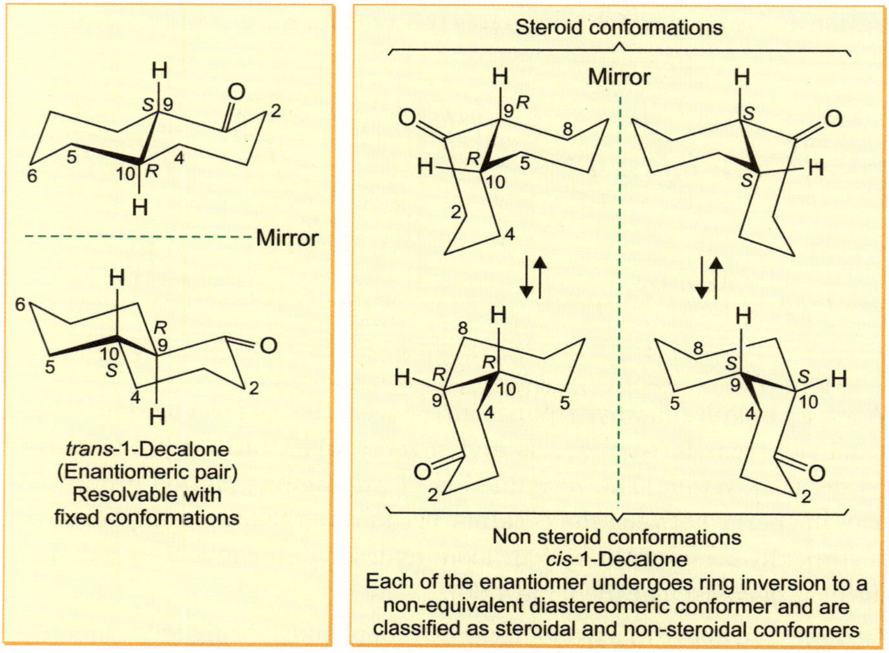

trans-1-Decalone (Enantiomeric pair) Resolvable with fixed conformations

cis-1-Decalone
Each of the enantiomer undergoes ring inversion to a non-equivalent diastereomeric conformer and are classified as steroidal and non-steroidal conformers

SCHEME 4.61

Both *cis*- as well as *trans* 1-decal-one can be easily equilibrated on treatment with base (Scheme 4.61*a*) in which the *trans*-isomer being more stable predominates.

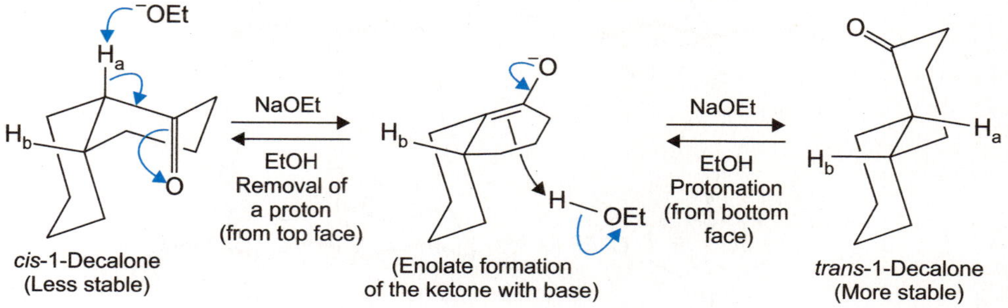

SCHEME 4.61a

(F) Fused Rings: The Hydrindanes (Bicyclo [4.3.0] Nonane)

Hydrindane represents as the rings C and D of steroids. The dissimilarly of the two rings makes the ring-fused carbon atoms chirality centers. A consideration of planar structures clearly shows the presence of two equivalent stereocenters and predicts the *trans*-isomer as a (±)-pair *i.e.*, two stable *trans*-isomers (Scheme 4.62) while *cis*-isomer as a *meso*-compound. However, *cis* isomer in which one axial and one equatorial bond of cyclohexane is used to fuse the cyclopentane ring is chiral and just like *cis*-decalin exists as a (±)-pair, the two enantiomers being interconvertible by ring inversion.

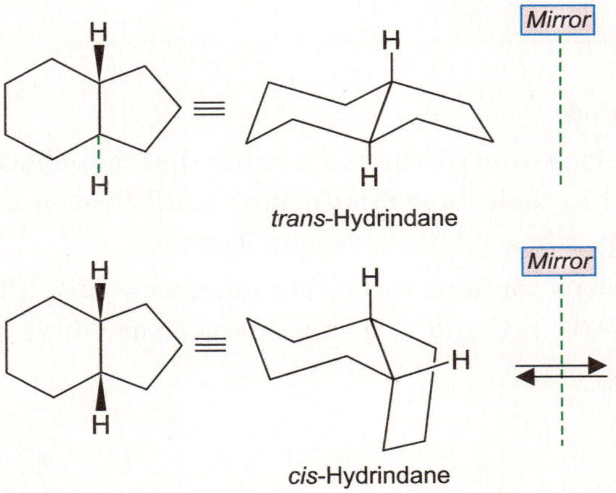

SCHEME 4.62

As with the decalins, *trans*-hydrindanes are conformationally rigid molecules, while *cis*-hydrindane is flexible and exists as a (*dl*)-pair of conformers. The free energy of activation of this conformational inversion is considerably smaller than the inversion barrier of *cis*-decalin. Moreover, the *cis* and *trans* -diastereomers of hydrindane have an energy difference around 2.9 kJ/mol (0.7 kcal/mol), a significantly smaller value as compared with that of decalin diastereoisomers. This is due to the more planar structure of the five-membered ring and as a result there is less strain in the fused system.

(G) Fused Rings: The Steroid Nucleus, Bridged Alkaloids

The tetracyclic ring system of perhydrocyclopentanophenanthrene is found in steroids and is composed of three fused 6-membered rings (perhydrophenanthrene itself exists in ten stereoisomeric forms; four enantiomeric pairs and two *meso* isomers). The prefixes *cis*- and *trans*-refer to the stereochemistry at the bonds of fusion between rings while *syn*- and *anti* (Scheme 4.63) have reference to the orientation of terminal rings with respect to each other. The most stable of these is the *trans-anti-trans* form, an all-chair conformation. The *trans-syn-trans* form is the least stable as this structure can be realized only when the middle ring has the energetically unfavorable boat conformation.

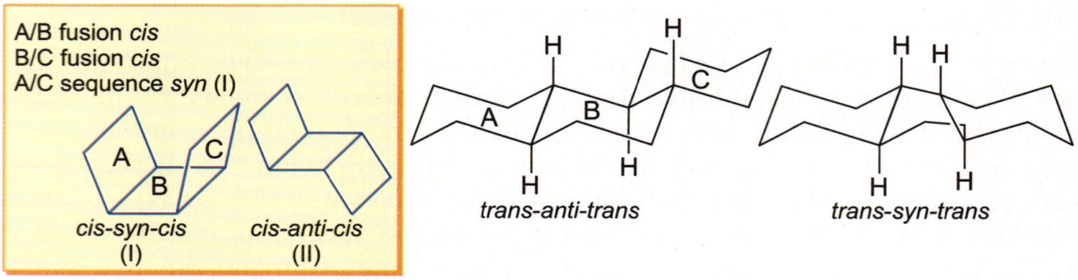

SCHEME 4.63

(*i*) Steroids

The following points may be noted:

- Steroids can broadly belong to one of the two families that resemble the decalins at the A and B rings. The cholestane family has an A/B *trans*-decalin system and the coprostane series has a A/B *cis*-decalin system.
- In fact the steroid nucleus contains three ring junctures (A/B, B/C and C/D). In cholestane, all the A/B, B/C and C/D ring junctions are more stable *trans* (Scheme 4.64).

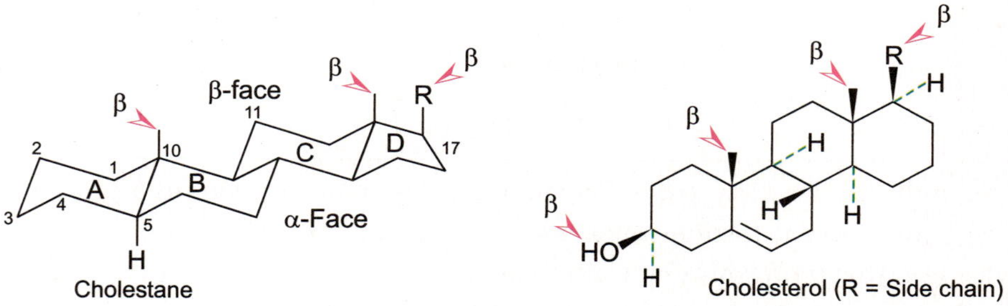

SCHEME 4.64

- The two methyl groups at ring junctions are called angular methyl groups.
- A group below the plane of the steriod ring system *trans*-to the angular methyl groups is called an α-group (corresponding face is called α) while the one above the plane *cis*-to the angular methyl groups is called the β-group (corresponding face is called β face).

(*ii*) Alkaloids

In the two-dimensional structure of strychnine (Scheme 4.65) six chirality centres are shown starred. Thus, strychnine is theoretically capable of existing in $2^6 = 64$ stereoisomers. A more realistic conformational formula helps to deduce and make the stereochemistry look comparatively simpler by considering the following points:

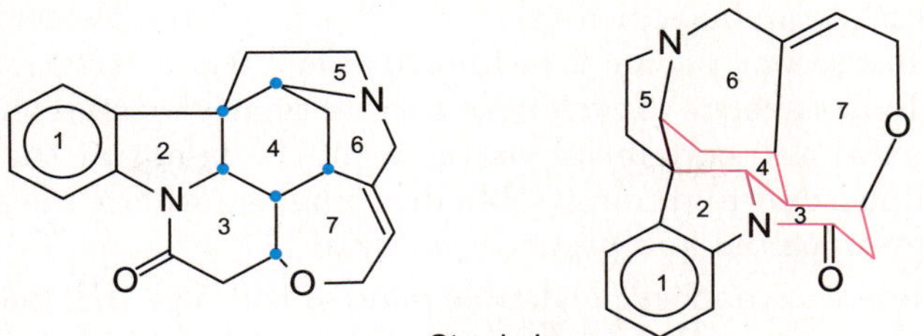

Strychnine

SCHEME 4.65

- Ring 4 is in chair conformation thus, forcing the bridging of ring 6 with it *via* and *cis*-diaxial bonds.
- This makes the fusion of ring 5 with 6 *via* one axial and one equatorial bond, *i.e., cis* fusion as it is geometrically impossible to bridge 1, 2-diaxial *trans* positions with a ring of ordinary size.
- Ring 2 is also *cis* fused with ring 4 *i.e.*, its bond with nitrogen is equatorial. Similarly, ring 7 is also *cis* fused and the juncture of rings 3 and 4 is obviously *trans*.

Similarly the conformational formula of morphine is drawn (Scheme 4.65*a*). Recall that the enantiomer of morphine can be written by inverting the stereochemistry at every chirality center.

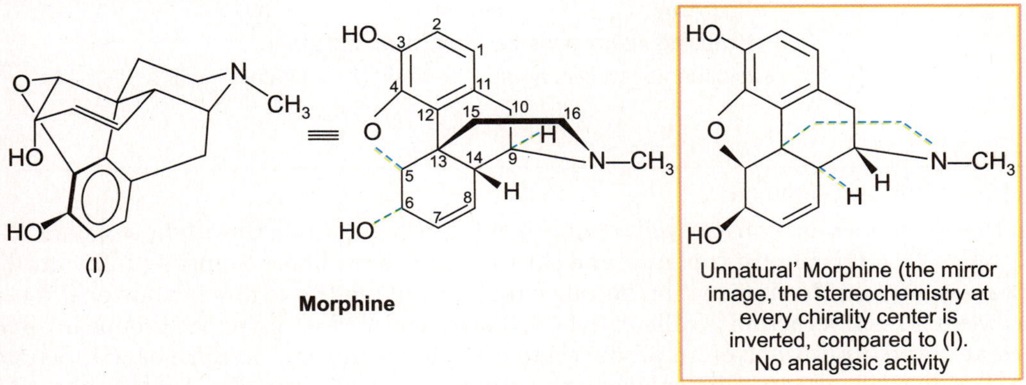

SCHEME 4.65a

4.17 CYCLOHEXENE AND CYCLOHEXANONES

(*i*) Cyclohexene (Allylic 1, 2- and 1, 3-strain) and Related Compounds

The four carbon atoms in the fragment C—C=C—C being almost coplanar, cyclohexene cannot adopt either a chair or a boat from and its preferred conformation is the half-chair shown in different forms (Scheme 4.67). The C1, C2, C3 and C6 atoms constitute a plane as are the two alkene hydrogens. The other two ring carbons C4 and C5 are disposed up or down or (down and up) alternatively with respect to this plane. The molecule has a C_2 axis which bisects the double bond (no σ plane, see structure I, Scheme 4.66). The compound thus is chiral belonging to point group C_2. The two enantiomers (I and II, Scheme 4.66) however, undergo fast interconversion by ring inversion and consequently constitute an inseparate (±)-pair. This fast interconversion of enantiomers by ring inversion has recently come to be known as enantio-merization (see also Scheme 4.33). The homoallylic carbons of cyclohexene have nearly normal axial and equatorial bonds while the axial and equatorial character of the bonds at the allylic carbons is significantly modified due to different dihedral angles (other than 60°). These bonds are thus termed pseudoaxial and pseudoequatorial.

A cyclohexene with endocyclic double bond of the type (III, Scheme 4.66) suffers from allylic 1, 2-strain also called $A^{1,2}$ -strain. When *R* and *R′* are bulky, they interfere to an extent that the axial conformer will predominate. This is so *e.g.,* when in (III, Scheme 4.66, *R* = Ph and *R′* = *t*-butyl), the conformer (IV) is predominant.

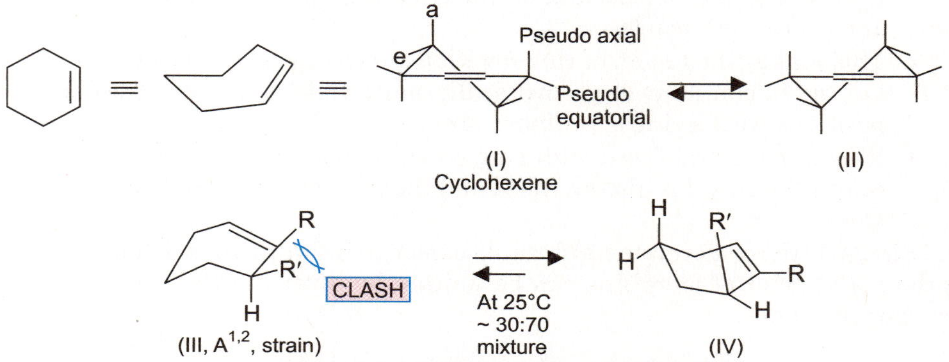

Eclipsing interactions between vinyl and allylic substituents result in allylic 1, 2-strain ($A^{1,2}$–strain)

SCHEME 4.66

(*ii*) Alkylidenecyclohexanes

In these compounds with exocyclic double bond (I, Scheme 4.67), the allylic segment R'—$C_\alpha = C_\beta$—C_γ—R is almost coplanar and thus *R* and *R′* are almost eclipsed (unfavourable interaction) and this interaction becomes further unfavourable due to shorter C = C in between. This interaction is called allylic 1, 3-strain or $A^{1,3}$ -strain as the groups involved are at 1 and 3 positions of an allylic system. As a result axial conformer (II, Scheme 4.67) may predominate particularly if the groups *R* and *R′* are bulky. Both $A^{1,2}$ and $A^{1,3}$ type strains play an important role to determine the stereochemical outcome of several important reactions involving cyclic and acyclic systems, *e.g.*, alkylation of enamines, alkylation and protonation of enolates (see problem 4.17 and Schemes 1.99, 6.52*c*, 6.91).

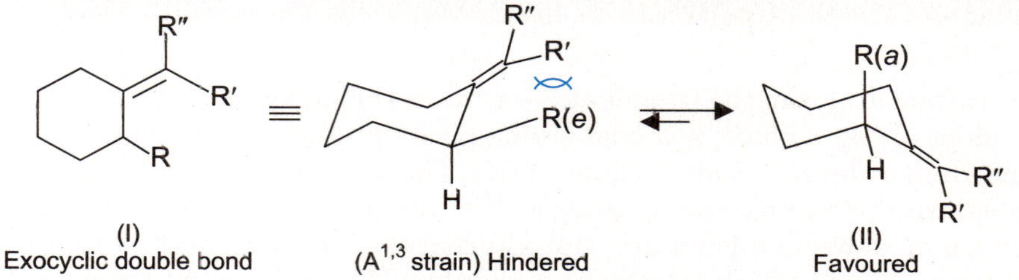

SCHEME 4.67

BOX 4.6
STEREOCHEMICAL PRINCIPLES AND $A^{1,2}$ AND $A^{1,3}$ STRAIN

The enamine is essentially a "nitrogen enolate, closely related to the enolate derived from ketones. Enamines are readily formed by the acid catalysed condensation reaction of a secondary amine with a ketone (Scheme B 4.7). Enamines behave as bidentate necleophiles, with both nitrogen and carbon functioning as nucleophiles. Alkylation at nitrogen is a problem with reactive halides such as methyl iodide or allyl bromide and when the groups on nitrogen are small.

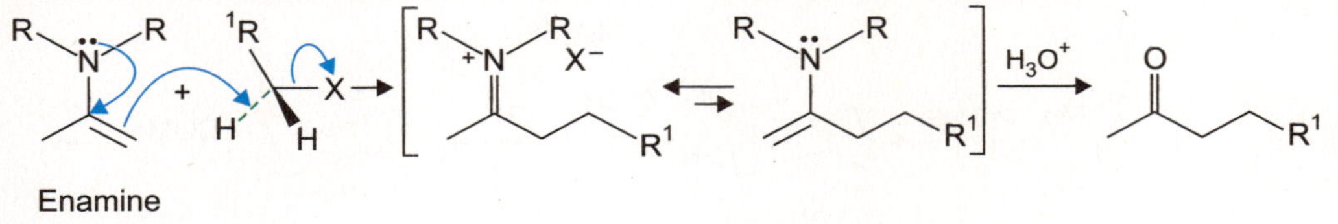

SCHEME B 4.7

Generally the enamine from an unsymmetrical ketone which can react at two sites is predominantly the less substituted enamine, due to reduced steric hindrance (Scheme B 4.8).

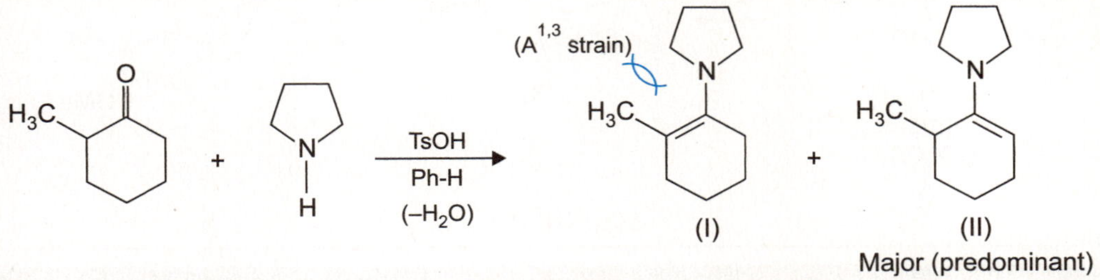

SCHEME B 4.8

An enamine is stabilized by interaction of the alkene π-system with the unshared electron-pair in a p orbital on nitrogen. This requires coplanarity of the bonds on the unsaturated carbon atoms and those to nitrogen, which is possible in (II, Scheme B 4.8) but, owing to steric repulsion, is not possible in (1). This can be usefully applied in synthesis. Thus in (II, Scheme B 4.8) to avoid $A^{1,2}$-strain (see, Scheme 4.66), the C-alkyl group occupies a pseudoaxial position (III, Scheme B 4.9). Alkylation, therefore, takes place at C-6 and there is no possibility of dialkylation.

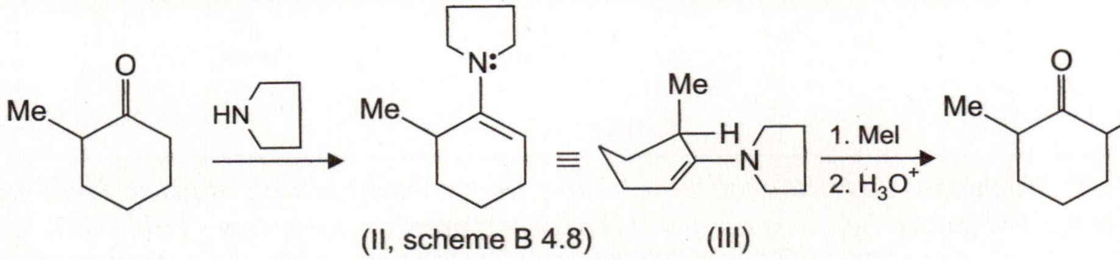

SCHEME B 4.9

Continued...

CHAPTER 4

1-Acetylcyclohexene (I, Scheme B 4.10) undergoes 1, 4-addition with phenylmagnesium bromide to give the bromomagnesio complex (II) which exists in two diastereomeric forms, (Z) and (E) differing in the configuration of the double bond.

The following points may be noted:

- *In this 1, 4-addition, axial conformer is preferred since now $A^{1,3}$-strain is avoided.*
- *Protonation on (II, Scheme 4.10) occurs from axial side to yield a cis-ketone as two conformers (III and IV)*
- *Conformer (III) enolises with acid to the enol (IV) with Ph axial in order to avoid again $A^{1,3}$-strain.*

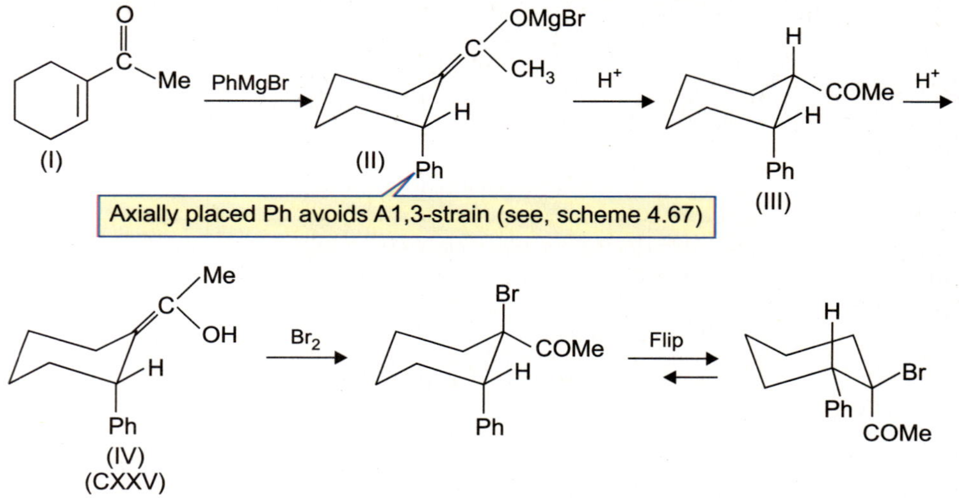

SCHEME B 4.10

EXERCISE 4.10

Explain the stereochemical outcome of the reaction (Scheme 4.67a).

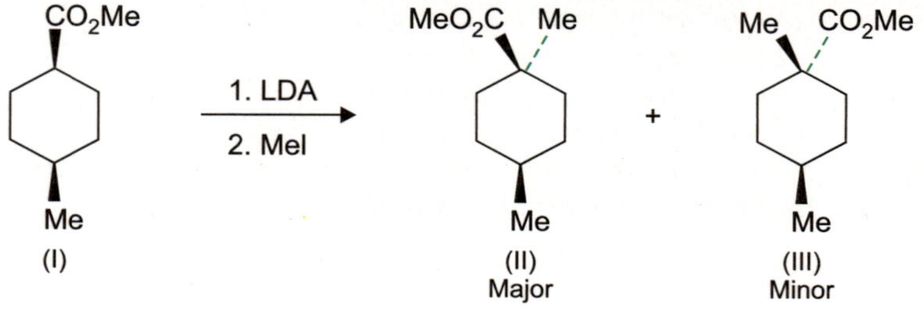

SCHEME 4.67a

ANSWER. *First consider the conformation of the reactant which is given (Scheme 4.67b), a methyl group behaves as if it were larger than an ester (see, Table 4.2).*

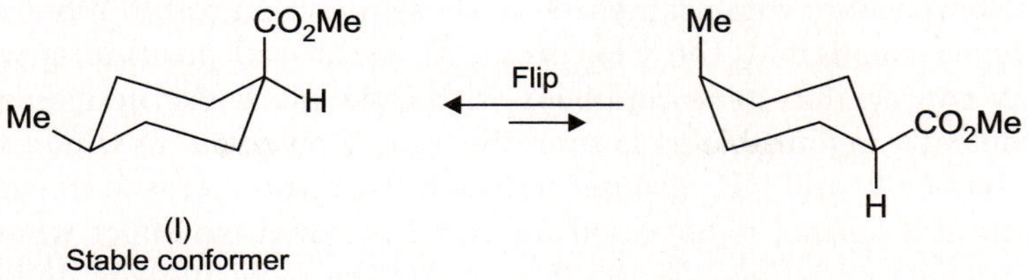

SCHEME 4.67b

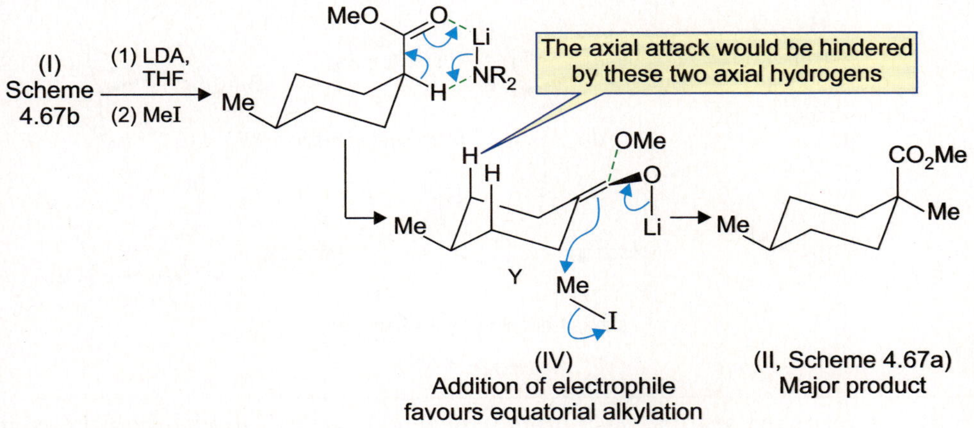

SCHEME 4.67c

In the 1, 4-asymmetric system (I, Scheme 4.67c), reaction with LDA generated the alkylidene enolate (IV) which on alkylation with iodomethane proceeded via path b (Scheme 4.67*d*), to give the major product. Interaction with the 1, 3-diaxial hydrogens in (V, Scheme 4.67*d*) destabilizes attack along path a to only give the minor product.

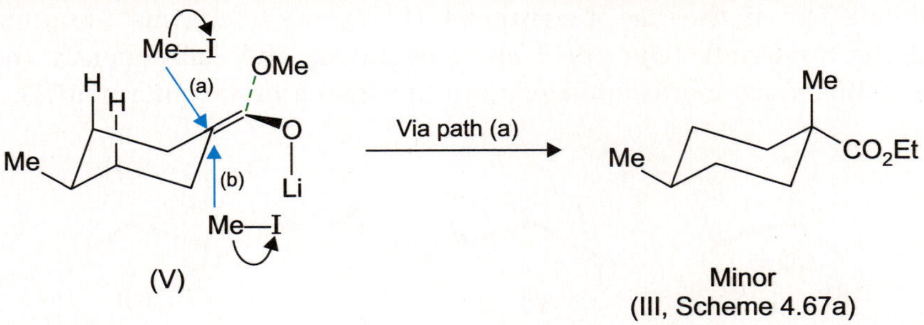

SCHEME 4.67d

(*iii*) Cyclohexanone Ring—Alkyl Ketone Effect

The equatorial hydrogens on the α-carbon (C2) are almost eclipsed with the carbonyl oxygen (I, Scheme 4.68). An equatorial substituent may be destabilized to some extent due to steric repulsions. This decreases the energy difference between the axial and the equatorial conformers when compared to those in cyclohexane. When $R = CH_3$ this value is, however, negligible, the CH_3 group in equatorial position is indeed far, to experience any appreciable steric repulsion with C=O, moreover, eclipsing of CH_3 with C=O is electronically favourable as is so in the case of propanal in which the preferred conformation has C=O and CH_3 groups eclipsed. With the increase in size of $R(C_2H_5)$ a value of around 3 kJ/mol is obtained for the 2-alkylketone effect when $R = i - Pr$ the value increases to 7 kJ/mol. When R = *t*-butyl the molecule adopts largely a boat conformation to avoid *t*-butyl/C=O eclipsing (III, Scheme 4.68). Compare these results with those in (Scheme 4.43).

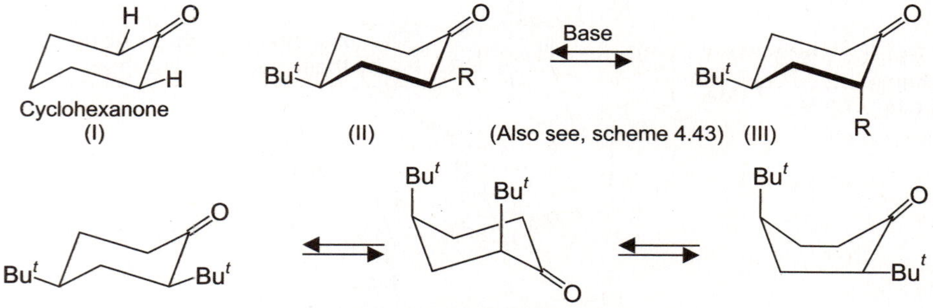

cis-2, 4-di-*t*-Butylcyclohexanone

SCHEME 4.68

Based on these arguments 2-bromocyclohexanone will predominantly exist as axial conformer which will have greater stability over the equatorial conformer. For these reasons *cis*-2, 4-di-*tert*-butylcyclohexanone adopts largely a boat conformation (Scheme 4.68).

In the case of a 3-alkylcyclohexanone (I, Scheme 4.68*a*) one 1, 3-interaction between the axial *R* group and axial *H* is missing. This is equivalent to one butane gauche interaction (3.75 kJ/mol) when $R = CH_3$. This decrease in $-\Delta G°$ value of *R* is termed 3-alkylketone effect. In the case of menthone (II, Scheme 4.68*a*), the 2-isopropylketone effect and the 3-methylketone effect are cooperating and consequently the diaxial conformer (III) is much more stabilized than the diequatorial conformer (II).

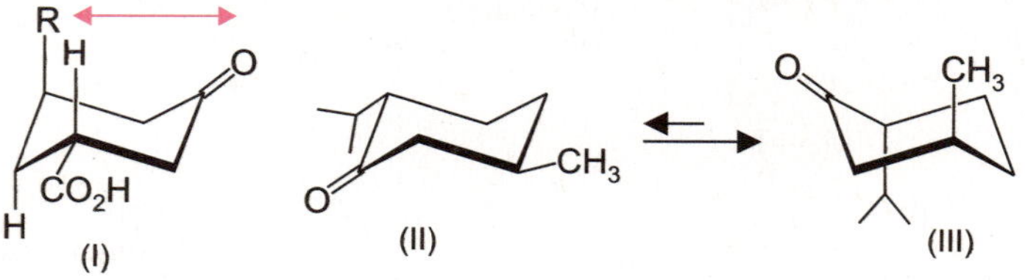

SCHEME 4.68a

(*iv*) Reactivity in Cyclohexanones a Comparison with Cyclopentanones

Cyclohexanone is reduced with hydride ion from borohydride to give cyclohexanol much faster than cyclopentanone. This difference in reactivity shows that unlike cyclopentanone cyclohexanone suffers from eclipsing strain (torsional strain) due to eclipsing of its oxygen atom by two α-hydrogens which are equatorial (see I, Scheme 4.68). This strain is relieved on the formation of cyclohexanol (Scheme 4.68*b*). However, on the formation of cyclopentanol there is an increase in the torsional strain.

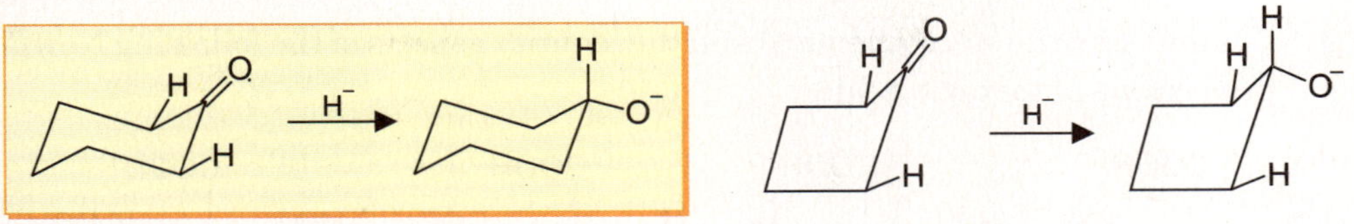

SCHEME 4.68b

The results may be compared with *e.g.*, S_N1 reactions where the reactivity is less in cyclohexyl derivatives compared to their cyclopentyl analogues. The formation of transition state in non-planar cyclohexane leads to increase in strain, while this is not so in cyclopentane which is almost flat leading to decrease in strain in the transition state (Scheme 4.68*c*).

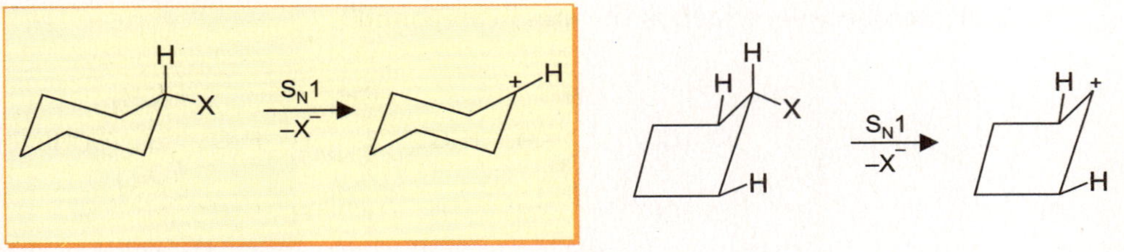

Scheme 4.68c

More on Cyclopentane Ring

The geometry of cyclopentane ring has been discussed as a planar and an envelope form (see, Scheme 4.15). The additional half chair form may have axial, equatorial or isoclinal bonds (i). In a monosubstituted cyclopentane a substituent is equatorial at the top of the ring (Scheme 4.68d).

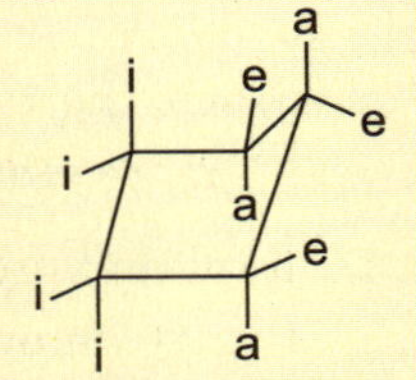

SCHEME 4.68d

PROBLEMS

4.1. Comment on the stereochemical and the identity aspects of the following Newman projections of tartaric acid:

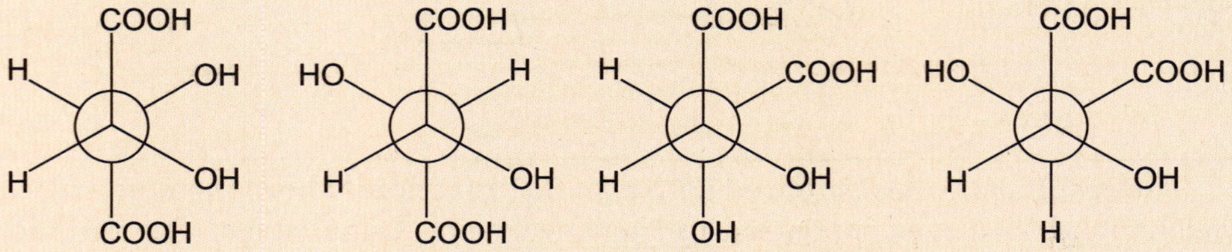

(I) (II) (III) (IV)

4.2. (*i*) Comment on the relationship among the following Newman projections of 2, 3-dichlorobutane, which is the *meso*-compound?

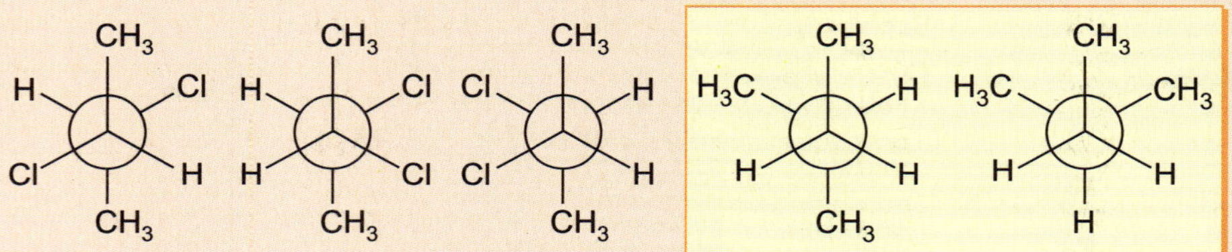

(I) (II) (III) (IV) (V)

(*ii*) Of the two staggered conformations (IV and V) of 2-methylbutane, one is more stable than the other. Explain.

4.3. Explain why only the *trans*-isomer of 2-bromocyclohexanol reacts with NaOH to give an epoxide?

4.4. Comment on the conformation and chirality of decalin.

4.5. Write a short note on *threo* and *erythro* nomenclature.

4.6. Explain the chirality of an adamantane derivative.

4.7. Comment on the configurational and conformational aspects of disubstituted cyclohexane derivatives. How their chirality depends on the nature of substituents *R* and *R'* ?

4.8. Write the most stable conformation of *trans*-1, 2-dimethylcyclohexane, is it chiral? Compare these results with its *cis*-isomer.

4.9. Write the conformations of 1, 3-dimethylcyclohexane. Discuss the chirality of *cis* and *trans*- isomers.

4.10. Draw the diastereomers of 1-bromo-1, 2-diphenylpropane, why one of these reacts as a faster rate than the other.

4.11. How the compound (A) is related to (B)?

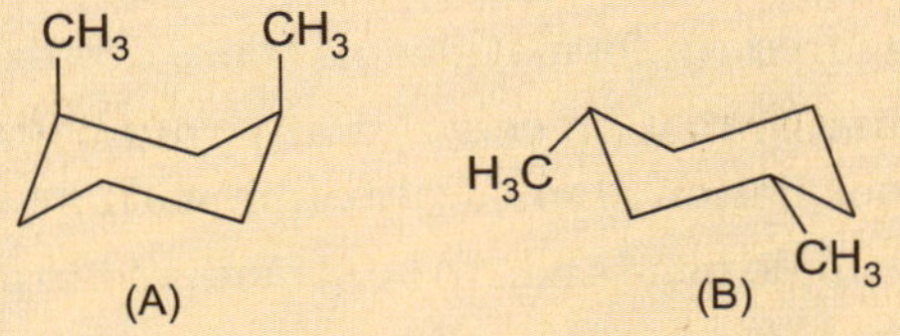

4.12. (*i*) Comment on the strain in the conformation of bicyclo [3.3.1] nonane. How it is related with adamantane ? How many bridgehead atom are present in adamantane ?

(*ii*) Comment on the conformation of compound (A).

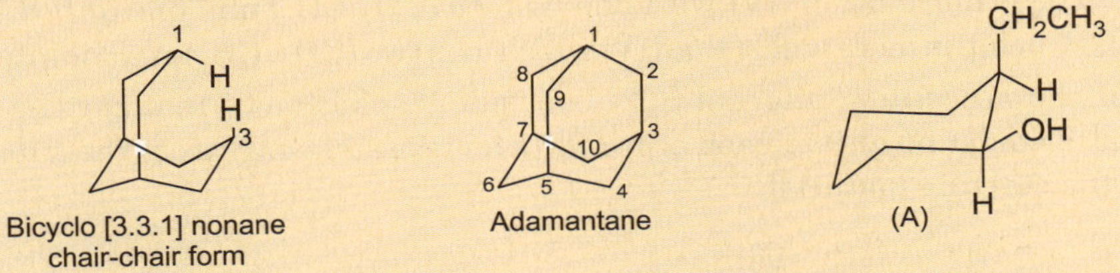

4.13. Compared to rotation about single bonds *e.g.*, in ethane (sp^3-sp^3 carbon-carbon bonds), what influences you expect on the rotational isomerism about sp^2-sp^3 hybridized carbon-carbon single bonds as for example in CH_3—CHO ?

4.14. When cyclopentane is reacted with excess chlorine at a high temperature, seven dichlorocyclopentanes are formed. Comment on their geometrical isomerism and chirality.

4.15. (*a*) Discuss the geometrical isomerism of 1-bromo-2-chlorocyclobutane. Give stereodescriptors to different chirality centers.

(*b*) Assign stereodescriptors to the two chirality centers in the compound (I).

(*c*) Assign nomenclature to compounds (II) and (III).

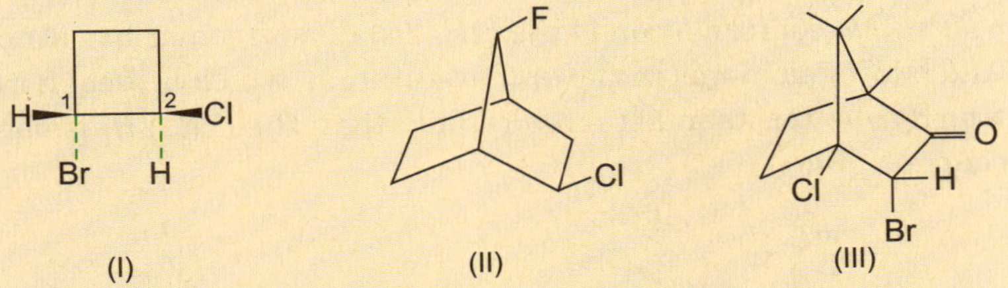

4.16. (*R*)-Camphor is reduced with lithium aluminium hydride to give a pair of alcohols. Give nomenclature to each chirality center in these. How these alcohols are stereochemically related ?

4.17. Considering $A^{1,3}$ strain, write the most stable rotamer in each case.

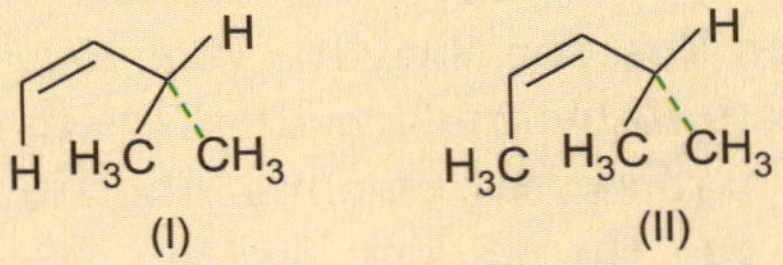

CHAPTER 4

ANSWERS TO SELECTED PROBLEMS

4.1. These are the staggered conformations as these are drawn keeping in mind:

(*a*) That a compound cannot be chiral if any rotamer is non-disymmetric or if each chiral conformation is in equilibrium with its enantiomer.

(*b*) Rotation about single bonds has to be considered to prove or disprove identity.

(*c*) While comparing stereochemical relationship of two compounds it is not sufficient to compare only one conformation of each compound. Rotation about single bonds is necessary to find the closest correspondence Newman projections (I and II) are enanatiomers while *e.g.*, (III) is a *meso*-compound; (I and IV) are identical.

4.2. (*ii*) The conformation (IV) with one *gauche* (CH_3,CH_3) and one *anti* (CH_3, CH_3) relationship is more stable than (V) which has two (CH_3, CH_3) *gauche* interactions.

4.3. For the necessary "internal S_N2" backside displacement to occur the alkoxide and Br must be *anti* to each other, which is possible only from the *trans* isomer of the starting material.

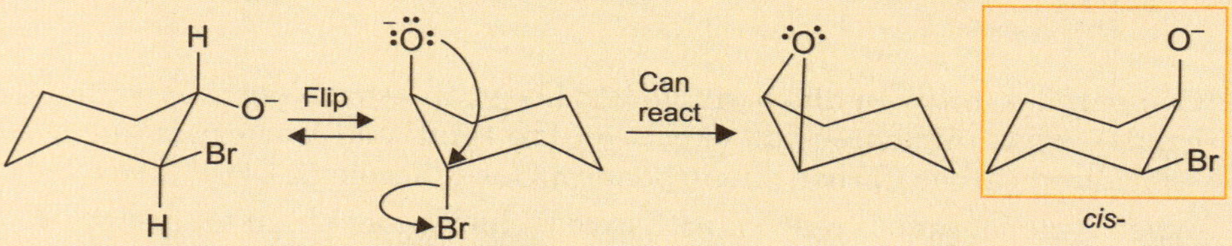

trans-biequatorial   Reactive conformation (trans, diaxial)

4.7. (*i*) 1, 1-Disubstituted cyclohexanes are achiral.

(*ii*) 1, 2-Disubstituted derivatives, *cis* chiral if R ≠ *R*′; achiral if R = *R*′, *trans* chiral.

(*iii*) 1, 3-Disubstituted derivatives, *cis* chiral if R ≠ *R*′; achiral if R = *R*′, *trans* chiral.

(*iv*) 1, 4-Disubstituted derivatives, *cis* achiral; *trans* achiral.

4.10. The most stable conformation of the (1*S*, 2*R*) diastereomer has already the bromine and hydrogen atoms in the *anti* orientation, which is ideal for the E2 reaction, which thus reacts much more rapidly than the (1*R*, 2*R*) isomer does (also see Scheme 5.19).

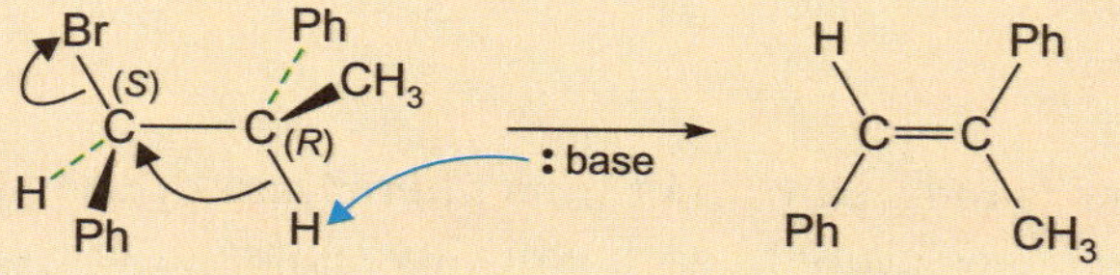

(Most stable conformation H and Br already anti)   (*E*)

The most stable conformation of (1R, 2R)-diastereomer has the H and Br atoms *gauche* to one another. Rotation of right carbon to bring them into *anti* arrangement generates an less stable conformation where now the phenyl groups are *gauche* placed. Thus in this case *anti*-elimination involves a conformation other than the most stable one, the reaction is slow.

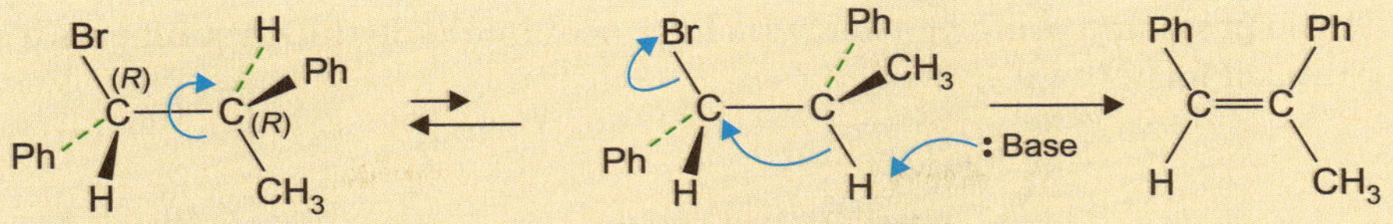

Rotation about the carbon-carbon bond (More stable conformation but H and Br are not *anti*)

(Less stable conformation H and Br are now *anti*)

(*Z*)

4.11.
- Ring flip (A) to (C), this operation does not effect the configuration of chirality centers.
- When (C) is rotated 180° it gives (B).
- Thus (A and B) as well as (C) are the same compound.
- This is confirmed from the same stereodescriptors at the chirality centers.

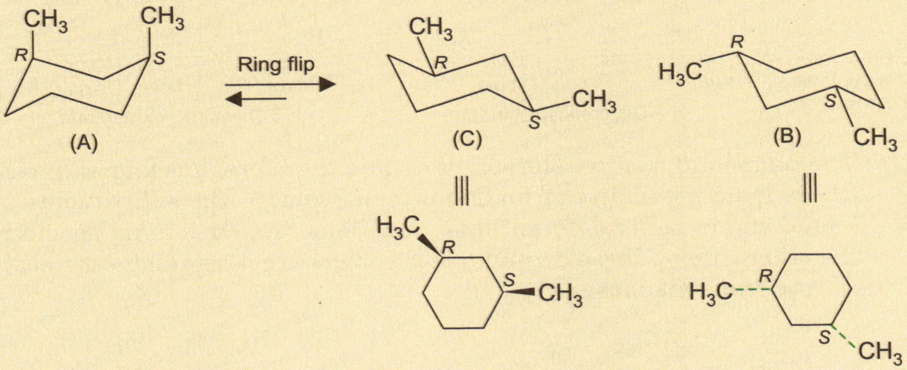

4.12. (*i*) Bicyclo [3.3.1] nonane constitutes a 1, 3 fusion of two cyclohexane chairs which are free of angle strain. However, the molecule in the conformation suffers from serious trans-annular interaction between the axial hydrogen at C-3 and C-7. The more attractive conformation is the one in which both chairs are somewhat stretched out to reduce this strain. Adamantane is the compound obtained by bridging the conflucting methylene groups, C-3 and C-7 in the chair-chair form. It has four bridgehead atoms (1, 3, 5 and 7).

(*ii*)

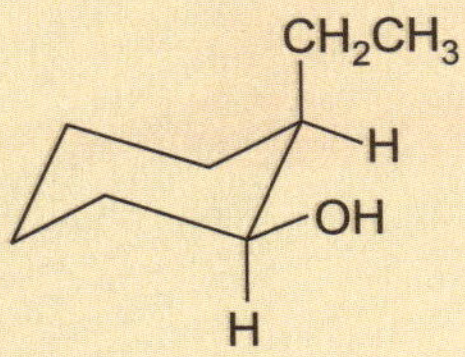

$\Delta G^\circ_{OH} - \Delta G^\circ_{ethyl} = -1.0 - (-1.8) = +0.8$

ΔG° is greater than zero; therefore the chair conformation after a ring-flip will be favored.

4.13. In saturated compounds generally the staggered conformation is preferred and the one with the groups at 180° *i.e.*, farthest from each other is the best staggered conformation. When however, a single bond is connected to a double bond, the double bond at one end of the single bond eclipses a hydrogen at the other end in the preferred conformation which for acetaldehyde thus, represents a slightly eclipsed form (by 9°).

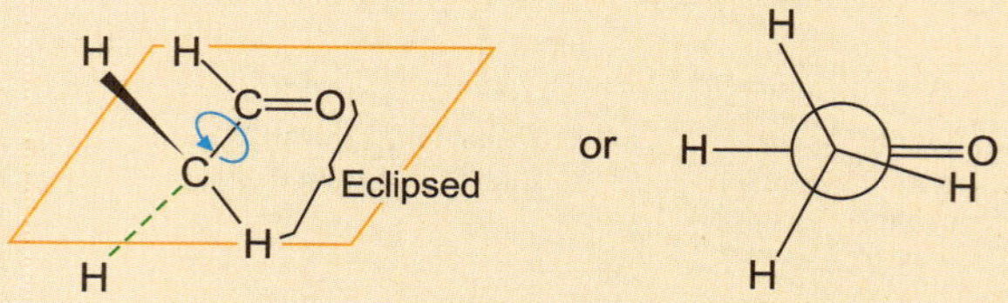

4.14. In common with other alicyclic rings the chirality can be discussed. The 1, 1-dichloro-derivative is achiral. The 1, 2 and 1, 3-compounds have two stereocenters. In each case *cis*-isomer is *meso* and the *trans* isomer exists as a pair of enantiomers.

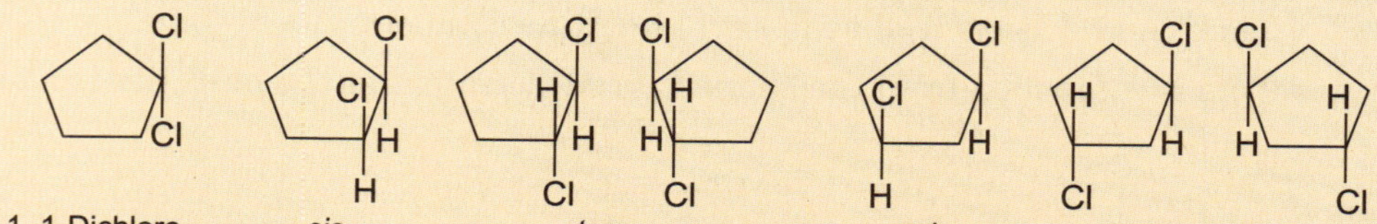

1, 1-Dichloro-cyclopentane (Achiral) | *cis* (*meso*) | *trans* (±) pair — 1, 2-Dichlorocyclopentane | *cis* (*meso*) | *trans* (±) pair — 1, 3-Dichlorocyclopentane

4.15. (*a*) The compound has two stereocenters and therefore, has four stereoisomers (there is no possibility of finding a *meso* isomer—the substituents on two stereocenters being different) one enantiomeric pair has *cis* geometry and the other *trans*-. These *cis* and *trans* isomers are diastereomers (recall from *cis* and *trans*-alkenes).

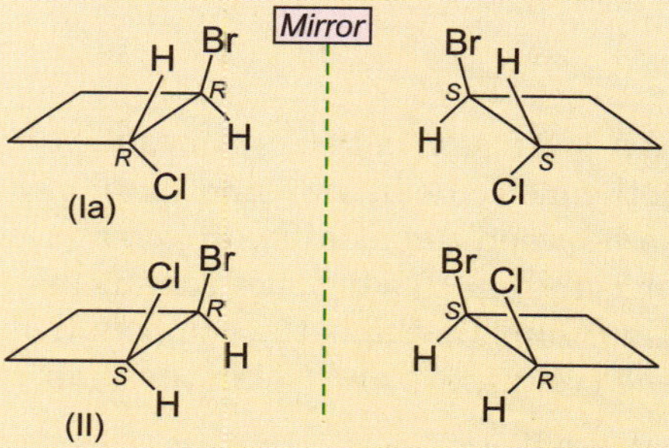

For assignment of R and S descriptors a geometrical figure e.g., (II) must be oriented properly in mind/make a model keeping the lowest priority group away (compare with α-D-Glucose (scheme 1.26)

Stereoisomers of 1-bromo-2-chlorocyclobutane

(*b*)

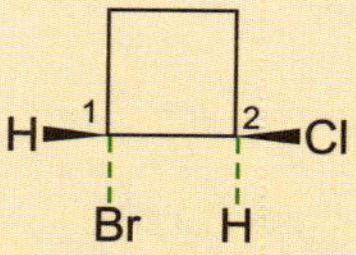

C1 = *R*; C2 = *R*

(I)

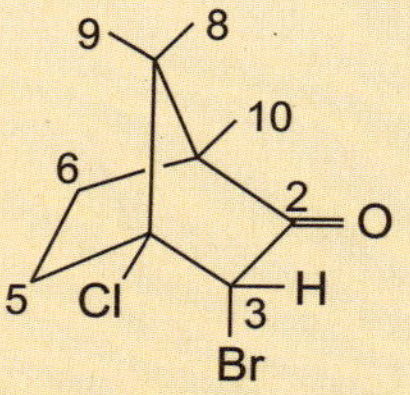

(*c*) II, 2-*exo*-chloro-7-*syn*-fluorobicyclo [2.2.1] heptane; III, 3-*endo-bromo*-4-chlorocamphor.

4.17. $A^{1,3}$ strain in acyclic systems leads to the following results:

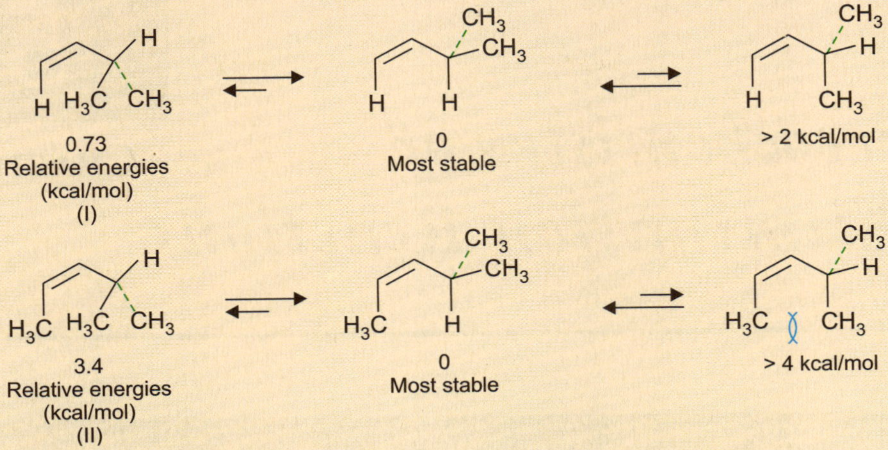

5

CHAPTER

Stereochemistry of β-Elimination Reactions

5.1 INTRODUCTION

The term elimination normally refers to the loss of two atoms or groups from a molecule and in the commonly known elimination reactions multiple bonds are formed through the loss of groups bonded to adjacent atoms. The process is usually termed 1, 2-elimination or β-elimination because the two departing ligands are on adjacent atoms. Often E1 and E2 reactions are referred to as β-eliminations (Scheme 5.1).

Most of the β-elimination reactions in this chapter are H/Het-eliminations *e.g.*, elimination of H/Br from an allyl bromide, H and OTs from a tosylate and so on. Other eliminations which are also β-eliminations Het[1]-Het[2] include Wittig and Peterson reaction.

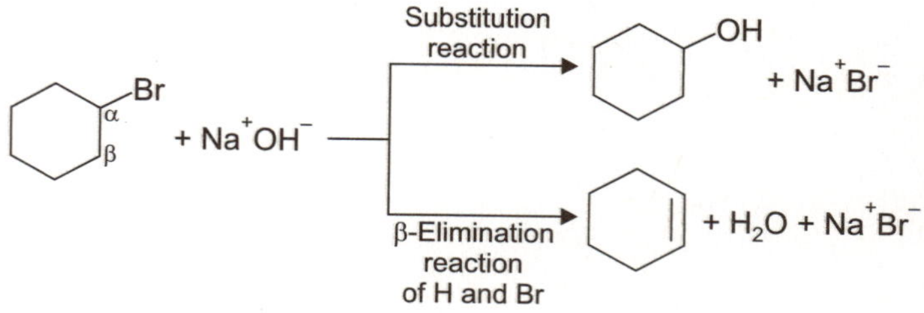

Generally in S_N1 and S_N2 reactions, substitution is accompanied by some elimination of H—L, where L is a leaving group to give an alkene.

SCHEME 5.1

Other elimination reactions may also occur in organic compounds, with the departing groups located at 1, 3 or more remote sites in the molecule to yield cyclic products. This type of elimination reaction is actually an intramolecular substitution (Scheme 5.2).

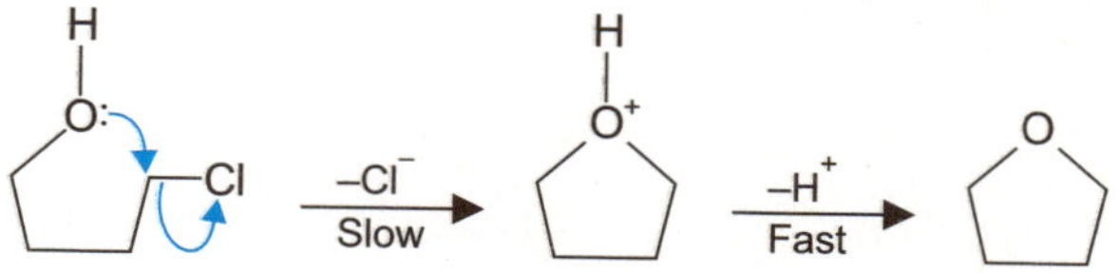

SCHEME 5.2

Still another mode of elimination may involve two groups departing from the same atom. Such 1,1-or α-eliminations are used for producing the very reactive species known as carbenes (Scheme 5.3).

Elimination reactions always face a competition from substitution reactions, since all nucleophiles are potential bases as all bases are potential nucleophiles.

Elimination reactions 1, 2 can occur by a variety of mechanisms and three mechanistic pathways are normally distinct routes for β-elimination reactions. Of these E2 and E1 are the most common. These two processes are closely related to the S_N2 and S_N1 mechanisms of substitution. A third mechanism is designated as ElcB, although less common, it provides one extreme in the possible elimination pathways (see Scheme 5.7).

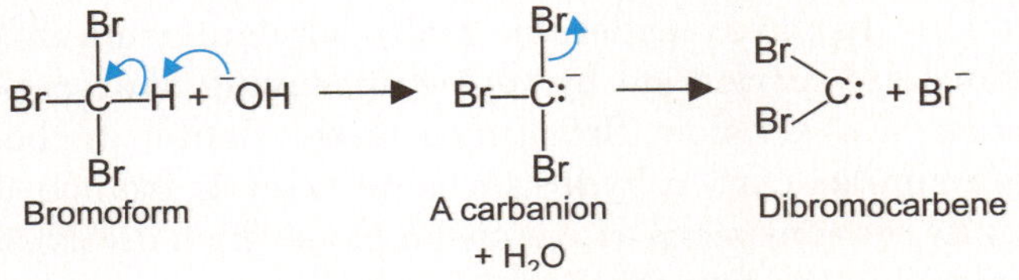

SCHEME 5.3

5.2 MECHANISMS E1, E2 AND ElcB

(A) Carbocation Mechanism (E1 Mechanism)

In these elimination reactions the initial step involves cleavage of the bond between the leaving group and its carbon with the generation of fully dissociated carbocation. These elimination reactions are called, E1 reactions (Elimination unimolecular).

E1 reactions almost always compete with S_N1 reactions as the initial step for both the reactions is similar. When *tert*-butyl bromide is treated with water, the reaction produces the substitution product as well as the elimination product (2-methyl-propene) (Scheme 5.4).

The initial step for both of these reactions is the formation of a *tert*-butyl cation (or an ion pair) and represents the rate-limiting step for both the reactions (Scheme 5.4).

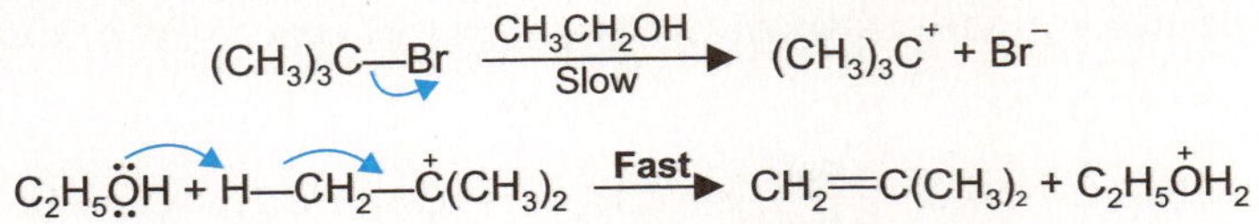

Rate = k_1 [$(CH_3)_3CBr$]

E1 Mechnism: Carbocation Intermediates

SCHEME 5.4

If a solvent molecule reacts as a nucleophile at the positive centre of the *tert*-butyl cation the product is *tert*-butyl alcohol (S_N1), however, if a solvent molecule acting as a base, accepts one of the β-protons, the product is 2-methylpropene (E1).

(B) The Concerted Reaction (E2 Mechanism)

Both the substrate and the nucleophile participate in the single step. Reaction kinetics are typically second-order. The nucleophile which is commonly a base abstracts a proton from the β-atom to the leaving group (Scheme 5.5).

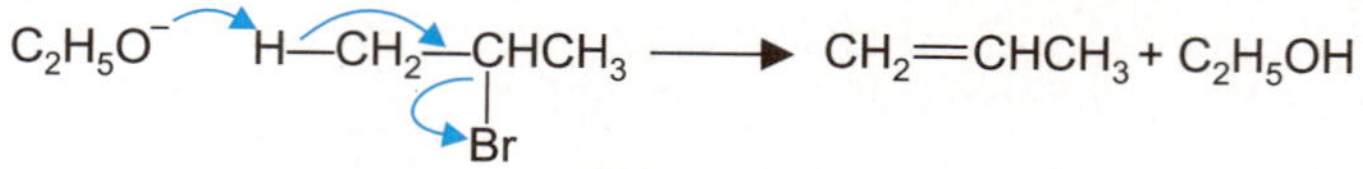

Rate = $k_2[(CH_3)_2CHBr]\,[C_2H_5O^-]$

E2 Mechanism: Synchronous Elimination

SCHEME 5.5

The following evidences support the existence of these general mechanistic representations:

- The rate expressions for the reactions are proper second order.
- The reactions display considerable hydrogen-deuterium isotope effect. When eliminations are carried out by replacing leaving hydrogen with deuterium, the reaction rates get slow. Breaking a carbon-deuterium bond is slower than cleaving a similar carbon-hydrogen bond (kinetic isotope effect). Therefore, these results are consistent with a mechanism which involves carbon-hydrogen bond cleavage in the transition state.
- The reactions also show a substantial element effect. The relative reactivity of 2-phenylethyl halides towards ethoxide ion is in the order:

  $$C_6H_5CH_2CH_2I > C_6H_5CH_2CH_2Br > C_6H_5CH_2CH_2Cl > C_6H_5CH_2CH_2F$$

  An element effect of this kind is again in keeping with a transition state involving carbon-halogen bond cleavage. Thus, 2-phenylethyl iodide reacts fastest because the carbon-iodine bond is weakest; 2-phenylethyl fluoride reacts slowest as the carbon-fluorine bond is strongest.
- E2 reactions, generally, are not accompanied by rearrangements. This is consistent with a mechanism which does not involve the formation of carbocations.
- The most compelling evidence for E2 mechanism is, however, the stereochemical outcome. This mechanism is stereospecific as illustrated (Scheme 5.19).

EXERCISE 5.1

Explain why the major product from the elimination (Scheme 5.6) contains deuterium?

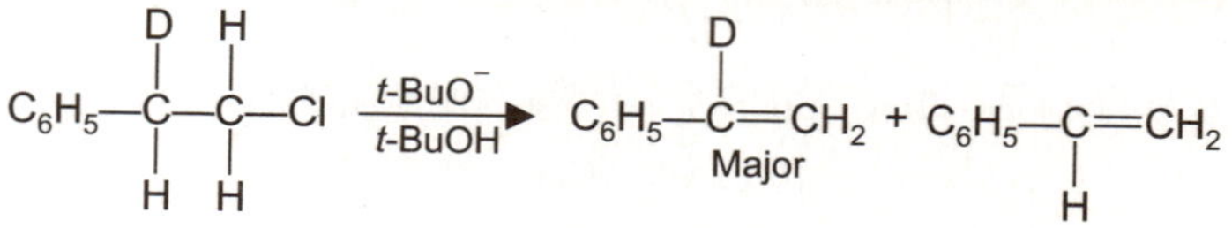

SCHEME 5.6

ANSWER. *The rate determining step requires the breaking of a C—H or C—D bond and C—H bond is easier to break.*

(C) The ElcB Mechanism

In the E1 mechanism the bond to the leaving group breaks during the first (slow) step to give the carbocation and the bond to the hydrogen breaks in a second step. In the E2 mechanism, both of these bonds break in a single step. There is a third possible mechanism in which the bond to the hydrogen breaks during the first step to give a carbanion and the bond to the leaving group breaks in a second step (Scheme 5.7).

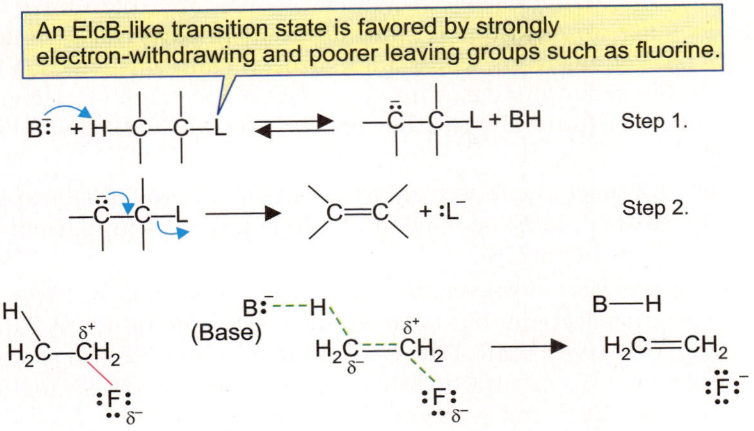

SCHEME 5.7

This third mechanism where base abstracts a β-hydrogen atom to form a carbanion in the first step, followed by the departure of the leaving group to give an alkene in the second step is called ElcB mechanism. The second step often determines the rate, since this step involves a unimolecular reaction of the conjugate base of the initial reactant, the mechanism is designated as elimination, unimolecular, conjugate base-or ElcB. This mechanism dominates by the presence of substituents that stabilize the intermediate carbanion. A poorer leaving group, which makes the second step less likely to be concerted with the first, also favors this mechanism as in alkyl fluorides.

BOX 5.1
BIOLOGICAL ElcB ELIMINATION REACTIONS

An important example of an elimination reaction in living organisms is the conversion of 2-phosphoglycerate to phosphoenolpyruvate during the metabolism of glucose (Scheme B 5.1).

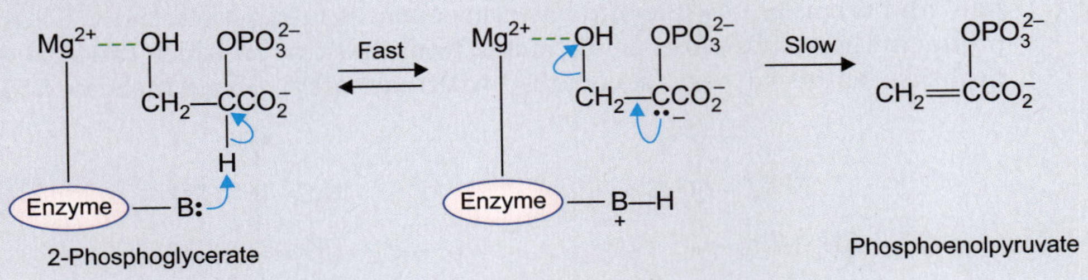

SCHEME B 5.1

This elimination is catalyzed by the enzyme enolase and follows an ElcB mechanism. The enzyme provides a base to remove the acidic proton and generate a carbanion in the first step. In addition, a Mg^{2+} cation in the enzyme acts as a Lewis acid and bonds to the hydroxy group, making it a better leaving group.

5.3 E1, E2 AND ElcB VARIABLES—THEIR MECHANISTIC SPECTRUM

The three most common elimination mechanism, E1, E2 and ElcB have been studied. These mechanisms are largely similar than different in each case, the leaving group departs with its pair of electrons while the hydrogen come off without them. These, however, differ in the order of steps, there is in fact a spectrum of mechanisms which range from one extreme (pure E1) to other extreme (pure ElcB). The following points may be considered:

- The pure E2 mechanism is somewhere in the middle of this spectrum when in the transition state the bonds to hydrogen and leaving group (X) are both breaking (II, Scheme 5.8).
- In the E2 reaction, however, the transition state may not be exactly in the middle of this spectrum but somewhere to one side or other. The C—H bond may begin to cleave ahead of the C—X bond as in the transition state (I Scheme 5.8). The reaction can still be bimolecular, however, the transition state will be "carbanion–like" and resemble the ElcB reaction.

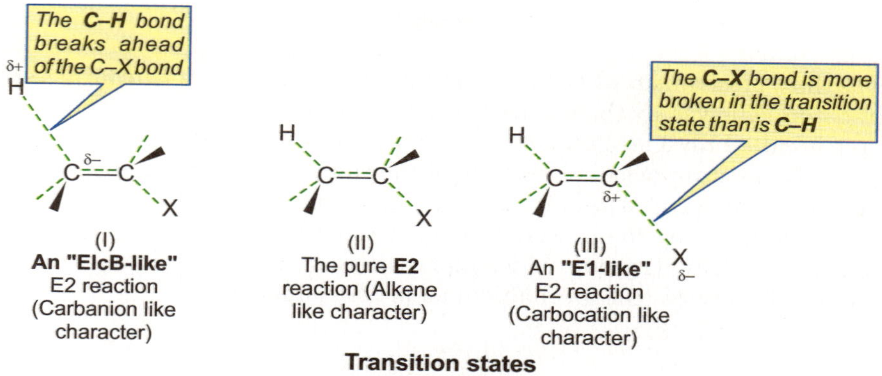

SCHEME 5.8

- As an example, the E2 reaction of quaternary ammonium ions and alkyl fluorides does not go by ElcB mechanism, however, the E2 transition state is itself "ElcB-like" and is carbanion–like and has consequences in regioselectivity. Thus the proton in these substrates is abstracted from that carbon which can hold on to negative charge the best *i.e.*, a carbon with least alkyl groups (Scheme 5.9).

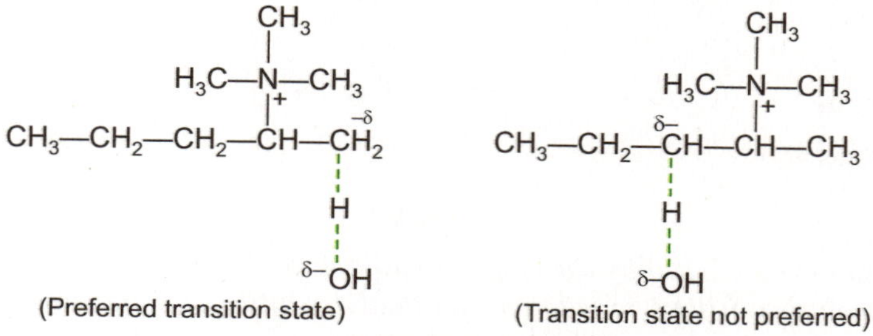

SCHEME 5.9

- Similarly the leaving group (X also called the nucleofuge) may depart just before the proton *i.e.*, the C—X cleavage is ahead of C—H cleavage. The situation is then expressed as an E2 reaction with a small amount of E1 character (see III, Scheme 5.8). In fact such a transition state leads to a carbocation intermediate *i.e.*, E1 elimination.

5.4 COMPETITION BETWEEN ELIMINATION AND SUBSTITUTION

Often one has to face a competition among four different mechanisms while carrying out an elimination reaction and these are S_N1, S_N2, E1 and E2. Thus one has to decide if the combination is going to be S_N2/E1 or E2 or S_N1/E1. Several factors influence this outcome and these are the structure of the substrate, the identity of the base or nucleophile, the solvent and the leaving group.

A particular substrate generally responds to changes in reaction conditions so as to favour one pathway over the other. Changes in the base, nucleophile generally have a marked effect on the elimination-substitution pathways. Smaller, but significant effect is also exerted by the reaction medium and temperature.

Significantly in the elimination-substitution competition, substitution is generally the predominant reaction. Substitution involves less bond reorganization (less bonds are broken and formed) and is therefore, generally more favourable energetically.

(a) Effect of Structure of the Substrate and Basicity Versus Nucleophilicity

(*i*) S_N2/E2 Condition: Substitution is Slowed by Steric Hindrance (Crowding) while Elimination is not

The following points may be considered:

- When the reactant is a primary alkyl halide, there is no possibility to form carbocations. Primary alkyl halides undergo only S_N2/E2 reactions (a high concentration of a good nucleophile/strong base). Thus primary substrates due to low steric hindrance are excellent for S_N2 reactions with almost any nucleophile (eqn. I, Scheme 5.10).

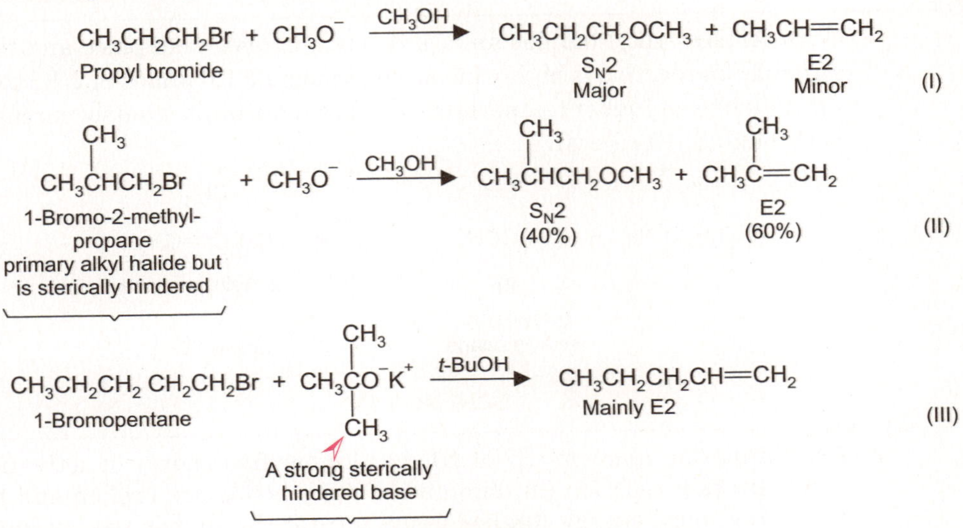

SCHEME 5.10

- When either the primary alkyl halide or the nucleophile/base is sterically hindered, the nucleophile will face difficulty to attack the back side of the α-carbon, thus elimination product will predominate (eqns. II and III, Scheme 5.10). Potassium *tert*-butoxide is a strong base and not a very good nucleophile due to its steric bulk, and is therefore, used for this purpose.
- When the reactant is a secondary halide, the elimination reaction is highly favored under S_N2/E2 conditions (strong base) *e.g.*, with the use of OH^- or OR^- (particularly hindered one) as shown (Scheme 5.11). Thus the elimination is the main reaction from 2-bromopropane.

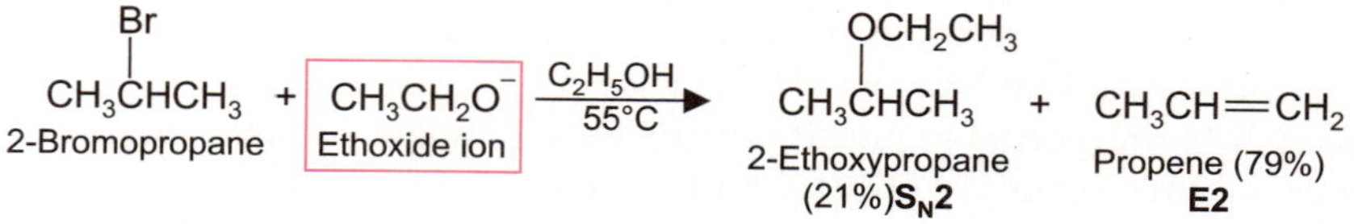

SCHEME 5.11

- Compared with the use of strongly basic ethoxide ion (Scheme 5.11), with weakly basic acetate ion only S_N2 substitution is observed (Scheme 5.12, acetic acid is a stronger acid *pKa* = 4.76 than ethanol (*pK*a = 15.9) to show that acetate ion is a weaker base than ethoxide ion).

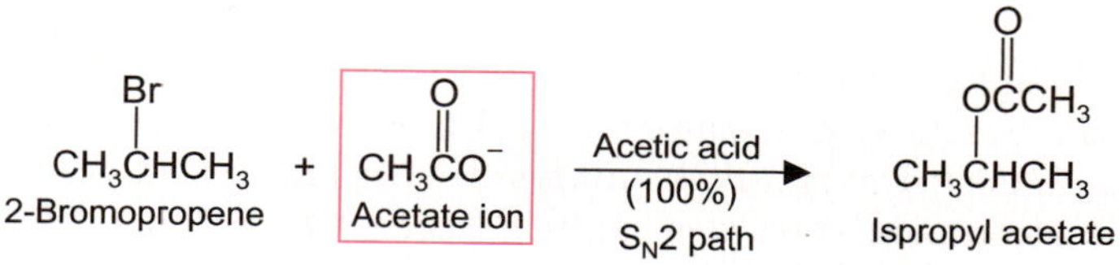

SCHEME 5.12

- With tertiary alkyl halides an S_N2 reaction cannot take place and therefore, elimination reaction is highly favoured (Scheme 5.13) under S_N2/E2 conditions (particularly at higher temperatures little substitution whatsoever may occur, is through S_N1 process).

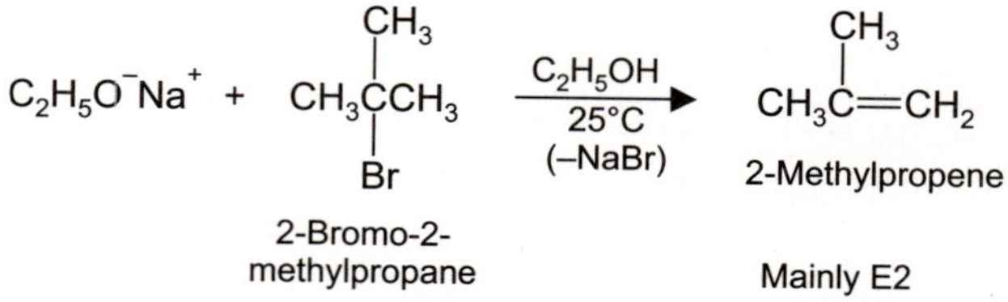

SCHEME 5.13

- An elimination reaction (E2 or E1) has higher free energy of activation than a substitution reaction (in elimination more bonds are broken and formed). By giving more energy (higher temperature) the higher energy barrier for elimination is surmounted.

(*ii*) Tertiray Halides – S_N1/E1 Conditions

- Both E1 and S_N1 reactions proceed through a common intermediate. E1 reactions are favored with substrates which can form stable carbocations *i.e.*, tertiary substrates and by the use of poor uncleophiles (weak bases). These are also generally favoured by the use of polar solvents. Increasing the temperature favors the reaction by the E1 mechanism.
- High yields of elimination product can be obtained by forcing E2 mechanism instead on a tertiary substrate *i.e.*, the use of a strong base (OH^- or OR^- see, Scheme 5.13).

(*iii*) Summary of the Products Expected in Substitution/Elimination Reactions

These are given in Table 5.1.

Table 5.1

RCH_2L *1°*	*R_2CHL* *2°*	*R_3CL* *3°*
Only S_N2/E2		*S_N1/E1 or E2*
Give mainly S_N2 products except with a hindered strong base [*e.g.*, $(CH_3)_3\,CH^-$] when mainly E2 pathway is followed.	Give mainly S_N2 products with weak bases [*e.g.*, I^-, CN^-, RCO_2^-) and mainly E2 with strong bases (*e.g.*, RO^-)	No S_N2 reactivity. In solvolysis gives S_N1/E1 at lower temperature S_N1 is favoured. When a strong base (*e.g.*, RO^-) is used E2 pathway predominates.

(b) Effect of Solvent

Primary alkyl halides undergo only E2 elimination reactions. E1 reactions are not possible because of the difficulty encountered in forming primary carbocations. *Secondary and tertiary* alkyl halides can undergo both E2 and E1 reactions. For these alkyl halides which can undergo both E2 and E1 reactions, the E2 reaction is favoured by the same factors that favour an S_N2 reaction and similarly the E1 reaction is again favoured by the same factors that favour an S_N1 reaction. *An E2 reaction is favoured by a high concentration of a strong base and an aprotic polar solvent* (DMSO, DMF). An E1 reaction is favoured by a weak base and a protic polar solvent (H_2O, ROH).

Further details are similar as already discussed (See, Schemes 3.18–3.21).

5.5 STEREOCHEMISTRY OF E2 ELIMINATION REACTIONS

In reactions like E2 elimination (it is a concerted reaction, the two substituents are eliminated in the same step) several bonds are made and broken simultaneously. This electron reorganization has strict requirements for the stereochemical relationship of these bonds (stereo-electronic requirement) as the reaction proceeds. These stereochemical factors occur since the bonds to the eliminated substituents (H and L) must be in the same plane since the sp^3 orbital of the carbon bonded to H and the sp^3 orbital

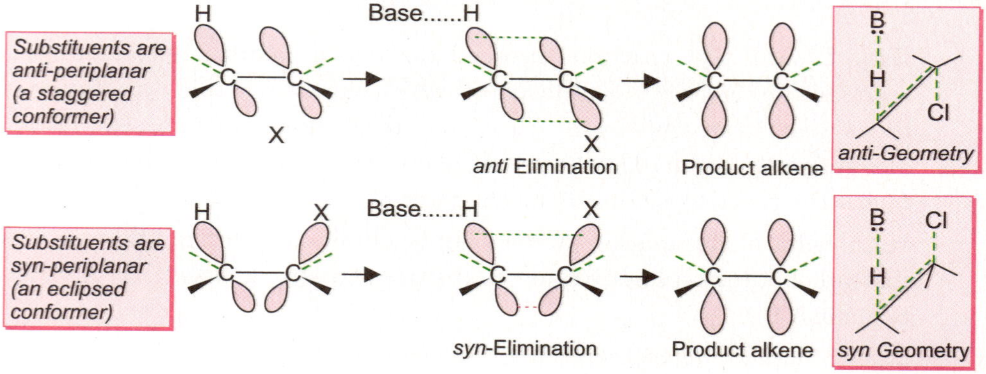

SCHEME 5.14

of the carbon bonded to leaving group L become overlapping *p* orbitals in the alkene. Thus the orbitals must overlap in the transition state (this overlap increases from the start of the reaction and provides significant stabilization in the transition state to help offset the energy cost of breaking other bonds). The overlap can only occur provided the orbitals are parallel (Scheme 5.14) and for to be parallel, these must lie in the same plane (*i.e.*, coplanar).

To be in the same plane the C—H and C—L bonds can either be parallel to one another on the same side of the molecule (*syn*-coplanar) or on the opposite side of the molecule (*anti*-coplanar). Thus when the eliminated groups are on the same side of the molecule, the reaction is termed *syn*-elimination and if these are eliminated from opposite sides, the reaction is called *anti*-elimination (Scheme 5.14). The following points may be noted:

- The *anti*-coplanar conformation is more stable (Scheme 5.14, the dihedral angle between H and L is 180°) since it is staggered when compared to *syn*-coplanar conformation in which all bonds are eclipsed.
- *Anti*-elimination is preferred in E2 reactions.
- *Syn*-elimination is however, less common but it does occur in substrates where structure is rigid and as a consequence, the hydrogen and leaving group are held *syn*-coplanar in an eclipsed or nearby eclipsed conformation (see, Scheme 5.26). Thus, E2 reactions require a dihedral angle of 180° (as in E2 *anti*-elimination) or perfect 0° (as in *syn* E2 elimination).
- Moreover, *anti*-elimination occurs more rapidly since the attacking electron rich base avoids the repulsion with electron-rich halogen on the same side of the molecule.
- The preference for *anti*-elimination has been proved always when the acyclic or cyclic substrate is able to achieve *anti*-periplanar geometry.

The following examples illustrate the above points. One must study the stereostructure of the substrate by considering the number of stereoisomers which may be imagined/drawn initially as Fischer projections and then translated into Newman or sawhorse projections or perspective drawings as already learnt to reach the results of elimination.

Example 1: ***E2-elimination from 1, 2-dibromo-1, 2-diphenylethane (sodium ethoxide-ethanol). Some E2 processes are stereospecific:***

- The compound has two chirality centers, since the group on each of the stereocenters are the same one expects only three stereoisomers, a *meso*-compound and a pair of enantiomers.
- It would be easy to first draw the *meso*-isomer (plane of symmetry) in the Fischer projection (Scheme 5.15). Redraw the Fischer projection as a Newman projection. Now rotate this projection to obtain the staggered conformer in which groups to be eliminated attain an *anti*-coplanar arrangement [there could be other staggered conformations as well which are not drawn here (see Scheme 5.19 for a related case) but only in the one shown the H and Br groups on the neighbouring carbons are in proper *anti*-coplanar arrangement].
- These two elimination reactions (Scheme 5.15) are stereospecific *i.e.*, *meso*-stereoisomer gives (*E*)-alkene having the phenyl groups *cis*, while the racemic dibromide give only (*Z*)-alkene.
- Thus elimination takes place most readily ($^{-}OCH_3/CH_3OH$) when H and Br are in *anti*-periplanar arrangement.
- In this example the phenyl group plays no role other than that of a stereochemical marker. The *meso*- and enantiomeric stereoisomers are properly namely by assigning *R* and *S* descriptors.

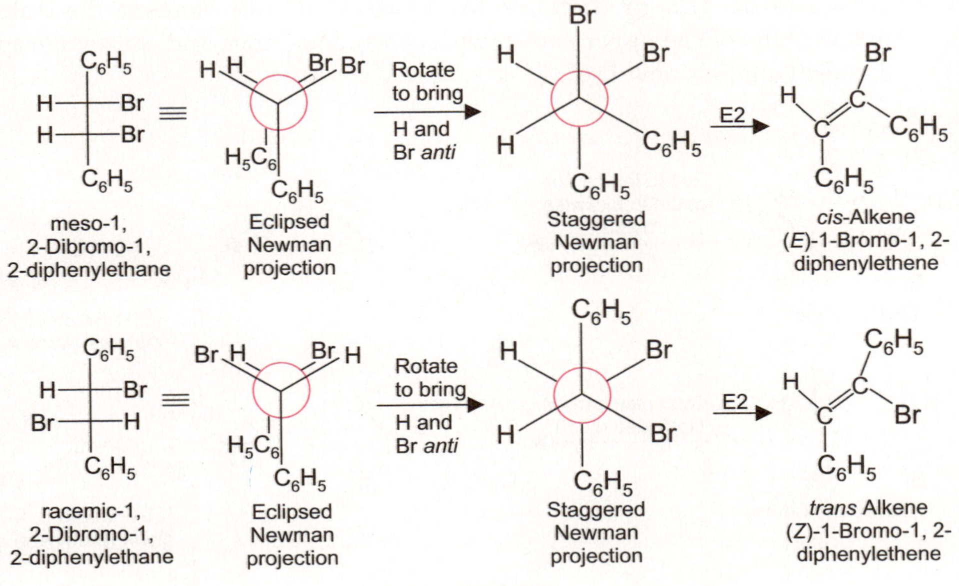

SCHEME 5.15

- The same result can be seen by drawing sawhorse projections (Scheme 5.16 in the case of racemic mixture the eclipsed forms are not drawn).

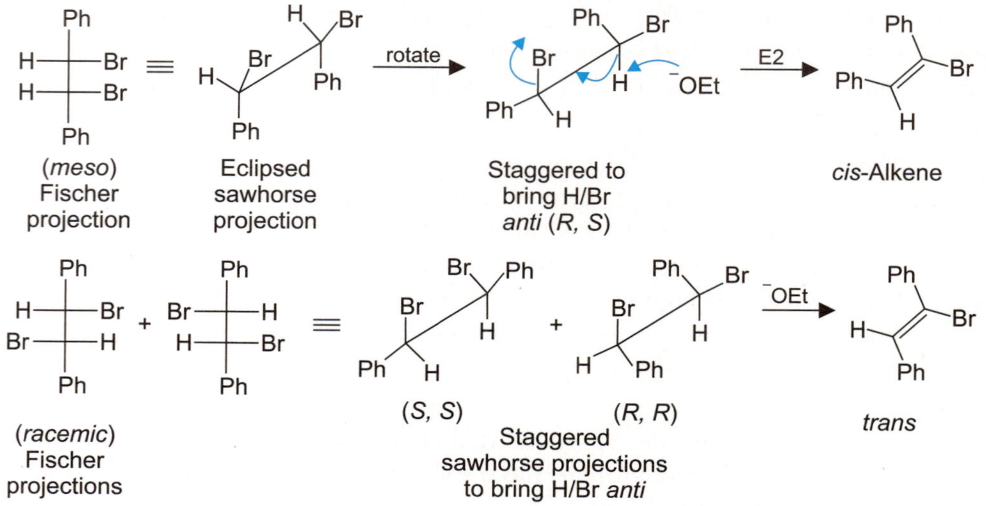

SCHEME 5.16

- Alternatively one may also redraw the Fischer projection (Scheme 5.15) on a wedge and dash structure, or start afresh by first drawing the *meso*-isomer on a wedge and dash structure to show the plane of symmetry (Scheme 5.17). The left carbon has to be rotated (only 60) to bring H and Br into *anti*-coplanar arrangement. This rotation now brings H—C—C—Br bonds of the molecule in the plane of the paper shown with continuous lines and consequently the dihedral angle is now 180°.

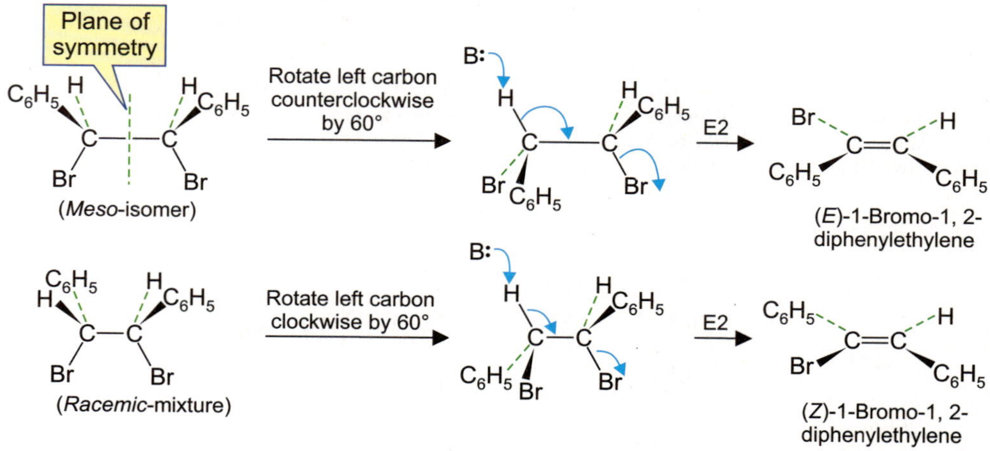

SCHEME 5.17

Example 2: ***E2-elimination from 1-chloro-1, 2-diphenylpropane***

- This compound has two chirality centers, thus it would exist in four stereoisomeric forms (as pairs of enantiomers).

- While drawing the Fischer projection for these stereoisomers *e.g.*, for *erythro*-diastereomer (Scheme 5.18), the carbon chain is to be kept vertical with the most oxidized carbon (CHCl) at the top (Scheme 5.18).

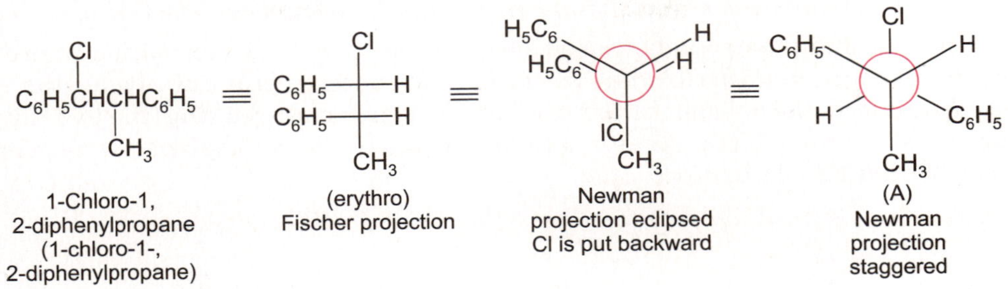

SCHEME 5.18

- The *erythro*-diastereomer is translated on to a Newman projection and staggered to give a conformer (*i.e.* rotamer A, Scheme 5.18) which however, is not suitable for *anti*-elimination, since H and Cl are not antiperiplanar.
- In this rotamer, the front carbon is rotated further appropriately so as to bring H and C1 in *anti*-periplanar relationship (I, Scheme 5.19), this conformation undergoes E2 elimination (NaOEt/EtOH) to give *Z*-alkene. Thus, out of three possible staggered rotamers of *erythro*-diastereomer, only rotamer (I, Scheme 5.19) is suitable for *anti*-elimination and that too to give *Z*-alkene.
- Similarly in the *threo*-diastereomer the rotamer (II) gives *E*-alkene.

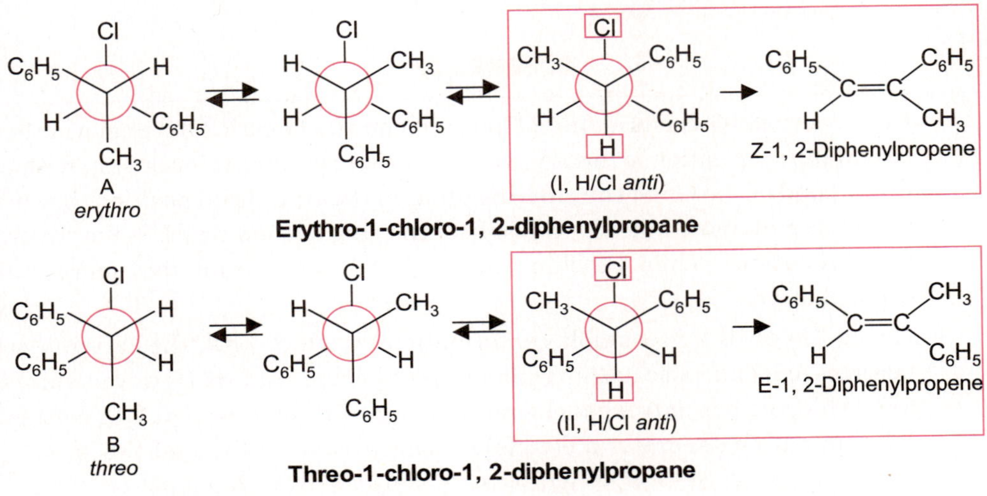

SCHEME 5.19

- The elimination reaction (E2) of each diastereomer proceeds in a stereospecific *anti*-manner *i.e.*, when the H atom and Cl are in the *anti*-periplanar arrangement.

- Recall that *erythro* diastereomer A can be converted into *threo*-diastereomer B by interchanging groups around one of the chiralitycenters *e.g.*, H and C_6H_5 on the rear carbon.

Example 3: *E2-elimination in cyclohexane rings—Conformation and reactivity.*

In open chain compounds a molecule can easily adopt a conformation where the leaving groups are *anti*-periplanar. In cyclic compounds, this is not always the case which thus effects the outcome of the reaction. In a six membered ring, the two leaving groups must be *trans*-diaxial so as to assume the requisite *anti*-coplanar arrangement prior concerted E2 elimination.

(i) E2-Elimination from Trans-1-Chloro-2-Methycyclohexane

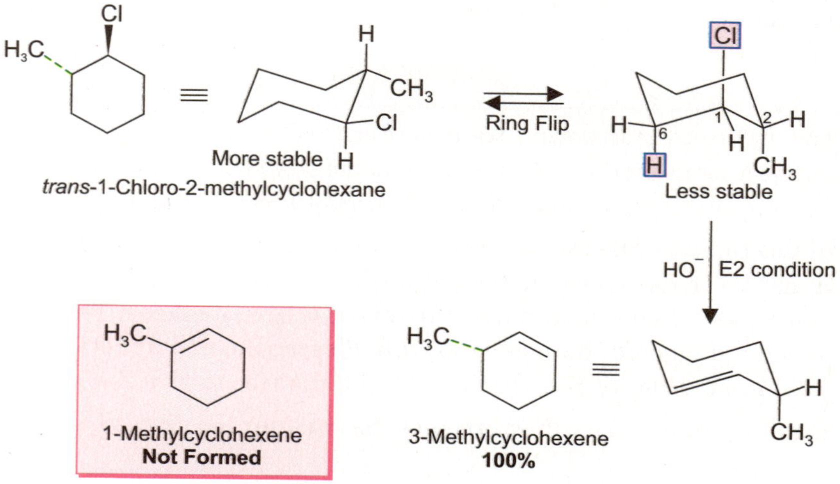

SCHEME 5.20

- In the more stable conformation of the compound (Scheme 5.20) the chlorine being in equatorial position, cannot have any *anti*-periplanar relationship with any C—H bond of the ring (The chlorine atom in the equatorial position, however, is *anti*-periplanar to the C—C bonds of the ring system which is not involved in elimination). Thus E2-elimination cannot occur from this more stable conformation.
- A ring flip gives a less stable conformation, in which now, the chlorine atom becomes axial and also *anti*-periplanar to a hydrogen at C-6. However, since the axial chlorine has only a gauche relationship with hydrogen at C-2, concerted E2 elimination generates a double bond only between C-1 and C-6.

(ii) E2-Elimination from Menthyl and Neomenthyl Chloride

Menthyl and neomenthyl chlorides (Schemes 5.22 and 5.23) are substituted cyclohexyl chlorides and are diastereomers, which differ only in the configuration of the chlorine substituent. These two compounds display dramatic behaviour when subjected to E2 conditions (use of strong base, ethoxide ion).

- In menthyl chloride all the substituents are equatorial. When it is treated with sodium ethoxide in ethanol, elimination occurs very slowly to produce the alkene 2-menthene as the only product (Scheme 5.21).

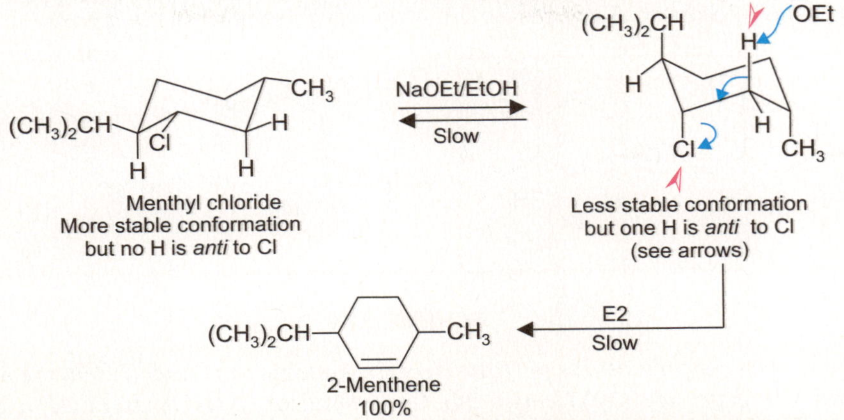

Slow E2 elimination but from less stable conformation of menthyl chloride. Note that in cyclic systems, there is a clash between Saytzeff's rule for regioselectivity and preference for *anti* elimination in E2 reactions. *anti*- elimination usually predominates.

SCHEME 5.21

- When, however, neomenthyl chloride is reacted under similar conditions it reacts rapidly to give a mixture of two alkenes, 2- and 3-menthenes (Scheme 5.22).
- The elimination must occur from a conformation where hydrogen and chlorine are *anti*-coplanar, thus in neomenthyl chloride the chlorine substituent must be axial (Scheme 5.22). Loss of either of two different *anti*-periplanar β-hydrogen atoms gives the mixture of alkenes (Scheme 5.22).

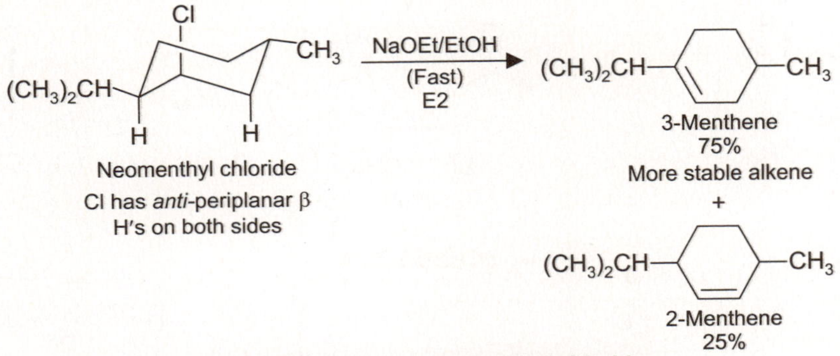

Fast elimination from neomenthyl chloride

SCHEME 5.22

- In menthyl chloride, the chlorine atom must be equatorial. The elimination, therefore, occurs from a less favourable conformation (see, Scheme 5.21) obtained by ring flip. At any one time only a small percentage of the molecules have this conformation. A low population of this reactive conformation leads to a slow elimination reaction.

EXERCISE 5.2

On attempted E2 reaction, (I) reacted much faster than (II) while (III) reacted much faster than (IV, Scheme 5.23). Write the product of reaction in each case.

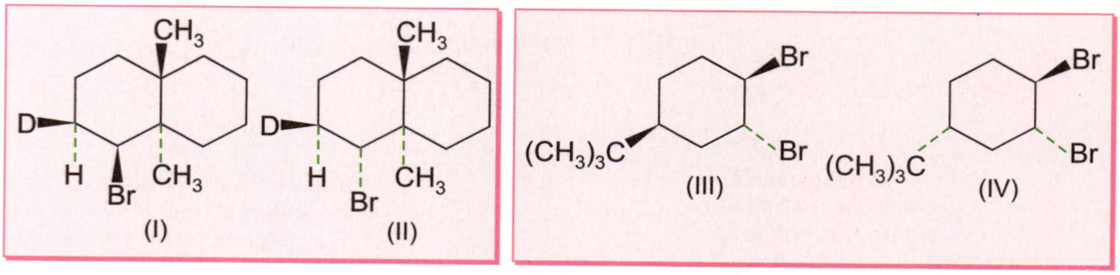

SCHEME 5.23

ANSWER. *The conformations of trans-decalin systems in (I and II Scheme 5.23) are in (V and VI) respectively (Scheme 5.24). Conformation of (III) is in (VIII) which can undergo E2 reaction while conformation of (IV) depicted in (IX) cannot do E2 unless it adopts a high energy boat conformation which could give the same product as from (VIII) via a slow E2 reaction. A ring flip in (IX, Scheme 5.25) is highly unlikely since, bulky t-butyl group is most stable in equatorial orientation.*

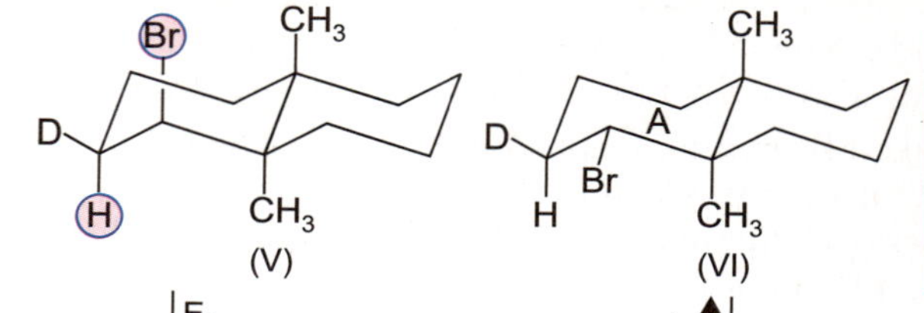

In (VI) ring flip is not possible (trans fusion). Attainment of a boat conformation in ring (A) brings D/Br anticoplanar for E2 elimination. Since this attainment of boat form requires energy, elimination in (VI) is very slow.

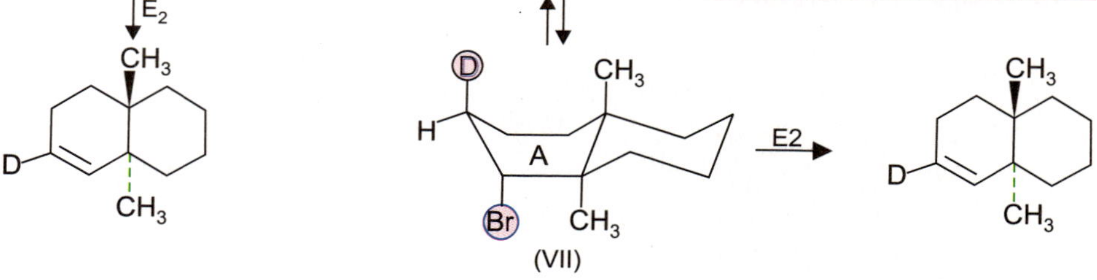

SCHEME 5.24

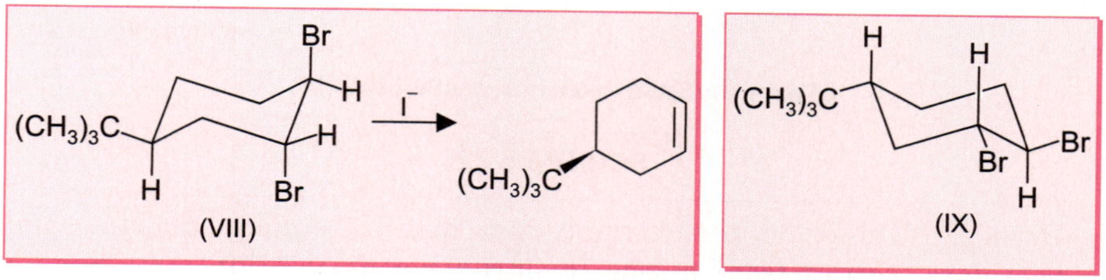

SCHEME 5.25

5.6 E2-*syn*-ELIMINATION

Some rigid molecules are not able to attain the favourable *anti*-periplanar conformation of the departing groups. Elimination is, thus, inhibited and may proceed by another mechanistic or stereochemical pathway.

It has been shown that arrangements in which the departing groups are coplanar on the same side of the molecule may undergo concerted elimination. The groups are termed to be *syn*-periplanar, and the process represents a *syn*-elimination. The deuterated norbornyl system (Scheme 5.26), gives almost 100% of the product with no deuterium. In this case the *exo* tasylate group cannot achieve *anti*-periplanarity *i.e.*, a dihedral angle of 180° with an *endo* β-hydrogen due to the rigid structure of the molecule, the dihedral angle being only 120°. In this situation, therefore, the leaving groups prefer to undergo a *syn*-elimination with a dihedral angle around 0° (*syn*-periplanarity) compared to *anti*-elimination when the angle available is only around 120° (Scheme 5.26).

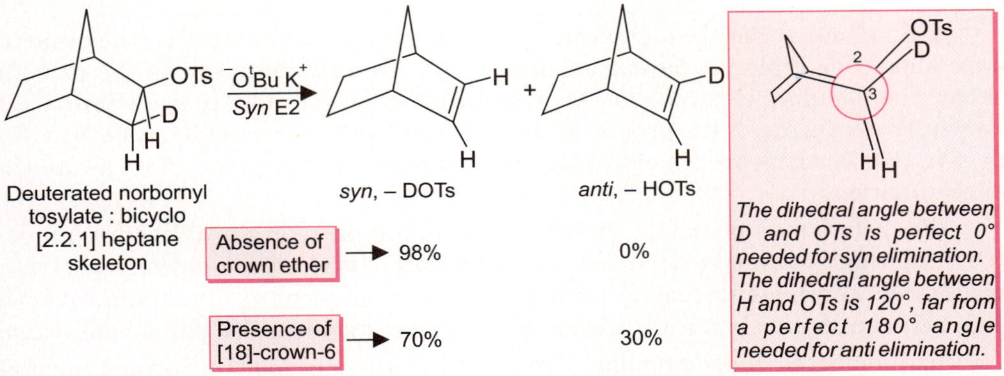

***syn*-Elimination at a dihedral angle of 0° is better than *anti*-elimination at 120°**

SCHEME 5.26

When suitable crown ethers are added to the systems where *syn*-elimination is exclusive or predominant, it is observed that substantial elimination then occurs via *anti*-pathway. A crown ether [18]-crown-6 can effectively bind K^+ ion which is then no longer available for ion pairing thus cutting *syn* mechanism.

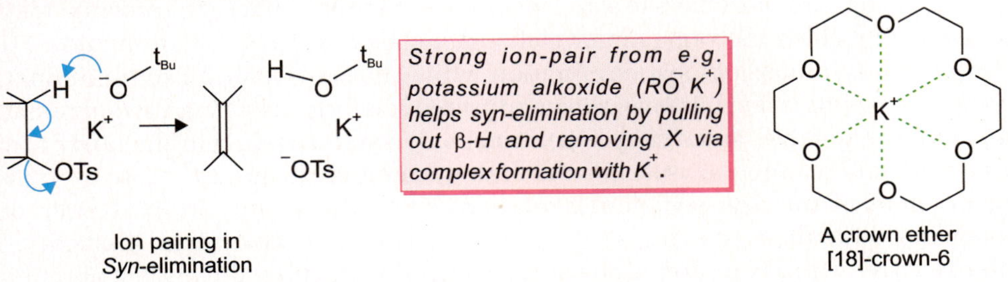

SCHEME 5.27

Syn-elimination is also observed in the Hofmann elimination of *N, N, N*-trimethylcyclobutylammonium hydroxide. Elimination of the deuterated analog shows that 90% of the elimination occurs by a *syn* pathway. A planar cyclobutane ring has eclipsed bonds. Although the actual geometry is slightly nonplanar to relieve some torsional strain, it is easy for the cyclobutane ring to attain the eclipsed geometry needed for *syn* elimination than to attain a nonplanar geometry to have the nitrogen and a hydrogen *anti*-planar (Scheme 5.28).

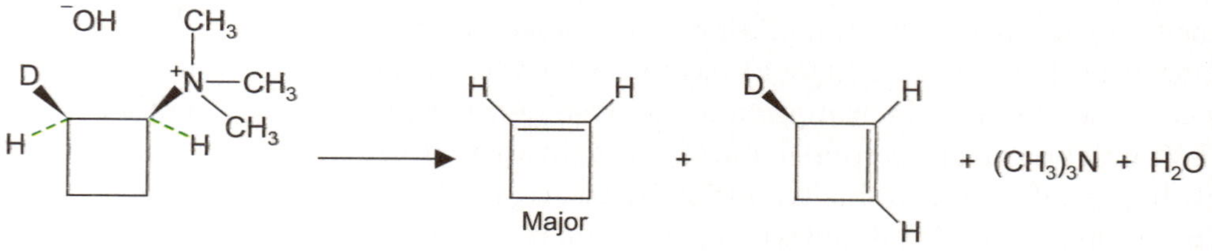

SCHEME 5.28

In E2-eliminations the significant condition for the reactions is that the departing groups should be coplanar no matter *anti* or *syn*. In *anti*-eliminations the incoming and outgoing bonding electrons are kept as far apart as possible. In *syn*-eliminations, however, the eclipsing of the groups in the reaction leads to a less favorable situation, thus, *syn*-eliminations are considerably less favorable, occuring at a slower rate than *anti*-eliminations.

Two points may be noted. Firstly *anti*-elimination requires a dihedral angle of 180°. If this angle cannot be achieved, *anti*-elimination is slowed or entirely prevented. Secondly, for the simple systems *syn*-elimination does not occur to any significant extent unless *anti*-elimination is greatly prevented by failure to achieve the 180° dihedral angle.

Six-membered rings singularly, from among rings of four to thirteen members present systems where strain-free *anti*-periplanar conformations can be achieved. It is, therefore, not surprising that *syn*-elimination is rare in six-membered rings. Cycloalkyltrimethylamonium hydroxides on elimination give the following precentages of *syn*-elimination products with ring sizes: four-membered 90%; five-membered 46%; Six-membered 4%; Seven-membered 31 to 37%. It may be noted that the $^{+}NMe_3$ group has a greater tendency for *syn*-elimination than other common leaving-groups such as OTs, Cl and Br.

The *cis*-and *trans*-isomers of a cyclopentane derivative 2-phenylcyclopentyl tosylate (Scheme 5.29) yielded the same major olefin (A, Scheme 5.29) by E2-elimination, thus, the *trans* isomer undergoes *syn*-elimination while the *cis*-isomer an *anti*-elimination. Moreover *anti*-elimination has been shown to proceed slightly faster than *syn*-elimination. This is understandable, since in cyclopentane, only small twisting of the fairly flexible ring is needed to attain a *syn*-periplanar arrangement of the groups *cis* to each other. As attainment of the *anti*-periplanar relationship of the groups *trans* to each other is somewhat difficult as the ring shall have to be bent well away from planarity. The greater reactivity of 2-H proton is one of the controlling factors, during the elimination of *trans* isomer (*syn*-elimination). The addition of crown ether (see, Scheme 5.27) reduces the percentage of (A, Scheme 5.29) by around 30. This again confirms the necessity of ion-pair formation during *syn*-elimination as already explained (see, Scheme 5.27).

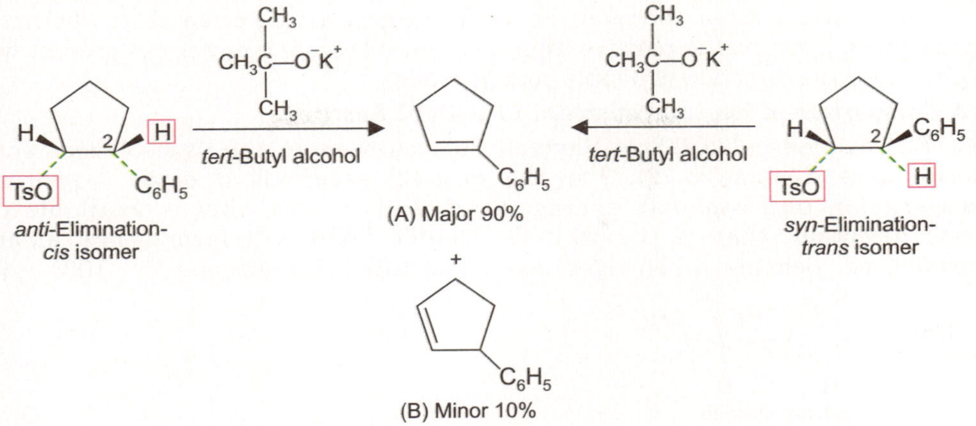

SCHEME 5.29

5.7 ORIENTATION OF THE DOUBLE BOND—REGIOCHEMISTRY OF THE ELIMINATION REACTIONS

While discussing the stereochemistry of substitution reactions, the concern has been stereogenic carbon bearing the leaving group. During elimination one is concerned with the regiochemistry *i.e.*, the production of structural isomers which have the double bond in different positions.

(a) Regiochemistry of E1 Reaction

When regioisomers are possible during E1 reaction the product distribution follows the Saytzeff rule—the double bond goes mainly toward the most highly substituted carbon. Put differently the major product is the one with more alkyl groups on the carbons of the double bond. This is so since in E1 mechanism the leaving group is gone to give a carbocation before the choice is made as to which direction the generated double bond will go. Thus the direction is determined almost exclusively by the relative stabilities of the olefins produced via elimination. From the heat of combustion data, it is known that olefin stability increases with alkyl substitution. Thus E1 reactions occur under thermodynamic control. The more stable product is formed preferentially. The eliminations which occur under thermodynamic control are said to give the Saytzeff product (Scheme 5.30).

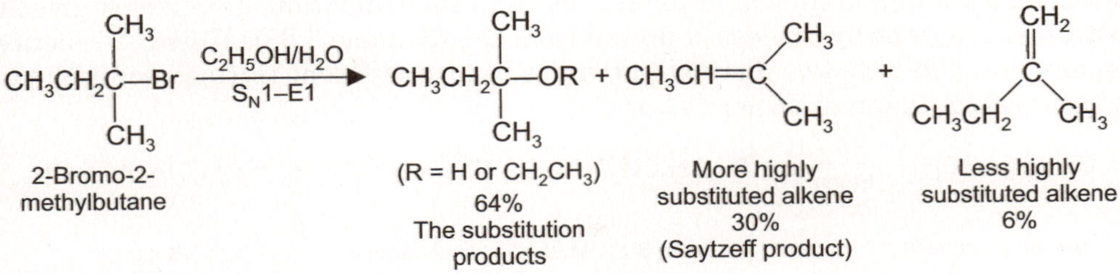

SCHEME 5.30

For E1 elimination, therefore, Saytzeff rule governs the orientation whether the leaving group is neutral or positive (this is so, since leaving group is not present when the choice for the direction of double bond is made).

(b) A Comparison of Regiochemistry of E1 with E2 Reaction

Most E2 reactions also follow Saytzeff rule as seen in the case of neomenthyl chloride (see Scheme 5.22). For the *anti*-E2 reactions a *trans* β proton is necessary (an *anti*-coplanar arrangement) and in case this is available only in one direction, that is the way the double bond will form. Thus menthyl chloride (see, Scheme 5.21) gives less substituted 2-menthene in 100% yield.

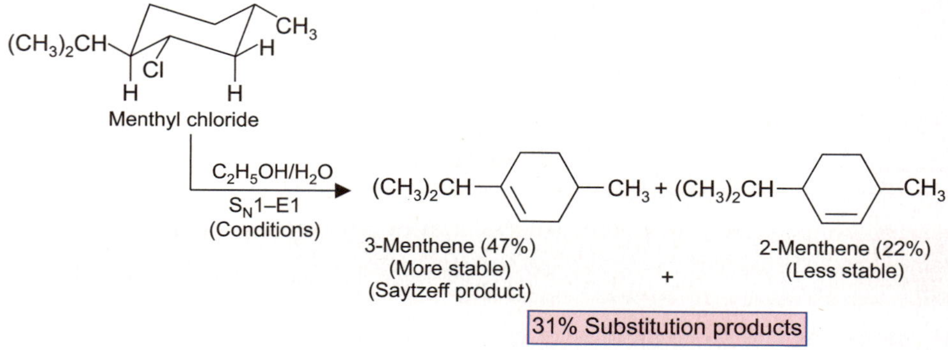

SCHEME 5.31

Interestingly in E1 reactions only a carbocation is formed and only at this point a C—H bond must be aligned parallel to the empty *p* orbital of the carbocation for the *pi* overlap to occur as the C—H bond beings to break. Thus *anti*-periplanar requirement is no longer required for E1 reactions in cyclohexane rings and under E1 conditions (Scheme 5.31) menthyl chloride gave Saytzeff product as the major alkene along with the less substituted alkene.

(c) E2 Reactions are both Regioselective and Stereoselective (Orientation of Double bond)

E2 reactions are both regioselective (more of one constitutional isomer is formed) and stereo selective as well since *e.g.*, during the elimination from *sec*-butyl bromide 2-butene contains more of (*E*)-2-butene than (*Z*)-2-butene. Unfortunately from synthetic viewpoint complete *trans* selectivity is rare in E2 reactions.

The regioselectivity of the E2 reaction is determined by the transition state which is stabilized by the same factors which stabilize the alkene (Scheme 5.33). The transition state for the formation of 2-butene (more substituted) by the loss of H from C-3 resembles it and is thus of lower energy than the transition state which gives less substituted 1-butene by the loss of proton from C-1 (Scheme 5.33). Thus E-2 reaction of 2-bromobutane follows two distinct pathways to give different regioisomers. The base can approach a proton at either C-1 or C-3.

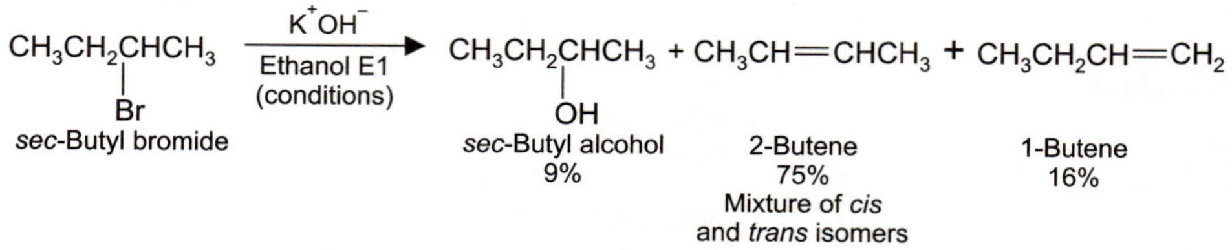

SCHEME 5.32

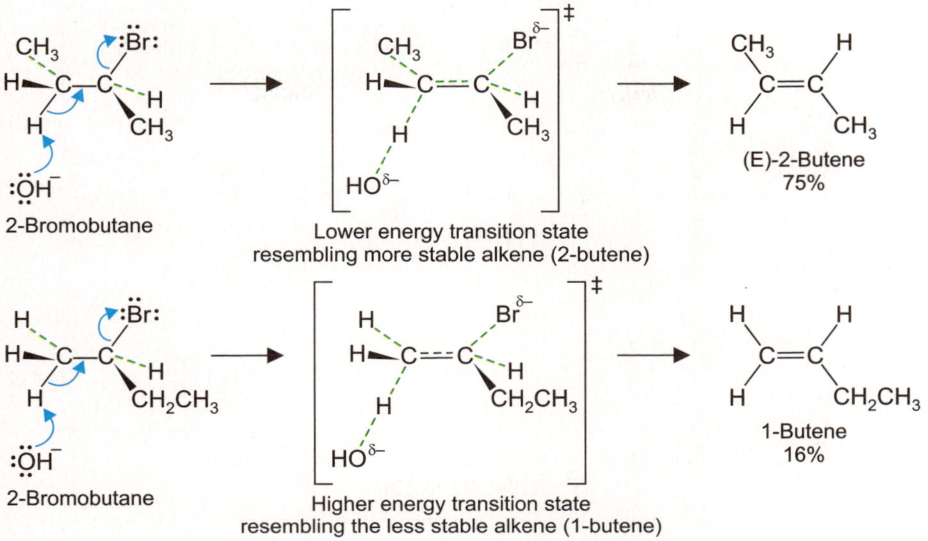

SCHEME 5.33

Saytzeff orientation predominates in E2 eliminations when the eliminated group is a halide ion like chlorine, bromine or iodine.

As another example of the involvement. of *anti* stereochemistry as well as the predominance of Saytzeff product during E2-elimination consider the alkene formation from stereoisomeric alkyl halides (I and II, Scheme 5.33a). The compound (I) leads to the exclusive formation of exo-olefin *via anti* pathway. In the case of (II) although there are two β-hydrogens which could be *anti* to the Br, but the one which leads to more substituted olefin is removed.

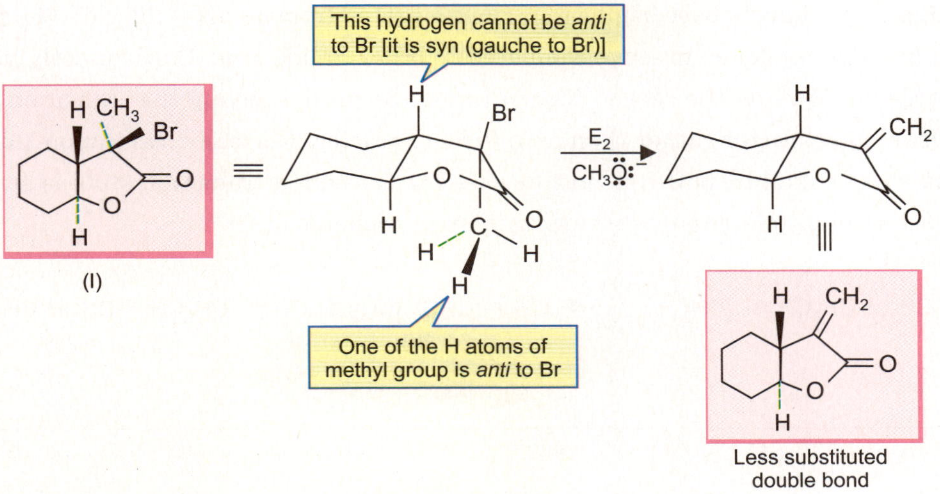

Continued...

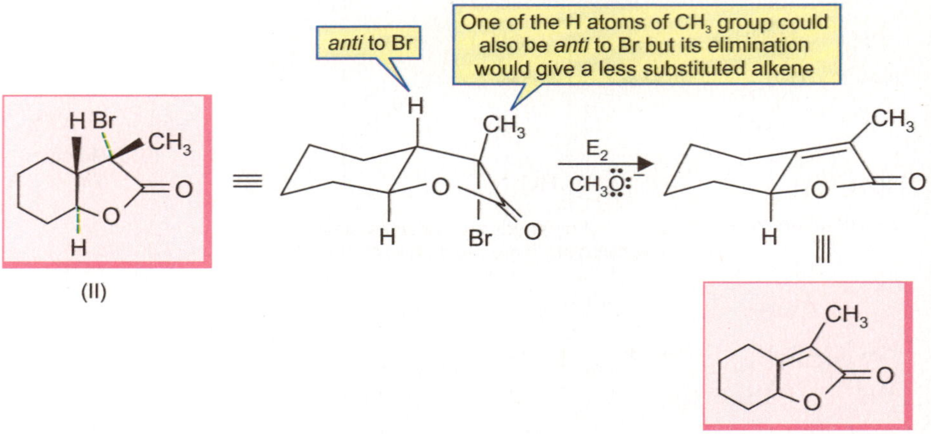

SCHEME 5.33a

Thus compared to the major product of E2 dehydrohalogenation of alkyl chlorides, alkyl bromides, and alkyl iodides (normally the most substituted alkene) the major product of the E2 dehydrohalogenation of alkyl fluorides is however, the least substituted alkene (Scheme 5.34). When a hydrogen and a chlorine, bromine, or iodine are eliminated from an alkyl halide, the halogen starts to leave as soon as the base begins to remove the proton (Scheme 5.33). Loss of the halogen and its bonding electrons prevents the buildup of negative charge on the carbon that is losing a proton, giving the transition state alkene character rather (see II, Scheme 5.10) than carbanion character. Of the halogen ions, the fluoride ion is the strongest base and therefore the poorest leaving group. Thus when a base begins to remove a proton from an alkyl fluoride, the fluoride ion has less tendency to leave compared to other halide ions. Consequently negative charge develops on the carbon that is losing the proton, giving the transition state a carbanion character rather than an alkene character. Normally, carbanion transition states are unstable, however, in this case the carbanion transition state is stabilized by the strongly electron-withdrawing fluorine (Scheme 5.35).

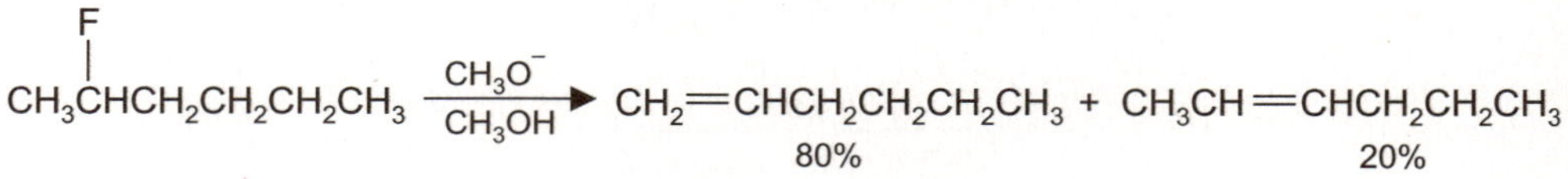

SCHEME 5.34

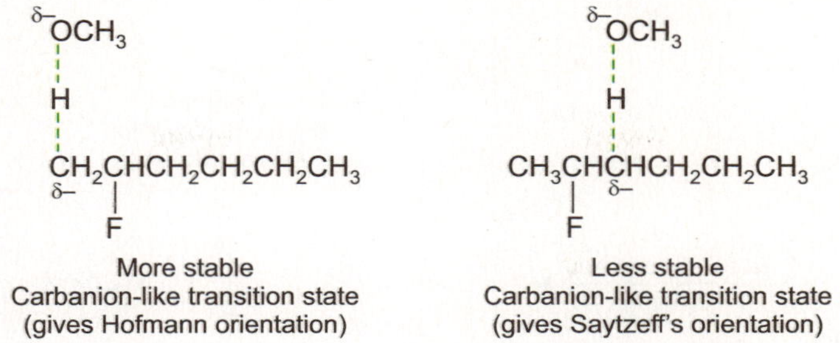

SCHEME 5.35

Positively charged carbocations are stabilized by electron-donating groups, while negatively charged carbanions are stabilized by *the fewer alkyl groups* that are bonded to a negatively charged carbon. In summary the tansition state for Hofmann elimination has a partial negative charge on a primary carbon (more stable) whereas the transition state for Saytzeff's elimination has a negative charge on a secondary carbon (less stable).

For similar reasons elimination from quaternary ammonium ions (Hofmann elimination), normally gives the less substituted alkene as the major product (Scheme 5.35*a*) due to the relatively poor leaving group property of an amine.

As seen above, exceptions to the Saytzeff rule occur during elimination from alkyl fluorides and quaternary ammonium ions (also during elimination from other charged leaving groups *e.g.*, SMe_2^+). Another exception is the use of a sterically hindered base (Scheme 5.35*b*).

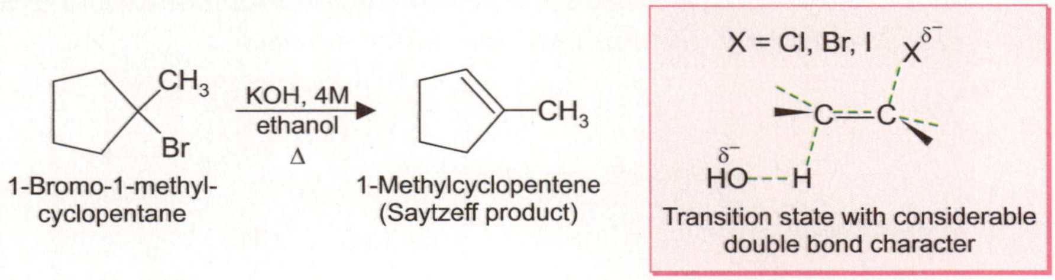

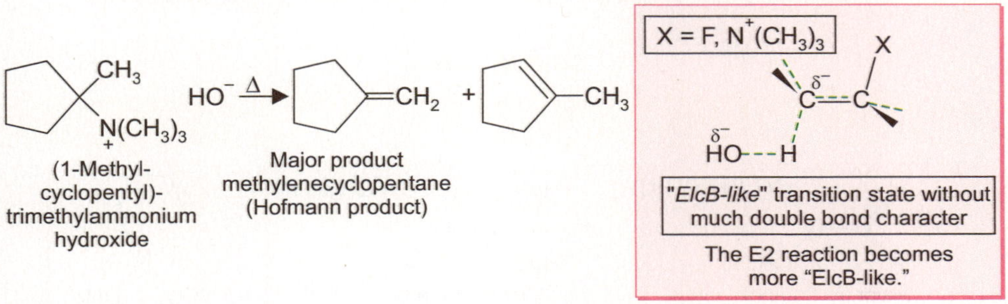

SCHEME 5.35a

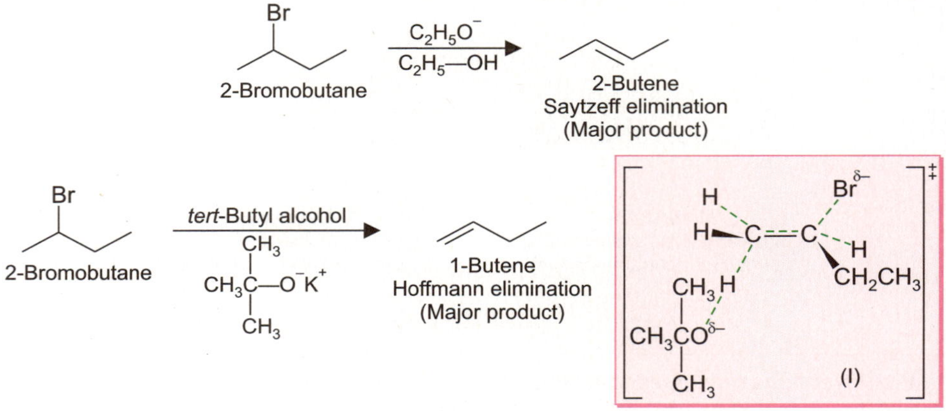

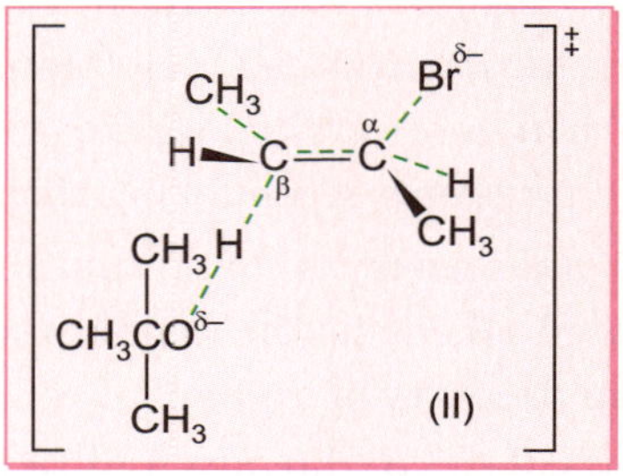

Formation of Saytzeff product (2-butene) from 2-bromobutane with a strong but sterically hindered base is not favored due to the involvement of a comparatively crowded transition state (II).

SCHEME 5.35b

EXERCISE 5.3

Write the major product from the eliminations (Scheme 5.35c).

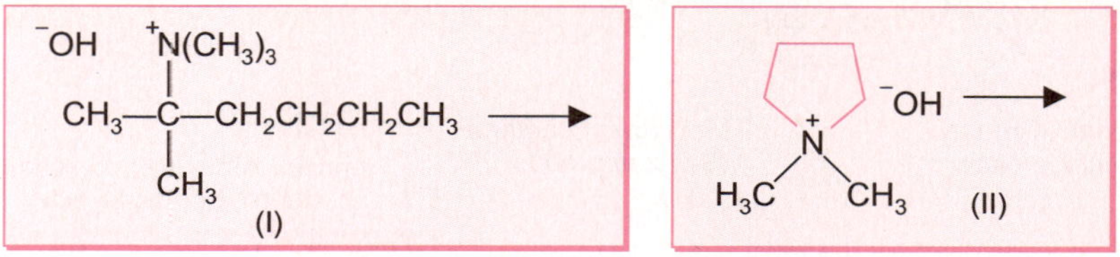

SCHEME 5.35c

ANSWER

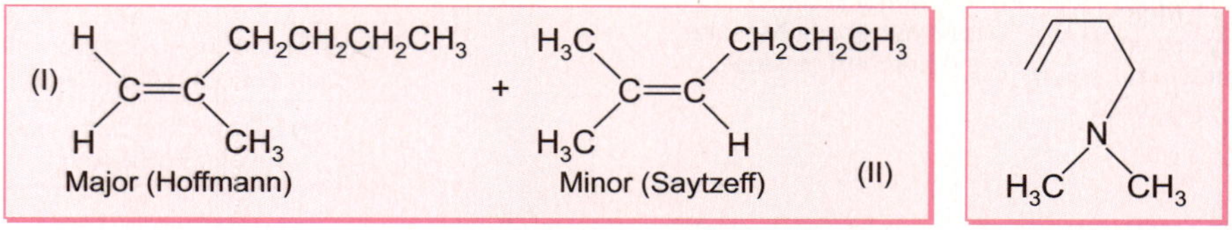

SCHEME 5.35d

Saytzeff and Hoffmann Eliminations
A Summary

Exceptions to Saytzeff rule occur under two situations. When the proton to be removed is in the sterically more hindered position or the use of a base whose basic center is also sterically hindered. These situations lead to the predominance of the less substituted alkene. When a transition state is more "Carbanion-like" but less alkene-like", then the alkyl substituents destabilize an adjacent negative charge. The proton is abstracted from a carbon which can accommodate the negative charge the best to give the less substituted alkene. In these cases the leaving group is poor [e.g., F or $N(CH_3)_3$] and thus C—H bond is relatively more fully broken.

In a Saytzeff elimination the transition state appears to have alkene-like character. A transition state may be "Carbocation–like" when the C—X bond is more broken in the transition state which leads to a carbocation and E1 elimination.

Another factor which affects the regiochemistry of E2 reactions is when the new double bond can be in conjugation with already present double bond (C═C or C═O) or an aromatic ring. No matter what the mechanism the conjugated product usually predominates (Scheme 5.36).

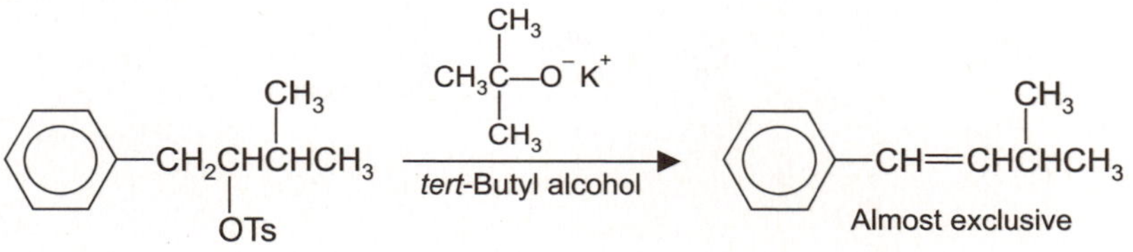

SCHEME 5.36

BOX 5.2

BIOLOGICAL E2 ELIMINATION REACTIONS

E2 reaction is very important in biological systems which occur during metabolism and the enzyme active site satisfy all the criteria for reactivity.

- *The substrate will have a good leaving group or a group readily converted to a leaving group.*
- *A base will be present to remove a proton from the carbon atom β to the leaving group.*
- *The substrate molecule will be bound at an enzyme active site in the orientation that favors anti elimination.*

Several amino acids in the protein backbone of an enzyme have side chains that are basic. The nitrogen atom in the imidazole ring of histidine (Scheme B 5.2) often functions as a base in living systems. The imidazole ring in the side chain of histidine is strong enough to remove a proton from water at pH 7, although the equilibrium lies to the left for this simple acid-base reaction.

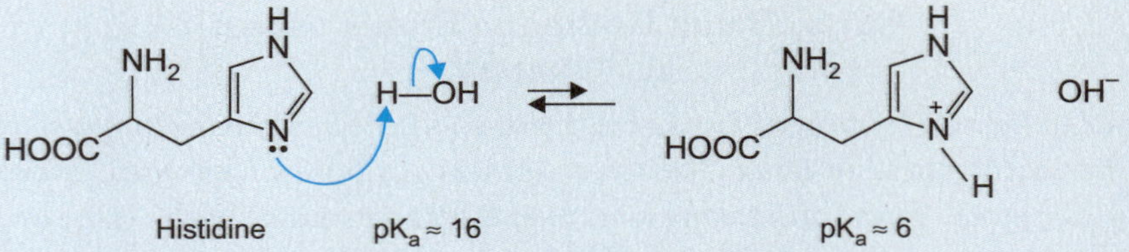

Scheme B 5.2

The enzyme can influence the properties of the leaving group in the substrate. Some of the substrates may have a good leaving group already—an ammonium ion and a phosphate ion. An amino acid with an acidic side chain, for example, aspartic acid may facilitate the loss of OH group as water.

Finally, the proper *anti* conformation of proton and leaving groups is ensured since a protein chain folds into a specific three-dimensional shape to create an enzyme active site.

These requirements as displayed during E2-elimination from the *citraconate hydrase* catalysed reaction are in (Scheme B 5.3)

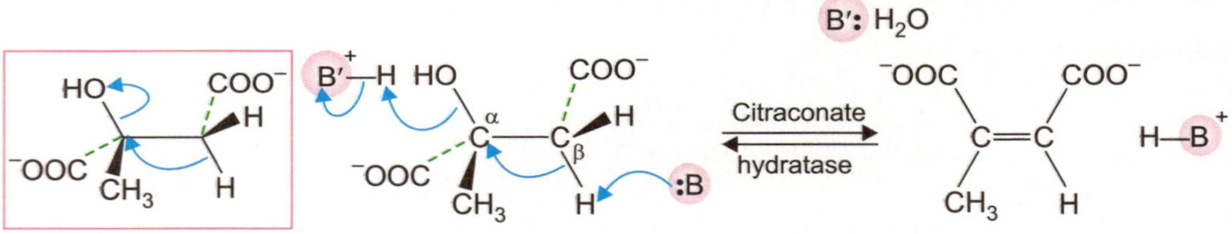

Scheme B 5.3

EXERCISE 5.3

The cis and trans isomers of 4-chlorocyclohexanol on reaction with OH⁻ give the products shown (Scheme 5.37). Explain.

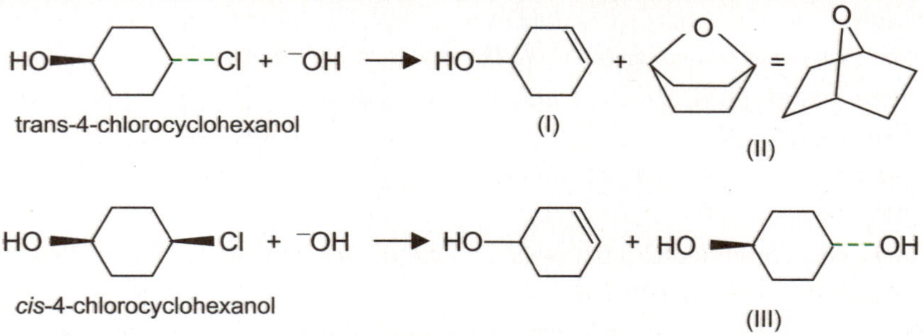

SCHEME 5.37

ANSWER

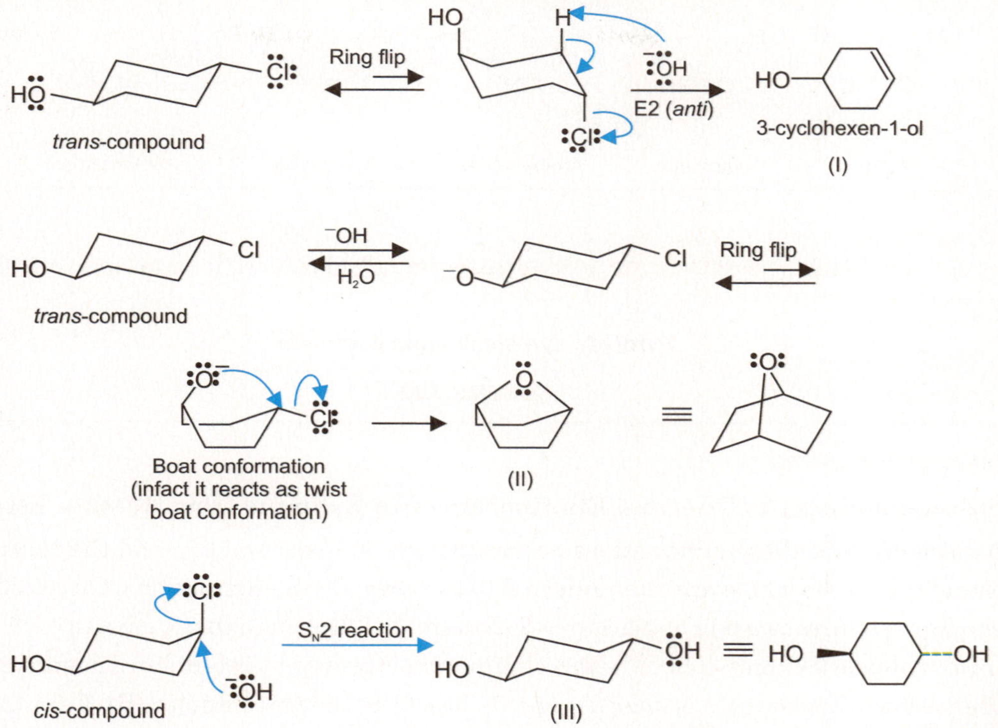

The *cis* compound could also undergo
E2 elimination to give (I).

SCHEME 5.37a

5.8 PYROLYTIC *syn* ELIMINATIONS (*Ei*-ELIMINATION INTERNAL)

These thermal eliminations occur within a small family of compounds, like acetate esters, methyl xanthate esters, tertiary amine oxides, sulphoxides and selenoxides which contain atleast one β-hydrogen atom with the formation of olefins. These eliminations have a common mechanistic feature: a concerted reaction *via* a cyclic transition state (Scheme 5.38) within which an intramolecular proton transfer is accompanied by *syn*-elimination to form a new carbon-carbon double bond. If more than one β-hydrogen is present, then mixtures of alkenes are generally formed. Since these reactions involve cyclic transition states, conformational effects play an important role in determining the composition of the alkene product.

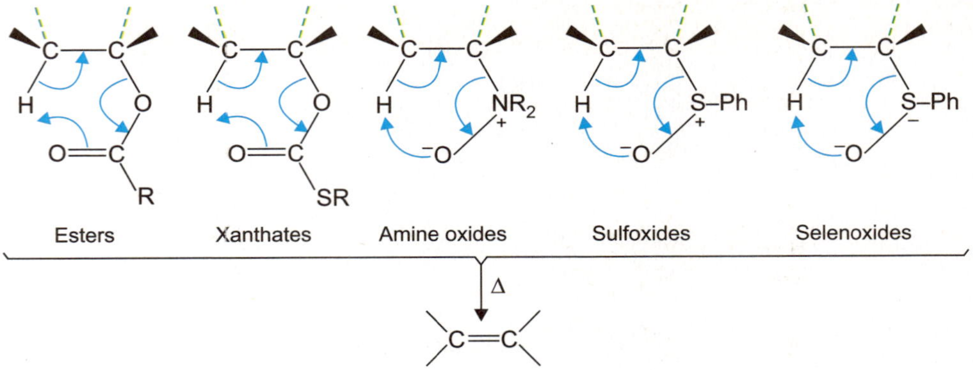

Pyrolytic *syn*-elimination reactions

SCHEME 5.38

(i) Acyclic Systems

Stereochemistry of Thermal Eliminations (An Example from Acetate Esters)

As already said these eliminations occur with *syn* stereochemistry and the reactions therefore, on the whole are often referred to as thermal *syn* eliminations. On pyrolysis, *erythro* and *threo* isomers of 1-acetoxy-2-deutero-1, 2-diphenylethane gave in each case *trans*-stilbene (Scheme 5.39). The *syn* elimination is proved by retention of deuterium in the product from *erythro* isomer (A) and its loss from the *threo* isomer (B). For this *syn* process either the hydrogen or the deuterium could be *syn* placed to the acetoxy group. However, the preferred conformations have the phenyl groups as far apart as possible. The molecule has two stereocenters and therefore, four stereoisomers are possible *i.e.*, a pair of enantiomers in each *erythro* and *threo* isomer. In eclipsed conformations (I and II, Scheme 5.39) Ph/Ph θ = 0°. The elimination, thus takes place from conformations with Ph/Ph far removed. In the transition state from A, a H is *syn* placed with the acetoxy group. In B now it is D which is *syn* placed with it.

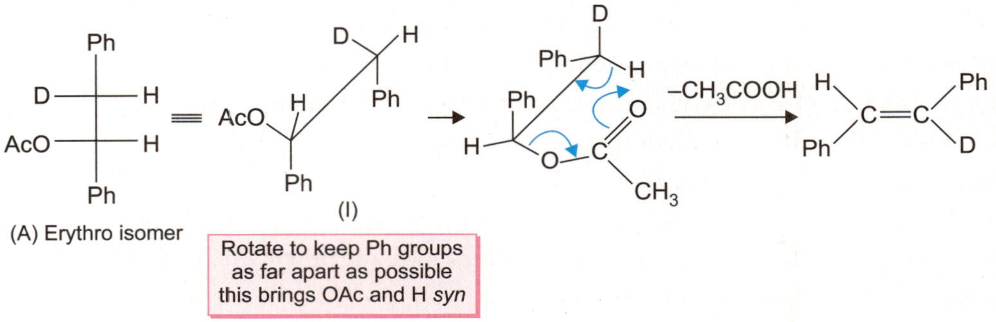

Continued...

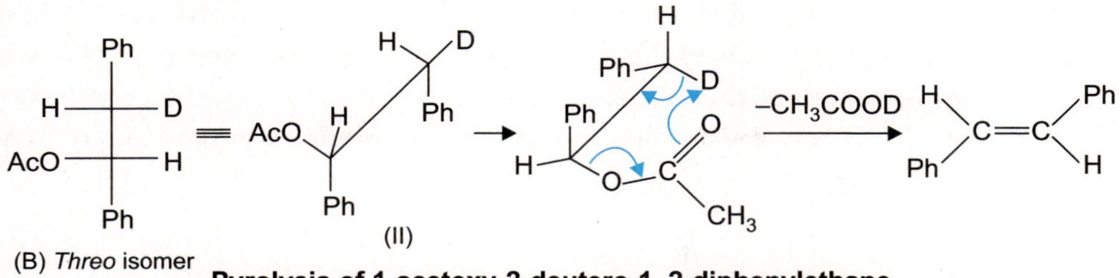

Pyrolysis of 1-acetoxy-2-deutero-1, 2-diphenylethane

SCHEME 5.39

If two or more *syn* periplanar arrangements are available, the one with least crowding is preferred, generally leading to the product in which the bulky groups are *trans*. Thus, pyrolysis of 2-butyl acetate affords a mixture of 1-butene, *cis*-2-butene, and *trans*-2-butene, with the *trans*/*cis* ratio around 2 (Scheme 5.40). The reaction provides a very good method to make pure terminal alkenes from primary acetates (A, Scheme 5.40).

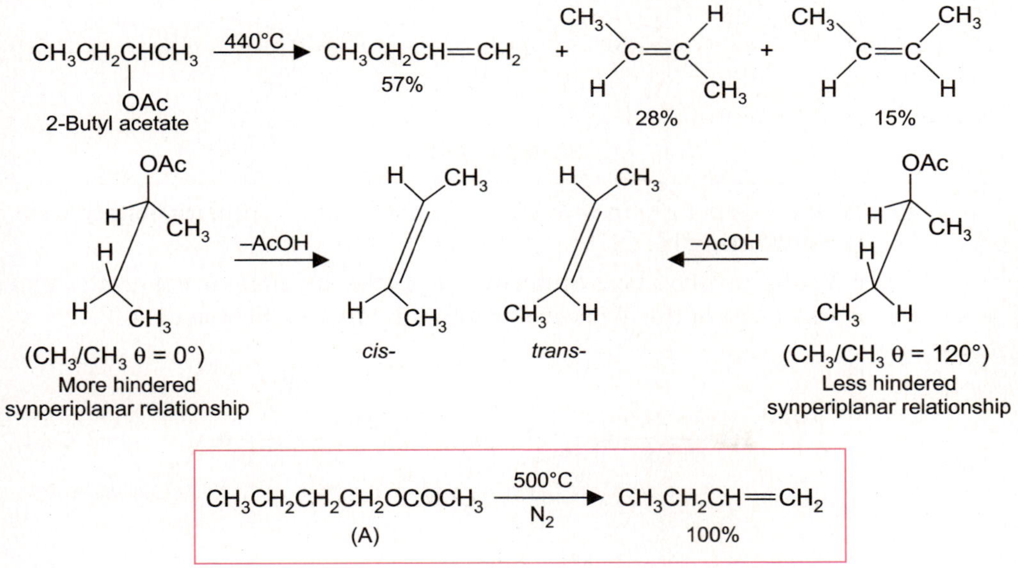

SCHEME 5.40

(ii) Alicyclic Systems

In these systems some restrictions are imposed by the conformations of the leaving groups and the necessity to form the cyclic intermediate.

The following points may be considered:

- In a cyclohexane system if the leaving group is axial then the hydrogen on the adjacent carbon must be equatorial *i.e.* a *cis* relationship between the leaving groups. Thus the cyclohexylacetate (I, Scheme 5.41) with the leaving group in axial position eliminates ethanoic acid to form only one alkene. The double bond in the direction of carbethoxyl group does not form even though it would be conjugated, since there is no equatorial hydrogen on that side, similarly (II, Scheme 5.41) provides another example of *syn*-elimination showing conformational effects.

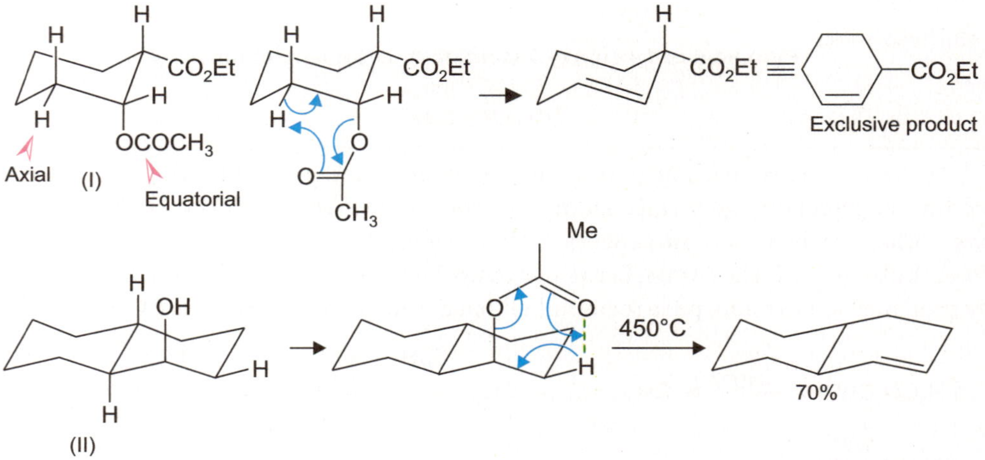

SCHEME 5.41

- Thus if the leaving group is axial, it cannot attain a transition state with an axial neighbouring H.
- One may also invoke a boat conformation as the alternate to generate a planar *cis* arrangement of the ester group with a hydrogen (Scheme 5.42).

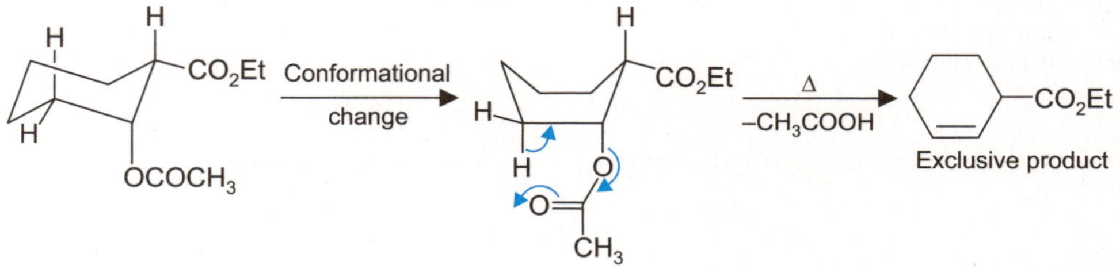

SCHEME 5.42

- In case the leaving group is equatorial, it can still form a transition state with a H atom on adjacent carbon that is either axial (therefore, *cis* relationship) or equatorial (therefore, *trans*). The six-membered transition states (unlike *e.g.*, in amine oxides where a five membered transition state is involved) need not be completely coplanar, thus an equatorial leaving group can equally well eliminate a neighbouring equatorial hydrogen. The methyl xanthate (Scheme 5.43) with an equatorial leaving group gives about 50% of each olefin since hydrogen atoms on both adjacent carbon atoms are accessible in this case.

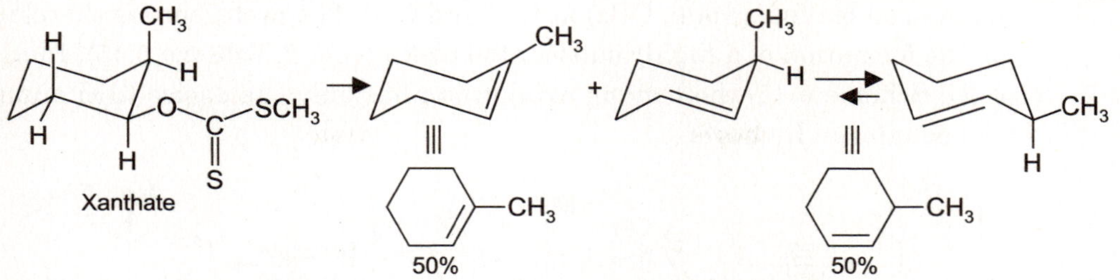

SCHEME 5.43

- More explanation with an equatorial leaving group eliminating an equatorial hydrogen (*trans*-arrangement) is the involvement of stable *trans*-decalin type transition state (Scheme 5.44).

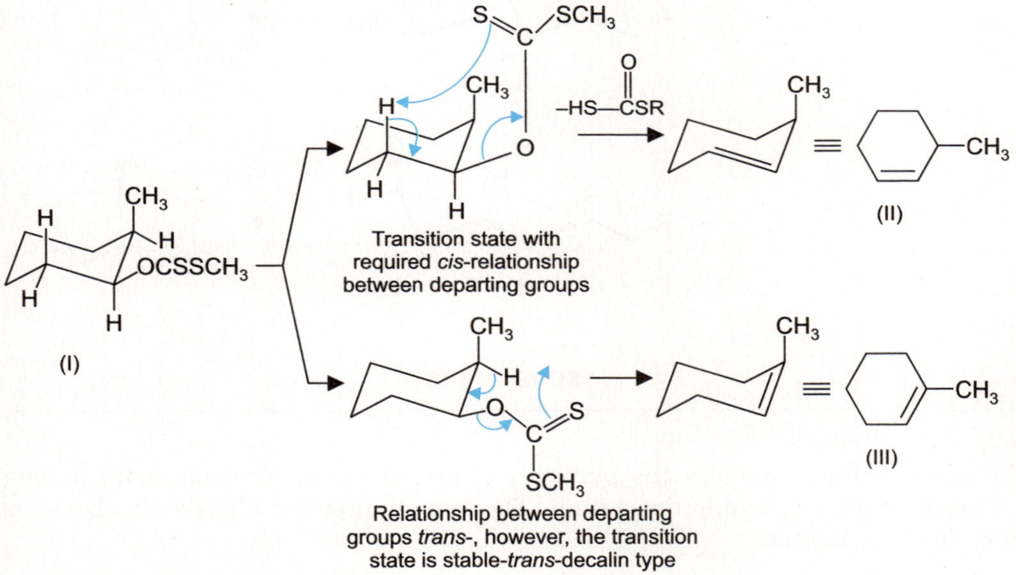

SCHEME 5.44

- The differing stereochemical preferences of E2 and pyrolytic eliminations make them complementary methods for creating specific olefinic centers. Either the alcohol is made to undergo E2 elimination or is first esterified and then subjected to pyrolysis. The 4-hydroxy compound in 5α-chloestane series on E2 elimination eliminates C–5 hydrogen to give more substituted olefin rather than C–3 hydrogen (eqn. 1, Scheme 5.45). When the C–4 hydroxy compound (I, Scheme 5.45) is first esterified (benzoate, OBz) and then heated the C4 ester group (axial leaving group) can only eliminate the available equatorial hydrogen at C–3 (eqn. 1, Scheme 5.45). In (II, Scheme 5.45) hydrogen atom is stereochemically suitably located on both sides to the departing group (equatorial leaving group, OBz) at C–6 and C–8. The preference is therefore, for the formation of a highly substituted olefin (eqn. 2, Scheme 5.45). In case of (II, Scheme 5.45) the leaving axial group has only one choice to eliminate C–6 equatorial hydrogen.

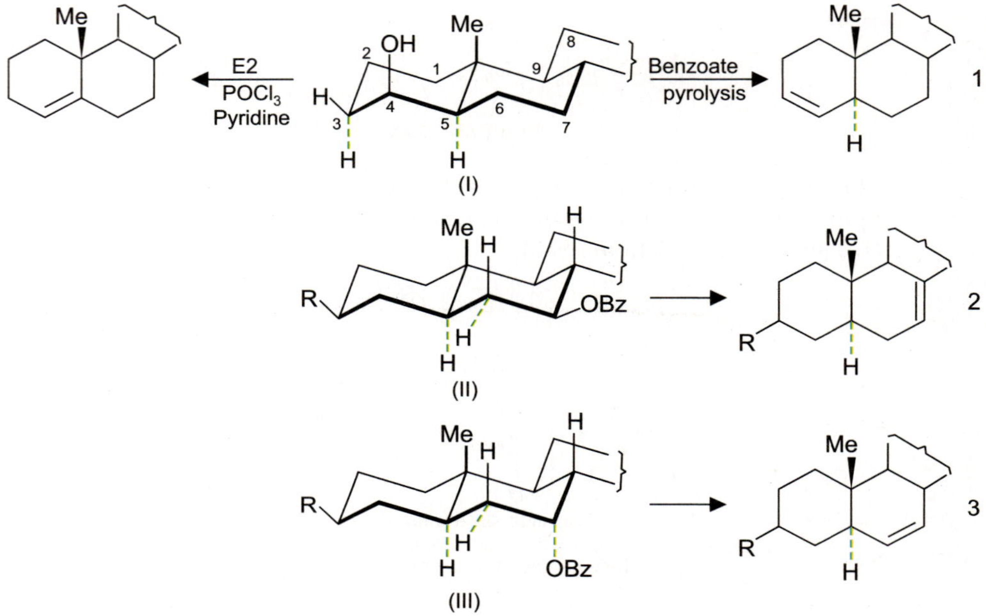

SCHEME 5.45

(iii) Cope and Related Reactions

The Cope reaction involves the pyrolysis of amine oxides (Scheme 5.46) having a hydrogen atom β to the amine group. The *syn*-elimination affords an alkene and dialkylhydroxylamine.

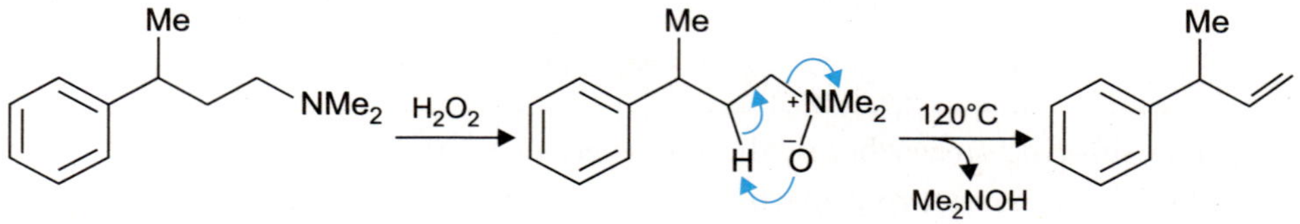

SCHEME 5.46

As already discussed in cyclohexane systems an axial equatorial alignment of the eliminating groups is required. Thus, (I, Scheme 5.47) gives the exclusive formation of (II) since only one axial H (circled) is available for elimination. In the case of (III) two axial protons are available for elimination, however the major product is formed by the exclusive loss of one of these protons to give a double bond in conjugation with the aromatic ring.

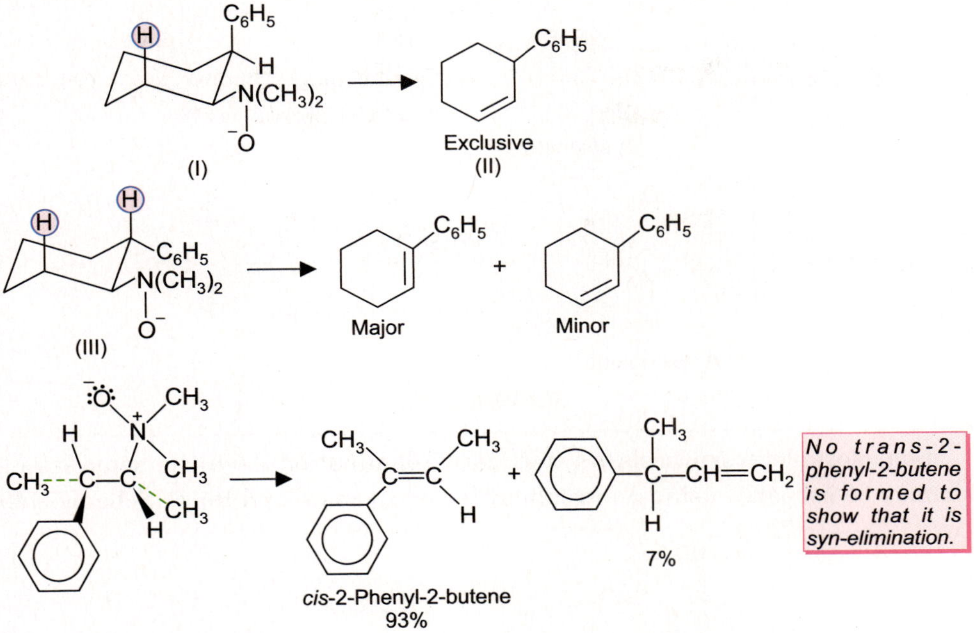

SCHEME 5.47

Moreover, as expected, less hindered elimination would be expected to give the major alkene (Scheme 5.48).

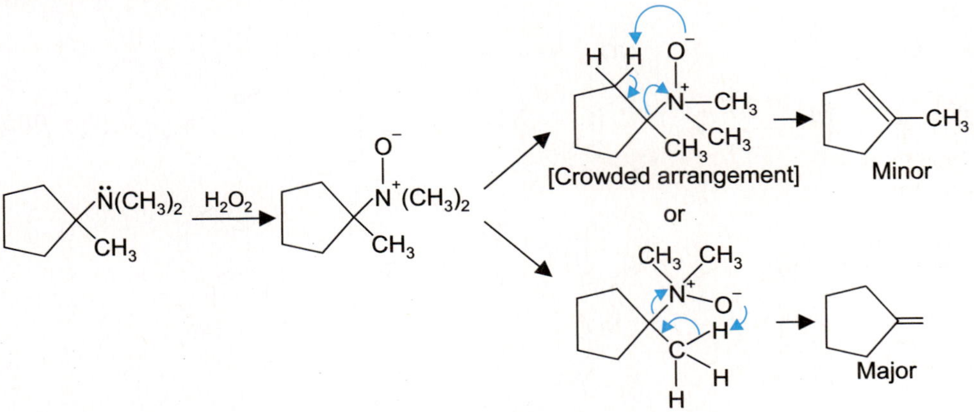

SCHEME 5.48

Lastly mention may be made of elimination from selenoxides. Selenoxides are more reactive and give better results toward β elimination and undergo thermal *syn*-elimination to form alkenes under milder conditions (room temperature). The selenoxides can be easily made from the corresponding selenides by oxidation with hydrogen peroxide. A selenide is formed from alkyl halides or tosylates (Scheme 5.48*a*) by reaction with phenyl selenide anion (Selenide anions are powerful nucleophiles to displace halides or tosylates and open epoxides).

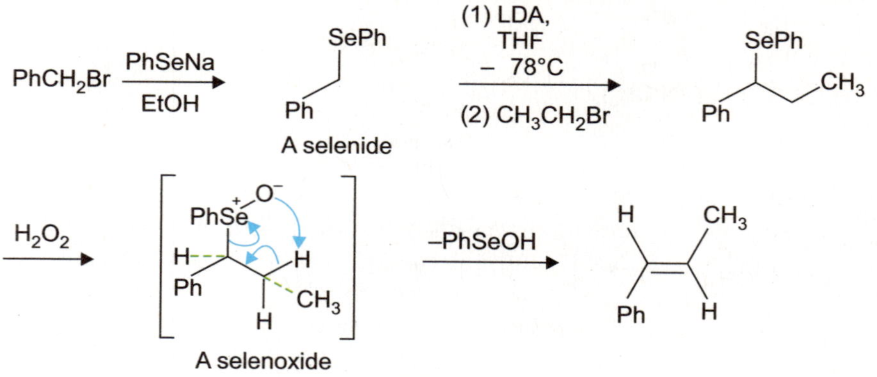

SCHEME 5.48a

Sharpless olefin synthesis is *syn*-oxidative elimination of *o*-nitrophenyl selenides, which may be prepared using *o*-nitrophenyl selenocyanate and Bu_3P (Scheme 5.49).

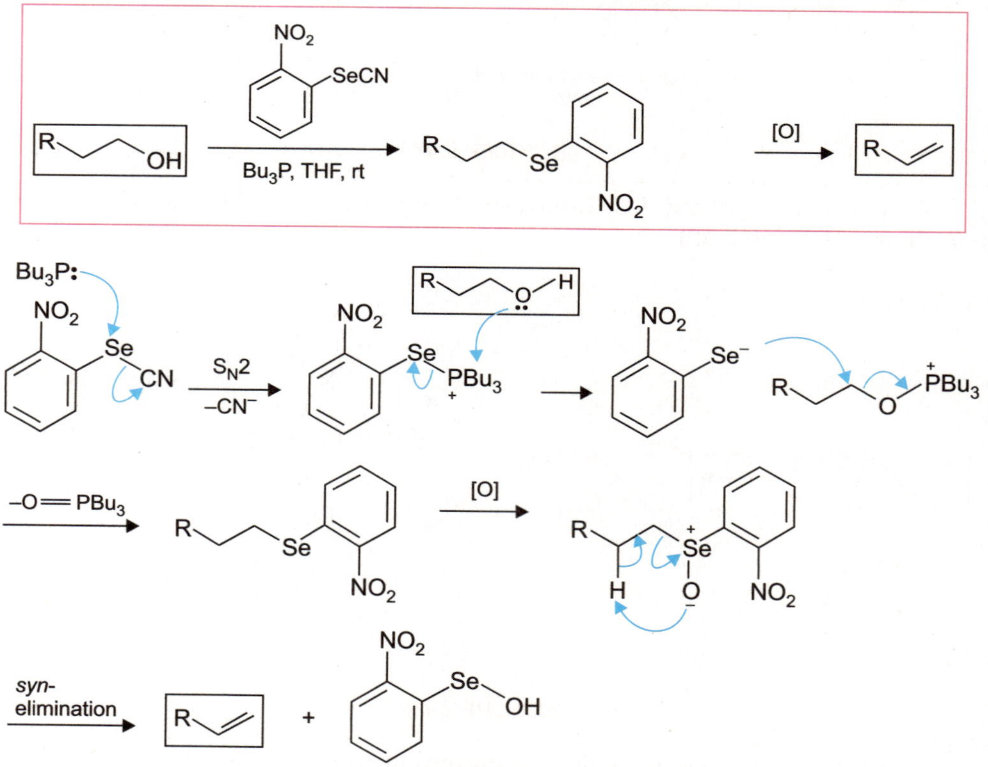

SCHEME 5.49

5.9 MOLECULAR REARRANGEMENT DURING ELIMINATION

During the dehydration of *e.g.*, *sec*-alcohols (formation of carbocations) a rearrangement may take place.

When the achievement of the *anti* conformation in cyclohexane systems is not possible, molecular rearrangement may precede elimination. Thus, in the *trans* decalin derivative (I, Scheme 5.50) the hydroxyl group is held in the equatorial position because the ring system cannot flip. Two of the ring residues (shown by arrows, Scheme 5.50) are in the appropriate *anti*-periplanar position with the OH group for rearrangement, and in the presence of acid, ring-contraction occurs. Interestingly, of the two ring residues that one migrates which leaves behind the more stable carbocation.

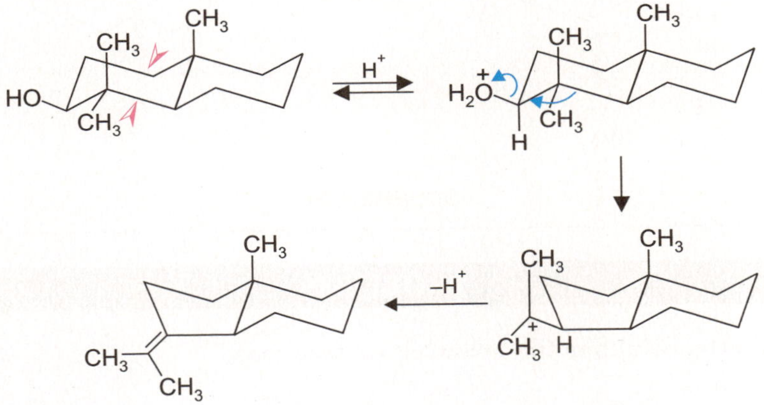

SCHEME 5.50

5.10 1, 4 ELIMINATION REACTIONS (FRAGMENTATION REACTIONS) GROB FRAGMENTATION

The stereoelectronic requirement of the ring formation, rearrangement, neighbouring group participation and fragmentation reaction is that the groups involved must be *anti*-periplanar. Generally, this involves diaxial relationship of the groups involved. The ring residue in a cyclohexane also has an *anti*-periplanar, relationship to an equatorial leaving group. This situation obtains in several fragmentation reactions (Scheme 5.51). Thus one may note that eliminations are not limited to the formation of a single bond in a reaction, stereospecific multiple eliminations (fragmentations) may take place if coplanar bonds are suitably aligned with leaving groups.

Consider 3-aminoalcohol derivatives in which the *p* orbital of N with a lone pair is *anti*-periplanar to 3-2 bond which in turn is *anti*-periplanar to the leaving group (OTs) can undergo a concerted 1, 4-elimination (I, Scheme 5.51) leading to the fragmentation of the molecule. The epimeric structure (II) cannot be so converted into the amino olefin to show that the stereoelectronic requirements in the reaction (known as Grob fragmentation) are very stringent.

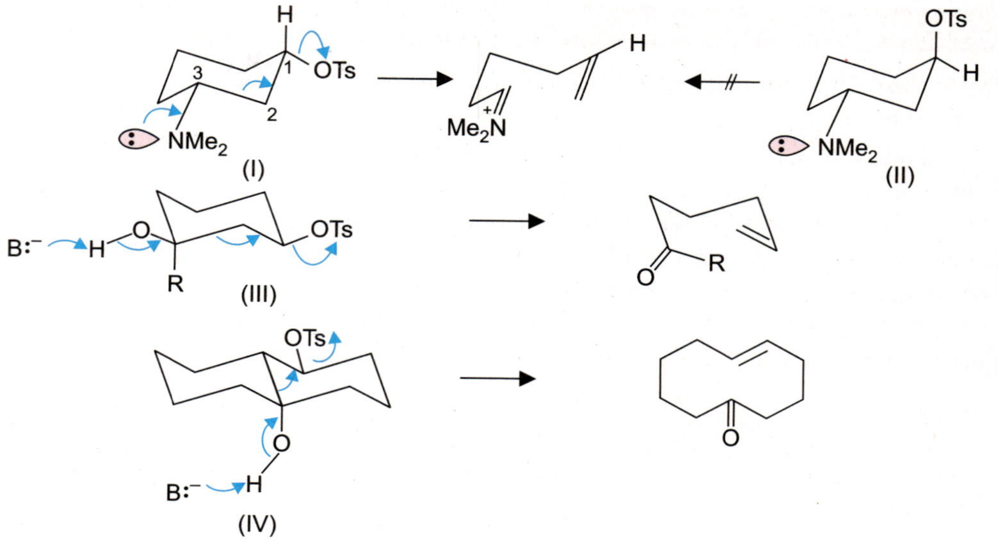

SCHEME 5.51

PROBLEMS

5.1. Explain the results in the following E2 reactions:

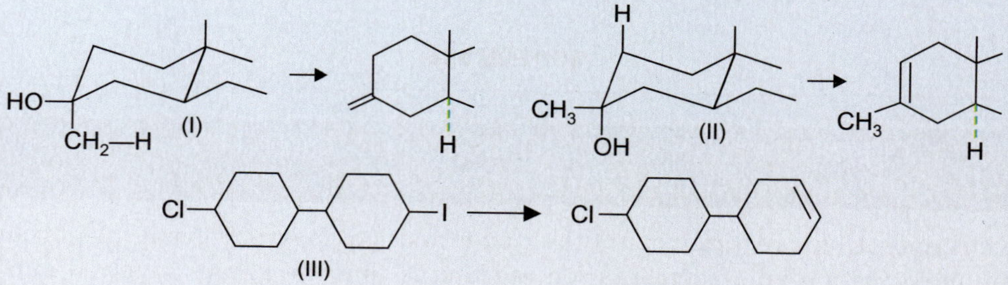

5.2. *Anti*-elimination occurs readily for *cis*-but nor for *trans*-1-bromo-4-(1, 1, dimethylethyl) cyclohexane. Suggest a reason for this difference.

5.3. (*a*) In the rigid *trans*-declain system (I, shown partly) of $\Delta 8$-lanosterol, the C-3 hydroxyl is equatorial, the compound, however, undergoes an easy E2 elimination. Explain.

(*b*) The norbornane *i.e.* [2.2.1] heptane system of compound (II) undergoes an E2 elimination giving only the product with methylenic double bond. Explain.

(*c*) The isomer (III) of 1, 2, 3, 4, 5, 6-hexachlorocyclohexane undergoes E2 elimination 7000 times more slowly than any of its stereoisomers. Explain.

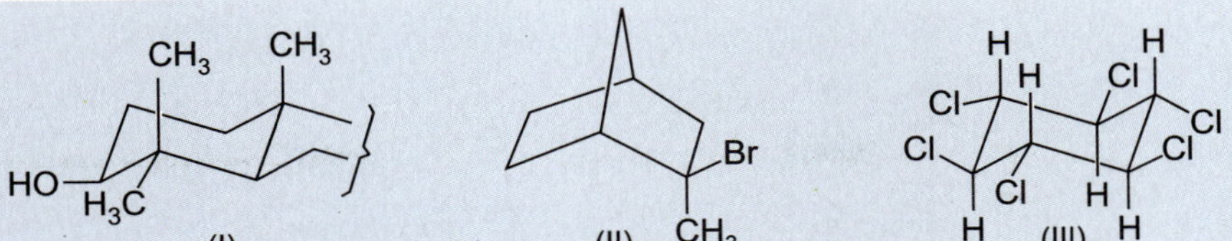

(I) (II) (III)

5.4. Predict the major product from the reaction.

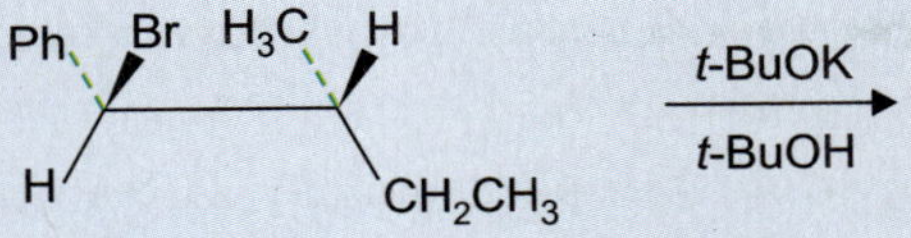

5.5. Write the mechanism of ElcB reaction. What are the structural features in the substrate which are essential for the operation of this mechanism?

5.6. What is aldol condensation? Which step resembles ElcB mechanism in this reaction?

5.7. Predict the product from the following reaction:

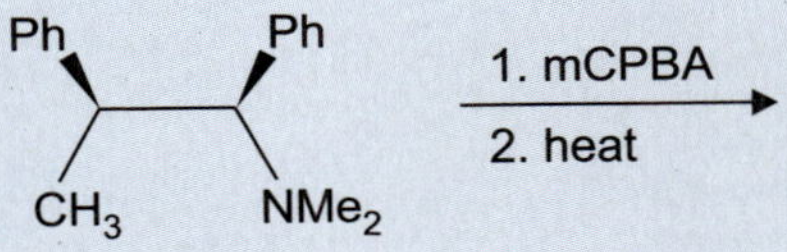

5.8. How one can bring about the following transformations?

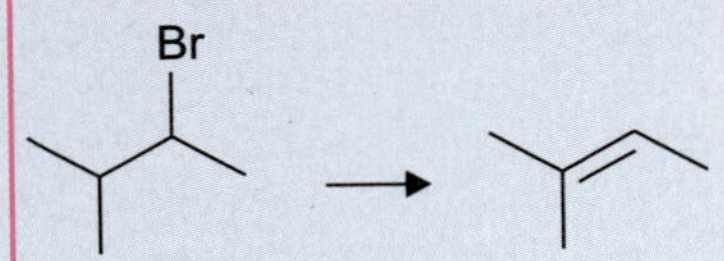

(I)

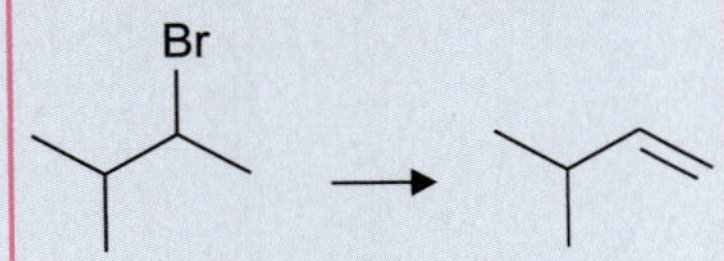

(II)

ANSWERS TO SELECTED PROBLEMS

5.1. The rigid *trans*-decalin systems in epimeric 3-methylcholestanols reflect the effect of configuration on the course of dehydration. In (I) only a hydrogen of the methyl group can attain an *anti*-relationship with the ring equatorial hydroxyl group. In (III) since I^- is a better leaving group, selective elimination of HI is allowed.

5.2. The ring is anchored by the bulky *t*-butyl group which occupies equatorial position. Of the two stereoisomers only in the *cis*-isomer *anti*-hydrogens are available.

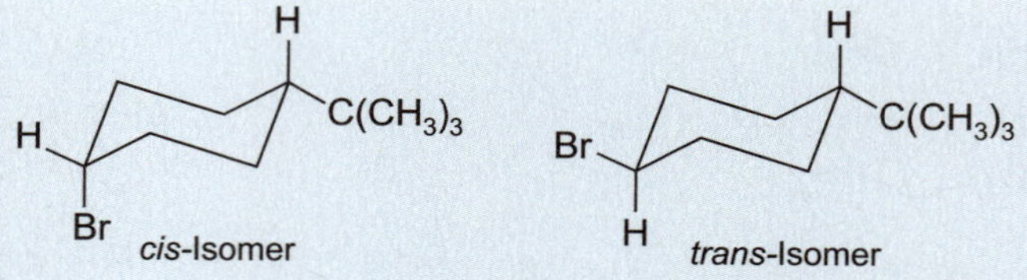

5.3. (*a*) An equatorial group on a cyclohexane ring can adopt an *anti*-arrangement with a neighbouring hydrogen by adopting a boat form. In the case of lanosterol derivative, the serve 1, 3-diaxial interaction between the methyl groups at the angular position and at C–4 is relieved in the ring *A* boat form.

(*b*) Since Br cannot attain *anti*-periplanarity with a ring hydrogen.

(*c*) All chlorines are equatorial and no *anti*-hydrogens are available.

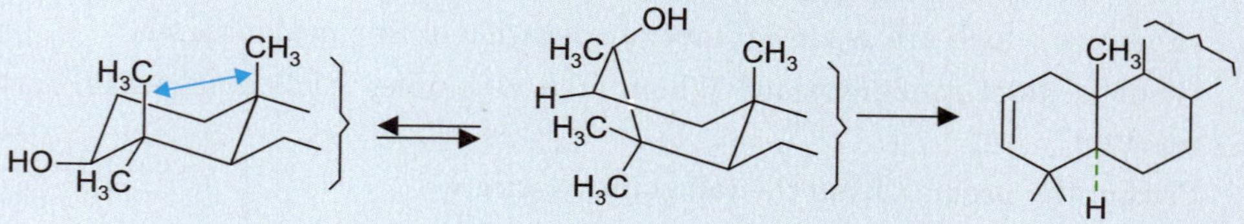

5.4.

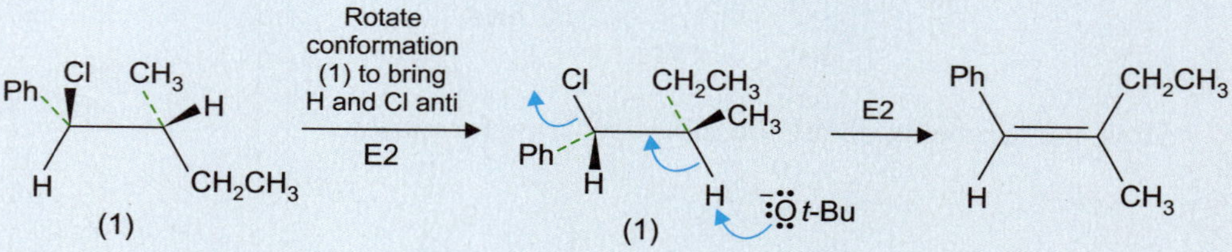

5.7.

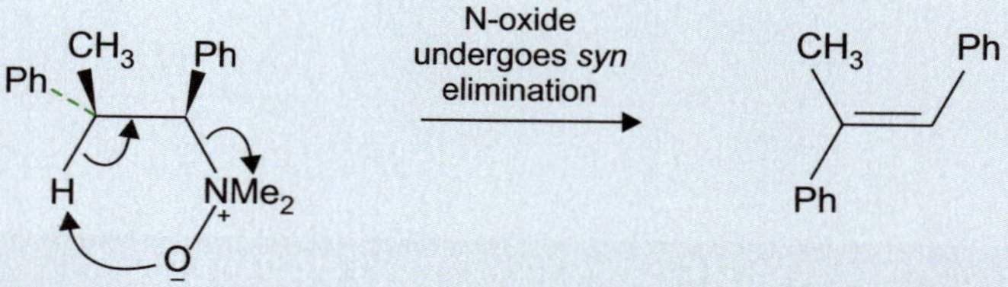

5.8.

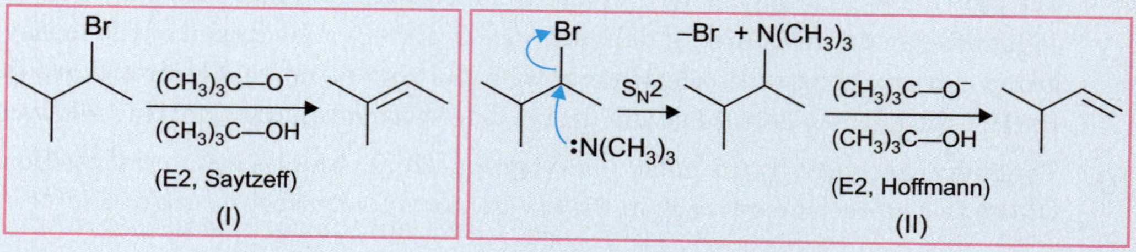

Stereochemistry and Mechanism of Additions to Carbon–Carbon Multiple Bonds and Hydroboration

6

CHAPTER

Addition to a Carbon–Carbon double or triple bond can take place in four different ways:

- By the initial attack of an electrophile (*e.g.*, bromination of a double bond).
- By the initial attack of a nucleophile (*e.g.*, addition of a polar reagent to an α, β-unsaturated carbonyl compound).
- By the initial attack of a free radical.
- In the fourth type the mechanism involves the attack at the two carbons of the double or triple bond simultaneously (*e.g.*, epoxidation of alkenes).

6.1 ADDITION REACTIONS OF ALKENES INVOLVING ELECTROPHILES

In an alkene the electron density of a *pi* bond is maximum below and above the plane of the double bond. The most favourable way for the attack of the electrophile based on stereoelectronic considerations is along these electron rich regions *i.e.*, perpendicular to the plane of the double bond. The attacking total reagent might be divided into an electrophilic and a nucleophilic portion (E-Nu). Each part of this reagent may add to the double bond of the reactant molecule from the same side (stereospecific *syn* addition) or from opposite side (stereospecific *anti* addition), or the reaction may be nonstereospecific.

(A) Addition of Bromine to an Alkene-Stereochemistry of Bromine Addition (*Anti* Addition)

Bromination of an alkene was discussed in some detail during the study of stereoselective and stereospecific reactions. Here a review is given and the following points may be noted:

- This mechanism involves the addition of electrophilic reagents in a two step process. An organic cation is formed in the first step which reacts with the nucleophile in the second. The organic cation may have a bridged structure or a carbocationic nature which may be viewed as in (Scheme 6.1).

- The organic cations are not simple cations but have bridged or otherwise complexed structures and the best understood of these is the bromonium ion which is a three-membered ring and is formed by donation of bromine lone-pair electrons to the vacant *p* orbital of the neighbouring carbocation. Although (Scheme 6.1) depicts bromonium ion formation as stepwise, this is done only for clarity. The bromonium ion is formed in a single step by interaction of the alkene with Br^+. Alkene π electrons attack bromine and push out bromide ion to give a bromo carbocation. The neighbouring bromo substituent stabilizes the positive charge by using two of its electrons to overlap the vacant carbon *p* orbital, to give a three-membered-ring bromonium ion.

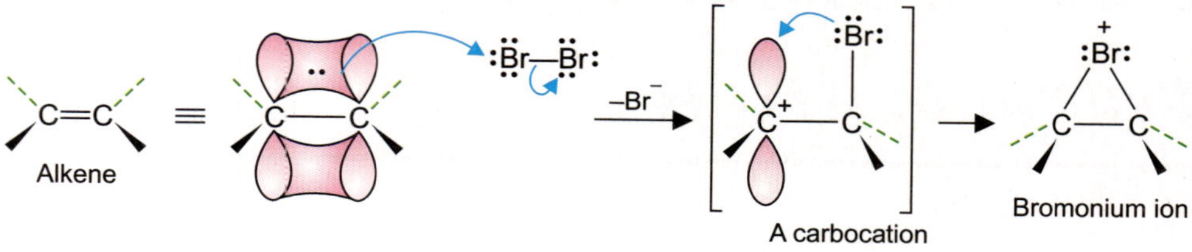

SCHEME 6.1

- The bromonium ion is often represented as a hybrid (Scheme 6.2) in which the structure (A), a tertiary carbocation is more important contributor to the hybrid compared to (B) a secondary carbocation.

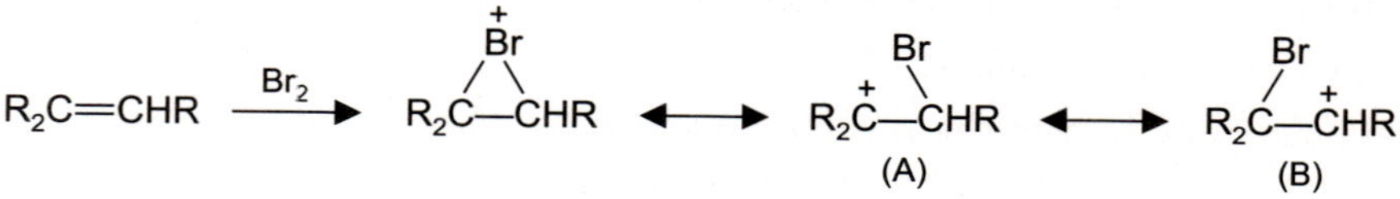

SCHEME 6.2

- When an alkene *e.g.*, cyclohexene reacts with bromine, *anti*-addition occurs. Bromonium ion is formed in the first step (Scheme 6.3) which is then attacked by the bromide ion from the opposite side of the bromonium ion in an S_N2 type reaction to result in *anti*-addition to give a *trans* product as a racemic mixture.

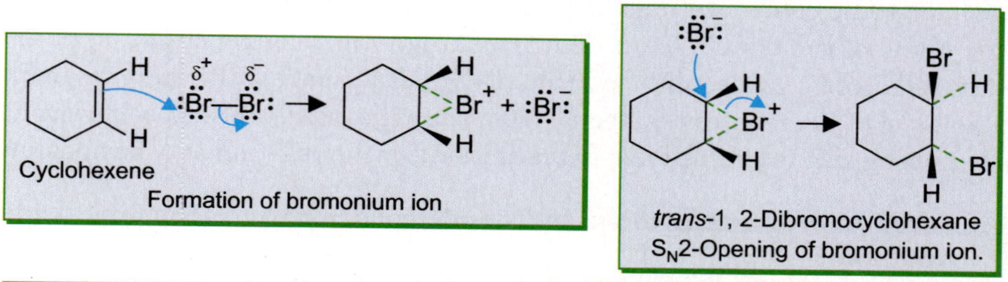

Bromine molecule is polarized by approaching the alkene. The cationic bromine atom (Br^+) acts as an electrophile and adds to the double bond to give a cyclic bromonium ion. In cyclohexene bromonium ion could be formed either at top face or bottom face (bromonium ion at bottom face is shown here). Br^- can attack (S_N2) either carbon of the bromonium ion. Attack at one carbon gives trans-1,2-dibromocyclohexane (one enantiomer). The attack at other carbon will give its enantiomer. (not shown here).

SCHEME 6.3

- Addition of bromine to cyclohexene is a stereoselective reaction (*cis*-1, 2-dibromo-cyclohexene is not formed).
- The nucleophilic attack by the bromide ion (S_N2) reaction on the bromonium ion causes an inversion of configuration at the carbon being attacked. This inversion of configuration at one carbon of the ring gives one enantiomer of *trans* 1, 2-dibromocyclohexane. The other enantiomer results from the equally probable attack of the bromide ion at the other carbon of bromonium ion.

The trans product formed from cyclohexene is drawn as a single enantiomer. Opening of the bromonium ion must take place in two equivalent ways to produce a racemic mixture i.e. both the enantiomers are formed in equal amounts. In such cases, however, it is the practice not to draw both the enantiomers, unless needed for clarity. Thus for clarity the anti bromination of cyclohexene may be shown to give a racemic product. (Scheme 6.4).

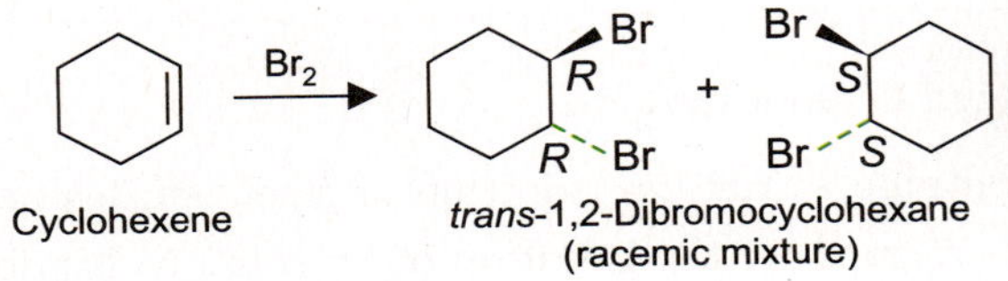

anti-bromination of cyclohexene shown to give trans-1,2-dibromocyclohexane as a racemic mixture (The hydrogens are omitted for clarity).

SCHEME 6.4

- If the addition of bromine to cyclohexene had involved a nucleophilic attack of the double bond by bromine with the resulting formation of a planar carbocation as an intermediate, some *cis*-1, 2-dibromo compound would be expected as a product (see, Scheme 1.102)
- During the addition of chlorine, the intermediate chloronium ion is believed to be less strongly bridged *i.e.*, the cyclic structure (see, Scheme 6.2) is of considerably less important in the hybrid.

Strong bridging ensures that the anti-product is likely to be the major if not exclusive. The bridged species formed by the attack of sulfur and selenium electrophiles have strong bridging features (Scheme 6.5a). Thus Se atom in CH_3SeCl (methylselenyl chloride) is electrophilic and has lone pairs with $CH_3Se^{\delta+}$ as the electrophile similarly in CH_3SCl, $CH_3S^{\delta+}$ is the electrophile. A strong bridged ion behave like an epoxide in an S_N2 reaction and the strength of bridged species therefore, determines both the regioselectivity and stereoselectivity of the addition reaction as seen (Scheme 6.5a). Thus the nucleophile attacks preferentially the less substituted carbon atom from the opposite side of the bridge. With these strongly bridged species, even if the approach of the nucleophile from the opposite side is hindered, anti-addition is still the preferred pathway as seen in addition to norbornene (Scheme 6.5).

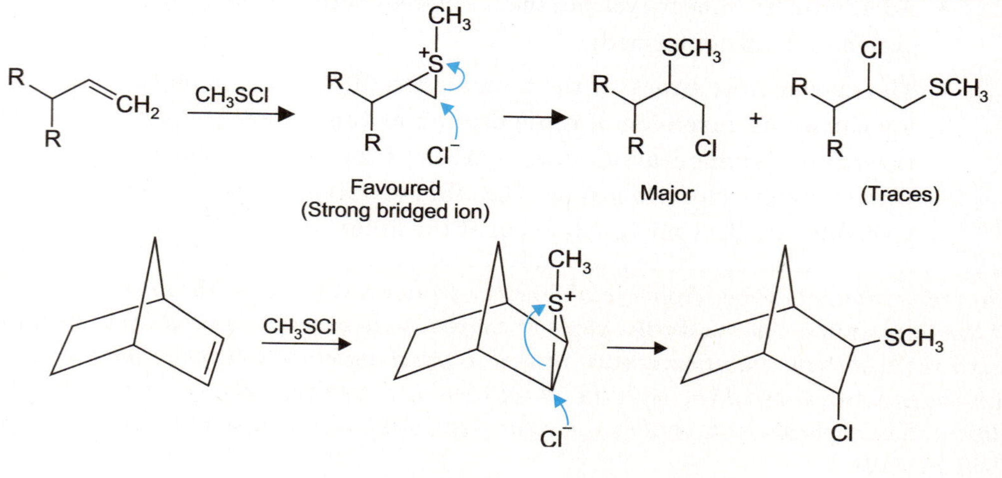

SCHEME 6.5

(B) Evidence for the Existance of Bridged Bromonium Ion

One case where a bromonium ion is stable to further addition of nucleophiles is (II, Scheme 6.5*a*). This bromonium ion is formed by the addition of bromine to hindered alkene (I, Scheme 6.5*a*). The bromonium ion (II) formed initially does not go to second stage to give a dibromo compound. Attack by Br^- on the intermediate is prevented by the cage like structure which shields each end of the original double bond from attack. An X-ray carystallographic study has confirmed the structure of the bromonium ion. Strong evidence which supports the involvement of a bromonium ion also comes from the work of G.Olah (Nobel Prize 1994) who prepared and studied stable solutions of cyclic bromonium ions in liquid SO_2 (Scheme 6.5*a*).

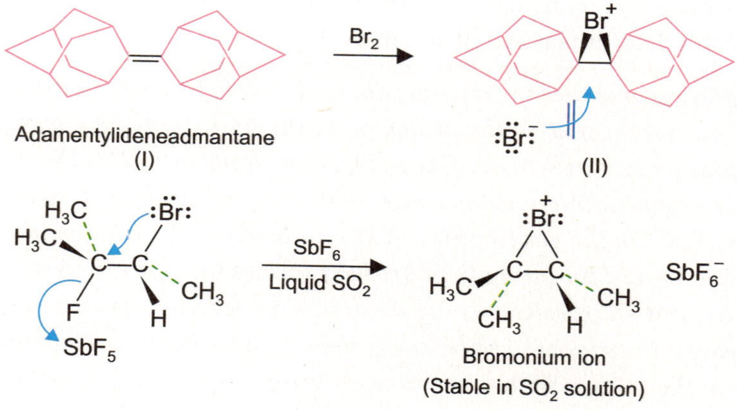

SCHEME 6.5a

- A symmetric alkene like cyclohexene can be approached by the electrophile either from top face (as in Scheme 1.102) or from bottom face (as in Scheme 6.3). However, the presence of substituents may hinder the approach of the electrophile from the side on which these are present. Thus (I, Scheme 6.6) will form the bromonium ion on the side opposite to the methyl group. In conformationally rigid systems the *trans* product formed is diaxial rather than diequatorial as shown in rigid-decalin structure (I, Scheme 6.6). In case ring flipping is possible, the diaxial compound formed initially converts into the diequatorial compound. This is so in cyclohexene (Scheme 6.6).
- The second example is of cholesterol a naturally occuring compound, whose top face (β-face) is hindered by the presence of two bulky angular methyl groups and the side chain at C–17 on that face. Thus cholesterol exclusively forms a bromonium ion from the bottom α-face), which is subsequently attacked by

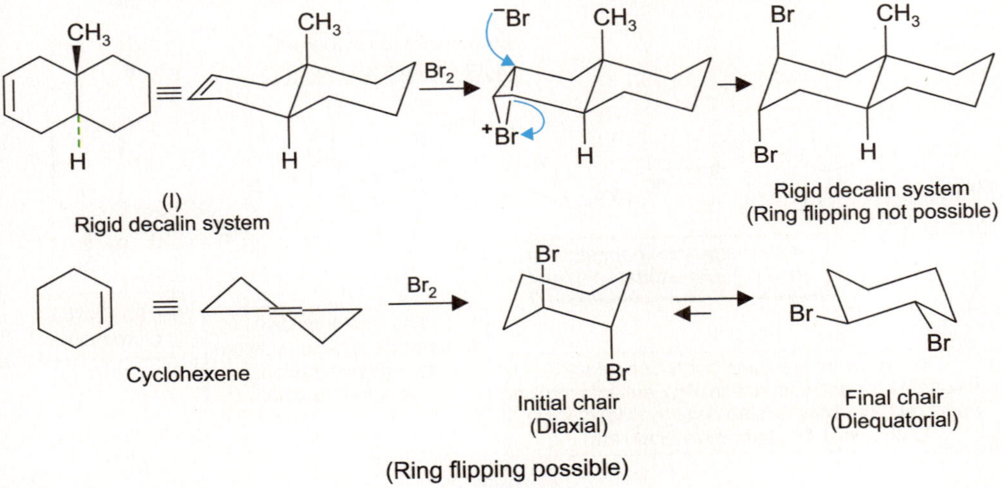

SCHEME 6.6

Br⁻ at (C–6 (S_N2 conditions) to form 5α, 6β-diaxial dibromide (Scheme 6.7). The 5α, 6β-diaxial bromide (I) undergoes a change when kept in chloroform solution to more stable diequatorial isomer (II). This is an example of a slow stereoisomerization and not a conformational change.

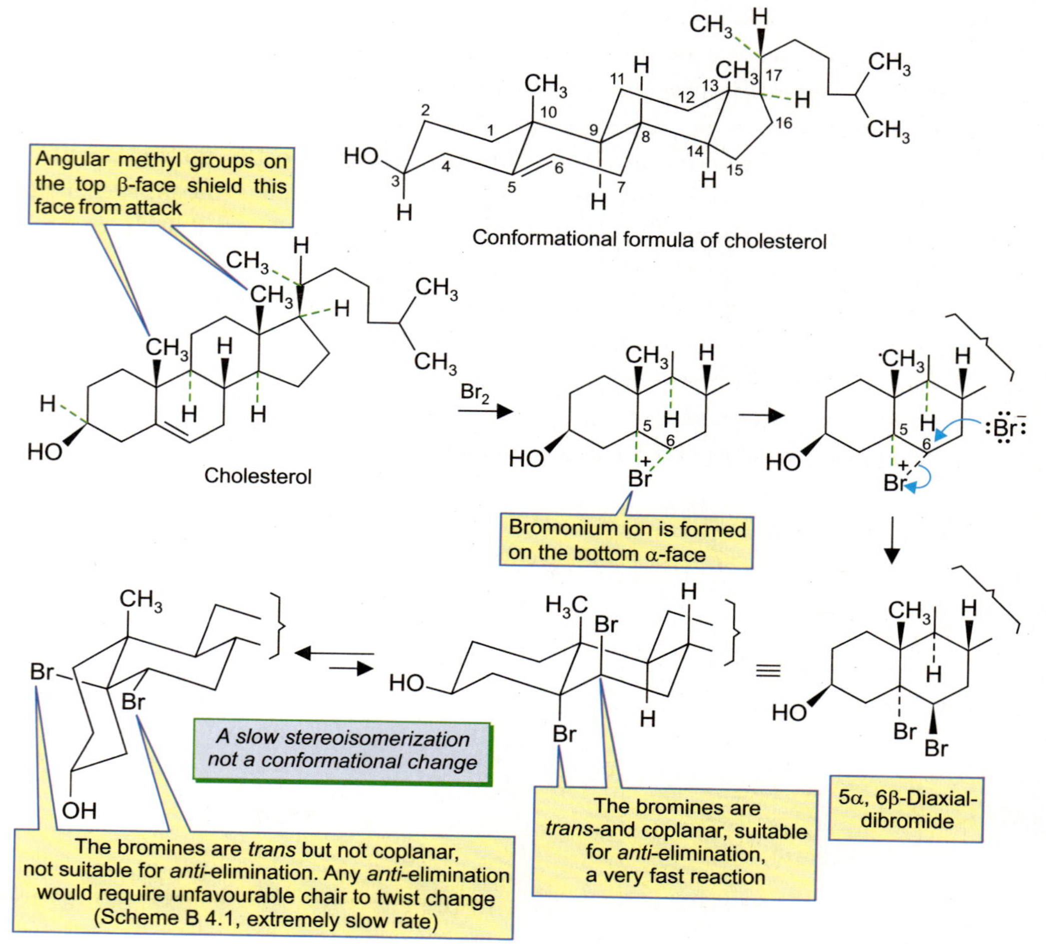

SCHEME 6.7

BOX 6.1
PURIFICATION OF CHOLESTEROL

Cholesterol isolated from natural sources e.g., gallstones contains small amounts (0.1–3%) of 3β-cholestanol, 7-cholesten-3β-ol and 5, 7-cholestadien-3β-ol along with the fourth component 24-hydroxycholesterol which is easily removed by crystallisation from alcohol. These are so very similar to cholesterol (Scheme B 6.1) in solubility that their removal by crystallization is not feasible. Purification can be achieved through the sparingly soluble dibromo derivative 5α, 6β-dibromocholestan-3β-ol (Scheme 6.7) 3β-Cholestanol is saturated and does not react with bromine; and remains in the mother liquor. 7-Cholesten-3β-ol and 5, 7-cholestadien-3β-ol are dehydrogenated by bromine to dienes and trienes, respectively, that likewise remain in the mother liquor and are eliminated along with colored by-products.

Continued...

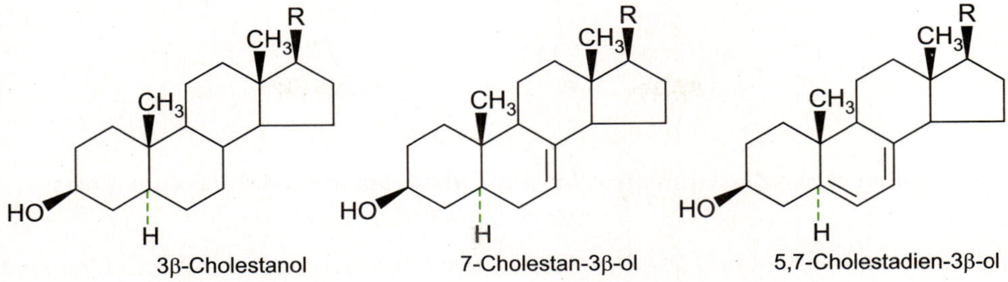

Impurities in cholesterol isolated from natural sources

SCHEME B 6.1

Purified cholesterol dibromide (Scheme 6.7) is then debrominated with I^- which induces anti-elimination to give pure cholesterol (Scheme B 6.2).

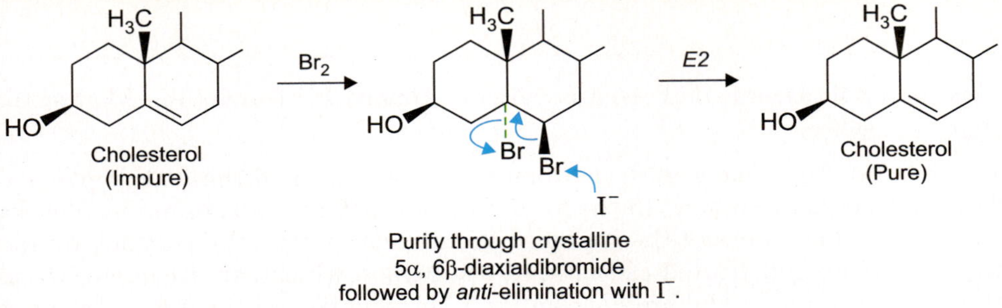

SCHEME B 6.2

EXERCISE 6.1

Explain the stereochemical outcome from each of the reactions (Scheme 6.7a).

SCHEME 6.7a

ANSWER. *(I). Recall that addition of bromine to a double bond is anti. Different compounds are formed (scheme 6.7b) from the reaction of bromide ion on the bromonium ion at two possible carbons see scheme 6.4. The starting alkene (I, scheme 6.7a) is chiral, each diastereomer is thus formed as a single enantiomer (the rings of a trans-decalin do not undergo the chair interconversion).*

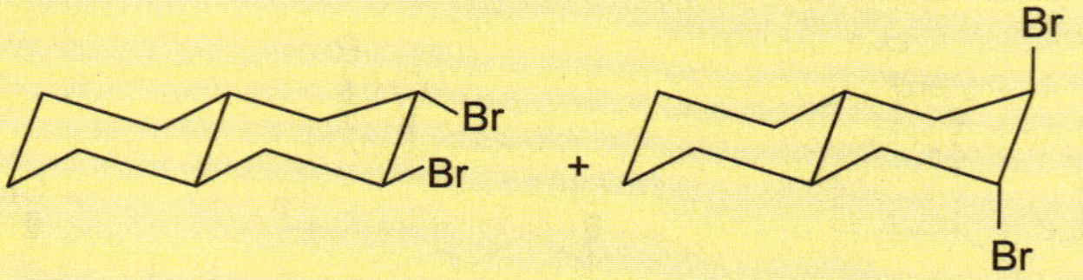

Formation of two diastereomeric *trans*-disubstituted *trans*-decalin derivatives

SCHEME 6.7b

(II). Two diastereomers of 1,2,4,5-tetrabromocyclohexane will be formed. One as an achiral meso while the other is the racemate (scheme 6.7c).

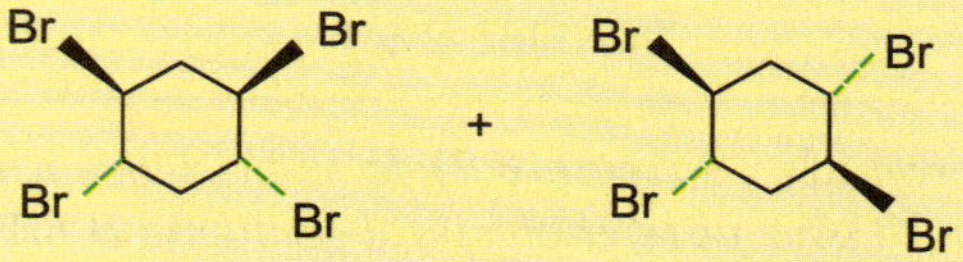

SCHEME 6.7c

(C) Addition of Bromine in the Presence of Water — Halohydrin Formation — Markovnikovs Regiochemistry

When a nucleophile reacts with a bromonium ion, the addition takes place with Markovnikovs regiochemistry. In the formation of bromohydrins, bromine adds to the least substituted carbon and the hydroxyl group (from nucleophilic attack by water) bonds to the more substituted carbon (*i.e.*, the carbon which can accommodate more of the positive charge in the bromonium ion.) Markovnikovs rule states: In the ionic addition of an unsymmetrical reagent an electrophile adds to the π bond so as to form a bond with the carbon atom that allows the formation of a most stable carbocation.

The addition of bromine to 1-methylcyclohexene in the presence of water gives the tertiary alcohol as the major product (Scheme 6.8). The following conclusions are drawn:

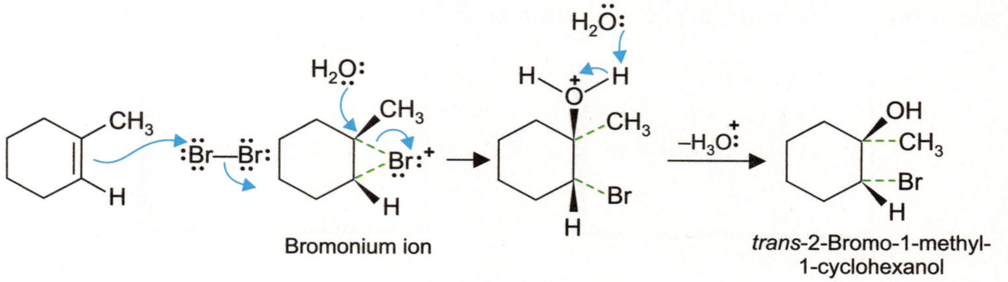

SCHEME 6.8

- The nucleophile (H_2O) attacks the more highly substituted carbon of the ring since this is more positively polarized than the other. The bromonium ion is below the ring and thus H_2O attacks from top of the ring.
- Since, tertiary alcohol is the major product, the reaction is regioselective. Nucleophile (H_2O) attacks at the more substituted carbon of bromonium ion since it has greater δ^+.
- Stereochemically the orientation of OH and Br is *trans* (*anti*-addition), thus the intermediate is a cyclic bromonium ion which is attacked by the nucleophile at the opposite face.
- A racemic mixture is obtained since there is no chiral influence in the reaction.
- In the first step of halohydrin formation, Br^+ is the only electrophile in the reaction mixture. The second step is fast and of the two nucleophiles Br^- and H_2O, H_2O is the solvent and is therefore, available in large concentration. The bromonium ion is thus more likely to collide with a molecule of water (than with Br^-) to give protonated bromohydrin, which immediately loses a proton.
- A strongly bridged bromonium ion shows reactivity like an epoxide in an S_N2 reaction (see Schemes 3.56*b* and 3.56*c*).

(D) The addition to acyclic alkenes follows the same mechanistic and stereochemical pathways. The stereospecific bromination of *cis* and *trans*-2-butene has been discussed (see, Schemes, 1.70, 1.18*e* and Problems 1.23 and 1.24).

(E) Stereochemistry of *Syn*-Addition of Bromine to an Alkene

An addition reaction which proceeds through a bromonium ion intermediate results in *anti*-addition (opening of the bromonium ion in S_N2 fashion) to give *trans*-products. However, a reaction which proceeds through a carbocation intermediate should result in a mixture of *syn*- and *anti*-addition since the planar carbocation can be attacked from either side by the incoming nucleophile. In these situations the alkene contains atleast one aryl group attached to the double bond. The stabilization of the positive charge in the intermediate (after electrophilic attack) is by delocalization on to the aromatic ring. As a result the effectiveness of bridging (formation of the bromonium ion) is reduced.

Acenaphthylene (Scheme 6.9) on reaction with bromine gives large amounts of the usual *trans*-dibromide (*anti*-addition) alongwith a large amount of *cis* dibromide (*syn*-addition). This is due to the formation of a resonance stabilized open carbocation which competes favourably with the formation of a bromonium ion, S_N2 reaction on cyclic bromonium ion will give only the *trans* product, while the open cation will give both *cis* and *trans* products (Scheme 6.9).

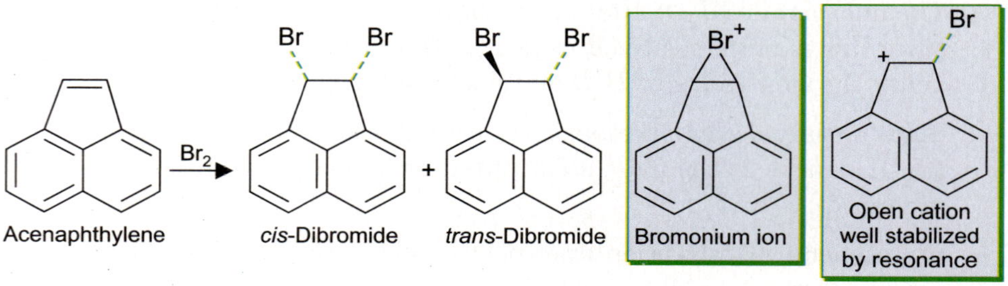

SCHEME 6.9

Similarly 1-phenylpropene $C_6H_5CH = CHCH_3$ (*cis* or *trans* isomer, Scheme 6.10) displays a nonstereospecific course (both *syn* and *anti* addition) during the addition of bromine. This may well be compared with the stereospecific addition of bromine to *E*-and *Z*-2 butene isomers (Scheme 1.100). Thus in alkenes in which atleast one aryl group is attached to the double bond, the selectivity is less since now the positive charged is instead stabilized in the intermediate by its delocalization on to the aromatic ring. As a result, effective bridging is reduced and rotation can occur (Scheme 6.10).

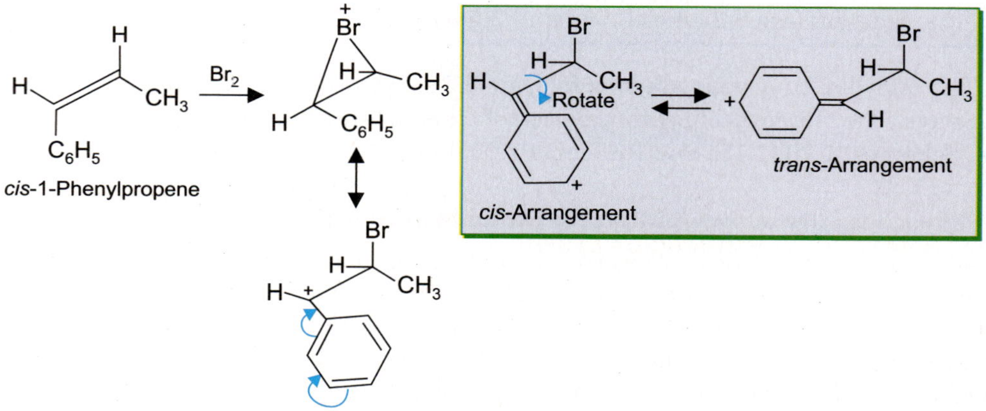

SCHEME 6.10

Additionally, increase in the polarity and ion-solvating ability of the solvent also stabilizes a carbocation than the bromium ion intermediate which as excepted leads to a decrease in *anti* stereoselectivity. This is seen in the addition of bromine to stilbene to give 95% *anti*-addition in solvents of low dielectric constants but is only 50% *anti* in a solvent with $\varepsilon = 35$.

However, irrespective of the fact, whether the intermediate is an open carbocation or a cyclic bromonium ion, the mechanism is called Ad_E2 (electrophilic addition bimolecular).

(F) Oxymercuration/Reduction of an Alkene (Markovnikovs Hydration of an Alkene). A Metal Acting as an Electrophile

This is a two step process, an alkene is first treated with mercuric acetate and the cyclic mercurinium ion is then reduced with sodium borohydride. The following points may be noted:

- Mercuric accetate gives electrophilic cationic mercury species which add to the double bond to form a cyclic (bridged) mercurinium ion (A, Scheme 6.11). Reaction can occur at either face.
- Oxymercuration is stereospecific *anti*-addition and highly regioselective, the mercurinium ion has a structure similar to that of a cyclic bromonium ion. (Scheme 6.2)
- Water that is present attacks the more substituted carbon (Markovnikov rule regioselectivity) to yield an alkylmercuric acetate intermediate.
- Replacement of mercury by hydrogen (demercuration) is carried out by sodium borohydride reduction.
- The alcohol obtained on demercuration is similar to the product of Markovnikov hydration of the substrate. It is however, a valuable alternative to acid-catalyzed hydration, since no carbocations are involved and thus no rearrangements are observed.
- Alkene (I, Scheme 6.11) gives two diastereomers on oxymercuration reduction, each as a racemic mixture.

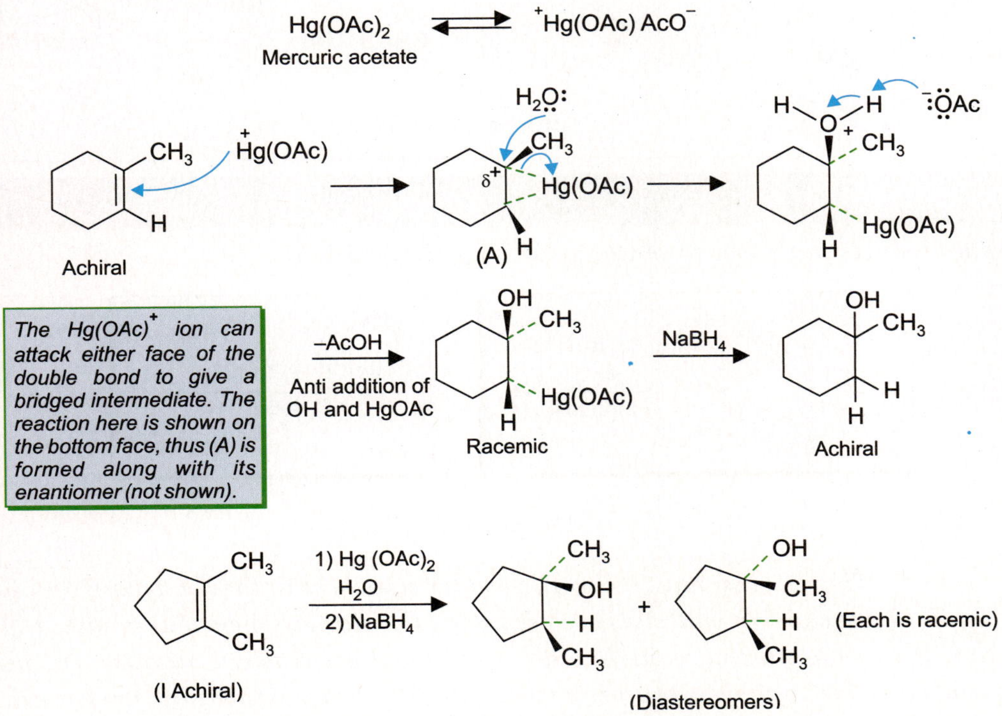

SCHEME 6.11

When mercuration is carried out in an alcohol solvent (solvomercuration) the product contains an alkoxy group (–OR) and the method converts alkenes to ethers *via* the addition of an alcohol across the double bond of the alkene (Scheme 6.12). Again the solvent attacks the mercurinium ion at the more highly substituted end of the double bond.

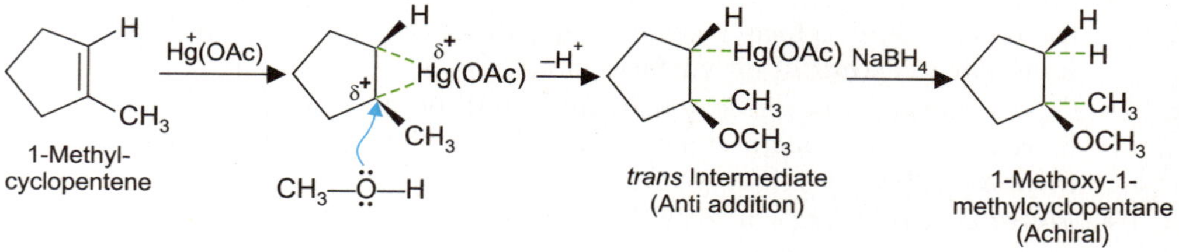

SCHEME 6.12

When a hydroxyl group is part of the reactant and geometrically suitably placed an intramolecular bond formation can take place (Scheme 6.12*a*).

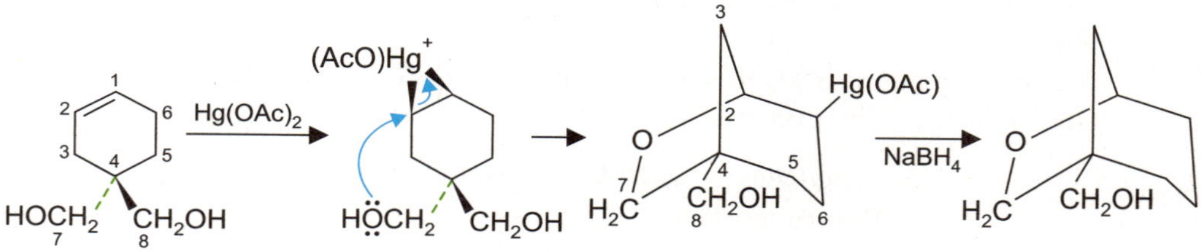

SCHEME 6.12a

EXERCISE 6.1a

What products you expect from the reactions (Scheme 6.12b)?

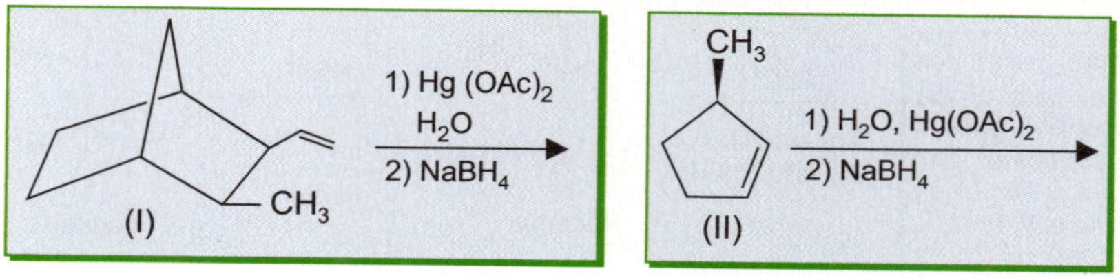

SCHEME 6.12b

ANSWER. *When olefins, have a chirality center elsewhere in the compound and if optically active starting materials are used, then the diastereomers will be optically active. If racemic starting materials are used, the diastereomeric mixture will be optically inactive. In either case different amounts of the two diastereomers, are formed. Alkene (I, Scheme 6.12b) gives, diastereomers. Alkene (II) gives two constitutional isomers, each of which is a mixture of two diastereomers.*

Continued...

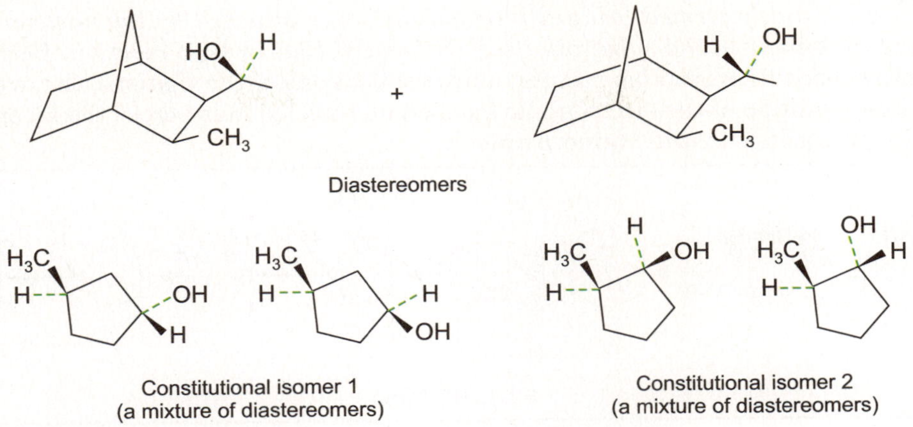

SCHEME 6.13

(G) Addition of HX-Hydrohalogenation and Hydration of Alkenes Markovnikovs Additions

The halogen acids *e.g.*, HCl add to alkenes (unsymmetrical) via the addition of a proton (electrophile) to give a carbocation. The orientation of addition follows the Markovnikov rule—The electrophile adds so as to form the most stable carbocation where the nucleophile then adds.

EXERCISE 6.2

Predict the products from reactions (I and II, Scheme 6.14).

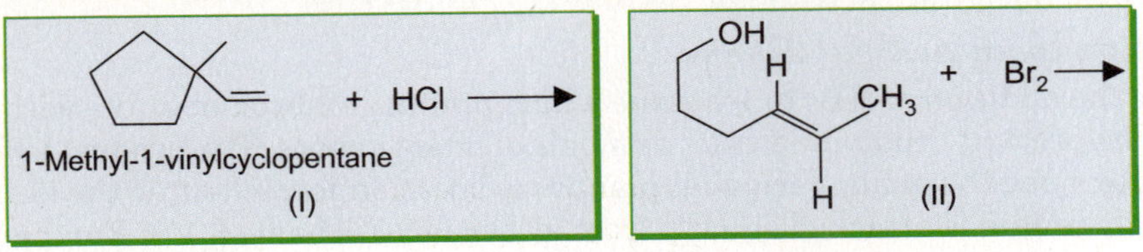

SCHEME 6.14

ANSWER. *In (I) the initially formed carbocation (secondary) rearranges to give a more stable carbocation (tertiary, Scheme 6.15a).*

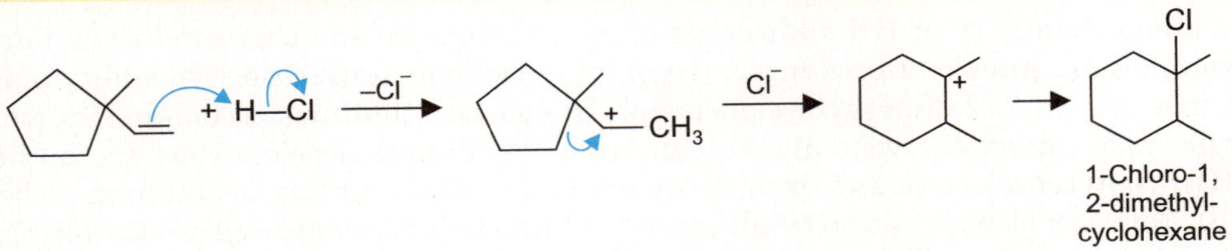

SCHEME 6.15

Continued...

CHAPTER 6

In (II) the initially formed bridged bromonium ion is attacked by the nucleophilic OH group present in the substrate itself (Scheme 6.15a) by S_N2 reaction. Because the S_N2 substitution reaction results in inversion at one of the C atoms, the overall reaction—addition of the electrophilic halogen and nucleophile across the π bond—usually proceeds with *anti* stereochemistry.

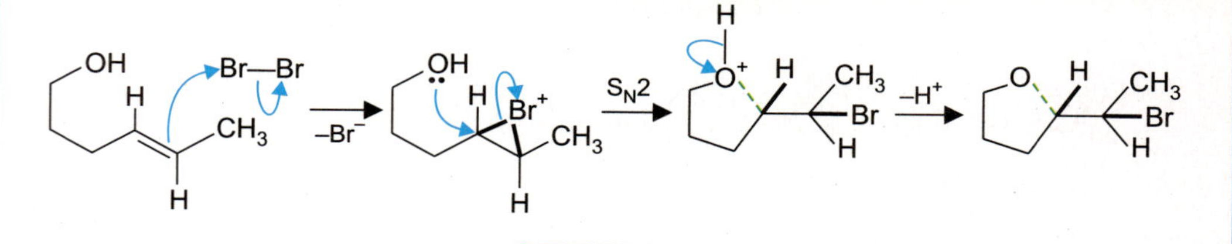

SCHEME 6.15a

As already discussed, the alkene (I, Scheme 6.16) with a chirality center adds bromine to give diastereomers.

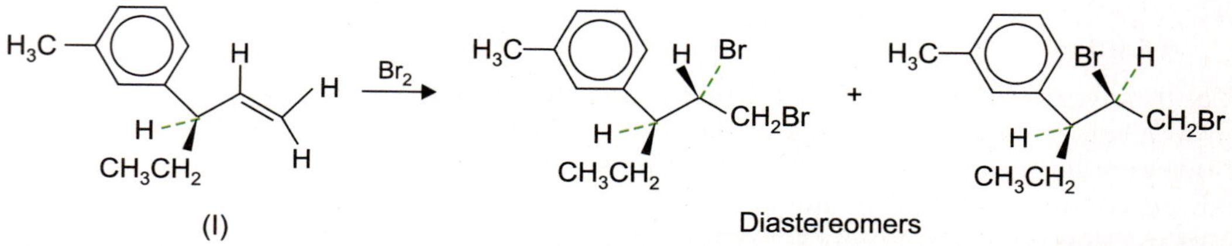

SCHEME 6.16

(H) Stereochemistry of Addition of HX and H_2O to Alkenes

1. *Addition to an Achiral Alkene*

Consider the addition of HBr to 1-butene which produces 2-bromobutane with a stereocenter. The product thus would exist as a pair of enantiomers. The product is formed as a racemate since the achiral trigonal planar carbocation formed after the electrophilic addition by proton is attacked equally from either face (Scheme 6.17). Further detailed discussion is given (see, Scheme 1.88).

2. *Addition of HX to a Chiral Alkene*

For detailed discussion (see, Scheme 1.95).

3. *Stereochemistry of HX Additions to an Alkene*

The stereochemistry of HX addition is varied, though mostly *anti* product is formed [Examples are known of predominant *syn, anti* and non-stereoselective addition as in the case of *e.g.,* 1, 2-dimethylcyclohexene]. In fact *anti* addition predominates with a simple nonconjugated cyclic alkene *e.g.,* with 1, 2-dimethylcyclohexene the addition of HBr is stereoselective and exclusively product of *anti* addition is obtained (Scheme 6.18). Both *syn* as well as *anti* products are obtained on many conjugated double bonds as in acenaphthylene (Scheme 6.18).

When proton is the electrophile the possibility of a cyclic intermediate is reduced (to account for stereoselectivity) since, hydrogen is not able to stabilize the positive charge as is done by *e.g.* bromine to form a bromonium ion. Hydration though regiospecific often proceeds by random *anti* and *syn* stereochemical pathway (Scheme 6.18*a*). Thus addition of water to 1, 2-dimethylcyclohexene gives almost equal amount of products of *syn* and *anti* addition.

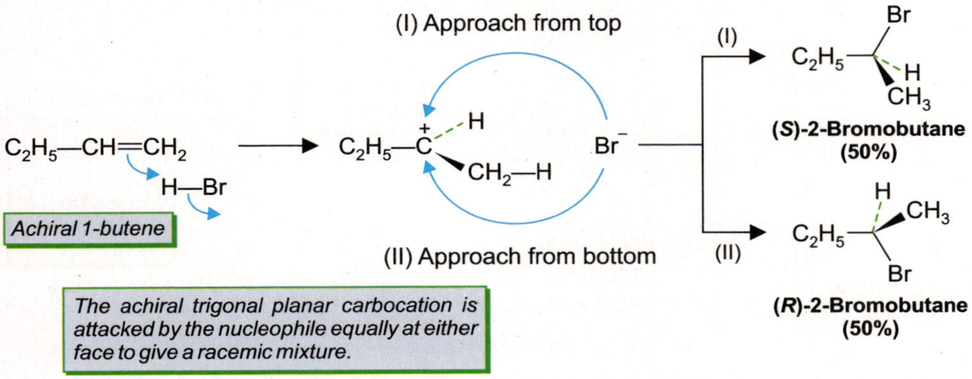

SCHEME 6.17

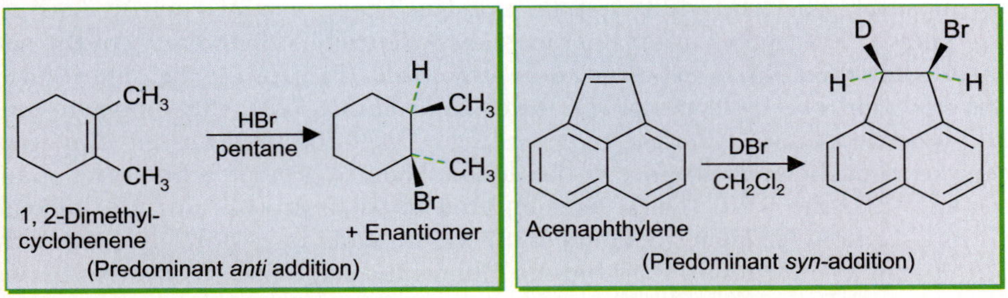

SCHEME 6.18

Consider the addition of DCl (deuterium chloride) to *Z*-1-phenyl-propene (Scheme 6.19). This is Ad_E2 addition and results in the formation of *erythro* product in greater amount (*syn*-addition) than the *threo* isomer (*anti*-addition). These results are clearly in contradiction to the formation of a free carbocation. In this case one may explain the predominant *syn*-addition by invoking the formation of a "tight ion-pair" (Scheme 6.19) in which the nucleophile is very closely associated with the carbocation that it attaches itself to the cationic centre before the bond can rotate to give a product of *syn*-addition. However, interestingly in other additions which involve carbocation intermediates, *anti* addition takes place to a greater extent and this fact further puts to doubt the true nature of the cation.

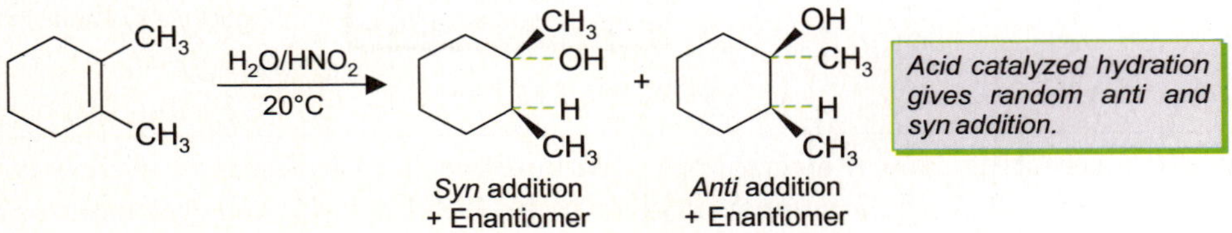

SCHEME 6.18a

CHAPTER 6

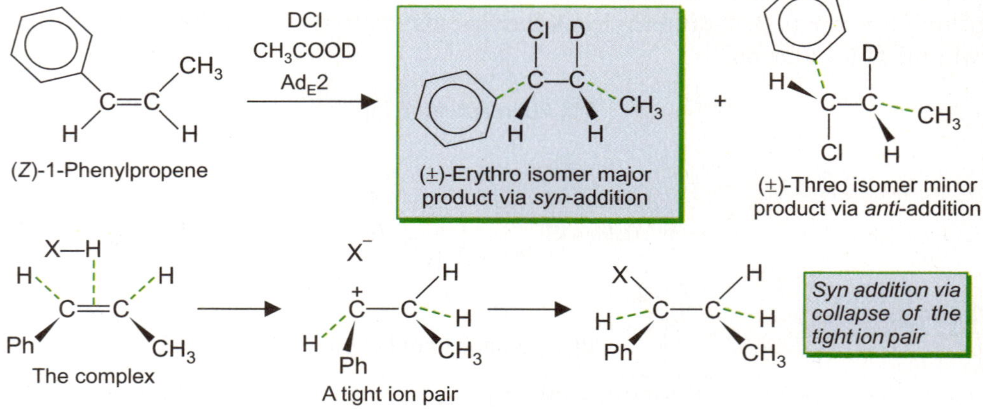

SCHEME 6.19

It is found that electrophilic addition of a halogen acid to an alkene from which relatively unstable carbocations are formed show a greater than first-order dependence on the concentration of the addition reagent E-Nu. These reactions are Ad_E3 Ad_E4 etc., and are known. The higher order reactions are favoured as the polarity of the solvent decreases and when HX is expected to be unionized. To account for *anti*-addition, it is suggested that one molecule of HX forms a π complex with the substrate and the second molecule attacks from the rear leading to *anti*-addition, *i.e.*, each molecule of HX attacks from the opposite face of the double bond to provide a nucleophile and an electrophile (Scheme 6.20). Thus, as compared to Ad_E2 process during the reaction of DCl to *Z*-1-phenylpropene (Scheme 6.19) which leads to *erythro* product in larger amount, a similar addition to *cis*-2 butene follows the Ad_E3 process leading instead to the formation of threo product in larger amount (Scheme 6.20).

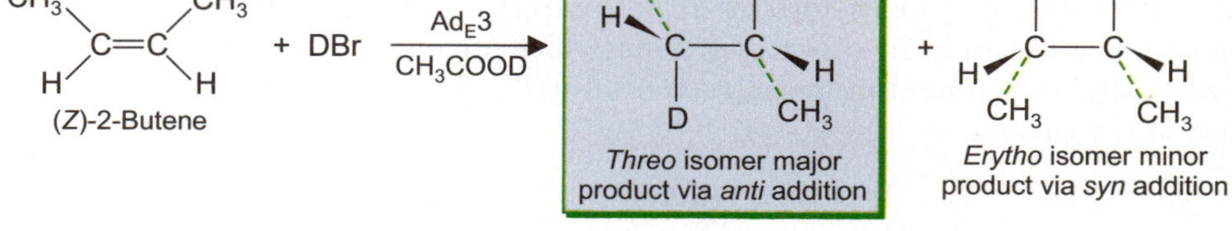

SCHEME 6.20

An interesting case of electrophilic *syn*-addition is also observed with *trans*-cyclooctene (Scheme 6.21). This compound gives *trans*-1, 2-disubstituted cyclooctanes. The preference for *syn*-addition is due to the inaccessibility of the rear face of the double bond which is shielded by the chain of methylene groups of the eight membered ring.

Addition of HX to norbornene presents an interesting example where steric factors and molecular rearrangement are involved. Addition of DBr to norbornene shows preference for *syn*-addition *i.e.*, *syn-exo* since *endo* side is sterically hindered. The *syn* addition leading to (I, Scheme 6.21) is via the collapse of ion pair. The presence of deuteriam at C–3 and C–7 position also indicate the involvement of a bridged ion (II, Scheme 6.21) in addition to ion pair formation.

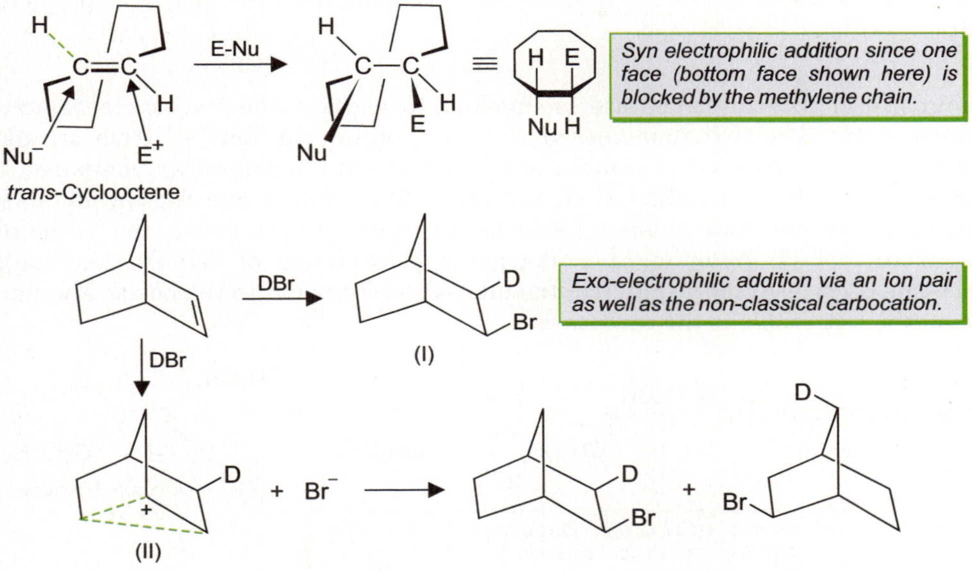

SCHEME 6.21

EXERCISE 6.3

What product one expects from reaction (Scheme 6.21a)? Depict the stereochemical outcome at every step.

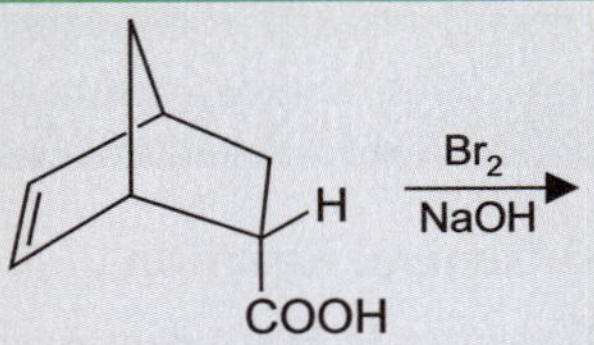

SCHEME 6.21a

ANSWER. *It is a typical reaction involving the formation of a cyclic bromonium ion on the less hindered exo-face of the molecule followed by intramolecular attack by COOH group (Scheme 6.21b)*

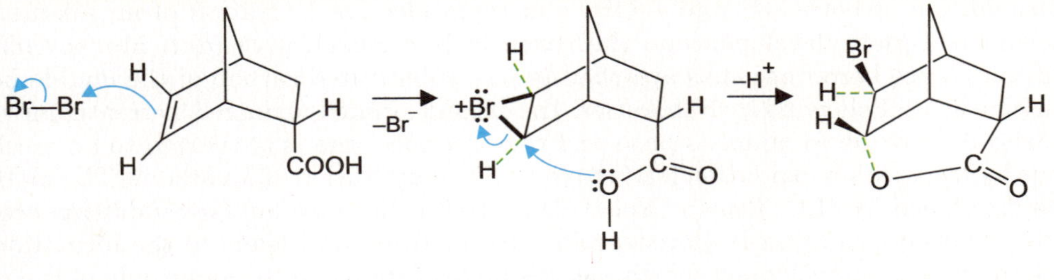

SCHEME 6.21b

6.2 ADDITION REACTIONS OF ALKYNES INVOLVING ELECTROPHILES

Alkynes are somewhat less reactive towards electrophiles. The additions follow Markovnikov rule. These reactions are closely similar to alkenes, however, stereoselectivity is lower. The addition of both bromine and chlorine give generally *anti*-addition which is in keeping with a bridged ion intermediate. The following two reactions prove these points.

(*i*) Bromination of Alkynes

A bromonium like intermediate is invoked to explain the *anti*-stereoselectivity (Scheme 6.22). The three membered ring bromonium ion derived from an alkyne (I, Scheme 6.22) has a full double bond and is thus strained as compared to a bromonium ion from an alkene (II, Scheme 6.22). This is one reason for alkynes to be less reactive than alkenes towards electrophiles. Moreover, the bromonium ion carbons of (I) have more *s* character than those of (II) further making (I, Scheme 6.22) less stable than (II), the more the *s* character in the positively charged carbon, the less stable is the carbocation.

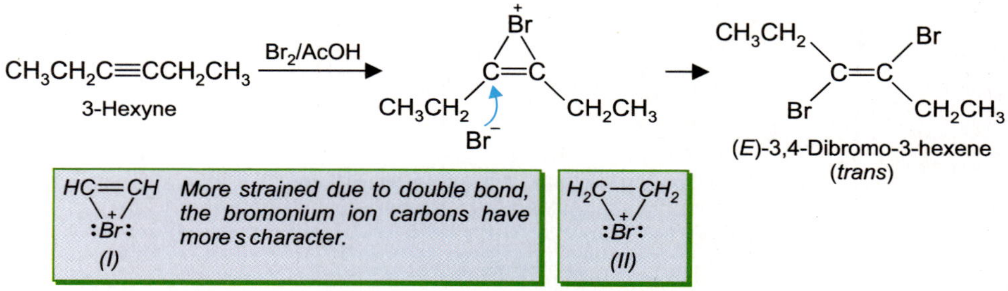

SCHEME 6.22

6.3 HYDROBORATION/OXIDATION HYDRATION OF ALKENES: A STEREOSPECIFIC ANTI-MARKOVNIKOV HYDRATION

(*a*) General Reaction

Diborane B_2H_6 (a dimer of borane, BH_3) adds readily to alkenes to givemorganoboranes. [Diborane is commercially available in the form of its complex with tetrahydrofuran (I, Scheme 6.23)]. Hydroboration oxidation is an useful method for both the stereospecific (*syn* addition of both —H and —OH) and regioselective hydration of an alkene. As shown for 1-methylcyclopentene the reaction is regioselective (*anti* Markovnikovs addition) since boron becomes attached to less substituted carbon of the double bond (Scheme 6.23). Following hydroboration, the organoborane is oxidized by treatment with hydrogen peroxide in aqueous base and the organoborane is converted to an alcohol. Thus hydroboration–oxidation leads to the overall hydration of an alkene. The method was developed by H.C. Brown (Nobel Prize 1979). Because only *syn*-addition occurs hydroboration–oxidation is stereospecific, *syn*-addition and leads to the formation of only the pair of enantiomers which has the added groups on the same side of the ring (see, Exercise 6.5).

(*b*) Regiochemistry of Addition and Mechanism

Borane, the monomer of diborane is very reactive electrophile since it has only a sextet of electrons. The addition process though concerted (no carbocation intermediate is involved) has some polar character in the transition state. Thus *e.g.*, consideration of transition states from the addition of borane to propene (Scheme 6.24), the transition state (I, Scheme 6.25) with boron at the primary carbon atom is more stable (The developing positive charge is on the secondary carbon) than the other transition state (II, Scheme 6.24) where the developing positive charge will be on primary carbon. Thus even though a carbocation intermediate is not formed a carbocation like transition state is formed (for this reason hydroboration/oxidation process occurs without rearrangement.

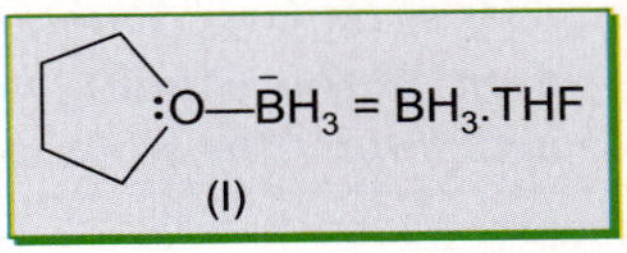

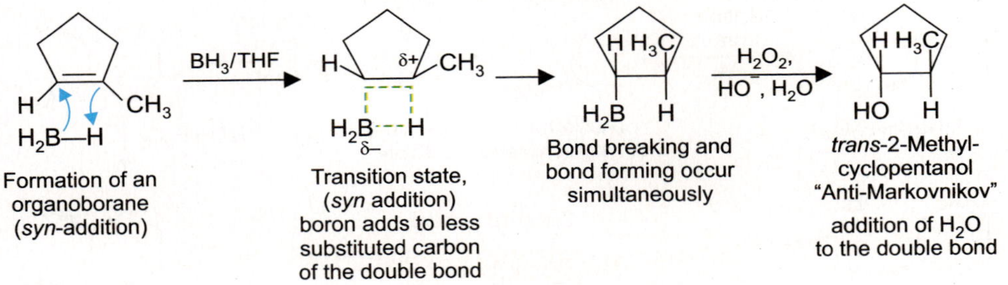

SCHEME 6.23

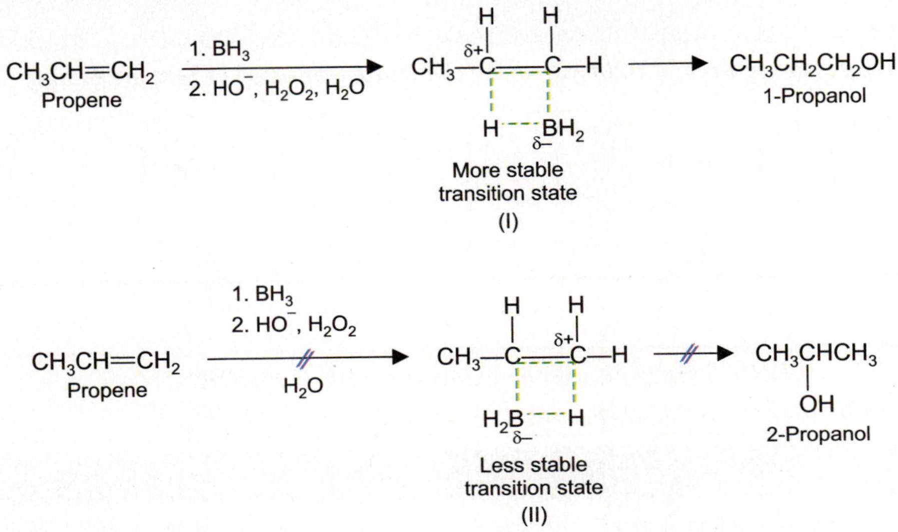

SCHEME 6.24

EXERCISE 6.4

Depict the addition of an achiral borane to an achiral trisubstituted alkene followed by oxidation/hydrolysis.

ANSWER. *The hydroboration of unsymmetrical alkenes yields monoalkylboranes (addition of H—BH_2), dialklylboranes (addition of H—BHR), or trialkylboranes (addition of H—BR_2), which are typical anti-Markovnikov products. Thus, the reaction sequence hydroboration/oxidation/hydrolysis brings about the anti-Markovnikov addition of H_2O to unsymmetrically substituted alkenes, as a syn-addition to give alcohols.*

Thus e.g., 9-BBN adds to both sides of its C=C double bond of 1-methylcyclohexene at the same reaction rate. The reason for this equality is that the transition states for both pathways are enantiotopic which have the same energy (just like enantiomers).

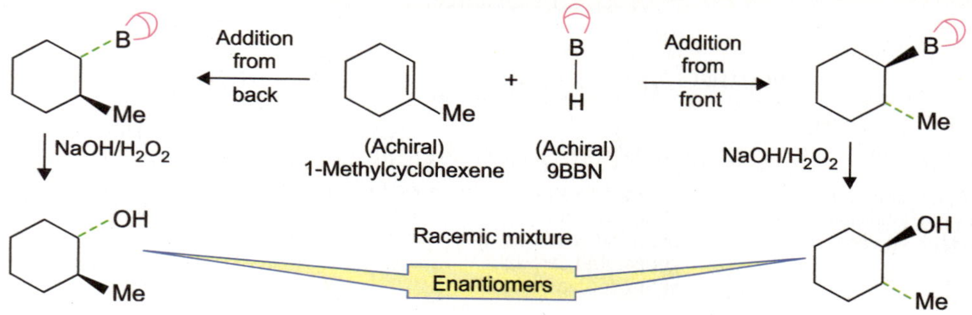

SCHEME 6.25

EXERCISE 6.5

What reagents are needed to bring about the transformations (Scheme 6.25a) ?

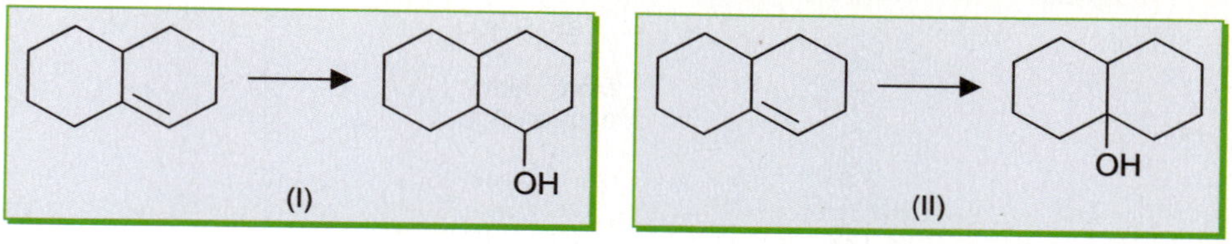

SCHEME 6.25a

ANSWER. *(i) BH_3 . THF, followed by oxidation with H_2O_2/OH^-*
(ii) Treatment with $Hg(OAc)_2$ followed by reduction with $NaBH_4$.

EXERCISE 6.6

How one can convert indene (I, Scheme 6.25b) into (II and III)? How ^{13}C NMR can be used to distinguish between them?

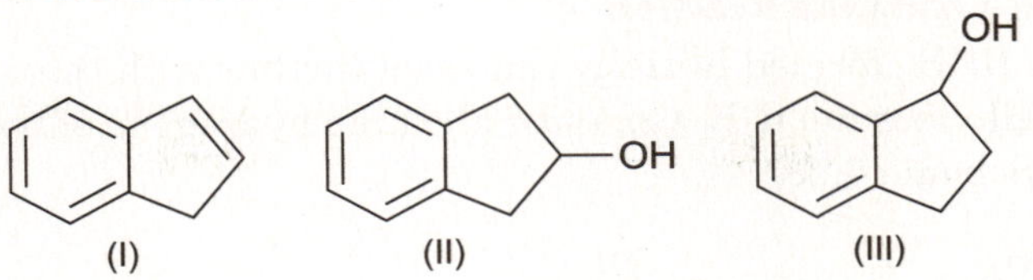

SCHEME 6.25b

ANSWER. *Compound (III) can be obtained by Markovnikov hydration (Scheme 6.25c) compound (II) can be made by hydroboration-oxidation which is an anti-Markovnikov addition (Scheme 6.25d). Compound (II, plane of symmetry) shows five ^{13}CNMR signals while compound (III) shows nine.*

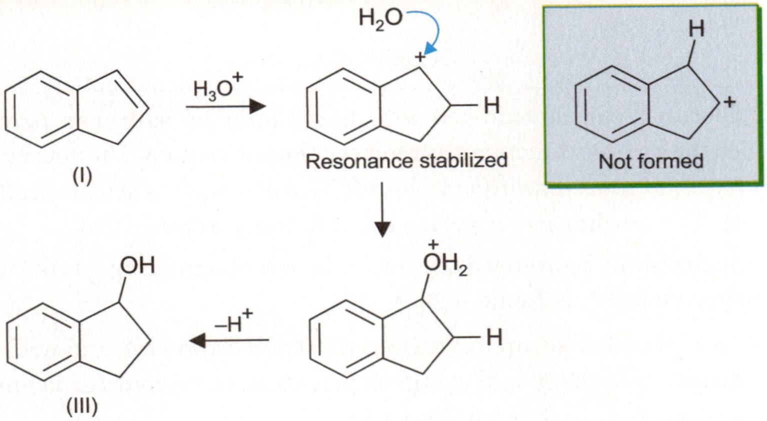

SCHEME 6.25c

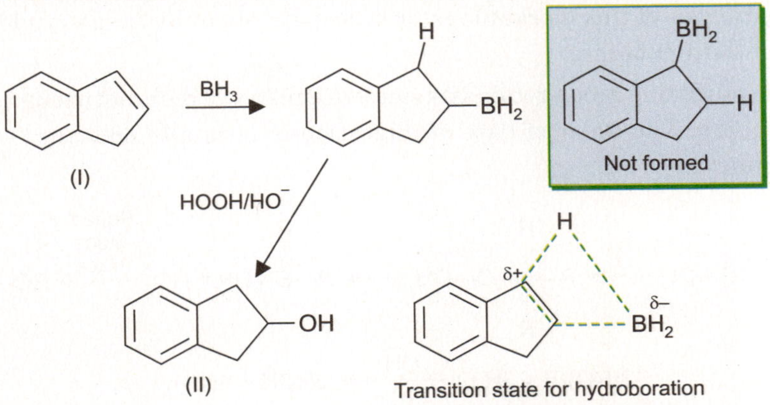

SCHEME 6.25d

The second tactic to explain the regiochemistry of hydroboration is steric effect. The borane containing group (BH_2) is much bulkier than hydrogen, thus boron attaches with the less hindered end (with more hydrogens) of the double bond.

(c) Hydroboration–Oxidation Reaction

An organoborane RBH_2 formed initially can react further with the alkene in stages to finally form a trialkylborane R_3B, *i.e.*, until all three hydrogens of borane are replaced by alkyl groups (Scheme 6.26).

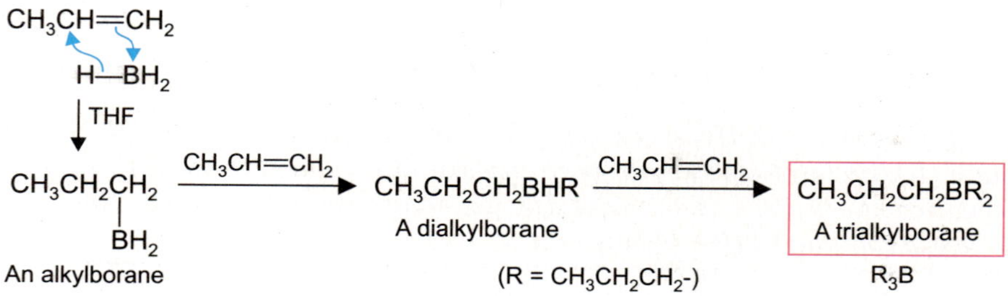

SCHEME 6.26

Trialkylboranes can be oxidized with basic aqueous hydrogen peroxide to give alcohols in which the hydroxy function has replaced the boron atom. The net result of this two step sequence *hydroboration-oxidation* is the addition of elements of water to a double bond (Scheme 6.26*a*). The mechanism involves the following steps:

- The nucleophilic hydroperoxide ion adds to boron making use of boron's vacant *2p*-orbital (step 1, Scheme 6.26*a*).
- In step 2, an alkyl group migrates with its electron pair and with retention of configuration to the neighboring oxygen atom and a hydroxide ion is expelled.
- The process is repeated till all the three alkyl groups get migrated to oxygen atoms to yield finally a trialkylborate (A, Scheme 6.26*a*).
- Hydrolysis of this inorganic ester is brought about by base to yield and alcohol and sodium borate.
- The migrating group retains its stereochemistry, if different groups are attached to boron. The order of ease of migration is normally primary > secondary > tertiary.

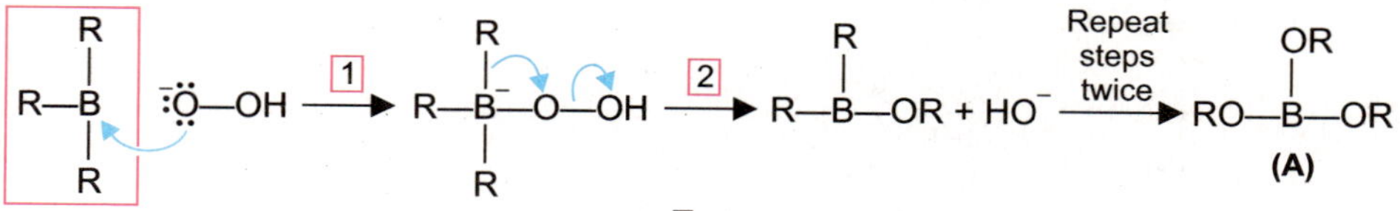

$$3(RO)_3B + 3H_2O \xrightarrow{\bar{O}H} 3ROH + B(OH)_3$$

(A)

The oxidation of an alkyl borane

SCHEME 6.26a

(*d*) Sterically Congested Boranes. Regioselective Hydroboration of Unsymmetrical Alenes

With sterically hindered alkenes reaction can be controlled and monoalkyl or dialkyl-boranes can be obtained (Scheme 6.27). These organoboranes with one or two B–H bonds, are of value as boronating agents, and sterically congested boranes like 9BBN and Sia_2 BH (Scheme 6.27)

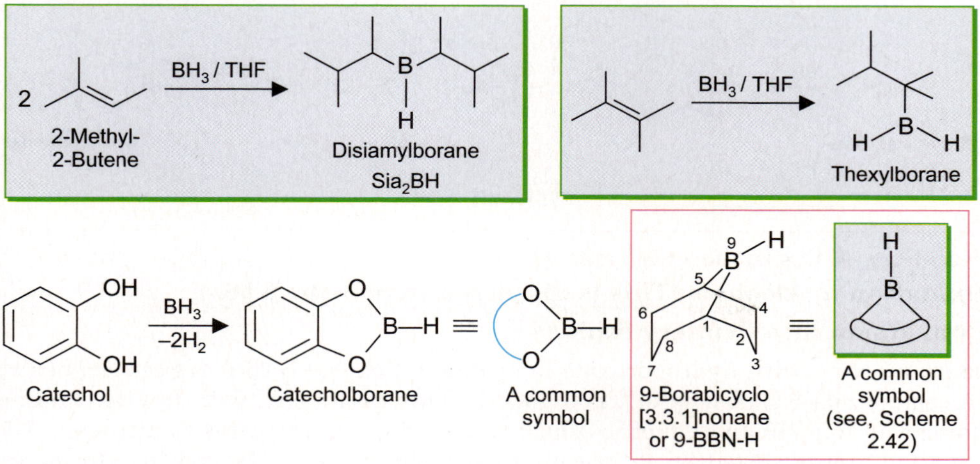

SCHEME 6.27

are very strongly regioselective due to steric bulk of the reagent. Catecholborane is another useful reagent derived from catechol.

The boron atom adds predominantly to the less substituted carbon atom. However, with 1, 2-disubstituted alkenes this selectivity is lost or is less pronounced (Scheme 6.28). By using a bulkyl reagent *e.g.* 9-BBN the boron atom almost exclusively adds to that carbon atom which has the less bulky of the two substituents, thus use of 9-BBN is almost a perfect solution to regioselectivity. The B-H bond of 9-BBN is much less readily accessible than the B-H bond in BH_3. Thus, the hydroboration with 9-BBN occurs much more slowly than the one with BH_3, though with much higher regioselectivity.

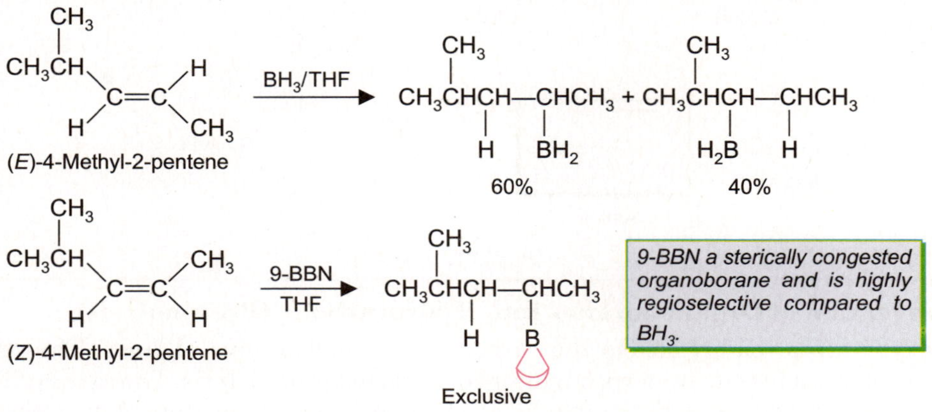

SCHEME 6.28

(e) Isomerization of Organoboranes

On heating organoboranes to about 150°C an isomerization occurs. Alkenes are formed and reconverted to organoboranes till the least substituted organoborane is formed (Scheme 6.29). By using this method a readily available internal alkene can be converted *e.g.*, into a primary alcohol.

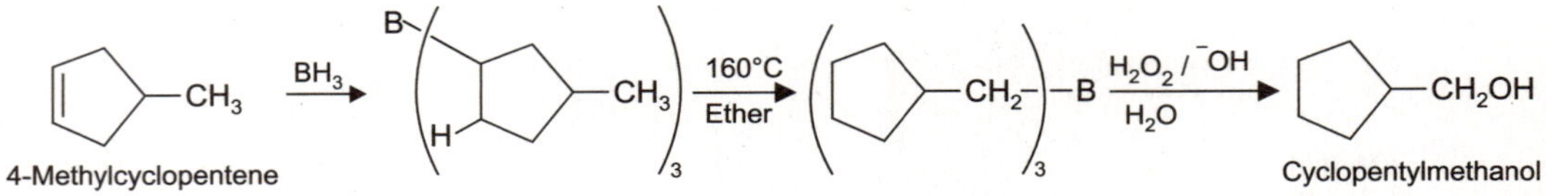

SCHEME 6.29

(f) Reactions of Organoboranes from Alkenes

1. Oxidation to Alcohols (This is discussed in Scheme 6.26a)

2. Conversion into Primary Amines

Like its reaction with hydroperoxide ion (Step 2, Scheme 6.26*a*) organoboranes react with compounds NH_2X where X is a good leaving group *e.g.*, with hydroxylamine-*O*-sulfonic acid to yield an adduct in which the alkyl group migrates to nitrogen. All the three alkyl groups, however, do not migrate in this case in contrast to hydroperoxide ion. Methyl groups attached to boron resist migration. Thus dimethylborane is used in the first step and only the alkyl group derived from the alkene reactant migrates in the reaction (Scheme 6.30, a good leaving group HSO_4^- initiates the migration of the alkyl group to nitrogen).

The nitrogen containing reagent initially reacts as a nucleophile by adding at boron and a boron to nitrogen rearrangement with the loss of chloride or sulphate ion follows.

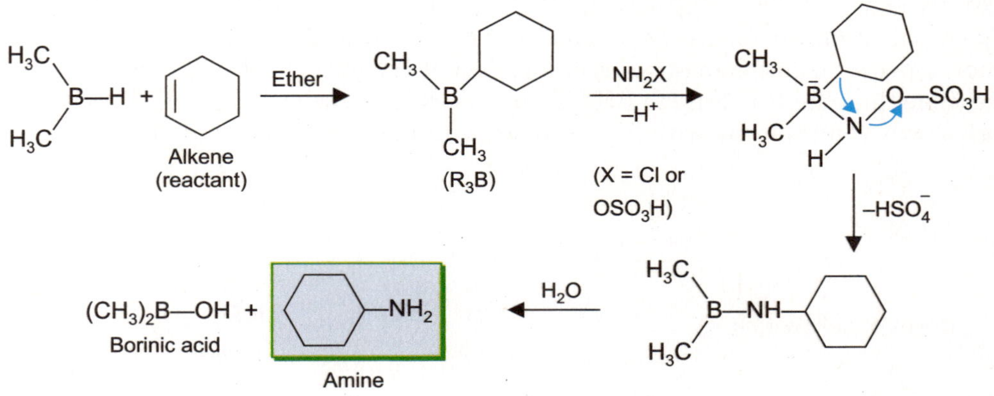

SCHEME 6.30

3. Conversion of Organoboranes into Hydrocarbons (Protonolysis)

Carboxylic acids readily cleave the carbon-boron bond to give alkanes. The process which replaces a hetero atom with hydrogen is termed protonolysis. The carboxylic acid firstly forms an adduct by reacting with electron deficient boron atom. A six membered transition state is involved in the reduction process (Scheme 6.31) which also accounts for complete retention of configuration at the site of reduction *i.e.*, the carbon-carbon bond.

In the case of dienes, it is possible to reduce selectively the less substituted alkene by using a hindered borane such as that from trimethylethylene, Sia_2BH (I, Scheme 6.31)

The protonolysis of carbon-carbon bond is used to introduce deuterium at a specific position in an organic compound (*e.g.* II, Scheme 6.31).

$$R_3B + 3R'CO_2H \longrightarrow 3RH + (R'CO_2)_3B$$

(Solvolysis of organoboranes with organic acids)

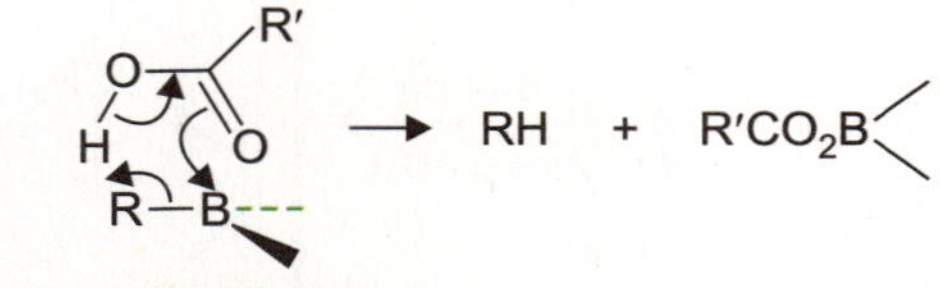

(The transition state is six membered)

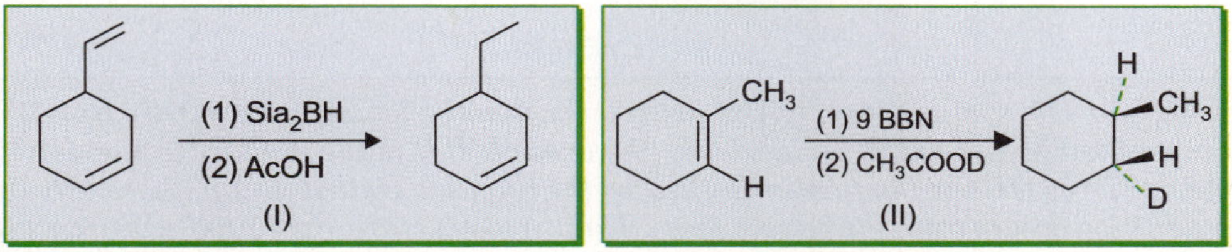

SCHEME 6.31

Stereospecific and Stereoselective Reactions Revisited

One may recall the definitions of stereospecific and stereoselective reactions (Scheme 1.99–1.106). If stereoisomers of a substrate react under identical reaction conditions to give products that are stereoisomers of each other, the reaction is called stereospecific. When the reactant is non-stereoisomeric but leads to stereoisomeric products in unequal amounts the reaction is said to be stereoselective.

More precisely stereospecific reactions have a mechanism which demands a specific stereochemical outcome. Stereoselective reactions offer alternative pathways so that the reaction proceeds via. (i) the most favourable pathway (kinetic control) or (ii) via a pathway which gives the most stable stereoisomer as the major product (thermodynamic control). Thus hydroboration-oxidation of α-pinene is stereoselective. Hydroboration–oxidation gives syn hydration of carbon–carbon double bonds with a regioselectivity opposite to Markovnikov's rule. Reagent attacks at the less hindered face of the double bond and only one alcohol is formed in high yield (Scheme 6.32).

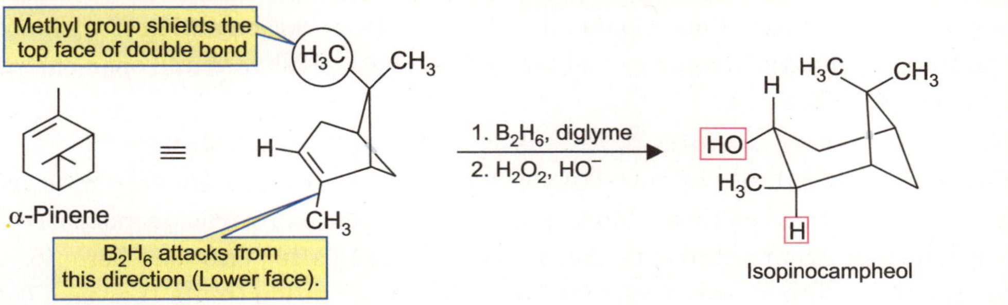

SCHEME 6.32

EXERCISE 6.7

Write the stereostructures of the products from reactions (Scheme 6.33).

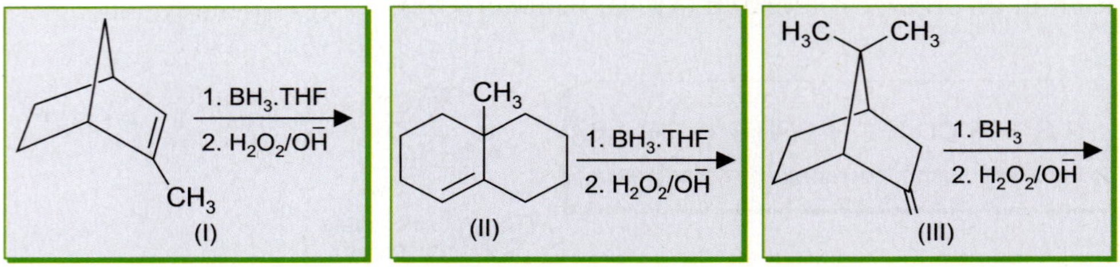

SCHEME 6.33

ANSWER. *The syn addition of BH_3 across the double bond occurs mostly from the more accessible face of the molecule. In the case of (I) it is the exo face of the double bond, while in (II) the top face is blocked by the angular methyl group, therefore, the syn addition occurs from the bottom faces. The stereostructure of the products are given (Scheme 6.33a). In the case of (III) the gem dimethyl group at C7 hinders the exo face.*

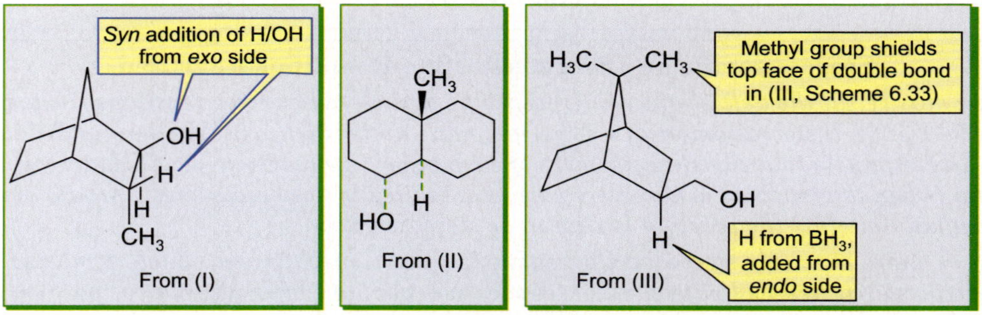

SCHEME 6.33a

4. Carbonylation of Organoboranes

Carbon monoxide forms Lewis acid-base complexes with organoboranes, and these adducts undergo boron-to-carbon migration of the boron substituents. The reaction can be controlled and therefore, directed in the migration of one, two or all the three substituents on boron. Conditions can be defined so that carbonylation of organoboranes can lead to primary, secondary and tertiary alcohols, aldehydes and open chain, cyclic and polycyclic ketones.

(*i*) Conversion of boranes to aldehydes and primary alcohols

When the carbonylation of the trialkylborane is done in the presence of a mild reducing agent *e.g.* lithium trimethoxyaluminium hydride or potassium triisopropoxyborohydride the reducing agent intercepts the intermediate (A) after only one boron-to-carbon migration has taken place (Scheme 6.34). This on oxidation (route I) gives aldehydes, or primary alcohols (route II, Scheme 6.38). An inspection of Scheme 6.38 shows that, the method as such has a disadvantage since only one of the three alkyl groups of the starting trialkylborane is converted into aldehyde, the others are wasted. This difficulty is solved by hydroboration of the alkene with 9-BBN. The B-alkyl derivative on reaction with CO in the presence of a reducing agent is attended with the preferential migration of the alkyl group (there is minimal tendency of the bicyclic ring to undergo migration) to give high yields of the aldehyde. While working with B-alkyl-9-BBN, since only the 9-alkyl group migrates, this method, thus converts (high yields) an alkene to a primary alcohol or an aldehyde containing one more carbon (Scheme 6.34). Significantly the use of chiral boranes gives products of retained configuration.

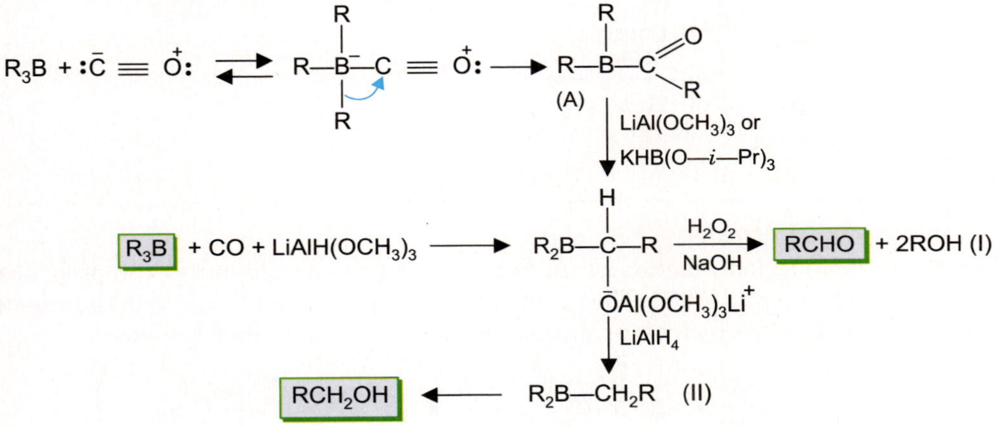

SCHEME 6.34

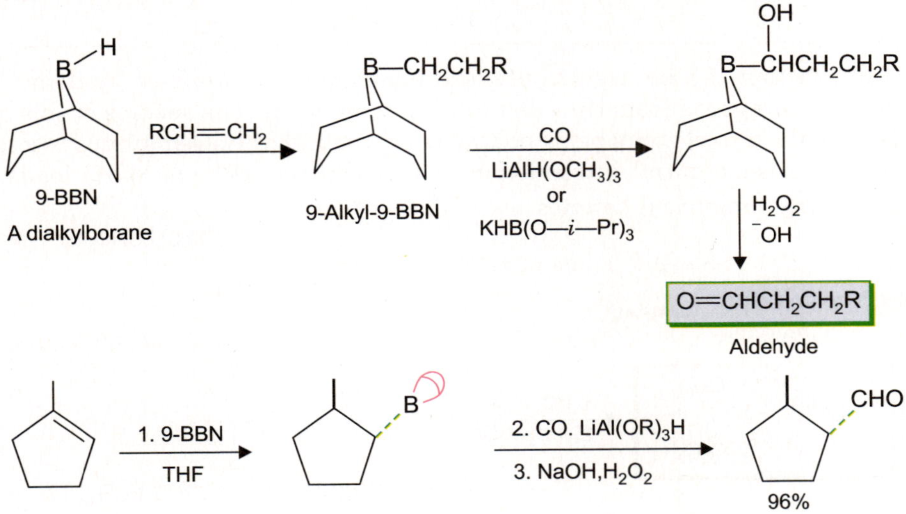

SCHEME 6.35

Another efficient method for the synthesis of aldehydes by boron homologation from alkenes involves boronic esters and their reaction with methoxy (phenyllithio) methyl lithium. The migration step is induced by mercuric ion (Scheme 6.35*a*). The use of chiral boronates makes the reaction enantioselective.

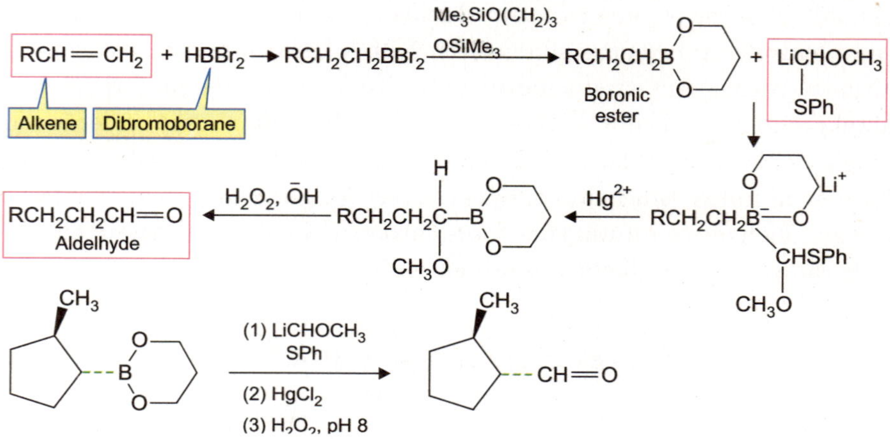

SCHEME 6.35a

(ii) Synthesis of Ketones

When the carbonylation reaction of (A, Scheme 6.34) is carried out in the presence of small amount of water at 100°C. The second alkyl group migrates from (A) to the adjacent carbon (Scheme 6.36), which on oxidation gives the corresponding ketone.

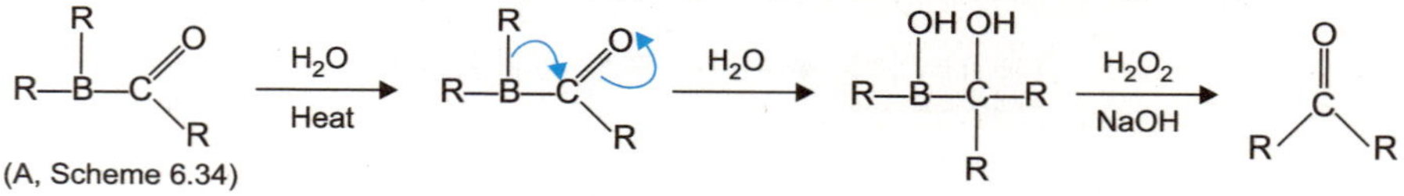

SCHEME 6.36

The reaction can be used to prepare unsymmetrical ketones by using 'mixed' organoboranes prepared from thexylborane (Scheme 6.37). The success of this method depends upon the thexyl group being noncomepetitive with the other groups in migration steps. Thus the sequenitial introduction of two alkenes (Scheme 6.37) leads to the synthesis of unsymmetrical ketones.

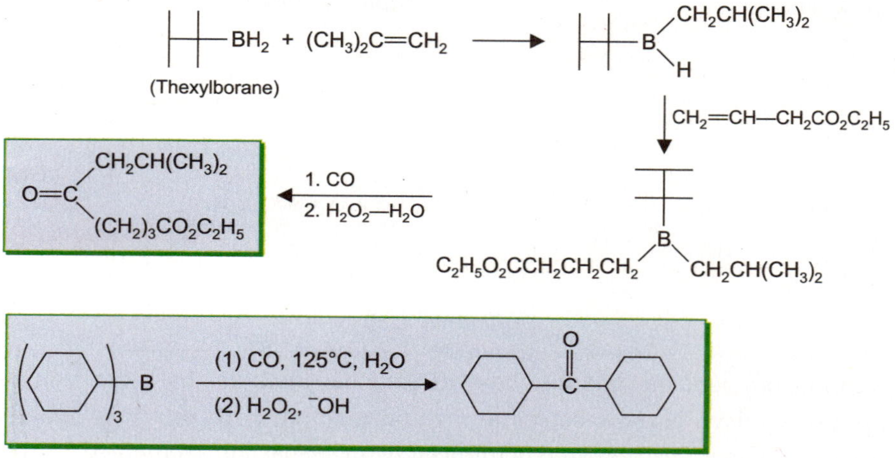

SCHEME 6.37

By working with appropriate dienes one can end up with cyclic or bicyclic ketones (Scheme 6.38).

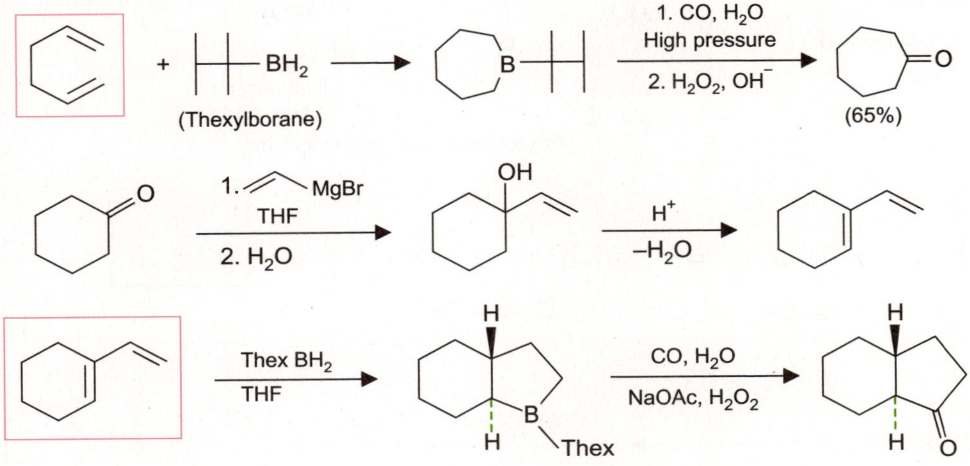

SCHEME 6.38

(*iii*) Synthesis of *tert*-alcohol

Carbonylation of trialkylboranes in the presence of ethylene glycol results in migration of a second and a third alkyl group to give, after oxidation, the corresponding *tert*-alcohols (Scheme 6.39).

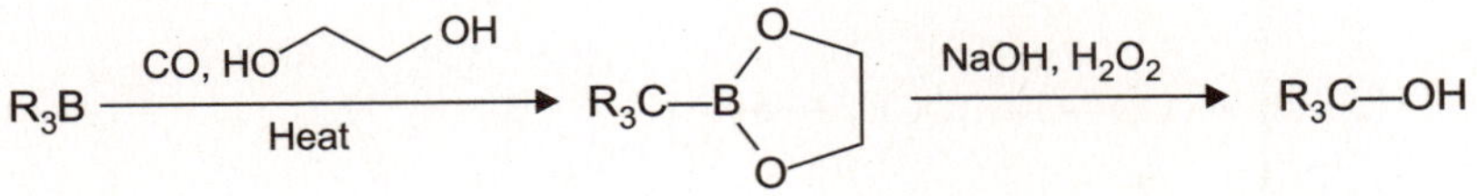

SCHEME 6.39

5. Cyanidation of Organoboranes

Cyanide ion is isoelectronic with carbon monoxide and a borane reacts similarly to initially form an adduct with cyanide ion, on reaction with sodium cyanide (Scheme 6.40). Thus this is another reagent besides carbon monoxide which serves as the electrophilic migration terminus. The nitrogen atom in the initial adduct is not sufficiently electron attracting to induce migration. This ability is increased by acylation of the cyano group with trifluoroacetic anhydride. Two alkyl groups are transferred (Scheme 6.40) at low temperature to give ketones by oxidation. The method has a merit over carbonylation due to low temperature conditions of this reaction. Moreover, as in the carbonylation reaction, one can avoid the wastage of the alkyl groups by working with thexylborane. Also, the thexyl group does not migrate and unsymmetrical ketones can be easily obtained.

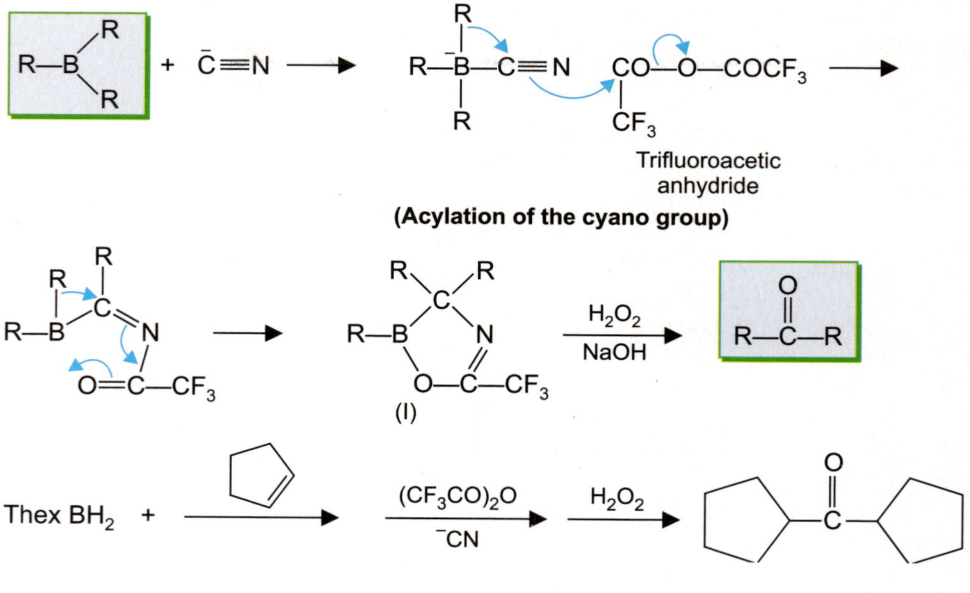

SCHEME 6.40

One can induce the migration of the third group also on (I, Scheme. 6.40) by using an excess of trifluoroacetic anhydride, to afford tertiary alcohols on oxidation (Scheme 6.41).

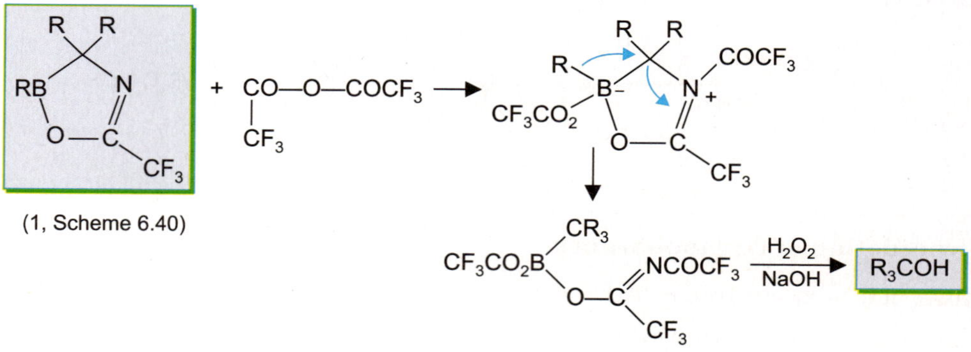

SCHEME 6.41

6. Synthesis of Esters

Excellent yields of esters can be obtained by reacting trialkylboranes with *e.g.*, enthylbromoacetate, (an α-haloester) in the presence of a base (*t*-BuOK, *t*-BuOH eqn. I, Scheme 6.42). The enolate adds to the organoborane to give a tetracoordinate boron intermediate on which the migration of an alkyl group occurs with displacement of bromide ion. Hydrolysis of the rearranged products gives an ester. Similar reactions are displayed by α-haloketones and α-halonitriles. The key step in this reaction is again the migration of an alkyl group from boron to adjacent carbon atom. Only one of the three alkyl groups of the trialkylborane is used in this reaction. This problem is solved as usual by using an alkyl derivative of 9-BBN instead of trialkylborane (eqn. II, Scheme 6.42). By following this method not only the alkyl group is fully utilized, but 9-BBN takes no part in alkylation. When β-alkenyl derivatives of 9-BBN are employed, one ends up with β-γ-unsaturated esters.

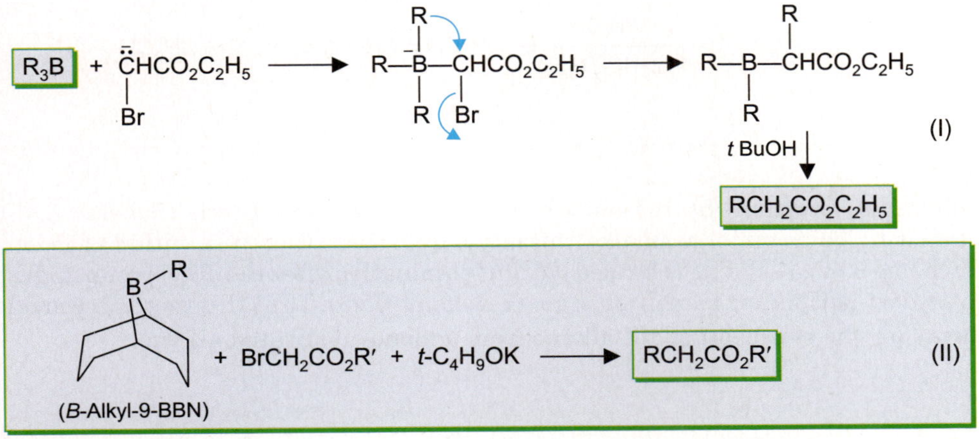

SCHEME 6.42

7. Hydroboration of Alkynes and Reaction of Derived Organoboranes

Borane adds to alkynes in a way similar to alkenes *i.e.,* one mole of BH_3 reacts with three moles of alkyne to form one mole of boron substituted alkene. On oxidation boron is replaced by an OH group and the enol rearranges to ketone (Scheme 6.43). For this synthesis of ketones, the reaction must stop at the alkene stage. For internal alkynes, the second addition is prevented by the bulky groups on the boron substituted alkene. In the case of a terminal alkyne, it is difficult to stop the addition at the alkene stage. Thus hindered boranes like, disiamylborane or 9-BBN-H (see Schemes 6.27 and 6.28) are used (infact for internal alkynes as well, hindered boranes are employed). By following this procedure internal alkynes give ketones while terminal alkynes give aldehydes (Scheme 6.43). This procedure is thus complementary to the direct hydration catalyzed by mercury (II) sulfate (Scheme 6.44, also see Scheme 6.23).

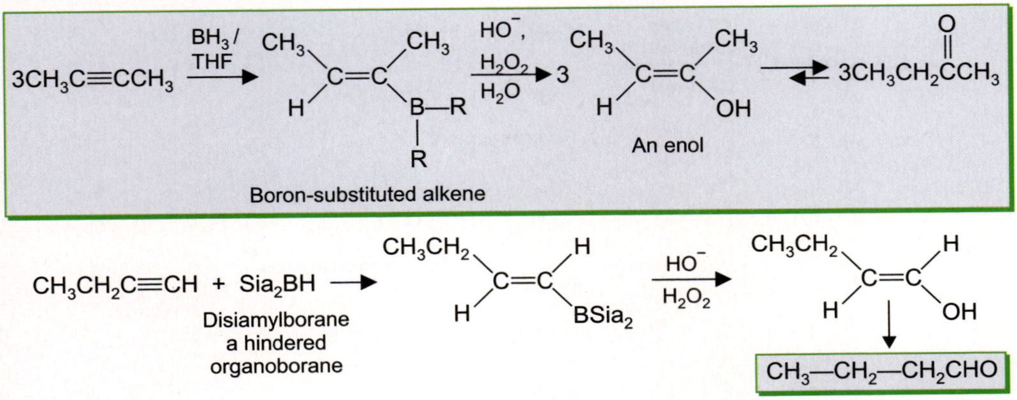

SCHEME 6.43

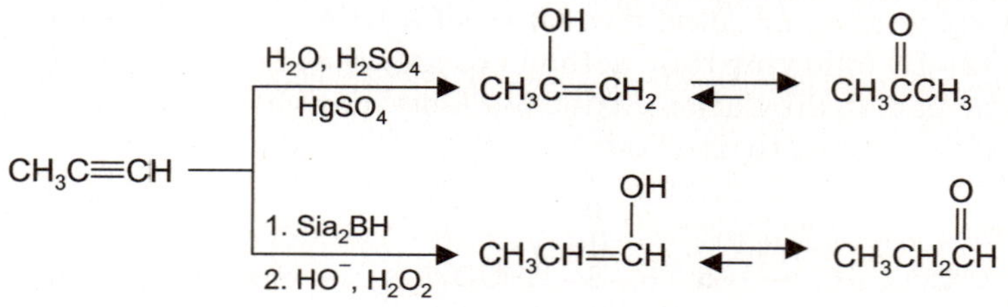

SCHEME 6.44

(*i*) Synthesis of Z-alkenes

Vinylboranes, prepared by the monohydroboration of triple bonds (Scheme 6.43) on reaction with iodine undergo an alkyl migration from boron to carbon, within an iodonium ion (Scheme 6.45). Z-alkene is formed *via anti* elimination after the alkyl group migration from an *anti*-periplanar transition state (I, Scheme 6.45). Thus this method, provides a pathway for the synthesis of a Z-alkene from a monosubstituted alkyne.

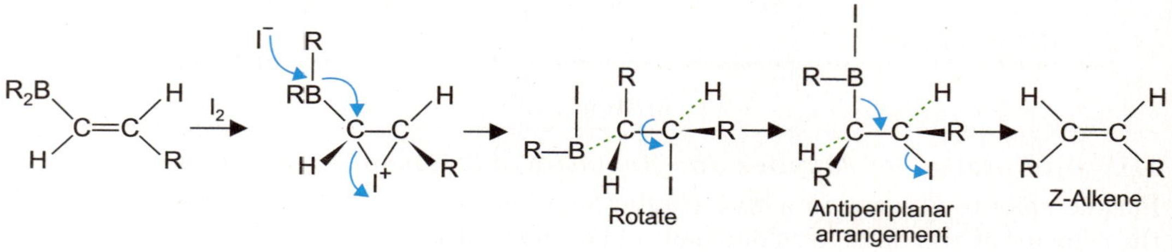

SCHEME 6.45

(*ii*) Synthesis of E-alkenes

For the synthesis of E-alkenes a 1-haloalkyne is used. The addition of a dialkylborane gives an α-halovinylborane (*i.e.* α-haloalkenylborane). On treatment with methoxide ion, this intermediate undergoes boron to carbon migration (Scheme 6.46) to afford an alkylated alkenylborane. Protonolysis gives an *E*-alkene.

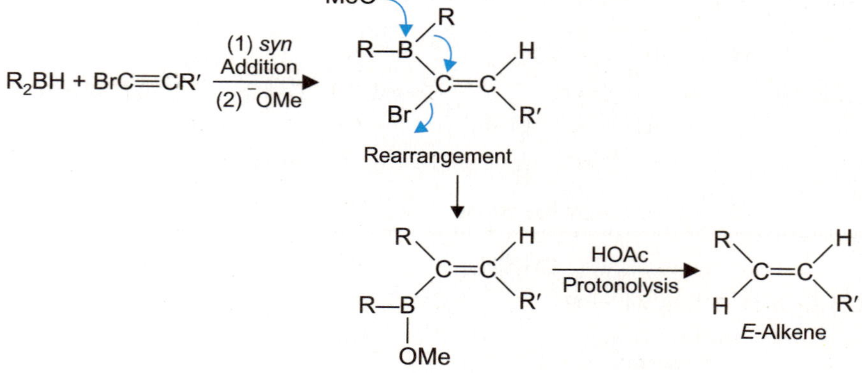

SCHEME 6.46

(*iii*) Synthesis of conjugated dienes

The principles used in the synthesis of *E*- and *Z*-alkenes can be applied for an appropriate geometry of the double bond(s) in a conjugated diene. Thus, *e.g.*, *E*, *E*-dienes can be made as shown (Scheme 6.47). Hydroboration of a 1-haloalkyne with thexylborane affords a thexyl-1-chloroalkyenylborane. This reacts with another alkyne to give thexyldialkenylborane. Reaction with methoxide ion inducts a rearrangement of alkenyl and protonolysis affords the diene.

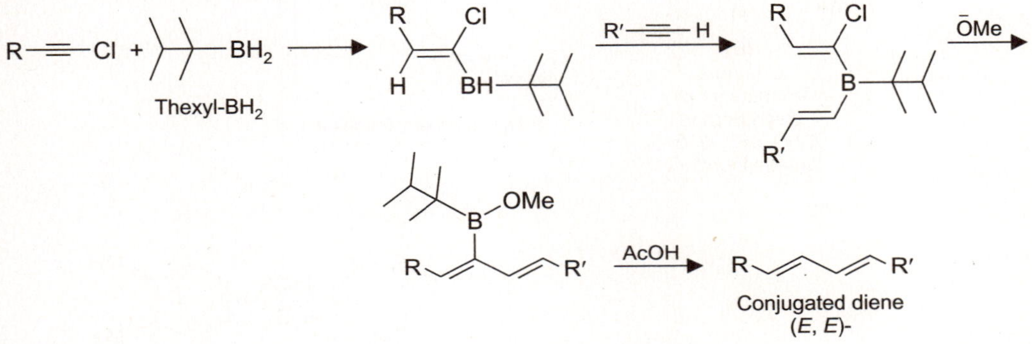

SCHEME 6.47

(*g*) Formation of Alkynes from Boranes and Acetylides

Organoboranes alkylate terminal acetylenes. Adduct formation occurs between a lithium acetylide and a trialkylborane. Reaction with iodine involves an electrophilic attack of iodine on the triple bond thereby inducing a migration of an alkyl group from boron to carbon (Scheme 6.48). This is followed by elimination of dialkyliodoboron.

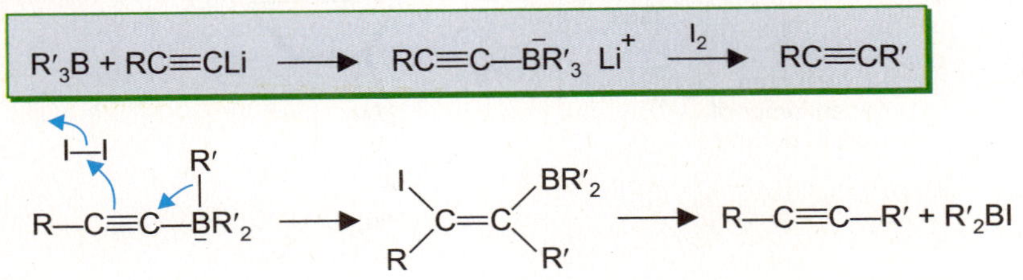

SCHEME 6.48

(*h*) Asymmetric Synthesis with Optically Active Alkylboranes or Sterically Hindered Achiral Alkenes – Enantioselectivity

Consider the outcome of hydroboration, of an achiral alkene and an achiral borane (Scheme 6.25) when a racemic mixture is obtained. The chiral borane however, chooses the face of the alkene where it adds (Scheme 6.49). Recall that the borane approaches the alkene in a manner so as to minimize steric interactions. Attack of chiral borane (B on the top face of alkene, A, Scheme 6.49) avoids steric interactions between the methyl substituent on the borane and the six membered ring. Thus the reagent here controls the stereoselectivity, and the mixture of the diastereomeric trialkylboranes C and D, both of which are pure enantiomers are formed with a high diastereoselectivity (*ds* =

97.8:2.2). After a 97.8:2.2 mixture of the enantiomeric *trans*-2-methyl cyclohexanols is formed (also see, Scheme 2.41*a*). The 1*R*, 2*R* alcohol will therefore have an *ee* value of 97.8% – 2.2% = 95.6%.

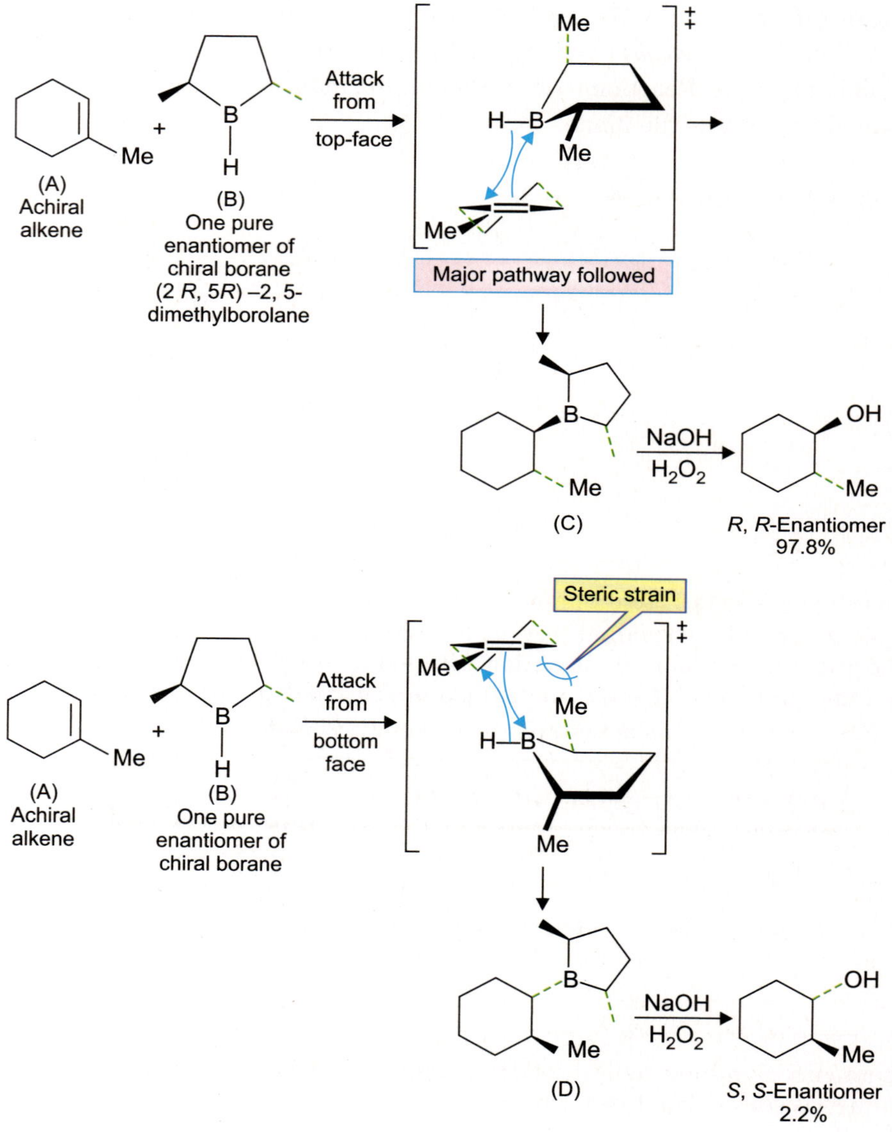

SCHEME 6.49

The hydroboration of enantiomerically pure α-pinene with achiral 9-BBN occurs only on the side of the C═C double bond from below *e.g.*, the α-face which is opposite the isopropylidene bridge. The addition is thus completely diastereoselective. Moreover, the trialkylborane obtained is a pure enantiomer, since the starting material is a pure enantiomer (Scheme 6.49*a*).

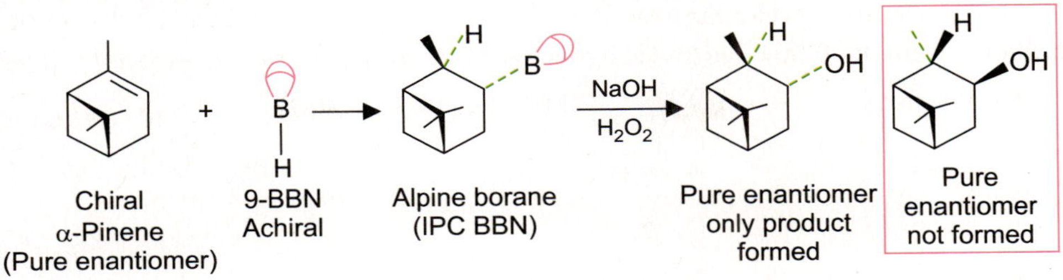

SCHEME 6.49a

- Asymmetric hydroboration with Ipc_2 BH of (*Z*)-1, 2-disubstituted alkenes followed by oxidation gives optically active secondary alcohols of high optical purity (see Scheme 2.40).
- For enantioselective reduction of carbonyl compounds with alpine borane (see Scheme 2.44).

Approach of a Borane to the Alkene

Significantly, hydroboration oxidation is a syn-addition of H/OH to the double bond. The stability of the product is not important since the thermodynamically more stable product is not always formed. An example is of β-pinene which gives the less stable cis-myrtanol (Scheme 6.49a). Thus, the reaction pathway is syn-addition, irrespective of direction of attack.

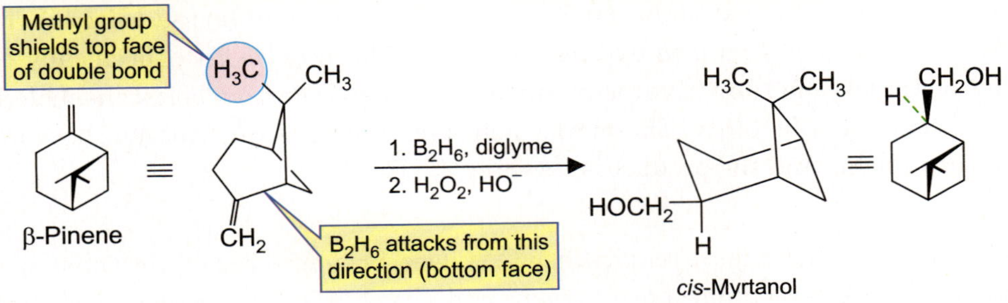

SCHEME 6.49b

CHAPTER 6

6.4 HYDROGENATION OF ALKENES AND ALKYNES

(A) Catalytic and Metal Ammonia Reductions

Hydrogenation of an olefin in the presence of a catalyst (*e.g.*, Ni, Pt, Pd) takes place under varying conditions of heat and pressure. The catalyst is believed to provide a surface to form a complex with the olefin and at the same time promote the activation of the adsorbed hydrogen molecule. Under the reaction conditions, the molecular hydrogen splits into hydrogen atoms which then add in a step wise way to the olefin and form two new C—H bonds on the same side of the ring. This type of addition is called *syn* addition and a *cis*-disubstituted cycloalkane will be obtained if the reactant is a 1, 2-disubstituted cycloalkene (Scheme 6.50). Catalytic hydrogenation is therefore, *stereospecific*. The two

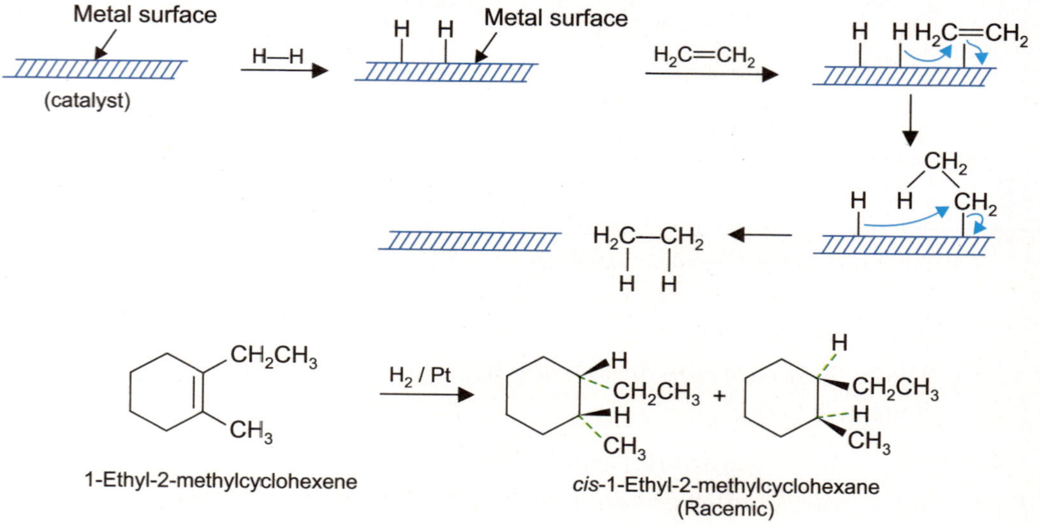

Two hydrogen atoms add to the same face of the *pi* bond (*syn* stereochemistry)

SCHEME 6.50

hydrogen atoms are added to the same face of the double bond **(*syn*-addition).** Thus, 1-ethyl-2-methylcyclohexane is hydrogenated over platinum to give specifically *cis*-1-ethyl-2-methylcyclohexane. Hydrogen can add from above or from below the molecular plane with equal probability. Therefore, each stereocenter is generated as both image and mirror image, and the product is racemic.

(*i*) Selectivity

Te C≡C bond is reduced more readily then C=C (see, Scheme 6.53*a*). Other unsaturated groupings with the exception of nitro groups and acid chlorides are reduced less readily. Thus catalytic hydrogenation is used for the selective reduction of C=C in the presence of aromatic rings and carbonyl groups.

(*ii*) Stereochemistry-Hydrogenation is Stereospecific

When addition of hydrogen to an alkene creates two chirality centers (Scheme 6.50), only two of the four stereoisomers are formed since only *syn*-addition can take place (The other two stereoisomers would arise *via anti*-addition). Which two stereoisomers would be formed will depend on the geometry of the reacting alkene *i.e.*, whether it is *cis* or *trans*. *cis*-Alkene on *syn*-addition will give *erythro*-pair of enantiomers (Scheme 6.51, when the groups attached to the two stereocenters are the same (R=R′), the *meso*-stereoisomer will be formed instead of *erythro* enantiomers (see, Scheme 6.57).

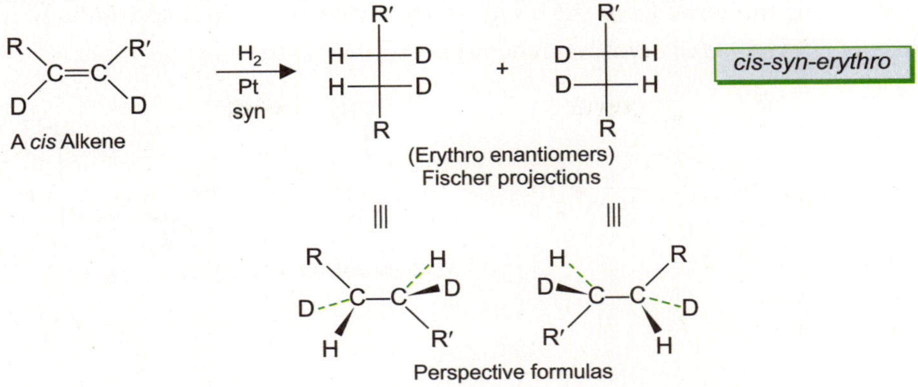

SCHEME 6.51

When the alkene is *trans*, the *syn*-addition of hydrogen will afford *threo* enantiomers. This shows that hydrogenation is a stereospecific reaction (Scheme 6.52).

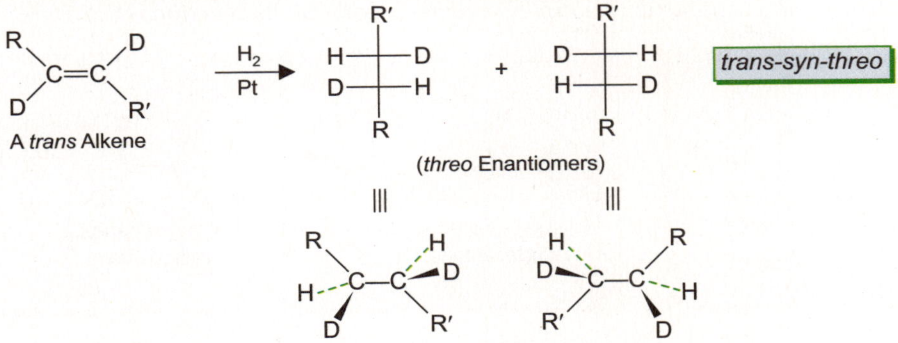

SCHEME 6.52

Problem Solving Hint

One can predict the outcome of reactions which generate two chirality centers by remembering the system cis, syn, erythro where every terms means to the same side. The stereochemical outcome can be predicted by changing any two of the terms in these systems at one time in a given reaction.

- *cis, syn, erythro (e.g.,* Scheme 6.51)
- *trans, syn, threo (e.g.,* Scheme 6.52)
- *cis, anti, threo (e.g.,* Scheme 6.20)

In rigid cyclic systems *e.g.*, cholesterol (Scheme 6.52*a*) the addition of H_2 is from the less hindered α side, thus cholesterol gives mainly *trans* ring-junction on reduction, since the angular methyl groups hinder the fit of the catalyst on the β side of the molecule. However, when the C-3 OH group is axial, (as in C-3 epi-cholesterol) the fit to the catalyst is hindered from the α side of the molecule as well and reduction gives a mixture of *cis* and *trans* decalins in equal amounts.

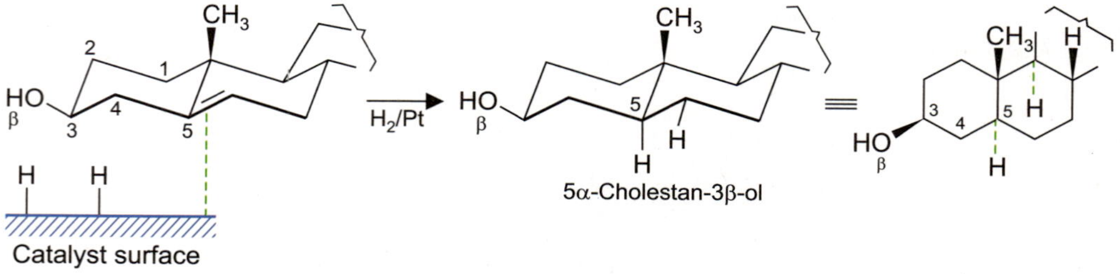

SCHEME 6.52a

EXERCISE 6.8

Predict the products along with stereochemical outcome of the reactions (Scheme 6.52b).

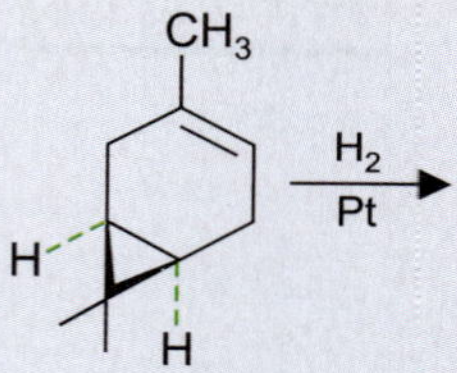

3-Carene
(A)

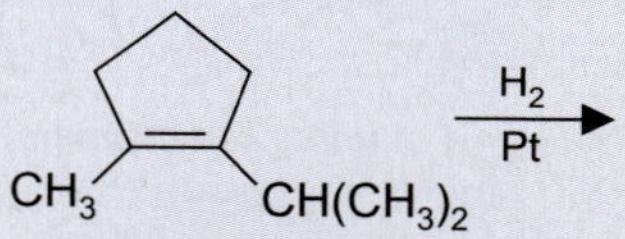

1-Isopropyl-2-methyl-cyclopentene
(B)

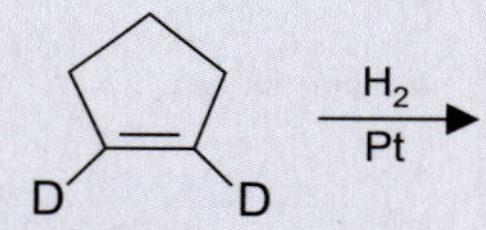

1, 2-Dideuterio-cyclopentene
(C)

SCHEME 6.52b

ANSWER. *(A) Steric hindrance completely inhibits hydrogenation on one side of the ring, thus hydrogen is added on the side opposite to the three membered ring to give (IV, Scheme 6.52c). (B) syn Addition of H_2 to a cyclic compound gives cis stereoisomers (V). (C) The product formed will be meso since the addition from either top or bottom face of the double bond gives only the same meso compound (VI).*

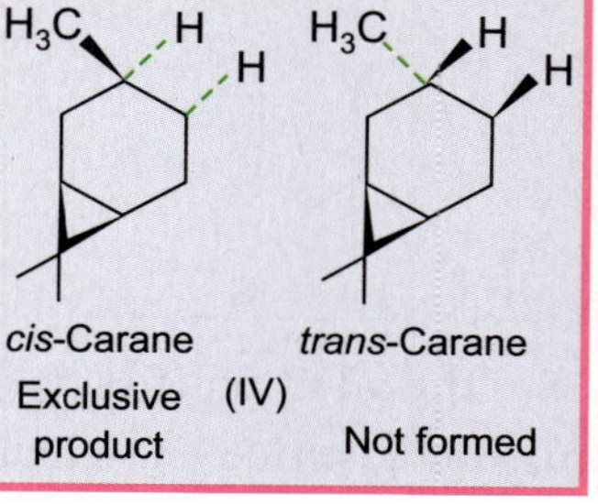

cis-Carane *trans*-Carane
Exclusive product (IV) Not formed

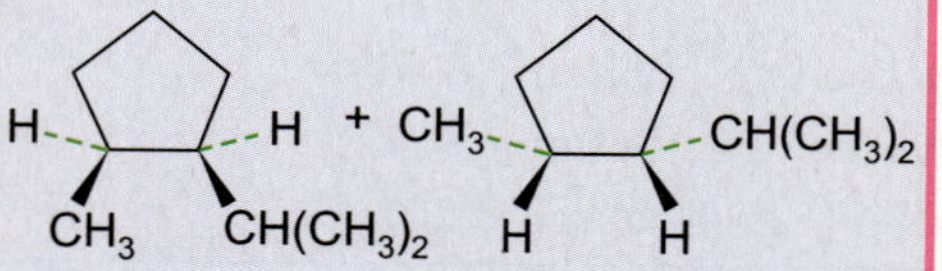

(V)
Mixture of enantiomers

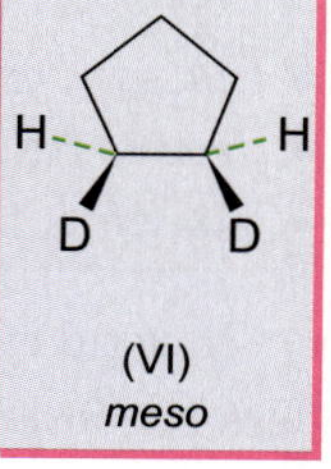

(VI)
meso

SCHEME 6.52c

If one considers that the major mode of hydrogen addition to the double bond is from the less hindered equatorial side the major product of hydrogen addition to (I, Scheme 6.52*d*) is (II, also see Scheme 4.67).

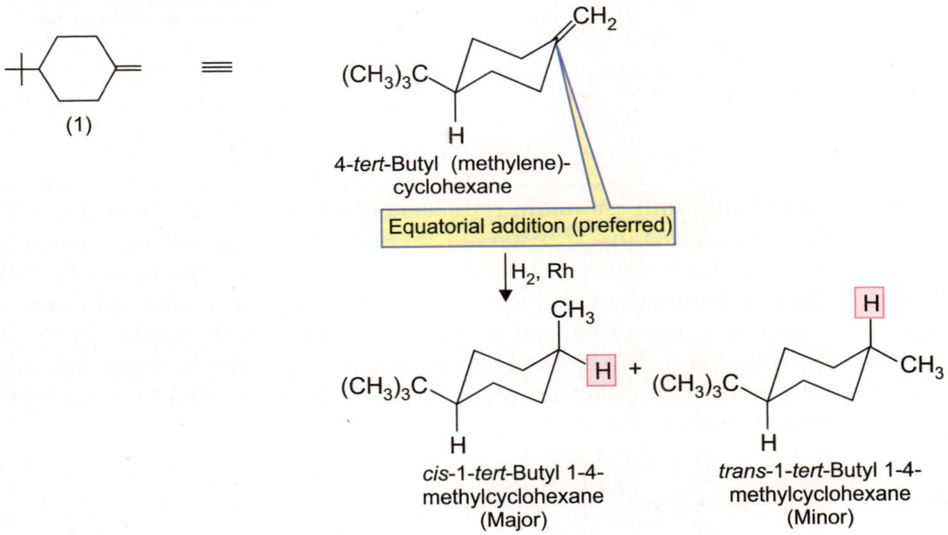

SCHEME 6.52d

Although *syn*-addition of H_2 is generally observed, *anti*-addition is sometimes observed, (*eq. a*, Scheme 6.53) if double bond isomerization occurs more rapidly than hydrogenation. In the case shown below, it has been established that isomerization occurs before reduction (Scheme 6.53). Palladium is particularly prone to catalyze double bond isomerization. Platinum, rhodium, or iridium should be used if isomerization is a problem.

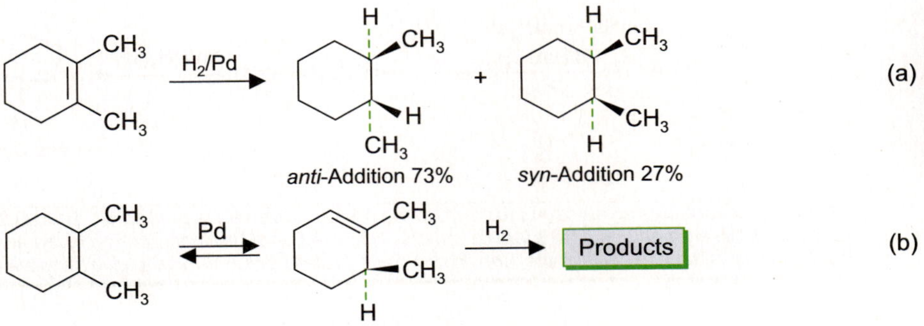

SCHEME 6.53

The catalytic hydrogenation of alkynes also leads to *syn*-addition of two H atoms and proceeds with *syn* stereochemistry (Scheme 6.53*a*). The hydrogenation of an alkyne can be stopped at the alkene stage by using a less reactive ("poisoned" catalyst). Lindlar's catalyst is one such poisoned palladium catalyst using barium sulphate poisoned with quinolene.

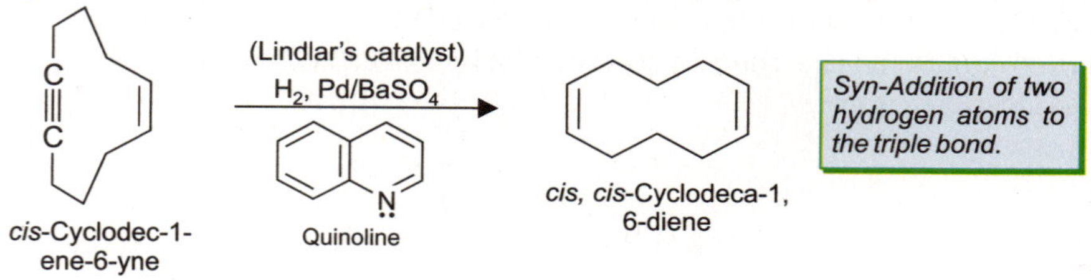

SCHEME 6.53a

Sodium metal in liquid ammonia reduces alkynes with *anti* stereochemistry to give *trans*-alkenes. Electrons are transferred more readily to acetylenes as compared to olefins (greater reactivity of acetylenes towards nucleophiles). The reagents of choice are the metal-amine and metal-ammonia systems. The reduction is stereoselective *anti*-addition since the vinyl radical formed during reduction is more stable in the *trans* configuration (Scheme 6.54). The method is complementary to the stereospecifically *syn* addition of hydrogen during catalytic methods. Thus, both *cis*- and *trans*-cyclooctene can be obtained from cyclooctyne.

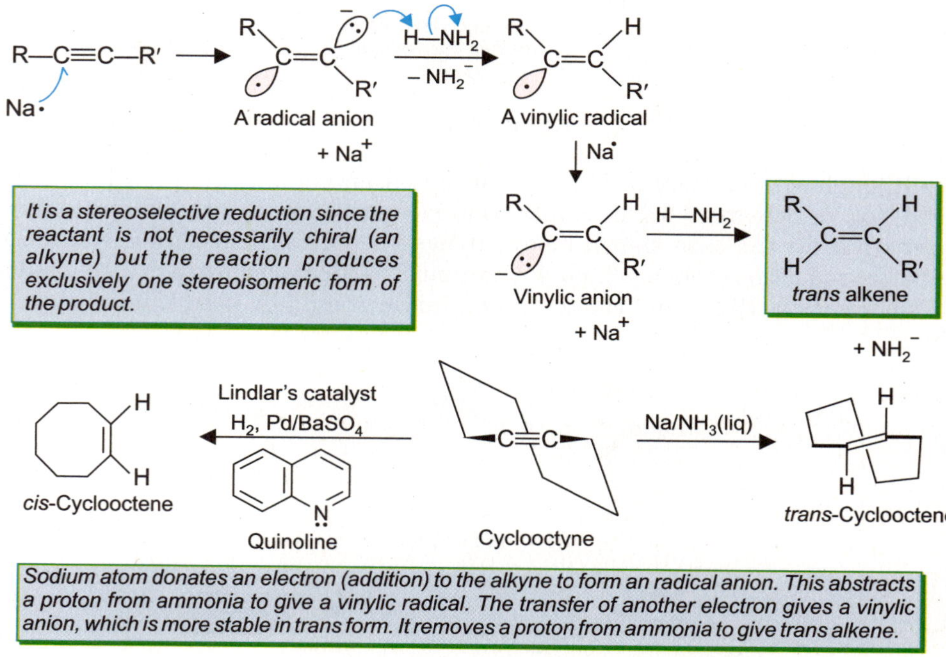

SCHEME 6.54

(B) Diimide Reductions

An interesting nonmetallic compound often used for the hydrogenation of carbon-carbon multiple bonds is diimide which is formed by oxidation of hydrazine (NH_2NH_2). Diimide is generated in the reaction mixture as needed.

Diimide is an unstable compound, it decompose to nitrogen and hydrogen (in the absence of additives), but when it is generated in the presence of an alkene, rapid reduction occurs. The driving force being the great stability of the nitrogen molecule compound to the —N=N-system (Scheme 6.55). Although both the *syn* and *anti* forms of diimide are produced by the oxidation of hydrazine, only the *syn* form reduces the double bonds by a cyclic mechanism. The addition is therefore, stereochemically *syn*.

Diimide is a highly selective reagent employed to reduce readily the symmetrical bonds like C≡C, C=C, N=N, and not the unsymmetrical and more polar bonds like C≡N, N=O, S=O, etc. Thus α-pinene gives almost exclusively *cis*-pinane which is an example of a stereoselective reaction since *trans*-pinane is not formed (also see Scheme 1.107).

Hydrogenation (catalytic) of α-pinene, occurs from the α-face of the double bond since the methyl groups on the bridge carbon block the top face, the hydrogenation in 100% stereoselective and leads to the formation of only *cis*-pinane.

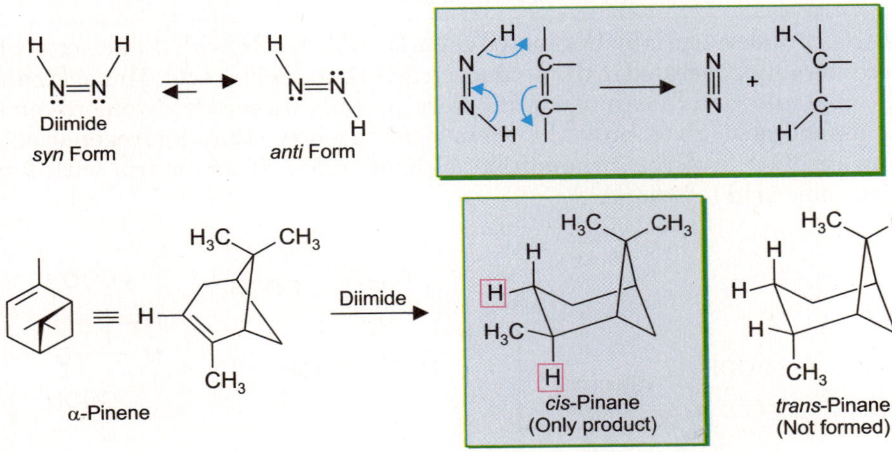

SCHEME 6.55

(C) Homogeneous Hydrogenation (Wilkinson's Catalyst)

In this method, hydrogen is activated not by chemisorption on the surface of the metal catalyst but by its incorporation into the co-ordination sphere around the metal atom, generally of the group VIII *e.g.*, rhodium, ruthenium. In the heterogeneous hydrogenation reactions two phases are involved, the solid phase of the catalyst (Pd, Pt, Ni etc.) containing the adsorbed hydrogen, while the liquid phase of the solution has the unsaturated reactant. When however, one uses a transition metal complex *e.g.*, tris (triphenylphosphine) chlororhodium, $Rh[(C_6H_5)_3P]_3Cl$ (called Wilkinson's catalyst), hydrogenation occurs in a single phase-in solution.

The Wilkinson's catalyst in solution probably exchanges reversibly one Ph_3P group for a loosely held solvent molecule (A, Scheme 6.56). The catalyst reacts with hydrogen to form a metal hydride (B, the dihydrido complex). The H—H bond is cleaved and each hydrogen is separately bonded to rhodium, and thus the metal uses one of its electron pairs and is oxidized to the rhodium (III) state. In the next step alkene reacts with the complex and replaces the solvent molecule. The alkene-metal bond formation is a result of the overlap of an empty orbital of the metal with π cloud of the alkene. Two hydrogen atoms are then transferred to the alkene one at a time and the net result is a *syn* addition.

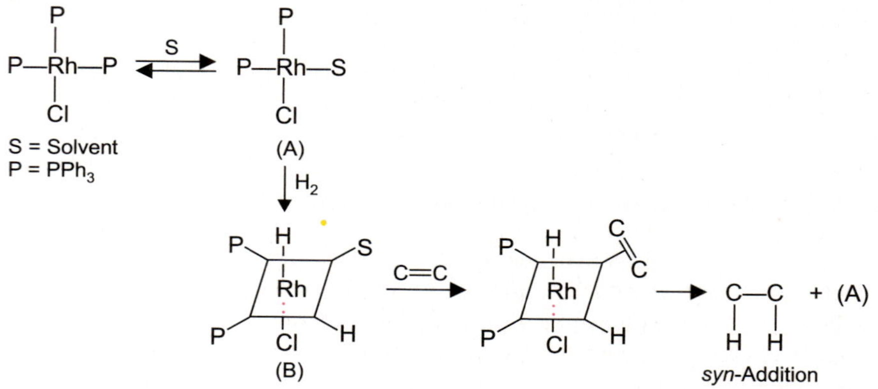

SCHEME 6.56

When *e.g.*, maleic acid is hydrogenated using D_2 in place of H_2 with Wilkinson's catalyst two stereocenters are generated in the product (2, 3-D_2) butanedioic acid. This acid can exist as a *meso* compound, or as a pair of enantiomers *i.e.*, only three stereoisomers can exist. However, maleic acid gives only the *meso* acid to show, that hydrogenation with Wilkinson's catalyst involves *syn*-addition (Scheme 6.57). If one starts with fumaric acid only racemic acid is formed.

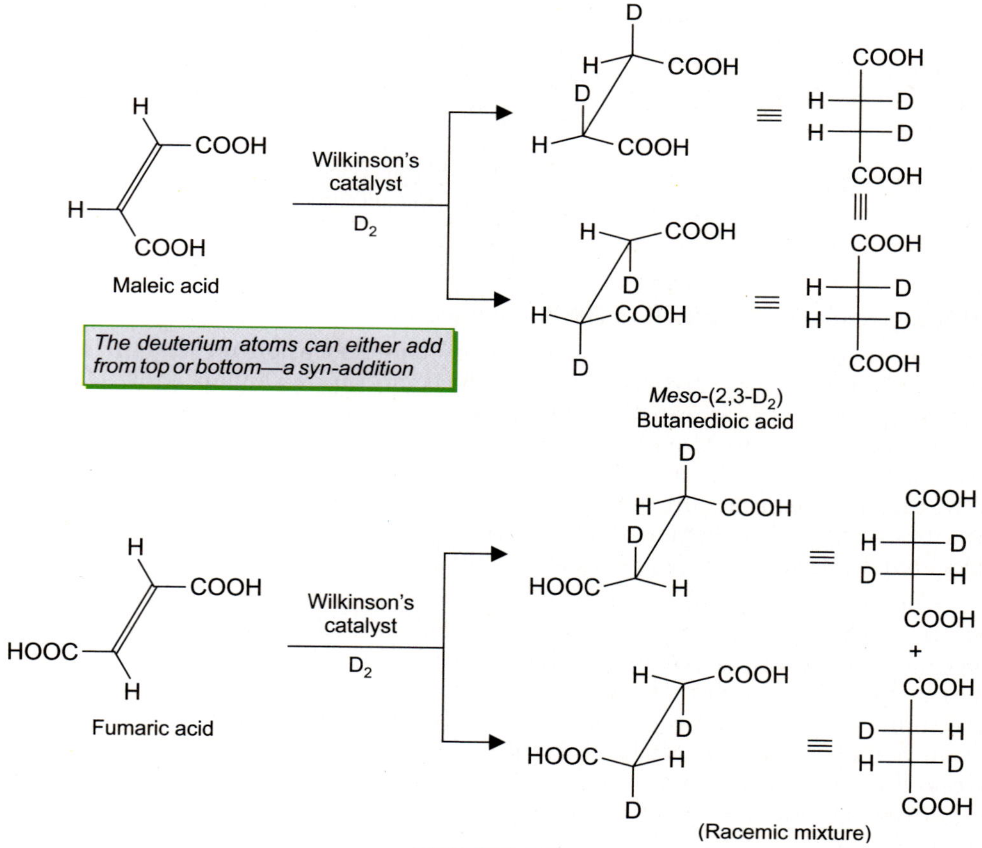

SCHEME 6.57

(D) Enantioselective Catalytic Hydrogenation with Chiral Wilkinson–Type Catalysts

The hydrogenation of an achiral starting material with *e.g.*, optically inactive Wilkinson catalyst will always produce a racemic modification. Thus, 2-acetylaminopropenoic acid (with enantiotopic faces) on hydrogenation with Wilkinson's catalyst give racemic acetylalanine since there is equal probability for hydrogen to add to the either face. Thus, here the situation is just the same as seen in the case of maleic or fumaric acids (Scheme 6.57). However, when the Wilkinson's catalyst is made chiral (optically active) by attaching optically active ligands, then the hydrogenation occurs in a chiral medium and can give optically active products. Several optically active ligands have been developed where the chirality is due to either stereogenic carbon or stereogenic phosphorus (Scheme 6.58).

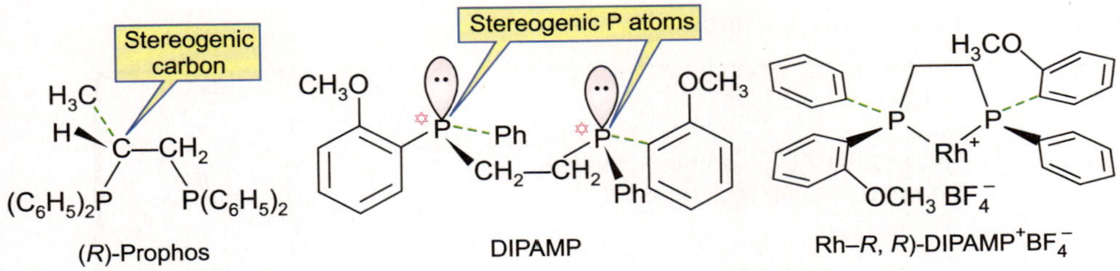

SCHEME 6.58

When the reduction of *e.g.*, 2-acetylaminopropenoic acid (an achiral compound with enantiotopic faces) is carried out with chiral Wilkinson's type catalyst, the two enantiotopic faces are differentiated and hydrogen is added *via syn* addition selectively only to one of the two faces to give one enantiomer (Scheme 6.59).

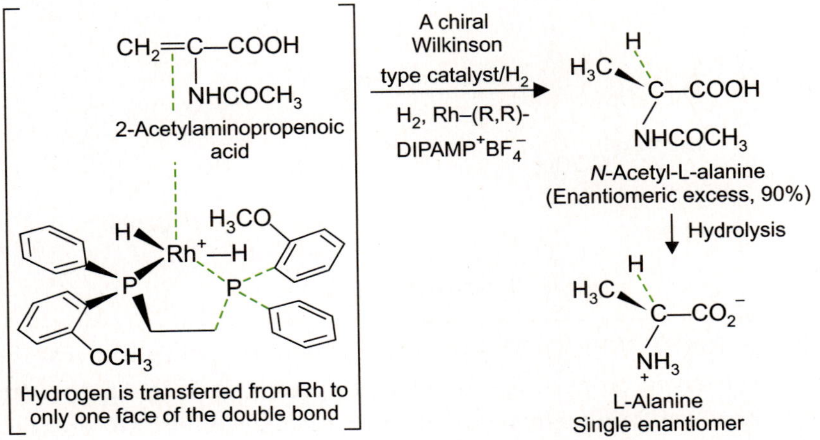

SCHEME 6.59

Similarly DIOP is a complex of rhodium with a chiral ligand (Scheme 6.59*a*) which has been used to synthesize biologically active (–) form of dopa. As the catalyst is chiral (and as in the above example), the transition states which would lead to two enantiomers are diastereo-topic, and have different energies, the transition state which leads to the (–) -enantiomer is of lower energy and is favoured.

> *Levodopa is used to cure Parkinson's disease to counteract a deficiency of dopamine (one of the neurotransmitters in the brain). Dopamine itself cannot be used as a drug, because it cannot cross the "blood-brain barrier".*

(–)-DOPA is an amino acid derived from tyrosine is capable of crossing the blood-brain-barrier to reach the site of action and is converted by brain enzymes into dopamine. Significantly only the (–) enantiomer of dopa is converted into dopamine while the other enantiomer is toxic to the patient.

Rhodium (DIOP)Cl_2—a chiral catalyst for enantioselective reduction =

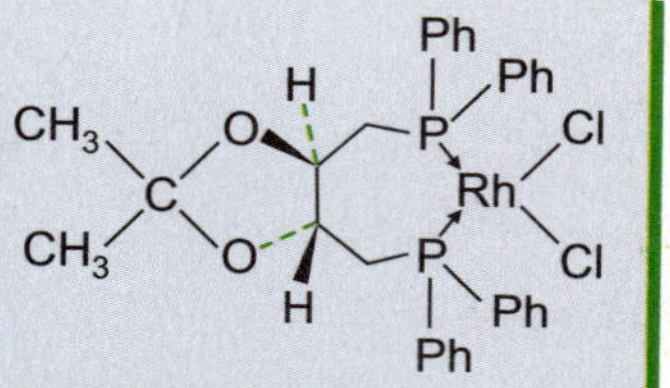

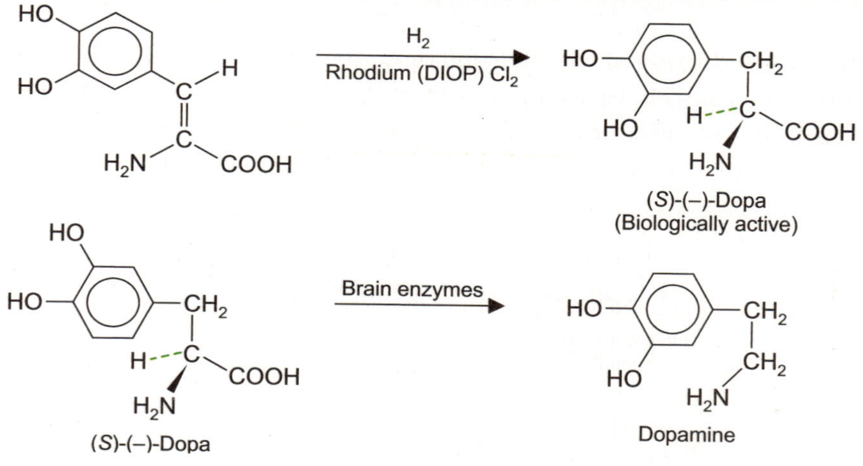

SCHEME 6.59a

Similarly Ru(BINAP)Cl_2 is a chiral ruthenium phosphine catalyst where the chirality is derived from 1, 1′-binaphthyl system. It is used as homogeneous hydrogenation catalyst for asymmetric reduction (Scheme 6.59*b*) *i.e.*, creation of a new stereocenter with the exclusive formation of one enantiomer. Thus, when the ligands in a Wilkinson's catalyst [$(Ph_3P)_3RhCl$] are chiral, optical activity is induced in the product of hydrogenation. An example of these ligands is binaphthyl-phosphine BINAP.

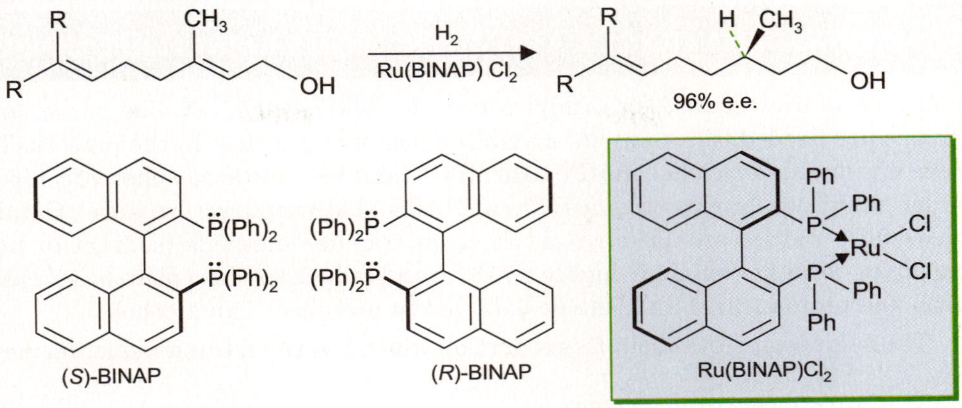

SCHEME 6.59b

(E) Advantages of Wilkinson's Catalyst (Homogeneous Catalytic Hydrogenation) over Hydrogenation on a Metal Catalyst

Only alkenes and alkynes are reduced, and groups like C=O, C=N and NO_2 remain intact. Moreover, mono- and disubstituted alkenes are far more reactive than tri- and tetrasubstituted double bonds. Thus a double bond may be selectively reduced as in carvone (Scheme 6.60) which is reduced to the dihydroderivative with trisubstituted double bond intact.

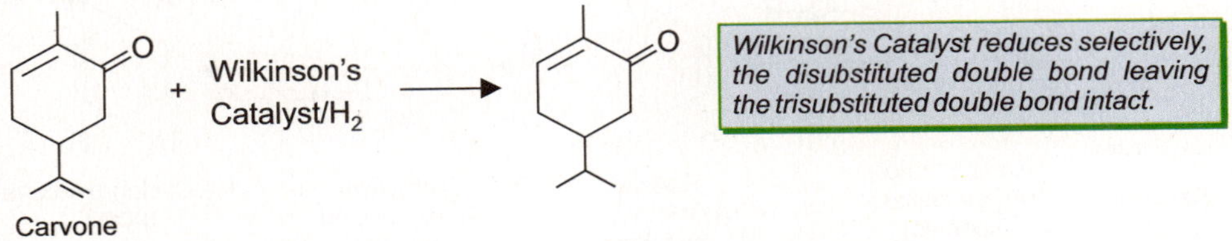

SCHEME 6.60

Moreover, hydrogenolysis is not observed and benzyl cinnamate gives its corresponding dihydro derivative (eqn. I, Scheme 6.61) without the cleavage of the *O*-benzyl bond. It may be mentioned that a benzylic group when attached to OH, OR, OCOR, NR_2, SR or halogen undergoes cleavage (hydrogenolysis) during catalytic hydrogenation (eqn. II, Scheme 6.61).

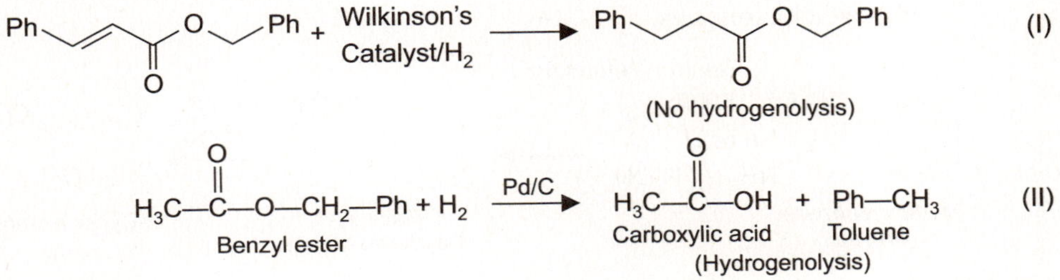

SCHEME 6.61

CHAPTER 6

6.5 HYDROGENATION OF AROMATIC RINGS—BIRCH REDUCTION

Aromatic rings are reduced by catalysts used for alkenes *e.g.*, Pt, but, more strong conditions are needed, since aromatic stabilization energy is lost in the process. Thus benzene is reduced to cyclohexane (Pt 10 hr, 100°C and 100–150 atmospheres) compared to milder conditions for most alkenes (1 hr at 20°C and atmospheric pressure). Catalytic methods fully reduce aromatic rings. Electron transfer methods (sodium in liquid ammonia) on the other hand are highly selective and reduction takes place by successive electron and proton transfers (Scheme 6.62) in the presence of an alcohol.

The following points about this reduction which is termed Birch reduction may be noted:

- Sodium in liquid ammonia is normally used as the reducing agent. An alcohol is used to trap the anion radical rapidly. Generally ethyl, isopropyl or *t*-butyl alcohol may be used. 1,4-Addition of hydrogen takes place to yield non-conjugated cyclohexadienes.
- The structure of the product is determined by the site of the first protonation. The anion thus formed reacts at the central carbon of the delocalized system to complete

$NH_3 + Na \rightleftharpoons NH_3 \cdot e^-$ (Deep blue solution) $+ Na^+$

Solvated electron

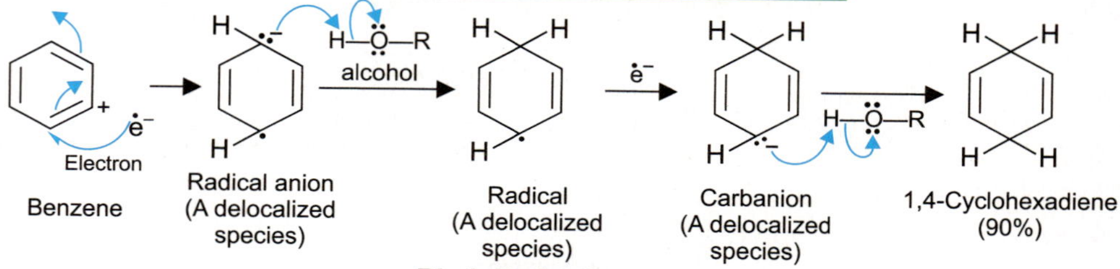

Birch Reduction

SCHEME 6.62

the 1,4-addition. Electron-releasing substituents direct the first protonation to the *ortho* position (Scheme 6.63) as in the Birch reduction of methoxybenzene (anisole) which gives 1-methoxy-1,4-cyclohexadiene, which can be hydrolyzed by dilute acid to 2-cyclohexenone, and this method provides a useful synthesis of 2-cyclohexenones.

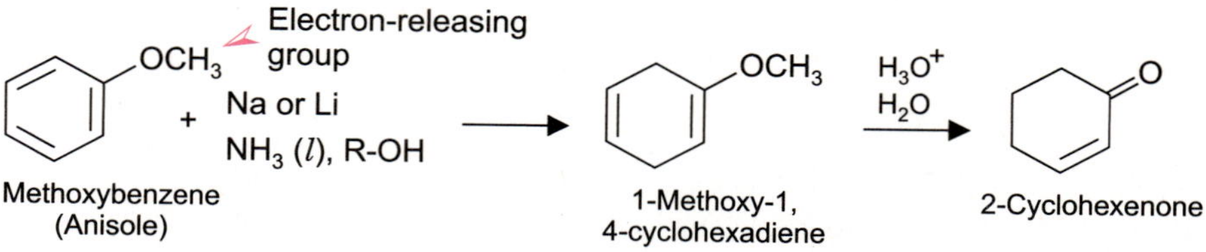

SCHEME 6.63

In contrast, the $-CO_2^-$ substituent directs protonation to the *para* position (Scheme 6.64).

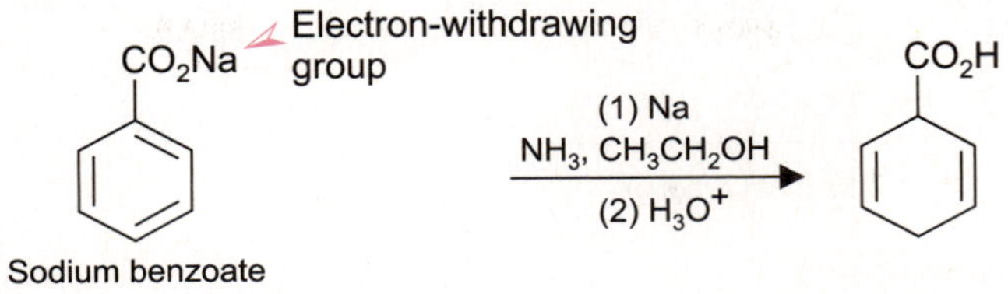

SCHEME 6.64

6.6 MICHAEL REACTION

Many different nucleophiles can add to the α, β-carbonyl compounds. When the anion is resonance stabilized enolate ion (*e.g.*, an anion of a diketone) and is used as a nucleophile in a conjugate addition, the reaction is termed a Michael addition. In fact the term Michael addition is often used rather loosely to include all conjugate addition reactions.

The Michael reaction takes place with a wide variety of α, β-unsaturated carbonyl compounds as well as α, β-unsaturated nitrites and nitro compounds. Effective nucleophiles may be derived from a β-diketone, β-keto-ester, β-ketonitrite or even an enamine (see box 4.5).

(i) Michael Reaction During Robinson Ring Annulation

The process fuses a cyclohexanone ring on the starting cyclic ketone and involves a Michael addition as well as an aldol condensation (Scheme 6.65). The reaction begins by the addition of the enolate anion of the ketone to methyl vinyl ketone in a conjugate sense to the β-carbon of the enone (Michael reaction) and a new carbon-carbon bond is formed. This is followed by another carbon-carbon bond forming reaction-Aldol condensation. In the last step, the aldol formed undergoes dehydration on heating to give the enone. (The reaction is made enantioselective under chiral influence, Scheme 2.56).

(ii) Chiral Enamines

Chiral enamines are prepared by condensation of ketones with enantioenriched *sec*-amines. The C_2-symmetric *trans*-2,5-dimethylpyrrolidine is often used chiral auxiliary to make enantiomeric enamines. Alkylation of chiral enamines followed by hydrolysis is an effective method for the enantioselective alkylation of ketones (Scheme 6.66).

(iii) Conjugate Addition with Gilman Reagents

Lithium diorganocopper reagents (Gilman reagents) add to α, β-unsaturated aldehydes and ketones to give products of 1,4-(*i.e.*, conjugate) addition (Scheme 6.67). The reaction is closely related with Michael reaction.

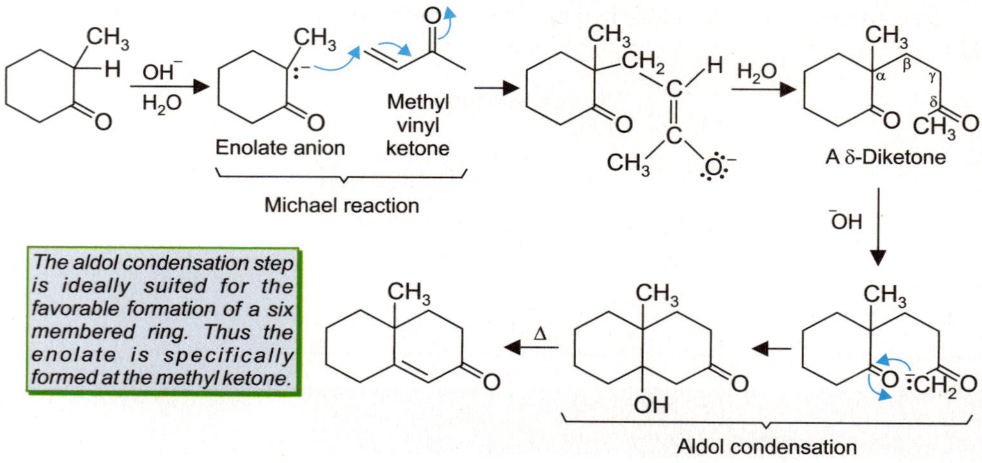

SCHEME 6.65

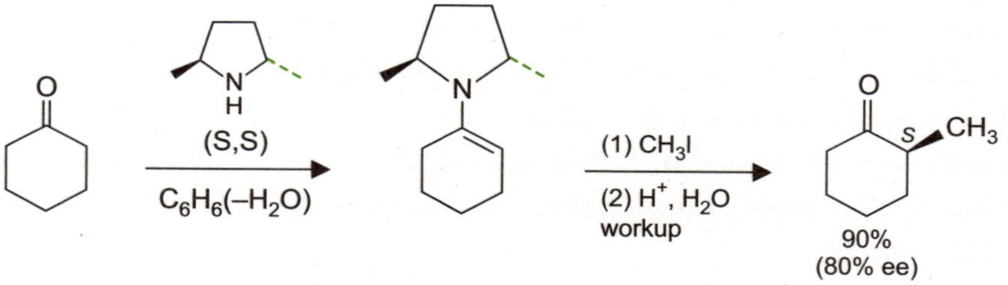

SCHEME 6.66

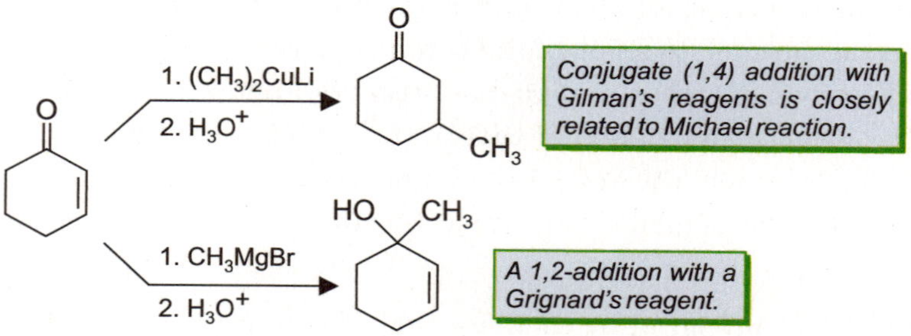

SCHEME 6.67

6.7 ADDITION OF FREE RADICALS TO π-SYSTEMS

The free radicals add readily to an alkene to give a new bond and a new free radical (Scheme 6.68). In the process, the π-bond is broken—the carbon–carbon σ bond which is formed is about 30 kcal stronger than the π-bond which is broken. The reactions are initiated by a peroxide or azoisobisbutyronitrile (AIBN) which on heating or irradiation give reactive radicals R• (Scheme 6.68).

The free radical reactions used in synthesis are chain reactions and a chain is terminated by the reaction of radicals among themselves.

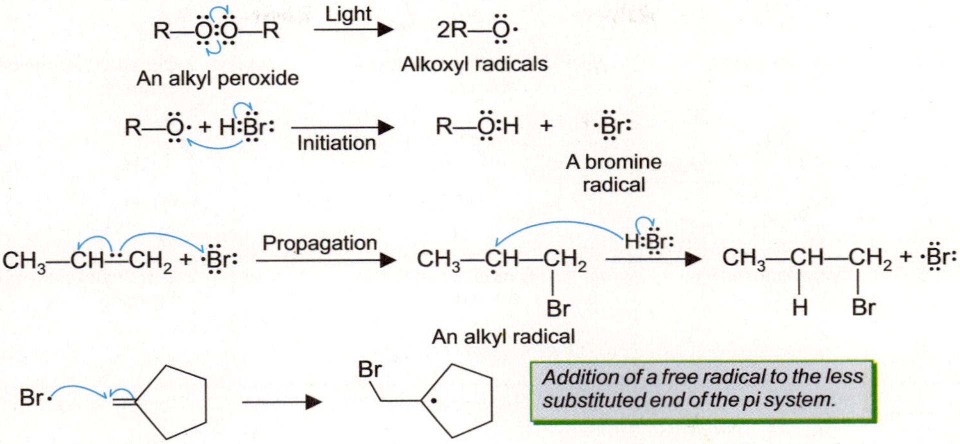

SCHEME 6.68

(i) Regioselectivity

A free radical addition occurs readily at the less substituted carbon atom of an unsymmetrical alkene like propene (Scheme 6.68). All substrates $CH_2{=}CHX$ are preferentially attacked at the CH_2 regardless of the nature of X or of the attacking radical and the major reason is therefore, steric. Hydrogen bromide is unique among the hydrogen halides in that it can add to alkenes either by electrophilic or free-radical addition. When one considers the free radical addition of HBr to an unsymmetrical alkene, photochemical conditions or in the presence of peroxides it comes out that the addition is *anti*-Markovnikov. This compared with the observed orientation of HBr (Markovinikov electrophilic) shows that in both situations the formation of the more stable secondary intermediate is involved (Scheme 6.69).

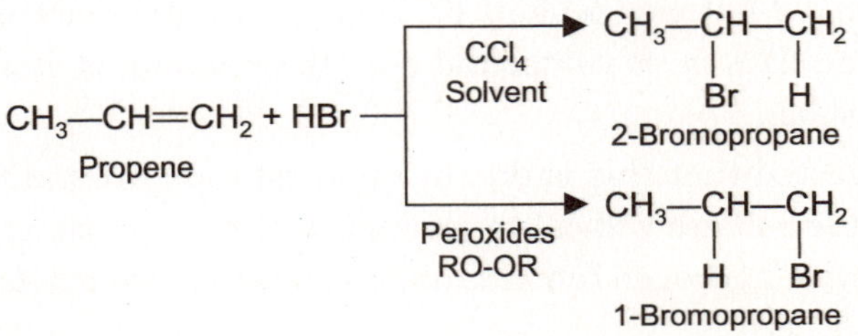

SCHEME 6.69

Thus 1, 6-dimethylcyclohexene and 1, 2-dimethylcyclohexene react with HBr in the dark (absence of peroxides) to give the same compounds (Scheme 6.69*a*).

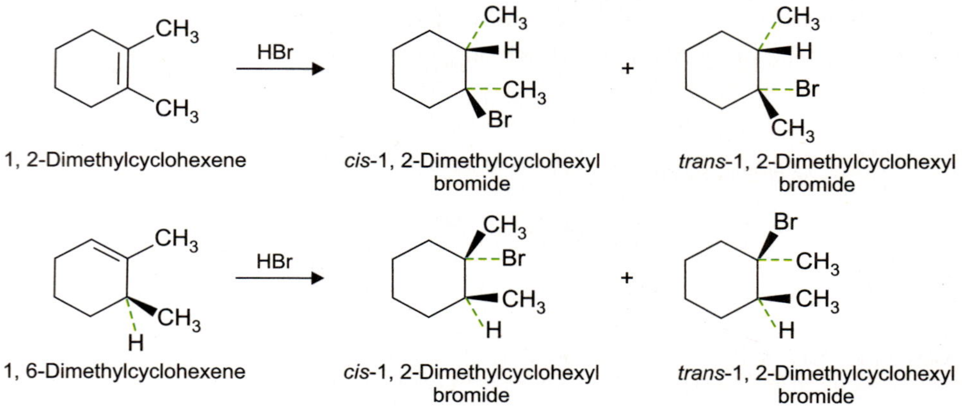

SCHEME 6.69a

(ii) Stereochemistry

Many free-radical additions are stereoselective, while others are not. Thus the addition of HBr to cyclohexenes gives *anti*-addition to give *trans*-diaxial addition products and is best explained on the basis of a bromine bridged intermediate (Scheme 6.70).

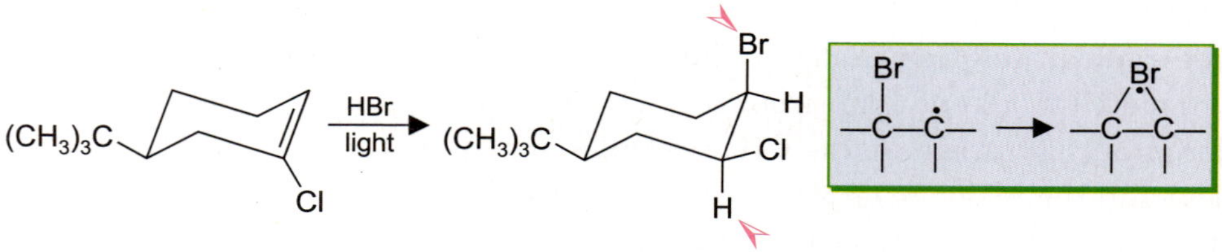

SCHEME 6.70

Consider the free radical addition of HBr to 2-bromo-2-butene and consider the following points:

- At room temperature this addition shows stereoselectivity since either of (*E*)- or (*Z*)-2-bromo-2-butene (Scheme 6.71) gave the same mixture of products 78% *dl* pair and 22% *meso* compound (*i.e.*, the reaction is stereoselective but not stereospecific).
- It is suggested that this is due to rapid rotation around the new C—C bond in the intermediate radical to give more of the less congested radical before reaction with hydrogen bromide occurs on the face opposite the new C-Br bond (Scheme 6.71).
- When the reaction is carried out at –80°C, the stereospecificity returns–the (*E*)–alkene gives only *meso* compound while the (*Z*) isomer gives only (±) product. It is suggested that at low temperature, the rotation in the intermediate radical is almost arrested and a bridged radical is instead formed which explains the observed stereospecificity.

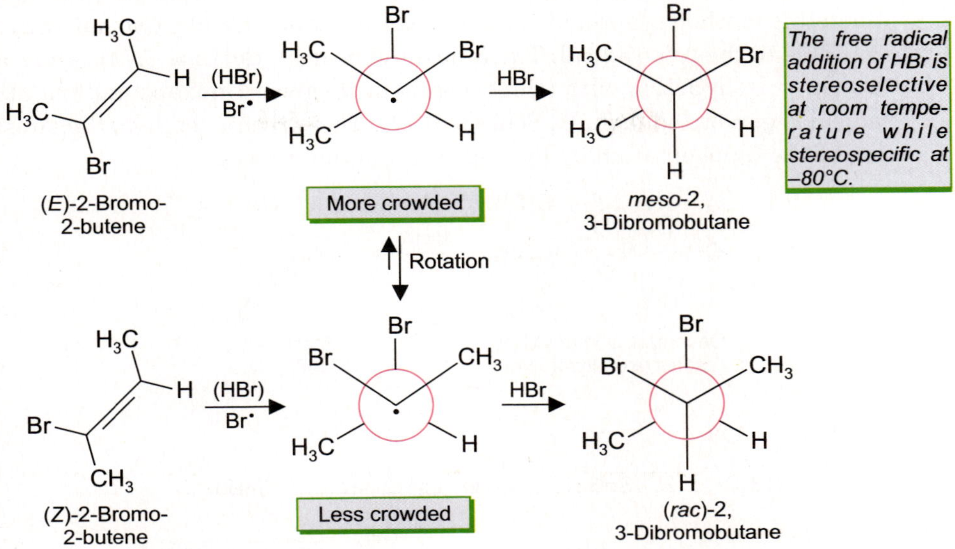

SCHEME 6.71

(iii) Regioselectivity of Intramolecular Addition of a Free Radical to a Double Bond—Free Radical Cyclization

Intramolecular olefin additions produce rings and these reactions are much faster than other reaction pathways. Intramolecular addition of radicals generated at a carbon having a 5, 6-double bond can give either a five membered ring or a six membered ring, but mostly a five membered ring is greatly preferred kinetically (Scheme 6.72, methylcyclopentane from 6-bromo-1-hexene). The mechanism and explanation for the formation of these products is as under:

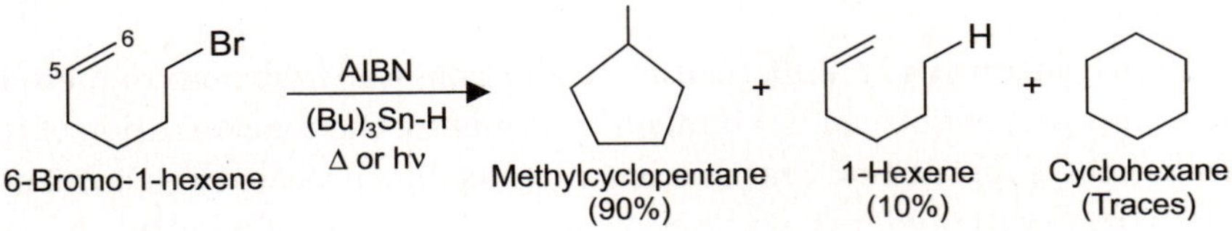

SCHEME 6.72

- Tributyltin hydride is a versatile reagent for the generation of a free radical at a specific carbon atom in a molecule. Tributyltin radical is generated by the abstraction of a hydrogen atom from tributyltin hydride with AIBN (Scheme 6.73) which is the initiation step. This radical abstracts efficiently the halogens (except fluorine), PhS and PhSe. Since the Sn-H bond is weak, it does not abstract hydrogen from an alkyl C—H group.
- The method thus provides a convenient way to convert an alkyl halide to the corresponding hydrocarbon (R—X→R—H) no matter what the structure of substrate.

- When this method is applied to an unsaturated alkyl halide, two pathways are available to the free radical. The 5-hexenyl radical (Scheme 6.74) can react in two ways. Its reaction with tributyltin hydride gives 1-hexene and tributyltin radical is generated (eqn. II, Scheme 6.74). This process leads to the substitution of the halogen atom by hydrogen (reduction process).

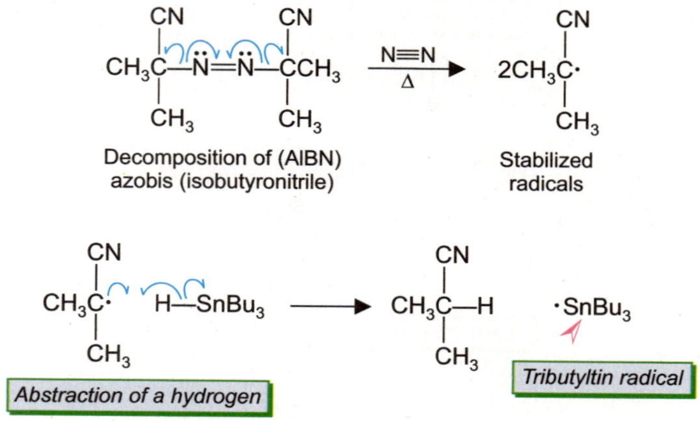

$$(Bu)_3Sn\cdot + \boxed{R\text{-}Br} \longrightarrow (Bu)_3SnBr + R\cdot$$

$$R\cdot + (Bu)_3Sn\text{-}H \longrightarrow \boxed{RH} + (Bu)_3Sn\cdot$$

SCHEME 6.73

- In the second pathway (eqn. I, Scheme 6.74) cyclization takes place by the addition of the alkyl radical to the π-bond generating a different alkyl radical. This cyclized radical then reacts with tributyltin hybride to give the saturated hydrocarbon and regenerating tributyltin radical to propogate the chain reaction.

These two pathways *i.e.* substitution (reduction) and addition compete with each other. The course of reaction can be changed by changing the concentration of tributyltin hydride. Since the substitution reaction is a bimolecular process high concentration of tributyltin hydride will favour it, while a lower concentration of tributyltin hydride will favour the cyclization process.

It is known that the rates of ring-forming free radical cyclizations are $5 > 6 > 7$. It was found that reaction (Scheme 6.72) gives the five-membered ring product exclusively. Thus the regioselectivity of ring formation is not controlled by thermodynamic considerations, but rather by kinetic control of the cyclization.

- It is found that bond formation between the radical and the alkene stereoelectronically requires an approach angle of about 110° between the free radical center and the olefinic plane. This is due to the fact that free radical addition results from donation of the unpaired electron on the radical into the π antibonding orbital of the olefin, which coincidentally makes an angle of about 110° with the olefinic plane (Scheme 6.75).

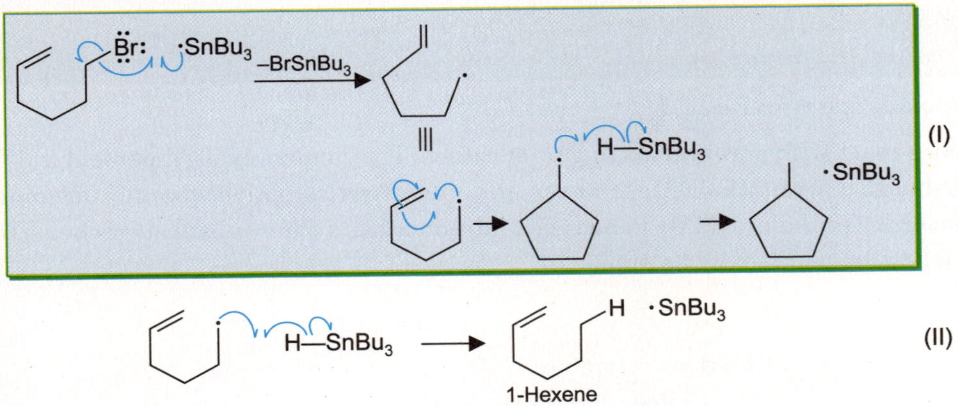

SCHEME 6.74

- This approach angle is easily achieved (during intramolecular cyclization) by attack on that end of the double bond which is closet to the radical center (favourable entropy factors) to result in a five membered ring (Scheme 6.75).
- If the attack was to occur on the other olefinic carbon (as would be required for formation of a six-membered ring) the radical shall have to reach across the double bond to achieve the proper approach angle. This indeed would be a higher energy path and is kinetically not favored.

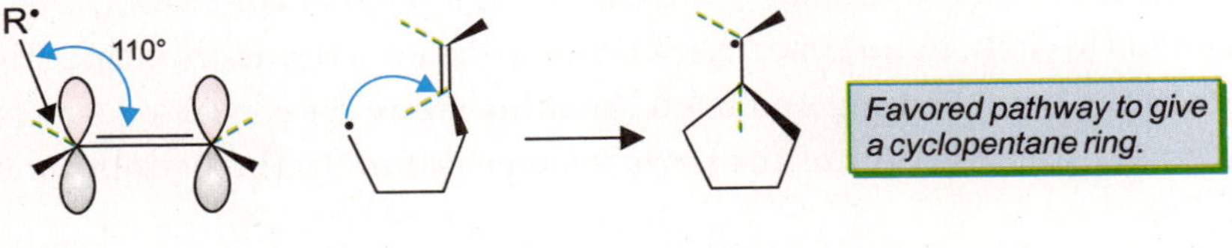

SCHEME 6.75

One may appreciate that while Diels Alder reaction is one of the methods to form fused six-membered rings, radical cyclization is superior to synthesize a fused ring system containing a five membered carbocycle. The power of the method for the synthesis of fused ring system with a five membered carbocycle is presented (Scheme 6.76). In this case the reaction is induced at a bridgehead position where radical reactions are normally difficult since the radical cannot achieve planar geometry.

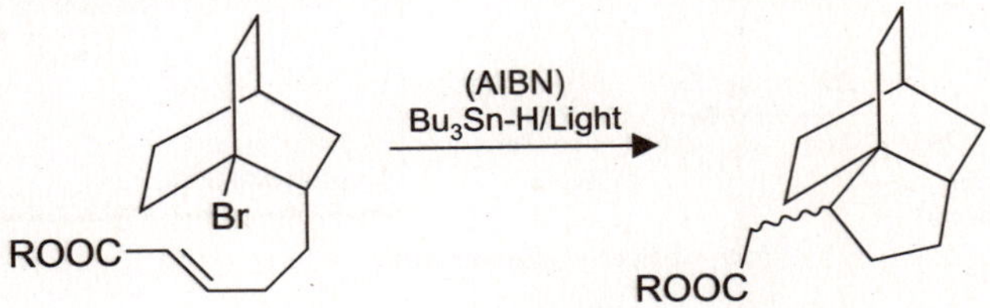

SCHEME 6.76

6.8 EPOXIDATION OF ALKENES

(a) Epoxidation with Peracids

Alkenes react with peroxyacids to give epoxides. The commonly used peracid is 3-chloroper-benzoic acid (MCPBA), however, it being shock sensitive, recently megnesium monoperoxy-phthalate (MMPP) has been introduced as a safer substitute (Scheme 6.77). The following points may be noted:

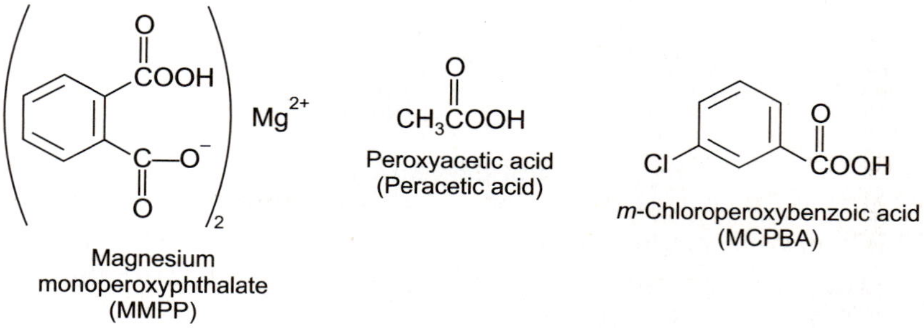

SCHEME 6.77

- The reaction is stereospecific *e.g.*, *cis*-2-butene gives only *cis*-epoxide while *trans*-2-butene gives the *trans*-product (see Scheme 1.99). Thus the configuration of the alkene is retained *i.e.* although the π bond of the carbon-carbon double bond is broken, at no time, there is free rotation around the remaining σ bond. Thus the bond making and bond breaking steps are concerted—epoxidation is a single step transfer of the peroxy oxygen atom to the double bond (Scheme 6.78).
- Epoxidation is a regioselective reaction, since the more electron rich double bond reacts when more than one double bond is present. This also shows that the peroxy acid behaves as an electrophile, thus 1, 2-dimethyl 1, 4-cyclohexadiene reacts selectively at the more highly substituted of the two double bonds (Scheme 6.78).

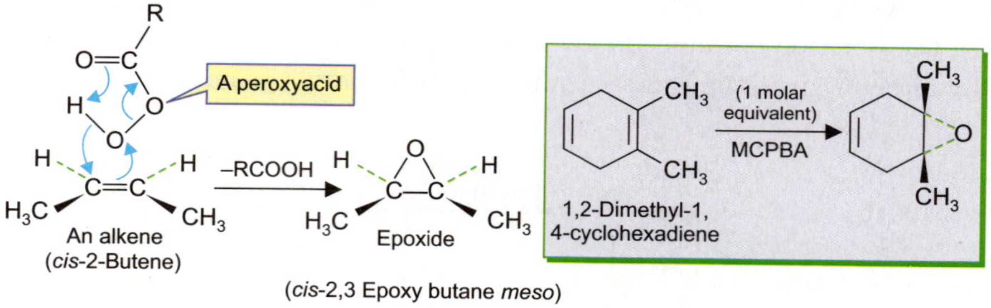

SCHEME 6.78

- The stereochemical outcome of epoxidation from *cis*-and *trans*-butene may be represented as in (scheme 6.79). The assignment of configuration to the chirality centers is as usual. Consider for example one of the enatiomers of *trans*-2,3-epoxybutane. Reorient the molecule so that the hydrogen at C-2 points away from one's eyes the order of decreasing precedence traces a clockwise path. The configuration C-2 is R, and that at C-1 is the same. In the *meso* form, the centers will have opposite configurations.

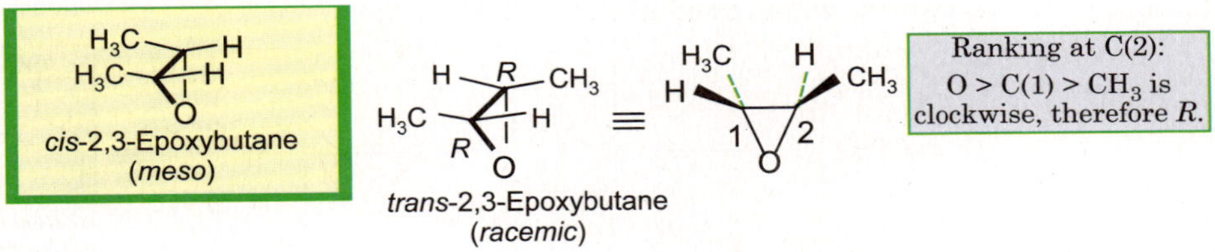

SCHEME 6.79

- Cyclic alkenes are attacked by the peracid predominantly from the less hindered side. Cholosterol molecule *e.g.*, presents extensive steric hindrance from its top face (β-face) by the angular methyl groups. Thus many reagents like a peroxyacid react preferentially at the relatively unhindered α-face (Scheme 6.80). Similar is the situation with epoxidation of 4-methylcyclopentene.

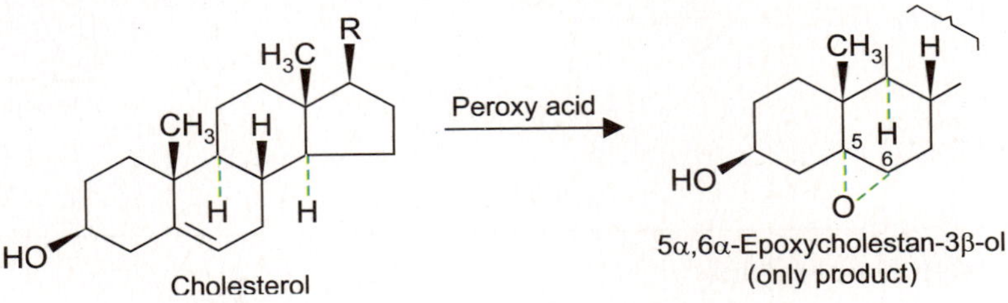

SCHEME 6.80

- α, β-Unsaturated carbonyl compounds can be epoxidized on reaction with alkaline hydrogen peroxide (Scheme 6.81). The nucleophilic addition of hydroperoxide anion to C=C is facilitated by the carbonyl group (compare with Michael reaction).

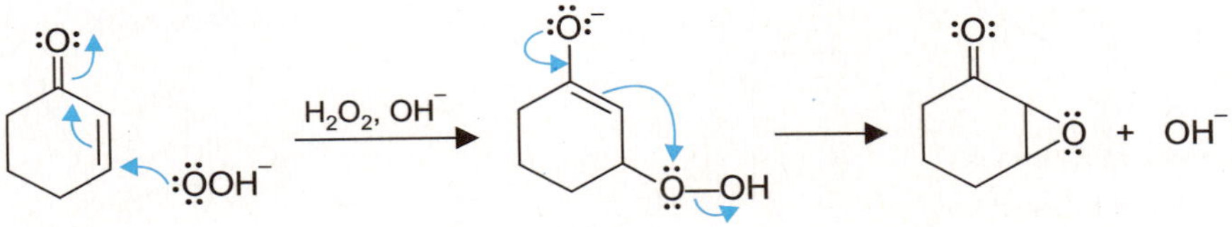

SCHEME 6.81

- Epoxides can also be made from ketones (see Scheme 7.34).

(b) Epoxidation of Allylic Alcohols

The epoxidation of (I, Scheme 6.82) a stereospecific *syn*-addition occurs on the opposite face of methoxy group for steric reasons (stereoselectivity) while hydroxy group guides it on the face of the double bond occupied by it (II, Scheme 6.82). Like peroxyacids, alkylhydroperoxides (ROOH) *e.g.*, *tert*-butylhydroperoxide bring about epoxidation of an alkene under catalysis by a transition metal like molybdenum, vanadium or titanium (compare these results with those in Scheme 6.101).

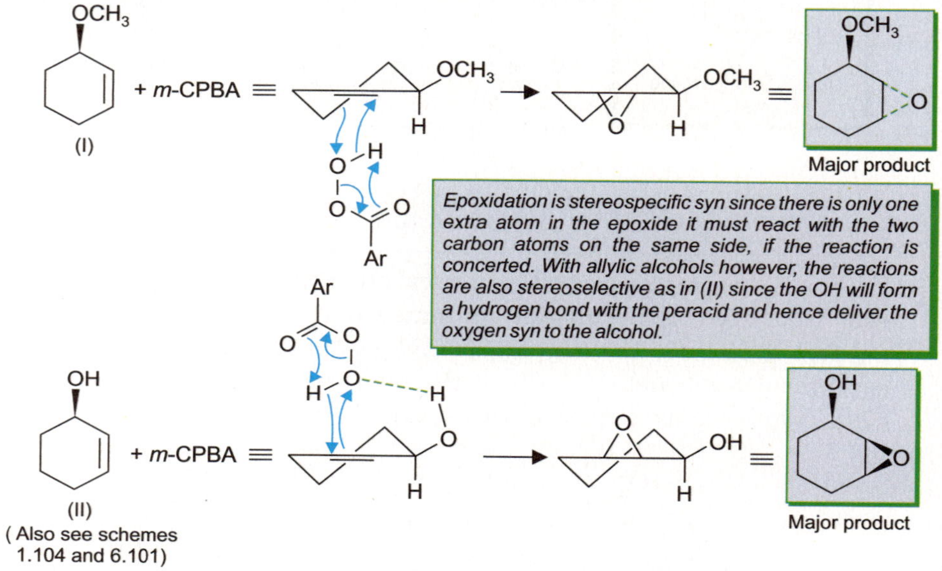

SCHEME 6.82

(c) Epoxides from Halohydrins

These are discussed (see Scheme 3.62*a*), a notable feature of the halohydrin route is that it affords epoxides with the stereochemistry opposite of that obtained using peroxy acids (Schemes, 6.83 and 6.84).

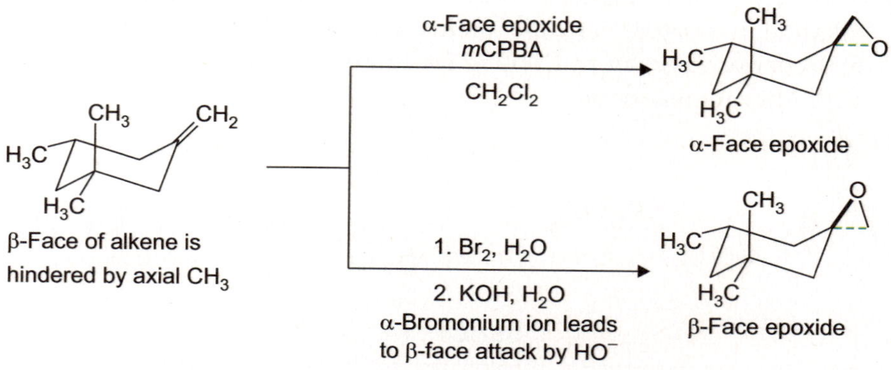

SCHEME 6.83

In aqueous solution chlorine and bromine convert an alkene to afford a halohydrin via the attack of the halonium ion intermediate by water (Scheme 6.84).

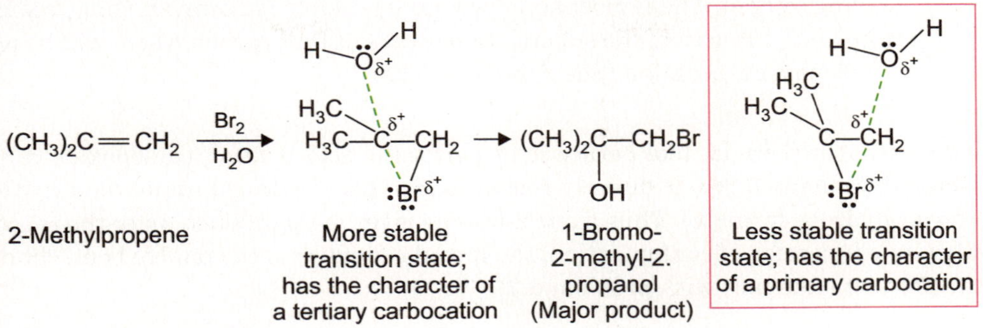

SCHEME 6.84

(d) Some Reactions of Epoxides

(i) Hydrolysis (Acidic or Basic Conditions) Gives 1,2-Diols

Acid or base catalyzed hydrolysis of an epoxide gives glycols with *anti*-stereochemistry (Scheme 6.85). The reactivity to the weak nucleophile (water) is enhanced by protonation of the ring oxygen. The protonated epoxide is then attacked from the side opposite the epoxy ring (S_N2 conditions) to give a *trans*-product as a racemic mixture, the mirror image structure is however, not drawn (Scheme 6.85). The following points may be noted:

- Other weak nucleophiles like alcohols and bromide ion are also effective when the reactivity of the epoxide is increased by protonation (See, Schemes 3.56*b* and 3.56*c*).
- In the case of an unsymmetrical epoxide different products are obtained under acid-catalyzed and neutral conditions (See, Scheme 3.56*b* and 3.56*c*).
- Under neutral (or basic) conditions S_N2 opening is observed *i.e.*, nucleophile attacks at the less hindered carbon of the epoxy ring.

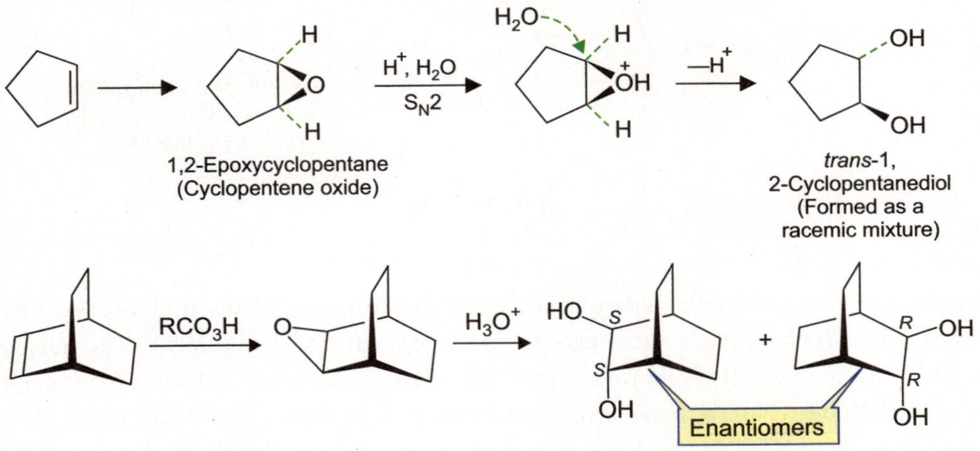

SCHEME 6.85

- Under acid catalyzed epoxide ring opening the stereochemistry is predicted as from an S_N2 reaction *i.e.*, nucleophile approaches from the side opposite the leaving oxygen. The regiochemistry where possible is however, that predicted from an S_N1 reaction, the substitution occurs at that carbon which will be more stable as a carbocation (See Scheme 3.56*c*).

(ii) Direct Hydroxylation

When a stronger peracid, like peracetic or performic acid is used, the epoxide cannot be isolated (Scheme 6.86). It directly reacts to give *anti* hydroxylation (via a hydroxy acetate or hydroxy formate). Thus *trans* 2-butene gives *meso* product, since the reaction involves one inversion of configuration, the initial geometry of the double bond changes in the product (compare with problem 6.2).

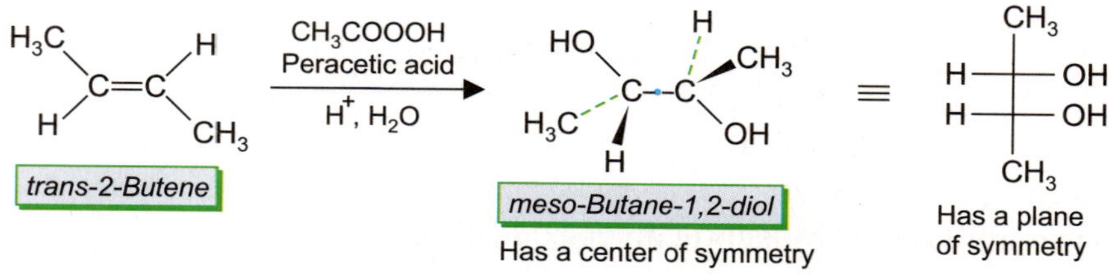

SCHEME 6.86

(iii) Formation of Trans-Diaxial Products

The ring opening of epoxides with base from a rigid cyclohexene devivative gives *trans* diaxial products (Scheme 6.87), the attack of OH^- at other carbon would have given diequatorial product which is not observed (also see problem 6.19).

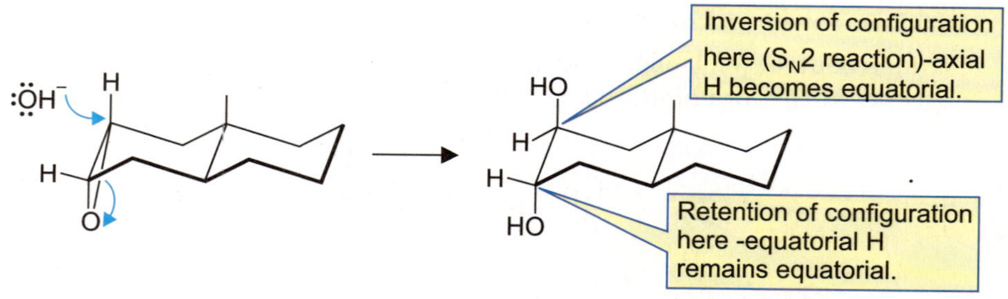

SCHEME 6.87

(iv) Reduction with Lithium Aluminium Hydride

Epoxides are reduced to alcohols with lithium aluminium hydride and as epoxides are derived from alkenes, the overall process is the hydration of an alkene. The method is complementary to the hydroboration method, since the hydride selectively attacks the less substituted carbon (S_N2 mechanism) of the epoxide to give more highly substituted alcohol (Scheme 6.87*a*). In rigid cyclohexane systems an axial alcohol is formed (Scheme 6.87*a*, also see scheme 3.56*b*).

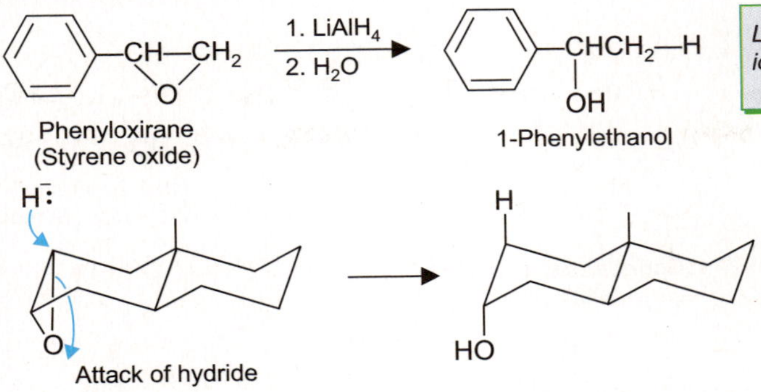

$LiAlH_4$ delivers the hydride ion to the less substituted

From rigid cyclohexane systems an axial alcohol is formed.

SCHEME 6.87a

EXERCISE 6.9

Write the stereostructure of the product from the reduction of cis-4-tert-butylcyclohexene oxide with $LiAlD_4$.

ANSWER. *The opening of the epoxide will give the trans-diaxial product (Scheme 6.88).*

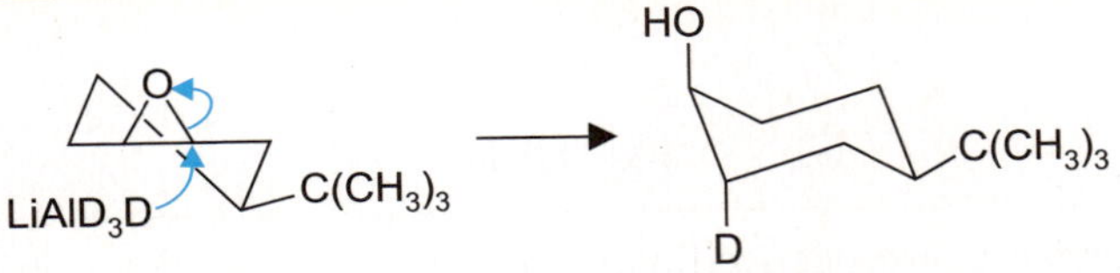

SCHEME 6.88

6.9 HYDROXYLATION OF ALKENES – DIOL FORMATION

(i) *Syn – Hydroxylation and Woodward Method*

Conversion of an epoxide into a *trans* 1, 2-diol has already been discussed. The process involves the delivery of oxygen *syn* to the double bond to give an epoxide and its subsequent acid catalysed hydrolysis (Scheme 6.85).

An alkene can be converted into a *cis*-1, 2-diol on reaction with OsO_4 or $KMnO_4$. With OsO_4 *e.g.*, a cyclic osmate ester is formed which is then hydrolyzed with sodium sulphite to give the addition product *i.e.*, a diol by the net addition of two hydroxyl groups to the same face of the alkene *i.e. syn* addition (Scheme 6.89). Pyridine complexes with osmium atom and is often added as a catalyst. The following points may be noted:

- Thus stereoselective formation of *trans*-1,2-diol from the oxirane of cyclopentene (Scheme 6.85) with acid in water is complementary to the oxidation of cyclopentene with potassium permanganate or osmium tetroxide which gives *cis*-1,2-cyclopentanediol. The reaction is considered as a concerted *syn*-addition of the reagent to the double bond, forming the cyclic osmate ester, which upon hydrolysis or reduction (H_2S, $NaHSO_3$, or Na_2SO_3) gives the *cis*-1,-2 diol. Stereochemistry of *cis*-diol is determined from the cyclic intermediate (Scheme 6.89).

CHAPTER 6

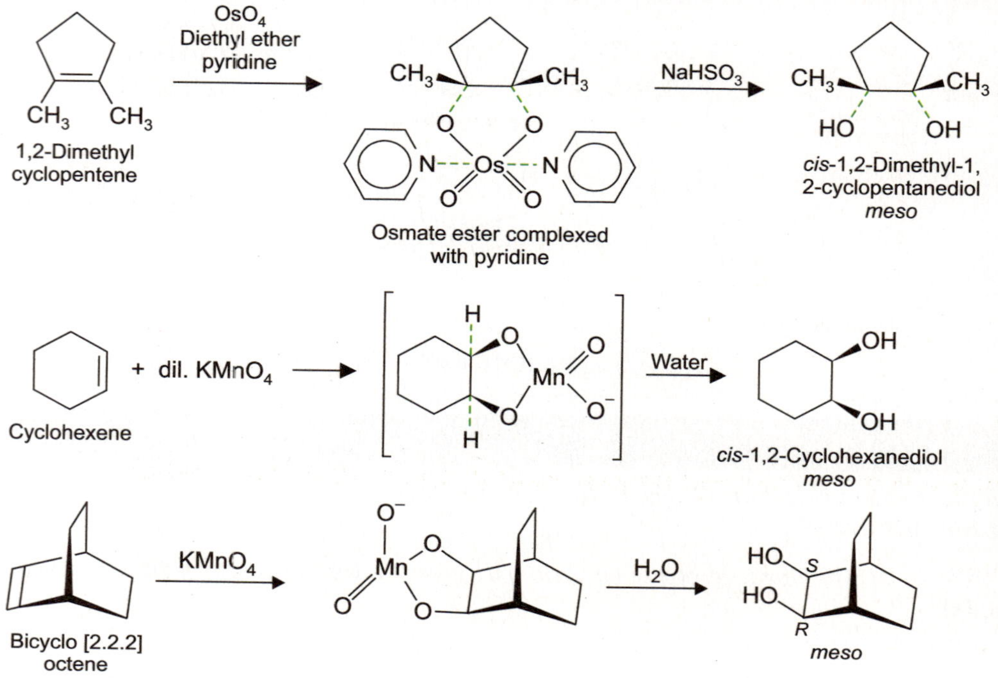

SCHEME 6.89

- The reaction is stereospecific *i.e.*, (*E*)-alkenes give the *syn*-1,2-diols, whereas (Z)-alkenes give the *anti*-1,2-diols (Scheme 6.90).

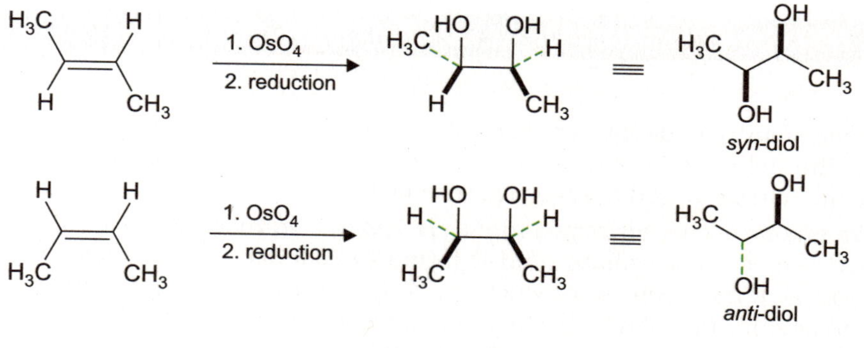

SCHEME 6.90

- The disadvantages associated with the high cost and safety in handling of OsO_4 are taken care of using catalytic amounts of the reagent together with a co-oxidant.

The most widely used co-oxidant is *N*-methylmorpholine *N*-oxide (NMO). Other possible co-oxidants are trimethylamine *N*-oxide (Me_3NO, H_2O) or *tert*-butyl hydroperoxide (*t*-BuOOH). In each case the reaction is a stereospecific *syn* addition and the hydroxy groups are introduced from the less hindered face of the double bond (Scheme 6.91). The reaction is also chemoselective and shows preference to react with more nucleophilic double bond in the presence of a triple bond.

- In rigid cyclic systems, OsO_4 attacks from the less hindered side, and the net result is the formation of the more stable of the two possible *cis* diols (Scheme 6.91), compare with the diol in Scheme 6.95).

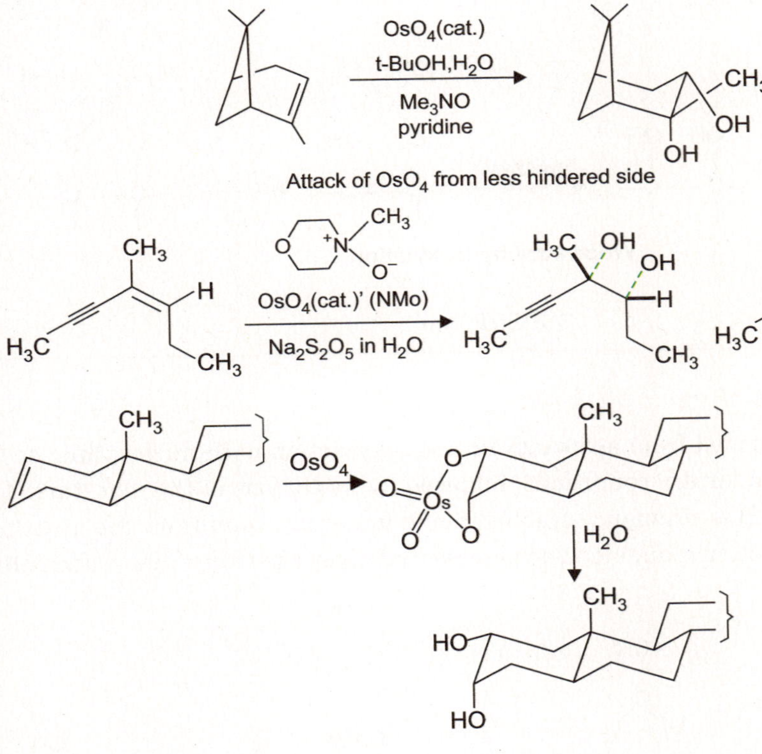

SCHEME 6.91

- In Woodward hydroxylation (iodine silver acetate wet *syn*-hydroxylation), an alkene on reaction with iodine, aqueous acetic acid and silver acetate also gives *cis* diols (Scheme 6.92). Iodine reacts with the double bond to give an iodonium ion which undergoes displacement by acetate in the S_N2 displacement, giving a *trans*-iodoacetate (I, Scheme 6.92). Anchimeric assistance by the acetate group, together with the powerful bonding capacity of silver ion for iodide gives a cyclic acetoxonium ion which reacts with water to give a *cis*-hydroxyacetate. Hydrolysis gives the *cis*-diol (Scheme 6.92).

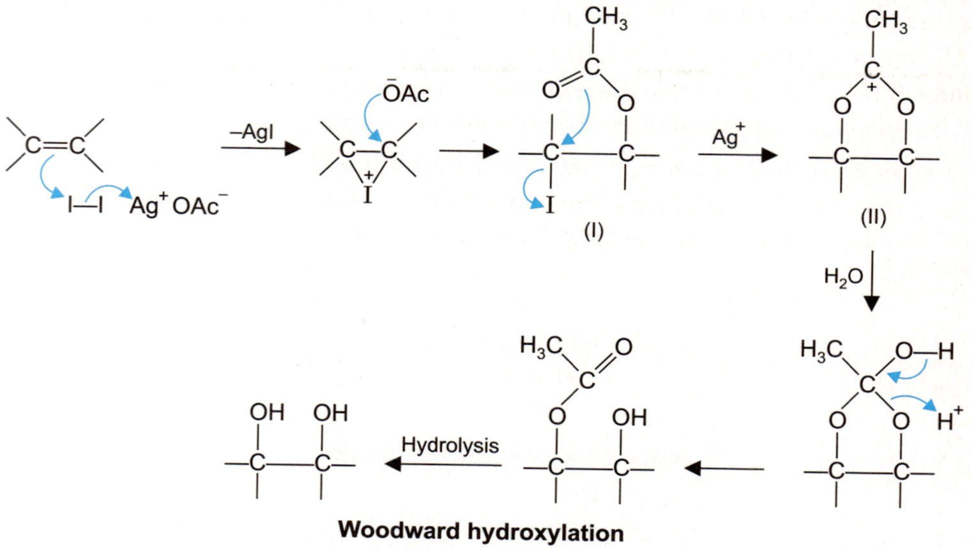

SCHEME 6.92

(ii) *Anti Hydroxylation (Prevost Method)*

The *trans*-diols are formed from acetoxonium ion by reaction with nucleophilic acetic acid (Scheme 6.93), under dry conditions, followed by hydrolysis of the diacetate. The cation (II, Scheme 6.92) is resonance stabilized and under dry conditions the attack on the cation by acetate ion in a bimolecular process and gives the *trans*- diacyl compound (Scheme 6.93).

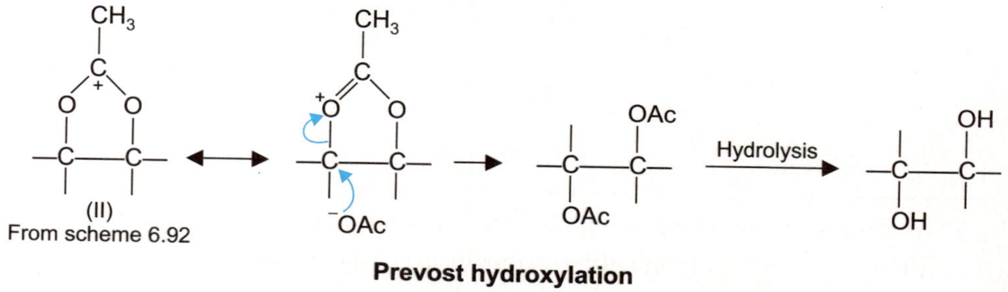

SCHEME 6.93

EXERCISE 6.10

What products will be formed from the steroid (Scheme 6.94) on reaction with (i) OsO_4 and (ii) with $I_2/CH_3COOAg/CH_3COOH/H_2O$?

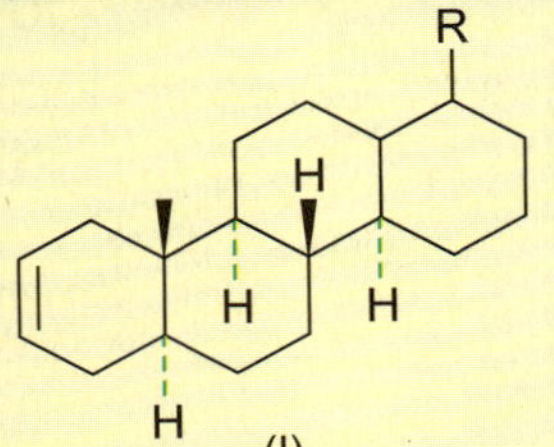

(I)

SCHEME 6.94

ANSWER. *With conformationally rigid molecules the cis diol formed by Woodward method may not have the same configuration to that obtained by OsO_4. Thus (I, Scheme 6.94) is attacked by OsO_4 at the less hindered side to give a less hindered diol (see Scheme 6.91) Woodward procedure gives the more hindered diol (via the less hindered iodonium ion, Scheme 6.95).*

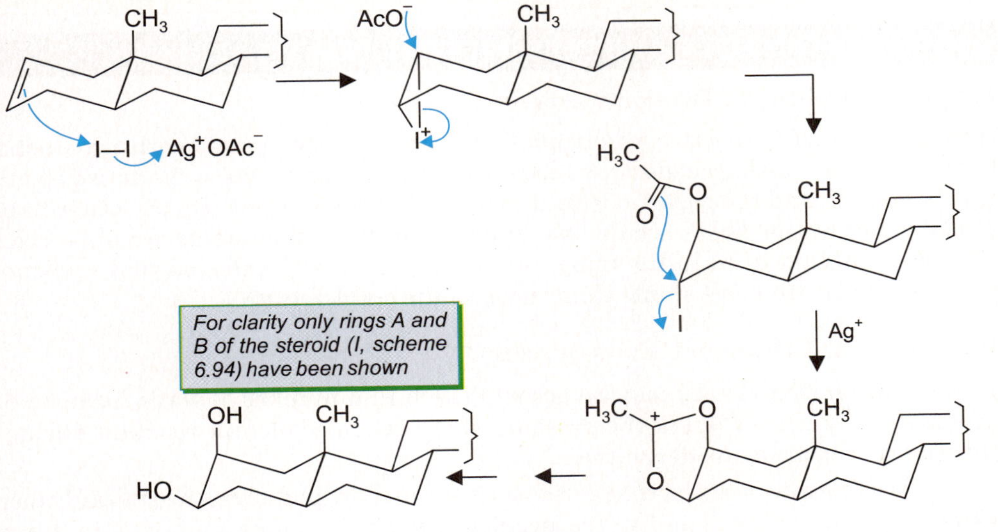

SCHEME 6.95

(iii) *Asymmetric Dihydroxylation of Alkenes with OsO_4 and Chiral Amine Ligands—Enantioselective cis Vicinal Dihydroxylation*

Hydroxylation of an alkene by OsO_4 (a *syn* addition) can lead to preferential formation of one enantiomer of the diol by incorporating selected alkaloid derivatives leading to almost total enantioface selectivity. The reaction is catalyzed by amines and derivatives of naturally occuring cinchona alkaloids. Dihydroquinidine-4-chlorobenzoate is one such chiral reagent which promotes *asymmetric bis*-hydroxylation reactions. Thus (*E*)-stilbene on hydroxylation with OsO_4 and an amine oxide in the presence of a chiral agent (I, Scheme 6.96) gives (*R*, *R*) diol in high yield and the (*S*, *S*) isomer is not formed (one may note that in the absence of the chiral agent a racemic mixture is formed).

Osmium tetroxide oxidations can be highly enantioselective in the presence of chiral ligands. The most highly developed ligands are derived from the cinchona alkaloids dihydroquinine (DHQ) and dihydroquinidine (DHQD see, Scheme 2.53*d*).

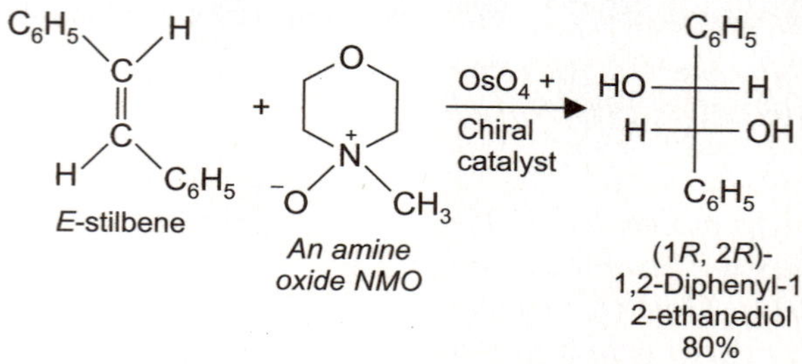

Total enantioface selectivity in osmium tetroxide oxidations

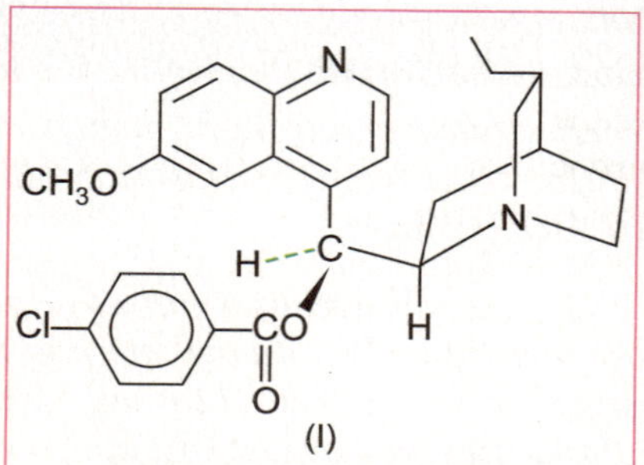

Dihydroquinidine 4-chlorobenzoate
a chiral agent used as a catalyst

SCHEME 6.96

6.10 ADDITION OF CARBENES (METHYLENE) TO ALKENES

(A) Spin Multiplicity of Divalent Carbenes

Carbenes are highly reactive, electrophilic divalent carbon species with two unshared electrons. If both unshared electrons occupy the same orbital, the spins of two electrons must be paired, and the carbene is said to be in the singlet state. On the other hand, if the two unshared electrons occupy two different orbitals; their spins are unpaired and the carbene is said to be in the triplet state. It is generally believed that carbenes in the triplet state are more stable than those in the singlet state.

(B) Generation of Divalent Carbon Species

The generation of carbenes may be accomplished in a number of ways (Scheme 6.97) but it is usually carried out in the presence of the intended olefinic reactant during the synthesis of cyclopropane derivatives.

It is generally believed that carbenes generated by photolysis are initially formed as singlet species and will add stereospecifically to liquid olefins. However, in the gaseous phase with low olefin concentration, such species may lose energy to become the more stable triplet species through collision with the container wall or with inert gas molecules (*e.g.*, N_2).

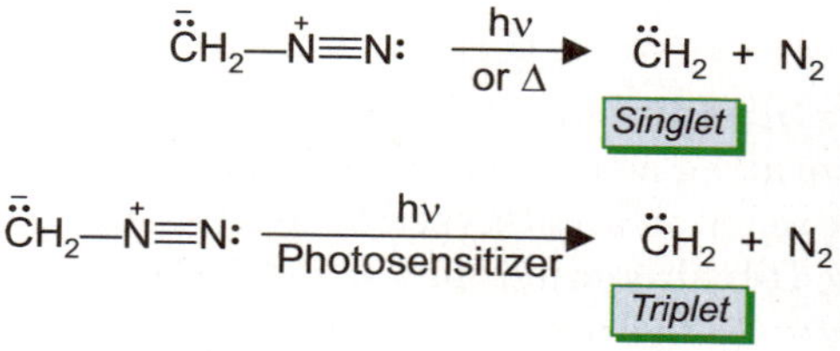

Carbenes (singlet species) can be generated by the decomposition of diazomethane (1,1-elimination) by heat or light. The triplet state is obtained by a special photochemical method using a photosensitizer.

SCHEME 6.97

(C) Addition of Singlet and Triplet Methylene to an Olefin

The triplet species add to a gaseous olefin under high dilution in a nonstereospecific manner. Singlet methylene reacts with an alkene stereospecifically, *e.g.*, *cis*-2-butene and its *trans* isomer react as shown in (Scheme 6.98). However, when triplet methylene reacts with either *cis*-or *trans*-2-butene the reaction is nonstereospecific, each isomer yielding a mixture of *cis*-and *trans*-1,2-dimethycyclopropanes. The difference in this stereochemical outcome is due to the fact that the mechanisms of the two reactions are not the same. Singlet methylene adds to the double bond in one step *i.e.*, reaction is concerted bond formation between the carbon of singlet methylene and the carbons of the alkene occurs at the same time. Thus, the stereochemistry of the alkene is preserved in the cyclopropane product (Scheme 6.98).

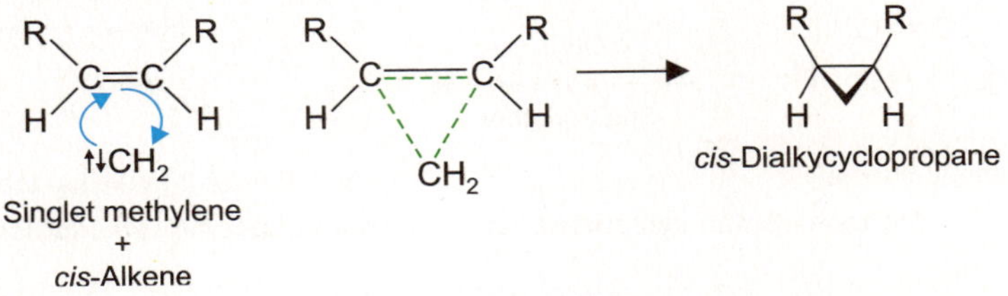

SCHEME 6.98

The electrons in triplet methylene are not paired, therefore, it reacts in a step-wise process. Triplet methylene, a diradical itself reacts with the alkene to afford an intermediate biradical in the form of conformation (I). The biradical (I) has sufficient lifetime to allow rotation of groups joined by single bonds to afford another conformation (II). The ring closure on (I) and (II) yields diastereomerically different cyclopropanes.

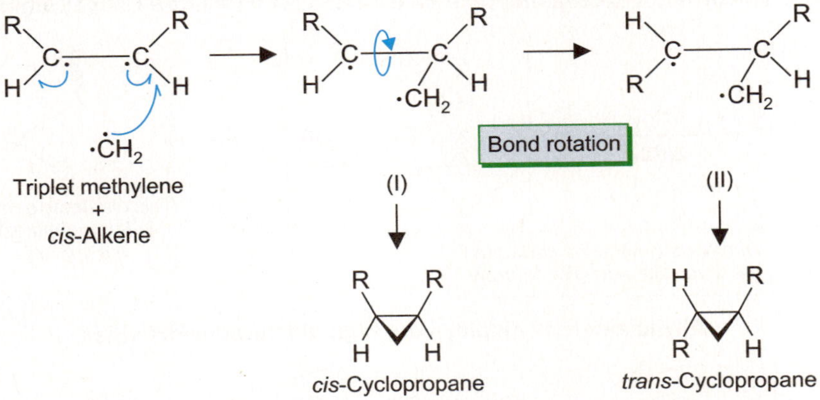

SCHEME 6.98a

(D) Simmons-Smith Cyclopropanation

Simmons and Smith (1958) reported that treatment of a zinc-copper couple with diiodomethane in ether forms a reagent that adds to alkenes to form cyclopropane (Scheme 6.99). The reaction displays stereospecificity *i.e.*, retention of olefin geometry to show that addition step is concerted (Scheme 6.100).

$$CH_2I_2 + Zn\ (Cu) \xrightarrow{Et_2O} \underset{\text{A carbenoid}}{ICH_2ZnI}$$

SCHEME 6.99

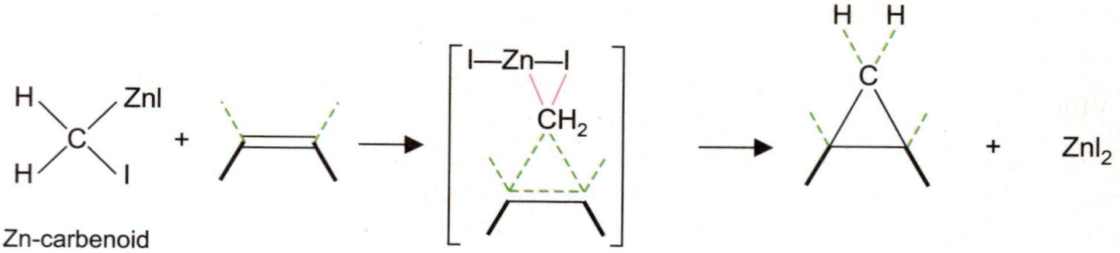

Zn-carbenoid

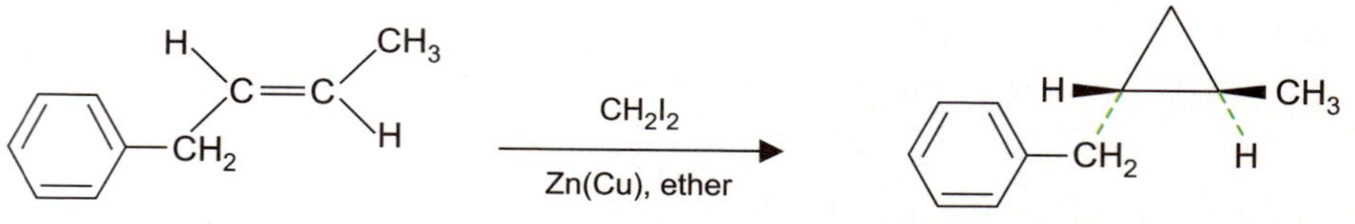

(*E*)-1-Phenyl-2-butene

trans-1-Benzyl-2-methylcyclopropane (50%)

Simmons-Smith cyclopropanation is a stereospecific *syn* addition reaction

SCHEME 6.100

Allylic alcohols also cyclopropanate almost 100 times faster than their unfunctionalized alkene equivalents. Coordination between the zinc atom and the hydroxyl group in the transition state explains both the stereoselectivity and the rate increase. This stereoelectronic control is displayed by proximal OH, OR groups, which favor cyclopropanation to occur from the same face of the double bond as the oxy substituents. The order of decreasing directive effects is OH > OR > C = O (Scheme 6.101).

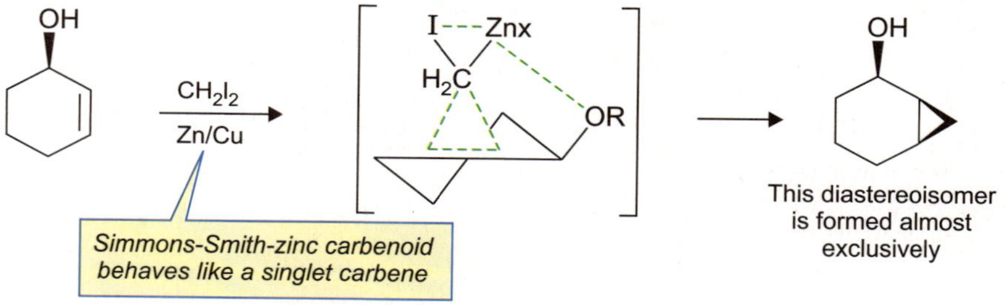

Allylic alcohols display very high diastereoselectivities

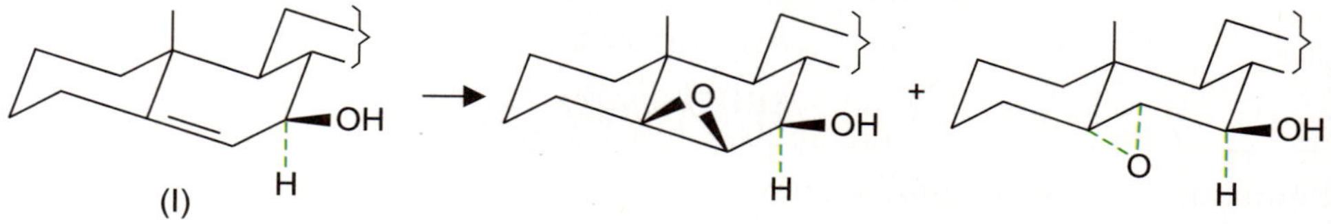

(I)

TBHP + V(acac)$_2$	100	:	~ 0
MCPBA	~ 70	:	~ 30

(Also see scheme 6.81) **Stereoselective epoxidation of a cyclic allylic alcohol resembles Simmons-Smith reaction.**

SCHEME 6.101

In the highly stereoselective epoxidation of *e.g.*, a steroidal allylic alcohol (I, Scheme 6.101) with *t*-butylhydroperoxide (TBHP) catalysed by vanadium or molybdenum, used as $V(acac)_2$ and $Mo(CO)_6$, vanadium coordinates with both the allylic OH and *t*-Bu-O-OH, and oxygen is transferred to the double bond completely from the side *cis* to allylic OH (proximofacial addition). A comparison with *m*-chloroperbenzoic acid (MCPBA) is also presented for comparison (also see, Scheme 6.81).

PROBLEMS

6.1. *meso*-Butane-1, 2-diol can be made from two diastereomeric alkenes. Give the mechanism of the two pathways and the geometry of the diastereomeric alkenes.

6.2. Predict the stereochemistry of the products of hydroxylation with osmium tetroxide followed by hydrolysis from *cis* and *trans*-crotonic acids. What product will be formed from crotonic acid on treatment with performic acid ?

6.3. Designate the two faces of 1-butene. Discuss the stereochemical consequences of its reaction with bromine.

6.4. Predict the outcome of *anti*-addition of bromine of *cis*-cyclodecene and its diastereomer. From the stereochemistry of the products formed in each case, is the reaction stereoselective or stereospecific?

6.5. How diastereomic *syn* and *anti* diol can be made from *E*-oct-4-ene?

6.6. Explain giving mechanism why the addition of bromine to 2-butene is a stereospecific reaction.

6.7. Addition of bromine to cyclohexene is a stereoselective reaction, explain giving mechanism.

6.8. Predict the product of epoxidation from the rigid cyclic alkenes.

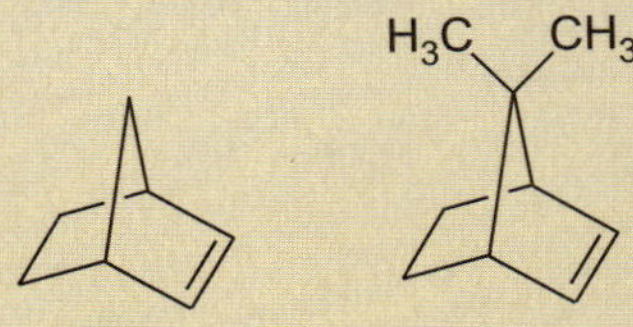

(I) (II)

6.9. Predict the product of reduction from *Z*, *E*-1, 5-cyclodecadiene.

6.10. Write the stereostructure of the product from the reaction of Br_2/H_2O with 1-methyl-cyclohexene. What features explain the regioselectivity and stereoselectivity of the reaction?

6.11. Predict the product stereochemistry from the reaction.

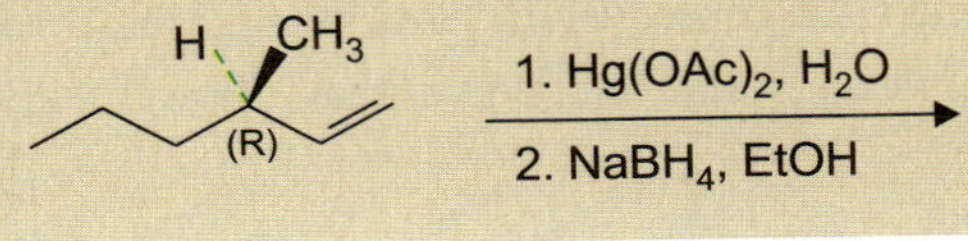

6.12. Depict the addition or bromine to *E*-2 butene. Detect the plane or center of symmetry in the product by writing suitable stereostructures.

6.13. Why the addition of bromine to 2-butene is termed as a stereospecific reaction? Explain briefly by drawing stereostructure of the products.

6.14. Why the bromonium ion from (*Z*)-2-butene is achiral while from (*E*)-2-butene is chiral?

6.15. Why *trans* cyclooctene displays electrophilic *syn* addition? Is cyclooctene chiral, if so, why?

6.16. Norbornene on electrophilic addition of DBr gives only the *exo-syn* product. How this product can arise?

6.17. Write the reagents for the following reactions:

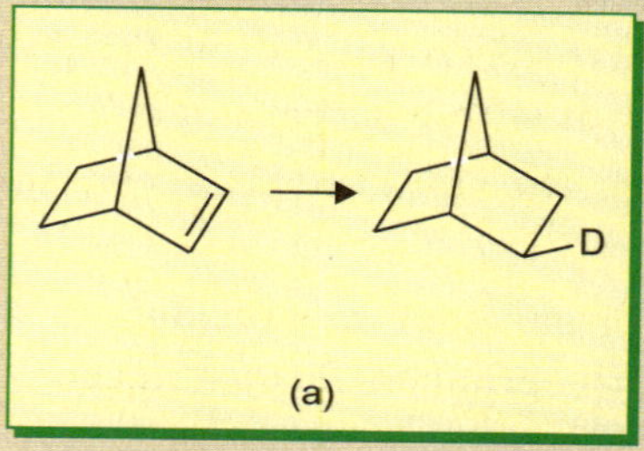

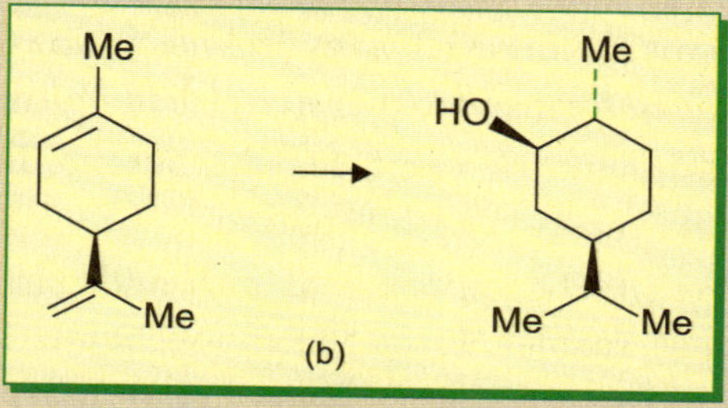

6.18. *trans*-2-Butene is treated with MCPBA and the isolated epoxide is subjected to acid catalyzed hydrolysis. Predict the stereochemistry of the product.

6.19. The following epoxide on acid catalyzed hydrolysis as expected gives a *trans* diol which could have either structure (I) or (II). Predict the correct product with reasoning.

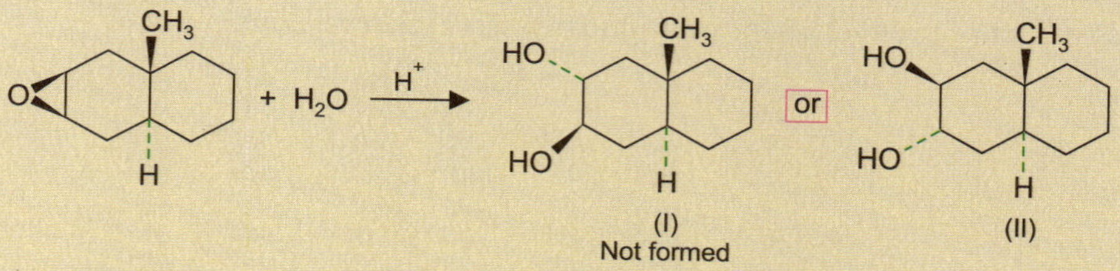

6.20. Predict the stereochemical outcome of hydroxylation of maleic and fumaric acids with osmium tetroxide.

6.21. Write the stereochemistry of the product of opening the following epoxide (1-phenylcyclohexene oxide) with hydroxide.

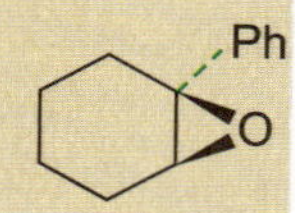

6.22. Opening of an epoxide in a rigid cyclohexane system leads to *trans*-diaxial products. An exception to this rule is the opening of the following 2β, 3β-epoxylanost-8-ene with HBr. Explain.

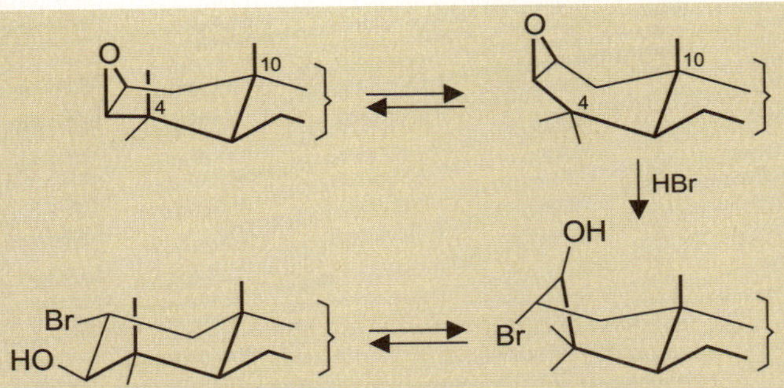

6.23. Write the Fischer projections of the products of bromination of *cis*- and *trans*-2-pentene. Write the perspective formulas of these products.

6.24. Write the Fischer projection of the product of addition to *trans*-2-butene, and transform it into its perspective drawing.

6.25. Predict the products from the reactions.

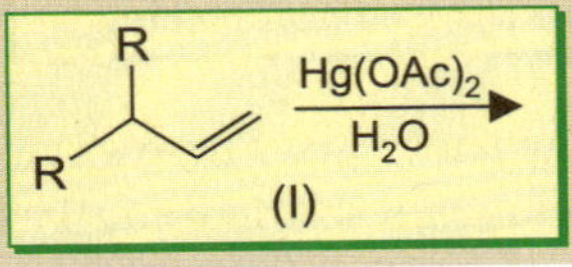

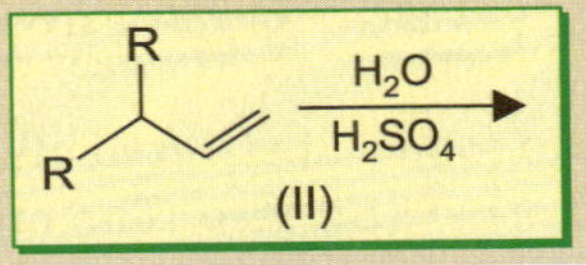

ANSWERS TO SELECTED PROBLEMS

6.1. *vicinal*-Dihydroxy compounds can be made from alkenes by two methods (*i*) epoxidation and S_N2 ring opening of the epoxide (*anti* addition) and (*ii*) via osmate ester formation and ring opening by treatment with $NaHSO_3$ (*syn* addition). The pathway in (I) is similar to bromination of an alkene and makes *trans*-2-butene the starting material (the inversion of configuration in the product forming step as *e.g.* in Scheme 6.85 will give the *meso* product as in Scheme 6.86). The pathway (II) does not involved inversion of configuration and the same *meso* product will be obtained from *cis*-2-butene.

6.2. Consult answer of problem 6.1. During hydroxylation with osmium tetroxide no inversion of configuration is involved at any stage. During hydroxylation with performic acid, the acid catalyzed ring opening of epoxide is attended with inversion of configuration.

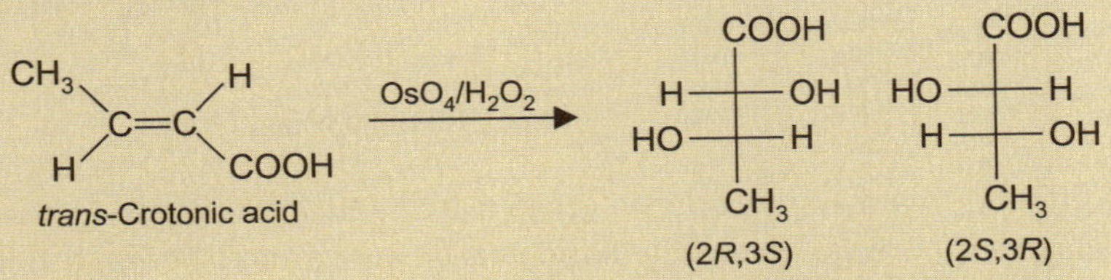

threo-2,3-Dihydroxybutanoic acid

No inversion of configuration at any stage of hydroxylation. The geometry of substrate is maintained in the product.

CHAPTER 6

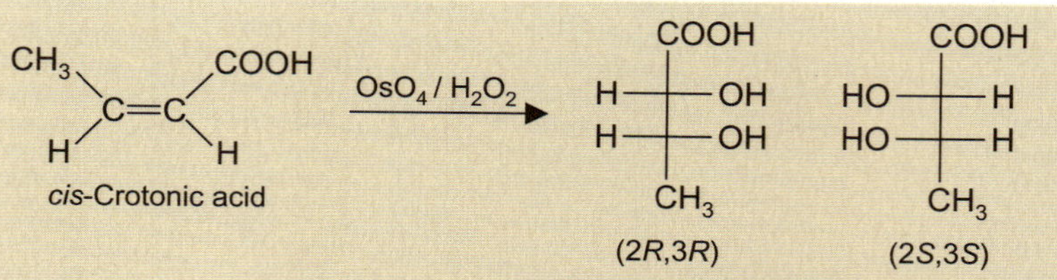

erythro-2,3-Dihydroxybutanoic acid

> *No inversion of configuration at any stage of hydroxylation. The geometry of substrate is maintained in the product.*

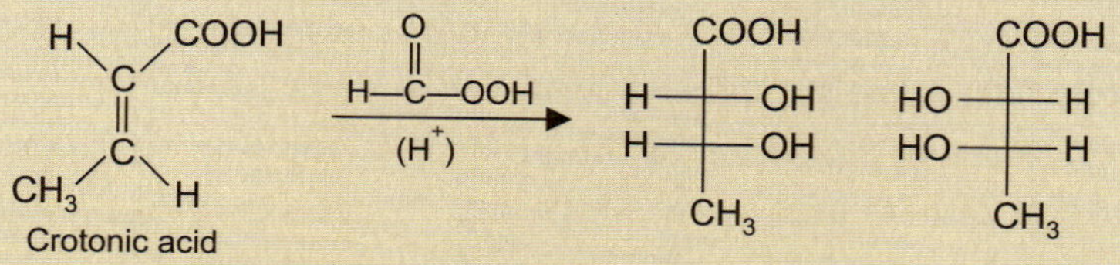

> *One inversion of configuration during hydroxylation. The initial geometry of the substrate changes in the product.*

6.3. In the case of a monosubstituted alkene there are always two enantiotopic faces one (*Re*) and other (*Si*) face. The bromonium ion is equally likely to be formed from either face to give bromonium ion (I or II). This is then attacked by Br^- (S_N2 fashion) on the more electrophilic C2 atom to give (*R*) and (*S*) 1, 2-dibromobutane as a racemic mixture. Any alternative attack at C1 will also be the same for both the bromonium ions and will still give the same racemic mixture.

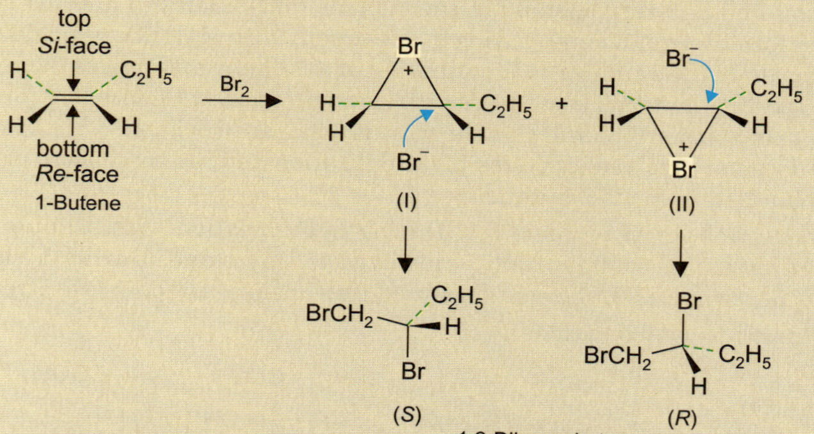

1,2-Dibromobutane

6.4. *anti*-Addition to a *cis*-cycloalkene gives a *trans* product while to *trans* cycloalkene a *cis* product. A *trans* double bond can be accommodated in a ten membered ring. The reacting alkenes are therefore, stereoisomers (diastereomers), which on bromination give different stereoisomers. The reaction is therefore, stereospecific (Note, *cis*-1,2-dibromo-cyclodecane has a plane of symmetry).

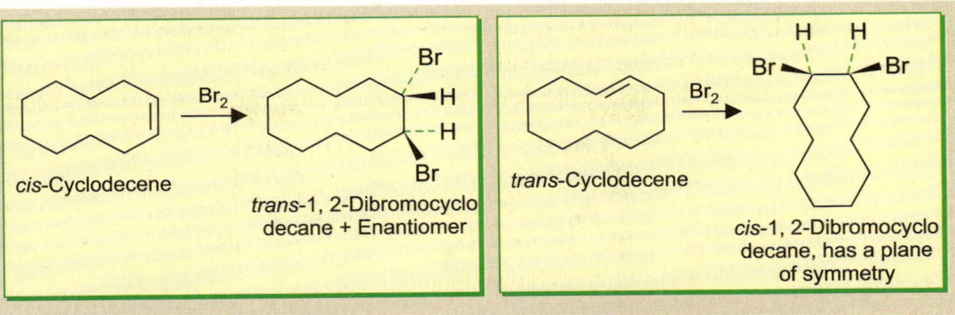

6.5.

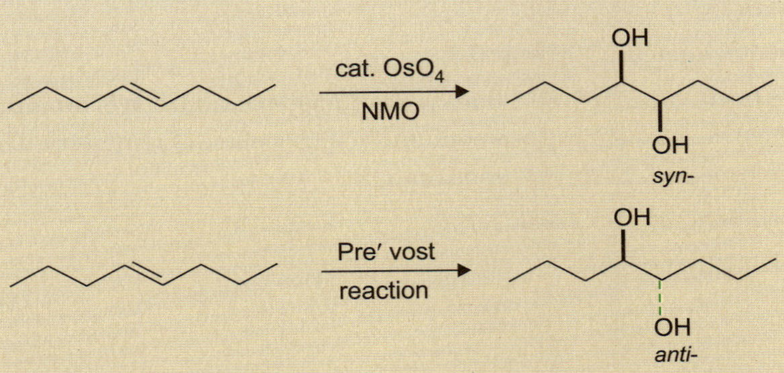

6.8. With conformationally rigid cyclic alkenes, the epoxidising reagent *e.g.*, *m*CPBA approaches the molecule (the double bond) from the less hindered side.

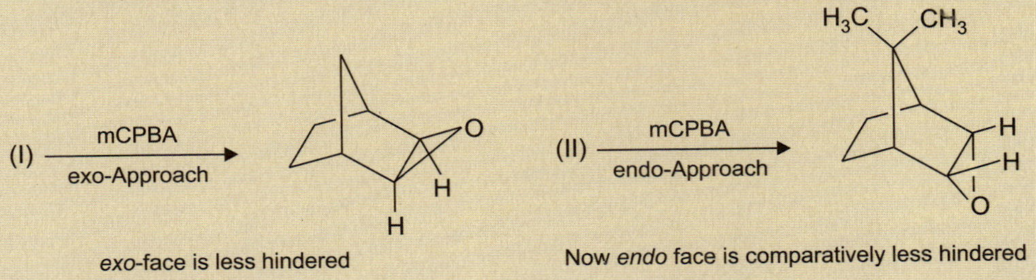

6.9. The rate of reaction with diimide is influenced by torsional and angle strain in the alkene. More strained double bonds react at accelerated rates. Thus the more strained *trans* double bond is selectively reduced in Z, *E*-1,5-cyclodecadiene. Diimide selectively reduces terminal over internal double bonds in polyunsaturated systems.

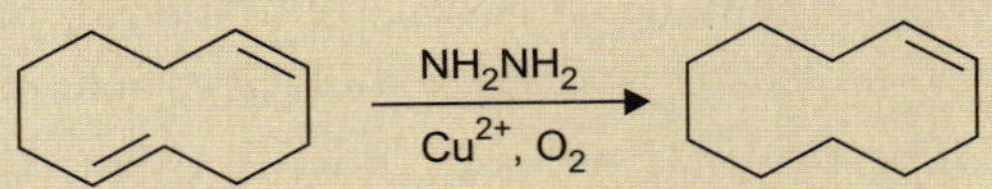

6.10. Hint (see Scheme 6.8), the cyclic bromonium ion character as well as the carbocationic character of the intermediate bridged bromonium ions is responsible for stereoselectivity as well as regioselectivity of the reaction.

CHAPTER 6

6.11. It is a Markovnikov hydration of an alkene. In this reaction a new chirality center is generated while the chirality center already present in the reactant (*R*) remains unchanged. The product is thus formed as a mixture of diastereomers.

Diastereomers

6.12. Hint (See Schemes 1.100).

6.13. Hint see Schemes 1.100 and 6.3.

6.14. Overall the two faces of (*Z*)-2-butene are homotopic, addition of bromine from either face gives the same achiral bromonium ion with a plane of symmetry. The two faces of (*E*)-2-butene are, however, enantiotopic.

6.15. Hint see Schemes 6.21 and 1.136.

6.16. Hint see Scheme 6.21.

6.17.

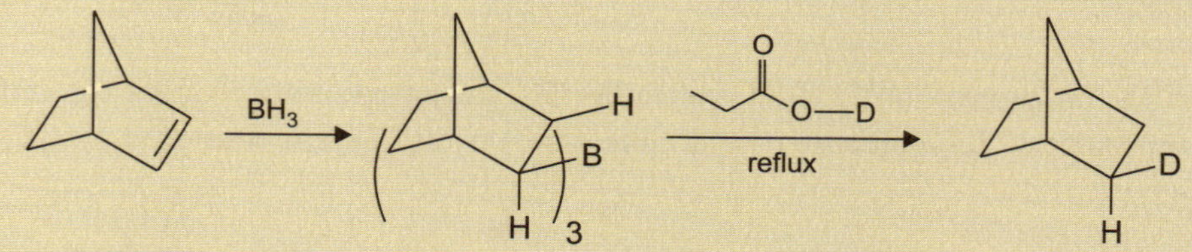

(a)

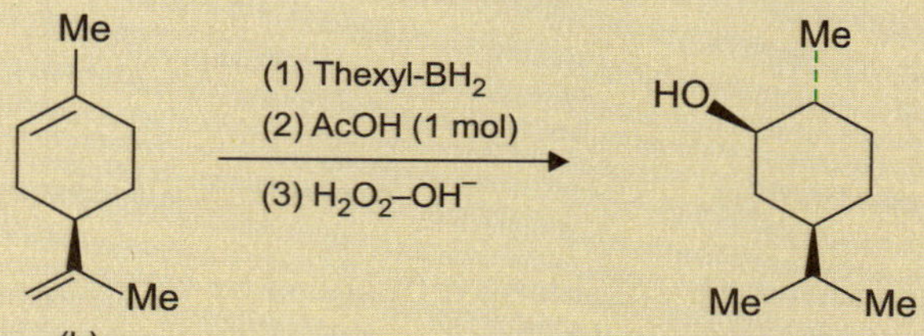

(b)

6.18. *trans*-2-Butene will give a *trans*-epoxide (See Scheme 1.103), just like the bromonium ion (See, Scheme 1.100). The protonated epoxide will be then attacked by water from the back side (S_N2) resulting in overall *anti* hydroxylation. The product formed will be *meso*-butane 1,2-diol (See, Scheme 6.86).

6.19. The glycol (II) will be the product, since only in this (considering its conformational formula) the two hydroxyl groups are *trans*-diaxial (see, Scheme 6.87).

6.20. It is a *syn*-hydroxylation. Adding both the hydroxyl groups from top or bottom face (*syn* addition) to maleic acid gives the same *meso* tartaric acid. The *syn* addition of both the hydroxyl groups to fumaric acid from top or from bottom gives the two enantiomers in the racemic form. Since during hydroxylation with OsO_4, there is no inversion of configuration at any stage, the initial geometry of the substrate is maintained in the product.

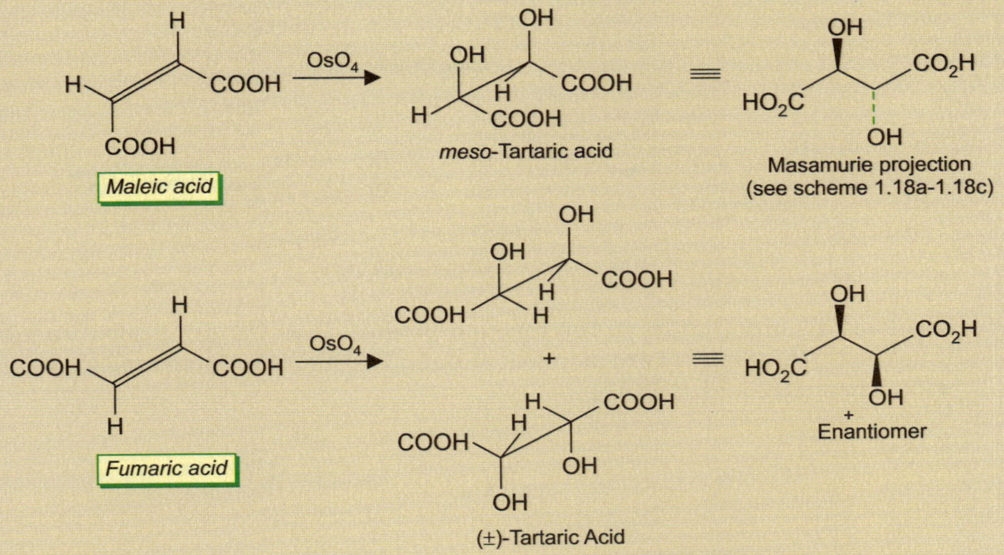

6.21. Opening of the epoxide with base proceeds via an S_N2—like transition state to give a single diastereomer.

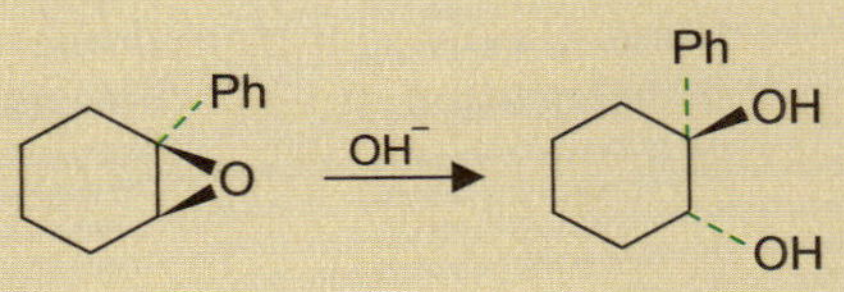

6.22. The chair conformation of ring A is destabilized by 1 : 3 type of non bonded interactions among the epoxy oxygen and the two β oriented methyl groups at C4 and C10. Therefore, ring A attains a boat conformation in which the diaxial addition takes place and a reversion to chair form leads to the diequatorial product.

6.23. When two chirality centers are generated via a bromonium ion intermediate, only one pair of enantiomers would be formed—addition of bromine is stereospecific (see, Scheme 1.100). Remember the stereochemical sequence *cis-syn-erythro* (all terms mean to the same side). One can change any two of the terms at a given situation to predict stereochemical outcome. Thus *cis*-2-pentene on *anti*-bromonation will give *threo*-enantiomers and *trans*-2-pentene on *anti* addition will give *erythro* enantiomers. Stereochemically, since after the formation of a bromonium ion there is one inversion of configuration during its opening, the initial geometry of the reactant changes in the substrate-*cis*-alkene gives *threo* product while *trans*-alkene gives *erythro*-product.

CHAPTER 6

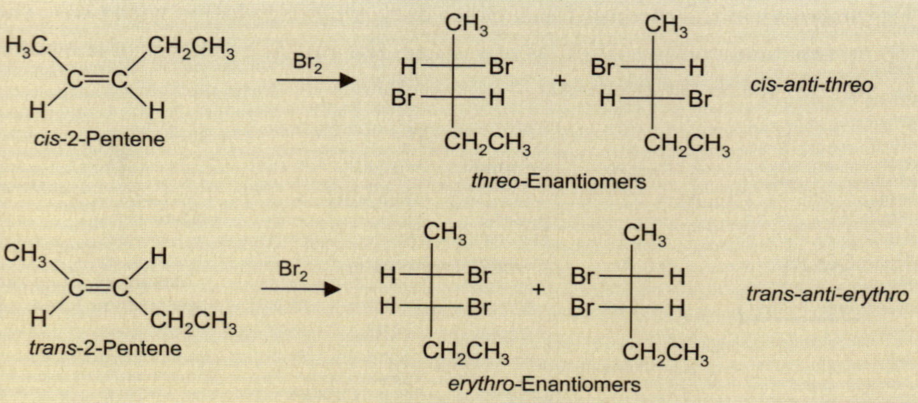

6.24. The two chirality centers in the reaction product have the same substituents, thus the *erythro* isomers are identical and represent a *meso* compound (compare with scheme 6.3 and 1.100).

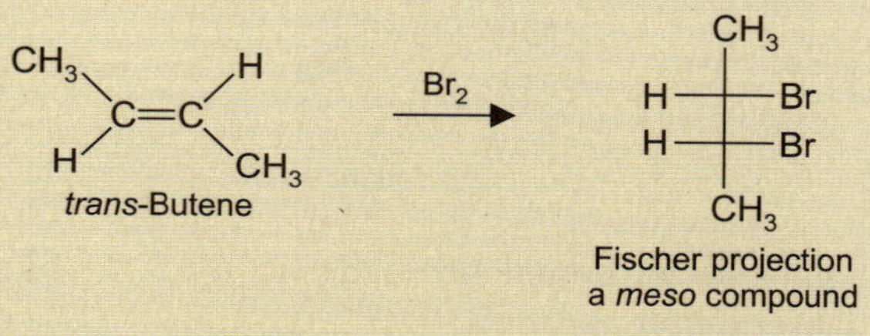

6.25. Both are hydration reactions. Oxymercuration-demercuration process does not involve the formation of a carbocation, thus in this case rearrangements are not commopn which, however is the case under acidic conditions.

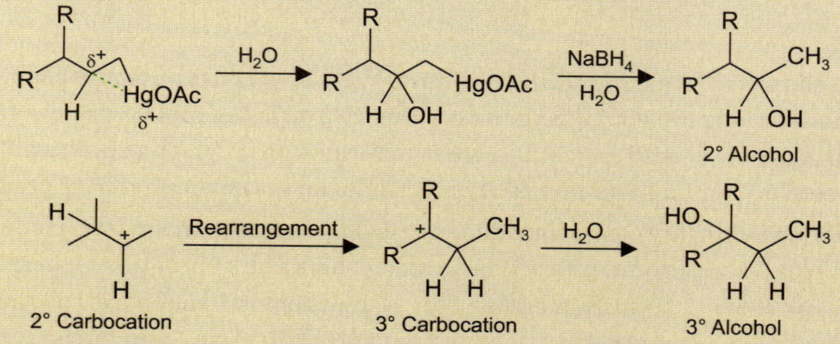

Addition to Carbon Hetero Multiple Bonds and Related Reactions – Stereochemistry Mechanism

7

CHAPTER

7.1 SOME GENERAL AND STEREOCHEMICAL ASPECTS OF ADDITION TO CARBONYL COMPOUNDS

The following points may be noted:

- The faces of aldehydes other than (formaldehye) and unsymmetrical ketones R—CO—R′ (R ≠ R′) are enantiotopic. In such carbonyl compounds, on nucleophilic addition a new stereocenter is generated. The product however, is optically inactive since equal mixture of enantiomers is formed (see Scheme 2.7*a*).
- When, however, there is chirality in R or R′ in a ketone R—CO—R′ (or in an aldehyde), the faces become diastereotopic. The racemic mixture will not be formed on nucleophilic addition and Cram's rule predicts the stereochemistry of the diastereomeric product (see Schemes 2.24—2.28).

7.2 MECHANISM AND STEREOCHEMISTRY OF METAL HYDRIDE REDUCTION

(a) Aldehydes and Ketones

(i) Mechanism of Reduction

Two hydride donating reagents which reduce aldehydes and ketones to alcohols are lithium aluminium hydride ($LiAlH_4$) and sodium borohydride ($NaBH_4$). The hydride ion ($H:^-$ nucleophile) adds to an aldehyde or a ketone to form an oxyanion which can then be protonated by adding water (eq. I, Scheme 7.1) [*The overall reaction adds H_2 to the carbonyl double bond. Each of the four hydrogen atoms in lithium aluminium hydride is available for transfer to the carbonyl group*. Lithium aluminium hydride is an extremely reactive reagent and reacts with most polar functional groups. It reacts, often with explosive violence, with any compound that has a relatively acidic hydrogen, such as an OH or NH_2. Therefore, reductions employing $LiAlH_4$ are usually conducted using ether or THF as the solvent. After addition of the hydride is complete, H_2O is carefully added to decompose any excess reagent and protonate the initially formed alkoxide anion.

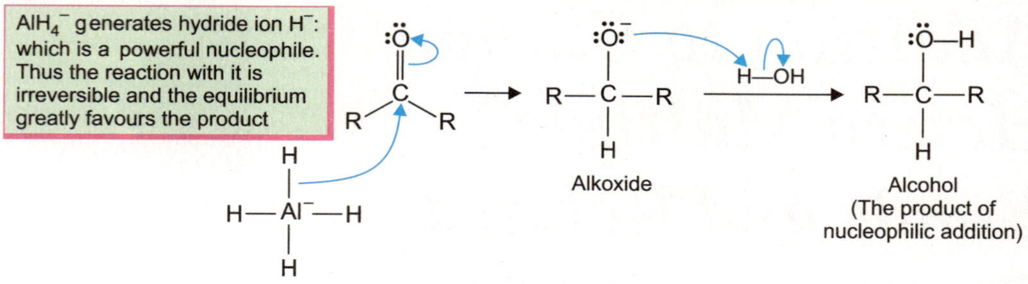

SCHEME 7.1

(ii) Stereochemistry of Reduction

The following points may be noted with different classes of carbonyl compounds:

(A) Stereochemistry/Stereoselectivity with cyclohexanones and other cyclic ketones

During the addition of nucleophiles to cyclohexanones, the nucleophile can approach from either the axial or the equatorial side, the carbonyl oxygen then moves into an equatorial or axial position respectively. With comparatively small hydride donors (sterically unhindered) like AlH_4^- and BH_4^- ions the more stable equatorial alcohol generally predominates (thermodynamic control). The nucleophile *i.e.*, the hydride ion is delivered axially (Scheme 7.2), and the reaction is highly stereoselective.

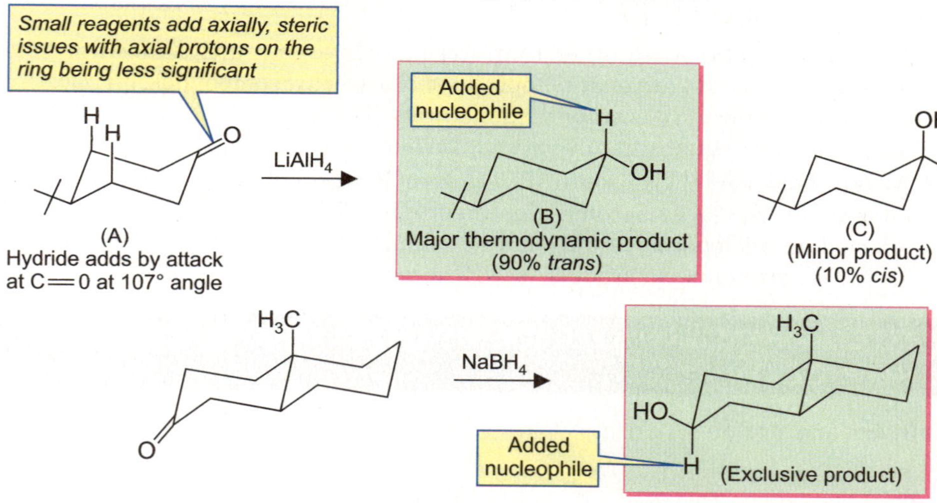

SCHEME 7.2

The approach route (trajectory of nucleophilic additions) to the carbonyl group directs the nucleophile over the smallest group in a nonperpendicular attack. The perpendicular attack allows a maximum overlap between the HOMO of the nucleophile and the *p*-orbital at the carbonyl carbon which makes up part of the LUMO π^*, however, there is a significant antibonding overlap with the other *p*-orbital on oxygen. Thus, the best compromise angle of attack is –107° (Bürgi-Dunitz trajectory Scheme 7.3).

$LiAlH_4$ reduction of ketone A (Scheme 7.2) gave a diastereoisomeric mixture of alcohols (ratio 90:10). There is clearly a choice in the reaction pathway; addition can either occur on the top or bottom face of the carbonyl group in (A) to give (B and C) respectively. Addition to the top face is preferred, so this reaction is highly stereoselective.

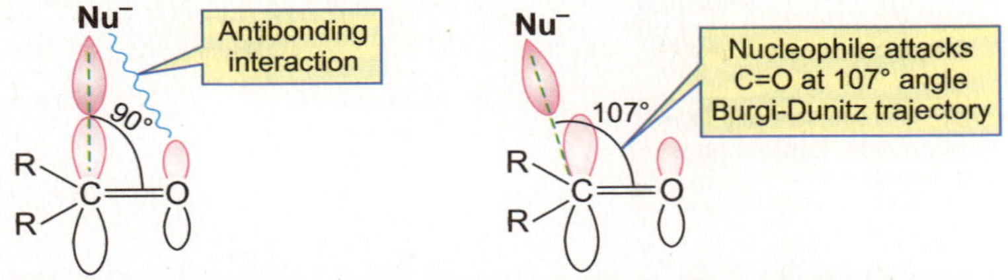

The angle of attack as the result of a compromise between maximum orbital overlap of the HOMO with π^* and minimum repulsion of the HOMO by the electron density in the carbonyl π bond, is 107°.

SCHEME 7.3

The hydride ion can be delivered to the carbonyl carbon of cyclohexanone: (1) *axial* attack with formation of the *equatorial* alcohol and (2) *equatorial* attack with formation of the *axial* alcohol. Two factors compete with each other: (1) *steric interaction* of the incoming "hydride" with the 3,5-diaxial hydrogens in the axial attack and (2) *torsional strain* of the incoming "hydride" with the 2,6-diaxial hydrogens in the equatorial attack (Scheme 7.4).

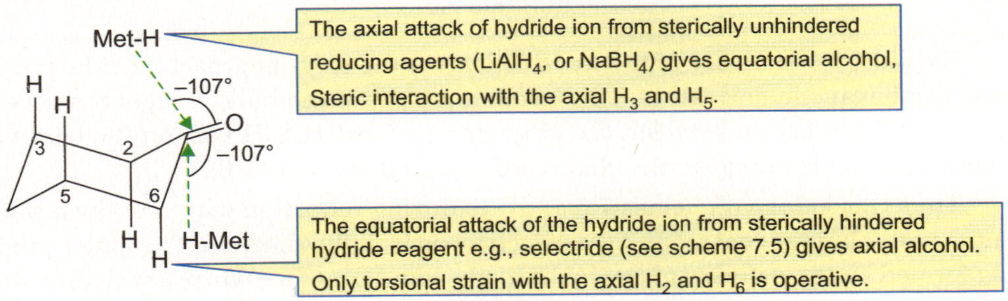

SCHEME 7.4

When the nucleophile is sterically hindered like a bulky hydride "selectride" (Scheme 7.8) then an axial alcohol formation predominates *i.e.*, now the nucleophile (H:⁻) is delivered from the equatorial side. Here one has a situation opposite to the one in (Scheme 7.2) *i.e.,* selectride is a nucleophilic hydride ion source like $NaBH_4$ but it is much bulkier than $NaBH_4$ or $LiAlH_4$.

The following points may be noted:

- This reduction with selectride like $NaBH_4$ reduction is under kinetic control.
- Approach path and stability of the transition state are important. The major product is formed by the delivery of the hydride *i.e.*, the approach of the reagent from the equatorial side (lower face) which is less crowded (Scheme 7.5).

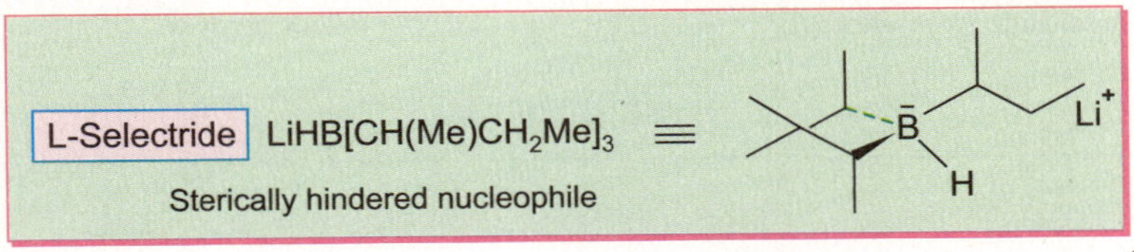

Continued...

CHAPTER 7

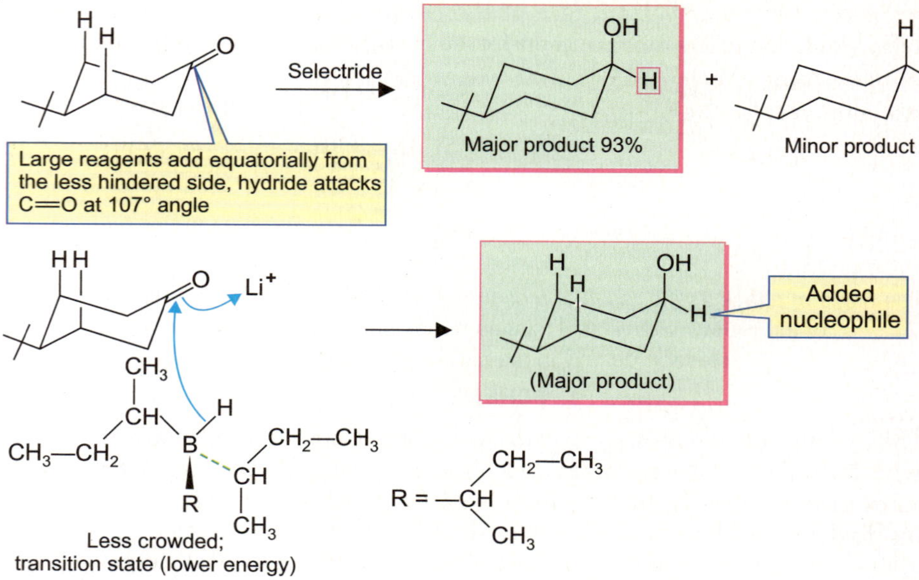

SCHEME 7.5

Consider the delivery of the hydride ion via the axial approach *i.e.*, the top face of the cyclohexanone derivative. This axial approach of the bulky reagent is severely restricted by interaction between the large groups $CH(CH_3)CH_2CH_3$ on the boron and the vertical axial H atoms on the chair conformation (Scheme 7.6).

Thus this ketone (Schemes 7.5 and 7.6) during reduction with a bulky reducing agent is under kinetic control and the selectride is facially selective. The major product being formed by the delivery of the hydride via equatorial face which involves a stable and less crowded transition state as compared to the one involved via axial approach by the bulky reducing agent.

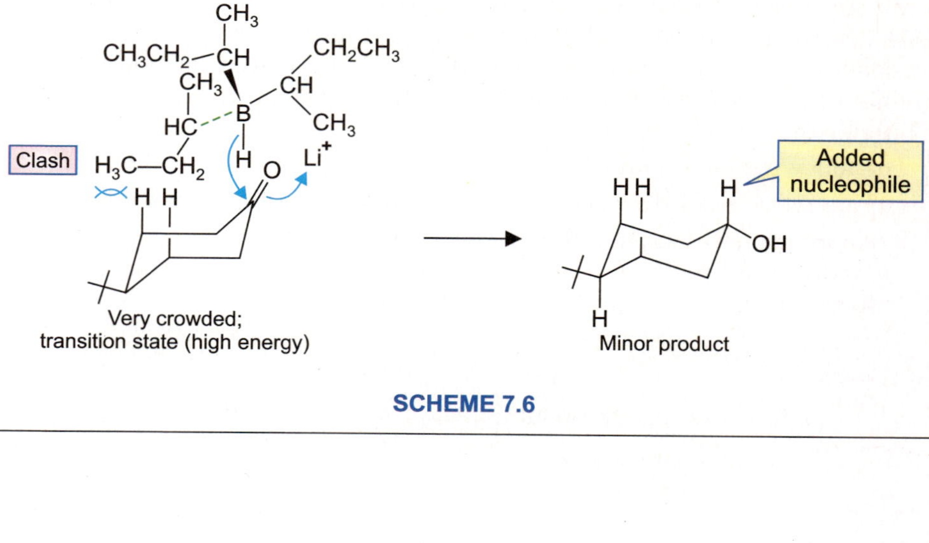

SCHEME 7.6

EXERCISE 7.1

Predict the outcome of the products from reactions (Scheme 7.7).

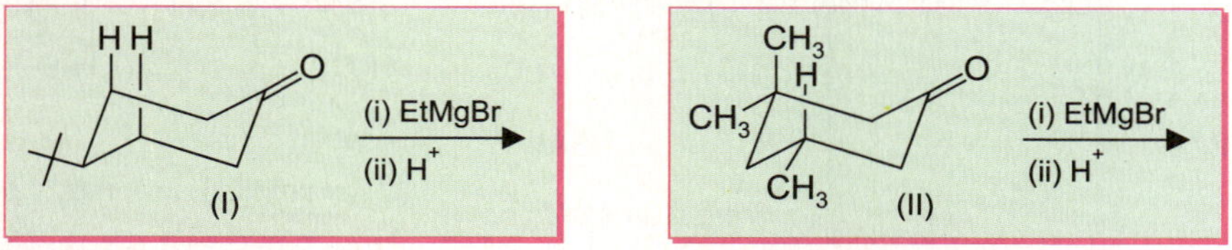

SCHEME 7.7

ANSWER. *In both reactions the nucleophile is comparatively bulky and thus will be delivered from the equatorial side. With the presence of an axial substituent at C3 position in the cyclohexanone substrate (II, Scheme 7.7) the nucleophile is exclusively delivered from the equatorial side and the stereoselectivity becomes more pronounced. The products are in scheme 7.8. Recall that axial approach by the nucleophile dominates with small nucleophiles like $LiAlH_4$* (compare Scheme 7.11).

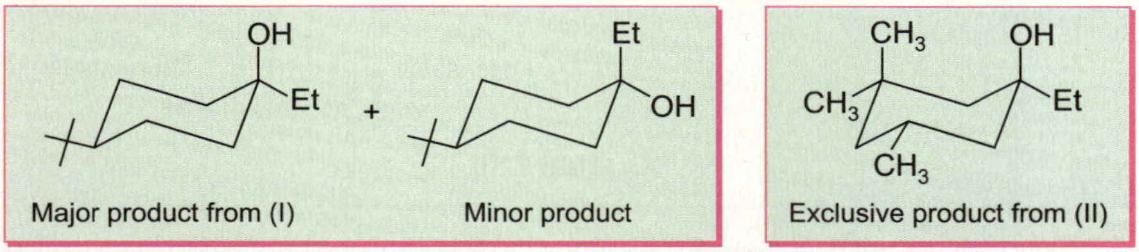

SCHEME 7.8

(B) *Stereochemistry (stereoselectivity) of addition to norbornane systems*

Consider the reduction of camphor. The two faces of the carbonyl group of camphor (*i.e., exo* and *endo* are diastereotopic) have different accessibility to nucleophilic reagents. The approach to the bottom (*endo*) face is hindered by the U-shaped cavity of the molecule, but the top (*exo*) face is strongly hindered by the overhanging C-7 methyl group. Thus reduction with $LiAlH_4$ occurs largely by approach of the nucleophile *e.g.,* AlH_4^- from, the *endo* side (Scheme 7.9). A more bulky reducing agent *e.g.,* L-selectride is more selective and gives almost exclusive delivery of hydride to the *endo* face. In the case of norcamphor, however, there is no hindrance from the *exo* side, but the hindrance to approach from the *endo* side remains. Thus, hydride reduction occurs by attack of the AlH_4^- now from the *exo* side (Scheme 7.10).

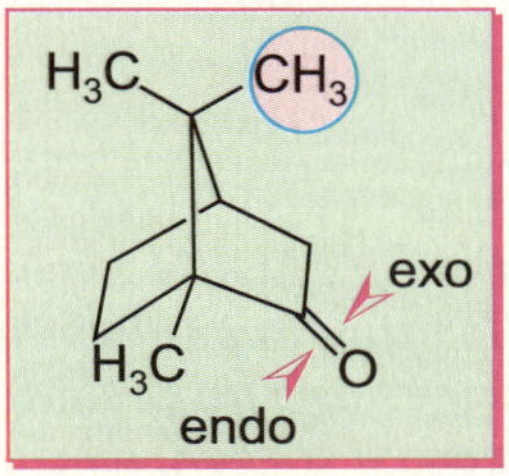

In camphor C=O (close to a stereocenter) has diastereotopic faces. Attack by the AlH_4^- from one face, endo in this case is more easy than exo. The faces refer to inside (endo) and outside (exo) of the boat shaped six-membered ring. Thus when the carbonyl group is embedded in a rigid system steric effects predominate, resulting in preferential transfer of the "hydride" from the less hindered side.

Continued...

CHAPTER 7

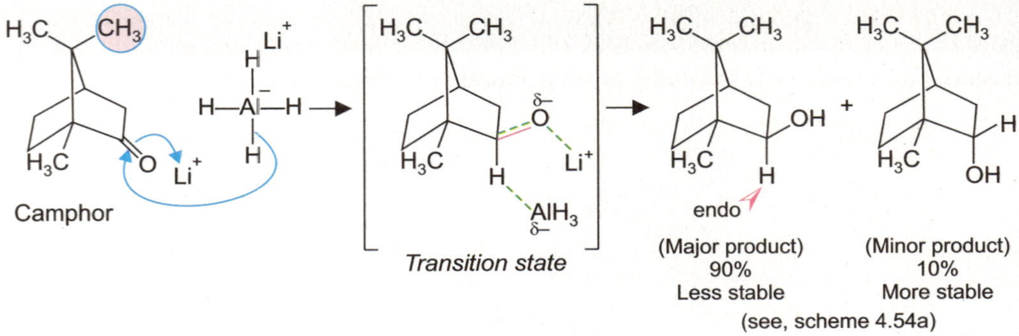

SCHEME 7.9

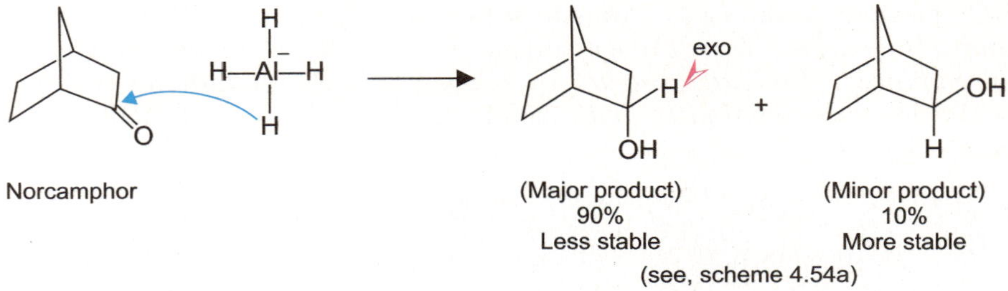

SCHEME 7.10

EXERCISE 7.2

The hydride reduction of cyclohexanone system (Scheme 7.2) gives 90% of equatorial alcohol. How one explains the reduction outcome when the cyclohexanone system has a bulky substituent [e.g., CH_3 at C(3)] in 3,3,5-trimethylcyclohexanone (Scheme 7.11).

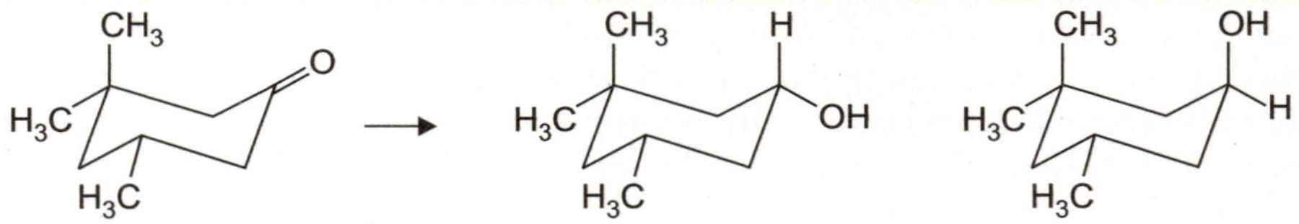

Reagent	***cis(%)***	***trans(%)***
$NaBH_4$, EtOH	33	65
$LiAlH_4$, THF	26	74
L-Selectride	–	100

SCHEME 7.11

ANSWER. *Using sterically unhindered reducing agents e.g., $LiAlH_4$ or $NaBH_4$ the effect of torsional strain prevails over the steric effect, leading to predominant axial attack to yield the equatorial alcohol. However, L-Selectride has a large steric demand resulting in the equatorial delivery of hydride to the carbonyl carbon, when there are bulky axial substituents at C3 [or and at C5 (as in 3,3,5-trimethylcyclo hexanone).*

One recent method for the preparation of equatorial alcohols from cyclohexanones in high excess is the reduction with *t*-butylmagnesium chloride using methylalumino derivative of *bis*-(2, 6-di-*t*-butyl-4-methylphenoxide) (Scheme 7.12).

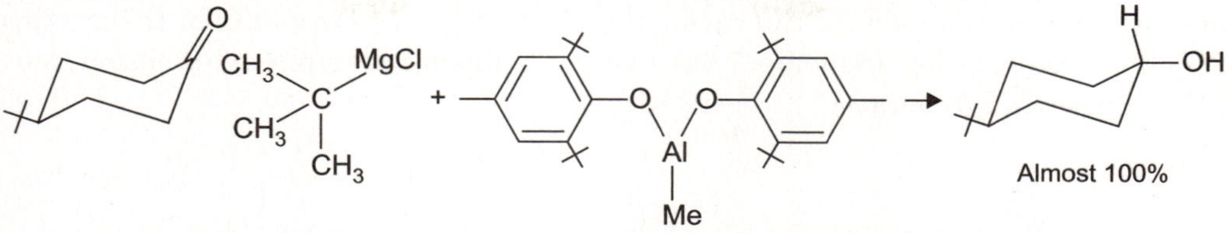

Formation of thermodynamically more stable equatorial alcohols

SCHEME 7.12

7.3 ADDITION OF ALLYLBORON COMPOUNDS TO A CARBONYL GROUP-SIMILARITY TO AN ALDOL CONDENSATION DIASTEREOSELECTIVE SYNTHESIS

Recently the role of allylboration has been considerably expanded, and these reagents add to carbonyl group in a way similar to aldol condensation. The reaction displays the same high diastereoselectivity as seen earlier (see, scheme 2.33), involves an allylic rearrangement. The reaction both with *E*- and *Z*-allyl derivatives is in (Scheme 7.13). Two modes of attacks are possible, between like (*Re-Re* or *Si-Si*) or the unlike faces (*Re-Si* or *Si-Re*) to involve transition states like (I or II, Scheme 7.13). For the *E*-allyls the two enantiomeric like approaches through transition state (as I) is favourable since R of the aldehyde occupies an equatorial position whereas in the unlike approaches (not shown), it would occupy an axial position and will give rise to a synaxial interaction with one of the ligands at boron, It thus fixes the relative stereochemistry at C-3 and C-4 as *anti* (see, Scheme 2.33). For the *Z*-allyls, a similar argument favours an approach to unlike faces and syn isomer is formed through the transition state (II).

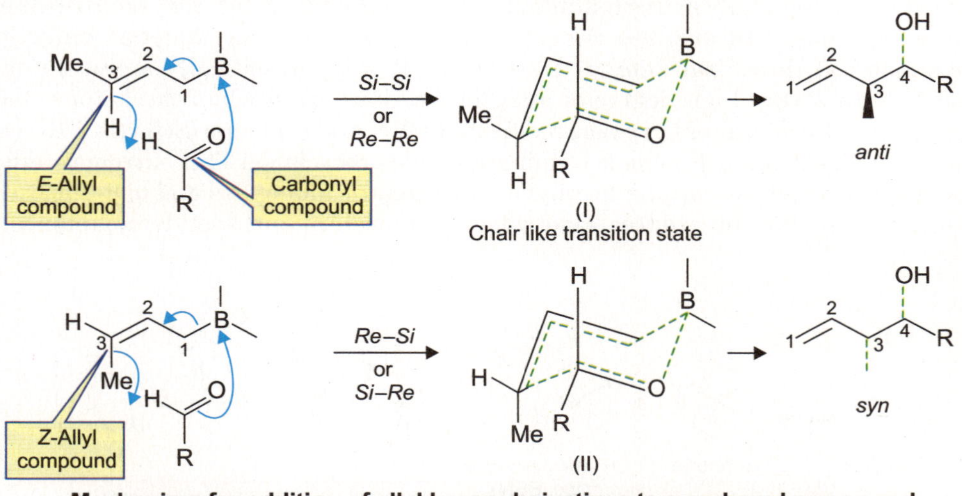

Mechanism for addition of allyl boron derivatives to a carbonyl compound

SCHEME 7.13

7.4 ENANTIOSELECTIVE SYNTHESIS

The use of allylboranes or allylboronic esters with chiral ligands is highly efficient to generate high asymmetry at the carbinol carbon. A good example is in the synthesis of (-)-artemisia alcohol (Scheme 7.14) using (–)-diisopinocamphenyl-(abbreviated as IPC)-(dimethylallyl) borane.

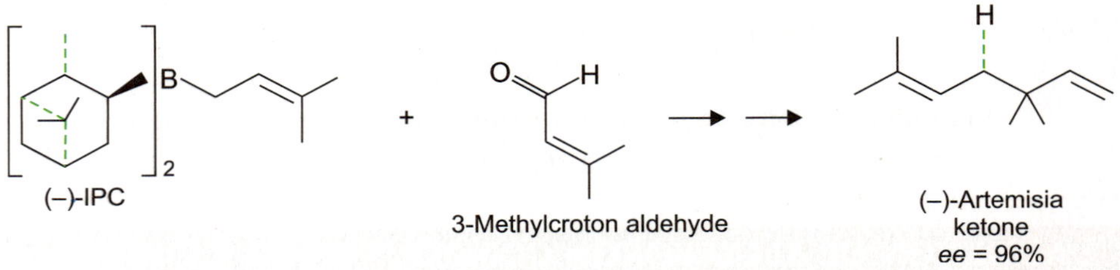

SCHEME 7.14

Chiral optically pure tartaric ester of allylboronic acid is useful due to its two faces being homotopic, asymmetry is induced in the same direction. Homoallyl alcohols (Scheme 7.15) are formed with 71-87% *ee*'s.

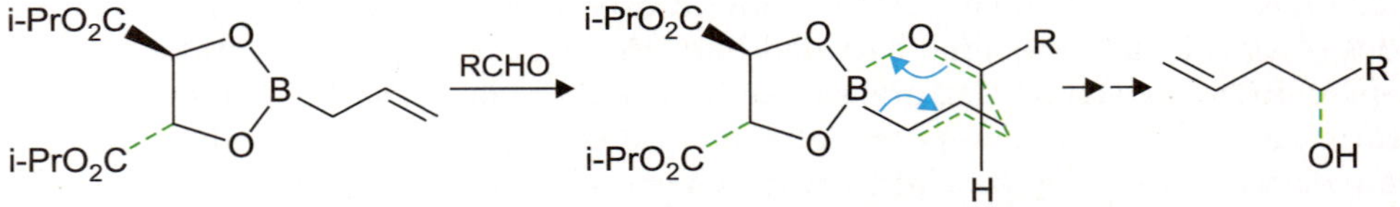

Synthesis of an enantiomerically pure homoallyl alcohol

SCHEME 7.15

A chiral, optically active (enantioselective) reducing agent, *e.g.*, Ipc.BBN (alpine-borane, (Scheme 7.16 also see Scheme 2.44) brings about asymmetric reduction of unsymmetrical (prochiral ketones R—CO—R′ R ≠ R′) to optically active secondary alcohols with a very high degree of enantioselectivity even when the ketone itself is achiral. The chiral reagent is prepared from 9-BBN and α-pinene (Scheme 7.16 see also Schemes 2.42—2.44). α-Pinene is eliminated and is recyclable. This extremely efficient reducing agent reduces, aliphalic, allylic and aromatic aldehydes and α, β-unsaturated ketones. Alpine borane reduces aldehydes, α-ketoesters and acetylenic ketones with

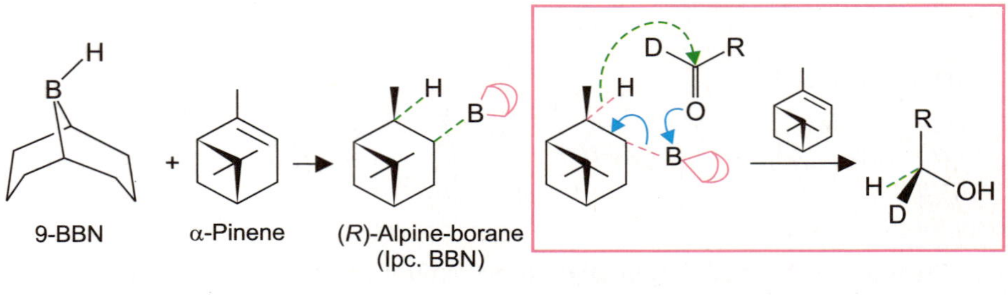

SCHEME 7.16

high enantioselectivity. The reduction proceeds through a cyclic transition state similar to MPV reduction.

The CBS (Corey–Bakshi–Shibata) reagent is a chiral catalyst derived from proline. Also known as Corey's oxazaborolidine, it is used in enantioselective borane reduction of ketones (Scheme 7.17)

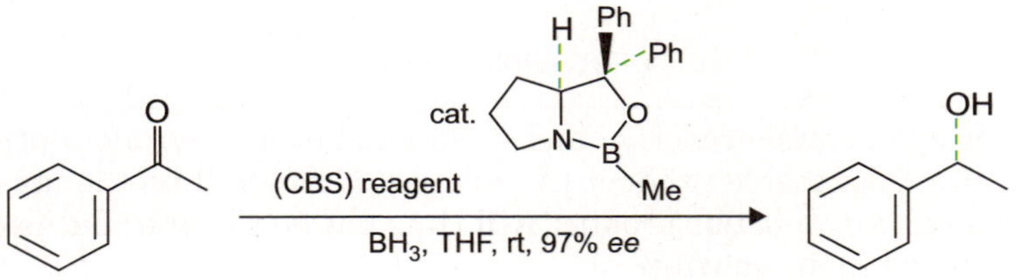

SCHEME 7.17

7.5 MEERWEIN – PONNDORF – VERLEY (MPV) REDUCTION – THE HYDRIDE TRANSFER REACTION

In the Meerwein–Ponndorf–Verley reduction, equilibrium is established between the carbonyl group to be reduced and isopropanol on the one hand and also between the product alcohol and acetone in the presence of aluminium isopropoxide. Since acetone is the lowest boiling component of the mixture, it is regularly distilled off so that the equilibrium is displaced to the right (Scheme 7.18).

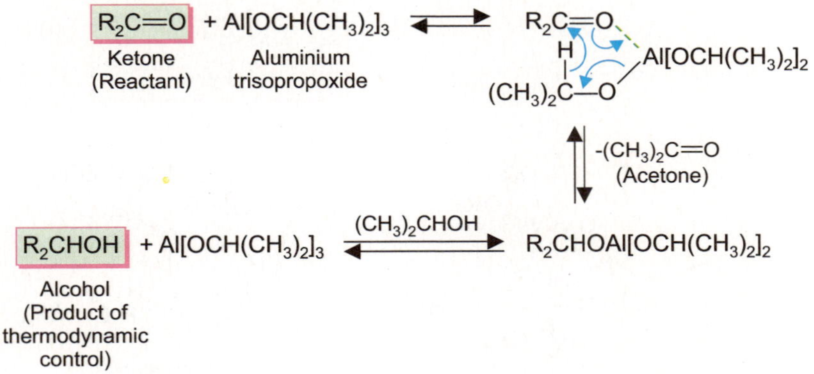

SCHEME 7.18

Consider the Meerwein–Ponndorf -Verley (MPV) reduction of 4-*tert*-butylcyclohexanone (Scheme 7.18*a*) and compare these results with $NaBH_4$, $LiAlH_4$ and bulky hydride reduction of the same cyclohexanone (Schemes 7.2, 7.5 and 7.6). Based on the accepted mechanism (Scheme 7.18*a*), aluminium alkoxide coordinates with the carbonyl oxygen of ketone and a hydride ion is transferred via a six membered transition state. The following points may be noted:

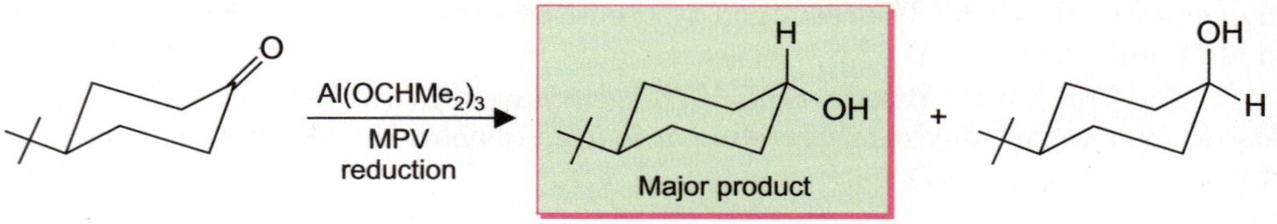

SCHEME 7.18a

- This is a reversible reaction and is under thermodynamic control and therefore, the most stable product (A, Scheme 7.18*b*) will be the major product at equilibrium (the major product will thus not be the one that is easily formed *i.e.*, under kinetic control).

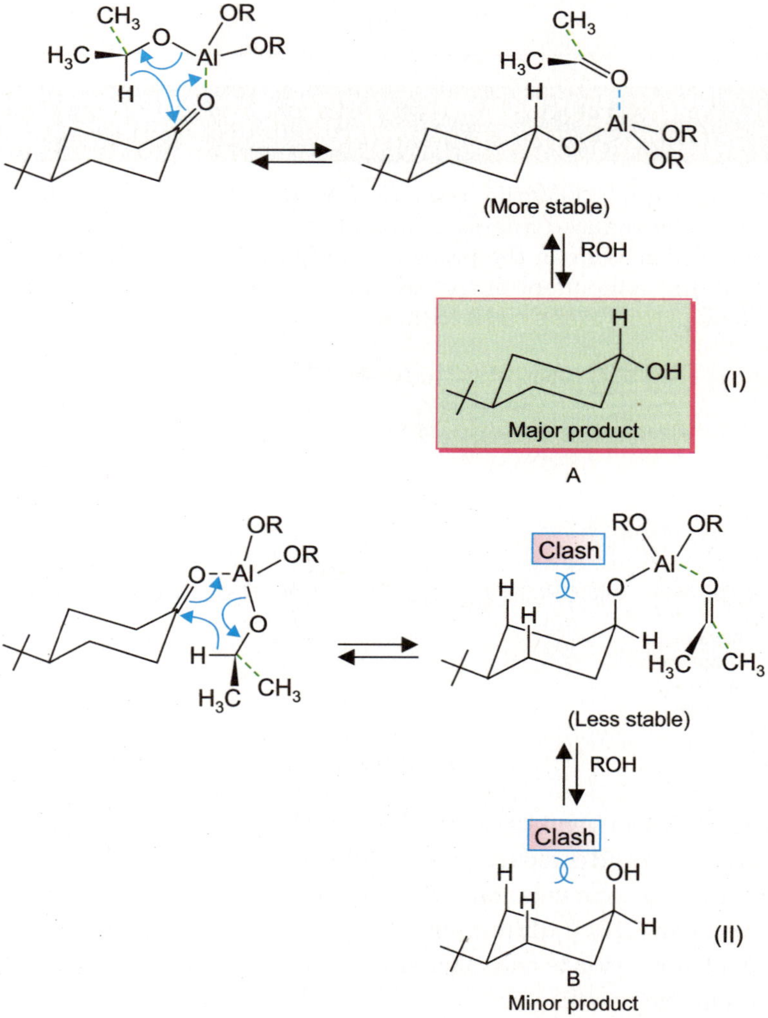

SCHEME 7.18b

- The attack is at one or the other face of the ketone at an angle of 109°. Attack from the top axial face yields the more stable alkoxide (eq. I, Scheme 7.18*b*) and the most stable of the alcohols.
- The attack from the equatorial side (eq. II) yields the less stable alkoxide and alcohol (B), since both the alkoxide and the corresponding alcohol suffer from unfavourable 1, 3-nonbonded interactions.
- One may think that the initially formed aluminium complex may be formed easily from the equatorial side, however this is irrelevant when the reaction is under equilibrium as is so in the present case (also see problem 7.1).

7.6 ADDITION OF NUCLEOPHILES TO CARBON – HETERO MULTIPLE BONDS

(a) Stereochemistry/Stereoselectivity of Addition to Aldehydes and Ketones

Groups or faces in molecules can be enantiotopic or diastereotopic and collectively these faces are termed as heterotopic. In carbonyl compounds when R^1 and R^2 are different the two faces of the trigonal centre then become different (enantiotopic). These faces become diastereotopic.

When R^1 and R^2 contain chirality centers or in other words when carbonyl group is close to a chirality center.

Crams rule is applied to predict the stereochemistry of the major product of addition to chiral aldehydes and ketones (see, Scheme 2.24).

When the two faces of the carbonyl group are differently hindered, the nucleophile (H:⁻) approaches from the less hindered face to give the major product (Scheme 7.19).

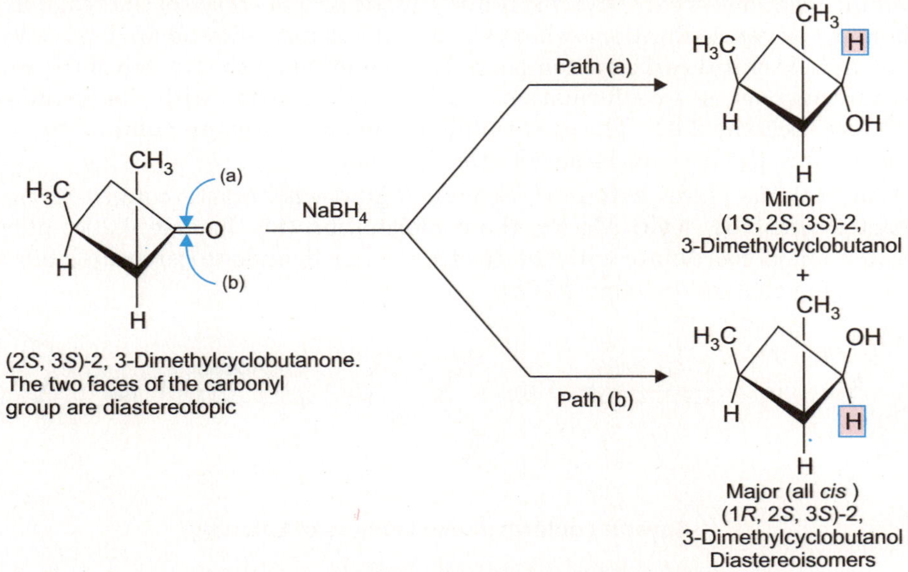

Path (a) *Re* face attack is hindered by the methyl groups whereas path (b) allows easy approach of H⁻.

SCHEME 7.19

Additionally during the reduction of a (2*R*)-2-substituted cyclopentanone with sodium borohydride (Scheme 7.20), the H^- ion cannot approach the carbonyl carbon atom at the *Re* face because its 109° angle of approach is prevented by the presence of a bulky substituent on that face. Thus, it approaches the *Si* face to give (*R*) alcohol.

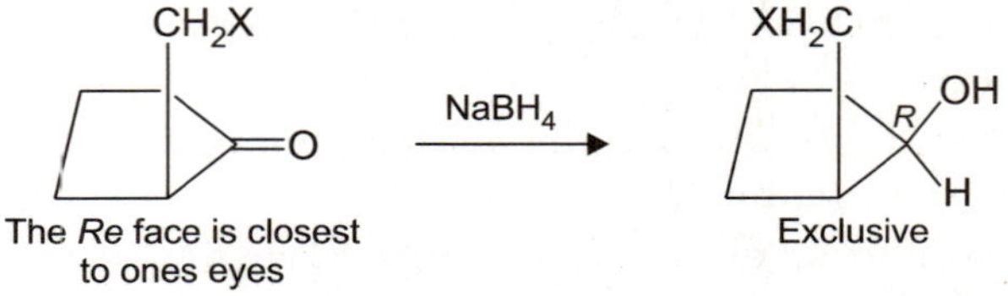

Nucleophilic attack on a 2-substituted cyclopentanone

SCHEME 7.20

Enantioselective Synthesis

Much effort has been spent to develop ways to perform the enantioselective reduction of ketones. Several drugs are chiral and have the alcohol functionality. In Chapter 2, methods are presented for making enantiomerically pure alcohols e.g., via hydroboration of an alkene (see Scheme 2.41). The enzyme catalyzed addition to a ketone gives only one enantiomer. Enzymes are chiral and these e.g., can position, the nucleophile in a way so that the C=O group can be attacked from only one side of the molecule (See, Scheme 2.22).

(b) Locking Effects in Nucleophilic Reactions at Carbonyl Groups

This subject has been discussed already, one can control the stereoselectivity of nucleophilic reactions of carbonyl compounds by the proper choice of the reagent. Recall that there are several situations where Crams rule is not followed during nucleophilic attack on acyclic chiral carbonyl compounds. Depending on the nature of the reactant, the reagent may act as a conformational lock by coordination with the ligands on the reactant (see, Scheme 2.28). The nucleophile is then delivered predominantly from the face which offers the least resistance to the approach.

Consider the chiral ketone (I, Scheme 7.20*a*) with diastereotopic faces. When it is reacted with CH_2 = CH Mg Br, the conformation (Ia, Scheme 7.20*a*) allows the magnesium ion to coordinate with the electron donor ligands to generate a planar five membered ring chelate (Scheme 7.20*b*).

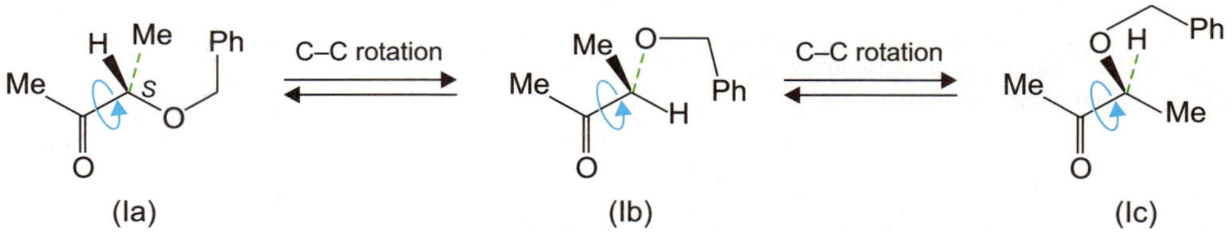

Several conformational isomers of ketone(I)

SCHEME 7.20a

This Grignard reagent (or a second CH_2 = CHMgBr coordinated to the first) then delivers the nucleophile (ethenyl group) faster to the less hindered face of the carbonyl group, with the smaller H remote from the larger methyl group. Thus the addition to the carbonyl group (Scheme 7.20*b*) is under kinetic control via Bürgi–Duntiz angle as expected (see, Scheme 2.26).

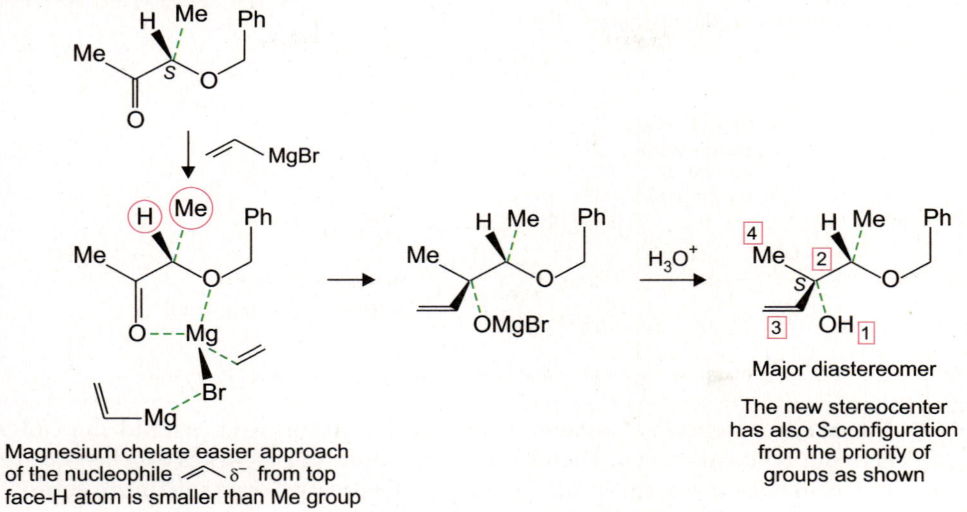

SCHEME 7.20b

More examples of the outcome of these reactions are in Scheme 7.21.

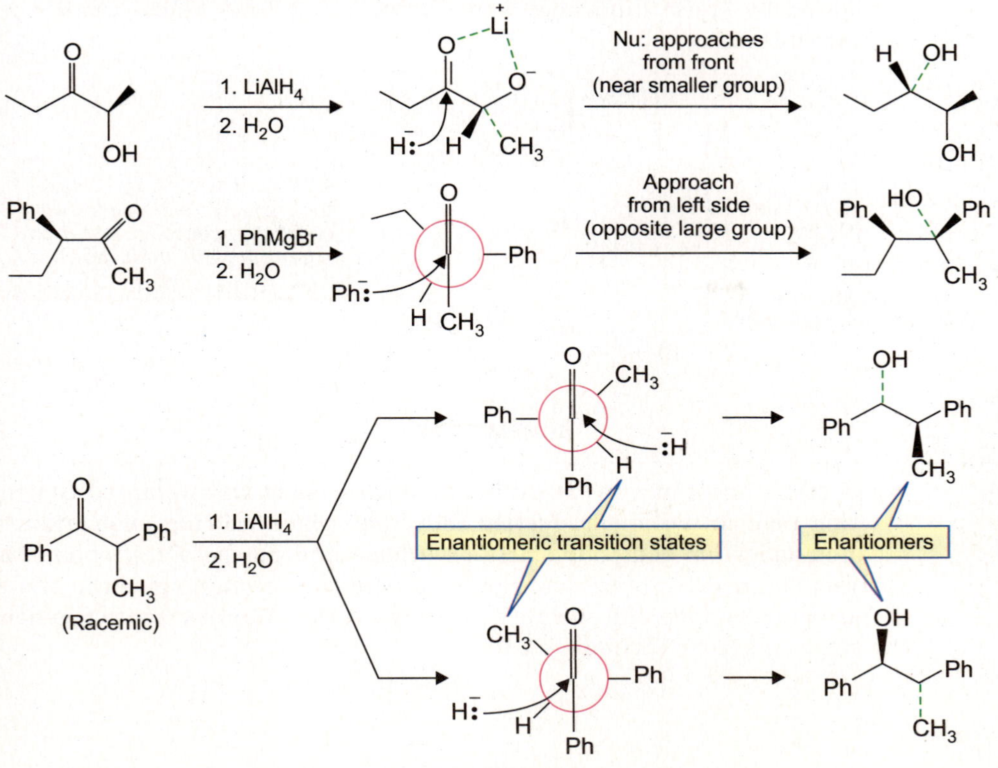

SCHEME 7.21

(i) Side Reactions During Sterically Congested Cases—Reaction with Ketones

- The ketones which contain very bulky groups *e.g.*, di-*t*-butyl ketone do not give tertiary alcohols with Grignard reagents but react successfully with alkyl-lithium reagents (Scheme 7.22).

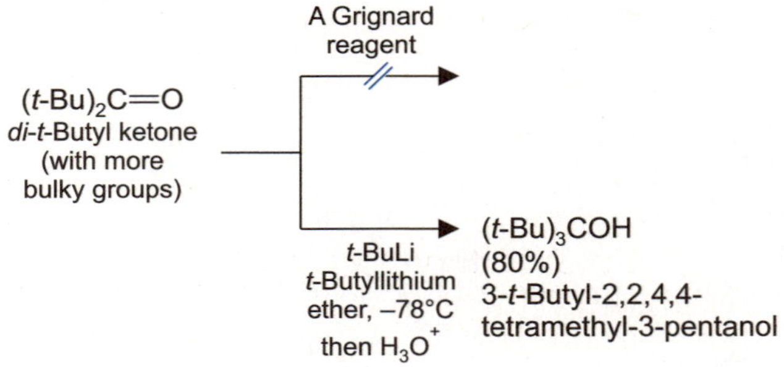

SCHEME 7.22

- When the ketone has somewhat less bulky groups and instead the Grignard reagent contains a branched alkyl group then the yields are either low or the reaction does not occur at all. However, the derived organolithium compounds from a branched alkyl group react. Thus diisopropyl ketone reacts with methylmagnesium bromide but not with isopropyl Grignard reagents to give tertiary alcohols (Scheme 7.23). However, organolithium compounds are less prevented by steric hindrance and diisopropyl ketone reacts readily with isopropyllithium.

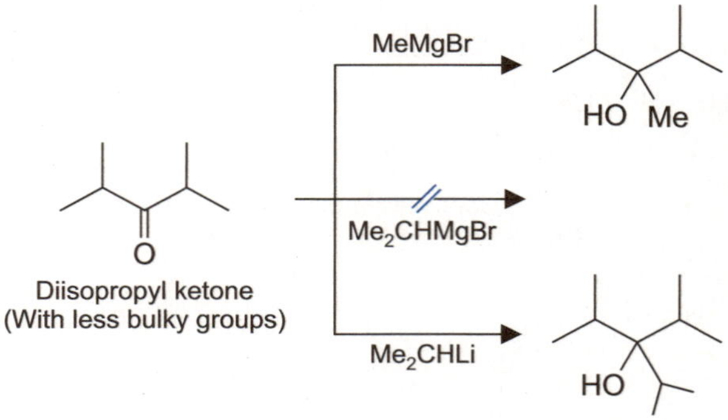

A Grignard reagent e.g., CH_3MgI does not react with ketones with very bulky groups, however, an organolithium undergoes a

SCHEME 7.23

- When steric hindrance is the problem, in the case of an attempted Grignard reaction enolization and reduction may take place. During enolization the organomagnesium compound acts as a base and not as a nucleophile and abstracts an α-proton (activated proton) from the ketone to give the enolate. When water is added, the enolate is hydrolyzed and tautomerizes to regenerate the starting ketone (Scheme 7.24).

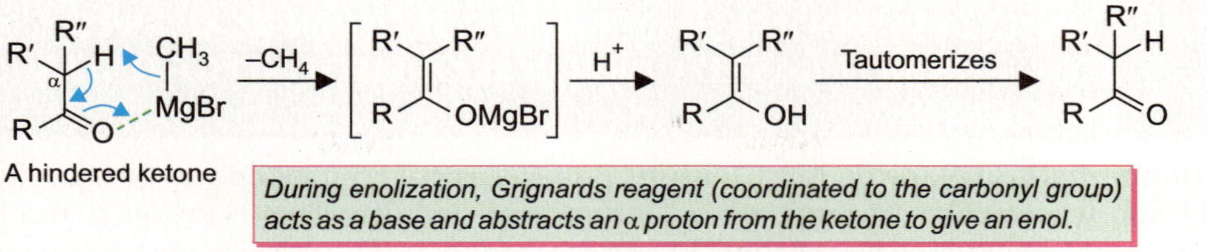

SCHEME 7.24

- Another important side reaction in hindered cases is reduction when the Grignard reagent has a β-hydrogen, the carbonyl group is reduced by hydride ion transfer. The process occurs in a six membered cyclic transition state (with Grignard reagent coordinated to the carbonyl group) and an alkene is formed (Scheme 7.25).

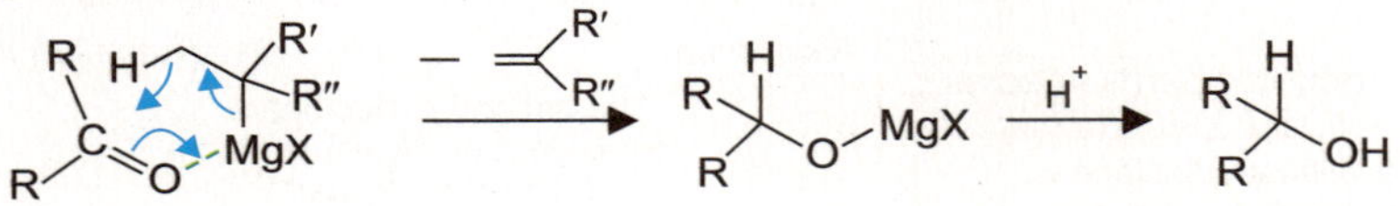

SCHEME 7.25

Only conjugate addition occurs when Gilman reagents (lithium dialkylcuprates (Scheme 7.27) react with α,β-unsaturated aldehydes and ketones. Grignard reagent should thus be used when one wants to add an alkyl group to the carbonyl carbon, whereas Gilman reagents should be used when one wants to add an alkyl group to the β-carbon.

(c) Reaction of Allylic Grignard Reagents with Carbonyl Compounds

Reaction of an allylic Grignard reagent with a carbonyl compound is attended with an allylic shift (shift of the double bond) and occurs through a six membered transition state (Scheme 7.26). One may note that allyl Grignard reagents are in rapid equilibrium, the major product of their reaction with an electrophile *e.g.*, carbon atom of C=O has less substituted double bond due to double bond shift.

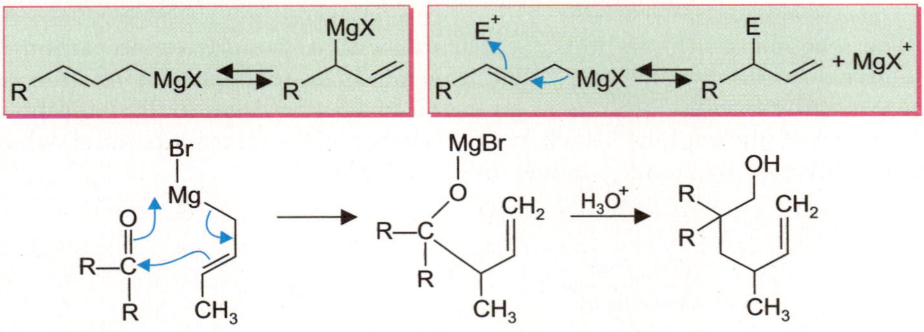

SCHEME 7.26

7.7 CONJUGATE ADDITION OF ORGANOCOPPER REAGENTS TO α, β-UNSATURATED CARBONYL COMPOUNDS-GILMAN REAGENTS

Lithium dialkylcuparates react with α, β-unsaturated compounds to give conjugate addition where these regents act as carbon nucleophiles. This reaction is the most prominent feature of these reagents in organic synthesis. These are made when a copper (I) halide reacts with two equivalents of an alkyllithium in diethyl ether or THF, the first equivalent forms an alkyl copper compound.

The driving force for this reaction is due to the preference of lithium, the more electropositive metal to exist as an ionic compound ($Li^+ I^-$). The copper of an alkyl copper reagent is a Lewis acid and therefore, the "alkyl anion" from the second equivalent of the organolithium reagent reacts with it to give lithium dialkyl cuparate (Scheme 7.27).

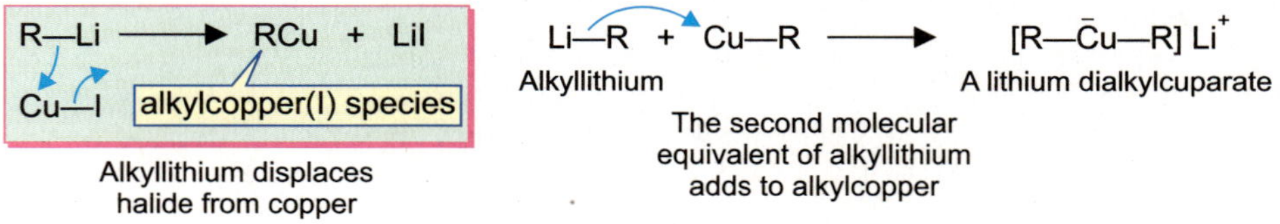

SCHEME 7.27

Lithium dialkylcuparates react some what like Grignard or lithium reagents; however since the "alkyl anion" is complexed by copper, a less electropositive element than lithium, the alkyl-copper bond has more covalent character (I, Scheme 7.27a) and these reagents are less reactive. Their reaction with acid chlorides and aldehydes is typical, and these react very slowly with ketones, and not at all with esters. The reaction of lithium dialkylcuparates with acid chlorides gives ketones in excellent yield.

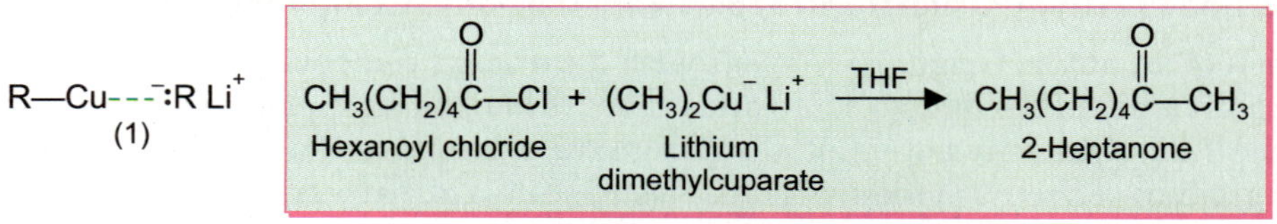

SCHEME 7.27a

The reaction of lithium dialkyl-cuparates with α, β-unsaturated carbonyl compounds gives excellent yields of 1, 4-addition products (Scheme 7.27*b*). One can introduce two different alkyl groups in an enone to afford vicinal dialkylated products. Initial product of nucleophilic attack at the β-carbon gives an enolate anion which can be further alkylated by an alkyl halide (Scheme 7.28).

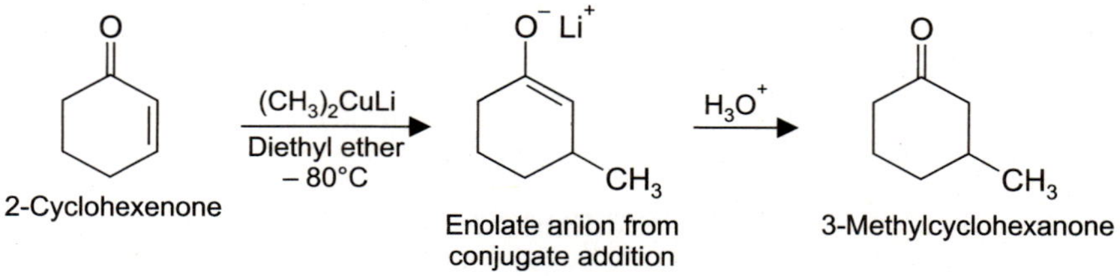

SCHEME 7.27b

Alkylation of an enolate proceeds via an early, hence a reactant like, transition state. Thus, steric factors in the ground state play a significant role.

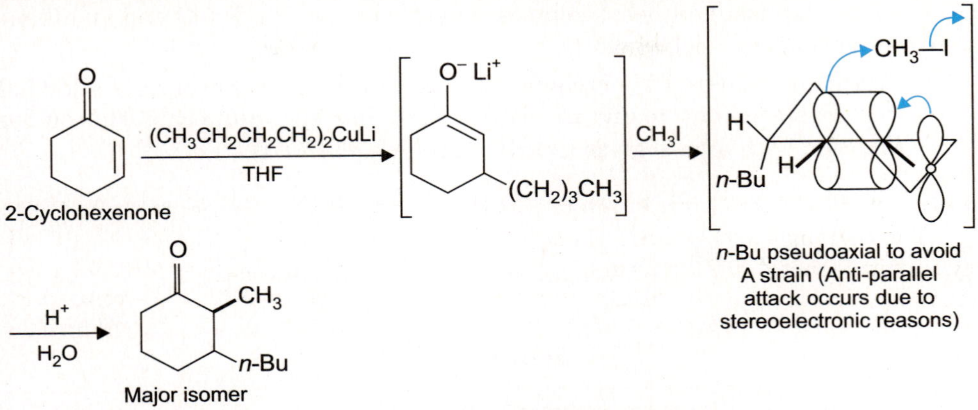

SCHEME 7.28

7.8 THE WITTIG AND RELATED REACTIONS ADDITION OF CARBANIONS TO CARBONYL CARBON OF AN ALDEHYDE OR A KETONE FOLLOWED BY β-ELIMINATION

(A) Introduction/Mechanism

The Wittig reaction is an regiospecific conversion of a carbonyl group to a alkene, when the >C═O bond of an aldehyde or a ketone is converted to a >C═C< bond (Scheme 7.29). Unlike Wittig reaction, reaction of *e.g.*, a ketone with Grignard reaction and dehydration of the tertiary alcohol thus formed instead gives an isomeric mixture of olefins.

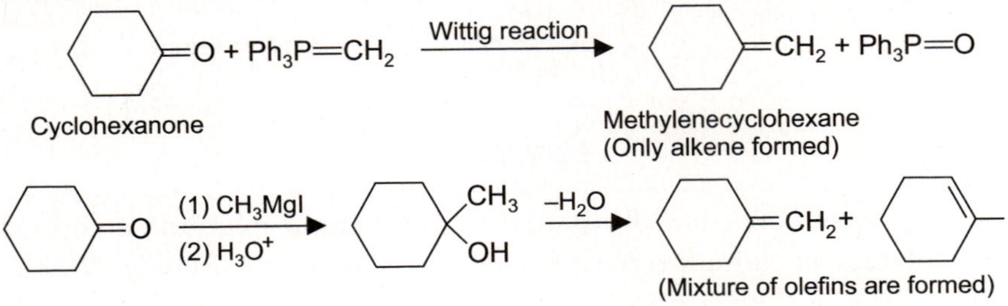

SCHEME 7.29

- The Wittig reaction is carried out by the reaction of a carbonyl compound with ylides. Ylides are special class of anions (Scheme 7.30).

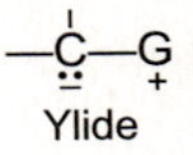

Ylides are carbanions, where G is appropriately substituted N, S or P. Wittig reaction involves phosphorus ylides—the negative charge on carbon is balanced by the positive charge on adjacent phosphorus.

SCHEME 7.30

CHAPTER 7

Reaction of phosphorus ylides with aldehydes and ketones first react in a [2 + 2] cycloaddition to give a heterocycle referred to as oxaphosphetane (see Schemes 7.32 and 7.34). The oxaphosphetanes decompose rapidly to give Ph_3 P = O and an alkene by a *syn* β elimination (see, Scheme 5.1).

- Ylides are prepared from triphenyl phosphine and an appropriate alkyl halide (S_N2 displacement) to give an alkyltriphenylphosphonium salt. This on deprotonation with a base gives a carbanion (Scheme 7.31).

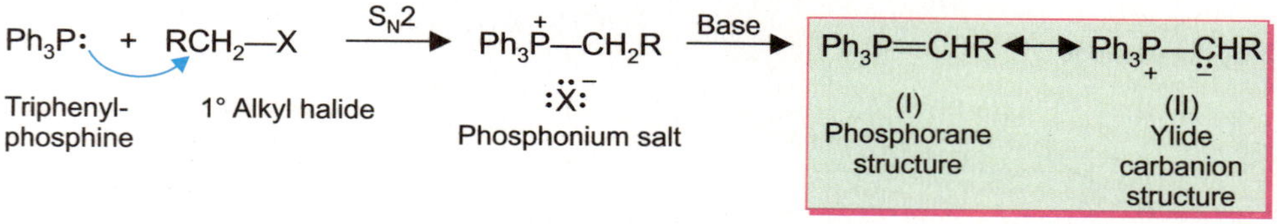

SCHEME 7.31

- The negative charge on carbon is balanced by the positive charge on the adjacent phosphorus and such zwitterions are often called ylides (II, Scheme 7.31). The other resonance structure is called (Phosphorane structure I). The ylides react as strong nucleophiles.
- Traditional mechanism of Wittig reaction is the nucleophilic addition of the ylide to an aldehyde or a ketone to give a diion (A, Scheme 7.32 called a *betaine*) as an intermediate. Its cyclization to the oxaphosphetane (B) and its elimination to triphenyl-phosphine oxide and an alkene completes the mechanism.

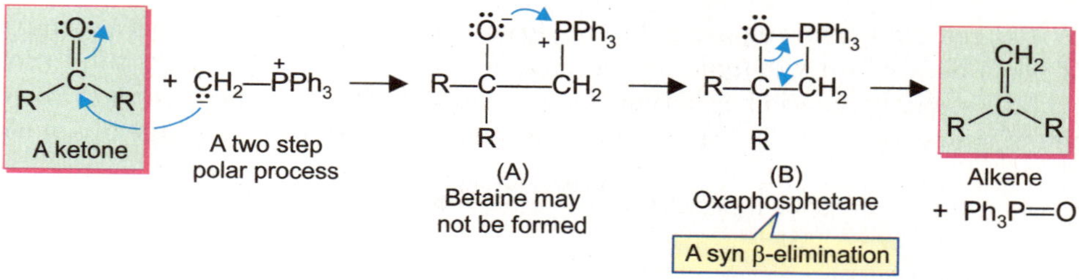

SCHEME 7.32

- However, evidence has accumulated that oxaphosphetane is formed directly by a [2+2] cycloaddition reaction [$\pi^{2s} + \pi^{2a}$] as shown (Scheme 7.33).
- The conversion of a carbonyl compound to an alkene using Wittig reaction is regiospecific unlike the use of Grignard reaction, followed by dehydration of the resultant tertiary alcohol which will yield mixture of isomeric alkenes.

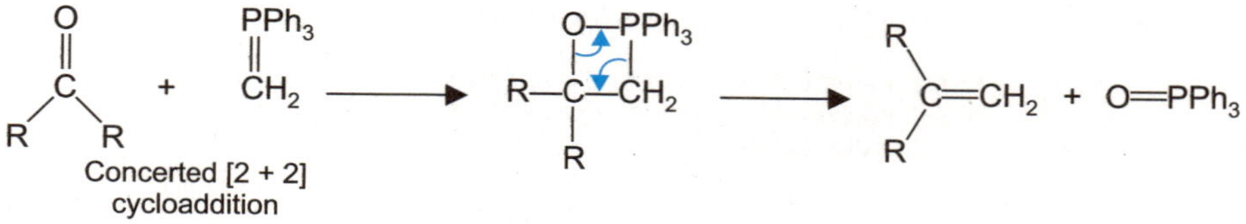

SCHEME 7.33

Apart from Wittig reaction where the carbanion is $Ph_3P^+—\bar{C}<$ there are two other olefination reactions in which a carbanion adds to a carbonyl group followed by elimination of the adduct thus formed. (Peterson, $Me_2Si^+—\bar{C}<$, Julia $PhSO_2—\bar{C}<$). Interstingly the reaction of the sulfur ylide with the carbonyl group of aldehydes, ketones, or enones gives a betaine intermediate, which decomposes by intramolecular displacement of Me_2S by the oxyanion to yield the corresponding epoxide (Scheme 7.34).

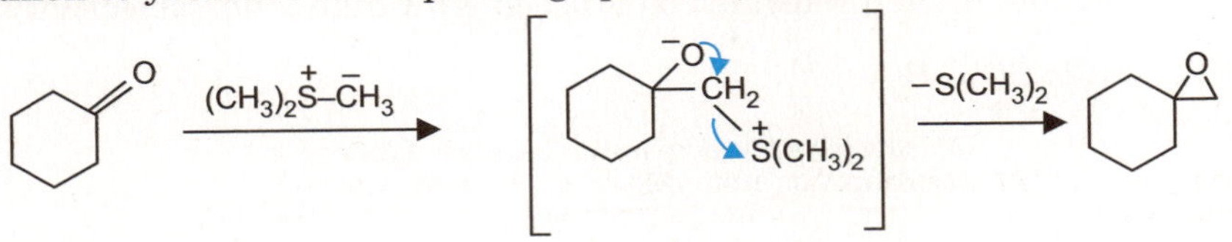

SCHEME 7.34

(B) Stereochemistry/Stereoselectivity—Wittig Reaction though Regiospecific is not Stereoselective—Reaction Conditions and Stability of the Ylides.

(i) Non-Stabilized Ylides/Stereoselectivity

When the groups attached to the original alkyl halide are hydrogens or simple alkyl groups (See Scheme 7.31. The α-hydrogen of the phosphonium salt is only weakly acidic and a very strong base *e.g.*, butyllithium is needed to give the ylide. This ylide is very reactive, unstable and highly nucleophilic. It gives the oxaphosphetane (irreversibly) which affords the alkene spontaneously. A mixture of both *E* and *Z* isomers is then obtained (Scheme 7.35), where generally the *Z* isomer predominates. The reaction is then called to proceed with good selectivity.

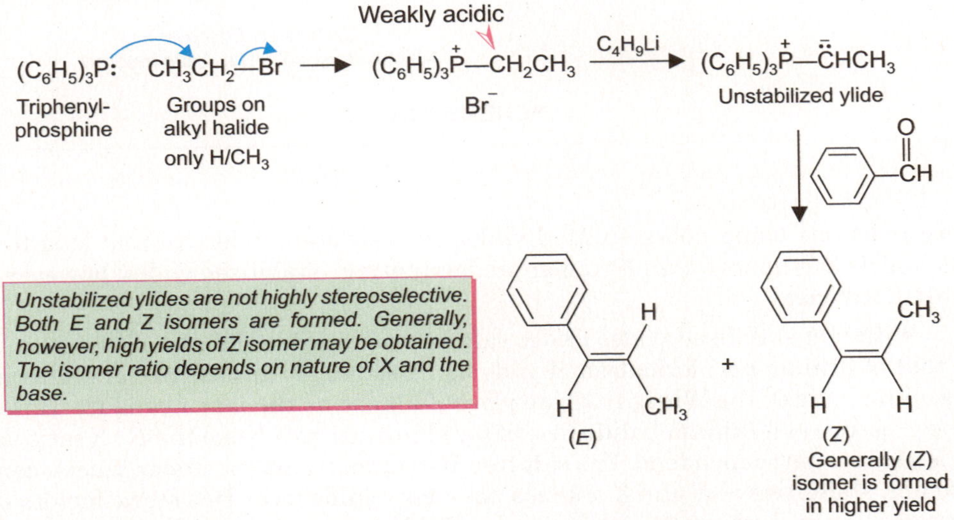

SCHEME 7.35

CHAPTER 7

(ii) Stabilized Ylides

When in the original alkyl halide a group of *-M* type *e.g.,* an ester is adjacent to an anionic carbon, the ylide can be formed using less strong bases. This type of a ylide is stable and can be isolated. The stability allows it to react *reversibly* with the carbonyl compound and at the end of the reaction the (*E*) isomer usually predominates (Scheme 7.36). This outcome is then reformed to proceed with high *trans*-selectivity.

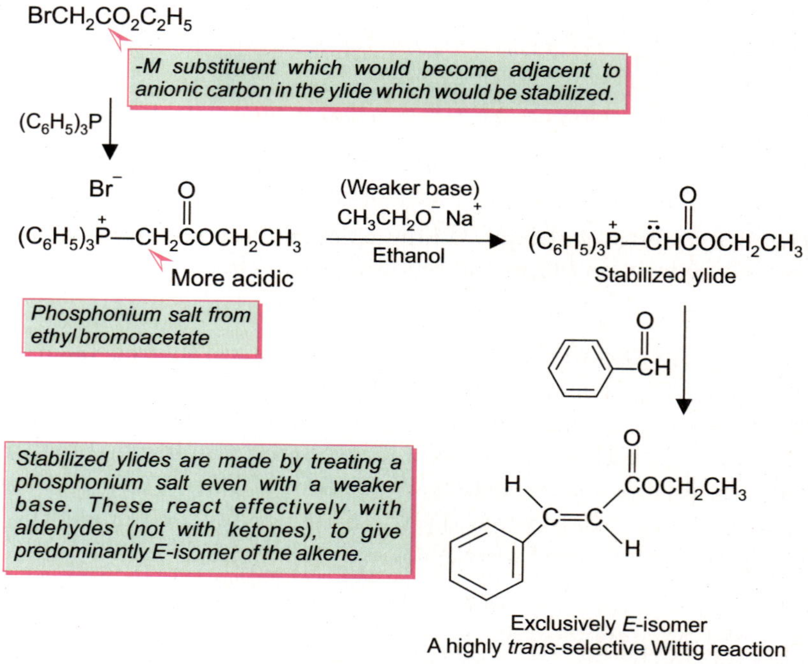

SCHEME 7.36

(iii) Wittig Reaction "Salt Free" or in the Presence of Salts and Steric Consideration—Finer Aspects of Mechanism

Wittig reactions using non-stabilized ylides are less stereoselective and lead to both *E*- as well as *Z*-isomers (often *Z*-isomer predominating), stabilized ylides, however, give mainly *E*-alkenes.

With non-stabilized ylides under salt free conditions (absence of inorganic salts like that of lithium *e.g.,* formation of ylide using sodamide as the base and filtering of the sodium halide), the Wittig reaction gives *Z*-alkene as the major product. However, in the presence of a lithium halide (use of butyl lithium as a base) the *E* : *Z* ratio varies and is also solvent dependent. The salt free Wittig reactions are under kinetic control, therefore, stabilities of *E* and *Z* alkenes have no significance. Two steric factors effect the proportion of *Z*-isomer, which increases if aldehyde contains an α-substituent (*e.g.,* $PhCH_2CR_2$ CHO) and decreases *e.g.,* when one of the *P*-phenyl groups is replaced by an isopropyl group (Scheme 7.37).

$$Ph_3\overset{+}{P}—\overset{-}{C}HCH_3 + PhCH_2CR_2CHO \longrightarrow PhCH_2CR_2CH{=}CHCH_3$$

If R = H, E: Z = 6 : 94; If R = CH_3, Z = almost 100%

SCHEME 7.37

The "salt free" Wittig reactions of nonstabilized ylides follow the cycloaddition mechanism (Scheme 7.34). For *Z*-alkene formation the spatial orientations are as in puckered four-membered transition state (Scheme 7.38) while the *E*-alkene formation involves an almost planar transition state not shown here.

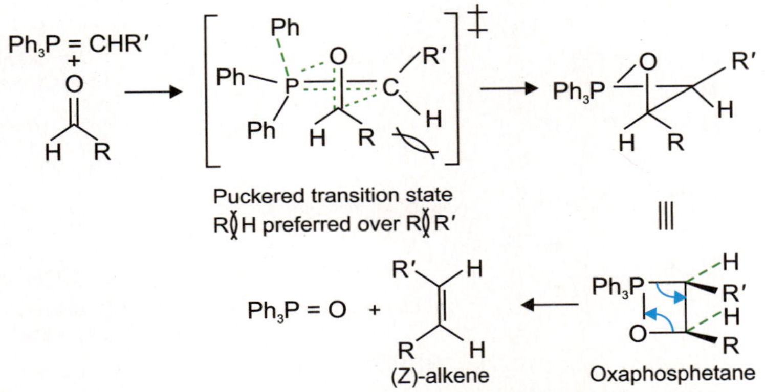

Mechanism of Wittig reaction ("salt free") with non-stabilized ylides.

SCHEME 7.38

A four-centered, puckered transition state (Scheme 7.38) is involved which keeps the substituent R' of the ylide, the Ph moieties of the phosphorus, and the R group of the aldehyde positioned in a way so as to minimize steric interactions and gives a (*cis*)-oxaphosphetane.

Thus the major considerations for stereoselectivity are based on steric effects which develop as the two reagents approach each other. In the presence of a polar solvent and lithium salts, the geometry of the oxaphosphetane which is formed depends on the steric approach of the ylide during its reaction with an aldehyde. The reaction yields more of *Z*-alkene. The first step in the addition of ylide (phosphorane) to the aldehyde is irreversible and the intermediate which is formed faster determines the major alkene, *i.e.*, *Z*-alkene when a reactive ylide interacts with an aldehyde (Scheme 7.39).

With stabilized phosphoranes, however, the formation of the intermediate is reversible (Scheme 7.39) and this allows the formation of the more stable stereoisomeric form of oxaphosphetane. Thus an interconversion to the more stable and thus more abundant isomeric form and a *syn*-elimination affords the *E*-alkene.

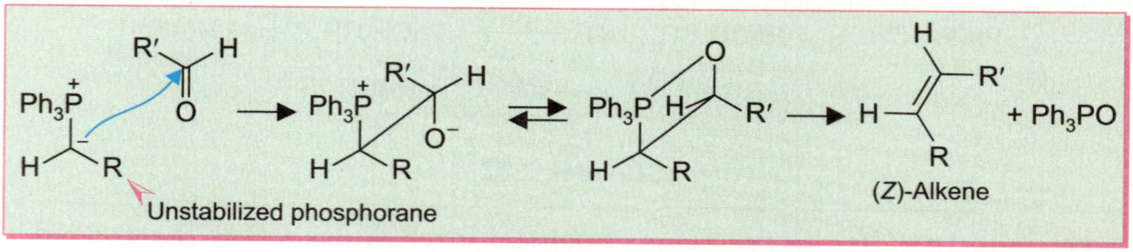

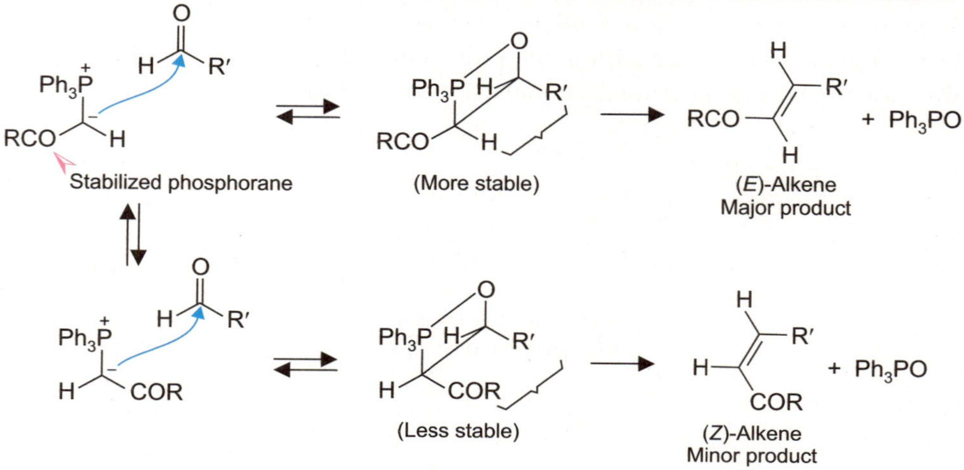

SCHEME 7.39

Thus in summary when lithium salts are present in the Wittig reaction both with stabilized and non-stabilized ylides alike, the stepwise mechanisms may be operating and the adducts are diastereomeric diionic species (I and II, Scheme 7.40) which eliminate to the *Z*- and *E*-alkenes respectively.

(iv) Schlosser Modification

Mainly E-alkenes are obtained from nonstabilized ylides by the Schlosser modification of the Wittig reaction (Wittig-Schlosser reaction) by working at low temperature (–78°C). Thus, treatement of the (*cis*)-oxaphosphetane intermediate (I, Scheme 7.40) with *n*-BuLi or PhLi at –78°C results in lithiation of the acidic proton adjacent to phosphorus to give the β-oxido phosphonium ylide (II). Protonation of II with *t*-BuOH gives the *trans*-1,2-disubstituted alkene (III).

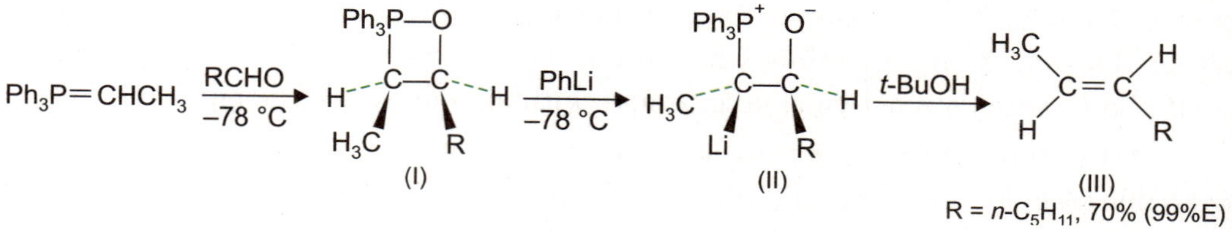

SCHEME 7.40

(v) Wadsworth-Emmons Reaction

The stabilized ylides, though useful, react only with aldehydes and these do not do so with ketones. A modification is introduced (Wadsworth-Emmons reaction (Scheme 7.41) in which triphenyl-phosphine is replaced by triethyl phosphite. Triethyl phosphite is reacted with the relevant halide and the resulting alkyl phosphonate is treated with base, like NaH. The product of reaction is mainly (*E*) alkene. The following points may be noted:

- Wadsworth-Emmons—Horner reaction involves the use of a phosphonate ester instead of a tripenylphosphonium salt of a Wittig reaction.

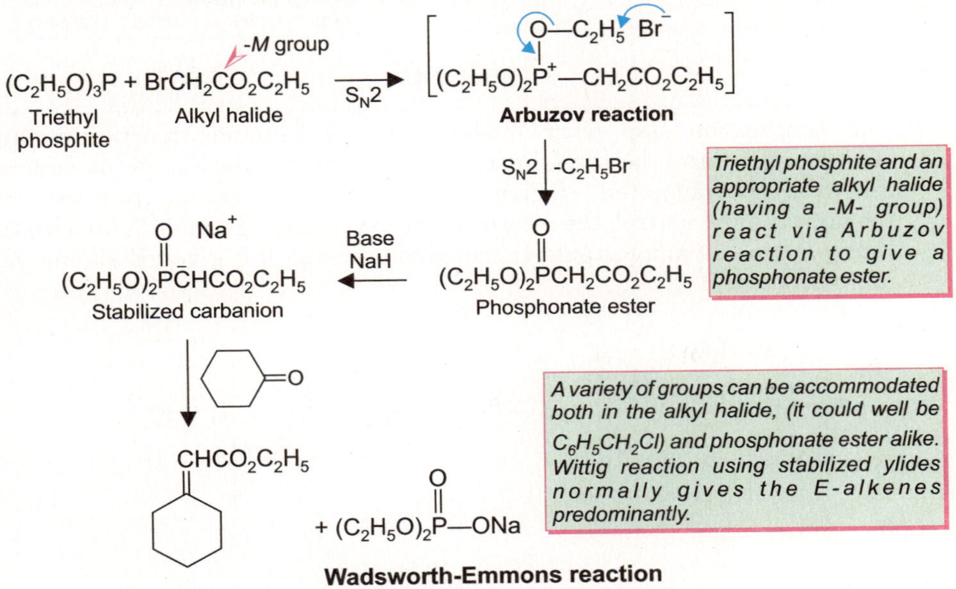

Wadsworth-Emmons reaction

SCHEME 7.41

- Non-stabilized ylides are highly reactive. These react with aldehydes and ketones to give mixture of *E* and *Z* isomers of alkenes.
- A stabilized ylide (—*M* group adjacent to carbanionic carbon are obtained using less basic conditions. Nucleophilic character and reactivity toward carbonyl group is decreased, since the negative charge is delocalized into —*M* group.
- A useful alternative uses phosphonate ester and the derived carbanions, (which are also stabilized Scheme, 7.41). A phosphonate ester is made by reacting triethyl phosphite with an appropriate halide *e.g.*, ethyl bromoacetate a process called Arbuzov reaction.
- Of the stabilized ylides, the one derived from a phosphonate ester is more nucleophilic (Scheme 7.42).

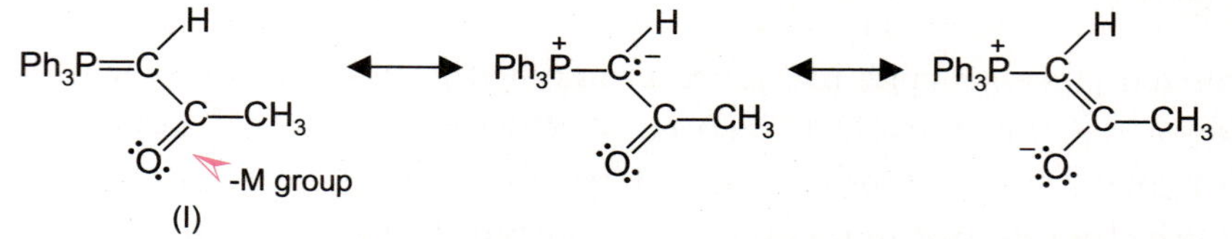

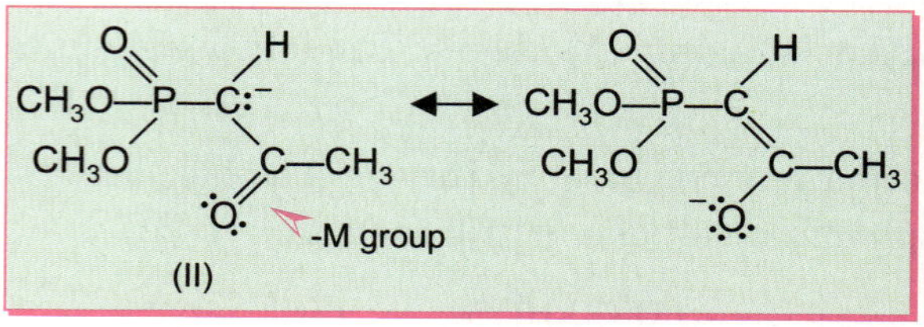

A familiar stabilized ylide (I, for ***Wittig reaction****) is less nucleophilic due to resonance. The stabilized ylide (II) from phosphonate ester (****Wadsworth-Emmons reaction****) is more nucleophilic (oxo and alkoxy groups on P ensure that negative charge is more on carbon).*

SCHEME 7.42

- The elimination step from Wadsworth and Emmons reaction is highly stereospecific and is thought to involve the formation of oxyanions of β-hydroxyphosphonates (Scheme 7.43). The reactions proceed under thermodynamic control, the intermediate oxyanion (I, Scheme 7.43) with *threo*-configuration is more stable thermodynamically and gives *E*-alkene as the main product.

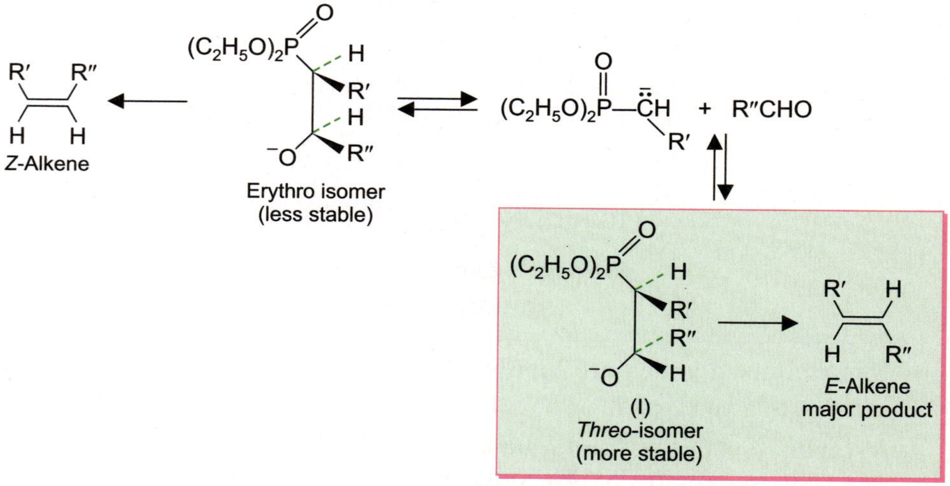

SCHEME 7.43

- The aldol product on heating can dehydrate to give an α, β unsaturated compound.

(vi) The Peterson Reaction

The reagent in this reaction is a trimethylsilylmethyl Grignard which adds to a carbonyl group to form a β-hydroxysilane. Elimination of the trialkylsilyl group and the hydroxyl group then yields the alkene. Which of the two mechanistic pathways this reaction follows depends upon whether the elimination is carried out under acidic or basic conditions (Scheme 7.44).

A merit of the Peterson reaction is in that a single diastereomer of a β-hydoxyalkylsilane *e.g.*, threo isomer can be used as a precursor for the stereospecific preparation of both (Z)- and (*E*)-alkenes, depending on whether the elimination is done under acidic H_2SO_4 or $BF_3.Et_2O$ in THF) or basic (KH in THF) conditions.

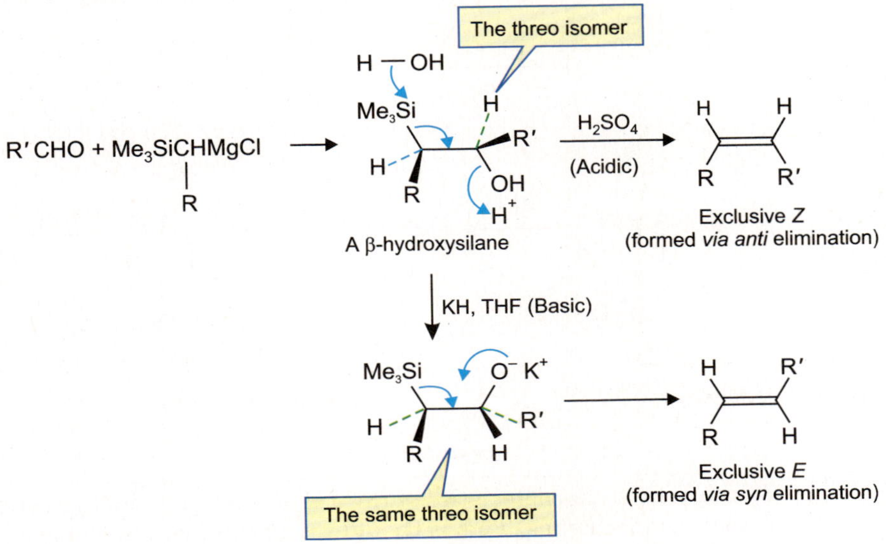

The Peterson silicon-based alkene synthesis

SCHEME 7.44

Silicon is more electropositive than carbon, thus the carbon–silicon bond is strongly polarized $Si^{\delta+}$—$C^{\delta-}$. Silicon stabilizes an α-carbanion (scheme 7.45). A further useful property is its ability to stabilize a β-carbocation.

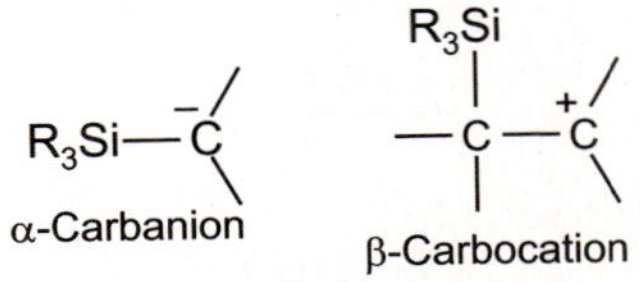

SCHEME 7.45

(vii) The Julia Olefination

The Julia olefination reaction is highly regioselective, (*E*)-stereoselective, and is a valuable alternative to the Schlosser reaction to make *trans*-disubstituted olefins. The reaction is a condensation of a metalated alkyl phenyl sulfone with an aldehyde to yield a β-hydroxysulfone (scheme 7.46) which is then subjected to a reductive elimination to produce the alkene.

In the first step, a primary alkylphenyl sulfone (I, scheme 7.46) is lithiated with *n*-BuLi in the position α to the sulfur atom. The resulting sulfonyl-stabilized organolithium compound (II) is then added to the carbonyl group of an aldehyde R^2–CHO. This gives a mixture of β-phenylsulfone alkoxides (III) as a mixture of diasteromers.

To cut out side reactions, the hydroxyl group is just converted to an acetate or tosylate and the *syn* : *anti* diastereomeric mixture acetates is then reduced with sodium amalgam (scheme 7.46)

CHAPTER 7

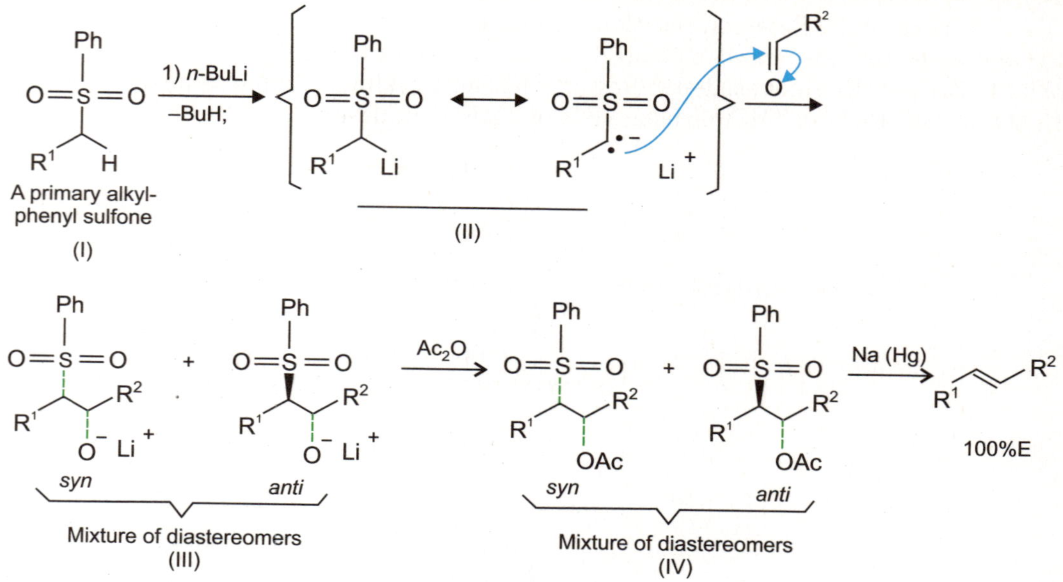

SCHEME 7.46

One of the important characteristics of the Julia reaction is its high stereoselectivity for (*E*)-disubstituted alkenes. This is thought to result from mediation of a carbanion in the reductive elimination which can rotate into the most stable configuration, corresponds to (*E*)-alkene, before eliminating acetate (Scheme 7.46*a*).

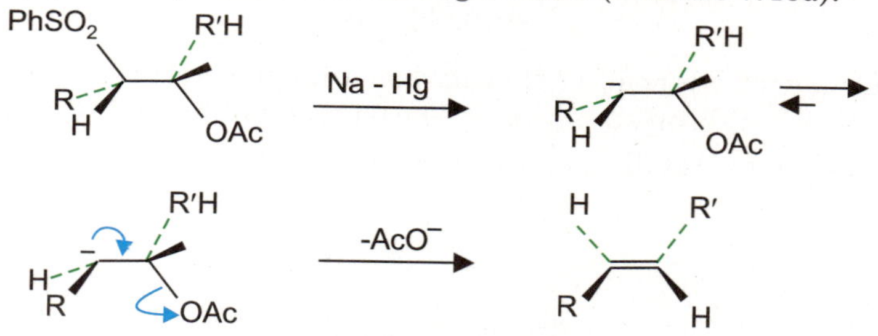

SCHEME 7.46a

EXERCISE 7.3

Write the stereochemistry of the product from (1) when treated with t BuO⁻ followed by reaction with ethylbromoacetate (scheme 7.46).

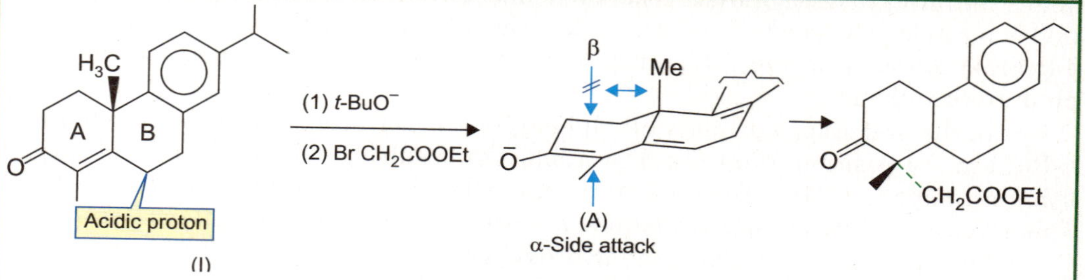

SCHEME 7.47

ANSWER. *The steric factor controlled the stereochemistry of alkylation of the dienolate (A, scheme 7.46) derived from a cyclohexenone. The approach of the alkylating agent (ethyl bromoacetate) from the β-side in a chair-like transition state is sterically blocked by a 1.3-synaxial interaction with the angular methyl group at C 10.*

PROBLEMS

7.1. Comment on the yields obtained from reduction of the following bridged bicyclic ketone.

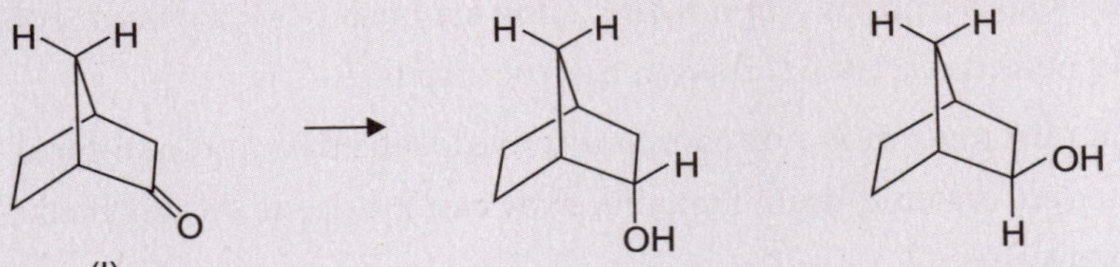

(I)

Reagent	*endo* (%)	*exo* (%)
Al(O*i*-Pr)$_3$, *i*-PrOH	20	80
$LiAlH_4$, THF	90	10

7.2. Compared to the reduction of (I, problem 7.1) bicyclo [2.2.1]-hept-2en-7-one also gives two products on hydride reduction with $LiAlH_4$ or $NaBH_4$. Explain.

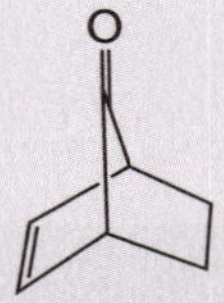

Bicyclo[2.2. 1]-hept-2-en-7-one

7.3. Comment briefly on the stereochemical outcome of the following Wittig reactions carried out under different conditions considering (*i*) nature of ylide, (*ii*) nature of the base, whether salt free and (*iii*) reaction conditions. Which mechanism operates in each case?

$$\underset{\text{(I)}}{C_2H_5\overset{+}{P}Ph_3\ Br^-} \xrightarrow[\text{(-NaBr)}]{NaNH_2} [CH_3\overset{-}{C}H\text{—}\overset{+}{P}Ph_3] \xrightarrow{PhCHO} \underset{\text{Mainly Z-alkene}}{PhCH{=}CHCH_3}$$

$$\underset{\text{(II)}}{C_2H_5\overset{+}{P}Ph_3\ Br^-} \xrightarrow{PhLi} CH_3\overset{-}{C}H\overset{+}{P}Ph_3 \xrightarrow[\text{(ii) PhLi; (iii) }\overset{+}{H}\text{ (Me}_3\text{COH)}]{\text{(i) PhCHO (– 78°)}} \underset{\text{Mainly E-alkene}}{PhCH{=}CHCH_3}$$

$$\underset{\text{(III)}}{Ph_3\overset{+}{P}CH_2CO_2C_2H_5\ Br^-} \xrightarrow{NaOH} Ph_3\overset{+}{P}\text{—}\overset{-}{C}HCO_2C_2H_5 \longrightarrow \underset{\text{Mainly }E\text{-alkene}}{PhCH{=}CHCO_2C_2H_5}$$

7.4. Complete the following equations by writing the products formed.

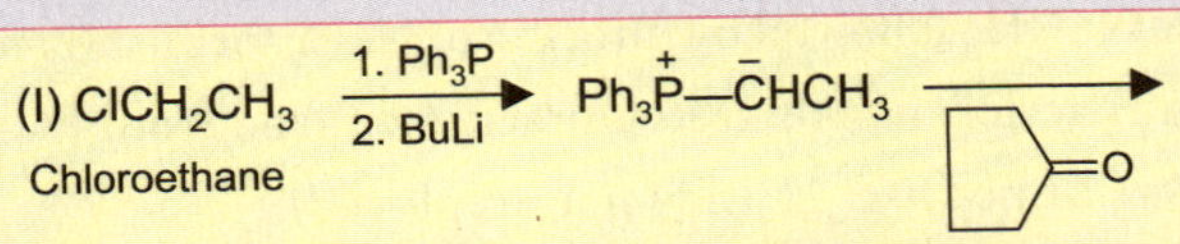

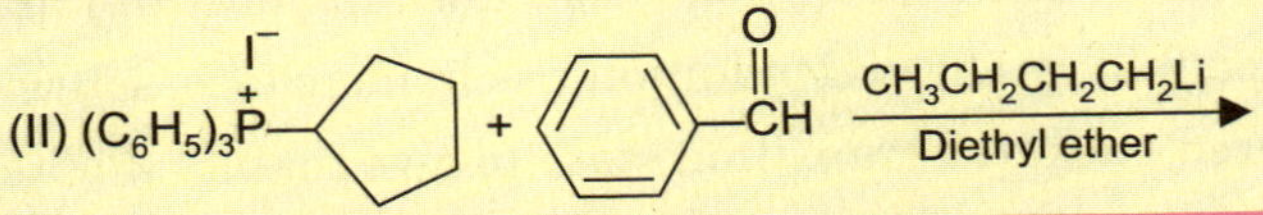

7.5. What reactions/reagents are needed in the Wittig reaction to synthesize the following compounds?

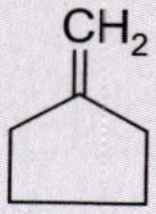

(I)

$C_6H_5CH{=}CHCH{=}CH_2$

(II)

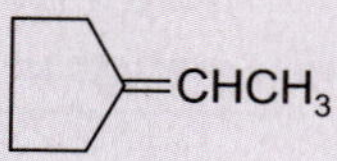

(III)

7.6. An organic compound (A) on reaction with C_6H_5MgBr followed by acidification gives a compound (B). The compound (B) on mild oxidation gives a compound C. Its mass spectrum $M^{+\bullet}$ $m/z = 106$, shows a base peak at $m/z = 105$ and another strong peak at $m/z = 51$. Assign a structure to A.

7.7. An E2 elimination is not completely regioselective and can form a mixture of alkenes *e.g.*, from 3-bromohexane. How can 2-hexene be synthesized completely regioselectively?

$$CH_3CH_2CH_2CH(Br)CH_2CH_3 \xrightarrow{HO^-} CH_3CH_2CH{=}CHCH_2CH_3 \text{ (3-Hexene)} + CH_3CH_2CH_2CH{=}CHCH_3 \text{ (2-Hexene)}$$

***E2 elimination** may not be completely regioselective.*

ANSWERS TO PROBLEMS

7.1. Meerwein-Ponndorf-Verley reduction is an equilibrium process which favours the most stable stereoisomer.

7.2. The hydroxyl group may be on the same side as the double bond or on the opposite side.

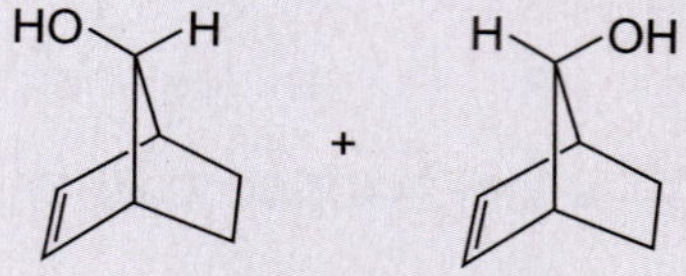

7.3. I (*i*) The reaction uses nonstabilized ylide prepared from CH_3CH_2Br.

(*ii*) The base $NaNH_2$ followed by filtering of NaBr formed (before reaction) shows that the reaction is salt free particularly free from lithium salts (since a base like PhLi or n-C_4H_9Li is not employed).

(*iii*) No special reaction conditions (*e.g.*, low temperature) are employed. This reaction follows the cycloaddition mechanism (see, scheme 7.38) involving a favored puckered transition state to give mainly *Z*-alkene.

II (*i*) The reaction again uses only non-stabilized ylide prepared from CH_3CH_2Br.

(*ii*) The base is PhLi–lithium salts are present.

(*iii*) Special reaction temperature conditions – 78° are used during adding of carbonyl compound. Thus adduct(s) though formed do not eliminate at this low temperature. A second molar addition of base leads to the formation of a new ylide which changes into a more stable *threo* diionic structure (see scheme 7.40) which eliminates to give mainly *E*-alkene.

III (*i*) The alkyl halide $BrCH_2COOC_2H_5$ contains a–*M* group, the ylide would be obtained with a comparatively weaker base. The ylide is stabilized to give *E*-alkene.

7.4.

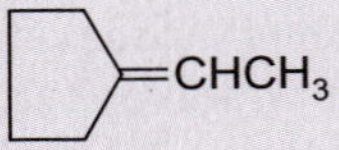

(I)

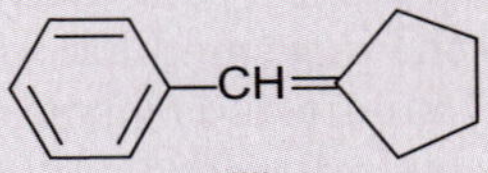

(II)

7.5. (*i*) CH_3I, $(Ph)_3P$, then strong base and finally reaction with cyclopentanon.

(*ii*) $CH_2{=}CHCH_2Br$, $(Ph)_3P$, then strong base and finally reaction with benzaldehyde.

(*iii*) Chlorocyclopentane, $(Ph)_3P$, then strong base and finally with acetaldehyde.

7.6. The mass spectrum of compound (C) is typical of an aromatic aldehyde (benzaldehyde). Phenylmagnesium bromide must react with formaldehyde to give a primary alcohol $C_6H_5CH_2OH$ which on mild oxidation gives C_6H_5CHO. The compound A is therefore formaldehyde HCHO.

7.7. Using Wittig reaction which is completely regioselective.

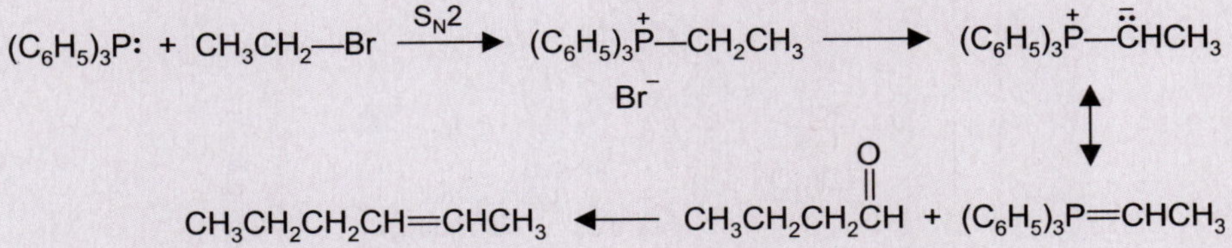

Pericyclic Reactions

8
CHAPTER

In a pericyclic reaction, there is a concerted bond reorganization and the essential bonding changes occur within a cyclic array of the participating atomic centers. These reactions do not involve the intermediate formation of either ions or radicals. Pericyclic reactions are also largely unaffected by polar reagents, solvent changes, radical initiators etc. These can however, be influenced only thermally *i.e.,* reactants are in their ground state or photochemically, *i.e.*, the excited state of a reactant is involved in the reaction.

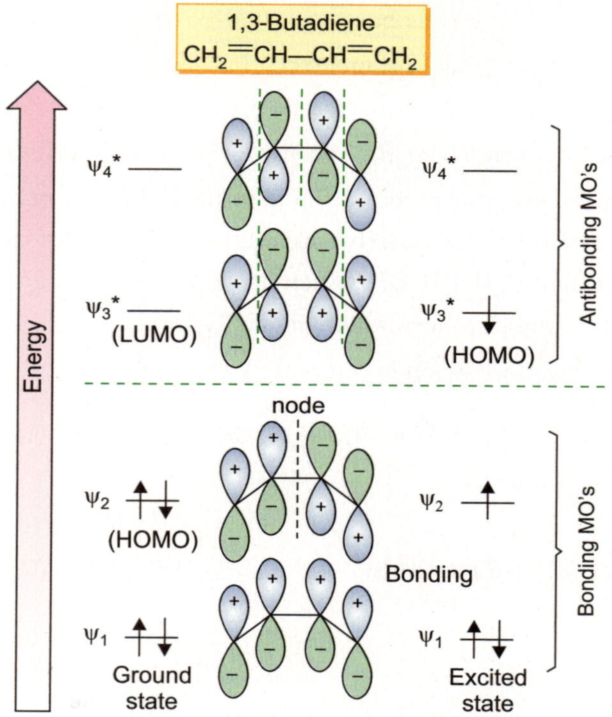

The four π molecular orbitals of 1,3-butadiene (Asterisk denotes an antibonding orbital)

SCHEME 8.1

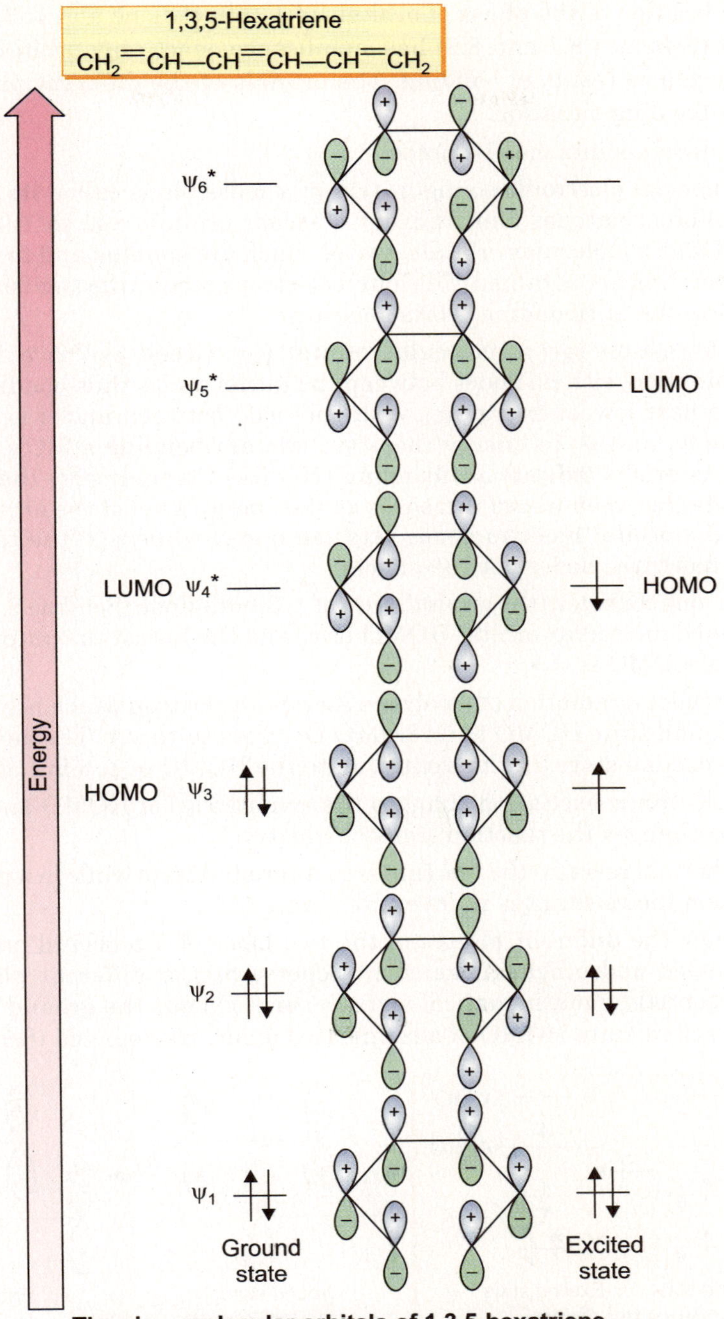

The six π molecular orbitals of 1,3,5-hexatriene

SCHEME 8.2

A consideration of the phase of orbitals of 1, 3-butadiene and 1, 3, 5-hexatriene are presented (Schemes 8.1 and 8.2) has significance, since only orbitals of the same phase will overlap to result in bonding. The orbitals of the different phase lead to a repulsive anti-bonding situation.

The following points may be noted:

- The normal electronic configuration of a molecule is called its ground state. When one combines *four* adjacent *p* atomic orbitals *e.g.,* in 1, 3-butadiene a set of four π molecular orbitals, two of which are bonding and two of which are antibonding are obtained. The four π electrons occupy the two bonding orbitals to leave the antibonding orbitals vacant.
- The lowest-energy π molecular orbital (ψ_1, Greek psi) is a fully additive combination with no nodes between the nuclei and is thus bonding. The π MO of the next lowest energy ψ_2, with one node between nuclei is also bonding. Above ψ_1 and ψ_2 in energy there are two antibonding π MO's, ψ_3* and ψ_4*. (The asterisks indicate antibonding orbitals.) One may note that the number of nodes between nuclei increases as the energy level of the orbital increases. The ψ_3* orbital has two nodes between nuclei while ψ_4*, the highest-energy MO, has three nodes between nuclei.
- When one considers the ground state of 1, 3-butadiene (Scheme 8.1) the highest occupied molecular orbital HOMO is ψ_2 and the lowest unoccupied molecular orbital LUMO is ψ_3*.
- Ultraviolet irradiation of a polyene excites an electron which is promoted from its ground state HOMO to its LUMO *i.e.,* from ψ_2 to ψ_3*. The molecule is then in an excited state. In the excited state the HOMO of 1, 3-butadiene is ψ_3*.
- The electronic excitation changes the symmetries of HOMO and LUMO and it also changes the reaction stereochemistry.
- In a thermal reaction the reactant is in its ground state while in a photochemical reaction the reactant is in its excited state.
- To show the different phases of the two lobes of a *p* orbital one phase may be shaded and other left as such (Scheme 8.1) or different phases may be represented by mathematical signs (+) or (–). Thus the ground state HOMO and excited state HOMO of a conjugated diene are depicted (Scheme 8.3).

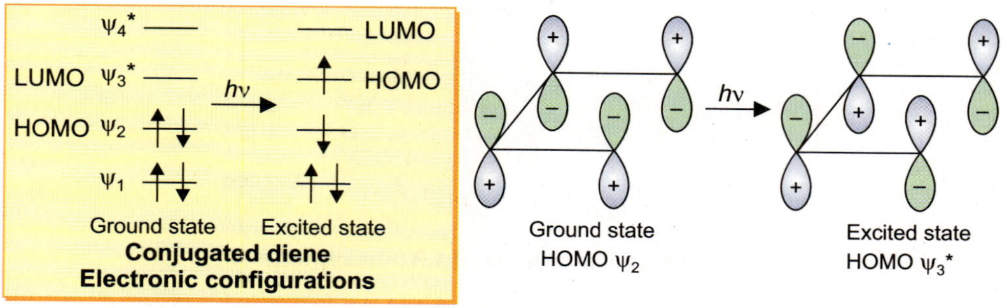

Ground state and excited state electronic configurations of a conjugated diene

SCHEME 8.3

A consideration of the six π molecular orbitals of 1, 3, 5-hexatriene (Scheme 8.2) shows that ψ_3 is the HOMO in the ground state and ψ_4* is the LUMO. In the excited state of 1, 3, 5-hexatriene ψ_4* is the HOMO and ψ_5* is the LUMO.

8.1 CONSERVATION OF MOLECULAR ORBITAL SYMMETRY

The resulting stereochemistry of the concerted pericyclic reactions depends on whether the reaction is thermal or photochemical. Woodward and Hoffmann in 1965 pointed out that symmetry of the molecular orbitals which participate in the chemical reaction determines the course of the reaction and they proposed the principle of the conservation of orbital symmetry in concerted reactions. A pericyclic reaction can take place only provided the symmetry of all reactant molecular orbitals is the same as the symmetry of the product molecular orbitals (symmetry allowed reaction). In other words the lobes of the reactant molecular orbitals must be of correct algebraic sign for bonding overlap to take place in the transition state to give the product.

In a concerted reaction, the symmetry present in the reactants is maintained during the course of the reaction and is present in the product as well (principle of conservation of orbital symmetry). For example, in the Diels-Alder reaction, the reactants, the diene and the dienophile each has a plane of symmetry which is maintained in the *transition* state as well as in the product cyclohexene (Scheme 8.4).

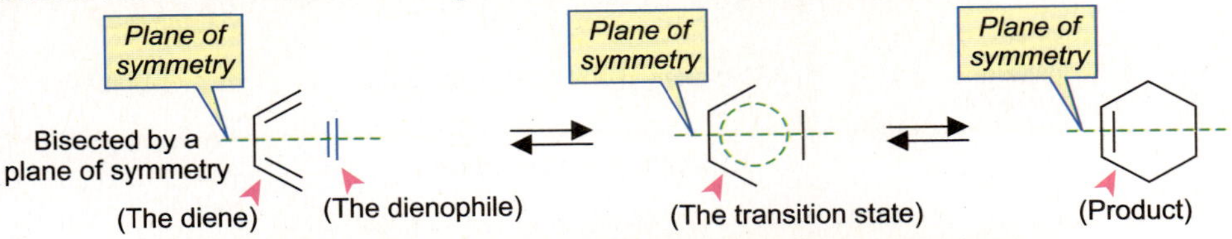

The conservation of symmetry *e.g.* in Diels-Alder reaction

SCHEME 8.4

8.2 METHODS TO EXPLAIN PERICYCLIC REACTIONS

Three approaches have been employed to explain the pericyclic reactions and these are:

1. Woodward Hoffman Rules – Correlation Diagrams

These rules require the smooth passage of the participating molecular orbitals of the reactants into the molecular orbitals of the product. The process is described by a correlation diagram. In case the conversion of the reactant orbitals into the product orbitals is favoured in terms of energy and if orbital symmetry is conserved in the process, the reaction is called symmetry allowed (The symmetries of the orbitals to be maintained are around the mirror plane and the two fold axis of rotation as a symmetry element). In case either of these conditions (energy and orbital symmetry) is not met the reaction is called symmetry-forbidden.

2. HOMO-LUMO Method-FMO Approach

The Fuki's method concentrates on the so called frontier molecular orbitals, the HOMO (highest occupied molecular orbital) and the LUMO (lowest unoccupied molecular orbital). In the ground state of 1, 3-butadiene ψ_2 is HOMO and ψ_3^* is the LUMO (Scheme 8.1). The Fuki's FMO approach examines as to how the orbitals of HOMO or in some cases, the orbitals of HOMO of one component and the LUMO of other overlap to form new bonds. If the overlaps are favourable (bonding overlaps) then the reaction is allowed and if not favourable (antibonding overlaps) then the reaction is forbidden.

3. Möbius-Hückel Analysis (PMO Method)

The idea behind this method is that a pericyclic reaction which proceeds through a *transition* state which has aromatic characteristics (electron interactions is energetically favourable) is allowed process. For system of $4n + 2$ electrons Hückel transition states are aromatic; for systems of $4n$ electrons Möbius transition states are aromatic.

8.3 ELECTROCYCLIC REACTIONS (FMO-APPROACH)

These are pericyclic reactions (intramolecular) which under the influence of heat or light involve either the formation of a ring, with the generation of one new sigma-bond and the consumption of one *pi*-bond or the reverse (Scheme 8.5). The reverse reaction, ring opening proceeds by the same mechanism, but in reverse.

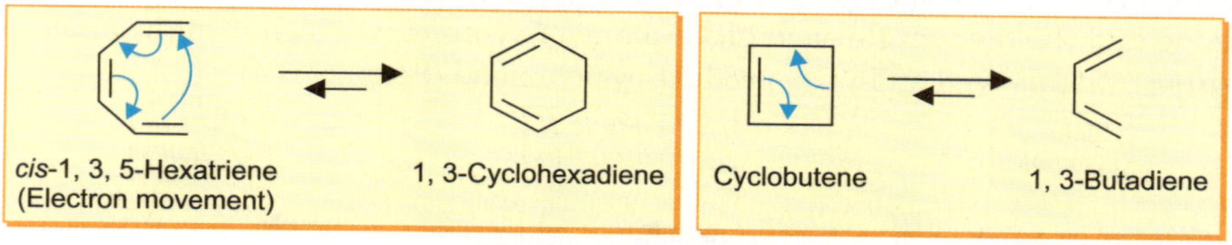

Electrocyclic reactions

SCHEME 8.5

Electrocyclic reactions are brought about by heat or light and are concerted and stereospecific. A symmetry-allowed pathway requires in-phase orbital overlap. When the HOMO is symmetric which has the end orbitals identical e.g., (II, Scheme 8.6) the rotation will have to be disrotatory to achieve the in-phase overalap (a symmetry allowed process). When however, the HOMO is asymmetric e.g., (I, Scheme 8.6) the rotation must be conrotatory to achieve an in-phase overlap. A symmetry-allowed pathway requires an in-phase orbital overlap.

The stereochemistry of electrocyclic reactions can be studied by using suitably substituted molecules. An intriguing feature of electrocyclic reaction is the stereochemistry of the product which depends on whether the reaction is thermally induced or photo-induced.

Regarding Electron Movement

In electrocyclic reactions conjugated polyenes close to give rings or rings open to polyenes shown by having electrons "chase each other's tails" around in a circle. These are among the conceptually simplest organic reactions. (It doesn't matter in which direction the curved arrows depicting electron motions are drawn in electrocyclic reactions, as long as they all move in the same direction (Scheme 8.5).

The following points may be noted:

- All the electrocyclic reactions are accounted for by orbital-symmetry arguments (FMO approach) by looking only at the symmetries of the two outermost lobes of the polyene. Thus the inner lobes of a reactant may not be shown (Scheme 8.6) and if shown, these may not be labelled +ve or –ve. The lobes of like sign can be either on the opposite side or on same side of the molecule (I and II respectively; Scheme 8.6). For bond formation the outermost lobes must rotate—a positive lobe overlapping a positive lobe or a negative lobe overlapping a negative lobe. When the two lobes of like sign are on the same side of the molecule (symmetric arrangement) the two orbitals (on the ends of the *pi* system) must rotate in different direction (clockwise and counterclockwise) and this motion is termed disrotatory (Scheme 8.7). When, however, the lobes of like sign are on opposite side of the molecule (asymmetric arrangement) both orbitals must rotate in the same direction (both clockwise or both counter-clockwise) and this motion is termed conrotatory.

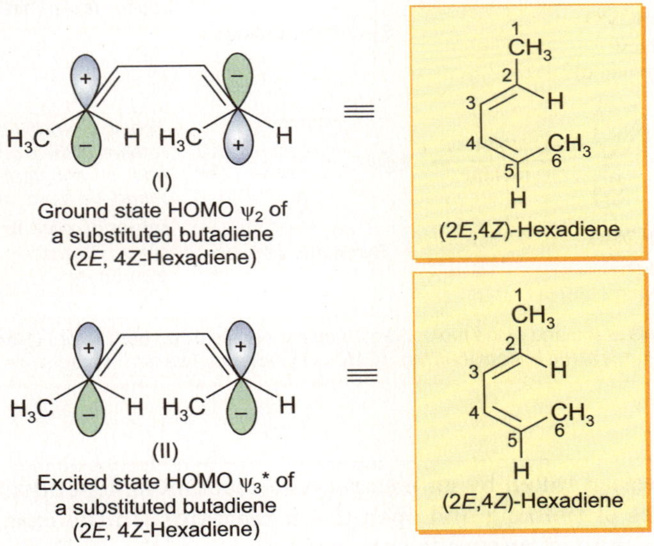

SCHEME 8.6

- The stereochemical outcome of an electrocyclic reaction depends on the number of double bonds in a polyene and on whether the reaction is thermal or photochemical. A thermal electrocyclic reaction involving $4n$ *pi* electrons (n = 1, 2, 3,...) proceeds with conrotatory motion (*i.e.*, a motion in which the bonds rotate in the same direction) while the photochemical reaction involves disrotatory motion (a motion, in which the bonds rotate in opposite directions). A thermal reaction involving $(4n + 2)$ *pi* electrons (where $n = 0, 1, 2,...$) proceeds with disrotatory motion while the photochemical reaction proceeds with conrotatory motion.

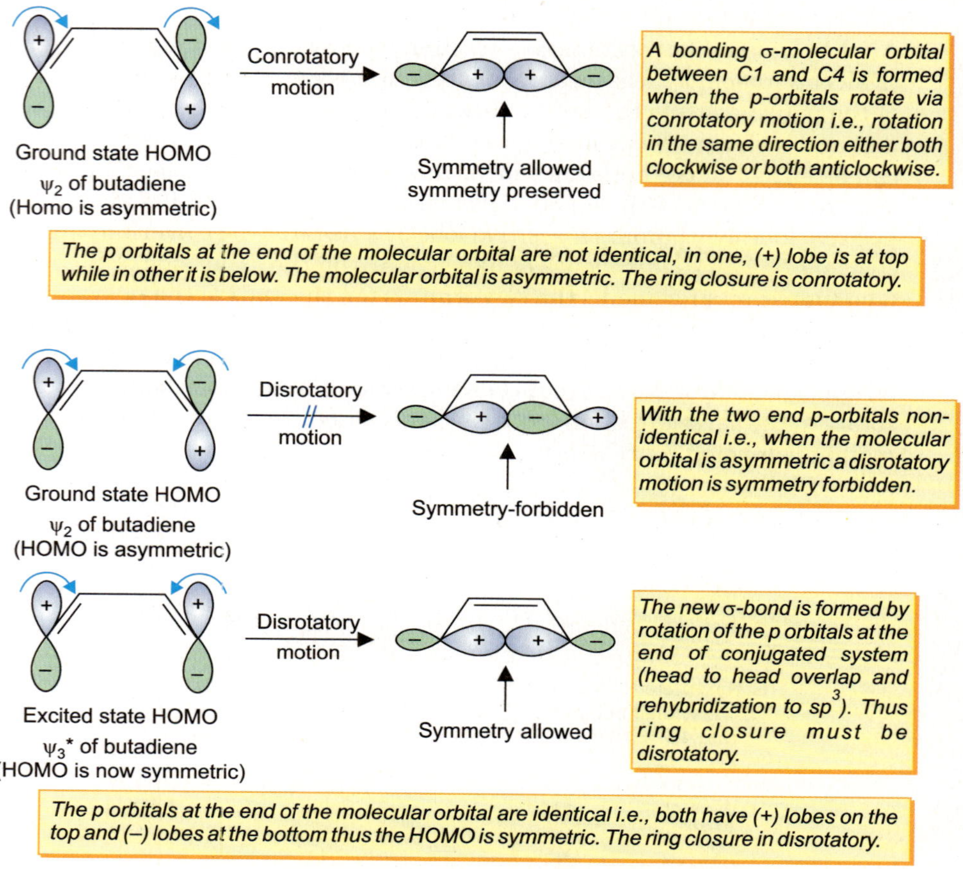

SCHEME 8.7

- The direction taken by an electrocyclic reaction is dependent on the relative stabilities of the ring and open-chain reactants. In the case of cyclobutanes the open chain structure is favoured because of the strain in the ring, during the thermal reaction.

(A) Thermally Induced Interconversion of a Conjugated Diene and a Cyclobutene

The following points may be noted:

- The symmetry of the HOMO of a conjugated polyene undergoing ring closure determines the outcome of the electrocyclic reaction. All conjugated polyenes with asymmetric HOMO's undergo conrotatory ring closure. The ground state HOMO of a conjugated compound with an even number of double bonds *e.g.*, in a conjugated diene is asymmetric and should undergo conrotatory ring closure (Scheme 8.8). Moreover the reaction is remarkably stereospecific (Scheme 8.8). One may recall that there are always two conrotatory modes clockwise and anticlockwise and both are equally probable. It is obvious that neither of the two possible disrotatory modes can be a favourable process each being an antibonding situation.

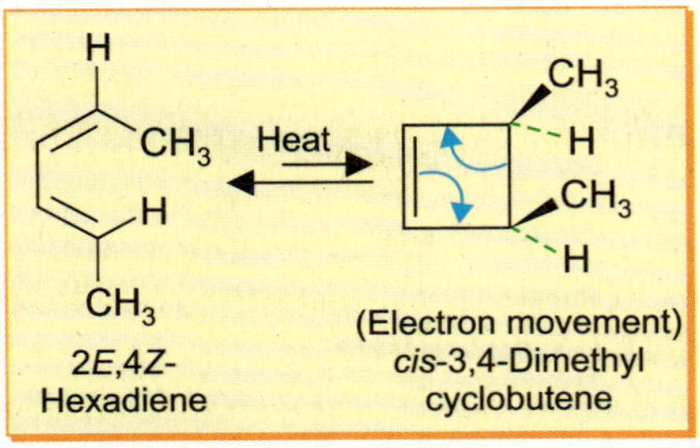

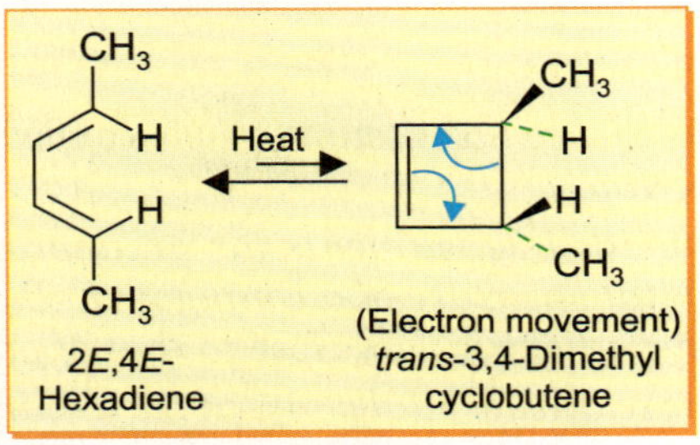

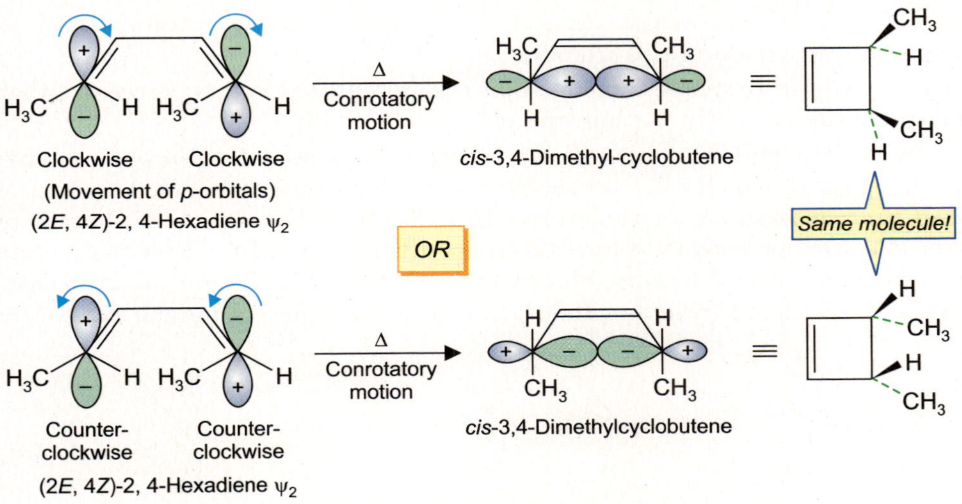

SCHEME 8.8

EXERCISE 8.1

Depict the conrotatory ring clasure of 2E, 4E-hexadiene and predict the stereochemistry of the product.

ANSWER. *Conrotatory ring closure is depicted in the counterclockwise fashion, the product is trans-3, 4-dimethylcyclobutene (Scheme 8.9). Same product would result by conrotation via clockwise motion.*

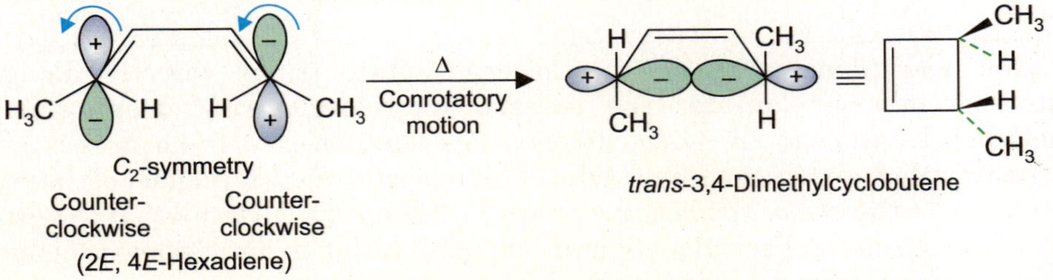

SCHEME 8.9

More on Electrocyclic Reactions

- *The diene must assume a s-cis-conformation for the terminal carbons p orbitals overlap.*
- *The alkyl substituents (as an approximation) do not effect the π molecular-orbital structure of a conjugated alkene. Thus the π molecular structure of 2, 4-hexadiene is very close to that of parent 1,3-butadiene.*
- *The relative orbital phase at the terminal carbon atoms of the HOMO (the orbital symmetry) is what that determines if the reaction would occur by conrotation or disrotation.*

(B) Thermal Ring Opening of Cyclobutenes

In keeping with the principle of microscopic reversibility the reverse process of thermal ring opening takes exactly the same path.

Due to conrotatory motion (Scheme 8.10). A σ bond will open so as to give the resulting *p* orbitals which will have the symmetry of the highest occupied π orbital of the product. Since in the case of cyclobutenes the HOMO of the product (*i.e.*, a butadiene) in the thermal reaction is ψ_2 therefore, the cyclobutene must open so that on one side the positive lobe lies above the plane, which on the other side it is below it. This process also forces the stereochemistry in the product from a substituted cyclobutene (Scheme 8.10).

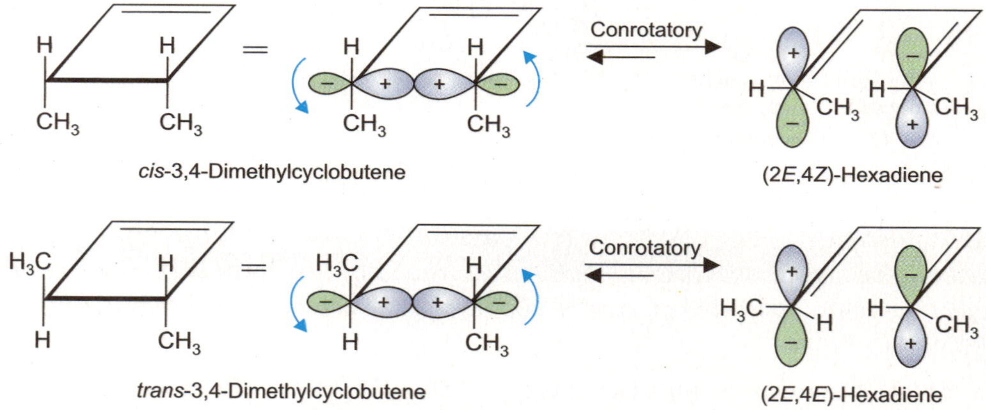

Thermal ring opening of *cis*- and *trans*-dimethylcyclobutene involves conrotatory motions

SCHEME 8.10

One may depict the opening of cyclobutane rings directly along with the stereochemistry. Consider the thermal electrocyclic ring opening in *cis*-3, 4-dimethylcyclobutene to 2*E*, 4*Z*-hexadiene. This is a concerted 4*n*-electron reaction in which two of electron pairs are involved; the sigma bond and the *pi* bond of the reactant are converted to the two *pi* bonds of the product (Scheme 8.11). Recall that a conrotation involving 4 electrons is thermally allowed. The conrotation in anticlockwise fashion will yield the same compound as already seen (see, Scheme 8.8).

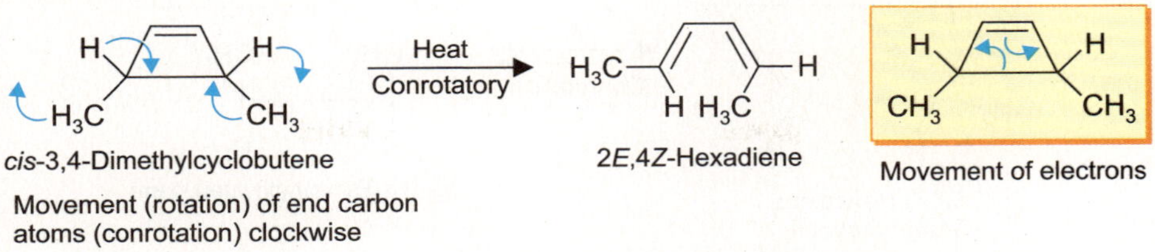

SCHEME 8.11

EXERCISE 8.2

Depict two possible conrotatory (thermal) modes of ring opening in trans-3, 4-dimethylcyclobutene.

ANSWER. *The two conrotatory modes lead to different products. (The two disrotatory modes on cis- and trans-disubstituted cyclobutenes are in Scheme 8.20a).*

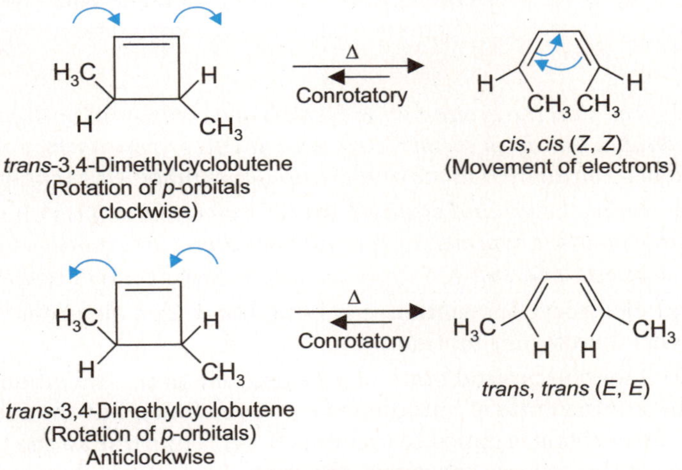

SCHEME 8.12

Thus one can also easily depict these conrotations (and also disrotations) directly without drawing molecular orbitals (Scheme 8.13). When a polyene undergoes an electrocyclic ring closure to give a cycloalkene, the terminal carbons of the polyene chain must rotate about 90° to convert the *p* orbitals on these carbons into the sp^3 orbitals forming the new σ bond. The substituents on these carbons are to be rotated into a plane which is approximately at right angles to the newly formed ring. Conversely, during the ring opening of a cycloalkene, the substituents on the atoms forming the bond undergoing cleavage will rotate into the plane of the new double bonds.

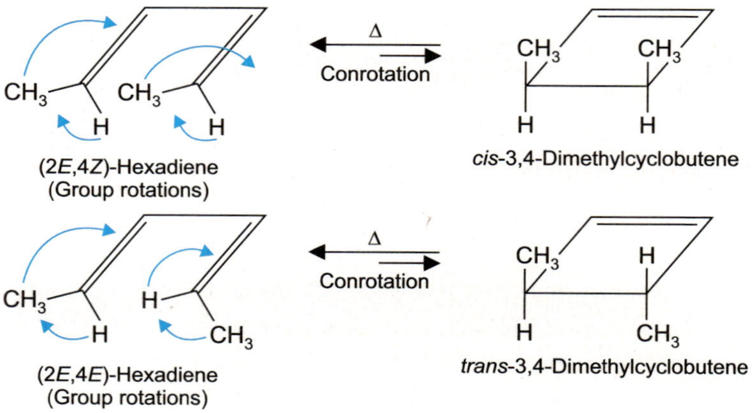

SCHEME 8.13

(C) Thermal Electrocyclic Interconversion of 1,3,5-Hexatriene and 1,3-Cyclohexadiene

Thermal Electrocyclic Ring Closure or Ring Opening–Conrotatory or Disrotatory Motions

One has already seen that the ground state HOMO of a compound with an even number of conjugated double bonds is asymmetric and involves conrotatory motion e.g., in interconversion of a conjugated diene and cyclobutene (involvement of 4n pi electrons).

On the other hand the ground state HOMO of a compound with an odd number of conjugated double bonds is symmetric and involves disrotatry motion e.g., in interconversion of 1, 3, 5-hexatriene and 1, 3-cyclohexadiene (involvement of 4n + 2 electrons).

A thermal electrocyclic reaction involving $(4n + 2)$ π electrons where $(n = 0, 1, 2, ...)$ proceeds with disrototary motion.

The HOMO for the ground state of a hexatriene is ψ_3 and when compared with the HOMO of the ground state of butadiene *i.e.*, ψ_2 one finds that the relative symmetry about the terminal carbons is opposite (Scheme 8.14). Thus unlike the thermal opening of a cyclobutene (or the reverse reaction—the ring closure) which requires conrotatory motion, in the thermal opening of a 1,3-cyclohexadiene and likewise the ring closure requires a disrotatory motion. Based on these arguments 2*E*, 4*Z*, 6*E*-octatriene gives specifically *cis*-5,6-dimethyl-cyclohexadiene (Scheme 8.15).

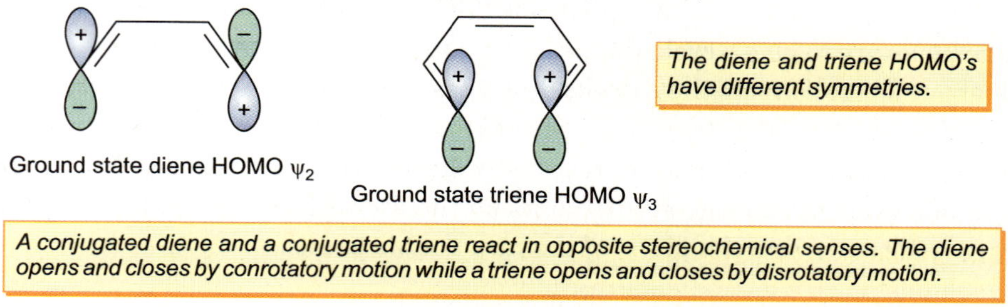

SCHEME 8.14

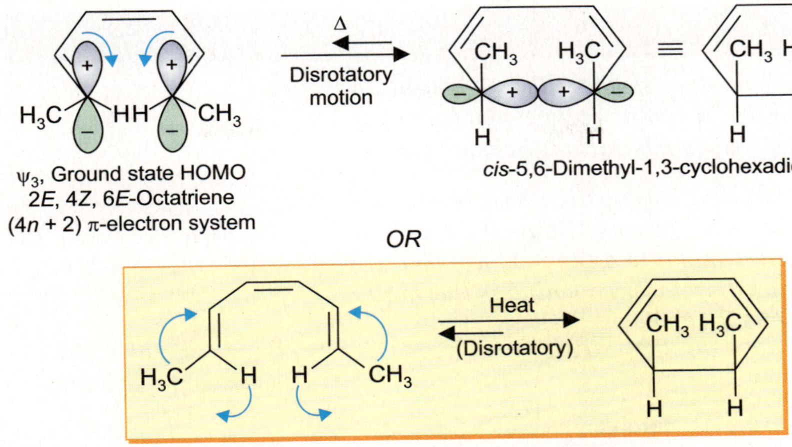

SCHEME 8.15

Similarly one finds that in the thermal cyclization of (2*E*, 4*Z*, 6*Z*)-octatriene as well the methyl groups rotate in the disrotatory fashion (Scheme 8.16).

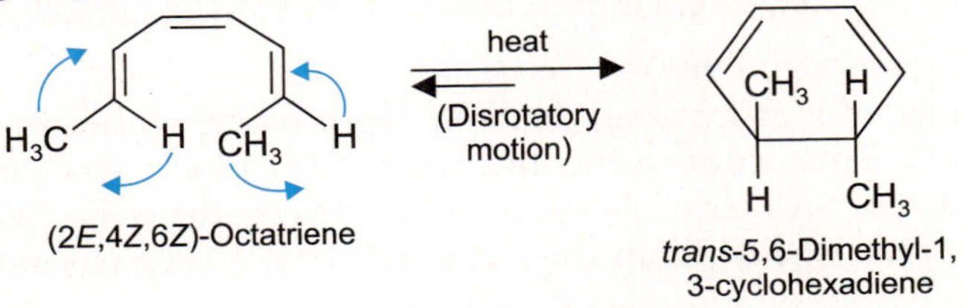

SCHEME 8.16

A thermal ring opening *e.g.*, in the case of *cis*-5, 6-dimethyl-1, 3-cyclohexadiene (see Scheme 8.15) must also be disrotatory. The ground state HOMO of the derived triene is to be ψ_3 (scheme 8.17).

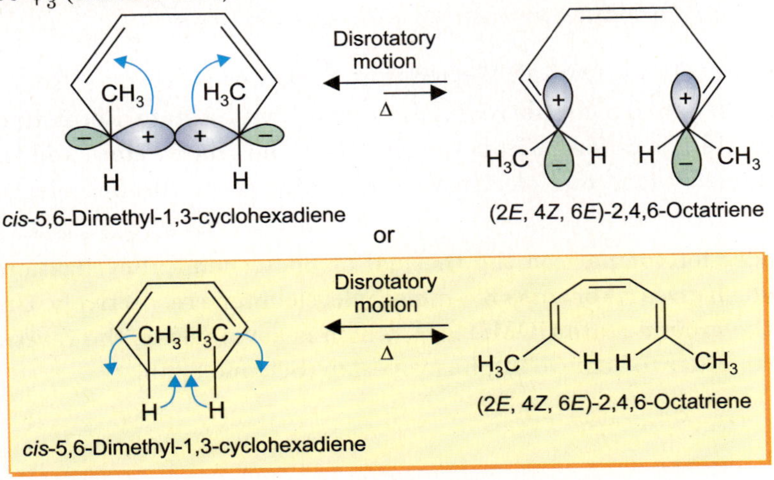

SCHEME 8.17

EXERCISE 8.3

What stereochemistry of the allyl carbocation is expected under thermal conditions from cis-dimethylcyclopropyl carbocation (Scheme 8.18)?

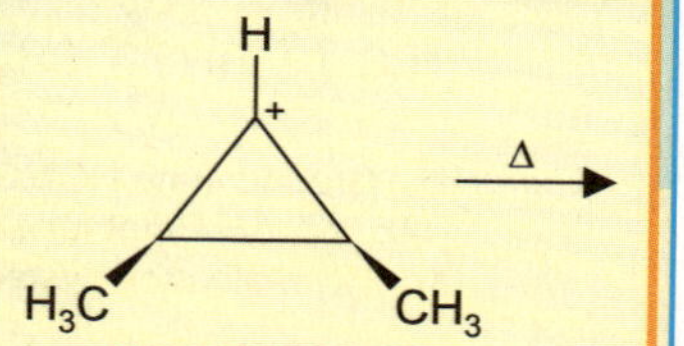

SCHEME 8.18

ANSWER. *The allyl carbocation will be formed by electrocyclic ring opening. Since the reactant involves two electrons it is a (4n + 2, n = 0, 1, ...) system and thus disrotatory opening is thermally allowed (Scheme 8.19).*

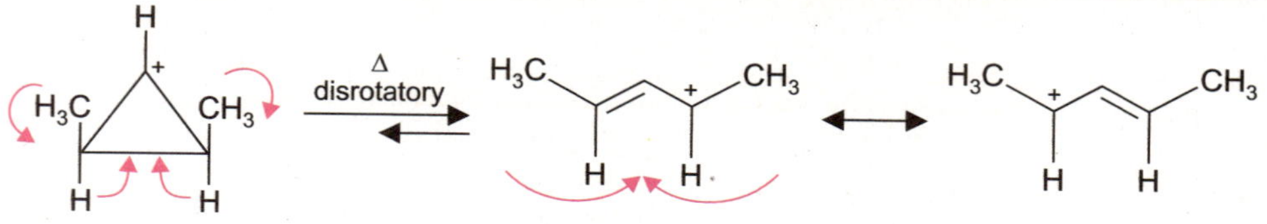

SCHEME 8.19

(D) Summary—Electrocyclic Thermal Reactions

Electrocyclic closure of a conjugated diene is conrotatory, while that of a conjugated triene is disrotatory. This is due to the difference is the phase relationships within the HOMO at the terminal carbons of these π systems. In the diene the HOMO has opposite phase at these two carbons while in the triene the HOMO has the same phase.

Conjugated alkenes with 4n π electrons (n = any integer) have antisymmetric HOMOs and undergo conrotatory ring closure while those with 4n + 2 π electrons have symmetric HOMOs and undergo disrotatory ring closure. Thus a conrotatory ring closure is allowed for systems with 4n π electrons and it is forbidden for systems with 4n + 2 π electrons. Conversely, disrotatory ring closure is allowed for systems with 4n + 2 π electrons and is forbidden for systems with 4n π electrons.

(E) Excited State (Photochemical) Electrocyclic Reactions

On absorption of light a molecule reacts through its excited state. Recall that the HOMO of the excited state is different from the HOMO of the ground state and therefore, has different symmetry (thermal electrocyclic reactions occur through electronic ground states).

Consider for comparison the thermal or photochemical cyclization of (2*E*, 4*Z*) hexadiene which gives *cis* or *trans* 3,4-dimethylcyclobutene respectively. Under thermal cyclization the ground state HOMO is ψ_2 which is asymmetric, the reaction has to be conrotatory in order to achieve in-phase overlap (Scheme 8.20).

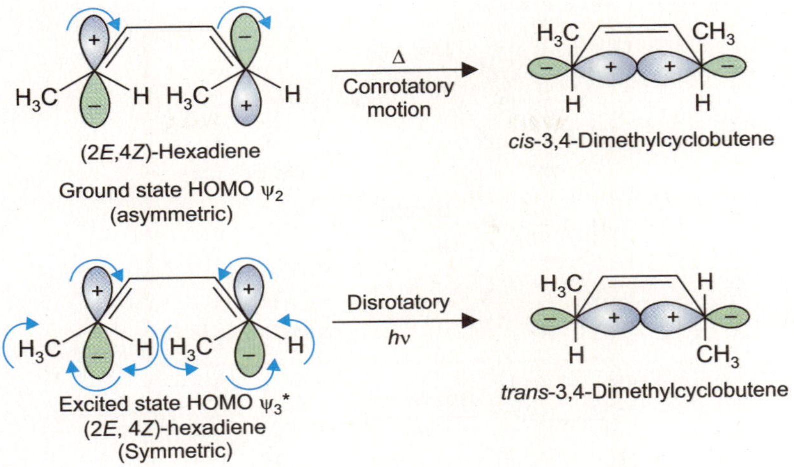

SCHEME 8.20

Similarly when (2E, 4E)-hexadiene on heating cyclizes to form *trans*-3, 4-dimethylcyclobutene, none of the *cis*-isomer is formed. In the reverse reaction the cyclobutene opens to produce only the (E, E)-isomer of the hexadiene. When the same isomer is photolyzed rather than heated, only *cis*-3, 4-dimethylcyclobutene is produced as the only product (Scheme 8.20*a*). Thus the reaction is completely stereospecific and involves conrotation under thermal conditions and disrotation on photochemical cyclization.

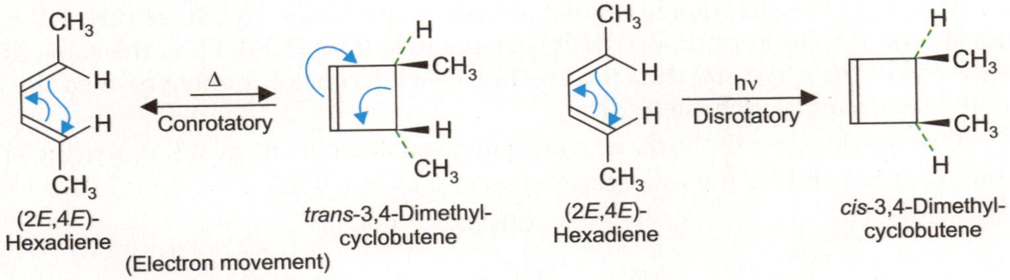

SCHEME 8.20a

EXERCISE 8.4

Depict the stereochemistry of the product from two disrotatory modes involved in the photochemical reaction of cis and trans-disubstituted cyclobutene.

ANSWER. *The products will be formed by disrotation, a cis-disubstituted cyclobutene opens to give cis, cis and trans, trans isomers of butadiene (Scheme 8.20b), while a trans-disubstituted cyclobutene gives the same product (in which there is one cis and one trans double bond).*

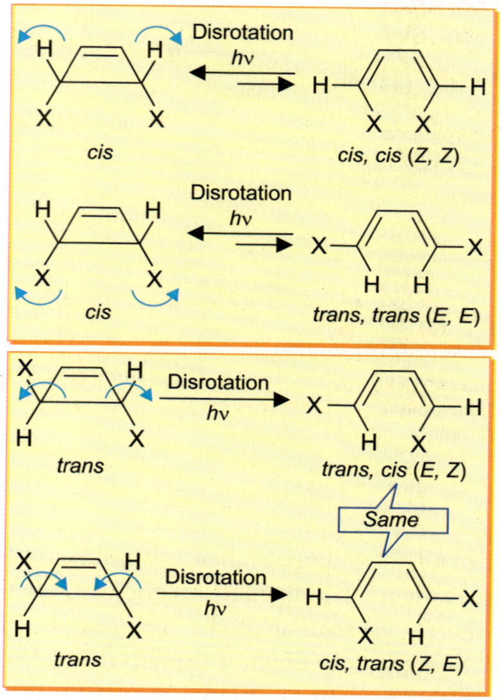

SCHEME 8.20b

When photocyclization is carried out the excited state HOMO of (2*E*, 4*Z*)-hexadiene is now ψ_3*, because this HOMO is symmetric (this HOMO has the same phase at each end of the π system) thus the bonding overlap can occur only provided the ring closure is disrotatory (Scheme 8.20).

One quickly gets to the same stereochemical outcome by avoiding writing of the *p* orbitals at the ends of the conjugated system (Scheme 8.21).

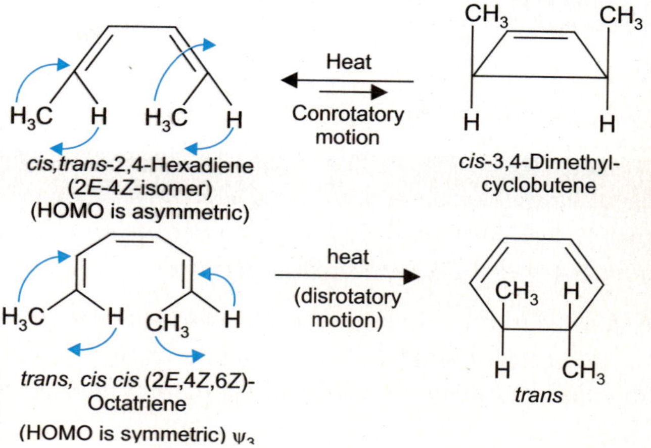

The ground state HOMO of a compound with an even number of conjugated double bonds is asymmetric (4n) π-electron system (conrotatory ring closure). The ground state HOMO of a compound with odd number of conjugated double bonds (4n + 2) π-electron system is symmetric (disrotatory ring closure).

Continued...

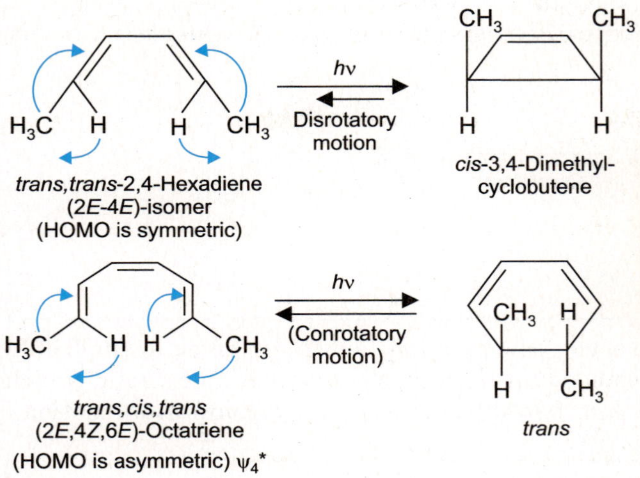

In reactions under photochemical conditions every thing is reversed. The ground state and excited state HOMO's have opposite symmetries. If the ground state HOMO is symmetric the excited state HOMO is asymmetric.

SCHEME 8.21

EXERCISE 8.5

Predict the stereochemical outcome from the electrocyclic reactions (Scheme 8.22).

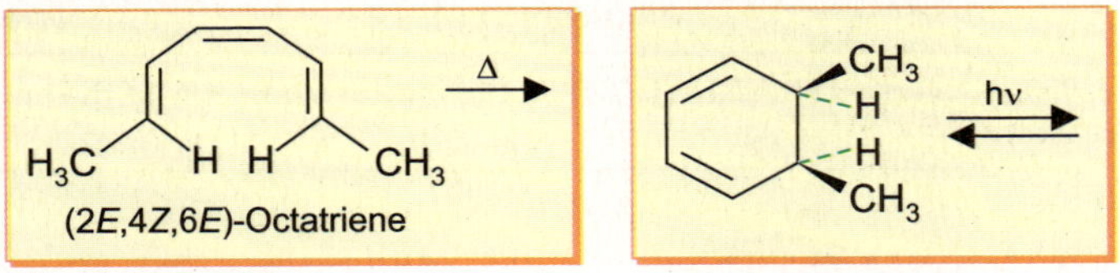

(I) (II)

SCHEME 8.22

ANSWER. (I) *Reactant has odd number of double bonds (4n + 2) π electron system, ground state HOMO of the triene (ψ_3) is symmetric. Ring closure is disrotatory.*

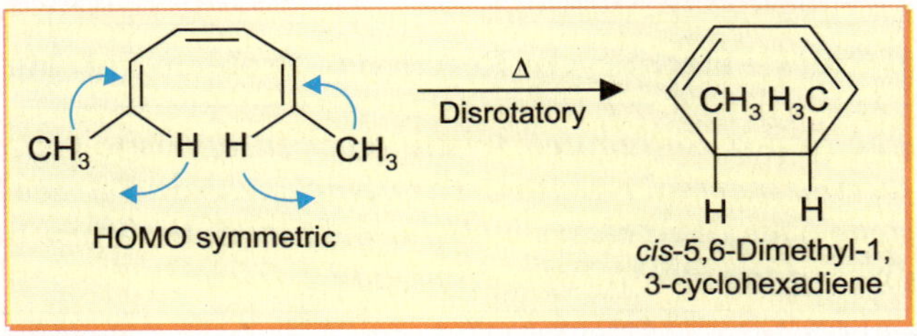

SCHEME 8.23

CHAPTER 8

(II) *Reactant is a (4n + 2) π electron system, photochemically it involves a conrotatory motion.*

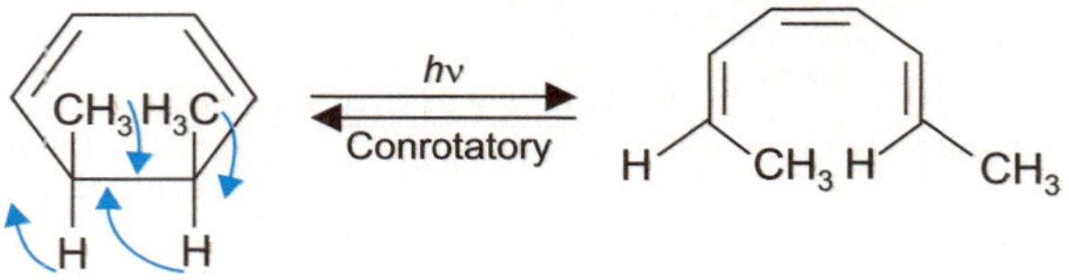

SCHEME 8.24

The selection rules for electrocyclic reactions which are given again (Table 8.1) will thus help to know the outcome of an electrocyclic reaction. Compound (I, Scheme 8.25) is a $4n$ π system, therefore, a thermal cyclization *via* a conrotatory motion is an allowed process.

Table 8.1: Selection Rules for Electrocyclic Reactions

Number of electrons*	Mode of activation	Allowed stereochemistry
$4n$	thermal photochemical	conrotatory disrotatory
$4n + 2$	thermal photochemical	disrotatory conrotatory

*n = an integer. These selection rules are based on the orbital symmetry of the open-chain (conjugated alkene) reactant.

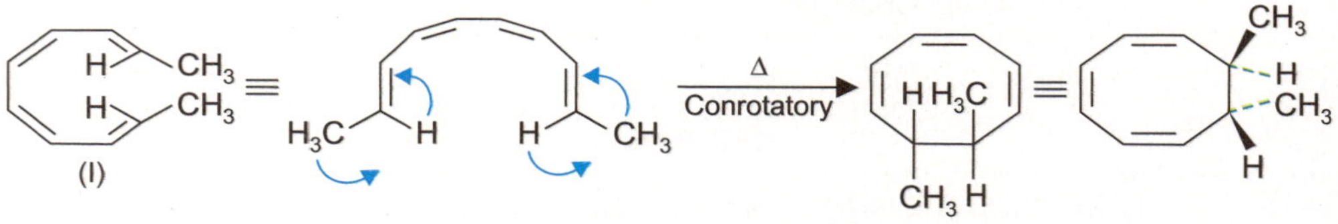

SCHEME 8.25

(F) Stereochemical Rules—Electrocyclic Reactions—Problem Solving Hint

A conjugated diene and a conjugated triene react in opposite (alternating) stereochemical senses during thermal reaction. The diene opens and closes by a conrotatory path while the triene opens and closes by a disrotatory path. These results are due to different symmetries of the HOMO of a diene and a triene.

On electronic excitation (photochemical reactions) the symmetries of HOMO and LUMO change and with it changes the reaction stereochemistry (which is reversed) and stereochemical rules for electrocyclic reactions are given (Table 8.1).

Problem Solving Hint 1

If in a reactant the bonds to the substituents point in opposite directions then these substituents will have cis stereochemistry in the product provided the motion is disrotatory (Scheme 8.25a)

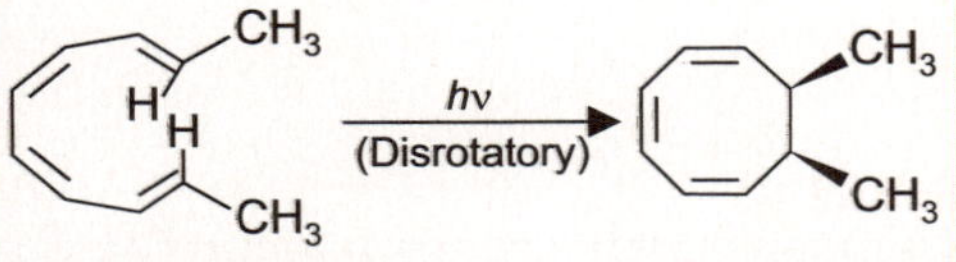

(4*n*) π-Electron system substituents point in opposite directions

cis-Stereochemistry

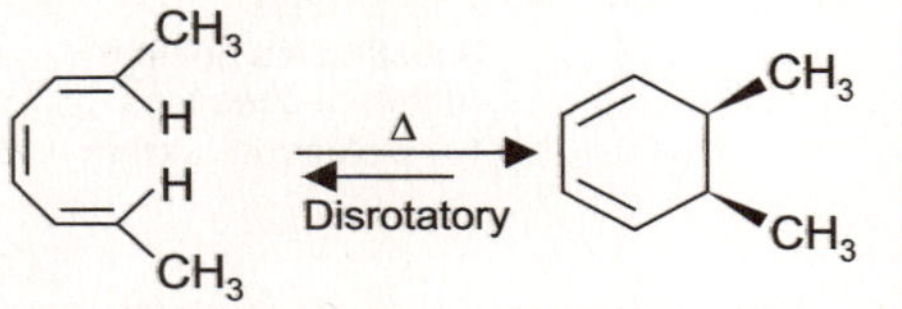

(4*n* + 2) π-Electron system substituents point in opposite directions

cis-Stereochemistry

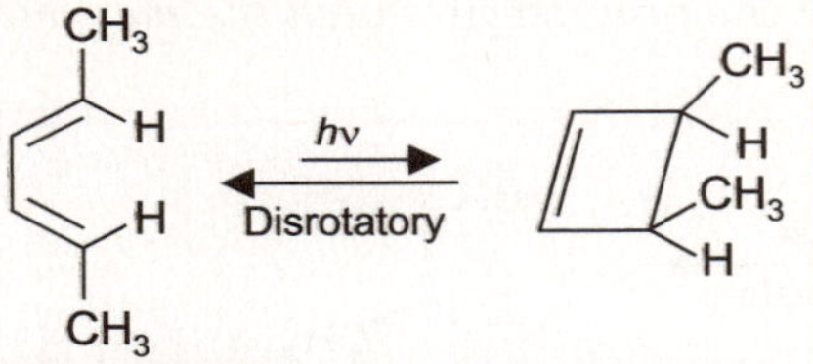

(4*n*) π-Electron system substituents point in opposite directions

cis-Stereochemistry

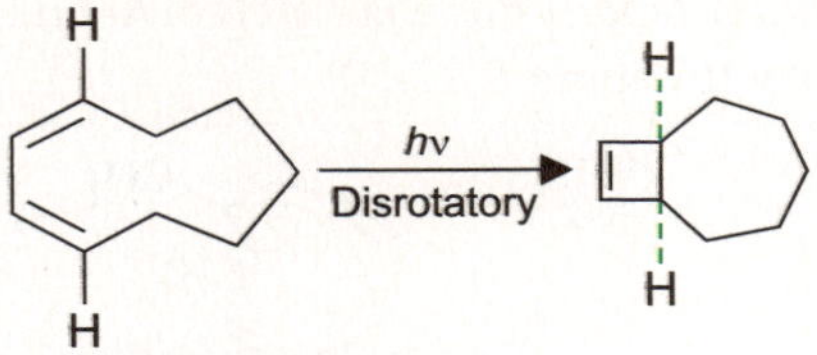

4*n* and (*Z*, *Z*) Substituents point in opposite directions

SCHEME 8.25a

Problem Solving Hint 2

If in a reactant the bonds to the substitutents point in opposite directions then these substituents will have trans stereochemistry in the product provided the motion is conrotatory (Scheme 8.26).

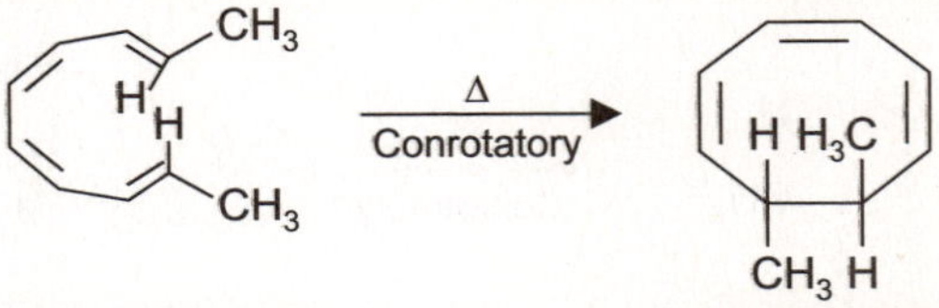

Substituents point in opposite directions (4*n*) π electron system

trans-Stereochemistry

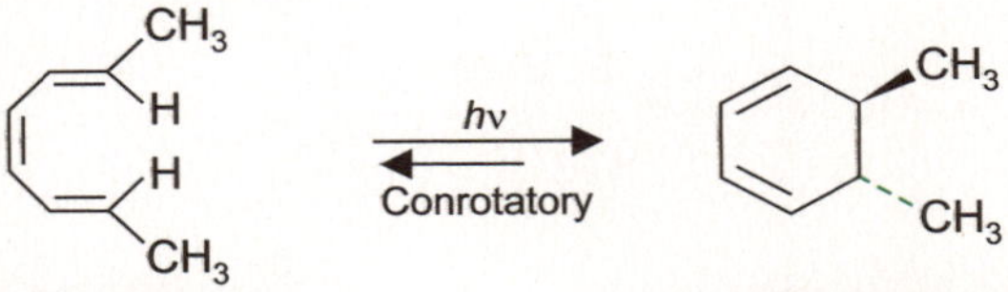

Substituents point in different directions (4*n* + 2) π electron system

trans-Stereochemistry

Continued...

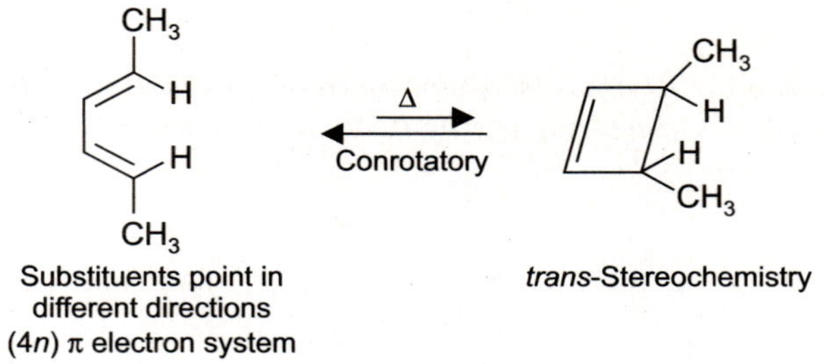

SCHEME 8.26

Problem Solving Hint 3

If in a reactant the bonds to the substituents point in the same direction then these substituents will have cis-stereochemistry in the product provided the motion is conrotatory (Scheme 8.27).

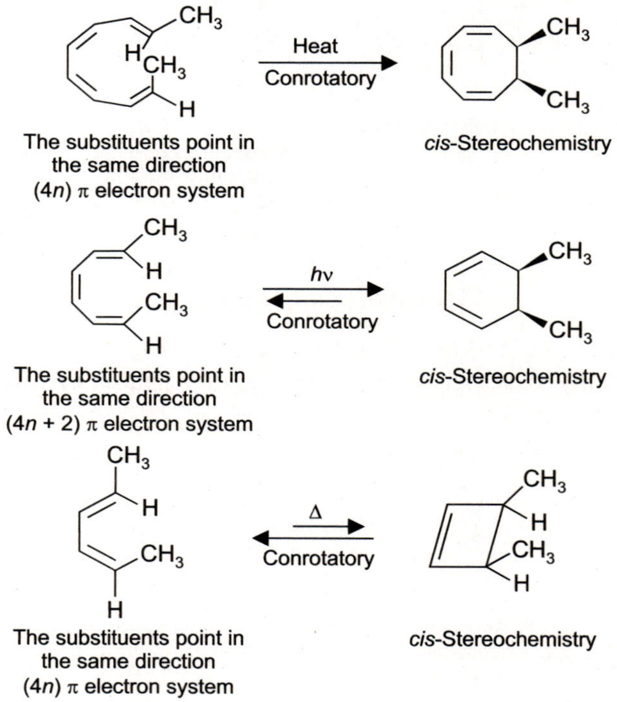

SCHEME 8.27

Problem Solving Hint 4

If in a reactant the bonds to substituents point in the same direction then the substituents will have trans-stereochemistry in the product provided the motion is disrotatory (Scheme 8.28).

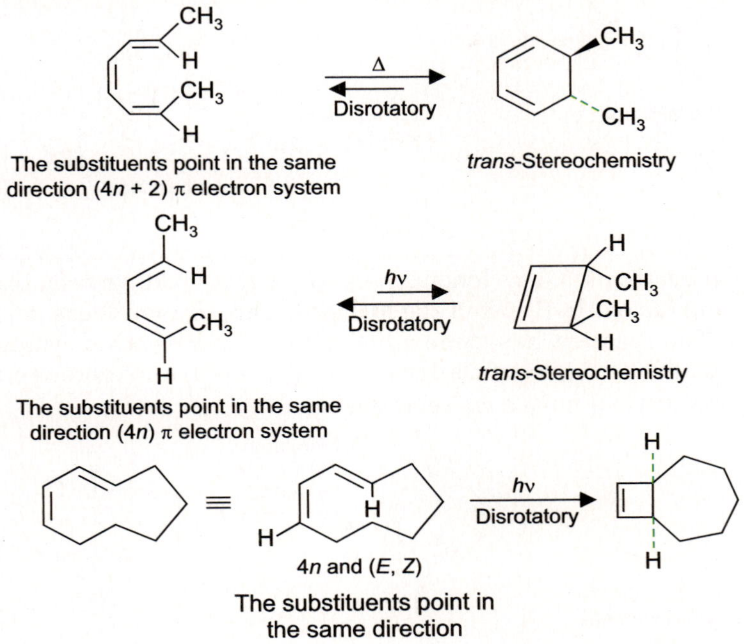

SCHEME 8.28

Example 1: Consider, firstly the photochemical ring closure of precalciferol which gives ergosterol and lumisterol (I and II respectively; Scheme 8.29) in which hydrogen and methyl substituents are *trans* to one another.

The reactant precalciferol has three conjugated π bonds (4*n* + 2) π electron system thus ring closure under photochemical conditions is conrotatory (see Table 8.1). The methyl and hydrogen at C-10 and C-9 respectively point in opposite directions in precalciferol (the H atom at C-9 is not shown which if drawn is in opposite direction to CH_3 at C-10, note that ring residue at C-9 and CH_3 substituent at C-10 are in the same direction).

Thus in precalciferol a conrotatory ring closure will cause the substituents which point in opposite directions in the reactant to be *trans* in the product. On disrotatory ring closure of precalciferol, however, these substituents will assume a *cis* relationship. The reason for the formation of two *cis* and two *trans* products is that *e.g.*, in the case of disrotatory ring closure (Scheme 8.29) two *cis* products arise due to the outward disrotatory or inward disrotatory motion.

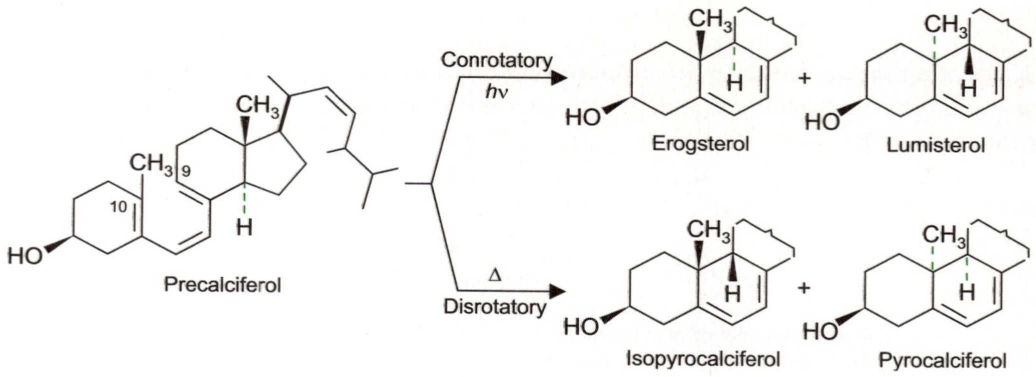

SCHEME 8.29

Example 2: On heating 1,3,5-cyclononatriene (I, Scheme 8.30) gives a bicyclic product (III) with *cis*-ring fusion. In (I) when the hydrogens are drawn, these point in opposite directions (II). The reactant has three π bonds (4*n* + 2) π electron system thus under thermal conditions the ring closure is disrotatory (table 8.1) and therefore, the hydrogen atoms in the product will have a *cis* relationship.

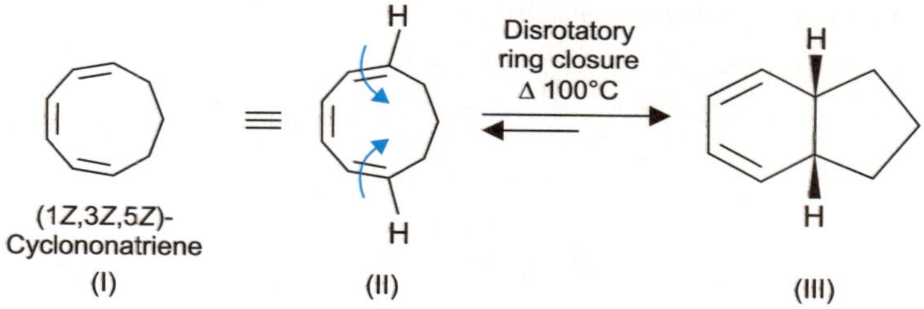

SCHEME 8.30

Example 3: Consider the photochemical ring opening electrocyclic reaction of (I, Scheme 8.31). The product (II) undergoing ring closure has three conjugated double bonds and thus under photochemical conditions (see Table 8.1) ring closure or ring opening is conrotatory. To get a product with *cis* hydrogens in (I, Scheme 8.31) the substituent hydrogens have to point in the same direction (II). Thus compared to 1,3,5,-cyclononatriene (I, Scheme 8.30) in which all the three double bonds have *Z* geometry in (II, Scheme 8.31) one of the double bonds has *E* configuration. However, (II) does undergo a disrotatory ring closure to give the expected *trans* product (III).

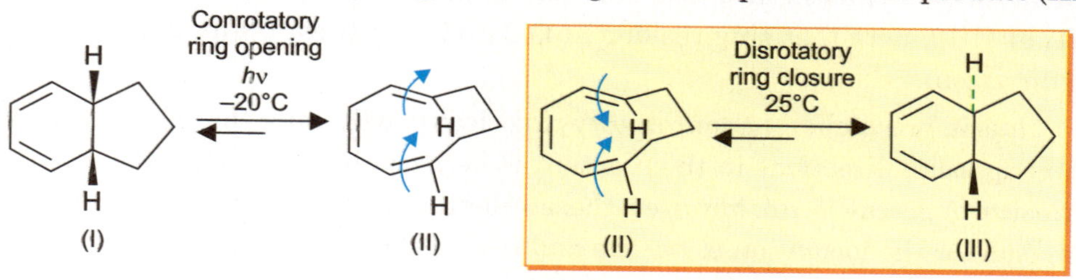

SCHEME 8.31

Example 4: 1,3-Cycloheptadiene (I, Scheme 8.32) closes to the cyclobutene by a disrotatory motion under photochemical conditions with expected *cis* ring fusion. A conrotatory opening of cyclobutene (II) is thermally allowed, since the hydrogens in (II, Scheme 8.32) are *cis*, these must point in the same direction in the ring opened product (III) *i.e.*, one of the double bonds in III must be *cis* and the other must be *trans*. However, a *trans*-double bond cannot be accommodated in a seven membered ring and thus (II, Scheme 8.32) is stable under thermal conditions. Similar arguments prove that compound (IV, Scheme 8.32) does not undergo ring opening under thermal conditions while (V) does.

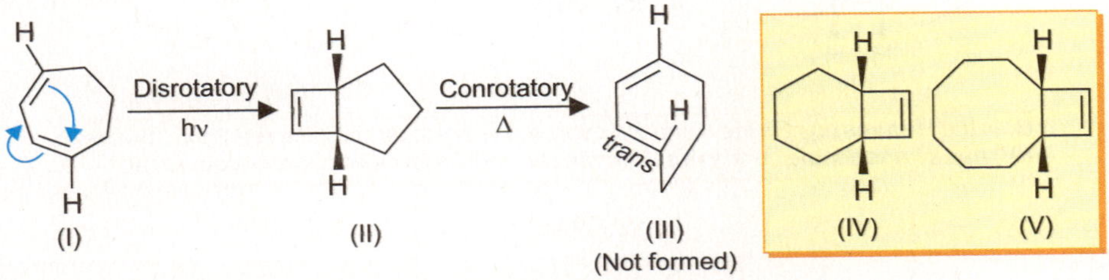

SCHEME 8.32

EXERCISE 8.6

Predict if the conversion shown (Scheme 8.33) is allowed or forbidden?

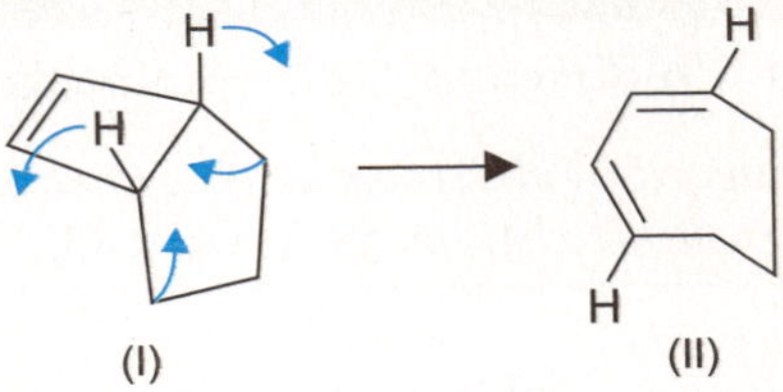

SCHEME 8.33

ANSWER. *It is a concerted 4n electron reaction. The ring opening shall have to be conrotatory (Table 8.1). Thermal process is allowed but the product is strained and is therefore, not formed (see Scheme 8.32). However, the conversion (Scheme 8.33) to unstrained all cis-diene is a disrotatory process as shown and is forbidden by the selection rules (Table 8.1).*

EXERCISE 8.7

One of the cyclobutane (Scheme 8.34) on heating reacts very fast while the other reacts at an extremely slow rate and at much higher temperature. Explain.

ANSWER. *Recall problem solving hints 1–4. Compound (I) reacts faster.*

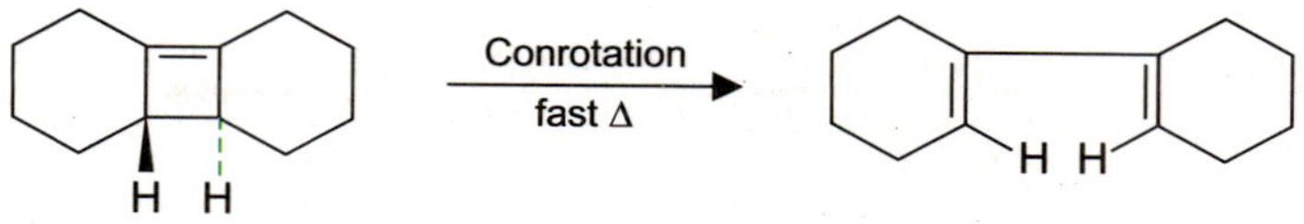

(4*n*) Electron system (I) (III)

Both the H atoms point in opposite directions, these are to be trans in the reactant.

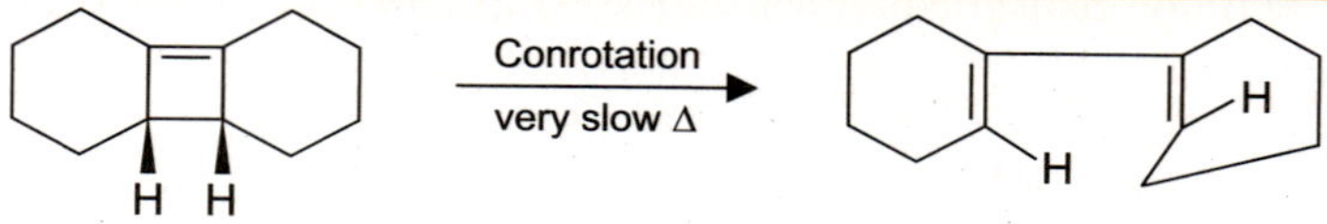

(II) (IV)

Both the H atoms point in the same direction, these are to be cis in the reactant, however (IV) has a trans double bond in a six membered ring. Its formation is very slow if at all.

SCHEME 8.34

The compound (II) can however, undergo an easy photochemical opening to the butadiene (III see Problem 8.9), involving disrotation. One may recall that e.g., on ring closure a butadiene under thermal or photochemical conditions gives cyclobutenes with opposite configurations. (see, Scheme 8.20a)

EXERCISE 8.8

How one can convert trans-5, 6-dimethyl-1, 3-cyclohexadiene into its cis isomer?

ANSWER. *This may be achieved by electrocyclic ring opening followed by electrocyclic ring closure under appropriate conditions (Scheme 8.35).*

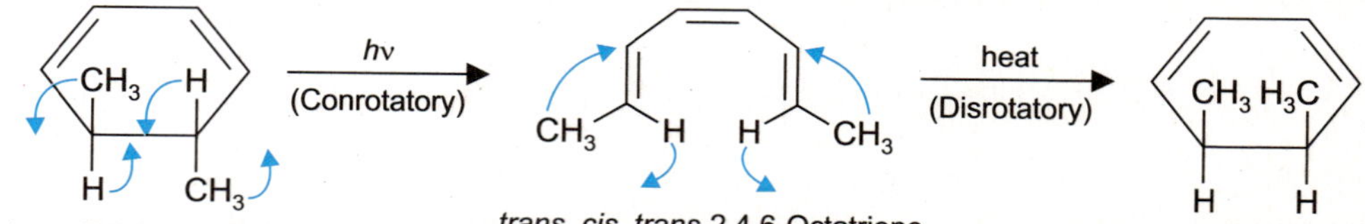

trans-5,6-Dimethyl-1, 3-cyclohexadiene — *trans, cis, trans*-2,4,6-Octatriene — *cis*-5,6-Dimethyl-1, 3-cyclohexadiene

SCHEME 8.35

Example 5: Dewar benzene has been synthesized and is stable at room temperature (at 25°C, the half life for its conversion to benzene is two days). It is much less stable than benzene due to angle strain and no stabilization due to aromaticity. Dewar benzene could therefore, easily isomerize to benzene. However, this conversion is an electrocyclic reaction (Scheme 8.36), it involves two pair of electrons one pair of π electrons and one pair of sigma electrons of the Dewar benzene, the third pair of electrons is located in exactly the same place both in the reactant and the product and is thus not involved in the reaction).

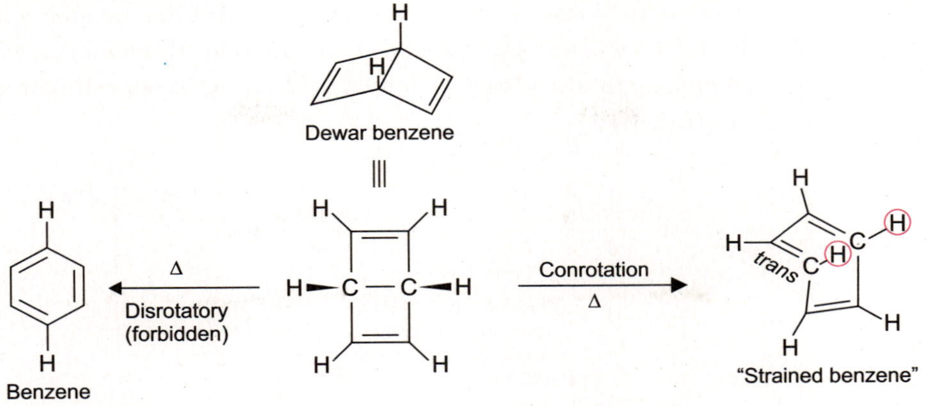

SCHEME 8.36

A thermally allowed electrocyclic reaction with two pairs of electrons ($4n$) π electron system must be conrotatory and the opening of the cyclobutene ring in Dewar-benezene (Scheme 8.37) would result in strained isomer of benzene ("strained benzene") with a *trans* double bond in six membered ring. (In strained benzene two hydrogens on the newly created diene moiety which point to the same side are circled. Thus the otherwise thermally allowed conversion of Dewar benzene into benzene is geometrically impossible since low energy pathway for this conversion is not available.

Example 6: The tetraene (Scheme 8.37) has an even number of π bonds ($4n$) π electron system and therefore, under thermal conditions (– 10°C) it will undergo conrotatory ring closure. Since the two methyl substituents on the tetraene point to opposite directions these will be *trans* in the ring closed cycloctatriene. In cycloctatriene one now has three double bonds in conjugation (an odd number, $4n + 2$ system) and therefore, the second thermal ring closure will now be disrotatory. Since the hydrogen substituents at the ends of the triene system (which are not drawn) are in opposite directions (Compare with Scheme 8.30), these must be *cis* in the final bicyclo product.

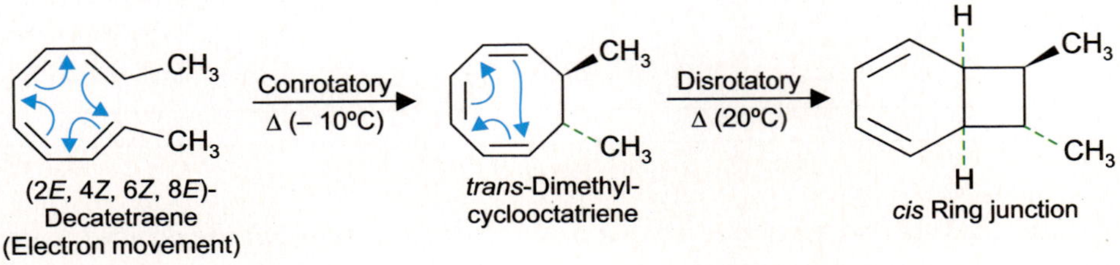

SCHEME 8.37

Example 7: A remarkable distinction between photochemical and thermal reactions is displayed by the known all *cis*-cyclodecapentaene (Scheme 8.38). One may consider its reacting system with three conjugated π bonds. A conrotatory ring closure under photochemical conditions with hydrogen substituents at the end of the considered triene system in opposite directions will end up in a *trans* ring junction. Thermal ring closure of a three π bond system is disrotatory, and with hydrogen atoms in opposite directions will end up in a *cis* ring junction.

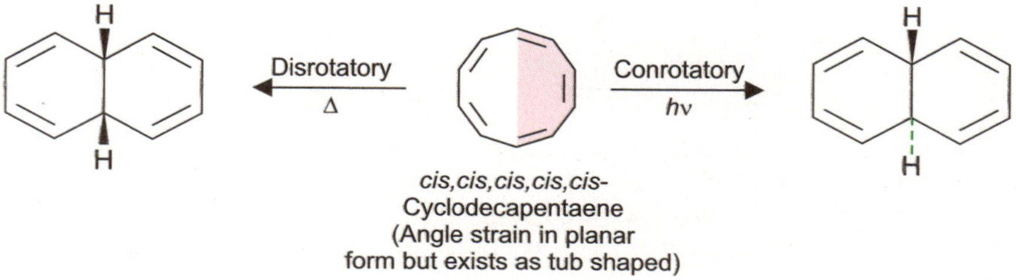

SCHEME 8.38

8.4 CYCLOADDITIONS (FMO-APPROACH)

The reactions of alkenes (the dienophiles) and polyenes (conjugated dienes) in which two molecules react to form a cyclic product, with π electrons being used to form two new σ bonds are called cycloaddition reactions. These reactions are classified on the basis of π electrons involved, in each component, the [4 + 2] cycloaddition being the well known Diels-Alder reaction (Scheme 8.39). The reaction of two alkenes to form a cyclobutane derivative is termed a [2 + 2] cycloaddition reaction (Scheme 8.39). A cycloaddition reaction requires only heat or light for initiation, radical and ionic intermediates are not involved.

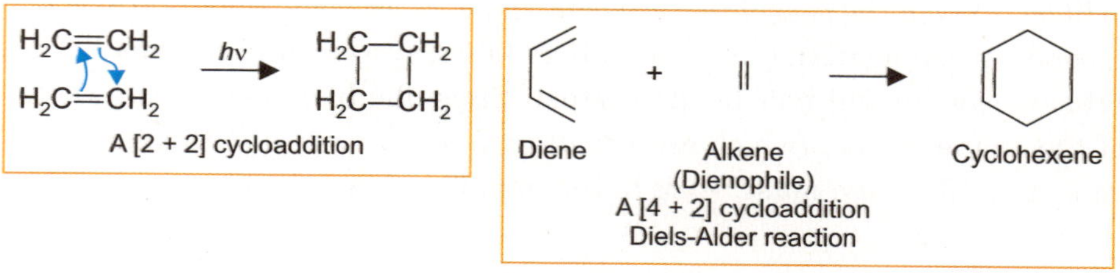

SCHEME 8.39

Molecular Orbitals of Ethylene

On heating ethylene its π electrons are not promoted, but remain in the ground state ψ_1. Irradiation with UV light excites an electron from ψ_1, the ground-state HOMO, to ψ_2^*, which becomes the excited-state HOMO. Interaction between the excited-state HOMO of one alkene and the LUMO of the second alkene indicates that a photochemical [2 + 2] cycloaddition reaction can occur by a suprafacial pathway (Scheme 8.40).

(Contd...)

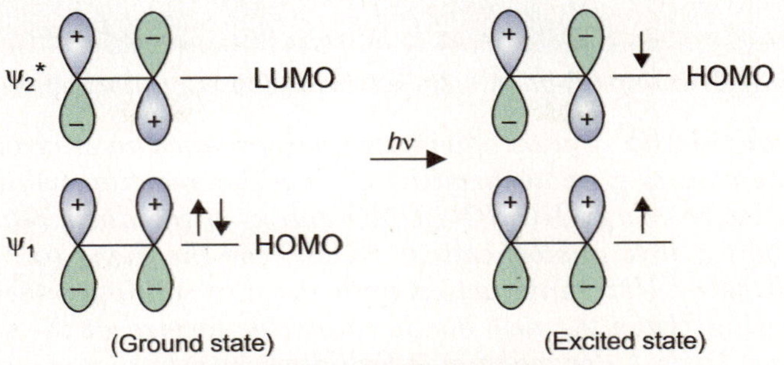

Molecular orbitals of ethylene

SCHEME 8.40

A. Diels-Alder Reaction—[4 + 2] Cycloadditions

These are concerted, thermal [4 + 2] cycloadditions. A consideration of orbital interactions (two combinations) accounts for this (Scheme 8.41), *i.e.*, the overlap can take place between the HOMO of one component and the LUMO of the other and *vice versa*.

As with electrocyclic reactions in cycloadditions as well, one is only concerned with the terminal lobes. The simplest [4 + 2] system involves the cycloaddition of 1, 3-butadiene (the diene) and ethylene (the dienophile) which is a thermally induced reaction. This thermally allowed reaction involves *e.g.*, the HOMO of 1, 3-butadiene (ψ_2) with the LUMO of ethylene ψ_2* (one could equally well use the diene LUMO and the alkene HOMO).

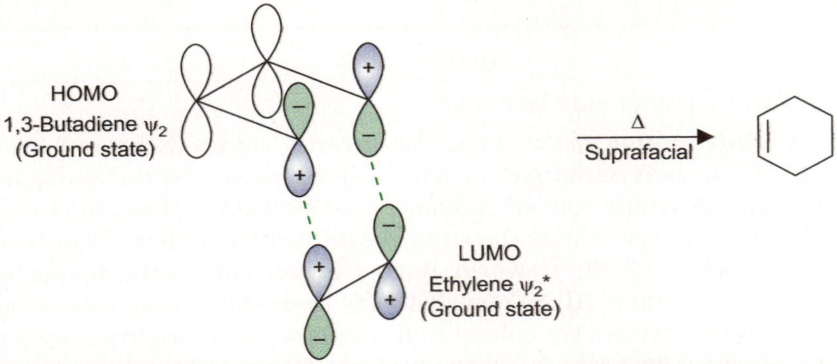

Symmetry-allowed thermal [4 + 2] cycloaddition : 1,3-butadiene and ethylene

SCHEME 8.41

In either case, the overlap brings together lobes of the same phase. Addition to the lobes on the same side of a π system is called suprafacial addition, while addition to lobes on opposite sides of a π system is termed antarafacial addition (for an example of antarafacial addition see Scheme 8.57). These modes of addition are identified by the symbols *s* and *a* respectively. Thus cycloaddition of two π bonds each reacting suprafacially would be called [$\pi^2s + \pi^2s$] reaction.

EXERCISE 8.9

Explain by orbital drawings that [4 + 2] cycloaddition is photochemically forbidden.

ANSWER. *Usually the absorption of a photon will promote an electron from HOMO to LUMO. In the case of a photochemical Diels-Alder reaction (which is the most uncommon) the lower energy HOMO–LUMO gap is in the diene partner. Thus on absorption of light a new photochemical HOMO for the diene (ψ_3*) is generated and now the HOMO–LUMO interaction with the dienophile partner involves one antibonding overlap. Thus the new bonds cannot be formed at the same time and the photochemical Diels-Alder reaction is forbidden by orbital symmetry (I, Scheme 8.42). However, one may note that photoinduced [4 + 2] cycloaddition cannot occur if either the diene or the dienophile is excited (II, Scheme 8.42).*

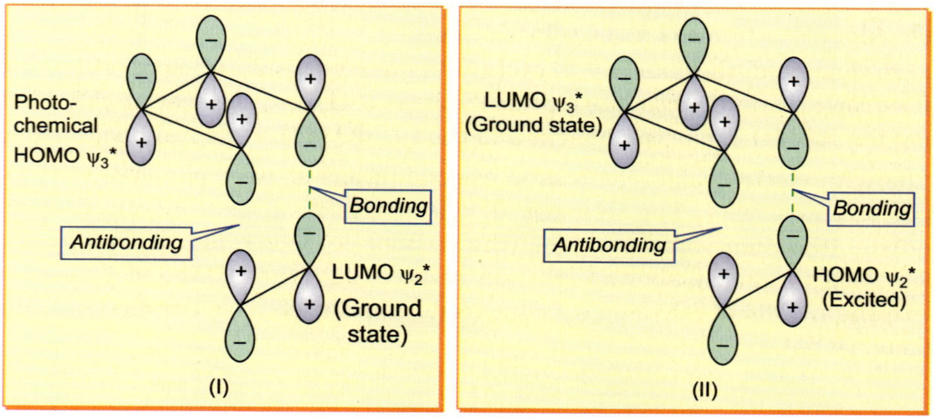

SCHEME 8.42

The following points may be noted:

- *s-cis Conformation of the Diene.* As correctly shown (Scheme 8.39) the diene reacts in the *s-cis* conformation, which allows the ends of the conjugated system to reach the doubly bonded carbons of the dienophile. That the *s-cis* geometry of the diene is essential is shown by the unreactive nature of the fixed *transoid* diene (I Scheme 8.43). However, dienes in which one of the double bonds is *cis* are poor substrates (II, Scheme 8.43) for Diels–Alder reactions because steric interactions between the substituents in the *s-cis* conformation are particularly severe, and dienes whose *s-trans* conformation is enforced do not undergo the Diels–Alder reaction.

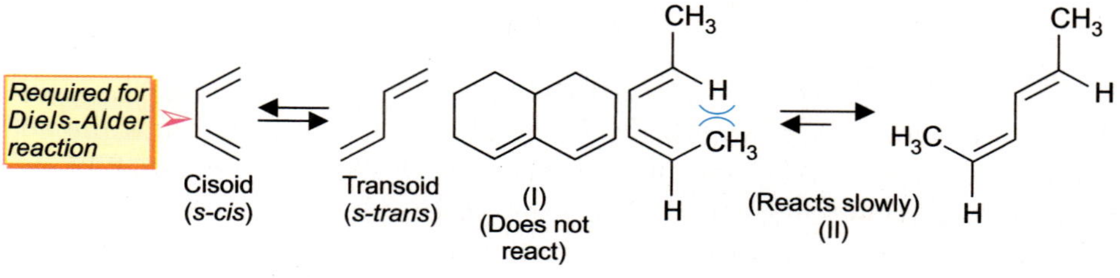

SCHEME 8.43

Cyclic dienes are among the useful dienes and particularly reactive in Diels-Alder reactions as the two double bonds are held in the *s-cis* conformation in five or six membered rings. Cyclopentadiene is highly reactive and forms a Diels-Alder adduct with itself. On heating, the commercially available dimer undergoes a *retro-Diels-Alder* reaction (The term *retro* means the reverse) to give cyclopentadiene (Scheme 8.44).

Another good diene *o*-xylylene can be generated by E2′ reaction when a π bond is formed between the two C atoms at which bond breaking occurs (Scheme 8.44). The base is F^-, and a Me_3Si group replaces the usual H.

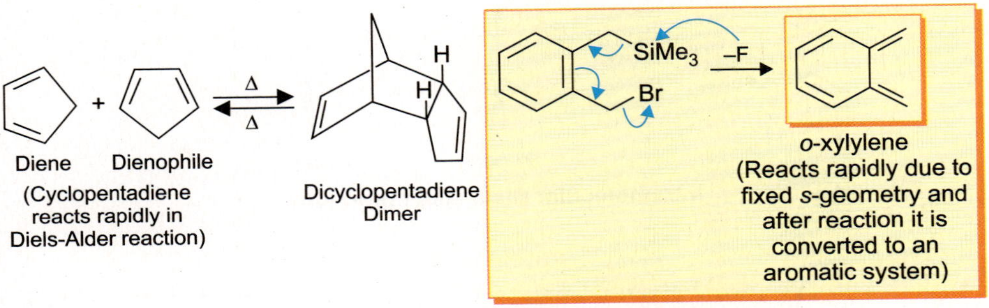

SCHEME 8.44

- When in the dienophile there is conjugation to a group of $-M$ type *e.g.*, carbonyl, nitro etc. the reaction occurs under milder conditions and gives good yields. The substituent lowers the energy of the LUMO of the dienophile so as to bring it closer in energy to the HOMO of the diene. Consequently the bonding interaction in the transition state increases. As expected, the reactivity is also increased by an electron releasing group in the diene. Conversely, when the diene contains an electron-withdrawing substituent the dienophile requires an electron-releasing substituent for ready reaction. In this situation the interaction is between diene's LUMO and the dienophile's HOMO. Thus, the bonding at the transition state is more effective when the HOMO of one reactant and the LUMO of other are more closely matched in energy.
- Thus many efficient Diels–Alder reactions occur when an *electron-rich diene* reacts with an *electron-poor dienophile.* The dienophile may be substituted with electron withdrawing group (carbonyl, CN, sulfonyl, NO_2). Dienophiles substituted *e.g.*, with two electron-withdrawing groups (diethyl fumarate, maleic anhydride, benzoquinone) are efficient substrates for the Diels-Alder reaction. The rate of a Diels-Alder reaction can be increased by making it intramolecular, which makes it possible to make bicyclic compounds, the activating effect of a carbonyl group conjugated to the dienophile is obvious (Scheme 8.45). The reaction becomes facile and favours the *cis*-product.

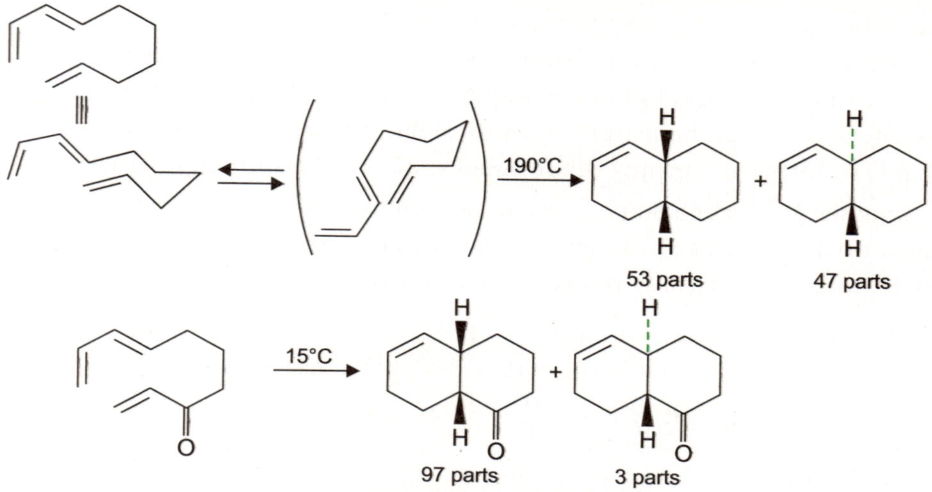

Intramolecular Diels-Alder reactions

SCHEME 8.45

Dienophiles having strained double bonds (benzyne, norbornadiene) are effective although these do not have any electron withdrawing group.

Among other reactive dienophiles are cyclic alkenes and alkynes with angular strain. The driving force for these reactions is lowering of angular strain associated with the transition state for the addition. Thus, cyclopropene reacts rapidly and stereoselectively with cyclopentadiene at room temperature to form the *endo* adduct in high yield, and butadiene gives norcarene in lower yield (Scheme 8.45*a*).

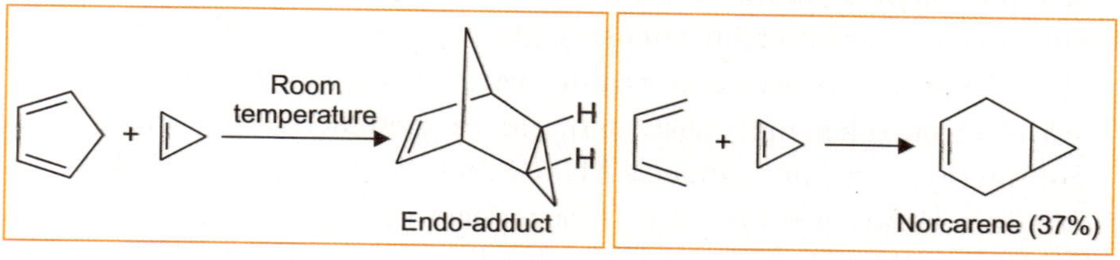

SCHEME 8.45a

- *syn-Stereochemistry.* That the Diels-Alder reaction is concerted (both the new bonds are formed in the same transition state) is shown by the fact, that it proceeds with retention of configuration of both the diene and the dienophile (Schemes 8.46 and 8.47 also see, Scheme 1.3) *i.e.*, it proceeds stereoselectively *syn* with respect to both the diene and the dienophile as expected of a concerted (supra, supra) mode of addition. One may note that if in a diene both groups *e.g.*, methyls are "outside", these ends up *cis* in the product (Scheme 8.47) and if one methyl is "inside" and one "outside" these end up *trans* in the product.

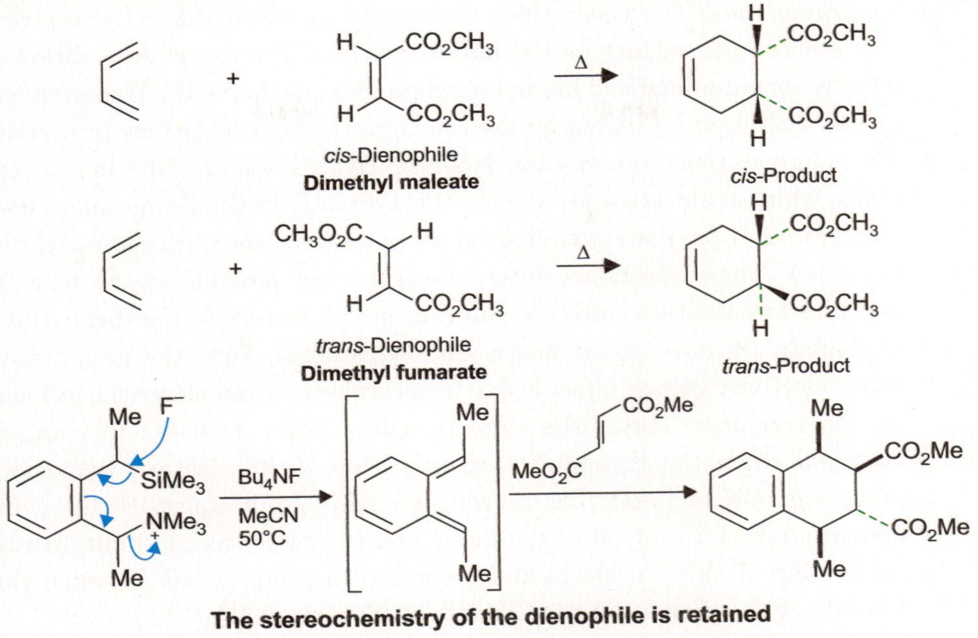

SCHEME 8.46

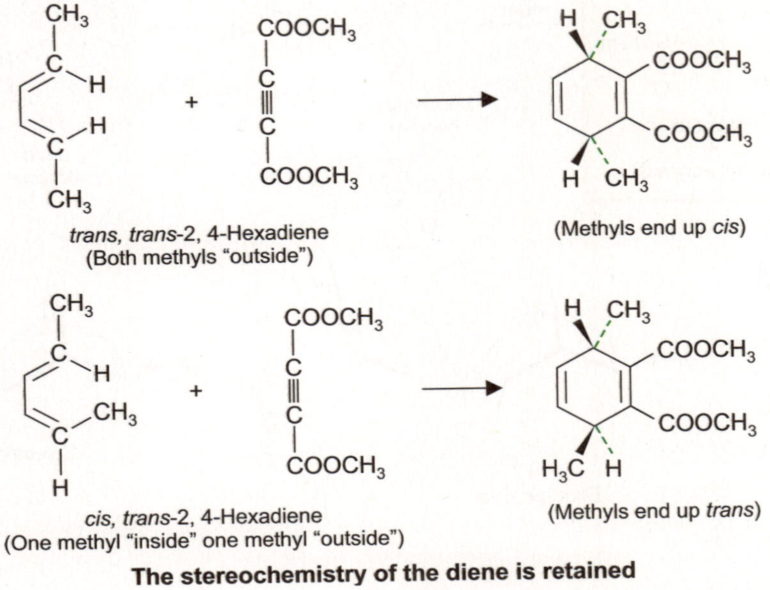

SCHEME 8.47

- *The Endo Rule.* The Diels-Alder reaction takes place generally to give the less stable *endo* adduct as the major product. For the *endo* addition *e.g.,* with cyclopentadiene and methyl acrylate (Scheme 8.48), the transition state can be stabilized (speeding up the reaction) through secondary interactions. These interactions involve the lobes of HOMO and LUMO of the same phase which themselves are not involved directly in the formation of bonds. One sees that for the *endo* addition the π-system lies more completely over the other. These secondary interactions are not possible in the transition state for *exo* addition since the relevant set of centers in the diene and the dienophile are now too far apart, from each other. Thus the preference for *endo* selectivity (which infact is due to several steric and electronic influences on the transition state—the *endo* transition state is lower in energy) is observed when the dienophile has a π bond in its electron withdrawing group *e.g.,* CN or C=O. The *p* orbitals of this group approach the central carbon atoms C2 and C3 of the diene and the resulting proximity leads to an overlap of the *p* orbitals and secondary overlap effects between the *p* orbitals of the diene and the dienophile (Scheme 8.48).

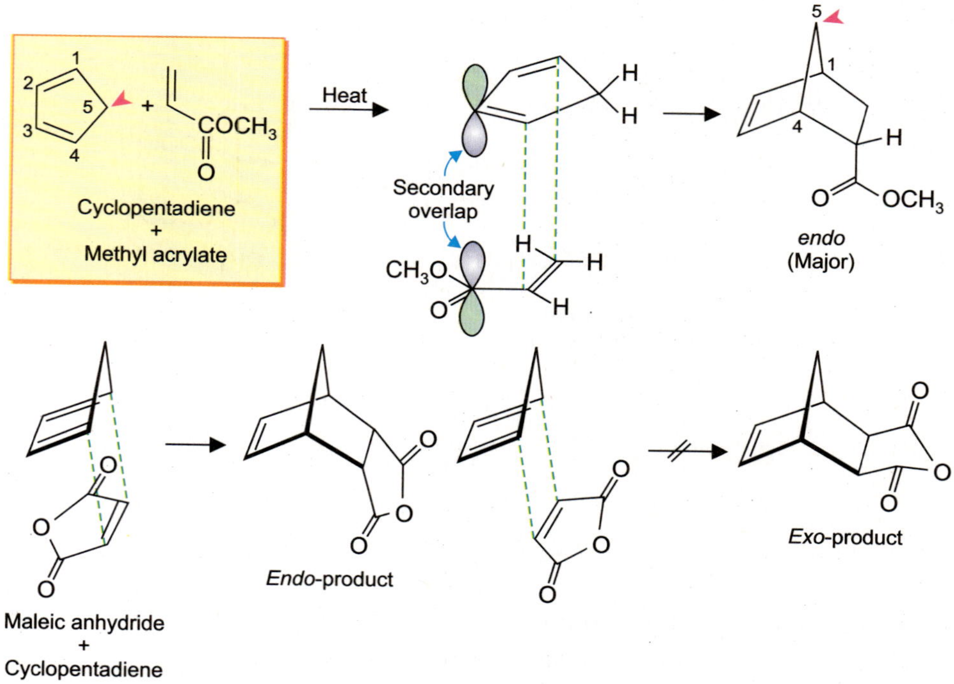

SCHEME 8.48

- *The endo Rule to Predict General Stereochemical Outcome.* One can use the *endo* rule to predict the stereochemical outcome of a reaction as detailed (Scheme 8.49). One can imagine the "inside" ligands of a diene to be the "CH_2" of cyclopentadiene and these will have a *cis*-relationship and shown on thick wedges. In keeping with the *endo* product formation when the product is a cyclohexene derivative the group on the dienophile (which is electron withdrawing) will be down (on a dotted wedge) due to it being inside the pocket of ring as a result of stability of the transition state (secondary overlap).

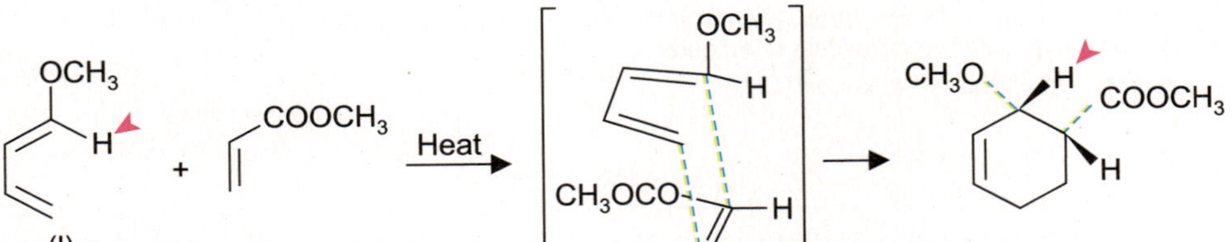

(I)

In (I) the H at C-1 ends "up" in the product (thick wedge) just like C5 methylene of cyclopentadiene (see, scheme 8.48) on its reaction with methyl acrylate. Based on endo rule $COOCH_3$ is below the

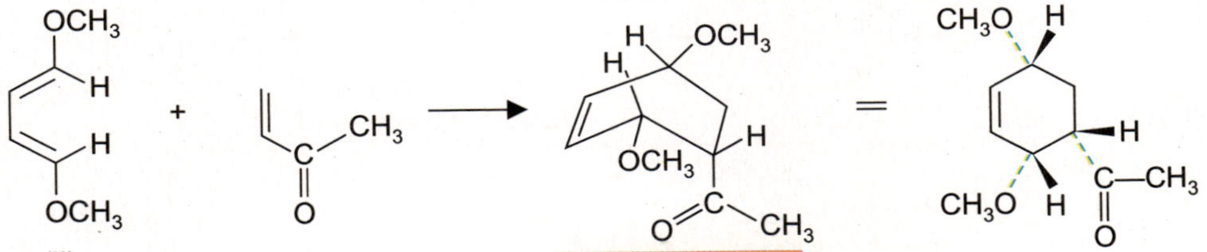

(II)

Relation with endo product formation

In (II) since both OCH_3 groups are outside these end up cis in the product. As in (I) both the inside H atoms end up on thick wedges. The $COCH_3$ group

SCHEME 8.49

Stereochemistry Solving Hint

Considering Schemes 8.47 and 8.49 one can evolve an useful stereochemistry solving hint of syn addition of concerted Diels-Alder reaction.

Two inner substituents always end up with a cis relationship in the product, and same is the case with two outer substituents. An inner substituent on one carbon ends up always trans to an outer substituent on the other.

When a ring junction is created then as expected the stereochemistry at the ring junction must be cis for a syn addition. Thus when maleic anhydride adds to trans, trans-2, 4-hexadine two diastereomeric syn addition products are possible (Scheme 8.49a).

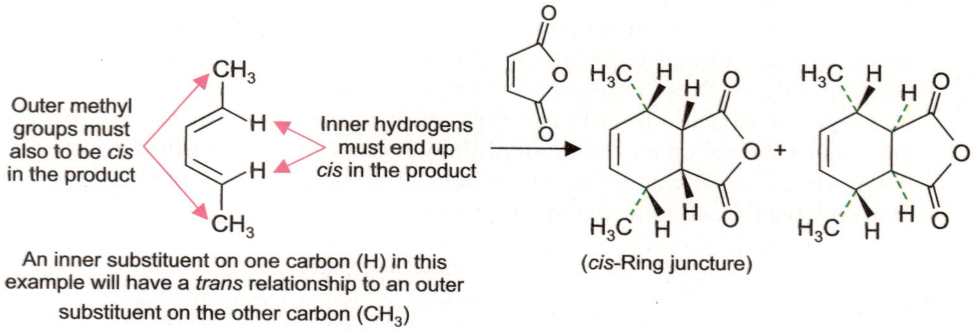

SCHEME 8.49a

- *Regioselectivity.* Cycloaddition of an unsymmetrically substituted diene and an unsymmetrically substituted dienophile can lead to regioisomers (Scheme 8.50).

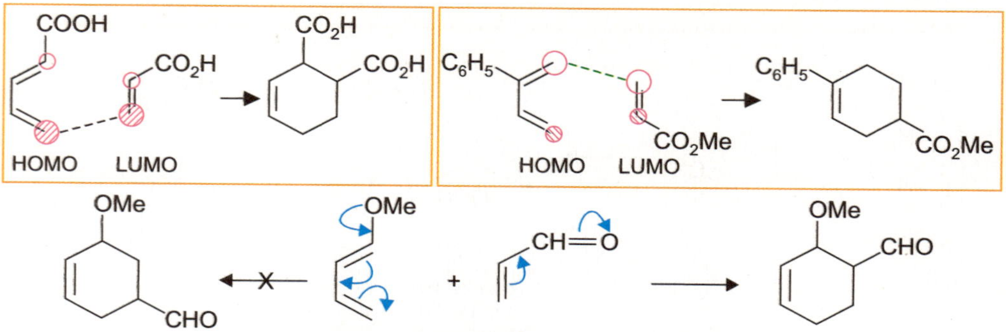

Regioselectivity in the Diels-Alder reaction

SCHEME 8.50

One has already seen that in a normal Diels-Alder reaction *i.e.,* between an electron rich diene and electron-deficient dienophile, the main interaction is between the HOMO of the diene and LUMO of the dienophile (In this situation these orbitals are more closely matched in energy, the better is the overlap and thus the reaction occurs more readily). However, the orientation of the product from an unsymmetrical diene and an unsymmetrical dienophile depends mostly on the atomic orbital coefficients at the reacting termini. The atoms with the larger terminal coefficients on each reactant, bond preferentially in the transition state, because of better orbital overlap. Consequently with 1-substituted butadienes the major product is 1,2 ("*ortho*") adduct while with 2-substituted butadienes, the major adduct is 1,4 ("*para*").

In several cases the regioselectivity can be determined by considering electrophilic and nucleophilic centers of the two components by looking at the curved arrows (Scheme 8.50).

In the case of butadiene-1 carboxylic acid and acrylic acid the frontier orbitals are polarized as shown (Scheme 8.50). The size of the circles as shown is roughly proportional to the size of the coefficients and an allowed reaction leads to 1,2-adduct. Similarly, now with 2-phenyl-butadiene and methyl acrylate the major product formed would be 1,4.

- *Lewis Acid Catalysts.* Some Diels-Alder reactions are catalyzed by Lewis acid catalysts. These catalysts form complexes with the polar groups on the dienophile which lower the energies of the frontier orbitals of the dienophile. Consequently, the energy difference between the HOMO of the diene and the LUMO of dienophile is reduced and the reaction becomes faster.

EXERCISE 8.10

Write the structure of the products from the reactions (Scheme 8.50a).

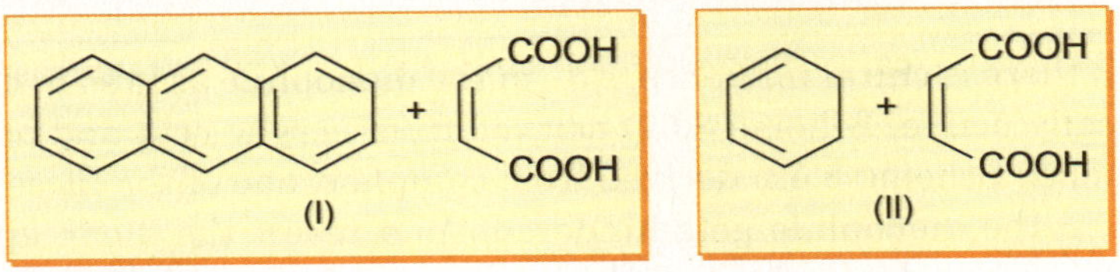

SCHEME 8.50a

ANSWER.

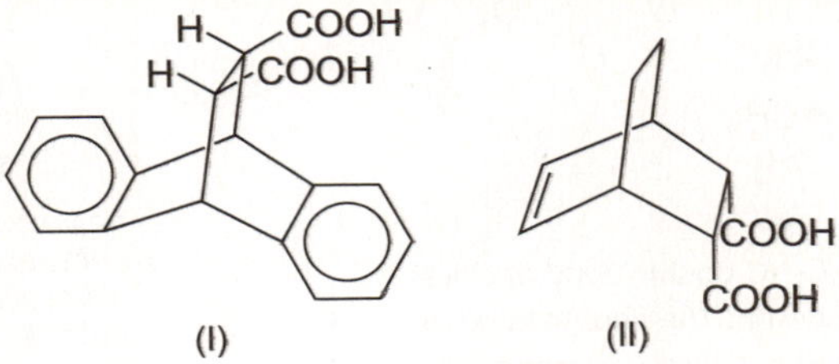

SCHEME 8.50b

(B) Asymmetric Diels-Alder Reaction

In asymmetric Diels-Alder reaction generally camphor based chiral auxiliaries are employed and this is due to the reason that both geometrically possible stereoisomeric forms of camphor are readily available. The following points may be noted:

- In the case of a monosubstituted alkene there are two enantiotopic faces *Re* or *Si*.
- A diene *e.g.,* buta-1,3-diene thus can add either to *Re* or *Si* face of a dienophile like acrylic acid to give a cyclohexene in which the stereocenter generated via this Diels-Alder reaction could be either *S* (*Re* approach of the diene) or *R* (*Si* approach of the diene, Scheme 8.51).
- Diels-Alder reaction, therefore, creates stereocenters and when both reactants are achiral and no other chiral influence is there, racemic mixtures are obtained.

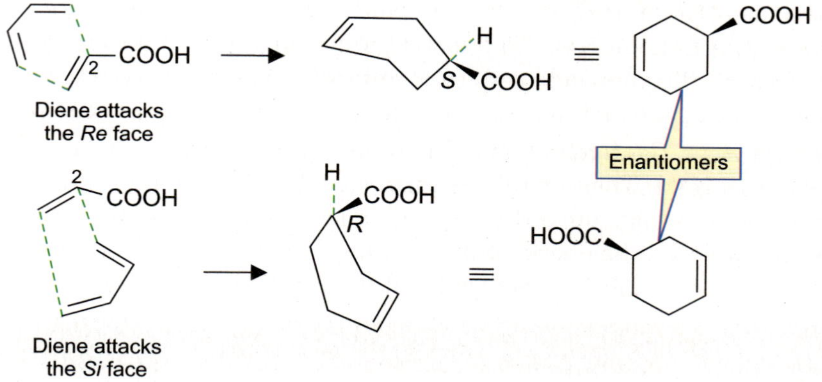

SCHEME 8.51

- When there is chiral influence, *e.g.*, in the dienophile as is so in chiral acrylate (optically active, Scheme 8.52) derived from acrylic acid and optically active alcohol (I, Scheme 8.52) derived from camphor, one of the faces of the double bond in the dienophile gets hindered. As a result the diene approaches the carbon-carbon double bond of the acrylate preferentially from one direction.

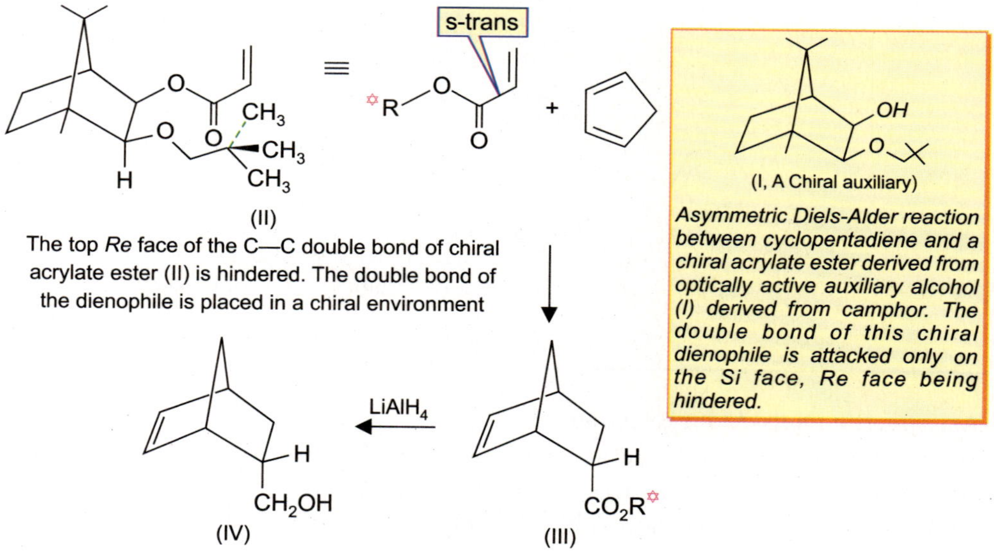

SCHEME 8.52

- The top *Re* face of the carbon-carbon double bond of the dienophile is hindered by the *t*-butyl group of neopentyl unit. This forces the addition to occur preferentially from the *Si*-face (back of the double bond to give (III) almost exclusively, to give only one of the enantiomers. The reaction is thought to occur in *s*-trans conformation in (II, Scheme 8.52).

- One knows that Diels-Alder cycloadditions follow the *endo*-rule.
- Reduction of the ester with lithium aluminium hydride gives the product (IV) in an optically pure form and regenerates the camphor derived chiral auxiliary.

Another example is in (Scheme 8.53). The dienophile (*E*)-methyl crotonate becomes chiral when optically active alkyldichloroborane (I) complexes with it to yield (II). Now the approach of diene from the rear face of the dienophile is blocked (naphthyl group). Attack occurs from front to give (IV) in optically pure state (*trans*-geometry of dienophile is preserved in IV).

(C) Hydrophobic Effects

It has been shown that some intermolecular Diels-Alder reactions are accelerated under hydrophobic effects in aqueous media. This was the case when cyclopentadiene reacted with methylvinyl ketone (Scheme 8.54) and it was observed that any additive which increased the hydrophobic effect also increased the rate *e.g.*, lithium chloride increases the hydrophobic effect by salting out nonpolar material.

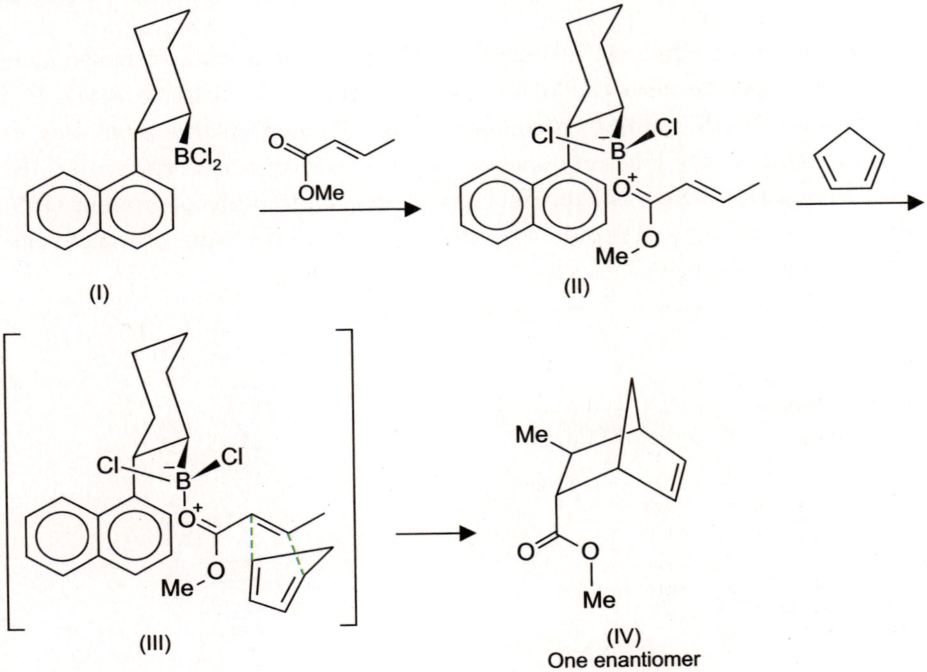

SCHEME 8.53

Hydrophilic and hydrophobic effects are water attractive and water repellant respectively. Soaps e.g., sodium oleate have a hydrophilic site (COO⁻) and a hydrophobic site (the hydrocarbon chain).

β-Cyclodextrin has a hydrophobic cavity and if the system of a particular Diels-Alder combination can fit within the cavity a significant rate enhancement is observed.

This is found in the case of methyl vinyl ketone and cyclopentadiene system in aqueous medium with β-cyclodextrin as additive. α-Cyclodextrin, however, has a smaller cavity which is not able to accommodate the reactive species and the rate is significantly diminished in its presence compared to that in β-cyclodextrin.

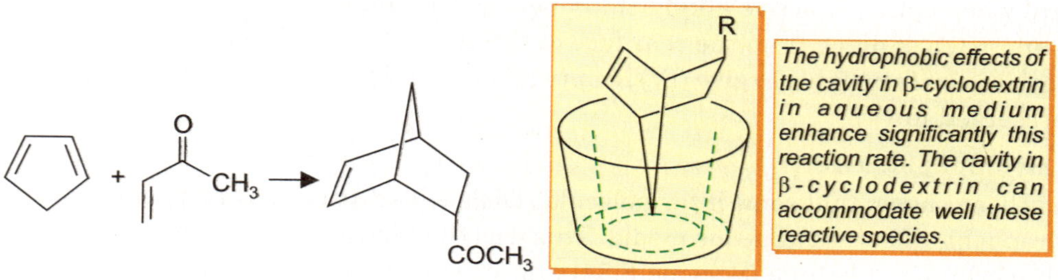

SCHEME 8.54

(D) [2+2] Cycloadditions

In the dimerization of ethylene, a thermal [2+2] cyclization would involve overlap of HOMO, of one molecule with the LUMO, of the other (see, Scheme 8.41). If in this concerted reaction both bonds to a component are formed on the same face *i.e.,* the process is suprafacial, the lobes of opposite phase would approach each other (Scheme 8.55). This interaction which is suprafacial with respect to both components $[\pi^{2s} + \pi^{2a}]$ is therefore, antibonding and repulsive and the concerted reaction, does not take place (symmetry forbidden process).

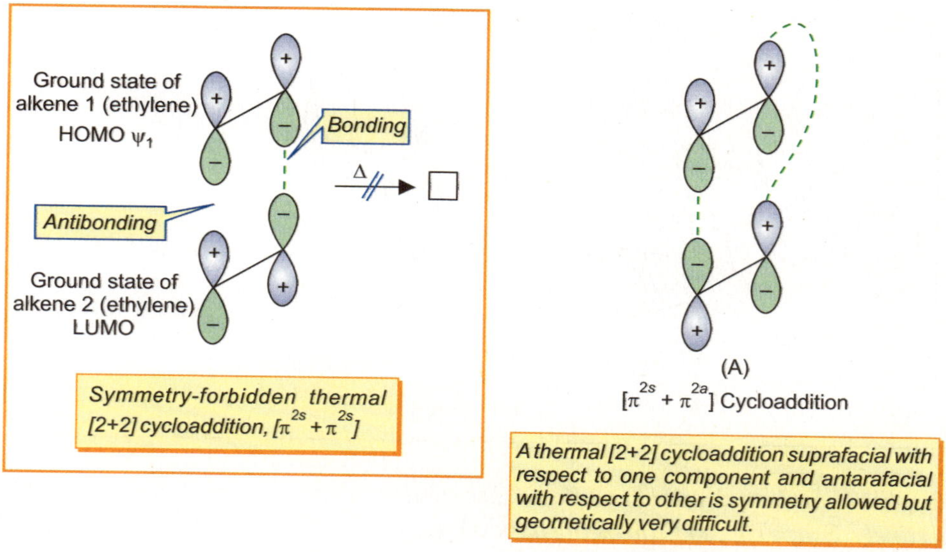

SCHEME 8.55

The photochemical [2+2] cycloadditions which are suprafacial with respect to both the components [$\pi^{2s} + \pi^{2a}$] will, however, permit a previously forbidden reaction to become a symmetry allowed process. During [2 + 2] cycloadditions, irradiation of an alkene with UV light excites an electron from ψ_1, the ground state HOMO to ψ_2* which now becomes the excited state HOMO. The interaction between the excited state HOMO of one alkene and LUMO of the second alkene is now a symmetry allowed process (Scheme 8.56).

The stereochemistry of the Diels-Alder reaction reveals that these are also $\pi^{4s} + \pi^{2s}$ processes. However, a thermal [2+2] cycloaddition could occur provided it is suprafacial with respect to one component and antarafacial with respect to the other *i.e.*, it is $\pi^{2s} + \pi^{2a}$ (A, Scheme 8.55). This process, through symmetry allowed is geometrically very difficult.

Thus the photochemical [2+2] cycloaddition reaction occurs smoothly and represents one of the best techniques to synthezise cyclobutane rings and cage compounds (Scheme 8.57). The CO double bond of an aldehyde or a ketone can act as one component in [2+2] cycloaddition with an alkene to form an oxetane (Scheme 8.57). The reaction can occur inter-or intramolecularly.

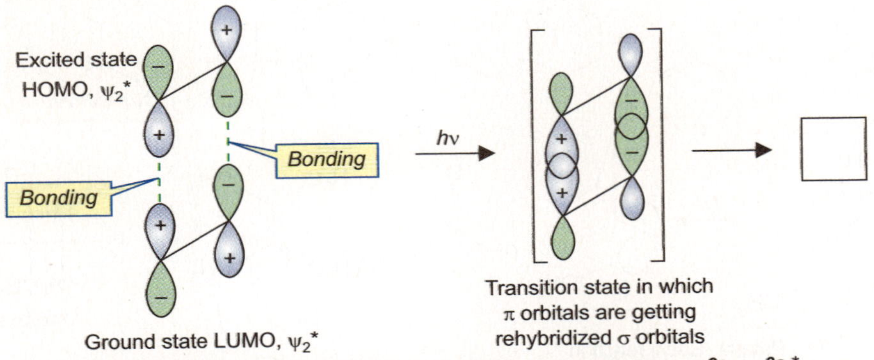

Symmetry-allowed photochemical [2+2] cycloaddition [$\pi^{2s} + \pi^{2a}$]*

SCHEME 8.56

Both thermal as well as photochemical cycloaddition reactions take place by opposite stereochemical pathways. As with electrocyclic reactions one can categorize cycloadditions according to the total number of electron pairs (double bonds) taking part in the rearrangement. Thus, a Diels-Alder [4+2] reaction between a diene and a dienophile involves an odd number (three) of electron pairs and takes place by a ground state (thermal) suprafacial pathway. A [2+2 thermal reaction between two alkenes involves an even number (two) of electron

Table 8.2: Cycloaddition Reactions Stereochemical Rules

Electron pairs (double bonds)	Thermal reaction	Photochemical reaction
Even number	Antarafacial	Suprafacial
Odd number	Suprafacial	Antarafacial

pairs and must takes place by an antarafacial pathway. However, it may be said that both suprafacial and antarafacial cycloaddition pathways are symmetry allowed. Only the geometric constraints inherent in twisting a conjugated *pi* electron system out of planarity make antarafacial reaction geometrically difficult in most of the cases. One may note that preferences for cycloadditions may be summarized further (Table 8.3) to quickly know the success of a particular cycloaddition and provides a rule of thumb.

Table 8.3: Cycloaddition Reactions

Number of Electron Pairs	Allowed Cycloaddition
Odd	Thermal
Even	Photochemical

As a last example of cycloaddition, cyclopentadiene reacts with cycloheptatriene system to give a product—which is [6+4] suprafacial process (Scheme 8.58, this would include an aromatic transition state with 10 electrons).

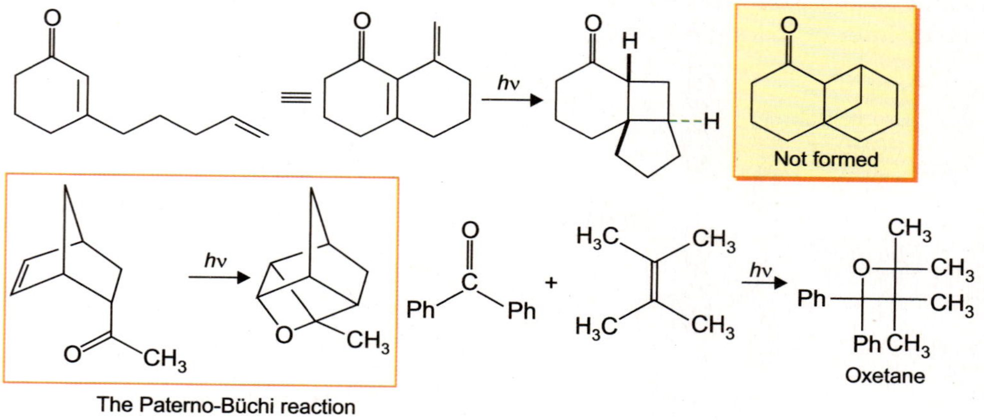

SCHEME 8.57

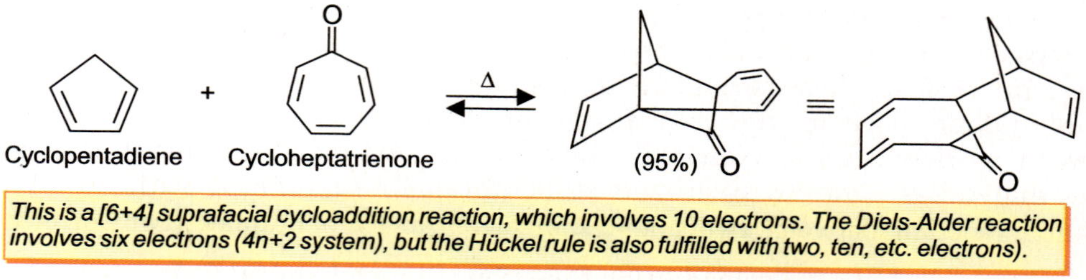

SCHEME 8.58

A general tendency for *exo*-type cyclisation to form five-membered ring is observed in case of β-(4-pentenyl) substituents. This involves initial bonding at the β-carbon of the enone and is termed "the rule of five".

EXERCISE 8.11

Indicate if the following reactions (Scheme 8.59) are allowed or forbidden.

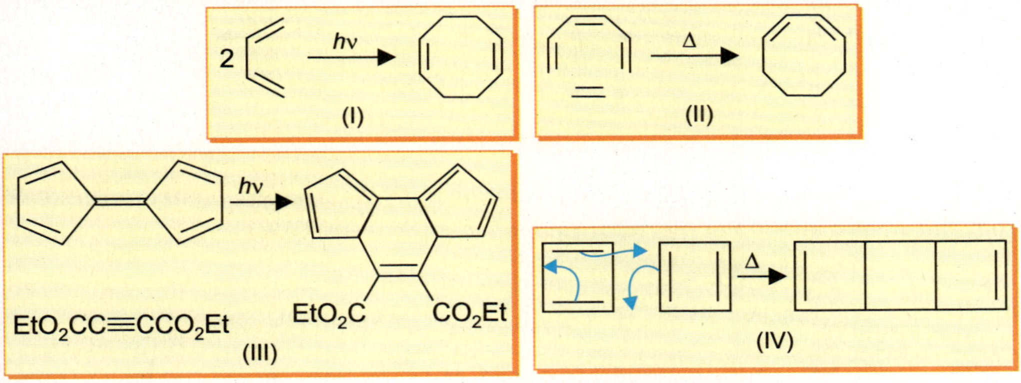

SCHEME 8.59

> **Hint.** *If in a reactant more π electrons are involved in the cyclization, the non-participating π electrons are not counted for the classification (see Table 8.3). A reaction involving odd number of electron pairs requires heat while a reaction involving even number of electron pairs requires light.*

ANSWER. *(I) A[4 + 4] cycloaddition, involves 4 electron pairs is photochemically allowed (see Table 8.3).*

(II) A[6 + 2] cycloaddition, involves 4 electron pairs is photochemically allowed and thermally forbidden.

(III) Since it is a [10 + 2] cycloaddition and involves 6 electron pairs, is photochemically allowed.

(IV) It is the allowed thermal dimerization [4 + 2].

EXERCISE 8.12

One of the reactions (Scheme 8.59a) requires heat and the other light. Which is which? Explain.

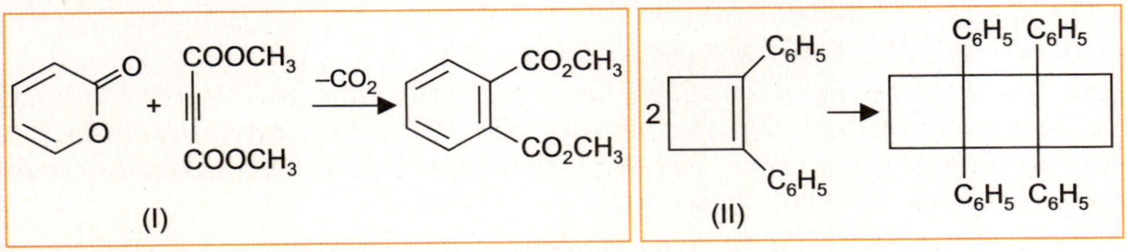

SCHEME 8.59a

ANSWER. *(I) This reaction is thermally allowed ; it being a [4 + 2] cycloaddition. The initially formed intermediate then undergoes a thermally allowed reverse Diels-Alder reaction (Scheme 8.59b).*

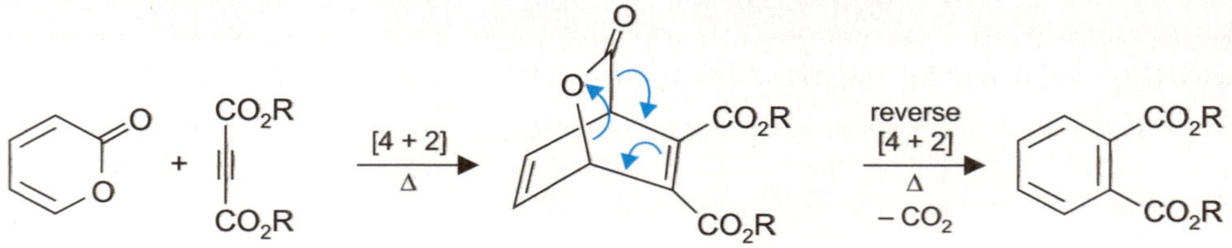

SCHEME 8.59b

(II) This is a photochemical [2 + 2] dimerization with the stereochemistry as dictated from end-to-end overlap of p-orbital components (Scheme 8.59c).

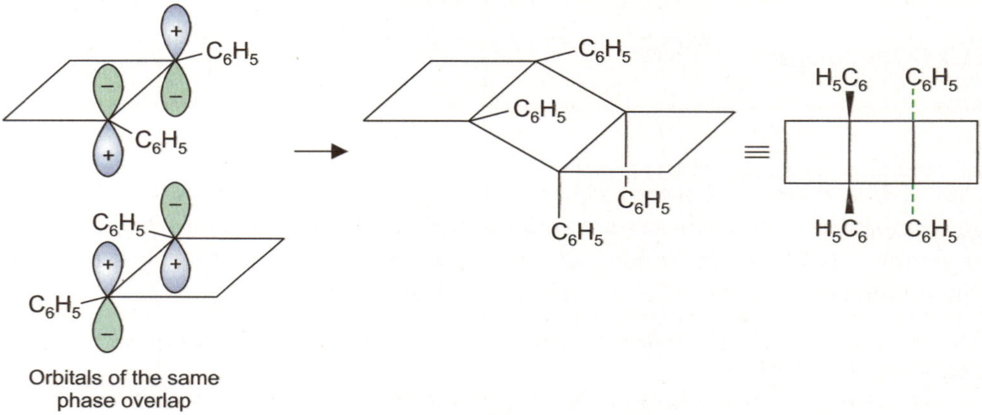

SCHEME 8.59c

BOX 8.1
THERMAL [2 + 2] CYCLOADDITION

Thermal [2 + 2] cycloaddition is not common due to antibonding interaction of HOMO-LUMO at one end (see, Scheme 8.55). Positive overlap between the orbitals at both terminal of the π systems can be obtained provided if one of the components reacts antarafacially. This orientation is geometrically very difficult and hence [2 + 2] cycloadditions do not normally undergo cycloaddition under thermal conditions.

The [2 + 2] cycloaddition of most alkenes and carbonyl compounds however, proceed only upon irradiation with light (see, Scheme 8.55).

Ketenes is a class of compounds which, however readily undergo [2 + 2] cycloaddition under thermal conditions. The two terminal of the C = C π bond of the ketene react from opposite faces of the π bond, and this leads to positive overlap between the orbitals at both terminal of the two π systems (Scheme B 8.1).

The antarafacial nature of the ketene has no stereochemical consequences, since there is no cis-trans relationship in the ketene to ends up in the product. The alkene component of the [2 + 2] cycloaddition with ketenes, however, reacts suprafacially, and its stereochemistry is retained in the product (see, Scheme B 8.2). Thus a ketene (and not the alkene) is able to react antarafacially. One may note that in a ketene one of the carbon atoms has an O atom on it which is sterically not demanding. In the case of alkenes sterically demanding substituents are present on both ends of the antarafacial component sterically inhibiting the reaction.

Thus ketenes and cumulenes with one or more sp hybridised carbon atoms which do not have a pair of interferring substituents at one of the reacting termini undergo [2 + 2] thermal cyclo-additions with olefins with ease.

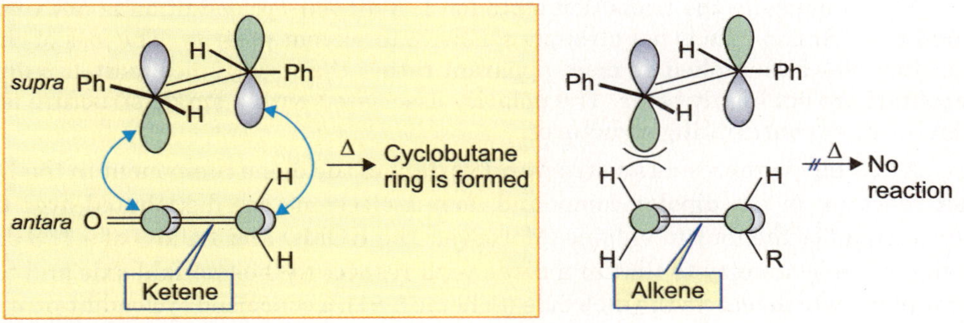

SCHEME B 8.1

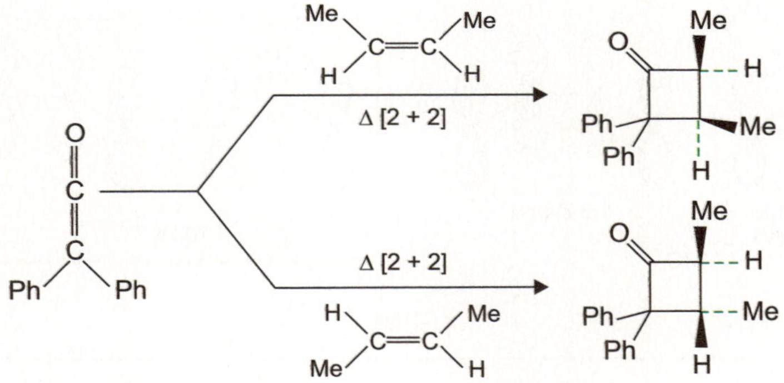

Ketene cycloaddition with an olefin is stereospecific

SCHEME B 8.2

The ketenes and cumulenes behave as antarafacial and olefins as suprafacial components.

Other cumulenes like isocyanates RN=C=O also undergo thermal [2 + 2] cycloadditions.

8.5 1,3-DIPOLAR CYCLOADDITIONS

These cycloadditions are analogous to the Diels-Alder reaction in that they are concerted [π^{4s} + π^{2s}] reactions. The 1,3-dipolar components are compounds whose representation requires ionic structures which include ones with charges on

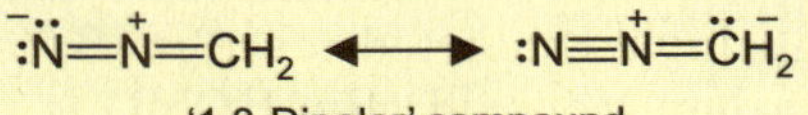

'1,3-Dipolar' compound diazomethane

SCHEME 8.60

atoms bearing 1,3-relationship, as in diazomethane (Scheme 8.60). These type of molecules which are called, 1,3-dipoles are isoelectronic with allyl anion. These have four π electron and each has at least one charge separated resonance structure with opposite charges in a 1,3 relationship. The other reactant (dipolarophile) in a dipolar cycloaddition has unsaturated bonds like, C=C, C ≡ C, C=O and C ≡ N. The 1,3-dipolar cycloadditions form useful reactions for the synthesis of five membered heterocyclic rings.

Mechanistically the transition state for 1, 3-dipolar cycloaddition is not very polar and the reaction rate is not strongly sensitive to solvent polarity. The loss of charge separation which is implied, is more apparent rather than real, since most 1, 3-dipolar compounds are not highly polar. The polarity associated with a single structure is balanced by other contributing structures.

A 1, 3-dipole represents a structural variant of the diene component in the Diels-Alder reaction; in the dipolar compound, four π-electrons are distributed over three atoms instead of the four in a diene. Moreover, the HOMO and LUMO of a 1, 3-dipole are similar in symmetry to that in a diene with respect to the two-fold axis and to the mirror plane which bisects the molecule (Scheme 8.61), a concerted cycloaddition *e.g.*, to an alkene is a symmetry allowed process. The reaction of an alkene with diazomethane to give a pyrazoline (Scheme 8.62 pyrazole derivative) belongs to this class.

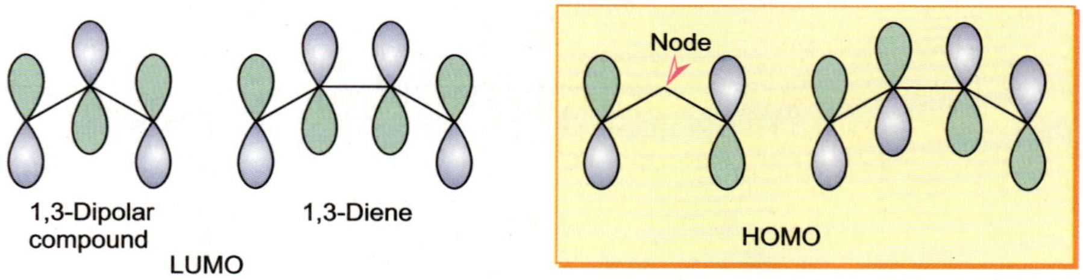

SCHEME 8.61

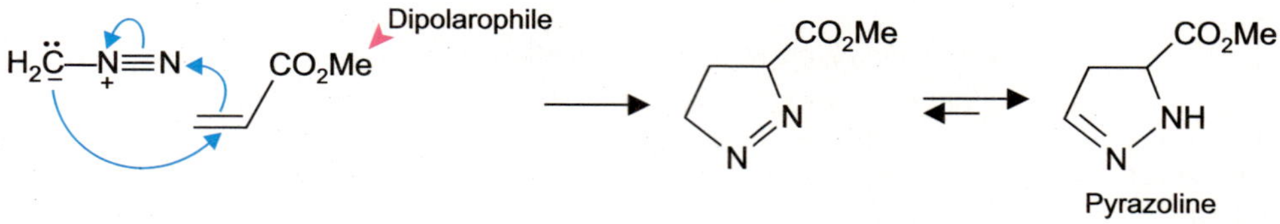

SCHEME 8.62

8.6 CHELETROPIC REACTIONS–A [4 + 2] CYCLOADDITION

In a cheletropic reaction two σ bonds that terminate at a single atom are made or broken during a concerted reaction (Scheme 8.64). In the case of molecules, sulfur dioxide or carbon monoxide the HOMO is that which has a lone pair of electrons in the plane having the atoms, while the LUMO represents the *p* orbital perpendicular to this plane (Scheme 8.63). Thus a cheletropic reaction is a cycloaddition in which one component interacts through a single atom (*i.e.*, C, N or S) which has a lone pair of electrons.

For a symmetry allowed cycloaddition of *e.g.*, SO_2 to a diene, the molecule of SO_2 must lie in a plane which bisects the *s-cis* conformation of the diene (Scheme 8.63). The interaction is suprafacial for diene and SO_2. In the transition state, the terminal carbon atoms of the diene must move in the disrotatory manner in order that the HOMO of SO_2 can interact with the LUMO of the diene, or the LUMO of SO_2 with the HOMO of the diene.

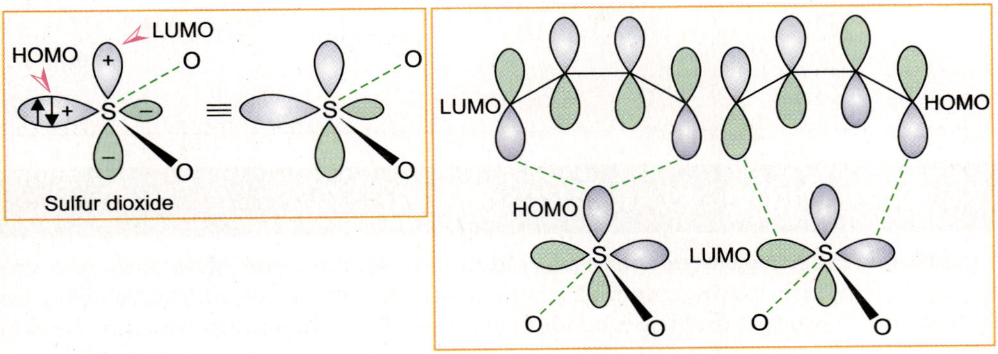

SCHEME 8.63

3-Sulpholene, a solid, is a convenient substitute for geseous butadiene. Butadiene is generated at high temperatures from 3-sulpholene in a reverse reaction (Scheme 8.64) and when a dienophile is present it is trapped in a Diels-Alder reaction. The Diels-Alder reaction in itself is usually reversible and has been used to protect double bonds.

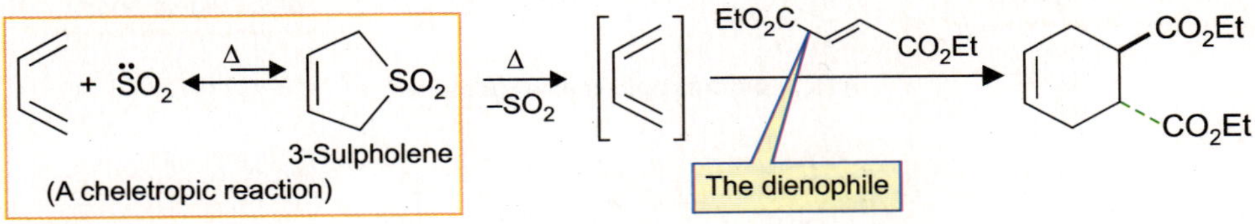

SCHEME 8.64

Cyclopentadienones undergo a facile Diels-Alder reactions due to their fixed s-*cis* conformation and antiaromaticity. The Diels-Alder reaction of a substituted cyclopentadienone and an alkyne is followed immediately by a retro-cycloaddition to generate CO and an aromatic system (Scheme 8.65).

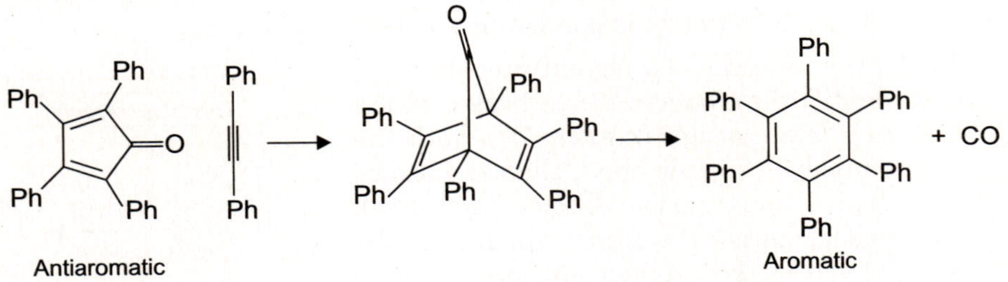

SCHEME 8.65

The cyclobutene derivative (Scheme 8.66) undergoes thermal conrotatory ring opening to generate the butadiene derivative which reacts with sulfur dioxide to give the cyclic sulfone derivative in quantitative yield.

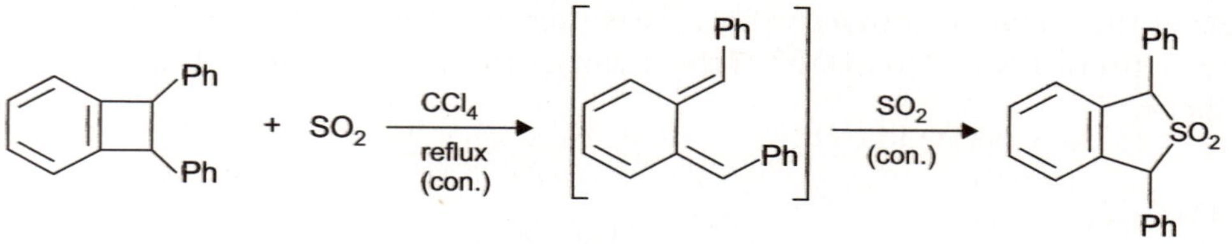

SCHEME 8.66

8.7 SIGMATROPIC REARRANGEMENTS

A sigmatropic rearrangement is a concerted intramolecular shift of an atom or a group of atoms. During this arrangement at σ bond is broken in the reactant and a new σ bond is formed in the product and the π bonds rearrange. The following points may be noted:

- The number of π bonds does not change, both the reactant and the product contain the same number of π bonds.
- The σ bond that cleaves can be in the middle of the π system or at end of the π system (Scheme 8.67).

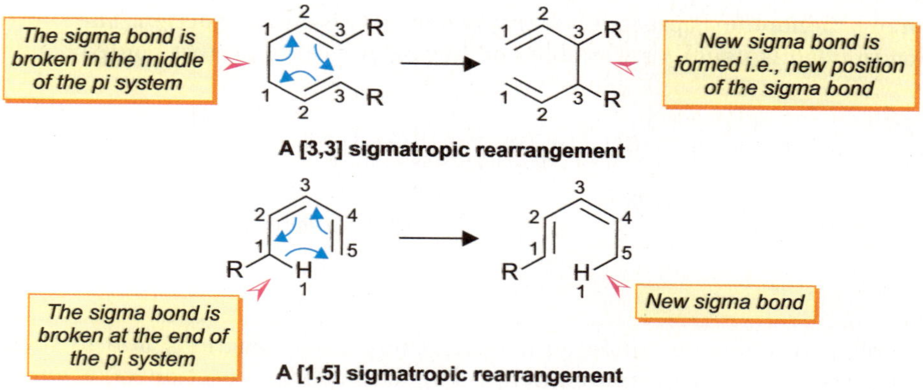

SCHEME 8.67

- The σ bond that breaks is bonded to an allylic carbon.
- To identify the order of a sigmatropic shift [*i*, *j*] first identify the σ bond which is broken in the reaction. Then assign number 1 to both the atoms involved in this bond, then the atoms in each direction from the bond being cleaved, upto and including the atoms which form the new σ bond in the product are numbered as atoms 2,3 and so on. The numbers assigned to

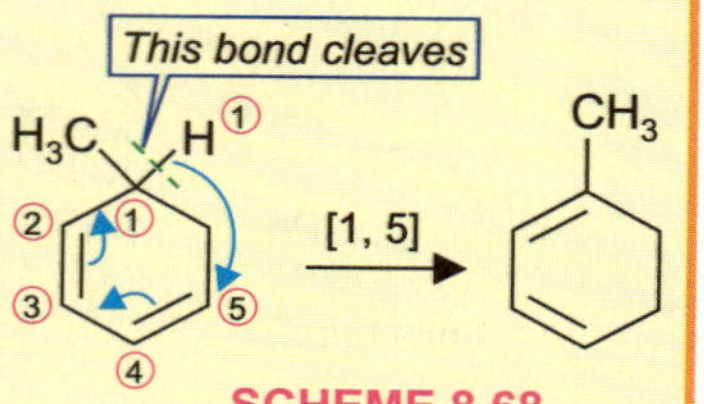

SCHEME 8.68

the atoms that form the new bond, separated by commas are put within the brackets to show the reaction order (Scheme 8.67). Similarly the migration of hydrogen (Scheme 8.68) is another example of [1,5] sigmatropic shift. The order [1,5] is not due to the fact that hydrogen migrates from C1 to C5 but since the hydrogen (one of the two atoms given the number 1) forms part of the new σ bond and had also formed part of the old σ bond. Only all the atoms taking part in the reaction have to be counted. Thus the rearrangement of cyclohexadiene (Scheme 8.68) cannot be labelled as [1,3] shift since the methylene group linking 1 and 5 is not involved in the reaction.

- Since in these reactions a change in the position of one σ bond takes place, Woodward and Hofmann coined the term "sigmatropic shifts".
- A [3,3] sigmatropic rearrangement of a 1,5-diene (when the six atoms involved are all carbons) is known as the Cope rearrangement (Scheme 8.69).
- The oxygen analog of the Cope rearrangement is called the Claisen rearrangement. Often one of the π bonds is part of an aromatic ring (II, Scheme 8.69). Allyl vinyl ethers also undergo Claisen rearrangement.

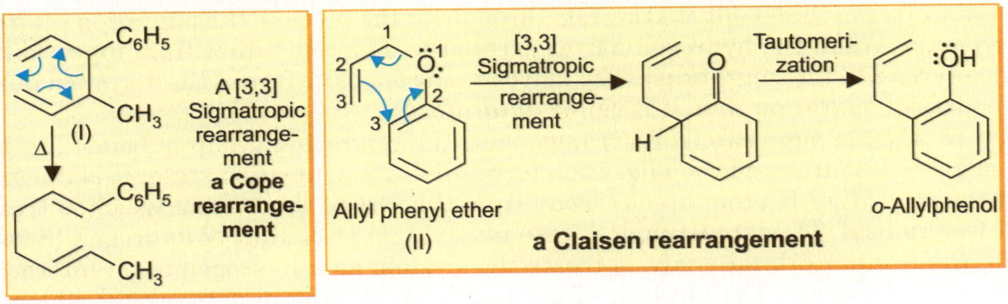

SCHEME 8.69

- More examples of sigmatropic rearrangements are in (Scheme 8.69*a*).

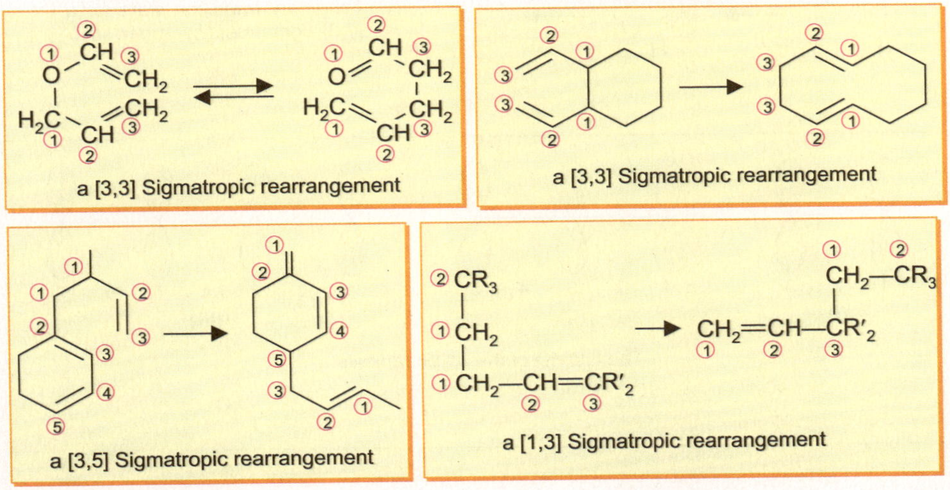

Continued...

- Two atoms (N, N) connect the old and new σ bonds in one fragment, and three atoms (C, C, C) connect the old and new σ bonds in the other fragment.

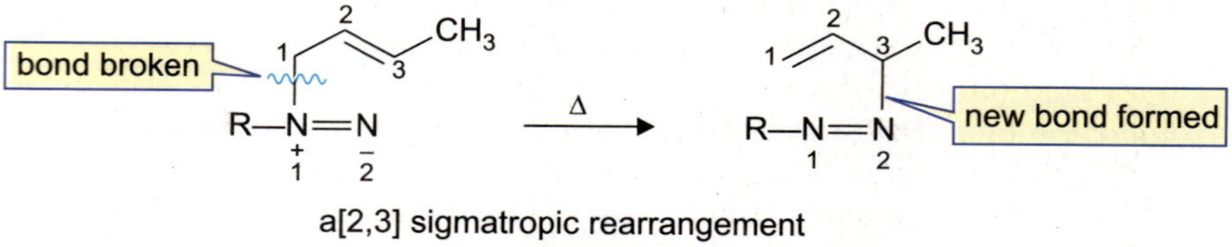

SCHEME 8.69a

(A) Sigmatropic Migration of Hydrogen

(i) Introduction

A hydrogen atom is reported to migrate from one end of a system of π bonds to the other, under thermal or photochemical rearrangements. In the transition state the hydrogen must be in contact with both ends of the chain at the same time. There are two distinct processes by which a sigmatropic migration can occur. If the hydrogen moves along the top or bottom face of the π-system *i.e.,* migrating group remains associated with same face of the conjugated system throughout the process, the migration is termed *suprafacial.* When the hydrogen moves across the π system either from top to bottom or vice versa *i.e.,* the migrating group moves to the opposite face of the π-system during the course of migration then it is called *antarafacial.*

In a given sigmatropic rearrangement, the migrating group is bonded to both the migration source and the migration terminii in the transition state. It is imagined that the migrating H atom breaks away from the rest of the system which is treated as a free radical. Thus in a simplest case involving a [1,3] shift of hydrogen (Scheme 8.70), the frontier orbital analysis treats this system as a hydrogen atom interacting with an allyl radical. The electron of the hydrogen atom is in a 1*s* orbital which has only one lobe. The HOMO of an allylic free radical depends on the number of carbons in the π-framework (Scheme 8.71).

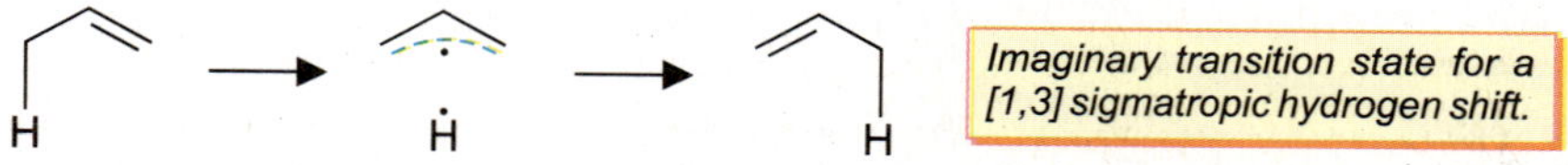

SCHEME 8.70

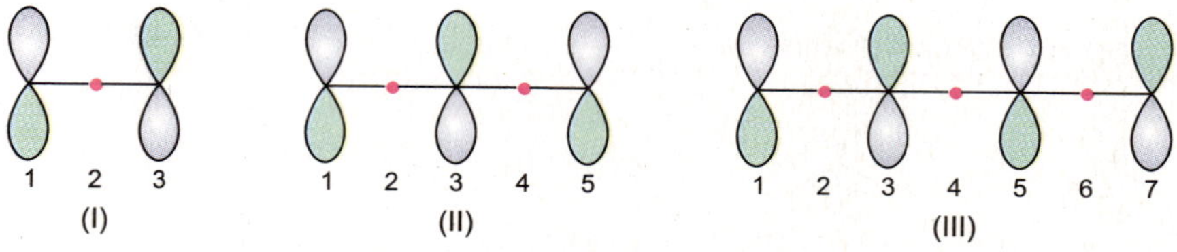

The HOMO of the *allylic* radicals

SCHEME 8.71

(ii) [1,3] Sigmatropic Rearrangement (Hydrogen Shift).

In the migration of hydrogen the H must move from a plus to plus or from minus to a minus lobe of the HOMO, it cannot move to a lobe of opposite sign.

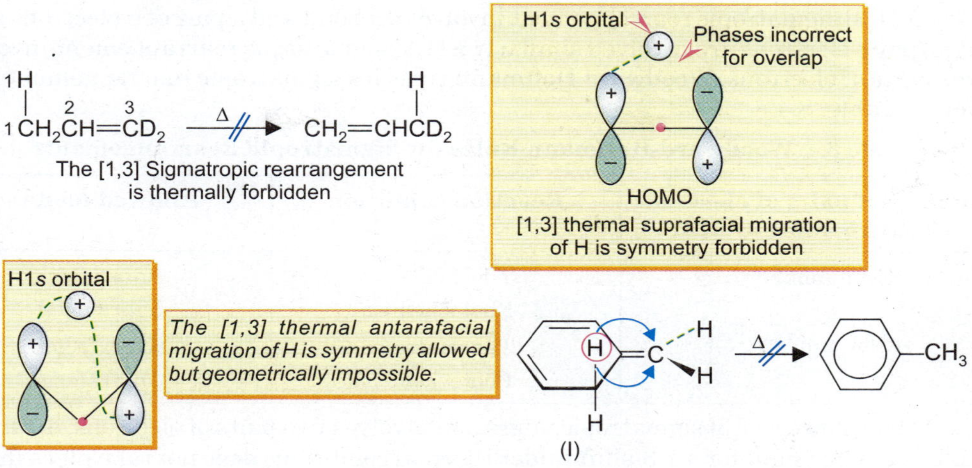

SCHEME 8.72

During a thermal [1,3] sigmatropic migration of a hydrogen, the overlap of the hydrogen 1*s* orbital with the HOMO of the allyl radical (I, Scheme 8.71, asymmetric) is bonding at one end and antibonding at the other end for the suprafacial migration (Scheme 8.72). Thus [1,3] sigmatropic suprafacial migration of hydrogen (under thermal conditions) is symmetry-forbidden (Scheme 8.72). However, in the antarafacial process (Scheme 8.72) the hydrogen atom shall have to cross over the *pi* system to the other face to form a four membered ring transition state, a geometrically very difficult situation. Thus over all thermal [1,3] sigmatropic rearrangements are rare. The stability of the triene (I, Scheme 8.72) which is not thermally isomerized to toluene, which is thermodynamically more stable is due to a symmetry-forbidden process (suprafacial H migration is symmetry forbidden, antarafacial H migration though symmetry allowed but sterically forbidden).

In a photochemical reaction promotion of an electron means that now (I, Scheme 8.73) becomes the HOMO. Suprafacial pathway for [1,3] shift now becomes an allowed process and antarafacial pathway forbidden. Thus, the compound (II, Scheme 8.73) displays a [1,3] hydrogen shift under photochemical conditions.

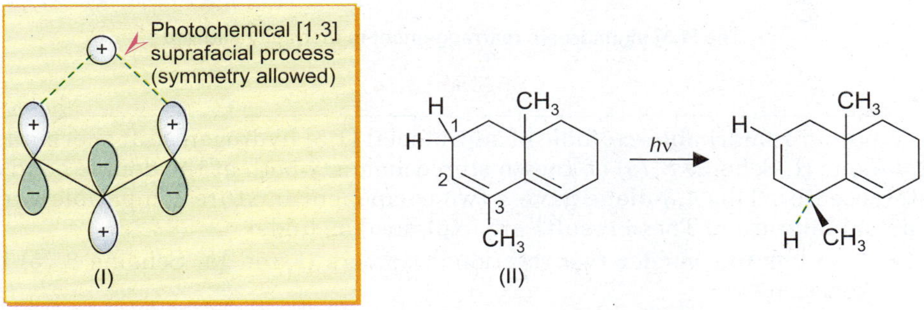

The [1,3] sigmatropic rearrangement is photochemically allowed

SCHEME 8.73

A [1,3] sigmatropic rearrangement involves a π bond and a pair of σ electrons so in all two pairs electrons are involved similarly a [1,5] sigmatorpic rearrangement involves three pairs of electrons. Woodward Hoffmann rules for sigmatropic rearrangements are given in Table. 8.4.

Table 8.4: Woodward-Hoffmann Rules for Sigmatropic Rearrangements

Number of pairs of electrons in the reacting system	Reaction condition	Allowed mode
Even number	Thermal	Antarafacial
	Photochemical	Suprafacial
Odd number	Thermal	Suprafacial
	Photochemical	Antarafacial

Thus, since a [1,3] sigmatropic migration involves two pairs of electrons, an antrafacial rearrangement for a 1,3-shift under thermal conditions does not take place due to geometrical constraints. 1,3-Shifts do take place photochemically [Table 8.4, moreover, since under photochemical conditions HOMO becomes symmetric (see, Scheme 8.73) hydrogen can migrate by suprafacial pathway].

(iii) [1,5] Sigmatropic Hydrogen Shift

The [1,5] sigmatropic shift of hydrogen or deuterium atoms is well known. These involve three pairs of electrons, thus these occurs via a suprafacial pathway under thermal conditions (see Table 8.4). These shifts can be analyzed by examining a hydrogen atom and a pentadienyl radical whose HOMO (see III, Scheme 8.71) is bonding at both the migration origin and the migration terminus. Thus the migration maintains orbital symmetry when the migrating group remains on the same side of the conjugated system (suprafacial process, Scheme 8.74).

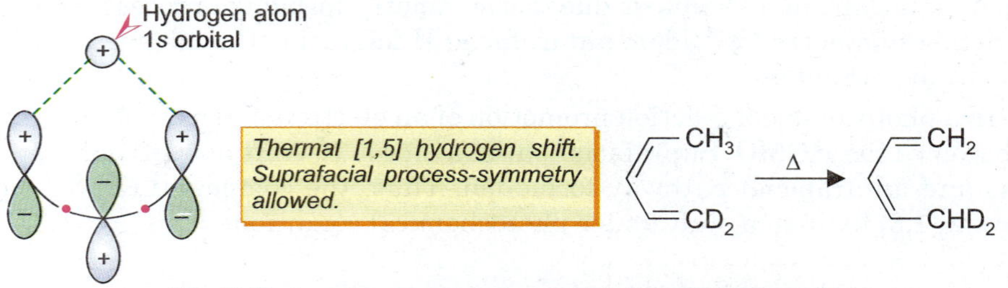

The [1,5] sigmatropic rearrangement is thermally allowed

SCHEME 8.74

Another remarkable example of suprafacial [1,5] hydrogen shift thermally, is in the 1,3-diene (I, Scheme 8.75) of known stereochemistry both at the double bond and at the stereocenter. This 1,3-diene gave a two component mixture compatible with only suprafacial migration. These results are explained as under:

- One has to consider two rotational isomers (I and I*a*, Scheme 8.75) for the reaction.
- Recall, a compound can have an infinite number of conformations but only one configuration.
- In (I) the methyl group is directed toward C4-C5 double bond while in (I*a*) it is now ethyl group that is directed toward C4-C5 bond.

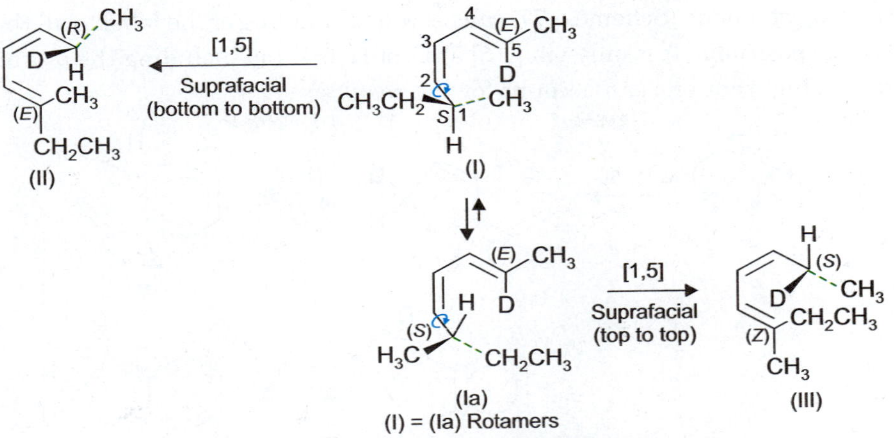

Stereochemistry of [1,5], superfacial H migration, chirality is transferred from C—1 to C—5.

SCHEME 8.75

- There are two suprafacial [1,5] pathways for the hydrogen in these two conformations (I and I*a* Scheme 8.75).
- Each of these suprafacial pathways gives a product with specific stereochemistry and both are formed.
- Considering the two stereogenic units in (I, Scheme 8.75) 4 stereoisomers could be considered. Two (II and III) are formed during suprafacial migration by the symmetry allowed pathway.
- If one considers, the antarafacial migration the hydrogen leaves from one side and is delivered from the other, and would lead to the formation of the remaining two stereoisomers (as a diastereomeric pair) (Scheme 8.76) which however, is not the case.

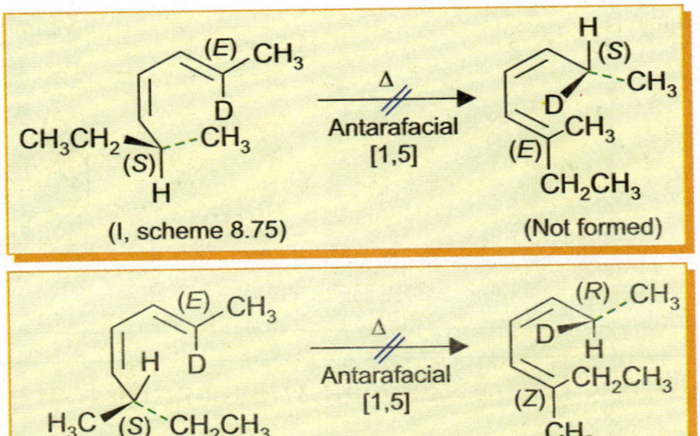

SCHEME 8.76

Heating of indene (Scheme 8.77) causes the scrambling of the label to all the three non-aromatic positions. It is only via [1,5] shift of H or D (by including the *p*-orbitals of the benzene ring) that one can account for the results.

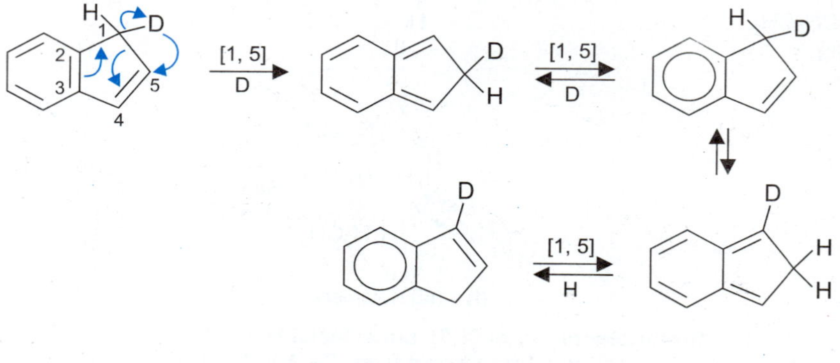

SCHEME 8.77

(iv) [1,7] Sigmatropic Hydrogen Shift

In the case of [1,7] hydrogen shifts, in a triene, the orbital symmetry rules (see III, Scheme 8.71) predict that the transfer of hydrogen must be antarafacial compared to [1, 3] shift, the transition state is not much strained and the shift is sterically feasible. This is seen in the thermal interconversion of vitamin D series (Scheme 8.78).

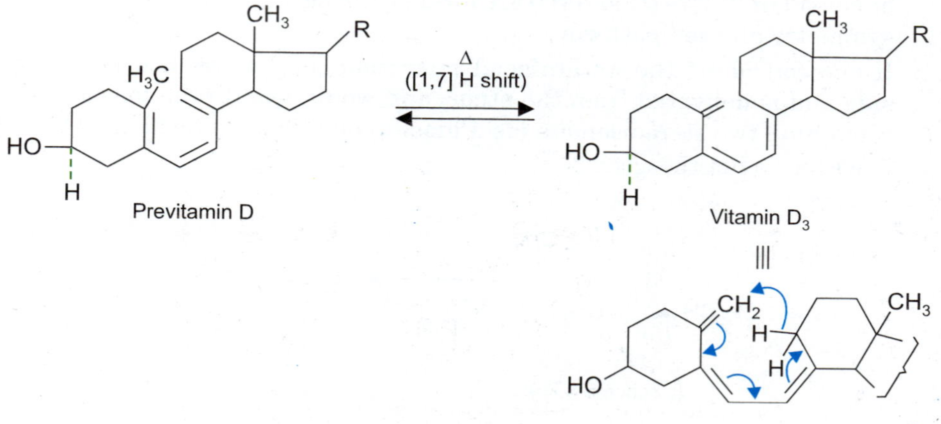

SCHEME 8.78

More on sigmatropic rearrangements (Degenerate rearrangements)

1, 3–Pentadiene on heating reagenerates itself via a [1, 5] shift of hydrogen, this kind of process in which a reactant rearranges to itself is termed degenerate rearrangement (Scheme 8.78a). Significantly as expected the [1, 3] shift is symmetry farbidden.

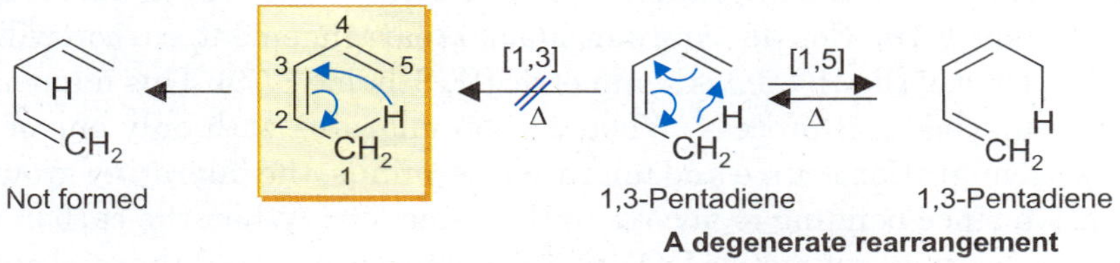

SCHEME 8.78a

The degenerate rearrangement can be estabilished either by isotopically labelled molecules or suitably substituted molecules (Scheme 8.78b).

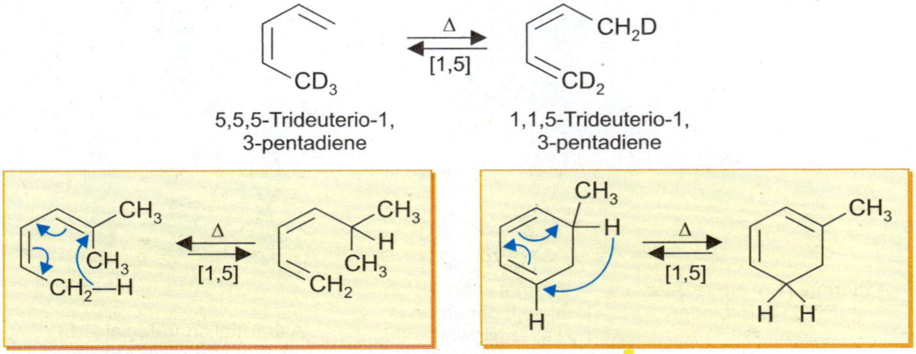

SCHEME 8.78b

Three membered rings can often play the role of a double bond and a [1, 5] H shift can take place just like in 1, 3-pentadiene and involves the opening of the cyclopropane (Scheme 8.78c)

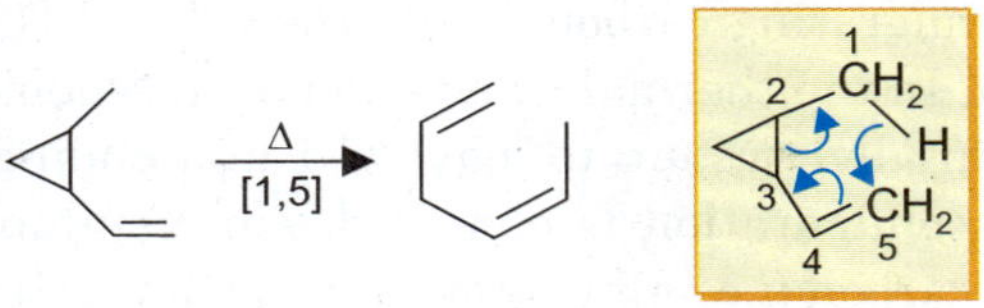

SCHEME 8.78c

(B) Sigmatropic Migrations of Carbon

As compared to a hydrogen atom which has its electrons in a 1*s* orbital that has only one lobe, a carbon free radical (free imaginary transition state) has its odd electron in a *p* orbital which has two lobes of opposite sign. Recall that a [1,3] sigmatropic suprafacial migration of hydrogen (thermally) is symmetry forbidden while an antarafacial reaction though allowed is geometrically improbable (see Schemes 8.72–8.74). Interestingly an additional possibility would exist if an alkyl group (carbon) rather than a hydrogen was potential migrator. A [1,3] shift can now be suprafacial migration (I, Scheme 8.79) if the migrating group does so antarafacially *i.e.*, it would result in inversion of configuration of the migrating group. Thus carbon can simultaneously interact with the migration source and the migration terminus using either one of its lobes or both of its lobes (Scheme 8.79). Considering suprafacial rearrangement, carbon will migrate using one of its lobes if the HOMO is symmetric (II, Scheme 8.79). This happens during a thermal suprafacial [1,5] process. When carbon migrates with only one of its lobes interacting with migration source and migration terminus, the migrating group retains its configuration since bonding is always to the same lobe. When the carbon migrates using both of its lobes. (asymmetric HOMO, I Scheme 8.79), a [1,3] thermal suprafacial migration would involve opposite lobes. Thus, if the migrating carbon was originally bonded *via* its positive lobe, it must now use its negative lobe to form the new C—C bond. The stereochemical outcome of such a process is the inversion of configuration in the migrating group.

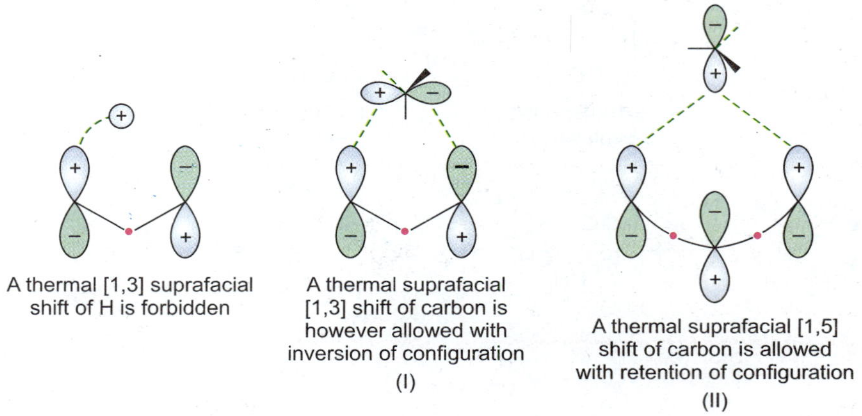

SCHEME 8.79

In summary, a suprafacial [1,5] thermal rearrangement proceeds with retention of configuration at the migrating carbon, while the related [1,3] suprafacial process proceeds with inversion. In the thermal conversion of (I, scheme 8.80) to (II) a carbon atom migrates across an allyl system to leave C-1 and ending up at C-3 via a [1,3] shift. The inversion of configuration is observed using suitable substrates (scheme 8.81). The [1,3] shifts of carbon (*i.e.*, alkyl groups) in such reactions involve expansions

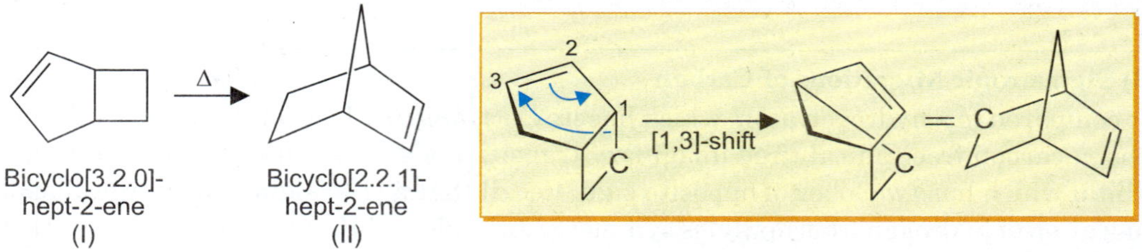

SCHEME 8.80

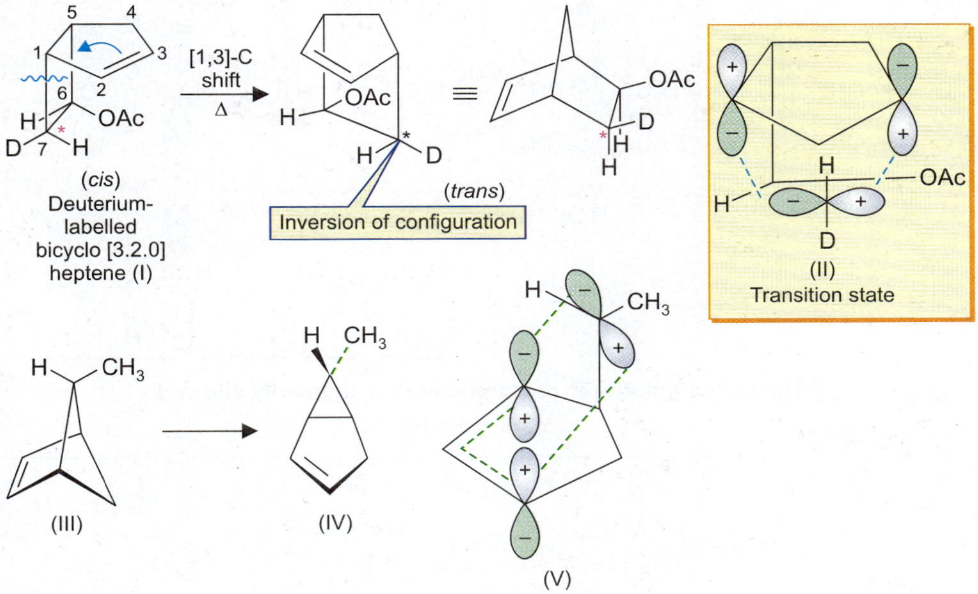

SCHEME 8.81

of strained three-or four-membered rings. As predicted by orbital symmetry conservation rules these reactions proceed almost entirely with inversion of configurations in the migrating group as in (I, Scheme 8.81). In this case, a label deuterium was placed at C7 which was *trans* to the acetoxy group. After the reaction, it was found to be exclusively *cis* due to inversion of configuration at C7. The transition state (II, Scheme 8.81) shows that it is a [1,3] sigmatropic shift of carbon. Similarly (III, Scheme 8.81) gives (IV) via the transition state (V) by a suprafacial [1,3] shift with inversion at the migrating carbon under thermal conditions.

(C) The Cope Rearrangement

A 1,5-diene on heating is rearranged to another 1,5-diene by concerted formation of a 1, 6-bond, breaking of the 3,4-bond and migration of both double bonds in a [3,3] sigmatropic rearrangement known as Cope rearrangement (see, Scheme 8.69). The compound rearranges by a [3,3] sigmatropic pathway and is also hypothetically pictured as split into two allyl radicals (Scheme 8.82). Interaction of the HOMO's of these allyl radicals is bonding at both ends, so the reaction is thermally allowed. The stereochemical outcome of this rearrangement is in keeping with their occurrence generally through the chair-shaped transition states (Scheme 8.83). *Meso* 3,4-dimethyl-1, 5-hexadiene gives *cis*, *trans*-2, 6-octadiene (in the starting compound the two methyl groups are having *cis*-relationship, in the chair form of a cyclohexane only 1, 2-axial, equatorial relationship is *cis*) while a boat shaped transition state would give *cis*, *cis*-product or *trans*, trans-product (Scheme 8.84).

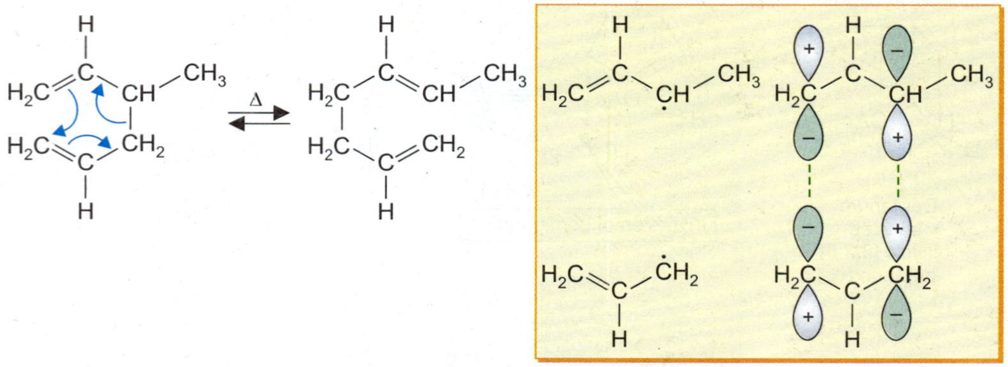

The [3,3] sigmatropic rearrangement is thermally allowed

SCHEME 8.82

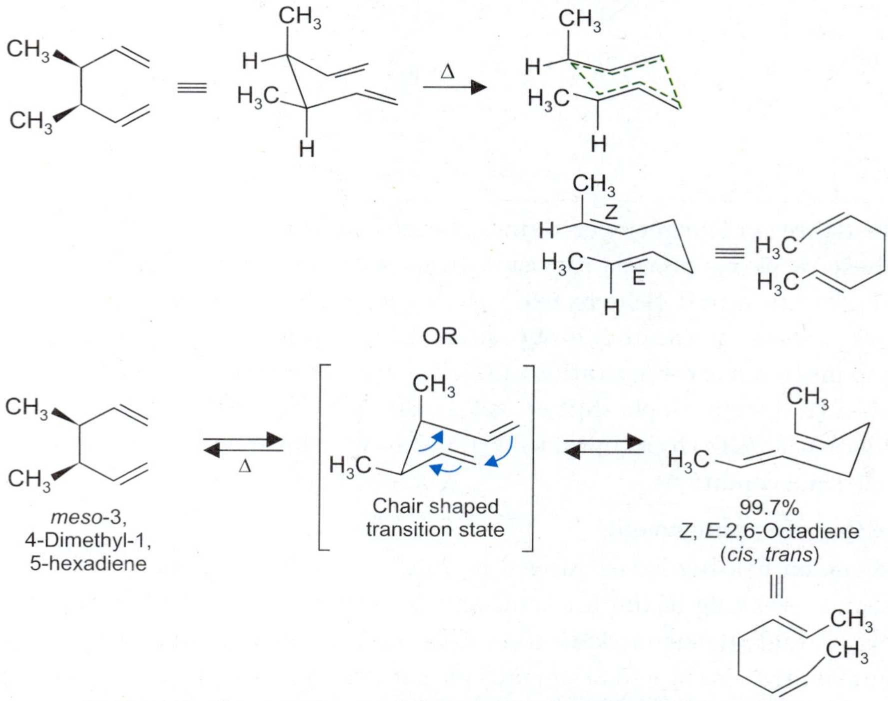

Cope rearrangement occurs via a chair shaped transition state

SCHEME 8.83

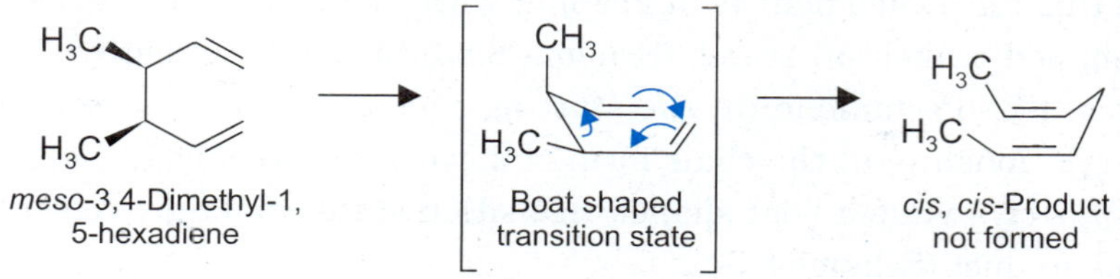

SCHEME 8.84

EXERCISE 8.13

Which diene you expect on pyrolysis of trans-3, 4-dimethylcyclohexadiene ?

ANSWER. *A Cope rearrangement occurs through a chair shaped transition state and the diene expected is E, E isomer of 2, 6-octadiene (Scheme 8.85).*

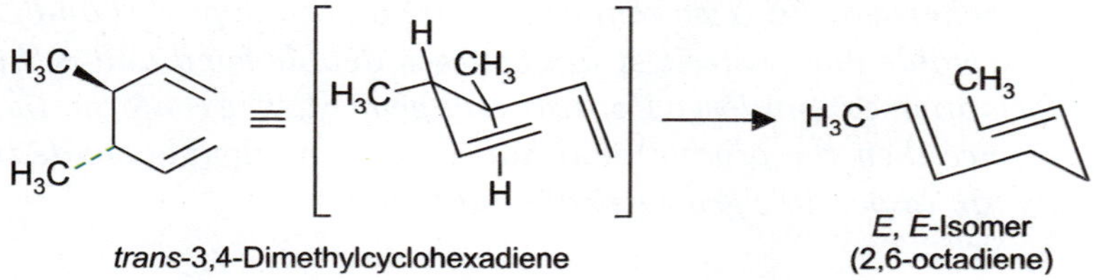

SCHEME 8.85

EXERCISE 8.14

Why Z, Z-2, 6-octadiene is not the product of pyrolysis of trans-3,4-dimethylcyclohexadiene ?

ANSWER. *The transition state (I, Scheme 8.86) with two pseudoequatorial groups is far more stable than (II) with two pseudoaxial groups. The Z, Z-isomer would arise from the less stable chair shaped transition state (II, Scheme 8.86).*

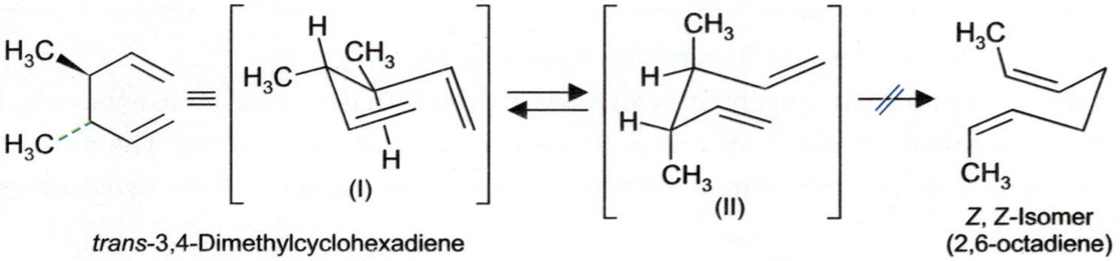

SCHEME 8.86

On introducing strain into the reactant, rate accelerations are observed and *cis*-divinylcyclopropane rapidly undergoes Cope rearrangement (Scheme 8.87). Similar reaction is however, not observed with *trans*-isomer where the reacting ends of the double bonds are too far apart. Thus the Cope rearrangement occurs at low temperatures in *cis*-1, 2-divinyl cyclopropane compared to Cope rearrangement of 1, 5-hexadiene itself which requires temperatures in the range of 200–300°C.

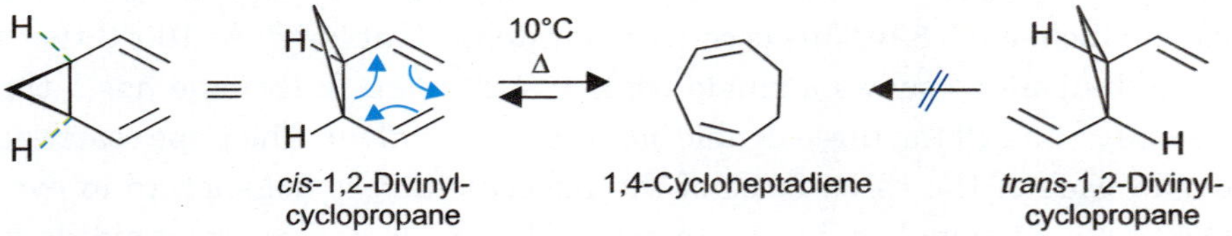

SCHEME 8.87

EXERCISE 8.15

Which two conformations of cis-1, 2-divinylcyclopropane can be considered for a possible Cope rearrangement ?

Which of these conformations is capable of undergoing this rearrangement ?

ANSWER. *See Scheme 8.87a. Two conformations can be adopted (I and II, Scheme 8.87a), (I) is less stable due to steric strain between double bond and a H atom and only this conformation can undergo Cope rearrangement. The conformation (II) will be unreactive since then the product will have two trans double bonds in a seven membered ring (an impossible geometrical situation).*

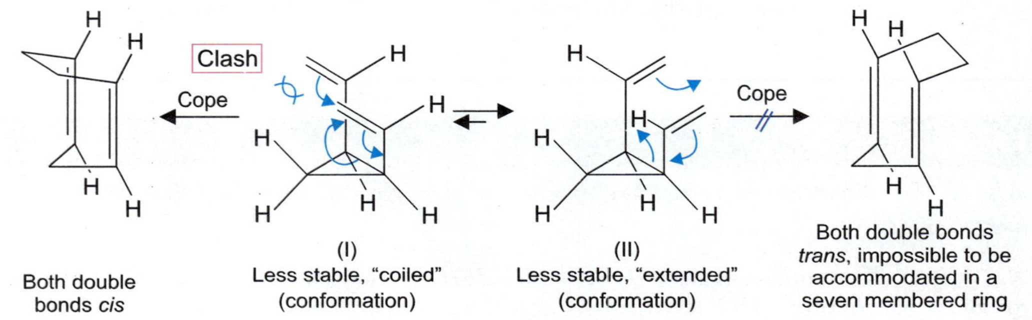

SCHEME 8.87a

(D) Fluxional Molecules — A Degenerate Rearrangement

Divinylcyclopropane rearrangements can proceed even with more ease in case the entropy of activation is made still negative by incorporating both vinyl groups into a ring. The system then becomes homotropilidene (Scheme 8.88) which undergoes a degenerate Cope rearrangement.

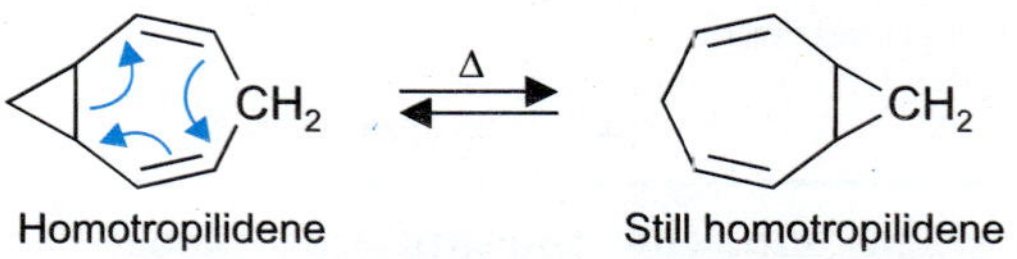

The *degenerate* Cope rearrangement of homotropilidene

SCHEME 8.88

A degenerate rearrangement leads to a product which is indistinguishable from the reactant. By bridging the two methylene groups in homotroplidene one gets a molecule of bullvalene (Scheme 8.89). This is converted into itself at 25°C. At 100°C the ^{1}HNMR spectrum of bullvalene shows a single peak at 4.22 ppm. Bullvalene has a three fold rotational axis; thus all the three double bonds are equivalent. The Cope rearrangement can occur in each of the three faces of the molecule and is degenerate in every case (Bullvalene is a fluxional molecule—a molecule which undergoes rapid degenerate rearrangement).

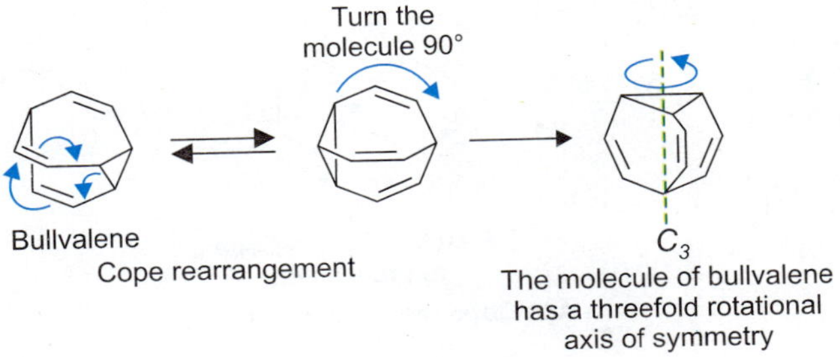

SCHEME 8.89

(E) Oxy-Cope Rearrangement

As seen in other pericyclic reactions Cope rearrangement is also reversible and the position of equilibrium depends on the relative stability of the isomers. This problem can be checked and the forward reaction can be made to predominate provided the product reacts further. This is so in oxy-Cope rearrangement when the reactant contains an oxygen group on C_3 or C_4 position. The alcohol variant of Cope rearrangement (Scheme 8.90) is called the oxy-Cope rearrangement and when the alkoxide derivative is used it is referred to as the "anionic oxy-Cope rearrangement. The rates of sigmatropic rearrangements are enhanced and the temperatures for the reactions are drammatically decreased compared to parent alcohols.

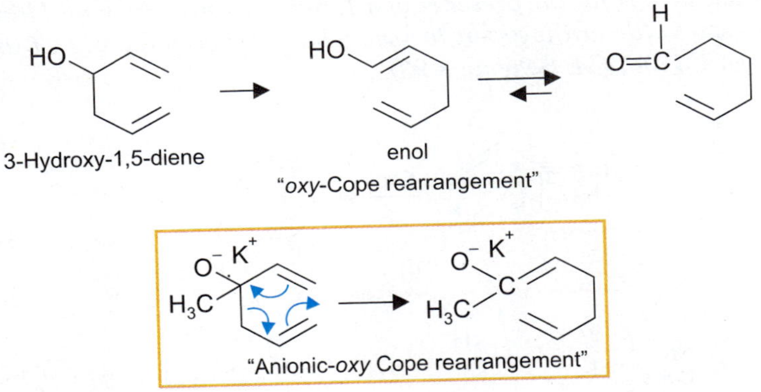

SCHEME 8.90

(F) Aza-Cope Rearrangement

It is well known that the presence of oxygen atom adjacent to the π bond accelerates the Cope rearrangement. Similarly a nitrogen usually as an minimum salt fragement in the diene also induces an aza-Cope rearrangement. Thus the reactant (II, Scheme 8.91) derived from (I) underwent a fast aza-Cope rearrangement at low temperature.

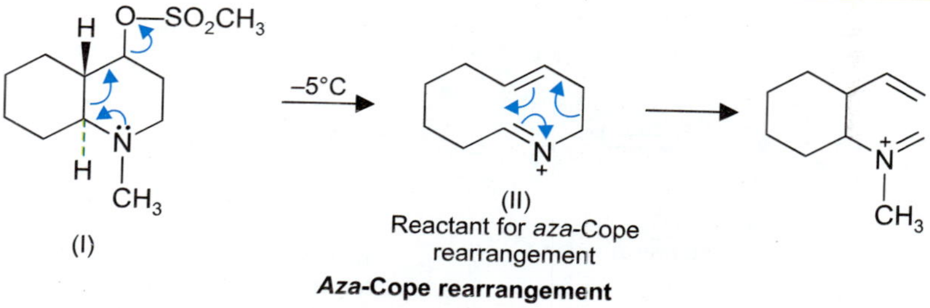

SCHEME 8.91

EXERCISE 8.16

Write the structure of products from the reactions (Scheme 8.92).

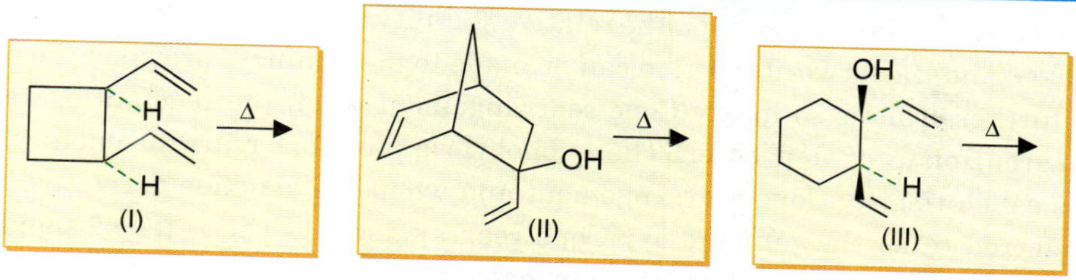

SCHEME 8.92

ANSWER. *Always look for the presence of a 1, 5-diene unit which will hint towards a Cope rearrangement (draw the arrow to form a bond between C1 and C6 and breaking a bond between C3 and C4, Scheme 8.93).*

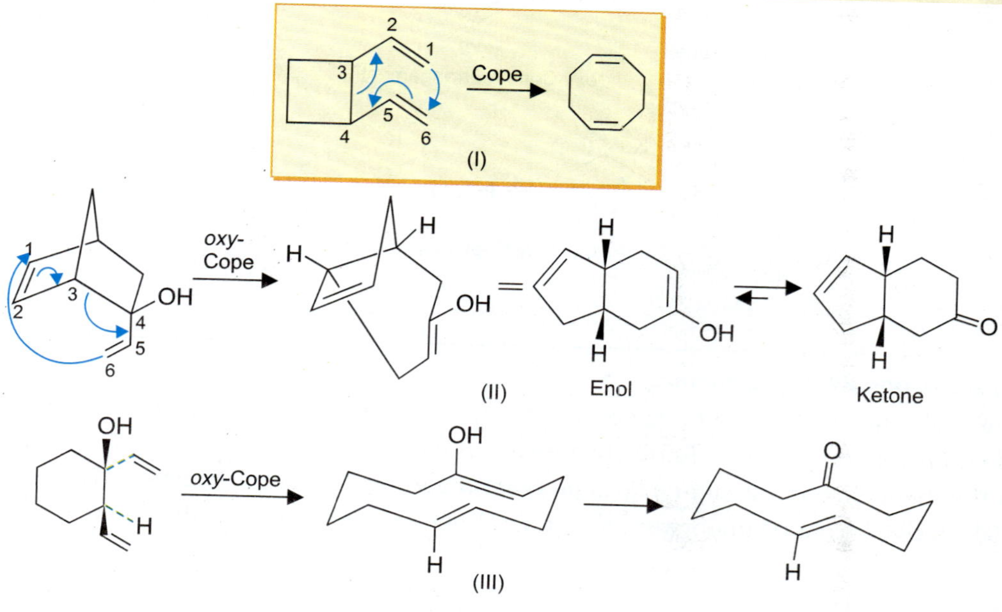

SCHEME 8.93

(G) The Claisen Rearrangement

Claisen rearrangement also involves a [3,3] sigmatropic pathway like Cope rearrangement, however, in Claisen rearrangement the substrates incorporate one or more heteroatoms in place of carbon in the 1,5-hexadiene system. The simplest example of Claisen rearrangement is the thermal conversion of allyl vinyl ether to γ, δ-unsaturated carbonyl compounds and allyl aryl ethers on heating rearrange to *o*-allyl phenols (Schemes 8.94 and 8.95).

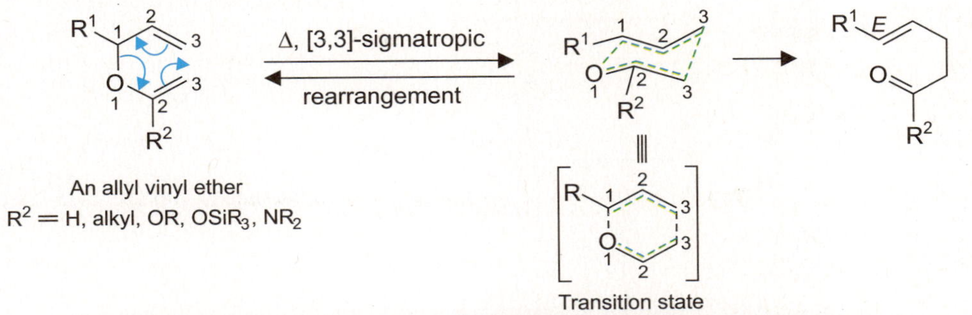

Claisen rearrangement involves a[3,3]-sigmatropic rearrangement

SCHEME 8.94

Studies using migrating groups labelled with ^{14}C or with substituents show that the allylic group is end-interchanged during the *ortho* rearrangement (Scheme 8.95). These and other results which show that the Claisen rearrangement is intramolecular provide strong support for a concerted mechanism. The reaction is highly stereoselective, leading predominantly to the *E*-configuration of the new double bond and to the controlled setreochemical disposition of substituents on the single bond. A chair conformation is preferred for the cyclic transition state with the substituent R^1 (Scheme 8.94) in the less-hindered pseudoequatorial position.

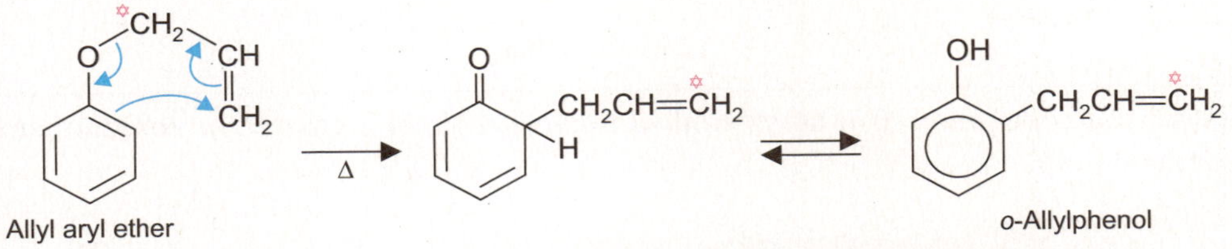

Allyl group normally migrates to the ortho position (the allylic group is end-interchanged)

SCHEME 8.95

Like Cope rearrangement reliable stereochemical predictions can be made from a chair-like transition state that minimizes 1, 3-diaxial interactions. Thus Claisen rearrangement provides a means of introducing functionality in a stereo-and regio-specific manner (Scheme 8.96). At the same time the geometry about the nearly carbon-carbon double bonds is also fixed. In (I) the methyl groups occupy pseudoequatorial positions in the transition state. Similarly in (II) the major product will have E configuration of the newly created double bond due to placement of the bulkier substituent in the pseudo equatorial position.

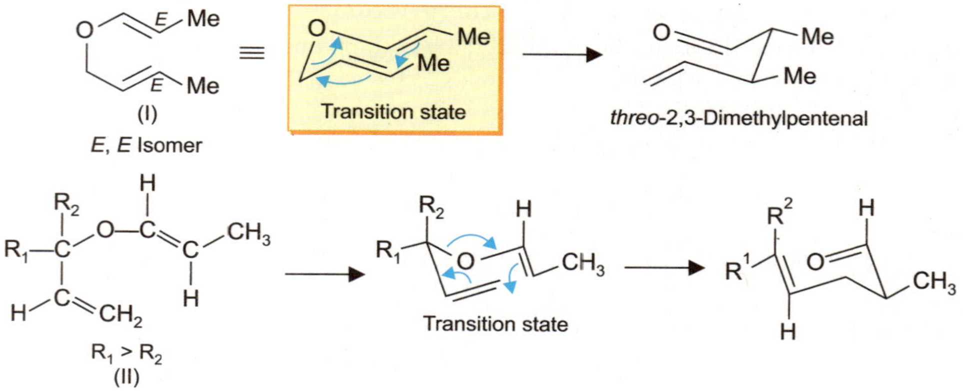

SCHEME 8.96

When both the *ortho*-positions are substituted, the allyl group undergoes a second migration *via* a concerted [3, 3] Cope rearrangement followed by enolization to give *para*-allylphenol (Scheme 8.96*a*). Sometimes, the *para*-product is obtained even if one or both *ortho*-positions are free. Migration to the *meta*-position has not been observed.

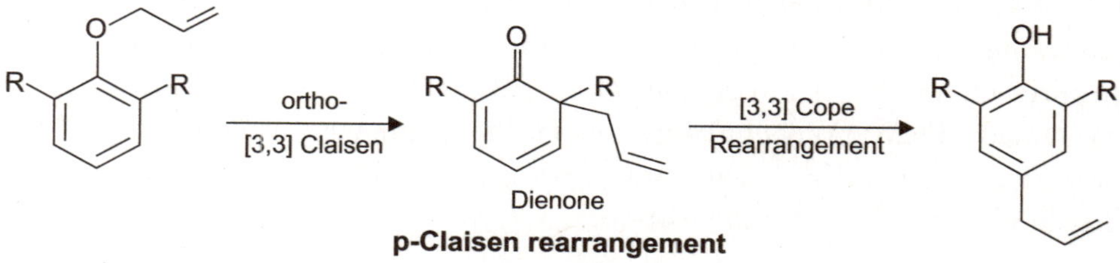

SCHEME 8.96a

This arrangement may also be viewed to involve a [3, 5] sigmatropic rearrangement (Scheme 8.96b).

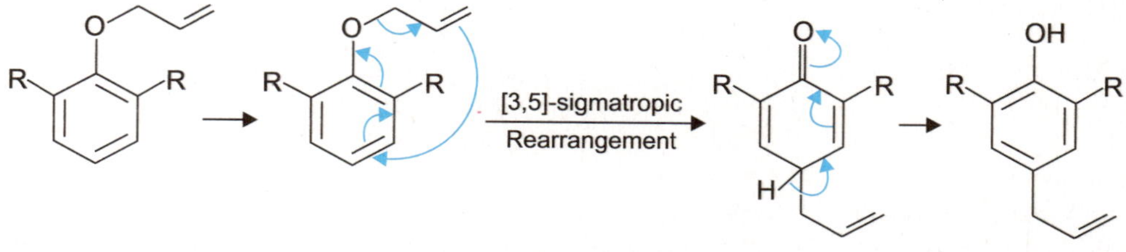

SCHEME 8.96b

BOX 8.2

ABNORMAL CLAISEN REARRANGMENT

In an abnormal Claisen rearrangement, the normal Claisen rearrangement product further rearranges with the β-carbon getting attached to the ring to give an isomeric *ortho*-allylphenol (Scheme B 8.3).

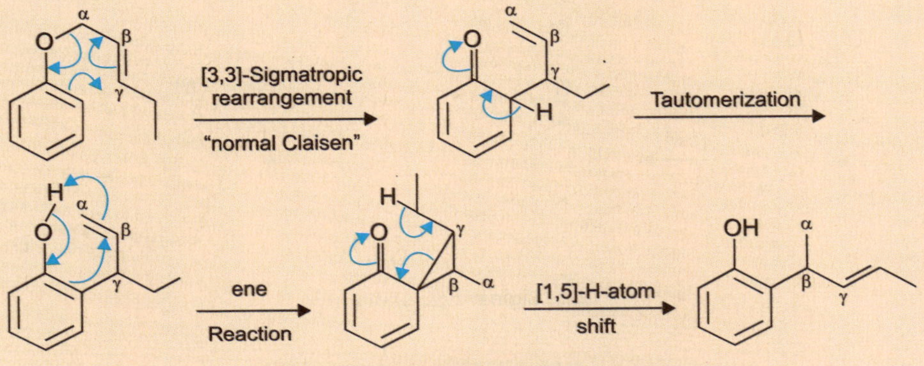

Abnormal Claisen rearrangement

SCHEME B 8.3

IRELAND-CLAISEN (SILYL KETENE ACETAL) REARRANGEMENT

The Ireland-Claisen rearrangement uses an allyl ester instead of the allyl ether used in the Claisen rearrangement. A base removes a proton from the α-carbon of the ester and the enolate ion is trapped as a trimethylsilyl ether. Mild heating results in a Claisen condensation (Scheme B 8.4).

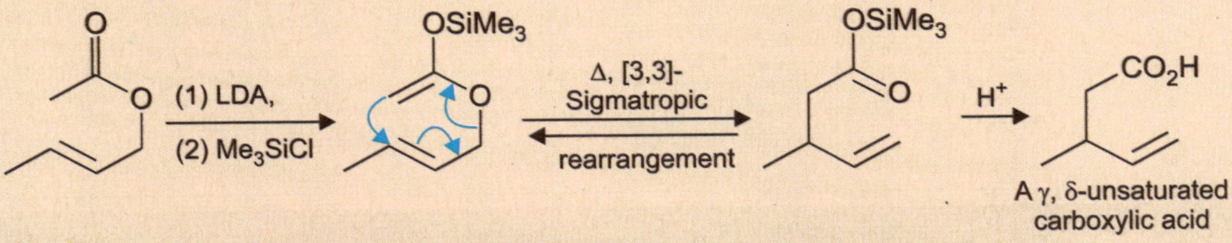

Ireland-Claisen (silyl ketene acetal) rearrangement

SCHEME B 8.4

(H) [5, 5] Sigmatropic Shifts

Thermal [5, 5] shifts are facile, however the compounds undergoing such rearrangement are not common. One type of substrates are pentadienyl ethers which give 4-pentadienylphenols as the major products along with minor products arising from *ortho*-Claisen rearrangement (Scheme 8.96*c*). With the help of deuterium labeling it has been shown that major products arise from direct [5, 5] sigmatropic shifts and not by two consecutive [3, 3] shifts.

It is proved that [5, 5] shifts occur very fast in negatively charged compounds. An interesting reaction is oxy-Cope rearrangement of (I, Scheme 8.96*d*) the arrangement does not proceed by a sequence of consecutive [3, 3] shifts, however it is indeed a result of [5, 5] shifts, (Scheme 8.96*d*).

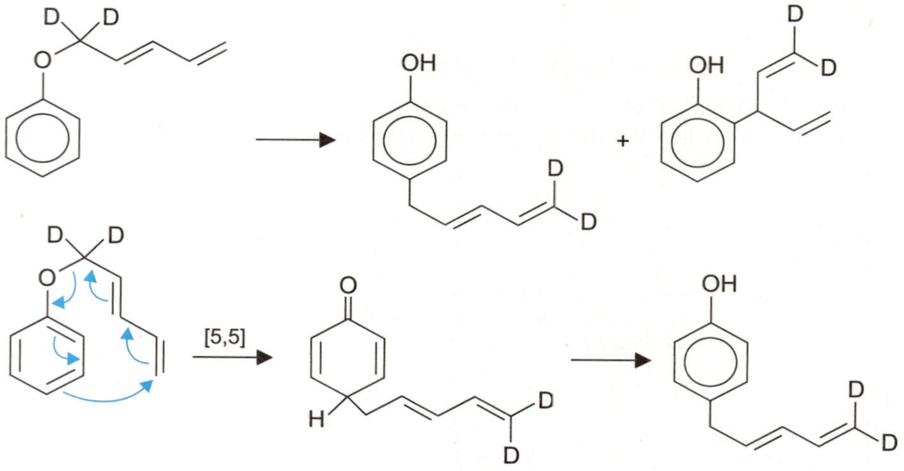

[5,5] Sigmatropic shifts

SCHEME 8.96c

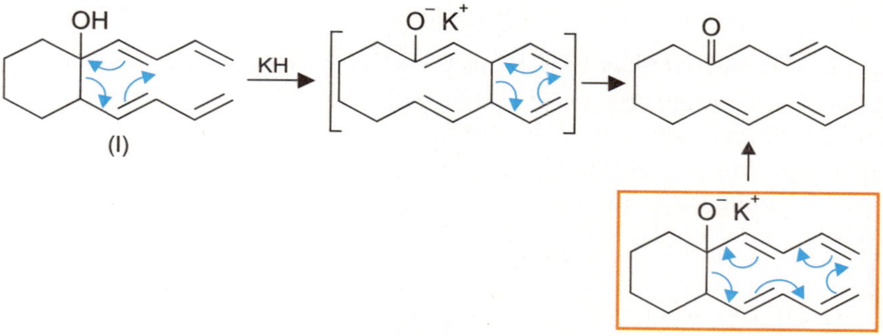

SCHEME 8.96d

8.8 THE ALDER ENE REACTION

In this reaction an alkene having an allylic hydrogen atom (the ene component) reacts thermally with a dienohile (C=C, C=O, N=N etc., called enophile) with the formation of a new σ bond to the terminal carbon of the allyl group by the migration of the allylic hydrogen and subsequent change in the position of allylic double bond. The *ene reaction* is always a six-electron reaction, with some characteristics with the [4 + 2] cycloaddition and the [1,5] sigmatropic rearrangement. Like the [4 + 2] cycloaddition, a four-atom component reacts with a two-atom component, but in the ene reaction, the four-atom component is a π bond and an allylic σ bond (usually to a H atom), not two π bonds. Like a [1, 5] sigmatropic rearrangement, a σ bond migrates, as in selenium dioxide oxidation of alkenes which is used to hydroxylate the alkenes at the allylic position (Scheme 8.97).

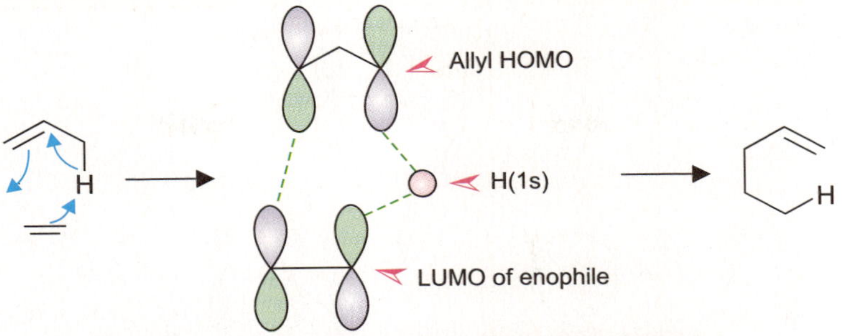

The Alder ene reaction is a symmetry allowed process

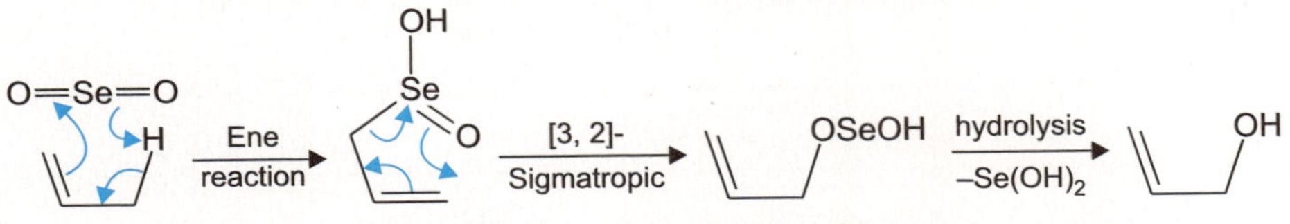

Selenium dioxide (SeO_2) hydroxylates alkenes at the allylic position

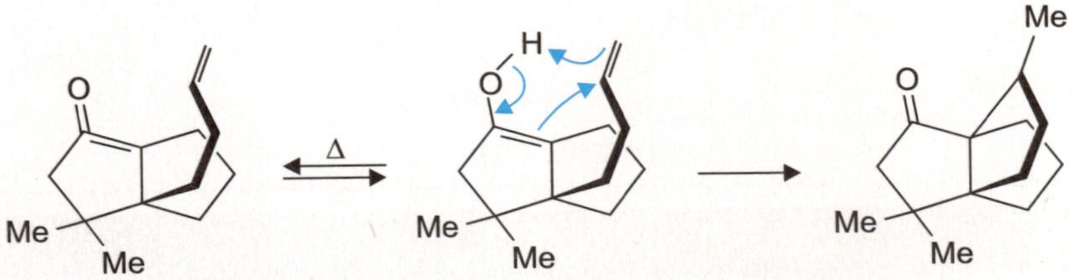

The enol serves as the ene component in the reaction

SCHEME 8.97

8.9 AROMATIC TRANSITION STATES

The reactions with aromatic transition states (2, 6, 10, 14 ... $4n + 2$) delocalized electrons are permitted, while the *anti*-aromatic systems (4, 8, 12 ... $4n$) delocalized electrons are the forbidden ones. It is clear that the Diels-Alder reaction has a 6π aromatic transition state isoelectronic with benzene while the forbidden cyclobutane formation has the unfavourable 4π transition state, isoelectronic with cyclobutadiene. This is the reason that although [4+2] reactions are common and general, the analogous concerted [2+2] and [4+4] thermal cycloaddition reactions generally do not occur, since the corresponding transition states involve 4 and 8 electrons respectively (Scheme 8.98). However, the other hand, cycloaddition reactions which involve 6 or 10 electrons occur readily (Scheme 8.99). It may be mentioned that *e.g.*, a [4+2] cycloaddition involves four π electrons of one system and two on another.

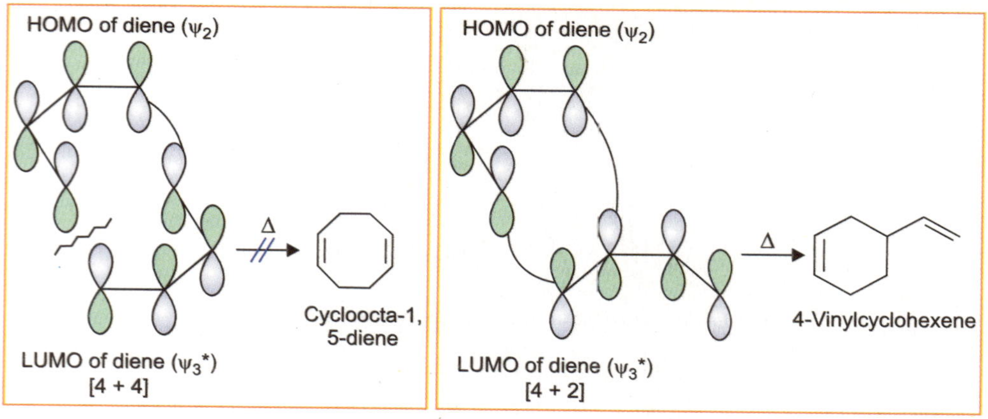

SCHEME 8.98

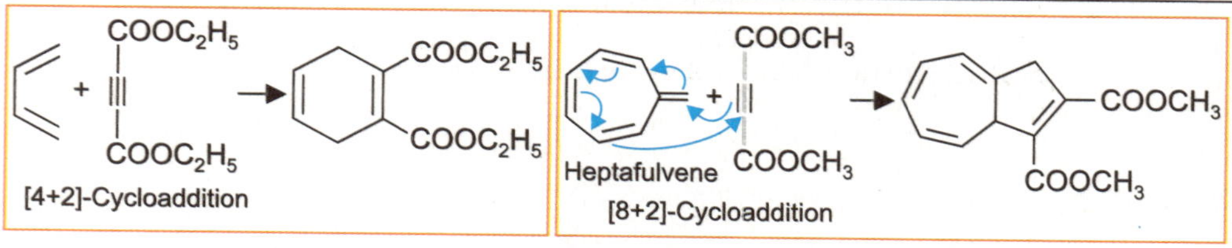

SCHEME 8.99

EXERCISE 8.17

How one can explain the outcome of the reactions (Scheme 8.100)?

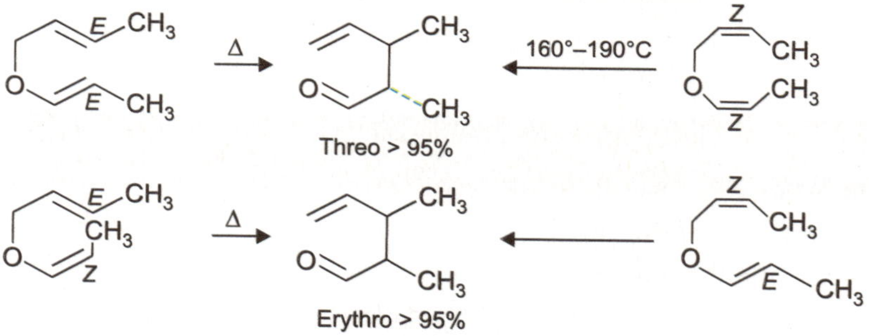

SCHEME 8.100

ANSWER. *This is the Claisen rearrangement of acyclic allyl vinyl ethers. Invoking a chair type transition state the outcome of stereocontrol is explained (see, Scheme 8.96).*

EXERCISE 8.18

The optically active diene (I, Scheme 8.101) undergoes a facile [3, 3] Cope rearrangement through a chair-like transition state (a symmetrical species). The asymmetry originally present in the starting diene (I) is lost in symmetrical transition state. The isomer (II) produced by the [3, 3] Cope rearrangement is the enantiomer of the starting compound. This facile racemisation occurs even at 50°C. Explain.

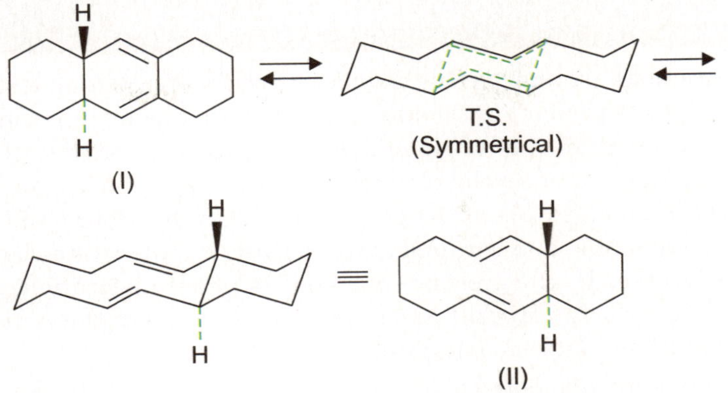

Racemisation during a [3, 3] cope rearrangement

SCHEME 8.101

EXERCISE 8.19

Why on thermal rearrangement optically active enol ether (I, Scheme 8.102) gives only the (S) product?

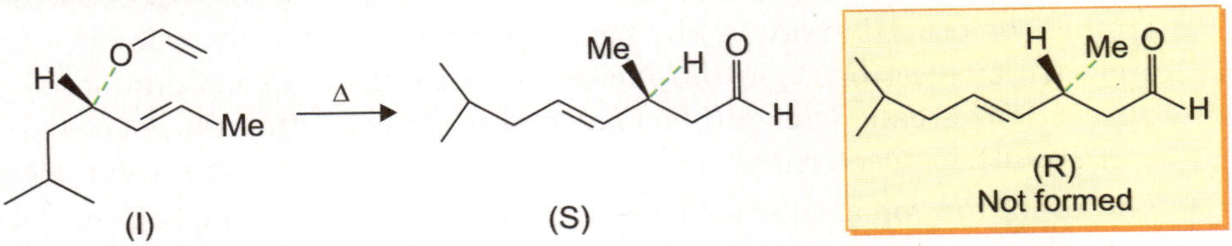

SCHEME 8.102

ANSWER. *Considering the chair like transition states, the formation of (R)-product would involve the axial orientation of the bulky sec-butyl group (Scheme 8.103).*

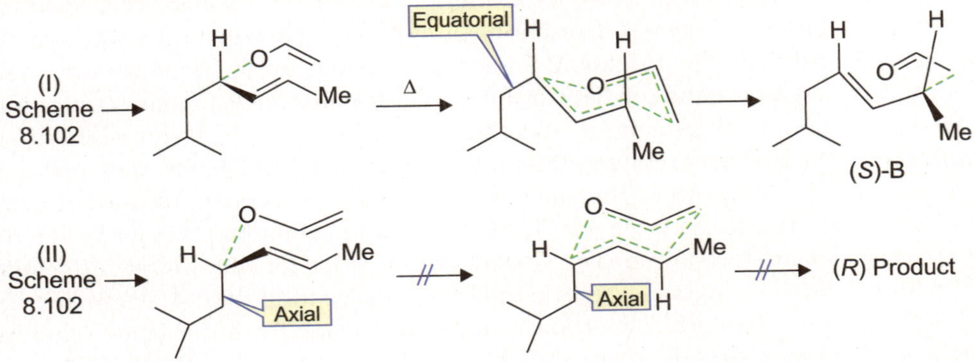

SCHEME 8.103

8.10 MÖBIUS-HÜCKEL ANALYSIS (PMO) APPROACH

The concerted reactions are analyzed by the classification of transition states as aromatic or antiaromatic. These predictions yield the same results as by other methods. In the previous section a mention was made of aromatic transition states. Hückel's rule of aromaticity states that a monocyclic planar conjugated system with $(4n + 2)\pi$ electrons is *aromatic* and therefore, stable in the ground state. On the other hand a system with $(4n)$ π electrons is unstable and is called antiaromatic. Further it has been shown that these rules are reversed by the presence of a node (a phase dislocation) in the array of atomic orbitals. The Möbius-Hückel concept is used to analyze the pericylic reactions without using the actual molecular orbitals.

The following points may be noted:

- Each atom of the interacting system is assigned a *p* orbital with one lobe black and one white (or some other designation). A hydrogen atom is represented by a circle of one color representing an *s* orbital.
- One draws each reactant with the black lobes on one side and the white on the other. Then one considers the transition state of a particular reaction, counts the number of electrons and the nodes in the array to reach the conclusion if that reaction is symmetry allowed or forbidden.
- A Hückel system has zero (or any even number) of nodes (phase changes) around the orbital array. A Hückel system with $4n + 2$ electrons is aromatic and with $4n$ electrons is antiaromatic. An array with an odd number of phase dislocations is called an *anti* Hückel system (Mobius system). An *anti*-Hückel system with $4n$ electrons is aromatic and with $4n + 2$ electrons is antiaromatic.
- The condition for aromaticity in *anti*-Hückel system is opposite to that for Hückel system.

1. Electrocyclic Reactions

(a) Thermal Ring Opening of Cyclobutenes—4n Systems

On the basis of FMO approach under thermal conditions the observed stereochemistry of the products indicates a conrotatory motion. Consider the thermal ring opening in *cis* 3,4-dimethyl-cyclobutene which on the basis of FMO method gives 2*E*, 4*Z*-hexadiene (Schemes 8.6 and 8.10). The Hückel-Möbius approach also predicts conrotatory motion under thermal conditions and predicts the formation of same diene. Consider the basis set for the starting butadiene (Scheme 8.104), the tilt at C-1 and C-4 as the butadiene system rotates toward the transition state is different for conrotatory and disrotatory modes. The transition state for conrotatory ring opening has one sign inversion (phase dislocation) and with four electrons it is aromatic. The conrotatory transition state for cyclobutene ring opening is therefore, aromatic. The transition state for disrotatory cyclobutene ring opening however, is anti-aromatic (no phase dislocation with 4 electrons). Thus the PMO approach like FMO method also predicts that for butadiene-cyclobutene interconversion the conrotatory transition state is the favoured aromatic transition state and thus thermal conrotatory ring opening in cyclobutenes is allowed and disrotatory opening is forbidden.

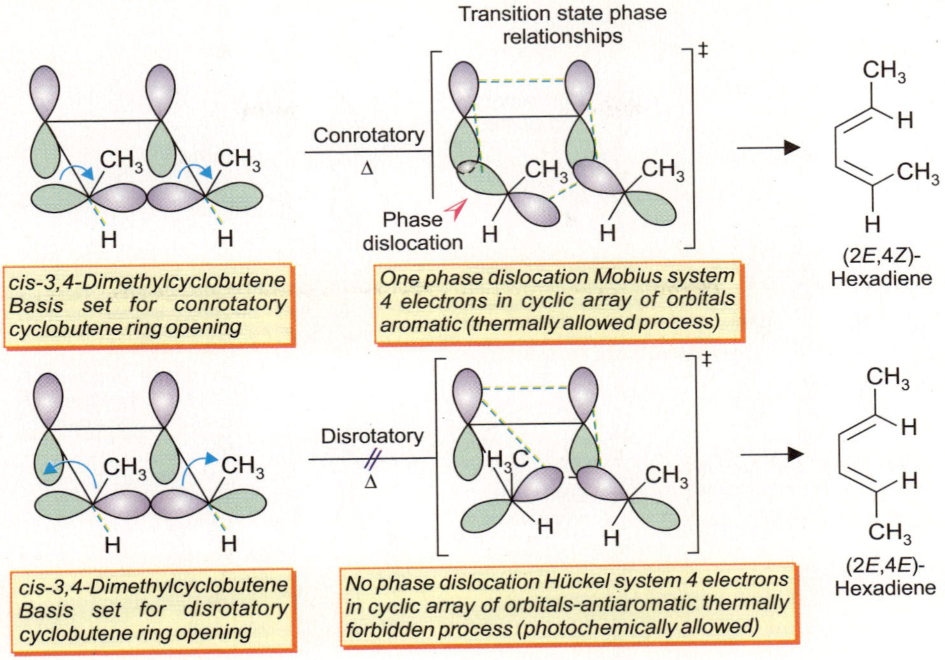

SCHEME 8.104

(b) Thermal Ring Closure of an Octatetraene—4n Systems

In this case also a conrotatory ring closure (Scheme 8.105) is allowed since in the transition state basis set there is one phase dislocation at the forming σ bond. The transition state is therefore, aromatic (8 electrons Möbius system). Disrotatory ring closure for the reaction (Scheme 8.105) will be however, antiaromatic eight electron Hückel system.

As with FMO approach the general statement is that thermal electrocyclic reactions in the conrotatory mode are allowed for 4*n* electron transition states.

The stereochemistry and other aspects are just the same as already discussed during FMO approach. Moreover, since Woodward-Hoffmann rules for photochemical reactions are always the reverse for thermal reactions, one can always predict the outcome of a reaction under photochemical conditions.

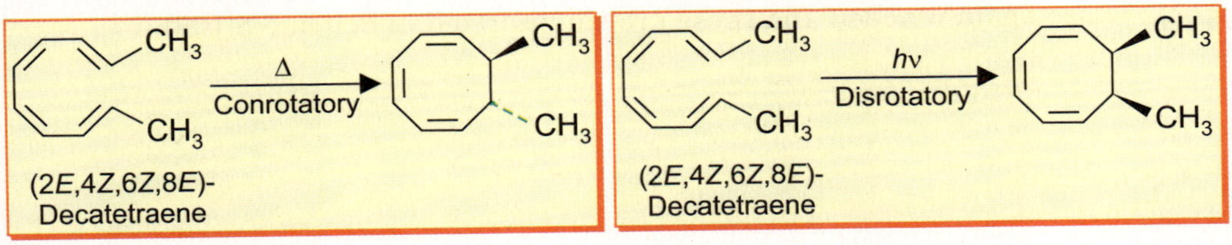

Continued...

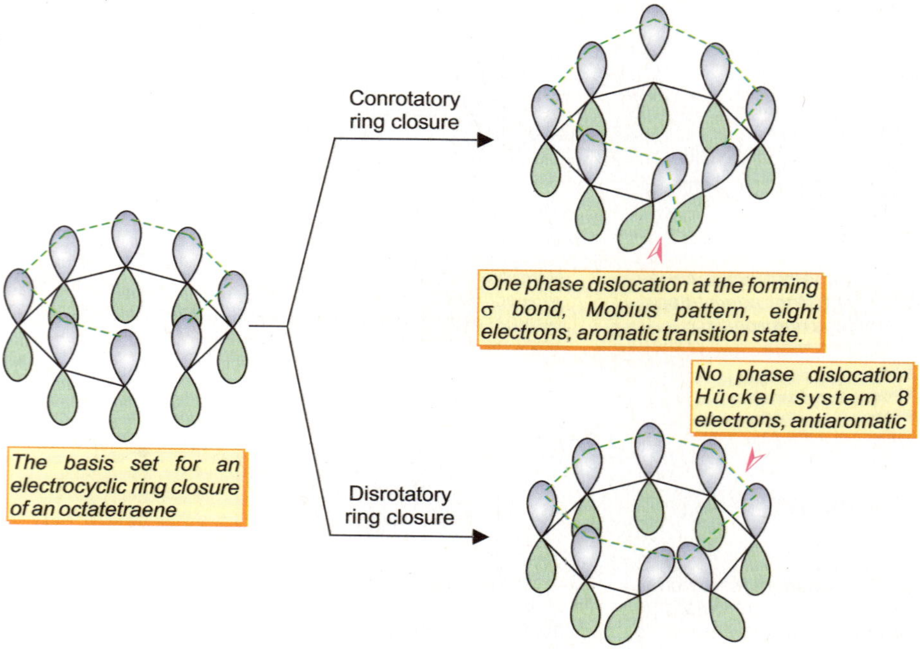

SCHEME 8.105

(c) Thermal Hexatriene Ring Closure—4n + 2 Systems

In FMO approach for $4n + 2$ electron transition states, in electrocyclic reactions, the disrotatory mode is allowed for thermal reactions and the conrotatory mode for photochemical reactions. In PMO method as well, this is found to be true. Starting from hexatriene basis set (I, Scheme 8.106), the transition state for disrotatory ring closure is a six-electron Hückel system and thus aromatic (II). The conrotatory ring closure proceeds through an antiaromatic anti-Hückel transition state. It is, therefore, correctly predicted that thermal ring closure of substituted hexatrienes should be disrotatory and the photochemical reaction should proceed via the opposite conrotatory path.

2. Diels-Alder Reaction

The selection rules may also be derived by the consideration of the aromaticity of the transition state of a Diels-Alder reaction. Same conclusions (as in the case of FMO approach) are again reached and these are summarized in Scheme 8.107).

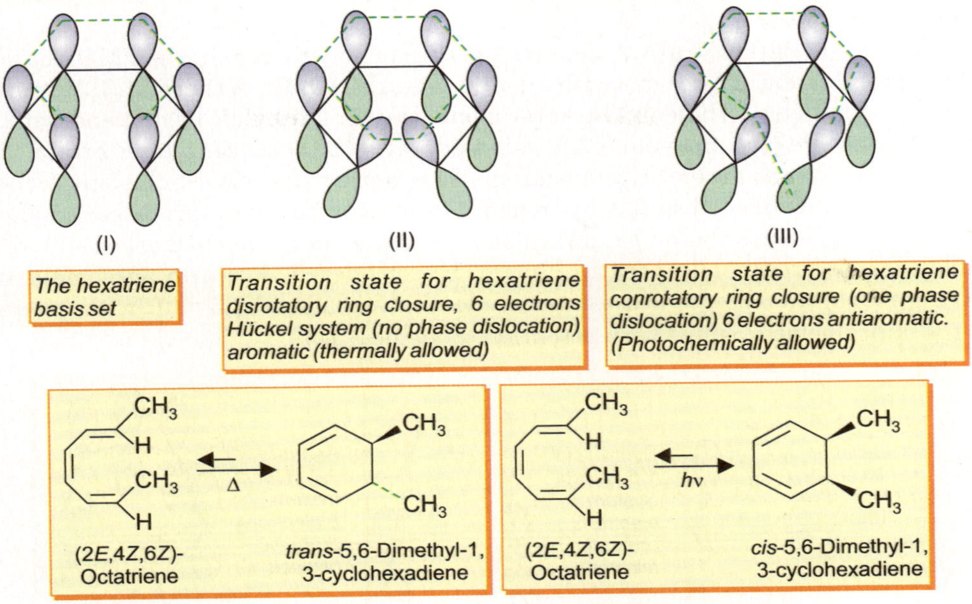

SCHEME 8.106

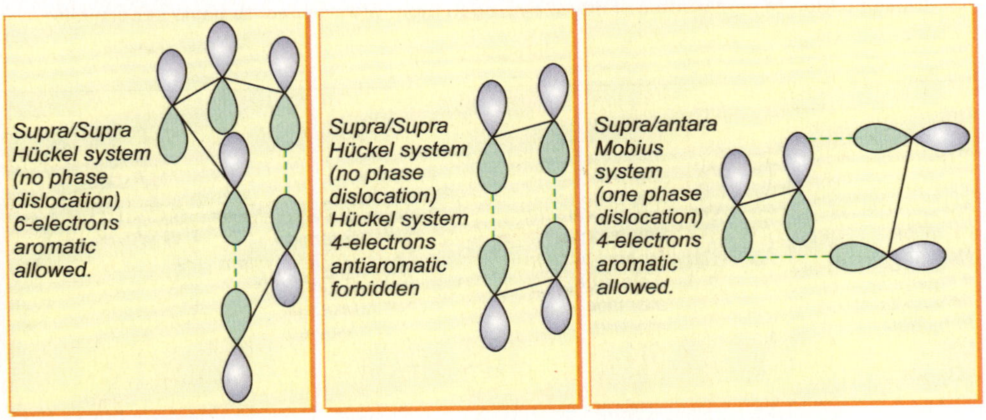

SCHEME 8.107

Antarafacial-Suprafacial cycloaddition is highly sterically hindered and is, therefore, less common. In case two ethylene molecules are brought together in such a way that a Mobius activated complex can be realized (see Scheme 8.107) the process become a suprafacial/antarafacial addition. This process should be allowed since it is a Mobius system with four electron and a node. For this process to be realized the ethylene molecules have to approach each other in a perpendicular geometry. The completion of this addition involves distortion of the carbon framework. The process, therefore, is difficult although allowed. For this reason simple alkenes do not display this addition.

3. Sigmatropic Shifts

Consider the simplest case of 1,3-sigmatropic shift of a hydrogen. In the FMO approach the hydrogen 1*s* orbital interacts with an allyl radicals HOMO. A thermal [1,3] suprafacial shift is symmetry forbidden, the antarafacial is symmetry allowed, but energetically very unfavourable (see, Scheme 8.72). A consideration of basis set atomic orbitals and their classification as aromatic or antiaromatic reaches the same conclusions (Scheme 8.108). The 1,3-suprafacial shift of hydrogen is forbidden, but the suprafacial 1,5-shift is allowed. The 1,7-shifts should be antarafacial, when an alkyl group (carbon) migrates, an additional stereochemical feature has to be considered. Again in agreement with FMO approach, the allowed processes include, the suprafacial 1,3-shift with inversion and the suprafacial 1,5-shift with retention (Scheme 8.109).

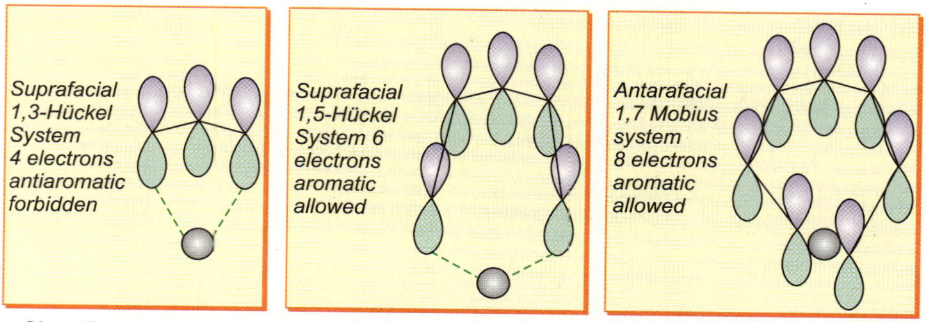

Classification of sigmatropic shifts of hydrogen atom with respect to basis set orbitals

SCHEME 8.108

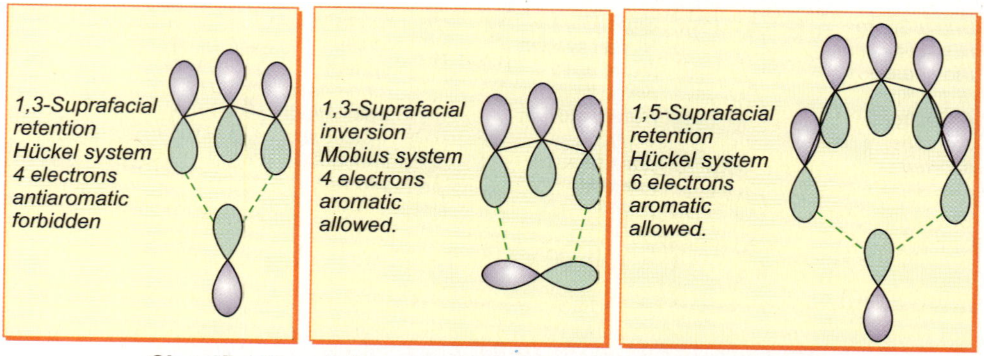

Classification of sigmatropic shifts of alkyl groups (Carbon) with respect to basis set orbitals

SCHEME 8.109

8.11 CORRELATION DIAGRAM METHOD

In this method a correlation of the geometrical symmetry of the orbitals between reactants and the products is involved. Based on this, a correlation diagram is developed which compares the symmetry characteristics from this comparison. A reaction can be easily predicted to be symmetry allowed or symmetry forbidden.

The following points may be noted:

- Molecular orbitals (of the reactant and the product) are either symmetric or antisymmetric around the mirror plane or around a two fold axis of rotation (C_2). Thus π orbital of ethylene in the ground state is symmetric (*S*) with respect to mirror plane (I, Scheme 8.110) while it is antisymmetric (A) with respect to rotational axis C_2 (II, Scheme 8.110). A two fold axis of rotation may be regarded as a pin at right angles passing in the middle. If a molecular orbital is spun 180° (*i.e.*, C_2 axis 360°/2), around this axis it would either yield an identical orbital *i.e.*, symmetric or an orbital with all signs the opposite of what these were originally *i.e.*, antisymmetric (A). Thus when the π orbital of ethylene is rotated it is found that the symmetry around the axis of rotation is not maintained. One would see that after the rotation the shaded lobes would be instead on the bottom rather than at the top (as was so in the original). Thus π orbital of ethylene is antisymmetric (A) in relation to its axis of rotation (II Scheme 8.110).

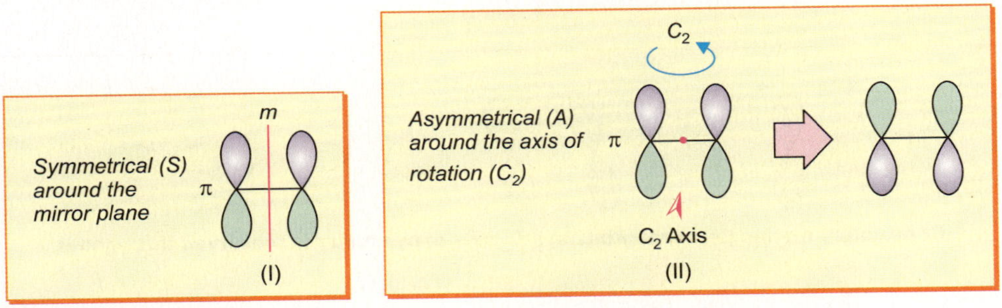

SCHEME 8.110

- On the other hand the same operations show that the antibonding π* orbital of ethylene is antisymmetric around the mirror plane but symmetric around the C_2 axis of rotation (Scheme 8.111).

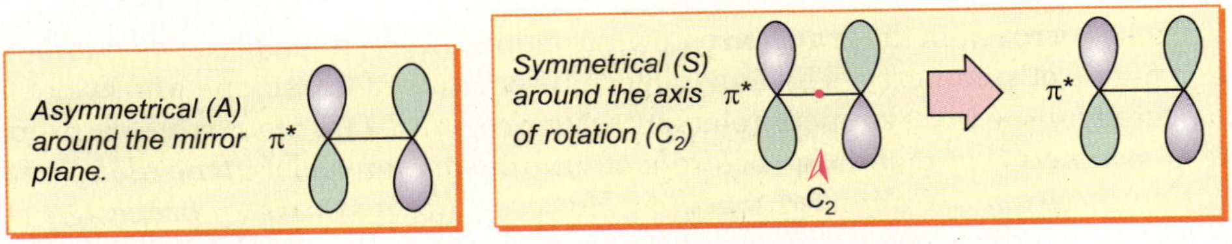

SCHEME 8.111

- Similarly the orbitals of the reactant and the product can be labelled. Thus the sigma orbital of a C-C covalent bond has a mirror plane of symmetry as well as C_2 symmetry, a σ* orbital is antisymmetric both with respect to plane of symmetry as well as rotational axis of symmetry (Scheme 8.112).

SCHEME 8.112

- During a disrotatory electrocyclic conversion, a plane of symmetry is maintained throughout (Scheme 8.113). If the reaction proceeds by a conrotatory motion a two fold axis of Symmetry (C_2) is preserved throughout (Scheme 8.114).

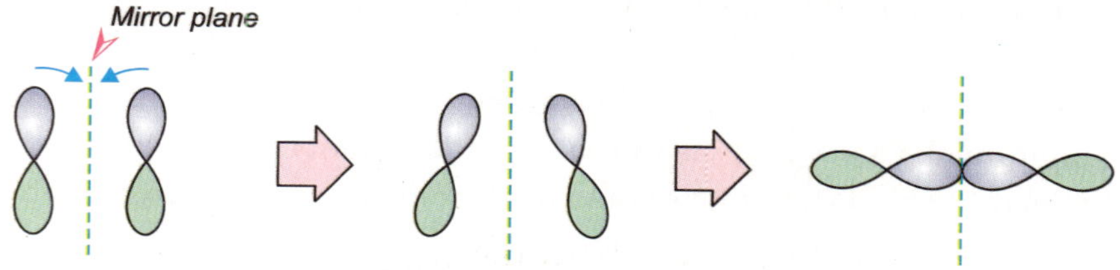

Disrotatory motion (terminal orbitals) a mirror plane of symmetry is maintained throughout

SCHEME 8.113

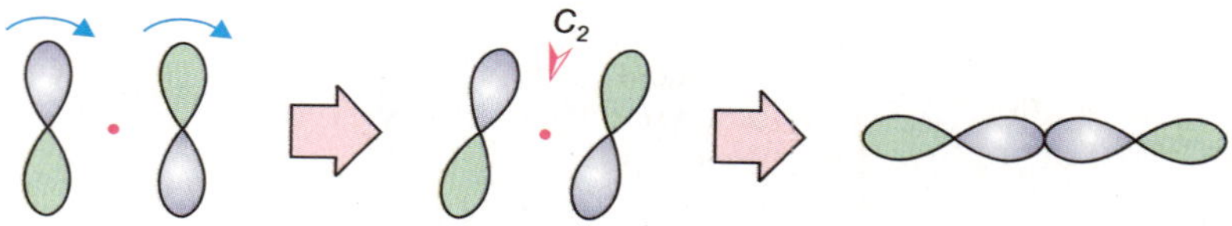

Conrotatory motion (terminal orbitals) a C_2 axis of symmetry is preserved throughout

SCHEME 8.114

- The most stable transition state is the one that conserves the symmetry of the reactant orbitals in passing to product orbitals—a symmetric (S) orbital in the reactant must transform to a symmetric orbital in the product, and an antisymmetric (A) orbital must transform to an antisymmetric orbital.

Correlation Diagrams For Electrocyclic Interconversion of 1,3-Butadiene and Cyclobutene

These correlation diagrams are now developed involving a plane of symmetry as well as an axis of symmetry. The four molecular orbitals of butadiene and cyclobutane are inspected for the two symmetry elements. [*In Scheme 8.115, the symmetric properties (plane of symmetry) of molecular orbitals of butadiene and cyclobutene along with the correlation diagram are depicted together. However, these symmetry properties can be translated onto a correlation diagram (as shown on the bottom of Scheme 8.115) for its study.*] Firstly one considers the correlation diagram for the disrotatory ring closure of 1,3-butadene to cyclobutene during which the mirror plane of symmetry is preserved (Scheme 8.115). The following points may be considered:

- ψ_1 can be converted to σ, however, ψ_2 cannot be converted to π which is the second-lowest orbital of cyclobutene.
- In order to conserve symmetry around the mirror plane, ψ_2 must instead be converted to π*, while it is ψ_3 which is converted to π.

- This symmetry correlation requires crossover between bonding and antibonding orbitals (Scheme 8.115). This is thermally an unfavourable energetic process (*i.e.*, symmetries of the molecular orbitals with respect to mirror plane do not show ground state correlation) and thus the disrotatory process is forbidden.
- When one considers the correlation diagram for the thermal conrotatory ring closure of 1,3-butadiene (Scheme 8.116) considering the symmetries of the orbitals in relation to the axis of rotation the following results arise.
- ψ_1 is now antisymmetric in relation to its axis of rotation (if the ψ_1 orbital is rotated 180° around the axis, the shaded lobes would come on the bottom instead of on top) similar is the case with π (Scheme 8.116).

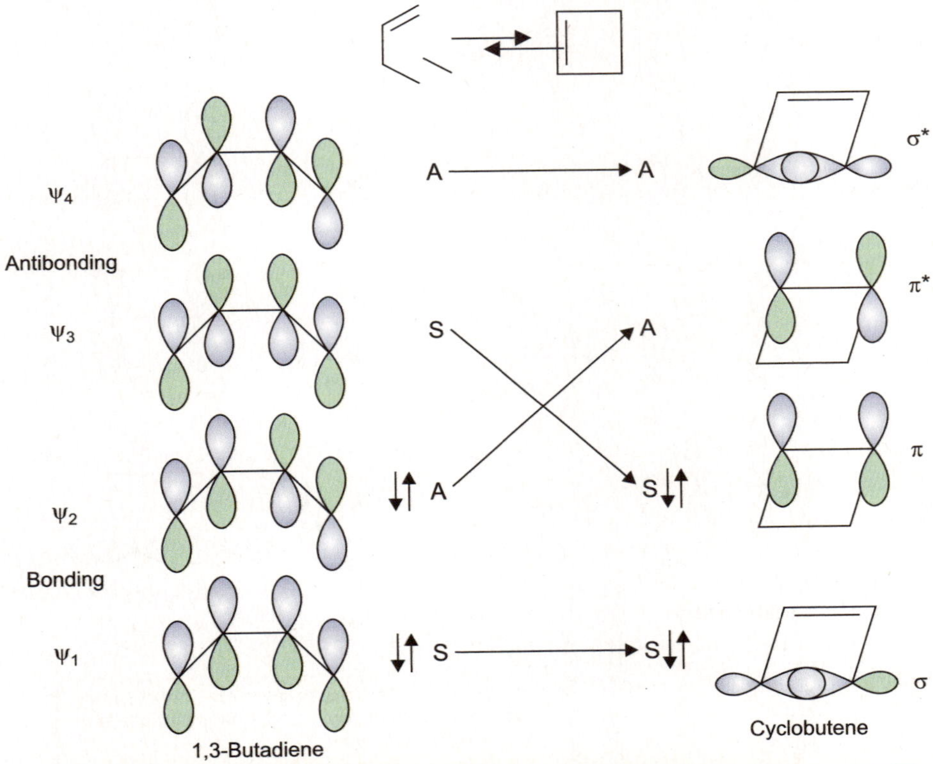

Symmetry properties of molecular orbitals and correlation diagram for disrotatory interconversion of butadiene-cyclobutene

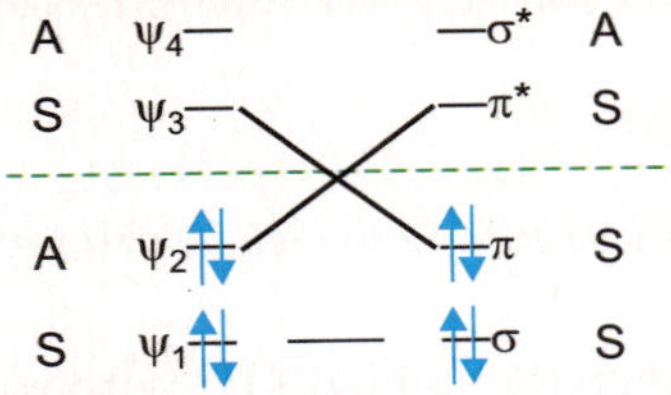

Correlation diagram for disrotatory interconversion of butadiene-cyclobutene (Mirror plane of symmetry is preserved)

Thermal disrotatory ring closure-1,3-butadiene (mirror symmetry maintained) symmetry forbidden

SCHEME 8.115

- However, ψ_2 and π^* are symmetric as in both the cases a 180° rotation would bring shaded lobes back to the top left and bottom right of the orbital.
- Similar operations reveal that σ is symmetric while σ^* is antisymmetric around the axis of rotation.
- These data show that correlation exists between the ground state bonding orbitals, therefore, a thermal conrotatory motion is symmetry allowed process.

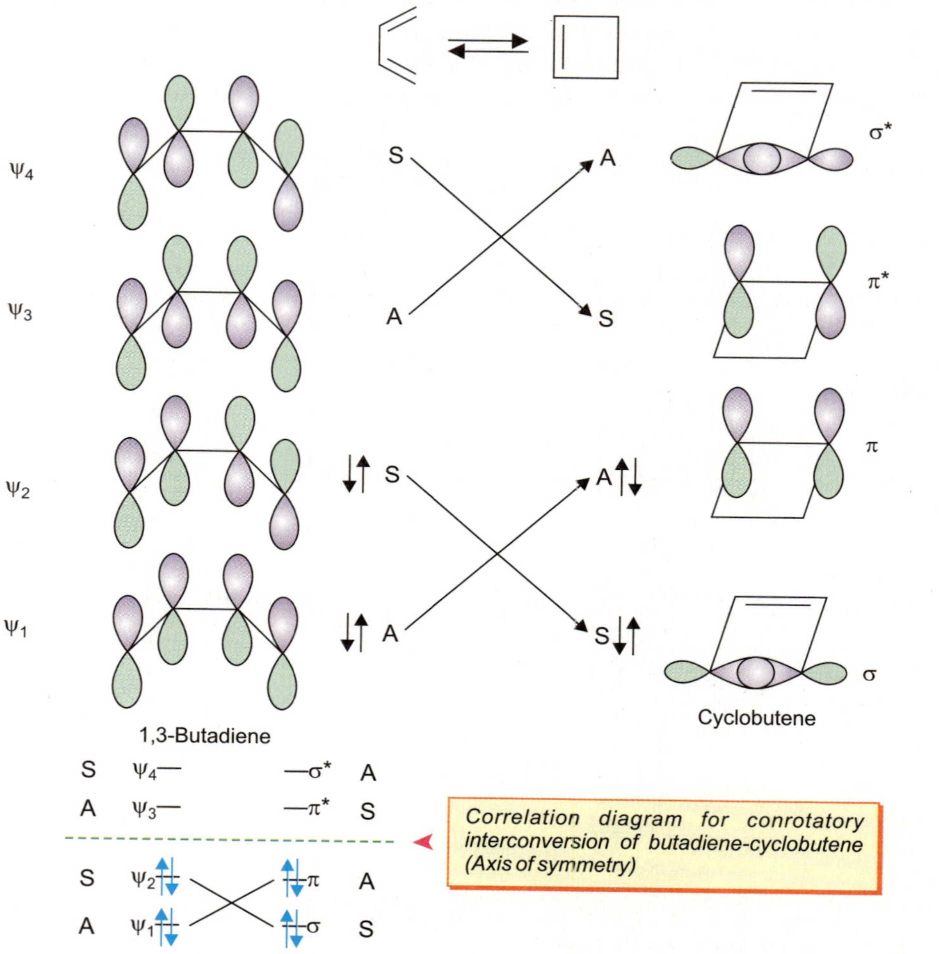

Thermal conrotatory ring closure-1, 3-butadiene (C_2 axis of symmetry maintained) symmetry allowed

SCHEME 8.116

Example 2: *[4 + 2] Cycloaddition of Ethylene to Butadiene to Give Cyclohexene Suprafacial Suprafacial Thermal Cycloaddition*

The orbital symmetry relationships are given (Scheme 8.117) with respect to the mirror plane of symmetry of the whole reacting system. The two σ-bonds of the product are considered as a symmetric and antisymmetric combination. After the classification of the orbitals with respect to symmetry these are arranged according to energy and the correlation lines can be drawn. (Scheme 8.117). It is found that all bonding levels of the reactants correlate with product ground state orbitals (orbital symmetry is conserved within the bonding orbitals and also within the antibonding set and no cross over between the two sets occurs). This therefore, is an allowed reaction.

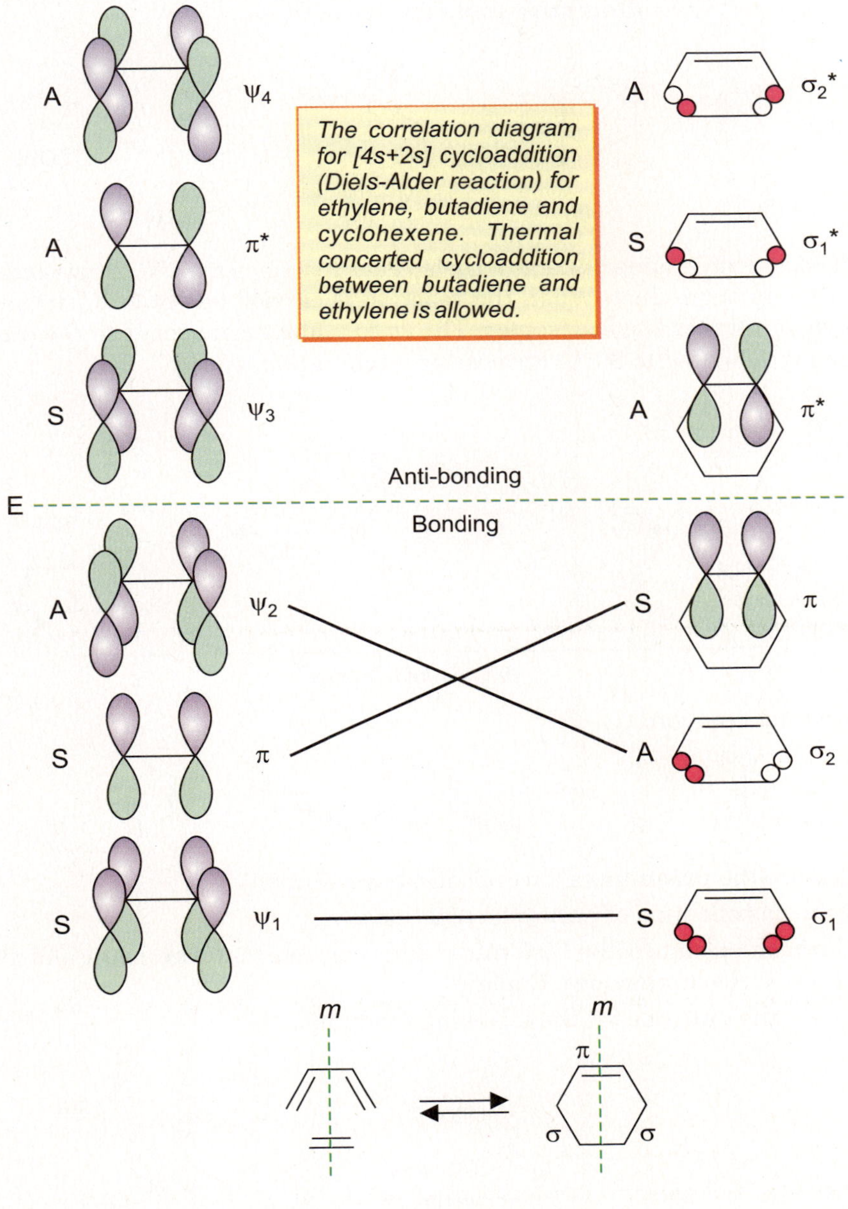

SCHEME 8.117

PROBLEMS

8.1. Write the stereostructure of the compound obtained by the Diels-Alder reaction of dimethyl maleate with butadiene.

8.2. Which diene and dienophile one would employ to synthesize the following compounds? Give alternative route for one of these.

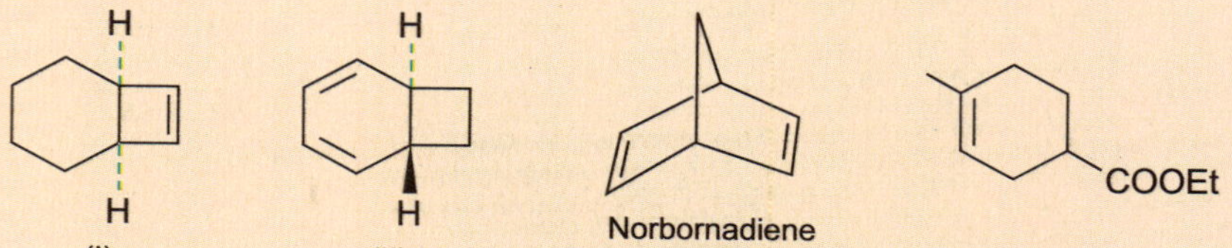

(I) (II) (III) (IV)

8.3. Furan and maleimide undergo a Diels-Alder reactions at 25°C to give *endo* adduct as the major product. When the reaction is carried out at 90°C, however, the major product is the *exo* isomer. The *endo* adduct isomerizes to the *exo* adduct when it is heated to 90°C. Propose an explanation.

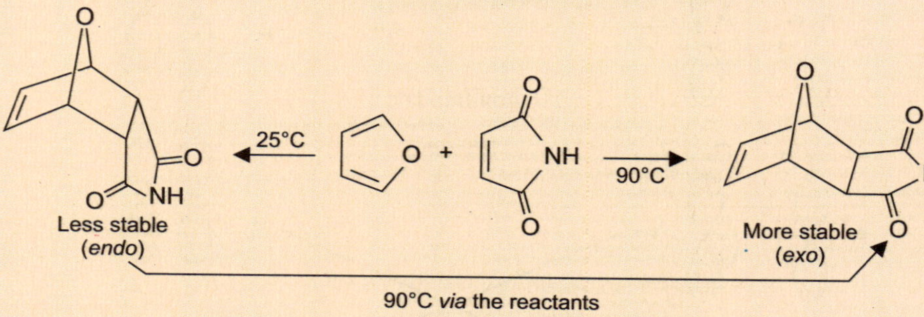

8.4. Name the reaction.

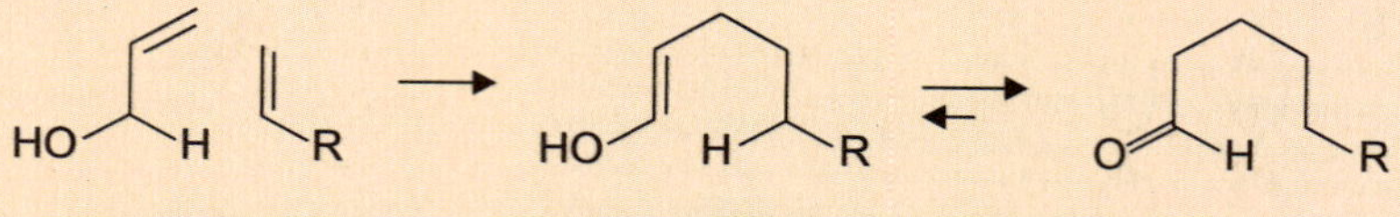

8.5. What are the preferences for cycloaddition reactions?

8.6. Give a classification of pericyclic reactions.

8.7. The transition state of the Diels-Alder pericyclic reaction is aromatic and compares with Cope rearrangement. Explain.

8.8. Explain the outcome for the following reaction:

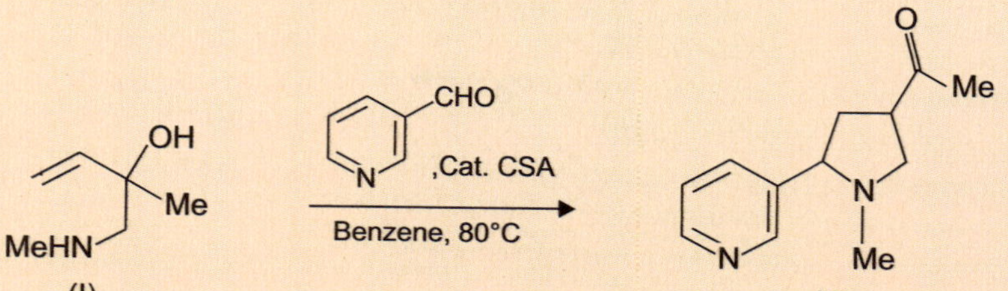

(I) (II)

8.9. Give the stereostructure of the products from the following electrocyclic reactions of (I and II) carried out under photochemical reactions. Discuss if each reaction takes place in a conrotatory or disrotatory fashion.

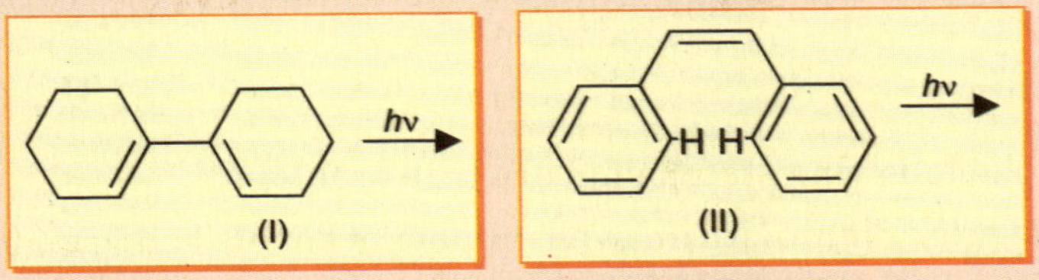

8.10. On thermal ring opening *cis* 3,4-dimethylcyclobutene gives two dienes (I and II). One of these is formed almost exclusively which is this diene and how it is formed?

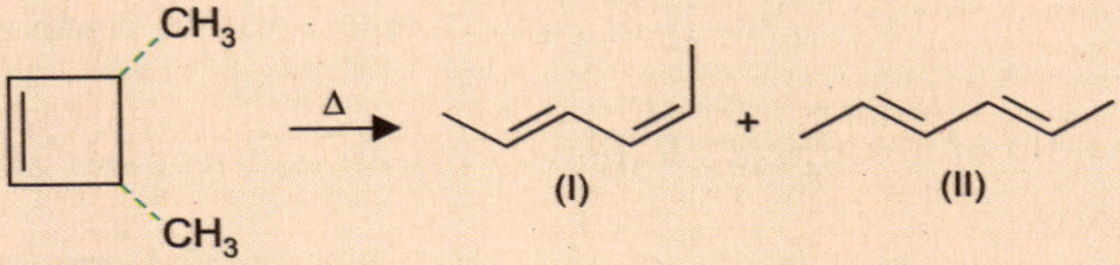

cis-3,4-Dimethyl-cyclobutene

8.11. Predict whether conrotatory or disrotatory motion will take place under the conditions mentioned against each compound. Write the structure of the product with stereo-chemistry in each case.

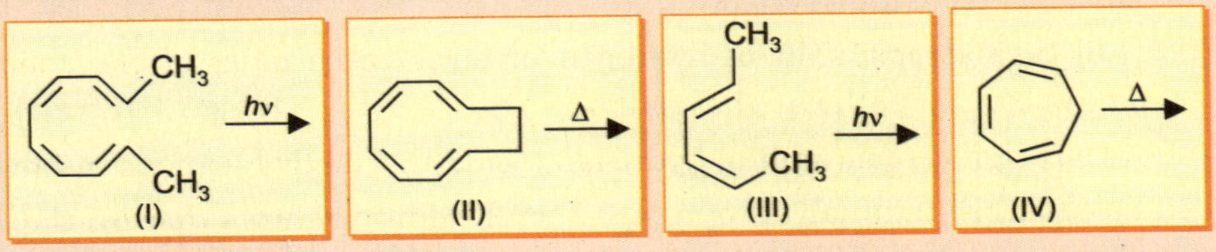

8.12. Explain briefly, taking one common example as to how FMO (frontier molecular orbital), method, PMO method and correlation diagram can be used for analyzing a pericyclic reaction.

8.13. Draw a correlation diagram for disrotatory conversion of butadiene to cyclobutene. Is the process allowed or forbidden? Explain.

8.14. A [3,3] sigmatropic rearrangement is thermally allowed via hypothetically formed allyl radicals. Explain by drawing appropriate bonding interactions.

8.15. Each of the following reactions (*a* and *b*) involve a sequence of two pericyclic reactions, what are these?

(*a*)

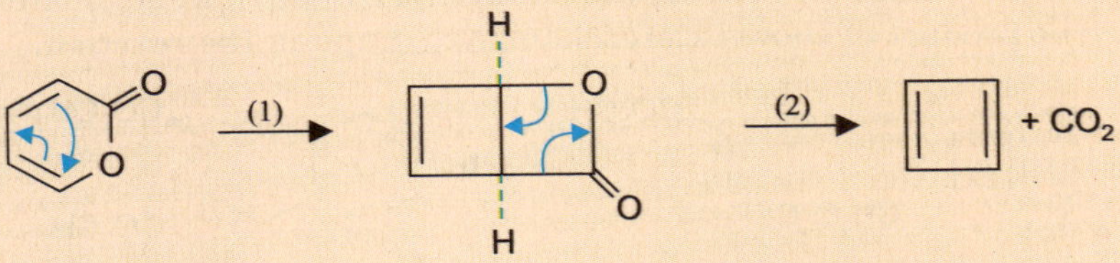

(*b*) 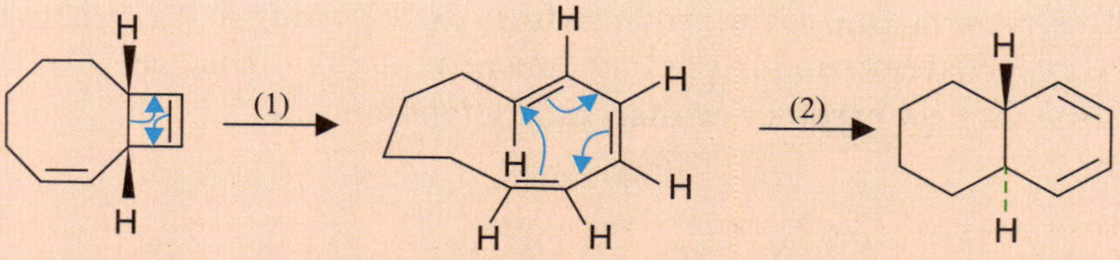

8.16. Which of the following statements are true or false?

(*i*) A conjugated diene with an even number of double bonds undergoes conrotatory ring closure under thermal conditions.

(*ii*) A conjugated diene with asymmetric HOMO undergoes conrotatory ring closure under thermal conditions.

(*iii*) A concerted antarafacial [1,3]-sigmatropic shift of hydrogen is thermally allowed.

(*iv*) The HOMO of a conjugated diene with an odd number of double bonds is symmetric.

(*v*) A [1,3] sigmatropic shift of carbon can occur under thermal conditions.

8.17. Fill in the blanks:

(*i*) A 1, 3-migration of carbon can take place thermally with ...... of configuration.

(*ii*) Pericyclic reactions are concerted, unaffected by catalysts or solvents and have ...... transition states.

(*iii*) [1,5] Sigmatropic shift of hydrogen involves three pairs of electrons and occurs by ...... pathway thermally.

(*iv*) A [1,7] sigmatropic shift of hydrogen occurs thermally by an ...... pathway.

(*v*) Frontier orbital analysis of a [4 + 2] cycloaddition shows that overlap of in phase orbitals to form new sigma bonds requires a ...... orbital overlap.

8.18. (*a*) Draw the transition states for suprafacial and antarafacial 1, 3 hydrogen shift by drawing the phase interactions in the basis sets. Show which process is thermally forbidden and which thermally allowed?

(*b*) Explain the results of the following photochemical reaction.

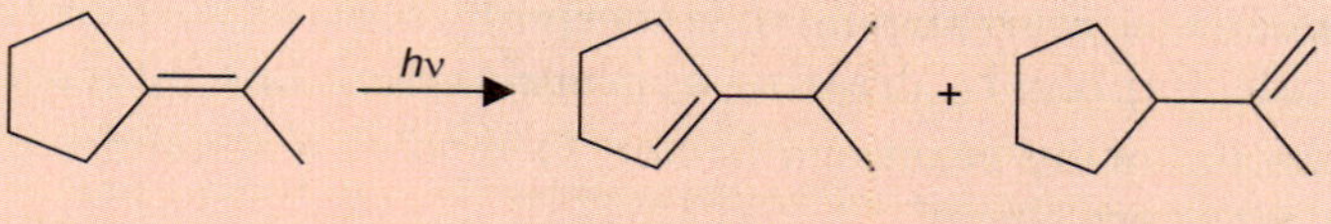

8.19. (*a*) Benzocyclobutene on heating with dimethyl *trans*-2-butene dioate (I) gives a bicyclic product of shown stereochemistry. Explain the reaction.

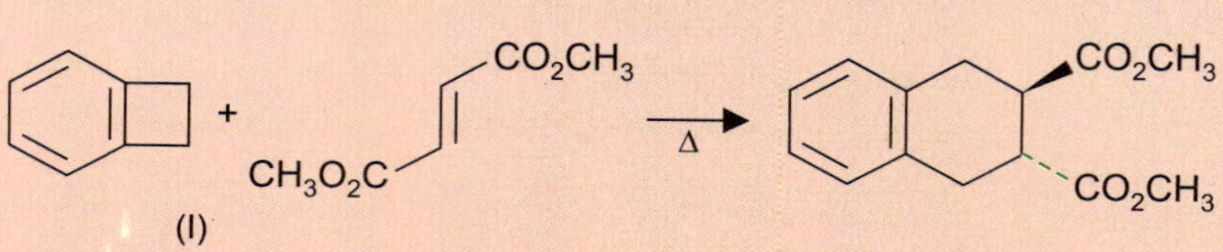

(I)

(*b*) Write the product with stereochemistry of the Diels-Alder reactions (II and III).

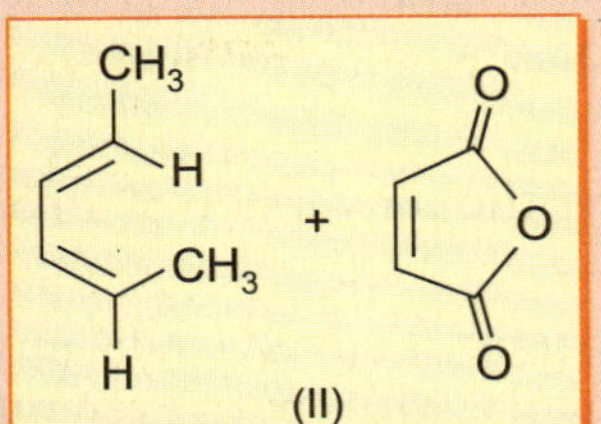

(II)

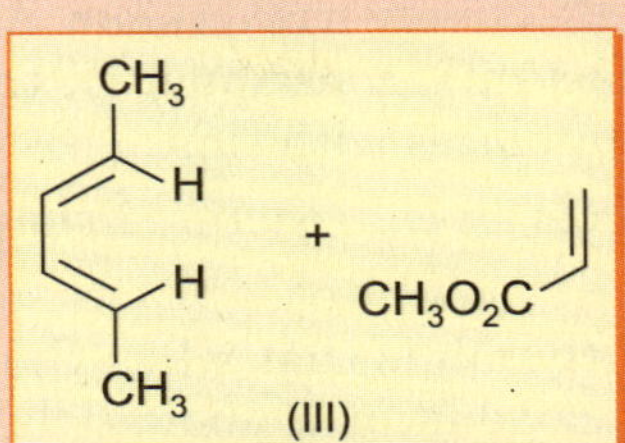

(III)

(*c*) Explain the reaction.

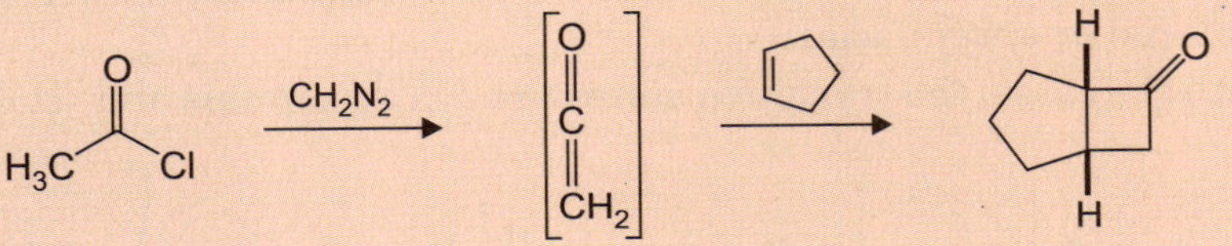

ANSWERS TO SELECTED PROBLEMS

8.1. The reaction follows a stereospecifically *syn* pathway. The product is *cis*-dimethyl cyclohexene-4, 5-dicarboxylate. Therefore, the groups which are *cis* in the olefin also occupy *cis*-positions in the cyclohexene ring.

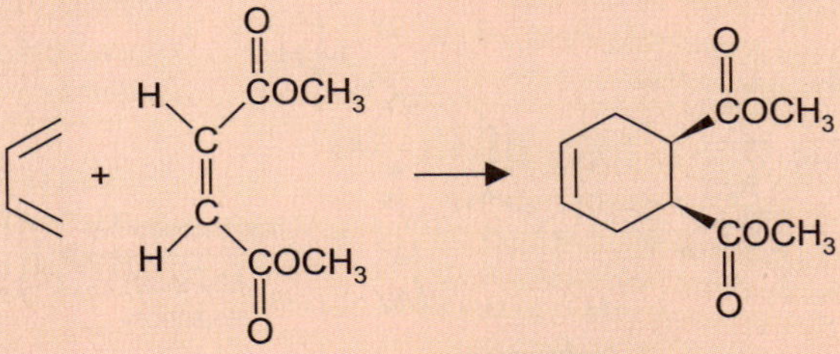

8.2. Cyclopentadiene and acetylene.

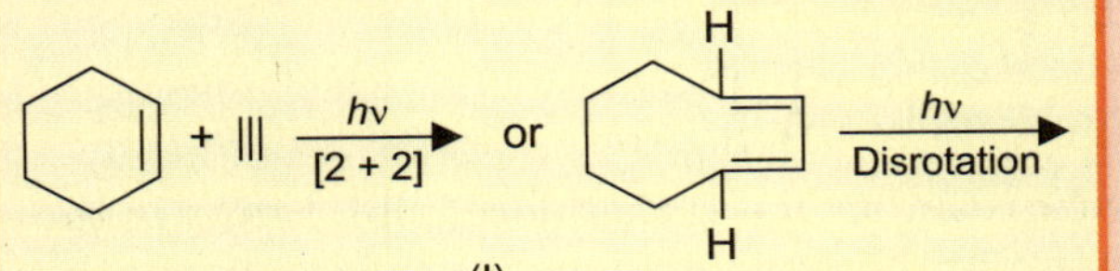

(I)

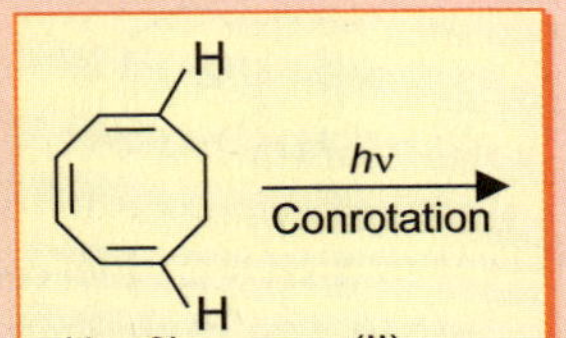

(4n+2) (II)

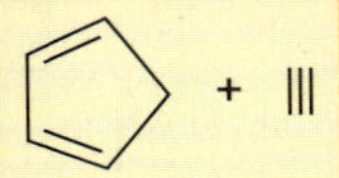

(III)

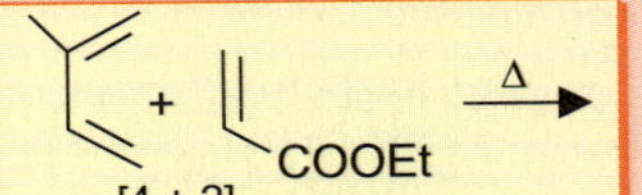

[4 + 2] (IV)

CHAPTER 8

8.3. The *exo* product is thermodynamically more stable. The less stable *endo* isomer (kinetically favoured adduct) is formed faster and predominates at 25°C, the reaction is effectively irreversible. At 90°C this product is in rapid equilibrium with the reactants, consequently, the less rapidly formed but more stable *exo* isomer accumulates with time.

8.4. The reaction is an *ene* reaction involving the conversion of an allyl alcohol to an aldehyde.

8.5. The rule of thumb is that when the reactants involve odd number of electron pairs, the cycloaddition is allowed thermally, while with even number of electron pairs, the cycloaddition is allowed photochemically.

8.6. Electrocyclic reactions are stereochemically classified as conrotatory and disrotatory, cycloadditions and sigmatropic rearrangements are classified as suprafacial or antarafacial.

8.7. In both the cases, the transition states involve six orbitals and six electrons.

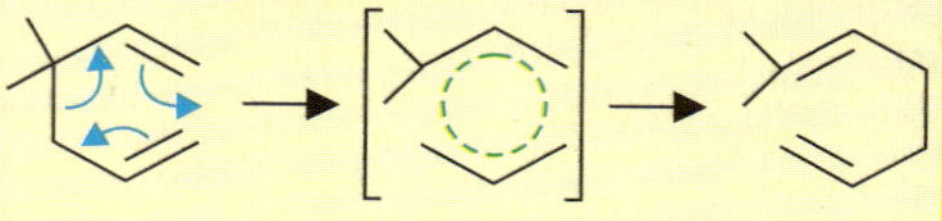

The **Cope rearrangement**

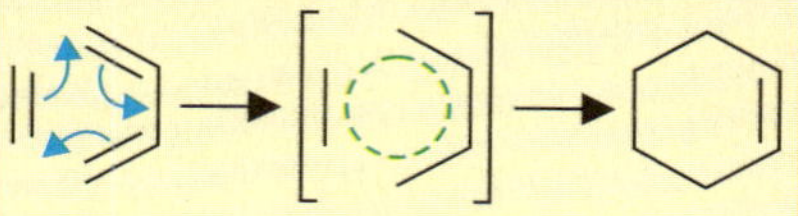

The **Diels-Alder reaction**

8.8. This is an [3, 3]-aza Cope rearrangement on the iminium ion [A, obtained from the butenyl amine (I) and the carbaldehyde with acid catalyst–camphor sulfonic acid CSA] to give (B) which undergoes intramolecular Mannich reaction.

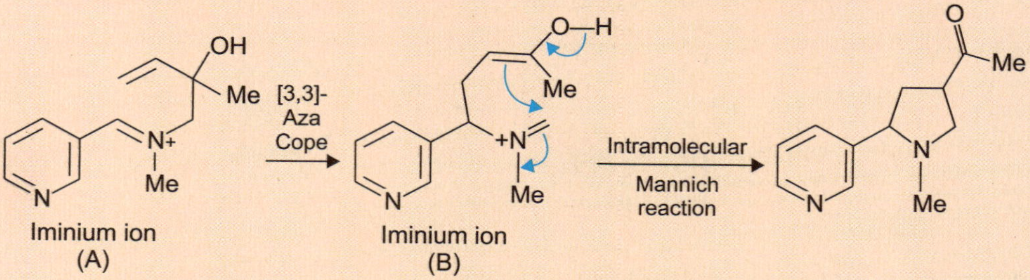

8.9. (I)

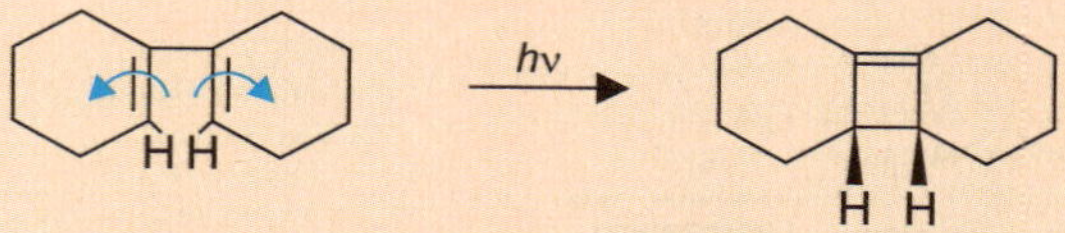

(II) It is a $4n + 2$ electrocyclic reaction; $n = 1$, therefore, conrotatory in the excited state, the hydrogen atoms at the ring junctions in dihydrophenanthrene will be *trans* to one another.

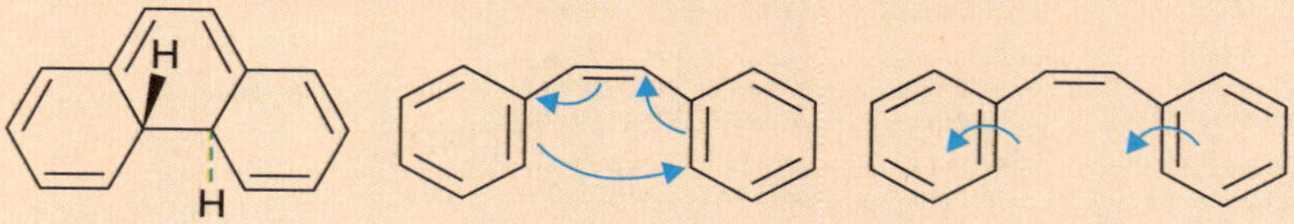

8.10. It is diene (I). Thermally a compound with two π bonds undergoes conrotatory ring closure. Since in the product the two methyls are *cis* placed, these point in the same direction in the reactant. In diene (I) the methyls point in the same direction.

8.11. The stereostructures of product in each case is presented (refer to Table 8.1).

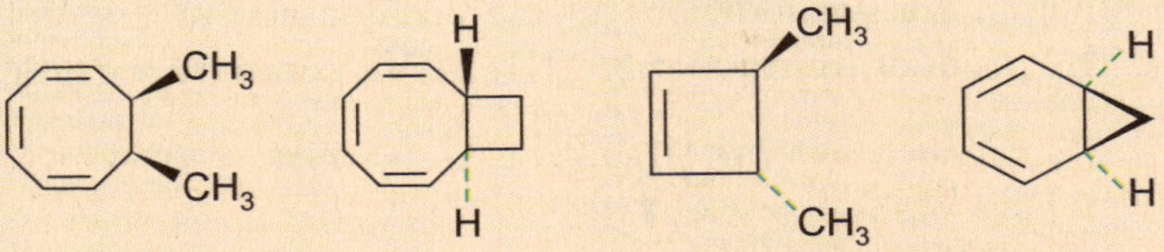

(I) (II) (III) (IV)

(I) The reactant being a (4*n*) π electron system will undergo disrotatory ring closure under photochemical conditions. Since the two methyl groups point in opposite directions these will be *cis* in the product.

(III) The reactant is (2*E*, 4*Z*) hexadiene (4*n*) π electron system. Photochemically it will undergo ring closure by disrotation. Since the methyl substituents point in the same direction, the product will have these in *trans* relationship.

(IV) The reactant 1,3,5-cycloheptatriene is a (4*n* + 2) π electron system. It will undergo disrotatory ring closure under thermal conditions, since in the triene, the hydrogen substituents (not shown) point in opposite directions these will be *cis* in the product norcaradiene. However, due to the strain, norcaradiene cannot be isolated and it reverts back to the starting material. In case the ring size of the starting triene is large as in 1,3,5-cyclononatriene (Scheme 8.30) the product is stable and is isolated as the exclusive product.

8.15. (*a*) Both (1 and 2) represent two successive photochemical disrotatory electrocylic reactions.

(*b*) The transformation (1) is a thermal controtory electrocylic ring opening, and the (2) transformation is a thermal distrototory electrocylic ring closure.

8.16. (*i*) True; (*ii*) true; (*iii*) true; (*iv*) true; (*v*) true.

8.17. (*i*) Inversion; (*ii*) cyclic; (*iii*) suprafacial; (*iv*) antarafacial; (*v*) suprafacial.

8.18. (*b*) This is a photochemical 1,3-hydrogen shift (see, Scheme 8.34). Two products are expected since two different allylic hydrogens can undergo this shift.

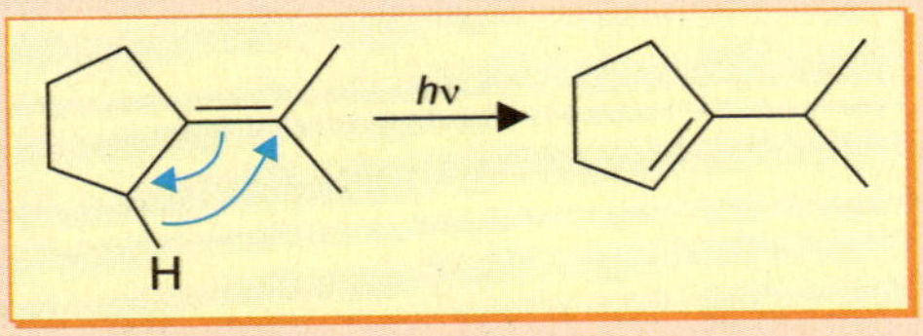

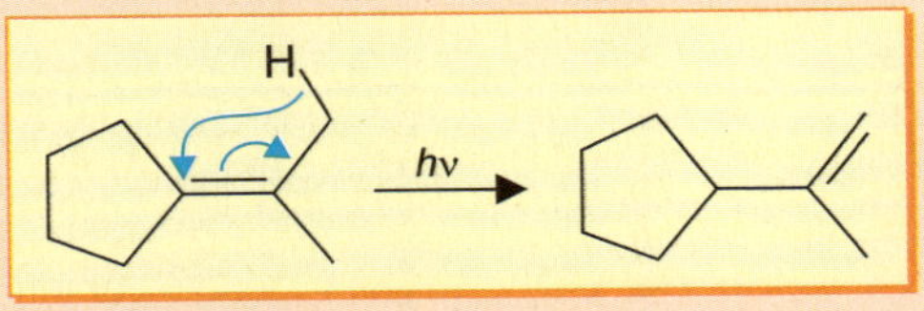

8.19. (*a*) It is a combination of electrocyclic reaction (to give IV) followed by Diels-Alder reaction.

(*b*) The reaction (II) gives the product (V) with the stereochemistry at centres other than ring junction as shown. This is deriveable from the discussion presented (Scheme 8.49*a*). Maleic anhydride is a *cis*-alkene, since the Diels-Alder reaction is a *syn* addition, the stereochemistry at the ring junction must be *cis*. The product from reaction (III) is (VI, compare with Scheme 8.49).

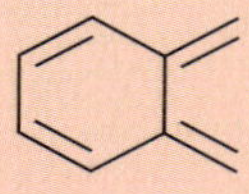

(IV)

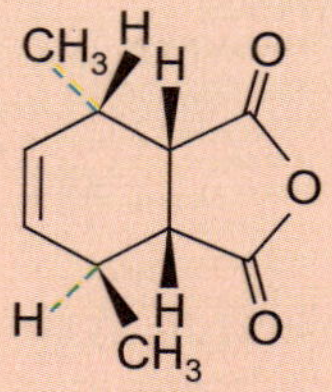

(V)

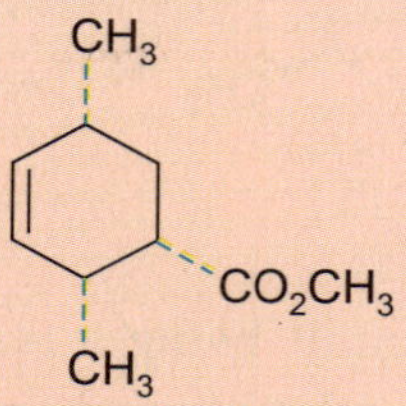

(VI)

(*c*) This is a ketene-alkene thermal cycloaddition to give cyclobutanones. Most ketenes are not kinetically stable, so they are usually generated in situ, either by E2 elimination of HCl from an acyl chloride or by a Wolff rearrangement of an α-diazoketone.

Index

Symbols

A

B

C

I

J

K

L

M

N

O

P

R

S

T

U

W

Z

NEW AGE
Publishing Globally
BEST SELLING BOOKS ON CHEMISTRY
A TEXTBOOK OF ORGANIC CHEMISTRY
RAJ K BANSAL
ISBN: 978-81-224-3967-0
PB MRP: ₹ 499.00
ORGANIC CHEMISTRY
PROBLEMS AND SOLUTIONS
ISBN: 978-81-224-3415-6
PB MRP: ₹ 399.00
ORGANIC CHEMISTRY
Volume I
ISBN: 978-81-224-2738-7
PB MRP: ₹ 599.00
ORGANIC CHEMISTRY
Volume II
ISBN: 978-81-224-3368-5
PB MRP: ₹ 499.00
ORGANIC CHEMISTRY
Volume III
ISBN: 978-81-224-3611-2
PB MRP: ₹ 450.00
ORGANIC REACTIONS AND THEIR MECHANISMS
P S KALSI
ISBN: 978-93-89802-08-5
PB MRP: ₹ 499.00
ORGANIC REACTIONS STEREOCHEMISTRY AND MECHANISM
Through Solved Problems
P S Kalsi • R S Oza
ISBN: 978-93-86418-52-4
PB MRP: ₹ 475.00
ORGANIC REACTION MECHANISMS
Raj K Bansal
ISBN: 978-81-224-3062-2
PB MRP: ₹ 399.00
ORGANOMETALLIC CHEMISTRY
A UNIFIED APPROACH
R C MEHROTRA
A SINGH
ISBN: 978-81-224-1258-1
PB MRP: ₹ 450.00
STEREOCHEMISTRY OF ORGANIC COMPOUNDS
Principles and Applications
D Nasipuri
ISBN: 978-93-89802-47-4
PB MRP: ₹ 499.00
STEREOCHEMISTRY WITH APPLICATIONS TO ORGANIC REACTIONS
ISBN: 978-93-88818-71-1
PB MRP: ₹ 299.00
BIOORGANIC, BIOINORGANIC AND SUPRAMOLECULAR CHEMISTRY
P S KALSI • J P KALSI • ASHU CHAUDHARY
ISBN: 978-93-88818-51-3
PB MRP: ₹ 499.00
BIOINORGANIC CHEMISTRY
K Hussain Reddy
ISBN: 978-93-87788-87-9
PB MRP: ₹ 399.00
PHOTOCHEMISTRY AND PERICYCLIC REACTIONS
JAGDAMBA SINGH
JAYA SINGH
ISBN: 978-93-86649-16-4
PB MRP: ₹ 400.00
SPECTROSCOPY OF ORGANIC COMPOUNDS
P S KALSI
ISBN: 978-81-943696-8-4
PB MRP: ₹ 450.00
SPECTROSCOPY OF INORGANIC COMPOUNDS
ISBN: 978-93-89802-17-7
PB MRP: ₹ 299.00
Physical Chemistry
ISBN: 978-81-224-2038-8
PB MRP: ₹ 295.00
PHYSICAL CHEMISTRY
Volume II
ISBN: 978-81-224-2403-4
PB MRP: ₹ 375.00
PHYSICAL CHEMISTRY THROUGH PROBLEMS
S K DOGRA
S DOGRA
ISBN: 978-81-224-3805-5
PB MRP: ₹ 699.00
BIOPHYSICAL CHEMISTRY
P S KALSI
N MAHANTA
ISBN: 978-93-86070-86-9
PB MRP: ₹ 299.00
A Textbook of PHYSICAL CHEMISTRY
THIRD EDITION
ISBN: 978-93-93159-79-3
PB MRP: ₹ 799.00
AN INTRODUCTION TO PHYSICAL CHEMISTRY
ISBN: 978-81-224-3221-3
PB MRP: ₹ 250.00
MOLECULAR STRUCTURE AND SYMMETRY
K VEERA REDDY
ISBN: 978-93-88818-52-0
HB MRP: ₹ 999.00
FUNDAMENTALS OF MOLECULAR SPECTROSCOPY
ISBN: 978-81-951755-1-2
PB MRP: ₹ 499.00
ELEMENTS OF MOLECULAR SPECTROSCOPY
ISBN: 978-81-224-7216-5
PB MRP: ₹ 350.00
Contact your nearest bookseller to buy these books. If you are unable to get it, you could send us Demand Draft in favour of NEW AGE INTERNATIONAL (P) LTD., payable at New Delhi with the total amount of book value at the address below:
FREE SHIPPING IN INDIA
NEW AGE INTERNATIONAL (P) LIMITED, PUBLISHERS
7/30 A, Daryaganj, New Delhi-110002
Visit us: www.newagepublishers.com
(CIN: U74899DL1966PTC004618)

NEW AGE
Publishing Globally
BEST SELLING BOOKS ON CHEMISTRY
Basic Concepts of ANALYTICAL CHEMISTRY
S M Khopkar
ISBN: 978-93-87477-79-7
PB MRP: ₹ 499.00
ANALYTICAL CHEMISTRY (Problems and Solutions)
ISBN: 978-81-224-2746-2
PB MRP: ₹ 275.00
QUANTUM CHEMISTRY
R K Prasad
ISBN: 978-93-93159-31-1
PB MRP: ₹ 550.00
THE ESSENCE OF HETEROCYCLIC CHEMISTRY
Arun R Parikh
Hansa Parikh
Ranjan Khunt
ISBN: 978-81-224-3436-1
HB MRP: ₹ 999.00
HETEROCYCLIC CHEMISTRY
SIXTH EDITION
Raj K Bansal
ISBN: 978-93-88818-56-8
PB MRP: ₹ 475.00
SOLID STATE CHEMISTRY
SECOND EDITION
D K CHAKRABARTY
ISBN: 978-81-224-4277-9
HB MRP: ₹ 999.00
SEPARATION CHEMISTRY
FOURTH EDITION
R P BUDHIRAJA
ISBN: 978-93-89802-25-2
HB MRP: ₹ 999.00
BASIC SEPARATION TECHNIQUES IN BIOCHEMISTRY
R O OKOTORE
ISBN: 978-81-224-4317-2
HB MRP: ₹ 499.00
Chemistry of Water
ALLA APPA RAO
ISBN: 978-81-224-2361-7
HB MRP: ₹ 495.00
WATER CHEMISTRY
INDUSTRIAL AND POWER STATION WATER TREATMENT
ISBN: 978-81-224-0876-8
PB MRP: ₹ 200.00
Essentials of NUCLEAR CHEMISTRY
H J ARNIKAR
ISBN: 978-81-224-3203-9
PB MRP: ₹ 450.00
CHEMISTRY OF ATMOSPHERE
SECOND EDITION
P S SINDHU
ISBN: 978-93-88818-66-7
HB MRP: ₹ 999.00
SYMMETRY AND GROUP THEORY IN CHEMISTRY
THIRD EDITION
RAMESHWAR AMETA
ISBN: 978-93-93159-60-1
HB MRP: ₹ 1199.00
SYMMETRY AND SPECTROSCOPY OF MOLECULES
K Veera Reddy
ISBN: 978-93-93159-69-4
PB MRP: ₹ 550.00
ASYMMETRIC SYNTHESIS
PRINCIPLES AND METHODOLOGY
SECOND EDITION
Y VATSALA
ISBN: 978-93-88818-74-2
HB MRP: ₹ 899.00
HETEROGENEOUS CATALYSIS
ISBN: 978-81-224-2197-2
PB MRP: ₹ 295.00
POLYMER SCIENCE
V R GOWARIKER
N V VISWANATHAN
JAYADEV SREEDHAR
ISBN: 978-93-89802-51-1
PB MRP: ₹ 450.00
EXPERIMENTS IN POLYMER SCIENCE
ISBN: 978-81-224-2388-4
PB MRP: ₹ 275.00
Introduction to POLYMER SCIENCE AND TECHNOLOGY
ISBN: 978-93-86286-53-6
PB MRP: ₹ 275.00
POLYMER COMPOSITE
SECOND EDITION
M C GUPTA
A P GUPTA
ISBN: 978-81-224-3244-2
HB MRP: ₹ 699.00
ENVIRONMENTAL POLLUTION ANALYSIS
S M KHOPKAR
ISBN: 978-81-224-2747-9
PB MRP: ₹ 299.00
ENVIRONMENTAL CHEMISTRY
ANIL KUMAR DE
ARNAB KUMAR DE
ISBN: 978-81-224-7227-1
PB MRP: ₹ 399.00
PERSPECTIVES IN ENVIRONMENTAL STUDIES
ISBN: 978-81-951755-2-9
PB MRP: ₹ 350.00
ENVIRONMENTAL CHEMISTRY
P S Sindhu
ISBN: 978-81-224-2731-8
PB MRP: ₹ 499.00
ENGINEERING CHEMISTRY AND ENVIRONMENTAL STUDIES
(Theory and Practical)
ALLA APPA RAO
ISBN: 978-81-224-2765-3
PB MRP: ₹ 125.00
Committed to Educate the Nation
Publishing Globally
IN INDIA
NEW AGE INTERNATIONAL (P) LIMITED, PUBLISHERS
22, Golden House, Ansari Road, Daryaganj, New Delhi-110 002 (India) • Tel.: +91 (11) 23262370, 23262368 • Fax : 43551305
E-mail: sales@newagepublishers.com • Visit us at www.newagepublishers.com
Bangalore (080) 26756823 • Chennai (044) 24353401 • Cochin (0484) 2377303 • Guwahati (0361) 2513881 • Hyderabad (040) 24652456 • Kolkata (033) 22273773 • Mumbai (022) 24927869